Microwave and Millimeter-Wave Vacuum Electron Devices

Inductive Output Tubes, Klystrons, Traveling-Wave Tubes, Magnetrons, Crossed-Field Amplifiers, and Gyrotrons

For the *Artech House Microwave Engineering Library*
go to the back of this book.

Microwave and Millimeter-Wave Vacuum Electron Devices

Inductive Output Tubes, Klystrons, Traveling-Wave Tubes, Magnetrons, Crossed-Field Amplifiers, and Gyrotrons

A. S. Gilmour, Jr.

ARTECH HOUSE
BOSTON | LONDON
artechhouse.com

Library of Congress Cataloging-in-Publication Data

A catalog record for this book is available from the U.S. Library of Congress.

British Library Cataloguing in Publication Data

A catalog record for this book is available from the British Library.

ISBN-13: 978-1-63081-728-2

Cover design by John Gomes

2020 Artech House
685 Canton Street
Norwood, MA 02062

10 9 8 7 6 5 4 3 2 1

Contents

Preface

In the beginning of the development of microwave and millimeter-wave vacuum electron devices, the devices were called microwave tubes because they operated at frequencies that were so high, the operating wavelength was a small fraction of normal radio signal wavelength. The devices were constructed using glass-receiving tube technology. Hence, they were called microwave tubes. Why are they now being called vacuum electron devices (VEDs)?

If we move forward about half a century, we find that there had been enormous advances in microwave tube technology, but solid-state technology was becoming a competitor. At the same time, MIL-HNDBK-217F was reporting tube failure rates that were over five times[1] higher than field experience with tubes. Tube people were becoming frantic at the thought of being put out of business by solid state. The cartoon in Figure P.1 became very popular. New systems were being based on the new solid-state technology rather than that old tube technology. Bob Symon's (Litton EDD) advice to potential customers was to use appropriate technology.

Figure P.1 Why use a thousand mice when one reliable horse can do the job? (Adapted from Rodney Vaughan, 1996 Monterey Power Tube Conference.)

A substantial effort was launched by the Navy to put a new face on microwave tubes by calling them VEDs. It's clear that the microwave tube industry got the word. There

[1] See A. S. Gilmour, Jr. *Principles of Traveling Wave Tubes,* Artech House, 1994.

was one report that one microwave tube manufacturer charged its engineers $1 for every use of the "word" tube rather than VED. In another report, an Air Force major insisted on using new VEDs rather than those old microwave tubes in a new system design! I don't know if he ever found out that they are the same thing.

Call them whatever you like, they are not going away! The primary purpose of this book is to describe tubes that are in use today, as well as some that have been phased out and those that may be needed for new applications. This book contains a balance of practical and theoretical material so that those new to vacuum electronic devices can benefit from their use.

I have made liberal use of color graphics to improve the clarity of the figures. To a large extent, each chapter builds on material from earlier chapters. For those who have microwave tube experience, most of the chapters can be used without referring to others. The appendices contain information that is relevant to tube operation but not critical to understanding fundamental operating mechanisms. For those who are interested in further study and research, the references to the works of over 500 authors dating from the invention of the microwave tube to the present time should be of value.

Contributions to this book have come from many sources. First of all, Thomas Bekker of the Port Hueneme Division of the Naval Surface Warfare Center provided the encouragement that led me to give my first 5-day course on microwave tubes in 1982 and to write *Microwave Tubes,* which was published by Artech House in 1986.

With the encouragement and support of Eugene (Joe) Dutkowski of the Crane Division of the Naval Surface Warfare Center, I developed an 8-day course on traveling-wave tubes (TWTs). That course was similar to a semester-long university course I taught 2 days per month for four months. At the end of each 2-day session, I assigned homework. The students completed the homework between sessions and then we reviewed it during the following 2-day session. That TWT course led to the publication in 1994 by Artech House of *Principles of Traveling Wave Tubes*.

Subsequently, I developed 8- and 10-day courses, which contained in-depth material on klystrons, crossed-field devices, and gyrotrons, as well as traveling wave tubes. I integrated the new material on these devices with revised material from my previous two books in *Klystrons, Traveling Wave Tubes, Magnetrons, Crossed-Field Amplifiers, and Gyrotrons,* which was published by Artech House in 2011. With support and encouragement from Brian Mitsdarffer of the Crane Division of the Naval Surface Warfare Center, I combined relevant material from my previous books with new material and color graphics into this current book.

I thank the anonymous reviewer who was assigned by Artech House and who made numerous suggestions for corrections and additions to the manuscript for this book.

Finally, I again thank my wife, Sally, for her patience and understanding.

Chapter 1

Introduction

1.1 THE DOMAIN OF RF VACUUM ELECTRON DEVICES

The blue colored part of the chart in Figure 1-1 represents the power/frequency domains that vacuum electron devices and solid-state devices dominate. The red area represents devices that are selected primarily on the basis of one or more significant attributes of the devices. A good example of devices that fall in this area is RF power amplifiers used on satellites for television broadcasting. TWTAs (traveling wave tube amplifiers) are frequently the amplifiers of choice because of their high efficiency (see Figure 1-8).

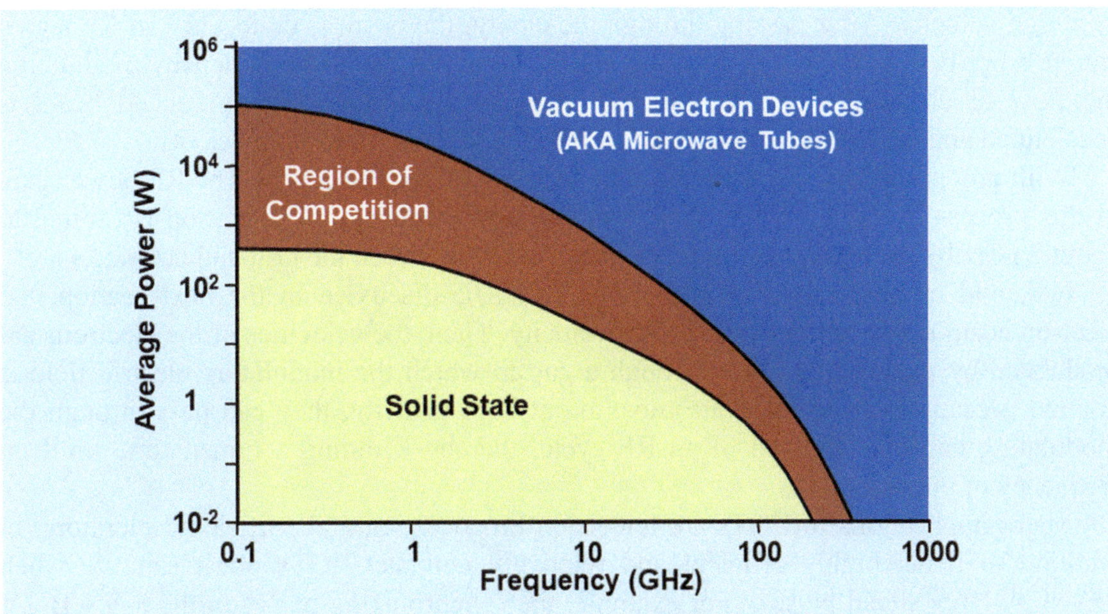

Figure 1-1 Ranges of application of RF vacuum electron devices and solid-state devices.

1.2 CLASSIFICATION OF RF VACUUM ELECTRON DEVICES

The chart in Figure 1-2 shows the four major categories of vacuum electron devices: inductive output tubes, linear-beam tubes, crossed-field tubes, and fast-wave tubes, which are usually referred to simply as gyrotrons. So that the major differences between these tube types can be understood, important characteristics of each tube type are described next.

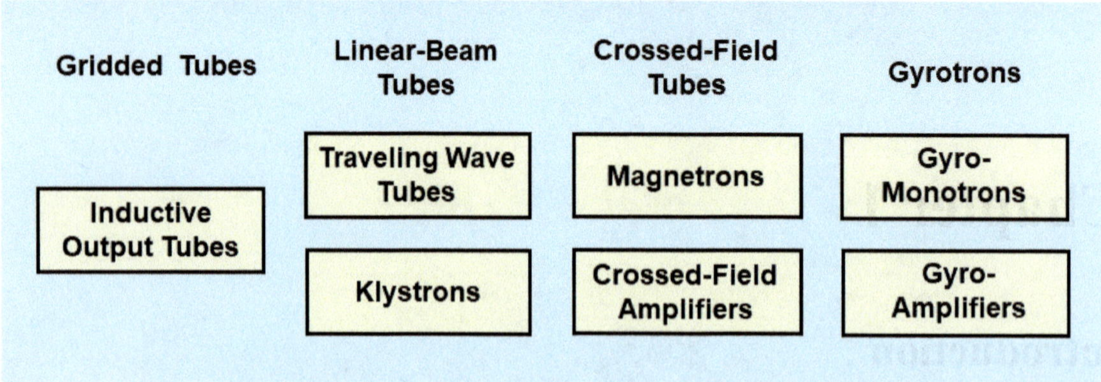

Figure 1-2 Major categories of vacuum electron devices.

1.2.1 Inductive Output Tubes

Inductive output tubes (IOTs) are the least complex of all of the tubes considered in this book. Some of the IOTs made by CPI are also called "Klystrodes," which is the CPI trademarked name for their IOTs. In its simplest form shown in Figure 1-3, the tube has a cathode, modulating grid, cavity, and a collector.

A high voltage applied between the cathode and cavity accelerates the electron beam to a high velocity. After passing through the cavity, the beam is collected. An RF signal can be amplified with the IOT by applying the signal to the grid and density modulating the flow of electrons from the cathode. Then, when the density-modulated beam is accelerated and passed through the cavity, RF current is induced in the cavity.

With proper loading, RF power can be extracted from the cavity. The RF power gain in this process is about 20 dB. The upper limit on frequency of operation is limited to about 2 GHz by the long transit time of the electrons between the grid and cathode.

It should be noted that, in a klystron, which is discussed in the next section, the electron beam is first accelerated to high velocity. Then, the velocities of the electrons are modulated by passing the beam through a gap to which the modulating electric field is applied. Because the electrons are traveling at high velocity, they can pass through the modulating gap in a fraction of an RF cycle, thereby avoiding a transit time limit on frequency of operation.

Major applications for IOTs are television broadcasting and particle accelerators. In addition to being highly efficient and relatively compact, IOTs don't saturate when driven by large signal pulses. For example, the synchronizing pulses in the 8-VWB TV modulation used in the United States are 6 dB larger than the video signal. Amplifiers that saturate have to be operated well below saturation to properly amplify these signals.

1.2.2 Linear Beam Devices

The family tree of linear-beam tubes given in Figure 1-4 shows klystrons and traveling wave tubes, and also indicates that there are hybrid devices that combine the klystron and TWT technologies.

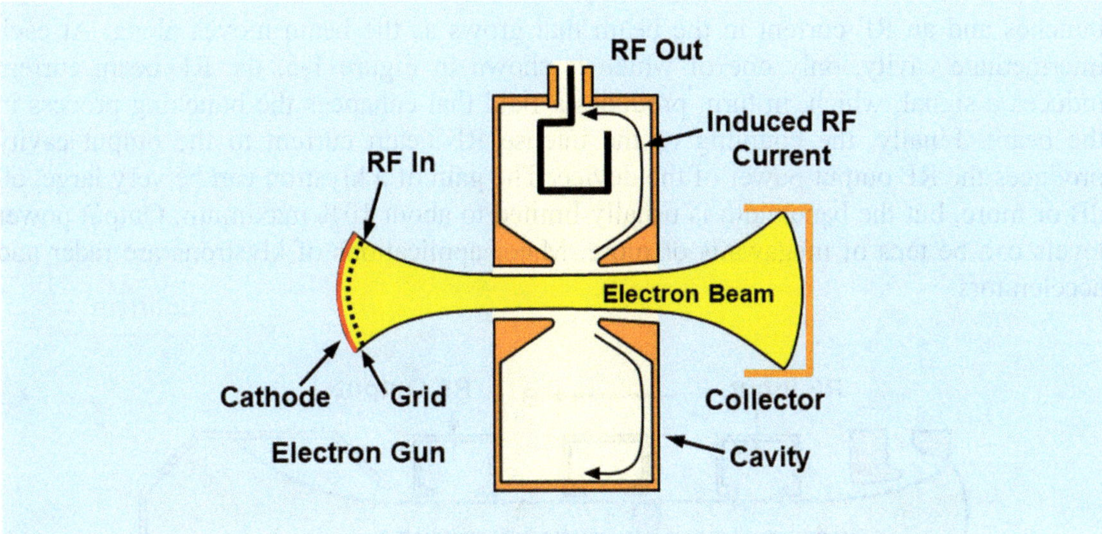

Figure 1-3 Parts of an IOT.

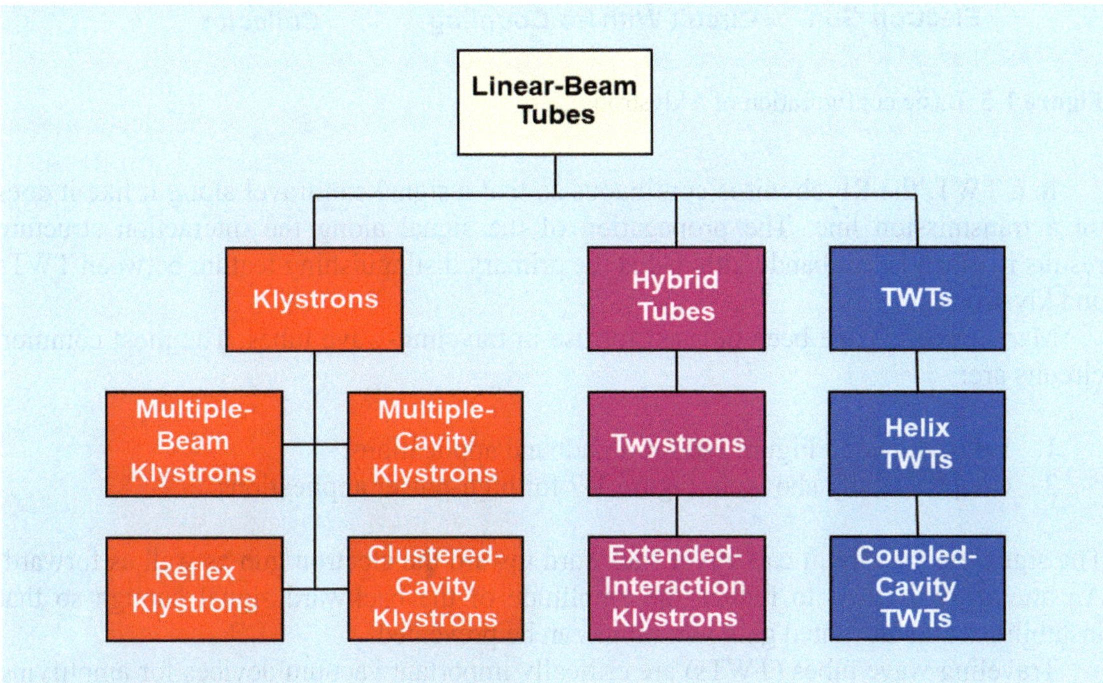

Figure 1-4 Family tree of linear-beam tubes.

Figures 1-5 through 1-8 indicate the differences between the klystron, TWT, and hybrid device technologies. In the klystron, interactions take place in discrete locations along the beam and there is no coupling of the signal between parts of the interaction structure. The RF signal is carried from cavity to cavity by the electron beam.

Figure 1-5 shows the basic configuration of a klystron. In a klystron, the RF circuit consists of resonant cavities. There is no electromagnetic coupling between cavities. The RF input signal, indicated by the small blue arrows, accelerates and decelerates electrons in the beam. As the beam drifts, faster electrons catch up with slower electrons to form bunches and an RF current in the beam that grows as the beam moves along. At each intermediate cavity, only one of which is shown in Figure 1-5, the RF beam current induces a signal, which, in turn, produces a field that enhances the bunching process in the beam. Finally, the coupling of the intense RF beam current to the output cavity produces the RF output power of the device. The gain of a klystron can be very large, 60 dB or more, but the bandwidth is usually limited to about 10% maximum. Output power levels can be tens of megawatts or more. Major applications of klystrons are radar and accelerators.

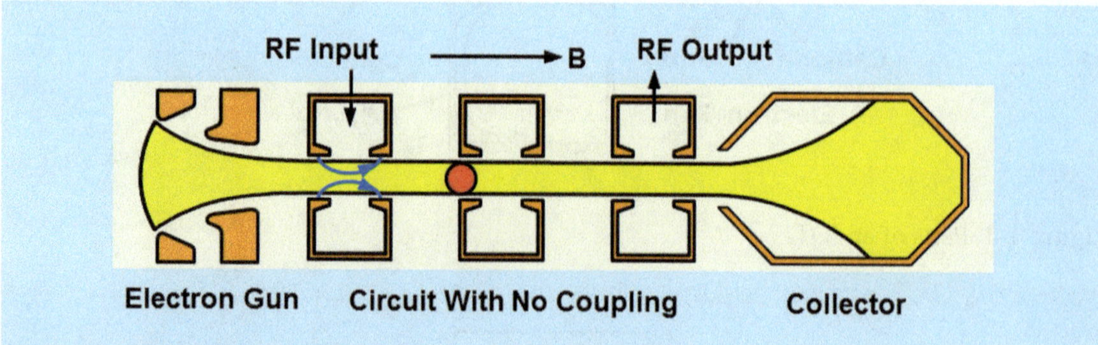

Figure 1-5 Basic configuration of a klystron.

In a TWT, the RF circuit is continuous so that a signal can travel along it like it does on a transmission line. The propagation of the signal along the interaction structure results in much larger bandwidth and is the primary distinguishing feature between TWTs and klystrons.

Many circuits have been devised for use in traveling-wave tubes. The most common circuits are:

1. Helix shown in Figure 1-6 for broadband applications;
2. Coupled cavity shown in Figure 1-7 for high-power applications.

The signal on the circuit can travel backward toward the electron gun as well as forward. An attenuator is used to reduce the amplitude of the backward signal enough so that instabilities and unwanted gain variations can be prevented.

Traveling wave tubes (TWTs) are critically important vacuum devices for amplifying radio frequencies over the entire microwave frequency range. Factors other than power are of significance for helix TWTs. Perhaps the most important are bandwidth, efficiency, and life. Helix TWTs have by far the largest bandwidth capability of all microwave tubes. This can exceed 2 octaves in applications such as electronic countermeasures systems. For satellite applications, efficiencies greater than 70% are now being achieved. This is remarkable because the basic electronic efficiency of a helix TWT is only about 15%. With sophisticated circuit designs, multistage depressed collectors, and great attention to detail in eliminating losses, these extremely high efficiencies can be obtained.

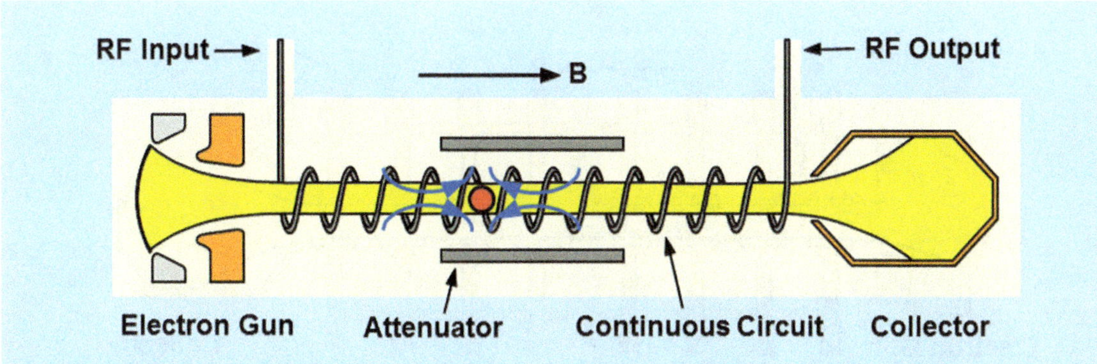

Figure 1-6 Helix TWT.

For space-based applications, the operational lifetimes now required are over 150,000 hours (18 years). Data for space-based traveling-wave tube amplifiers show an MTBF (mean time between failures) of 8 million hours. Of particular note is the performance of the X-band TWT on the Voyager I satellite, which was still in operation after 37 years in space. This TWT was the backup for the solid-state amplifier that failed early in the mission.

Helix TWTs are the amplifiers of choice for many satellite programs. Figure 1-8 shows the growth over a 15-year period in the number of helix TWT amplifiers installed on satellites. The number of SS (solid state) amplifiers installed per satellite steadily declined during that period.

Over 50% of the sales volume of all microwave tubes is for TWTs. Applications for TWTs are varied. They serve as the final amplifiers in most electronic warfare systems. In many radar systems, one or more TWTs are used as the high-power amplifier that generates the transmitted RF pulse. In others, a TWT may be used as the driver for some other device such as a crossed-field amplifier. TWTs are used as the power boosters in microwave power modules (MPMs) and millimeter MPMs (MMPMs). TWTs are the amplifiers of choice for many satellite applications.

In the coupled-cavity version of the TWT, cavities similar in appearance to klystron cavities are used. There are coupling apertures in the walls of the cavities so that the RF energy can propagate along the interaction structure in addition to being carried along by the electron beam.

There are two types of coupled cavity circuits. The one shown in Figure 1-7 is representative of the circuits capable of operating at power levels up to a few hundred kilowatts. The other, which is not shown here but is described in Chapter 15, is capable of megawatt operation. Other TWT circuits are mentioned in this book, but the majority of the discussions center on helix and coupled-cavity circuits.

The TWT circuit is designed so that the velocity of the signal traveling on it is near the velocity of the electron beam passing through it. Also, the circuit is designed so that the electric fields from a signal on the circuit penetrate into the beam as indicated by the blue arrows in the figures. Electron bunches (known as electron-density waves or space charge waves) are formed in the electron beam. These travel somewhat slower than the beam. Energy is extracted from beam electrons as they are slowed to the bunch velocity.

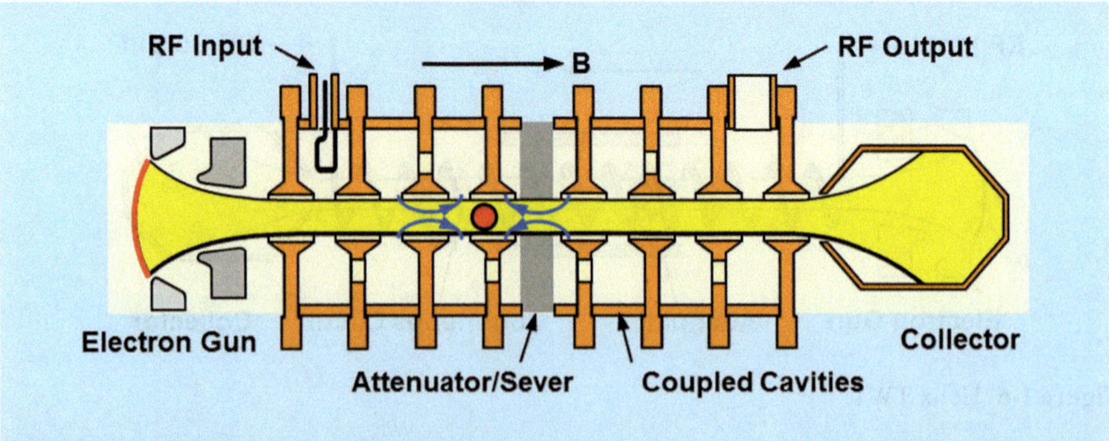

Figure 1-7 Coupled-cavity TWT (CC TWT).

As the bunches pass by the nearby circuit, they induce RF current in the circuit. This causes the amplitude of the RF field on the circuit to increase. The increased field, in turn, increases the bunching action in the beam. As a result, as the bunches in the beam and the field on the circuit travel along (at the same velocity), the bunching action in the beam increases. As the bunching becomes more intense, the RF current in the beam and the field on the circuit become larger. Eventually, the amplified signal is coupled off from the circuit. The resulting gain of a TWT is usually in the 30–50-dB range and the bandwidth can be from 20% to over 2 octaves. Output power levels range from tens of watts for helix TWTs to hundreds of kilowatts or even megawatts for coupled-cavity devices.

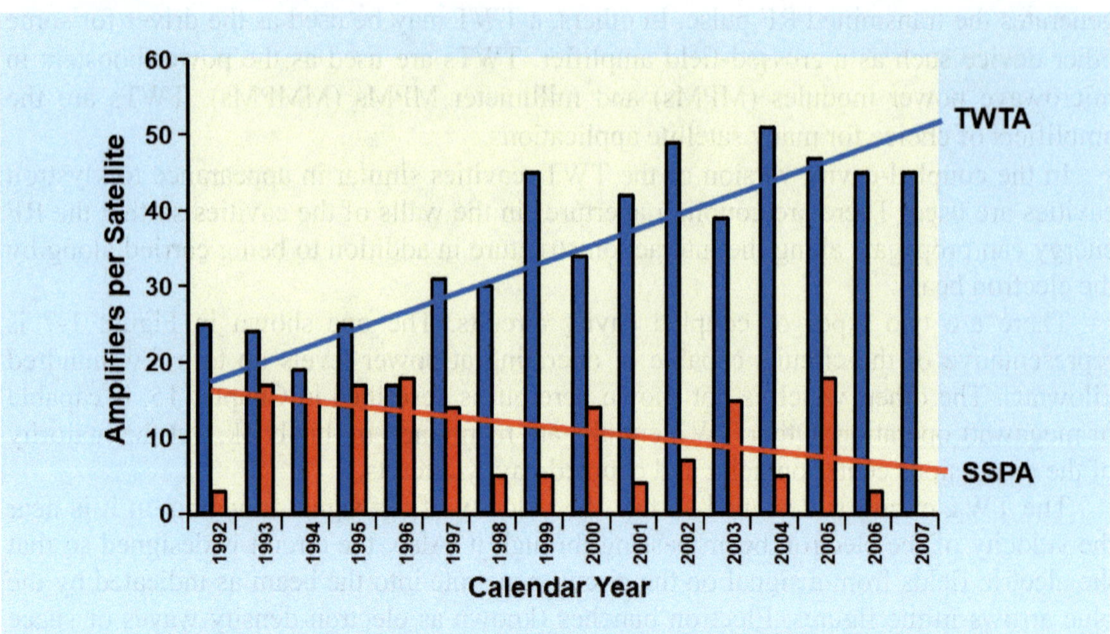

Figure 1-8 Geostationary communications satellite payload trends 1993-2007.

In the clustered-cavity klystron or the extended-interaction klystron in Figure 1-9, one or more of the single cavities of the conventional klystron is replaced by a structure containing two or more interaction gaps. Bandwidth is increased over that of the conventional klystron. In addition, at the output end of the tube, the voltage at each interaction gap is reduced, permitting operation at very high-power levels.

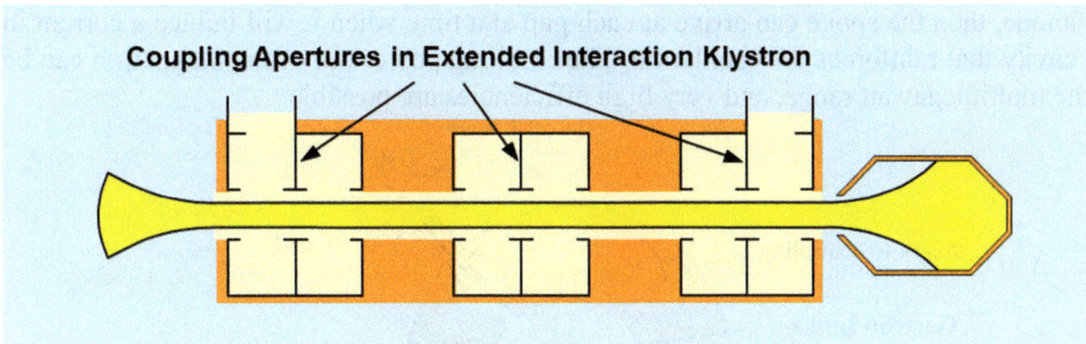

Figure 1-9 Clustered-cavity klystron and extended-interaction klystron.

The twystron, shown in Figure 1-10, consists of a klystron input section and a coupled-cavity output section. Twystrons are capable of several megawatts at bandwidths of 10%–15%.

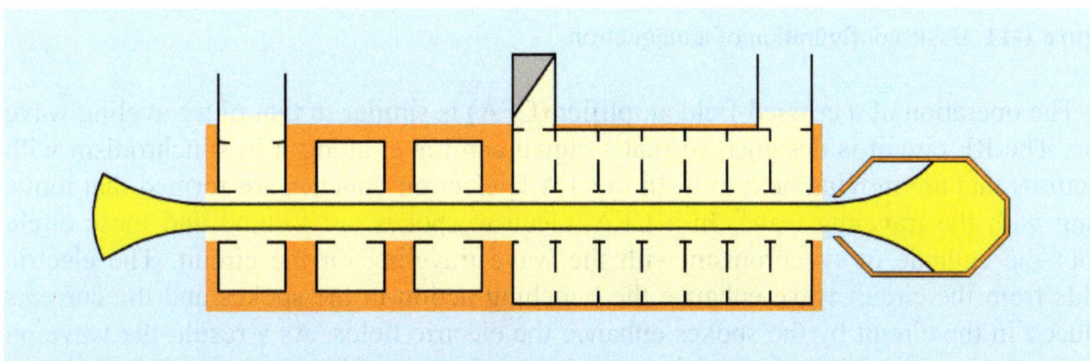

Figure 1-10 Twystron amplifier. (Adapted from: Armand Staprans et al., *Proc. IEEE*, March 1973. © 1973 IEEE.)

1.2.3 Crossed-Field Devices

The basic configuration of a magnetron is shown in Figure 1-11 and a crossed-field amplifier is in Figure 1-12. In these devices, the cathode is a cylindrical emitter in the center. In general, electron flow is radially outward toward the RF circuit, which also serves as the anode. Because a magnetic field is used, which is perpendicular to the cathode-to-anode electric field and to the direction of electron flow, electrons are forced to move in paths that circle about the cathode. When an RF field is present, the electrons are bunched into spoke-like configurations.

The magnetron is an oscillator in which the RF circuit consists of resonant cavities. The cavities are arranged so that the RF magnetic field generated in each cavity is coupled to adjacent cavities. Ideally, the entire cavity structure resonates at a single frequency with the RF electric field in adjacent cavities being 180° out of phase. As the RF field oscillates, the field pattern across cavity gaps rotates about the cathode. If an electron spoke circles about the cathode in synchronism with the rotating field pattern on the anode, then the spoke can arrive at each gap at a time when it will induce a current in the cavity that reinforces the oscillation. The resulting power out of the magnetron can be in the multimegawatt range, and very high efficiencies are possible.

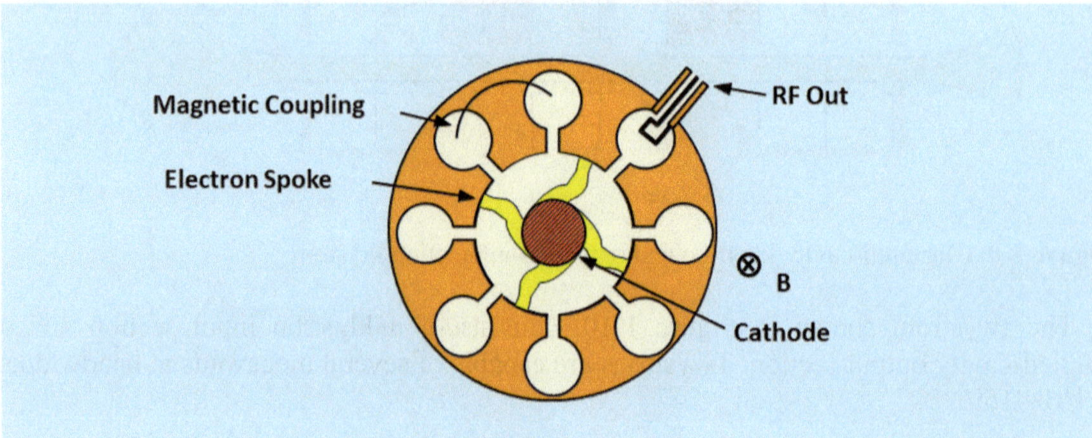

Figure 1-11 Basic configuration of a magnetron.

The operation of a crossed-field amplifier (CFA) is similar to that of a traveling wave tube. The RF circuit is designed so that a signal can travel along it in synchronism with electrons that are moving next to it. In the TWT, electron bunches are formed that move along with the traveling wave. In a CFA, electron spokes are formed and these circle about the cathode in synchronism with the wave traveling on the circuit. The electric fields from the circuit wave enhance the bunching action in the spokes and the currents induced in the circuit by the spokes enhance the electric fields. As a result, the wave on the circuit grows as it travels from input to output. In some CFAs, the circuit is designed so that a backward wave interaction can occur and produce amplification. The gain of CFAs is usually less than 20 dB, but power outputs of tens of megawatts are possible.

The crossed-field amplifier was invented by W. C. Brown, who called it an "amplitron." The amplitron can amplify a broad band of microwave frequencies and is a critical component of many military systems. Perhaps the most publicized amplitron was the device shown in Figure 1-13. Brown designed and built the amplitron which powered the reliable and efficient long-distance communications on the Apollo Moon missions, enabling the world to watch Buzz Aldrin step onto the moon and hear Neil Armstrong's immortal words, directly from the moon.

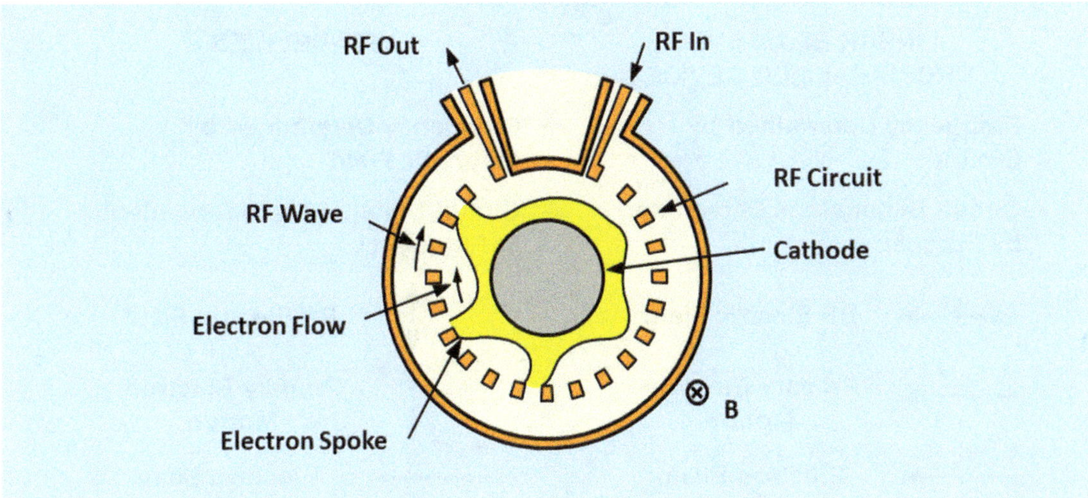

Figure 1-12 Crossed-field amplifier.

Figure 1-13 The QKS1300 amplitron that made it possible for the world to hear Neil Armstrong say, "That's one small step for a man, one giant leap for mankind," as Buzz Aldrin stepped on the moon, July 20, 1969.

1.2.4 Gyro-Devices

Gyro-devices, whether they are oscillators or amplifiers, are sometimes all referred to as gyrotrons. In gyro-devices, the beam-wave interaction mechanism is completely different than in linear beam and crossed-field devices. As Figure 1-14 indicates, in linear beam and crossed-field devices, the frequency of operation is determined by the RF circuit and so circuit dimensions scale with wavelength. As frequency is increased, power capability is decreased. In gyro-devices, the frequency of operation is determined by the cyclotron frequency that depends on the magnetic field. Frequency is independent of circuit dimensions, and very high powers can be generated at very high frequencies.

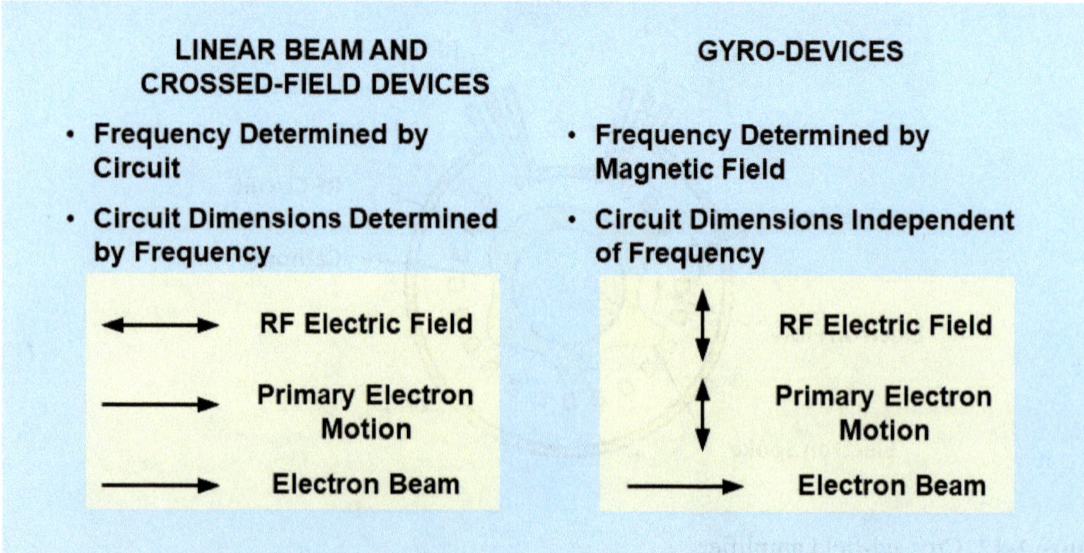

Figure 1-14 Comparison of RF electric field configurations and electron motions in linear beam, crossed-field, and gyro-devices.

Figure 1-15 shows the basic parts of a gyrotron oscillator. As this figure indicates, the electron beam in a gyro-device is hollow. Electrons in the beam spiral about magnetic flux lines in near synchronism with a transverse RF electric field. The transverse velocities of the electrons are normally 1.2 to 1.5 times the axial velocities. It is the transverse energy of the electrons that is used to amplify the electric field.

Gyrotrons are capable of enormous powers at extremely high frequencies. For example, gyrotrons being developed for ITER, the International Thermonuclear Experimental Reactor, will operate at 2MW CW at 170 GHz.

One of the uses for the Haystack radar facility operated by the MIT Lincoln Laboratory is tracking objects in space. In a major upgrade dubbed HUSIR (Haystack Ultrawideband Imaging Radar), the ability to operate at W-band was added to the facility. The upgrade made use of W-band gyro-twystrons driven by a gyro-TWT to increase the resolution of radar images. Figure 1-16 shows a comparison of an X-band image and a W-band image of a satellite obtained with HUSIR.

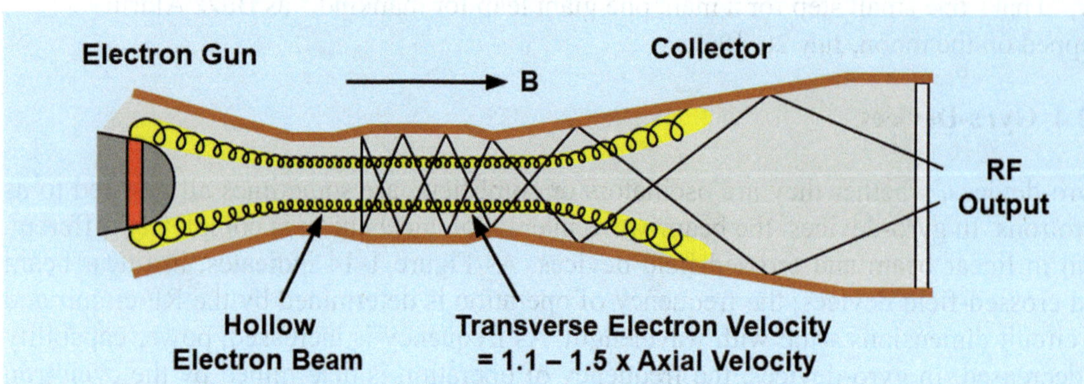

Figure 1-15 Basic configuration of a gyrotron oscillator.

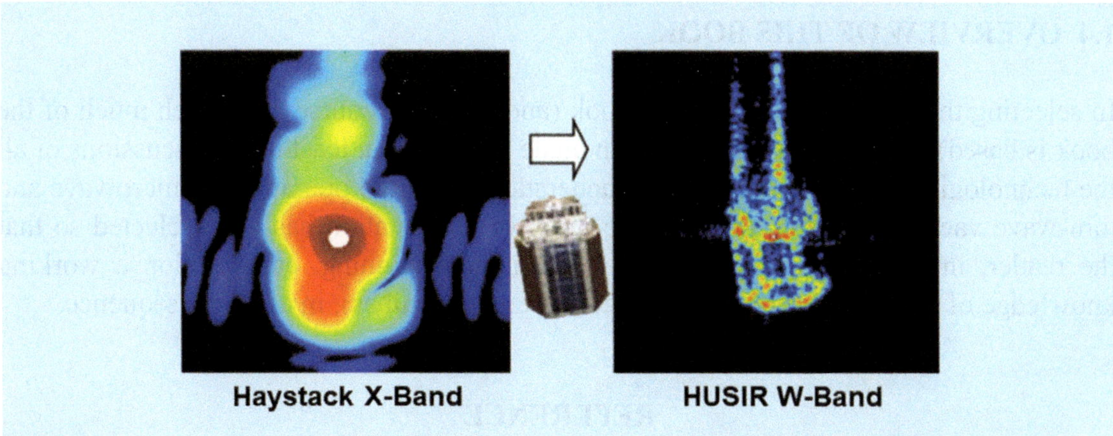

Figure 1-16 Comparison of X-band and W-band radar images of a satellite obtained with the MIT Lincoln Laboratory Haystack radar. HUSIR is the upgrade to the radar with W-band gyro-amplifiers.

1.3 CAPABILITIES OF RF DEVICES

A chart showing the relative power capabilities of various RF devices and the progress that has been made in the last several decades is shown in Figure 1-17 [1]. The figure of merit, $P f^2$, where P is average power in MW and f is frequency in GHz, indicates the ability of a device to generate RF power. Note in particular that $P f^2$ for helix TWTs has increased over three orders of magnitude in the past 50 years and is an order of magnitude greater than that for solid-state devices. Note also that the development of coupled cavity TWTs reached the state of the art in power capability several decades ago.

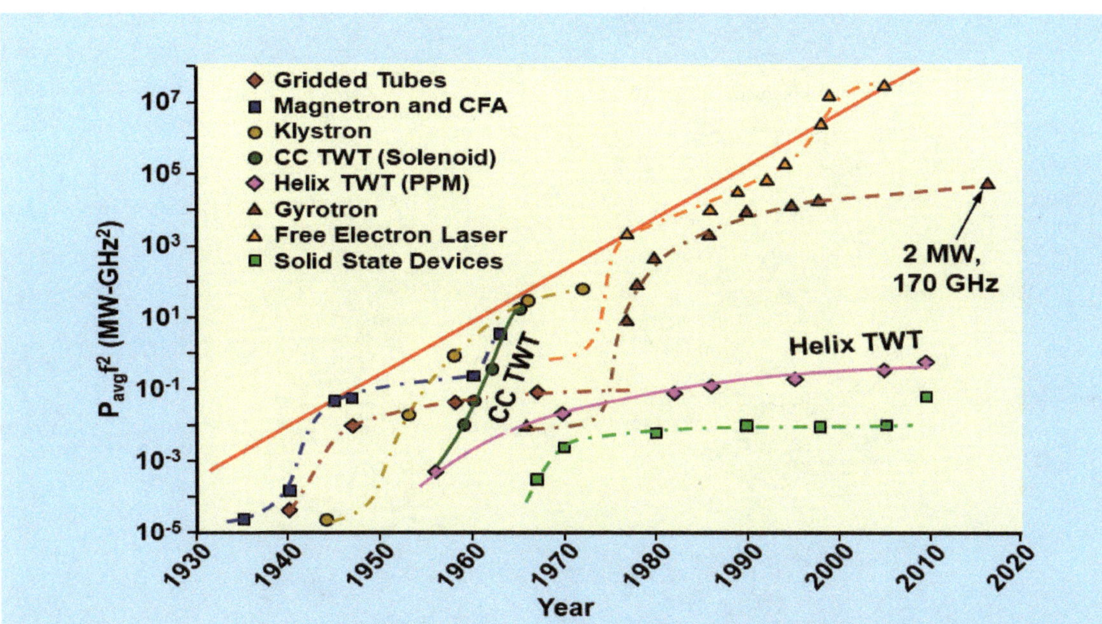

Figure 1-17 Vacuum tube amplifiers, nine decades of progress. The data point for the 2 MW, 170 GHz tube was added by the author.

1.4 OVERVIEW OF THIS BOOK

In selecting the subject matter for this book (and the short courses on which much of the book is based), a serious attempt has been made to include at least basic discussions of all the technologies that are essential to an understanding of the operation of microwave and mm-wave vacuum electron devices. The order of presentation has been selected so that the reader, in working through the book from the beginning, can develop a working knowledge of vacuum electron devices (aka microwave tubes), in a logical sequence.

REFERENCE

[1] Joe X Qui, Baruch Levush, et al., "Vacuum Tube Amplifiers," *IEEE Microwave Magazine*, December 2009, pp. 38–51.

Chapter 2

Basic Properties of Electromagnetic Waves

2.1 WAVE PROPAGATION

Electromagnetic waves propagate through free space at the speed of light, c, which is $3\text{x}10^8$ m/s (~ 1 foot per nanosecond). The relation between wavelength, λ, of a signal and its frequency, f, is

$$\lambda = \frac{c}{f} \tag{2-1}$$

The wavelengths of microwave signals are shown in Figure 2-1.

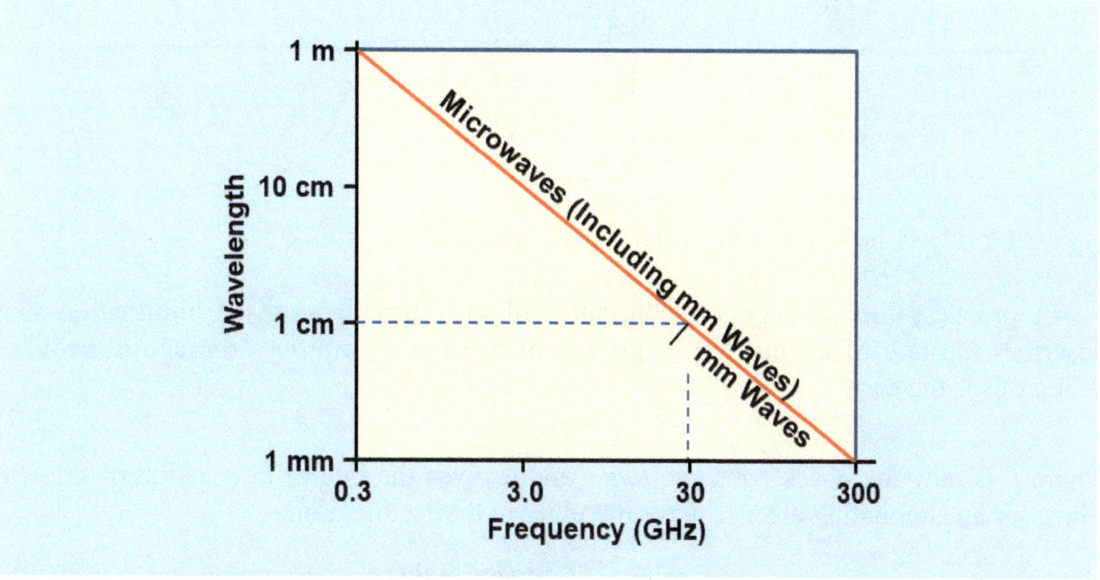

Figure 2-1 Wavelengths of signals in the microwave range.

The propagation of an electromagnetic wave can be expressed by

$$E = E_o \sin(\omega t \pm \beta z) \tag{2-2}$$

where E_o is the peak value of the electric field, ω is the angular frequency, β is the propagation constant, t and z are time and distance, and wave amplitude is constant.

To show how this equation represents the propagation of a wave, a constant value of the electric field, E, may be chosen. For the field to remain constant, at some value, E_1, the phase angle of the sine wave ($\omega t \pm \beta z$) must remain constant at some value, ϕ_o, so

$$\phi_O = \omega t \pm \beta z \tag{2-3}$$

If the minus sign is chosen first, then,

$$z = \frac{\omega t - \phi_o}{\beta} \tag{2-4}$$

Thus, as t increases, z must increase, and so a wave moving in the $+z$ direction is represented as shown in Figure 2-2. Similarly, when the plus sign is chosen, a wave moving in the -z direction is described.

The velocity of the wave, u, can be determined from (2-4) taking the derivative with respect to time, which is

$$u = \frac{dz}{dt} = \frac{\omega}{\beta} \tag{2-5}$$

Therefore, the propagation constant, β, is

$$\beta = \frac{\omega}{u} \tag{2-6}$$

and is a measure of the phase characteristic of a wave moving through a medium. For free space, $\beta = \omega/c$.

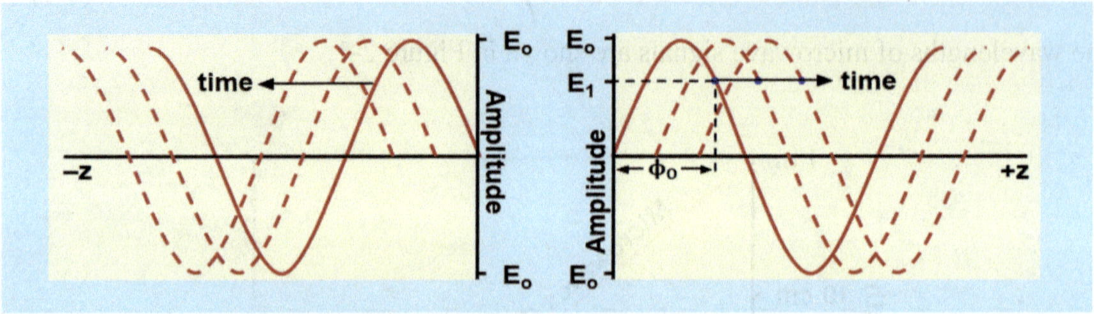

Figure 2-2 Waves propagating in $+z$ and in the $-z$ direction.

In practice, the propagation constant will also usually contain information that describes the rate of attenuation or growth of a wave as it moves through a medium. When this is the case

$$\gamma = \beta + j\alpha \tag{2-7}$$

where β is now the phase constant, ω/u , and α gives the rate of attenuation or growth. Thus, for an attenuating electric wave traveling in the +z direction,

$$E = E_o e^{-\alpha z} \sin(\omega t - \beta z) \tag{2-8}$$

and appears as shown in Figure 2-3.

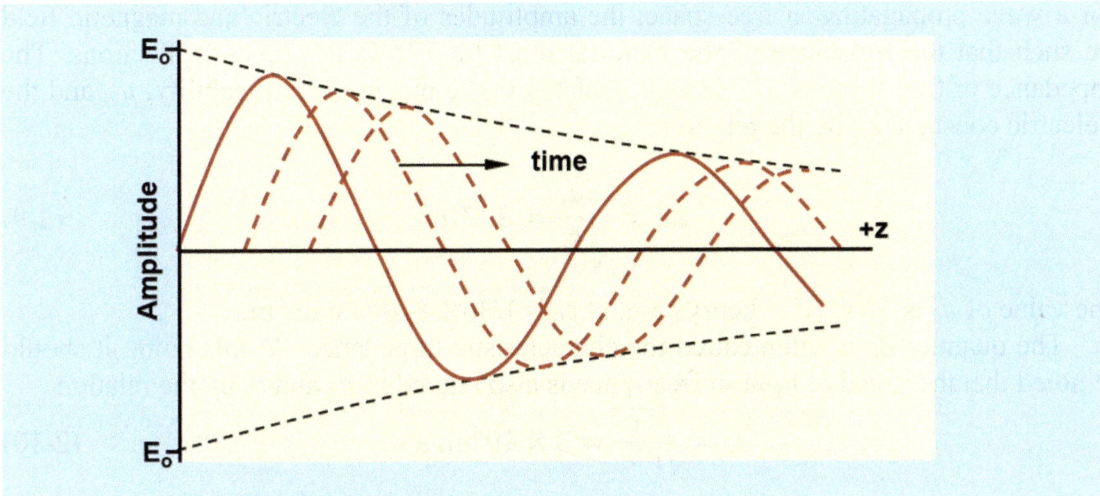

Figure 2-3 Attenuating electric wave traveling in the +*z* direction.

An electromagnetic wave has both electric and magnetic components. The variations of amplitude and phase with position can be very complex. A particularly simple case is the plane electromagnetic wave in a lossless medium. At one instance in time, the magnitudes of the electric and magnetic fields are constant in any plane perpendicular to the direction of travel of the wave. The relation of the electric field to the magnetic field for a plane wave is indicated in Figure 2-4. Both the electric and magnetic fields are sine waves with the same propagation constant. The direction of the magnetic field is at a right angle to the direction of the electric field. When an electromagnetic wave, such as that shown in Figure 2-4, impinges on a sheet of lossy material, voltages and currents are induced in the material. If the resistance of the material is such that all of the power in the wave is absorbed and none is reflected, then there is a perfect impedance match.

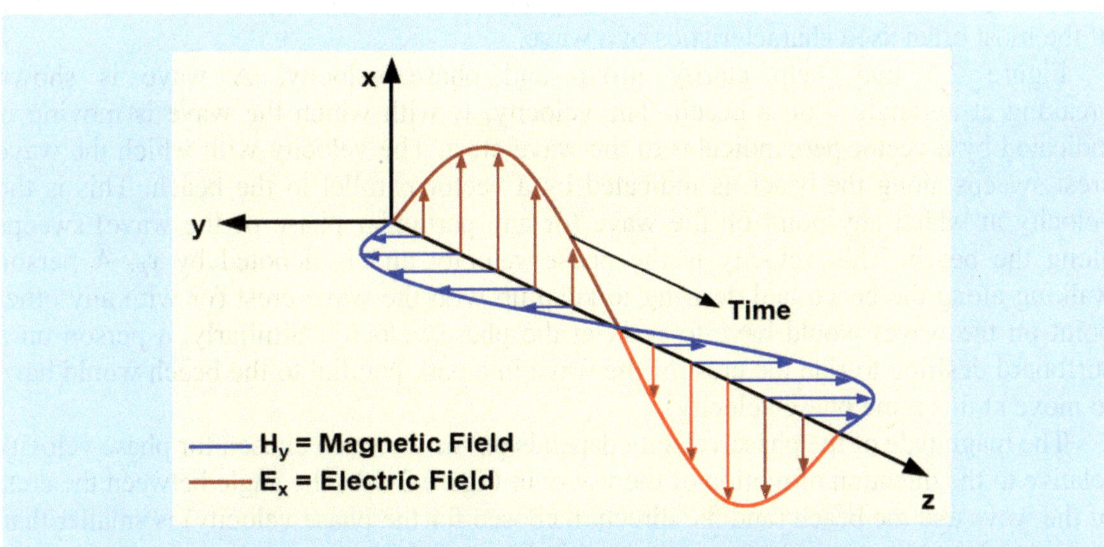

Figure 2-4 Electric and magnetic fields in a plane electromagnetic wave.

For a wave propagating in free space, the amplitudes of the electric and magnetic field are such that the resistance of the material must be 377 Ω to prevent reflections. The impedance of free space is 377 Ω and is related to the magnetic permeability, μ_o, and the dielectric constant, ε_o, by the relation

$$Z_o = \sqrt{\frac{\mu_o}{\varepsilon_o}} = 377\Omega \tag{2-9}$$

The value of μ_o is $4\pi \times 10^{-7}$ henrys/m and ε_o is $1/36\pi \times 10^{-9}$ farads/m.

The quantity Zo is often called the characteristic impedance. At this point, it should be noted that the speed of light in free space is also related to μ_o and ε_o by the relation

$$c = \sqrt{\frac{1}{\mu_o \varepsilon_o}} = 3 \times 10^8 \text{ m/s} \tag{2-10}$$

For values of μ or ε other than those for free space, the impedance and speed change accordingly. For example, for aluminum oxide, a material commonly used in microwave components, the dielectric constant is about nine times that of free space. As a result, the characteristic impedance and speed of light are one-third of the values for free space.

A circuit in which the wave velocity varies with frequency is said to have dispersion or to be dispersive. Before discussing dispersion in detail, it is important to examine wave velocities and, in particular, the concepts of the group and phase velocity.

2.2 GROUP AND PHASE VELOCITY

In microwave tube work, it is rare that we deal directly with plane electromagnetic waves traveling at the free-space speed of light. Instead, the waves are constrained to move in complex paths. We often find that it is convenient to use one or more characteristics of a wave to explain the effect of the wave. The group velocity and the phase velocity are two of the most often used characteristics of a wave.

Figure 2-5 may help clarify group and phase velocity. A wave is shown breaking at an angle θ on a beach. The velocity, v, with which the wave is moving is indicated by a vector perpendicular to the wave crest. The velocity with which the wave crest sweeps along the beach is indicated by a vector parallel to the beach. This is the velocity at which any point on the wave (or any particular phase of the wave) sweeps along the beach. This velocity is the phase velocity and is denoted by v_p. A person walking along the beach and desiring to keep up with the wave crest (or with any other point on the wave) would have to walk at the phase velocity. Similarly, a person on a surfboard desiring to ride the crest of the wave in a path parallel to the beach would have to move at this same phase velocity.

The magnitude of the phase velocity depends on the direction chosen for phase velocity relative to the direction of motion of the wave. In Figure 2-5(b) the angle between the crest of the wave and the beach (and the direction chosen for the phase velocity) is smaller than in Figure 2-5(a). As a result, the wave crest in Figure 2-5(b) sweeps along the beach more rapidly than in Figure 2-5(a). Thus, the phase velocity has increased.

The group velocity, v_g, is the velocity of the energy of the wave in the direction of interest. For example, in Figure 2-5(a), as the energy in the wave at point A moves to point B, the distance that energy moves parallel to the beach (and to the phase velocity) is from A to C. Thus, in this case, the group velocity is the component of the wave velocity in the direction of the phase velocity. Like the phase velocity, the magnitude of the group velocity depends on direction. This can be seen by comparing Figure 2-5(a) to Figure 2-5(b). The group velocity in Figure 2-5(b) is smaller than that in Figure 2-5(a).

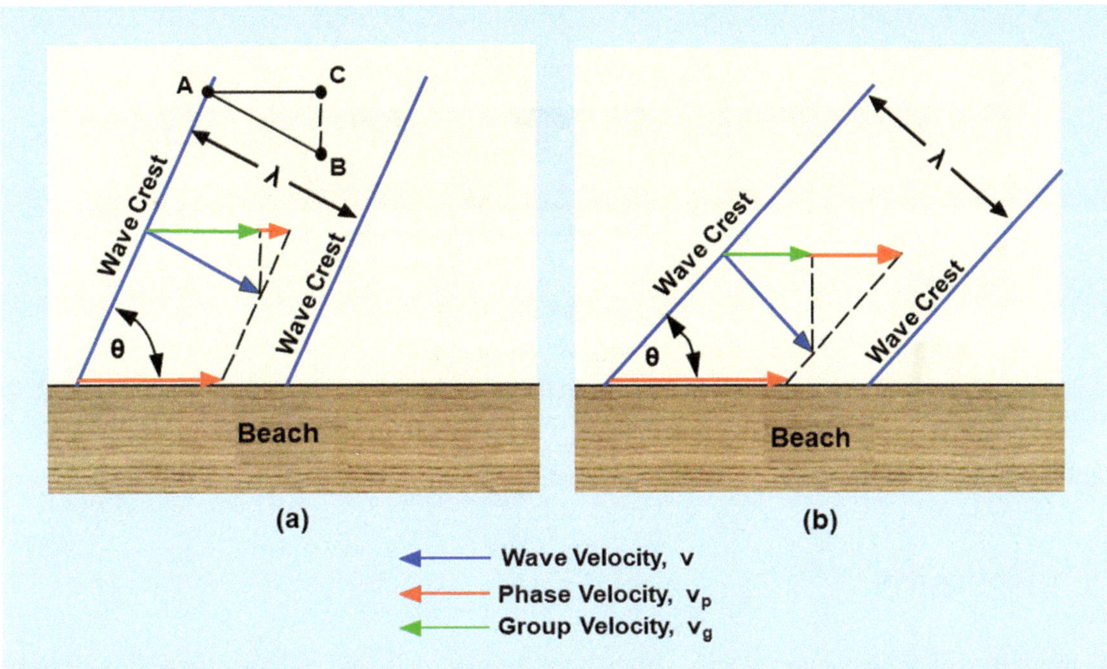

Figure 2-5 The magnitudes of phase velocity and group velocity depend on the direction of wave travel. In (b), phase velocity is greater and group velocity is smaller than in (a).

Group velocity and phase velocity can have opposite directions. Consider, for example, Figure 2-6. The waves are moving to the left and the group velocity is to the left. It is assumed that there is a seawall with equally spaced openings along the shoreline. It is also assumed that a person walking along the beach would be able to see the waves only when passing one of the openings in the seawall. The person could walk from right to left and pass successive openings at the same time that one of the wave crests did.

Alternatively, the person could walk from left to right (from 1 to 2, for example) and pass the openings in synchronism with successive wave crests moving from right to left (from 3 to 2, for example). In this case, the phase velocity is to the right while the group velocity is to the left. For a person on a surfboard to move to the right and continue to be accelerated by the wave crests (which are moving to the left), that person would have to jump from wave crest to wave crest.

When the group velocity and the phase velocity are in opposite directions, interaction with the wave is referred to as a *backward wave interaction.* An important aspect of this interaction, which was evident in the previous discussion, is that the interaction is periodic rather than continuous as it is when the group velocity and the phase velocity are in the same directions.

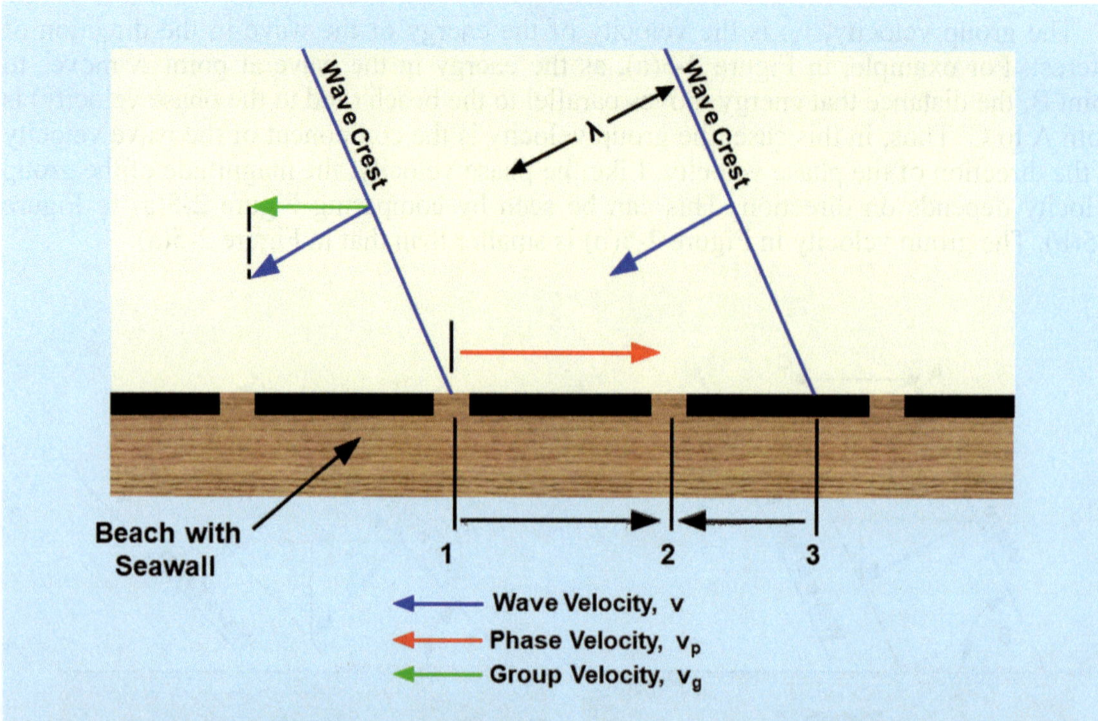

Figure 2-6 Illustration of backward wave interaction.

2.3 DISPERSION

A circuit in which the wave velocity varies with frequency is said to have dispersion or to be dispersive. Two ways are commonly used to illustrate the dispersion characteristics of a circuit. One, shown in Figure 2-7, is straightforward. The vertical axis represents phase velocity and the horizontal axis represents frequency. The plot for a circuit with no dispersion is simply a horizontal line. For a circuit with dispersion, the line is not horizontal.

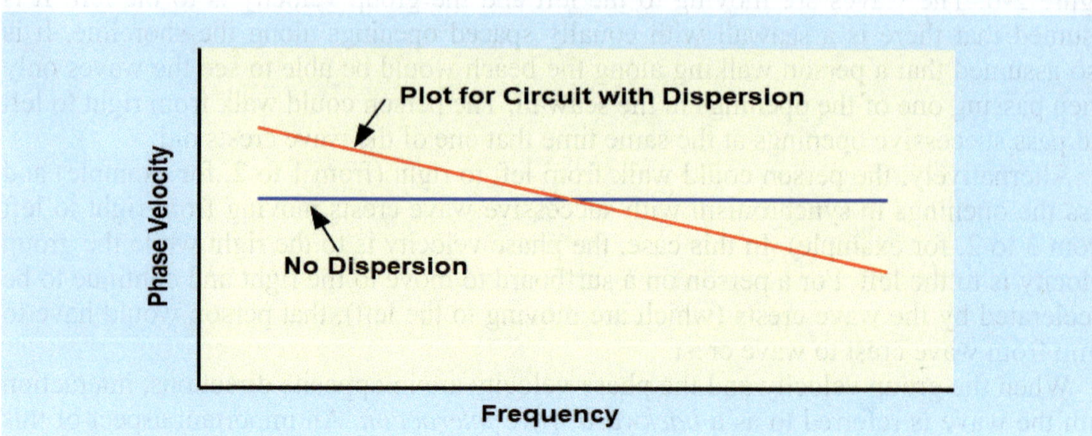

Figure 2-7 Straightforward representation of dispersion.

The second way to show dispersion characteristics is with a Brillouin diagram, which is often referred to as an ω-β diagram. Here, ω is angular frequency and β is the wave propagation constant. This diagram is not as straightforward as the first (phase velocity versus frequency). However, its use permits the display of considerably more information than the first. Because of the importance and widespread use of ω-β diagrams in the microwave component industry, their development is considered here.

A convenient way to develop an understanding of ω-β diagrams is to first consider simple microwave circuits and how their characteristics are represented. For this purpose, we will use a coaxial transmission line, which is not dispersive, and waveguides, which are dispersive. Following these two examples, the more complicated example of a periodically loaded waveguide is examined.

2.3.1 Coaxial Line

In the fundamental TEM mode in a vacuum insulated coaxial line shown in Figure 2-8, the wave velocity and the electric and magnetic field lines are in planes perpendicular to the axis of the transmission line. The group velocity and the phase velocity are the same and are the velocity of light, c, for all frequencies. Thus, the coaxial line has no dispersion.

The plot of ω *versus* β for the coaxial line is given in Figure 2-9. The plot is a straight line with slope c. In general, those portions of an ω-β diagram with constant slope represent frequency regions of operation of a circuit in which phase velocity is constant, so there is no dispersion.

2.3.2 Rectangular Waveguide

Waveguides are used for carrying microwave power to and from microwave tubes. The waveguides used for this purpose are metallic pipes, which are usually made from copper and have either rectangular or circular cross sections.

The phase velocity of a wave in a rectangular waveguide varies with frequency so the waveguide is dispersive. To understand why phase velocity varies with frequency and how phase and group velocity are shown on an ω-β diagram, it is useful to examine the configuration of waves in a rectangular waveguide.

A convenient place to start is with the consideration of two waves of equal amplitude traveling in free space, as indicated in Figure 2-10. The direction of propagation of one is at an angle $+\alpha$ with respect to the z axis and the other is at an angle $-\alpha$. The solid lines represent positions of wave crests (maximum value of E field out of the figure) and the dashed lines represent positions of wave troughs (minimum value of E field into the figure).

Wherever wave crests and troughs (solid and dashed lines) intersect, the electric field is zero and only a magnetic field exists. In fact, all along a plane through zero-field points and parallel to the y-z plane, the electric field is zero. The E field is zero not only for the instance of time implied in Figure 2-10, but also at other times for waves propagating at the angles shown. Thus, conductors can be inserted along these planes without disturbing the E fields as long as the currents in the conductors correspond to those that would be induced by the magnetic fields.

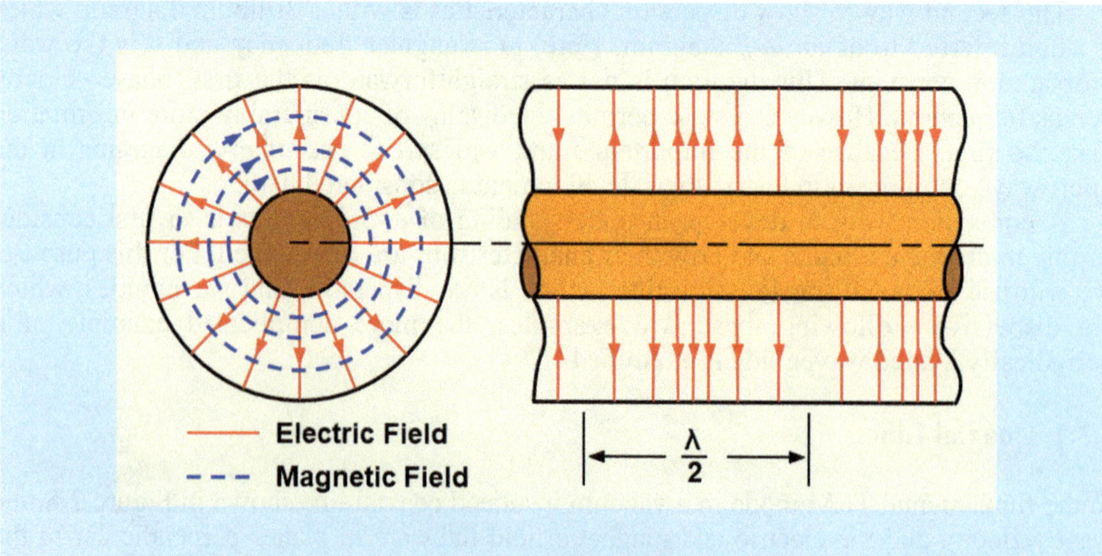

Figure 2-8 The TEM mode in a coaxial line has no dispersion.

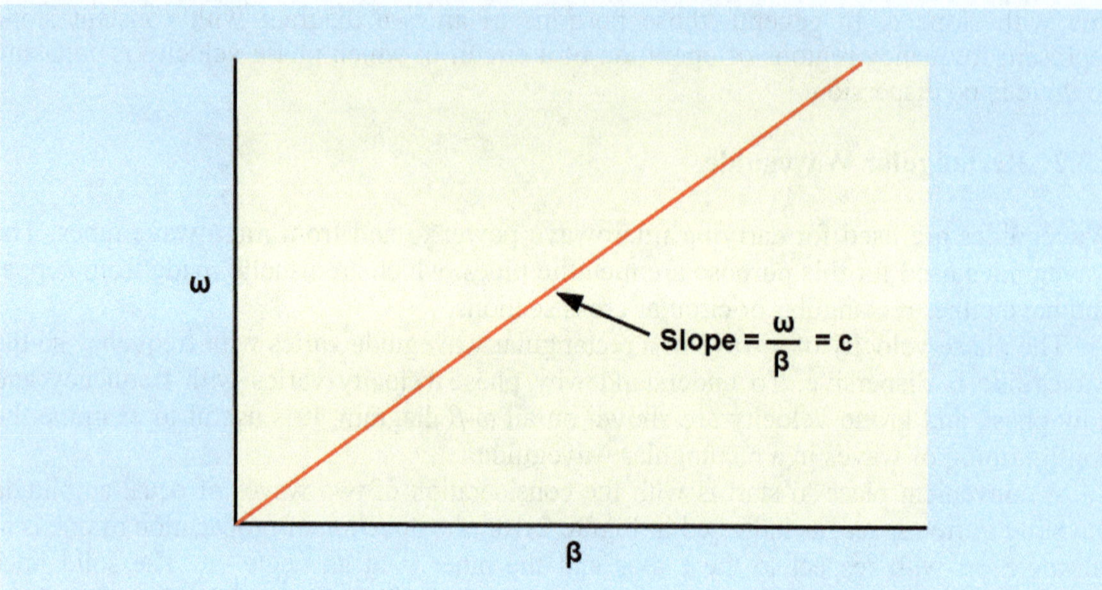

Figure 2-9 The ω-β diagram for the TEM mode in a coaxial transmission line.

The magnetic fields are also parallel to any *x-z* plane, and the electric fields are perpendicular to these planes. Thus, conductors with currents and charges corresponding to those that would be induced by the fields can be placed in the *x-z* planes.

Finally, two conductors in the *y-z* plane can be joined to the conductors in the *x-z* plane to form a rectangular pipe (waveguide) containing portions of the two waves. These two waves can propagate as though they were in free space, except that, in reality, they are continually reflected back and forth from one side of the guide to the other. In fact, there is only one wave because each wave is a reflection of the other.

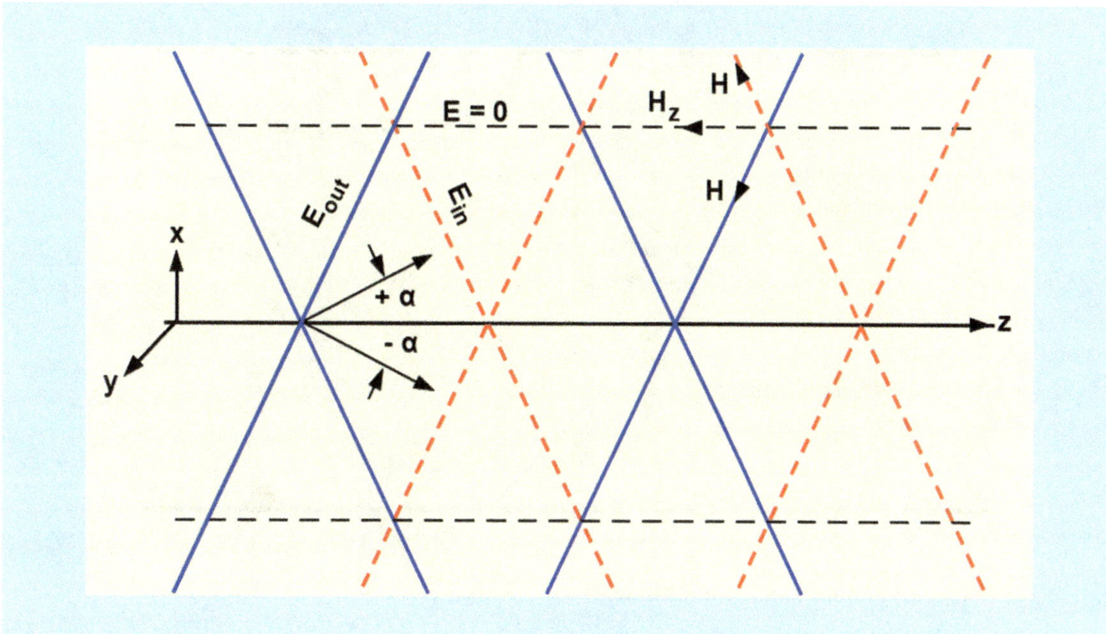

Figure 2-10 Plane waves of equal amplitude propagating at angles of $\pm\alpha$ with respect to the z direction.

The velocity with which the wave energy propagates along the waveguide (the group velocity) can easily be determined with the aid of Figure 2-11. Each component of the wave travels at the speed of light, c, at an angle α with respect to the axis of the guide. As a result, the group velocity is smaller than c and is

$$v_g = c\cos\alpha \tag{2-11}$$

The phase velocity of the wave (the velocity of a particular phase or amplitude of the wave) is indicated by v_p. Thus, as the crest of the wave propagates at angle α a distance proportional to the velocity of light, c, the crest sweeps along the wall of the waveguide a distance proportional to higher velocity, v_p, which is

$$v_p = \frac{c}{\cos\alpha} \tag{2-12}$$

If an electron traveling parallel to the wall of the waveguide was to ride the crest of the wave and continually be deflected into the figure by the electric field, it would have to move at velocity v_p. Of course, this is not possible because, in this example, $v_p > c$, which characterizes a fast wave. TWT circuits are designed to be "slow-wave" circuits so that $v_p < c$ and so electrons can move at a velocity equal to the phase velocity.

To examine the dispersion of waveguide, it is necessary to consider the change in wave configuration that takes place in a waveguide when frequency is varied. Shown in Figure 2-12 are wave configurations for three different frequencies $f_1 > f_2 > f_3$. Group velocities and phase velocities are also indicated.

The ω-β relation for the rectangular waveguide may be derived by examining the wave velocities as a function of frequency (or wavelength) with the aid of Figure 2-13.

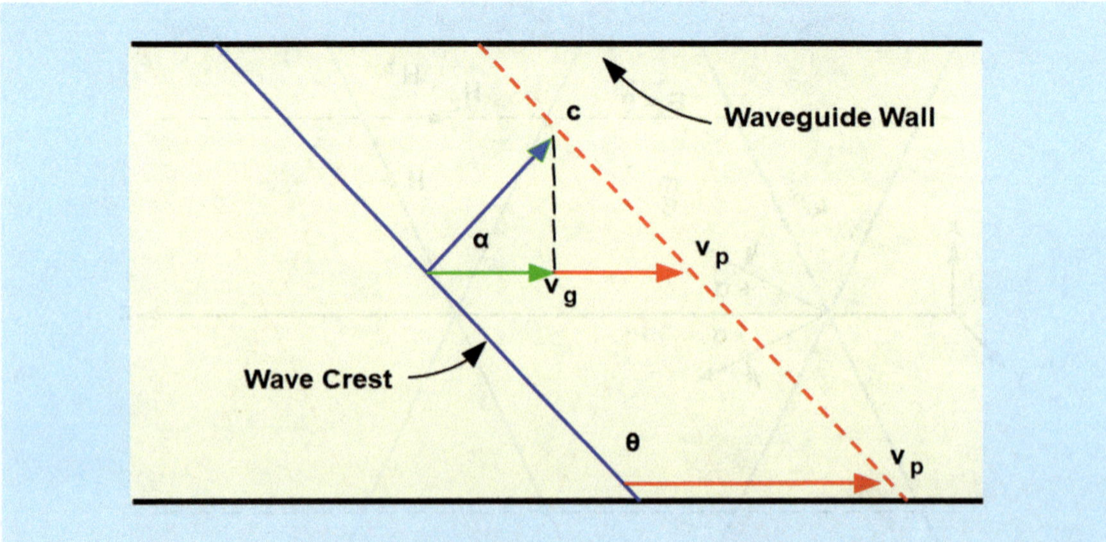

Figure 2-11 Wave velocities in a waveguide.

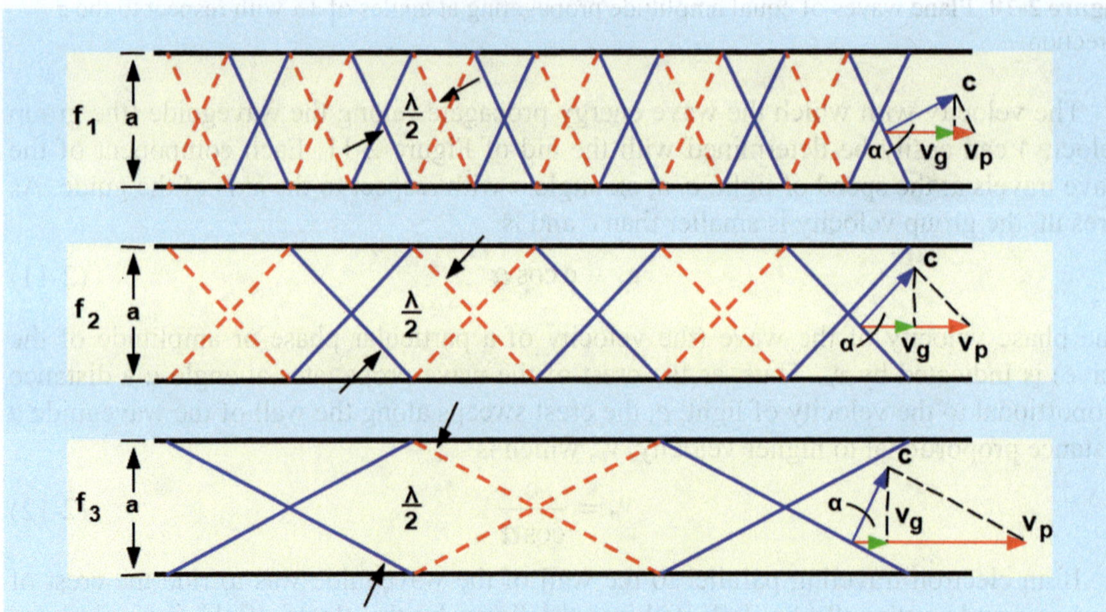

Figure 2-12 Wave configurations for frequencies $f_1 > f_2 > f_3$.

Thus,

$$\left(\frac{\lambda_c}{2}\right)^2 = \left(\frac{\lambda}{2}\right)^2 + \left(\frac{\lambda_c}{2}\cos\alpha\right)^2$$

or

$$1 = \left(\frac{\lambda}{\lambda_c}\right)^2 + \cos^2\alpha . \tag{2-13}$$

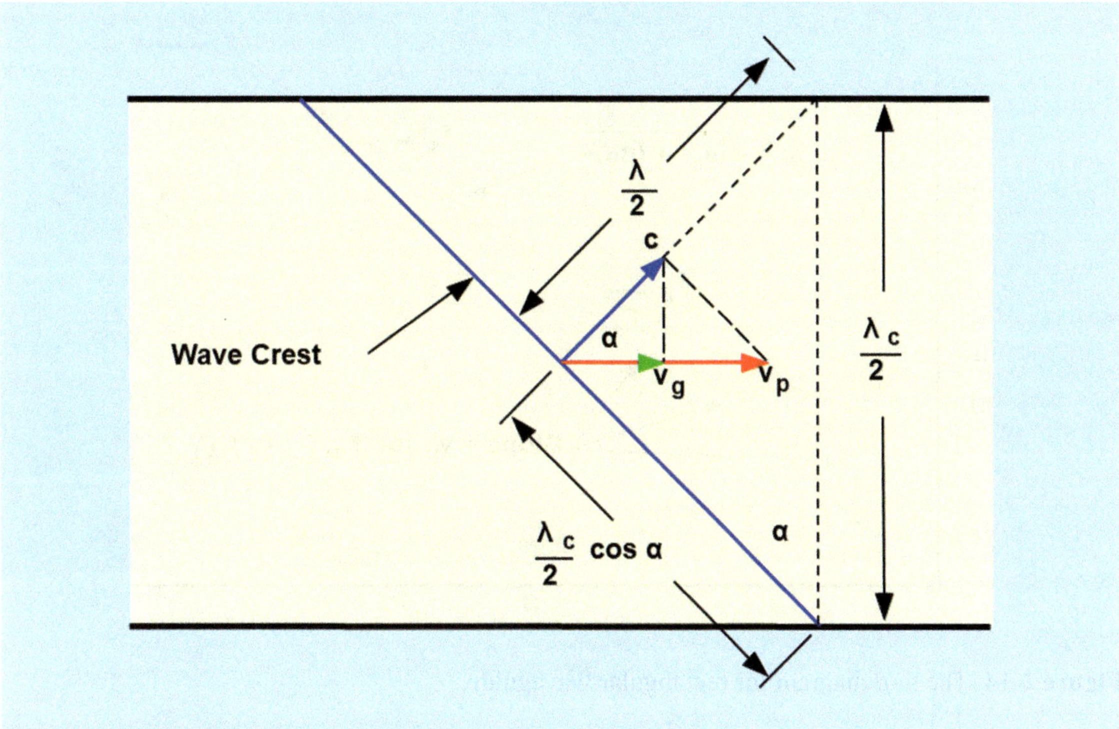

Figure 2-13 Quantities used in derivation of an ω-β relation.

Also,

$$\cos\alpha = \frac{c}{v_p} \tag{2-14}$$

but

$$\frac{\lambda}{\lambda_c} = \frac{c}{f}\frac{f_c}{c} = \frac{\omega_c}{\omega} \tag{2-15}$$

and

$$\frac{c}{v_p} = \frac{\omega}{v_p}\frac{c}{\omega} = \beta\frac{c}{\omega} \tag{2-16}$$

so

$$1 = \left(\frac{\omega_c}{\omega}\right)^2 + \beta\left(\frac{c}{\omega}\right)^2 \tag{2-17}$$

and

$$\omega = \sqrt{\omega_c^2 + (\beta c)^2} \tag{2-18}$$

which is plotted in Figure 2-14. Since $\beta = \omega/v_P$, v_P at any ω and β is simply the slope of a line drawn from the corresponding point on the ω-β curve through the origin. The phase velocities for the frequencies used in Figure 2-12 are indicated in Figure 2-14.

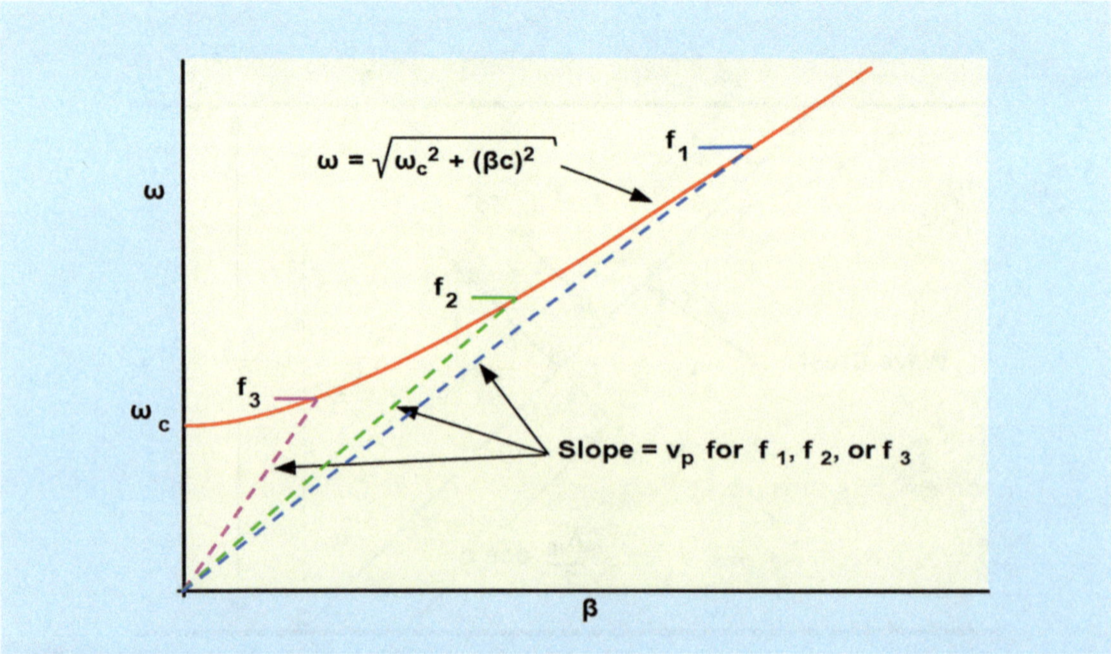

Figure 2-14 The ω-β diagram for rectangular waveguide.

The group velocity of the wave may also be displayed on an ω-β diagram. To determine how this is done, consider Figure 2-15, which shows the propagation constants k and β in vector form and the changes that occur (Δk and $\Delta\beta$) when the frequency is varied by $\Delta\omega$. First, from Figure 2-13,

$$\cos\alpha = \frac{c}{v_p} = \frac{c/\omega}{v_p/\omega} = \frac{1/k}{1/\beta} = \frac{\beta}{k} \tag{2-19}$$

This is why the relations of β, k, and α are as indicated in Figure 2-15. When ω is reduced by $\Delta\omega$, α increases by $\Delta\alpha$, k decreases by Δk, and β decreases by $\Delta\beta$. However, for a small $\Delta\omega$ and a small $\Delta\alpha$,

$$\frac{\Delta k}{\Delta\beta} \approx \cos\alpha \tag{2-20}$$

and, from Figure 2-13,

$$v_g = c\ \cos\alpha \tag{2-21}$$

so

$$v_g = \frac{\Delta\omega}{\Delta\beta} \tag{2-22}$$

Thus, group velocity can be shown on the ω-β diagram as indicated in Figure 2-16. The group velocity is the slope of the ω-β curve. Group velocities for the various wave configurations in Figure 2-8 along with the cutoff condition are also indicated in Figure 2-16.

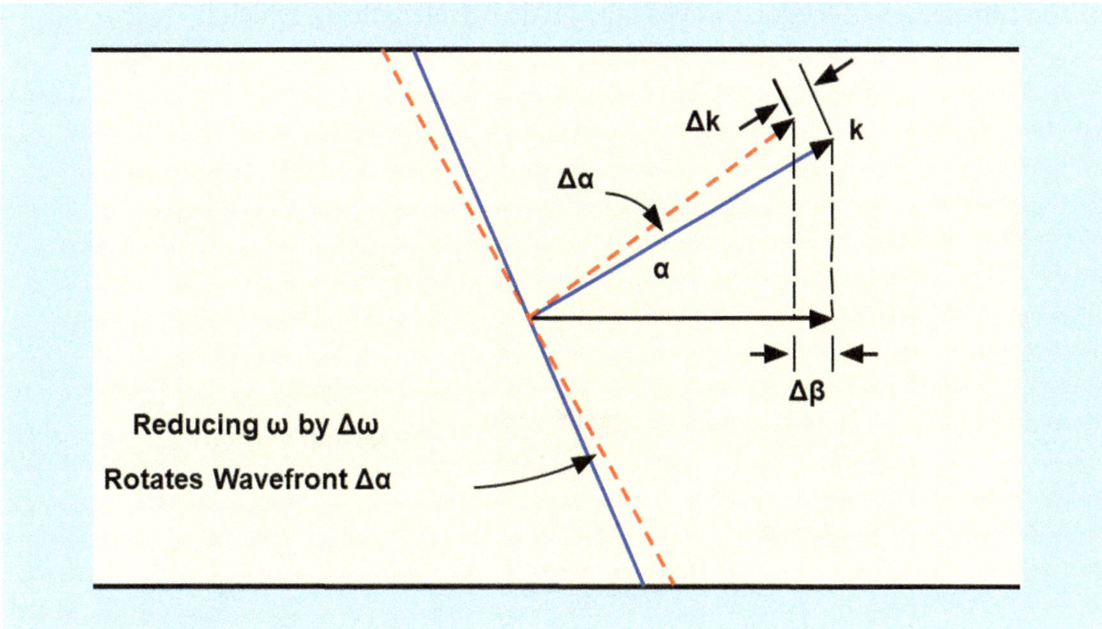

Figure 2-15 Changes in the propagation constant as frequency are varied.

In summary,

1. Phase velocity is the ratio of ω to β;
2. Group velocity is the slope of ω-β curve;
3. For a waveguide, as $\omega \rightarrow \infty$, $\beta \rightarrow k$, $v_p \rightarrow$ c, and $v_g \rightarrow$ c.

The guide wavelength is longer than the free-space wavelength, λ, at all normal operating frequencies. In applications of waveguides, only the guide wavelength is of value. The free-space wavelength was an artifice of use in explaining waveguide behavior but plays no role in the use of waveguide. Thus, in the following discussions on wavelength in the waveguide, it is to be assumed that guide wavelength is the quantity in question even though the term *wavelength* is used.

Finally, in considering the ω-β diagram for a waveguide, it is important to note that waves can propagate with identical characteristics in either direction and so the ω-β diagram is symmetrical about the ω axis as shown in Figure 2-17. At higher frequencies, in either direction, the group and phase velocities approach the velocity of light.

In this section, we have treated the wave in a rectangular waveguide as though it were composed of two waves. As noted previously, there is only one wave, because each is a reflection of the other. The amplitude of the resulting electric field as functions of axial and transverse positions in the waveguide is shown in Figure 2-18. It should be noted that this is the fundamental TE_{01} mode of propagation. TE refers to the configuration of the electric field (E), which is transverse (T) to the axis of the guide. There are no variations of field across the narrow dimension of the guide (thus, the subscript 0) and one-half sinusoidal variation in the broad dimension (subscript 1). At higher frequencies, other modes not of interest at this point in the discussion can propagate.

The wavelength of the total wave is the guide wavelength, λ_g, which is

$$\lambda_g = \frac{v_p}{f} \tag{2-23}$$

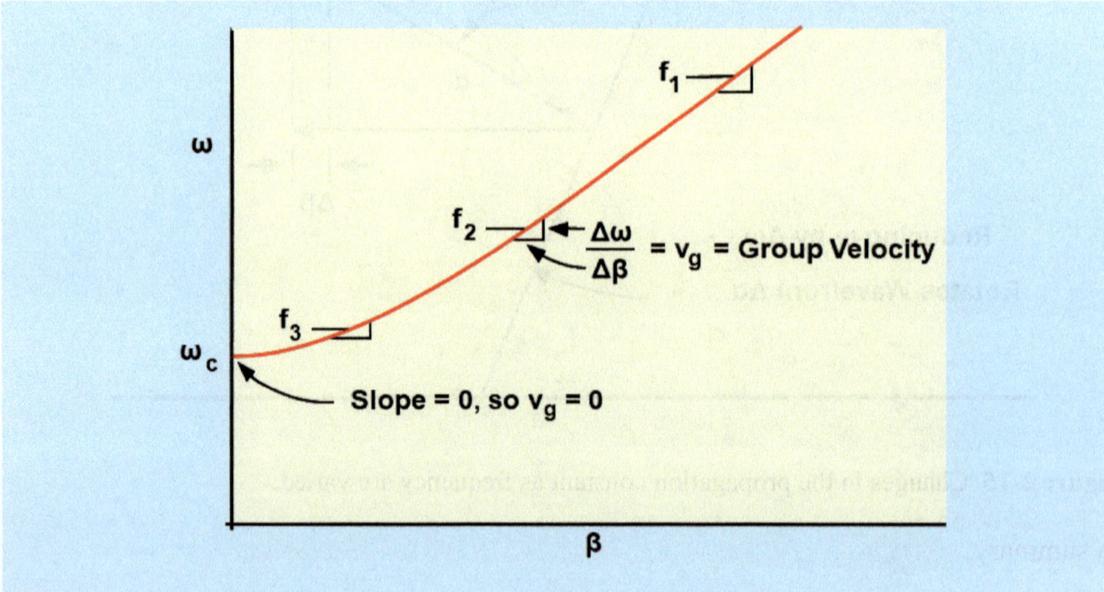

Figure 2-16 Group velocity is determined by the slope of the ω-β curve.

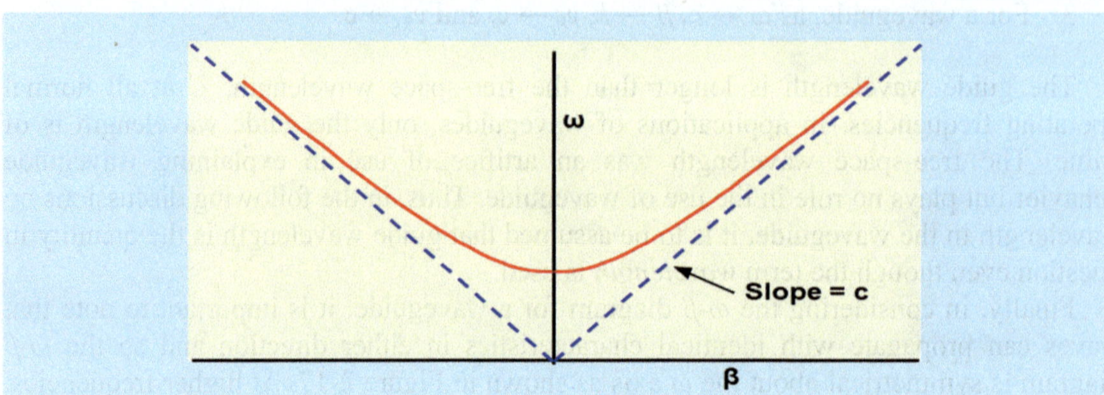

Figure 2-17 The ω-β diagram for waves propagating in either direction in a waveguide.

2.4 MODES

2.4.1 Modes in Rectangular Waveguide

The fundamental mode of propagation in rectangular waveguide has been used in the previous sections for defining group and phase velocity. The configuration of the electric field lines in this mode is shown in Figure 2-18.

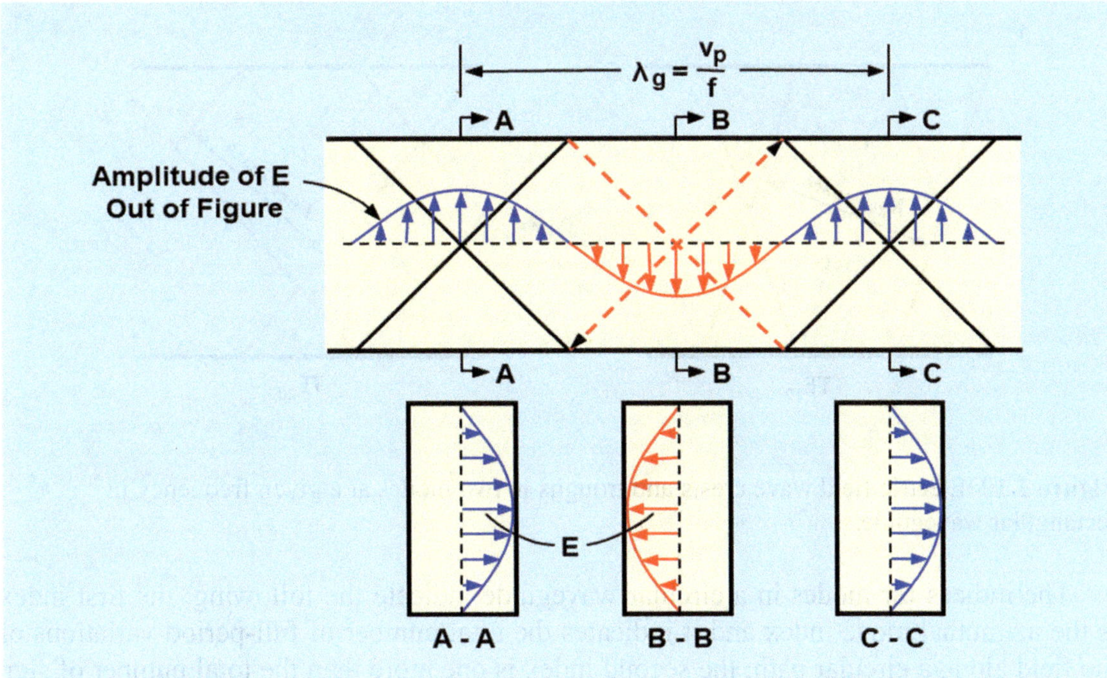

Figure 2-18 Electric field distributions in fundamental TE_{10} mode in a rectangular waveguide.

Up to this point, only dominant modes have been discussed. These are the lowest frequency modes of operation for coaxial lines and waveguides and are the modes usually used in their operation. For the coaxial line, this is the TEM mode; for the rectangular waveguide, it is the TE_{10} mode; and, for the circular waveguide, it is the TEu mode. At higher frequencies, other modes can exist. For example, the electric field wave crests and troughs for two modes at a given frequency are shown in Figure 2.19. One of these is the dominant TE_{10} mode that has already been discussed. The other is also a transverse electric mode, but has two half-wave variations of the electric field in the wide dimension of the waveguide. The cutoff frequency for this mode occurs when the free-space wavelength is the same as the width of the guide.

2.4.2 Modes in a Circular Waveguide

Circular waveguides are used in conjunction with microwave tubes for many reasons. For example, the ceramic disks used as windows through which power is propagated out of a tube are frequently circular and are brazed into a section of a circular waveguide. As another example, in some fast-wave tubes such as the gyrotron, parts of the microwave circuit and the output waveguide are circular waveguides. The dominant mode in a circular waveguide is the TEu mode illustrated in Figure 2.20 [1]. The field configuration corresponds to the dominant TE_{10} mode in a rectangular waveguide.

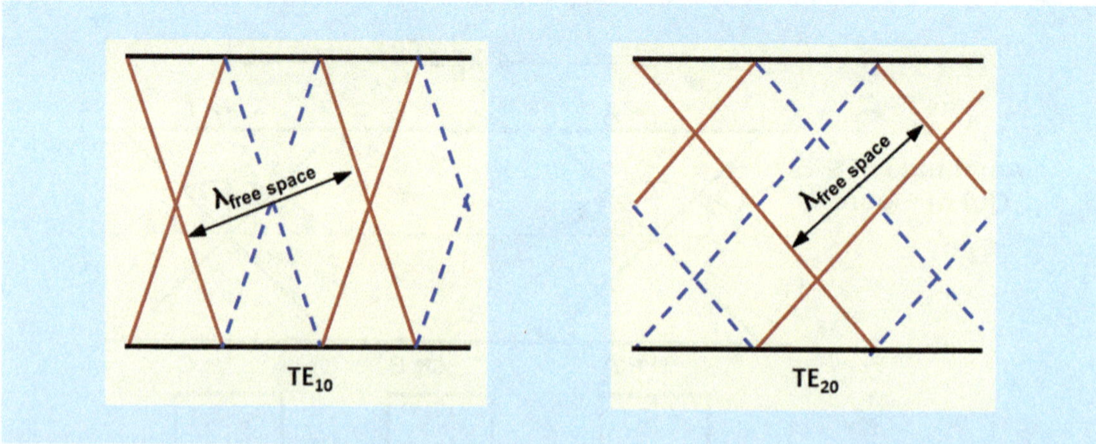

Figure 2.19 Electric field wave crests and troughs in two modes at a given frequency in rectangular waveguide.

The indices for modes in a circular waveguide indicate the following: the first index is the azimuthal mode index and it indicates the total number of full-period variations of the field along a circular path; the second index is one more than the total number of sign reversals along a radial path.

Other than the dominant TEu mode, the TE_{01} mode is of particular interest. The attenuation for this mode decreases indefinitely as frequency increases. As a result, this mode is of importance in millimeter devices where a large waveguide is needed for operation at high power levels and low attenuation is desired.

In coaxial lines, when the frequency becomes high enough so that the circumference halfway between the conductors is approximately a wavelength, the first higher-order mode can propagate. Thus, even though it is desirable to use lines that are as large in diameter as possible to minimize losses, size is limited by the first higher-order mode, for which losses are above those for the dominant mode.

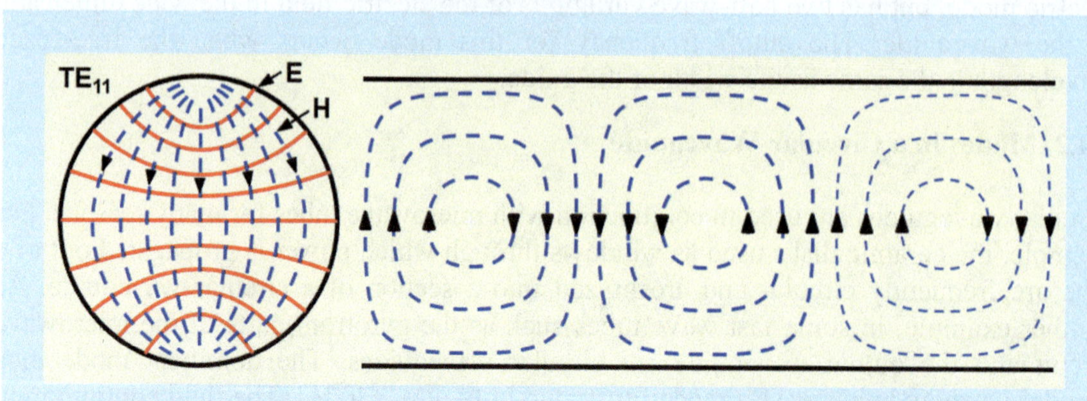

Figure 2.20 Field configurations for dominant TEu mode in circular waveguide.

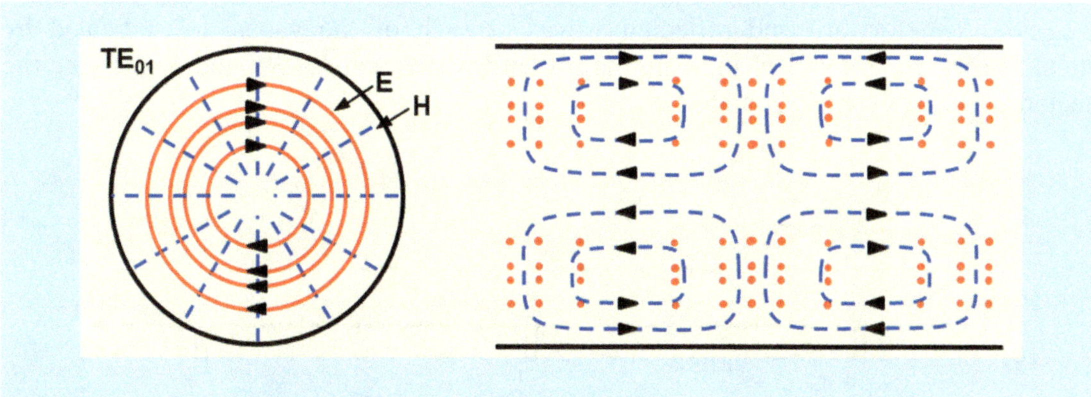

Figure 2.21 TE_{01} mode in circular waveguide.

2.5 PERIODICALLY LOADED WAVEGUIDE

An obstacle, such as a metal vane in a waveguide, produces a change in impedance and is said to load the waveguide. This loading causes part of a propagating wave to be reflected. If equally spaced obstacles are used, the waveguide is said to be periodically loaded. In an infinitely long waveguide, if the reflections from the obstacles add in phase, complete reflection of a wave will eventually occur. Thus, the wave is stopped and, in a frequency range where complete reflection occurs, a stopband is produced.

An example of a periodically loaded waveguide is shown in Figure 2-22. The obstacles are metal vanes attached to the top and bottom of the waveguide. The impedance produced by these vanes is capacitive. The phase shift that occurs in the electric field reflected from each vane is 180°. At frequencies near cutoff, the phase velocity is high and the wavelength is much longer than the spacing between vanes, *L*. Wave reflections from vanes do not add in phase, so there is little effect on wave propagation.

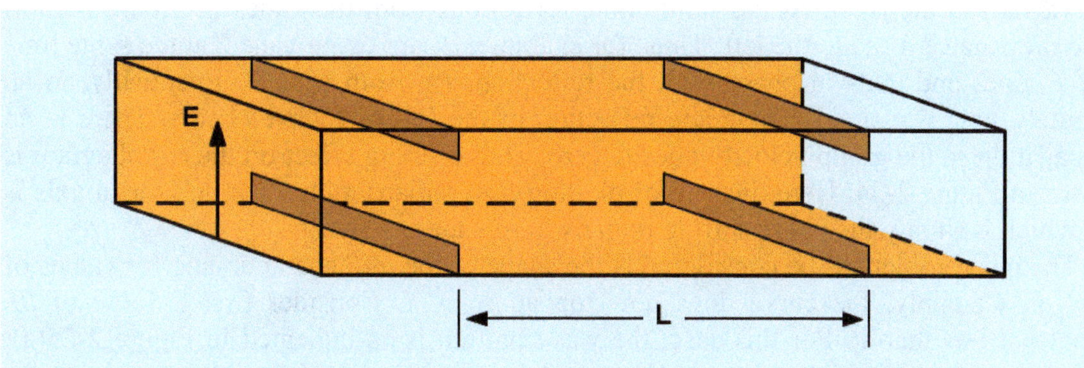

Figure 2-22 Use of vanes to periodically load waveguide.

As frequency is increased, the wavelength decreases and approaches $2L$ so that the phase shift between vanes is 180°. Then reflections from the vanes add in phase, as is illustrated in Figure 2-23.

In Figure 2-23(a), the sinusoid indicates the amplitude of the electric field in the waveguide. The forward and reflected waves at each of the vanes are indicated by vertical arrows. F_a, F_b, and F_c indicate the forward waves and R_a, R_b, and R_c indicate the reflected waves at vanes 1, 2, and 3.

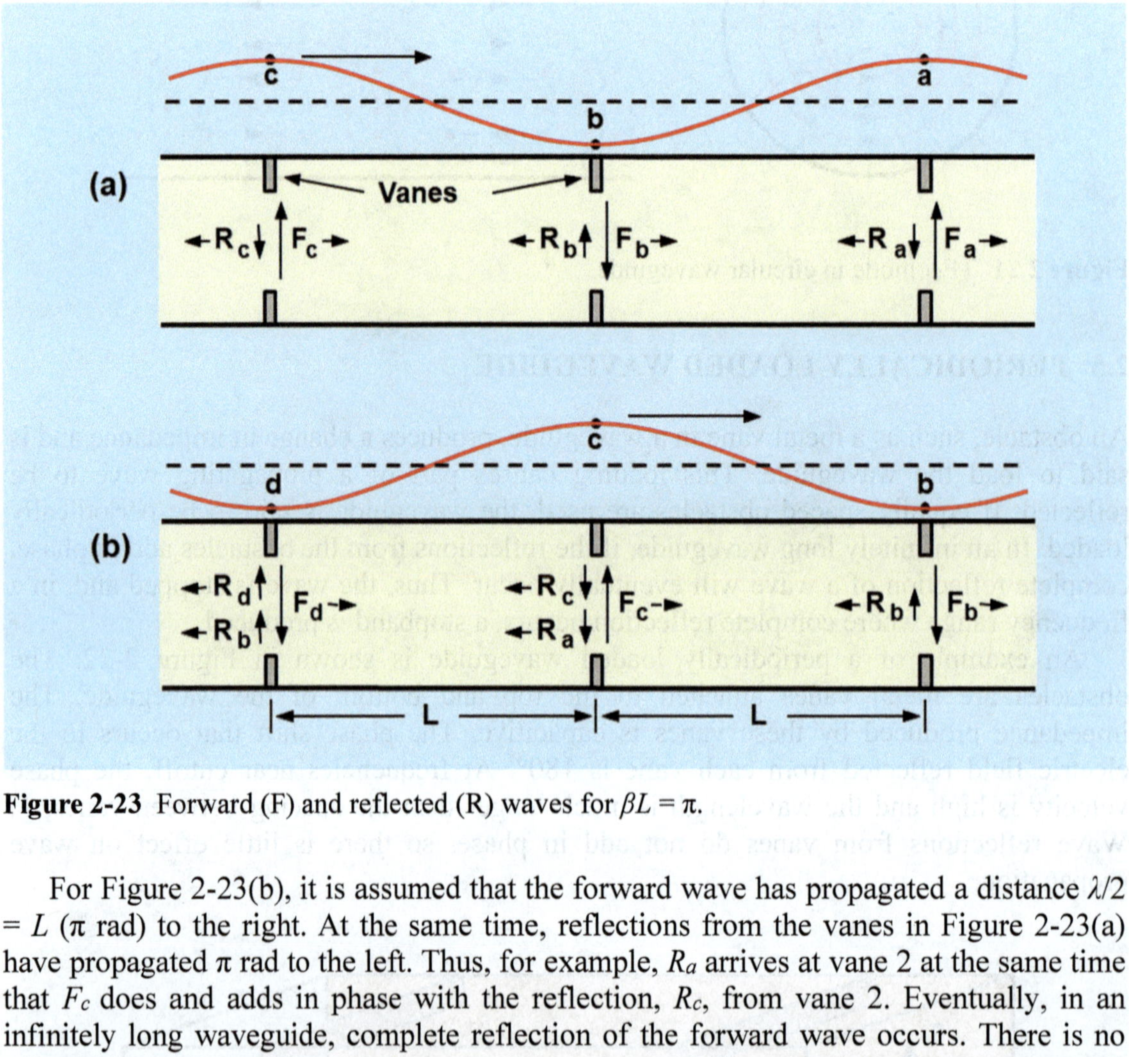

Figure 2-23 Forward (F) and reflected (R) waves for $\beta L = \pi$.

For Figure 2-23(b), it is assumed that the forward wave has propagated a distance $\lambda/2 = L$ (π rad) to the right. At the same time, reflections from the vanes in Figure 2-23(a) have propagated π rad to the left. Thus, for example, R_a arrives at vane 2 at the same time that F_c does and adds in phase with the reflection, R_c, from vane 2. Eventually, in an infinitely long waveguide, complete reflection of the forward wave occurs. There is no propagation so the group velocity goes to zero. The resulting effect on the ω-β diagram is shown in Figure 2-24. Here, instead of plotting β in radians/meter, the horizontal axis is βL, which is simply the phase shift in radians from vane to vane.

The next problem to be considered is the shape of the ω-β characteristic for values of $\beta L > \pi$. Certainly, the curve does not stop at $\beta L = \pi$. Consider first a value of βL somewhat less than π. For this case, the wave pattern is as indicated in Figure 2-25(a). The phase velocity is higher than at $\beta L = \pi$ and the wavelength is longer.

It is useful to consider an electron moving along the axis of the guide. If the electron moves at the phase velocity of the wave, then of course the electron will move from vane to vane in the same time that it takes the wave crest to move from vane to vane. The electron is said to maintain synchronism with the wave.

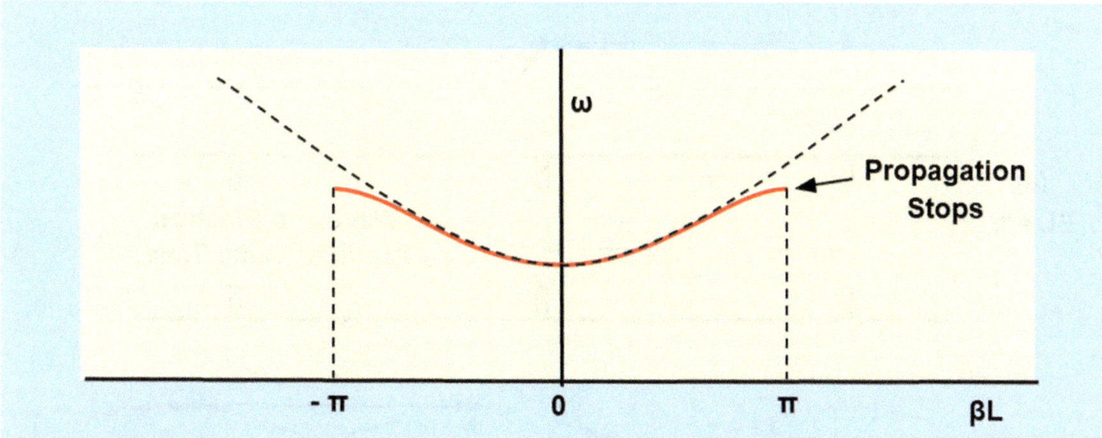

Figure 2-24 Effect of periodically loaded waveguide on an ω-β diagram.

To consider a value of $\beta L > \pi$, assume, as indicated in Figure 2-25(b), that a wave with the same frequency as that for Figure 2-25(a) is traveling in the backward direction. The wavelength and phase velocity (in the backward direction) are the same as those for the forward wave in Figure 2-25(a). An electron moving to the right with the same velocity as in Figure 2-25(a) will arrive at the next vane before the next backward-moving wave crest does. The velocity of the electron must be reduced for the electron to arrive at the next vane when the next backward-moving wave crest does (for the electron to be synchronous with the backward wave). Thus, the phase velocity of the wave as seen by the electron in Figure 2-25(b) is lower than the phase velocity of the forward wave in Figure 2-25(a).

As a result of these considerations, it is seen that as βL increases above π, a backward wave exists and the ω-β curve must turn downward as shown in Figure 2-26. In fact, for values of βL in the range $\pi < \beta L < 2\pi$, the wave is a backward wave and the ω-β characteristic is the same as for $-\pi < \beta L < 0$.

As βL approaches 2π, the backward wave approaches the cutoff condition with wavelength and phase velocity (in the reverse direction) approaching infinity. For an electron velocity (and phase velocity in the forward direction) such that $\beta L = 2\pi$, the electron moves from vane to vane in synchronism with wave crests of the backward wave that are an infinite distance apart but moving with infinite velocity.

Finally, as βL increases to values greater than 2π, a forward wave again exists and the ω-β diagram turns upward again. The shape of the curve from $\beta L = 2\pi$ to $\beta L = 3\pi$ is the same as for 0 □ βL □ π. The part of the ω-β diagram shown in Figure 2-27 is referred to as a *passband* because waves can propagate. This band will be of primary interest when coupled-cavity TWTs are considered.

Before coupled-cavity structures can be considered (in Chapter 15), it should be noted that there are other passbands for periodically loaded waveguides. There are bands of frequencies in which reflections from the vanes do not add in phase and wave propagation takes place. However, whenever the frequency approaches a condition where the distance between vanes is equal to several half wavelengths, a cutoff condition occurs and this places a limit on the upper or lower frequency of the band. As a result, the ω-β diagram contains an infinite set of passbands, two of which are shown in Figure 2-27.

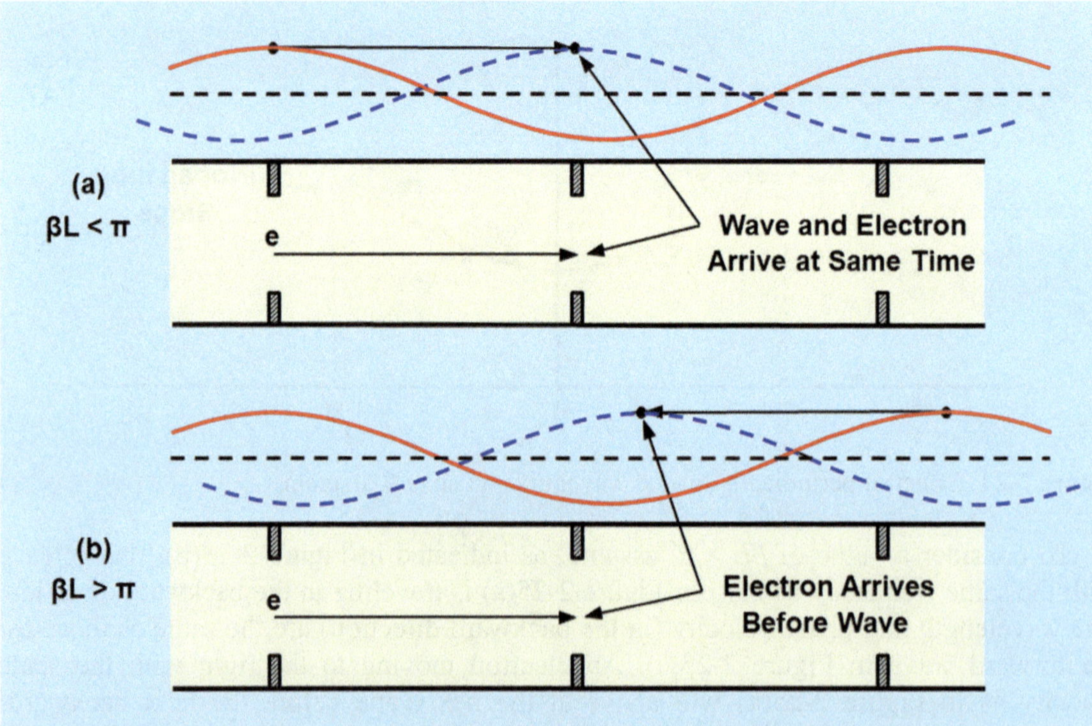

Figure 2-25 (a) Forward and (b) backward waves for βL near π.

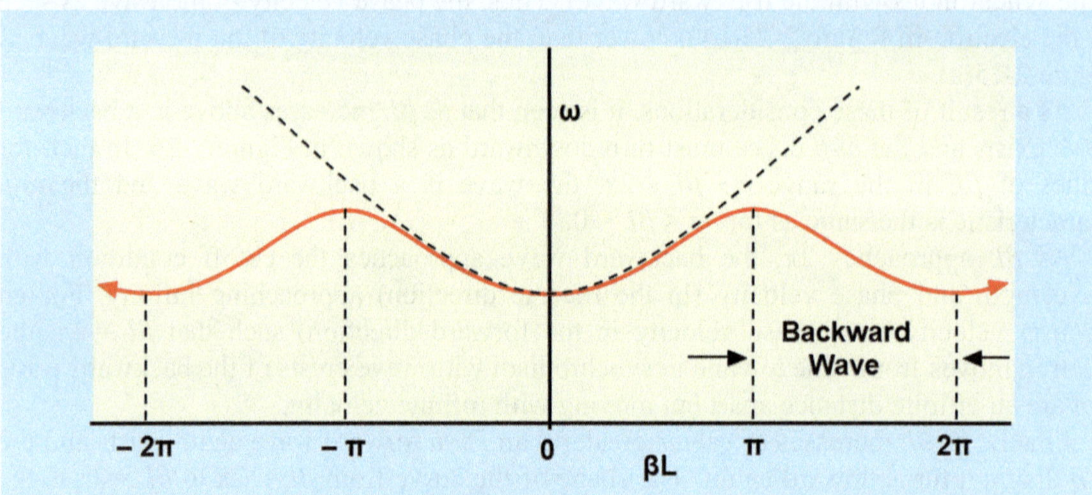

Figure 2-26 Extension of ω-β curve to $\pi < \beta L < 2\pi$.

2.6 EFFECT OF SKIN DEPTH

When an electromagnetic wave impinges on a good conductor, most of the energy is reflected. This is because of the impedance mismatch between the surface of the conductor, which acts essentially as a short circuit, and the air. The impinging wave causes current to flow along the surface of the conductor. The current density is a minimum at the surface and decreases exponentially with depth into the conductor.

The depth at which the current density has dropped to 1/e of the value at the surface is called the skin depth [1]. The skin depth is usually denoted by δ and is given by (2-24).

$$\delta = \frac{1}{\sqrt{\pi f \rho \mu_r}} \tag{2-24}$$

where f is the frequency of the electromagnetic wave in gigahertz, ρ is the resistivity in ohm-cm, and μ_r is the relative permeability (relative to the free-space value) of the conductor.

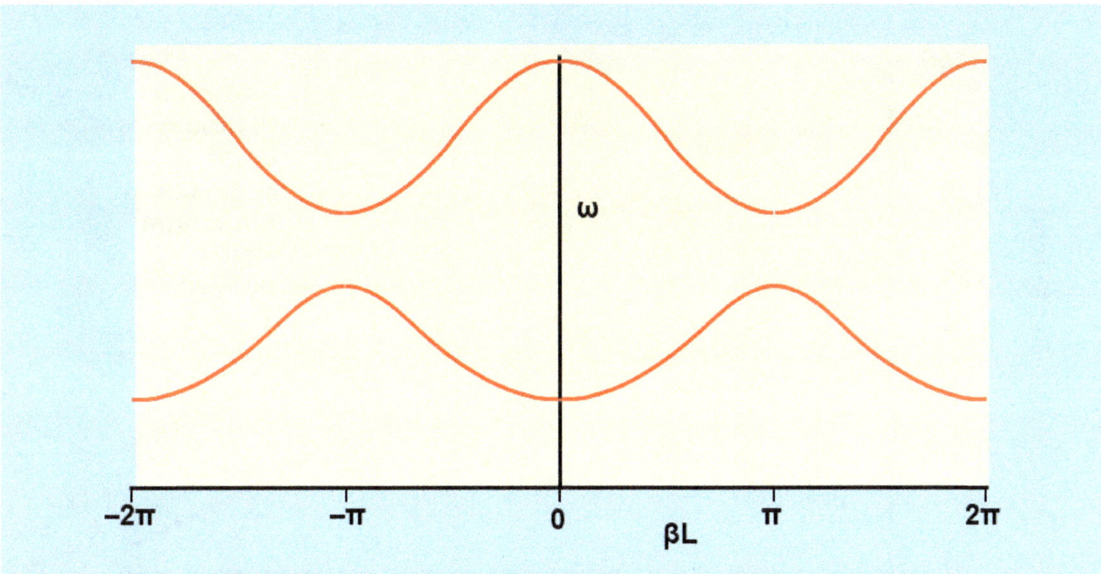

Figure 2-27 The ω-β diagram for a vane-loaded waveguide.

At microwave frequencies, the skin depth is very small as shown in Figure 2.6. Notice, for example, in the case of copper for which ρ is 1.72×10^{-6} Ω-cm and μ_r is unity, the skin depth is only 2.1×10^{-4} cm at 1GHz. This is approximately one-tenth of one-thousandth of an inch (0.1 mil). For magnetic materials, skin depth may be much smaller. For example, in Figure 2-28, skin depth for a material with a μ_r of 5,000 is shown. The resistivity used for this material was ten times the value for copper and typical of the resistivity for a magnetic material.

Caution must be exercised in attempting to determine the skin depth of magnetic materials. In the first place, permeability is strongly dependent on magnetic field level. Thus, not only does the permeability at the surface of the conductor vary with the intensity of the impinging field, it also varies with distance into the metal as the field decays. Second, the hysteresis effects and the hysteresis loss are not included in (2-24).

A small skin depth is beneficial when a conductor is being used to shield a region from electromagnetic waves. However, for conductors used in transmission lines for electromagnetic waves, the small skin depth leads to high transmission-line losses. The losses are the same as if the total current were of uniform density to a depth δ. The resistance of a conductor resulting from the flow of current near the surface is given by

$$R_s = \frac{\rho}{\delta} = \sqrt{\pi f \rho \mu_r} \tag{2-25}$$

This is surface resistance and has units of ohms per unit width per unit length, or simply ohms per square. Thus, resistance increases with frequency and resistivity. The problem of loss resulting from a small skin depth is compounded by the fact that conductor surfaces are often not as smooth as they should be. Irregularities with dimensions comparable to the skin depth are often present because of the machining and processing techniques used. Because the current must flow over these irregularities, the current path is lengthened and so the loss is increased.

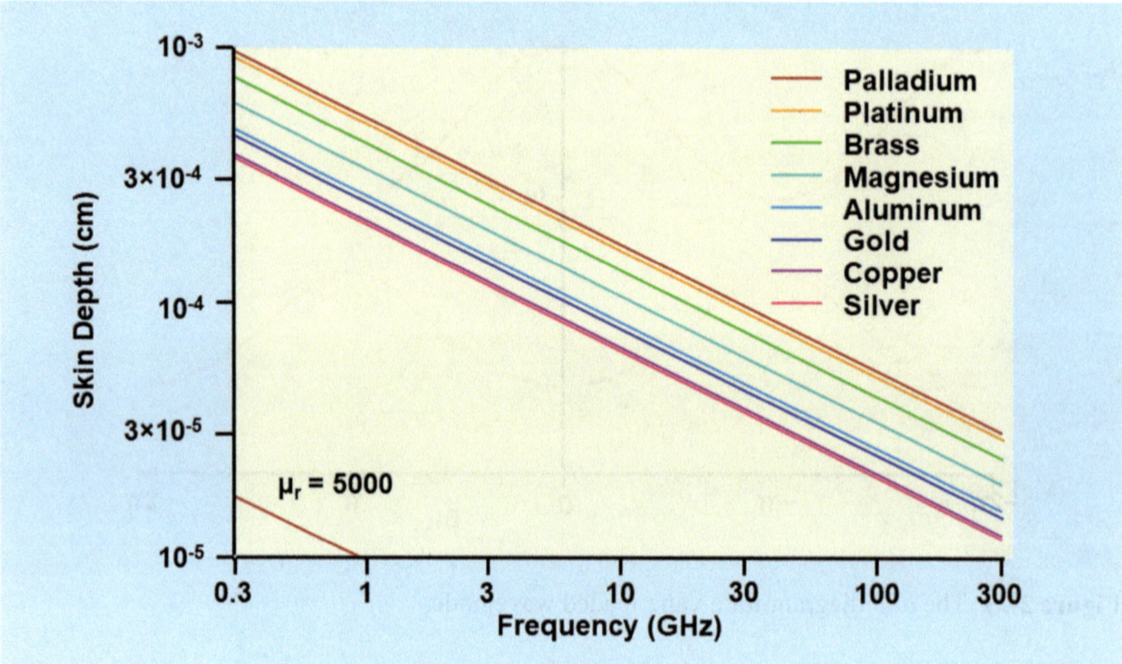

Figure 2-28 Skin depth for several commonly used metals.

REFERENCE

[1] S. Ramo, J. R. Whinery and T. Van Duzer, *Fields and Waves in Communication Electronics,* John Wiley and Sons, New York, 1965.

Chapter 3

Vacuum Requirement

Before discussing the operation of vacuum electron devices in detail, it is important to examine the harsh environment that must exist in tubes if they are to operate properly. First and most important, an ultrahigh vacuum must be maintained for the life of a tube. In this chapter, the reasons for the vacuum requirement are discussed. In Appendix B, the techniques, processes, and materials used to achieve the required vacuum levels are described.

3.1 VACUUM LEVEL

Many years ago, the most commonly used unit of pressure was the millimeter of mercury (mmHg). Atmospheric pressure is 760 mmHg. Now, the accepted units of pressure are the torr and the newton/m^2 (pascal). The torr is used almost universally by the microwave tube industry. The relation between these units is

$$1 \text{ torr} \approx 1 \text{ mmHg} \approx 1.33 \times 10^2 \text{ Pa}$$

Atmospheric pressure at sea level is 760 torr. For many types of microwave tubes to operate properly, the pressure must be ~10^{-8} torr or lower. This is 11 orders of magnitude (one hundred billionth) below atmospheric pressure. Some characteristics of gas at the pressure of 10^{-8} torr are shown in Figure 3-1 and are

- Density ~ $3\text{x}10^8$ molecules/cm^3;
- Monolayer formation time ~ 3 min;
- Mean free path length ~ 3000 m;
- Viscous and molecular flow.

In the following sections, these characteristics and their impact on microwave tube operation are discussed.

3.1.1 Gas Density

At atmospheric pressure, the gas density is about 2.7×10^{19} molecules/cm^3. As shown in Figure 3-1, density drops linearly with decreasing pressure and, at 10^{-8} torr, is about 3×10^8 molecules/cm^3. This density is about 3 orders of magnitude below the electron density in a typical klystron, TWT, or gyrotron. It is low enough to prevent excessive positive ion generation (resulting from electron collisions) during tube operation. Ions can cause many problems including the following:

- Sidebands on power output;
- Amplitude and phase distortions;
- Degraded beam focusing;
- Cathode degradation;
- Attenuator degradation.

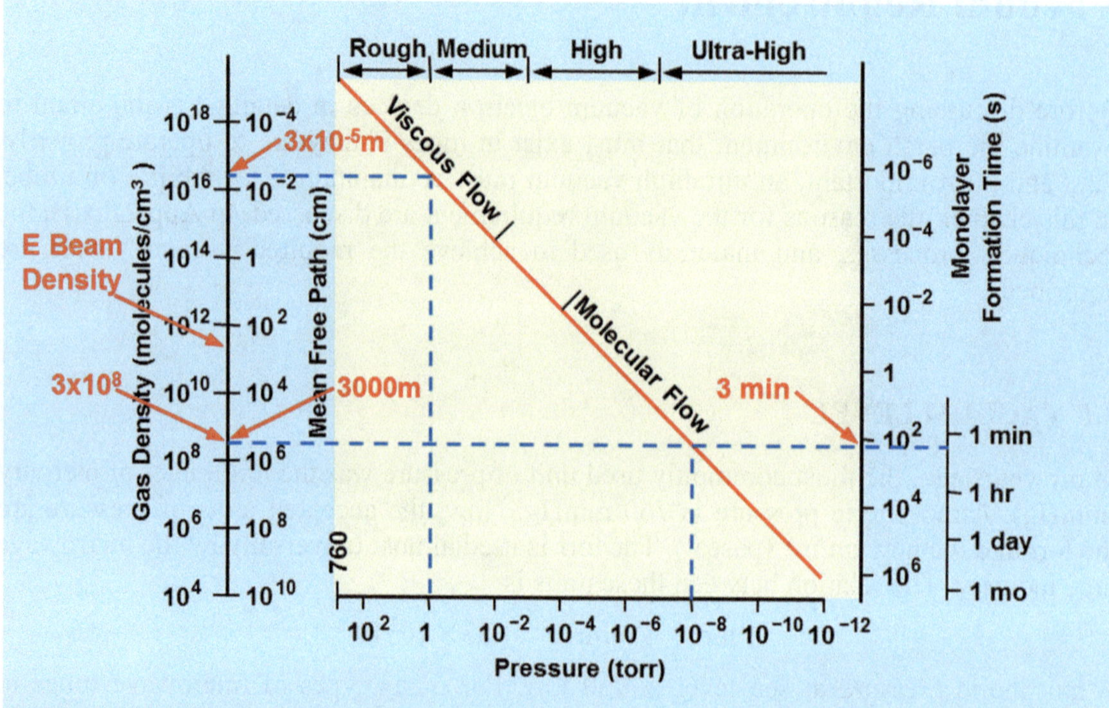

Figure 3-1 Vacuum quantities.

Examples of the deleterious effects of ions on CW coupled-cavity TWTs are shown in Figure 3-2. The TWTs from which these results were obtained were 10 kW, X-band devices that were 10,000 electrical degrees long. It is thought that these 1° phase fluctuations, which are 0.01% of the total phase length, were caused by 0.01% fluctuation in beam velocity, which, in turn, was caused by 0.02% changes in beam voltage. The actual voltage applied to the beam is the power supply voltage minus the beam-generated potential depression. This potential depression is a trap for ions. The charging and discharging of trapped ions causes the phase fluctuation in Figure 3-2.

Several elements of the grid structure in Figure 3-4 were damaged by one or more arcs that may have been initiated by ions or gas bursts. Arcing is a common occurrence in a TWT and so protection circuitry is usually provided to limit the energy into an arc to a few joules. From the extensive damage that this cathode and grid assembly incurred, it is clear that the TWT, in which this assembly was used was not properly protected. The energy required to melt the grid elements must have been well over 100 J.

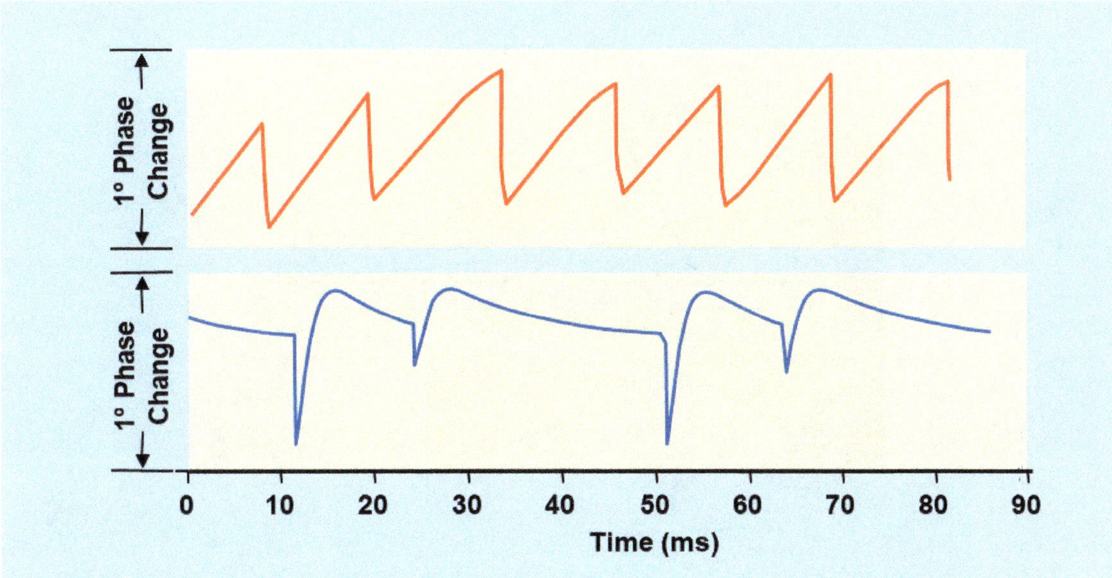

Figure 3-2 Phase fluctuations induced by ions in CW, confined-flow, coupled-cavity TWTs. (From: A. S. Gilmour, Jr., et al., *IEEE Trans. Plasma Science*, June 2004. © 2004 IEEE.)

Another illustration of effects of ions is shown in Figure 3-3. Ions formed in the body of the tube are focused to a spot on the cathode. The ion burn in Figure 3-4 was caused by an ion beam that was generated by electron impact with gas molecules in the tube.

When the ions strike the cathode, they sputter material from the cathode surface. The discoloration of the grid elements surrounding the ion burn in Figure 3-4 probably resulted from the deposition of sputtered material. The fact that the ion burn is not centered in the grid indicates that the gun of this tube was not precisely aligned when the tube was built.

Several elements of the grid structure in Figure 3-4 were also damaged by one or more arcs that may have been initiated by ions or gas bursts. Arcing is a common occurrence in a TWT, and so protection circuitry is usually provided to limit the energy into an arc to a few joules. From the extensive damage that this cathode and grid assembly incurred, it is clear that the TWT in which this assembly was used was not properly protected. The energy required to melt the grid elements must have been well over 100 J.

Appendix D contains additional information on ion effects in electron beams.

3.1.2 Monolayer Formation Time

The time required to form a monolayer is shown in Figure 3-1. At a pressure of 10^{-8} torr, a monolayer is formed in about 3 minutes, so all surfaces in a TWT are covered with one or more monolayers of gas.

This gas can cause problems if the electron beam is suddenly deflected and strikes a gas-covered surface. This could happen because the magnetic field used to focus the electron beam was distorted (by a magnetic tool, for example) or a transient in the power supply voltage for the tube occurred. The removal of a monolayer of gas from 1 cm^2 into a volume of 1 cm^3 will cause the pressure to rise from 10^{-8} torr to 10^{-4} to 10^{-3} torr. At that pressure, it is likely that electrical breakdown will occur.

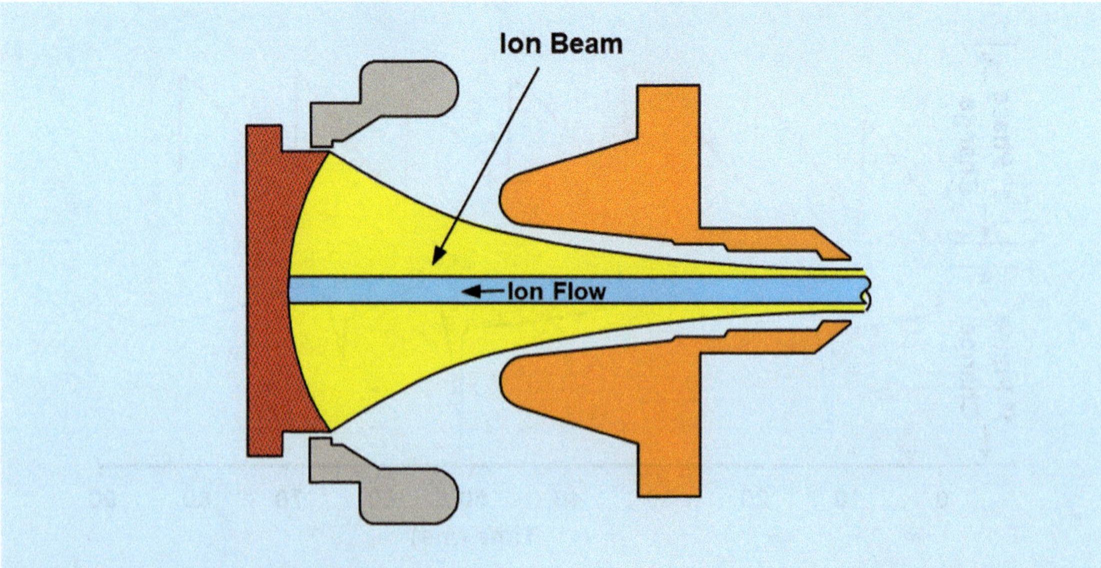

Figure 3-3 Ion flow into an electron gun.

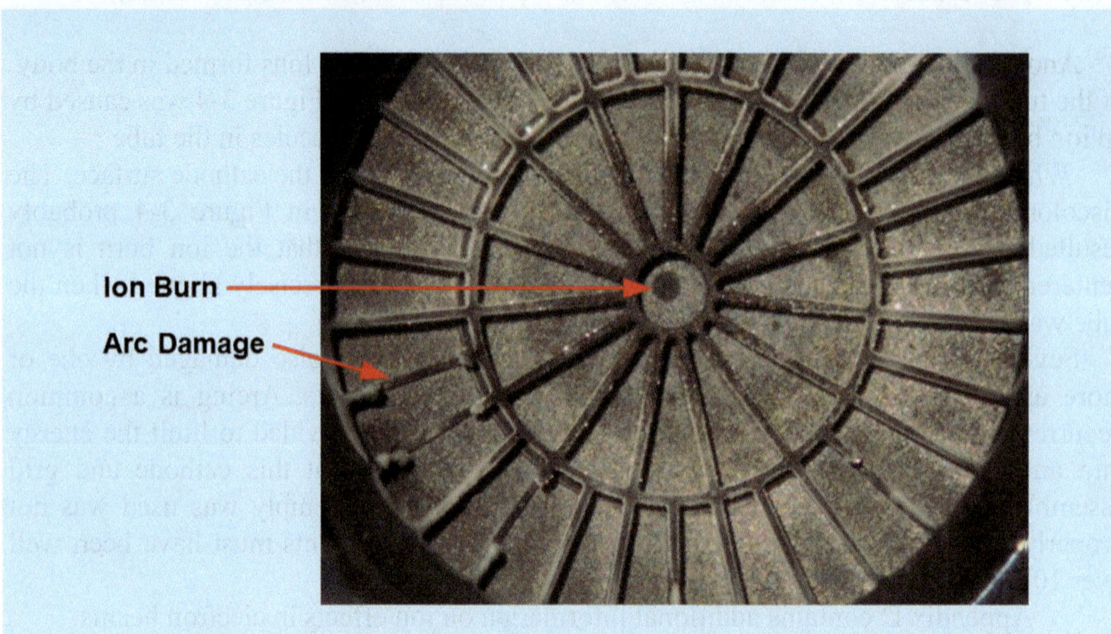

Figure 3-4 Ion and arc damage to a cathode and grid assembly.

3.1.3 Viscous and Molecular Flow

The effects of viscous and molecular flow in vacuum system operation are illustrated in Figure 3-5. In the viscous flow region where the mean free path length is less than a fraction of a centimeter, molecules collide frequently with other molecules. Molecules gradually migrate in the direction of the pump because of the reduced pressure, ΔP.

With molecular flow, the mean free path length is large enough that molecules rarely collide with other molecules. Instead, molecules collide only with the walls of the chamber and only occasionally are directed toward the pump. As a result, the pump simply accepts and traps molecules that happen to enter it. A large pump orifice is required if a high pump speed is to be achieved. For example, if 1-inch (2.54-cm) tubulation was used to connect the vacuum system to the pump, the speed would be limited to a few liters per second. If a 6-inch (15.24-cm) diameter orifice was used, the speed could be well over 1,000 liters per second.

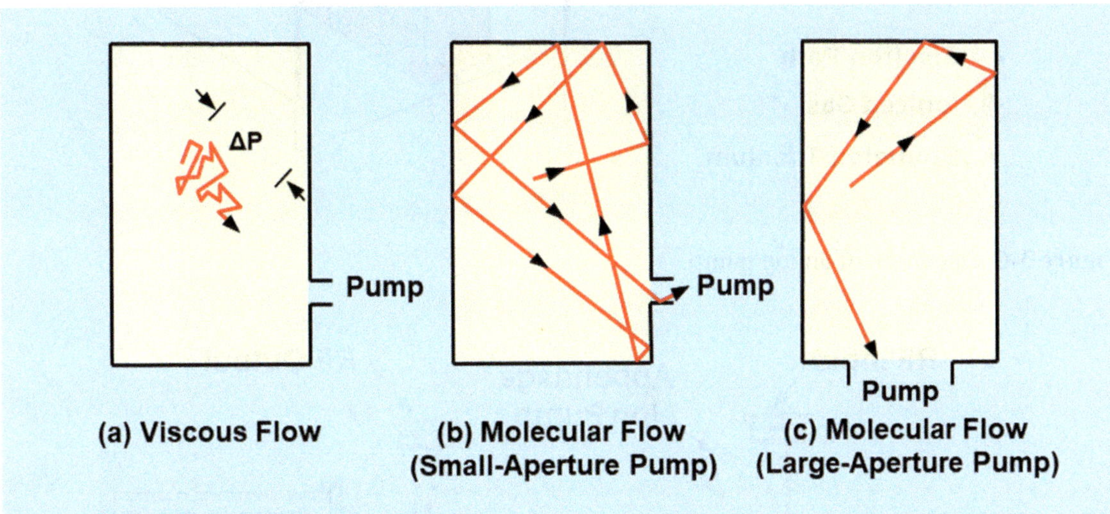

Figure 3-5 Viscous and molecular flow.

Thermionic cathodes operate at ~ 1000°C, and gas is continuously released from electron gun parts. An appendage ion pump or a getter is often used to help maintain the required vacuum level in the tube for the duration of its life.

The operation of an ion pump is indicated with the drawing in Figure 3-6. Depending on the pumping capacity required, the pump has one or more cells in an anode that looks like an egg crate. The negative electrodes are made of titanium. A magnetic field causes free electrons to spin in the anode cells and ionize gas molecules. The positive ions that are formed are attracted to the negative electrodes where they sputter the titanium and are trapped. The sputtered titanium deposits on the anode and combines with gas molecules that come in contact with it. An example of a TWT with a permanently attached appendage ion pump is shown in Figure 3-7.

One example of an effect of molecular flow is the positioning of the appendage pump. Because the pressure in the tube is in the high to ultra-high range, gas flow is in the molecular range, gas conductance is low, and pumping speed is severely limited if the gas flow path is long. As a result, the pump should be located as close to the gas source (the electron gun) as possible. However, the pumps require magnets and, because of the stray magnetic field from the magnet and its possible impact on beam focusing, the pumps are often located far from the cathode as shown in Figure 3-7. As a result, pumping speed is severely limited and the pressure in the gas source (the gun) may be one to two orders of magnitude above the pressure in the pump.

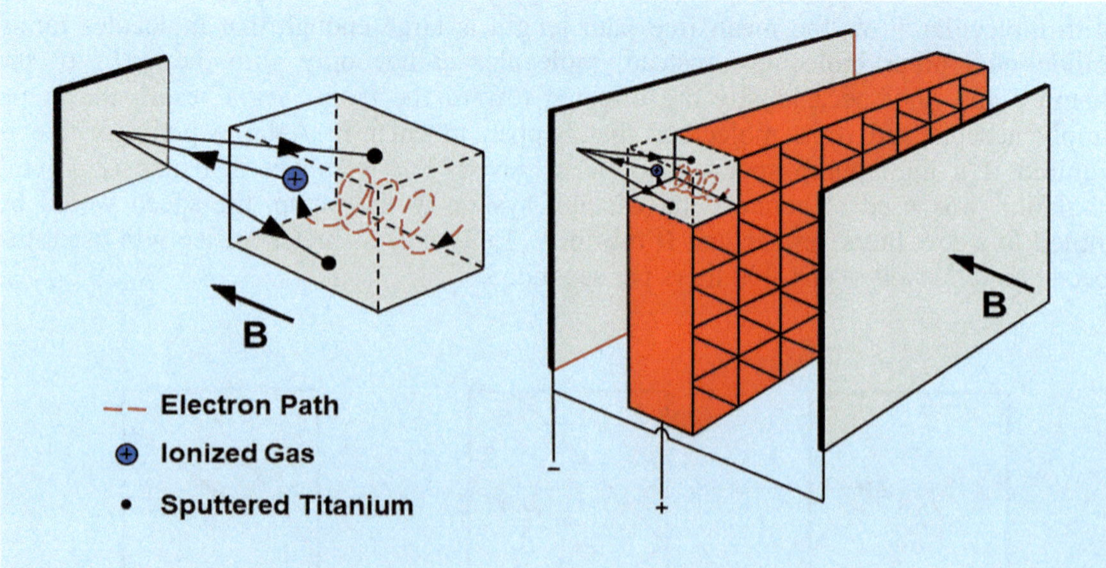

Figure 3-6 Operation of an ion pump.

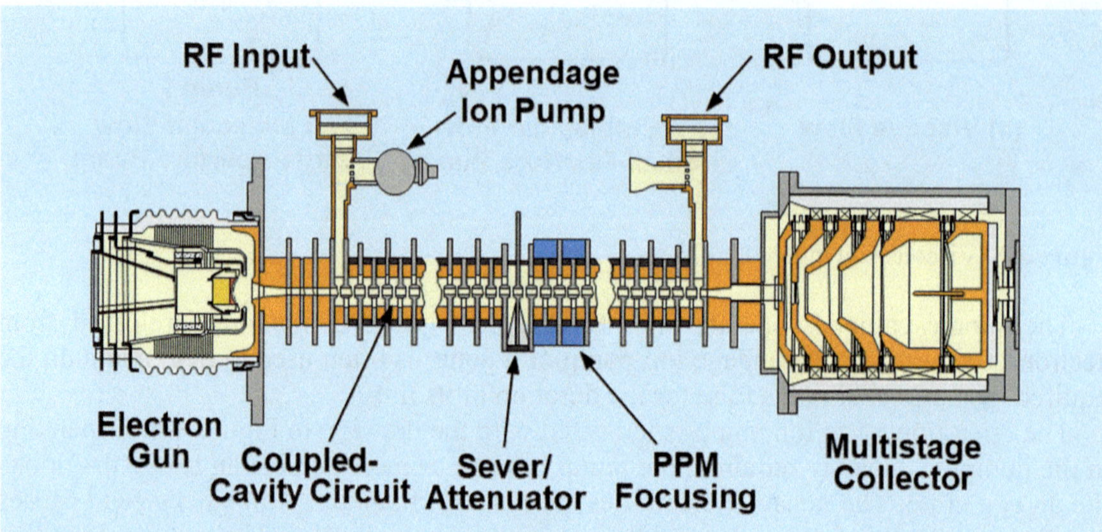

Figure 3-7 Coupled-cavity TWT with an appendage pump.

An alternative to the appendage ion pump is a getter. A getter placed in the gun near the gas source can be far more effective than an appendage pump in removing gas. The phenomenon of gettering has long been used in vacuum tubes. In conventional getters, a chemically active metal is evaporated to form a film on the inner surface of the vacuum envelope. Gases that react with the metal film are said to be gettered. Conventional getters have not found application in high-power microwave tubes primarily because the evaporated film may cause electrical leakage or breakdown in high-voltage structures.

Nonevaporable getters are now commonly used in microwave tubes. Getters available from SAES Group are particularly attractive for use. Some of these getters utilize a mixture of titanium and molybdenum powders in a weight ratio of 9:1. The material is lightly pressed and sintered to form a structure that is porous and has good mechanical stability. It is possible to use these getters in containers as is indicated in Figure 3-8.

Some of the microwave tube experience with nonevaporable getters installed in an electron gun is as follows:

1. Appendage ion pump current, which might sometimes be high, is virtually zero;
2. Arcing in the electron gun is greatly reduced;
3. After extended periods of storage, voltage can be applied immediately to a tube without gassing or arcing;
4. The getters are recoverable. After activation and then exposure to air, getters can be reactivated.

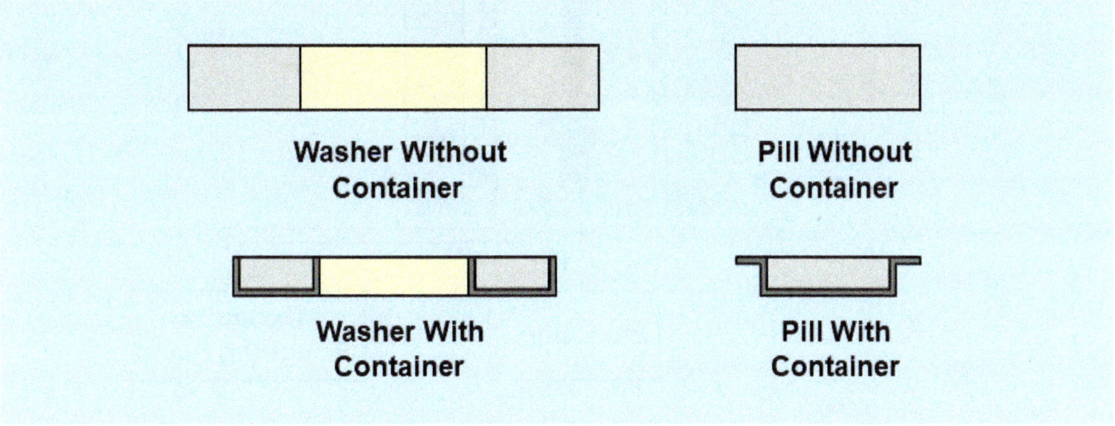

Figure 3-8 Getters without and with nichrome containers. (Courtesy of SAES Group, Inc.)

3.1.4 Bakeout

Evacuation is facilitated by heating (baking) a tube at a high temperature for an extended period. The bakeout temperature is ~500°C–525°C. At this temperature, the tube glows dull red. The bakeout time depends on the size of the tube and can be as much as a week for very large tubes.. Oxidation of the tube envelope is prevented by providing a partial vacuum outside of the tube.

3.1.5 Materials Selection

To withstand the bakeout process and to obtain and maintain an ultrahigh vacuum, a tube must be fabricated using only low vapor pressure materials. Vapor pressures of selected materials are in Figure 3-10.

Some of the metals that have a low enough vapor pressure to be used in a microwave tubes are those to the right of the dashed bakeout line. These are molybdenum (Mo), osmium (Os), tantalum (Ta), tungsten (W), copper (Cu), gold (Au), and iron (Fe).

The accidental use of the wrong material, for example brass, which contains zinc (Zn), can lead to disastrous results. At bakeout (>500°C), zinc vapor would diffuse throughout the TWT. The use of cadmium (Cd) plated screws would be even worse since its vapor pressure is an order of magnitude larger than that of zinc.

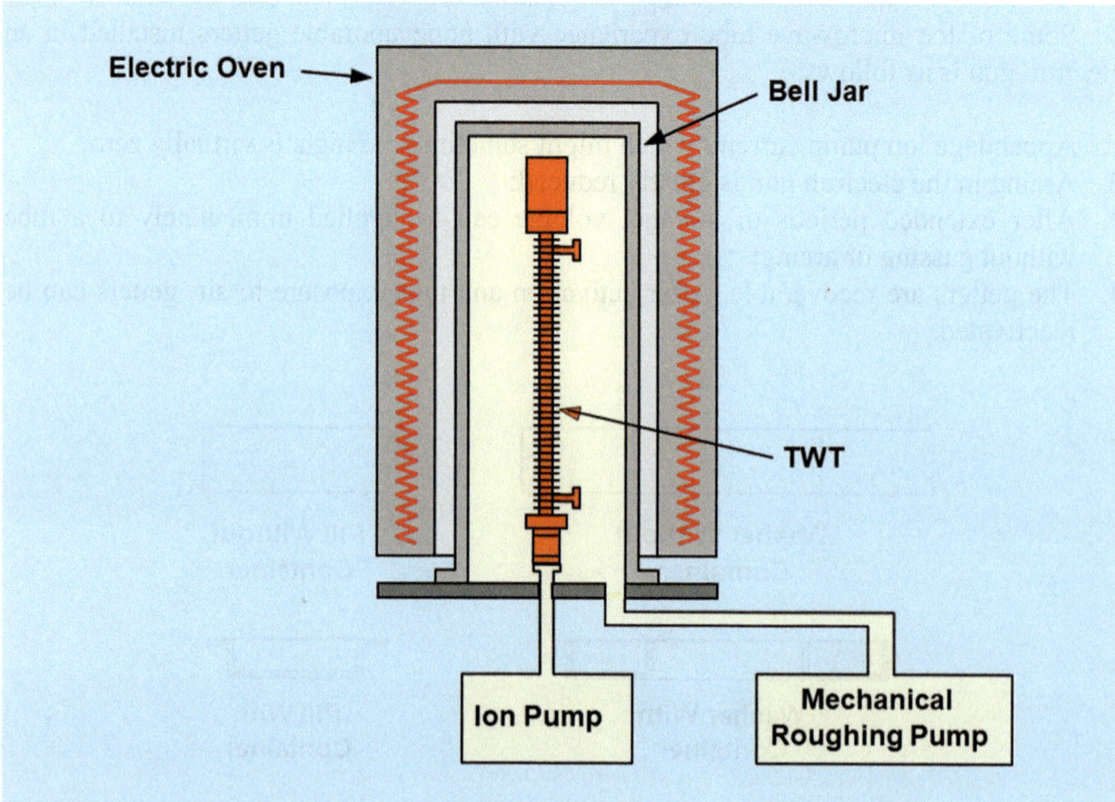

Figure 3-9 Equipment for baking out a TWT.

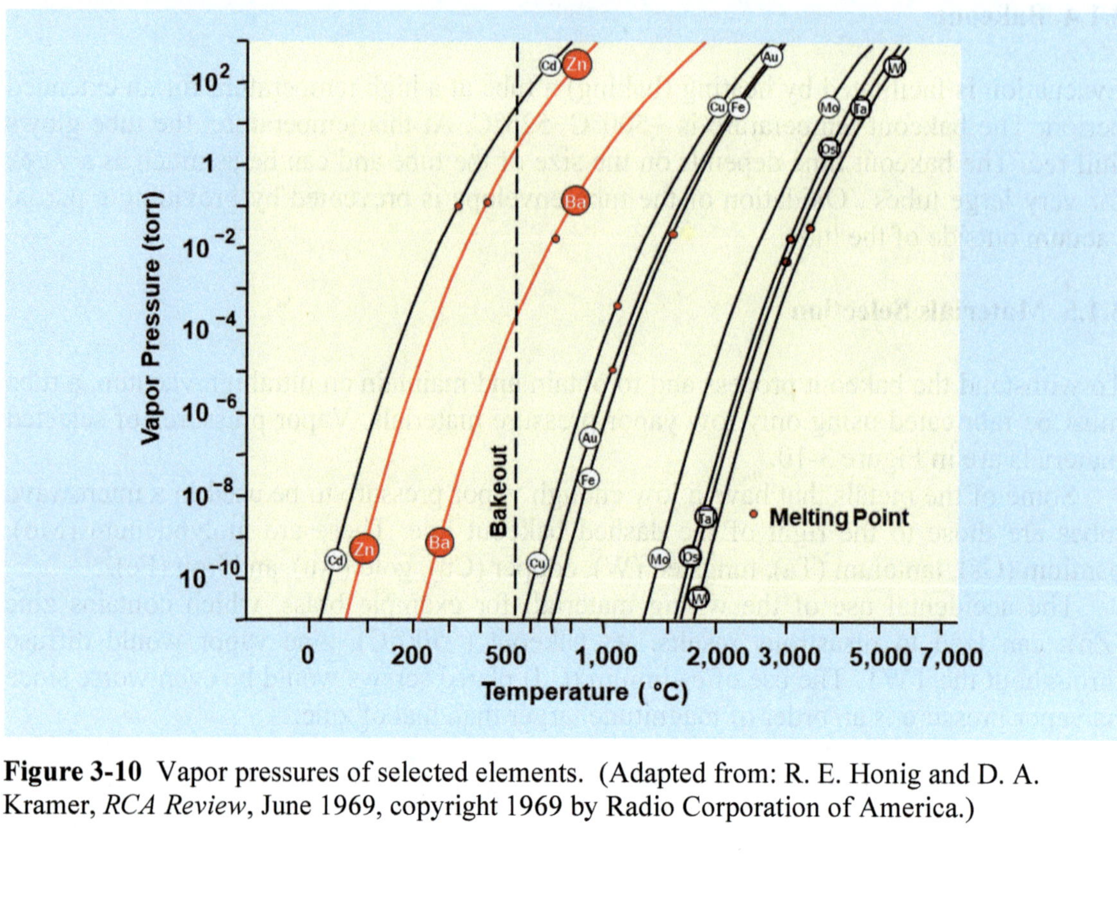

Figure 3-10 Vapor pressures of selected elements. (Adapted from: R. E. Honig and D. A. Kramer, *RCA Review*, June 1969, copyright 1969 by Radio Corporation of America.)

Barium (BA) is highlighted in Figure 3-10 because it is a critical element used in most dispenser cathodes. At a cathode operating temperature of 1,000°C, the vapor pressure is ~ 1 torr, which is high enough to cause severe contamination. Fortunately, mechanisms occurring at the cathode surface provide a force that attracts barium to the surface and reduces the rate of evaporation by orders of magnitude.

Chapter 4

Thermionic Cathodes

4.1 THERMIONIC EMISSION

The cathode is the source of electrons for the electron beam in every microwave tube. The current density of electron emission from the cathode ranges from milliamperes to tens of amperes per square centimeter of cathode area [1].

Emission mechanisms that are in use in conventional microwave tubes are:

1. Thermionic emission is used in linear beam tubes;
2. Secondary emission and thermionic emission are both used in crossed-field tubes;
3. Field emission has long been of interest but very little success has been reported.

Thermionic cathodes are discussed in this chapter. Secondary emission cathodes and field emission cathodes are described in Chapter 5.

J. R. Pierce [2] listed the primary characteristics that an ideal cathode should have:

1. Emits electrons freely, without any form of persuasion such as heating or bombardment (electrons would leak off from it into vacuum as easily as they pass from one metal to another);
2. Emits copiously, supplying an unlimited current density;
3. Lasts forever, its electron emission continuing unimpaired as long as it is needed;
4. Emits electrons uniformly, traveling at practically zero velocity.

Of course, real cathodes have none of these ideal characteristics. For example, real cathodes must be heated to a temperature near 1,000°C to produce appreciable electron emission. At that temperature, the current density is not unlimited, but instead is on the order of a few tens of amperes per square centimeter at most. Because of the necessity for a high operating temperature, some of the key constituents of a cathode evaporate, which leads to their depletion and eventually to the end of the life of the cathode.

The reason for the fourth characteristic of an ideal cathode is not as obvious as the other three. First, it should be noted that, in real cathodes, there are random microscopic fluctuations in the rate at which electrons are emitted. Second, electrons are emitted with finite random velocities in random directions relative to the cathode surface. There are at least two important reasons why an ideal cathode should emit electrons uniformly, traveling at practically zero velocity:

1. Fluctuations in electron emission and variations in electron emission velocities lead to noise currents in the electron beam, which produce noise on the output from the tube.
2. Variations in emission velocities and in the directions with which electrons leave the cathode surface lead to problems in focusing the electrons into a well-defined beam. This problem is particularly severe in very small, high-density beams.

While real cathodes have none of the characteristics that Pierce's ideal cathode would have, there has been enormous progress in cathode capabilities since Pierce listed his ideal characteristics. This progress is indicated in Figure 4-1. Note, in particular, the large increase in emission capability that was achieved in the 1970s and 1980s. As pointed out by Thomas et al., this progress was due largely to the advent of modern surface analytical capabilities that made possible a better understanding of the physics and chemistry of emitter surfaces [3]. In this chapter, we attempt to provide sufficient background material so that, at least qualitatively, the operation of modern cathodes, with the exception of the scandate cathode, can be understood. The operation of the scandate cathode is the subject of active research at the present time.

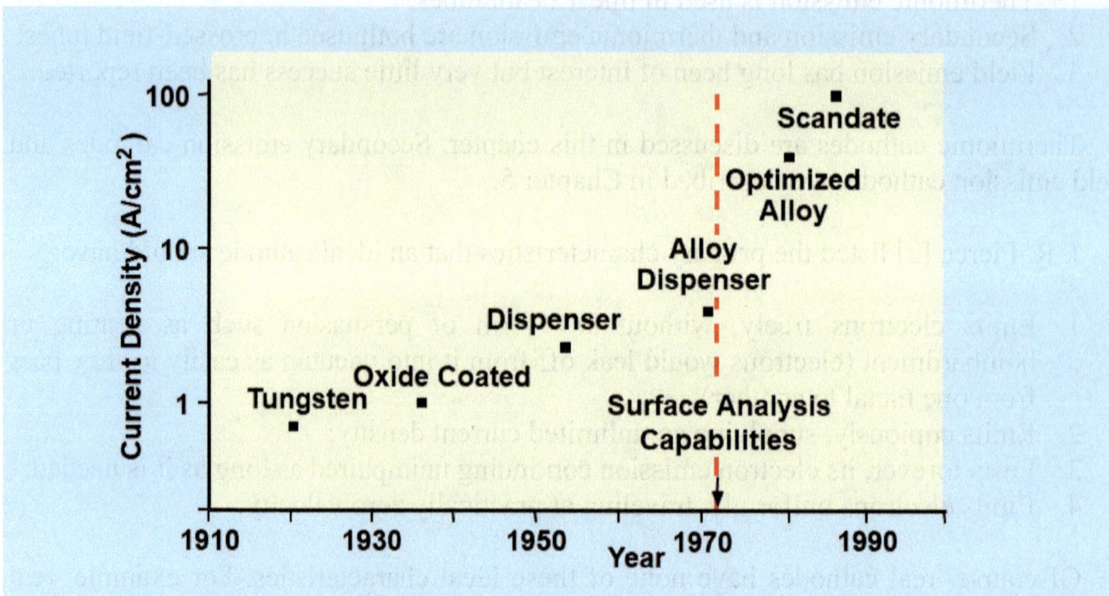

Figure 4-1 Historical perspective of thermionic cathode emission capabilities. (From: R. E. Thomas et al., *IEEE Trans. Electron Devices*, March 1990. © 1990 IEEE.)

The chapter starts with the basic theory of thermionic emission. This theory is straightforward and gives a reasonably accurate prediction of the maximum emission that can be expected from a cathode at a given temperature, in the absence of field effects, if the work function is known. Next, the enhancement of thermionic emission that results from the application of an electric field to the surface of the cathode (the Schottky effect) is discussed. The emission regulating effect of the electron cloud adjacent to the surface of the cathode (space charge limited emission) is of extreme importance in the use of thermionic cathodes in linear-beam tubes. For the person who must use a cathode in a linear-beam tube, the concept of space charge limited emission may be the most important material discussed in this chapter.

To maintain space charge limited emission, the cathode work function must remain low enough so that adequate emission is obtained at a given temperature during the life of a cathode. The work function of the cathode is critically dependent on the materials and condition of the surface of the cathode. Thus, a large portion of this chapter is devoted to a discussion of cathode materials and their impact on life of the cathode and the tube in which the cathode is used.

The physical arrangement of the cathode and the elements that provide the power necessary to heat it to the required temperature in the time permitted by the application are described. Conventional heaters are discussed along with fast warm-up assemblies that are required in some special applications of tubes.

The magnetic field of the filament can have a significant impact on the operation of an electron gun. Electron trajectories near the cathode surface are perturbed by this field with the result that beam generation and focusing are degraded. If an ac filament current is used, the beam shape may be modulated at the ac frequency and this, in turn, modulates the power output of the tube. These effects are discussed at the end of the chapter.

4.1 Thermionic Emission

At temperatures above absolute zero, some electrons have sufficient energy to escape from a cathode surface. As temperature is increased, the number of electrons with sufficient energy to escape increases. In addition to temperature, the nature of the surface of the cathode has an extremely strong effect on the rate at which electrons are emitted. Electron emission resulting from the heating of a surface is referred to as thermionic emission [3, 4].

The fundamental aspects of thermionic emission may be understood with the aid of Figure 4-2, which is the classical energy level diagram for electrons near the interface between a cathode and vacuum. In the cathode, the parabolic curves represent electron energy levels adjacent to atoms. The energy levels merge to form the conduction band. At a temperature of absolute zero (0K), no electrons have energy greater than E_o, which is the top of the conduction band and is called the *Fermi level*. The difference in energy between the top of the conduction band in the cathode and the vacuum level adjacent to the cathode is known as the *work function*, commonly denoted by $e\varphi$.

At temperatures above absolute zero, some electrons have energy greater than E_o. Emission of the electrons can occur if their energy is $E_o + e\varphi$ or greater. It is necessary to keep in mind, however, that the electrons move in random directions within the cathode. Those moving toward the surface have the highest probability of being emitted.

For those electrons that are emitted, the current density is

$$J = \rho u_x = e n_e u_x \tag{4-1}$$

where n_e is the number density of emitted electrons and u_x is the x-directed velocity component (at the emitting surface) of those that are emitted. This velocity is readily found from the energy relation

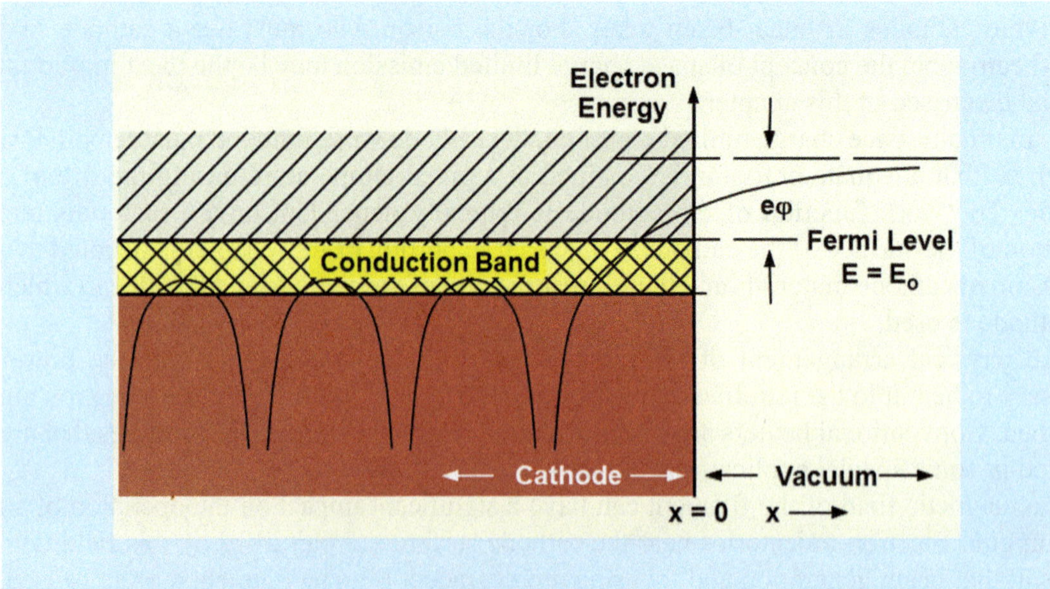

Figure 4-2 Energy level near the surface of a metal.

$$\frac{1}{2}mu_x^2 = E_o + e\varphi \tag{4-2}$$

The number density being emitted, n_e, is not so easy to find [1]. This is computed from the density of energy states in the cathode and the probability that they are occupied. When the analysis is carried out, it is found that

$$J = A_o T^2 e^{-\frac{e\varphi}{kT}} \tag{4-3}$$

This is the Richardson-Dushman equation for thermionic emission. A_o is a universal constant and has the value

$$A_o = 1.20 \times 10^6 \ \mathrm{A/m^2 deg^2} \tag{4-4}$$

The most significant aspect of the Richardson-Dushman equation is the exponential variation of current density with work function and temperature. The variation of current density with T^2 is negligible by comparison. This can be readily demonstrated by considering the relative effects of a small change in temperature. For example, for a typical operating temperature of 1,000°C, a 1% change in T produces roughly a 70% change in the value of the exponential term, but only a 2% change in T^2.

To compare theory with experiment, the Richardson-Dushman equation is first written in the following form:

$$\ln\frac{J}{T^2} = \ln A_o - \frac{e\varphi}{kT} \tag{4-5}$$

This suggests that a plot of $\ln J/T^2$ versus $1/T$ should be a straight line. The work function, $e\varphi$, should be the slope of the line and ln A_o should be the intercept of the extrapolation of the line to infinite temperature ($1/T = 0$). The data is plotted in the form shown in Figure 4-3, which is known as a Richardson plot.

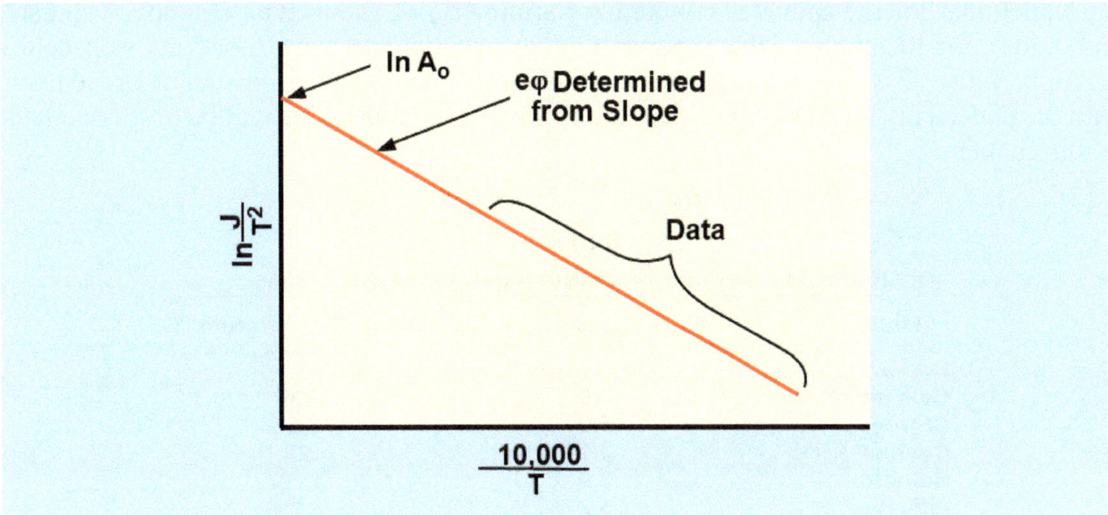

Figure 4-3 Richardson plot for determining work function and A_o.

In practice, the experimental value of A_o turns out to be on the order of one-quarter to one-half of the predicted value. Two reasons for this are:

1. The work function has been assumed to be independent of temperature. In fact, $e\varphi$ may vary with temperature.
2. The same value of $e\varphi$ has been assumed for all parts of the emitting surface. In fact, $e\varphi$ may be different for different crystal surfaces. Suppose, for example, that half of the emitter area has work function $e\varphi_1$ and the other half has work function $e\varphi_2$ where $e\varphi_2 > e\varphi_1$. Then, because of the exponential dependence of emission on work function, practically all of the emission will be from the $e\varphi_1$ areas. The emitting area is, therefore, only half of the nominal cathode area, and the measured value of A_o is only half of the theoretical value.

From the Richardson-Dushman equation, it is clear that to obtain high values of emission, the work function must be low and the temperature must be high. Unfortunately, as shown in Table 4-1, when pure metals are considered, materials with low work functions also have low melting temperatures. Cesium, with a work function of 2.1 eV and a melting temperature of 28°C, is an extreme example. Barium also has a reasonably low work function, but has a moderate melting temperature (725°C). It turns out that barium is commonly used in thermionic cathodes, but not because it has a low work function. Instead, barium is used because of its capability (accompanied by other elements including oxygen) to reduce the work function of the underlying metal surface. Thus, even though they have high work functions, it is possible to use high melting temperature metals such as tungsten and osmium by introducing barium and other elements to reduce the work function of the surface.

Figure 4-4 shows theoretical emission densities as functions of temperature and work function (dashed lines). Also shown are measured emission densities for several cathodes, most of which contain metals listed in Table 4-1. For W and ThW, the experimental curves are located where expected based on the values of $e\varphi$ in Table 4-1. (For ThW, thorium diffuses to the surface of the tungsten and so the work function is approximately that of thorium.)

Notice that for the cathodes containing barium (BaW, BaNi, type B, alloy, scandate, and oxide), the locations of the experimental curves indicate work functions well below that of barium (~ 2.7 eV). In most cases it is the interaction of the barium and other elements with the underlying metal that produces the low work function. This will be discussed later in this chapter.

Table 4.1

Room Temperature Work Functions of Various Metals

Metal	Work Function eV*	Melting Temperature °C
Barium	2.7	725
Calcium	2.9	839
Carbon	5.0	~3,550
Cesium	2.1	28
Hafnium	3.9	2,227
Iridium	5.2	2,410
Molybdenum	4.5	2,620
Nickel	5.2	1,455
Osmium	5.4	3,045
Platinum	5.3	1,773
Rhenium	5.1	3,180
Scandium	3.5	1,539
Sodium	2.7	97
Strontium	2.6	769
Tantalum	4.2	2,996
Thorium	3.4	1,750
Tungsten	4.6	3,410
Zirconium	4.1	1,852

*Work function is dependent on crystalline orientation. The values listed are estimates of average values.

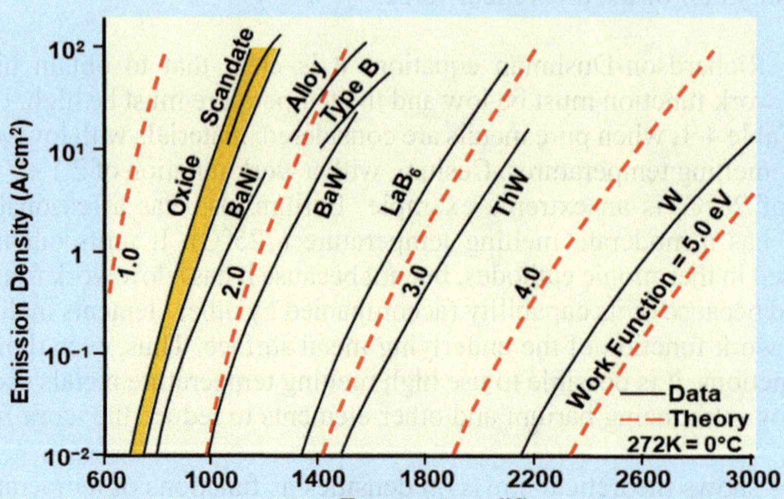

Figure 4-4 Emission characteristics of several cathodes. (From: *Methods of Experimental Physics*, Vol. 4, by G. A. Haas, copyright 1967 by Academic Press.)

4.1.1 Schottky Effect

In the derivation of the Richardson-Dushman equation, the effect of an electric field at the cathode surface was not taken into consideration and so neither voltage nor electric field appears in the equation. In fact, when an electric field is applied to the surface of a cathode, it is found that the emission increases slightly.

The reason for this effect may be explained with the aid of Figure 4-5. As was shown in Figure 4-2 and again in Figure 4-5(a), an electron must be emitted with energy $e\varphi$ or greater to overcome the work function barrier at the surface of the cathode. When a positive voltage exists on an anode near the cathode, then an electric field, E_a, is applied to the region between the cathode and anode. Near the cathode surface, as shown in Figure 4-5(b), this field reduces the barrier that electrons must overcome to be emitted so that more electrons are emitted. The resulting increase in emission current is called the *Schottky effect*. The amount by which the work function is reduced is $e\Delta\varphi$ where

$$\Delta\phi = \left(\frac{e\,E_a}{4\pi\varepsilon_o}\right)^{1/2} \tag{4-6}$$

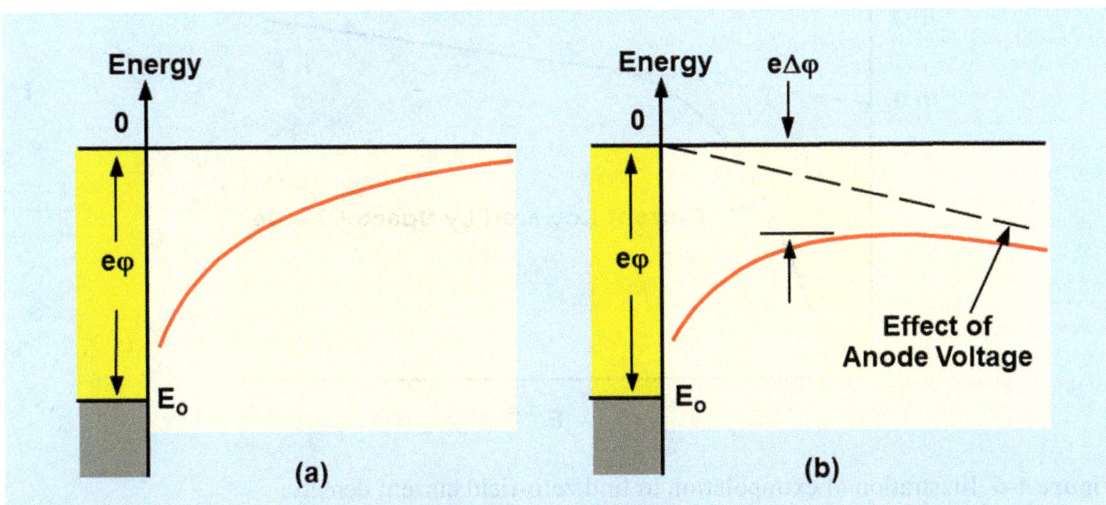

Figure 4-5 Potential energy as a function of distance from the surface of the cathode.

Using this $\Delta\varphi$, the Richardson-Dushman equation can be modified as follows:

$$J = A_o T^2 e^{-\frac{e}{kT}\left[\phi - \left(\frac{e\,E_a}{4\pi\varepsilon_o}\right)^{1/2}\right]} \tag{4-7}$$

This can be written as

$$J = J_o e^{\frac{e}{kT}\left(\frac{e\,E_a}{4\pi\varepsilon_o}\right)^{1/2}} \tag{4-8}$$

where J_o is the Richardson-Dushman current density and is sometimes referred to as the zero-field current density or the saturation current density [4]. J_o is a useful parameter for comparing performances of different cathodes.

In practice, to determine J_o experimentally, it is necessary to extrapolate from values measured at relatively high electric fields. This is because, at low electric field levels, electrons near the cathode surface alter the applied electric field. This is known as the electron space charge effect and will be discussed in the following section.

To determine J_o, proceed as follows:

$$\ln J = \ln J_o + \frac{e}{kT}\left(\frac{e E_a}{4\pi\varepsilon_o}\right)^{1/2} \tag{4-9}$$

This indicates that a plot of ln J versus $E^{1/2}$ should be a straight line with an intercept, when extrapolated to $E = 0$, with the value of ln J_o. This is illustrated in Figure 4-6. Note that, at low values of electric field, the measured value of current density falls below the extrapolated value because of the effect of the charge of the electrons in the region near the cathode surface. Some error in J_o may result from the change of the true work function by the applied electric field.

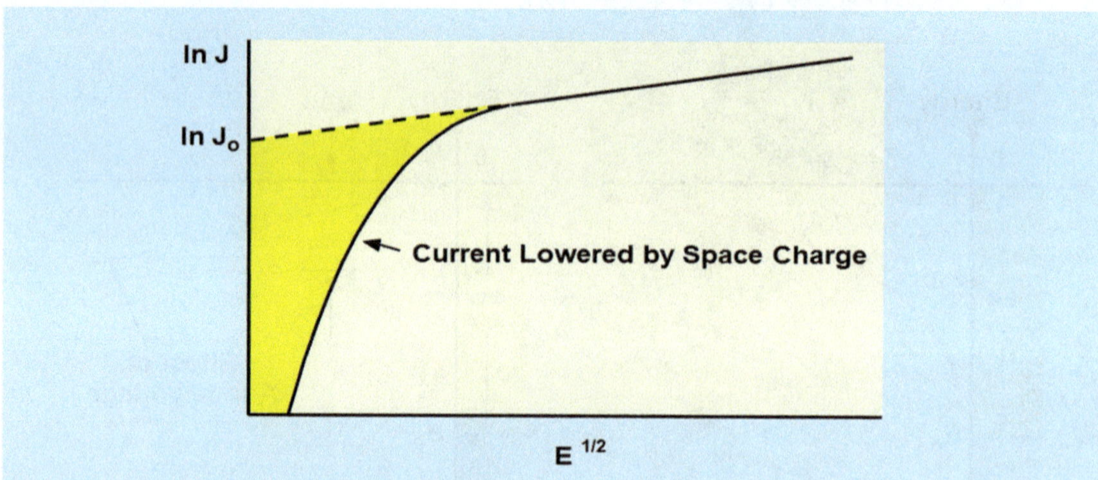

Figure 4-6 Illustration of extrapolation to find zero-field current density.

4.1.2 Space-Charge Limitation

The effect of the negative charge of an electron is to reduce the potential that is present in the absence of the electron [5]. Near an emitting cathode where many electrons are present, the reduction in potential can be appreciable. This is shown for a parallel-plane diode in Figure 4-7. With no electrons, the potential increases linearly from the cathode to the anode. With electrons, the potential profile is depressed as indicated by the dashed line.

As the electron emission rate is increased (by increasing temperature, for example), the potential is further decreased. As indicated by Figure 4-8, there is a limit to how far the electron emission rate can be increased. This limit occurs when the density of electrons adjacent to the cathode surface becomes so large that it tends to depress the potential below zero. If the potential were to become negative, electrons would be forced back to the cathode.

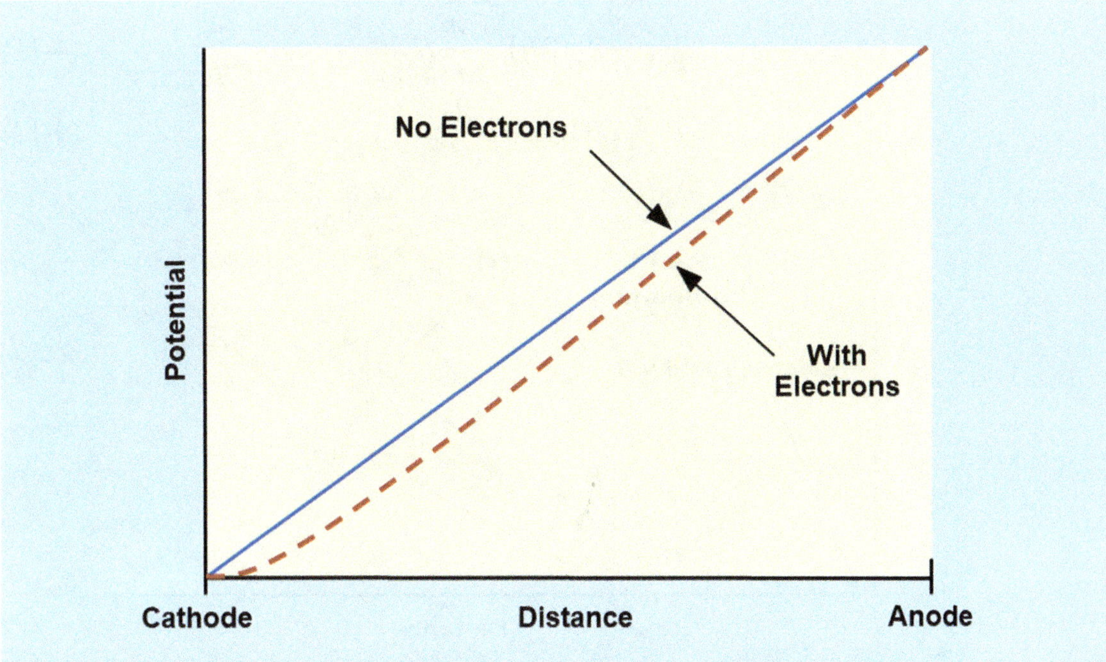

Figure 4-7 Potential distribution with and without electrons in a parallel-plane diode.

In Figure 4-8(a), the electric field at the cathode surface causes all emitted electrons to leave the cathode, thereby depressing the potential near the surface. In Figure 4-8(b) the electric field at the cathode surface forces electrons back to the cathode surface and increases the potential. An equilibrium condition occurs when, as shown in Figure 4-8(c), the potential adjacent to the cathode surface is zero; that is, when the electric field at the cathode surface is zero. If the potential tends to become positive, more electrons flow from the cathode, thereby depressing the potential. If the potential tends to become negative, fewer electrons flow from the cathode, thereby raising the potential.

When the electric field at the cathode surface is forced to zero by the electron cloud near the cathode surface, the emission is said to be space charge limited. When a sufficient number of electrons are present so that emission is space charge limited, then cathode temperature and surface condition have no effect on emission. This is indeed a very fortunate situation because it eliminates the necessity for extreme uniformity of temperature across the surface of a cathode. It also eliminates the necessity for precise control of the voltage and current to the heater in the cathode. When the voltage is increased in a diode with a space-charge-limited cathode, the potential at all points between the cathode and anode tends to increase. As a result, additional electrons flow from the cathode to cause the electric field at the cathode surface to remain at zero. It is evident, therefore, that diode current is dependent on anode voltage.

The Child-Langmuir law is the relation between voltage and current in a space-charge-limited diode [1]. This relation is

$$I = PV^{3/2} \tag{4-10}$$

where P is known as perveance and, for a parallel plane diode, is

$$P = \frac{4}{9}\varepsilon_o (2\eta)^{1/2} \frac{A}{d^2} \quad (4\text{-}11)$$

or

$$P = 2.33\times10^{-6} \frac{A}{d^2} \quad (4\text{-}12)$$

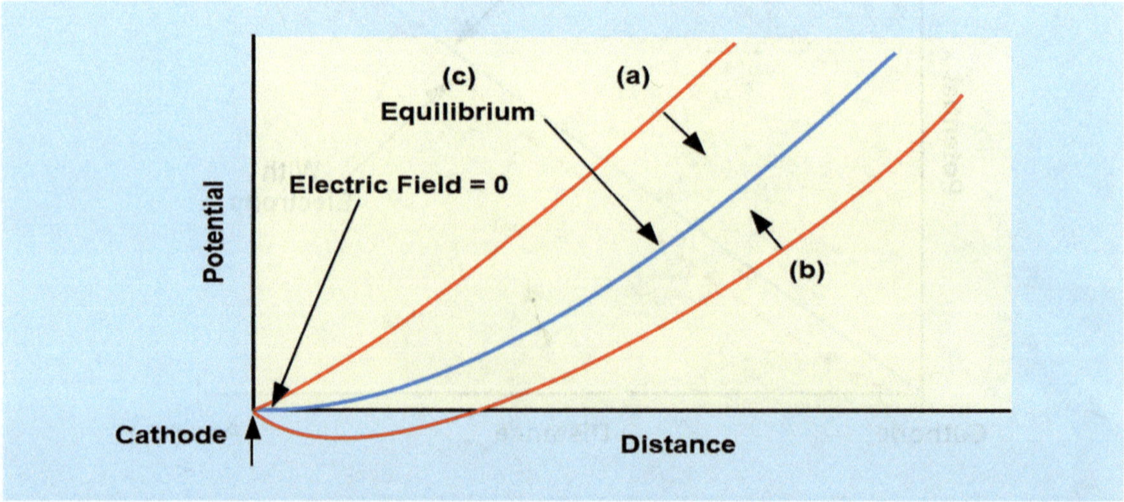

Figure 4-8 Potential profiles near the surface of a cathode.

The area of the diode is A and the spacing between electrodes is d. Note that perveance is a function only of geometry. Space-charge-limited diodes with geometries other than planar, such as cylindrical and spherical, all obey $I = PV^{3/2}$, where the expression for perveance is the one appropriate for the geometry used.

In the analysis, it is assumed that the velocities of the electrons are zero when they are emitted. In fact, electrons are emitted with small initial velocities. If the electric field was zero at the cathode surface, all of the emitted electrons would be able to leave the surface and the saturation emission, J_o (total thermionic emission capability of the cathode), would flow. In practice, space-charge-limited emission is much less than saturated emission because the presence of large quantities of electrons near the surface causes a potential minimum to be formed just outside the cathode surface. The voltage at the minimum is a small fraction of a volt below cathode potential. Only the most energetic electrons can overcome this potential minimum and pass on to the anode. The less energetic electrons are returned to the cathode.

The effect of increasing the anode voltage so that more current flows is to decrease the depth of the potential minimum and move it closer to the cathode. As J approaches the saturation current density, J_o, the electric field approaches zero at the cathode surface. If the voltage is increased further, current will be limited to the saturation current density and space-charge-limited operation will cease.

In practice, space-charge-limited diodes behave as predicted. This is readily demonstrated by making a two-thirds power plot or a log plot of current versus voltage. As shown in Figure 4-9, the resulting plot is a straight line for space-charge-limited operation. The slope of the line is the perveance.

When the voltage is increased to a value high enough so that the cathode cannot supply sufficient current to maintain the space-charge-limited condition, the cathode is said to be *TL* (temperature-limited). The TL region is readily apparent because the plotted results depart from a straight line.

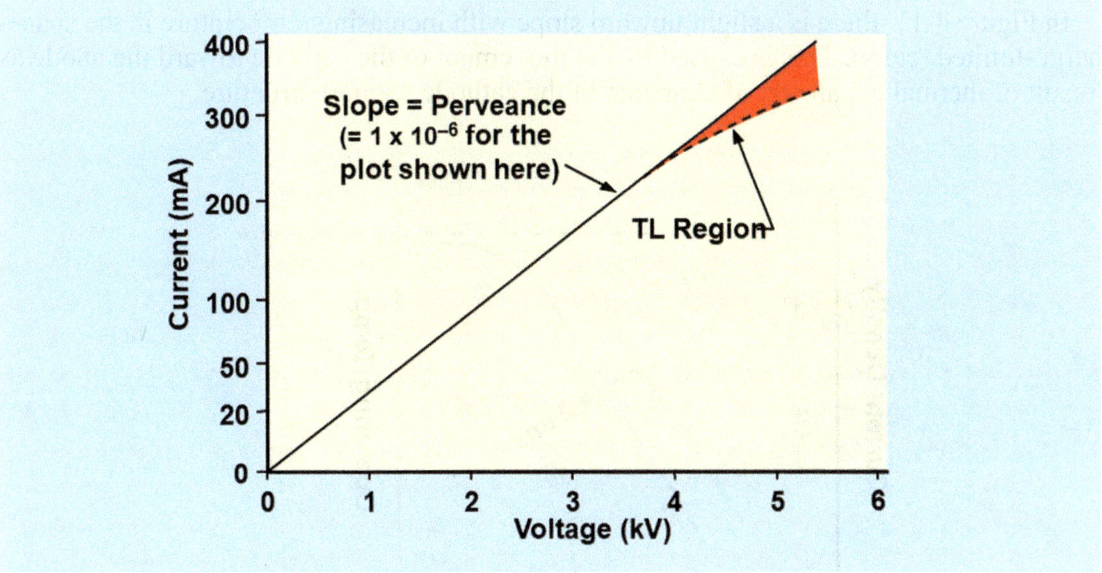

Figure 4-9 Current versus voltage plotted on two-thirds power graph paper.

In Figure 4-10, current is shown plotted as a function of temperature for a thermionic cathode. In the region of TL operation, current increases with temperature. At high temperatures, current is independent of temperature, indicating space-charge-limited operation. As space-charge-limited voltage is increased, the temperature at which the transition from TL to space-charge-limited operation occurs is increased.

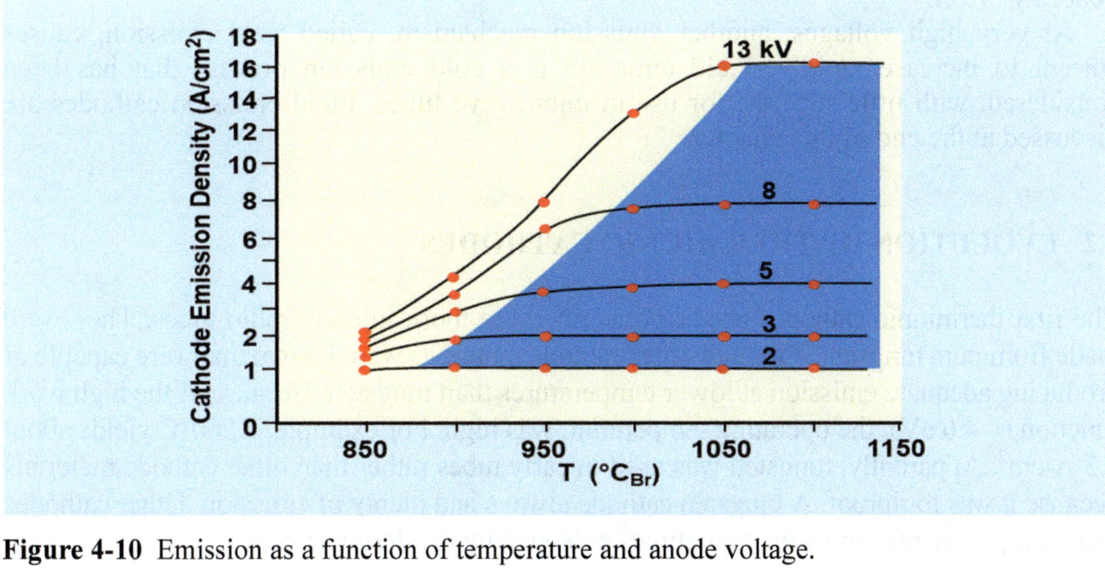

Figure 4-10 Emission as a function of temperature and anode voltage.

As indicated in Figure 4-11, the TL portion of a current versus temperature curve moves to higher temperatures as time increases. The reason for this will be discussed later in this chapter. It is necessary to select an operating temperature for a cathode that remains above the knee of the curve for the life of the cathode. The rate at which the TL curve moves increases with increasing temperature, so an excessive temperature will reduce cathode life.

In Figure 4-11, there is a slight upward slope with increasing temperature in the space-charge-limited region. This is caused by the movement of the cathode toward the anode as a result of thermal expansion of elements of the cathode support structure.

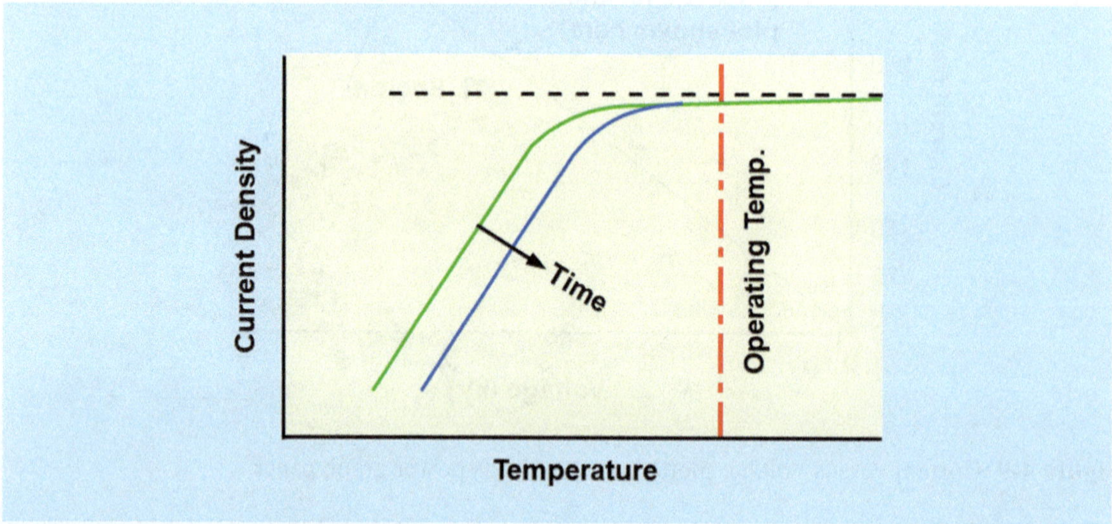

Figure 4-11 Movement of the TL curve with time.

The mechanisms dominating current flow in a thermionic diode are summarized in Figure 4-12. At low voltages, current is space charge limited. As voltage is increased, the emission limit of the cathode is reached and current becomes dependent on cathode temperature. Current continues to increase slowly as voltage is increased because of the Schottky effect.

At very high voltages, another emission mechanism, called field emission, causes current to increase rapidly. Field emission is a cold emission process that has been considered, with little success, for use in microwave tubes. Field emission cathodes are discussed at the end of this chapter.

4.2 EVOLUTION OF THERMIONIC CATHODES

The first thermionic cathodes used in quantity were those in early radio tubes. They were made from pure tungsten, although other cathode materials were known that were capable of producing adequate emission at lower temperatures than tungsten. Because of the high work function (~ 4.6 eV), the operating temperature was high. For example, 2,200°C yields about 0.3 A/cm^2. Apparently, tungsten was used in early tubes rather than other cathode materials because it was foolproof. A tungsten cathode always had plenty of emission. Other cathodes were temperamental and were sometimes poisoned by mysterious agents.

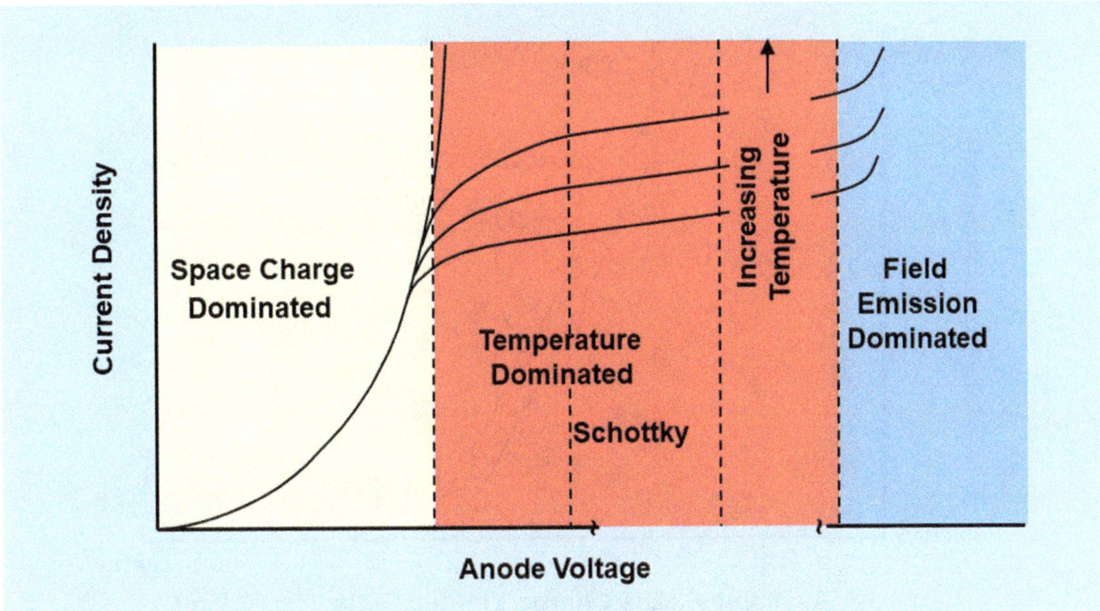

Figure 4-12 Summary of mechanisms dominating current flow in a thermionic diode.

For some applications, thorium oxide in a concentration of about 2% is added to tungsten in the process of making the tungsten wire for the cathode. By processing, thorium is brought to the surface of the tungsten. The work function of thorium is well below that of tungsten, so the emission is about 1,000 times higher than for tungsten.

The cathode surface is sometimes carburized; that is, heated in a vapor of a hydrocarbon such as alcohol, so that a layer of tungsten carbide is formed on the surface. This increases cathode life by reducing the rate of evaporation of thorium and protecting the cathode from ion bombardment.

The life expectancy of high-power switch tubes containing thoriated tungsten cathodes is shown in Figure 4-13. Thoriated tungsten cathodes are in common use in high-power switch tubes, transmitter tubes, and microwave oven magnetrons.

Early cathodes were all directly heated; that is, the filament, which was heated by passing a current through it, was also the cathode. Thus, as a result of the heater current, there was a voltage variation along the surface of the cathode. Because batteries were used as power sources, the filament voltage was dc, so there was no modulation of the emission current.

The development of ac radios and the need to heat the filaments with ac current brought about the need for a cathode, the surface of which was all at the same potential. The indirectly heated cathode resulted, whereby the heater was separate from the electron emitting surface. Directly and indirectly heated cathodes are illustrated in Figure 4-14.

In indirectly heated cathodes, tungsten is used for the filament. Because of heat losses and the relatively large cathode area, the cathode surface had to operate at a temperature substantially below the heater temperature. This provided the requirement for a low work function emitting surface.

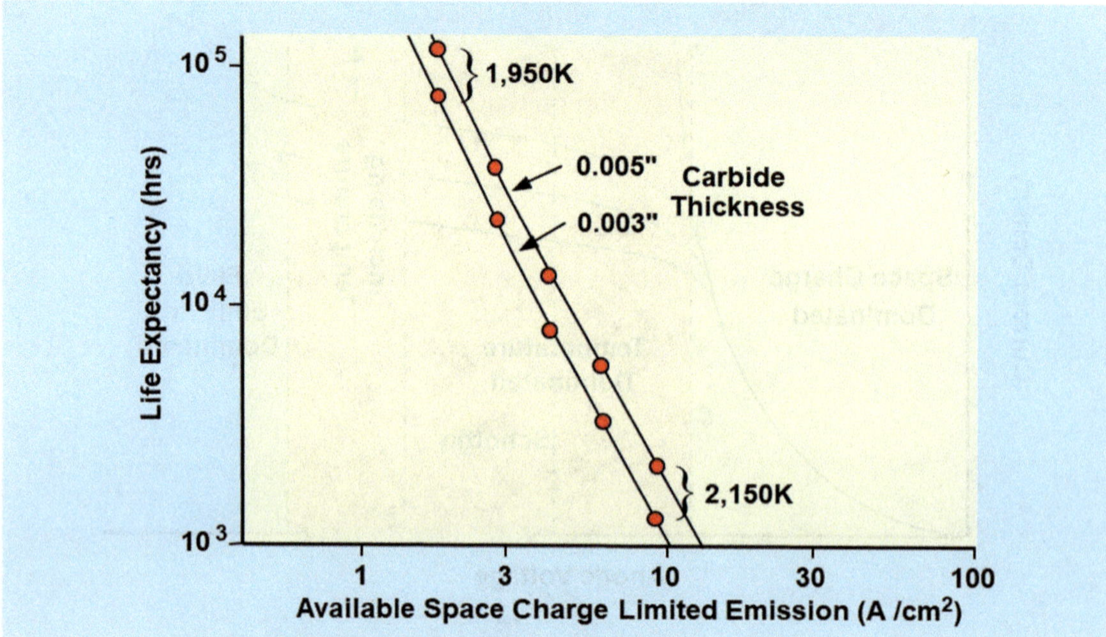

Figure 4-13 Life expectancy for a thoriated tungsten cathode (actual life □ 1/2 of life expectancy). (From: T. E. Yingst et al., *Proc. IEEE*, March 1973. © 1973 IEEE.)

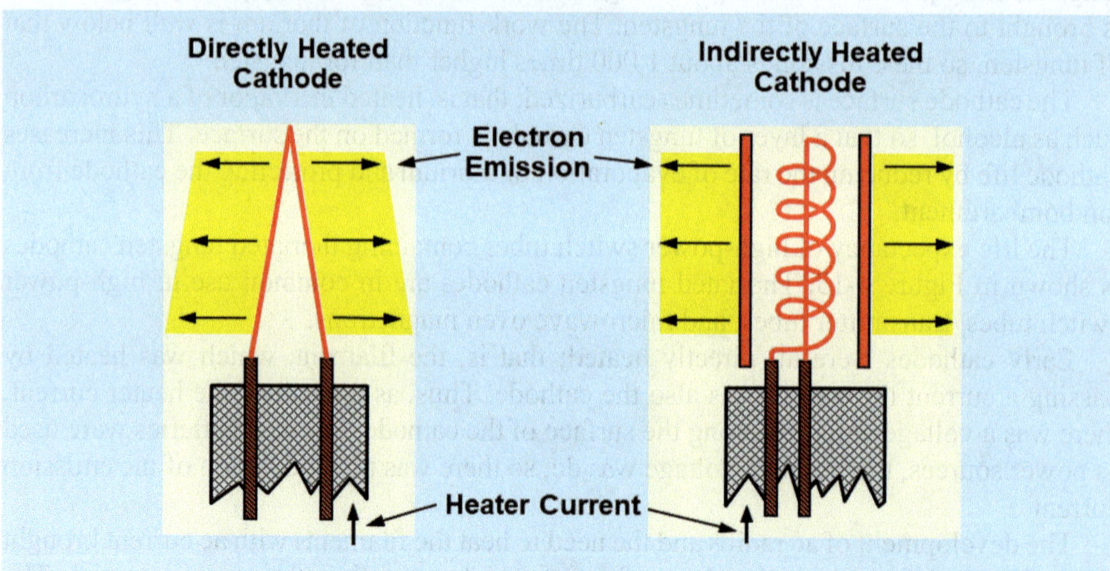

Figure 4-14 Comparison of directly and indirectly heated cathodes.

The oxide-coated cathode, which was discovered by Wehnelt in 1904, long before radio tubes, provided the solution to the problem because of its very low work function (~1.5 eV) and resulting low operating temperature [6]. Oxide-coated cathodes served the industry well from the 1920s through World War II and into the early 1950s. They are still prevalent today in applications requiring relatively modest cathode emission and long life. Early oxide-coated cathodes were described by J. R. Pierce [2] as follows:

The early history of oxide coated cathodes seemed without rhyme or reason. It was found that if a wire intended to emit electrons was heated and rubbed with sealing wax and the sealing wax then burned off a truly phenomenal cathode was produced. Further work traced the emission to alkaline earth oxides, and the compound from which they came was humorously described as a barium resinate. The sealing wax technique died hard, and for several years the process of coating filaments consisted in stringing fine platinum ribbon out in long lengths, heating it with an electric current, and rubbing a resinous compound containing barium and strontium over it repeatedly. The narrow ribbon was then cut into lengths, put into a crude vacuum tube of the day, the tube was pumped and baked until a "hard vacuum" was achieved, and the filament was activated. This activation was a wishful process consisting of glowing the filament brightly, applying voltage, and drawing current to the plate until, in happy circumstances the plate got red-hot, giving off enough gas to produce a blue glow and further increase the activity and the current. After a lucky period of this treatment the filament became active; that is, it would emit electrons at a lower and reasonable temperature, at which it would last for a number of thousands of hours.

The art and science of oxide cathodes are now highly developed. A mixture of barium, strontium, and calcium carbonates is applied to a metal base (usually nickel) and then activated in vacuum. Activation consists of heating and reducing the carbonates to oxides.

As oxide-coated cathodes were improved, dc emission increased from mA/cm^2 to hundreds of mA/cm^2. Pulsed emission densities are dependent on pulse length and can be as high as tens of A/cm^2. Oxide-coated cathodes are not capable of high dc emission densities because of the high resistivity of the coating and the resistive heating that occurs at high currents. They are also easily poisoned.

The search for better cathodes than oxide-coated cathodes resulted from their low dc emission and susceptibility to damage. In 1951, to circumvent the problems with oxide-coated cathodes, Varian introduced a klystron for television service that used a replaceable tantalum cathode that was heated by electron bombardment on the back side. The source of the electrons was a tungsten filamentary emitter. Varian claimed 10,000 hours of continuous operation with the easy replacement of the cathode assembly after that. The life of the cathode assembly, even under severe operating conditions, was limited by the life of the tungsten emitter rather than the tantalum cathode [7].

The L cathode, illustrated in Figure 4-15, was developed by Philips about 1950. In this cathode, the emissive material is sandwiched between the heater and a porous tungsten pellet. During operation, emissive material diffuses through the porous tungsten and forms a low work function emitting surface.

Cathodes like the L cathode are referred to as *reservoir cathodes* for the obvious reason that a reservoir of emissive material is used. Early problems have been overcome and structures like the L cathode are now in use. Some of these are referred to as *MK cathodes*.

A reservoir cathode, developed by the Naval Research Laboratory, is known as the *controlled porosity dispenser* (CPD) cathode. The reservoir contains BaO, CaO, and Al_2O_3, as well as particles of tungsten, which act as a reducing agent that frees barium from the BaO. The CPD cathode uses a thin sheet of tungsten or alloy (to be discussed next) in which pores have been etched or drilled with a laser. The CPD approach to cathode construction offers the advantages of improved emission uniformity and improved reproducibility and performance. At present, the difficulties of fabricating CPD cathodes, especially for concave cathodes, limit their usefulness.

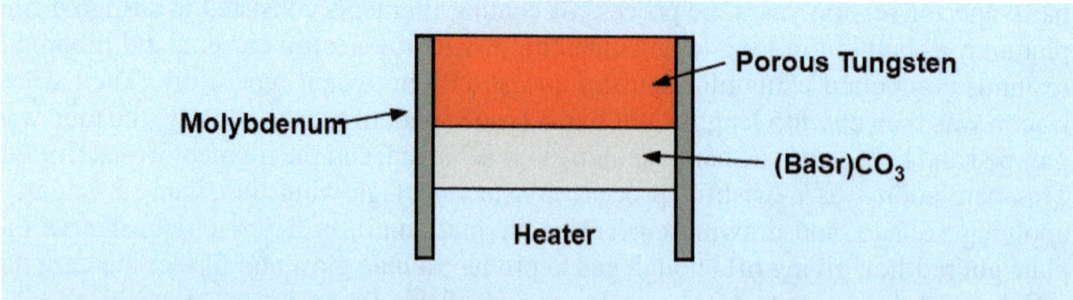

Figure 4-15 Essential features of the Philips L cathode.

From 1950 to 1955, the evolution of the impregnated dispenser cathode illustrated in Figure 4-16 occurred. The Philips B-type cathode developed by Levi is still in use by the microwave tube industry. The pores of a porous tungsten pellet are impregnated with a compound of BaO, CaO, and Al_2O_3. Barium is released when the impregnant reacts with the tungsten. The freed barium migrates to the surface of the porous tungsten to form the emitting surface.

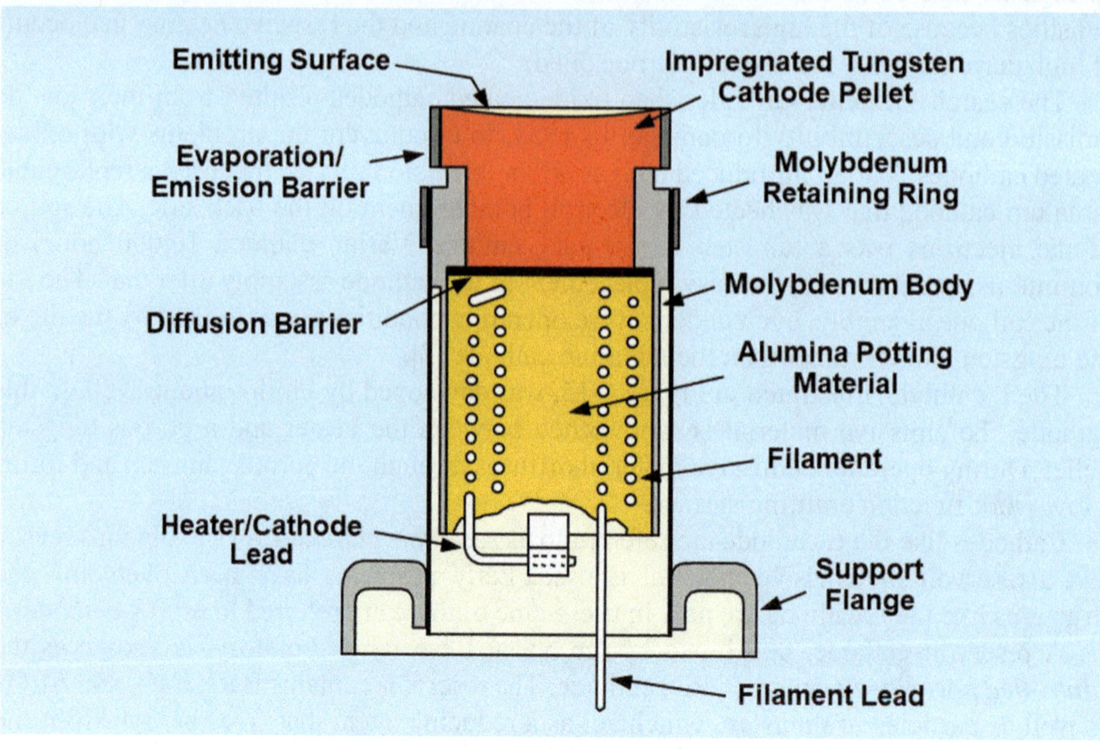

Figure 4-16 Typical B-type impregnated dispenser cathode.

A common emissive mix used in the B-type cathode consists of five parts BaO, three parts CaO, and two parts Al_2O_3 and is frequently referred to simply as 5:3:2. Other mixes are 6:1:2 and 4:1:1. The 4:1:1 cathode is sometimes referred to as an *S-type cathode*. The choice of mix depends on the application of the cathode [8]. For example, the 5:3:2 cathode is often used in large tubes at low emission densities. The 6:1:2 cathode has very good poisoning resistance and a low barium evaporation rate and is recommended for most applications. The 4:1:1 cathode is very active and poison-resistant and can usually be operated 30°C cooler than other types.

The B-type cathode is rugged and relatively impervious to damage and can readily provide emission densities of several A/cm^2. At these high densities, the operating temperature must be 1,100°C or higher because of the relatively high work functions of these cathodes. As a result, life is limited to a few thousand hours.

The search for a cathode capable of an emission density comparable to the B-type cathode, but with a lower operating temperature and longer life, led to the development of the M-type cathode in the 1960s. In its simplest form, the M-type is a B-type with a film several thousand angstroms thick of osmium, iridium, or rhenium applied to the surface. Compared to a B cathode, the effect of the film is to reduce the work function ~ 0.2 eV and the cathode operating temperature ~ 90°C (dependent on film metal).

A problem early in the development of M cathodes was degradation of the enhancing film. At normal operating temperatures, the tungsten substrate and coating metal would interdiffuse, and an alloy surface was formed with a composition that changed with time. The lowest work function occurs when the alloy is about 50/50 tungsten and coating metal. Thus, during the life of a cathode, emission would actually increase at first, then degrade, and then eventually revert to that of a B cathode as the surface became tungsten.

Several fabrication techniques for controlling the properties of the alloy formed on the surface of an M cathode have been examined. In one cathode, the mixed metal matrix (MMM) cathode, the cathode pellet contains particles of the enhancing metal (usually iridium). Fabrication is difficult because the size distribution of the tungsten and iridium particles must be carefully controlled. In the controlled doping (CD) cathode developed by Varian, multiple layers of alloy material, having different compositions, are deposited. The composition of the layers is adjusted to minimize interdiffusion of the tungsten and the enhancing metal. A few percent of the enhancing metal can be added to the tungsten matrix prior to coating and this reduces the diffusion of tungsten into the coatings.

The performance characteristics of the cathodes just described are summarized in Figure 4-17. The increase in emission capability over B cathodes of cathodes containing osmium, iridium, or rhenium is clearly shown.

4.3 DISPENSER CATHODES

4.3.1 Fabrication

At present, the most commonly used cathodes in the microwave tube industry are impregnated dispenser cathodes. These are fabricated from small grains of tungsten that have been pressed together under high pressure and sintered at a temperature of over 2,000°C for 1 to 2 hours to form porous billets. The pores are then filled with a plastic material to facilitate rough machining to the desired shape. After machining, the plastic is removed and the porous matrix is filled with a compound containing barium oxide, calcium oxide, and aluminum oxide.

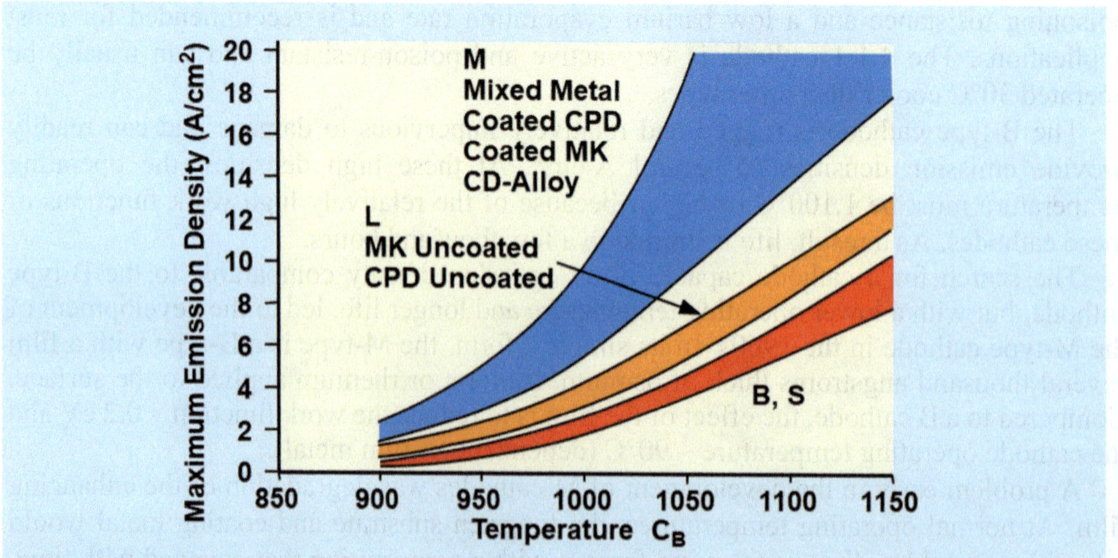

Figure 4-17 Performance comparison for various dispenser cathodes. (From: L. R. Falce, *Technical Record*, 1983 IEDM. © 1983 IEEE.)

Usually, there is then an additional machining operation of the emitter surface to the final dimensions. This is a difficult procedure because the cathode is a porous structure. Care must be taken to avoid excess smearing of the tungsten over the pore openings. If the pore openings become plugged, then, during cathode operation, the emissive compound will be prevented from diffusing to the cathode surface. Figure 4-18 shows the surface and interior of a cathode after final machining. The differences in grain sizes and pore distributions between the surface and the interior show that significant smearing occurred during machining.

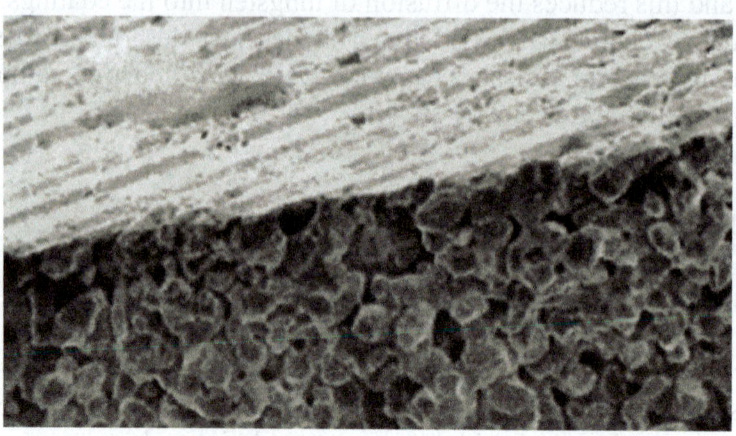

Figure 4-18 SEM image showing surface and interior of a porous tungsten cathode. (From: R. L. Ives et al., *IEEE Trans. Electron Devices*, December 2005. © 2005 IEEE.)

Before a new dispenser cathode is installed in an electron gun, the cathode surface is usually processed to remove excess impregnant. One technique is to place the cathode in a vacuum chamber (not the tube in which it will eventually be installed) and heat it to a high temperature. This causes the excess impregnant to evaporate from the cathode surface. Another technique is a liquid cleaning process, referred to as SRLV or EDTA cleaning [9]. This process removes a shallow layer of impregnant from the tungsten surface.

After final installation, a cathode must be activated before it can be used. This process consists of heating the cathode to a temperature ~100°C to200°C above final operating temperature to leave the cathode free from excessive aluminates or poisoning agents such as sulfur or phosphorous. See [10] for an example.

Figure 4-19 shows the scale of the features of the surface (without smearing) of a typical B-type cathode [11]. A coated cathode (M type, for example) has a very thin film of enhancing metal on the surfaces of the tungsten grains. Note that all dimensions are typically a few microns. Note also that there is a wide distribution in the sizes of the tungsten grains and that they are irregularly shaped. This results in there being a lack of control over the porosity of the cathode surface and this contributes to the variability of the work function of the cathode (to be discussed next).

There have been many attempts to control the porosity of a cathode surface. The work of the Naval Research Laboratory is mentioned earlier in this chapter. None of these attempts have found application for any one of several reasons [12].

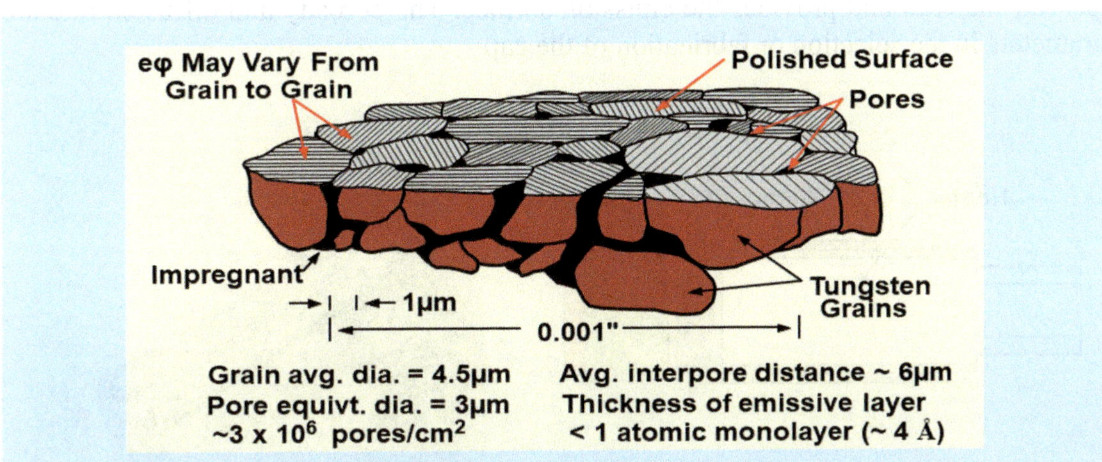

Figure 4-19 Scale of features of an impregnated cathode. (From: M. C. Green, *Technical Report,* RADC-TR-81-211, July 1981.)

4.3.2 Controlled Porosity Reservoir Cathodes

Ives et al. [13] have developed a novel technique for controlling porosity that overcomes many of the shortcomings of previous techniques. An emitter is fabricated from tungsten wires, the diameter of which is in the micron range. These wires are bundled together, compressed, and then sintered. The resultant material is essentially solid tungsten with a uniform array of pores corresponding to the gaps between the wires. Figure 4-20 shows a cross section of sintered wire material.

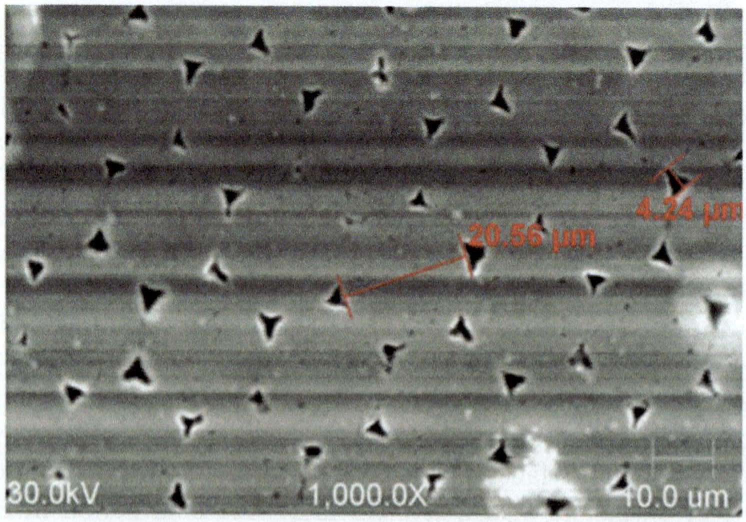

Figure 4-20 SEM photograph of sintered tungsten wires. (From: R. L. Ives et al., *Terahertz Science and Technology*, December 2011, © 2011 IEEE.)

The basic configuration of a reservoir cathode is like that of the Philips L cathode (Figure 4-15) and is shown in Figure 4-21. The thin porous tungsten disc shown here is the porous element described in the previous paragraph. The disc serves as the cap on the reservoir material and provides the emission surface. The porosity and diffusion rate are parameters in the selection or fabrication of the cap.

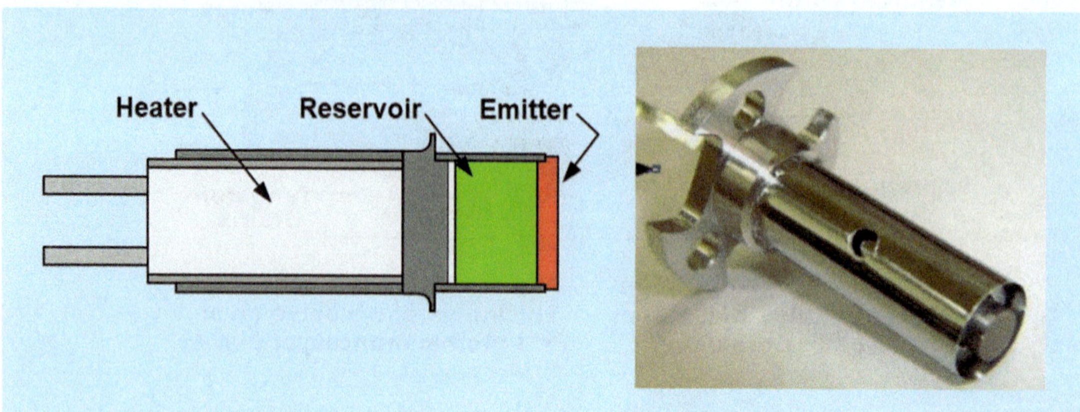

Figure 4-21 Controlled porosity reservoir cathode.

The starting point for the design of the cap is the emission density required of the cathode. That, in turn, makes it possible to specify the operating temperature of the cathode. (Complete coverage of the emitter surface with emissive material (i.e., barium) is assumed so that the work function can be specified.) Knowing the operating temperature, the evaporation rate of the emissive material from the cathode surface can be found. For optimum use of reservoir material in maintaining the monolayer, the cap must be designed so that the diffusion rate of the emissive material through it to the emitter surface matches the evaporation rate of emissive material from the cap.

The diffusion rate depends on the thickness of the tungsten cap and on the size of the pores between wires. With a given piece of tungsten cap, thickness can be trimmed to provide the desired diffusion rate. An apparatus has been developed for measuring diffusion rate. An example of the result of a diffusion test is shown in Figure 4-22.

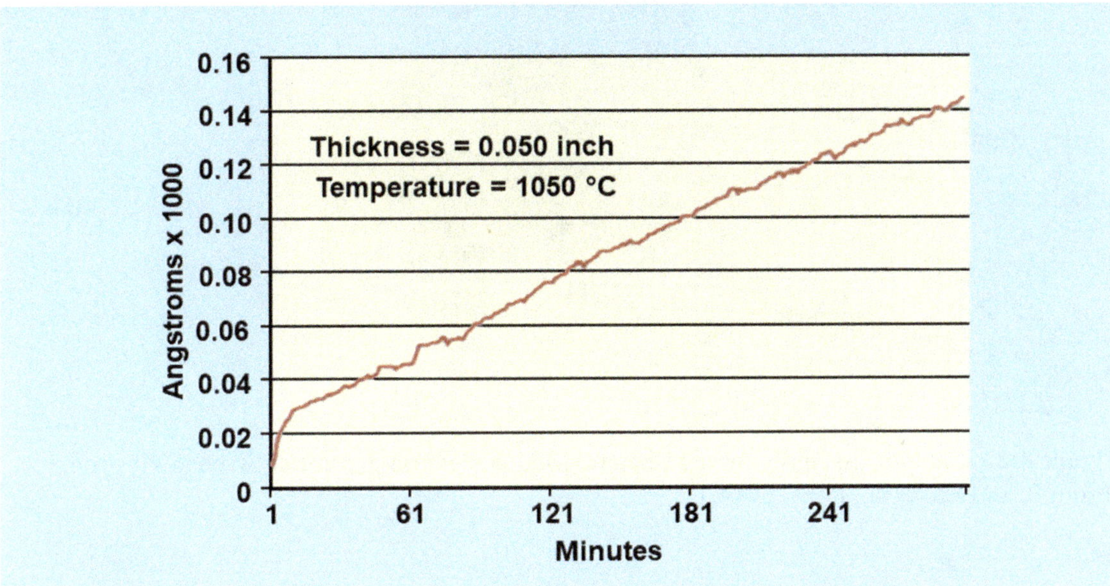

Figure 4-22 Diffusion measurement for 0.050 inch thick emitter at 1050°C. (From R. L. Ives, et al., IVEC 2014.)

The results of testing of early controlled porosity reservoir (CPR) cathodes were very positive and this led to attempts to use cathodes like those shown in Figure 4-23 in several high-power tubes. Unfortunately, the tests with high-power tubes that have been reported to date have been unsuccessful. In some cases, the failure (cracked insulator or grid short) does not appear to be related to the cathodes. In one case, however, the failure appears to have resulted from arcing as well as a defect in cathode construction. Arcing made it necessary to reduce the gun voltage from 25 kV to 19.5 kV. It was possible to operate the tube at this reduced voltage for about 15 hours before the vacuum was lost.

Using the perveance for which the gun was designed, the total beam current was calculated to be 15A. However, as shown in Figure 4-24, the current barely reached 12A during the 15-hour test and would not have exceeded about 13A if the test time could have been longer. Also, the gun was very slow to turn on. Some delay is expected because of the time required to fill the tungsten matrix with emissive material, but the delay shown in Figure 4-24 was excessive.

It was determined that significant differences occurred between the wire packing density of the cathodes used in the klystron and the cathodes for which the emissive material flow rate was determined. As a result, the rate of flow of emissive material to the cathode surface was inadequate and the cathodes were not operating space charge limited.

Figure 4-23 Cathode assembly for a 15-beam, gridded electron gun for an X-band klystron. (From R. L. Ives, et al., IVEC 2014.)

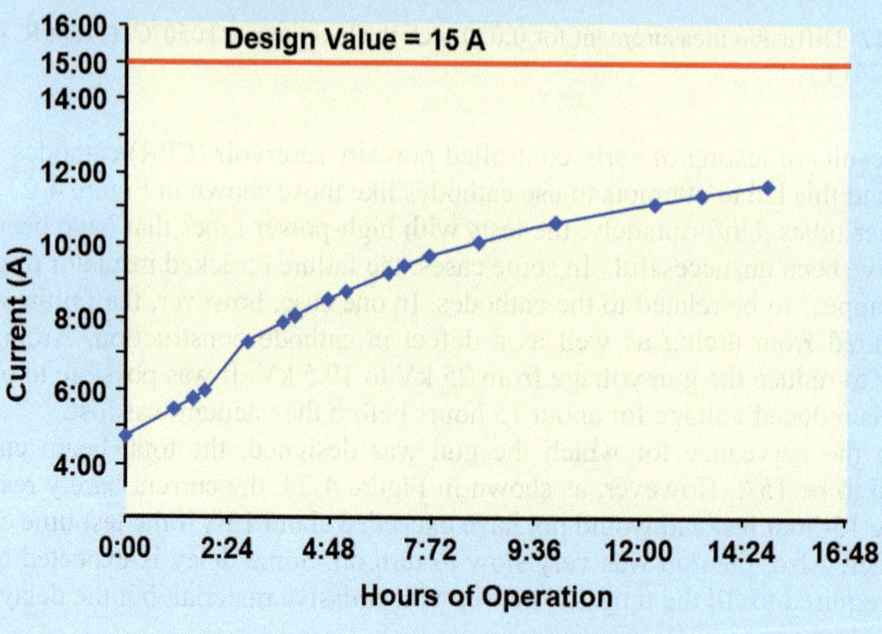

Figure 4-24 Electron gun current as a function of processing time. (From R. L. Ives, et al.)

4.3.3 Operation of Dispenser Cathodes

When a cathode is heated to the operating temperature, barium is released from the interior of the tungsten matrix by reactions of the impregnant with the tungsten. The freed barium migrates through the pores to the surface of the cathode. Once there, the barium migrates across the surface and eventually evaporates. The rate of diffusion of barium to the surface, along with the energy of desorption of barium atoms from the surface (which controls the evaporation rate), determines the surface coverage.

It is important to avoid operating a cathode at a temperature that is higher than necessary because the rate of evaporation of barium is highly temperature-dependent. Clearly, it is also important for the surface to be heated to a uniform temperature. The hottest portions of the cathode become depleted of barium soonest and this leads to reduced emission and the end of cathode life. It is desirable that the temperature should not vary more than a few degrees centigrade over the entire surface of the cathode.

The work function of the cathode depends on the fraction of the cathode surface that is covered by barium along with the work function of the metal substrate. The reason for this will be discussed later in this chapter. An example of the variation of work function with surface coverage is shown in Figure 4-25. For optimum coverage, which turns out to be nearly a monolayer, the work function is a minimum. With time, as a cathode operates, the pores become depleted and the rate of diffusion of barium to the cathode surface decreases. As a result, the coverage of the surface decreases, the work function increases, the rate of emission decreases, and the space-charge cloud may disappear. If that happens, emission becomes temperature-limited. By increasing the cathode temperature, the rate of diffusion of barium to the surface can be increased and the rate of emission can be increased, and it may be possible to return to space-charge-limited operation. Of course, the increased temperature causes the rate of evaporation to increase and this hastens the return of temperature-limited operation.

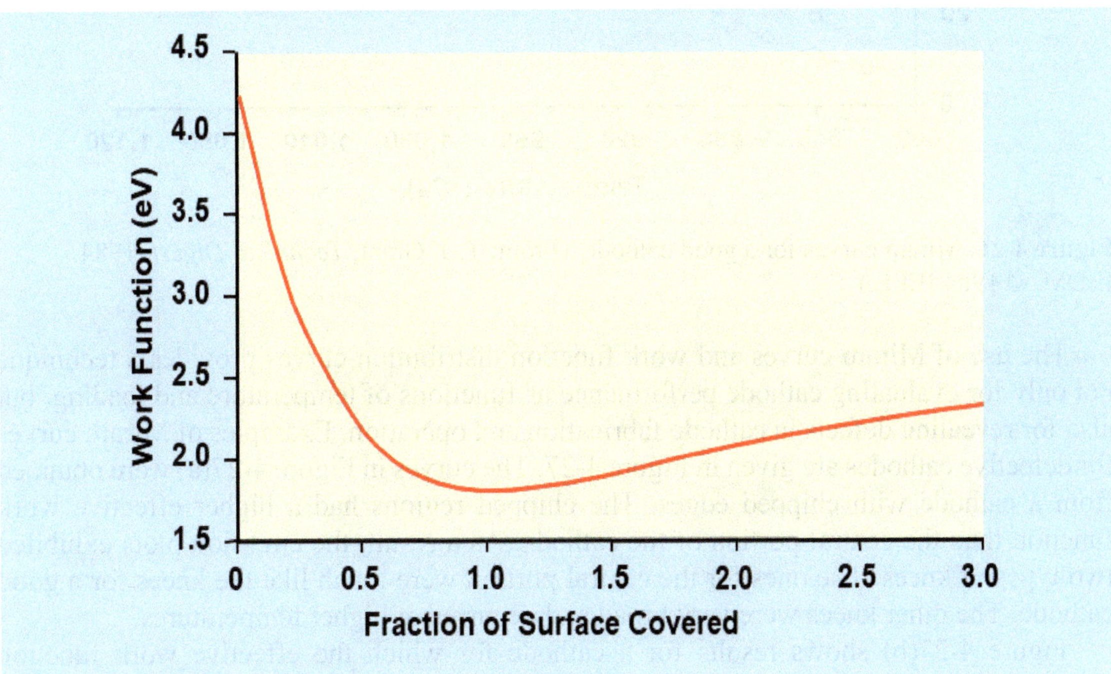

Figure 4-25 Variation of work function with a fraction of surface covered by barium. (From: L. Falce and L. Garbini, Fifth IEEE IVEC. © IEEE 2004.)

4.3.4 Miram Curves

Cattelino et al. developed a technique for presenting emission data that provides a powerful tool for comparing cathodes, for diagnosing their performance, and for helping to understand their operation [14]. The basic concepts may be understood with the aid of Figure 4-26. Values of emission current are plotted as percentages of the fully space-charge-limited (FSCL) current. Curves plotted in this manner are often referred to as *Miram curves*. Several curves are shown for TL operation. With the Miram technique, the curves are plotted for FSCL current densities that are separated by a factor of 2. From the Richardson-Dushman equation, the temperature difference required to produce an increase in current density by a factor of 2 is about 40°C, and, in fact, the TL curves are separated by about 40°C.

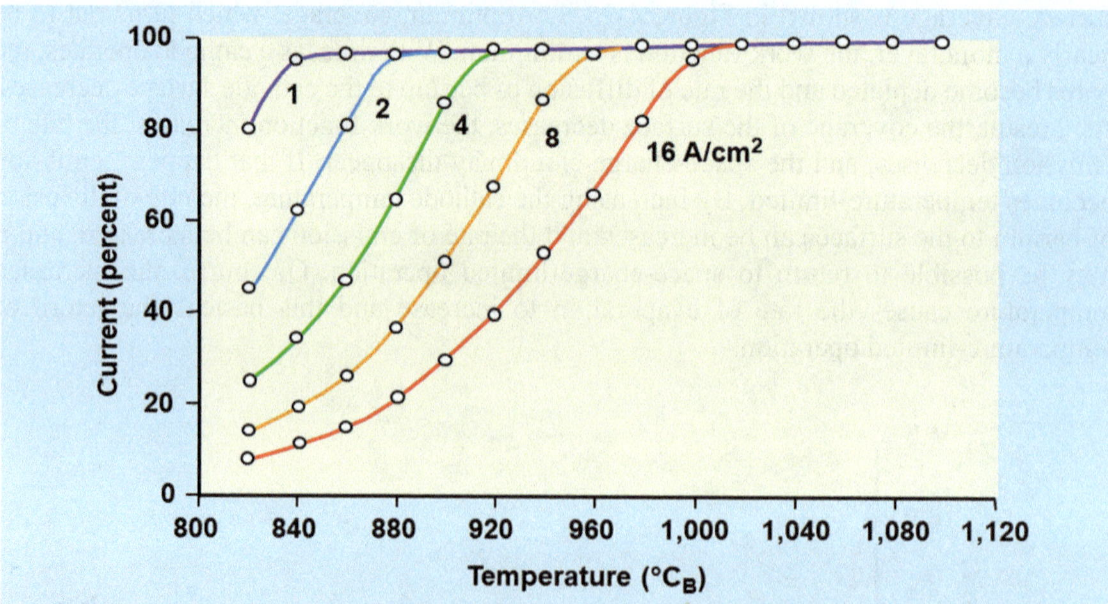

Figure 4-26 Miram curves for a good cathode. (From: T. J. Grant, *Technical Digest*, 1984 IEDM. © 1984 IEEE.)

The use of Miram curves and work function distribution curves provides a technique not only for evaluating cathode performance as functions of temperature and loading, but also for revealing defects in cathode fabrication and operation. Examples of Miram curves for defective cathodes are given in Figure 4-27. The curves in Figure 4-27(a) were obtained from a cathode with chipped edges. The chipped regions had a higher effective work function than the central portion of the cathode. As a result, the emission plots exhibited two types of knees. The ones for the central portion were much like the knees for a good cathode. The other knees were very broad and occurred at higher temperatures.

Figure 4-27(b) shows results for a cathode for which the effective work function increased as the temperature was raised above 940°C. Note that above 940°C, the separation between the 8-A/cm^2 curve and the 4-A/cm^2 curve is greater than the separations between the 4-and 2-A/cm^2 curves or the 2- and 1-A/cm^2 curves. This large change in separation indicates an increase in the work function.

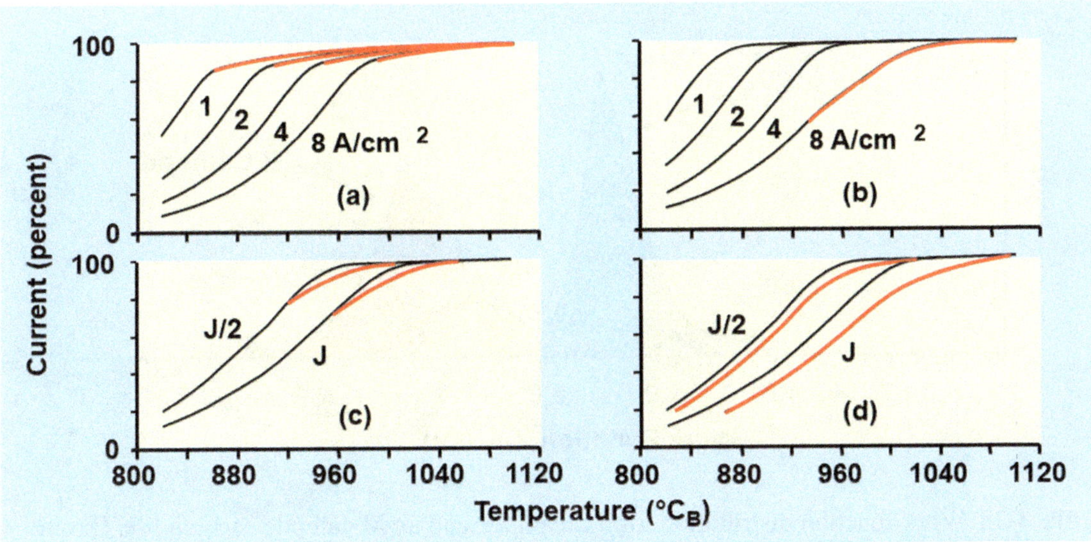

Figure 4-27 Miram plots showing evidence of cathode and grid defects. (From: M. J. Cattelino et al. and T. J. Grant, *Technical Digests*, 1982 and 1984 IEDMs. © 1982 and 1984 IEEE.)

In the plots shown in Figure 4-27(c), the knees of the curves have changed, as indicated by the red lines. The change may have occurred as a function of time, or the change may represent the difference between a cathode assembly that is known to be operating properly and another assembly that is supposed to be identical to the first. If only the knees of the curves have changed as shown in Figure 4-27(c), then it is likely that something has caused a nonuniformity in the electric field near the cathode surface. With a gridded electron gun, this could be movement or misalignment of one of the grids.

If both the spacing and shapes of the knees have changed as shown in Figure 4-27(d), then there is a cathode emission problem and possibly a misalignment of gun electrodes.

4.3.5 Work Function Distribution

Because the energy of desorption and the work function vary from grain to grain, the work function of a cathode is not a single-valued quantity but instead has a distribution of values. For example, the B cathode can have twice as many crystalline surface orientations as a coated cathode, causing a broader work function distribution as shown in Figure 4-28.

To understand the overall emission characteristic of a cathode, it is necessary to understand the distributed nature of the work function. An important consideration is the thickness of the electron space-charge cloud at the surface of the cathode during space-charge-limited operation. Because of the finite velocities of emission of electrons, there is a small depression in the potential (~0.2V) near the surface of the cathode. The location of the position where the potential is a minimum is a measure of the thickness of the space-charge cloud. Figure 4-29 shows the distance between the potential minimum and the cathode surface. Note that at emission densities of a few A/cm^2, which are typical for coated dispenser cathodes, the position of the potential minimum is comparable to the tungsten grain size (see Figure 4-29) and thus the emission patch size, so we can conclude that the various patches emit independently at rates determined by their individual work functions.

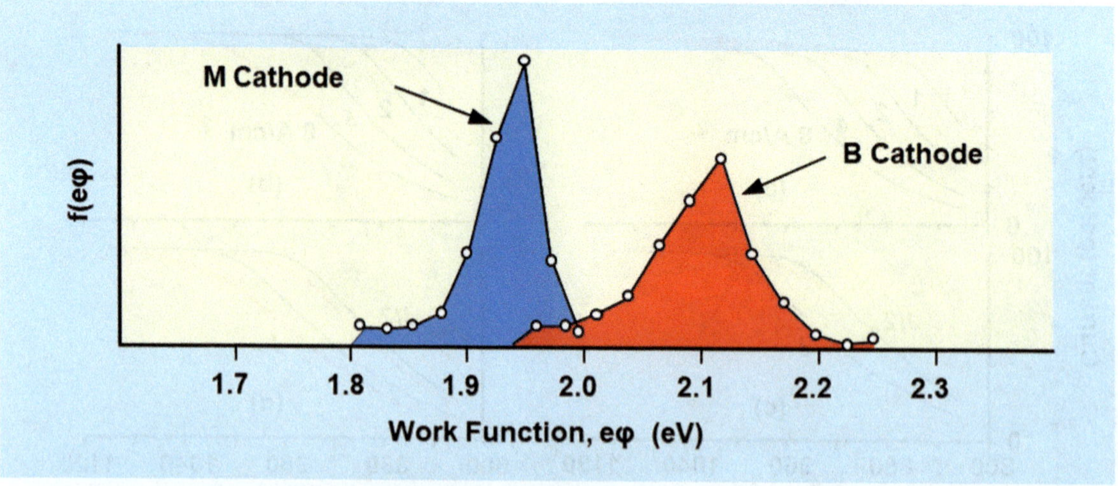

Figure 4-28 Work function distribution for a B cathode and an M cathode early in life. (From: T. J. Grant, *Technical Digest*, 1986 IEDM. © 1986 IEEE.)

Because the various patches operate independently, the surface of the cathode behaves like an array of diodes with varying work functions that are connected together in parallel. Thus, the emission characteristic for the cathode can be modeled as a parallel combination of the emission characteristics of the diodes (patches). In the TL region of operation of each patch, the Richardson-Dushman equation applies. In the space-charge-limited region (indicated by *FSCL* to show that the calculation is based on an operation that is fully space-charge-limited; that is, no part of a patch is operating temperature-limited), the Child-Langmuir law applies.

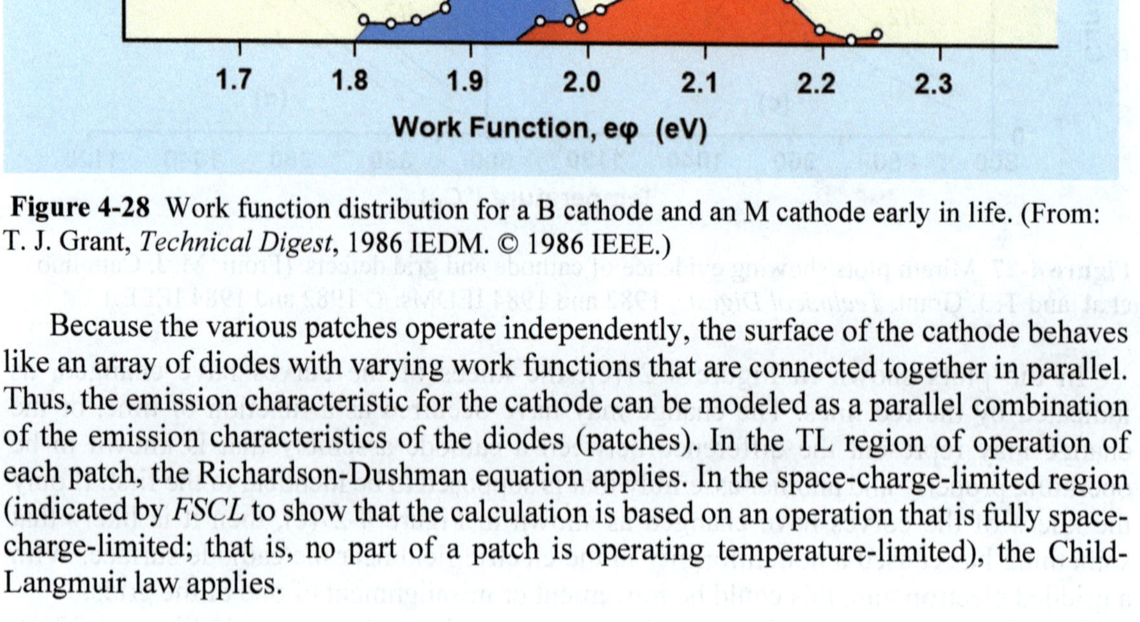

Figure 4-29 Distance between potential minimum and cathode surface.

Figure 4-30 shows, conceptually, how the TL and FSCL curves can be combined to obtain the overall emission characteristic (Miram curve), which is often called a *roll-off curve*. Notice that below the FSCL current, emission consists of a combination of currents from patches that are operating TL and from other patches that are operating space-charge-limited (SCL). In Figure 4-30, all currents are plotted as percentages of the FSCL values. It has been assumed that all emission patches operate at the same temperature and that the electric field applied by the anode is the same for all emission patches.

Drawing current from a thermionic cathode cools the cathode because each electron takes away, at a minimum, energy equal to the work function times the electronic charge. If the work function is uniform across the cathode surface, the power removed is equal to the work function times the average current. With a pulsed cathode, the cooling effect is dependent on the duty cycle. When the cathode current density is high, the cooling effect can be significant and the duty cycle can affect the cathode current.

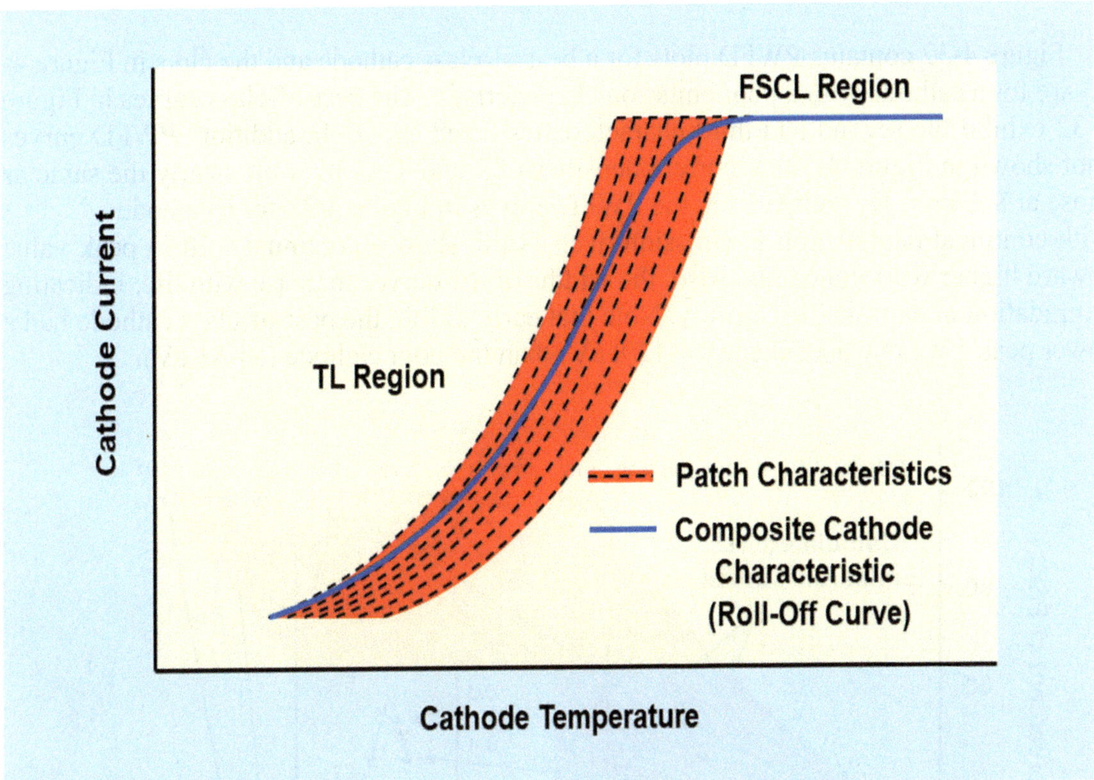

Figure 4-30 Emission curves for several patches combine to form the overall emission characteristic (roll-off curve) for a cathode.

In an ideal electron gun, the emission capability of the cathode far exceeds the current drawn. The current is for space charge limited and is independent of small changes in cathode temperature. In reality, at high cathode current densities this is not necessarily the case. Because of the distributed nature of the work function, a small change in temperature may cause some high-work-function grains of the cathode surface to become temperature limited. Then, a change in duty cycle, which affects cathode temperature, will affect cathode current.

Cattelino and Miram have used the *practical work function distribution* (PWFD) as a tool for evaluating cathode quality and predicting life [15]. Figure 4-31 shows how the PWFD plot is constructed. First, the Miram curve is overlaid on a set of TL emission curves calculated using the Richardson-Dushman equation. Currents up to the 0.5 FSCL value are normalized with respect to the FSCL current and plotted against the work function. Currents above the 0.5 FSCL value are subtracted from the FSCL current, normalized, and plotted against the work function.

Cattelino and Miram used PWFD to compare cathodes. The best of class is the cathode of a particular type such as the M-type that has the best emission performance. The PWFD curves for best-of-class cathodes have the following characteristics:

1. The curves are nearly the same for all current densities;
2. The mean value and the width of the curves are nearly constant throughout life;
3. The curves are narrow.

Figure 4-32 contains PWFD plots for a best-of-class cathode and the plots in Figure 4-33 are for a cathode with a poor emission characteristic. The best-of-class curves in Figure 4-32 exhibit the second and third characteristics listed above. In addition, PWFD curves (not shown in Figure 4-32) at current densities of 2 and 4 A/cm^2 were nearly the same as those at 8 A/cm^2. By comparison, the PWFD curves in Figure 4-33 for a cathode with continual degradation in emission over its life show a continual shift in peak value toward higher work functions. Also, the widths of the curves increase with life, indicating degradation in emission uniformity. Note that early in life, the best-of-class cathode had a lower peak PWFD work function (~1.78 eV) than the poor cathode (~1.81 eV).

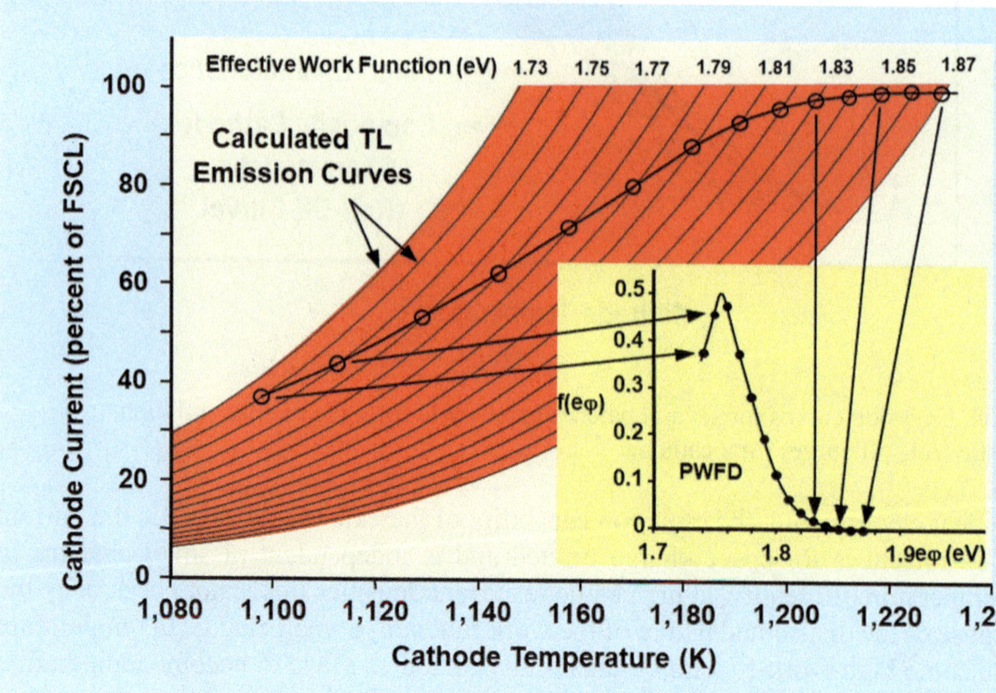

Figure 4-31 Determination of the PWFD. (From: Cattelino and Miram, *Applied Surface Science*, Vol. 111, 1997.)

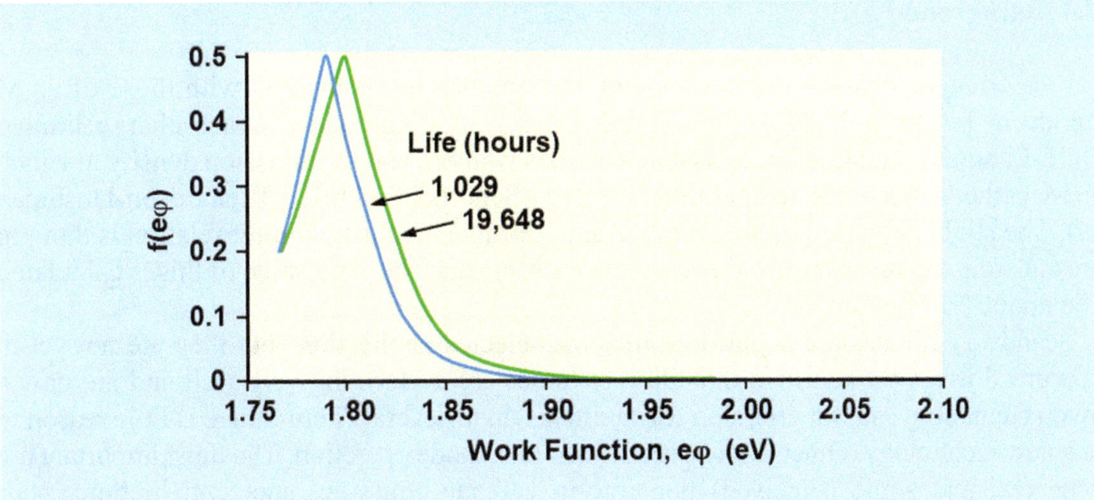

Figure 4-32 PWFD for best-of-class alloy-coated cathode at 8 A/cm^2. (Adapted from: Cattelino and Miram, *Applied Surface Science*, Vol. 111, 1997.)

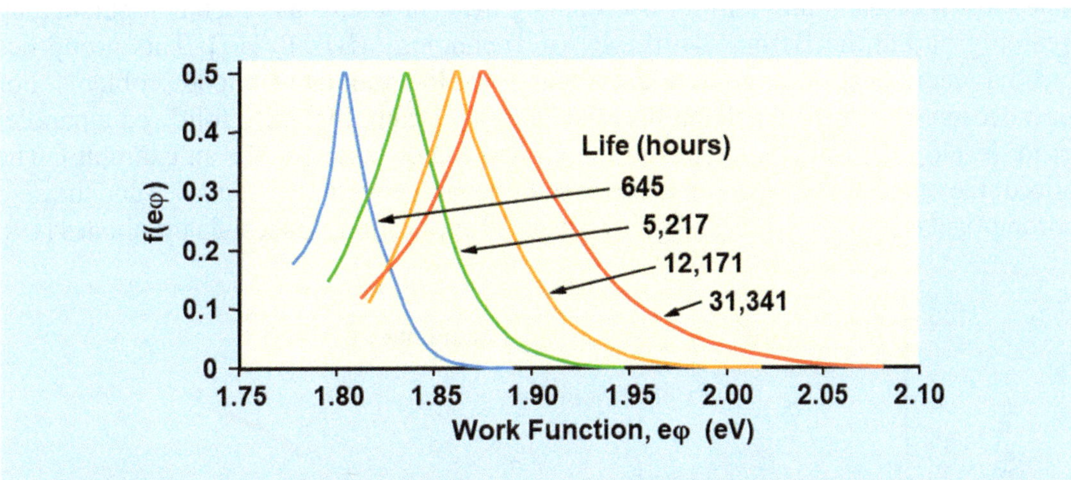

Figure 4-33 PWFD for an alloy-coated cathode at 8 A/cm^2 with poor emission and life. (From: Cattelino and Miram, *Applied Surface Science*, 1997.)

4.4 SCANDATE CATHODES

Scandate cathodes represent the next big step beyond alloy cathodes toward Pierce's vision of an ideal cathode (Section 4.1.). For over 40 years, there has been considerable interest in scandate cathodes because of their very low work function and the possibility of very long life. A scandate cathode is a dispenser cathode that has been modified by introducing a small amount of scandium oxide (Sc_2O_3 or scandia) into the porous tungsten pellet or into the impregnant along with the barium calcium aluminate. The scandia reduces the work function of the emitting surface to ~1.5 eV or lower. As a result, emission densities of tens of A/cm^2 at temperatures < 900°C can be attained. The mechanism by which scandium lowers the work function is not known.

4.4.1 Background

The performance characteristics of a scandate cathode are compared with those of an M cathode in Figure 4-34 [16]. Notice that the scandate cathode is space charge limited (Child-Langmuir equation) at emission densities well beyond the emission density at which the M cathode becomes temperature limited (Schottky modified Richardson-Dushman law). The slight deviation of the scandate data from the Child-Langmuir relation is thought to result from the resistance of the scandate emitting surface. The value of this is calculated to be about 2.5 ohm/cm^2.

Scandate cathodes are being used in some selected applications but they are not yet in widespread use. One reason is that alloy cathodes are performing very well and are now a proven technology, so, for emission requirements up to several A/cm^2, there is little reason to use a new technology. Other reasons are related to cathode operation. The most important are (some real and some perceived) nonuniform cathode emission, uncertain lifetime, poor reproducibility, and possible vulnerability to poisoning and to arc and ion damage. These problems are compounded by the lack of understanding of the role played by the scandium in enhancing emission.

The most successful program for overcoming these problems has been that pursued by a research group at the Beijing University of Technology (BJUT) [17]. This group has followed a methodical program to address and solve the scandate cathode problems. For the past decade, Bernard Vancil and his associates at eBeam, Inc. have followed a parallel effort to develop the BJUT cathodes in the United States (see [18] for an example). The benefits of the scandate cathodes are recognized by several government agencies and they are supporting fundamental research as well as advanced applications for scandate cathodes [19].

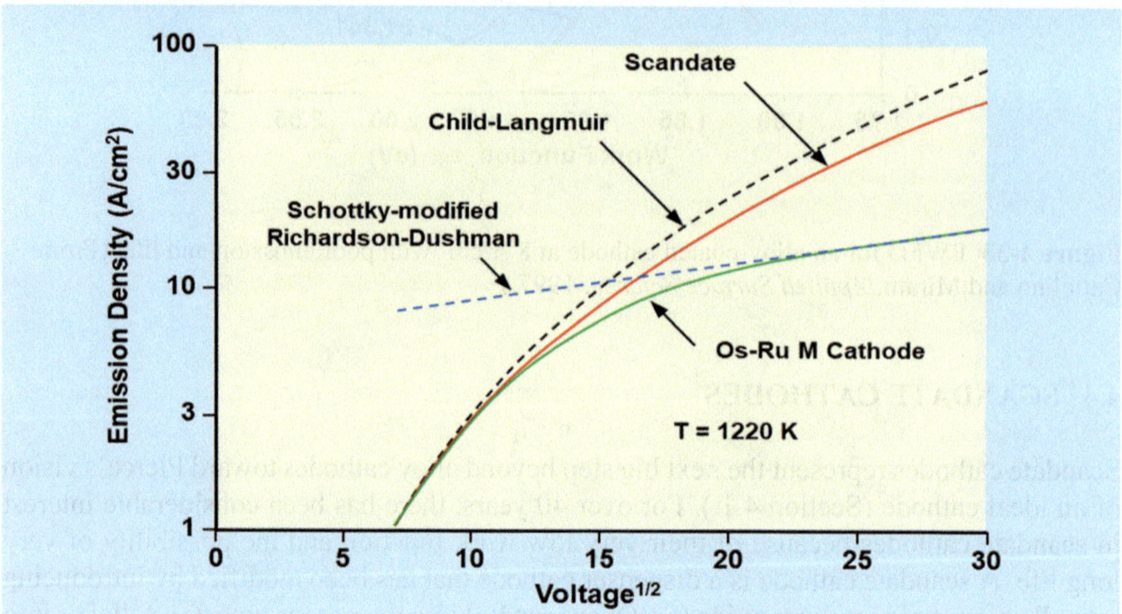

Figure 4-34 Justification given by Philips Research Lab for interest in scandate cathodes. (Adapted from J. Hasker et al., App. Surface Science, 26, 1986.)

One example of an immediate application is for tubes operating at very high frequencies where beams are small enough so that thermal effects govern beam size. With a higher emission capability, the cathode in one of these tubes could be smaller so the area compression ratio of the gun and the effective beam temperature could be reduced. Beam focusing could be improved and beam noise could be reduced.

Another example is in multibeam devices such as multibeam klystrons (MBKs) which require several properly focused electron beams. Each beam is formed in a Pierce gun with a beam area convergence ratio that makes it possible to use a relatively large cathode so as to limit cathode emission density and ensure long cathode life. The size of the resulting gun complex on a multibeam device is proportional to cathode size and, therefore, inversely proportional to the cathode emission density.

It would also be very desirable to be able to use cathodes with reduced operating temperatures in many existing applications. With these cathodes, the rate of evaporation of emissive material would be reduced. For example, a 100°C reduction in temperature would result in an order of magnitude reduction in evaporation rate. If depletion of emissive material was the primary reason for failure, then life would be increased by as much as an order of magnitude. Reduced evaporation would also reduce tube failure rates resulting from grid shorts, electrical leakage, and breakdown. The traveling-wave tube with a life of only about 5,000 hours in the Navy's SPQ-9B radar is an example of a tube that would benefit from the use of an improved cathode.

4.4.2 Structure and Fabrication

There are, basically, three types of scandate cathodes:

1. Scandia-added impregnated cathodes are similar to conventional impregnated cathodes except that scandium oxide is added to the impregnant compound containing barium, calcium, and aluminum oxides.
2. Scandia-doped tungsten matrix cathodes in which pellets are made by pressing and sintering tungsten powder:
   - To which scandium oxide has been added in a doping process,
   - Which are then impregnated with barium and calcium aluminates [17].
3. A top-layer cathode consists of a porous tungsten pellet impregnated with a compound of barium, calcium, and aluminum that is overcoated with a porous layer consisting of tungsten and scandium oxide particles [16].

Porous tungsten pellets are used in these scandate cathodes just as in other dispenser cathodes. Emission densities of tens of A/cm^2 have been demonstrated with each of these cathodes. The best and most consistent operation of scandate cathodes has been by using the scandia-doped, pressed-cathode process with ~ 1 μm tungsten powder.

As shown in Figure 4-35, the emitting surface of the scandate cathode is very different than the surface of a conventional dispenser cathode. The entire surface of the grain in the M cathode emits, whereas it appears that only 100-nm diameter areas produced by activation of 50-nm diameter scandate crystals emit [21].

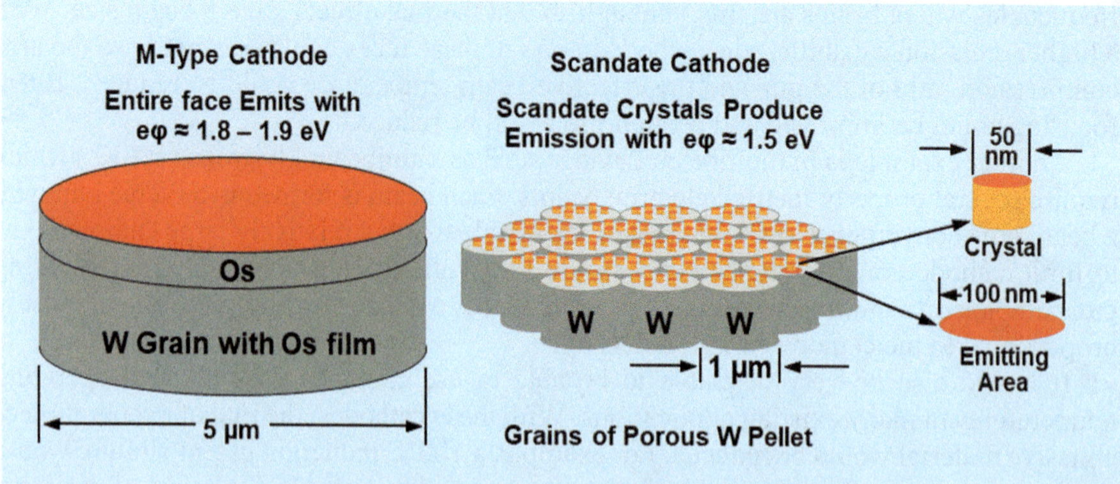

Figure 4-35 Comparison of surface features of M-type and scandate cathodes.

Figure 4-36 supports the validity of the model of the scandate cathode shown in Figure 4-35. The electron sources show up as bright spots with an average diameter of 100 nm [21]. It is presumed that each emission spot is produced by the activation of a scandate crystal.

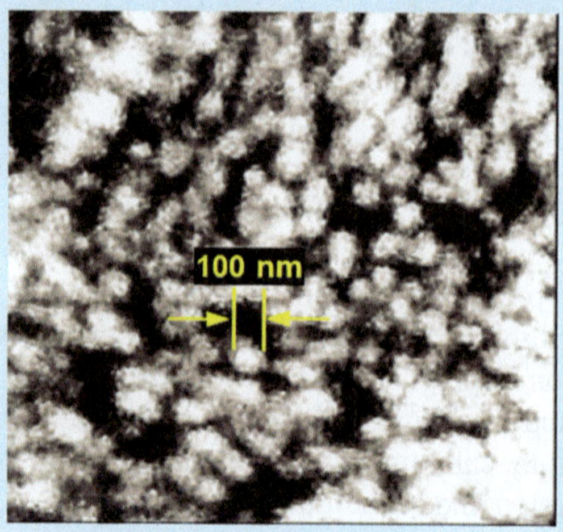

Figure 4-36 Image of scandate cathode surface. Bright spots are emitting sites. The average diameter of an emission spot is 100 nm. (From I. Brodie, B. Vancil, IVEC 2016.)

The surface features of scandia-doped matrix cathodes have been studied in detail in an attempt to explain the operation of these cathodes. Figure 4-37 from Yiman Wang et al shows depth profiles of the elements on the surface of a scandate cathode and an M cathode. SiO_2/Si was used as a sputter rate standard [17]. On the M cathode, the emitting layer of barium and oxygen is only about 10 nm thick, which corresponds to the well-accepted monolayer thickness for an M cathode. The emitting layer on the BJUT scandate cathode is far more complex and is about 100 nm thick.

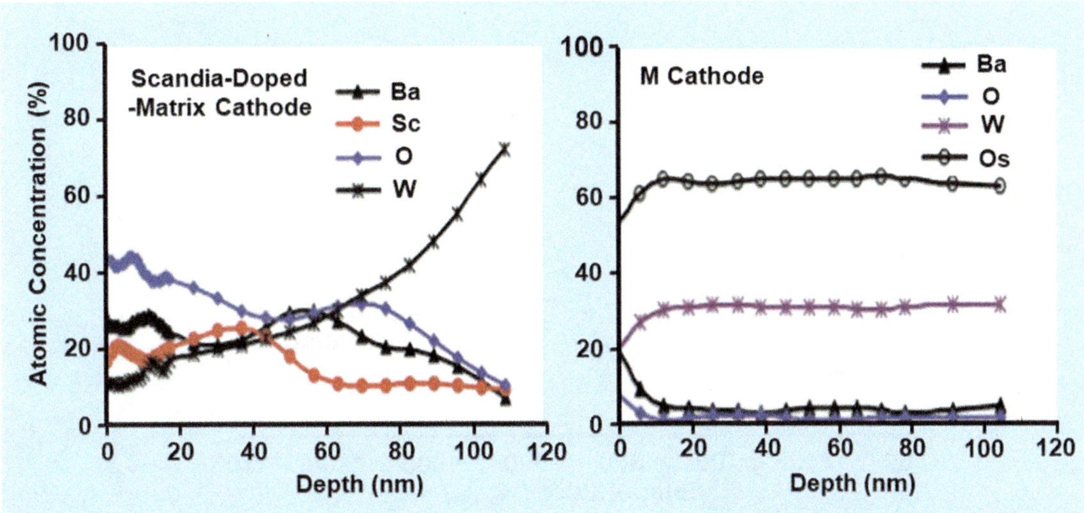

Figure 4-37 Surface elements for a Scandia-doped-matrix impregnated cathode and for an M cathode. (From Y. Wang et al., *IEEE Trans. Electron Devices*, Vol. 54, No. 5, May 2007.)

There are significant differences between the operation of a scandate cathode and a conventional dispenser cathode. The emitting surface of an M cathode or an alloy cathode is treated as a monolayer and is described very well by the Richardson-Dushman equation for pure metal emission.

The emitting surface of a scandate cathode is a semiconductor layer. It has characteristics of emission from a pure metal combined with semiconductor characteristics. One of these characteristics is the resistance of the semiconductor layer. An example of an effect of resistance was given in Figure 4-34. As we observed in Figure 4-34 and as pointed out by Brodie [22], the voltage applied to the scandate cathode is reduced by the increment of voltage dropped across the resistance of the scandate semiconductors. Brodie assumes that this resistance decreases exponentially with increasing temperature. He uses this effect to explain the poorly defined knee observed on the roll-off curves for scandate cathodes with insufficient Scandia doping.

The results of experiments reported by Miram (Figure 4-47), by Yang (Figure 4-48), and by Hasker (Figure 4-34), and others support the suggestion that the resistance of the scandate crystals must be considered.

Brodie and Vancil have stated that "For optimum emission at zero field at a given temperature, the number of crystallites per unit cathode area should be as high as possible to the point of complete coverage" [23]. The data published by Vancil and displayed in Figure 4-38 shows a transition comparable to that for an M cathode [18]. With an operating temperature ~100°C below an M cathode and ~50°C below an alloy cathode, this cathode should have a life in excess of 100,000 hours. This data was presumably obtained for a cathode with an extremely high density of scandate crystals.

Tungsten grain size is found to be very important in maximizing emission current density as well as optimizing emission uniformity in scandate cathodes. Early pressed scandate cathodes were made with traditional tungsten powder having ~5 μm grains. Typical grain dimensions are shown in Figure 4-19.

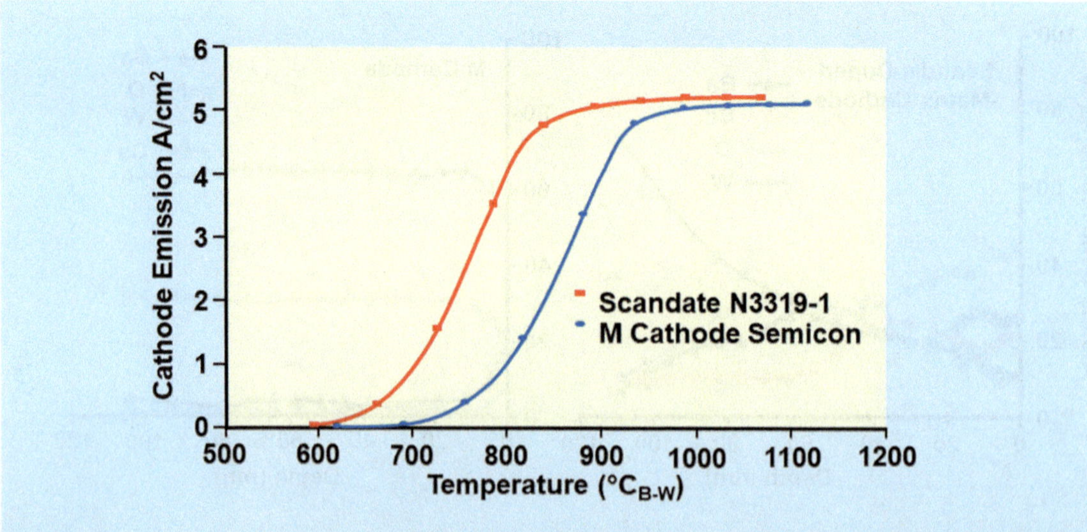

Figure 4-38 Scandate cathode and M cathode roll-off curves. (Adapted from: B. Vancil et al., IVEC 2013.)

Wang et al. performed a series of experiments with scandia-doped tungsten matrix pressed cathodes made with varying grain sizes and different mixing methods shown in Figure 4-39 [17]. Emission results are shown in Figure 4-40. The cathodes made with the small 200- to 400-nm and 600- to800-nm grains produced emission densities 2 to 2.5 times larger than the cathodes made with larger grains. The cathodes made from the smallest grains had insufficient porosity or fractured easily so further development was not pursued.

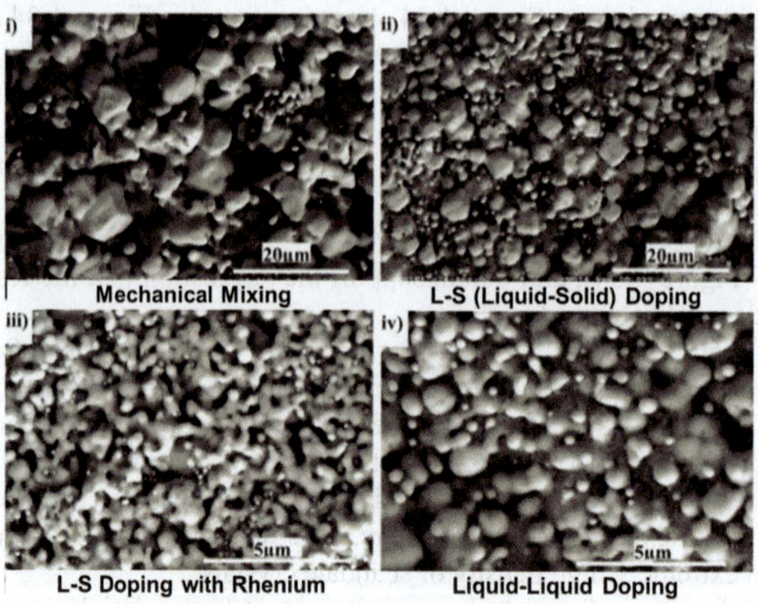

Figure 4-39 Scandia doped tungsten powders used for emission tests. (From: Y. Wang et al, *IEEE Trans. Electron Devices*, Vol. 54, No.56, June 2007.)

At the completion of these tests, Wang concluded [22]

> Based on the above results, it was recognized that the key factor is the sub micrometer microstructure of the matrices, with nanometer-size particles of Scandia uniformly distributed throughout the interior of the matrix. This microstructure plays a dominant role in the improved emission capability and uniformity of such cathodes.

The slopes listed in Figure 4-40 are said by the authors to be measures of the uniformity of the cathode. Alternatively, this may be a series resistance effect. For an ideal diode with a fully space-charge-limited cathode, $I \propto V^{3/2}$ so the slope of the I-V characteristic is 1.5. Scandium oxide is a highly resistive material so the emitting surface of the cathode must be like the sketch in Figure 4-41. That is, it functions as the parallel connection of space-charge-limited diodes, each of which has a resistor in series. The calculated resistance, R, that would produce each of the slopes listed in Figure 4- 40 is listed with the corresponding slope. The resistance of the cathodes with large tungsten grains is three times the resistance for those with small grains. This may imply that there is a factor of three fewer emitting particles in the cathode with large grains than for the cathode with small grains.

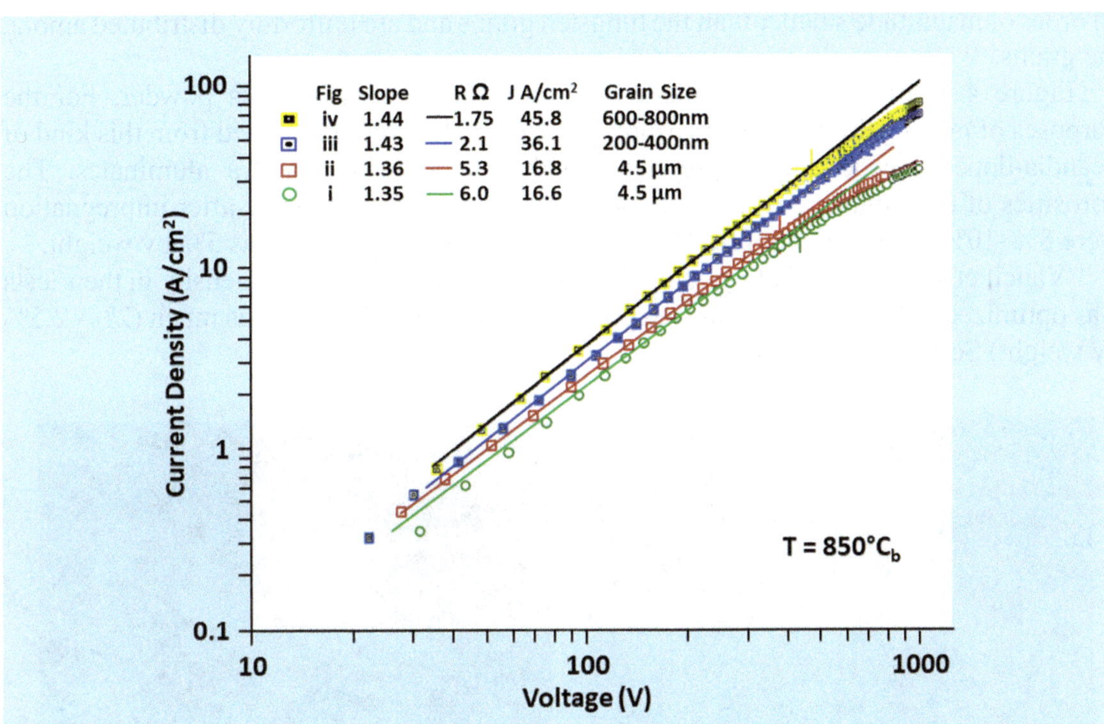

Figure 4-40 Emission densities obtained for cathode pellets fabricated from tungsten powders in Figure 4-39. (From: Y. Wang et al., *IEEE Trans. Electron Devices*, Vol. 54, No.56, June 2007.)

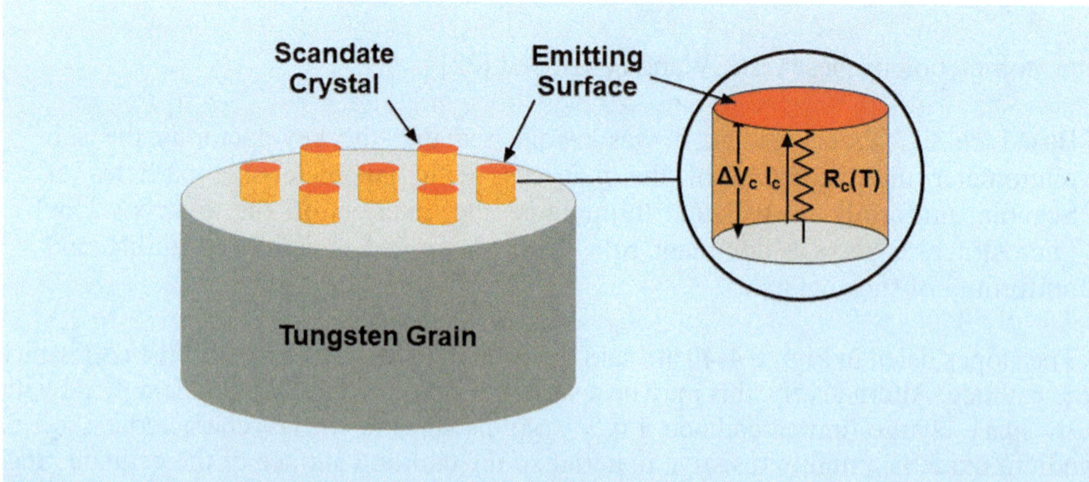

Figure 4-41 Model used in explaining the effect of the resistances of scandate crystals.

Additional studies and experiments by Wang [24] resulted in the development of the ~1-μm tungsten powder that was mixed with scandium oxide particles using the spray-drying procedure. That powder is shown in Figure 4-42(a). Notice that the tungsten grains are very uniform in size with dimensions of about 1 μm. The scandium oxide crystals are an order of magnitude smaller than the tungsten grains and are uniformly distributed among the grains.

Figure 4-42(b) shows the cathode structure after processing this powder. For the purposes of making dc emission measurements, cathodes were fabricated from this kind of Scandia-doped powder and impregnated with S type (411) Ba, Ca aluminates. The porosities of the cathode pellets were 24%–28% and the weight gains after impregnation were 8%–10%. The amount of Sc_2O_3 added to the tungsten powder was 5% by weight.

Vancil et al. reported that by using their doping process, emission density in their tests was optimized with the use of tungsten powders containing about half as much (2%– 2.5% by weight) Sc_2O_3 as Wang used [25].

Figure 4-42 Scandia-doped tungsten powder (a) and sintered and pressed cathode pellet (b). The tungsten grains are quasispherical in shape and the distribution of scandium oxide particle sizes is quite uniform with a size of approximately 50 nm. (From Y. Wang et al. 2014 IVEC.)

4.4.3 Operation

Test results for the powder in Figure 4-42 are shown in Figure 4-43. The current densities listed are the densities at the first departure of the measured current from the linear plot. The current densities at 950°Cb are for pulsed operation as well as for DC operation. Pulsed current densities of 140 A/cm^2 and dc current densities of 70 A/cm^2 were obtained at 950°Cb [20].

As shown in Figure 4-44, a knee temperature of 900°CB-Mo was measured from Miram curve data at an initial pulsed current density of 58 A/cm^2. A peak work function of 1.41 eV was obtained from the PWFD plot.

The PWFD plot for this scandate cathode is far better than many scandate cathodes but it is not as well defined as that for an alloy cathode as shown in Figure 4-44. A narrow PWFD like that for the alloy cathode is desirable because it makes it possible to set the operating temperature of the cathode closer to the knee of the roll-off curve than is possible with a broad PWFD. This minimizes cathode temperature and maximizes cathode life.

Cathodes fabricated using the tungsten powder shown in Figure 4-42 have been life tested at eBeam Inc. The diode used for the tests was a Pierce gun with a drift tube that simulated the beam environment in a TWT or klystron. A 1,000-hour life test was performed at 50 A/cm^2 with a pulse width of 40 μs and a duty cycle of 0.05%. The cathode face temperature was 1005°CB-W. The emission was steady for nearly the entire test period.

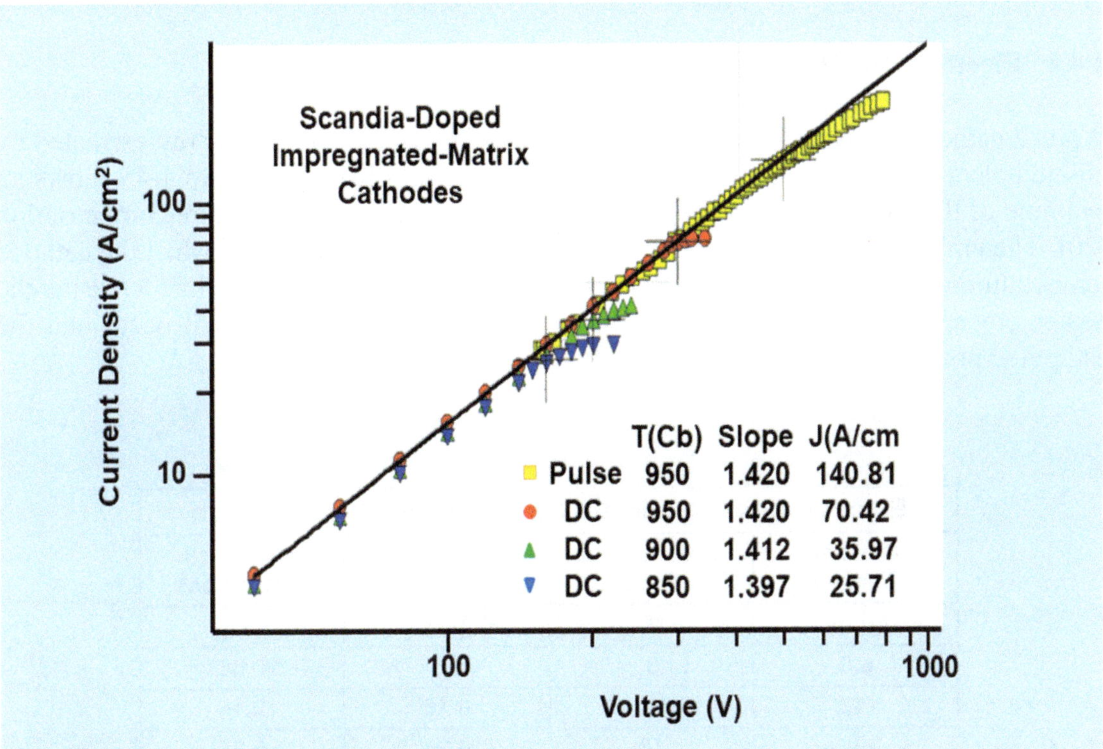

Figure 4-43 DC and pulsed emission densities for scandia-doped, impregnated matrix cathodes with temperature corrections for electron cooling. (From Wenlong Liang et al. *IEEE Trans Electron Devices*, Vol. 61, No. 6, June 2014.)

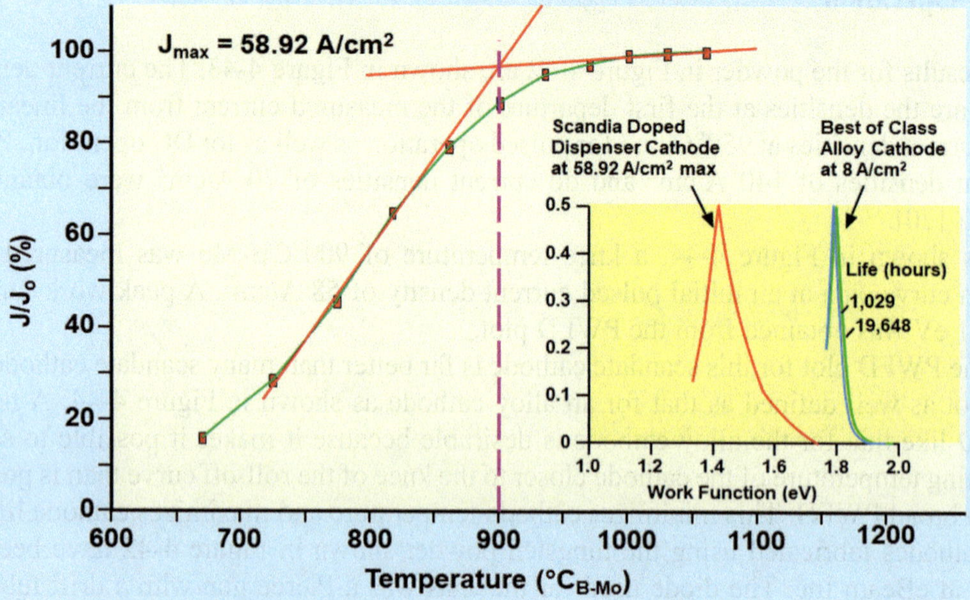

Figure 4-44 Miram curve for cathode fabricated with scandia-doped tungsten powder. The PWFD is compared with the PWFD of an alloy cathode in the insert. (Adapted from: Y. Wang et al., IEEE IVEC 2014, pp. 51–52 and combined with Figure 4-32.)

4.4.4 Electron Cooling

When a cathode is operating at a high average or high dc current, the energy extracted by emitted electrons is high enough to cool the cathode significantly. Figure 4-45 shows an example of the change in emission density that can occur with temperature compensation [20]. The emission density increased from 17.75 A/cm^2 to 27.48 A/cm^2 when the cathode temperature was brought back to the zero emission value. This was done by increasing the heater power. Table 4-2 contains a listing of the temperature drops that occur and the estimated power lost for various emission densities.

Table 4-2

Electron Cooling Effect & Heater Power Compensation

Current Density	Temperature Drop	Heater Power Comp.	Estimated Power Loss
A/cm²	°C	(W)	(W)
6.0	3	0	0.08
11.5	9	0.16	0.15
29.5	18	0.24	0.25
35.4	28	0.44	0.45
43.1	32	0.49	0.55

From W. Liang et al., *IEEE Trans. Electron Devices*, Vol. 61, No. 6, June 2014.

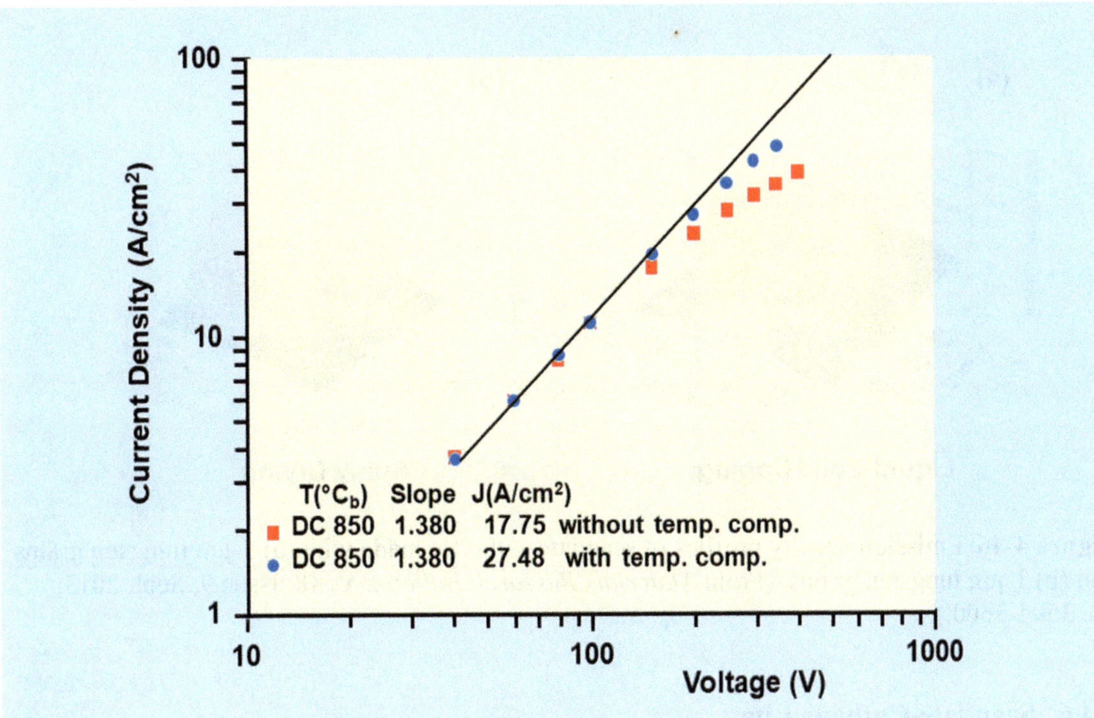

Figure 4-45 Electron cooling becomes significant at high emission densities. (From W. Liang et al., *IEEE Trans. Electron Devices*, Vol. 61, No. 6, June 2014.)

4.4.5 Emission Uniformity

Until recently, the lack of uniformity of the emission from scandate cathodes was a severe problem. Among other things, nonuniform cathode emission contributes to difficulties in beam focusing and control. The reason for nonuniform emission is clear from Figure 4-34. That is, the density of emitting crystals was not uniform.

In Figure 4-46(a), the emission density is shown for a cathode made with 5 μm tungsten grains that were mechanically mixed with scandate crystals. The nonuniformity of emission in this plot is thought to be representative of the nonuniform emission from most scandate cathodes. The plot on the right in Figure 4-46(b) shows the dramatic improvement made in emission uniformity with the BJUT cathodes.

There are two fundamental differences between the processes used to make the cathodes in Figures 4-46(a) and 4-46(b). First, the size of the tungsten grains in (b) was nearly an order of magnitude smaller than in (a) and the grains were much more uniform in size and were quasispherical. The total surface area of the tungsten grains in (b) was much larger than in (a) and so far more scandate crystals, which attach to the tungsten grains, could be accommodated in (b) than in (a).

Second, the process for preparing the mix of tungsten grains with scandate crystals was different. For the cathode used in Figure 4-46(a), the tungsten powder was mixed mechanically with scandate crystals. The cathode for the plot on the right was made with nano-sized, scandium-doped tungsten powder having a grain size of about 1 μm. That powder was made by processing a liquid containing tungsten and scandium [24].

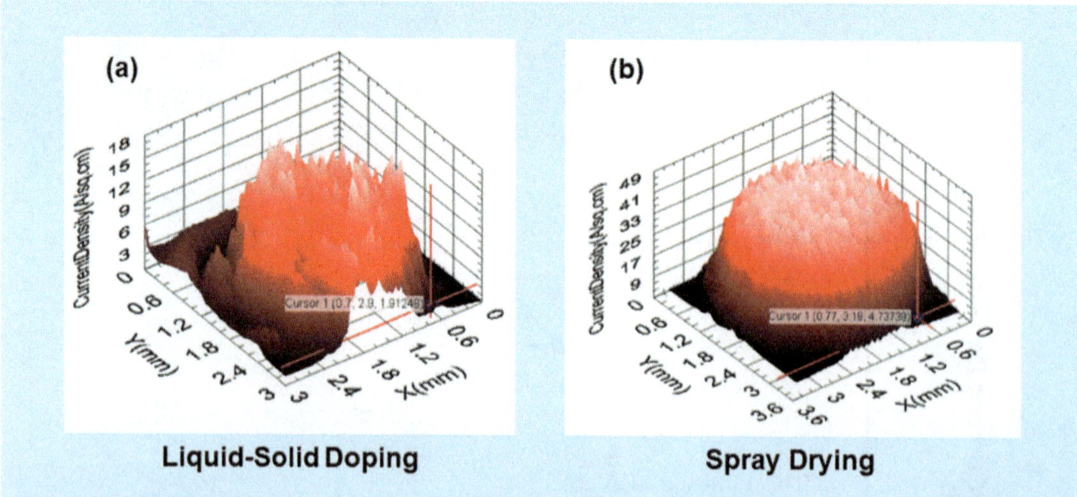

Figure 4-46 Emission density profiles of scandate cathodes made with (a) 5-μm tungsten grains and (b) 1-μm tungsten grains. (From *Materials Research Bulletin*, V. 48, Issue 9, Sept. 2013, pp. 3594-3600.)

4.4.6 Scandate-Cathode Life

The life of existing dispenser B and M cathodes is well defined by the supply of barium and the operating temperature. If barium depletion like that in B and M cathodes was the only reason for failure of a scandate cathode, then life prediction would be straightforward. However, as Figure 4-37 shows, the surface of the scandate cathode may be very different than the surface of a B or M cathode. Also, the role played by the scandium in reducing work function is not known and that certainly has an effect on life.

There are many referrals to life tests in the references but the majority of the tests are a few hundred to a few thousand hours long. As an example, cathodes fabricated using the tungsten powder shown in Figure 4-42 have been life tested for several thousand hours with a dc load of 20- 40 A/cm^2 at eBeam Inc. The diode used for the tests was a Pierce gun with a drift tube that was supposed to simulate the beam environment in a TWT or klystron. Presumably, the cathode would be exposed to ion bombardment similar to that encountered in a linear-beam tube [26].

Paff et al. mention a longer test but leave the reader wondering about the value of the test by stating that "The temperature and knee point data appear to show a moderate change only over ~ 20,000 hours" [27].

4.4.7 Robustness

The emission capability of the nano-sized tungsten grain cathodes increases as tungsten grain size decreases. However, when grain size is made too small, for example ~0.1–0.2 μm, the pressing and sintering process used to make the pellets fails to make pellets that are strong enough to withstand machining operations. A compromise is reached with slightly less emission being obtained with pellets fabricated with ~1 μm grains [17]. It has been demonstrated that cathodes fabricated using 1 μm tungsten grains can be machined to form the curved surface required in a Pierce electron gun [28].

Emission properties after air exposure and gas poisoning have been studied. It was found that emission could be fully restored after exposure of the cathode to air or poisoning gases [29]. The time required for reactivation was much shorter than the initial activation time.

It was demonstrated that pressure thresholds for poisoning of the SDD cathode by residual gases are about half to one order of magnitude higher than for an M cathode. The comparison was made with the cathodes at the same temperature but with higher emission level for the SDD cathode than for the M cathode. The SDD cathode is easier to recover after poisoning than the M cathode and can operate stably in a nonideal environment.

The effects of ion bombardment on a top layer cathode and on an impregnated cathode have been described by Zhang et al. [30]. Unfortunately, the tungsten matrix is not described and so the conclusion reached by Zhang that the top layer cathode shows better resistance to ion bombardment than the impregnated cathode may not apply to the BJUT cathodes.

4.4.8 Anomalous Behavior

In Figure 4-47, roll-off curves for a scandate cathode and for an M-type dispenser cathode are shown. The temperature-limited and fully-space-charge-limited portions of the scandate curve appear to be able to be explained by the Richardson-Dushman and Child-Langmuir relations. However, between 800°C and 1,050°C, current increases linearly with temperature. To operate at the 100% fully space-charge-limited current, the temperature of this cathode would have to be at least 1,100°C. This abnormal behavior is illustrated again by Ping Yang in Figure 4-48.

Brodie has suggested that the abnormal transition from TL operation to FSCL operation is caused by the resistance of the scandate coating. He assumes that the coating is a semiconductor. The resistance of a semiconductor decreases exponentially with increasing temperature. He uses this effect to explain the poorly defined knees often observed on the roll-off curves for scandate cathodes.

The Yang results in Figure 4-48 appear to support Brodie's suggestion. From 70% to 100% of the maximum current, the currents for emission densities from 2 A/cm^2 to 16 A/cm^2 are constant indicating space-charge-limited operation. However, the space-charge-limited current increases with increasing temperature. This can happen if there is resistance in series that decreases with increasing temperature.

When operating temperature-limited, scandate cathodes exhibit another anomalous behavior. Because there is no space-charge cloud, an electric field is present at the emitter surface. For a metallic cathode, the Richardson-Dushman equation with the Schottky modification describes the temperature-limited emission very well. Then, in theory, Miram plots for current as a function of temperature are separated by ~40°C when current is doubled. However, as shown in Figures 4-48 and 4-49, for the scandate cathode the separation is only ~20°C. Y. Wang attributes this (anomalous Schottky) 20°C behavior to field enhanced thermionic emission resulting from the penetration of the electric field into the semiconductor [31]. This anomalous behavior is of concern primarily because it is an aspect of scandate cathode operation that is not completely understood.

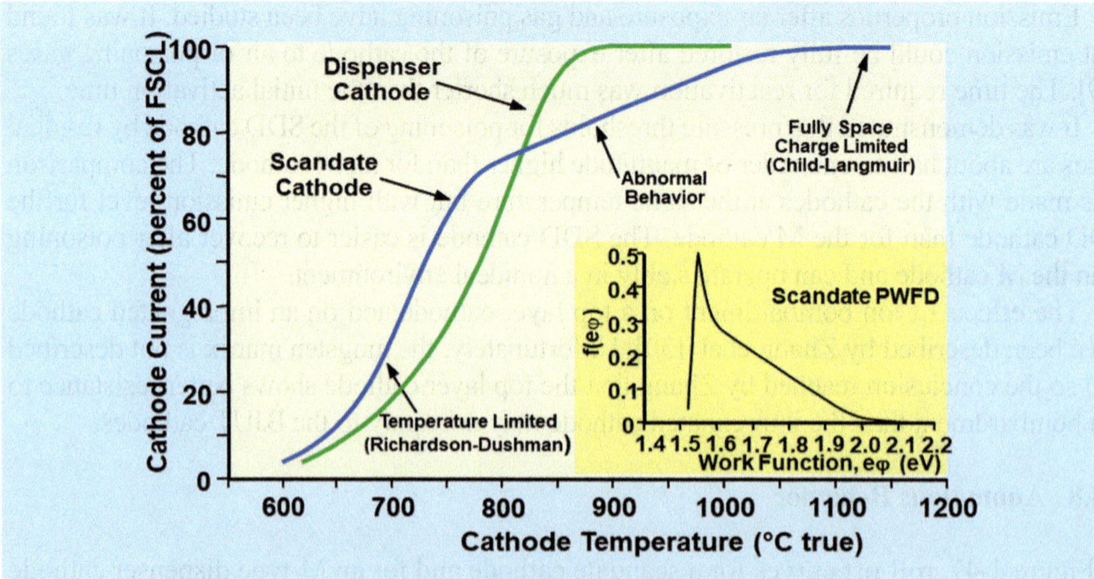

Figure 4-47 Miram curves comparing performances of dispenser and scandate cathodes. (Adapted from: G. V. Miram et al., Fifth IEEE IVEC, © IEEE 2004.)

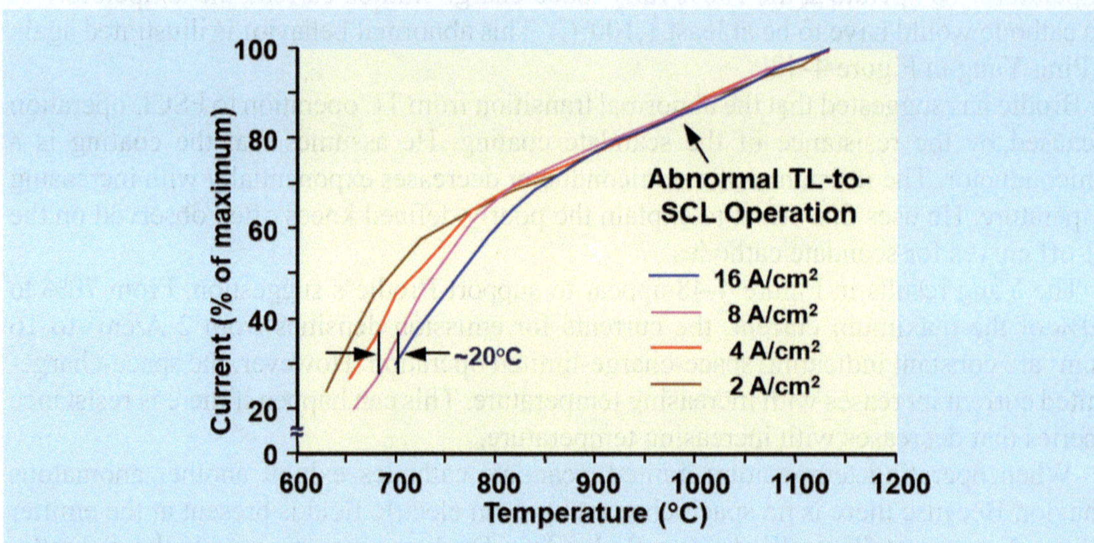

Figure 4-48 Yang data showing anomalous transition from temperature limited to space charge limited operation for a traditional 4:1:1 dispenser cathode coated with a Sc_2O_3-W film ~2200A thick (Adapted from: P. Yang et al., 2009 IVEC, © IEEE 2009.)

4.5 LIFE CONSIDERATIONS

Long-life requirements often range from 10,000 hours to over 100,000 hours. While life testing is sometimes done, it is more common to predict cathode life on the basis of known cathode degradation mechanisms. Grant identified three distinguishable types of emission degradation mechanisms [32]:

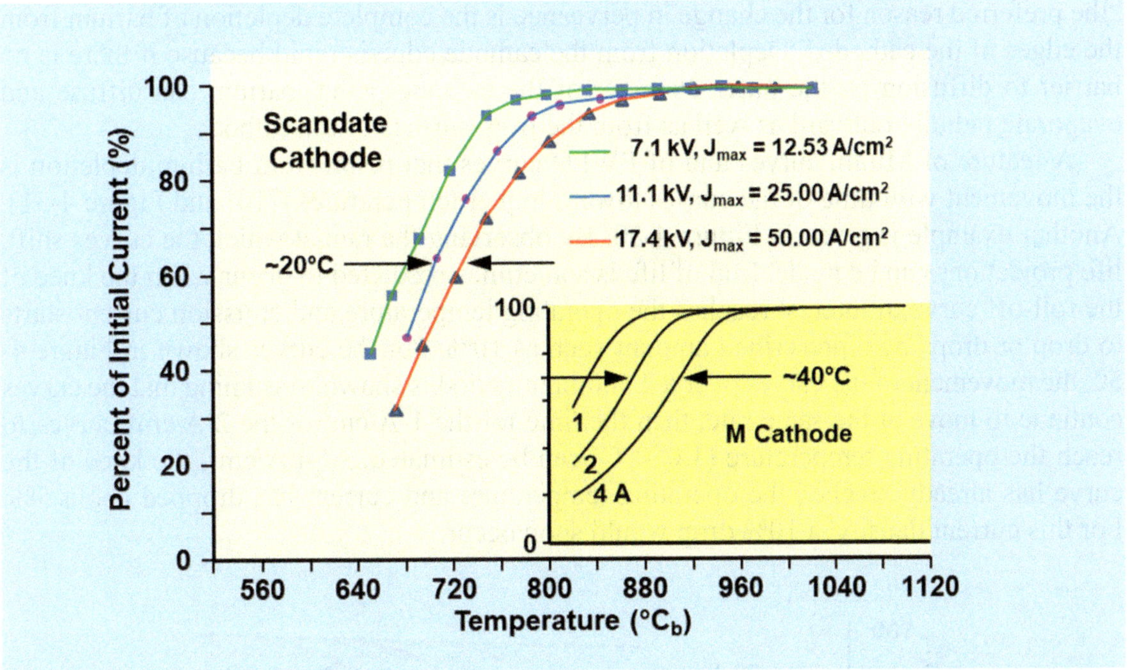

Figure 4-49 Richardson plots for current increase of 2X are separated by ~20°C for the scandate cathode rather than the predicted ~40°C (shown for an M cathode). (Adapted from: Y. Wang et al., IVEC 2009.)

Type I Gradual decrease in perveance; that is, FSCL emission decreases with time;
Type II Change in work function distribution with time and without dependence on changes in cathode temperature;
Type III Change in work function distribution with temperature as well as with time.

The type II mechanism results from a change in the base metal work function with time. This mechanism applies to coated cathodes and is attributed to the diffusion of tungsten through the coating. Grant pointed out that the work function distribution can move to lower as well as higher work functions early in life depending on coating composition, thickness, and interdiffusion rate.

When the type III mechanism occurs, the work function distribution changes in a reversible manner as temperature changes. That is, if temperature is increased or decreased, the distribution increases or decreases and the measurements are repeatable. This can occur early in life and is attributed to an insufficient supply of barium, a change in sticking coefficient of the barium to the cathode surface, or to external poisoning.

The reason given for the type I degradation is usually the gradual depletion of the barium supply in the impregnated tungsten pellet. Thus, the rate at which barium diffuses to the cathode surface decreases, so the barium coverage of the surface decreases and work function increases.

Grant noted that the Miram curves do not always support the conclusion that barium coverage of the entire cathode depletes with time. In some cases, a reduction in current persists even if cathode temperature is increased 150°C above the point where the cathode is 99.5% FSCL. If depletion of the entire cathode had been the reason for the reduced current, the increase in temperature would have increased the current. Grant stated that "the preferred reason for the change in perveance is the complete depletion of barium from the edges of the cathode." Depletion from the cathode edge is rapid because if there is no barrier to diffusion on the outer diameter of the cathode pellet, barium can diffuse and evaporate radially outward as well as from the front surface of the cathode.

A feature of Miram curves and of PWFD curves that result from barium depletion is the movement with time of the curves toward higher temperatures ([10] and Figure 4-31). Another example is given in Figure 4-50. By observing the rate at which the curves shift, life projections can be made. End of life is sometimes predicted to occur when the knee of the roll-off curve of interest reaches the operating temperature and emission current starts to drop or drops by a prescribed amount such as 10%. For the curves shown in Figure 4-50, the movement of the curves over a 2,500-hour period is shown. Assuming that the curves continue to move at the same rate, then the time for the 1-A/cm^2 or the 2-A/cm^2 curves to reach the operating temperature (1,050°C) can be estimated. At 4 A/cm^2, the knee of the curve has already reached the operating temperature and current has dropped about 5%. For this current density, a 10% drop would soon occur.

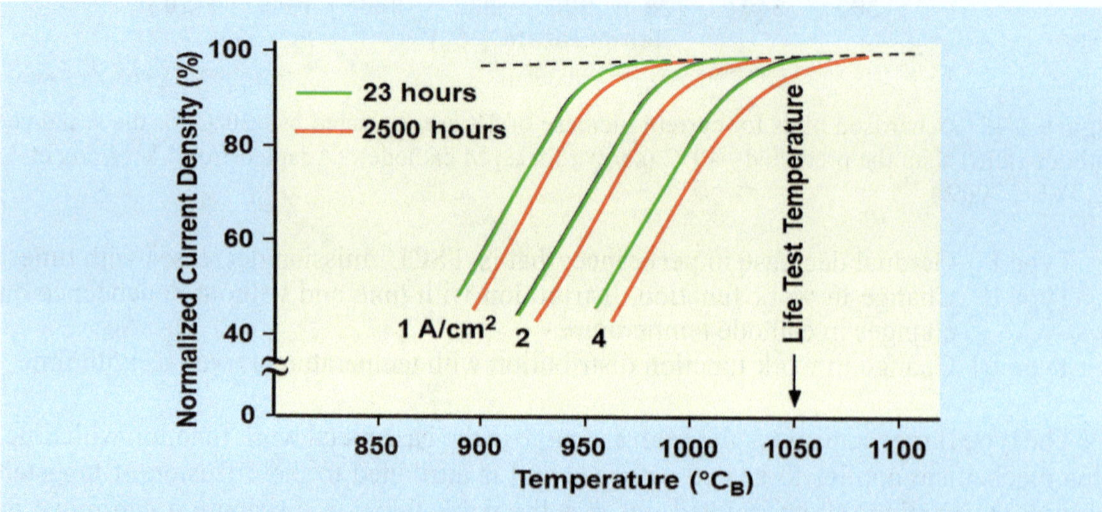

Figure 4-50 Miram plot for B-type cathode. (From: M.C. Green, *Technical Report* RADC-TR-81-211, July 1981.)

Under ideal conditions, the selection of the temperature at which a cathode should be operated to maximize life would be straightforward. For example, consider the roll-off curve shown in Figure 4-49. A cathode could have this emission characteristic if

1. All areas of the emitting surface had the same work function;
2. There was no variation in temperature across the emitting surface;
3. The electric field near the emitting surface was the same everywhere.

End of life would occur when the knee of the curve reached the operating temperature and suddenly there was a precipitous drop in emission current. The operating temperature could be selected by using the life prediction model of Grant and Falce [33]. Life predictions for B and M cathodes are shown in Figure 4-52 and in Figure 4-53.

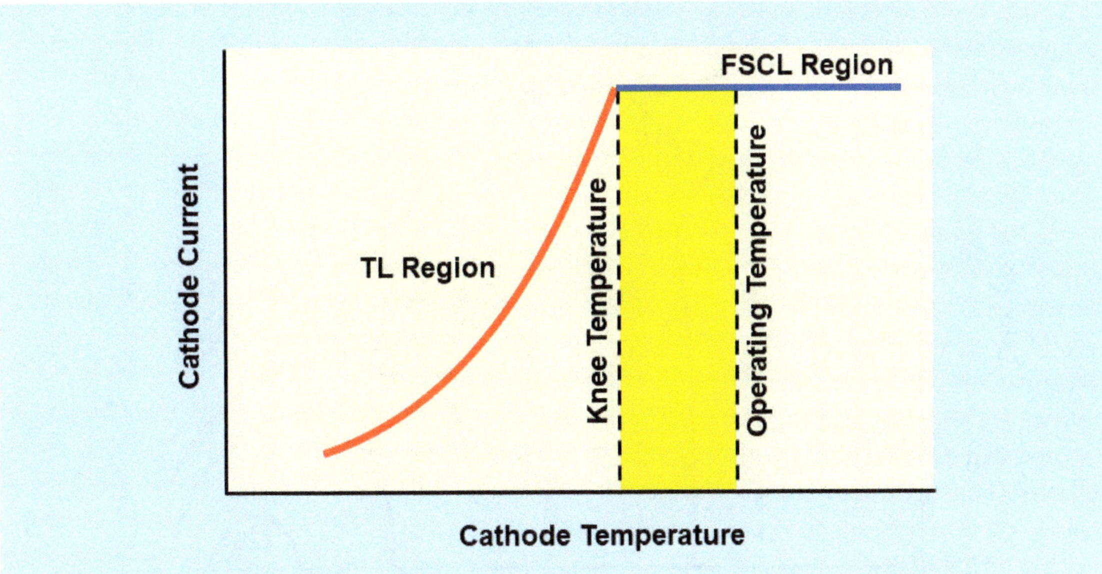

Figure 4-51 Idealized emission characteristic assuming no variations in work function, temperature, or electric field near the cathode surface.

4.5.1 Grant and Falce Life Prediction Model

The model used by Grant and Falce to predict life is based on the type I and type II degradation mechanisms (gradual depletion and base metal work function change) [33]. The curves in Figures 4-52 and 4-53 show results for B- and M-type cathodes. The model is valid for both, the only difference being the variation with time of the base metal work function in the M-type alloy coatings. The time of the onset of temperature-limited operation was defined as the life of the cathode.

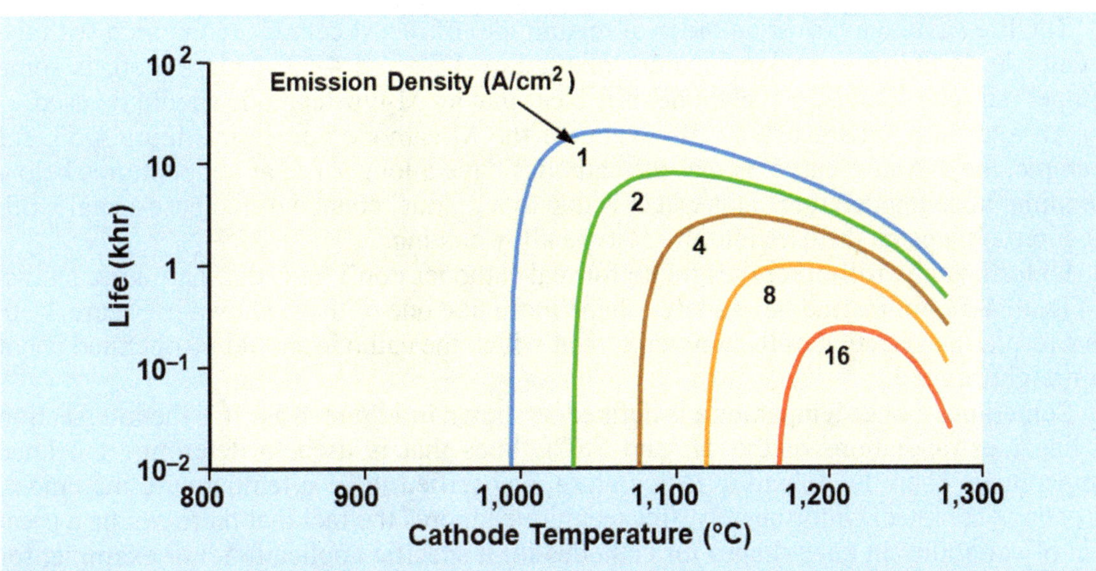

Figure 4-52 Life of a B-type cathode as a function of operating temperature. (From: T. J. Grant and L. R. Falce, Third IEEE IVEC, April 2002, © 2002 IEEE.)

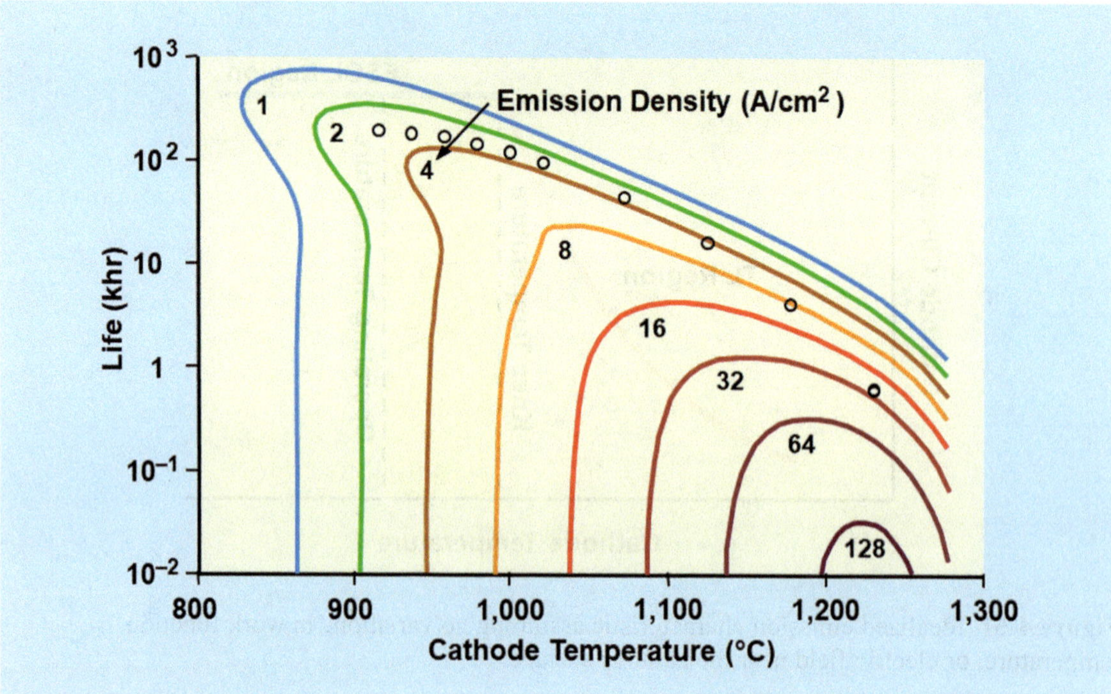

Figure 4-53 Life of an M-type cathode as a function of operating temperature. (From: T. J. Grant and L. R. Falce, Third IEEE IVEC, April 2002. © 2002 IEEE.)

As an example of the use of these curves for a B-type cathode having the emission characteristic in Figure 4-50, assume that life is to be maximized for operation at 1 A/cm^2. The knee temperature in Figure 4-50 is slightly below 1,000°C. Life is maximized with an operating temperature of ~1,050°C or ~50°C–60°C above the knee temperature. It should be noted, however, that there is very little sacrifice in life if the operating temperature is only ~40°C or as much as ~100°C above the knee temperature.

The life advantage (over an order of magnitude) of the M cathode, compared with the B cathode, is the most striking feature of Figures 4-52 and 4-53. Unless there is some compelling reason to use a B cathode, it is clear that an M-type cathode should be used.

An interesting characteristic of several of the M-cathode curves in Figure 4-53, for example, the 1-A/cm^2 curve, is that the cathodes have a longer life at temperatures below the initial knee temperature. This effect is due to a gradual change of the base-metal work function as tungsten diffuses into the M-type alloy coating.

Unfortunately, roll-off curves found for real cathodes don't have the sharp knee shown in Figure 4-51 but instead have a knee shape more like one of those shown in Figure 4-50. As a result, the selection of the temperature at which the cathode should be operated is not so straightforward.

Sometimes a knee temperature is defined as shown in Figure 4-54. It is the intersection of linear extrapolations of the TL and FSCL lines that is used to determine the knee temperature. Then the operating temperature is specified to be a temperature increment, ΔT, above the knee. Unfortunately, this technique ignores the fact that there can be a great deal of variability in knee shapes for cathodes for a specific application. For example, for the cathode in Figure 4-54(a), FSCL operation would be possible for a time period of ~$\Delta T/2$, whereas for the cathode in Figure 4-54(b), the onset of TL operation and reduced current would occur much sooner.

Ideally, the operating temperature should be chosen so that end of life is determined when the cathode current in a tube has decreased by an amount that causes one of the operating parameters of the tube to fail to meet specification. In this case the "knee" temperature is the temperature that produces that cathode current. Then the operating temperature is specified to be a ΔT such as 50°C above that knee temperature.

It is not as easy to determine end-of-life temperature in this manner as it would appear to be. This is because there are several factors that can cause irregularities in the shape of the roll-off curve in the FSCL region. Perhaps the most important factor is the thermal expansion characteristic of the cathode support structure. This causes the cathode-to-anode spacing and thus perveance to vary with temperature. This effect has to be compensated to find the true FSCL current. Other factors that complicate the determination of FSCL current include grid movement and scatter in measured data.

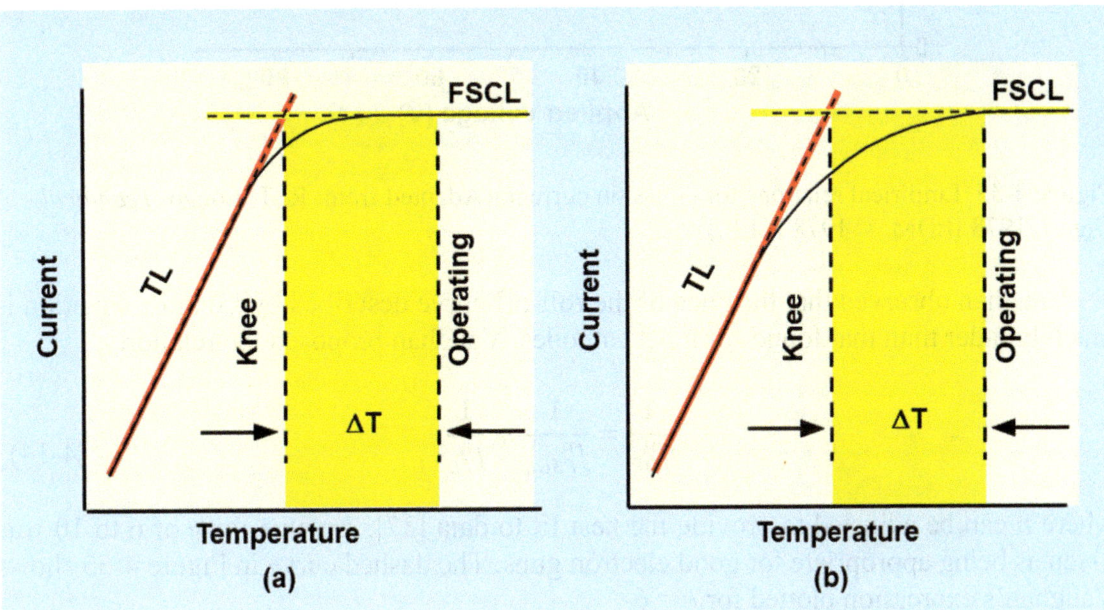

Figure 4-54 A technique for defining the knee temperature for a real cathode.

4.5.2 Longo Life Prediction Model

Another life prediction based on the depletion model is by Longo [34–36]. He observed that the following simple relation described the emission from an impregnated cathode over a wide range of voltages and temperatures:

$$\frac{1}{J_{OP}} = \frac{1}{J_{FSCL}} + \frac{1}{J_{TL}} \tag{4-13}$$

In this equation, J_{OP} is the observed current density, J_{FSCL} is the current density that would be observed if the entire cathode was operating space-charge-limited, and J_{TL} is the current density that would be observed if the entire cathode was operating temperature-limited. Shown in Figure 4-55 is a comparison of Longo's equation with measured results.

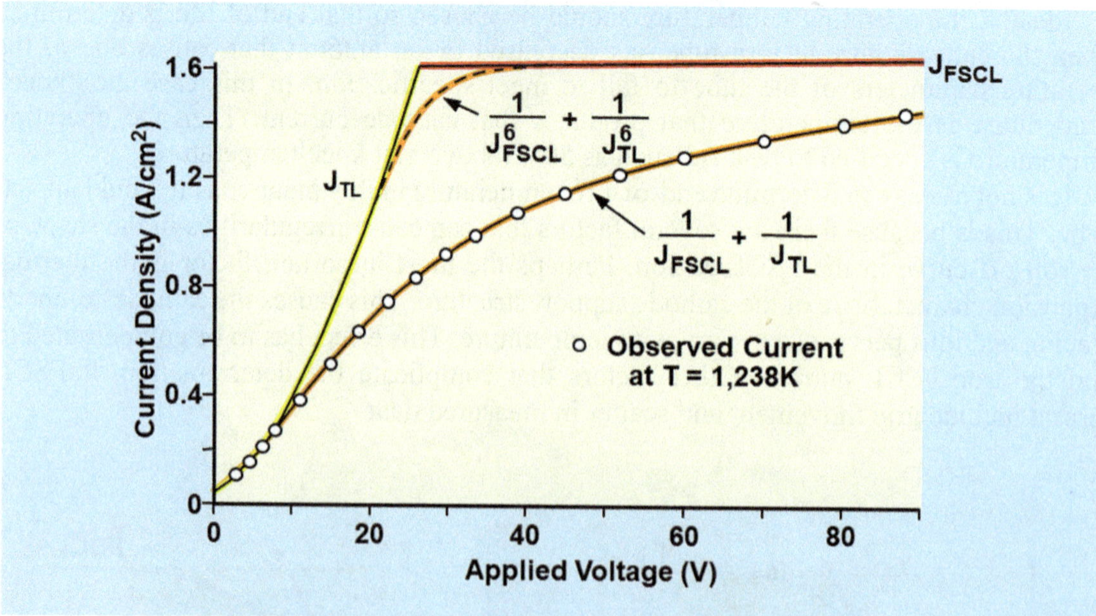

Figure 4-55 Empirical relations for emission current. (Adapted from: R. T. Longo, *Technical Digest*, 1978 IEDM. © 1978 IEEE.)

Vaughan observed that the knee of the roll-off curve described by Longo's equation is much broader than that found for most cathodes. Vaughan proposed the relation

$$\frac{1}{J_{OP}^{n}} = \frac{1}{J_{FSCL}^{n}} + \frac{1}{J_{TL}^{n}} \tag{4-14}$$

where n can be adjusted to provide the best fit to data [37]. A value for n of 6 to 10 was given as being appropriate for good electron guns. The dashed curve in Figure 4-55 shows Vaughan's expression plotted for $n = 6$.

Longo used his relation to predict emission current density as a function of time by assuming that the work function in J_{TL} (the Richardson-Dushman current) varies with time. The time dependence of work function is assumed to result from the variation with time of the fraction of the cathode surface that is covered by barium and by how much barium is on the covered surface. Surface coverage is determined by considering losses by evaporation and increases due to the evolution of material in the porous tungsten and the migration of this material to the surface by gas flow under conditions known as *Knudsen flow*.

To use the prediction model for an M-type cathode, Longo took into account the diffusion of tungsten into the osmium coating and the resulting variation with time of the base metal work function and of the minimum cathode work function. He calibrated his model against the results of accelerated life tests as shown in Figure 4-56(a) and then used the model to predict the operating lifetime at normal operating temperatures as shown in Figure 4-56(b). In this figure, $J(0)$ is the initial cathode current density and $J(t)$ is the current density at a later time. Note that lifetimes over 100,000 hours are predicted with only small decreases in emission current.

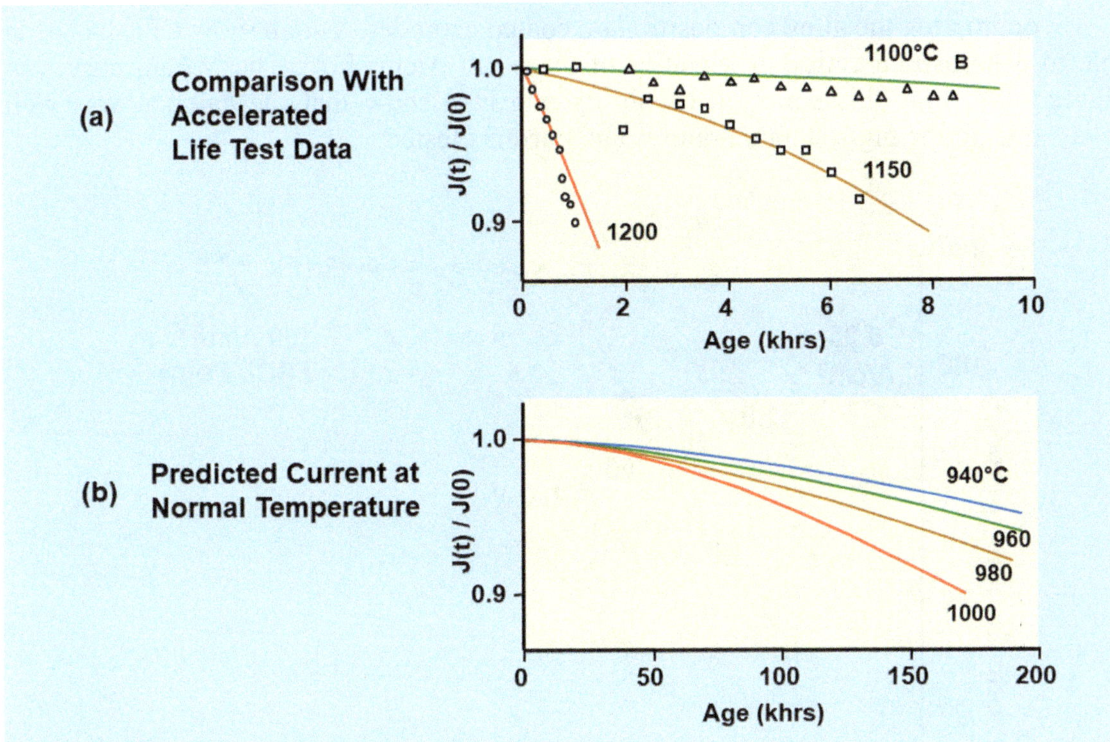

Figure 4-56 Results of Longo's life prediction model. (From: R. T. Longo et al., *Technical Digest*, 1984 IEDM. © 1984 IEEE.)

Since the first use of cathodes many decades ago, there has been a continuing requirement for increased emission capability along with long life. At present, for microwave tubes, these requirements are driven primarily by millimeter-wave tubes where small, high-current density cathodes are required for acceptable beam formation. There are also requirements for high-current density cathodes for technologies other than microwave tubes. One of these is for free electron lasers where a high-current density cathode (commonly referred to as a *high brightness cathode*) is needed to help limit the divergence of the beam generated by the electron gun.

There is a practical limit to the current density that can be drawn from a cathode, regardless of its emission capability. This limit results from the electrical breakdown characteristics of vacuum. For a 1-cm gap with a uniform field, the breakdown field is about 10^5 V/cm, so breakdown across the 1-cm gap would be expected for a voltage of about 100 kV. For a gap with hot electrodes (such as a thermionically emitting cathode) and filled with a high electron density, the breakdown voltage would be below 100 kV.

For a planar SCL diode with an anode-cathode separation of 1 cm, a voltage of about 120 kV is required for a current density of 100 A/cm^2. While the average field would be 120 kV/cm, the field at the SCL cathode would be zero and at the anode would be above the average field. Even if breakdown did not occur within the current flow region, it is probable that it would occur between the focus electrode and the anode where the spacing would be less than 1 cm. Thus, it appears that the practical upper limit to emission current densities is less than 100 A/cm^2. Higher-current densities are possible at very short pulse lengths because the breakdown field strength increases as pulse length becomes shorter.

By optimizing the alloy composition on coated cathodes, Varian (now CPI) has been able to demonstrate emission densities of over 100 A/cm^2 [38]. The Miram curves in Figure 4-57 show this emission capability and also show that the cathode was well behaved up to the highest temperatures and currents tested.

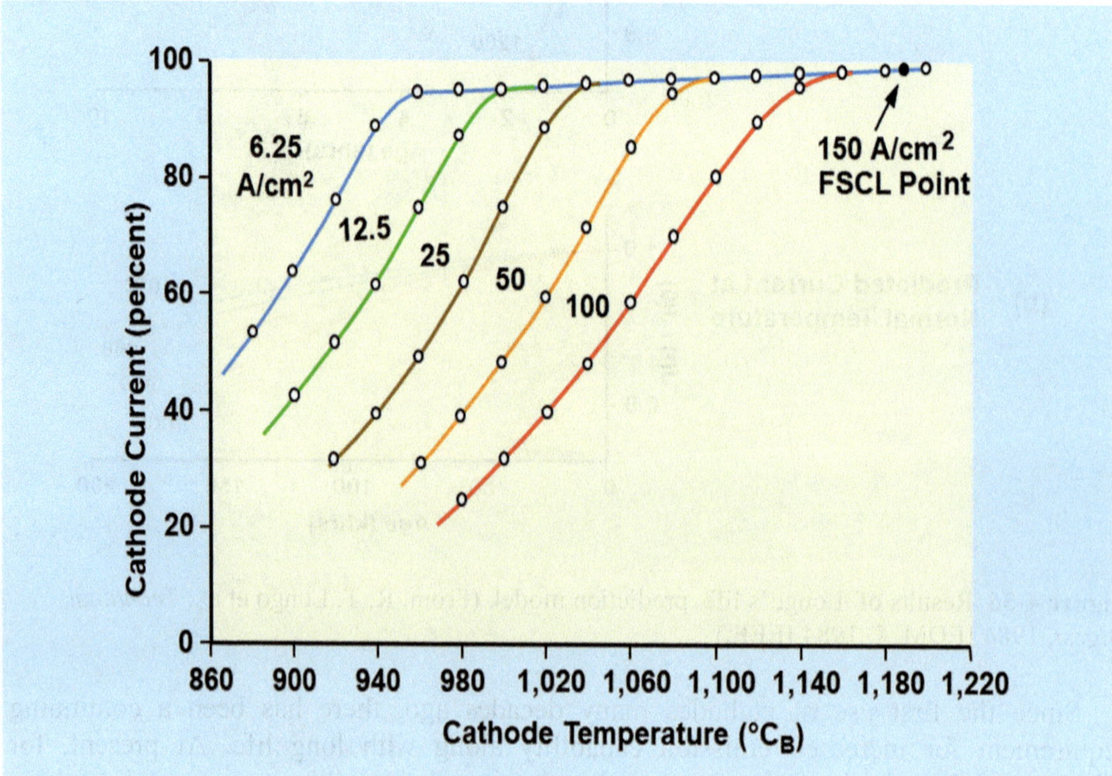

Figure 4-57 100 A/cm^2 dispenser cathode test data. (From: G. V. Miram et al., *Vacuum Electronics and Display Conference*, 1989.)

Thomas et al. used barium depletion calculations to predict cathode life at high emission densities [3]. The criterion used for end of life was when the temperature-limited emission dropped to half the initial level. Evaporation rate data for B-type cathodes was used to predict life as a function of temperature for each cathode type. Thomas et al. noted that the approximation is crude because of differences in impregnant composition, pore structure, and surface composition, all of which affect the net rate of evaporation. Still, a B-type cathode tested at LLNL and an M type tested at NRL, both at about 40 A/cm^2, are in reasonable agreement with predicted life. Thus, the results as shown in Figure 4-58 appear to be valid within a factor of 2 or 3 uncertainty.

The life of the scandate cathode was also assumed to be limited by barium depletion. Thomas et al. noted that at the time the paper was written, it was not clear what other mechanisms may limit the life of this cathode. Also, for all of the cathodes shown in Figure 4-58, it is often found that a higher temperature than expected (with resulting shortened life) is required to maintain emission because of the effects of poisoning gases.

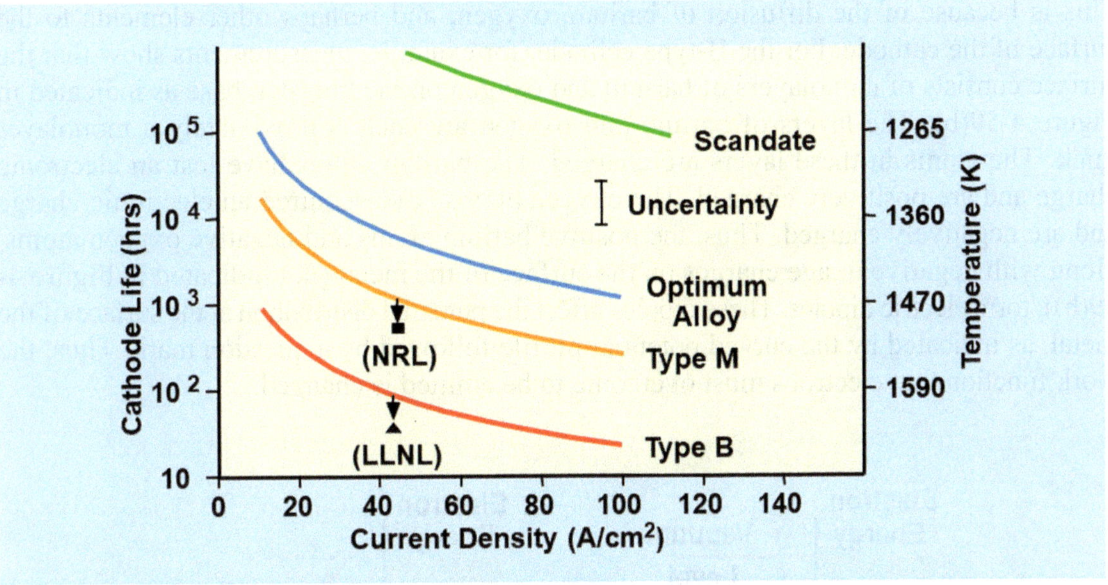

Figure 4-58 Predicted cathode life versus current density. (From: R. E. Thomas et al., *IEEE Trans. Electron Devices,* March 1990. © 1990 IEEE.)

4.6 DISPENSER CATHODE SURFACE PHYSICS

The basic theory of thermionic emission discussed at the beginning of this chapter has been well established and understood for several decades. What has not been well understood until recently has been the work functions of surfaces with adsorbed atoms of various materials. As a result, cathodes have historically been developed more by trying various combinations of materials and techniques rather than by design based on a thorough knowledge of surface physics. Fortunately, however, the modern state of cathode theories is significantly advanced from that given by Pierce in 1946 [2]:

> It is clear that as far as science goes, the state of the oxide coated cathode is confused indeed. Not only does theory fail to help us make good cathodes; it seems not to fit all the evidence. Maybe cathodes work sometimes one way and sometimes another, which isn't as ridiculous as it may sound. At any rate, while some physicists are uninterested and others confused, cathode making continues to be an art practiced, it seems, by artists with a good deal of witch doctor in their make-up. When something goes wrong, this or that demon is exorcised, and eventually things go right again. And, unlikely as it seems, cathodes continue to improve.

While it is not yet possible in theory to design a new type of cathode having a predetermined emission density, operating temperature, and life, present cathode theories do explain very well how most dispenser cathodes operate. The following section contains an overview of the present theories of dispenser cathode operation.

In the fundamental theory of thermionic emission, an energy level diagram for electrons in and near the metal surface, like that shown in Figure 4-59(a), was used. In dispenser cathodes, the emitting surface is more complex than that shown in Figure 4-59(a).

This is because of the diffusion of barium, oxygen, and perhaps other elements to the surface of the cathode. For the B-type cathode, for example, measurements show that the surface consists of monolayers of barium and oxygen on the tungsten base as indicated in Figure 4-59(b). The layers of barium and oxygen are each approximately a monolayer thick. The atoms in these layers are charged. The barium atoms have lost an electronic charge and are positively charged. The oxygen atoms have acquired an electronic charge and are negatively charged. Thus, the positive barium atoms and negative oxygen atoms, along with negative image charges on the surface of the metal (not indicated in Figure 4-59(b)), form electric dipoles. These dipoles affect the potential distribution at the surface of the metal, as indicated by the curved potential profile followed by a question mark. Thus, the work function that electrons must overcome to be emitted is changed.

.

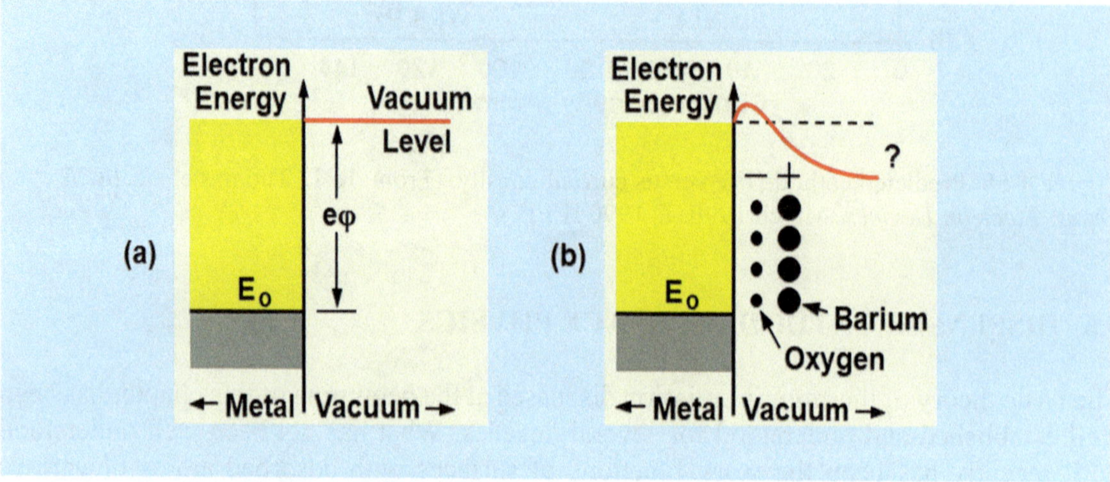

Figure 4-59 Effect of oxygen and barium on the energy level diagram.

The positive charge of the barium atoms also results in electrostatic forces that attract the atoms to the surface of the metal. This causes the rate of evaporation of the barium atoms from the tungsten surface to be far below (by over an order of magnitude) the rate of evaporation of barium atoms from a barium surface.

Before attempting to explain the effect of two (or more) layers of charged atoms of different elements on the work function of a surface, it is helpful to consider a single layer of atoms of one element. Following the work of Zalm and van Stratum [39], who were early developers of the M-type cathode, consider again the energy level diagram for a metal surface in Figure 4-60(a). With no foreign atoms near the surface, the energy levels are as previously discussed. If a small number (far fewer than required to form a monolayer) of atoms that can easily give up electrons (have a low ionization potential) are brought into the electric field near the surface of the metal, it is probable that some of the atoms will become ionized and give up electrons to the metal surface (see Figure 4-60(b)). The presence of the resulting positive charge on the surface of the metal produces a reduction, $\Delta e\varphi$, in the work function, which, in turn, increases the probability of electron emission. The influence of the positive ions near the surface of the metal is similar to that of a microscopic grid at a positive potential. Such a grid would reduce the potential barrier at the surface of the metal and permit increased electron emission.

As additional atoms are brought near the surface, some become ionized and further reduce the work function (Figure 4-60(c)). For small numbers of atoms (much fewer than required for a monolayer), the amount by which the work function is reduced, $\Delta e\varphi$, increases linearly with the density of atoms on the metal surface.

As more atoms are brought near the surface and more ions are formed, two primary effects reduce the rate of ion formation:

1. The reduction in electric field near the surface that the ions produce results in a reduced probability of ionization. Eventually, the field is reduced enough that the probability of ionization approaches zero. The addition of more atoms will not produce additional reductions in the work function.
2. As the number of ions increases, the distance between ions decreases. As this distance decreases, the repulsion forces between ions increases, and this causes more ions to be ejected from the metal surface.

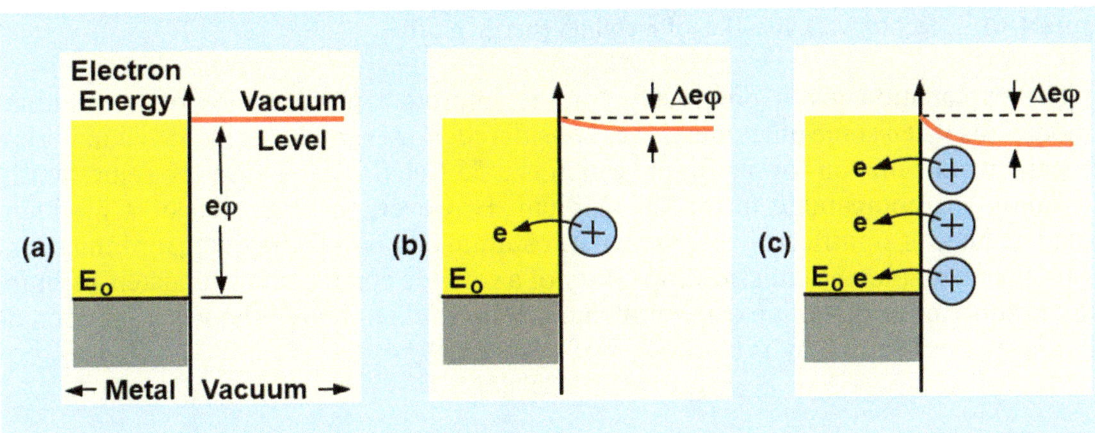

Figure 4-60 Energy level diagrams for a metal surface with (a) no foreign ions, (b) a few foreign ions, and (c) many foreign ions on the surface.

As still more atoms are brought near the surface and the film thickness grows to over a monolayer, the work function gradually changes to that of the atoms on the surface.

Using the concepts outlined in the preceding discussion, it is possible to understand the reasoning that led Zalm and van Stratum to using osmium and other high work function metals in cathodes. For example, consider osmium with a work function of about 5.4 eV and tungsten with a work function of about 4.6 eV. The energy level diagram in Figure 4-61 indicates why, with an adsorbed layer of ions, Zalm and van Stratum expected the work function for osmium to be lower than for tungsten. In each case, as more atoms are brought near the surface, ionization will occur until the electric field has dropped to the point where the probability of ionization approaches zero (P_i □ 0). This field will be approximately the same for the two metals.

Because the work function for osmium is higher than for tungsten, the field at the surface is higher, and more ions can be produced than for tungsten, so the resulting reduction in work function for osmium is higher than for tungsten. In fact, the reduction for osmium is high enough that the overall work function of the metal plus adsorbed ion layer is lower than for tungsten. This results in a higher electron emission from osmium (with an adsorbed layer) than for tungsten.

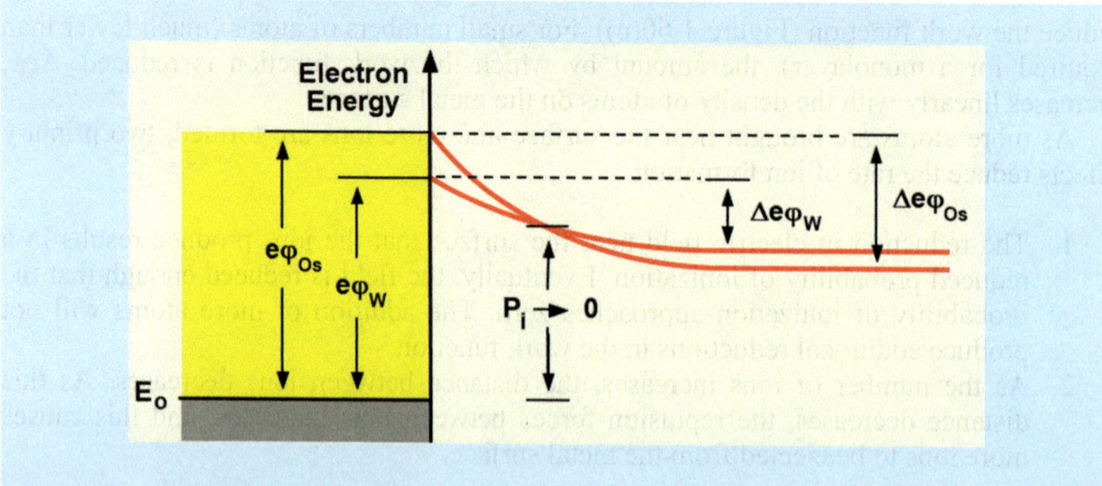

Figure 4-61 Effect of work function of metal on energy profiles.

The explanation given previously for the improved performance of an osmium cathode fails when some other metals are considered. For example, as indicated in Figure 4-62, the work function for platinum (which is a 5d transition metal in the same family as osmium) is comparable to that of osmium. However, the emission of a platinum coated cathode is significantly below that of osmium and is even below that of tungsten. As another example, the emission capability of a cathode coated with a tungsten-osmium alloy is superior to that of an osmium-coated cathode even though the work function of the tungsten-osmium alloy is below that of osmium.

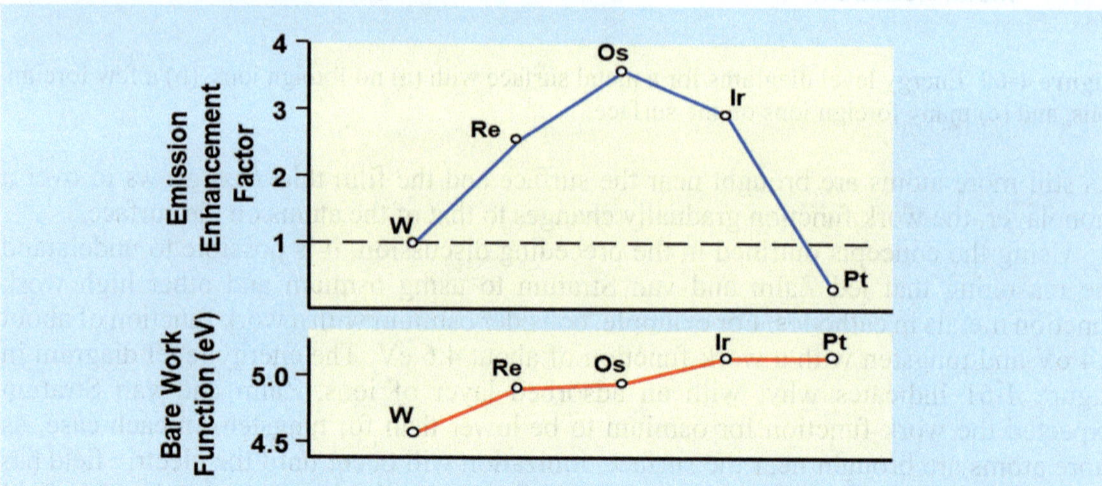

Figure 4-62 Emission enhancement and work function for dispenser cathode metals. (From: M. C. Green, *Technical Report,* RADC-TR-81-211, July 1981.)

To understand the differences in performance between cathodes coated with a tungsten-osmium alloy, osmium, or platinum, it is necessary to examine the atomic structure of the metal cathode surface and its relation to adsorbed oxygen and barium atoms. The work of Müller [40–42] has led to calculations of work functions for dispenser cathodes that are in excellent agreement with experimental results.

In Müller's work, a microscopic portion of the surface of a cathode is modeled as a cluster of atoms. The models are chosen on the basis of crystal structures found on actual cathodes. An example is shown in Figure 4-63. Here, the B-type cathode is modeled as a cluster consisting of 19 tungsten atoms in a body-centered cubic configuration and three groups of BaO molecules. The tungsten substrate in an actual B-type cathode is polycrystalline with many crystal faces present. The 100-face orientation was chosen because of its simple structure. The clusters used for modeling the osmium and osmium-tungsten alloy substrates represented the hexagonal-close-packed lattice. For platinum, the face-centered cubic lattice was used.

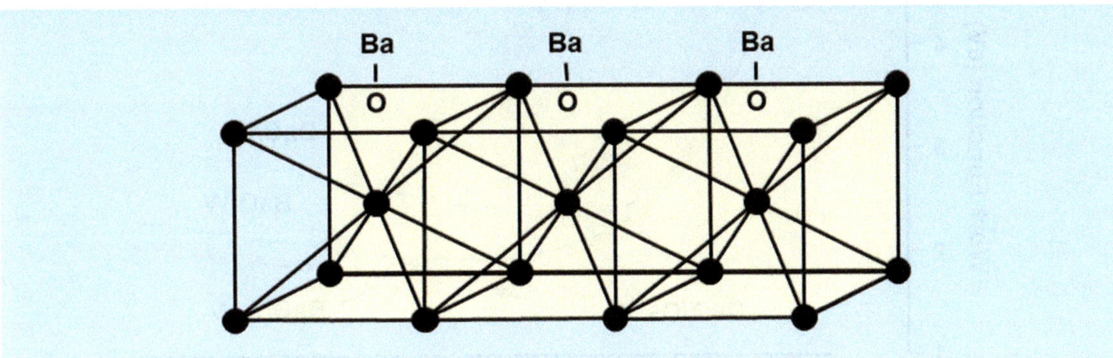

Figure 4-63 Body-centered cubic cluster (100 face) assumed for BaO/W. (From: W. Müller, *Technical Record*, 1991 IEDM. © 1991 IEEE.)

The calculations were performed using large computer codes capable of analyzing the interactions between the relatively large numbers of atoms in the clusters. Calculated properties included electronic energies, densities of states, electron densities, charge distributions, surface electric dipoles, and barium binding energies on different substrates.

Only the work function change that occurs with the sorption of barium and oxygen on the metal surface is obtained from the calculated properties. Thus, the minimum work functions depend critically on the bare-metal work functions of the various crystal faces considered. Because accurate values of the bare surface work functions are not known in most cases, some assumptions and approximations had to be made. In addition, empirical temperature dependencies of polycrystalline work functions for pure metals were used to estimate work functions at 1,200K. The initial bare-metal work functions affect the results by their magnitude and temperature dependence. A high value leads to a large charge transfer from barium to oxygen and the metal substrate. Alloying with W, in general, reduces the work functions for Re, Os, Ir, and Pt.

The work-function change that is induced by the sorption of BaO is found to be the largest for the hexagonal structures (Os, Ir, Os-W). The larger change more than offsets the initially higher bare-metal work functions of the alloy substrates and, as shown in Figure 4-64, leads to work functions that are up to 0.2 eV lower than for BaO on tungsten.

The calculated barium binding energies for high surface coverages are 5.4 eV for BaO/W and ~6.0 eV for the alloy substrates. This increase of 0.6 eV is in excellent agreement with the observed increase in the energy of desorption for Ba from B- and M-type cathodes. This, along with a reduced operating temperature for a given emission current, is the reason for the long life of the M-type cathode. For BaO on platinum, the calculated binding energy drops to 3.6 eV. Thus, barium evaporates at a higher rate from platinum than from the other metals considered, and the resulting reduced coverage of the surface leads to a higher minimum work function.

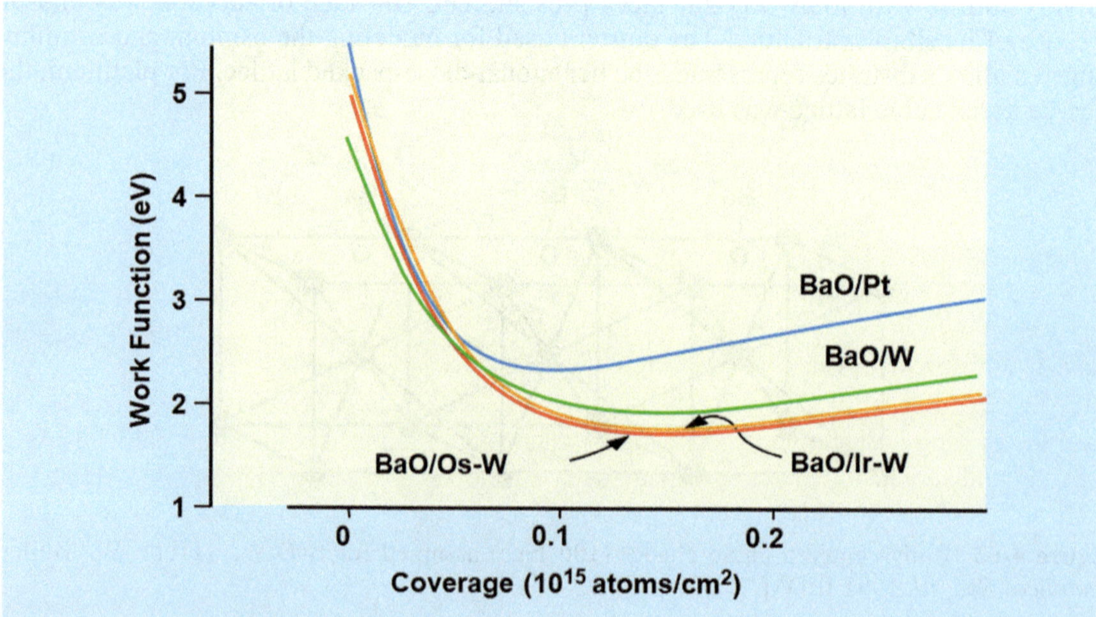

Figure 4-64 Calculated work functions for B- and M-type cathodes. (From: W. Müller, 1992 Tri-Service/NASA Cathode Workshop.)

4.7 HEATERS

4.7.1 Conventional Heater Assemblies

Typical oxide coated cathodes operate near 850°C and B-type cathodes operate near 1,100°C [43]. To raise a cathode to this temperature, the heater must be very hot. The materials of choice for heaters are pure tungsten or tungsten with a small amount of rhenium added. The rhenium makes the tungsten easier to work with.

After being formed, the heaters are coated with an insulating layer 0.002–0.003 inch thick of aluminum oxide. The oxide can be applied by spraying or by cataphoretic coating (an electrolytic process). The aluminum oxide insulation is sintered at high temperature to drive off organic binders and to form a hard, scuff-resistant coating.

Heaters for oxide-coated cathodes generally operate at a temperature near 1,100°C. This is not excessive for long life in tungsten or tungsten-rhenium heaters. Oxide cathode heaters are generally mounted in a fairly simple cathode structure with no special steps taken to improve the transfer of heat from the heater to the cathode. Two typical structures are shown in Figure 4-65.

One side of the heater is often connected to the cathode so that no large voltage difference exists between the cathode and any part of the heater. The aluminum oxide insulation on the heater is thin and will not withstand much voltage without breakdown.

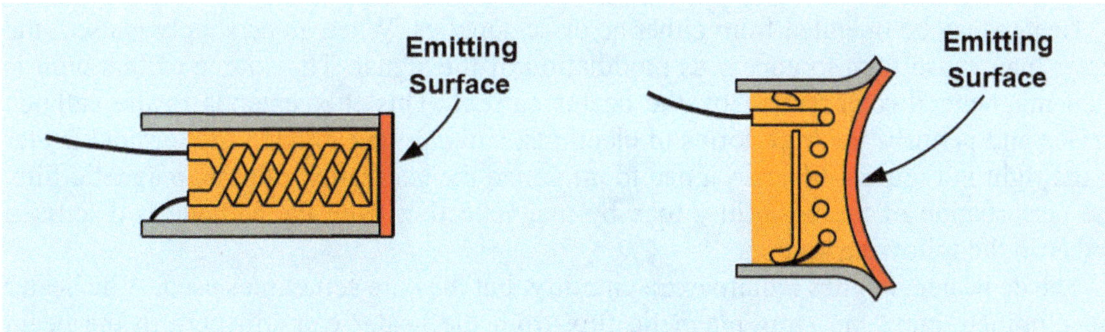

Figure 4-65 Typical heater assemblies for oxide-coated cathodes.

Heaters for B-type dispenser cathodes present a more severe problem because of the 1,100–1,150°C operating temperature of the cathode. The usual solution is to embed the heater in an aluminum oxide pellet at the back of the cathode as shown in Figure 4-66. Heat transfer between the heater and the cathode is much improved by this embedding or potting, so the heater operating temperature needs to be only 150°C–200°C higher than the cathode. Also, the potting gives the heater mechanical support, making it resistant to damage from vibration and shock. The heater potting material is added as a powder, possibly compacted under pressure and then sintered at high temperature.

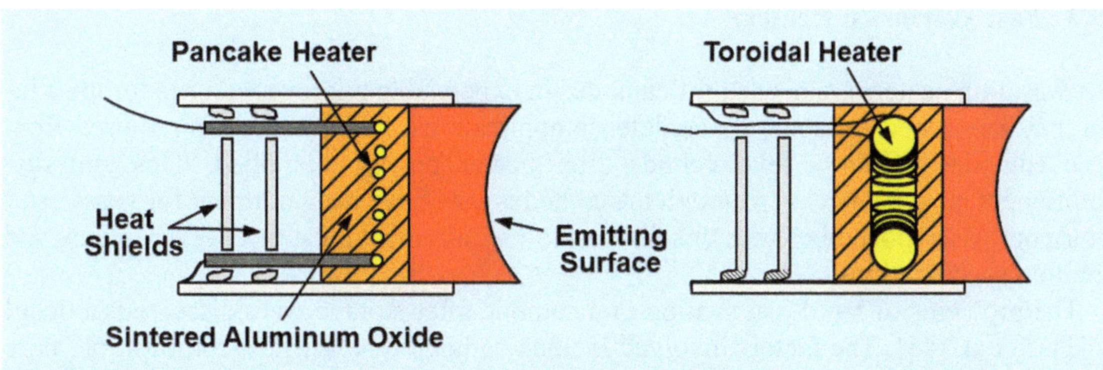

Figure 4-66 Heater assemblies for dispenser cathodes.

Potted heaters have one disadvantage. The extra mass of the potting material increases the warm-up time over an unpotted version in a similar cathode. In most applications this warm-up time penalty (typically 2–3 minutes) is accepted as the price for improving heater reliability. Some applications, in missiles, for example, require fast warm-up. (Fast warm-up heaters are discussed in the following section.) Fortunately, such applications do not require long life, so the heater can be designed for low mass and high temperature operation.

Dispenser cathodes require special steps to conserve heater power. Heat shields are used extensively to minimize radiation losses to the side and back. Support members are made with as thin a cross section as possible to minimize conduction losses. Finally, considerable care must be taken by the manufacturer to ensure consistency in the heater potting operation. Variations in contact between the heater, the potting material, and the cathode can lead to dangerous heater hot spots and cathode cold spots.

Heaters can be operated from either ac or dc supplies. When an ac supply is used, the heater may cause hum to appear as modulation of the signal. The source of this hum is often magnetic flux generated by the heater current. This flux extends to the cathode surface and perturbs the trajectories of electrons, leaving the cathode. The toroidal heater on the right in Figure 4-66 is designed to minimize the generation of stray magnetic flux. The perturbation of the beam in a tube by magnetic flux from the cathode is discussed further in the following section.

The dc heater supplies require extra circuitry, but they are sometimes used. A dc heater will eliminate hum, but stray magnetic flux from the heater can still perturb the beam emerging from the gun and cause increased interception current. The rectifier circuitry for a dc heater supply adapts easily to voltage regulation, which is of value when ac line voltage variations are excessive.

When a dc supply is used, the polarity of the supply can be important. If one side of the heater is connected to the cathode, that side should go to the positive side of the heater supply. The heater is then negative with respect to the cathode. This precaution is taken to avoid emission current from the back of the cathode to the heater. If the heater were positive with respect to the cathode, it would attract any electrons unintentionally emitted from the back of the cathode. The voltage and current involved are small, but over a long period of time this current can damage the heater.

4.7.2 Fast Warm-Up Heaters

Fast warm-up heaters are needed for cathodes in expendable microwave tubes for missiles and may be needed for some space defense applications. Full emission is required from these cathodes within a few seconds after heater power is applied. This emission requirement must be met even though the cathodes may have been dormant for years prior to turn on. Also, most tubes with this fast turn-on requirement must meet severe shock and vibration specifications.

The problems of rapid reactivation of a cathode after storage were discussed in detail by Haas et al. [44]. The factors involved include cathode type, surface composition, pore spacing, Ba impregnant concentration, Ba-O bonding, exposure to poisoning gases, and heating time and temperature. Haas et al. pointed out that it is the elimination of the poisoning products of ambient gases from the surface along with the arrival of new Ba surface atoms via the impregnant diffusing out of the pores that constitutes the reactivation process of the cathode surface. Some turn-on tests indicate that essentially no delay occurs. This could have been due to temperature overshoot as the cathode is very quickly brought to the operating temperature or to the passivating effect of other gases such as hydrogen. Other tests have correlated turn-on delays with measured pressures of oxidizing gases.

Shown in Figure 4-67 are the temperature characteristics of some fast warm-up heater-cathode assemblies [45]. The cathode diameter in each case was 0.2 inch (0.79 mm). In all cases, the fast warm-up times were achieved by using low thermal mass components. The 3-second design used an integral tungsten-3% rhenium heater cataphoretically coated with alumina. A voltage surge as high as twice the normal operating voltage was used.

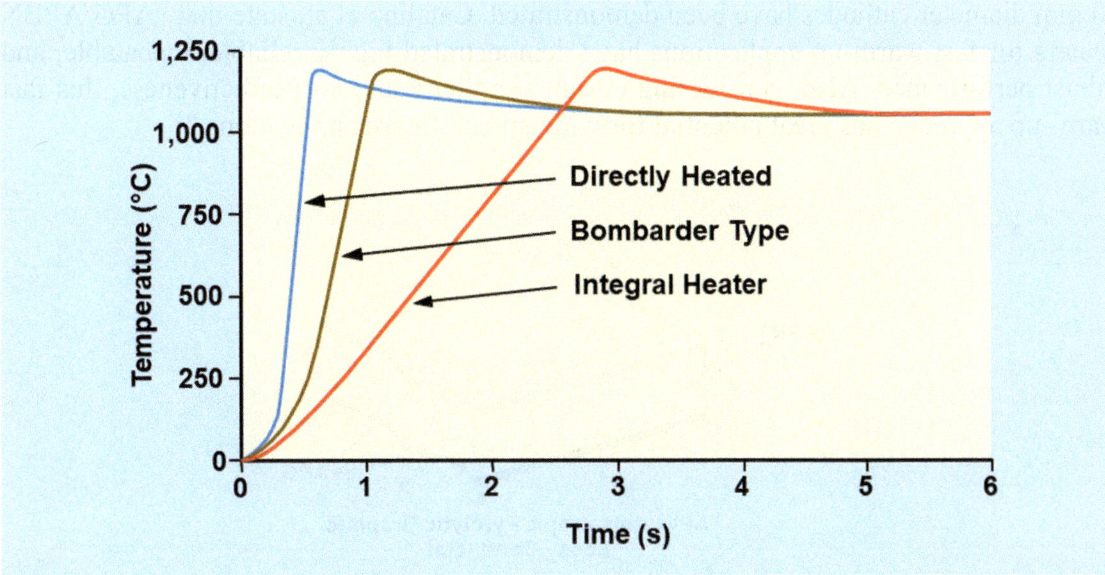

Figure 4-67 Temperature characteristics for fast warm-up heater-cathode assemblies. (From: G. V. Miram et al., *Vacuum Electronics and Display Conference*, 1989.)

The bombarder-type assembly, which is shown in Figure 4-68, achieved a warm-up time of 1 second [46]. The heater is made of thoriated tungsten and designed to be heated to about 1,700°C in a fraction of a second by a large overvoltage. At that temperature the heater is also an emitter. Several hundred volts are applied between the heater and the cathode. Bombardment of the back of the cathode by electrons from the heater raises the cathode to the operating temperature.

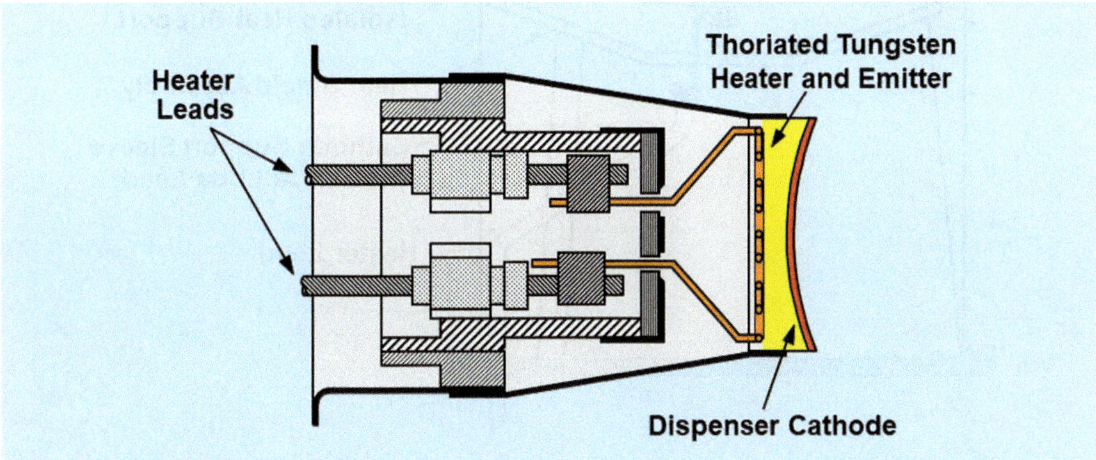

Figure 4-68 Bombarder-type fast warm-up heater. (From: R. C. Treseder et al., *Technical Digest*, 1983 IEDM. © 1983 IEEE.)

Cattelino et al. described fast warm-up heaters for large diameter cathodes [47]. As Figure 4-69 shows, the heater consists of an anisotropic boron nitride wafer with anisotropic pyrolytic graphite deposited on both surfaces. Both planar and spherical heaters can be made. A spherical cathode-heater assembly is shown in Figure 4-70. Warm-up times of 1 to 2 seconds for 4–6.4 mm diameter cathodes and less than 20 seconds for

60 mm diameter cathodes have been demonstrated. Catalino et al. state that "APG/APBN heaters for fast warm-up applications have demonstrated highly reliable, repeatable, and robust performance. Also, considering design simplicity and cost effectiveness, this fast warm-up approach has great potential for wide-spread future applications."

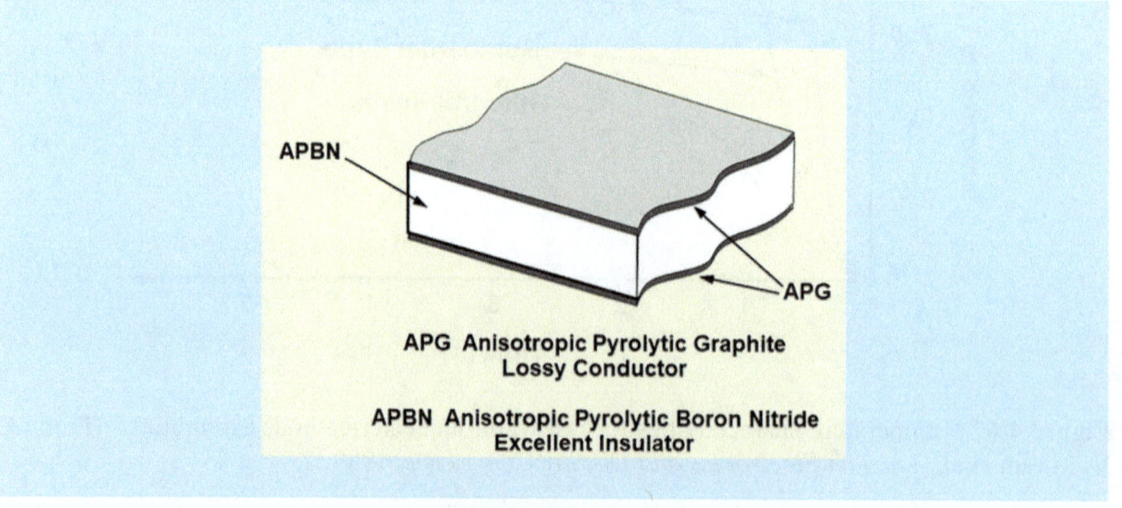

Figure 4-69 APG – APBN Heater Assembly. (From M. J. Cattelino et al., *IEEE Trans. Electron Devices*, Vol. ED-38, No. 10, October 1991, pp. 2239–2243.)

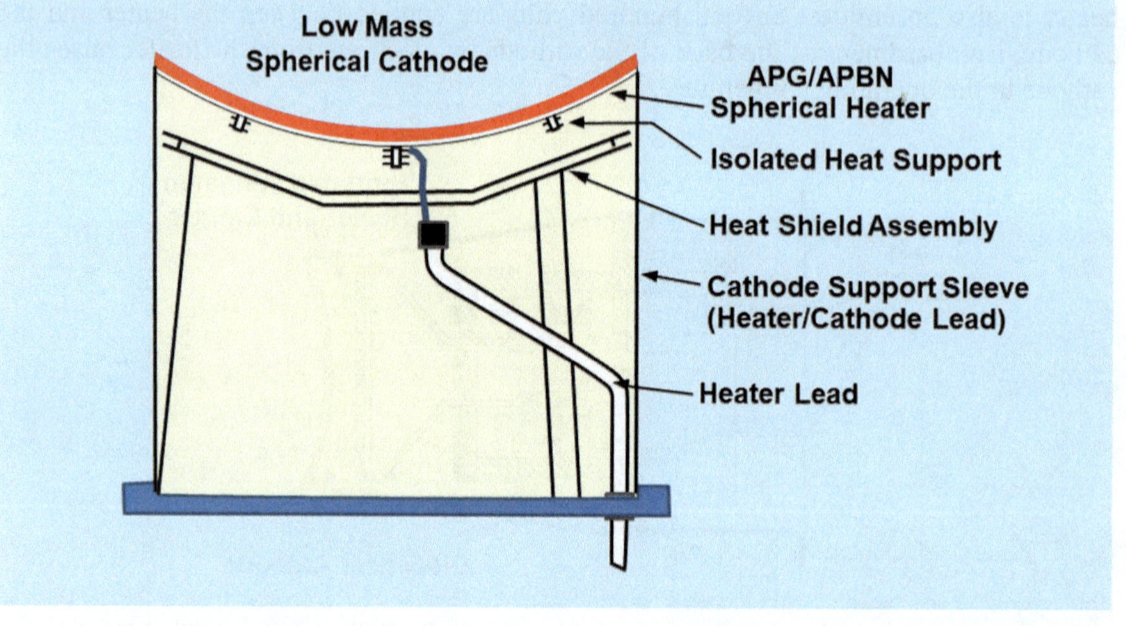

Figure 4-70 APG – APBN Spherical Heater Assembly. (From M. J. Cattelino et al., *IEEE Trans. Electron Devices*, Vol. ED-38, No. 10, October 1991, pp. 2239–2243.)

4.7.3 Heater Testing

By examining heater current and voltage, it is possible to determine if a heater is operating properly and usually if the cathode is at or near the proper operating temperature. It is important to realize that heater resistance varies rapidly with heater temperature. For the tungsten-3% rhenium wire often used for heaters, resistance increases by a factor of 6 as the wire is heated from room temperature to approximately 1,550°C. Thus, a plot of heater current versus heater resistance, as shown in Figure 4-71, can reveal malfunctions of the heater assembly.

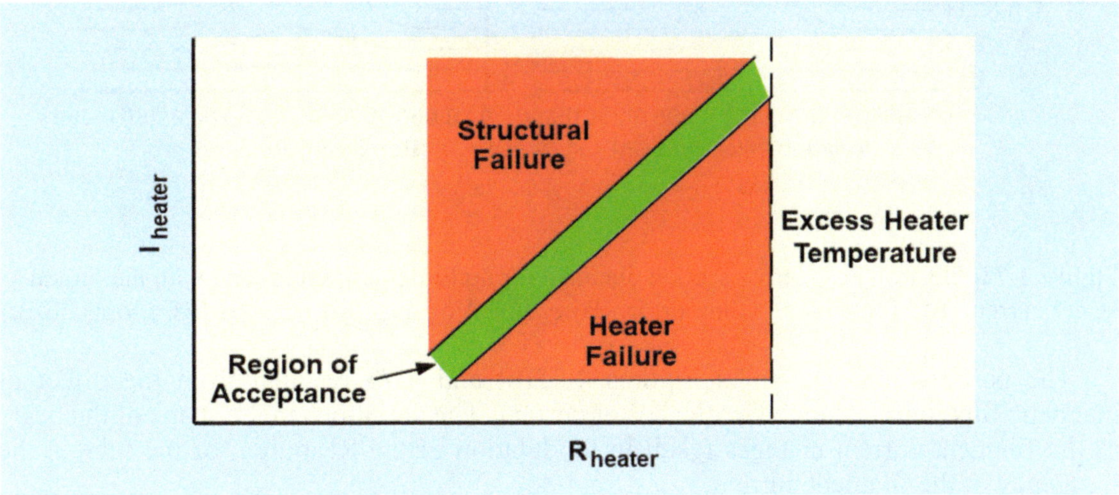

Figure 4-71 Heater performance. (Adapted from: T. J. Grant, *Technical Digest*, 1984 IEDM. © 1984 IEEE.)

To provide meaningful information, the region of acceptance must be determined for a particular class of heater assemblies that are known to operate at the desired temperature. Then Figure 4-71 becomes the benchmark against which other heater assemblies can be compared. Excessive heater current implies that the heater assembly is at a lower temperature than it should be (for that current, the heater resistance is too low so the temperature is too low). Possible problems are that the tube is gassy (gas cools the heater) or that a heat shield is touching the heater or cathode assembly and carrying heat away. Operation below the region of acceptance indicates excessive heater wire temperature and possible heater failure.

4.7.4 Effect of Filament Magnetic Field

Detailed measurements made on the electron beam from a magnetically shielded Pierce gun (discussed in the following chapters) illustrate the effect that an ac heater can have on a beam. The current used in the cathode heater was ac, and the beam voltage was pulsed on at various times during the heater current cycle [48].

When the heater current was zero, the electron beam cross section of Figure 4-72(a) was obtained. When the heater current was at a maximum value, cross sections like that of Figure 4-72(b) were obtained. Under very specialized conditions, an image of the spiral-wound heater could be detected in the beam, as shown in Figure 4-73.

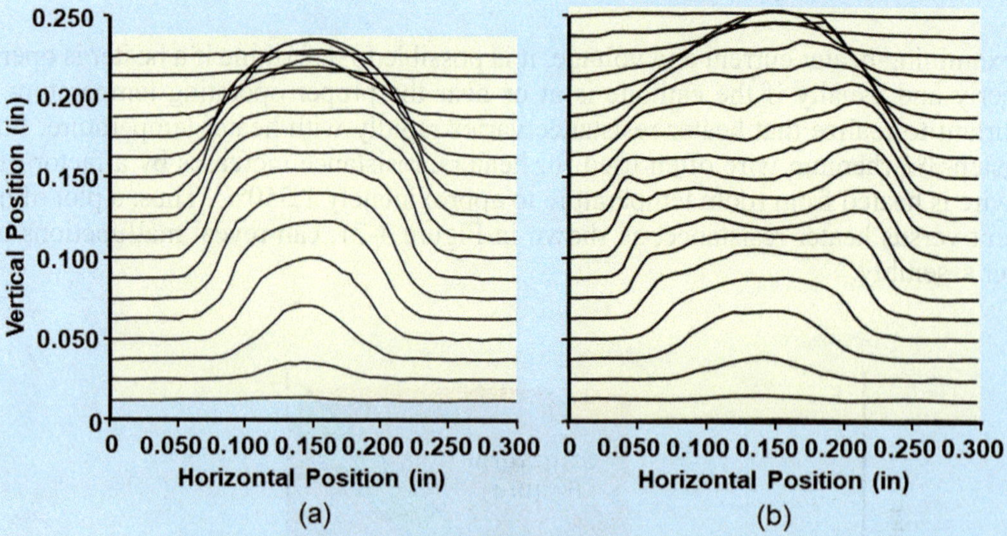

Figure 4-72 Electron beam cross section obtained for no heater current (a) and with maximum heater current (b). (From: A. S. Gilmour, Jr., *Proc. IRE*, Vol. 49, 1961. © 1961 IRE (now IEEE).)

The perturbations shown in Figures 4-72(b) and 4-73 resulted from the effect of magnetic flux lines generated by the heater current. The variations in the shape of the beam as the filament current changes result in modulation of the RF output of the tube at the frequency of the filament current.

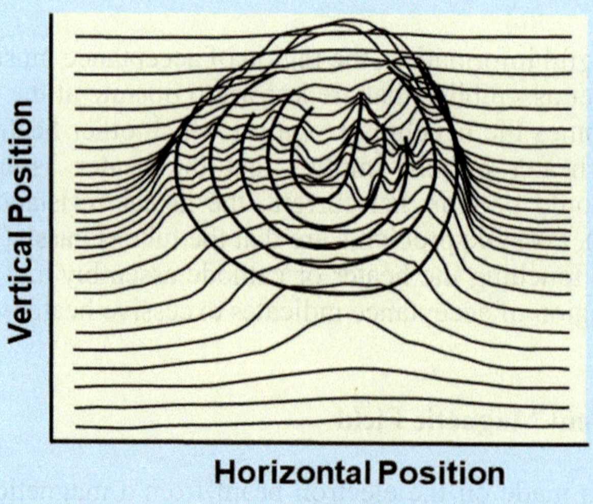

Figure 4-73 Image of spiral-wound cathode heater in electron beam cross section. (From: A. S. Gilmour, Jr., *Microwave Tubes*. Copyright 1986 by Artech House.)

REFERENCES

[1] Please refer to A. S. Gilmour, Jr., *Principles of Traveling Wave Tubes*, Boston: Artech House, 1994 for details.

[2] J. J. Coupling, a pseudonym of J. R. Pierce, *Astounding Science Fiction*, November 1946.

[3] R. E. Thomas et al., "Thermionic sources for high-brightness electron beams," *IEEE Trans. Electron Devices*, Vol. 37, No. 3, March 1990. pp. 850–861.

[4] J. L. Cronin, "Modern dispenser cathodes," *IEE Proc*, Vol. 128, pt. 1, No. 1, February 1981, pp. 19–32.

[5] G. Herrman, and P. S. Wagener, *The Oxide-Coated Cathode*, London: Chapman and Hall, 1951

[6] C. L. Hemenway, R. W. Henry, and M. Caulton, *Physical Electronics*, New York: John Wiley and Sons, 1967.

[7] Emilio Ciardiello, "Velocity modulated tubes: The Klystron," Radiomuseum.org.

[8] J. O. Tarter, and J. J. Farrell, "Selection of dispenser cathode impregnant types," *Technical Digest*, Fifth IVEC, 2004, pp. 301–302.

[9] J. J. Farrell, S. Conkwright, and J. O. Tarter, "The effects of chemical cleaning on impregnant removal as a function of impregnant type," *Technical Digest*, Fifth IVEC, 2004, pp. 297–298.

[10] A. M. Shroff, and G. Firmain, "Long-life travelling-wave tubes. Use of M-type cathode. Life prediction model," *Technical Digest*, IEDM, 1985, pp. 346–349.

[11] M. C. Green, "Dispenser cathode physics," *Final Technical Report*, RADC TR 81 211, July 1981.

[12] R. L. Ives et al., "Controlled porosity cathodes from sintered tungsten wires," *IEEE Trans. Electron Devices*, Vol. 52, No. 12, December 2005, pp. 2800–2805.

[13] R. L. Ives et al.," Advanced CPR cathode research," IEEE IVEC 2013.

[14] M. J. Cattelino, G. V. Miram, and W. R. Ayers, "A diagnostic technique for evaluation of cathode emission performance and defects in vehicle assembly," *Technical Digest*, IEDM, 1982, pp. 36–39.

[15] M. Cattelino, and G. Miram, "Predicting cathode life expectancy and emission quality from PWFD measurements," *Applied Surface Science*, Vol. 111, 1997, pp. 90–95.

[16] J. Hasker, J. van Esdonk and J. E. Crombeen, "Properties and Manufacture of Top-Layer Scandate Cathodes," *Applied Surface Science* 26, pp. 173-195, 1986.

[17] Y. Wang, J. Wang, W. Liu, K. Zhang, and J. Li, "Development of high current-density cathodes with Scandia-doped tungsten powders," *IEEE Trans. Electron Devices*, Vol. 54, No. 5, pp. 1061–1070, May 2007.

[18] B. Vancil, et al., "Scandate dispenser cathodes with sharp transition and their application in microwave tubes," *IEEE Trans. Electron Devices*, Vol. 61, No. 6, pp. 1754–1759, June 2014.

[19] D. M. Kirkwood, et al., "Frontiers in thermionic cathode Research," *IEEE Trans. Electron Devices*, Vol. 65, No.64, pp. 2061–2071, June 2018.

[20] W. Liang et al., "DC emission characteristics of nanosized-scandia doped impregnated dispenser cathodes," *IEEE Trans. Electron Devices*, Vol. 61, No. 6, pp. 1749–1753, June 2014.

[21] I. Brodie and Bernard Vancil, "Electron emission microscopy of electron emitting sites on scandate cathodes," IVEC 2016.

[22] I. Brodie, "A new model for the mechanism of operation of scandate and refractory oxide cathodes," *IEEE Trans. Electron Devices*, Vol. 58, No. 4, pp. 1247–1254, April 2011.

[23] I. Brodie and B. Vancil, "The nature of the emitting surface of scandate cathodes," *Technical Digest*, IEEE IVEC, 2014, pp. 53–54.

[24] Y. Wang, et al., "Recent progresses on nanosized-scandia-doped dispenser cathodes," *Technical Digest*, IEEE IVEC, 2014, pp. 51–52.

[25] B. Vancil, et al., "New findings on powder synthesis for scandate cathode matrices, " *IEEE Trans. Electron Devices*, Vol. 65, No. 6, pp. 2077–2082, June 2018

[26] B. Vancil, et al., "Scandate cathode performance comparisons," IVEC 2016.

[27] J. Paff, G. Rudd and B. Vancil, "Extended test results from scandia-doped tungsten dispenser cathodes," IVEC 2016.

[28] W. Liu, et al., "Investigation on robustness of nanosized-scandia doped dispenser cathodes," *Technical Digest*, IEEE IVEC, 2016.

[29] Y. Yang, et al., "Robustness investigation on nano-sized-scandia-doped dispenser cathodes," *IEEE Trans. Electron Devices*, Accepted for inclusion in future edition.

[30] M. Zhang et al., "Investigation of recovery characteristics after ion bombardment on scandate cathode," *Technical Digest*, IEEE IVEC, 2004, pp. 153–155.

[31] Y. Wang, et al., "Correlation between emission behavior and surface features of scandate cathodes, " *IEEE Trans. Electron Devices*, Vol. 56, No. 5, pp. 776–785, May 2009.

[32] T. J. Grant, "Emission degradation characteristics of coated dispenser cathodes," *Technical Digest*, IEDM, 1986, pp. 700–703.

[33] T. J. Grant, and L. R. Falce, "Life prediction model for vacuum electron devices using barium calcium aluminate impregnated tungsten cathodes as electron sources," *Technical Digest*, Third IEEE IVEC, 2002, pp. 300–301.

[34] R. T. Longo, "Long life, high current density cathodes," *Technical Digest*, IEDM, 1978, pp. 152–158.

[35] R. T. Longo, "A study of thermionic emitters in the regime of practical operation," *Technical Digest*, IEDM, 1980, pp. 467–470.

[36] R. T. Longo, E. A. Adler, and L. R. Falce, "Dispenser cathode life prediction model," *Technical Digest*, IEDM, 1984, pp. 318–321.

[37] J. R. M. Vaughan, "A synthesis of the Longo and Eng cathode emission models," *IEEE Trans. Electron Devices*, Vol. ED-33, No. 11, November 1986, pp. 1925–1927.

[38] G. Miram, M. Green, and M. Cattelino, "Gridded gun design considerations for pulse and CW applications," *Vacuum Electronics and Display Conference*, Garmisch Partenkirchen, Austria, 1989.

[39] P. Zalm, and A. J. A. v. Stratum, "Osmium dispenser cathodes," *Philips Technical Review*, Vol. 27, No. 3/4, pp. 69–75, 1966.

[40] W. Müller, "Theoretical study of cathode surface," *Final Technical Report*, NASA CR-182166, July 1989.

[41] W. Müller, "Computational modeling of dispenser cathode emission properties," *Technical Digest*, International Electron Devices Meeting, 1991, pp. 399–402.

[42] W. Müller, "Mechanism of emission enhancement in barium dispenser cathodes," *Technical Digest*, Tri-Service/NASA Cathode Workshop, Greenbelt, MD, 1992.

[43] A. T. Raczinski, "TWT principles," TWT training course presented at NSWC, Crane, IN, 1982.

[44] G. A. Haas et al., "Rapid turn-on of shelf stored tubes: An update," *IEEE Trans. Electron Devices,* Vol. ED-38, No. 10, October 1991, pp. 2244–2251.

[45] G. V. Miram et al., *Vacuum Electronics and Display Conference*, 1989.

[46] R. C. Treseder, T. J. Grant, and G. V. Miram, "Design of quick start, high current density electron guns," *Technical Digest*, IEDM, 1983, pp. 453–455.

[47] M. J. Cattelino, G. V. Miram, and B. Smith, "Fast-and super-fast-warm-up cathodes using novel APG/APBN heaters," *IEEE Trans. Electron Devices*, Vol. ED-38, No. 10, October 1991, pp. 2239–2243.

[48] A. S. Gilmour, Jr., "Effect of filament magnetic field on the electron beam from a Pierce gun," *Proc. IRE,* Vol. 49, 1961, p. 976.

Chapter 5

Cold Cathodes

5.1 SECONDARY EMISSION CATHODES

In crossed-field tubes, electrons are emitted from the cathode by secondary electron emission, which may be combined with thermionic emission. In this section, the secondary emission process and its role in the operation of crossed-field devices are discussed.

5.1.1 Introduction

As shown in Figure 5-1, when electrons bombard a surface, they may cause other electrons to be emitted from that surface. The bombarding electrons are called primary electrons and the emitted electrons are called secondary electrons. Secondary emission is of extreme importance in the operation of cathodes of crossed-field tubes where the electric and magnetic field configurations cause primary electrons to bombard the cathode. The way in which this occurs is indicated in Figure 5-2, which shows the RF field distribution in a magnetron.

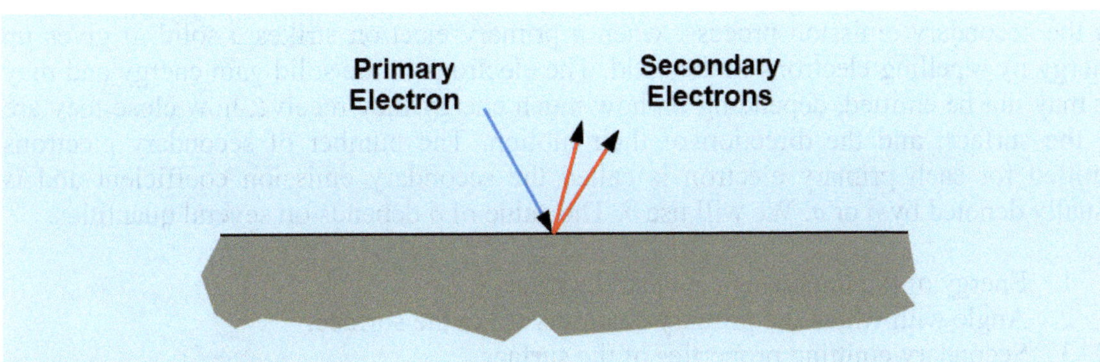

Figure 5-1 Secondary emission.

The path of an electron emitted at point A on the cathode is shown. After emission from the cathode, the electron is accelerated toward the anode, which, in crossed-field devices, contains an RF circuit (eight resonant cavities in Figure 5-2). Because a magnetic field is present (the direction is into the paper), the electron moves in a curved path in the clockwise direction. The direction of the RF field is such that it accelerates this particular electron and drives it back into the cathode surface. If the energy with which the electron strikes the cathode surface is high enough, then it can cause as many as several secondary electrons to be emitted.

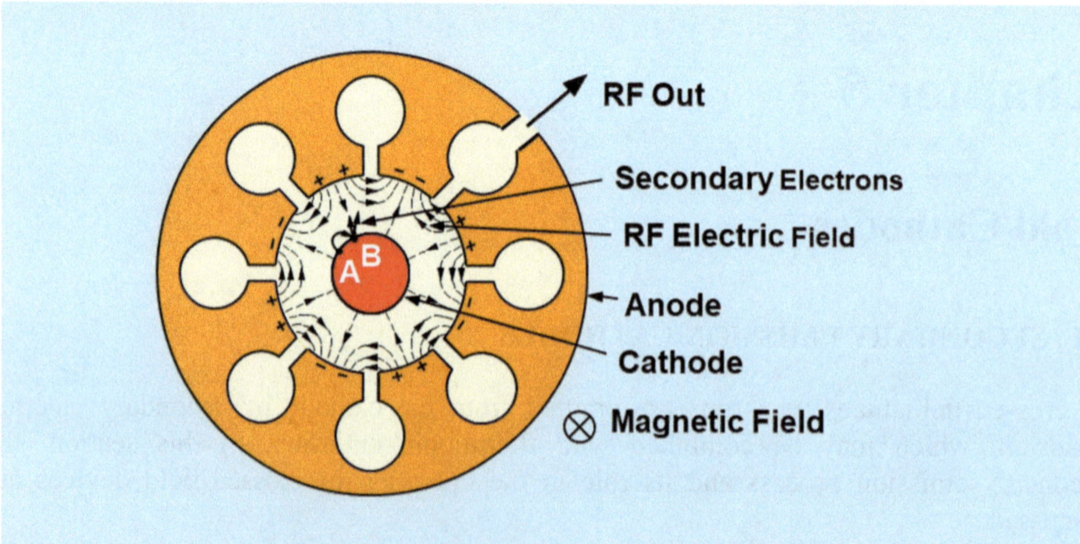

Figure 5-2 Secondary emission from the cathode in a crossed-field device.

Often, all or at least a major portion of the electrons from the cathode of a crossed-field device are emitted by secondary emission and the remainder are thermionically emitted. An example is the Philips YJ 1183 spin-tuned magnetron in which it is estimated that thermionic emission produces 3 A/cm^2 and secondary emission produces 23 A/cm^2 for a total cathode emission of 26 A/cm^2.

5.1.2 Characteristics of Secondary Emission

In the secondary emission process, when a primary electron strikes a solid, it gives up energy by repelling electrons in the solid. The electrons in the solid gain energy and may or may not be emitted, depending on how much energy they receive, how close they are to the surface, and the direction of their motion. The number of secondary electrons emitted for each primary electron is called the secondary emission coefficient and is usually denoted by δ or σ. We will use δ. The value of δ depends on several quantities:

1. Energy of the impacting primary electrons;
2. Angle with which the primary electrons strike the surface;
3. Secondary emitting properties of the surface.

5.1.3 Energy of Impacting Primary Electrons

The typical variation of δ with primary electron energy is shown in Figure 5-3. At very low energy levels, δ approaches zero. As primary electron energy is increased, δ increases, crosses unity, and reaches a maximum. Then, as primary electron energy is increased beyond this point, δ decreases and becomes less than unity again. The reason for this decrease in δ is that the primary electrons penetrate so deeply into the surface that their energy is dissipated in electrons that cannot reach the surface.

There are four quantities associated with the secondary emission property of a surface that are usually of significance. These are:

1. First crossover energy, E_I (53V in Figure 5-3);
2. Maximum secondary emission coefficient, δ_{max} (2.0 in Figure 5-3);
3. Primary electron energy, E_{max}, at which δ_{max} occurs (350V in Figure 5-3);
4. Second crossover energy, E_{II} (1,710V in Figure 5-3).

The first crossover energy, E_I, is the lowest primary electron energy that produces a δ of unity. The maximum secondary emission coefficient, δ_{max}, is a convenient way of characterizing the secondary emissivity of a surface. The second crossover energy, E_{II}, is the highest electron energy that produces a δ of unity.

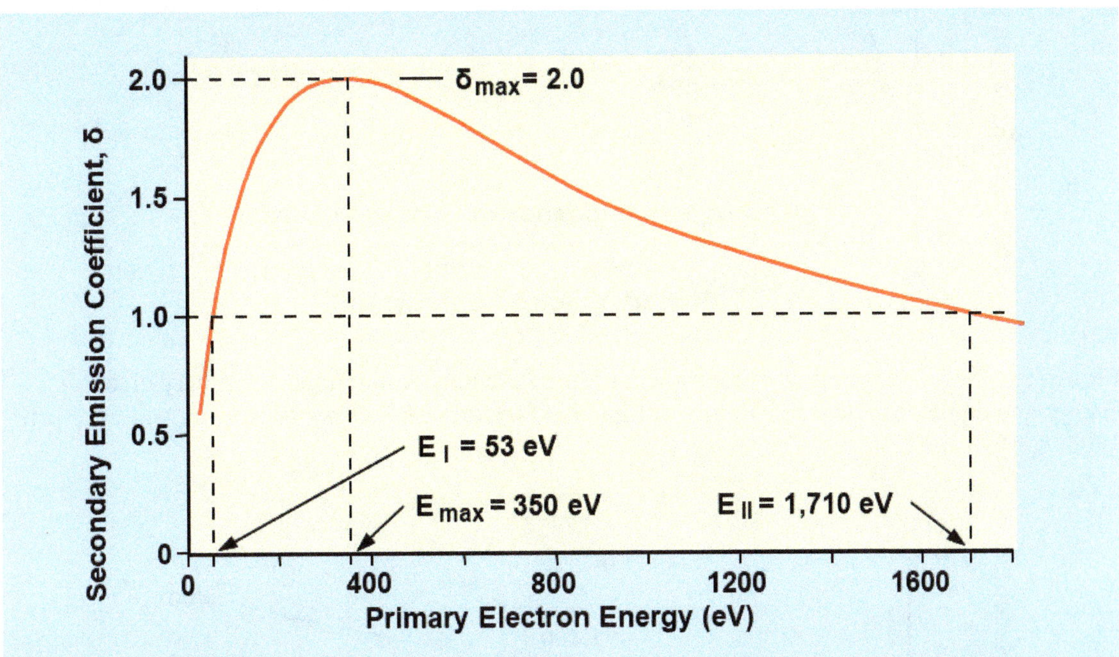

Figure 5-3 Secondary emission ratio as a function of primary electron energy.

5.1.4 Angle of Incidence of Primary Electrons

The angle of incidence, θ, of the primary electrons has a strong effect on the secondary emission coefficient of a surface. This dependency is shown in Figures 5-4 and 5-5 for molybdenum [1]. Note that δ_{max} increases from <1.4 to about 1.9 or about 40% as θ increases from 0° to 60°. The reason for the increase in δ with θ is that, as θ increases, primary electrons remain closer to the surface and so they give up their energy to electrons closer to the surface of the secondary emitter. As a result, there is a greater probability that the excited electrons will be emitted from the surface.

It is interesting that the first crossover energy of about 100 eV is nearly independent of θ. Also, it is clear, although not shown in Figure 5-4, that the second crossover energy increases significantly as θ increases.

5.1.5 Secondary Emitting Properties of Surfaces

The secondary emitting properties of surfaces vary widely from material to material and are critically dependent on the conditions of the surface. Examples are shown in Figures 5-6 through 5-10. In each case, δ is shown for an incidence angle of 0°. A material with an extremely low δ ($\delta_{max} \approx 0.25$) is shown in Figure 5-6. This material is textured carbon on a copper surface and was developed by NASA for possible use in multistage depressed collectors where it is important to inhibit secondary electron emission.

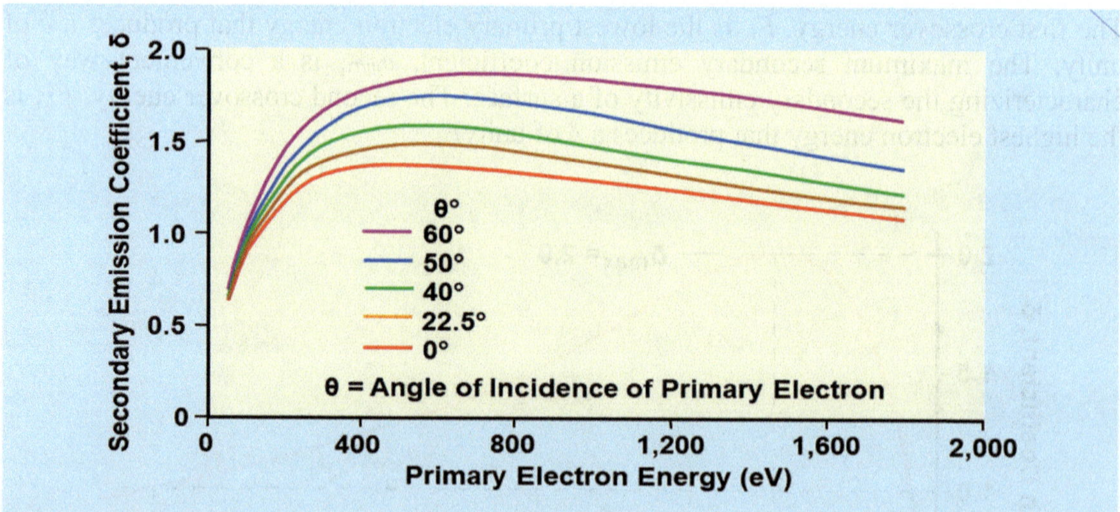

Figure 5-4 Secondary emission coefficient of molybdenum as functions of primary electron energy and angle of incidence. (From: A. Shih and C. Hor, *IEEE Trans. Electron Devices*, April 1993. © 1993 IEEE.)

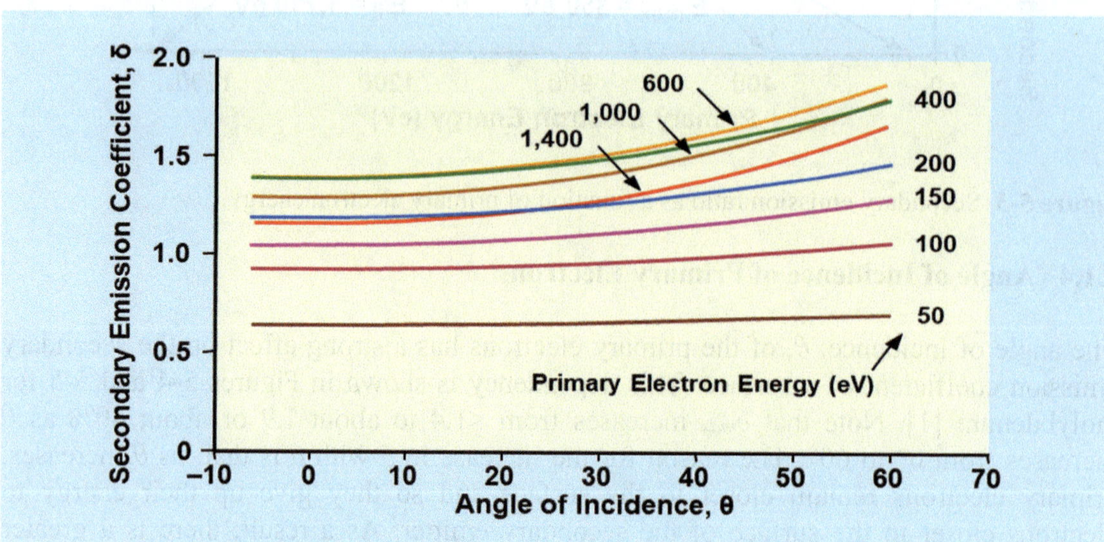

Figure 5-5 Secondary emission coefficient of molybdenum as functions of angle of incidence and primary electron energy. (From: A. Shih and C. Hor, *IEEE Trans. Electron Devices*, April 1793. © 1993 IEEE.)

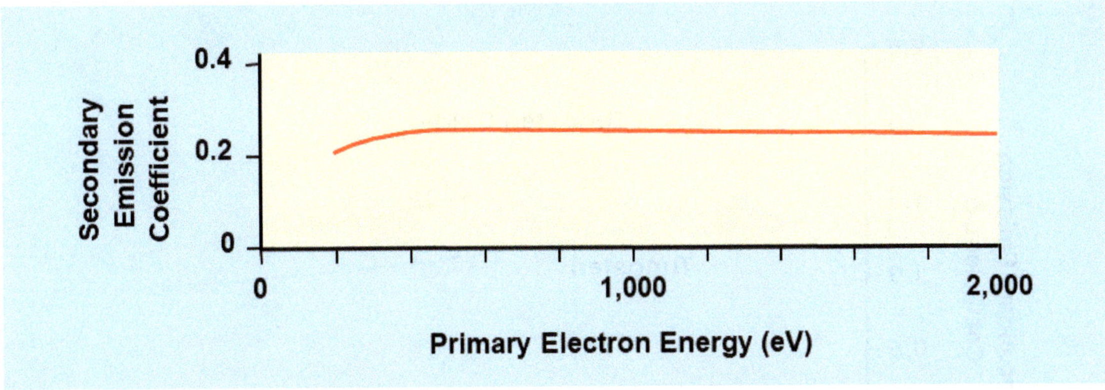

Figure 5-6 Secondary emission coefficient of textured carbon on a copper substrate. (From: A. N. Curren, *IEEE Trans. Electron Devices*, November 1986. © 1993 IEEE.)

In Figure 5-7, the extremely high δ ($\delta_{max} > 20$) of a diamond film on a molybdenum substrate is shown. When this film was subjected to ion bombardment so that the surface was converted to carbon, the value of δ_{max} dropped to about unity. Diamond films have been considered for use in the cathodes of crossed-field amplifiers, but the degrading effects of ion and electron bombardment make their use questionable.

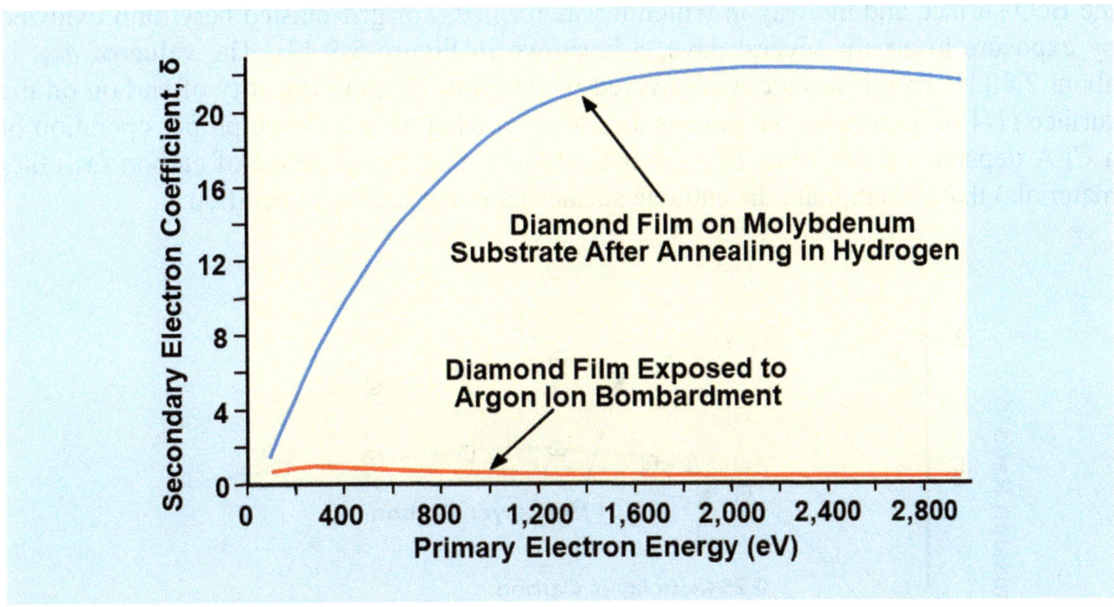

Figure 5-7 Secondary emission characteristics of a diamond film on a molybdenum substrate. (Adapted from: T. L. Bekker et al., *Final Report*, NSWC/Crane Division Diamond Film Study, August 1993.)

In Figure 5-8, the secondary emission characteristics of tungsten are shown [private communication from A. Dallos et al., Raytheon]. The value of δ_{max} is about 1.4. The presence of oxygen on the surface of the tungsten increases δ_{max} to about 2.3. This clearly illustrates the necessity for careful surface preparation if the true secondary emission coefficient of the surface is to be determined.

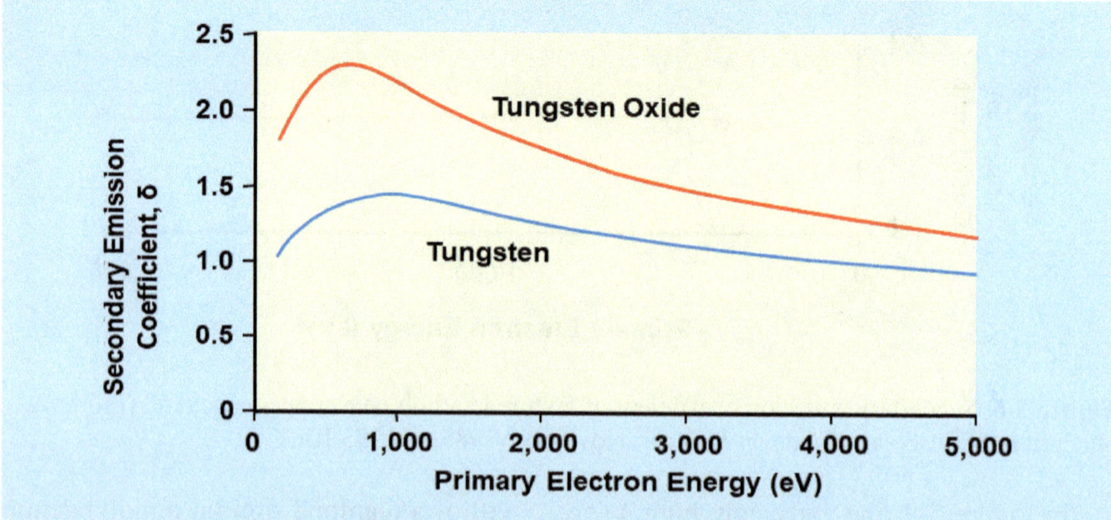

Figure 5-8 Secondary emission characteristics of tungsten and tungsten oxide.

Beryllium oxide (BeO) is frequently used as a cathode material in crossed-field amplifiers. The secondary emission coefficient for this material depends on the nature of the BeO surface and the way in which it was formed. For grit-blasted beryllium oxidized by exposure to air for several days, δ is shown in Figure 5-9 [2]. The value of δ_{max} is about 2.8 (1% of the surface was covered by carbon). A small quantity of carbon on the surface (1/4 of a monolayer) reduces δ_{max} to somewhat over 2. Since proper operation of a CFA depends on the value of δ_{max}, it is obvious that the presence of carbon (or other materials) that contaminate the cathode surface can degrade tube operation.

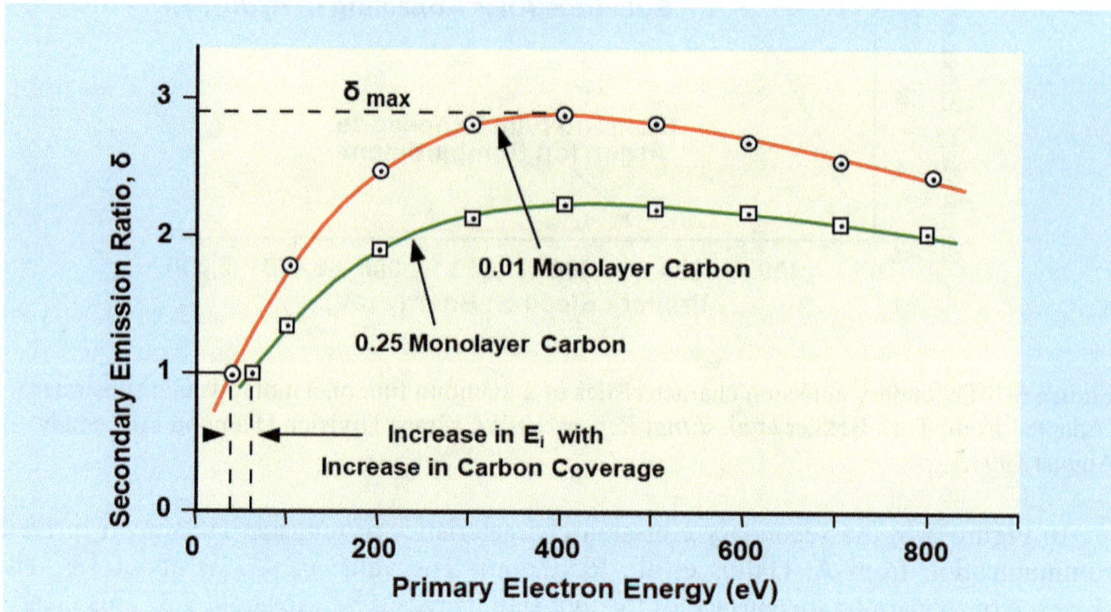

Figure 5-9 Secondary emission coefficient of oxidized beryllium with and without carbon contamination. (From: V. H. Ritz et al., *Surface and Interface Analysis*, November 1988.)

Figure 5-9 also shows that the first crossover energy, E_I, increases rapidly with carbon contamination. This is important in the operation of CFAs that use cold cathodes because there is no thermionic emission and tube operation depends entirely on secondary emission. Electron emission is initiated by the interaction of the RF input signal with stray electrons. E_I is important because the lower E_I is, the lower the RF input signal level has to be to start the tube properly. Figure 5-10 shows the dependence of δ_{max} and E_I on carbon. It was noted by Ritz et al. that a 0.03 difference in carbon atomic fraction will change δ_{max} by 0.1 and shift E_I by 7 volts. Again, this illustrates the strong dependency on carbon of the secondary emitting properties of a cathode most important to CFA operation.

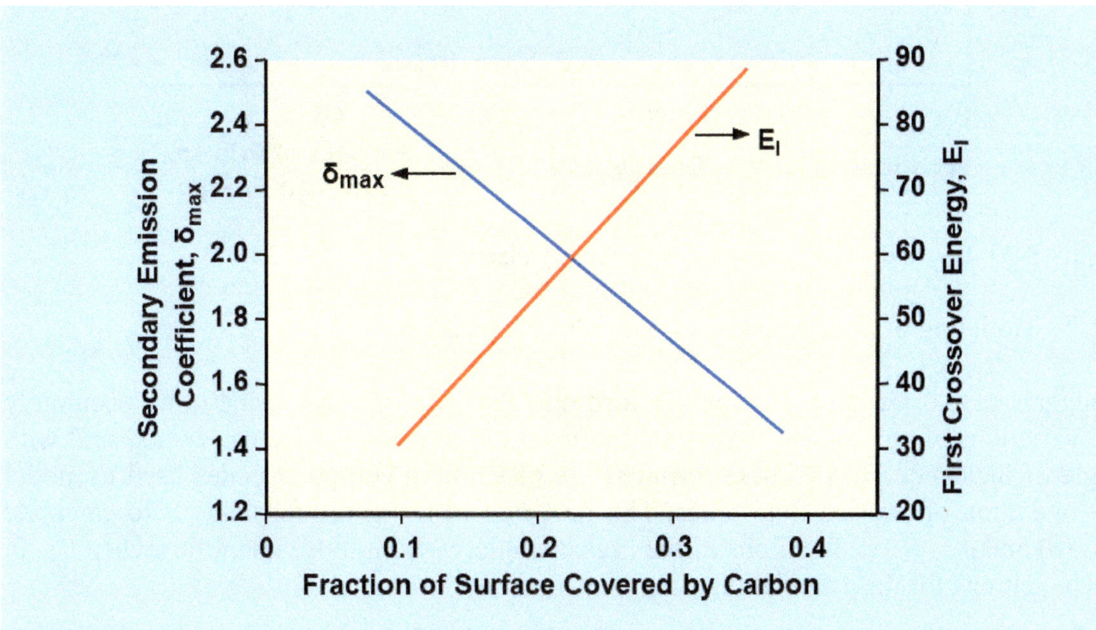

Figure 5-10 Effect of carbon coverage on beryllium oxide. (Adapted from: V. H. Ritz et al., *Surface and Interface Analysis*, November 1988.)

5.1.6 Energy Distribution of Secondary Electrons

The energy distribution of secondary electrons is shown in Figure 5-11. The majority of the electrons are emitted with energies of a few tens of electron volts. These are the electrons in the solid that acquire energy as the result of collisions with the primary electrons. A second peak in the secondary emission plot occurs at energy levels that are near the primary electron energy (near 100%). The electrons in this peak are those that experience elastic collisions at the solid surface and are reflected from that surface with an energy that is at or near the primary electron energy.

Perhaps the most significant aspect of the secondary electron energy distribution is the high values of electron energies (compared with thermionically emitted electrons). Thermionic electrons have energies of a small fraction of an electron volt. Secondary electrons have energies two orders of magnitude higher. This is part of the cause of the high noise output of crossed-field devices.

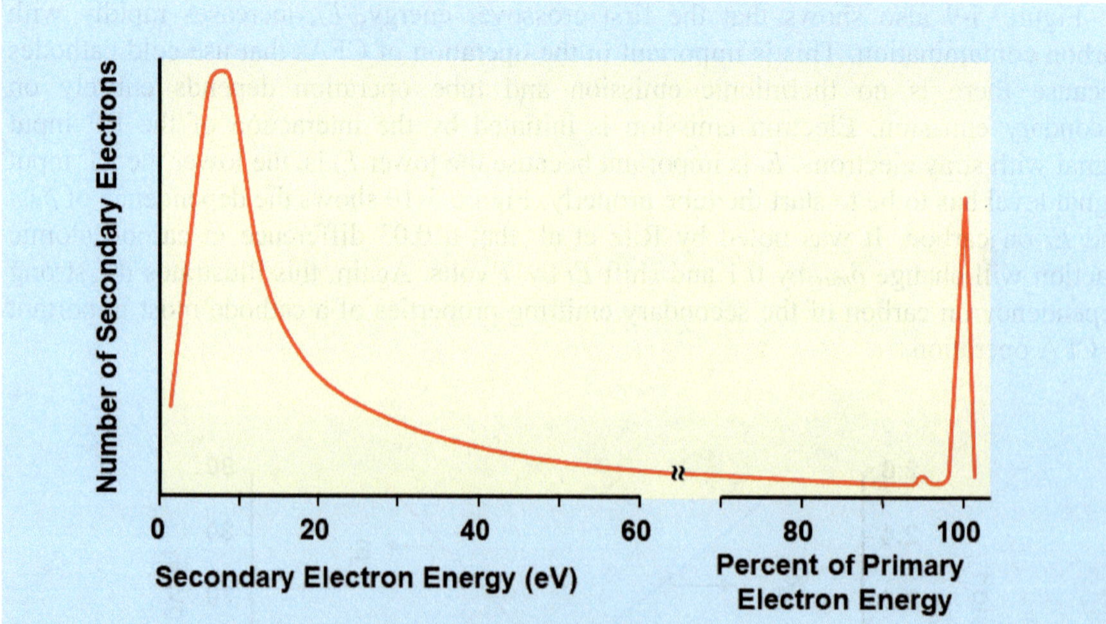

Figure 5-11 Distributions of energies of secondary electrons.

5.1.7 Modeling of Secondary Emission Characteristics

Vaughan developed simple empirical formulas that appear to describe quite accurately the variations of a secondary emission coefficient with primary electron energy and with angle of incidence [3, 4]. These formulas are of value in computer codes used to model the operation of crossed-field tubes. The first step in using the formulas is to calculate $E_{max}(\theta)$ and $\delta_{max}(\theta)$ as functions of the impact angle, θ, from those quantities when $\theta = 0$. The equations for this calculation are

$$E_{max}(\theta) = E_{max}(0)\left(1+k_{sE}\,\theta^2/2\pi\right) \tag{5-1}$$

and

$$\delta_{max}(\theta) = \delta_{max}(0)\left(1+k_{s\delta}\,\theta^2/2\pi\right) \tag{5-2}$$

where k_{sE} and $k_{s\delta}$ are factors, the values of which depend on the smoothness of the surface being considered. They should each be assigned a default value of 1.0, which Vaughan said is appropriate for typical dull surfaces in a working tube environment. Lower values would be appropriate for deliberately roughened surfaces, down to zero for textured carbon. Higher values up to about 2.0 will be appropriate for exceptionally smooth, clean, oxide-free surfaces.

The values of E_{max} and δ_{max} are then used to find δ as a function of E using

$$v = \frac{E-E_0}{E_{max}-E_0} \tag{5-3}$$

and

$$\frac{\delta}{\delta_{\max}} = \left(v\, e^{\,1-v}\right)^{k} \tag{5-4}$$

where

$$k = 0.56 \qquad \text{for } v < 1 \tag{5-5}$$

$$k = 0.25 \qquad \text{for } 1 < v \le 3.6 \tag{5-6}$$

and

$$\frac{\delta}{\delta_{max}} = 1.125/v^{\,0.35} \qquad \text{for } v > 3.6 \tag{5-7}$$

The quantity E_o is the minimum impact energy at which any secondary electrons at all are generated and is taken as 12.5 eV.

Shih and Hor [1] showed that Vaughan's formulas agree extremely well with their experimental results for molybdenum. Careful measurements similar to those of Shih and Hor are needed for other materials to provide a more stringent test of Vaughan's formulas.

5.1.8 Operation of Cathodes in Crossed-Field Devices

Some crossed-field devices use hot cathodes and others use cold cathodes. The cold cathode devices are crossed-field amplifiers. Cathode emission is initiated by the interaction of the RF input signal with a few stray electrons (at least 10 "good" electrons according to one CFA engineer). The source of these stray electrons is not clear. Field emission and the multipactor effect are both possible sources. The RF input signal drives these electrons into the cathode surface, causing secondary emission. If the secondary emission coefficient of the cathode is high enough, then emission current quickly (in nanoseconds) builds up to the operating current level.

In either hot-cathode or cold-cathode devices, the environment for the cathode is extremely harsh and many degradation mechanisms exist. First of all, as shown in Figure 5-12(a), there is intense electron bombardment of the surface. This causes heating and, if the temperature is not properly controlled, excessive evaporation of cathode constituents. In addition, electron bombardment leads to dissociation of emissive materials, such as barium oxide, on the surface of the cathode. The only cathodes free from the deleterious effects of electron bombardment are those fabricated from pure metal or from carburized thoriated tungsten. Electron bombardment also decomposes carbon monoxide and leaves a deposit of carbon on the cathode surface.

When arcs occur in crossed-field tubes, arc products are deposited on the cathode surface as is indicated in Figure 5-12(b). The arc products contaminate and reduce the secondary emission capability of the cathode. As a result, the noise figure and jitter of the tube increase. There must be some built-in mechanism in the tube to restore emission. In some tubes, it is the residual gas that provides ions during tube operation that clean the cathode surface. In tubes with BeO cathodes, an atmosphere of oxygen at a pressure of about 10^{-6} torr is used to maintain the oxidized surface of the cathode. This oxygen also becomes ionized and the positive ions bombard and clean the cathode surface. Also, the oxygen serves to oxidize carbon on the cathode surface and this goes off as CO or CO_2. In some tubes with platinum cathodes, the anode vane tips are also coated with platinum. This material is eroded from the anode during tube operation and is deposited on the cathode so that the platinum surface is maintained.

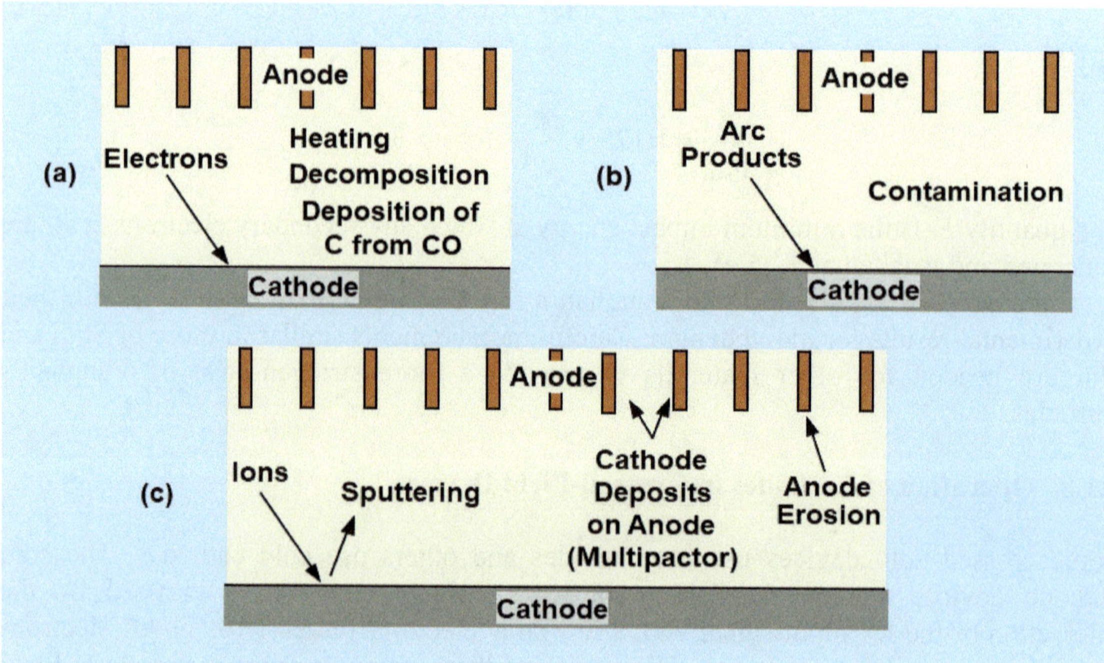

Figure 5-12 Cathode degradation mechanisms.

Ions serve to clean the cathode surface, but they also sputter emissive material from the surface as indicated in Figure 5-12 (c). Not only does this degrade emission, but it also results in the deposition of material on the anode. The secondary emission coefficient of that material is high so multipactor discharges may occur between elements of the RF circuit on the anode.

Another degradation mechanism indicated in Figure 5-12(c) is the erosion of the anode and this can be produced by decomposition products from the cathode. For example, Vaughan reported magnetron anode erosion in the presence of cathodes containing oxides [5]. He cited an example in which a thoria-molybdenum cathode was replaced by a tungsten-thorium cathode and this completely eliminated a severe anode erosion problem. It was not clear whether it was O or CO that was causing the erosion.

5.2 FIELD EMISSION CATHODES

5.2.1 Field Emission

As the electric field applied to the surface of a cathode is increased to the $10^9 - 10^{10}$ V/m (0.1–1.0 V/Å) level, it is found that electron emission increases very rapidly. Furthermore, the increase in emission is almost independent of cathode temperature.

The energy profile at the surface of the metal that an electron must overcome to be emitted is shown in Figure 5-13. This profile is like that for the Schottky effect (Section 4.1.1), except that the applied field is much larger and the width of the potential barrier is much narrower. Because of the wave nature of the electron, there is a finite probability that an electron can exist on the vacuum side of the barrier even though it does not have sufficient kinetic energy to overcome the barrier. This is known as the *tunneling effect.*

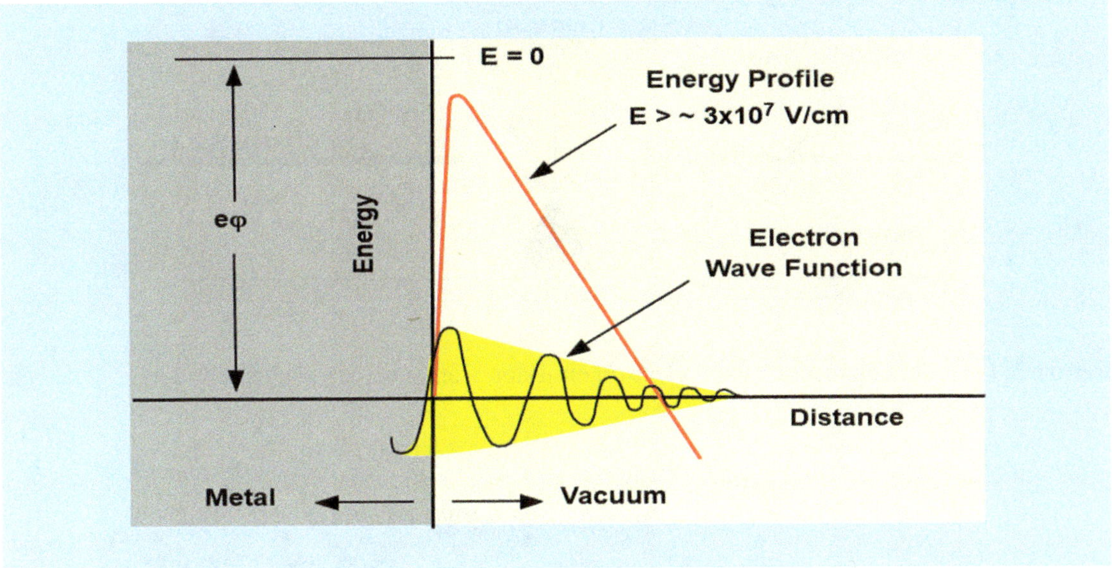

Figure 5-13 Energy profile with a very strong electric field.

5.2.2 Cathode Assembly

There are many reasons for the strong interest in field emission cathodes and these will become clear after the characteristics of the cathodes have been discussed [4].

The relation for current density as a function of electric field when field emission is occurring is the Fowler-Nordheim equation and is

$$J_o = C_1 E^2 e^{-C_2/E} \tag{5-8}$$

where C_1 and C_2 are constants that depend on work function and E is the electric field.

The electric field is $\beta V/d$ where β is the field enhancement factor (see Appendix E), V is the voltage between electrodes and d is the electrode spacing. The current density is I/A where I is the emission current and A is the area assigned to the emitter. We will be working with arrays of field emission emitters and I will be the total current from the array and A will be the area of the array. Making the substitutions for E and J, the Fowler-Nordheim equation can be written as

$$I = AC_1(\beta V/d)^2 \, e^{-C_2(d/\beta V)} \tag{5-9}$$

From (5-9) it is clear that large values of β and V and small spacings are required for substantial emission. To achieve a high field enhancement factor, it is necessary to use a needle-shaped emitter as is shown in Figure 5-14. The needle shape is used to concentrate the equipotential surfaces and enhance the electric field so that emission occurs at the tip of the emitter. The requirements for high β and small spacing are met in the emitter developed by C. Spindt [7] and shown in Figure 5-15 [8]. The I-V characteristics of the emitter are shown in Figure 5-16. Figure 5-17 shows that all dimensions are on the order of 1μm or less. A field emission cathode consists of a large array of Spindt emitters.

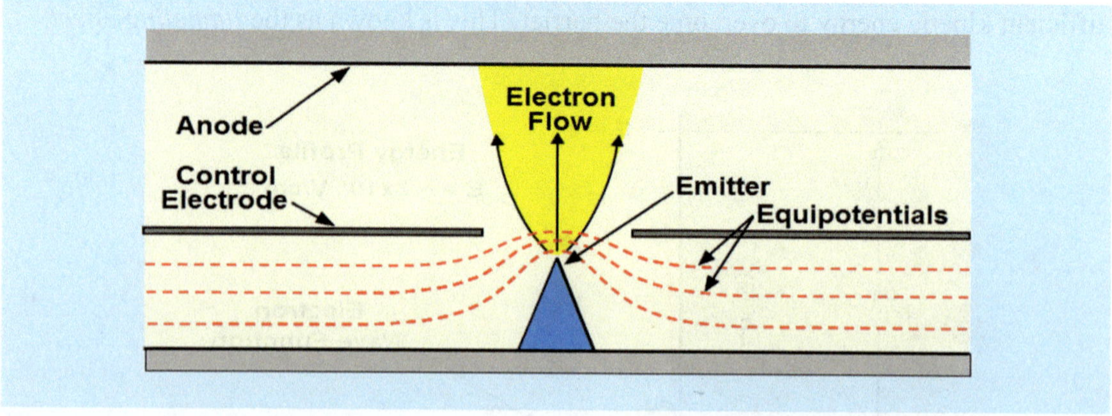

Figure 5-14 Field enhancement and electron emission from a needle-shaped emitter.

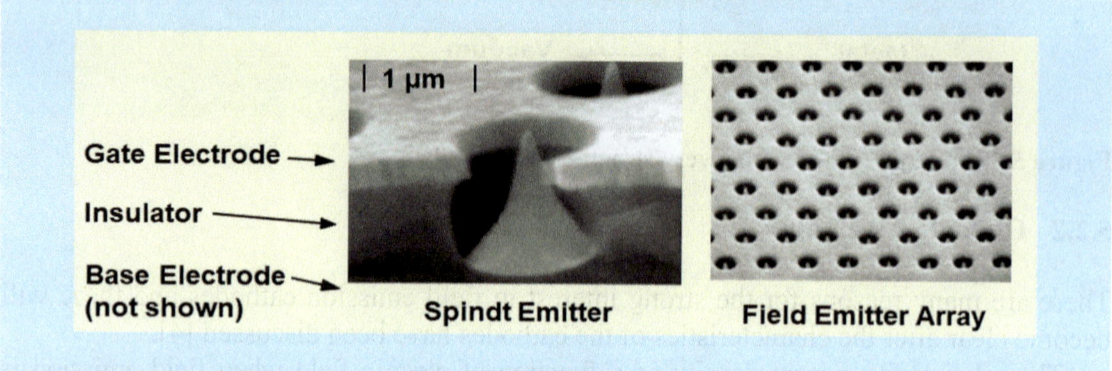

Figure 5-15 Elements of a field emission cathode.

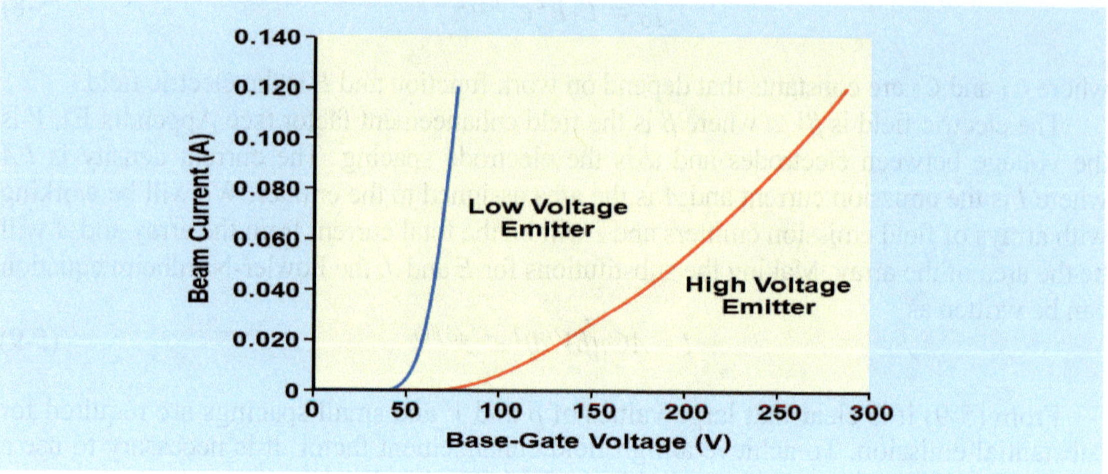

Figure 5-16 I-V characteristics for Spindt emitters.

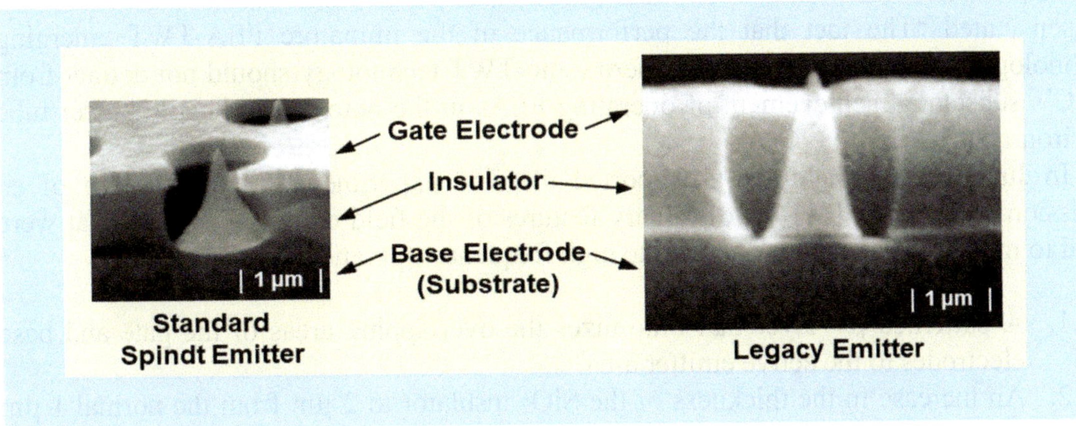

Figure 5-17 Comparison of standard emitter with emitter used in emission-gated TWT experiments (legacy emitter).

Now that the operation of field-emitter-array cathodes have been pointed out, we can understand the reasons for the strong interest in the development of these cathodes:

1. Room temperature operation and elimination of power loss due to operation of a heater;
2. No inherent cathode depletion mechanism;
3. High current density operation;
4. No time delay for heating.

Many of the other projected advantages of using a gated field emitter cathode in a TWT are direct spin-offs from the ability to control the source current at high frequencies without controlling the beam voltage. For example [9]:

1. TWT length can be reduced by 1/3 if the beam is RF density bunched at the
2. emitter. This is because the launching loss (see Chapter 13), which results from the use of only 1/3 of the input signal to initiate bunching in conventional TWTs is eliminated. As a result, TWT size and weight are reduced.
3. Improved AM/AM linearity because peak emission current is independent of beam voltage and can be controlled at microwave frequencies.
4. Improved beam control at MM wave frequencies and above because elimination of area compression of gun current more than offsets the emittance of the field emission cathode.

5.2.3 Proof of Principle Tests

There have been many attempts to use ungated field emission cathodes as replacements for the thermionic cathodes in TWTs. Perhaps the most successful results were obtained by NEC [10] when they achieved 27.5W output power, 19 dB gain at 10.5 GHz with a 40-mA beam current, and 82% beam transmission. A lateral resistive layer under the emitter tips was used to limit current. In a second TWT, NEC achieved 8W output power and 22 dB gain at 11.5 GHz with a 20 mA beam current and 98% beam transmission. A vertical resistor embedded just below the emitter cone was used to limit current. In [6],

Jensen stated "The fact that the performance of the immature FEA-TWT emerging technology lags the well-developed thermionic-TWT technology should not detract from NEC's substantial achievement of operating FEAs in the notoriously hostile power tube environment."

In June 2002, Whaley et al. reported the first experimental demonstration of an emission-gated TWT [11]. Two primary features of the field emission cathode that were used to minimize capacitance and make high frequency operation possible are

1. A patterned p^{++} layer that minimizes the overlapping areas of the gate and base electrodes to the active emitter area.
2. An increase in the thickness of the SiO_2 insulator to 2 μm from the normal 1 μm used in a standard low-frequency cathode. This made it necessary to grow emitter cones to twice the height and aspect ratio of the standard cathode. One of these cones is shown along with a standard cone in Figure 5-17. The emitter with this cone is referred to as a legacy emitter to distinguish it from more recent emitters.

The cathode was 300 μm in diameter and consisted of 10,000 molybdenum emitter tips with tip-to-tip spacing of 2.7 μm.

5.2.4 Cathode Failure Mechanisms and Failure Prevention

Because of the advantages just described and many other advantages, many workers have investigated field emission cathodes. While the performance has been exciting, the cathodes have been found to be very unreliable. There are several possible reasons for this. Ion blunting of emitter tips is one; however, this has been essentially eliminated by using a shield voltage in the electron gun. Li et al. [12] stated "We believe one main reason for Spindt cathode failure in initial testing is metal particle adhesion produced in cone formation process. It is also a potential failure reason in high current applications." They reported that "Preliminary experiments show megahertz ultrasonic cleaning is effective in removing the particles but further work is still needed."

It has been thought that when operating at high emission levels, poor uniformity between emitter tips in an array leads to some emitters producing excessive emission, overheating, and vaporizing thereby generating an arc that shorts out the structure. However, recently (in the history of Spindt emitters) it was found that the cause of damaging arcs at high emission levels is not emitter tip vaporization, but rather spurious emission from the triple junction formed by the silicon, silicon dioxide, and vacuum at the base of the cathode cavities that triggers a flashover across the cavity's oxide walls [13]. (Breakdown initiated at a triple junction is discussed in Appendix E.) This problem has been alleviated by changing the gate and inserting a silicon nitride wafer shield between the gate and the SiO_2 insulator. Also, by using a nickel post between the emitter and base, the distance between the shield and the base can be increased without changing the operating voltage. As a result, the field on the insulator surface can be reduced. Finally, the capacitance can also be reduced.

Whaley et al. reported the results of a detailed analysis of the emittance of a beam produced by emitter tip radii from 25 to 150Å operating at voltages ranging from 60 to 190V. They stated that "The technological hurdle for field emitter cathode developers is to reduce the beam divergence of the individual emitter tips so that the emittance of the composite beam originating from many thousands of these emitters, can be reduced from its current level" [14]. For a field emitter cathode to work effectively in a 100W, 100 GHz TWT, v_r would have to be reduced by a factor of 2. To do this, the gate voltage and/or the rms angle of emission would have to be reduced.

REFERENCES

[1] A. Shih, and C. Hor, "Secondary emission properties as a function of the electron incidence angle," *IEEE Trans. Electron Devices*, Vol. 40, No. 4, April 1993, pp. 824–829.

[2] V. H. Ritz et al., "Secondary electron emission characteristics of oxidized beryllium cathodes," *Surface and Interface Analysis*, Vol. 11, No. 6–7, April 1988, pp. 389–397.

[3] J. R. M. Vaughan, "A new formula for secondary emission yield," *IEEE Trans. Electron Devices,* Vol. ED-36, No. 9 pt. 2, September 1989, pp. 1963–1967.

[4] J. R. M. Vaughan, "Secondary emission formulas," *IEEE Trans. Electron Devices*, Vol. 40, No. 4, April 1993, p. 830.

[5] J. R. M. Vaughan, "Magnetron anode erosion in the presence of cathodes containing oxides," *IEEE Trans. Electron Devices*, Vol. ED-17, No. 4, April 1970, p. 377.

[6] K. L. Jensen, "Field emitter arrays for plasma and microwave source applications," *Phys. Plasmas*, Vol. 6, pp. 2241–2253, 1999.

[7] C. A. Spindt, *J. Appl. Phys*, Vol. 39, No. 3504, 1968.

[8] V. M. Aguero and R.C. Adamo, "Space applications of Spindt cathode field emission arrays," 6th Spacecraft Charging Technology Conference, AFRL-VS-TR-20001578, September 1, 2000, pp. 347-352.

[9] D. R. Whaley and B. M. Gannon, et al., "Application of field emitter arrays to microwave power amplifiers, " *IEEE Transactions on Plasma Science*, Vol. 28, No. 3, June 2000, pp. 727–747.

[10] H. Takemura and Y. Tomihari, et al., *Technical Digest of the 1997 IEEE- IEDM,* IEEE, Piscataway, NJ, 1997, p. 709.

[11] D. R. Whaley and B. M. Gannon, et al., "Experimental demonstration of an emission-gated traveling-wave tube amplifier," *IEEE Transactions on Plasma Science,* Vol. 30 No. 3, June 2002, pp. 998-1008.

[12] Xinghui Li, Guodong Bai, et al., "A Potential Failure Reason for Spindt Cathodes in High Current Applications," Conference Record, 2014 IVEC.

[13] C. Spindt, C. E. Holland, and P. R. Schwoebel, "An improved, reliable Spindt cathode design for high-current applications," in Proc. 11th Int. Vac. Electron. Conf., Monterey, CA, May 2010, pp. 201–202.

[14] D. R. Whaley, "Practical Design of Emittance Dominated Linear Beams for RF Amplifiers." *IEEE Trans. Electron Devices*, Vol. 61, No. 6, June 2014, pp 1726–1734.

Chapter 6

Electron Guns for Linear-Beam Tubes

The electron guns for all linear-beam tubes have operating principles that are nearly the same. The only major differences are in size and operating voltage and current, which come from differences in power requirements. In this chapter, the operating principles of the electron guns used on nearly all linear-beam tubes are discussed. Most of these guns are designed using guidelines set forth by J. R. Pierce and are known as Pierce guns [1].

6.1 PIERCE GUNS

The electron gun in a microwave tube is used to form the electrons from the cathode into a beam suitable for interaction with a microwave circuit. Electron gun designers must cope with the following two basic problems:

1. Electrostatic repulsion forces between electrons tend to cause the beam to diverge;
2. The current density required in the electron beam is normally far greater than the emission density that the cathode can supply with an acceptable life expectancy.

The solution to these problems, described by Pierce, is summarized in Figure 6-1. It is assumed that there is no magnetic field in the gun. Thermal velocities of electrons are ignored and laminar electron flow is assumed. The gun is divided into three regions. In the first region, assuming that a beam with a circular cross section is desired, the shape of the cathode disc (referred to as the *cathode button*) is spherical. (This may not be the optimum shape as pointed out by Brewer [2].) The focus electrode is designed to produce equipotential surfaces in region 1 that are nearly spherical, with the same center of curvature as the cathode. As a result, electrons flow toward the center of curvature of the cathode.

In region 2, because the anode must contain a hole to let the electron beam pass through, equipotential surfaces bow into the anode aperture. As a result, a divergent electrostatic lens exists that produces a defocusing action on the electron beam.

Finally, in region 3, the electrons have escaped from the accelerating field of the cathode-to-anode regions and are drifting under the influence of space-charge forces. Thus, the electrons in the beam follow universal beam-spread trajectories.

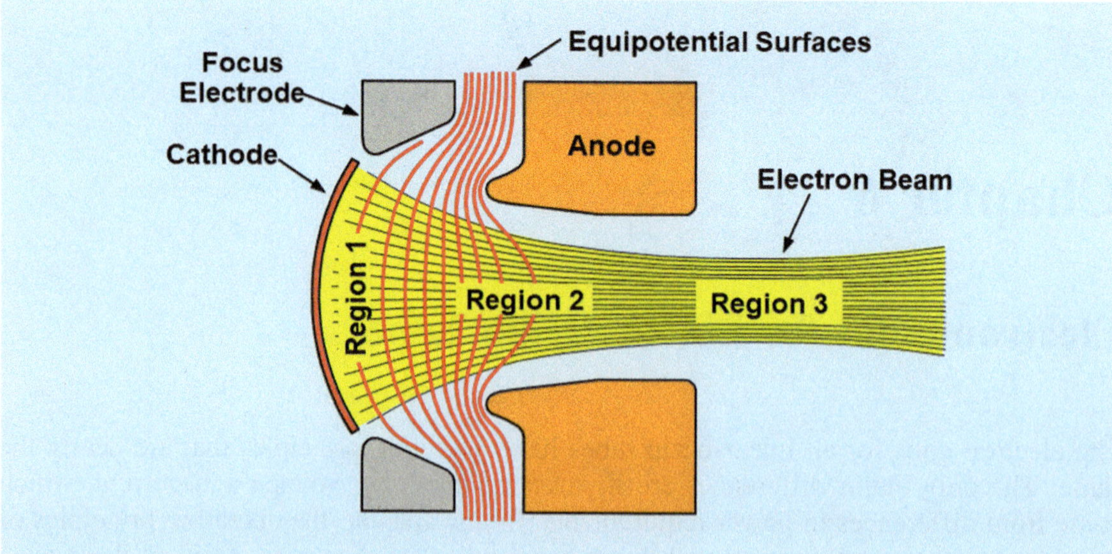

Figure 6-1 Overview of a Pierce gun.

6.1.1 Focus Electrodes for Parallel Flow

Dealing first with the repulsion force problem as electrons leave the cathode and move toward the anode, it is instructive to examine the problem of attempting to cause electrons to travel in parallel paths as they leave a cathode as indicated in Figure 6-2.

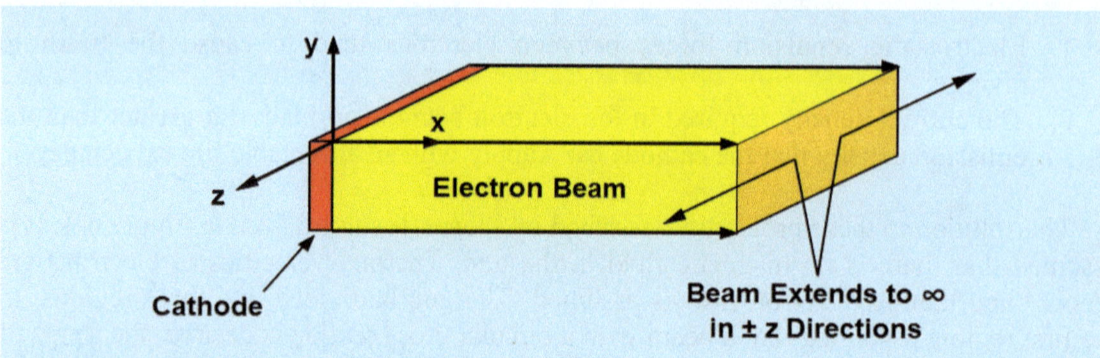

Figure 6-2 Parallel electron flow from a planar cathode.

When no electrons are present, the equipotential profiles in a parallel-plane diode are parallel and spaced equal distances apart for equal voltage increments. This is shown by the dashed lines in Figure 6-3. When electrons are present, the equipotential profiles are deflected to the right as indicated by the solid lines. Electrons tend to move perpendicularly to the equipotential profiles, so the electron trajectories diverge as the electrons move away from the cathode. This is simply another way of saying that the electrons are pushing on each other, and this causes their trajectories to diverge. If an electrode arrangement could be devised that would straighten the equipotential profiles even though electron space charge forces are present, then the electrons would travel in parallel paths.

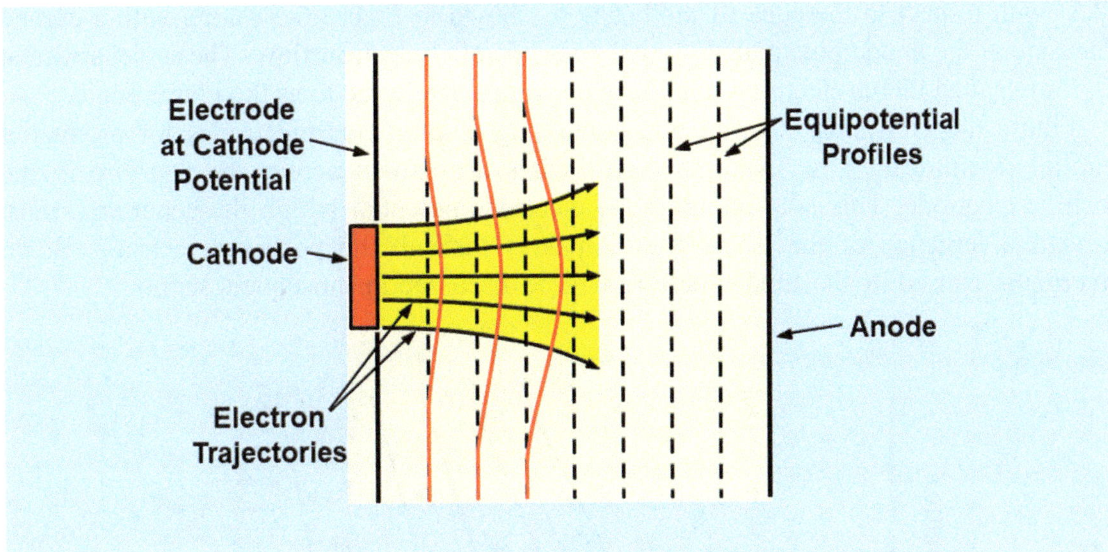

Figure 6-3 Equipotential profiles without (dashed lines) and with (red lines) electrons.

Intuitively, it can be seen that, if the electrodes adjacent to the cathode are tipped toward the anode, as indicated in Figure 6-4, then the equipotential profiles will be straightened. In the absence of electrons, the profiles curve toward the cathode. The electron charge causes them to move toward the anode.

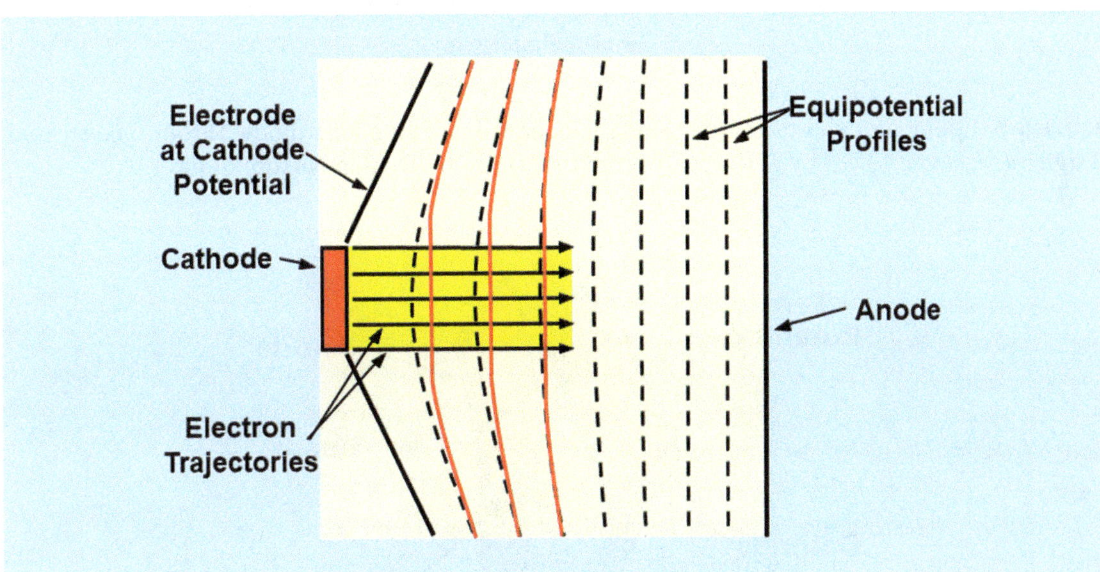

Figure 6-4 Straightening of equipotential profiles in an electron beam.

The mathematical solution that leads to a definition of the electrode configuration for parallel electron flow is similar to the solution for the problem of space-charge-limited flow discussed in Chapter 4. The main difference now is that the equations must be solved inside and outside the beam, and the solutions must match at the edge of the beam. Pierce showed that planar electrodes at the potential of the cathode placed at an angle of

67.5° with respect to the edge of the beam, as shown in Figure 6-5 (along with a curved electrode at the anode potential, V_a), will cause parallel electron flow. These electrodes at 67.5° are called Pierce electrodes and the 67.5° angle is referred to as the Pierce angle.

Figure 6-6 shows an electron gun containing 67.5° Pierce electrodes for producing parallel (rectilinear) flow. Note that grid wires are shown across the aperture in the positive electrode. This is to maintain the electrode potential within the beam and, thus, prevent divergence of the beam after it passes through the positive electrode. Beam divergence caused by the anode aperture will be discussed in subsequent sections.

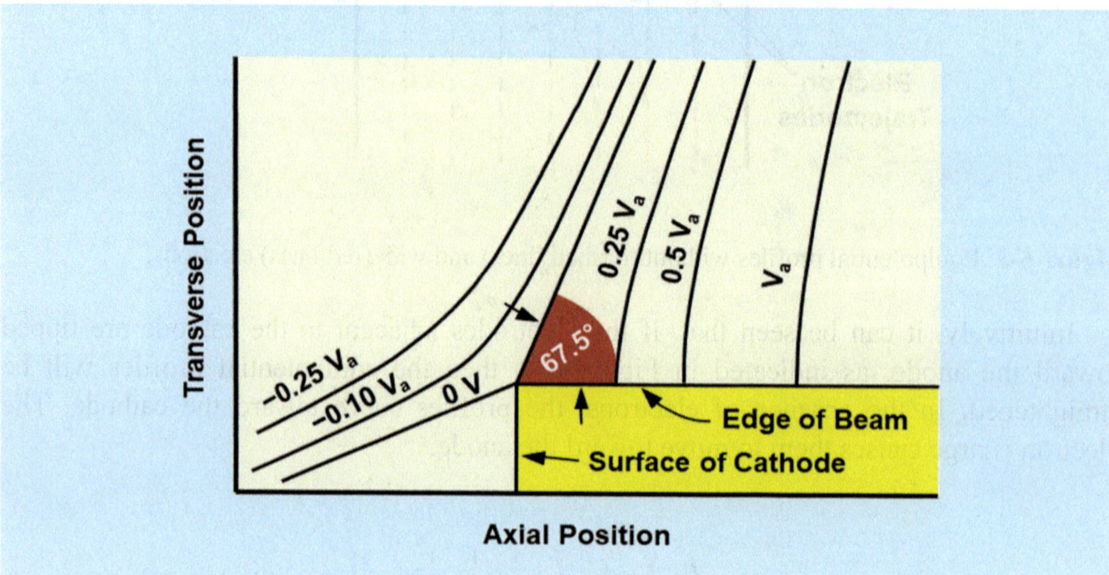

Figure 6-5 Equipotential profiles for parallel electron flow in a planar diode. (From: *Theory and Design of Electron Beams* by J. R. Pierce, copyright 1954 by D. Van Nostrand, Inc.)

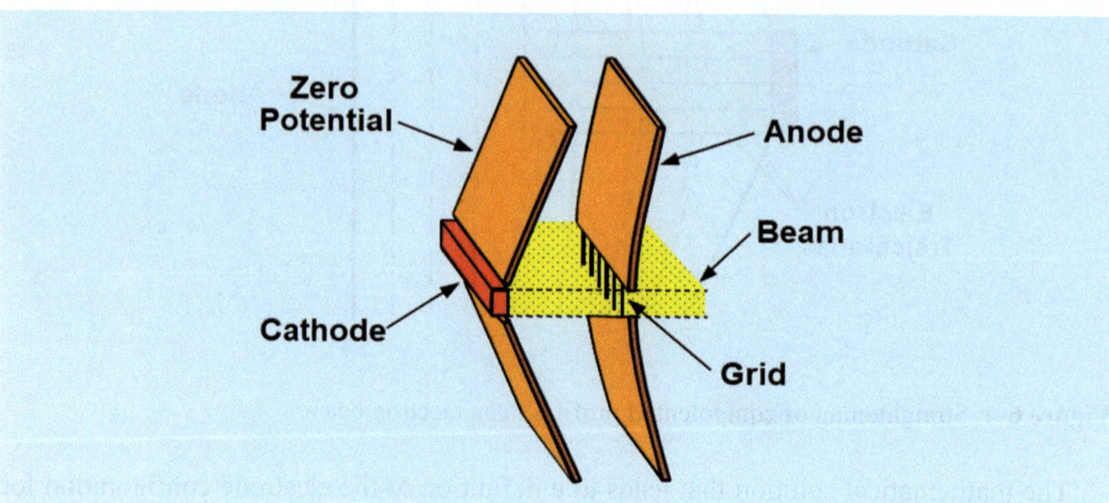

Figure 6-6 Electron gun for producing rectilinear (parallel) electron flow. (Adapted from: *Theory and Design of Electron Beams* by J. R. Pierce, copyright 1954 by D. Van Nostrand, Inc.)

6.1.2 Focus Electrodes for Convergent Flow

The current density required in the electron beam in a microwave tube is normally far greater than the emission density capability of the cathode. As a result, it is necessary to use a cathode having a relatively large area and then compress the flow of electrons as the beam is formed. The starting point in Pierce's analysis is a spherical diode. The inner surface of the outer sphere is the cathode and the inner sphere is the anode. A conical section of the diode is used as shown in Figure 6-7. Space-charge-limited emission is assumed.

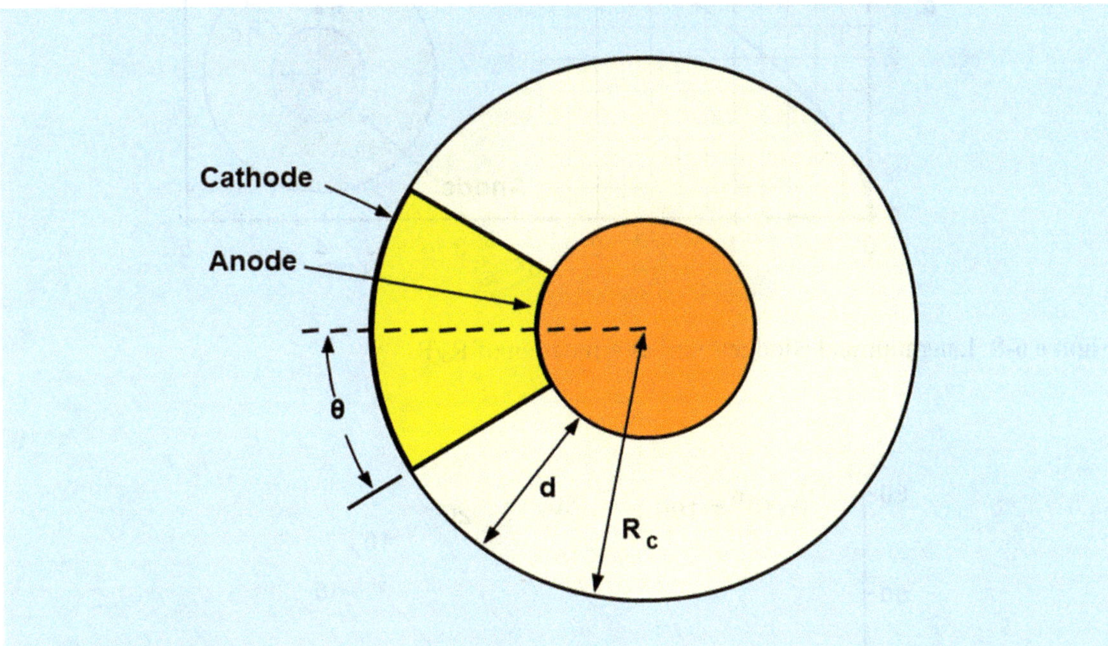

Figure 6-7 Conical diode used in a Pierce gun.

When the analysis is carried out, the perveance in the conical diode is found to be related to the half angle of the cone, θ, by

$$P = 14.67 \times 10^{-6}(1-\cos\theta)/\alpha^2 \tag{6-1}$$

The quantity α^2 used here was defined by Langmuir and Blodgett [3] and is a function of the dimensions of the diode as is shown in Figure 6-8.

Figure 6-9 shows perveance as a function of cone angle and normalized cathode-anode spacing. It will be shown that the extremely high values of perveance indicated in Figure 6-9 cannot be achieved in real electron guns when anode effects are considered.

The problem with the conical diode is that space-charge forces will cause the beam to diverge just as they did for the planar diode. Equipotential contours, which should be spherical in shape, are deflected by the electron space-charge potential as indicated in Figure 6-10. To retain radial electron flow when the remainder of the sphere is removed, it is necessary, as was the case for rectilinear flow in a parallel plane diode, to use electrodes that cause the equipotential profiles to have the shape they had in the absence of the electrons (spherical in this case).

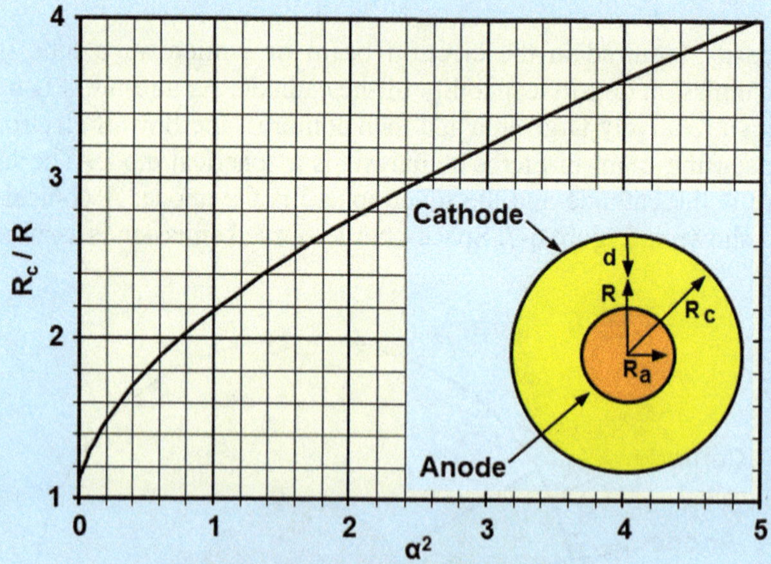

Figure 6-8 Langmuir and Blodgett's α^2 as a function of R_c/R.

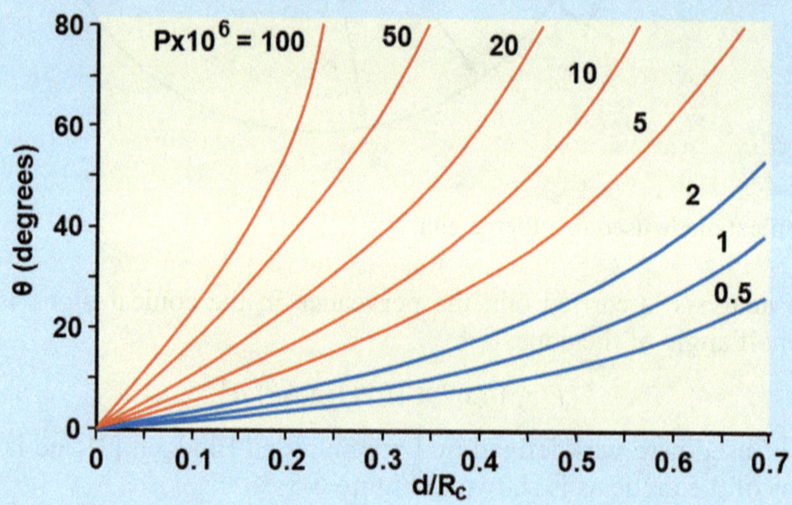

Figure 6-9 Perveance as functions of cone half angle and cathode-anode spacing.

Conceptually, as for the planar diode, it can be seen that by tipping the electrodes adjacent to the cathode forward, the equipotential contours can be made more nearly spherical. In fact, as indicated in Figure 6-11(a), close to the edge of the cathode, electron flow appears to originate from a planar portion of the cathode with electron flow perpendicular to that portion. Thus, an electrode placed at a 67.5° angle with respect to the edge of the beam will have the desired effect on the equipotential profiles near the edge of the cathode. For small cone half angles (small θ), this effect extends to the center of the cathode so no further consideration of the shape of the focus electrode is necessary.

For large cone angles, as indicated in Figure 6-11(b), the potential profiles will still not be spherical. As a result, as the electrons leave the cathode region, the electron trajectories will not be focused toward the center of curvature of the cathode.

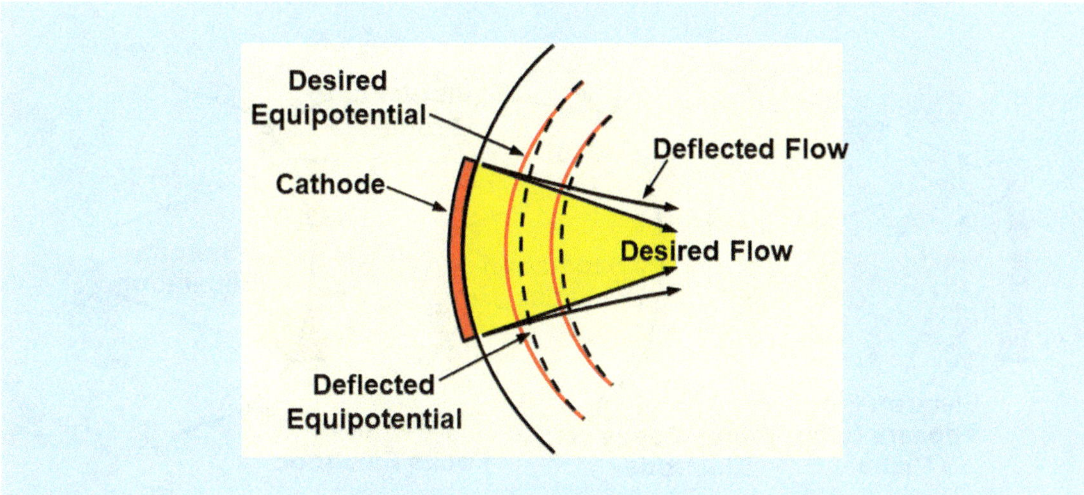

Figure 6-10 Deflection of equipotential contours in a conical diode by electron space-charge potential.

A more important problem may be that nonuniform emission from the cathode will occur. For high perveance guns in which the cathodes are large in diameter, this problem can be severe. Assume, for example, that the separation between the cathode and one of the equipotential surfaces near the cathode is twice as large at the cathode center as it is at the edge. Since the equipotential and the cathode are, in effect, a diode, the emission current density from the edge of the cathode surface will be four times that at the center. (Remember that current in a diode varies inversely with the square of cathode-anode spacing.) This will compound the focusing problem of the electron gun and will also shorten the cathode life because the entire cathode will have to be at a temperature high enough to provide the emission required at the outer edge.

To correct these problems, the shape of the focusing electrode must be modified so that the equipotential profiles near the cathode surface become more nearly spherical. Researchers have experimented with many possible shapes, one of which is shown in Figure 6-11(c) [4]. In effect, the modified focus electrodes lift the equipotential profiles away from the edges of the cathode. There is little effect on the shapes of the profiles in the center region of the cathode because the electron space-charge fields dominate in controlling the potentials there.

6.1.3 Defocusing Effect of Anode Aperture

To remove electrons from the conical diode, it is necessary to cut a hole in the anode. The resulting distortion of the equipotential profiles near the anode aperture is illustrated in Figure 6-12 [5]. Notice that at a low perveance as in Figure 6-12(a), the distortion of the profiles is limited to the region very close to the anode aperture. As perveance is increased, as in Figure 6-12(b, c), the size of the anode aperture increases and the anode to cathode separation decreases. At a perveance of 3×10^{-6} the distortion of the profiles extends to the cathode surface. As a result, the electron trajectories in the cathode-to-anode region are seriously perturbed and emission from the cathode is very nonuniform.

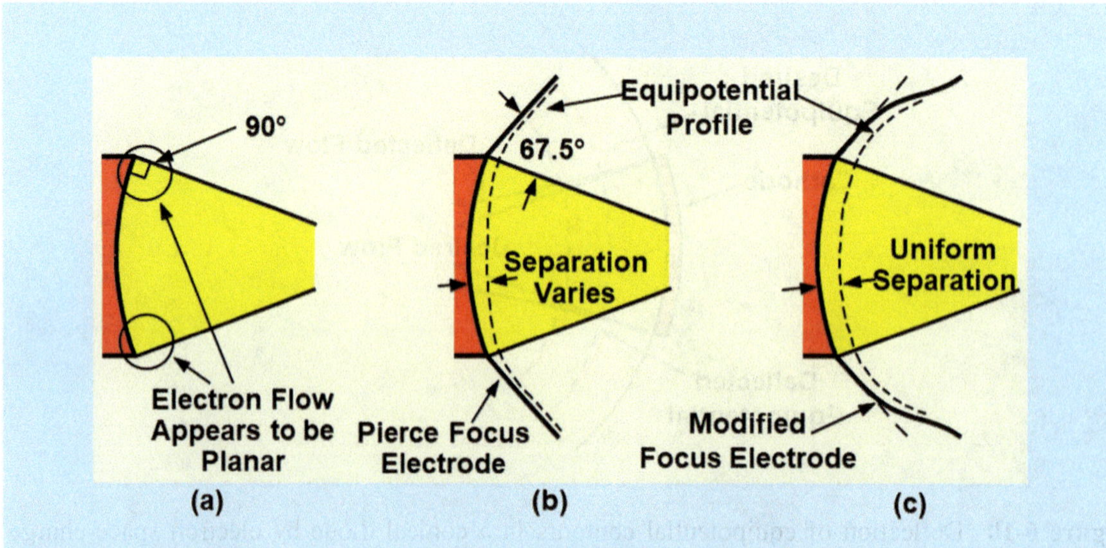

Figure 6-11 Modification of Pierce electrodes to produce spherical equipotentials.

In [5], Brewer described the modification of the focus electrode as shown in Figure 6-12(d) to reduce distortion of the equipotentials and improve electron focusing and cathode emission uniformity. The modified focus electrode has the same effect as that shown in Figure 6-11(c). That is, the equipotentials are lifted away from the regions near the edge of the cathode. Although focusing and emission uniformity are improved, perveance decreases because of the reduced electric field near the cathode edges.

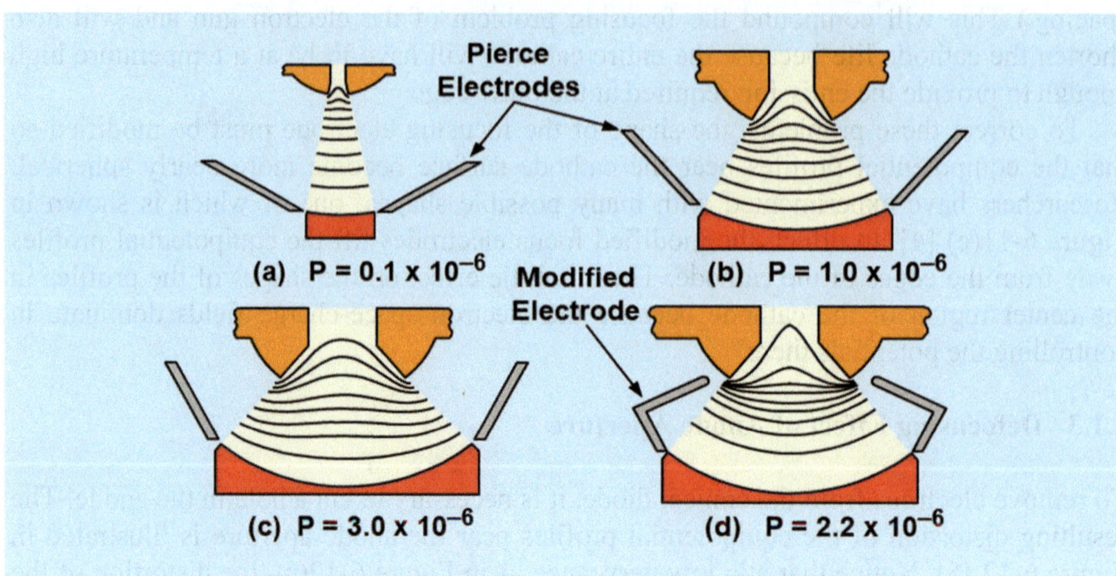

Figure 6-12 Effect of anode aperture on equipotential profiles. (G. R. Brewer, *Jour. App. Phys.*, January 1957 and *Focusing of Charged Particles*, copyright 1967 by Academic Press.)

The distortion of the equipotential profiles near the anode aperture results in the formation of a divergent lens. This lens changes the focus of the electron flow to a point beyond the center of the sphere, as shown in Figure 6-13.

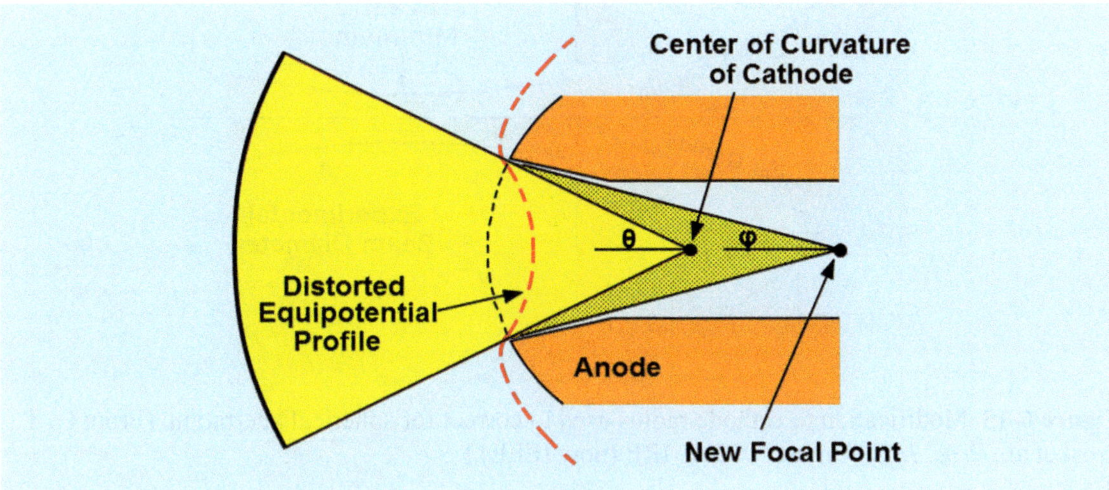

Figure 6-13 Defocusing effect of the anode aperture.

The lens formed by the distorted equipotential profiles is nowhere near a perfect lens because it suffers from spherical aberration. As is shown in Figure 6-14, this aberration results in a focal length that varies with the radial position in the lens. Notice that the trajectory of the electron that leaves the cathode at position 1 is focused to a point to the left of the trajectory of the electron that leaves the cathode at position 2.

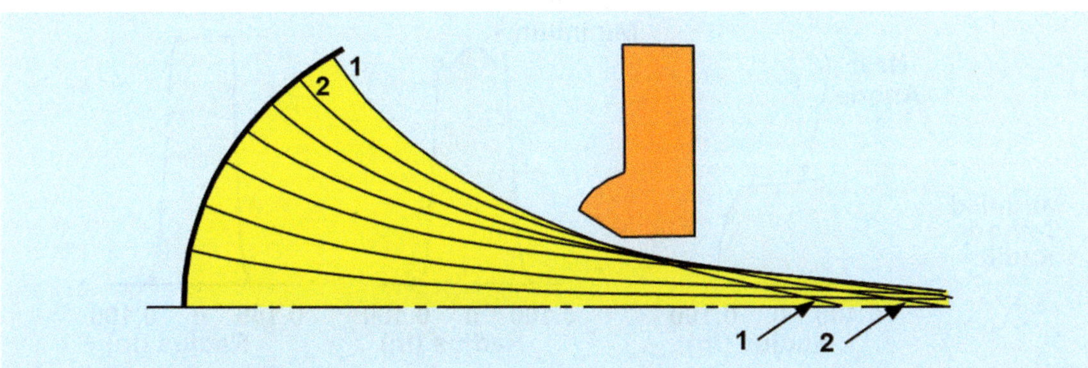

Figure 6-14 Effect on the focal length of spherical aberration. (Adapted from: G. R. Brewer, U.S. Patent 3,139,552, June 30, 1964.)

Frost, Purl, and Johnson [6] found that the measured beam from a high perveance gun was larger in diameter and less uniform in current density than expected from theory. By modifying the spherical radius of the cathode, as shown in Figure 6-15, they were able to correct for the lens aberration. The resulting improvements in beam characteristics are shown in Figure 6-16. The beam-current profiles at the top left of the figure were obtained before the correction was made. Note the severe nonuniformity in current as a function of radial position in the beam.

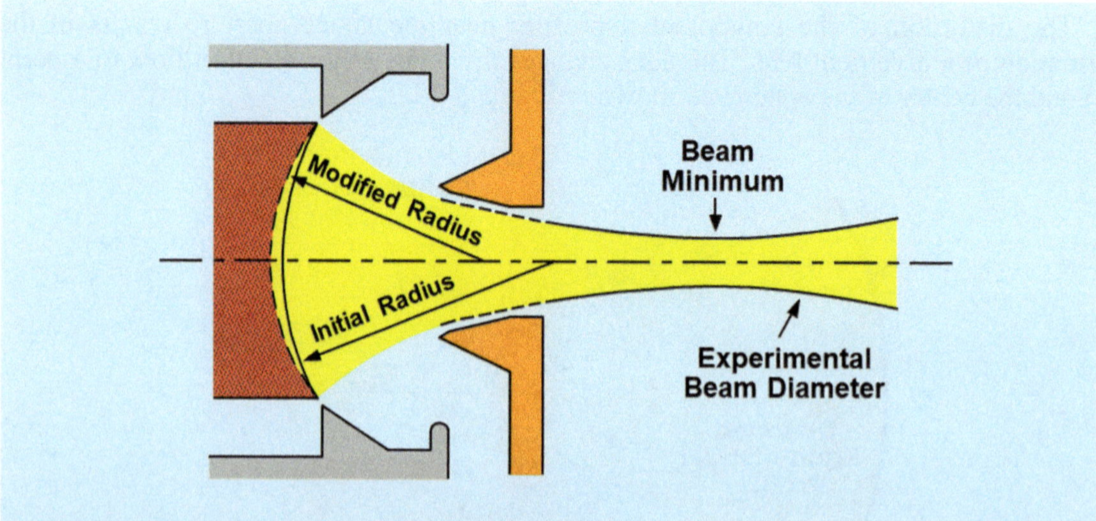

Figure 6-15 Modification to cathode radius used to correct for spherical aberration. (From O. T. Frost et al., *Proc. IRE*, August 1962. © IRE (now IEEE).)

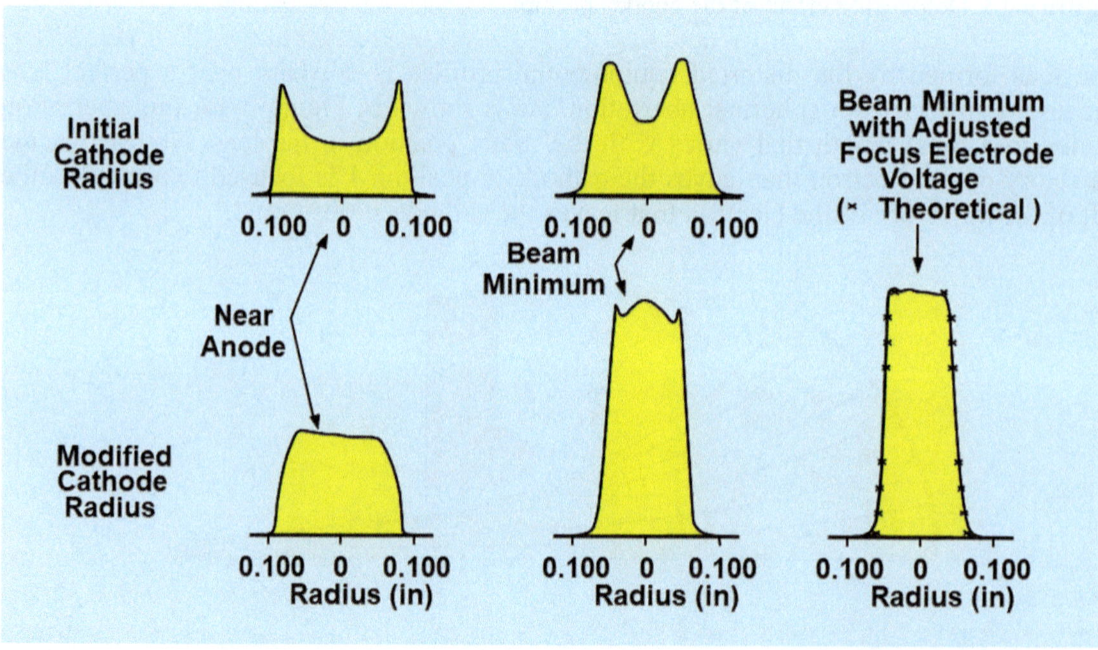

Figure 6-16 Beam-current profiles measured before and after correcting for spherical aberration. (Adapted from: R. D. Frost et al., *Proc. IRE*, August 1962. © 1962 IRE (now IEEE).)

After modifying the cathode radius, the profiles at the bottom left of Figure 6-16 were obtained. Not only was the uniformity of the current improved, but the radius of the beam at the position of the beam minimum was also reduced. After an additional small change in the focus electrode voltage, the profile to the right was obtained. As indicated by the ×s, the shape of this profile was very close to that predicted by theory (the theories of Pierce and of Cutler and Hines [7], who predicted thermal velocity effects).

Figure 6-17 shows Brewer's concept for a nonspherical cathode, which was shaped to correct for spherical aberration.

As reported by Vaughan [8], True, in an unpublished work, defined the values of the gun angle, θ_T, and the radius of curvature of the cathode, R_{cT}, corrected for spherical aberration by

$$\frac{\sin\theta}{\sin\theta_T} = k \tag{6-2}$$

and

$$\frac{R_{cT}}{R_c} = k \tag{6-3}$$

where k is an empirical constant less than unity. For diode guns, $k = 0.905$ yields results that agree well with practice.

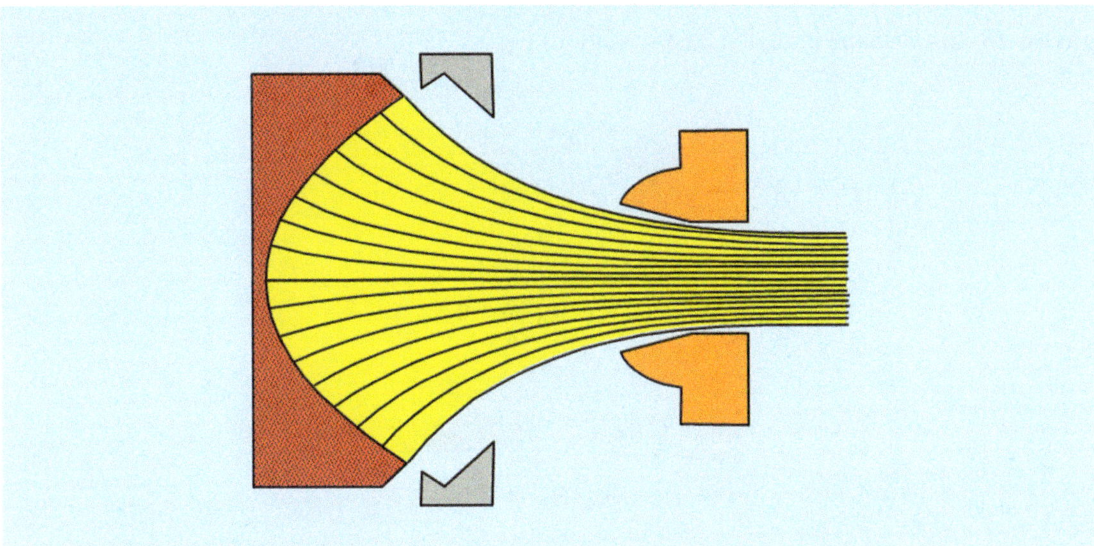

Figure 6-17 A nonspherical cathode that corrects for spherical aberration. (From: G. R. Brewer, U.S. Patent 3,139,552, June 30, 1964.)

6.1.4 Formation of Minimum Beam Diameter

After the beam leaves the anode aperture and enters region 3 (see Figure 6-1), the axial electric field is nearly zero and the only forces on the electrons are those resulting from the space charge. As a result, the shape of the beam is defined by the universal beam spread curve, shown in Figure 6-18. From the analysis for this curve (Chapter 3), the minimum beam radius, b_m, and the axial position of the minimum radius, z_m, can be found as functions of the entrance conditions into region 3. It is essential to have accurate knowledge of these quantities to properly inject the beam into the magnetic focusing structure.

It is important to realize that the values of b_m and z_m are very sensitive to R_c/R_a (the ratio of the cathode and anode radii of curvature). This is illustrated in Figure 6-19, where beam profiles are shown for values of R_c/R_a that differ by 25%. As R_c/R_a increases from 1.6 to 2.0, significant changes occur in both the minimum beam diameter and its axial position. As R_c/R_a increases to 2.5, the minimum beam diameter is again reduced by a significant amount. This figure clearly illustrates the importance of precisely locating the cathode with respect to the anode.

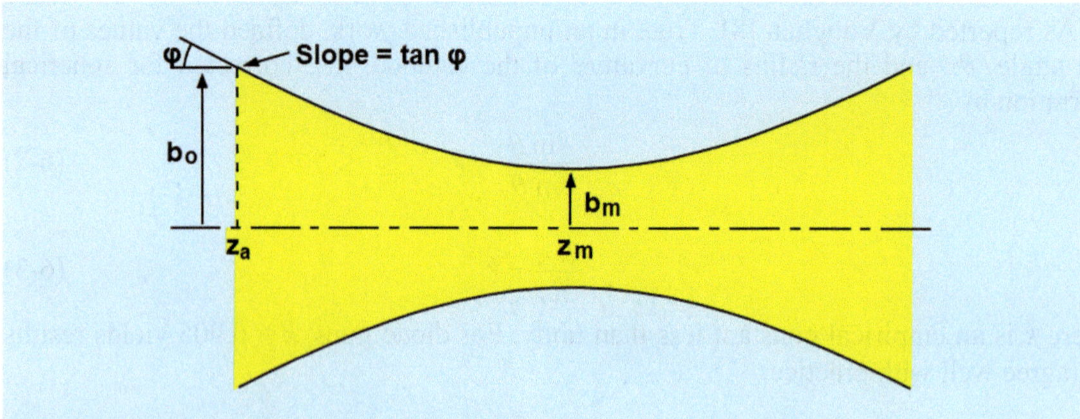

Figure 6-18 Beam shape in region 3 (drift region).

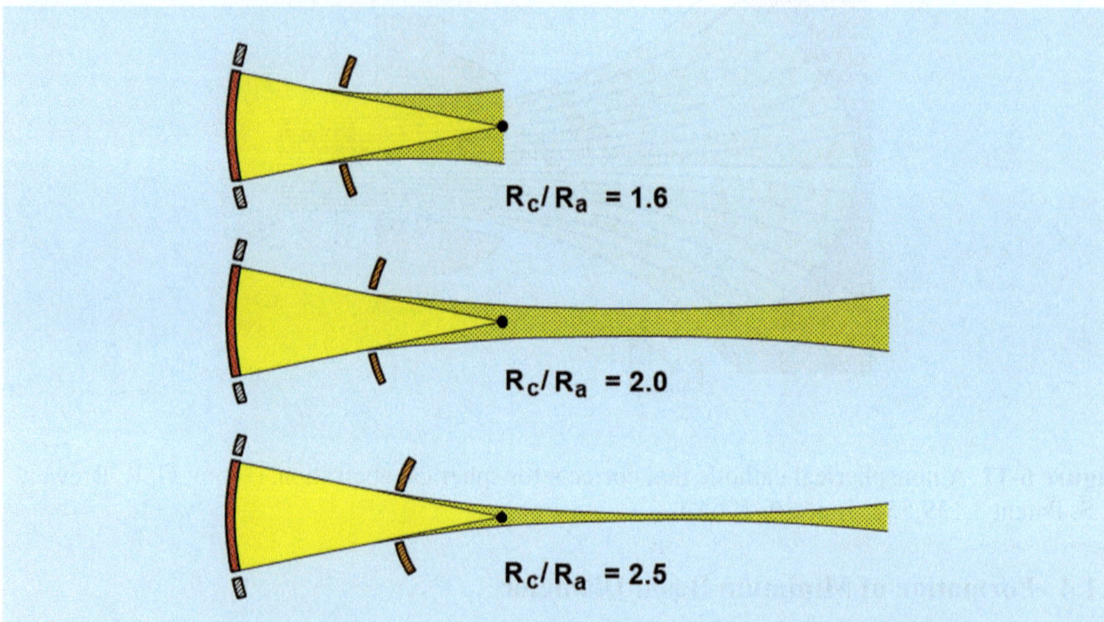

Figure 6-19 Emerging electron stream for various ratios of cathode radius to anode radius. (Adapted from: *Theory and Design of Electron Beams* by J. R. Pierce, copyright 1954 by D. Van Nostrand, Inc.)

Vaughan described a simple technique for synthesizing the design of a Pierce gun [8]. A design example is given in [9]. The technique provides a way to quickly check the design of an electron gun to see if dimensions and parameters are reasonable.

6.1.5 Thermal Velocity Effects

In this section we will consider the effects of the thermal velocities of electrons on the beam produced by an electron gun. It is assumed that the magnetic field is zero. The nature of electron emission from a thermionic cathode is discussed early in Chapter 4. The Maxwellian velocity distribution of the emitted electrons is of importance here. Consider a small spot on the cathode of an electron gun. Electrons are emitted in all directions, and the beamlet of electrons from the spot grows in diameter and the current density decreases as the beamlet is accelerated toward the anode. The current has a Maxwellian distribution with a normal deviation, σ, as shown in Figure 6-20.

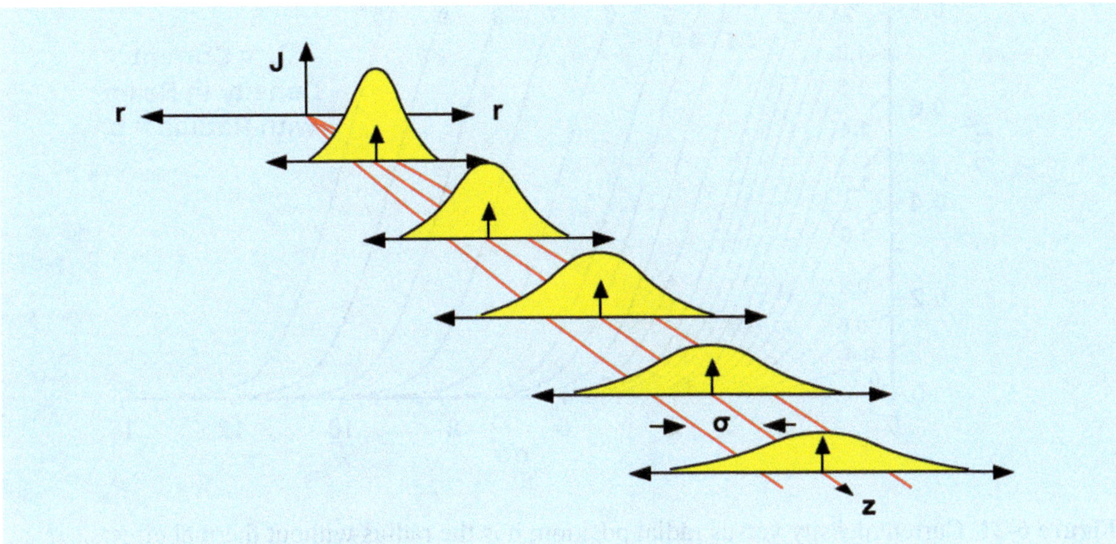

Figure 6-20 Effect of thermal velocities on a beamlet emitted by a spot on a cathode.

Cutler and Hines [7] analyzed the effects of thermal velocities on electron flow in a Pierce gun. They define the normal deviation by

$$\sigma = \frac{\Delta r}{(dr/dt)_c}\left(\frac{kT}{m}\right)^{1/2} \tag{6-4}$$

where Δr is the trajectory deviation of an electron with an initial transverse velocity at the cathode $(dr/dt)_c$. For a 10-kV, 1-A, 0.1-cm radius beam from a cathode at 1,000°C with a radius of 1 cm, the value of σ is ~0.01 cm.

Cutler and Hines summed the currents in beamlets from the entire cathode surface to find an expression for the total current distribution in the beam leaving the gun. The result is shown in Figure 6-21. Here, r is the radial position in the beam, b is the nominal beam radius (the radius in the absence of thermal effects), and J_o is the current density in a beam having the nominal radius.

For $\sigma = 0.01$ cm and a beam radius of 0.1 cm ($b/\sigma = 10$), the curve on the right in Figure 6-21 indicates the beam shape. Figure 6-22 shows the percentage of the total beam current to be found within any given radius of the beam. About 92% of the current would be contained within the 0.1-cm beam radius.

When very small beams are required, thermal effects are severe. An example from Danielson et al. [10] is given in Figure 6-23. Ignoring thermal effects, the minimum beam radius was calculated to be less than 2×10^{-3} in. The experimental radius enclosing 95% of the beam current (r_{95}) was over four times larger (about 8×10^{-3} inch). The minimum r_{95} radius calculated with the Cutler-Hines theory agreed well with experiment, but the variation of the radius with distance was quite different.

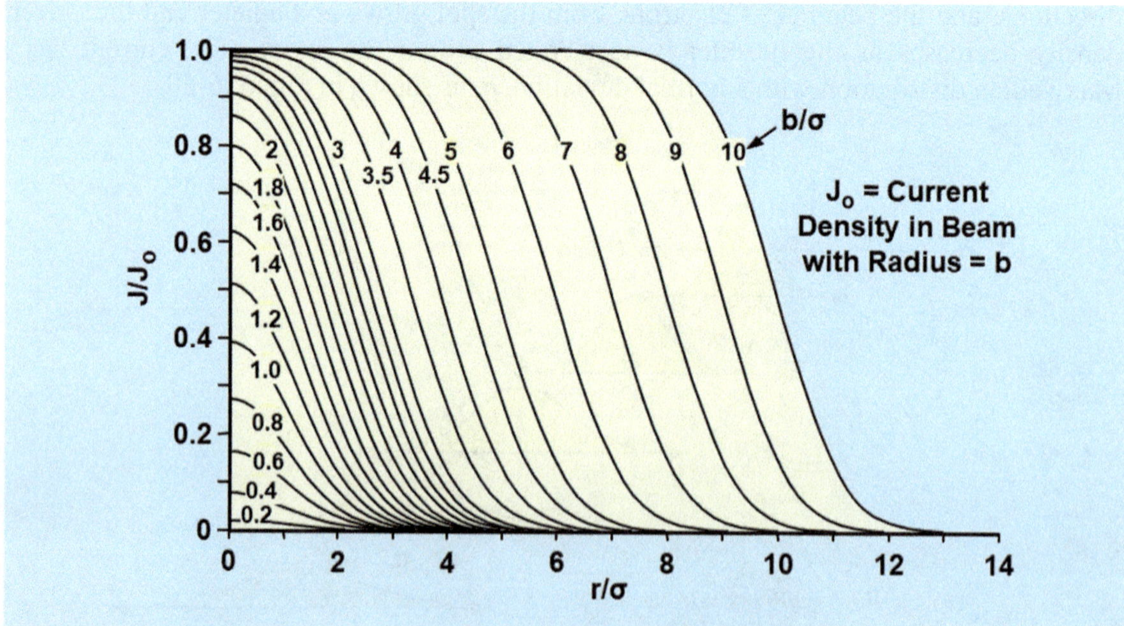

Figure 6-21 Current density versus radial position; b is the radius without thermal effects. (From: C. C. Cutler and M. E. Hines, *Proc. IRE*, March 1955. © 1955 IRE (now IEEE).)

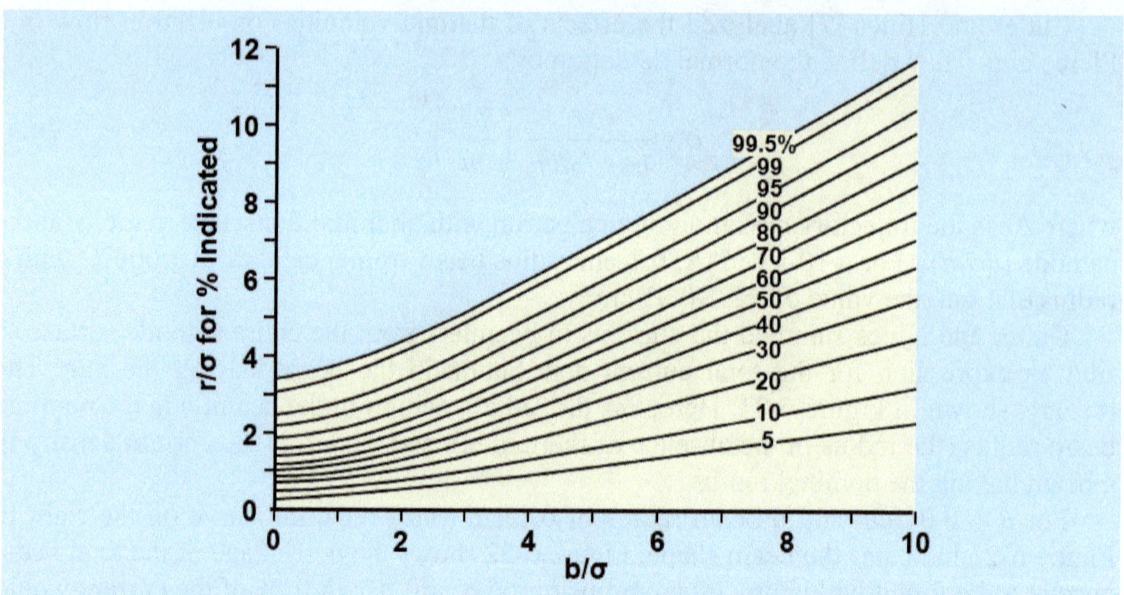

Figure 6-22 Percentage of total beam current within any radius. (From: Cutler and Hines, *Proc. IRE,* March 1955. © 1955 IRE (now IEEE).)

Danielson et al. corrected the Cutler-Hines theory by including the correction to the Davisson-Calbick equation and by using the space-charge field in the thermal beam rather than that in the nominal beam. The minimum radius that they predicted also agrees well with experiment, and the variation of radius with distance is closer to experiment than that predicted by Cutler and Hines.

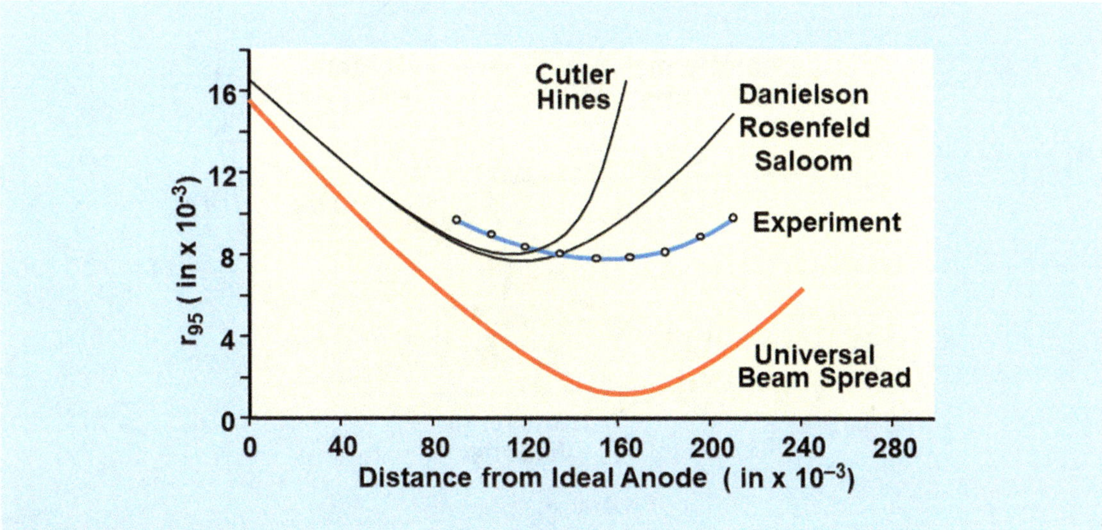

Figure 6-23 Computed and experimental profiles for a small beam dispersed by thermal velocities. (Adapted from: Danielson et al., *Bell Sys. Tech. Jour.*, March 1956.)

Before leaving the discussion of thermal velocities, it is important to consider the effect of beam convergence in the electron gun on the transverse velocity distribution of the electrons. Figure 6-24 shows the trajectories of several nonthermal electrons, which are convergent on point F. The trajectory of one thermal electron is shown crossing the trajectories of the nonthermal electrons. Electric fields are assumed to be negligible. It is clear that the angle that the trajectory of the thermal electron makes with the trajectories of the nonthermal electrons increases as the beam converges. Thus, the component of thermal-electron velocity perpendicular to the nonthermal-electron trajectories increases as the beam converges.

Using Figure 6-24, it is easy to calculate the increase in the transverse component of the velocity of the thermal electron. If v is the velocity of the thermal electron, then at the positions where the radius is r_1 and r_2, the transverse velocity components are

$$v_{t1} = v \sin \alpha_1 \quad \text{and} \quad v_{t2} = v \sin \alpha_2 \tag{6-5}$$

so

$$\frac{v_{t2}}{v_{t1}} = \frac{\sin \alpha_2}{\sin \alpha_1} \tag{6-6}$$

but

$$\frac{\sin \alpha_2}{\sin \alpha_1} = \frac{a / b_2}{a / b_1} = \frac{b_1}{b_2} \tag{6-7}$$

and

$$\frac{r_1}{b_1} = \frac{r_2}{b_2} \quad \text{so} \quad \frac{r_1}{r_2} = \frac{b_1}{b_2} \tag{6-8}$$

As a result,

$$\frac{v_{t2}}{v_{t1}} = \frac{r_1}{r_2} \tag{6-9}$$

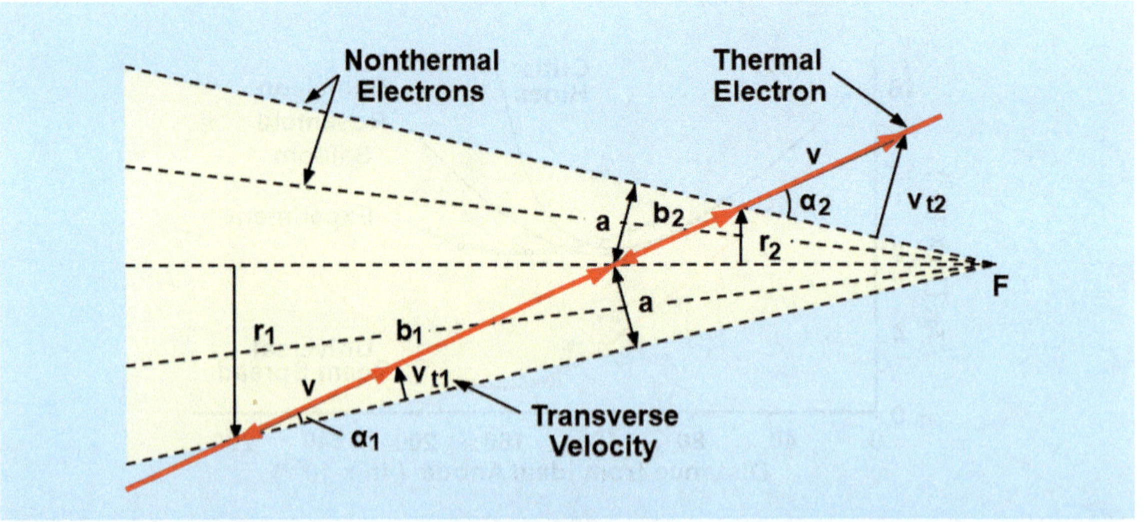

Figure 6-24 Increase in transverse velocity of thermal electron in convergent flow.

That is, the transverse velocity increase is inversely proportional to the radius compression ratio. Thus, to limit the growth of transverse velocities in an electron gun, the amount by which the beam is compressed should be held to a minimum. A cathode with a high emission capability helps in this regard because, for a given beam size and current from a gun, the cathode can be smaller, so the radius compression ratio is smaller than for a cathode with a low emission capability.

6.1.6 Effects of Patchy Emission and Cathode Roughness

The thermal velocity effects on beam shape discussed in the previous section are frequently overshadowed by the relatively large transverse electron velocities produced by patchy emission and cathode roughness. The surface of a dispenser cathode is made up of grains of tungsten that have various crystalline orientations and that may have various work functions. Because the thickness of the space-charge cloud at the cathode surface is comparable to the dimensions of the grains and the pores between grains, each grain emits independently of the others and the emission is said to be patchy. For those grains that are temperature-limited, emission varies with grain work function and most of the emission comes from the grains with low work functions.

There are at least two ways by which the patchy nature of emission produces transverse velocity components to electrons leaving the cathode. These are indicated in Figure 6-25(a, b). In Figure 6-25(a) we note that space-charge forces within the beamlet emitted from a patch cause spreading of the beamlet. In Figure 6-25(b) we show that equipotential surfaces near the cathode surface are rippled because of the varying space-charge density near the patches. This produces a spreading of the beamlets.

As explained in Chapter 4, coated dispenser cathodes have a narrower distribution of work functions than uncoated cathodes, so patchy emission is less pronounced for coated than for uncoated cathodes. If the work function distribution of a cathode increases with time, then the effects of patchy emission become worse. This may cause the beam diameter to increase so that current intercepted by the circuit increases.

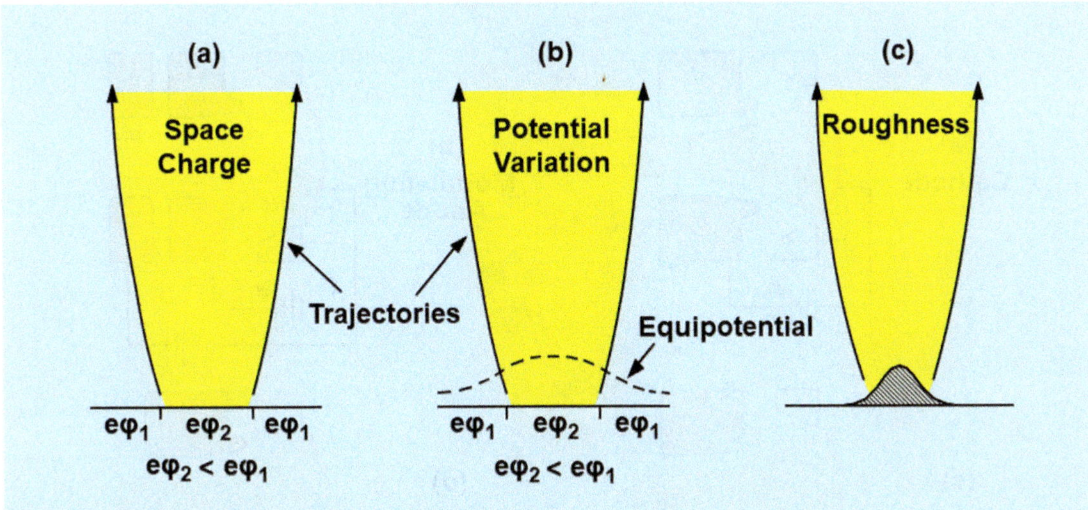

Figure 6-25 Spreading effects of patchy emission and cathode roughness on electron trajectories.

If the surface of the cathode is rough, as indicated in Figure 6-25(c), then local transverse electric fields are present, which produce transverse electron velocities. A problem sometimes encountered with coated dispenser cathodes is that the coating may develop microscopic cracks and peel away from the tungsten matrix. Then a cathode that was originally very smooth can become rougher as it is operated. If this happens, transverse velocities increase, beam size increases, and interception current may increase.

6.2 BEAM CONTROL TECHNIQUES

In all applications of microwave tubes, beam current must be controlled. Usually, the beam current must be switched on and off. Sometimes, continuous control is required. The brute-force way to control beam current is with the cathode-to-anode voltage as indicated in Figure 6-26(a). A second anode (called a *modulating anode*) may be placed between the cathode and final anode for current control as shown in Figure 6-26(b). In some guns, current can be controlled with the focus electrode as shown in Figure 6-26(c). Finally, in the majority of applications, one or more grids is placed near the cathode surface, as indicated in Figure 6-26(d), for current control. In this section, these control techniques are discussed. The primary emphasis is on grid control because of its wide use in TWTs.

6.2.1 Cathode Pulsing

Conceptually, cathode pulsing is the simplest technique for switching a microwave tube on and off. In virtually all applications, the anode of the gun and the body of the tube are at ground potential. To turn the beam on, the cathode is pulsed negative. To turn the beam off, the cathode is pulsed back to ground potential. The primary problem with cathode pulsing is that the power supply (the power modulator) must switch *all* of the beam power so the power modulator must be large, heavy, and usually inefficient.

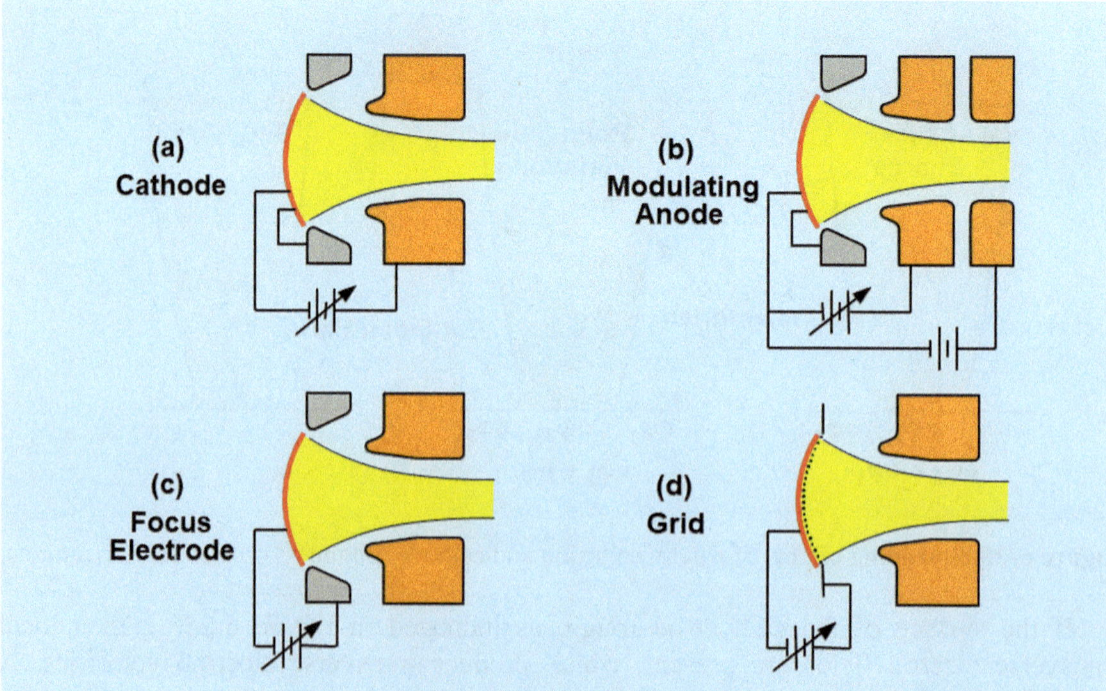

Figure 6-26 Techniques for current control in an electron gun.

6.2.2 Modulating Anode

In the modulating anode gun, the anode of the electron gun (the modulating anode) is separated from the body of the tube, a part of which is a second (final) anode. The configuration of a modulating anode gun is shown in Figure 6-27.

Although the voltage range over which the modulating anode must be varied to vary beam current is large, the power required to drive the modulating anode is small because the current intercepted by this anode is small. At intermediate voltage levels, perveance may be controlled with the modulating anode without severely affecting beam focusing.

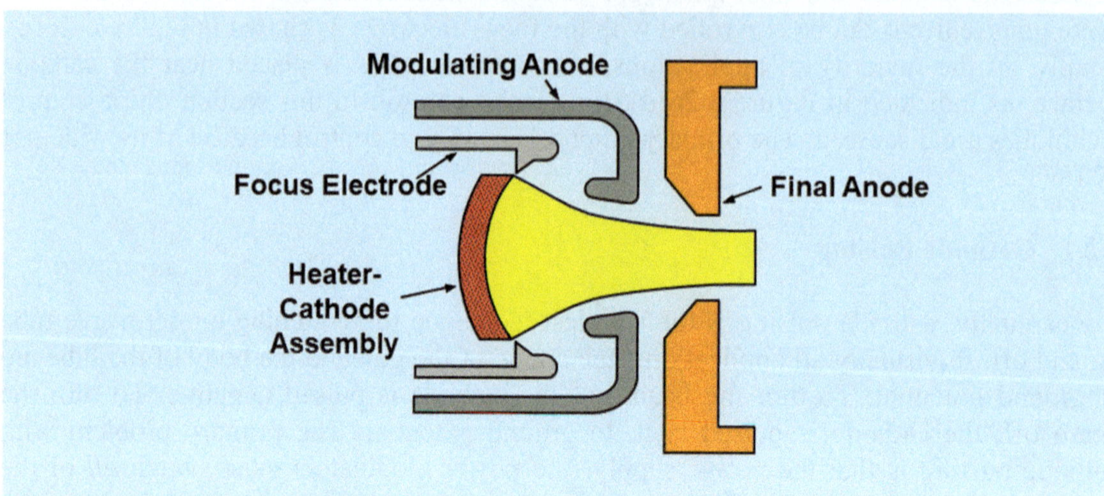

Figure 6-27 Modulating anode gun.

6.2.3 Control Focus Electrodes

The focus electrode (FE) can be used to turn the beam on and off. By using the FE for controlling current, it's not necessary to put anything (such as a grid, to be discussed next) in the electron beam. Any electrode that is inserted in the beam increases beam noise and also disturbs the trajectories of the electrons leaving the cathode so that beam focusing is degraded. As well, the voltage required for turning the beam off may be less than that required if a mod anode is used. Disadvantages of using the FE electrode are that breakdown may occur between the cathode and FE, an excessively high voltage may be required, and normally only beam-on or beam-off states are possible because gun focusing at voltages between the on and off state voltages is poor.

True [11, 12] has addressed the problem of high FE voltages and has shown that there is a linear relationship between the focus FE voltage, V_{fe}, required for turn-off and the perveance of the gun. This relationship is shown in Figure 6-28, which is a plot of the ratio of V_{fe} to anode voltage, V_a, as a function of microperveance. Notice that this ratio is nearly unity for a very large perveance. To understand why this variation occurs, it is necessary to first understand how the focus electrode can turn off the beam.

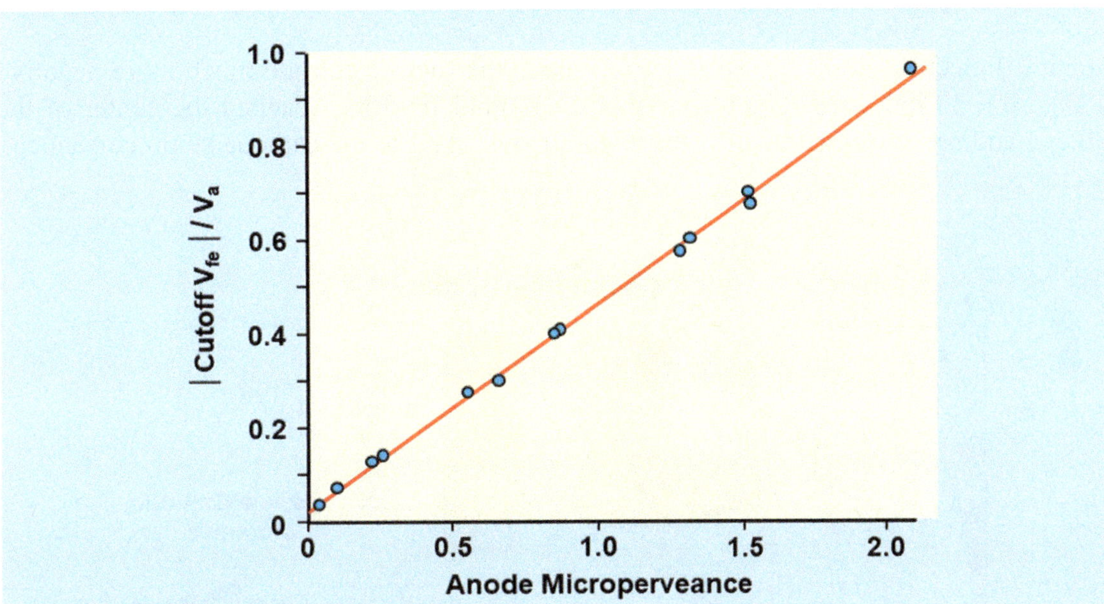

Figure 6-28 Relation of cut-off voltage to perveance for beam control with the focus electrode. (From: R. B. True, IVEC 2012. © 2012 IEEE.)

The electrode configuration in Figure 6-29 is representative of that for a gun with a microperveance near unity. Equipotential surfaces are shown in Figure 6-29(a) for the FE at the cathode voltage. In Figure 6-29(b), The FE voltage is below cathode potential but not far enough to cut off beam current. The blue lines represent equipotential surfaces that are negative with respect to the cathode. Notice that

1. Emission from the outer regions of the cathode is inhibited;
2. The electron beam is badly focused and severe interception by the anode occurs.

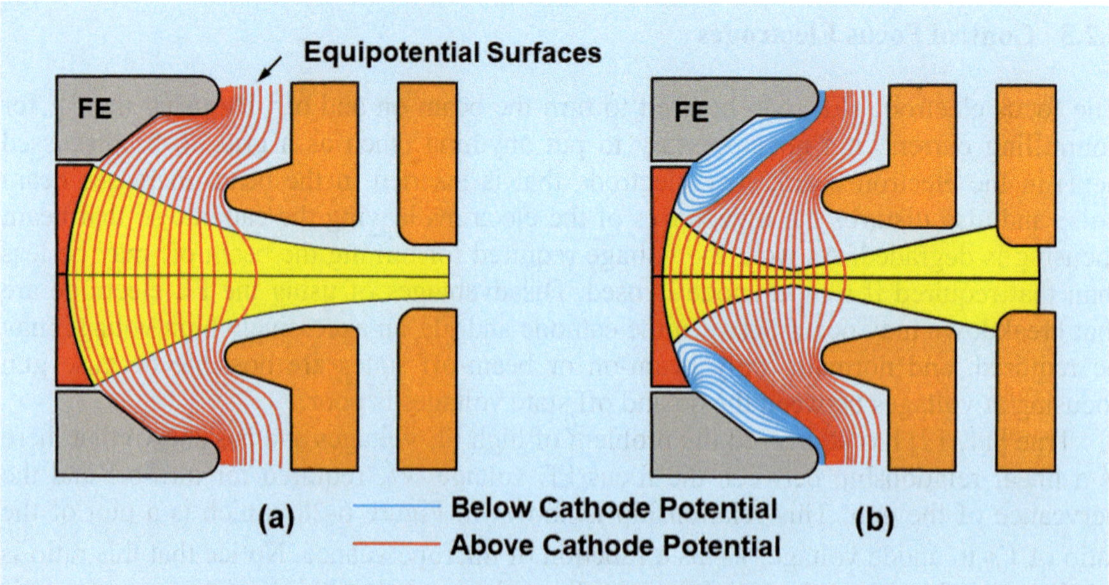

Figure 6-29 Equipotential surfaces (red and blue) for FE voltage: (a) at cathode potential and (b) below cathode potential but not cut off. (From: R. B. True, U. S. Patent 8492978 B2.)

To cut off electron flow, it is necessary to make the focus electrode much more negative as Figure 6-30 indicates. The negative equipotential field has reached the center of the cathode and prevents the anode voltage from penetrating to the cathode so no current can be drawn from the cathode.

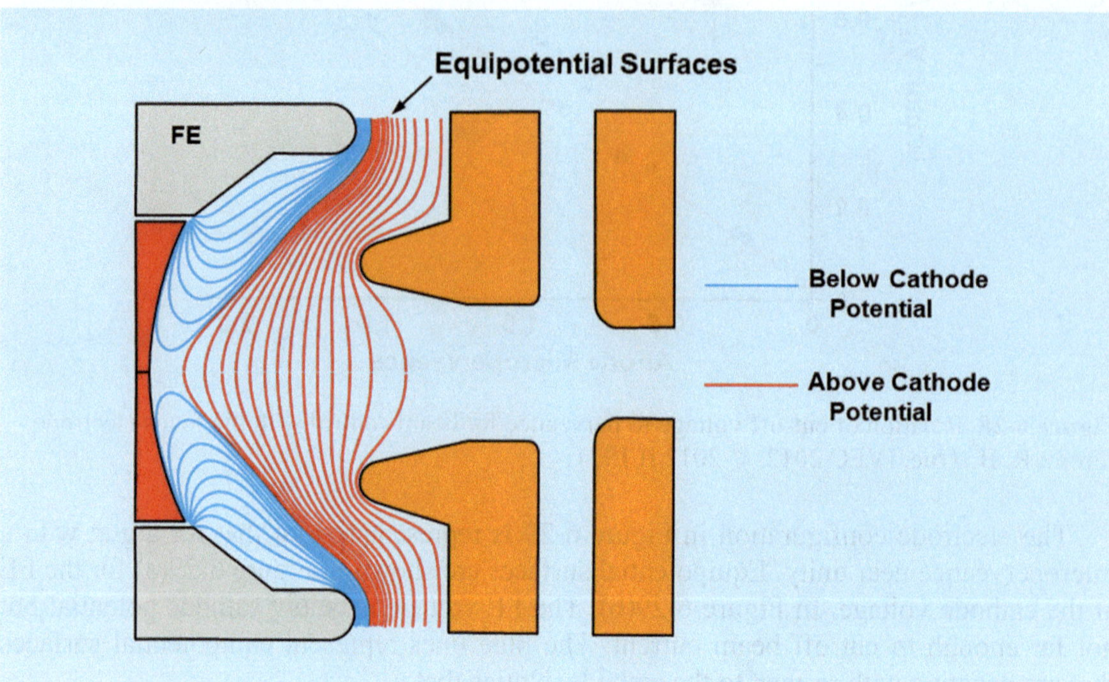

Figure 6-30 Equipotential surfaces for FE voltage adjusted for cutoff. (From: R. B. True, U. S. Patent 8492978 B2.)

Now, returning to the question of why V_{fe} for cutoff varies with microperveance, we must first remember that perveance varies with cathode area. The distance that the equipotential field from the FE penetrates across the cathode surface depends primarily on the strength of the anode field and not on the cathode size. So, for example, assume that the FE field is just adequate to cut off current flow from a cathode with a disc radius of 1 cm. Next, consider the same FE voltage and the same anode field and a cathode with a disc radius of 2 cm. The FE field will shield the outer 1 cm of the cathode from the anode field but current will still flow from the unshielded center portion. The FE voltage would have to be increased to cut off the center portion.

So we have conflicting requirements when we consider FE control of current in high perveance guns. In the first place, we want to use FE control because beam focusing is excellent. We want to use high perveance to minimize anode voltage. But, at the same time, high perveance cutoff requires high FE voltage (91% of the anode voltage for microperveance of 2). There are at least two problems in using a high FE voltage. First, the FE is close to the cathode and so breakdown may occur. Second, the switching power supply for the FE must operate at voltages beyond the capabilities of standard high-voltage switches.

In his patent entitled "Dual Element Switched Electron Gun," True describes circuitry and procedures that can be used to alleviate the requirement for high FE voltages for high perveance guns [12]. The concept is illustrated in Figure 6-31. A basic gun with a microperveance of 2 and an anode voltage of 7.5 kV has been selected. When only the FE is switched, a voltage of 6.9 kV (0.91x7.5 kV) below the cathode voltage is required to turn off the beam. When two elements are switched (FE and MA), the voltage required if equal voltages are used is 3.6 kV. This is because when the mod anode voltage is -3.6 kV, the MA to cathode voltage is reduced to 3.9 kV and so the FE voltage must be 0.91 x 3.9 or about 3.6 kV below cathode potential. Finally, if the cathode voltage is increased by 1.8 kV at the same time that the FE and MA voltages are reduced by 1.8 kV, cutoff can be achieved.

Notice in this example that with two or three switched elements, the FE to cathode voltage is 3.6 kV. While this is a factor of nearly two below that voltage with a single switched element, it is high enough to be of concern. Also, the complexity of the switching power supplies cannot be ignored. Still, an excellent beam can be generated and this helps in maximizing the RF performance of the tube.

The voltage needed to cut off beam current can be reduced by placing an additional electrode in the center of the cathode (Figure 6-32). Staprans [13] reported the use by McCune at Varian of this electrode (referred to as a modulating focus electrode) in a 250-kW average output power X-band klystron. Staprans stated "It is therefore evident that the beam optics of this gun are excellent."

6.2.4 Grids

Under normal operating conditions, the guns with modulating anodes and those with control electrodes operate as Pierce-designed guns. When grids are used, however, electron trajectories near the cathode are affected, and so beam focusing is degraded and beam dynamics are affected. Still, grid control is very widely used because of the small modulator size that it makes possible.

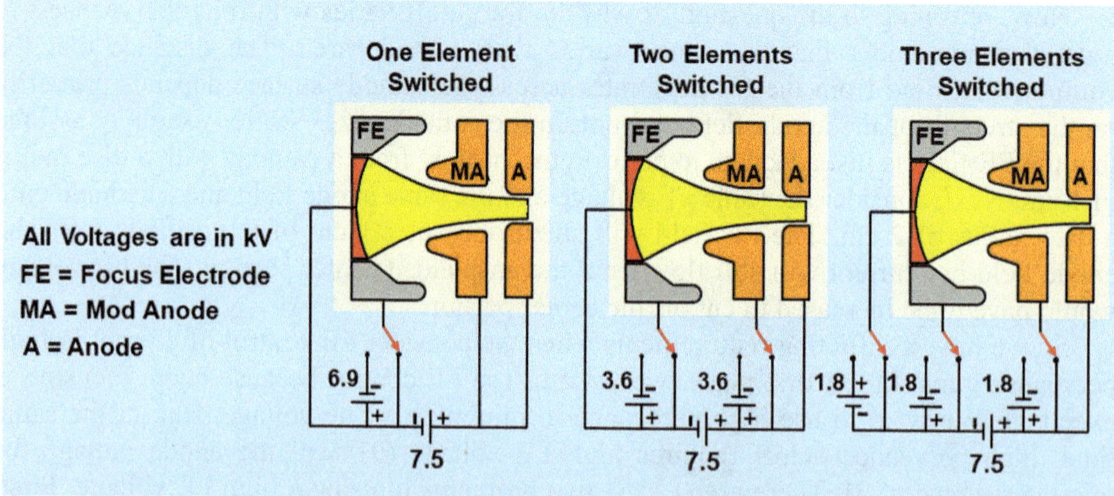

Figure 6-31 Focus electrode control of current. (From: R. B. True, U. S. Patent 8492978 B2.)

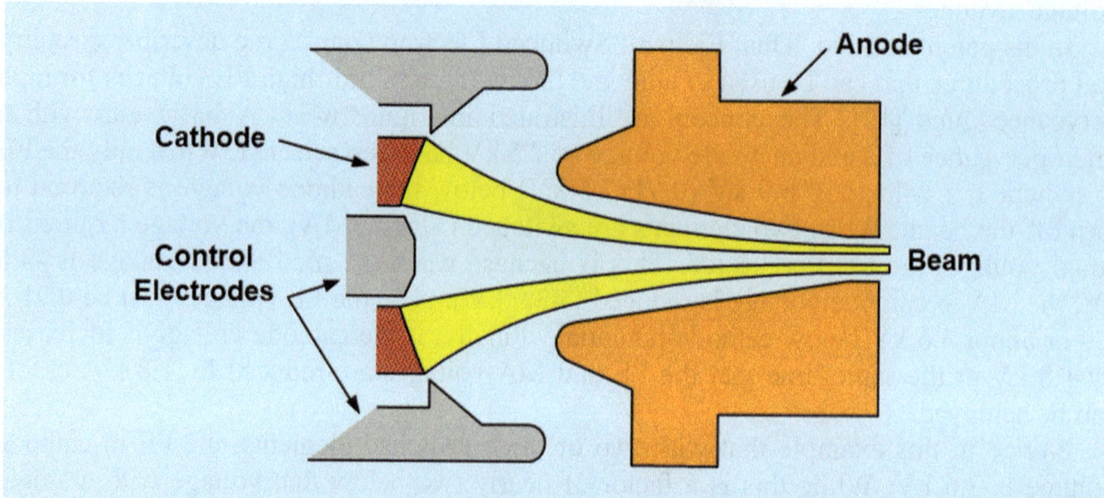

Figure 6-32 Control focus electrode gun. (From: Armand Staprans et al., *Proc. IEEE*, March 1973. © 1973 IEEE.)

Several types of grid structures are illustrated in Figure 6-33. The simplest structure has a single grid element. A computer simulation of electron flow in a Pierce gun containing a single grid operating at a voltage positive with respect to the cathode is shown in Figure 6-34. Thermal velocities are ignored here, and because the grid is positive, electron flow is very laminar. The problem with a single grid that is positive with respect to the cathode is that the cathode current intercepted by the grid can be very high. In some cases, as much as 15% or more of the cathode current is intercepted. The results of the interception are excessive heating of the grid at high duty cycles and the need for a relatively powerful grid modulator.

Many grid structures contain a shadow grid between the control grid and the cathode surface. The shadow grid operates at or near cathode potential so that heating by intercepted current is small. The elements of the shadow grid are aligned with the elements of the control grid so that the control grid is shielded and intercepts very little current. Several shadow-grid structures are shown in Figure 6-33.

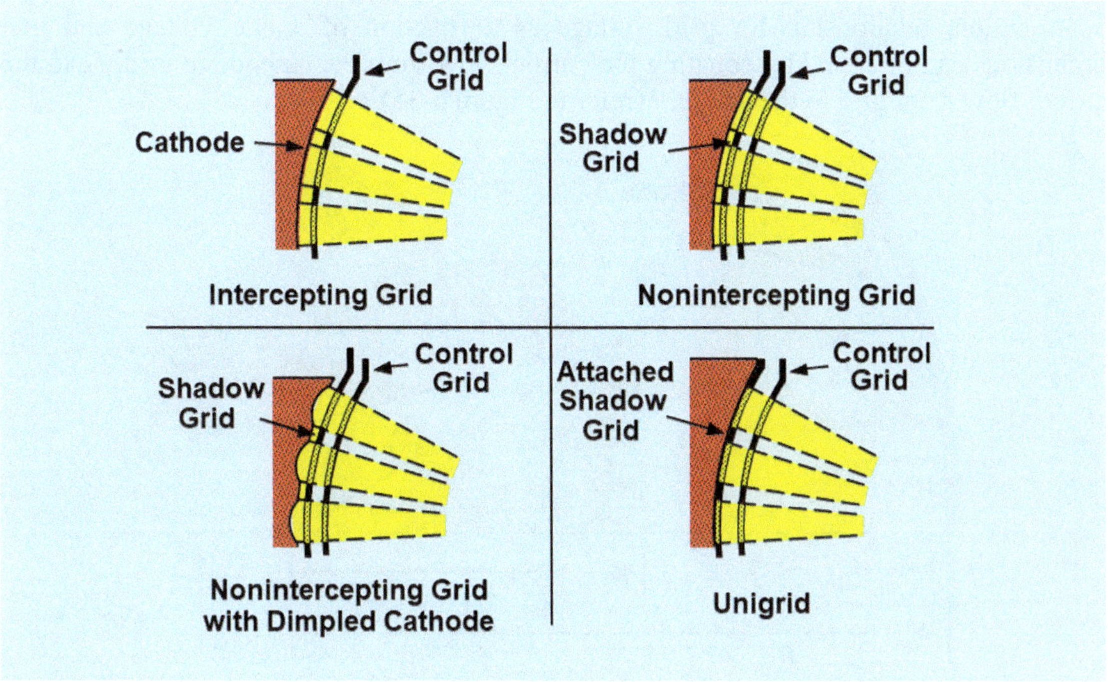

Figure 6-33 Various grid structures used in Pierce guns for current control. (From: *Microwave Tube Manual* by Varian, Air Force Publication No. T.O.00-25-251, October 1979.)

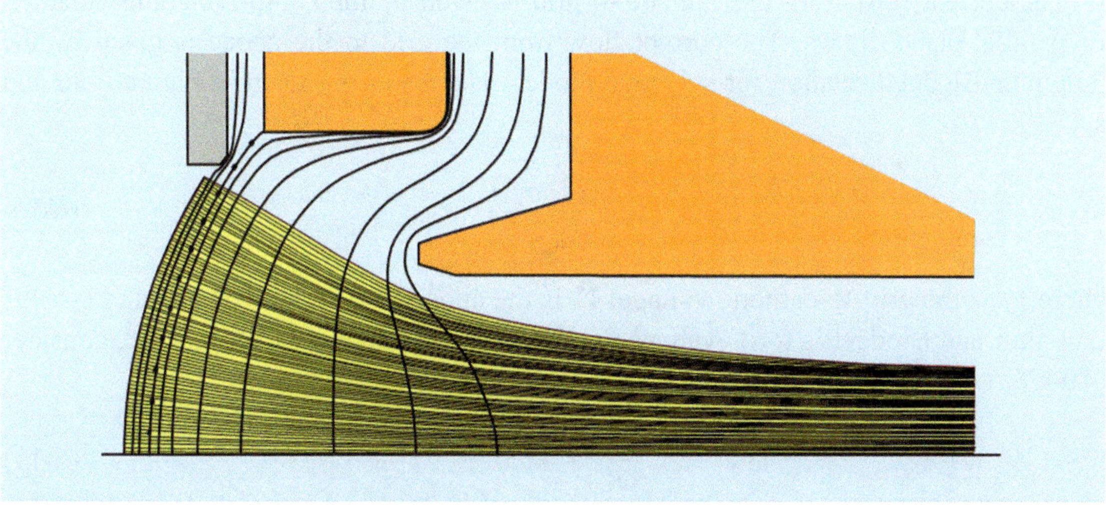

Figure 6-34 Computer simulation of electron flow in a Pierce gun with a single grid. (Adapted from: R. True, *Technical Digest*, 1985 IEDM. © 1985 IEEE.)

Grid control of current in electron tubes has been used since the invention of the vacuum tube. In 1960, Wolkstein discussed methods of grid placement and associated design considerations for convergent-flow Pierce guns [14]. The basic concept is to form the grid so that its shape is identical to that of an equipotential contour near the cathode surface (usually spherical). Then, if the grid is at the potential of that contour and intercepts negligible current, electron flow will not be disturbed. To interrupt the current flow, the grid voltage is switched negative with respect to the cathode.

A simple relationship for grid voltage as a function of anode voltage and gun dimensions can be found by equating the current flow from the cathode to grid, I_c, to the current flow from grid to the anode, I (refer to Figure 6-35).

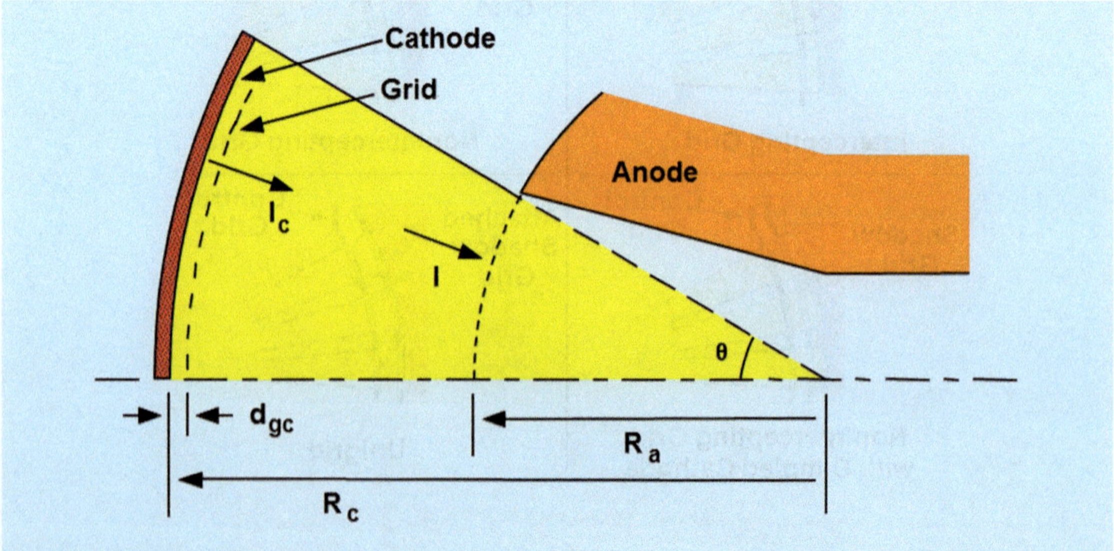

Figure 6-35 Gun parameters for determining grid-cathode spacing.

The current flow from the cathode to grid is given by the Child-Langmuir equation for parallel-plane diodes. The current flow from the grid to the anode is given by the Langmuir-Blodgett relation for spherical diodes. The result is shown in Figure 6-36 and is

$$\frac{V_a}{V_g} = \left(\frac{(-\alpha_a)}{d_{gc} / R_c} \right)^{4/3} \tag{6-10}$$

where V_g is the grid-to-cathode voltage, V_a is the anode-to-cathode voltage, and $(-\alpha_a)$ is Langmuir and Blodgett's $(-\alpha)$ evaluated at the anode. As an example of its use, consider a 10-kV, 1A gun design with $R_c/R_a = 2.25$. For a grid voltage of 100V,

$$\frac{d_{gc}}{R_c} \approx 0.032 \tag{6-11}$$

If we assume that $R_c = 1$ cm, then

$$d_{gc} \approx 0.032 \text{ cm} \approx 0.013 \text{ inch} \tag{6-12}$$

So we see that the grid must be very close to the cathode if the grid voltage is to be low.

Next, we will examine the configuration of the grid elements. Examples of control grids for Pierce guns are shown in Figure 6-37. The grid configuration on the left is often used in guns with intercepting grids and that on the right is used in shadow-gridded guns.

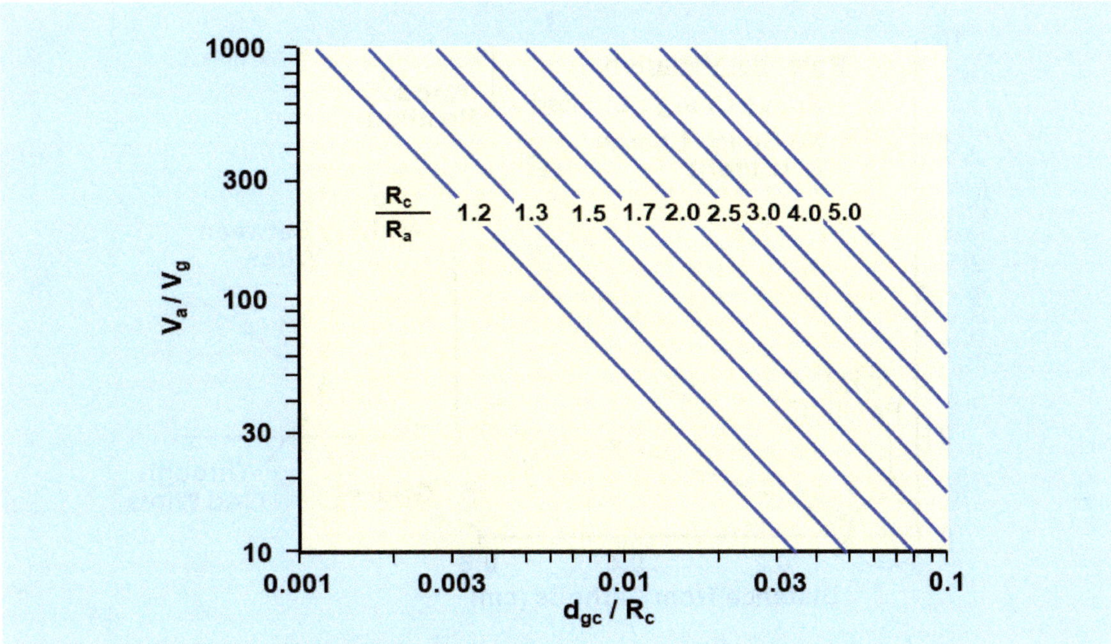

Figure 6-36 Relation of grid and anode voltages to grid, cathode, and anode positions. (Adapted from: H. J. Wolkstein, *RCA Review*, September 1960.)

To stop (cut off) electron flow from the cathode, as shown in Figure 6-38, a negative voltage (with respect to the cathode) is applied to the grid. In Figure 6-38, the dashed curves show the potential variation between the cathode and anode for the 10-kV, 1A electron gun with the grid 100V above the cathode potential. The solid curves show the potential variation when the grid is sufficiently negative with respect to the cathode to cut off electron flow. Note in the expanded drawing of the grid region that two solid curves are shown. One is the potential variation through a grid wire and the other is the potential variation midway between grid wires. The net effect of these two potential variations must be that the potential near the cathode is zero or negative if current flow is to be prevented.

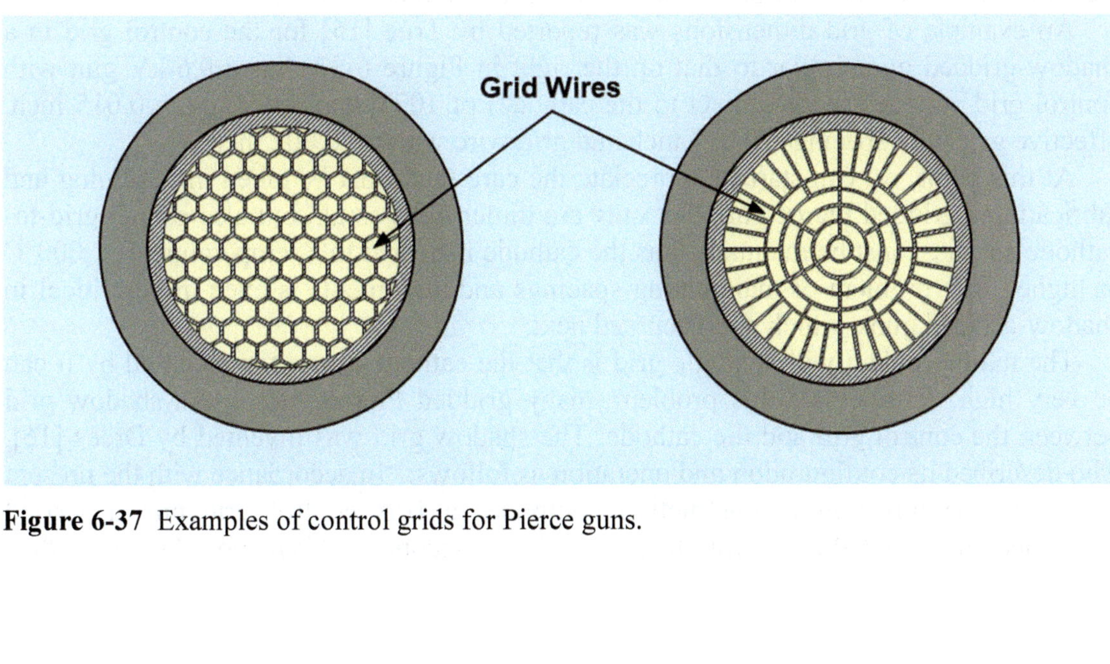

Figure 6-37 Examples of control grids for Pierce guns.

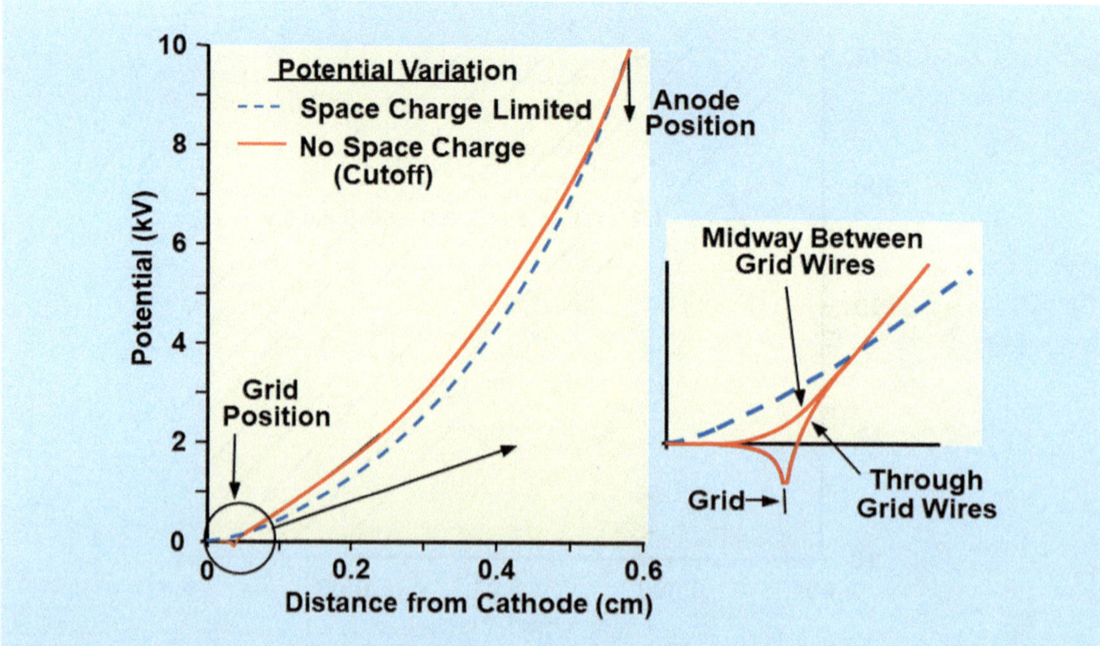

Figure 6-38 Potential variations in a gridded 10-kV Pierce gun.

To avoid electrical breakdown between the grid and the cathode, the grid voltage used for cutoff must not be excessive. The cutoff grid voltage is controlled largely by the spacing between grid wires. Clearly, for large spacings, the cutoff grid voltage must be large because of the strong tendency for the anode field to penetrate the regions between grid wires. Thus, for relatively small cutoff grid voltages (and low probability of grid-cathode breakdown), the grid wire spacing must be small.

In addition to small grid wire spacing, the area of the grid wires (and thus the width of the grid wires) must be small compared with the cathode area if the current intercepted by the grid is to be small. The ratio of the grid area to the cathode area is the screening fraction, which is the fraction of the cathode current intercepted by the grid. Thus, it is clear that for small interception, the screening fraction must be small.

An example of grid dimensions was reported by True [15] for the control grid in a shadow-gridded gun similar to that on the right in Figure 6-37. For a 9.6-kV gun with control grid voltages (with respect to the cathode) of 102V and –86V, d_{gc} = 0.015 inch, effective grid wire spacing = 0.023 inch and grid wire width = 0.002 inch.

At this point, we can start to appreciate the care that must be taken in designing and fabricating a gridded Pierce gun. Not only are dimensions small, but also proper grid-to-cathode spacing must be maintained as the cathode is heated to a temperature of 1,000°C or higher. The problem of maintaining spacings and alignments is even more critical in shadow-gridded guns, which are discussed next.

The main problem with a single grid is that the cathode current intercepted by it can be very high. To alleviate this problem, many gridded Pierce guns use a shadow grid between the control grid and the cathode. The shadow grid was invented by Drees [16], who described its configuration and operation as follows: "In accordance with the present invention, the novel grid construction comprises a first control grid and a second grid spaced from and aligned with the first grid. The second grid is disposed between the control grid and the cathode. In this arrangement, with the second grid preferably being held at cathode potential, the second grid acts as a (shadow) grid in that it shadows electrons from the control grid so that they are not intercepted by the control grid."

The effects on electron trajectories of the shadow grid, its location, and its voltage are illustrated in Figure 6-39 [17]. Each of these plots shows computer-generated electron trajectories from a cathode through a pair of grids. In an ideal shadow gridded gun, both of the grids would be at voltages that coincide with the equipotential surfaces that exist in the absence of the grids. Thus, except for interception and the finite sizes of the grid wires, electron flow would not be disturbed. The plot in Figure 6-39(d) shows trajectories with grid voltages adjusted to approximate the ideal shadow-gridded gun voltages.

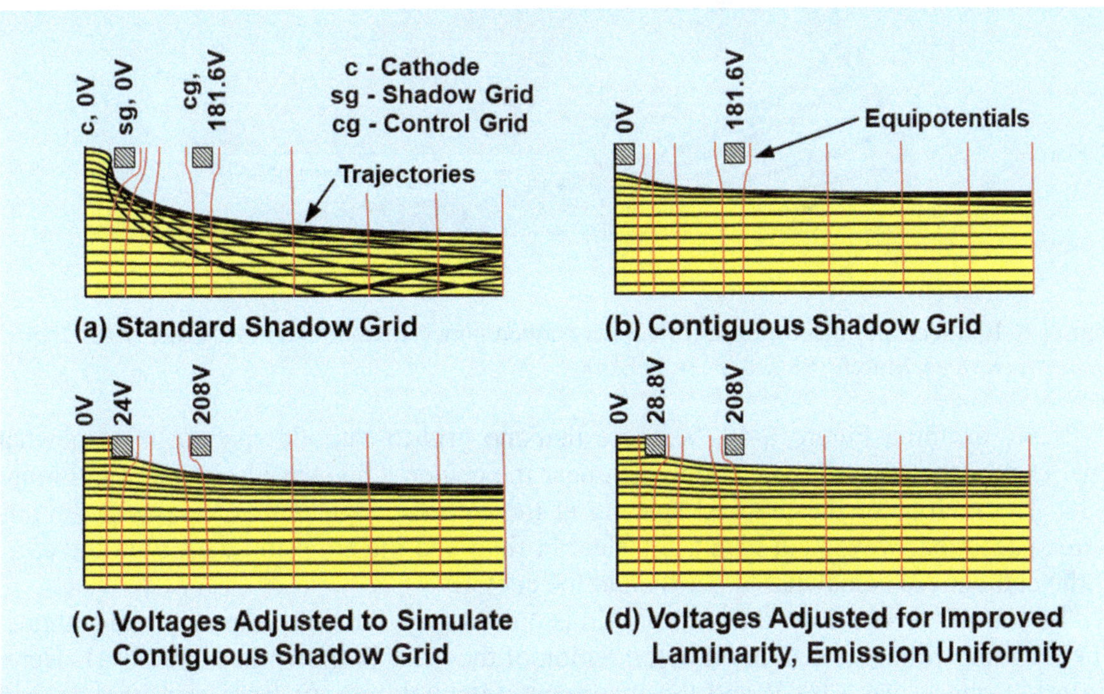

Figure 6-39 Electron trajectories for shadow-gridded structures (planar model). (From: R. True, 1979 IEDM. © 1979 IEEE.)

For the standard shadow-gridded gun (Figure 6-39(a)), the shadow grid is at cathode potential but is placed a short distance (usually a few thousandths of an inch) from the cathode surface. Thus, a lens action occurs near each wire of the shadow grid. This is evident from the bending of the equipotential surfaces near the shadow grid. As a result of the lens action, electron trajectories are severely distorted.

When the shadow grid is attached to or placed on the cathode surface, as in Figure 6-39(b), the distortion of the trajectories is significantly reduced. Unfortunately, of course, the shadow grid must operate at the cathode temperature, which can cause difficulties with grid emission and grid alignment.

An advantage of placing the shadow grid at cathode potential, as in Figure 6-30(a, b), is that only one grid power supply is required. If a separate power supply for the shadow grid can be used, then, as shown in Figure 6-39(c, d), the beam laminarity of the contiguous shadow grid can be equaled as in Figure 6-39(c) or improved upon as in Figure 6-39(d).

Shown in Figure 6-40 are plots of electron trajectories in a shadow-gridded gun with the standard shadow grid [18]. The beam radius containing 95% of the current, r_{95}, is indicated by dots. If the trajectories outside the r_{95} radius are followed back to the cathode, it is seen that they are the trajectories of electrons originating close to the grid wires.

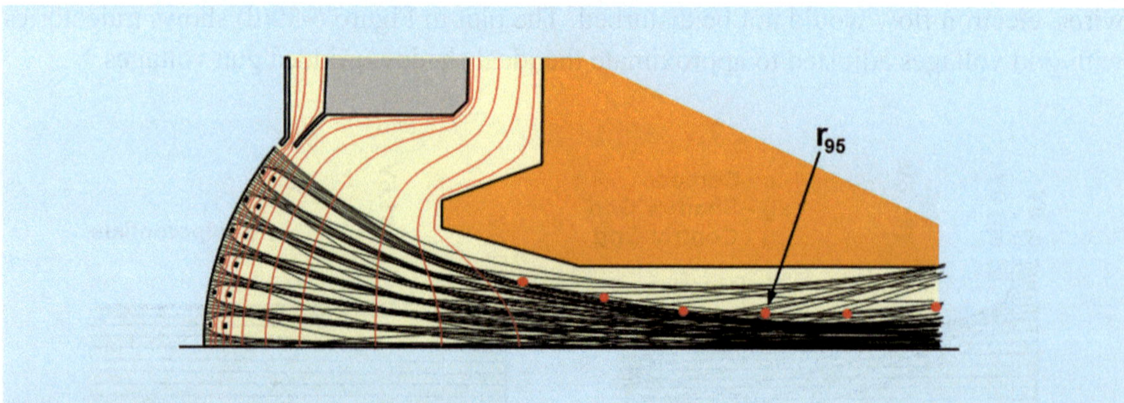

Figure 6-40 Electron trajectories in a shadow-gridded gun. (Adapted from: R. True, *IEEE Trans. Electron Devices*, March 1984. © 1984 IEEE.)

If we examine Figure 6-40, we note that the grid-to-cathode spacing is somewhat smaller near the edge of the cathode than near the center. This was also evident in Figure 6-34. The reason for the reduced spacing at the cathode edge is that the equipotentials were spaced that way. As a result, the electric field and the emission density at the edge of the cathode are somewhat larger than at the center.

To cut off current flow, the control grid is pulsed negative with respect to the cathode. The resulting voltage distribution in the region of the grids is shown in Figure 6-41. Here, the grid voltage was –160V and the innermost potential contour from grid wire to grid wire was at –10V, so the grid voltage was slightly below that required for cutoff.

We have noted that the electric field is somewhat higher at the edge of the cathode than at the center. Also, when grids like that shown on the right in Figure 6-37 are used, the spacing between grid wires increases with increasing radius. Thus, for negative control grid voltages, the anode field has a greater tendency to penetrate the regions between grid wires near the edge of the cathode than near the center. This means that as control grid voltage is made negative with respect to the cathode, the last part of the cathode to be cut off is that near the outer edge. To compensate for this effect, True and Laycock tapered the widths of the grid wires so that they increased in width near the cathode edge [19]. As a result, cutoff voltage was reduced by a factor of almost 2.

As the control grid voltage is varied, cathode current varies as shown in Figure 6-38. In addition, the beam current intercepted (the body current), mostly by the RF structure, varies. This is because, as control grid voltage and cathode current are changed while cathode voltage remains fixed, the gun perveance changes and focusing conditions change.

As a result, the entrance conditions of the beam from the gun into the magnetic focusing field change. This causes the beam shape and body current to vary.

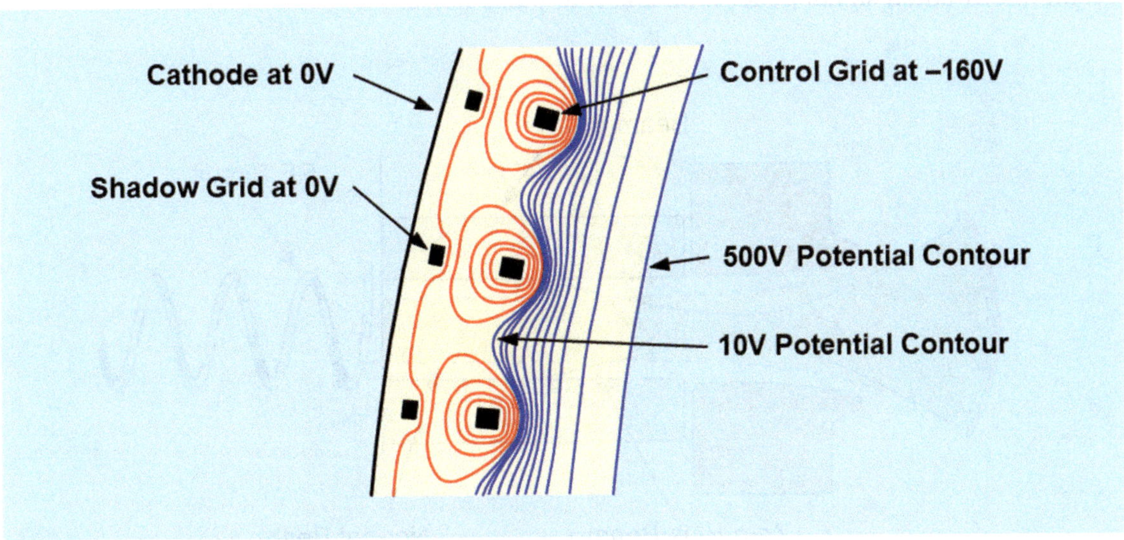

Figure 6-41 Voltage distribution near the grids at a control grid voltage slightly below cutoff. (Adapted from: R. True, *Technical Digest*, 1989 IEDM. © 1989 IEEE.)

A particularly disturbing aspect of the body current characteristic shown in Figure 6-42 is that body current tends to become maximum when grid voltage is zero. This means that if grid bias is lost (perhaps because of a power supply failure), the RF circuit may be damaged because of excess body current. To prevent this damage, a beam shaver like that shown in Figure 6-43 is often placed between the gun and the RF circuit.

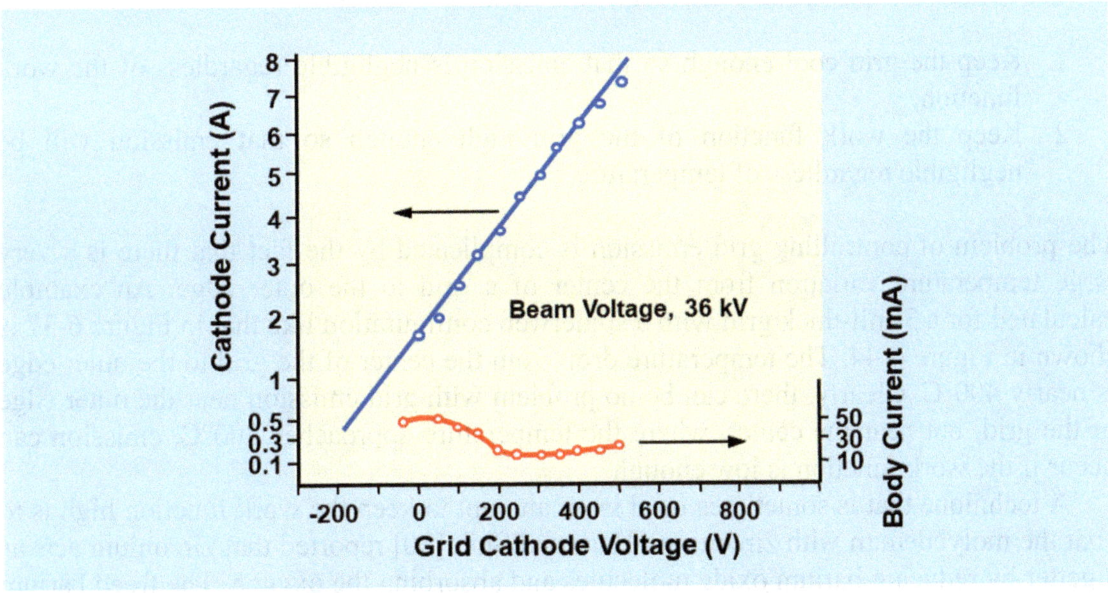

Figure 6-42 Shadow-gridded electron gun characteristics. (From: A. Staprans et al., *Proc. IEEE*, March 1973. © 1973 IEEE.)

When the control grid is pulsed negative to cut off the beam current, complete cutoff may not occur if the grid is emitting electrons. The resulting small beam current can result in unwanted signal generation and interpulse noise.

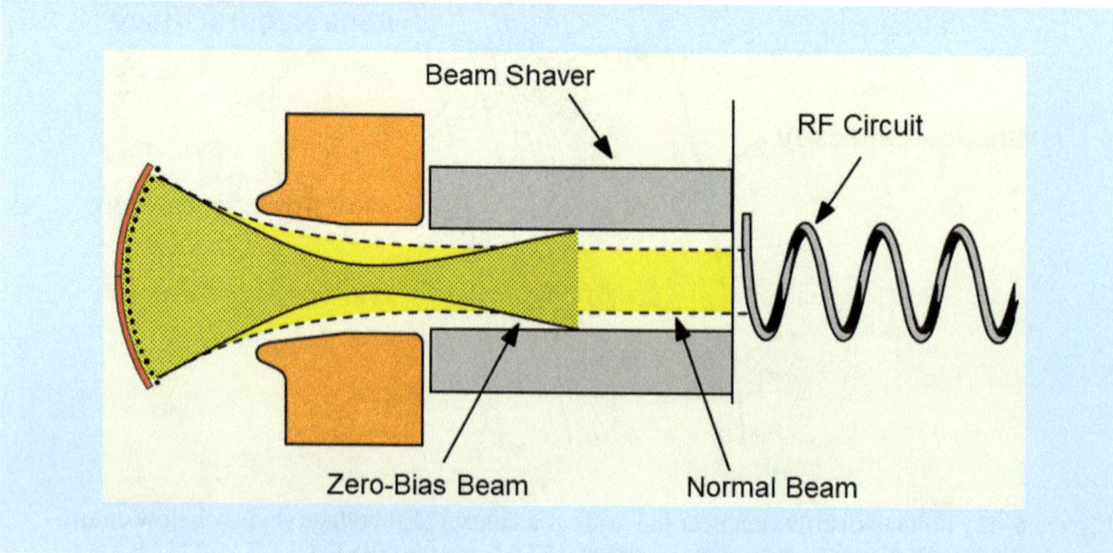

Figure 6-43 Placement of beam shaver to protect the RF circuit.

Grids are often fabricated from molybdenum. Because of its high work function, grid emission from the molybdenum is insignificant even at high grid temperatures. However, when barium and barium oxide vapors from the cathode collect on the molybdenum, the work function can decrease to the point where grid emission occurs.

There are two obvious options for preventing grid emission:

1. Keep the grid cool enough so that emission is negligible regardless of the work function;
2. Keep the work function of the grid high enough so that emission will be negligible regardless of temperature.

The problem of controlling grid emission is complicated by the fact that there is a very large temperature variation from the center of a grid to the outer edge. An example calculated for a 3-mil-thick grid with a spiderweb configuration like that in Figure 6-37 is shown in Figure 6-44. The temperature drop from the center of the grid to the outer edge is nearly 400°C. Clearly, there can be no problem with grid emission near the outer edge of the grid, but near the center, where the temperature approaches 800°C, emission can occur if the work function is low enough.

A technique that is sometimes used in an attempt to keep the work function high is to coat the molybdenum with zirconium. Haas and Shih [20] reported that zirconium acts as a getter by reducing barium oxide molecules and absorbing the oxygen. The freed barium then immediately vaporizes from the zirconium surface so that the work function is not reduced.

Unfortunately, as the oxygen content of the zirconium increases, its effectiveness decreases and eventually excessive grid emission can occur. Green and Goeser [21] reported that a thin layer of tungsten deposited on the molybdenum prior to zirconium deposition serves to improve the adherence and extend the life of the zirconium film. Thus, the zirconium/tungsten/molybdenum combination is found to outperform pure zirconium grids.

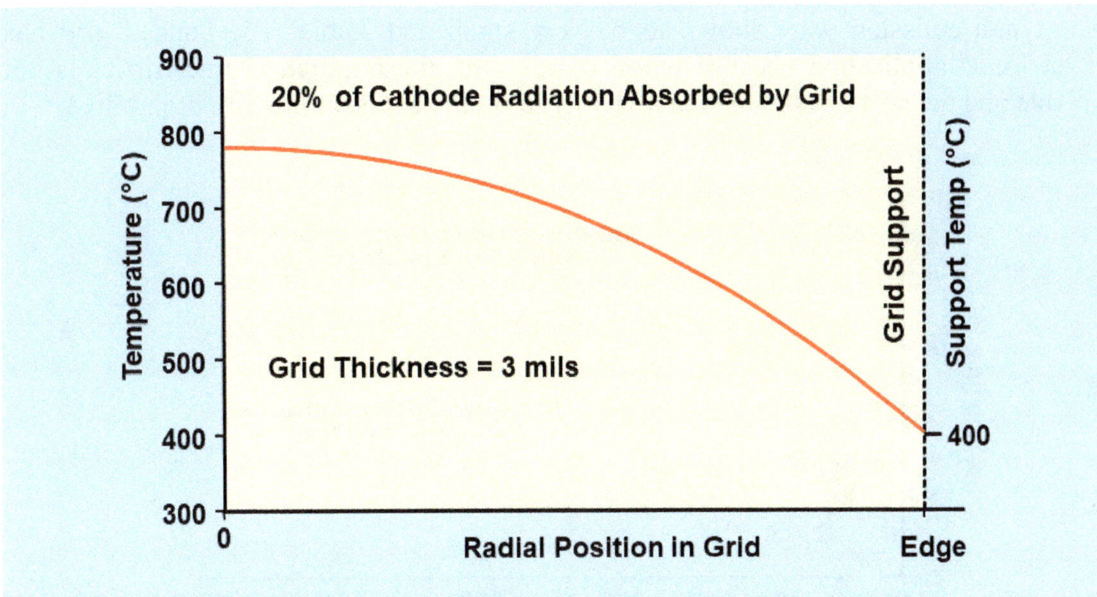

Figure 6-44 Grid temperature as a function of radial position.

The best way to prevent excessive control grid emission is to ensure that the temperature of the center of the grid is less than about 600°C. Then, even if the grid work function is as low as 2.2 eV (highly unlikely), emission will be negligible. To do this, the grid support must be properly cooled and the radial thermal conduction of the grid must be as high as possible.

The fabrication of shadow grid structures is difficult, especially in high-frequency tubes, where dimensions are small. Alignment of the elements of the control and shadow grids is critical. Alignment must be maintained despite the fact that the shadow grid operates at a higher temperature than the control grid because it is closer to or perhaps in contact with the cathode.

An example of the grid motions that occur as a cathode warms up is given in Figure 6-45. These results were computed using the thermal expansion characteristics of the grids and their support structures [22]. Note that at one point, the shadow grid moved to within about 0.001 inch of the cathode surface. This small spacing exceeded the total tolerance on the position of the grid. Thus, the shadow grid could touch the cathode, which, in turn, could lead to deformation of the shadow grid [23]. The author of [22] stated that this phenomenon had also been noticed experimentally and corrective action was undertaken.

In an effort to overcome the problems of maintaining proper grid spacings in shadow-gridded guns, Miram et al. developed the bonded grid structure shown in Figure 6-46 [24–26]. The grid elements were pyrolytic graphite deposited on thin boron nitride spacers. To suppress grid emission, the control grid was overcoated with a thin boron nitride film. Technical feasibility was demonstrated in a 2,000-hour life test in which grid leakage and emission were shown to be very small and stable. The bonded grid has not yet found application because boron nitride with the required characteristics is not available and the cost of establishing a facility to produce it is, at present, prohibitive.

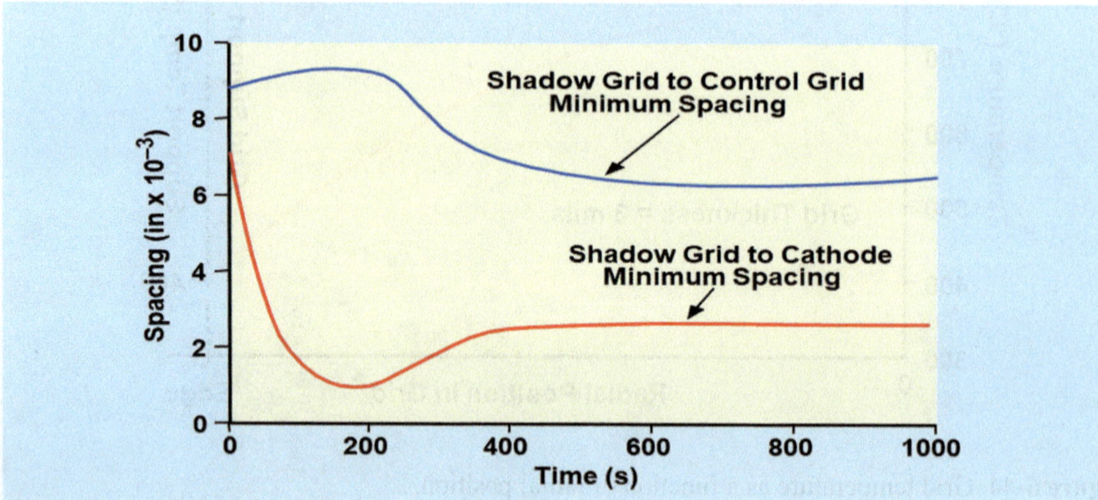

Figure 6-45 Relative spacing between shadow grid, control grid, and cathode during gun warm-up. (From: R. Sawicki, *Technical Digest*, 1978 IEDM. © 1978 IEEE.)

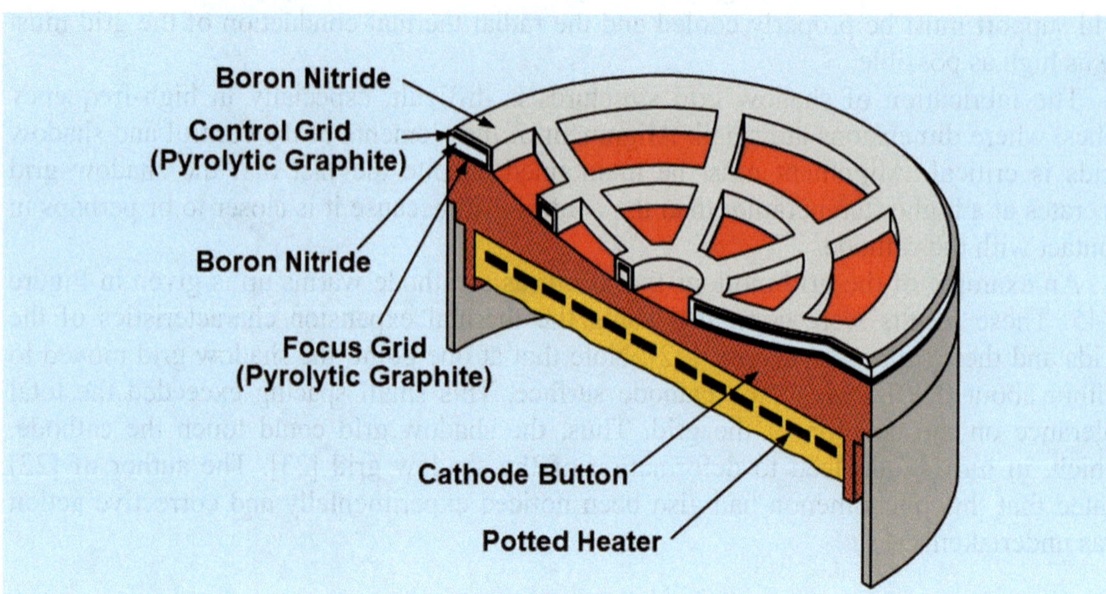

Figure 6-46 Bonded grid and cathode assembly. (From: T. J. Grant et al., *Technical Digest*, 1983 IEDM. © 1983 IEEE.)

Very sophisticated grid structures have been developed for some microwave tube applications. In the example in Figure 6-47, two control grids are used along with a shadow grid. This grid structure was developed for a tube that could operate efficiently at two different power levels. This is referred to as dual-mode operation. The grid closest to the anode controls emission from the central part of the cathode. The next grid controls emission from the outer portion of the cathode. Thus, beam current and diameter can be switched to change the beam power level while maintaining constant beam voltage and proper focusing conditions.

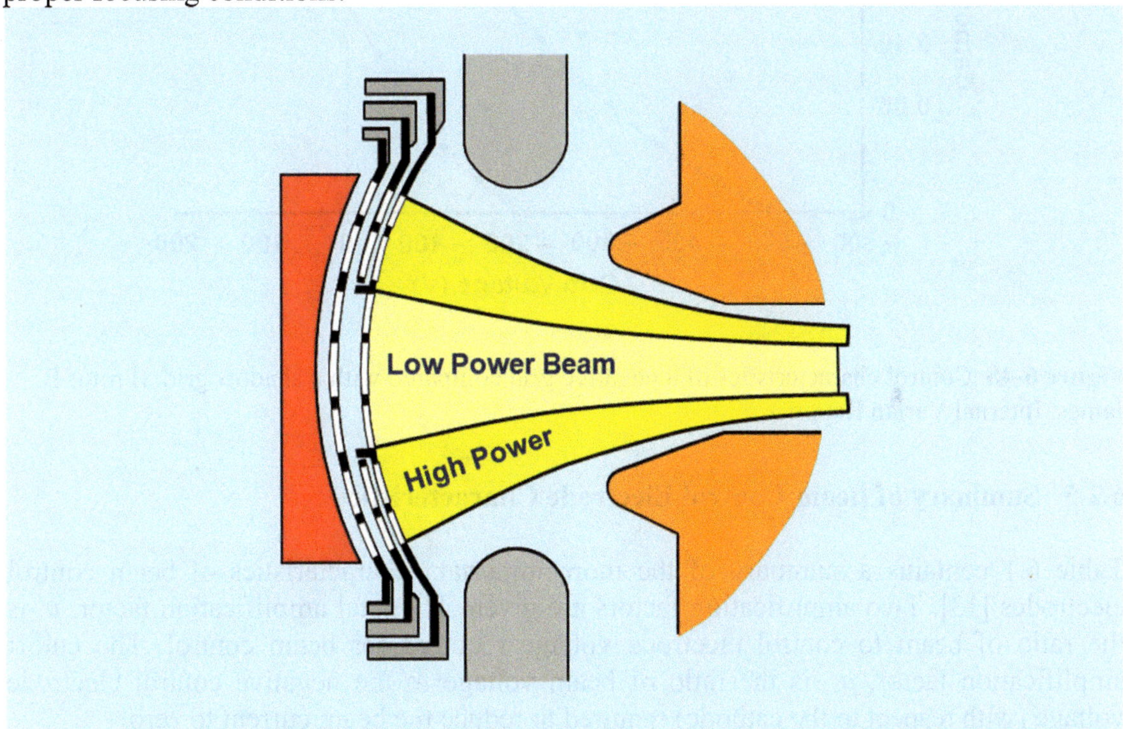

Figure 6-47 Dual-mode gun. (From: *Microwave Tube Manual* by Varian, Air Force Publication No. T.O.00-25-251, October 1979.)

In [27], True described a dual-mode structure using two grids rather than three. The grid farthest from the cathode is a control grid like that on the right in Figure 6-47. The grid next to the cathode is like a shadow grid except that the outer openings contain closely spaced wires. When this grid is pulsed positive with respect to the cathode (+36V in [27]), current flows from the entire cathode. When the grid is negative (–36V in [27]), the outer portion of the cathode is cut off and current flows only from the center portion.

The use of a shadow grid makes it possible to reduce interception on the control grid to about 0.1% of the cathode current. Unfortunately, at high frequencies where gun dimensions are very small, the fabrication of shadow grid structures becomes very difficult. To facilitate gun construction, negative grids are often used. With these grids, only a single grid is required and there is no grid interception. The beam optics are comparable to those of a shadow-gridded gun. The current-voltage characteristic of a negative grid is compared with a unigrid (see Figure 6-33) in Figure 6-48.

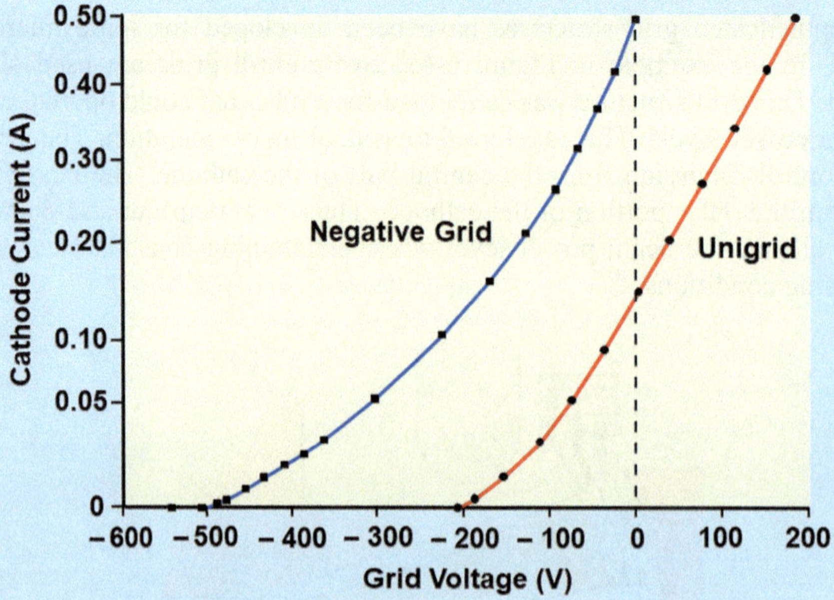

Figure 6-48 Control characteristics of a negative grid compared with a shadow grid. (From: B. James, Internal Varian Report.)

6.2.5 Summary of Beam Control Electrode Characteristics

Table 6.1 contains a summary of the more important characteristics of beam control electrodes [13]. Two amplification factors are given. The total amplification factor, μ, is the ratio of beam to control electrode voltage required for beam control. The cutoff amplification factor, μ_c, is the ratio of beam voltage to the negative control electrode voltage (with respect to the cathode) required to reduce the beam current to zero.

Table 6.1
Typical Characteristics of Control Electrodes

Type	μ	μ_c	Capacitance	Grid Interception	Focusing at Low Voltage
Modulating Anode	1 to 3		50 pF	0	Good
Control Focus Electrode	2 to 10	2 to 10	100 pF	0	Poor
Intercepting Grid	50	100	50 pF	15%	Fair
Shadow Grid	30	300	50 pF	0.1%	Fair

Source: A. Staprans et al., *Proc. IEEE*, Vol. 61, No. 3, March 1973, pp. 299–330.

As would be expected, the gridded guns have by far the highest amplification factors and the modulating anode gun has the lowest amplification factor. The capacitances of all of the control electrode structures to the surrounding electrodes are essentially the same. Capacitance is important, of course, in the design of the modulator that must be used to drive the control electrode structure. Finally, as would be expected, the interception of the intercepting grid structure is much higher than the other structures. Only the modulating anode gun has good focusing because the design of that gun is essentially the design of an ideal Pierce gun.

REFERENCES

[1] J. R. Pierce, *Theory and Design of Electron Guns*, 2nd ed., New York: Van Nostrand, 1954.

[2] G. R. Brewer, "Charged Particle Gun with Non Spherical Emissive Surface," U.S. Patent 3,139,552, 1964.

[3] I. L. Langmuir, and K. Blodgett, "Currents limited by space charge between concentric spheres," *Phys. Rev.,* Vol. 24, July 1924, p. 53.

[4] R. Helm, K. Spangenberg, and L. M. Field, "Cathode-design procedure for electron-beam tubes," *Elec. Comm.,* Vol. 24, March 1947, pp. 101–107.

[5] G. R. Brewer, "Formation of high-density electron beams," *Jour. App. Phys.,* Vol. 28, No. 1, January 1957, pp. 7–15.

[6] R. D. Frost, O. T. Purl, and H. R. Johnson, "Electron guns for forming solid beams of high perveance and high convergence," *Proc. IRE,* August 1962, pp. 1800–1807.

[7] C. C. Cutler, and M. E. Hines, "Thermal velocity effects in electron guns," *Proc. IRE,* March 1955, pp. 307–315.

[8] J. R. M. Vaughan, "Synthesis of the Pierce gun," *IEEE Trans. Electron Devices,* Vol. ED-28, January 1981, pp. 37–41.

[9] A. S. Gilmour, Jr., *Principles of Traveling Wave Tubes*, Norwood, MA: Artech House, 1994.

[10] W. E. Danielson, J. L. Rosenfeld, and J. A. Saloom, "A detailed analysis of beam formation with electron guns of the Pierce type," *Bell Sys. Tech. Jour.*, March 1956, pp. 375–420.

[11] R. True, "Dual-Element switching in Pierce guns," IVEC 2012.

[12] R. True, U.S. Patent 8492978 B2.

[13] A. Staprans, E. W. McCune, and J. A. Ruetz, "High-power linear-beam tubes," *Proc. IEEE,* Vol. 61, No. 3, March 1973, pp. 299–330.

[14] H. J. Wolkstein, "Design considerations for grid-controlled electron guns for pulsed traveling wave tubes," *RCA Review*, September 1960, pp. 389–413.

[15] R. True, "Calculation and design of grids in Pierce guns," *Technical Digest*, IEDM, 1989, pp. 215–218.

[16] J. M. Drees, "Non-intercepting grid structure for an electron tube," U.S. Patent 3,484,645, December 16, 1969.

[17] R. True, "An ultra-laminar tetrode gun for high duty cycle applications," *Technical Digest*, IEDM, 1979, pp. 286–289.

[18] R. True, "A theory for coupling gridded gun design with PPM focusing," *IEEE Trans. Electron Devices,* Vol. ED-31, No. 3, March 1984, pp. 353–362.

[19] R. B. True, and D. E. Laycock, "Gridded electron gun," U.S. Patent 4,737,680, April 12, 1988.

[20] G. A. Haas, and A. Shih, "Study of high work function materials needed for close-spaced grid applications," *Applications of Surface Science,* Vol. 4, 1980, pp. 104–126.

[21] M. Green, and G. Goeser, "Deposition techniques for suppressed emission," *Final Report, Contract No. N00014-84-C-2270*, October 1988.

[22] R. Sawicki, "Analytical determination of the thermal/mechanical performance of traveling wave tube electron guns," *Technical Digest*, IEDM, 1978, pp. 160–163.

[23] H. Y. Yeh, "The role of structural stability in the reliability of microwave vacuum devices," *IEEE Trans. Electron Devices,* Vol. ED-34, No. 2, February 1987, pp. 494–497.

[24] G. V. Miram, and E. L. Lien, "Convergent electron gun with bonded nonintercepting control grid," *Technical Digest*, IEDM, 1978, pp. 164–167.

[25] G. V. Miram, and E. L. Lien, "Convergent electron gun with bonded nonintercepting control grid," *Technical Digest*, IEDM, 1979, pp. 290–292.

[26] T. J. Grant, R. Garcia, and G. V. Miram, "Bonded grid electron gun for 95 GHz extended interaction amplifier (EIA)," *Technical Digest*, IEDM, 1983, pp. 141–143.

[27] R. True, "Dual-mode electron gun," U.S. Patent 4,593,230, June 3, 1986.

Chapter 7

Electron Beams from Pierce Guns

In Chapter 6, Pierce electron guns used in linear-beam tubes were discussed. In this chapter, the magnetic focusing of a beam from a Pierce gun is examined ([1–3] are recommended for further reading). Before proceeding with the discussion of magnetically focused beams, it is important to point out that numerous electrostatic focusing techniques have been examined in the more than 70 years that microwave tubes have been in existence. With the exception of a very few special cases, only magnetic focusing is used now.

The primary problem in focusing, as illustrated in Figure 7-1, is that space charge forces are so high in the beams in microwave tubes that, without focusing forces, rapid divergence of a beam occurs. With constraining (focusing) forces, beam divergence can be prevented.

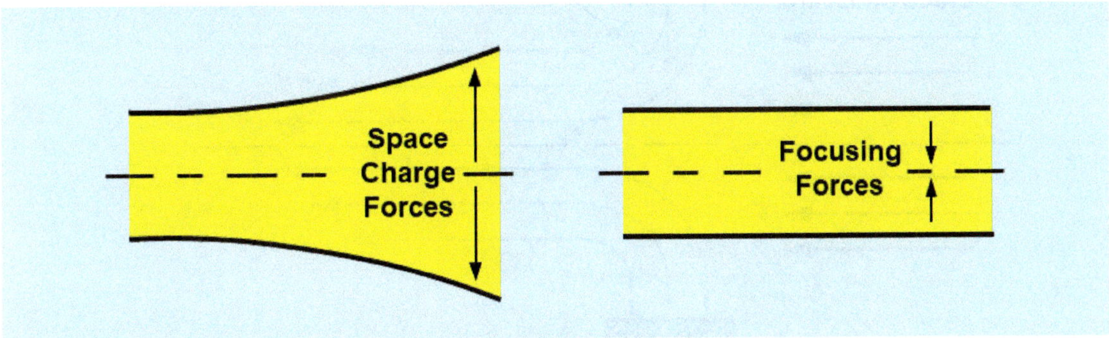

Figure 7-1 Illustration of the effects of space-charge forces and focusing forces on an electron beam.

In linear-beam tubes, the focusing forces are provided by a magnetic field aligned with the axis of the electron beam. In this chapter, techniques for applying magnetic focusing fields are described. One technique is by using a solenoid as indicated in Figure 7-2.

7.1 UNIFORM-FIELD FOCUSING AND LAMINAR FLOW

To simplify the description of the focusing process, it is initially assumed that there is no magnetic flux through the cathode and that the beam first encounters magnetic flux lines as it enters the solenoid. It is also assumed that there are no transverse velocity components (from thermal or grid effects, nonuniform cathode emission, and so forth). Magnetic focusing of the beam is initiated as the electron beam crosses radially directed components of magnetic flux lines (illustrated in Figure 7-3).

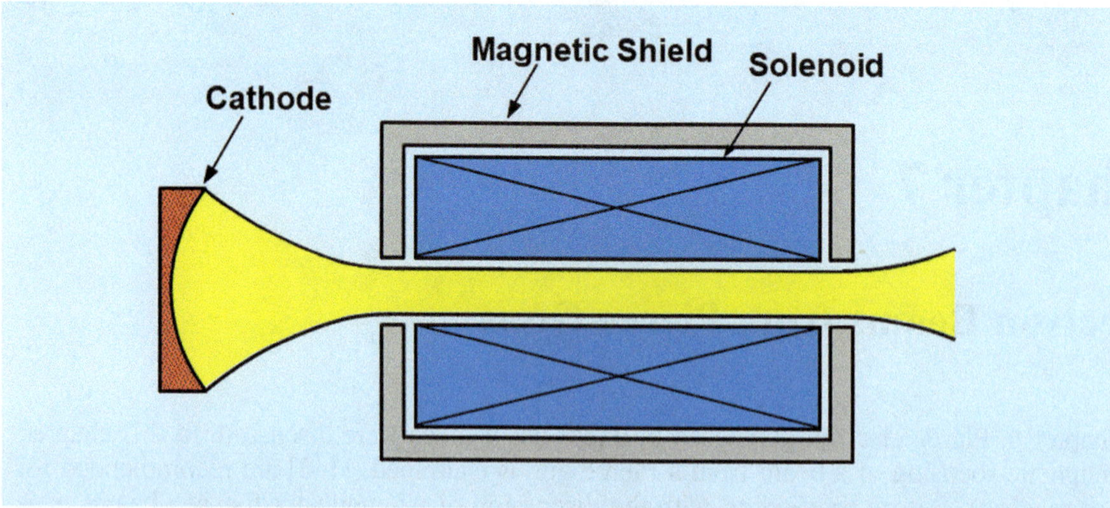

Figure 7-2 Illustration of the use of a solenoid to focus the beam in a TWT or klystron.

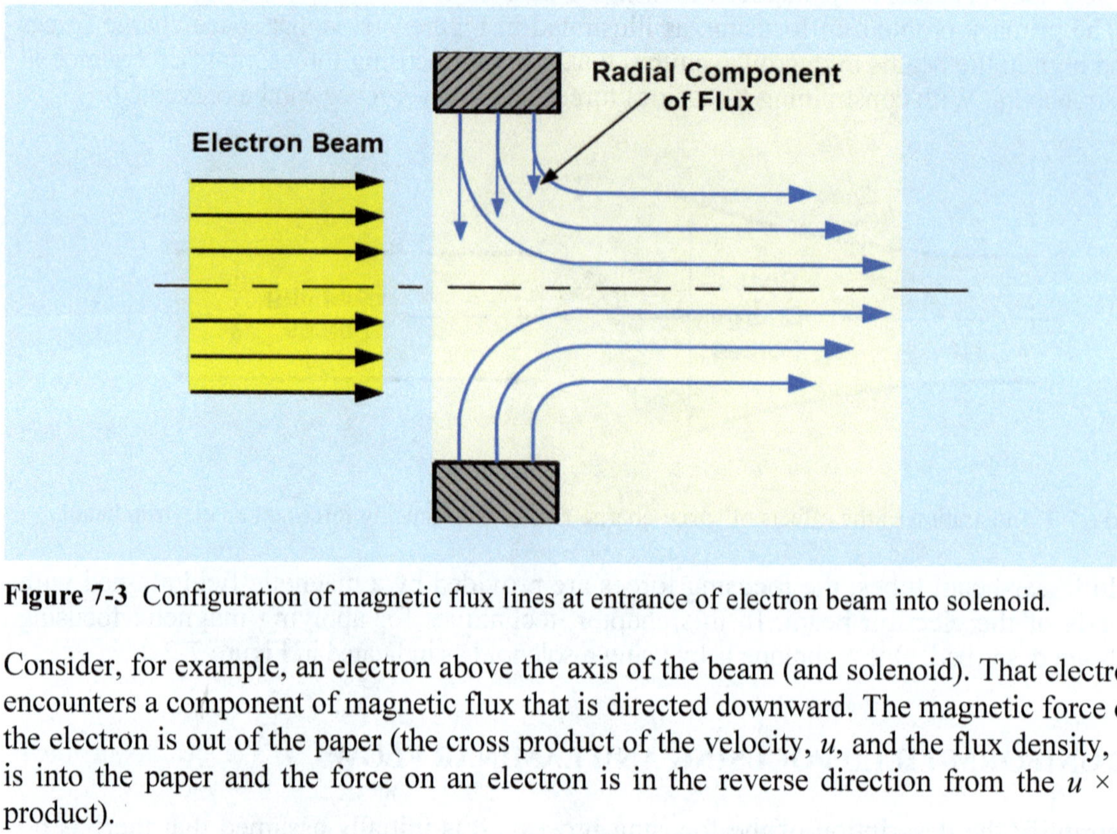

Figure 7-3 Configuration of magnetic flux lines at entrance of electron beam into solenoid.

Consider, for example, an electron above the axis of the beam (and solenoid). That electron encounters a component of magnetic flux that is directed downward. The magnetic force on the electron is out of the paper (the cross product of the velocity, u, and the flux density, B, is into the paper and the force on an electron is in the reverse direction from the $u \times B$ product).

Consider next an electron below the axis. That electron encounters a component of magnetic flux that is directed upward. The magnetic force on that electron is into the paper.

When the magnetic forces on all electrons are considered, the net effect is that with the flux lines in the directions shown in Figure 7-3, the electron beam starts rotating in the clockwise direction as it enters the field. If the directions of the flux lines were reversed, the direction of rotation of the beam would be counterclockwise.

As the electron beam moves further into the magnetic field, it encounters stronger axial components of magnetic flux. Consider again an electron above the beam axis. As a result of crossing radial components of flux, this electron has a velocity component out of the paper. The cross product of that velocity component with the axial magnetic flux is a vector away from the axis so that the force on the electron is toward the axis. Similar considerations for other electrons show that they all experience a magnetic force toward the axis. Thus, it is the rotation of the beam (established as the beam crosses radially directed flux lines) interacting with the axial components of flux that produces the magnetic focusing force. As each electron in the beam is followed, it is found that a helical trajectory about the axis of the solenoid is traversed as indicated in Figure 7-4.

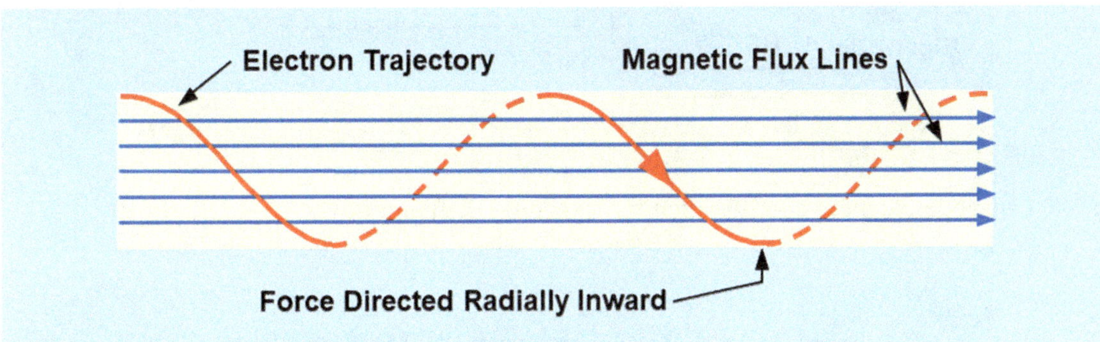

Figure 7-4 Electron trajectory in an axial magnetic field.

7.1.1 Brillouin Flow

If the space-charge forces (plus centrifugal forces resulting from beam rotation) tending to cause the beam to diverge are just balanced by equal and opposite magnetic forces tending to compress the beam, and if the electrons enter the magnetic field with no radial velocity components, then the beam diameter will remain constant as the beam rotates and moves through the magnetic field.

For a beam with uniform charge density throughout and no magnetic flux through the cathode, the magnetic flux level that produces a magnetic force that exactly balances the space charge and centrifugal forces is called the *Brillouin flux level*, commonly denoted by B_B. The resulting electron motion in the form of a rotating beam with constant diameter is called *Brillouin flow*. The configuration of the electron gun and magnetic pole piece for launching Brillouin flow are shown in Figures 7-5 and 7-6.

When the beam exits from the magnetic field at the collector end of the tube, the sequence of interactions is just the opposite of that at the entrance of the beam into the field. That is, as the beam leaves the magnetic field, the radial compressive force disappears, rotation stops, and the beam expands under the influence of space-charge forces as it enters the collector.

Now, as we have said, for Brillouin flow, the beam must have a uniform charge density and must enter the magnetic field on axis with no radial electron motion and with a radius, b_o, equal to the equilibrium radius, a (the radius at which space-charge and centrifugal forces are balanced by magnetic forces). This situation is shown in Figure 7-7.

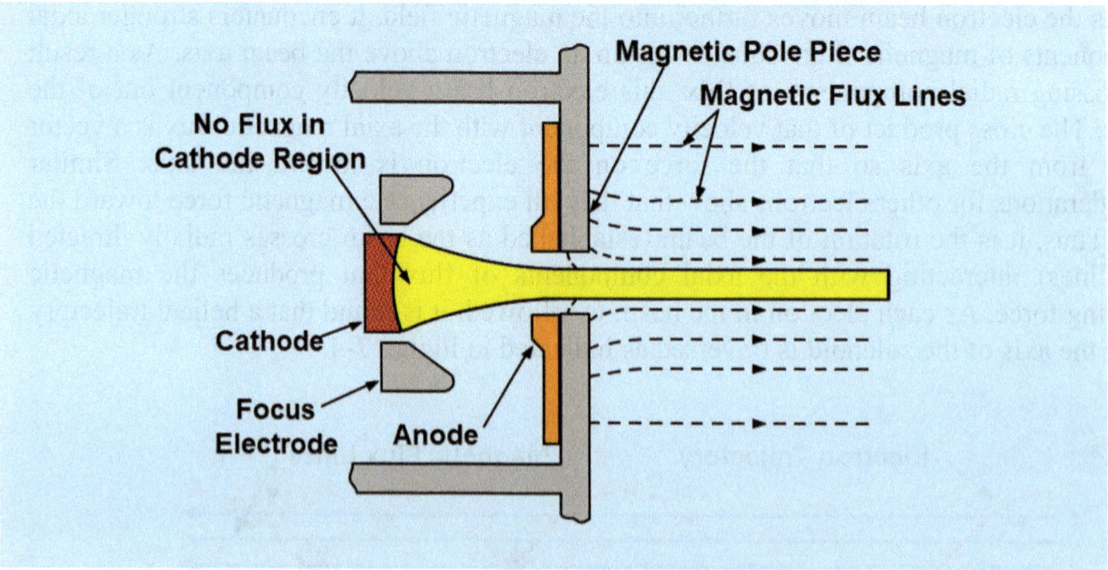

Figure 7-5 Gun and pole piece configuration for launching Brillouin flow.

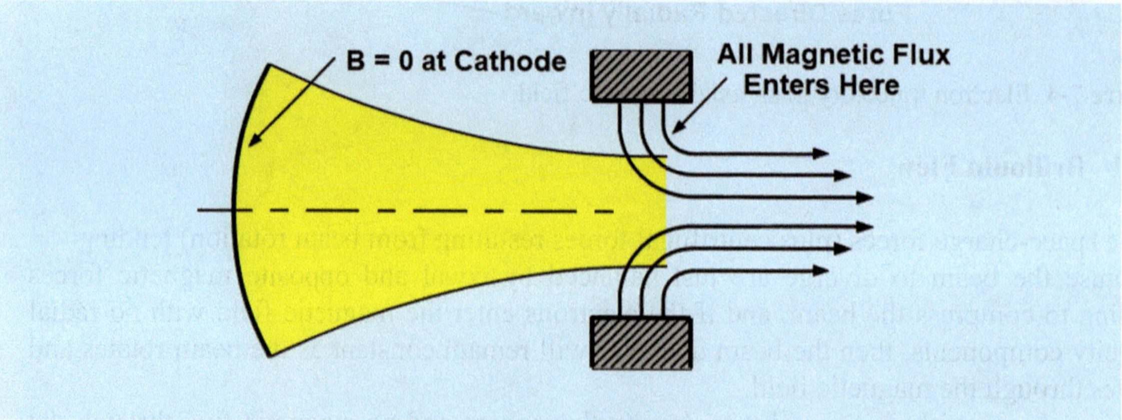

Figure 7-6 Configuration of the magnetic field at the entrance to the focusing structure for Brillouin flow.

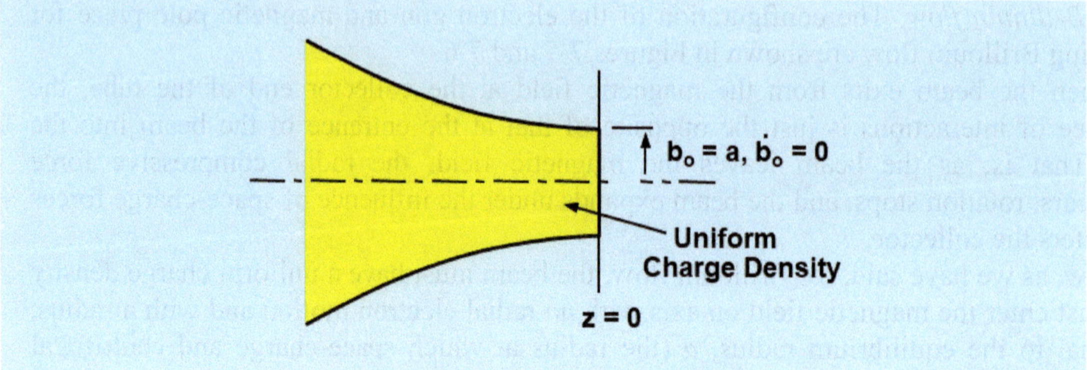

Figure 7-7 Entrance conditions for Brillouin flow.

The equilibrium radius for a Brillouin beam is given by

$$\alpha \sim I^{1/2}/B_B \tag{7-1}$$

where B_B is the Brillouin magnetic flux density, I is beam current. As would be expected, to make the equilibrium radius smaller for a given current and velocity, B_B must increase. To make the equilibrium radius larger for a given current and velocity, B_B must decrease.

If the equilibrium radius, α, is known (because of a specific TWT design), then B_B can be found from

$$B_B = 0.83 \times 10^{-3} \frac{I^{1/2}}{a V^{1/4}} \quad \text{T} \tag{7-2}$$

7.1.2 Scalloping

When the electron gun is shielded from magnetic flux, it is extremely difficult to achieve the perfectly balanced force condition that is required for Brillouin flow. For example, consider the case when the electron beam enters the magnetic field with a radius smaller than the equilibrium radius shown in Figure 7-8.

At plane z_1, where the beam enters the magnetic field, the space-charge force is larger than the magnetic focusing force and so the beam starts to expand. When the beam reaches plane z_2, the radial forces are equal, but the electrons are moving radially outward. As a result, the electrons overshoot the equilibrium radius. By the time the electrons have reached plane z_3, the radial motion of the electrons has been stopped because the space-charge and centrifugal forces have become smaller than the magnetic forces. In fact, this force imbalance causes the beam to become smaller again and, by plane z_4, the electrons are again crossing the equilibrium radius. Because the electrons are moving radially inward at plane z_4, they overshoot the equilibrium radius. By the time the beam reaches plane z_5, the inward radial motion has been stopped because the space-charge and centrifugal forces exceed the magnetic forces.

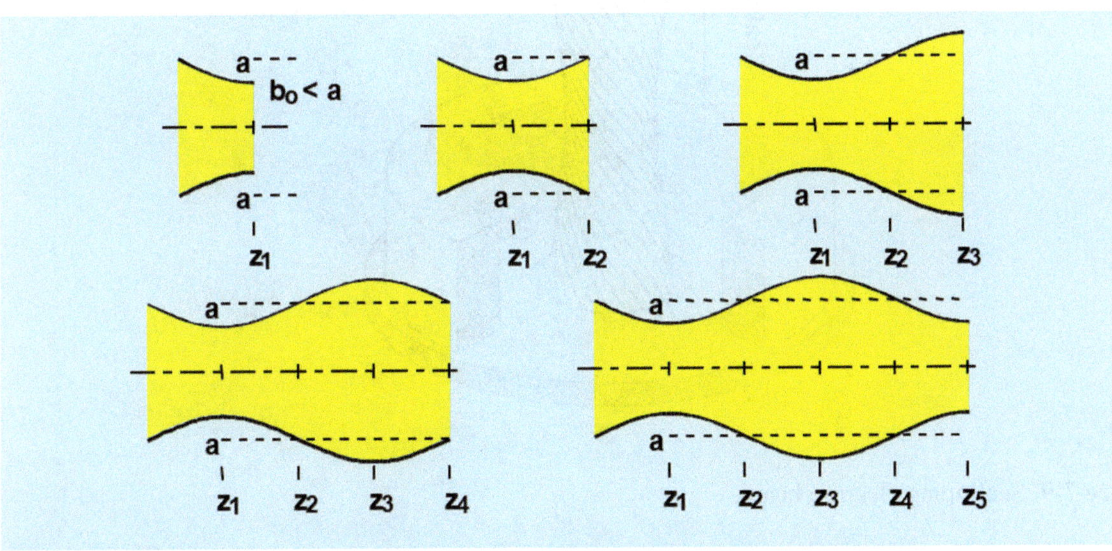

Figure 7-8 Beam dynamics resulting from the beam entrance into magnetic field at less than the equilibrium radius.

In a uniform axial field, the beam radius at plane z_5 is the same as it was at plane z_1 and the entire process that took place between planes z_1 and z_5 starts over again. As a result, the beam radius continues to oscillate as the beam moves in the axial direction. This periodic variation of beam radius is referred to as *scalloping.* A scalloping beam is shown in the drawing in Figure 7-9. The scalloping frequency and wavelength are

$$\omega_s = \sqrt{2}\,\omega_L \tag{7-3}$$

$$\lambda_s = 0.030 \times 10^{-3} \frac{V^{1/2}}{B} \text{ m} \tag{7-4}$$

where ω_L is the Larmor frequency[1], V is the beam voltage, and B is the flux density in Tesla.

The way in which scallop wavelength and beam shape vary with B is shown in Figure 7-10. Next consider the case when the beam enters the focusing field with $b = a$ but db/dz either greater or smaller than zero. That is, the beam is either expanding or contracting. When the beam is expanding, the beam radius varies with B as shown in Figure 7-11.

When a beam traveling parallel to the axis but having a radius smaller than the equilibrium radius enters the magnetic field as is indicated in Figure 7-12 (a), it is possible to adjust the magnetic field level so that the beam radius is the equilibrium radius. Then scalloping can be eliminated.

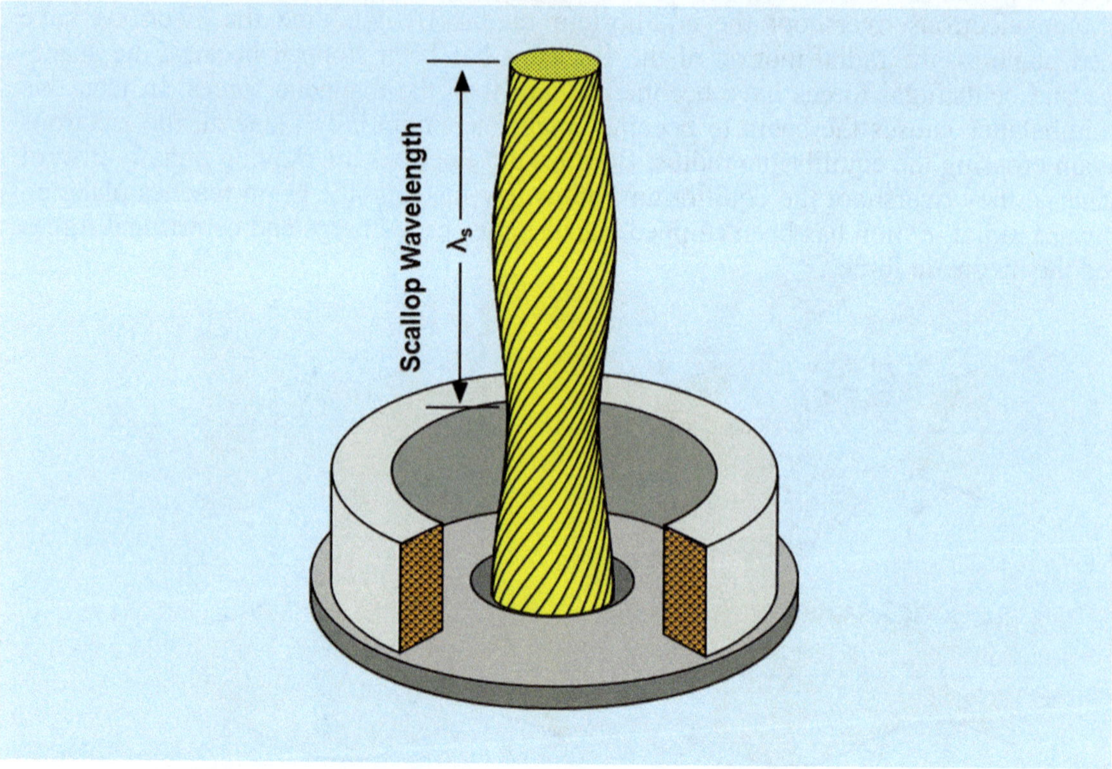

Figure 7-9 Scalloping electron beam.

[1] Appendix D contains a discussion of ω_L, ω_s, and ω_c.

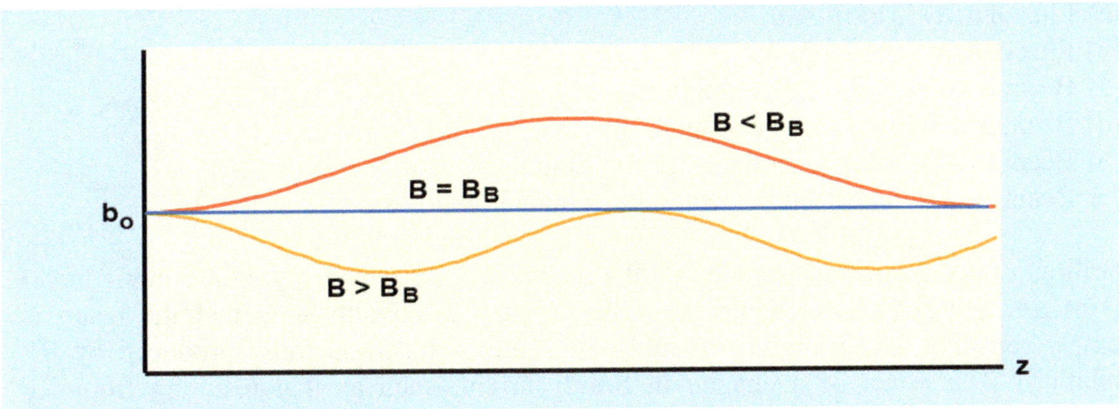

Figure 7-10 Beam shape as a function of the magnetic flux density.

A similar situation occurs (Figure 7-12 (b)) when a beam traveling parallel to the axis but having a radius larger than the equilibrium radius enters the magnetic focusing field. Then the beam scallops inward because the focusing forces are larger than the space-charge and centrifugal forces. Again, it is possible to adjust the magnetic field level to eliminate scalloping.

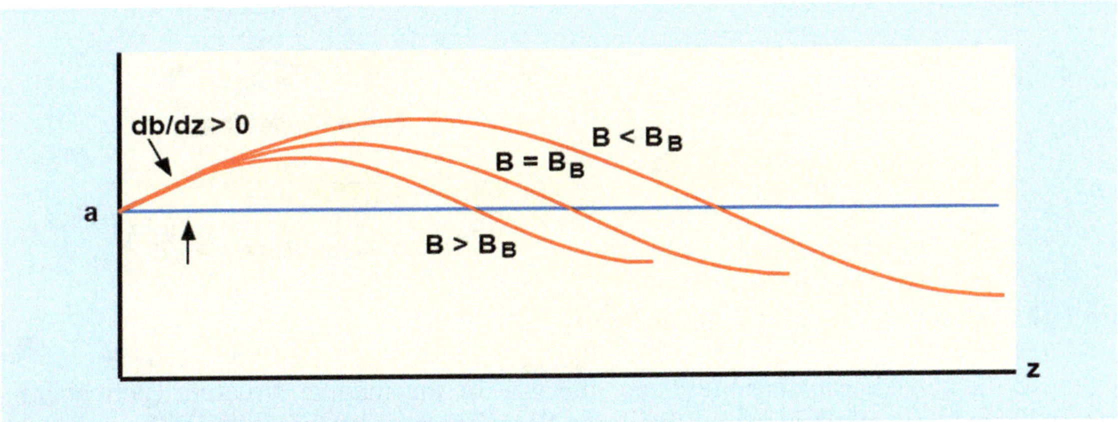

Figure 7-11 Beam shapes for $db/dz > 0$.

If, as indicated in Figure 7-12 (c) and (d), the electron gun is at the wrong axial position so that the beam is either diverging or converging as it enters the magnetic field, then no adjustment of the magnetic field can be made to eliminate the resulting scalloping.

At this point it is appropriate to comment on the beam behavior that occurs if the axis of the electron beam is not aligned with the axis of the magnetic field. This could happen if the electron gun was misaligned during tube fabrication. The gun axis could be either off center or tilted with respect to the axis of the tube. Other causes for misalignment include errors in cathode, grid, or control electrode positioning or nonuniformities in magnetic pole piece material. The result of beam misalignment is that the entire beam spirals about the magnetic field axis at the cyclotron frequency.[2]

Figure 7-12 contains a summary of the problems that could be encountered in attempting to form a Brillouin beam flowing parallel to the axis. These problems are;

[2] Appendix D contains a discussion of ω_L, ω_s, and ω_c.

(a) Flux density is too small;
(b) Flux density is too large;
(c) Beam is converging at the entrance;
(d) Beam is diverging at the entrance;
(e) Beam axis is offset from the magnetic field axis;
(f) Beam axis is tilted with respect to the magnetic field axis.

Brillouin flow has the advantage that the magnetic field required for focusing is lower than for any other focusing technique. The primary disadvantage is that the beam is extremely sensitive to misalignments and perturbations (such as those produced by RF modulation). The effect of a change in beam current, such as that resulting from RF bunching of electrons, is shown in Figure 7-13. Note that an increase in current of 50% results in an increase in beam diameter of over 40%.

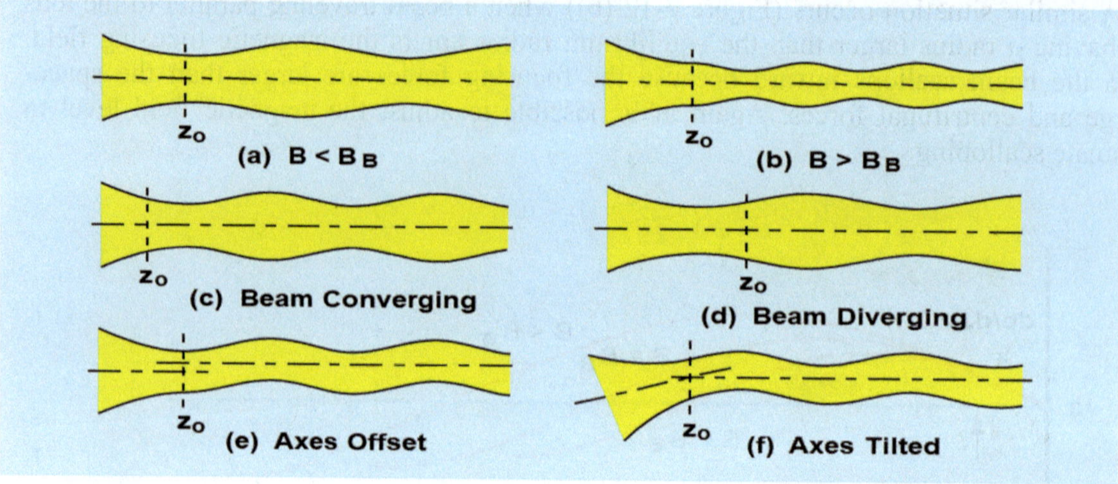

Figure 7-12 Problems in forming a Brillouin beam.

Because the flux density is minimized, the size of the magnet structure (permanent magnet or solenoid) is minimized with Brillouin flow. The penalty for using Brillouin flow, as shown in the following discussion, is that the beam is very sensitive to misalignments, incorrect magnetic fields, RF drive, and so forth.

Experiments show that Brillouin-focused electron beams can behave very nearly as predicted. Results obtained by scanning a current probe across the center of a beam at equally spaced axial positions are shown in Figure 7-14 [4, 5]. The inward scalloping of the beam when $B > B_B$ is clearly shown. When $B = B_B$, the beam diameter was nearly constant.

Ideally, each of the current profiles in Figure 7-14 should be rectangular. At least three factors prevented the current from rising abruptly from zero to a constant value at the edges of the beam profiles. One factor was the finite diameter of the aperture in the current probe (0.010 inch (0.25 mm)) used for current sampling. The second factor was the thermal velocities of the electrons, and the third factor was the patchiness and roughness of the oxide-coated cathode.

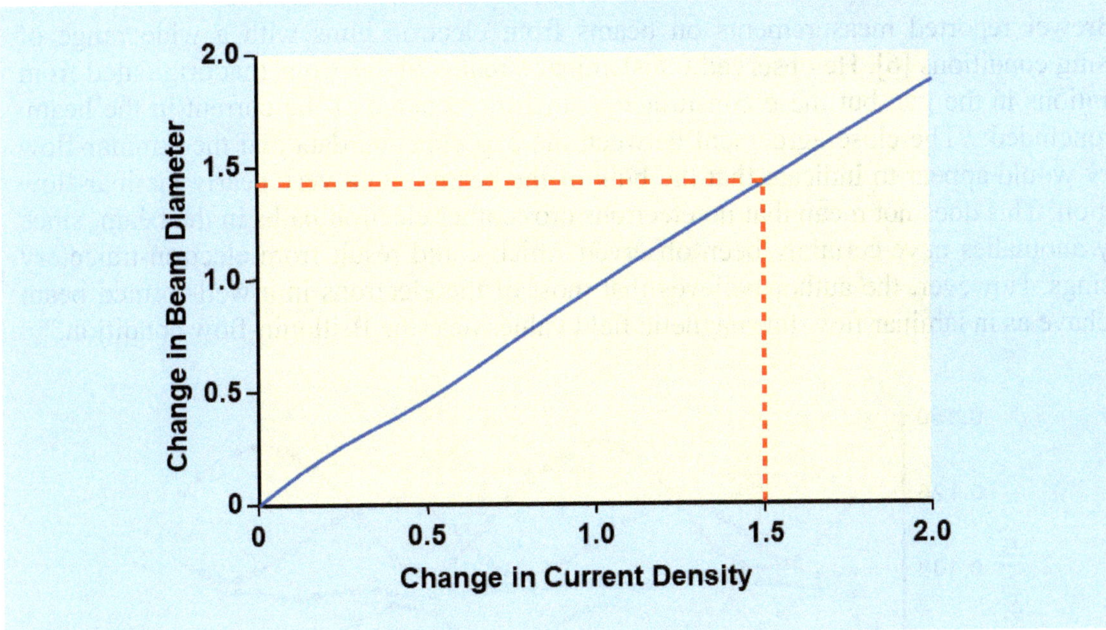

Figure 7-13 Change in beam diameter resulting from a change in current density for a Brillouin beam. (From: *Power Travelling Wave Tubes* by J. F. Gittins, published 1965 by American Elsevier.)

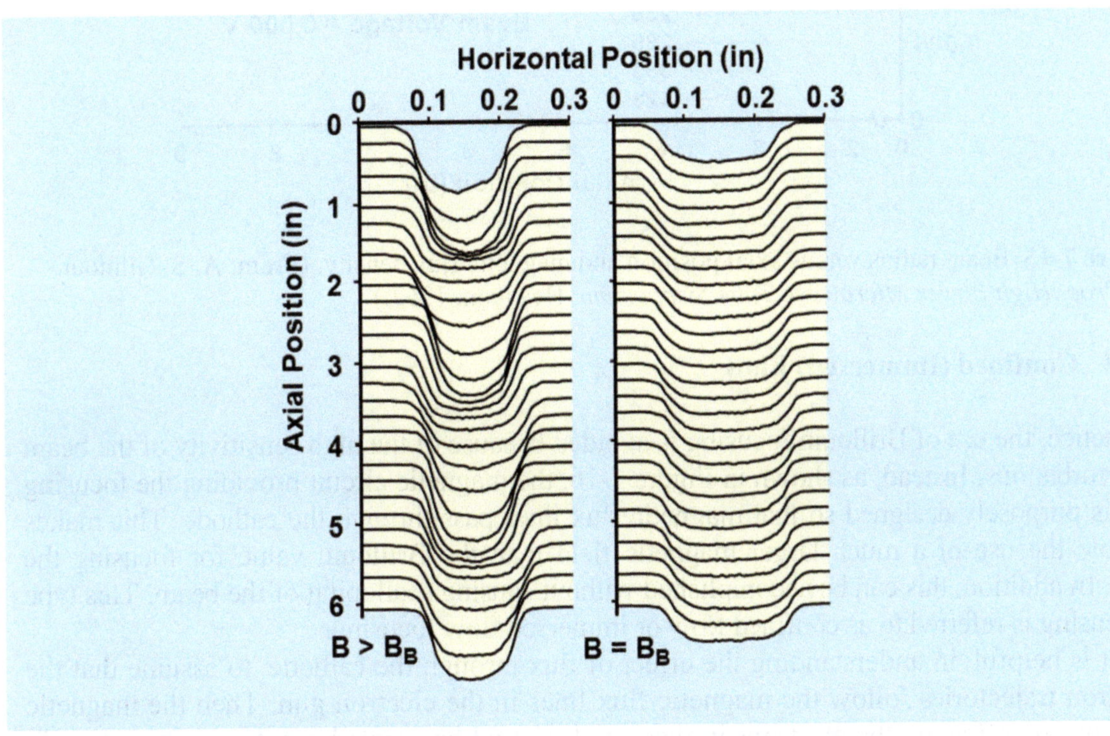

Figure 7-14 Current density profiles for a beam from a shielded gun. (From: A. S. Gilmour, Jr., *Proc. High Power Microwave Tube Symposium*, Hexagon, 1962.)

From data like that in Figure 7-14, the beam radius as functions of the axial position and the magnetic flux density were determined and are shown in Figure 7-15. At the predicted Brillouin flux density of 285G, the variation in beam diameter was less than 2%.

Brewer reported measurements on beams from electron guns with a wide range of focusing conditions [6]. He observed translaminar streams of electrons that originated from aberrations in the gun but these constituted a small percentage of the current in the beam. He concluded: "The close agreement between the experimental data and the laminar-flow theory would appear to indicate that the bulk of the beam is in a very nearly laminar-flow situation. This does not mean that no electrons cross other electron paths in the beam, since many anomalies have certainly been observed which could result from electron-trajectory crossings. However, the author believes that most of the electrons in a well-formed beam do behave as in laminar flow for magnetic field values near the Brillouin-flow condition."

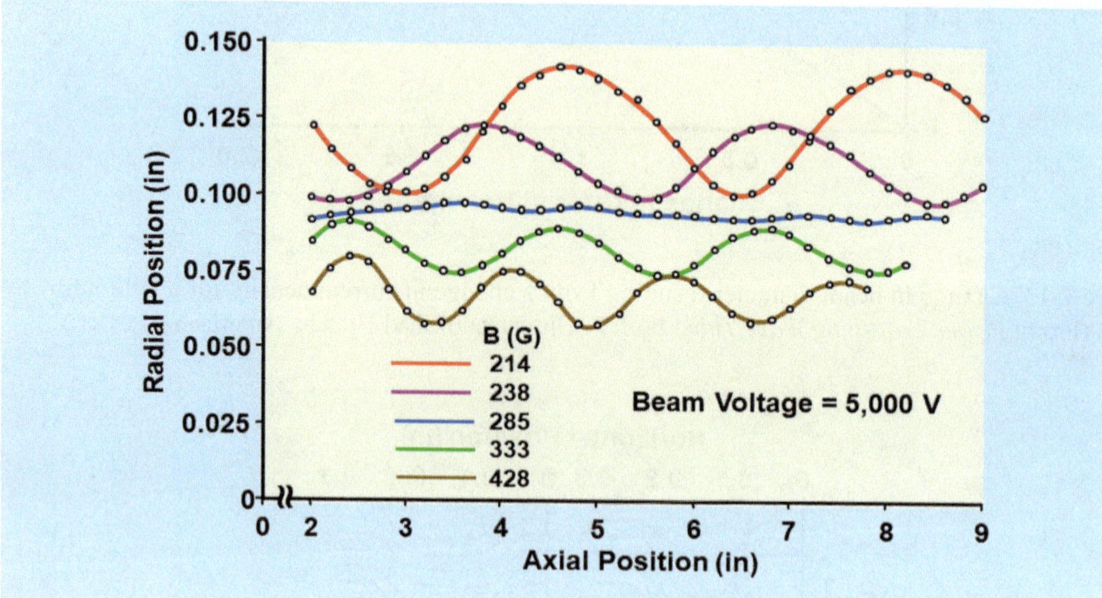

Figure 7-15 Beam radius versus axial position and magnetic flux density. (From: A. S. Gilmour, Jr., *Proc. High Power Microwave Tube Symposium*, Hexagon, 1962.)

7.1.3 Confined (Immersed) Flow

In practice, the use of Brillouin focusing is avoided because of the high sensitivity of the beam to perturbations. Instead, as shown in Figure 7-16, the magnetic circuit providing the focusing field is purposely designed so that magnetic flux lines pass through the cathode. This makes possible the use of a much larger magnetic field than the Brillouin value for focusing the beam. In addition, this can be accomplished without causing scalloping of the beam. This type of focusing is referred to as confined flow or immersed flow focusing.

It is helpful, in understanding the effect of flux through the cathode, to assume that the electron trajectories follow the magnetic flux lines in the electron gun. Then the magnetic field has no effect on the electrons in the gun. It should be pointed out that electrons do, in fact, cross flux lines in the gun (as is indicated in Figure 7-16). The resulting behavior of the beam is essentially the same as if it is assumed that the electrons do not cross flux lines.

As the electron beam leaves the gun and enters the main part of the focusing field, the increase in flux density must be sufficient to produce a focusing force that counterbalances the space charge and centrifugal forces. The focusing force results, just as it did for Brillouin focusing, from the interaction of the beam rotation with the axial magnetic field. In this case, however, the axial field is larger than it was for Brillouin flow so the beam rotates more slowly than for the Brillouin case. Also, since the beam rotates more slowly, the centrifugal force is reduced and this further reduces the rotational rate required for equilibrium.

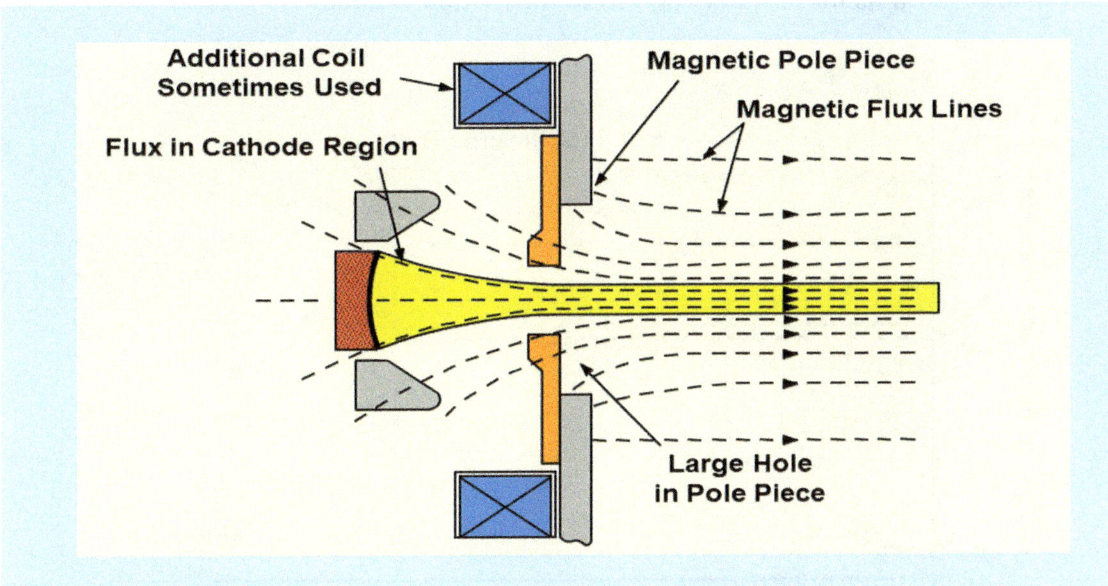

Figure 7-16 Magnetic circuit for launching confined (immersed) flow.

Next, consider a perturbation to the beam that causes the electrons to move radially outward. This could result from electron bunching caused by the interaction of the electrons with an RF field. As the outward moving electrons cross flux lines of the very large confined flow focusing field (often two to three times the Brillouin value), the rate of rotation of the beam increases rapidly. The interaction of this increased rotation with the large focusing field produces a focusing force far larger than that which occurs with Brillouin focusing. As a result, the radial expansion of the beam is much less than for Brillouin focusing.

The primary benefit of confined flow focusing is that beam control is significantly better than for Brillouin focusing. This is illustrated in Figure 7-17, which is like Figure 7-13 except that curves for field levels two times (m = 2) and three times (m = 3) the Brillouin value are shown. For an increase in beam current of 50%, the change in beam diameter is decreased from over 40% for Brillouin focusing to about 8% for m = 2 and less than 4% for m = 3. The primary disadvantage of confined flow focusing is that the magnet(s) must be larger and heavier than those for Brillouin focusing to produce the required field. Also, if electromagnets are used, the power and cooling requirements are larger than those for Brillouin flow.

Using Busch's theorem[3], it is possible to more clearly understand the details of confined flow focusing. For example, the rate of rotation of the beam, $\dot{\theta}$, compared to the rate of rotation of the Brillouin focused beam, $\dot{\theta}_B$, can be found to be

---

[3] Busch's theorem is discussed in Appendix D.

$$\frac{\dot{\theta}}{\dot{\theta}_B} = m - (m^2 - 1)^{1/2} \tag{7-5}$$

This relation is plotted in Figure 7-18. Note that as m increases from unity (the Brillouin value) to the range of 2 to 3 commonly used for confined flow focusing, the rate of rotation of the beam drops by a factor of 4 to over 5 from the Brillouin value.

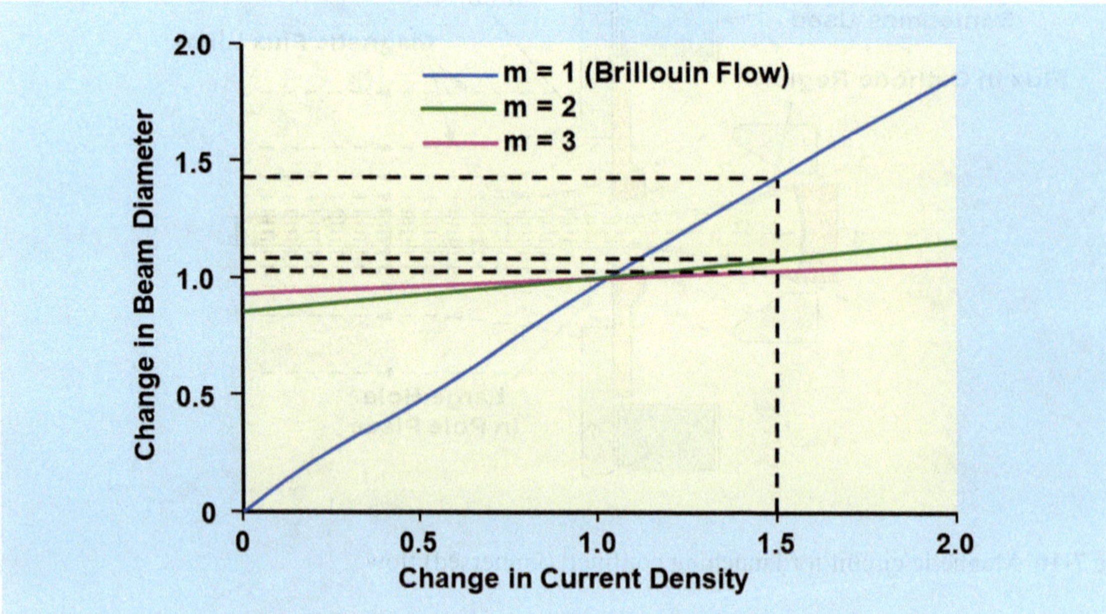

Figure 7-17 Effect of current density on beam diameter for fields greater than Brillouin. (From: *Power Travelling Wave Tubes* by J. F. Gittins, published 1965 by American Elsevier Publishing.)

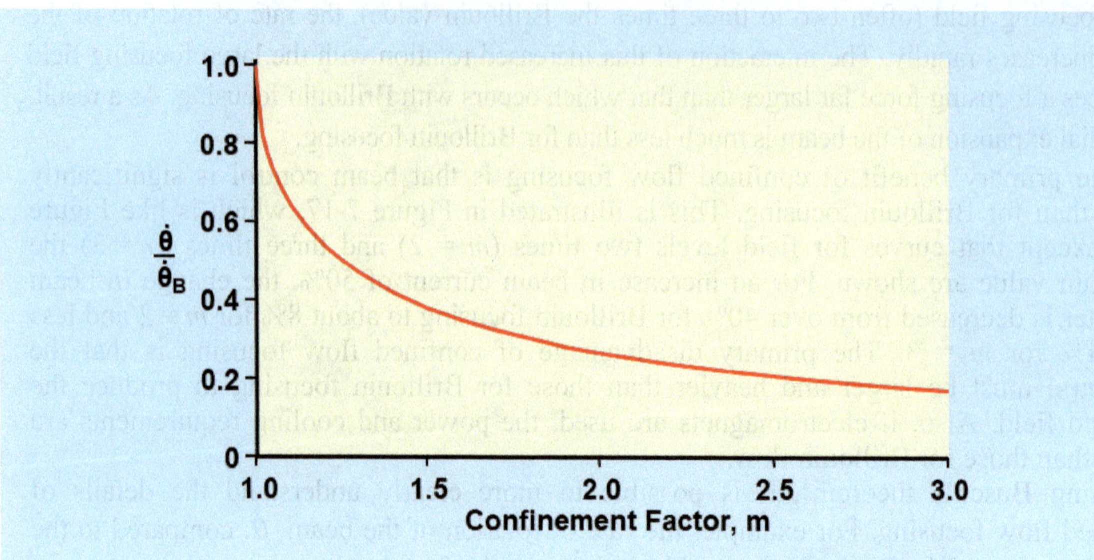

Figure 7-18 Ratio of rotation rates of confined and Brillouin beams as a function of the confinement factor.

Because the beam rotates so slowly with confined flow focusing, the centrifugal force, which is equal to the space-charge force for Brillouin focusing, almost vanishes. Nearly all of the focusing force produced by the slowly rotating beam in the strong focusing field is used to offset space-charge forces.

It is of value to compare the beam dynamics for confined focusing to those for Brillouin focusing. First, assume that the beam enters the focusing field at the equilibrium radius but with a radial component of velocity. Then the resulting scallop amplitude, C, compared with the amplitude of the scallop on a Brillouin beam, C_B, for the same entrance conditions is [7]

$$\frac{C}{C_B} = \frac{1}{\left(2m^2 - 1\right)^{1/2}} \tag{7-6}$$

This result is shown in Figure 7-19. Note that a significant decrease in scallop amplitude occurs as the flux density is increased.

Next consider the effect on beam radius of a change in current in the beam. Here it will be assumed that the beam velocity remains constant. Current may change because of the use of a control electrode, such as a grid, in the electron gun to modulate the electron flow. Also, current changes in localized regions when the beam interacts with the RF field on a nearby circuit to produce bunching of the electrons.

When a change in current occurs, let the change in the beam radius for confined flow be δ and for Brillouin flow be δ_B. The ratio of δ to δ_B can be shown to be [7]

$$\frac{\delta}{\delta_B} = \frac{1}{2m^2 - 1} \tag{7-7}$$

The ratio, δ/δ_B, is plotted in Figure 7-19 as a function of the confinement factor m. It is clear that the effect of confined flow on beam perturbations resulting from current changes is significant and that the use of confined flow is very desirable to reduce these perturbations. The penalty is that the increase in flux density required for confined flow necessitates an increase in the size of the magnets used and, if solenoid focusing is employed, an increase in solenoid power.

In his paper on magnetically focused electron beams, Brewer clearly showed the confining effect of magnetic flux through the cathode [6]. As the cathode flux density was increased, beam scalloping decreased and variations in current density with distance were reduced. Brewer also noted that cathode flux prevents translaminar electrons from crossing or coming close to the beam axis, so cathode flux improves the laminarity of the beam.

7.2 UNIFORM-FIELD FOCUSING AND NONLAMINAR FLOW

In Chapter 4 on cathodes and Chapter 5 on guns, several mechanisms were identified that produce transverse velocity components to electron trajectories:

1. Thermal effects;
2. Patchy emission;
3. Cathode roughness;

4. Filament magnetic field;
5. Field distortion by grids;
6. Gun aberrations.

As a result of the transverse velocity components, electron trajectories cross and the electron flow is said to be nonlaminar.

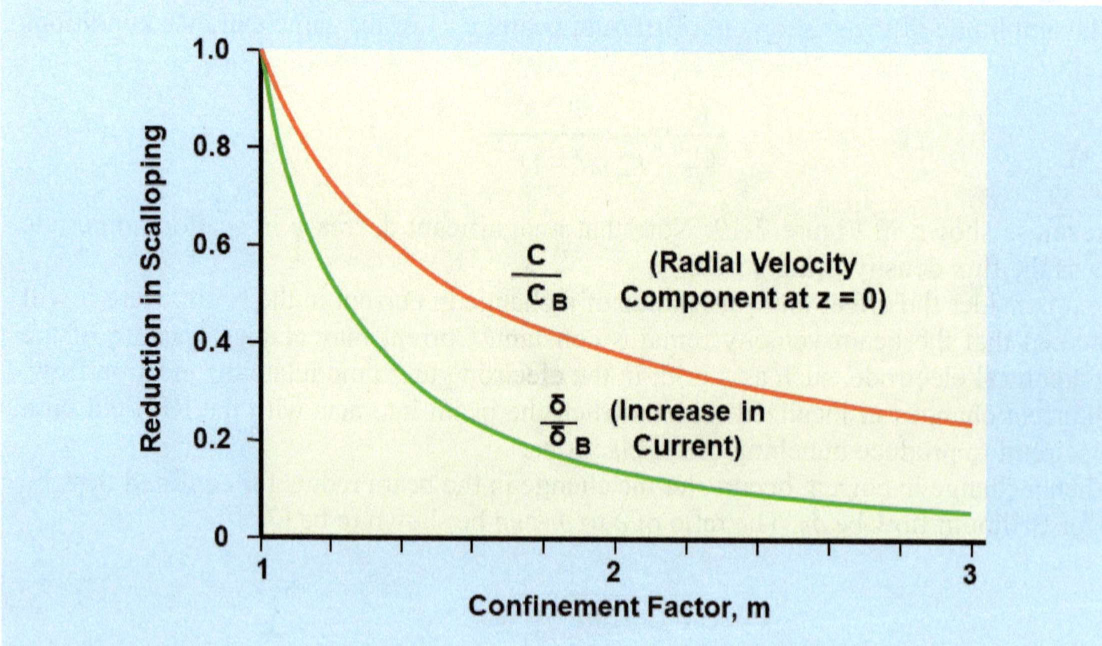

Figure 7-19 Scallop amplitudes for confined flow, C and δ, relative to those for Brillouin flow, C_B and δ_B, as functions of the confinement factor.

There is no single theory for treating all forms of nonlaminar flow. In fact, the only way for determining, with reasonable accuracy, electron behavior in nonlaminar flow is by tracing trajectories using numerical techniques. Results of this approach will be shown in a following section, which is concerned with PPM-focused beams from gridded guns.

Several theories exist for special cases of nonlaminar flow. One of the first was that by Pierce and Walker for the effects of thermal velocities on Brillouin flow [8]. This theory was based on the assumption of a Maxwellian distribution of charge density within the beam. The beam was assumed to be in thermal equilibrium with a temperature that was suggested to be the cathode temperature multiplied by the area compression ratio of the gun. Pierce and Walker computed the percentage of the beam current to be found outside the nominal Brillouin radius.

Brewer compared his experimental beam profiles with the predictions of Pierce and Walker [8]. The temperature used was the cathode temperature multiplied by the area compression ratio of the gun. Brewer showed that this was reasonable by examining the radial spread of electrons passing through an aperture located on the axis of the beam. The observed temperature was within 20% of the calculated temperature. Brewer's measurement of the current profile for the beam edge compared favorably with the Pierce and Walker theory near the nominal beam radius. At larger radii, there was a low current-density tail on the current profile that was far from the prediction of the theory. This tail was thought to have resulted from the translaminar streams of electrons produced by gun aberrations.

Gilmour et al. [4, 5] and Brewer [6] showed that beams focused for near Brillouin flow conditions do behave very nearly as predicted by laminar flow theory. However, Ashkin [9] and Harker [10] reported that electron flow in beams, which had been thought to be laminar, was actually very nonlaminar. Following the works of Ashkin and Harker, Herrmann presented an optical theory for thermal velocity effects that discarded the assumption of laminarity [11].

Herrmann's theory treats beams as perfect optical systems. The thermal velocities of the electrons emitted in all directions by the cathode are assumed to have a Maxwellian distribution. Electron trajectories are assumed to obey the laws of geometrical optics of ideal optical systems. A comparison of electron trajectories in Herrmann's optical model with those for a laminar flow model is shown in Figure 7-20. For laminar flow, as discussed earlier in this chapter, the electrons move side by side without ever crossing each other's paths. For imbalanced space-charge and magnetic forces, the beam radius oscillates with a scallop wavelength, λ_s.

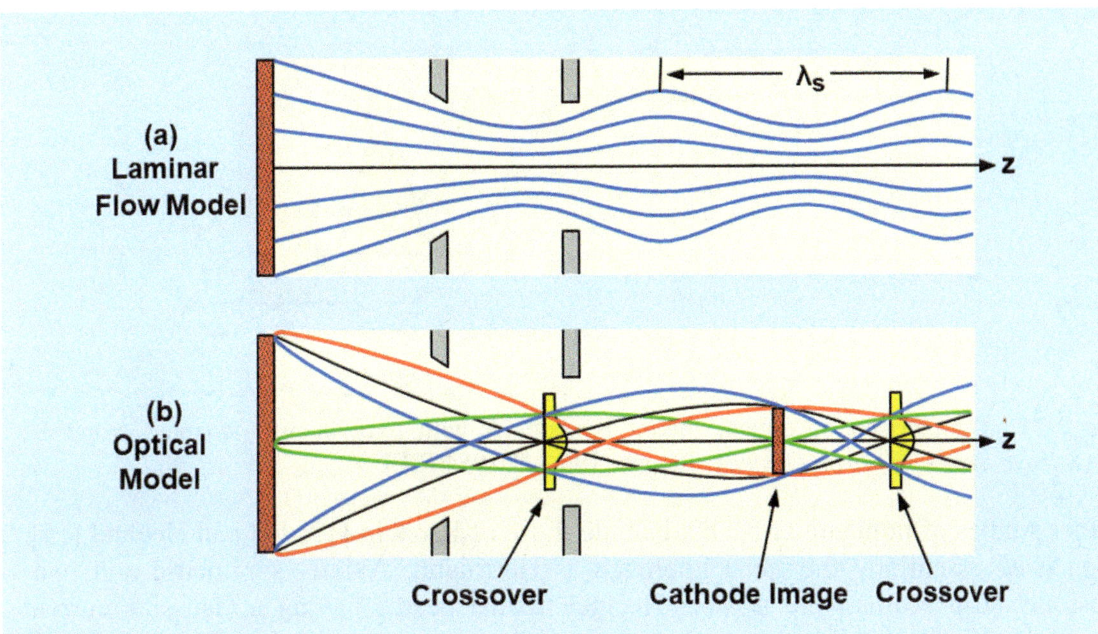

Figure 7-20 Comparison of (a) laminar and (b) optical models of electron flow. (Adapted from: Gabriel Herrmann, *Jour. App. Phys.*, February 1958.)

With the optical theory, electrons emitted from a point on the cathode focus periodically to form cathode images. As explained by Herrmann [11]: "One finds along the axis a succession of cathode images alternating with crossovers which impart to the beam a distinct structure which depends primarily on the internal motion of the electrons rather than on the shape of a boundary electron, as is the case in laminar theory."

Amboss compared the predictions of Herrmann with measurements on highly convergent, low-perveance (0.015×10^{-6}) beams focused by a flux density over an order of magnitude above the Brillouin value [12]. As shown in Figure 7-21, he found excellent agreement with the predictions of the optical theory. In this figure, λ_I is the distance between cathode images (the first and second images are shown). Note that the current profile changes from rectangular to Gaussian in the crossover region between images.

The Amboss results show the validity of the optical theory under conditions far from normal beam focusing conditions. For fields closer to the Brillouin level, Brewer found that the behavior of the translaminar electrons, which were produced by gun aberrations, was similar to that predicted by optical theory. In Gilmour's beam studies, the only evidence of optical-theory behavior was when an image of the heater for the cathode appeared in the electron beam. Magnetic flux lines generated by the heater current perturbed the flow of electrons from the cathode and apparently led to the optical behavior of the subsequent electron trajectories.

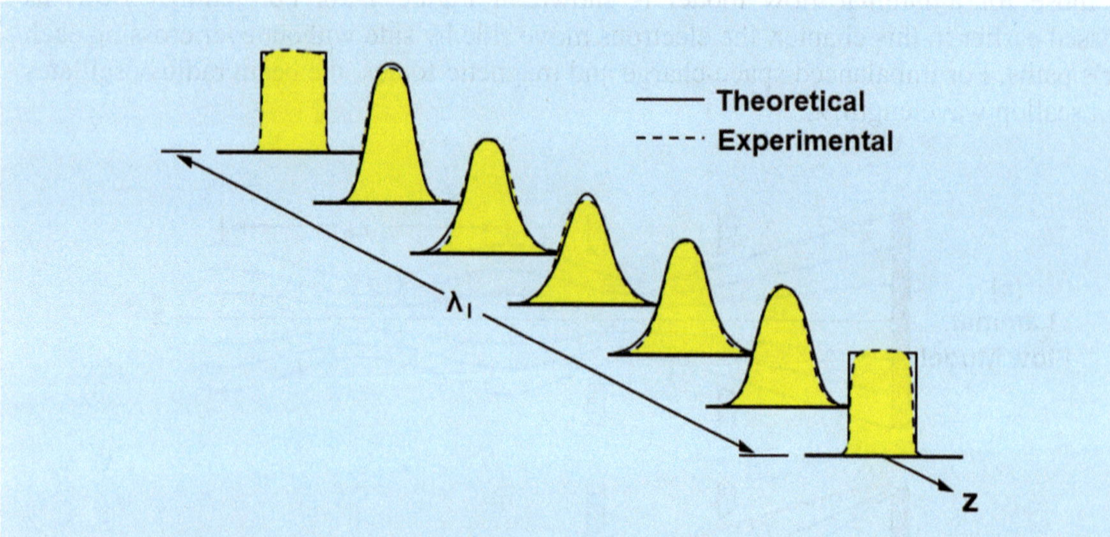

Figure 7-21 Comparison of optical theory predictions with experiment. (Adapted from: K. Amboss, *IEEE Trans. Electron Devices*, October 1964. © 1964 IEEE.)

Other studies of nonlaminar beams include those of Kirstein [13, 14] and Hechtel [15]. Kirstein used essentially the same approach as Herrmann. Axially symmetric and two-dimensional strip beams were treated. Hechtel treated beams having a Gaussian current distribution in the radial direction at the exit from the gun. The reason for considering this type of beam is that at millimeter-wave frequencies where beams are very small, thermal velocity effects in the gun result in a beam having a current distribution that is nearly Gaussian. Expressions are given for the magnetic field required to achieve various percentages of beam transmission through a tunnel.

7.3 FOCUSING WITH PERMANENT MAGNETS

7.3.1 Overview

A permanent magnet that produces a unidirectional field can be used in place of a solenoid to focus an electron beam. This eliminates the requirement for a power supply and ensures that the magnetic focusing field is always present. Disadvantages are size and weight. Also, in some applications, a disadvantage is that it is not possible to adjust the magnetic field to optimized tube performance.

As shown in Appendix C, the weight reduction of a periodic permanent magnet (PPM) focusing system, when compared with a solenoid or a unidirectional magnet, can be one to two orders of magnitude. A PPM focusing system is shown in Figure 7-22.

The basic principle of PPM focusing is straightforward. As a beam enters the field of a magnet section, the forces on the electrons act in exactly the same way as was described for Brillouin focusing. That is, the beam starts to rotate and the interaction of the rotational motion with the axial field produces a radial force that compresses the beam.

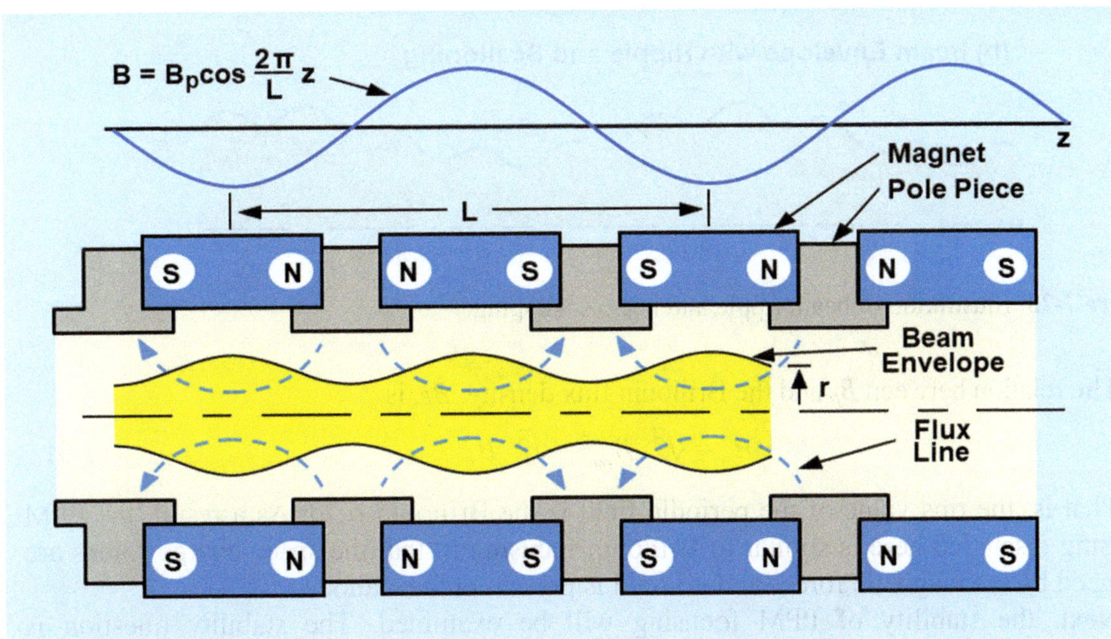

Figure 7-22 Periodic focusing of an electron beam with permanent magnets.

As the beam leaves the magnet section, rotation stops, focusing forces vanish, and the beam expands under the influence of space-charge forces. The beam then enters another magnet section with the field in the opposite direction from the previous one. Then the beam rotates in the opposite direction, but is focused just as it was in the previous section.

The overall result, as the beam traverses the alternating field of the PPM structure, is that its rotation oscillates back and forth, producing alternating periods of magnetic focusing and beam expansion and beam ripple. The difference between beam ripple, which results from the periodicity of focusing, and beam scalloping, which is the oscillation of a beam that is not in equilibrium, is indicated in Figure 7-23.

7.3.2 Laminar Flow, No Cathode Flux

The variation of the axially directed magnetic field along the axis of a PPM stack is very nearly sinusoidal; that is

$$B_z = B_p \cos \frac{2\pi z}{L} \tag{7-8}$$

where B_p is the peak field and L is the magnet period, or twice the distance between adjacent pole pieces as shown in Figure 7-22.

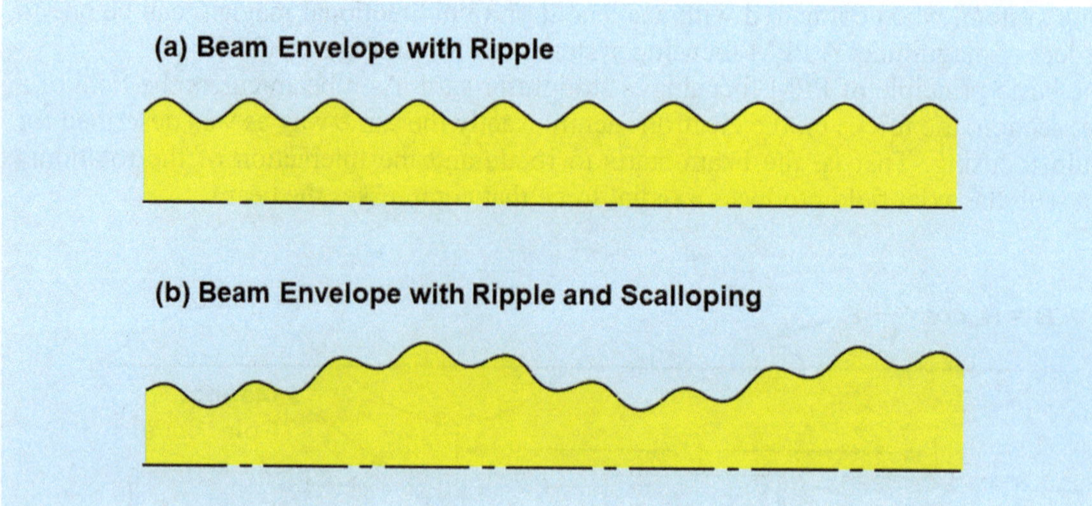

Figure 7-23 Illustration of beam ripple and beam scalloping.

The relation between B_p and the Brillouin flux density, B_B, is

$$B_p = \sqrt{2}\, B_{rms} = \sqrt{2}\, B_B \tag{7-9}$$

That is, the rms value of the periodic field is the Brillouin field. As a result, the PPM focusing discussed here is similar to Brillouin focusing in that the space-charge forces are balanced by the magnetic forces and no field is present at the cathode.

Next, the stability of PPM focusing will be examined. The stability question is important because there are ranges of values of flux density, magnet period, L, beam current, and beam voltage for which the beam diverges so that no beam current is transmitted through the PPM structure [16, 17].

Define α, the magnetic field coefficient, as

$$\alpha = \frac{1}{2}\left(\frac{\omega_{Lp}}{\omega}\right)^2 \tag{7-10}$$

and β, the space charge coefficient, as

$$\beta = \frac{1}{2}\left(\frac{\omega_p}{\omega}\right)^2 \tag{7-11}$$

where ω_{Lp} is the peak value of the Larmor frequency, $\eta B_p/2$, ω_p is the electron plasma frequency and

$$\omega = \frac{2\pi u_o}{L} \tag{7-12}$$

is the frequency at which the field appears to the moving electrons to alternate.

Typical beam envelope curves for three values of the magnetic field are shown in Figure 7-24. It was assumed that the electrons were injected into the magnetic field with no radial velocity at $z = 0$, which was where the magnetic field was a maximum. Both scalloping and ripple are present in Figure 7-24(a, c) when the magnetic field is not adjusted for optimum focusing.

Curves similar to those in Figure 7-24 were obtained by Mendel et al. for a wide range of values of β and from these curves the universal plot for optimum focusing given in Figure 7-25 was constructed [16]. Note that optimum focusing occurs when

$$\alpha = \beta \tag{7-13}$$

or when

$$\omega_{Lp} = \omega_p \tag{7-14}$$

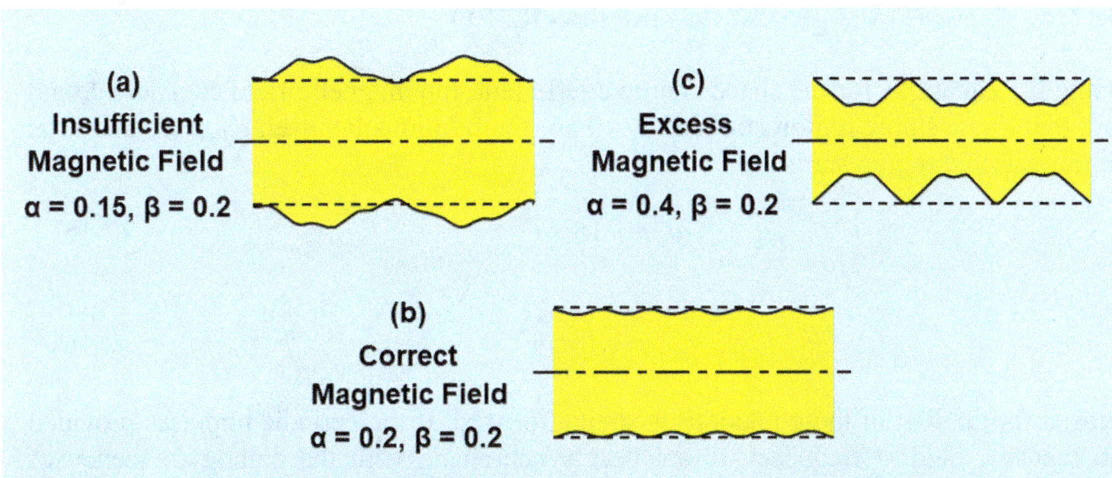

Figure 7-24 Analog computer plots of beam contours. (From: J. T. Mendel et al., *Proc. IRE*, May 1954. © 1954 IRE (now IEEE).)

When substitutions are made for rms flux density (B_B), charge density, and velocity, this relation is seen to be that for the equilibrium radius for Brillouin flow. This is not surprising since, on average, the space-charge forces are the same as for the Brillouin case and the rms field is the Brillouin field so the centrifugal and focusing forces are the same as for the Brillouin case.

The minimum ripple for optimum focusing is also shown in Figure 7-25. For values of β below about 0.35, the ripple amplitude is reasonably small. As β increases above 0.35, the ripple amplitude becomes very large and, in fact, for $\beta > 0.66$, the beam diverges.

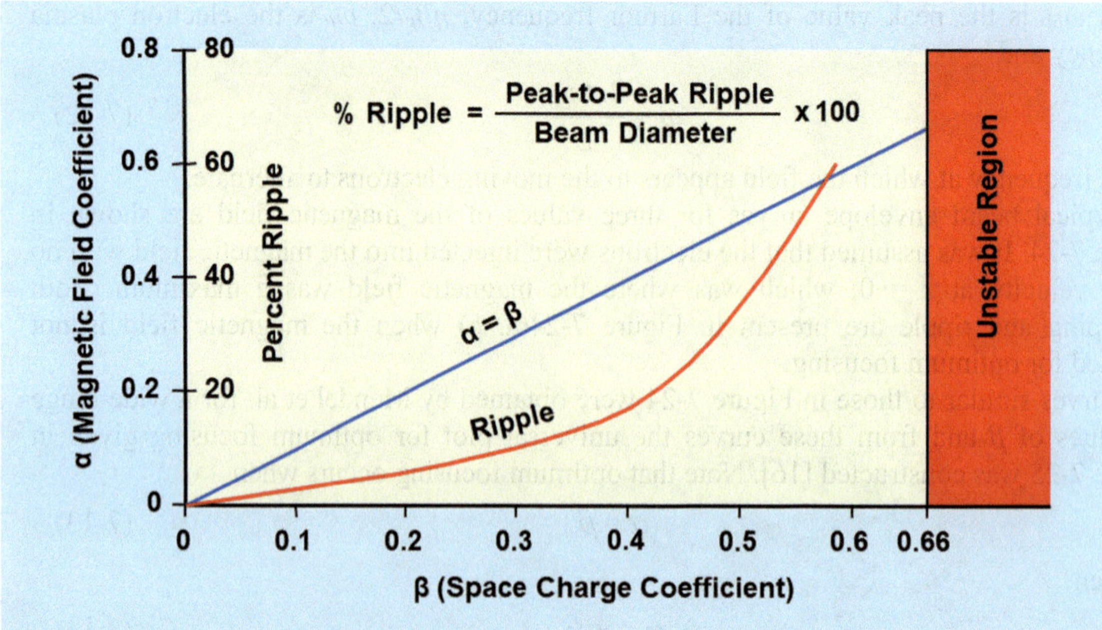

Figure 7-25 Optimum focusing conditions and minimum ripple amplitude. (Adapted from: J. T. Mendel et al., *Proc. IRE*, May 1954. © 1954 IRE (now IEEE).)

Using the equations for the space charge coefficient, the magnetic field coefficient, and the fact that the unstable region starts at $\alpha = 0.66$, the relations between ω_p, ω_{Lp}, and ω at which the unstable region starts are

$$\omega_p = 1.15\ \omega \tag{7-15}$$

and

$$\omega_{Lp} = 1.15\ \omega \tag{7-16}$$

The reason for the first of these relations is straightforward. If the periodic impulses provided by the focusing field at frequency ω are near synchronism with the natural frequency of oscillation of the electrons, ω_p, the beam diverges. The reason for the second relation can be understood by remembering that PPM magnets form a lens system that periodically focuses the electron beam as is indicated in Figure 7-26. The strength of each lens is proportional to B^2 and, therefore, to ω_{Lp}^2 and the focal length is inversely proportional to ω_{Lp}^2. Thus, if the lenses are too strong or are too far apart as indicated qualitatively by analogy to optical lenses in Figure 7-27, the beam diverges. In Figure 7-27(a), the lens strength and separation are such that the focal length, *f*, is one half of the lens separation. The ray, which is analogous to an electron trajectory, remains focused throughout the lens system. In Figure 7-27(b), the focal length is much shorter than in Figure 7-27(a) but the ray remains focused. In Figure 7-27(c), the focal length has been reduced slightly compared with Figure 7-27(b) and the ray diverges as it travels through the lens system.

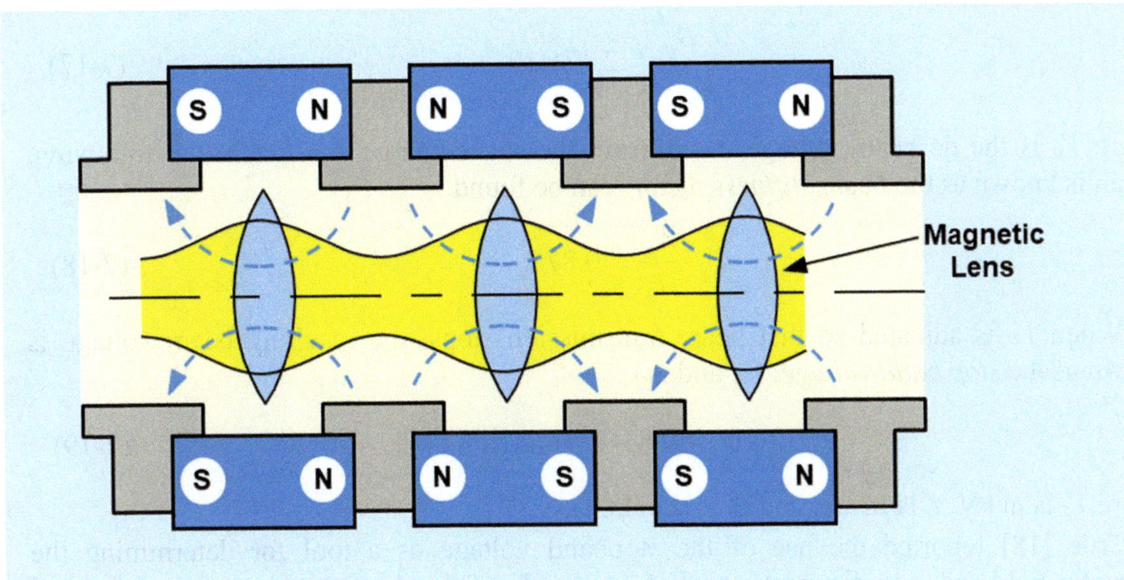

Figure 7-26 PPM focusing indicated by a series of convergent lenses.

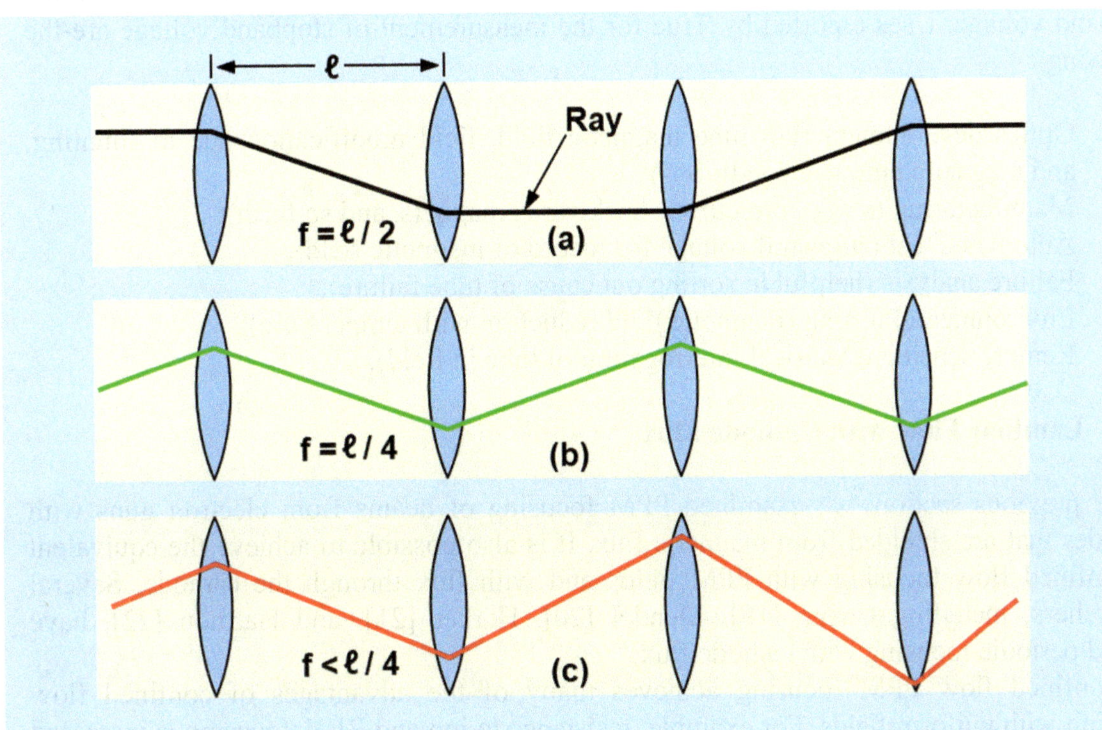

Figure 7-27 Examples of optical rays (analogous to electron trajectories) in a series of lenses. Focusing instability occurs for f < ℓ/4.

In practice, PPM structures are designed to operate well below the unstable region. As a result, when the parameters contained in ω_{Lp} and ω of the magnetic field coefficient are used, it is found that

$$\frac{L^2 B_p^2}{V_o} < 2.37\times10^{-9} \quad (7\text{-}17)$$

where V_o is the dc beam voltage. Also, from the space-charge coefficient the following, which is known as the *beam stiffness factor,* can be found

$$\frac{\lambda_p}{L} > 0.87 \quad (7\text{-}18)$$

When V_o is adjusted so that beam transmission stops, the resulting beam voltage is known as the *stop band voltage*, V_s, and is

$$V_s = 0.844L^2 (B_{mrs})^2 \quad (7\text{-}19)$$

where V_s is in kV, L is in cm, and B_{rms} is in kGauss.

True [18] reported the use of the stopband voltage as a tool for determining the magnetic field and as a diagnostic tool. An example of the determination of the stopband voltage is shown in Figure 7-28. Beam transmission is plotted as a function of beam voltage with perveance held constant. True reported that the magnetic field strength measured using a gaussmeter was in excellent agreement with the field calculated from the stopband voltage. Uses reported by True for the measurement of stopband voltage are the following:

- Optics development (hot tube magnetic field, field modification due to shunting, and magnetic entrance conditions);
- Manufacturing (stack correctness, backwards magnets, and so forth);
- Automated test (stopband voltage for record of magnetic field);
- Failure analysis (helpful in sorting out cause of tube failure);
- Environmental testing (magnetic field reduction with temperature);
- Remote sensing (electrical interrogation of tube in field).

7.3.3 Laminar Flow with Cathode Flux

In the previous section, we examined PPM focusing of beams from electron guns with cathodes that are shielded from magnetic flux. It is also possible to achieve the equivalent of confined flow focusing with PPM fields and with flux through the cathode. Several researchers, including Chang [19], Mendel [20], Harker [21], and Harmon [22], have treated periodic focusing with cathode flux.

Confined flow PPM focusing achieves many of the advantages of confined flow focusing with uniform fields. For example, resistance to ion and RF defocusing is increased and the control of beams from gridded guns is improved. A deficiency in the PPM case that is not present in uniform-field focusing is that the rotational energy of the beam increases as the beam becomes more confined. As a result, the energy in the axial motion of the beam and the efficiency decrease.

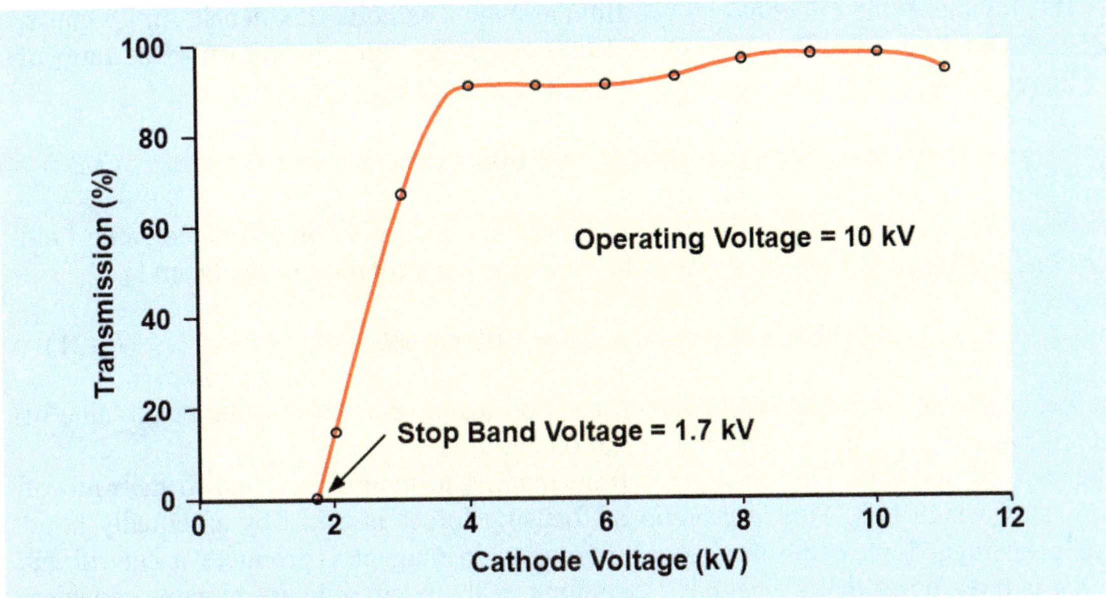

Figure 7-28 Beam transmission as a function of cathode voltage in a PPM-focused TWT with a shadow-gridded gun. (From: R. B. True, 1988 Microwave Power Tube Conference.)

The reason for the increase of rotational energy with the cathode magnetic field can be understood with the aid of Figure 7-29. Consider first the rotation of the beam within the first magnet. From Busch's theorem, we know that the rate of rotation is determined by the *change* in flux density encountered by the beam as it enters the first magnet. However, with cathode flux as shown in Figure 7-29 in the opposite direction to that in the first magnet, the *change* in flux density is larger than it is with no flux through the cathode. Thus, the rate of rotation of the beam in the first magnet is larger than it is with no flux through the cathode.

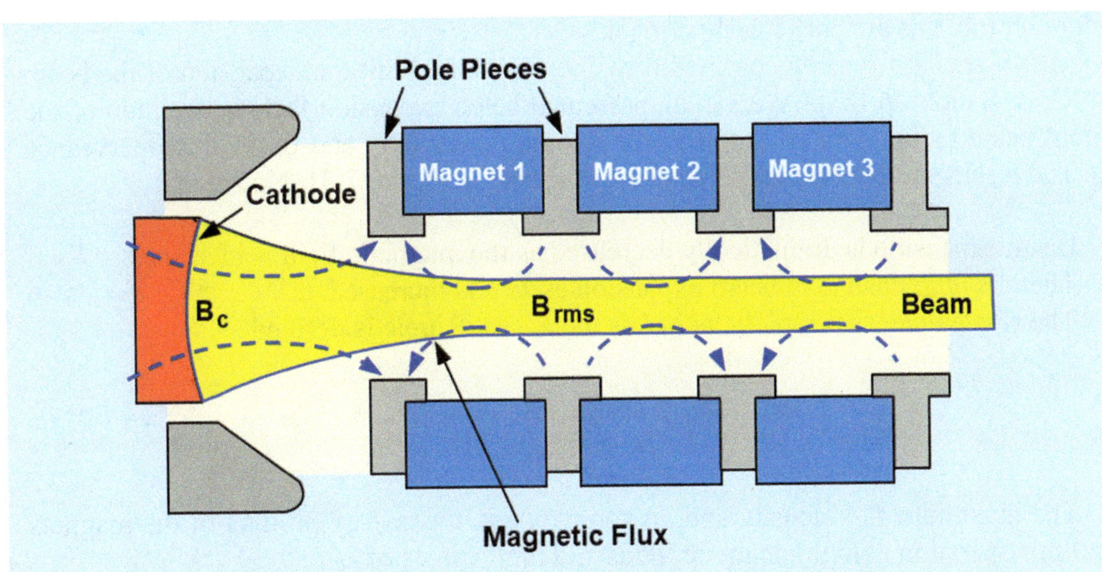

Figure 7-29 Magnetic field configuration for confined flow PPM focusing.

As an example of the effect of cathode flux, assume a cathode flux density of 25 gauss, a peak flux density in the first magnet of 1,000 gauss, and an area convergence ratio of 20:1. Then the peak rate of beam rotation in the first magnet is

$$\dot{\theta} \propto B_p + 20 \times B_c = 1{,}500 \text{ gauss} \tag{7-20}$$

As the beam moves from the first magnet to the second, the direction of the magnetic field reverses and is aligned with the cathode field. Then the peak rotation of the beam is

$$\dot{\theta} \propto 1000 - 20 \times 25 = 500 \text{ gauss} \tag{7-21}$$

As the beam moves on to the third magnet, rotation again reverses and the peak value of the rate of rotation is again proportional to 1,500 gauss.

The ratio of magnetic focusing forces from magnet to magnet is equal to the ratio of rotation rates, which is 3. This large ratio of focusing forces is offset by an equally large ratio of centrifugal forces (the large rate of rotation in magnet 1 produces a centrifugal force that is three times that in magnet 2). As long as the beam velocity remains constant, the ripple amplitude from magnet to magnet remains constant.

It should be observed that because there is more energy in rotation in magnet 1 than in magnet 2, the axial velocity of the beam in magnet 1 is lower than in magnet 2. As a result, the beam density in magnet 1 is larger than in magnet 2 and this produces a fluctuation in the radius of the beam that is similar to beam scalloping.

Figure 7-30 shows the variation of peak rotational beam energy with magnetic flux density and with microperveance (perveance $\times 10^6$) [23]. Note that for reasonable values of confinement, a large fraction of the beam energy is in rotation for microperveances above ~0.5. As a result, efficiency is degraded. Thus, confined flow PPM focusing is limited to low-perveance designs. Hence, it is particularly suitable for millimeter-wave tubes for which perveance is usually limited by cathode loading considerations and in which beam interception problems are particularly critical.

A primary reason for using confined flow focusing is to limit the expansion of the beam during RF operation. In [23], Ayers et al. presented beam expansion factors (the ratio of the maximum beam radius with RF drive to the dc radius) for 0.45 and 0.036 microperveance TWTs. The results for 0.036 microperveance are shown in Figure 7-31. Notice that

1. Beam expansion is dramatically decreased as the magnetic field is increased.
2. There is little change in beam expansion as λ_p/L is increased.
3. The ripple decreases rapidly as λ_p/L is increased. Ripple is defined as

$$\%\,\text{ripple} = \frac{r_{\max} - r_{\min}}{r_{\max} + r_{\min}} \times 100 \tag{7-22}$$

1. The maximum flux density and, to some extent, the energy product of the magnets limit operation at high magnetic fields and high values of λ_p/L.

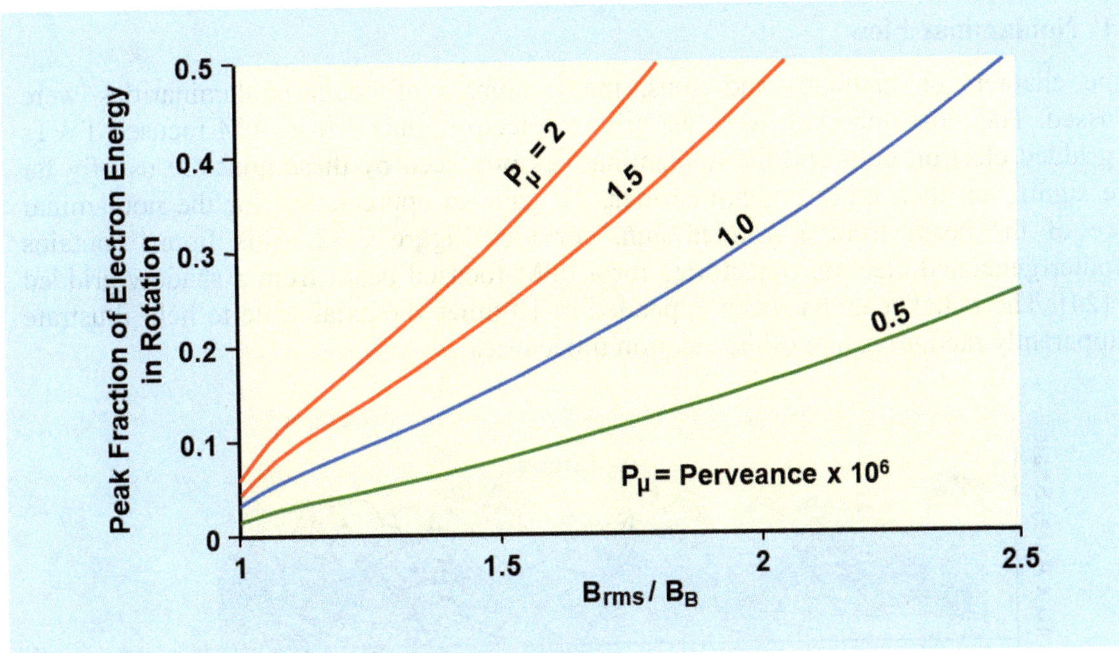

Figure 7-30 Dependence of peak transverse energy on magnetic field and perveance. (From: W. R. Ayers et al., *Technical Digest*, 1985 IEDM. © 1985 IEEE.)

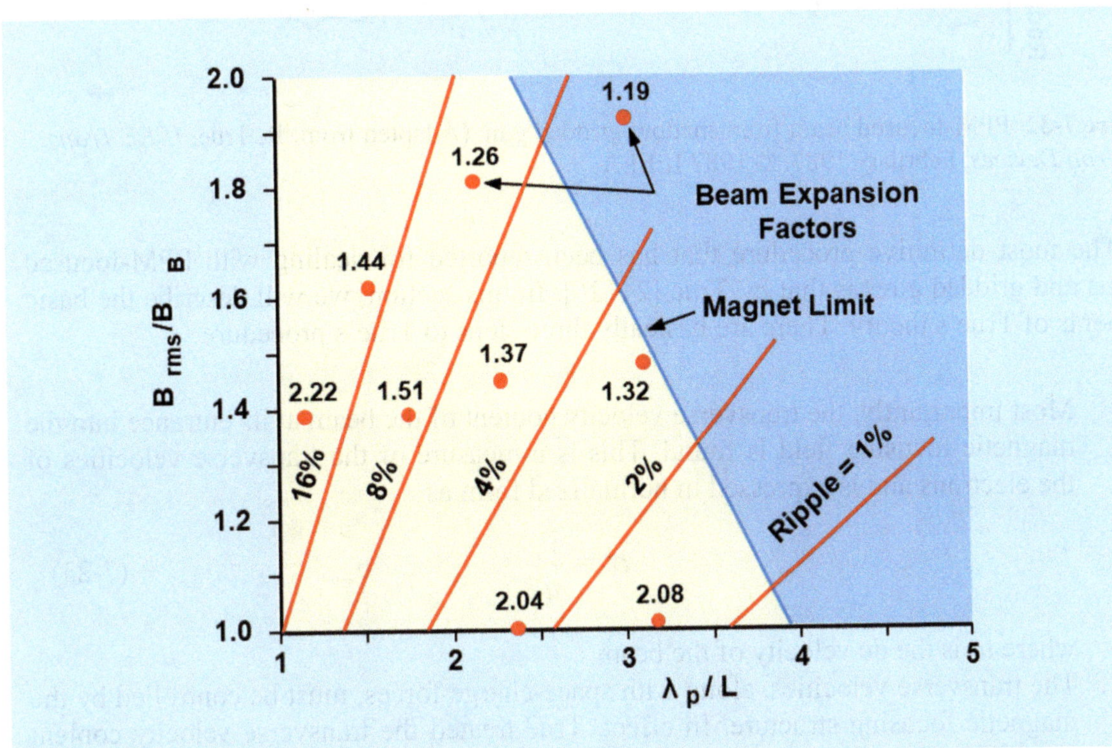

Figure 7-31 Beam expansion factors for 0.036 microperveance TWT. (From: W. R. Ayers et al., *Technical Digest*, 1985 IEDM. © 1985 IEEE.)

7.3.4 Nonlaminar Flow

In the chapters on cathodes and guns, many sources of beam nonlaminarities were discussed. The most important were the grids in electron guns. Most PPM-focused TWTs use gridded electron guns and the nonlaminarities produced by these guns are usually far more significant than other nonlaminarities. To gain an appreciation for the nonlaminar nature of the beam from a gridded gun, consider Figure 7-32. This figure contains computer-generated electron trajectories for a PPM-focused beam from a shadow-gridded gun [24]. The radial scale has been expanded to 10 times the axial scale to help illustrate the apparently random nature of the electron trajectories.

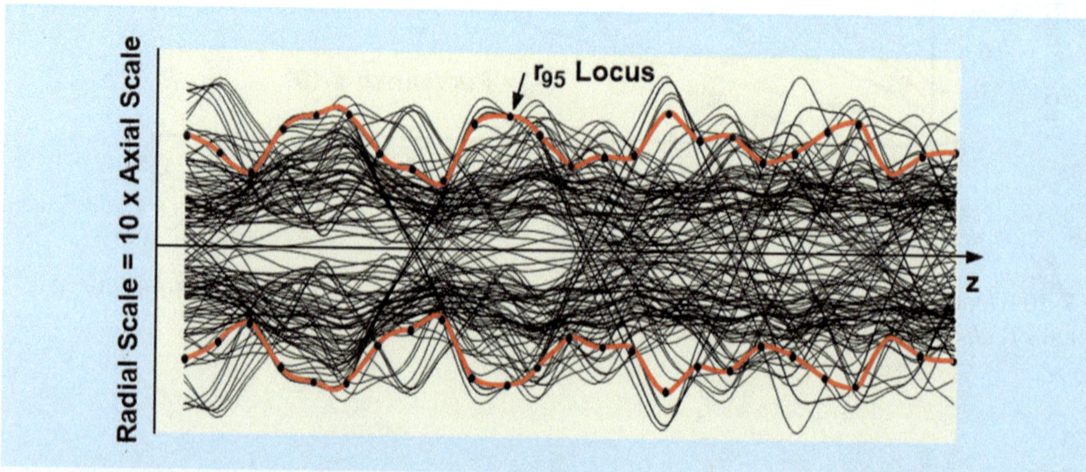

Figure 7-32 PPM-focused beam from shadow-gridded gun. (Adapted from: R. True, *IEEE Trans. Electron Devices*, February 1987. © 1987 IEEE.)

The most definitive procedure that has been reported for dealing with PPM-focused beams and gridded guns is that by True [24–26]. In this section, we will describe the basic elements of True's theory. There are basically three steps to True's procedure:

1. Most importantly, the transverse velocity content of the beam at its entrance into the magnetic focusing field is found. This is a measure of the transverse velocities of the electrons and is expressed in normalized form as

$$\sigma = \frac{\overline{v}_r}{u_o} \tag{7-23}$$

where u_o is the dc velocity of the beam.

2. The transverse velocities, along with space-charge forces, must be controlled by the magnetic focusing structure. In effect, True treated the transverse velocity content as an additional force (in addition to the space-charge force) that must be focused. This is done by finding an effective microperveance, P_μ^+, for the beam (microperveance equals perveance multiplied by 10^6), which is the sum of the regular beam microperveance, P_μ, and a scattering microperveance, $P_{\mu s}$, that results from the transverse velocity content. That is,

$$P_\mu^+ = P_\mu + P_{\mu s} \tag{7-24}$$

3. In the absence of magnetic flux through the cathode, the field level required to focus an ideal beam is the Brillouin field, and this is proportional to the square root of perveance. True assumed that the increased focusing field, B^+, made necessary by transverse electron velocities should be at least

$$\frac{B^+}{B_B} = \left(\frac{P_\mu^+}{P_\mu}\right)^{1/2} \tag{7-25}$$

The ratio B^+/B_B is referred to as the magnetic field intensification ratio. True finds that the perveance ratio, P_μ^+/P_μ, must be multiplied by a factor of 1.65 to minimize the beam ripple in actual magnet designs.

The normalized transverse velocity content of the beam, σ, is composed of two parts. The first part, σ_T, results from the thermal velocity content of the beam, given by

$$\sigma_T = \frac{r_c}{r_{95}}\left(\frac{k\,T}{2\,e\,V_o}\right)^{1/2} \tag{7-26}$$

where r_c is the disc radius of the cathode and r_{95} is the beam radius containing 95% of the beam current. The other part of σ is the nonthermal part, σ_{NT}, which results from gun aberrations and scattering by grid wires. This part can be found by a computer simulation that includes grid wire details. True found that a good approximation for the combined thermal and nonthermal parts is

$$\sigma = \left(\sigma_{NT}^2 + \sigma_T^2\right)^{1/2} \tag{7-27}$$

For gridded guns, $\sigma_{NT} >> \sigma_T$ so $\sigma \approx \sigma_{NT}$. In the remainder of this section, we will treat only gridded guns and we will ignore thermal effects. We will also substitute σ for σ_{NT}.

The value for σ is found at or near the beam minimum from the weighted standard deviation of the slopes of the electron trajectories. Although σ is defined in terms of slopes, it is essentially equal to v_r/u_o. True found, as shown in Figure 7-33, that σ can be approximated by

$$\sigma = 0.0015\;\theta \tag{7-28}$$

where θ is the half angle of the electron gun. Note that guns that have small half angles also have low values of σ. The reason was given in Chapter 6 when an electron crossing a convergent electron stream was considered. The transverse velocity increase was found to be inversely proportional to the radius compression ratio.

The next step in the analysis is to incorporate the effect of transverse velocities on the spreading of the electron beam. As pointed out earlier in this section, True treated the transverse velocity content as a force that, along with the space-charge force, must be focused by the magnetic field. This is done by finding an effective microperveance, which is $P_\mu^+ = P_\mu + P_{\mu s}$. The scattering microperveance, $P_{\mu s}$, is found to be related to σ by

$$P_{\mu s} = 197.6\,\sigma^2 \qquad (7\text{-}29)$$

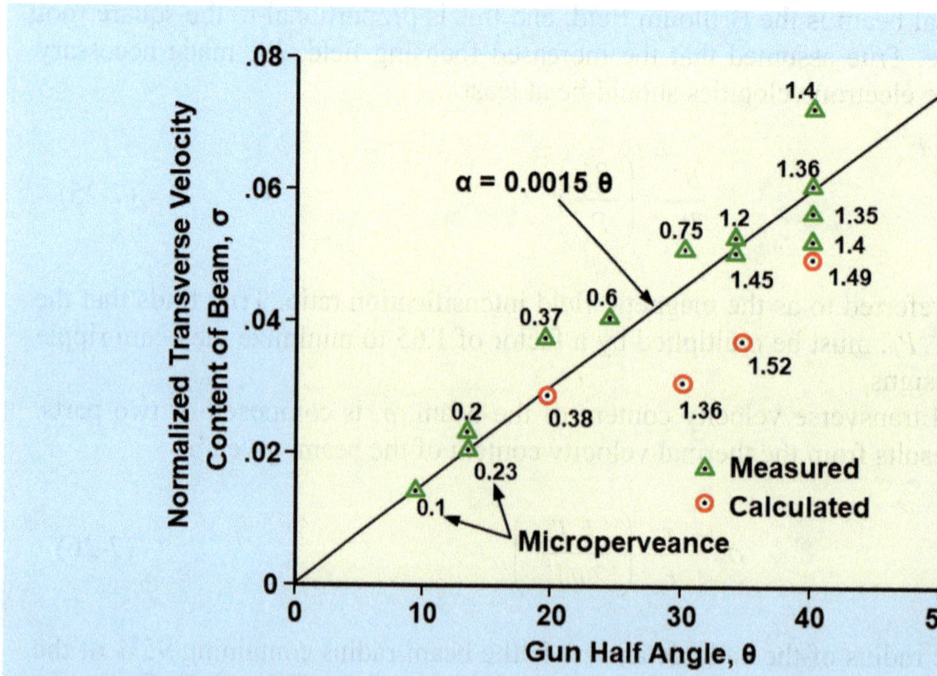

Figure 7-33 Justification for $\sigma = 0.0015\theta$. (From: R. B. True, *IEEE Trans on Electron Devices*, March 1984. © 1984 IEEE)

This relation is derived by considering electron behavior and the rate of beam spread near the beam minimum. Thus, the effective microperveance is

$$P_\mu^+ = P_\mu + 197.6\sigma^2 \qquad (7\text{-}30)$$

factor, B^+/B_B. The magnetic field is proportional to the square root of perveance. Also, allowance must be made for effects such as spherical aberration, nonideal entrance conditions, and gun-design imperfections, so True assumed that the intensification factor should take the form

$$\frac{B^+}{B_B} = \left(F\frac{P_\mu^+}{P_\mu}\right)^{1/2} \qquad (7\text{-}31)$$

where a value for F of 1.65 is chosen on the basis of numerical and experimental data. Now, substituting the relation for effective microperveance, the intensification factor becomes

$$\frac{B^+}{B_B} = \left[F\left(1 + \frac{197.6\sigma^2}{P_\mu}\right)\right]^{1/2} \qquad (7\text{-}32)$$

To make use of this equation, the relation between σ and P_μ is needed. Equation (7-28) was an empirical relation between σ and the gun half angle, θ. The exact relation between perveance and θ was discussed in Chapter 6. For our purposes here, a much simpler equation developed by True can be used. That is,

$$P_\mu = 0.006\,\theta^2 \frac{r_{95}}{r_c} \tag{7-33}$$

By substituting this equation and that for σ into the intensification factor,

$$\frac{B^+}{B_B} = \left[1.65\left(1 + 0.0741\,\frac{r_c}{r_{95}}\right)\right]^{1/2} \tag{7-34}$$

This result is plotted in Figure 7-34. It is interesting that the intensification factor is a function of only the radius compression ratio. We now see why beams from high convergence guns require field levels that are significantly higher above the Brillouin level than do beams from low convergence guns.

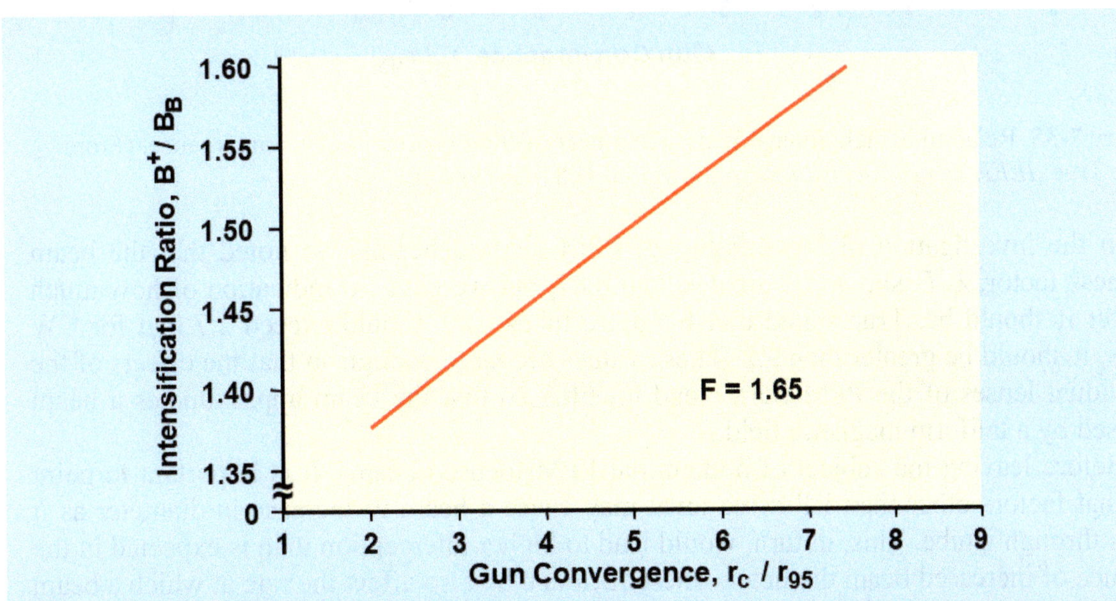

Figure 7-34 PPM field intensification ratio versus gun convergence. (From: R. True, *IEEE Trans. Electron Devices*, March 1984. © 1984 IEEE.)

The magnetic field intensification factor may be combined with r_{95}/a, where a is the tunnel radius (so that r_{95}/a is the fraction of the tunnel filled by the beam) to obtain a normalized focusing factor, denoted by True as FOC F_1 and given by

$$\text{FOC } F_1 = \frac{B^+ / B_B}{r_{95} / a} \tag{7-35}$$

Acceptable limits for this parameter for good tube design are shown in Figure 7-35.

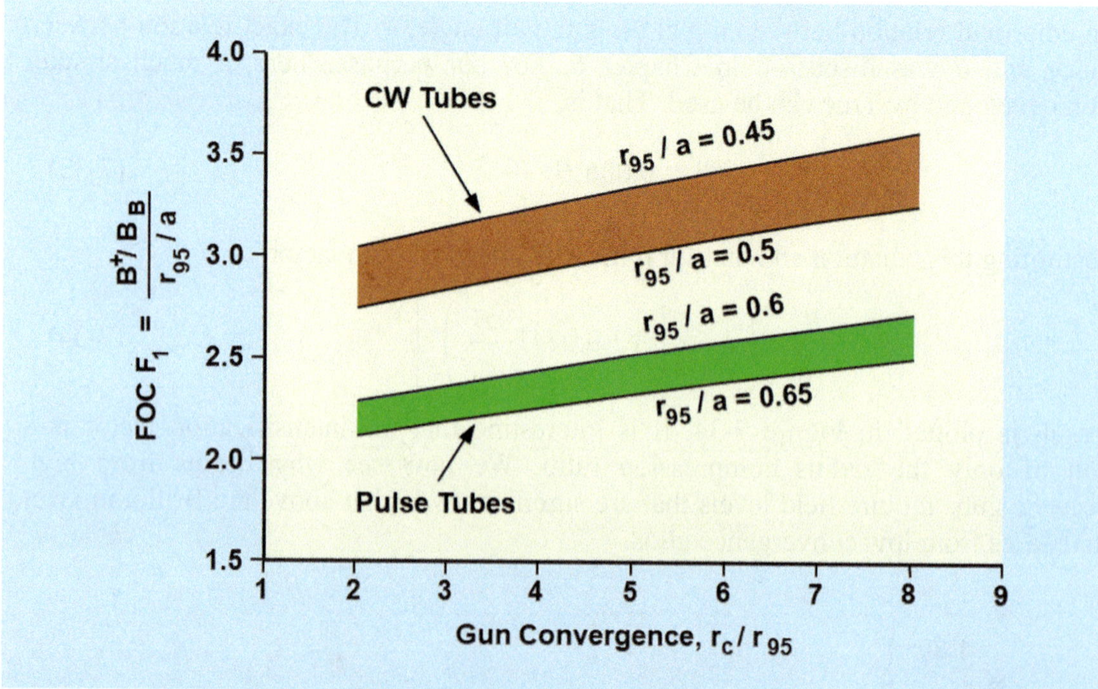

Figure 7-35 Relation of field intensification and beam-filling factors to gun convergence. (From: R. B. True, *IEEE Trans. Electron Devices*, March 1984. © 1984 IEEE.)

In the investigation of the stability of PPM-focused beams, we noted that the beam stiffness factor, λ_p/L, should be greater than 0.87, but we gave no indication of how much greater it should be. True stated that for pulse tubes, λ_p/L should exceed 2.7 and for CW tubes, it should be greater than 3.2. These values are large enough so that the effects of the individual lenses of the PPM field blend together so that the beam approximates a beam focused by a uniform magnetic field.

Before leaving the subject of nonlaminar PPM-focused beams, it is important to point out that factors other than RF drive level may cause a beam to increase in diameter as it drifts through a tube. This, in turn, would lead to higher interception than is expected in the absence of increased beam diameter. These factors may also affect the rate at which a beam expands as it enters the collector and this, in turn, affects the collector design. To deal with these phenomena, True [26] defined the term *tunnel emittance*, which is given by

$$\text{Tunnel emittance} = \sigma \frac{r_{95}}{a} \qquad (7\text{-}36)$$

Figure 7-36 shows the growth of tunnel emittance with distance. This growth is due partly to an increase in normalized transverse velocity (σ) and partly to a small increase in beam size.

One important factor that affects tunnel emittance growth is the current profile of the beam leaving the electron gun. The ideal beam shape is shown in Figure 7-37(a). This profile is shown slightly dished to match the radial variation in the magnetic field. The profiles shown in Figure 7-37(b, c) are less desirable shapes because of the concentration of charge in the center of the Gaussian distribution or in the ears of the "rabbit ears" of the hollow distribution. These charge concentrations contain potential energy that is converted to increased transverse electron velocities as the beam flows through the tube.

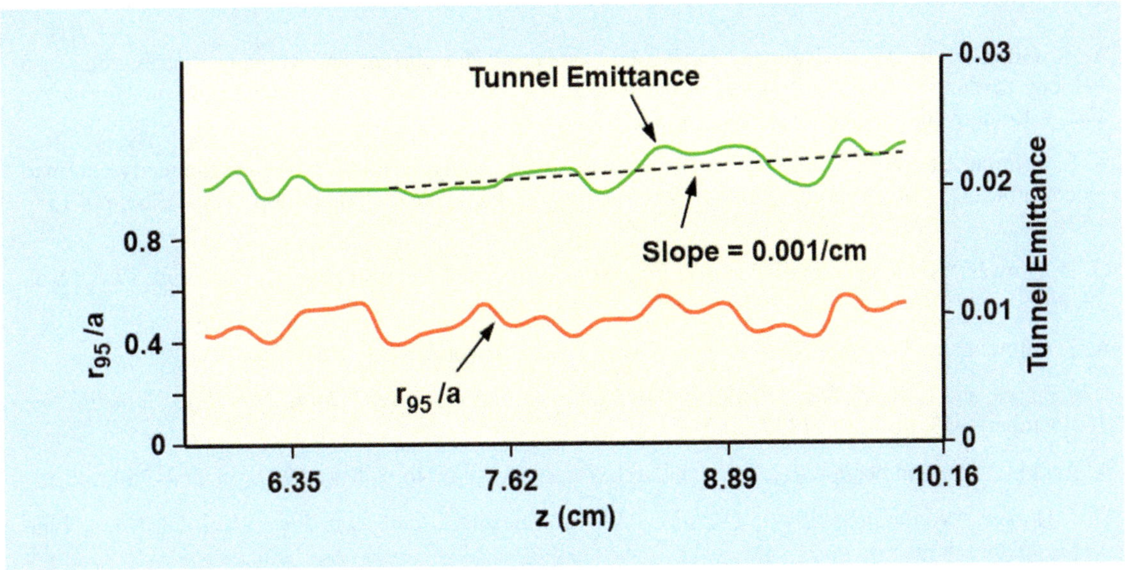

Figure 7-36 Growth of tunnel emittance with distance. (From: R. B. True, *Technical Digest*, 1985 IEDM. © 1985 IEEE.)

Another factor that affects tunnel emittance growth is the entrance point of the beam into the magnetic field. True considers the beam minimum position (the beam waist) to be the point where tunnel emittance is at a minimum. The axial position of the waist should be at the point where the first magnetic field peak has risen to $1/\sqrt{2}$ of its maximum value.

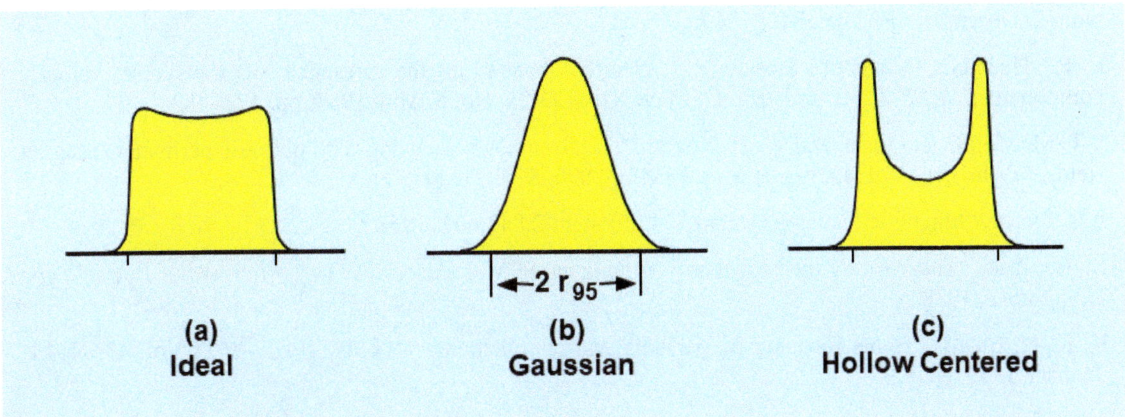

Figure 7-37 Current density versus radius at beam minimum. (From: R. B. True, *IEEE Trans. Electron Devices*, February 1987. © 1987 IEEE.)

REFERENCES

[1] M. Chodorow, and C. Susskind, *Fundamentals of Microwave Electronics*, New York: McGraw-Hill, 1964.

[2] J. F. Gittins, *Power Travelling Wave Tubes*, New York: American Elsevier, 1965.

[3] W. J. Kleen, *Electronics of Microwave Tubes*, New York: Academic Press, 1958.

[4] A. S. Gilmour, Jr., G. C. Dalman, and L. F. Eastman, "The current and velocity distributions in a velocity modulated Brillouin focused electron beam," *Proc. 4th International Congress on Microwave Tubes*, Eindhoven, the Netherlands: Centrex, 1963.

[5] A. S. Gilmour, Jr., and D. Hallock, "A demountable beam analyzer for studying magnetically confined electron beams," *Advances in Electron Tube Techniques*, New York: Pergammon Press, 1963, pp. 153–157.

[6] G. R. Brewer, "Some characteristics of a magnetically focused electron beam," *Jour. App. Phys.*, Vol. 30, No. 7, July 1959, pp. 1022–1038.

[7] A. S. Gilmour, Jr., *Principles of Traveling Wave Tubes*, Norwood, MA: Artech House, 1994.

[8] J. R. Pierce, and L. R. Walker, "Brillouin flow" with thermal velocities," *Jour. App. Phys.*, Vol. 24, No. 10, October 1953, pp. 1328–1330.

[9] A. Ashkin, "Electron beam analyzer," *Jour. App. Phys.*, Vol. 28, No. 5, May 1957, pp. 564–569.

[10] K. J. Harker, "Nonlaminar flow in cylindrical electron beams," *Jour. App. Phys.*, Vol. 28, No. 6, June 1957, pp. 645–650.

[11] G. Herrmann, "Optical theory of thermal velocity effects in cylindrical electron beams," *Jour. App. Phys.*, Vol. 29, No. 2, February 1958, pp. 127–136.

[12] K. Amboss, "Verification and use of Herrmann's optical theory of thermal velocity effects in electron beams in the low perveance regime," *IEEE Trans. Electron Devices*, Vol. 11, No. 10, October 1964, pp. 479–485.

[13] P. T. Kirstein, "On the effects of thermal velocities in two-dimensional and axially symmetric beams," *IEEE Trans. Electron Devices*, Vol. 10, No. 2, March 1963, pp. 69–80.

[14] P. T. Kirstein, "Thermal velocity effects in axially symmetric solid beams," *Jour. App. Phys.*, Vol. 34, No. 12, December 1963, pp. 3479–3490.

[15] J. R. Hechtel, "Magnetic focusing of electron beams in the presence of transverse velocity components," *IEEE Trans. Electron Devices,* Vol. ED-28, No. 5, May 1981, pp. 473–482.

[16] J. T. Mendel, C. F. Quate, and W. H. Yocom, "Electron beam focusing with periodic permanent magnet fields," *Proc. IRE*, Vol. 42, No. 5, May 1954 pp. 800–810.

[17] J. R. Pierce, "Magnetic focusing system," U.S. Patent 2,847,607, 1958.

[18] R. B. True, "The role of the stop band voltage in PPM focused TWTs," *Microwave Power Tube Conference*, 1988.

[19] K. K. N. Chang, "Beam focusing by periodic and complementary fields," *Proc. IRE*, Vol. 43, No. 1, January 1955, pp. 62–71.

[20] J. T. Mendel, "Magnetic focusing of electron beams," *Proc. IRE*, Vol. 43, No. 3, March 1955, pp. 800-810.

[21] K. J. Harker, "Periodic focusing of beams from partially shielded cathodes," *IEEE Trans. Electron Devices*, Vol. 2, No. 4, October 1955, pp. 13–19.

[22] W. A. Harmon, *PPM Focusing Concepts*, Hughes Electron Dynamics Division Technical Report No. 70, 1987.

[23] W. R. Ayers, et al., "RF beam spread in PPM confined-flow TWTs," *Technical Digest,* IEDM, 1985, pp. 357–360.

[24] R. True, "A theory for coupling gridded gun design with PPM focusing," *IEEE Trans. Electron Devices*, Vol. ED-31, No. 3, March 1984, pp. 353–362.

[25] R. True, "Emittance and the design of beam formation, transport, and collection systems in periodically focused TWTs," *IEEE Trans. Electron Devices*, Vol. ED-34, No. 2, February 1987, pp. 473–485.

[26] R. True, "Tunnel emittance growth in PPM focused TWTs," *Technical Digest,* IEDM, 1985, pp. 192–195.

Chapter 8

Beam Modulation in Linear Beam Tubes

In all linear beam microwave tubes, two interactions of the electron beam and the RF circuit are responsible for the operation of the tube. These are indicated in Figure 8-1 and are:

1. A signal is placed on the electron beam by RF modulating the velocities of the electrons (alternately accelerating and decelerating electrons).
2. As accelerated electrons overtake decelerated electrons, the beam becomes RF density modulated and can induce RF current in nearby metallic surfaces. This current, in turn, interacts with the impedance of the circuit to produce a voltage that enhances the modulation of the beam until RF power is removed.

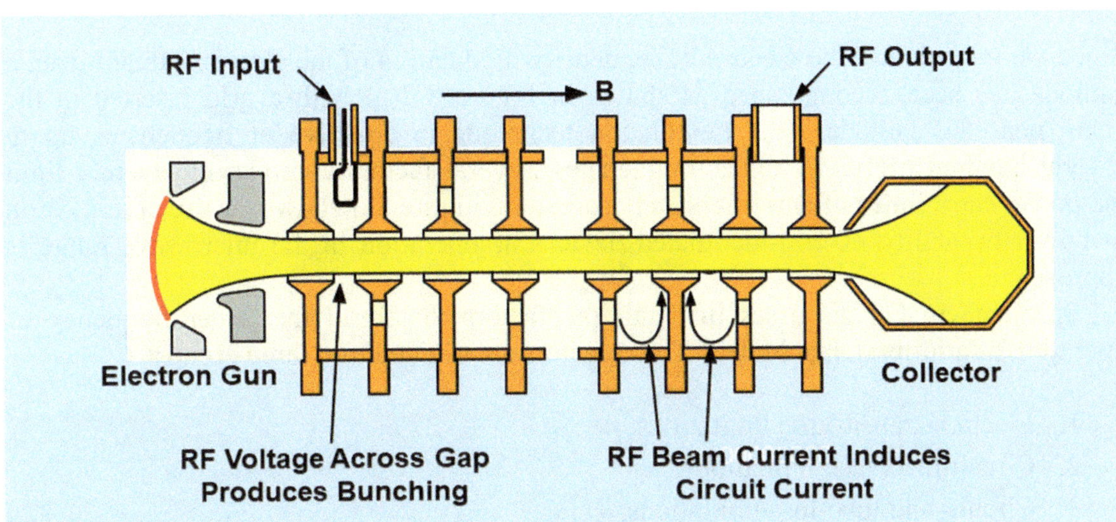

Figure 8-1 Illustration of locations where gap interactions take place in a coupled-cavity TWT.

8.1 BEAM MODULATION

There are two ways to modulate electron flow from a cathode. These are indicated in Figure 8-2 and are density modulation and velocity modulation. Density modulation is accomplished at low electron velocities by controlling the emission from a cathode. Velocity modulation is accomplished at high electron velocities by passing the electrons through an RF electric field that modulates the velocities of electrons.

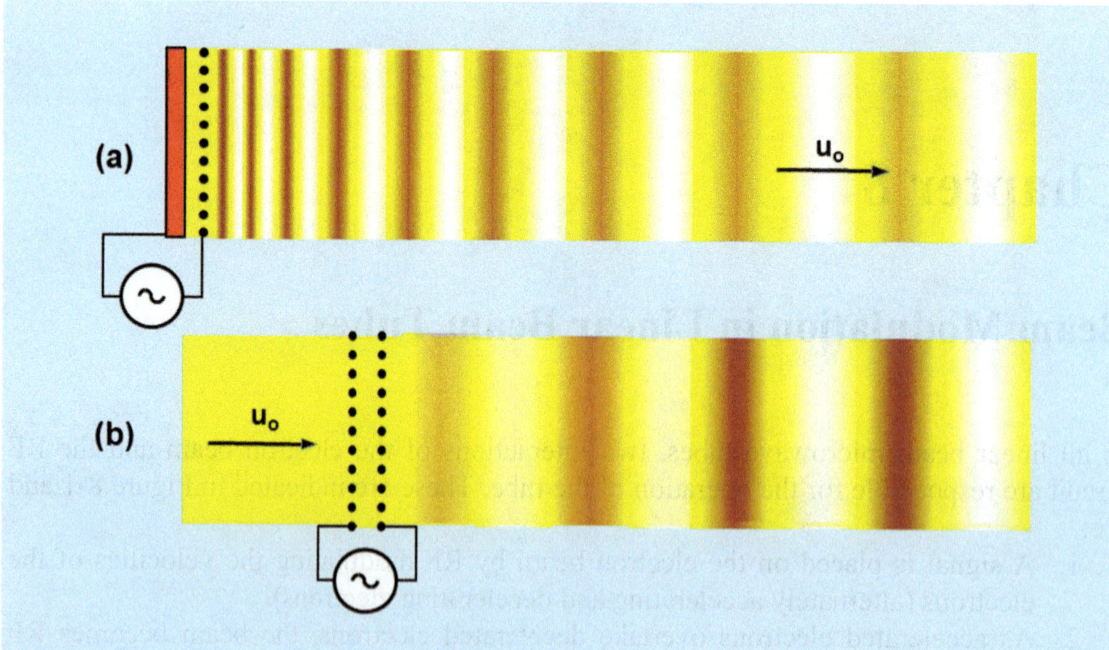

Figure 8-2 Density modulation (a) and velocity modulation (b).

8.1.1 Density Modulation

Since the invention of the electron tube, density modulation of the electron stream from a cathode has been accomplished as shown in Figure 8-2(a) with a grid inserted in the beam near the cathode. This modulation technique is effective at frequencies up to several hundred MHz. At higher frequencies, several factors combine to severely limit the power capabilities of amplifiers and oscillators. Figure 8-3 shows the precipitous drop in power capability of grid-modulated devices as operation in the microwave range is approached.

Spangenberg [1] discusses in detail the factors that limit the upper frequency of operation of grid-modulated tubes. These factors are divided into three groups:

1. Electron-transit-time limitations;
2. Circuit-reactance limitations;
3. Circuit- and tube-loss limitations.

The electron-transit-time limitation results from the time required for electrons to travel from the cathode to the grid. This is because the grid voltage changes polarity in the time that it takes for the electrons to travel from the cathode to the grid so the electrons are returned to the cathode. This problem is aggravated by the space-charge cloud near the cathode surface, which reduces electron velocities in the grid-cathode region.

The transit time problem can be alleviated by moving the grid closer to the cathode but the dimensions become smaller and smaller as frequency is increased so that the power capability vanishes. For example, at 3 GHz, the grid-cathode spacing in a 1-W triode is ~ 0.001 inch.

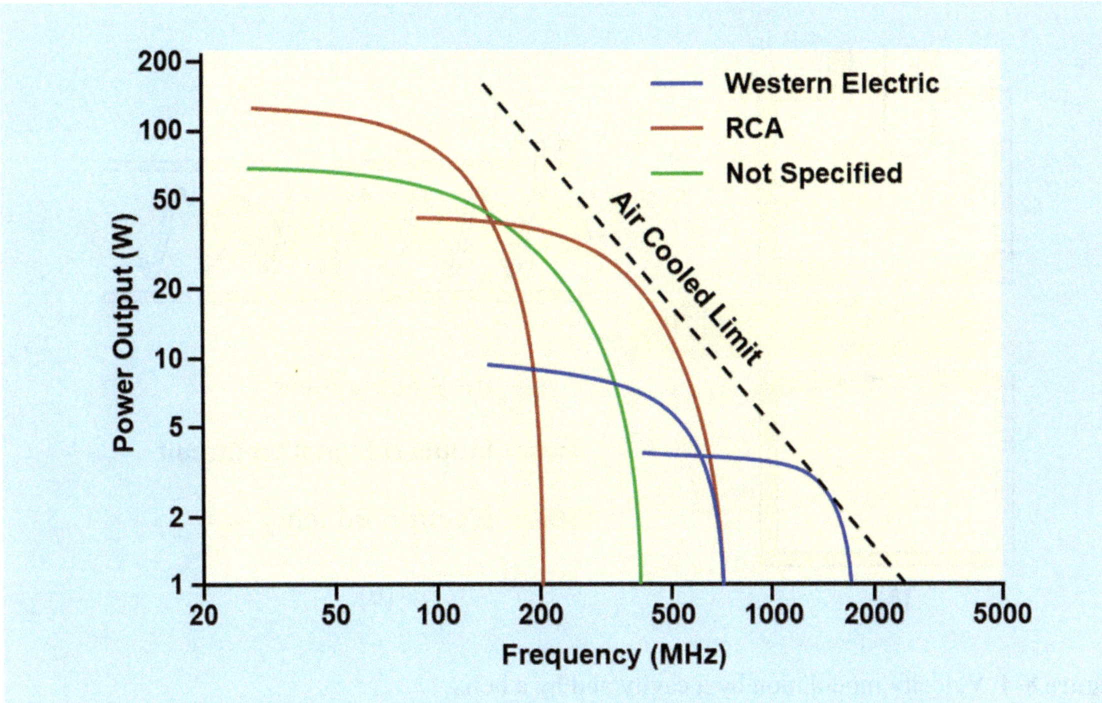

Figure 8-3 Power limitations of gridded transmitter tubes at low microwave frequencies [1].

One of the promising features of the field emitter arrays described in Chapter 5 is that the gate-to-emitter distance is two orders of magnitude smaller than the grid-cathode spacing in the 3-GHz triode. This means that it should be possible to modulate the emission current from a field emitter array at frequencies as high as tens of GHz. Whaley et al. have demonstrated the feasibility of gate modulation of a field emitter array at ~5 GHz and state that gate modulation of that array at frequencies up to 20 GHz should be possible [2]. However, progress in developing reliable field emission cathodes for microwave tubes has been very slow and it is not at all clear when or if they will be available for use in microwave tubes.

Even if the transit time problem is alleviated, there are the circuit impedance problems (item 2 on Spangenberg's list) and RF heating (Item 3 on Spangenberg's list) to contend with.

8.1.2 Velocity Modulation

Velocity modulation is accomplished as is shown in Figure 8-2(b) by first accelerating a beam to high velocity and then passing the beam through a region (gridded gap in Figure 8-2(b)) where an R- cavity TWT the modulating field exists between cavity drift tube tips indicated in Figures 8-1 and 8-4(a). In a helix TWT, the input signal velocity modulates the electrons and initiates the bunching process Figure 8-4(b). However, the bunches that develop induce a signal on the circuit that quickly dominates the beam-circuit interaction process.

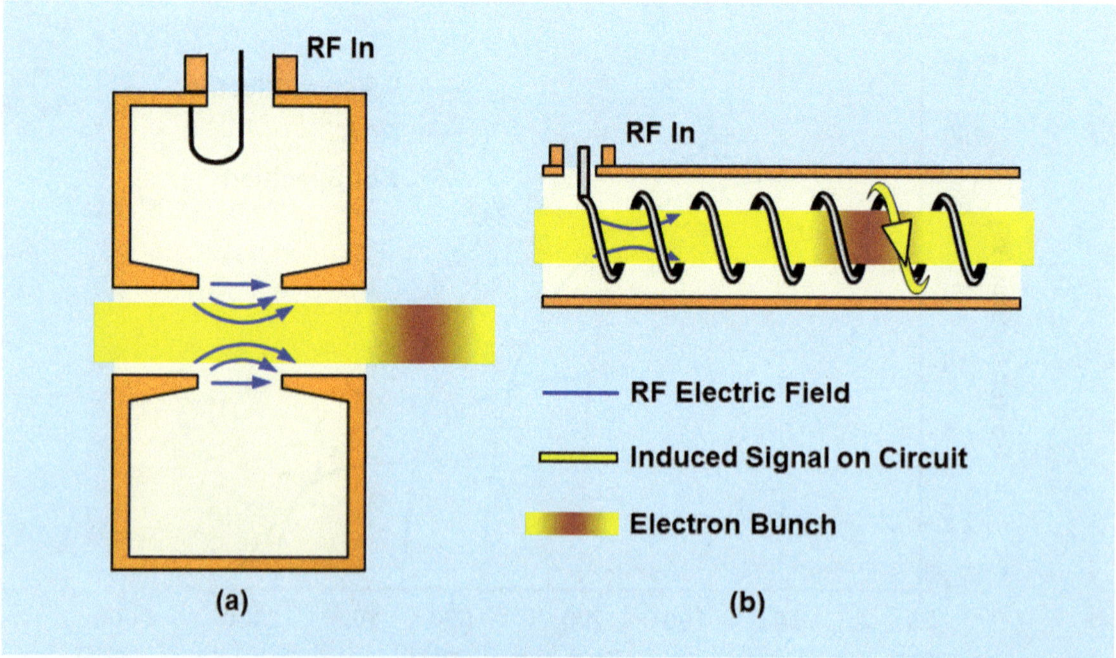

Figure 8-4 Velocity modulation by a cavity and by a helix.

A straightforward way to examine velocity modulation is to analyze the interaction between an electron beam and a gridded gap. In the first klystrons, the beam was modulated by passing it through a gap between two parallel grids across which RF voltage was applied. Beam power was very limited because of grid heating by electron impact. With the use of gridless gaps, beam power can be orders of magnitude above that possible with gridded gaps. Modulation by gridded and by gridless gaps is described in the following sections.

8.1.3 Gridded (Planar) Gaps

Consider the interaction of an electron beam with the electric field in a gridded gap as shown in Figure 8-5. The velocity of the beam entering the gap is u_o and is modified by an amount Δu as the beam passes through the gap. The time and position at the entrance to the gap are t_o and $z = 0$. The width of the gap is ℓ and the voltage applied to the gap is $V_1 \sin \omega t$. The equation of motion for electrons in the gap is

$$\frac{d^2z}{dt^2} = -\eta E = \eta \frac{V_1}{d} \sin\omega t \tag{8-1}$$

and so the velocity of the electrons leaving the gap is

$$u = u_o \left(1 + \frac{\alpha M_p}{2} \sin \omega t\right) \tag{8-2}$$

where α is called the *depth of modulation* and is defined as follows:

$$\alpha = \frac{V_1}{V_o} \tag{8-3}$$

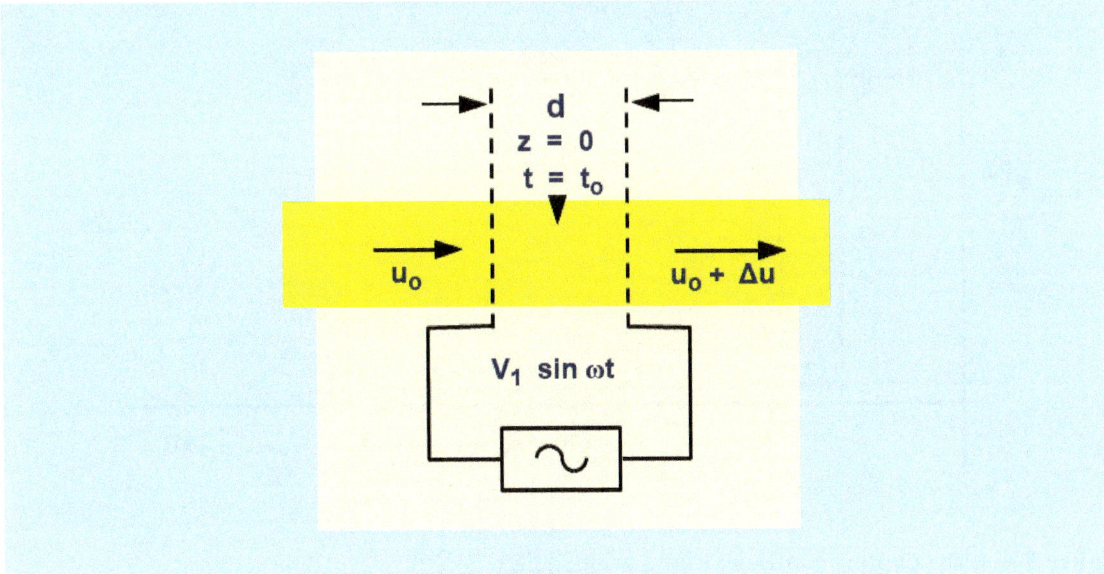

Figure 8-5 Velocity modulation with a gridded gap.

The dc acceleration voltage for the electrons in the electron gun is V_o. The quantity M_p, which is plotted in Figure 8-6, is the *modulation coefficient* or the *gap coupling coefficient* and is

$$M_p = \frac{\sin\frac{\varphi_o}{2}}{\frac{\varphi_o}{2}} \tag{8-4}$$

Here, φ_o is the dc transit angle through the gap and is defined as follows:

$$\varphi_o = \frac{\omega}{u_o} d \tag{8-5}$$

For a very narrow gap, φ_o is small so an electron sees a nearly constant voltage as it traverses the gap. For this case, the gap coupling coefficient is maximum and is nearly unity. As the gap is widened, the voltage varies as an electron traverses the gap and so the effective gap voltage (V_{eff} in Figure 8-7) is less than the maximum gap voltage (V_{max}). The effectiveness of the gap in accelerating the electron decreases (i.e., the gap coupling coefficient goes down). If the gap width is increased to the point where $\varphi_o = 2\pi$, the applied voltage varies through one complete cycle as an electron traverses the gap and the net acceleration of the electron is zero. In this case, $M_p = 0$. Klystron gap transit angles are usually 1.0 – 1.5 radians so the gap coupling coefficient (if a gridded gap was used) would be from 0.67 to 0.85. For gridless gaps, which are discussed next, Mp is somewhat smaller.

The value of the gap coupling coefficient is that, in the analysis, a gap with a finite width ℓ can be replaced with a gap having an infinitesimal width and voltage reduced by M_p without affecting the velocity. In addition, the reference plane where $z = 0$ can be shifted from the first grid of the gap to the center of the gap where $z = d/2$. This has been done in (8-2).

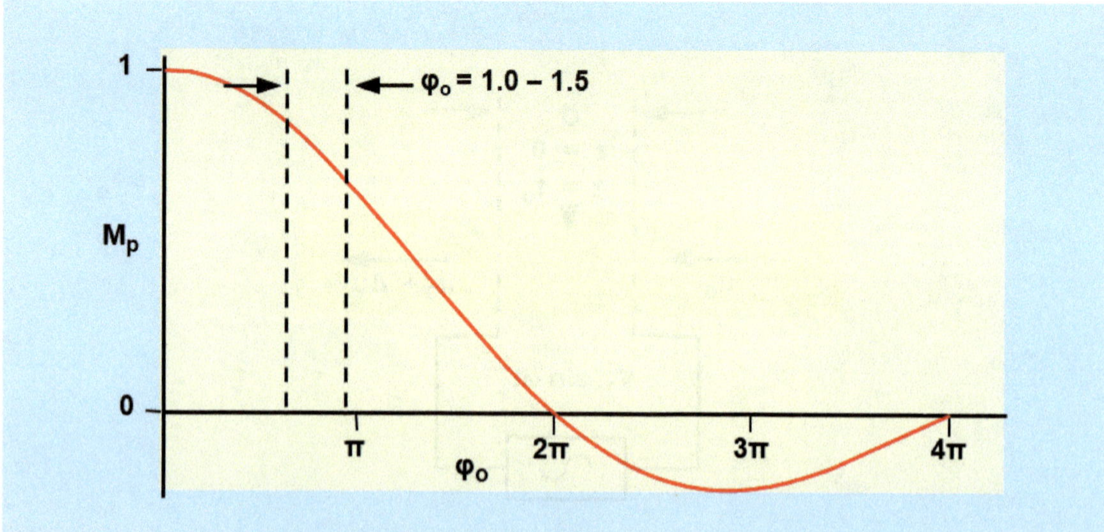

Figure 8-6 Gap coupling coefficient for a gridded gap.

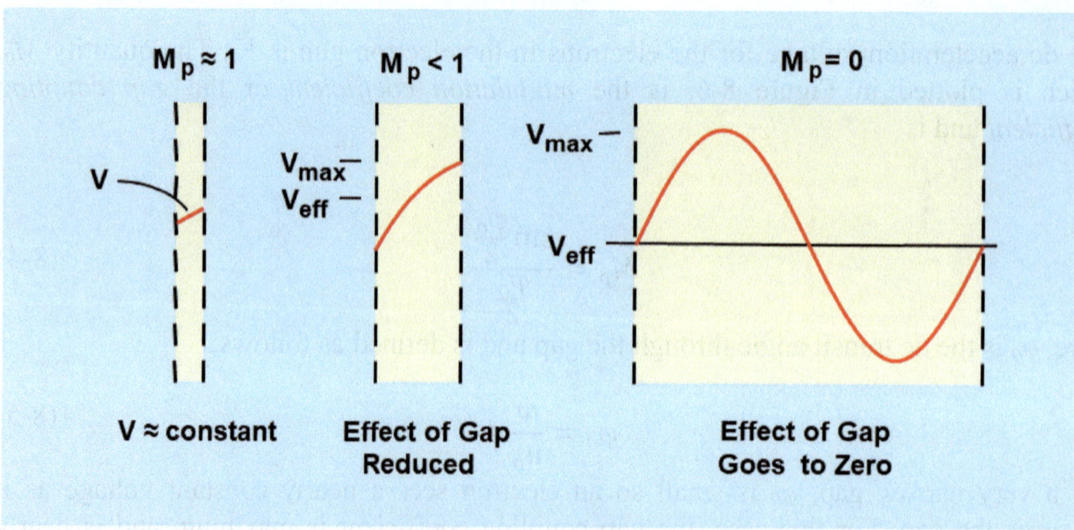

Figure 8-7 Effect of gap width on gap coupling coefficient.

8.1.4 Gridless (Nonplanar) Gaps

Gridless gaps are used for velocity modulation in virtually all microwave tubes. These gaps are formed by two facing electrodes in a coupled-cavity TWT and by adjacent helix turns in a TWT. For gaps between cylinders, the field produced is somewhat like that shown in Figure 8-8.

The electric field extends over a large region inside the gap and has variations in both the axial and radial directions. When a beam passes through the gap, electrons near the axis must travel further to see the full gap voltage than electrons near the gap. The field on axis may go through a large portion of a cycle while the field near the gap varies through a small portion as the electrons pass through, so electrons on axis receive a smaller net velocity change than those near the gap. Thus, the gap coupling coefficient varies with radius.

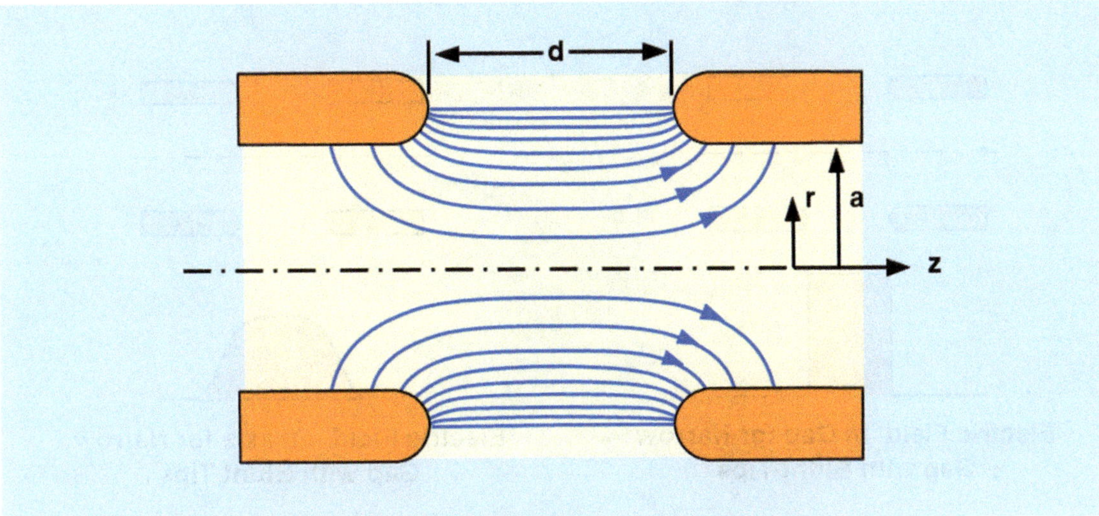

Figure 8-8 Electric field distribution in a gridless (nonplanar) gap.

The analysis of the interaction of the beam in the gridless gap is significantly more complicated than in the gridded gap. Still, under small signal conditions, the beam velocity can be expressed quite simply as

$$u = u_o \left(1 + \frac{\alpha M_{np}}{2} \sin \omega t\right) \tag{8-6}$$

where M_{np}, which is a function of radius, is the gap coupling coefficient for the nonplanar gap. The magnitude of M_{np} depends on the electric field distribution within the gap. The field, in turn, depends on several factors including the gap width, the cylinder radius, and the shapes of the cylinder tips.

For example, as shown in Figure 8-9, if the gap is narrow compared with the cylinder radius and the cylinder tips are blunt, then the field within the gap will be nearly constant. A straightforward way to model the gap is that described by Vaughan [3]. He used the gap coupling coefficient found for a uniform gap field and then devised an empirical correction factor.

With a uniform field in the gap, the field experienced by the electron traveling close to the walls of the cylinders is the same as the field in the gridded gap; that is, constant, and so the electron velocity leaving the gap (at $z = d/2$) is given by (8-2). The expression found for M_{np} as a function of radius is shown in Figure 8-10 and is

$$M_{np} = M_p \frac{I_o(\beta_e r)}{I_o(\beta_e a)} \tag{8-7}$$

where M_p is the modulation coefficient for the planar (gridded) gap (8-4), $\beta_e = \omega/u_o$ is the propagation constant for the electron beam, and a is the cylinder radius. Thus, the modulation coefficient for the gridless gap is M_p modified by the I_o Bessel function.

As expected, the modulation coefficient is smaller in the beam center than at the beam edge. It is usual practice to limit $\beta_e a$ to $\approx$ 1 to 1.5 radians to ensure adequate interaction with the electron beam.

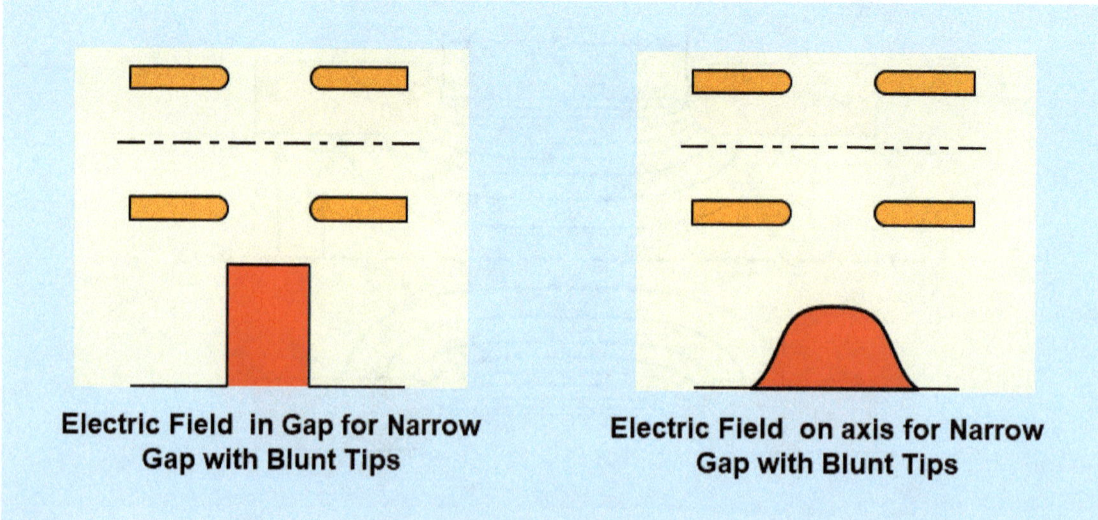

Figure 8-9 Electric field amplitude between tips for a narrow gap with blunt tips.

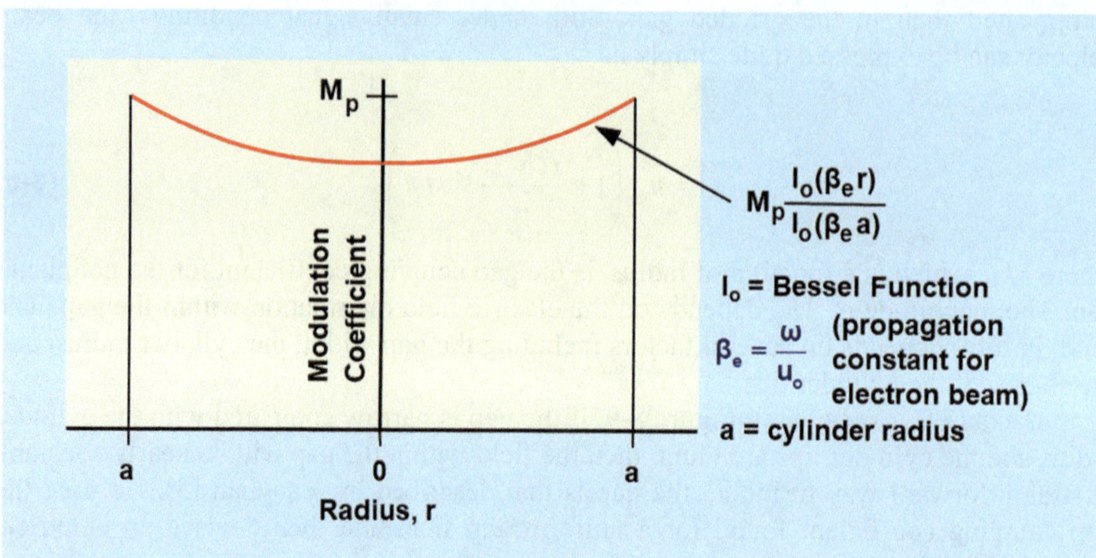

Figure 8-10 Dependence of gap coupling coefficient on gap radius. M_p is the coefficient for a gridded gap.

8.2 BEAM LOADING

The first question to be addressed is: What is beam loading? It takes power to modulate an electron beam and so the beam loads the signal source. The amount of loading is expressed as a conductance and a susceptance. During the modulation process, signal power is expended whenever electrons are accelerated and recovered when electrons are decelerated. More energy is expended in accelerating electrons than is recovered in decelerating electrons because electron energy is proportional to the square of electron velocity. The beam also provides a reactive component of admittance. The presence of electrons in the capacitive portion of the circuit (the gap) has an effect on capacitance and so the resonant frequency of the circuit is affected.

Figure 8-11 illustrates the conductive part of beam loading. One cycle of velocity modulation is indicated by the dashed blue line. Electron energy, which is proportional to velocity squared, is indicated by the red line. The power supplied to accelerated electrons is represented by the green-colored area. The power recovered from decelerated electrons is represented by the blue-colored area. The net power supplied by the source is the difference between supplied and recovered power.

Admittances for gridded and gridless gaps are shown in Figures 8-12 and 8-13 [4]. G_o is the dc beam conductance, I_o/V_o, G_B is the loading conductance, and B_B is the loading susceptance. In general, G_B can be positive or negative. However, for klystron cavities, a/d is normally from 0.5 to 1.0. Then, G_B is positive and modulation power is being delivered to the beam. When G_B is positive and the load is a resonator, the quality factor, Q, of the resonator is reduced and the bandwidth is broadened. In Figures 8-12 and 8-13, the usual range of transit angles for klystron cavities of 1.0 to 1.5 radians is highlighted.

Figure 8-13 shows beam loading conductance for gap widths ranging from four times the drift tube radius to ½ of the drift tube radius. The beam loading conductance varies from small values for short gaps to a maximum of about $0.2G_o$ for a gap somewhat longer than π radians. For a 10,000-V, 1-A beam, $G_o = 1\times10^{-4}$ mhos ($R_o = 1\times10^4$ ohms) and so $G_B = 2\times10^{-5}$ mhos ($R_B = 5\times10^4$ ohms). At small signal levels, this beam loading resistance is comparable to the cavity resistance. In addition to reducing the Q of a resonator, the beam-loading admittance affects the resonant frequency of the resonator.

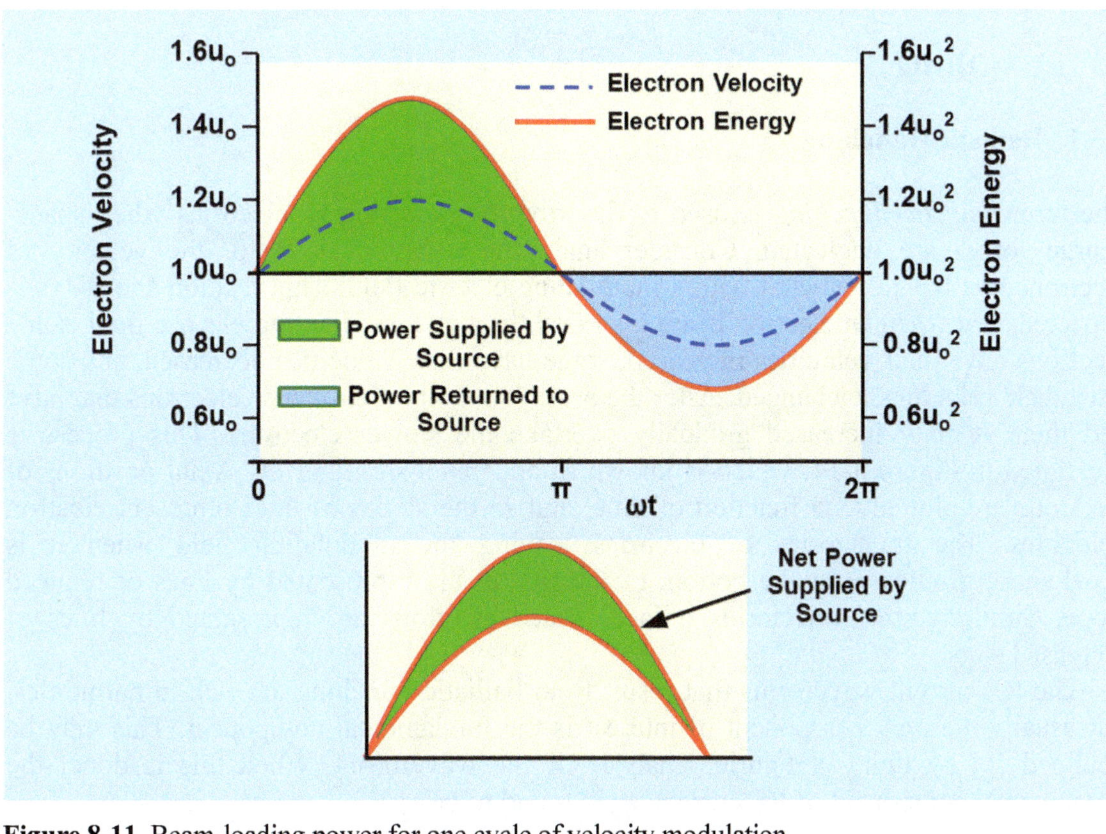

Figure 8-11 Beam-loading power for one cycle of velocity modulation.

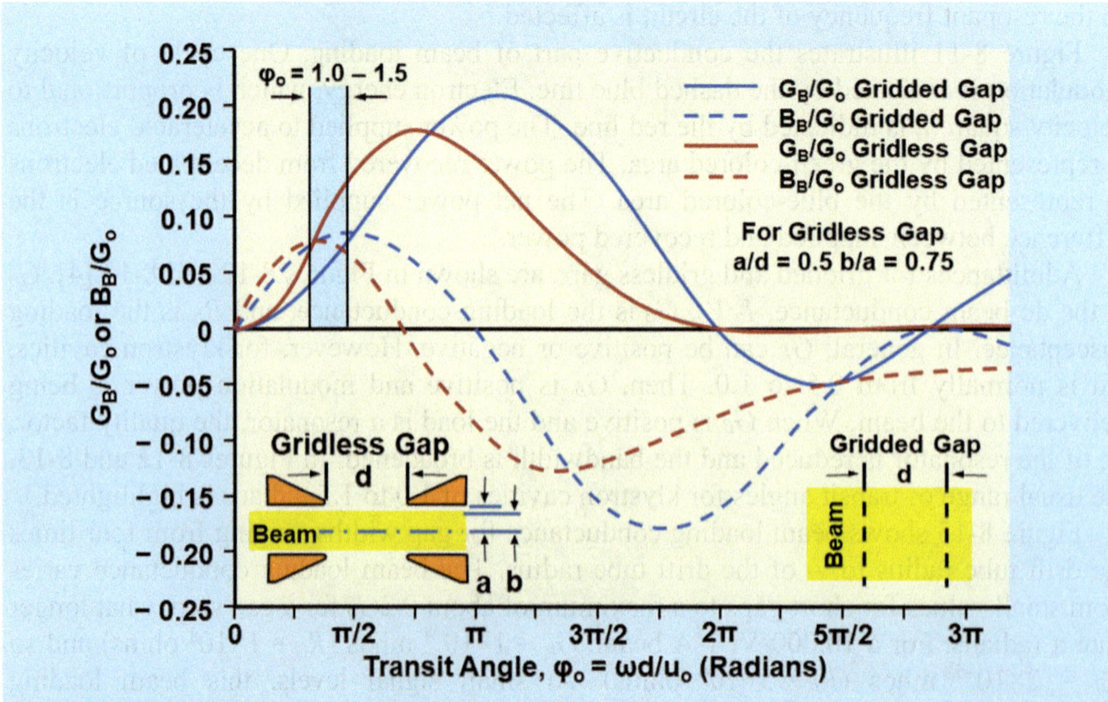

Figure 8-12 Beam-loading admittance for a gridless gap, with *a/d* = 0.5 and b/a = 0.75 and for a gridded gap.

8.3 BUNCHING

8.3.1 Ballistic Bunching

The term *ballistic bunching* is used to describe the bunching of electrons when space-charge forces are neglected. Consider again the equation (8-2) for the velocity of electrons that have interacted with a modulating electric field. This relation shows, as is expected, that when an electron beam is passed through a modulating electric field, some electrons have their velocities increased, some have their velocities decreased, and some have their velocities unchanged. After the beam leaves the field, those electrons that have had their velocity increased gradually overtake the slower electrons. This process is illustrated in Figure 8-14, which is known as an *Applegate diagram.* Axial positions of electrons are plotted as a function of time, and so the slopes of lines represent electron velocities. The trajectories of electrons leaving the modulating field when it is decelerating (indicated at the bottom of the figure) are represented by lines of reduced slope. Similarly, the trajectories of accelerated electrons are represented by lines of increased slope.

The RF current waveforms that result from ballistic bunching are rich in harmonics, but usually the only component of interest is the fundamental component. This may be obtained [1] by doing a Fourier analysis of the waveforms. When this is done, the fundamental component of RF current, i_f, is found to be

$$i_f = 2I_o J_1(X) \tag{8-8}$$

where J_1 is a Bessel function of the first kind. This relation is plotted in Figure 8-15. Note that the peak value of i_f is larger than the dc current.

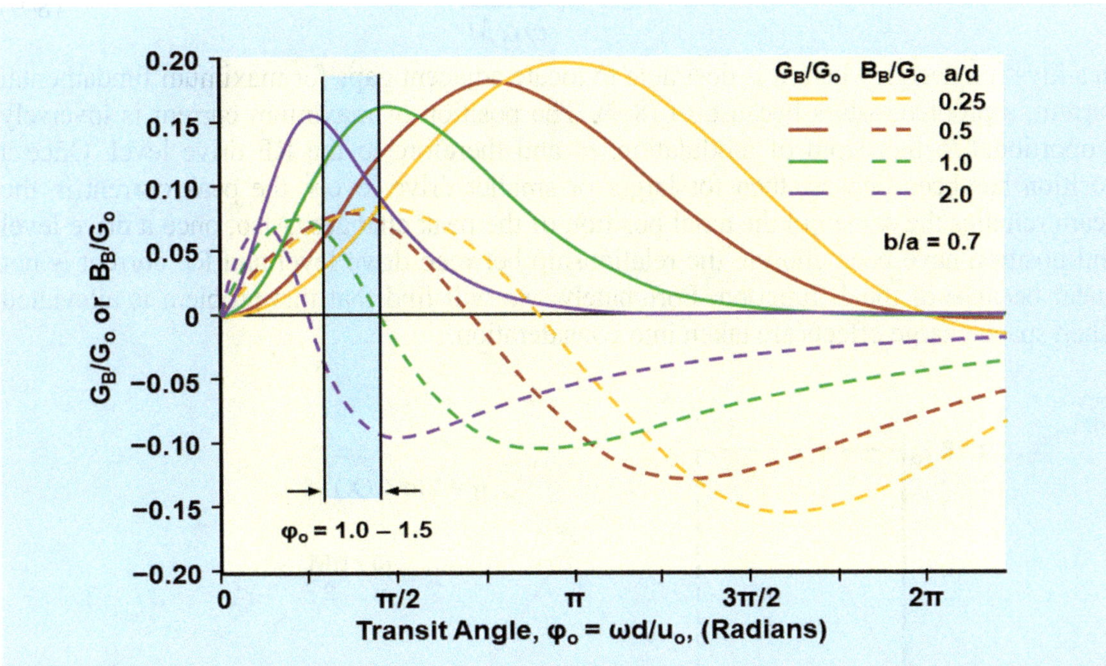

Figure 8-13 Beam-loading admittances for gridless gaps.

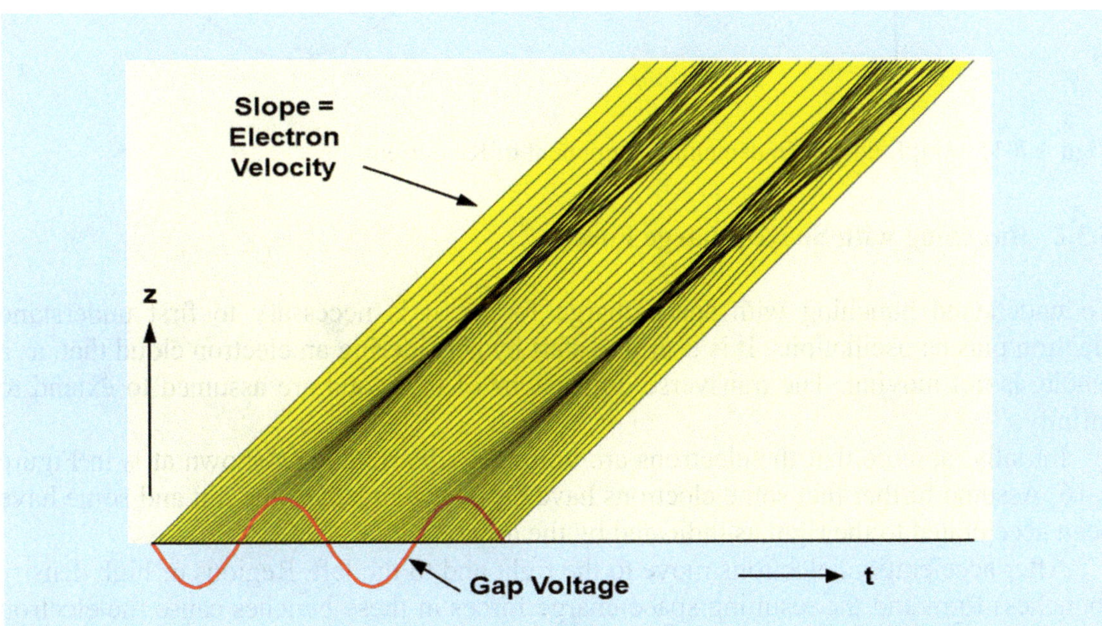

Figure 8-14 The classic Applegate diagram used for illustrating ballistic electron bunching.

The value of X for which i_f is maximum is 1.84, and so the axial position, z, at which the peak value occurs is

$$z = 1.84 \frac{2u_o}{\omega \alpha M} \tag{8-9}$$

In a klystron design where it is desirable to locate adjacent gaps for maximum fundamental current, a problem arises because of (8-9). The position of maximum current is inversely proportional to the depth of modulation, α, and therefore to the RF drive level. Once a position has been chosen, then for larger or smaller drive levels, the peak current in the beam remains the same but the axial position of the peak changes. Also, once a drive level and position have been chosen, the relationship between drive level and RF current is not linear because of the J_1 function. Fortunately, we will find that this problem is alleviated when space-charge effects are taken into consideration.

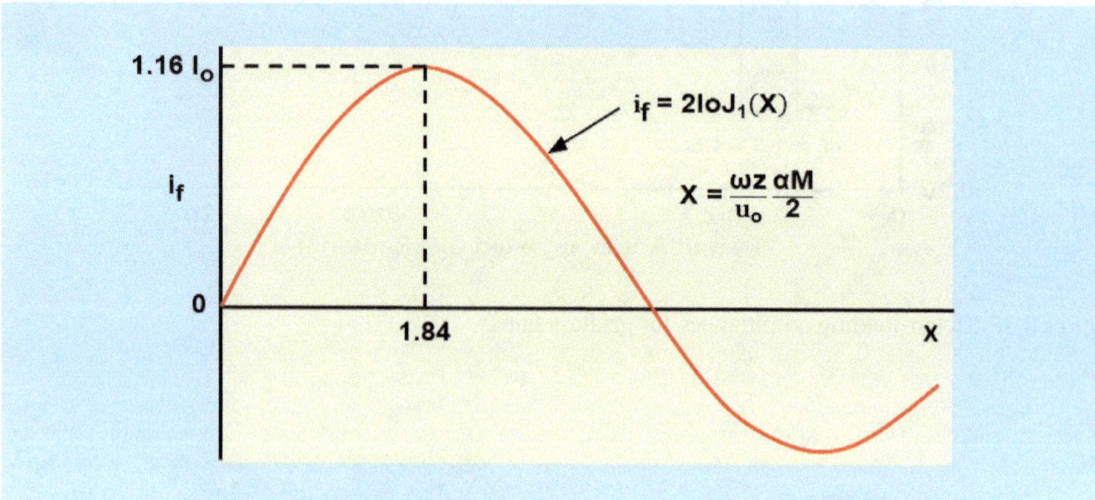

Figure 8-15 Amplitude of fundamental component of RF current.

8.3.2 Bunching with Space-Charge Forces

To understand bunching with space-charge forces, it is necessary to first understand electron plasma oscillations. It is useful to start by considering an electron cloud that, as a whole, is not moving. The transverse dimensions of the cloud are assumed to extend to infinity.

Initially, assume that the electrons are uniformly distributed as shown at t_0 in Figure 8-16. Assume further that some electrons have been accelerated to the left and some have been accelerated to the right as indicated by the arrows.

After acceleration, electrons move to the right and to the left. Regions of high density (bunches) form and the resulting space-charge forces in these bunches cause the electron movement to stop (at t_1) and then reverse directions. Next, at t_2, the electrons become uniformly distributed again but are moving in directions opposite to those shown at t_0. As the electrons continue to move, new bunches form at t_3. Again, the space-charge forces that build up in the bunches cause the electron movement to stop and reverse directions. After that, the distribution shown at t_4, which is the same as t_0, returns and the process starts over again. Thus, the electron cloud oscillates.

The frequency with which the cloud oscillates is called the *electron plasma frequency.* It turns out that when wall effects are ignored, this frequency is dependent *only* on the electron density. The electron plasma frequency is an extremely important quantity in the analysis and understanding of wave motion in a TWT.

Now we can return to the consideration of bunching with space-charge forces. As electrons converge in the bunching process after the beam leaves the modulating gap, space-charge forces build up, impeding the rate at which fast electrons overtake slow electrons. In addition, slow electrons have their velocities increased because of the bunch that forms behind them (toward the modulating cavity). The resulting electron trajectories are illustrated in Figure 8-17, which is a portion of an Applegate diagram modified to show the effects of space-charge forces.

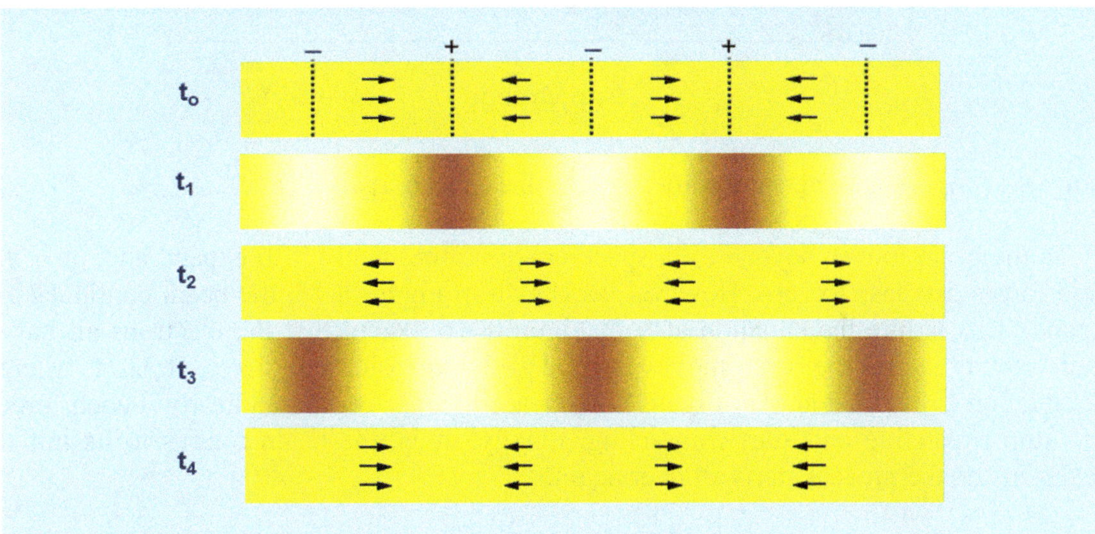

Figure 8-16 Electron distributions during plasma oscillations. Arrows indicate directions of electron motions.

The distance, λ_p, is the electron plasma wavelength. The origin of this wavelength is the natural frequency of oscillation (the electron plasma frequency) of the cloud of electrons coupled with the velocity with which the cloud of electrons is moving.

At the axial position denoted by $\lambda_p/4$, fast electrons have been slowed to the dc velocity of the beam and slow electrons have been accelerated to the dc beam velocity. At $\lambda_p/4$, all electrons have the same velocity. Also, at $\lambda_p/4$, the RF electron density and the RF current reach maximum values. For small to medium RF signals, the RF current is nearly sinusoidal.

A very important characteristic of the bunching process with space-charge forces is that *all* electrons are either speeded up or slowed down to the same velocity (the dc beam velocity) at the same axial position ($\lambda_p/4$). In addition, even if the amplitude of the modulating field is changed so that initial electron velocities are changed, the axial position of the bunch remains the same.

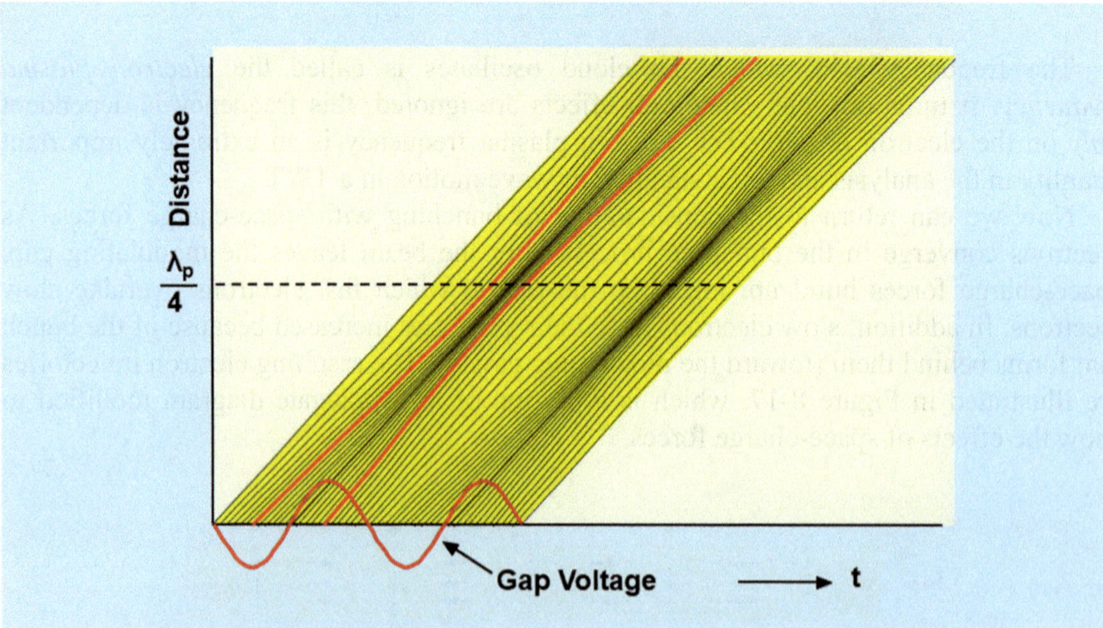

Figure 8-17 Applegate diagram modified to show the effect of space charge.

As the beam moves beyond $\lambda_p/4$, space-charge forces cause it to expand and, at $\lambda_p/2$, these forces decrease to zero. However, as shown in Figure 8-18, the beam continues to expand. This is like the situation at t_2 in Figure 8-16 except that the electrons all have axial velocity u_o. As the beam moves beyond $\lambda_p/2$, expansion continues, at $3\lambda_p/4$, where the situation is like that at t_3 in Figure 8-16, space-charge forces in a nearby bunch force it to stop expanding and start contracting. Finally, at λ_p, the beam returns to its initial condition and the process starts all over again.

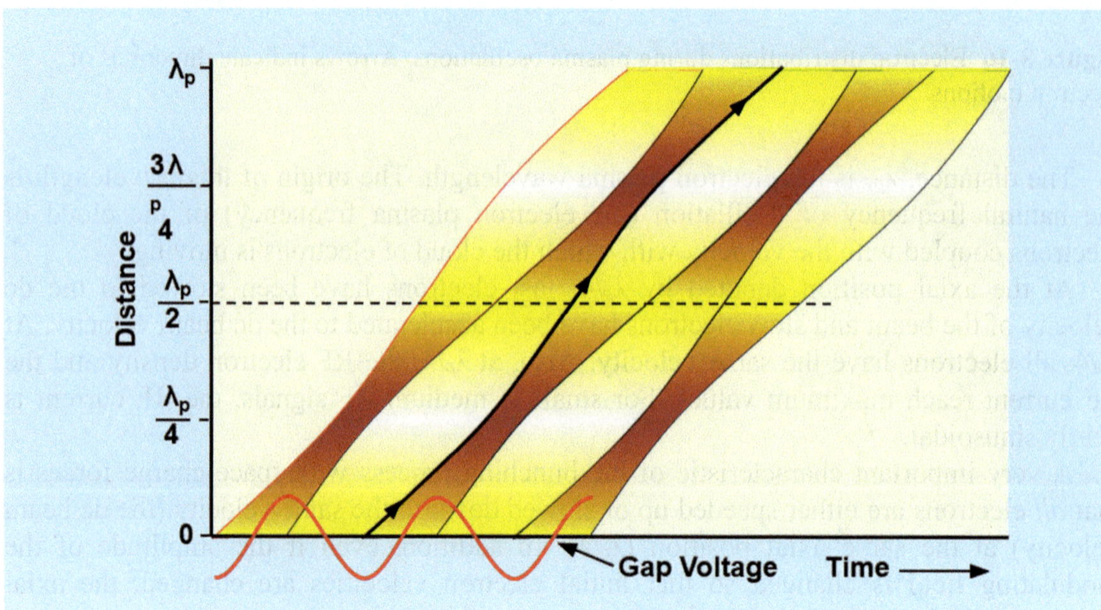

Figure 8-18 Electrons oscillate between bunches as the beam moves axially.

Figure 8-19 depicts the motion of electrons as seen by an observer who is traveling at beam velocity. If we choose the phase of the RF when maximum acceleration occurs in the modulating gap, then the ac velocity, u_z, of the electrons appears as shown by curve 1. From (8-6), the peak velocity of the electrons is $\alpha M/2$. If we choose a phase such that maximum deceleration occurs, then u_z appears as shown by curve 2. At other times, u_z lies between curves 1 and 2 as indicated by the dashed lines. Note that regardless of the starting phase, u_z is zero for all electrons at $\lambda_p/4$.

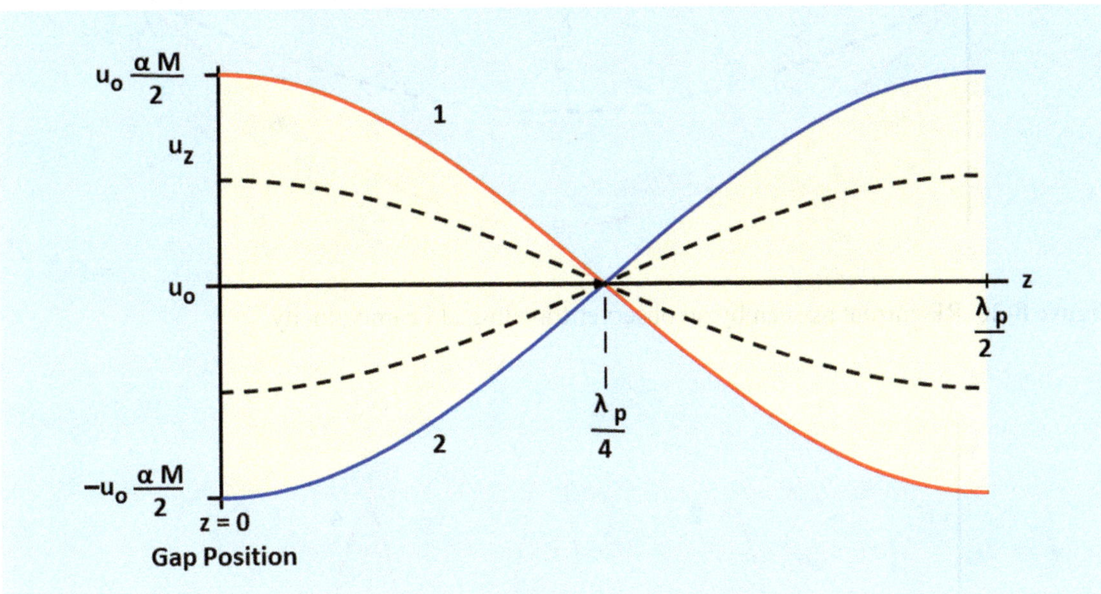

Figure 8-19 Electron velocity for various starting conditions as seen by an observer traveling at beam velocity.

As the fast electrons approach the slow ones at $z = \lambda_p/4$, a bunch is formed. Furthermore, the bunching action causes the fast electrons to slow to the dc velocity and the slow electrons to speed up to the dc velocity. As electrons drift beyond $\lambda_p/4$, the space-charge forces within the bunch cause the electrons in the front of the bunch to speed up and those in the rear of the bunch to slow down.

The RF current in the beam varies sinusoidally with distance as shown in Figure 8-20. The maximum current occurs when $z = \lambda_p/4$, which corresponds to the position where the ac velocity goes to zero. If the electrons drift beyond $\lambda_p/4$, then the bunch disperses and the current density decreases. The bunch and current disappear at $\lambda_p/2$ and then reappear as the beam passes $\lambda_p/2$.

The variations of current shown in Figure 8-20 are those that an observer moving at the beam velocity would see. If, instead, at an instant in time, a snapshot of current is considered, it would appear as shown in Figure 8-21. The maximum current is represented by the current peak labeled 3. The current peaks labeled 1 and 2 result from bunches of electrons that are in the process of forming. Beyond $\lambda_p/4$, the bunches disperse and the current peaks (4 and 5) decrease.

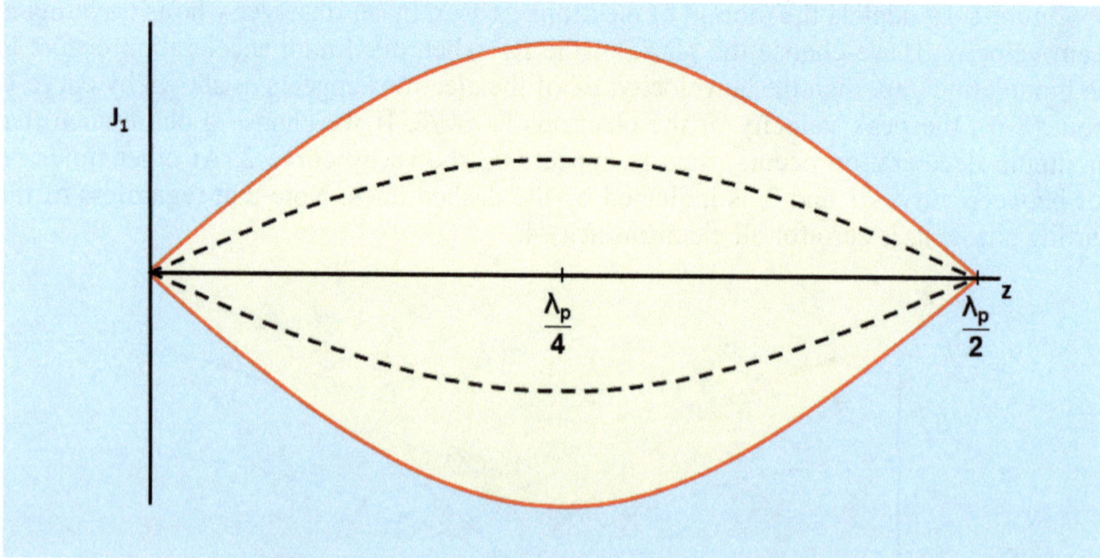

Figure 8-20 RF current as seen by an observer traveling at beam velocity.

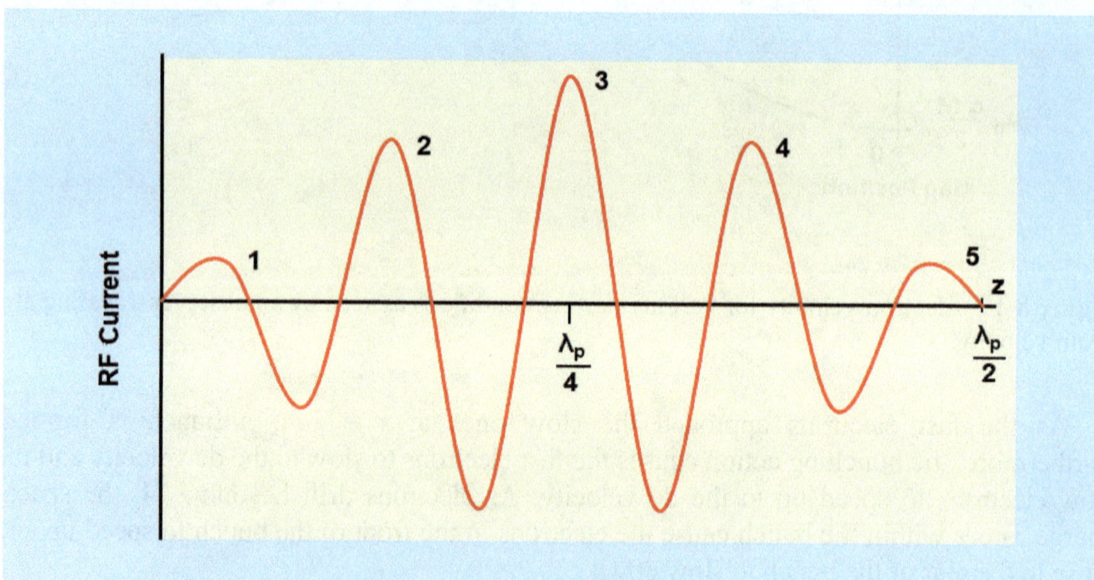

Figure 8-21 Snapshot of current as a function of distance.

8.4 TRANSITION TO LARGE SIGNAL BUNCHING

8.4.1 Bunching with and without Space-Charge Effects

At the beginning of this chapter, we examined bunching with no space-charge forces. That ballistic bunching produced an RF current that depended only on the bunching parameter, *X*, which was proportional to $\alpha\, z$ (α = depth of modulation and z = distance along the electron beam). The maximum RF current was 1.16 times the dc beam current. This same maximum RF current could be obtained at any drive level (any α) simply by choosing the correct axial position in the beam. Thus, current as a function of distance and drive level is as shown in Figure 8-22.

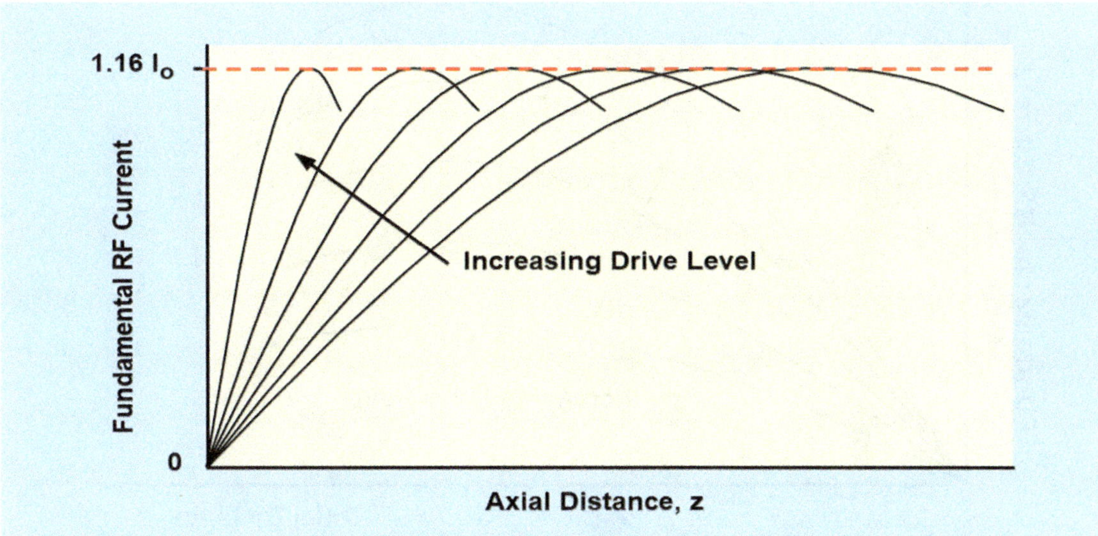

Figure 8-22 RF current as a function of axial position and drive level for ballistic bunching.

When the effects of space charge on bunching were considered, the variation of RF current with α and z was found to be very different than for the ballistic case. At small signal levels, the maximum RF current always occurred at the same distance from the modulating cavity, $\lambda_p/4$. As the drive level was varied, the amplitude of the current varied and the current as a function of distance and drive level is as shown in Figure 8-23.

8.4.2 Mihran's Experimental Results

The transition from space-charge wave to ballistic behavior was perhaps best illustrated by Mihran [5]. He used the experimental apparatus in Figure 8-24 to measure RF currents in a beam modulated by a klystron cavity. Mihran's apparatus was a two-cavity klystron with the following additional features:

1. The output cavity was movable allowing drift length to be variable.
2. A control grid allowed variation of beam current at fixed beam voltage.
3. Apertures could be placed over the control grid to vary the beam diameter.
4. The output cavity was tuned to a frequency, f_2, which was sufficiently far from the input frequency to ensure that the RF output voltage would always be much lower than the beam voltage. This guaranteed that small signal conditions would prevail in the output gap, thus giving direct information on the electron bunching in the beam.

A set of experimental results is shown in Figure 8-25. These curves show the transition from space-charge wave to ballistic behavior as signal level was increased. The dashed lines show the current maxima predicted by large signal theories.

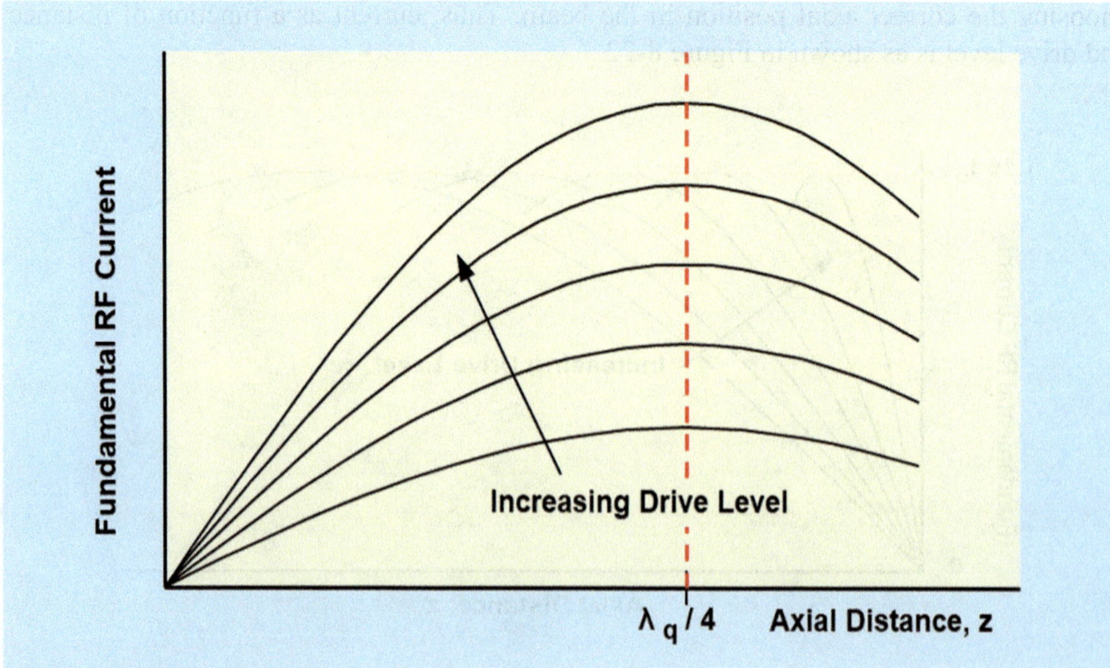

Figure 8-23 RF current as a function of axial position and drive level for bunching with space-charge forces.

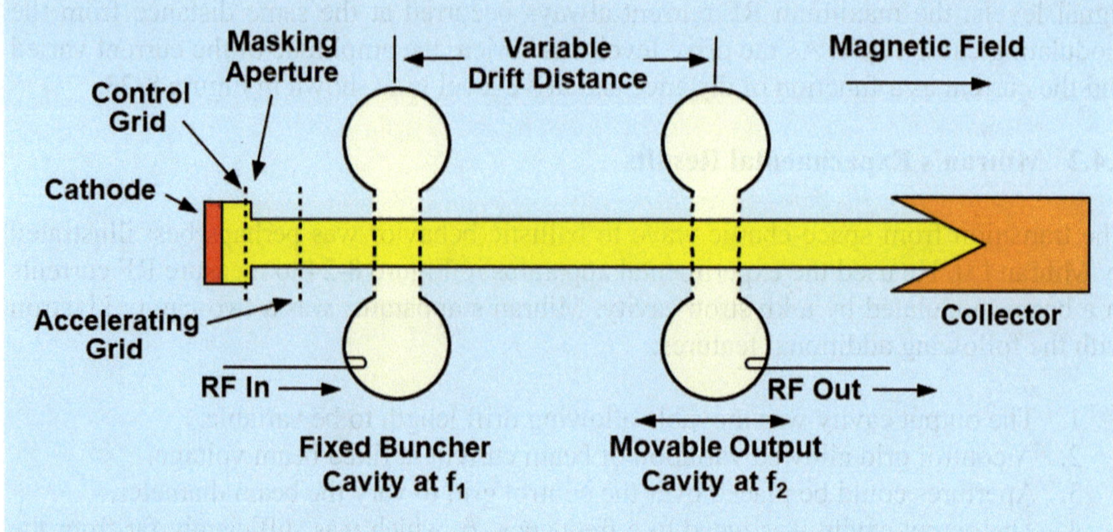

Figure 8-24 Mihran's experimental two-cavity klystron setup.

The most significant result of Mihran's work was demonstrating the amount by which the distance between cavities should be below $\lambda_q/4$ at large signal levels. Mihran gave the optimum distances between cavities of a multicavity klystron as being approximately

$$90°: 90°: \cdot\cdot\cdot : 65° : 45°: 30°$$

that is, the spacing between cavities should vary from $\lambda_q/4$ to 1/3 $\lambda_q/4$ as the signal increases from the input level to saturation. Mihran was the first to specify the amount by which drift lengths should be reduced in the output region of a klystron. These reduced spacings resulted in improved efficiencies. This will be discussed in detail in following chapters.

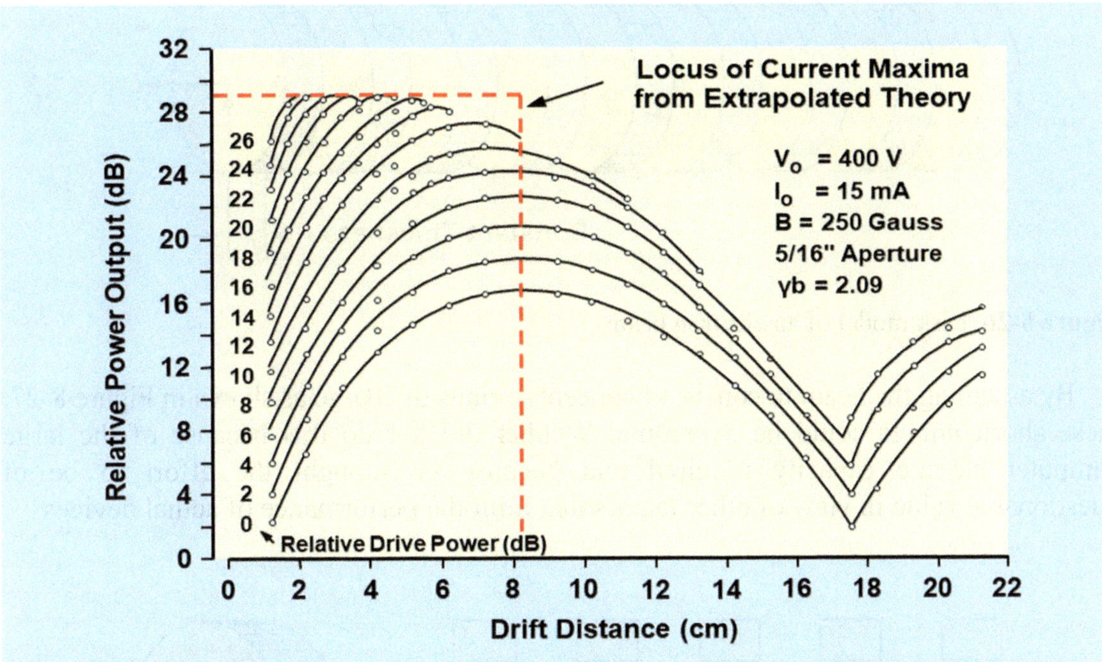

Figure 8-25 Mihran's results showing the transition from space-charge wave to ballistic behavior. (From: T. G. Mihran, *IRE Trans.on Electron Devices*, January 1959. ©1959 IRE (now IEEE).)

8.4.3 Webber's Disk and Ring Model Results

Large signal theories failed to predict the drift tube shortening that Mihran demonstrated was necessary to optimize efficiency. However, Webber [6] was able to use a disk electron model like that shown in Figure 8-26 and compute curves for current versus distance. In Webber's model, 16 disks per wavelength were used for most calculations. Charge was uniformly distributed throughout the disks and the disks were invariant throughout the calculation. The space-charge force was determined by first calculating the electric field due to a single disk and then obtaining the average axial force on another disk. The total force on a disk was obtained by integrating this force expression over all other disks in the beam.

The disks were assumed to behave ballistically, like individual electrons, in that each was given an initial velocity and/or phase position and then was allowed to bunch accordingly. Disks were allowed to cross over or pass through one another. Some shortcomings of Webber's model are:

1. Only axial motions of electrons or disks were considered.
2. The disk approximation did not allow any variation of velocity or density modulation across the beam.

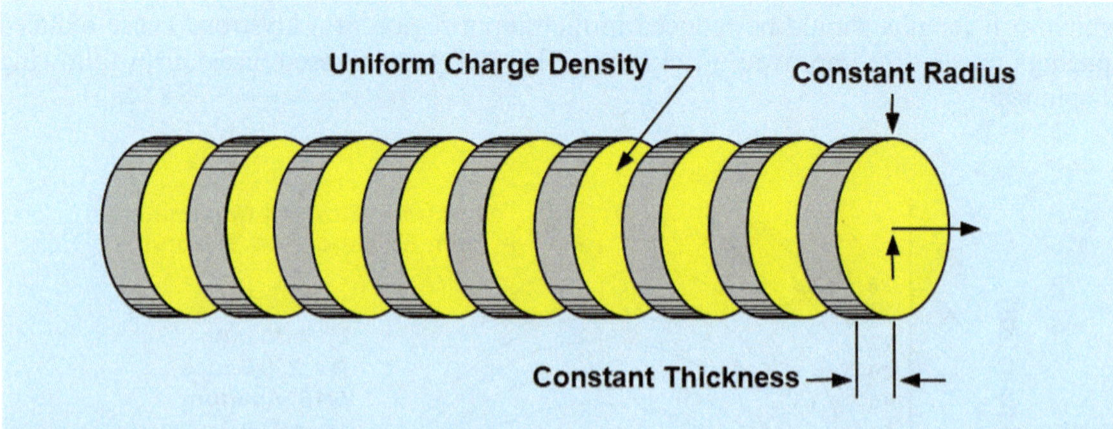

Figure 8-26 Disk model of an electron beam.

By assuming the beam to consist of concentric rings of charge as shown in Figure 8-27, these shortcomings could be overcome. Webber did not do this because of the large computer storage capacity required and because he thought the effort to be of questionable value in view of other factors that limit the performance of actual devices.

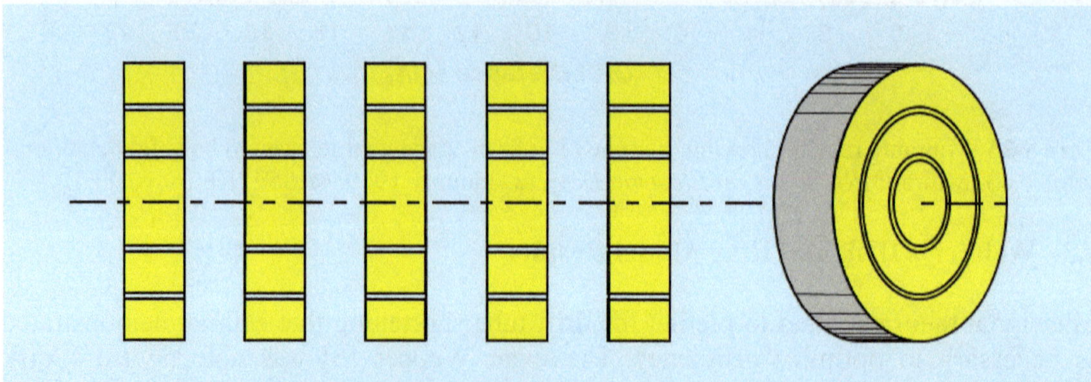

Figure 8-27 Ring model of an electron beam.

A set of Webber's curves showing current versus distance is shown in Figure 8-28. The quantity α contained in $\omega_q/\omega\alpha$ is the depth of modulation so decreasing values of $\omega_q/\omega\alpha$ correspond to increasing drive levels. In general, there is good agreement with Mihran's experimental results. The primary difference is in the transition from space-charge-limited behavior to ballistic behavior. Mihran's results show a more gradual transition than do Webber's calculations. Mihran points out that the disagreement is probably attributable to the inaccuracy of the disk electron assumption for large γb where $\gamma = \omega/u_o$. For $\gamma b = 2$ (Mihran's results are for $\gamma b = 2.09$ and Webber's calculations are for $\gamma b = 1.6$), there is a 2 to 1 variation of fields from the beam edge to the beam center. This would warp the electron disks assumed in Webber's calculations.

Another example of the results of Webber's calculations is shown in Figure 8-29. Here $\gamma b = 0.4$ and the beam was considerably smaller in diameter than for the example shown in Figure 8-28. Notice that for very large drive levels, the current exceeds the ballistic value of 1.16 I_o by a substantial amount. The reason for this is that space charge is actually enhancing the bunching process. Webber showed that electrons near the center of the bunch, which would normally cross over (fast electrons would pass slow ones) if moving ballistically, are stopped and reversed relative to the center of the bunch before crossing takes place. The result is that the fundamental component of current in the beam is enhanced. Webber showed that the charge densities at which this effect is optimized correspond to relatively low beam perveances. We will see in a later chapter when high-efficiency klystrons are discussed that a relatively low perveance (substantially less than 1×10^{-6}) is usually used for high efficiency.

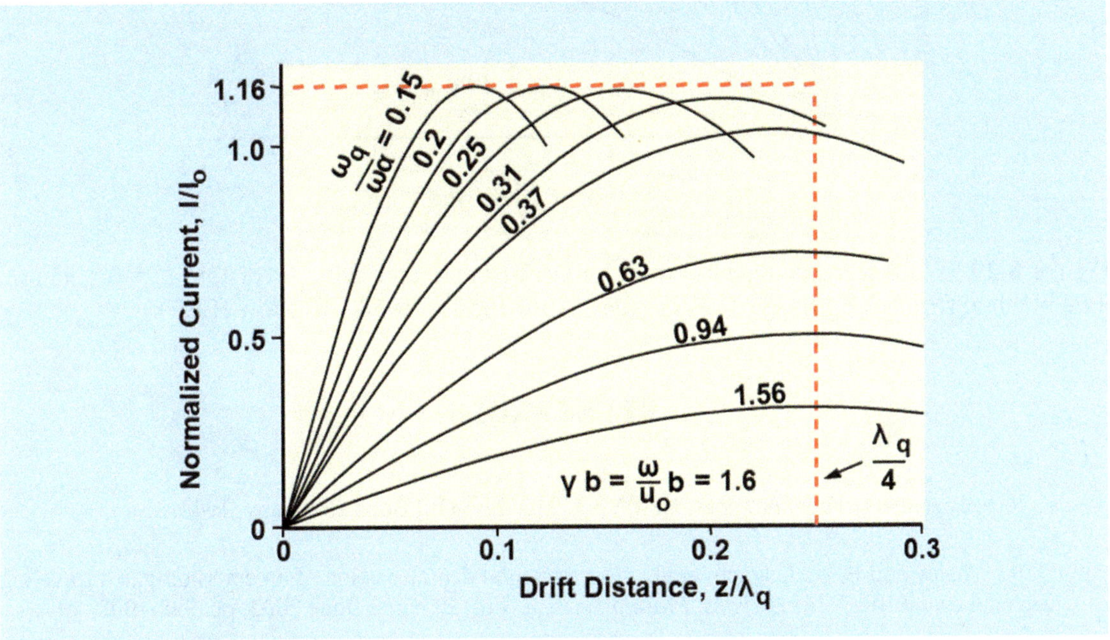

Figure 8-28 Webber's calculated current versus distance and drive level for γb = 1.6. (From: S. E. Webber, *IRE Trans. Electron Devices*, April 1958. ©1958 IRE (now IEEE).)

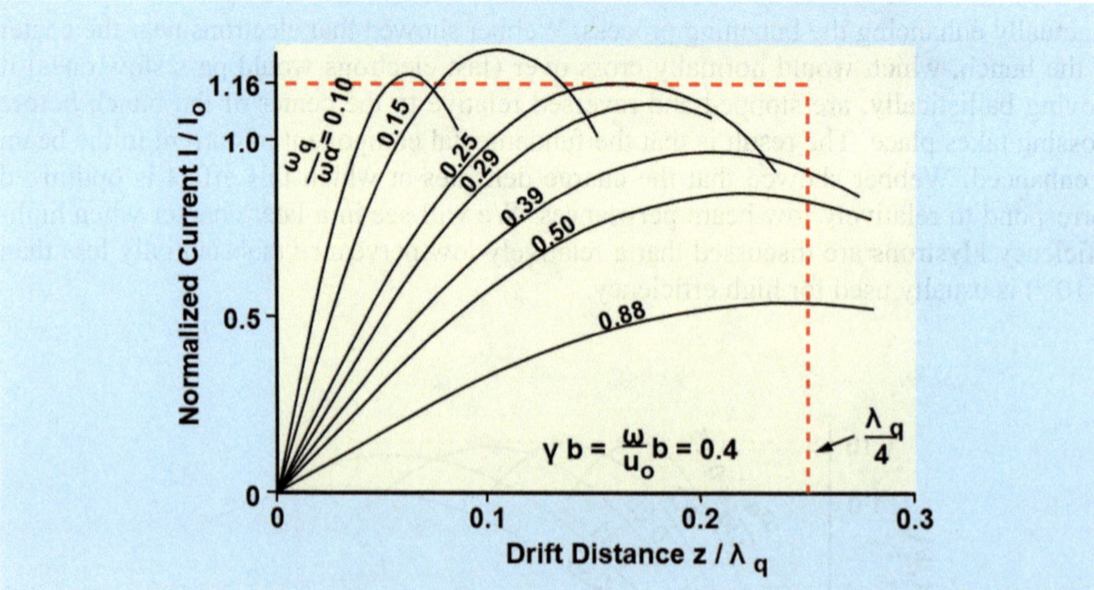

Figure 8-29 Webber's calculated current versus distance and drive level for γb = 0.4. (From: S.E. Webber, *IRE Trans. on Electron Devices*, April 1958. ©1958 IRE (now IEEE).)

REFERENCES

[1] K. R. Spangenberg, *Vacuum Tubes*, New York, McGraw-Hill Book Company, 1948.

[2] D. R. Whaley and B. M. Gannon, et al., "Experimental demonstration of an emission-gated traveling-wave tube amplifier," *IEEE Trans. Plasma Science,* Vol. 30 No. 3, June 2002, pp. 998-1008.

[3] J. R. M. Vaughan, "A model for the klystron cavity gap," *IEEE Trans. Electron Devices*, Vol. ED-32, No. 11, November 1985, pp. 2482–2484.

[4] E. J. Craig, "The beam loading admittance of gridless klystron gaps," *IEEE Trans. Electron Devices*, Vol. ED-14, No. 5, May 1967, pp. 273–278.

[5] T. G. Mihran, "The effect of space charge on bunching in a two-cavity klystron," *IRE Trans. Electron Devices*, Vol. ED-6, No. 1, January 1959, pp. 54–64.

[6] S. E. Webber, "Ballistic analysis of a two-cavity finite beam klystron," *IRE Trans. Electron Devices*, Vol. 5, No. 2, April 1958, pp. 98–108.

Chapter 9

Current Induction and Circuit Response

In this chapter, the interaction of a modulated linear electron beam with a circuit is discussed. When a bunch of electrons passes through a cavity as indicated in Figure 9-1, it repels electrons on nearby surfaces and causes current to flow in the cavity walls. The questions to be addressed here are: What is the magnitude of the induced current? What is the waveform shape of the induced current? How does the circuit respond to the induced current?

Figure 9-1 An electron bunch passing through a cavity induces current flow in the cavity walls.

9.1 CURRENT INDUCTION

9.1.1 Ramo's Theorem

Consider first a sheet of electron charge passing between two planar parallel grids as shown in Figure 9-2. The surface charge density on the charge sheet is assumed to be σ_e coulombs per unit area. Positive charge densities σ_L and σ_R are induced on the left and right grids, and as the charge sheet moves, electron current I_i flows in the circuit connecting the grids.

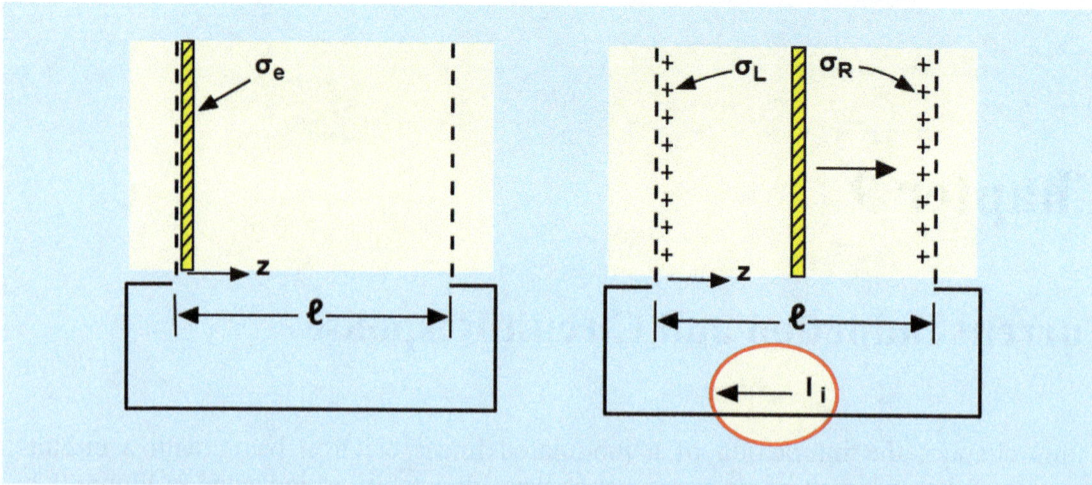

Figure 9-2 Current induction by a moving sheet of electrons.

Neglecting fringing fields at the edges of the grids and charge sheet, the charge density induced on the left grid is

$$\sigma_L = \sigma_e\left(1 - \frac{z}{\ell}\right) \tag{9-1}$$

and that induced on the right grid is

$$\sigma_R = \sigma_e \frac{z}{\ell} \tag{9-2}$$

As the position of the charge sheet changes, σ_L and σ_R vary as shown in Figure 9-3. The electron current density, J_i, induced in the grids is

$$J_i = \frac{\sigma_e}{\ell} u \tag{9-3}$$

where u is the velocity of the charge sheet. The circuit current is $I_i = A\, J_i$, where A is the area of a grid.

The preceding result is a form of Ramo's theorem. The importance is that it shows that immediately upon the entrance of a charge into a region, current starts to flow in nearby electrodes. The charge can be emitted from one of the electrodes and collected by the other or can enter and leave the region through grids. This analysis can be applied to RF gaps by assuming that RF magnetic fields are negligible (as in conventional microwave tubes) and that field propagation effects are negligible (valid if dimensions are small compared with a wavelength).

9.1.2 Induced Current Waveform

There is much to be learned from this simple relationship (9-3). First and most importantly, the current waveform induced in the grids and circuit may be very different than the current waveform of the charge sheet. Since the thickness of the charge sheet has been assumed to be infinitesimal, the volumetric charge density, ρ_e, must be infinite. The current density of the sheet, $J_e = \rho_e u$, is therefore infinite.

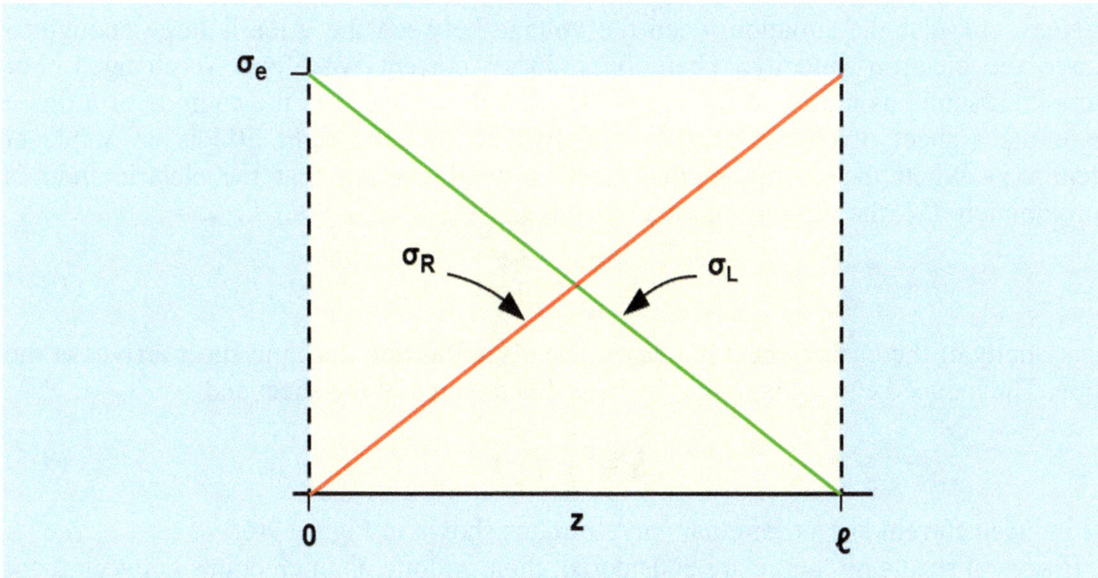

Figure 9-3 Variation of charge densities induced on the left and right grids by charge sheet at position z.

On the other hand, for constant velocity, the current density in the grids is constant as the sheet of charge traverses the gap. Thus, J_i has a rectangular waveform with a duration $\tau = d/u$, which is the transit time of the charge sheet across the gap. The waveforms for J_i and J_e are illustrated in Figure 9-4 for the charge sheet located at position z in the gap.

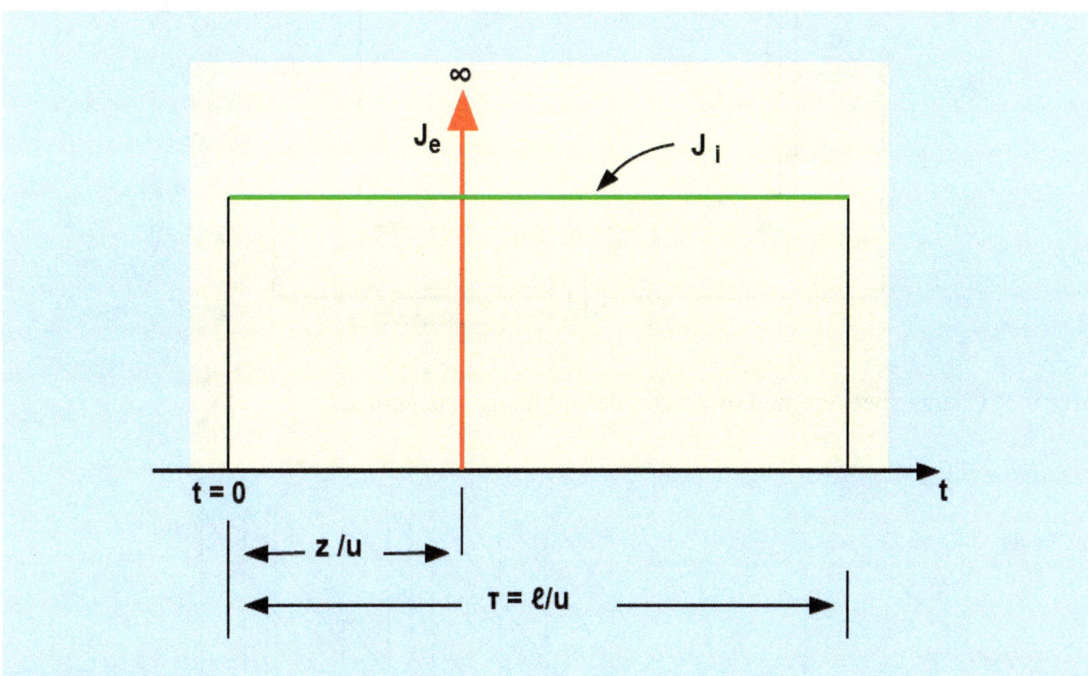

Figure 9-4 Current waveform induced in the circuit (J_i) and current waveform of the charge sheet (J_e).

Next, consider the situation when the voltage between the grids is large enough to change the electron velocity. Then the induced current waveform is changed. For example, assume, as indicated in Figure 9-5, that the left grid is the cathode of a diode that emits a sheet of charge at zero velocity and that the right grid is an anode at potential V. When the charge in the sheet is small enough that the electric field is approximately V/ℓ, the velocity of the electrons is

$$u = \frac{dz}{dt} = \eta \frac{V}{\ell} t \tag{9-4}$$

The velocity of the charge sheet increases linearly with time until the sheet arrives at the anode. The induced current is $I_i = A\,J_i$ where A is the area of the sheet and

$$J_i = \frac{\sigma_e}{\ell} \eta \frac{V}{\ell} t \tag{9-5}$$

The induced current has a triangular waveform, as shown in Figure 9-6.

If several sheets of charge are considered, then, as long as interactions between them are ignored, the current waveforms produced are superimposed. For the case of several sheets of electrons emitted at random times and leaving the cathode with zero velocity and being accelerated across the gap to an anode, the current pulses might appear as shown in Figure 9-6. This situation is fundamental to the analysis of shot noise in the current emitted by a cathode.

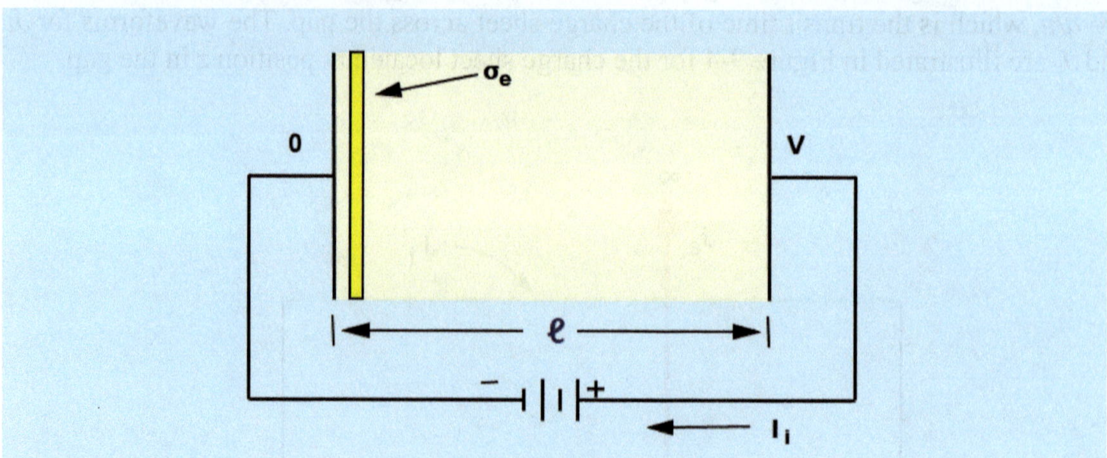

Figure 9-5 Charge sheet emitted by a cathode and being accelerated.

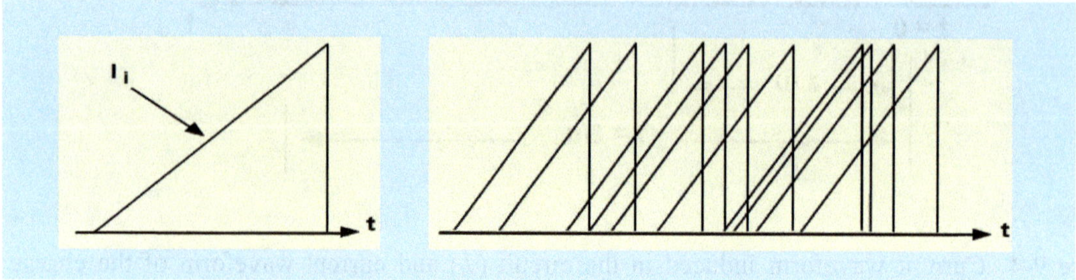

Figure 9-6 Current waveforms induced by accelerating charge.

9.1.3 RF Current Induced by an Electron Beam

To determine the current induced by an electron beam, the beam can be divided into slices with a thickness Δz, each one of which induces the current density, ΔJ_i, into the circuit. This is indicated in Figure 9-7. Then all of the increments of current can be added to obtain $I_i = A\,J_i$ where

$$J_i = \frac{1}{\ell}\int_0^{\ell} J_e dz \tag{9-6}$$

As an example of the application of (9-6), assume (as is the case in the small signal analysis of the modulation of electron beams) that the beam current contains an RF current component that varies sinusoidally with time and distance so that the total beam current density is

$$J_e = J_o + J_1 \sin\left(\omega t - \frac{\omega}{u_o} z\right) \tag{9-7}$$

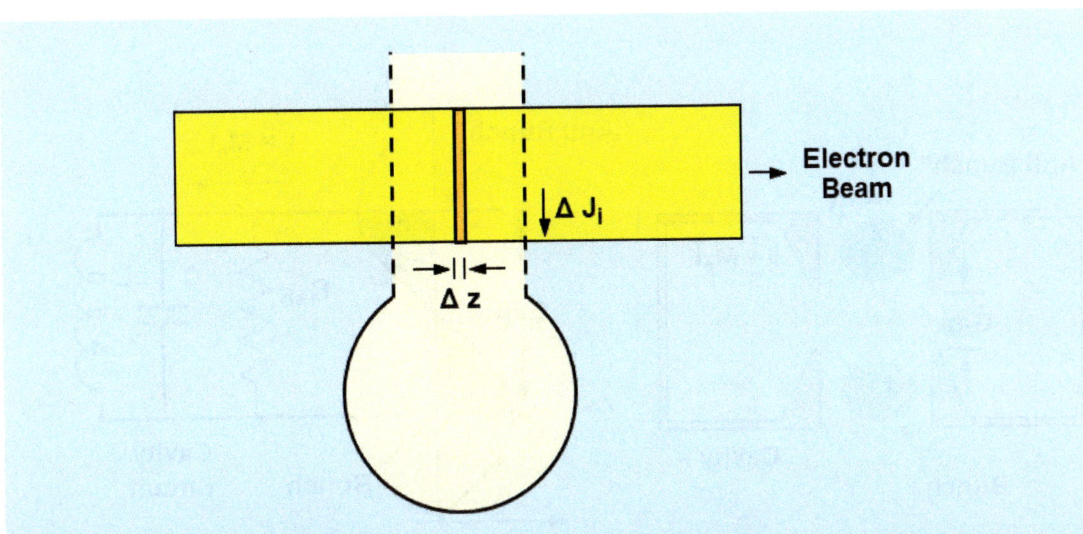

Figure 9-7 Situation assumed for determining the current induced by an electron beam.

When this current is used in (9-6), the RF electron current induced in the circuit and referenced to the center of the gap is found to be

$$I_i = J_i A M_p \sin\omega t \tag{9-8}$$

where M_p is the gap coupling coefficient for the planar gap (8-4) and A is the cross-sectional area of the electron beam.

The RF beam current density produced by a modulating cavity, J_1, is proportional to the velocity modulation; that is,

$$J_1 \propto u_1 \tag{9-9}$$

Since $u_1 \propto M_p$ (8-2), then

$$J_1 \propto M_p \tag{9-10}$$

This means that the induced current must contain M_p^2. (This assumes that the modulating gap and the current induction gap are the same. If the two gaps are different, then the induced current contains the product of the modulation coefficients.) Thus,

$$I_1 \propto M_p^2 \sin \omega t \tag{9-11}$$

For wide gaps, therefore, the induced current is much smaller than for narrow gaps. This effect is even more pronounced in gridless gaps, where M is a function of the radial as well as the axial dimensions of the gaps.

A cavity, its circuit model, and the electron bunch and the antibunch to be considered here are shown in Figure 9-8. The bunches are generated by the velocity modulation process described in Chapter 8. The antibunches are the regions from which electrons were taken to form the bunches.

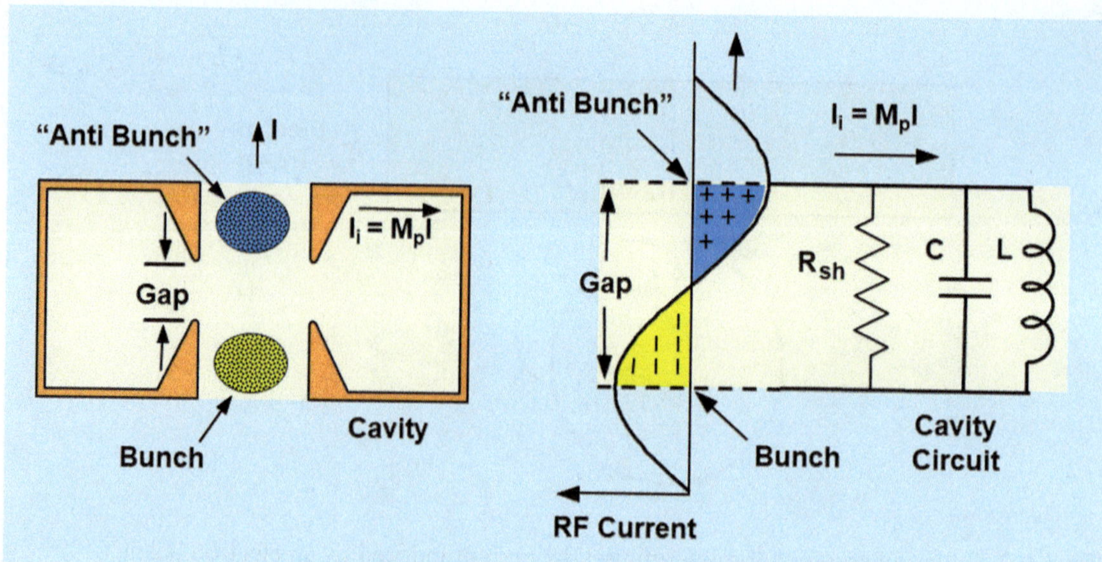

Figure 9-8 Cavity and circuit models used for RF current induction discussions.

9.2 CIRCUIT RESPONSE

The process by which the sinusoidal component of beam current is induced in a circuit is shown graphically in Figure 9-9. Here it is assumed that the induced current is in phase with the RF beam current. When discussing TWTs in Chapter 13 it will be shown that, in that case, there is a 90° phase shift between the RF beam current and the RF circuit current.

At time t_o, the position of the beam is such that the increase in beam charge density (electron bunch) is equal and opposite to the adjacent decrease in density (sometimes called an antibunch and denoted here by plus signs), so the electron current, I_i, induced in the circuit at t_o is zero. At time, t_1, only an electron bunch is between the grids and the induced current is maximum. The situation at t_2, is similar to that at t_o in that the net beam charge between the grids is zero and the induced current is zero. At t_3, only an antibunch is between the grids and the circuit current is minimum.

In these discussions of current induction, as long as the gap voltage is small compared to the beam voltage, the induced current depends only on the geometry of the gap. However, the interaction of the induced current with the impedance of the circuit attached to the gap does produce a voltage across the gap. That voltage, in turn, does produce effects on the beam.

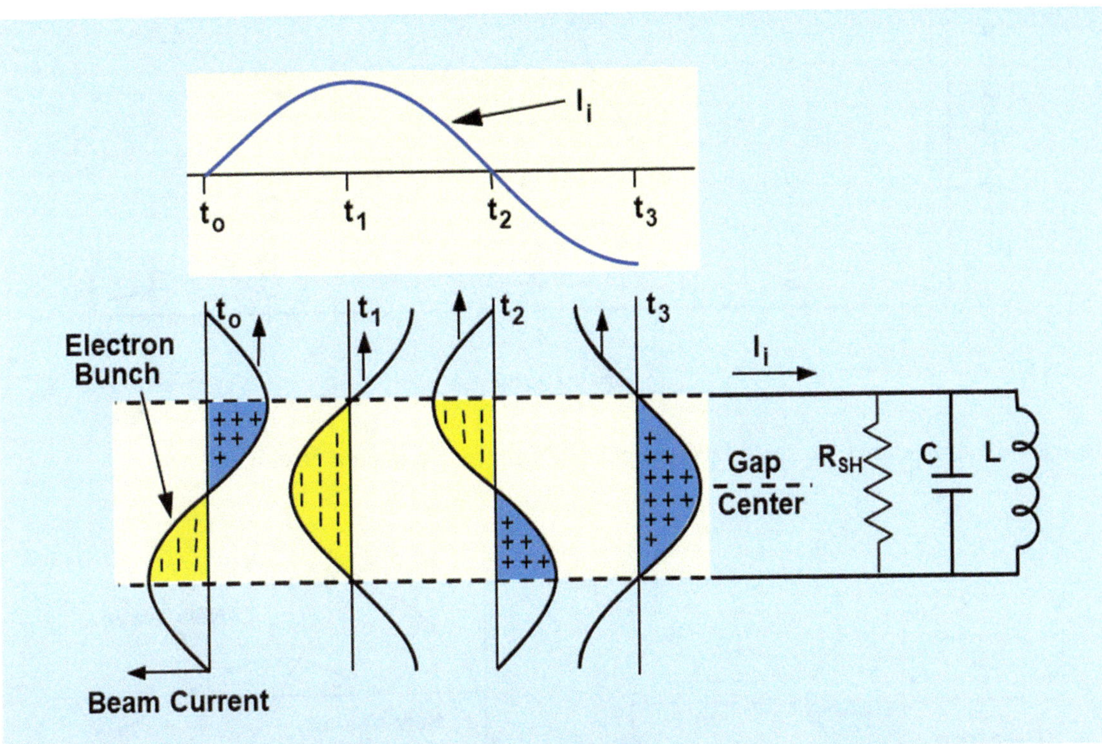

Figure 9-9 Current induction by a beam passing through a cavity.

9.2.1 Parallel RLC Circuit

First, it is important to remember that as a function of frequency, the impedance of a parallel RLC circuit is as shown in Figure 9-10. At resonance, the inductive and capacitive impedances are equal and opposite so they cancel. The impedance of a circuit driven at a frequency below the resonant frequency is inductive. At a frequency above resonance, the circuit is capacitive.

9.2.2 Resistive Load

When an electron beam modulated at frequency f_o passes through the gap, the induced current flows through the resistive portion of the circuit and produces the voltage shown in Figure 9-11. The phase of the induced voltage is such that it slows the bunch (slightly at small signal levels). At the same time, a new signal is launched on the beam, which is 90° out of phase with I_i. Part of the energy extracted from the bunch is dissipated in R_{sh} and part is used to launch the new signal.

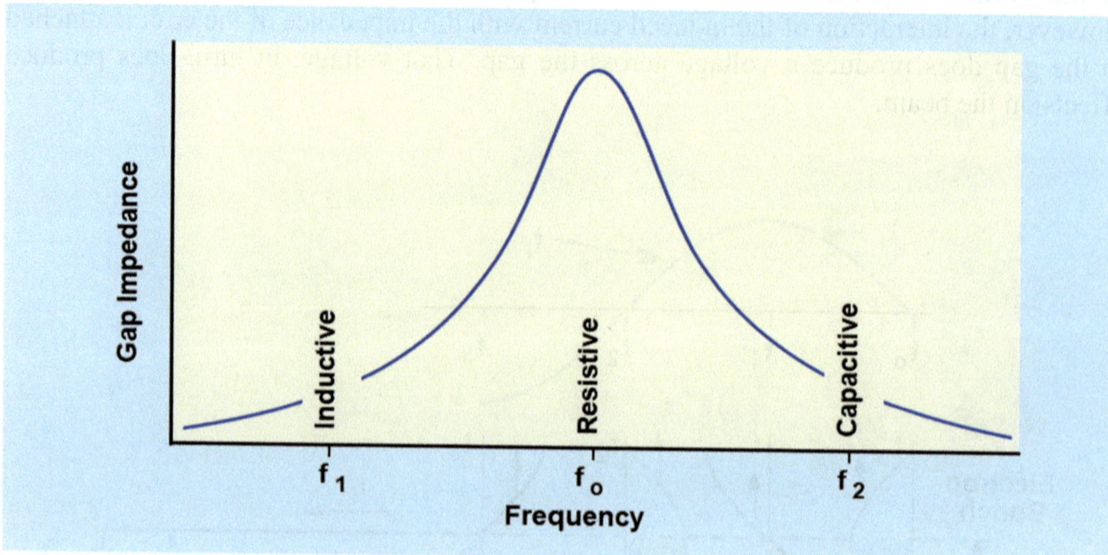

Figure 9-10 Impedance presented to an electron beam by a resonant circuit.

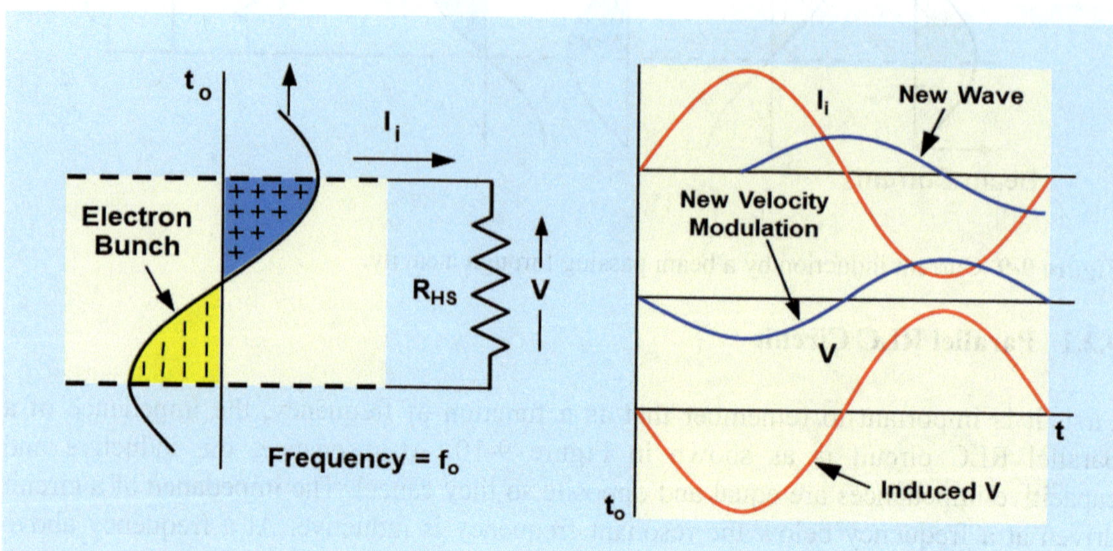

Figure 9-11 When the circuit is resistive (because the electron beam is modulated at f_o), the gap voltage generated is phased 180° with respect to the induced current. Energy is extracted from the bunch.

9.2.3 Inductive Load

When the electron beam is modulated at $f_1 < f_o$, the impedance presented to the beam is primarily inductive. The phase of the induced voltage with respect to I_i is as shown in Figure 9-12. (Remember that with conventional current in an inductive circuit, voltage leads current. With electron current, voltage lags current.) The voltage across the gap at t_o decelerates electrons in and near the leading edge of the bunch and enhances bunching. When the voltage has reversed 180° later, it accelerates electrons in and near the trailing edge of the bunch and again enhances bunching. Thus, at small signal levels, the signal launched on a beam by an inductive circuit voltage is in phase with the signal on the beam. At large signal levels near the output circuit of a klystron, an inductively tuned resonator is frequently used to enhance bunching by sweeping electrons from regions between bunches (decelerating electrons ahead of a bunch and accelerating those behind).

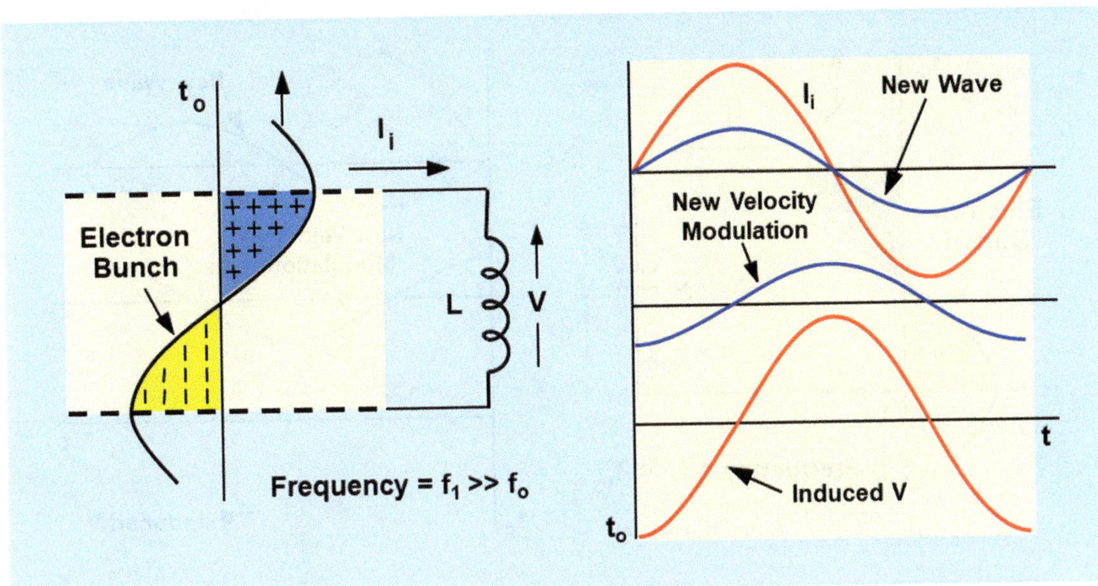

Figure 9-12 When the circuit is inductive, the induced voltage lags the electron current by 90°.

9.2.4 Capacitive Load

When the electron beam is modulated at $f_2 > f_o$, the impedance of the resonant circuit is primarily capacitive as shown in Figure 9-13. In this case, the induced electron current lags the voltage by 90°. The resulting effect on the electron bunches in the beam is the opposite of that which takes place with an inductive circuit. Thus, the voltage accelerates electrons away from the leading and trailing edges of a bunch, which tends to destroy the bunch. At small signal levels, the signal launched on the beam by a capacitive circuit is 180° out of phase with the signal on the beam.

9.2.5 Total Cavity Response

Because we're dealing with a velocity modulated beam, the total response of a cavity as evidenced by the effect of the cavity on the beam must be viewed from the position on the beam where the beam is to be used. Thus, the total response depends not only on the cavity itself but also on the lengths of the drift tubes on either side. This is because the gain depends on the ratio of the RF current produced by a cavity to the RF current entering the cavity. However, these RF currents depend on the velocity modulation on the beam and on the distances that the beam must travel for the velocity modulation to become density modulation. Figure 9-14 shows the gain pole that occurs near resonance along with a gain dip at a frequency that becomes higher as the drift tubes are made longer [1].

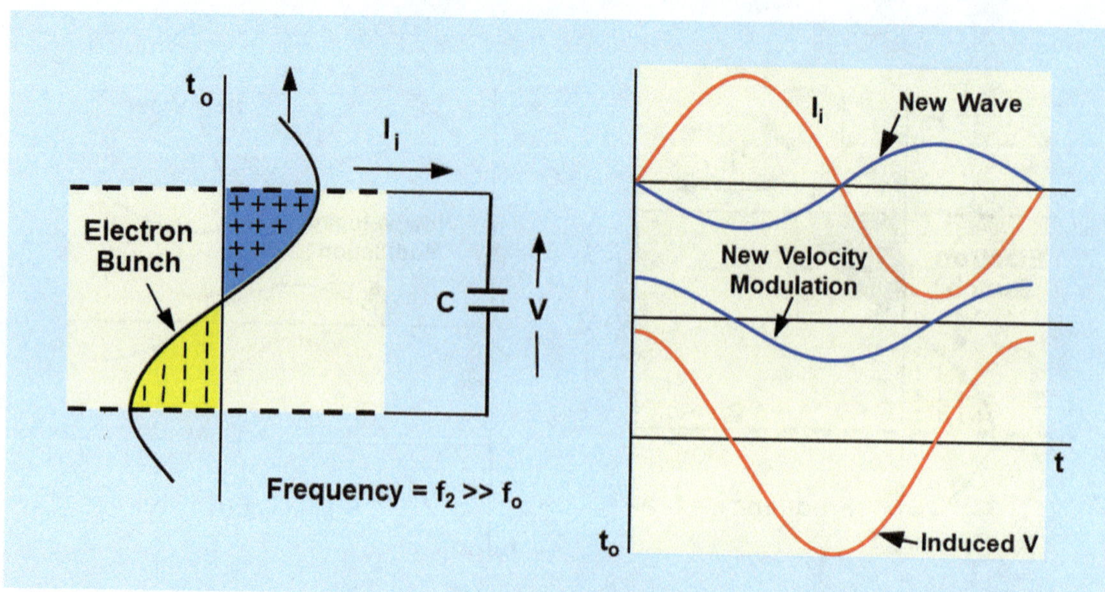

Figure 9-13 For a capacitive circuit, the induced electron current lags the voltage by 90°.

The reason for the gain dip can be understood from the discussion of capacitive operation in the previous section. When the frequency at which the cavity is being driven is well above the resonant frequency of the cavity, the impedance of the cavity is capacitive and the RF current generated by the cavity is phase shifted 180° with respect to the current exciting the cavity and so the residual driving current leaving the cavity offsets the current generated by the cavity. When almost complete cancellation occurs, the cavity gain becomes negligible and a zero occurs. This effect increases with drift tube length up to 90° because the peak of the driving current occurs at $\lambda q/4$ from the current source as velocity modulation becomes density modulation.

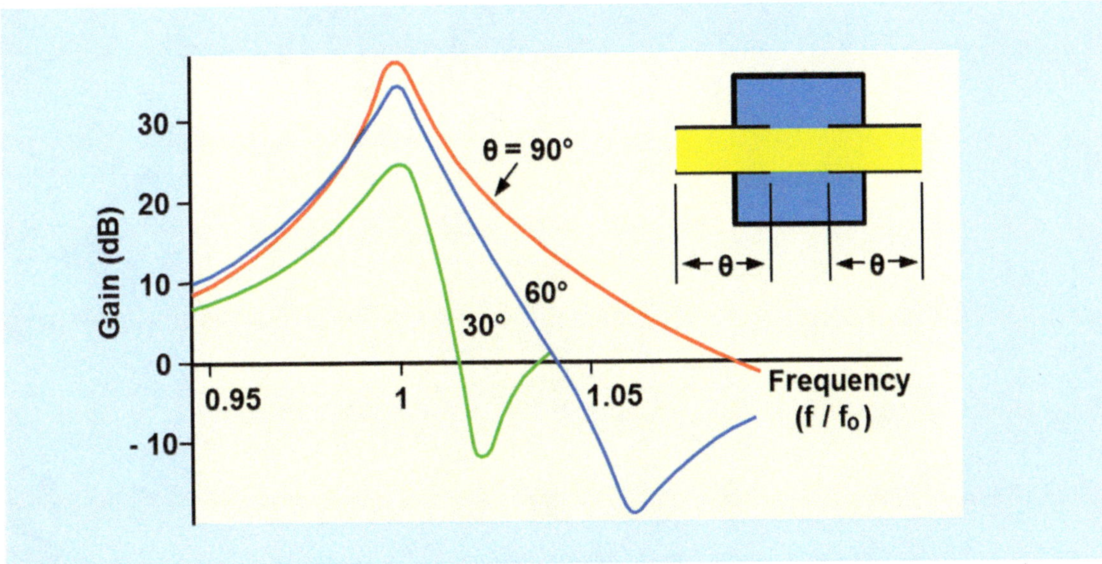

Figure 9-14 Example of the pole and zero of a single cavity as a function of drift tube lengths. (Adapted from: G. Faillon, 1985 IEDM. © 1985 IEEE.)

REFERENCE

[1] G. Faillon, "A 200 kW S band klystron with TWT bandwidth capability," *Technical Digest,* IEDM, 1985.

Chapter 10

Inductive Output Tubes

10.1 INTRODUCTION

Inductive output tubes (IOTs) are linear beam amplifiers that operate in the UHF-to-low-microwave-frequency range. The operation of an IOT can be explained by first considering a tetrode. In the tetrode, as Figure 10-1 shows, the functions of electron beam collection and RF power extraction are performed by a single electrode, which is the plate (anode). The size of this plate scales inversely with frequency of operation. At high frequencies, the small plate size is one of the factors that limits the power capability of tetrodes.

In an IOT, Figure 10-1 shows that beam collection and RF extraction are performed by separate electrodes. Most importantly, RF power is removed from the electron beam, not by collecting beam current, but by inductively coupling the RF current in the beam to a nearby circuit (usually a cavity). The beam current with the RF signal removed is then recovered by a collector. The size of the collector is determined by the dc beam power rather than the frequency of tube operation.

The IOT is sometimes referred to as a klystrode because it uses features of a klystron and a tetrode. Klystrode is Varian's trademarked name for the IOT.

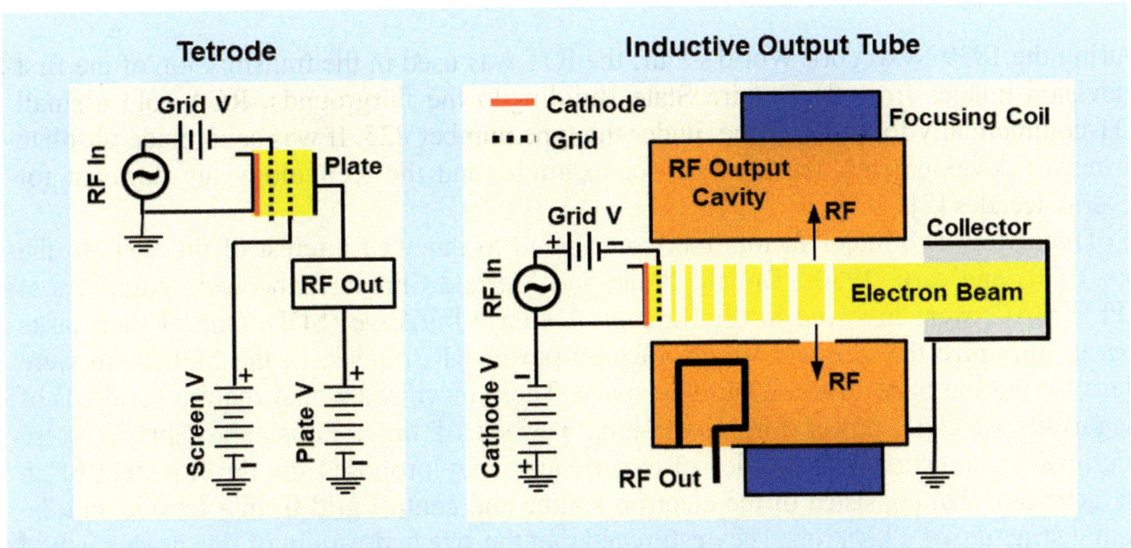

Figure 10-1 Comparison of output parts of a tetrode and an inductive output tube.

10.2 HISTORY AND BASIC OPERATION OF THE IOT

The IOT was invented in 1939 by Andrew V. Haeff. A patent [1] was issued for the IOT to Haeff and assigned to the Radio Corporation of America (RCA). In his patent, Haeff showed several output cavity arrangements for IOTs. One of the drawings is in Figure 10-2. The primary IOT components are identified as follows:

41 Cathode	36 Tail pipe
42 Grid	43 Collector
47, 48 Lecher line signal RF input to grid	52, 53 Output to load
38, 39 Output cavity and anode	

The functions of these components are described in the following paragraphs. The most important feature by far of Haeff's invention is the use of the output cavity to extract RF power from the beam without intercepting the beam. The beam is then intercepted by the collector.

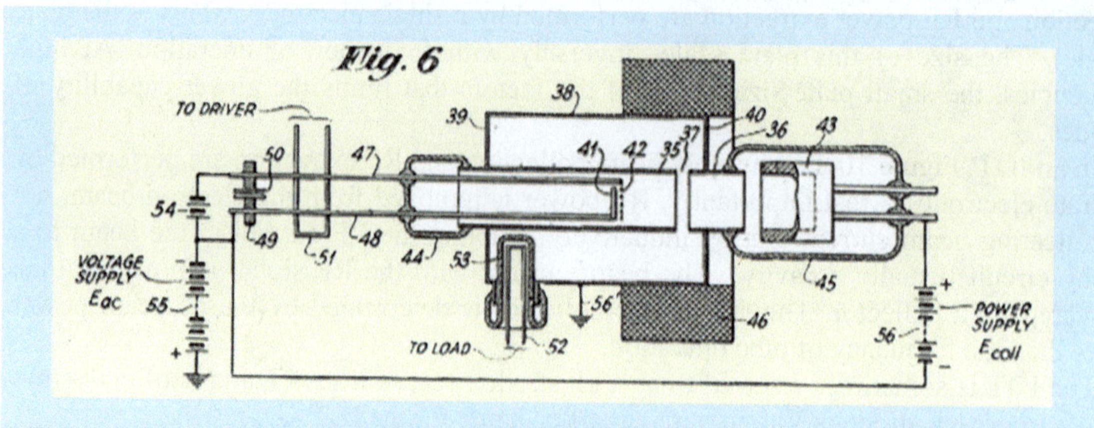

Figure 10-2 One of Haeff's inductive output tube figures.

During the 1939 New York World's Fair, the IOT was used in the transmission of the first television images from the Empire State Building to the fairgrounds. RCA sold a small IOT commercially for a short time, under the type number 825. It was soon made obsolete by newer developments, the klystron for example, and the technology lay dormant for several decades [2].

There were two major factors that contributed to renewed interest of the IOT. In the late 1970s and early 1980s, Donald Preist and Merrald Shrader, who were engineers at CPI/EIMAC, were involved in the Strategic Defense Initiative (SDI). One of their tasks was to minimize the size and weight of high-power RF sources. In the SDI, there were plans for putting particle accelerators in space. These machines would require hundreds of megawatts of UHF power for minute-long periods of time. Preist and Shrader were microwave engineers with decades of experience. They proposed the development of an RF generator that consisted of the electron source and control grid from a tetrode and the output structure of a klystron. They estimated that the size and weight of this device would be an order of magnitude smaller than for a klystron. They named this device the klystrode.

The 1981 paper by Preist and Shrader, "A high power klystrode with potential for space application" was their announcement for the new device [3]. That paper was soon followed by another: "The klystrode—an unusual transmitting tube with potential for UHF-TV" [4]. It wasn't long after the publication of their papers that Preist and Shrader learned about Andrew Haeff's invention of the IOT four decades earlier. Now, Haeff's name for the tube IOT is the name usually used for these devices. Klystrode is the CPI/EIMAC trademarked name for their version of the IOT.

Two important characteristics of IOTs that were highlighted by Preist and Shrader were efficiency and linearity. These characteristics led to the immediate application of the IOT to television broadcasting. Within 10 years (1992), a large percentage of TV broadcasting stations had replaced their existing amplifier with an IOT. A similar transition is taking place at a much slower pace in the particle accelerator industry.

Operation of an IOT is straightforward. An RF signal is applied between the cathode and the grid of a Pierce-like electron gun. This signal modulates the current from the gun. This process is referred to as density modulation. A small negative bias voltage applied to the grid prevents all but a small quiescent current from flowing when there is no input signal.

The IOT operates with the anode and RF output cavity at ground potential and a negative voltage of several tens of kilovolts on the cathode. The density-modulated beam is accelerated to the output cavity where it interacts with the cavity just as the beam in a klystron does. The RF power is removed from the beam and the spent beam is sent to the collector where its remaining power is dissipated or recovered.

10.3 ELECTRON GUN

As shown in Figure 10-3, the cathode and grid assembly is configured like that in a Pierce gun so that an electron beam can be formed. The beam is usually confined by a magnetic field. The grid is biased so that current flows only on positive (with respect to the cathode) portions of the signal input. As a result, the current leaves the cathode in discrete bunches.

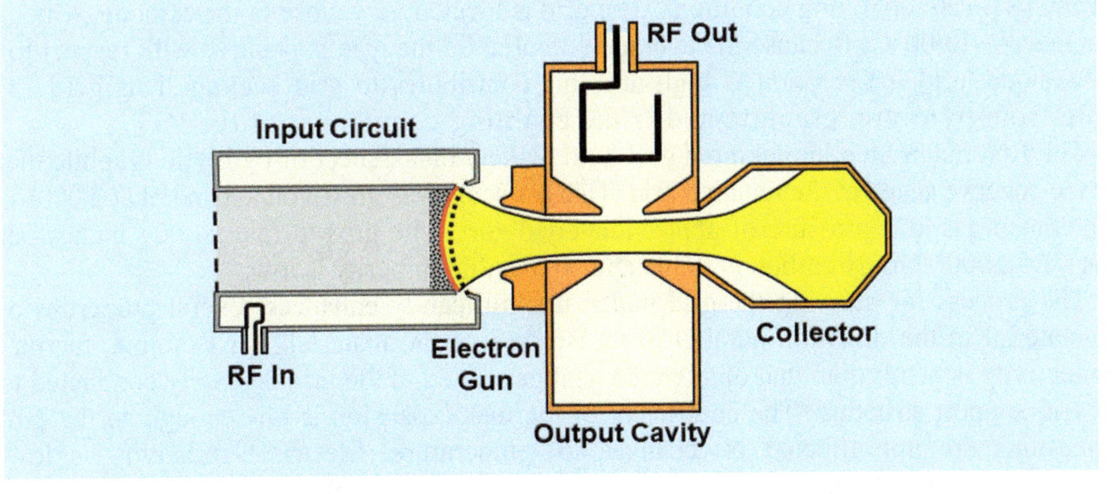

Figure 10-3 Drawing of inductive output tube (klystrode).

10.3.1 Grid

First, we have to remember that because the frequency of the operation of the IOT is in the UHF to low microwave range, proper RF connections to the grid and cathode are critical. In IOTs, it is common to use the quarter-wavelength resonant circuit shown in Figure 10-4 to couple the RF input signal to the grid.

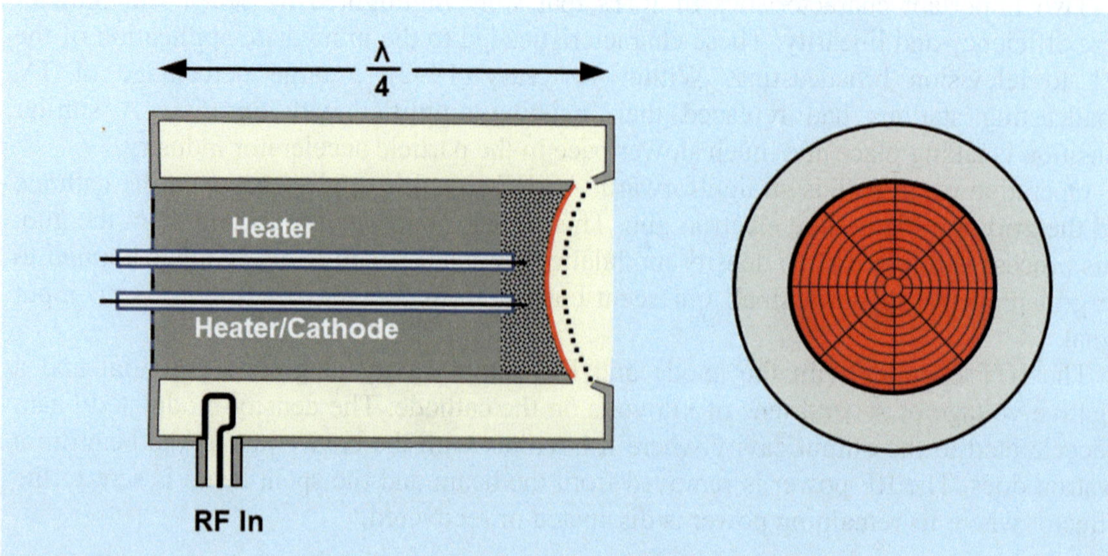

Figure 10-4 Use of λ/4 resonator to couple grid signal to grid.

The input circuit is complex only because it is necessary to drive the grid with RF at frequencies up to 1 GHz or higher while applying tens of thousands of volts negative to the cathode and an additional small bias voltage to the grid. In addition, it is necessary to maintain the outer surface of the enclosure for the input circuit and the frequency tuning mechanism at ground potential.

The control grid for an IOT is shown in Figure 10-5 [5]. This grid is subjected to extremely severe operating conditions. The grid is located very close to the cathode, which operates at ~1,000°C. Because the average potential on the grid is positive with respect to the cathode, grid interception is high and this contributes to grid heating. This grid is made from pyrolytic graphite and is the enabling component of the IOT.

The first major step in making a grid is to make a blank sheet of pyrolytic graphite that has the concave shape of the required grid. This blank sheet is grown on a mandrel at 2,000° C in methane gas at a pressure of approximately 1 torr. The growth rate is .001 inches per hour. At 2,000°C, only carbon is stable and so metallic impurity is low.

The process for growing the grid material significantly enhances several properties of the material in the direction parallel to the surface of the material. For example, thermal conductivity is better than that of silver so heat generated in the grid is readily conducted to the grid support structure. The coefficient of thermal expansion is low enough so the grid dimensions are not affected by changes in temperature. Electrical resistivity is low, comparable to that of copper.

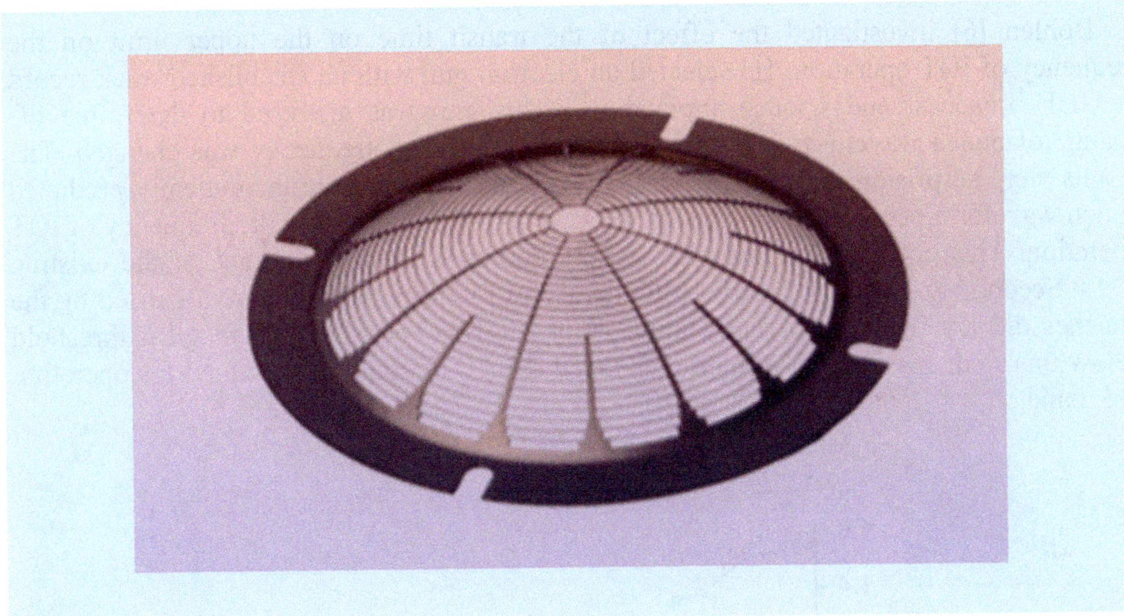

Figure 10-5 Pyrolytic grid used in an IOT. (From: M. Boyle, 2010 IEEE IVEC.)

10.3.2 Density Modulation

In an IOT, the RF input signal is applied between the cathode and the grid. A negative (with respect to the cathode) DC bias voltage is applied to the grid so that a small quiescent current flows with no RF drive. When a signal is applied to the grid, it is driven positive with respect to the cathode so that current flows. The electron gun electrodes shape this current flow into an electron beam just as they do in the electron gun of a klystron or TWT. The resulting beam current is proportional to the magnitude of the grid pulse voltage. The beam current is said to be density modulated.

There are two characteristics of density modulation that make the operation of the IOT superior to the operation a klystron or TWT, which use velocity-modulated beams:

1. Beam current only flows when it is needed rather than 100% of the time;
2. Beam current is limited by cathode emission rather than geometry (perveance).

The electron beam is thus density-modulated within the gun. The cathode is held at the negative beam potential of tens of kilovolts and the density-modulated beam is accelerated toward the anode and output cavity.

10.3.3 Transit Time, Frequency Limit

IOTs operate at frequencies in the upper part of the UHF band to L band. The maximum frequency at which IOTs are capable of operating is limited by the transit time of electrons from the cathode to the grid. If this transit time is comparable to the period of the RF signal, then emitted electrons may be returned to the cathode. This problem is aggravated by the space-charge cloud near the cathode surface, which reduces electron velocities.

Bohlen [6] investigated the effect of the transit time on the upper limit on the frequency of IOT operation. He selected an electron gun with an established track record in UHF broadcast and science applications. This gun was analyzed to determine the change of bunch waveform and fundamental RF current as frequency was changed. The results were surprising for two reasons. First, deviations from the theoretically predicted bunch waveform occurred at about 100 MHz, which was well below the frequency of IOT operation. This occurred without ever having degraded the performance of the existing IOTs. Second, as shown in Figure 10-6, the useful fundamental RF current carried by the bunches did not drop significantly until about 2 GHz. There was no magical threshold below this frequency and so Bohlen concluded that IOTs could be designed for operation in L band.

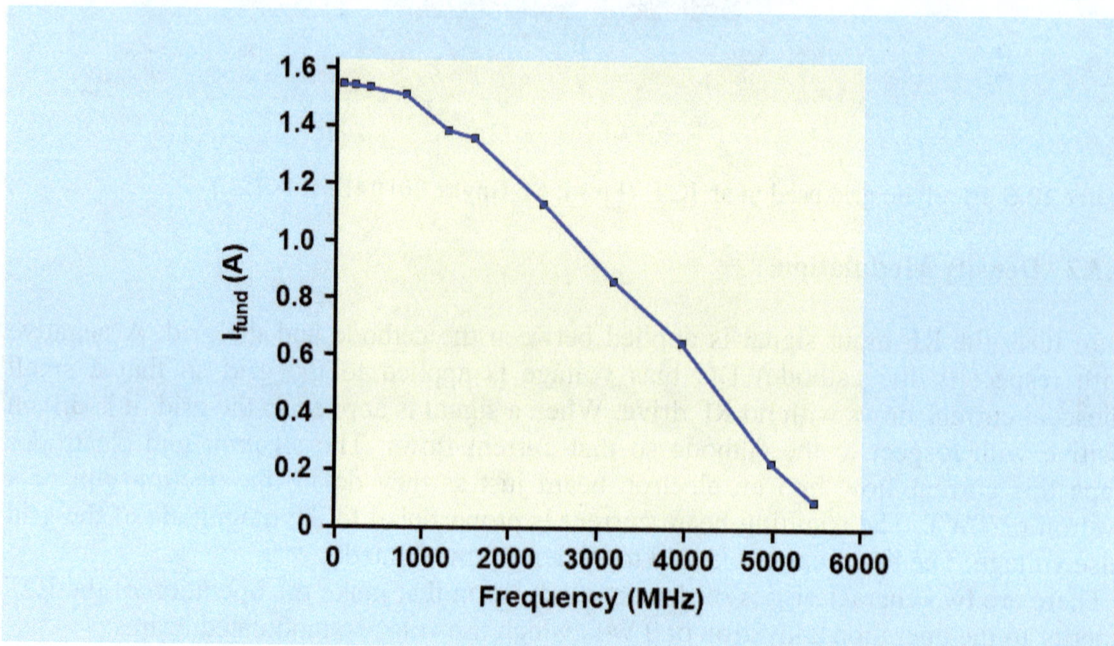

Figure 10-6 Calculated effect of transit- time limit. (From: H. Bohlen, IEEE IVEC 2003.)

10.3.4 Linearity

Transfer curves (power out vs. power in) for an IOT and a klystron are shown in Figure 10-7. The power levels are normalized with respect to the point of maximum efficiency of the klystron. Notice that when the klystron saturates, no change in output power occurs when the input power is changed. However, at that same output power, the IOT power change is nearly proportional to the input power change. Thus the IOT is far more linear than a klystron.

As in a klystron, field-free drift tube regions just before and after the output cavity gap serve as waveguides beyond cutoff to ensure that the output interaction space is isolated from the input space (grid-anode) and also from the collector.

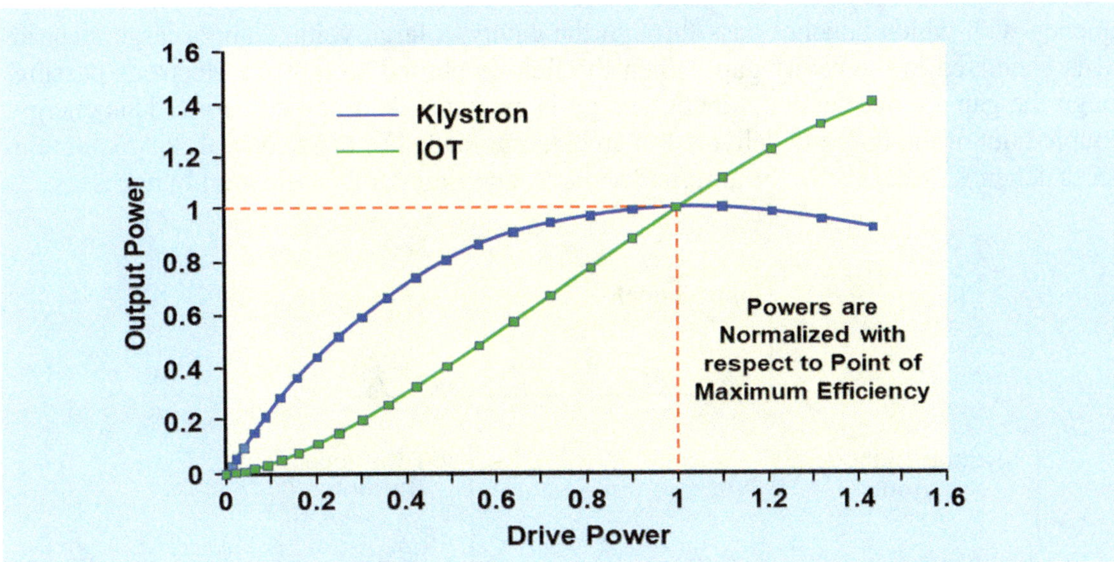

Figure 10-7 Transfer curves for a klystron and an IOT. (From: H. Bohlen, IEEE IVEC 2003.)

In Figure 10-8, the effect on the current of amplification of an RF input signal in a klystron and an IOT is shown. In a klystron, for a given anode voltage, the current is controlled by the perveance of the electron gun, which depends only on the geometry of the gun. As Figure 10-8(b) indicates, there is no effect of the input signal on cathode current. The RF beam current is generated by the velocity modulation process, which amounts to moving the phase positions of the electrons. There is no mechanism for adding electrons to the bunches. The result, shown in Figure 10-8(c), is that when the signal is very large saturation occurs and the signal cannot be sufficiently amplified. The saturation curve for a klystron is shown in Figure 10-7.

As Figure 10-8(d) shows, the IOT passes a small idle current continuously but the primary conduction occurs when a grid signal is present. Assuming that the cathode is operating space-charge-limited, the amount of current is controlled by the amplitude of the grid signal. When a signal with a large amplitude is processed, the grid voltage is made large enough to produce the current required to amplify the signal. The linearity of the transfer curve is a reason why the IOT is particularly suited for the transmission of digital-television and high-definition digital-television signals with the 8-VWB TV modulation that is used in the United States. The amplitude of the sync pulses is 6 dB (4×) the amplitude of the video signal. In order to process this signal, it is necessary to operate a klystron at a power level well below saturation. The IOT, on the other hand, can be operated at the klystron saturation power level and still accommodate the 6-dB sync pulses.

10.3.5 Efficiency

After being accelerated by the anode voltage the bunches pass through the output cavity. The interaction of the bunches with the cavity is identical to the interaction that occurs in the output cavity of a klystron. The RF current in the density-modulated beam induces RF current in the nearby cavity walls. If the resonant frequency of the cavity is near the frequency with which bunches pass through the cavity, a large voltage and a large electric field is generated in the cavity gap. When the field is phased so that the electrons passing through the gap are decelerated, kinetic energy is extracted from the electrons. This energy is coupled out of the IOT and delivered to an external load. The efficiency of this extraction process depends primarily on the amplitude, shape, and timing of the electron bunches.

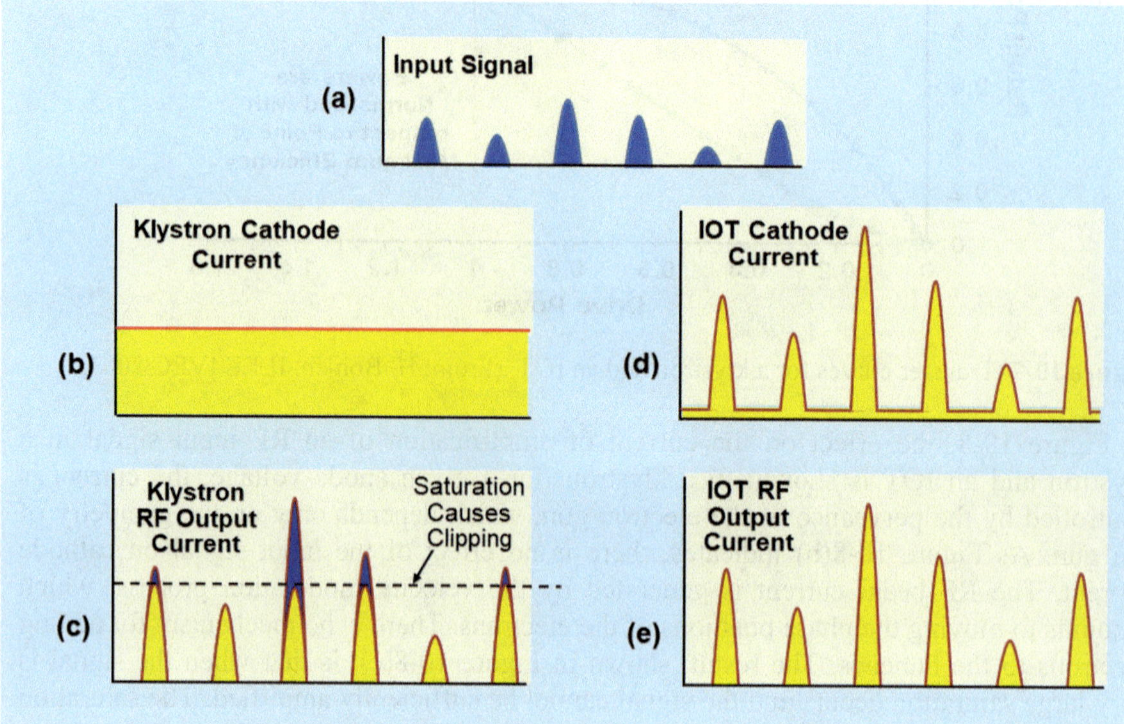

Figure 10-8 Comparison of IOT and klystron currents.

As with klystrons and TWTs, there are many factors that must be considered when specifying the efficiency of an IOT. These factors are all discussed in detail in Chapter 16. First, there is the overall efficiency, which is the efficiency that determines the price that has to be paid to operate the IOT. This includes power to the focusing solenoid, beam power, power to operate the cooling system, and filament power.

The electronic efficiency is the efficiency with which RF beam power is extracted from the electron beam during interaction with the RF circuit. The circuit efficiency is the efficiency with which the extracted power is delivered to the output port of the tube. Finally, the collector efficiency is the efficiency with which the collector recovers energy from the spent electron beam.

Electronic efficiencies approaching 80% are possible in class C operation. One of the reasons for high efficiency is that, as shown in Figure 10-8, current flows in the IOT when the grid signal turns the beam from the cathode on so the beam consists of bunches of electrons. If the bunch characteristics (primarily current, duration, and timing of arrival at the output gap) are properly controlled, then maximum energy will be extracted from the bunch when it passes through the output gap. Then electronic efficiency will be maximized. For most accelerator applications, a single dedicated frequency of operation is used and so the IOTs have high electronic efficiency.

In IOTs used in television service, the first order of business is to amplify the video and audio signals with an acceptable level of distortion. After that has been accomplished, excessive energy remains in the spent electron beam. If that energy can be recovered and returned to the power source, efficiency of the IOT can be improved. Energy can be recovered by sorting the spent electron beam into energy classes and then slowing the electrons in each class nearly to zero with an electric field. The energy thereby recovered is then sent to the cathode of the IOT so the power input to the IOT from the power supply is lower than without recovery. This energy sorting and recovery process is accomplished with a collector on an IOT that has several electrodes operating at the appropriate voltages. This collector is called a multistage depressed collector (MSDC).

All modern IOTs used for television service use MSDCs. CPI reported the results in Figure 10-9 for one of their MSDCs.

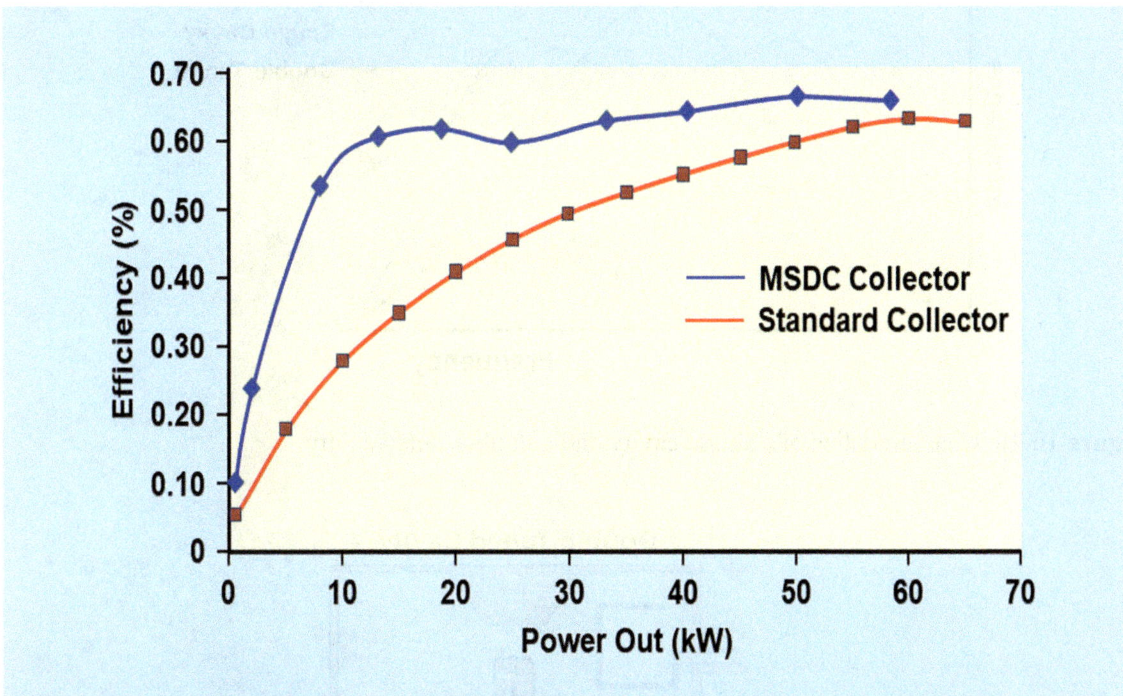

Figure 10-9 IOT efficiency improvement with an MSDC. (From C. Yates, IEEE IVEC 2010.)

10.4 OUTPUT CIRCUIT

10.4.1 Bandwidth

The bandwidth of an IOT with a single output cavity is very limited and is on the order of 1% or less. The bandwidth required for television is from 6 to 8 dB depending on the modulation technique. Bandwidth can be increased by modifying the RF output circuit. For example, Figure 10-10 shows the improvement in bandwidth that can be obtained by using a double-tuned output cavity. Figure 10-11 is a drawing of an IOT with a double-tuned output cavity.

10.5 COLLECTOR

The collector for an IOT is either a simple copper bucket or a sophisticated multielectrode component, which is usually referred to as an MSDC. The difference between the bucket and the MSDC is the way in which the power remaining in the spent electron beam is handled. In the bucket collector, the beam impinges on the metallic surface of the collector and all of the energy in the beam is dissipated as heat. In the MSDC, the electrons in the beam are sorted according to their energy content and much of the energy is recovered. IOTs with MSDCs are used for applications such as TV broadcasting where efficiency is important. Chapter 16 contains detailed information concerning collectors.

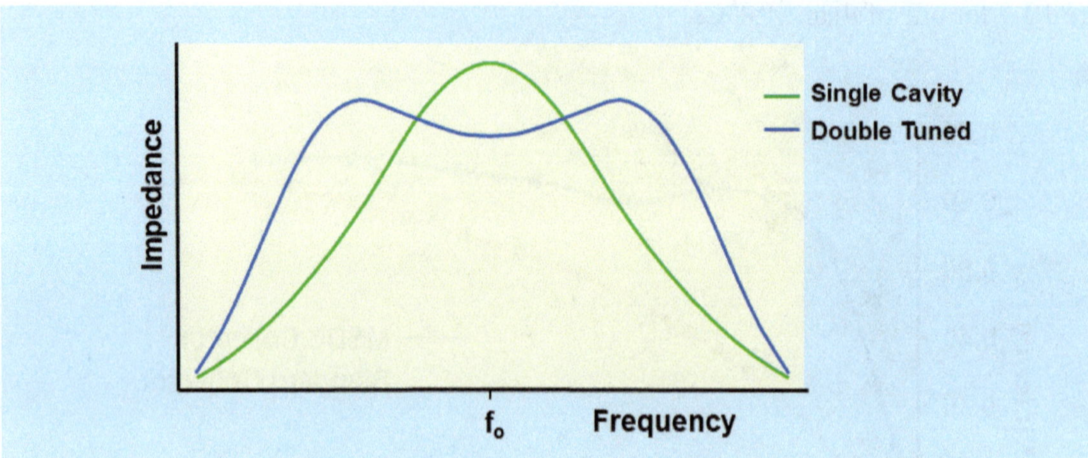

Figure 10-10 Characteristics of a single cavity and a double-tuned cavity.

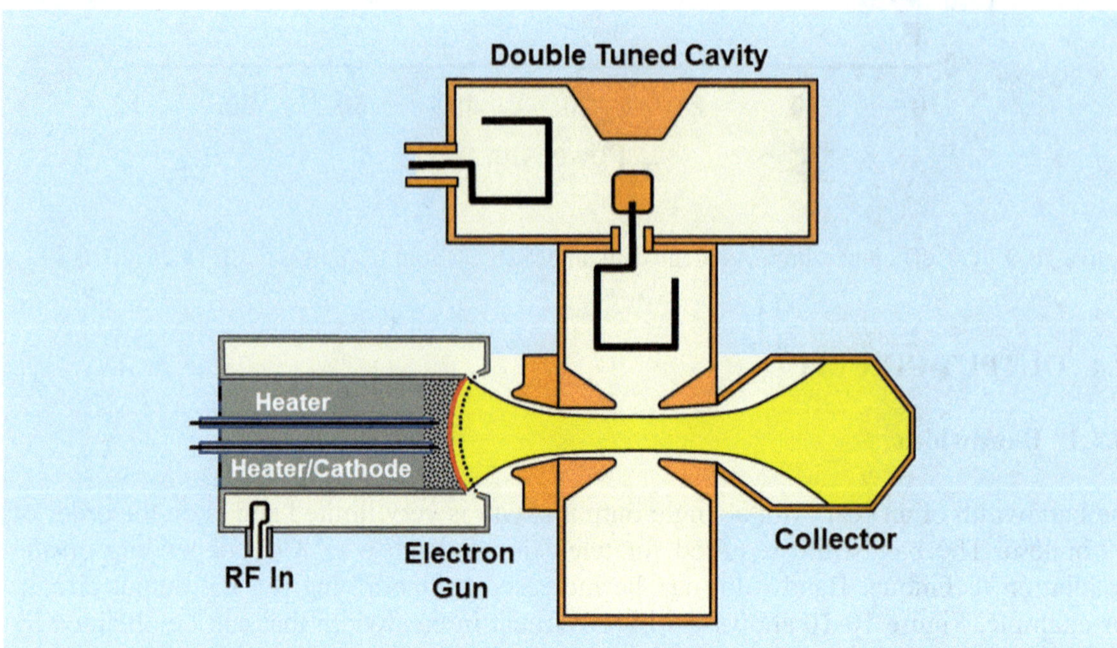

Figure 10-11 IOT with double-tuned output cavity.

10.6 POWER

The power capability reported for most IOTS is less than 100 kW. One exception is the Chalk River tube, which operated at 67 kV and produced 250 kW CW at 267 MHz [7].

A power level of over 1 MW was required for the European Spallation Source (ESS) accelerator. To meet that requirement, multibeam IOTs were developed. These tubes consist of multiple modestly powered IOTs operating in parallel. Each beam assembly in one of these tubes is based on the design of IOTs with proven track records.

The primary problem with trying to build single-beam devices operating at the megawatt level is that, to achieve long life, the cathode emission density must be limited. So, to obtain high power, a large cathode area is required. But a grid must span the entire cathode area with uniform spacing to the cathode. The control of this spacing becomes more and more difficult as cathode area increases.

10.7 SIZE AND WEIGHT

Figure 10-12 shows the primary reason why IOTs are smaller and weigh less than klystrons. In a klystron, the beam is velocity-modulated at an RF input gap. After the electrons leave the input gap, fast electrons catch up with slow electrons as they pass through a long drift space and so electron bunches are formed. This process is repeated several times before the bunches are sufficiently dense to produce the desired power output.

In an IOT, on the other hand, the RF input signal is applied between the cathode and a grid that is positioned close to and in front of the cathode. The electron beam is thus density modulated so that bunches are produced in the gun. The bunches then leave the gun and are immediately accelerated to high velocity by the cathode-to-anode voltage that is tens of kilovolts. The bunches then pass through and give up their RF energy to the output gap.

Thus the IOT requires one cavity next to the gun to bunch the beam whereas the klystron normally requires at least five widely spaced cavities, each one of which is approximately the size of the IOT cavity, to do the same thing.

10.8 APPLICATIONS

10.8.1 IOTs for Television

The primary application for IOTs is now television service. Because of their linearity, efficiency, and small size, they have replaced the klystrons that were previously used. MSDCs, which are described in detail in Chapter 16, are used to enhance efficiency.

Figure 10-13 shows an Eimac MSDC IOT that has a three-stage collector. This tube is installed in the hardware set, which contains the driver circuitry and the double-tuned output external cavity. This tube delivers 130 kW peak and 30 kW average.

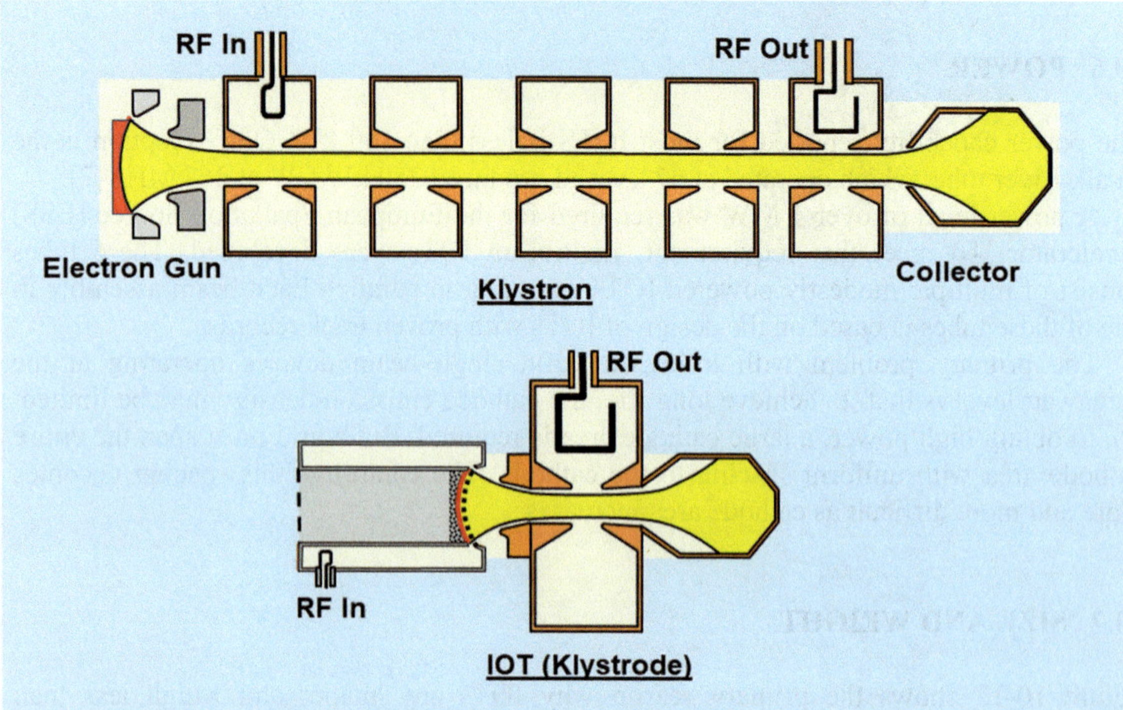

Figure 10-12 Comparison of IOT and klystron sizes.

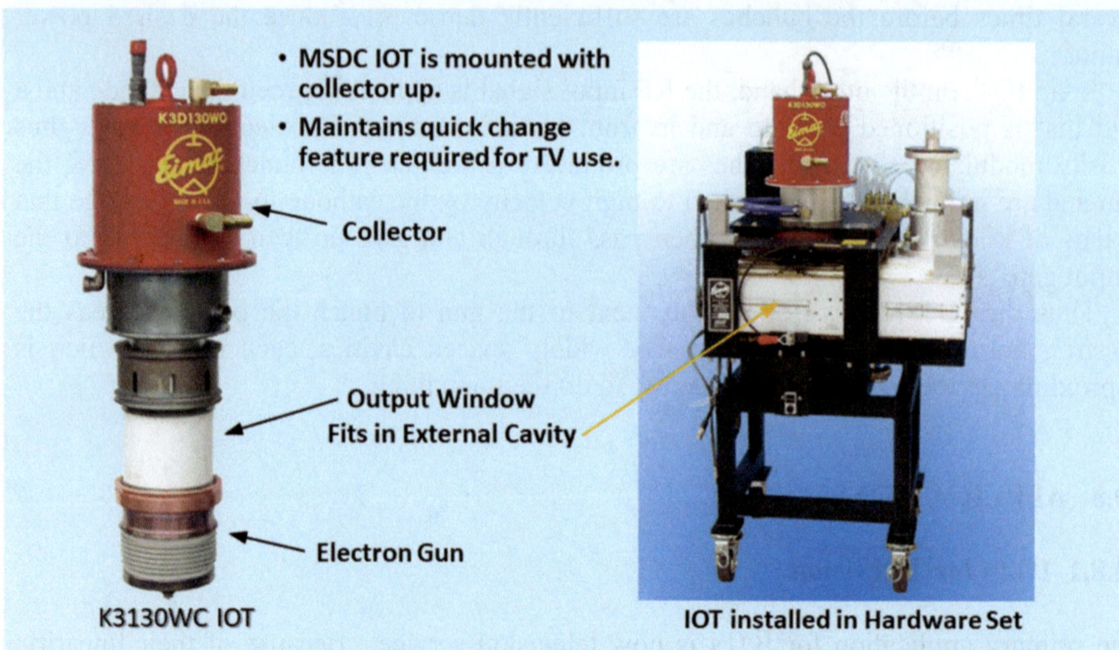

Figure 10-13 CPI/EIMAC IOT with a multistage depressed collector. (From H. Bohlen et al., LINAC 2004, August 16-20, 2004.)

An L-3EDD IOT and its hardware set are shown in Figure 10-14. This IOT is called a constant efficiency amplifier (CEA) by L-3 [9]. The tube uses an MSDC to produce a maximum efficiency of nearly 65%, as indicated in Figure 10-15. Notice that the CEA efficiency is substantially higher than the efficiency of the solid-state amplifier (yellow plot).

Figure 10-14 L-3 constant efficiency amplifier (IOT). (From M. Boyle, IEEE IVEC 2010.)

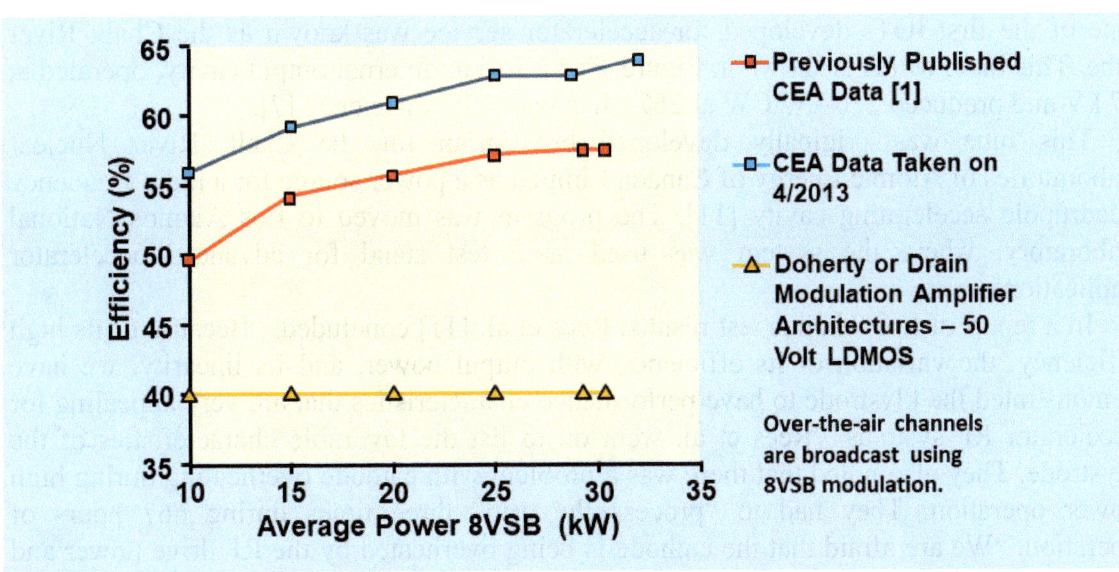

Figure 10-15 Efficiency of L-3 CEA. (From M. Boyle, IEEE IVEC 2010.)

Figure 10-16 contains information concerning the life of L-3 IOTs. In commenting on Figure 10-16, Boyle stated "When L-3 started manufacturing IOTs in 1998 the expected lifetime for an IOT was thought to be on the order of 4 years. Since that time we have been monitoring the installation of over 500 IOTs and MSDC IOTs." This duration can be seen in Figure 10-16. "With many IOTs still in service after 10 years, the average life can now be expected to be more than double the original estimates" [10].

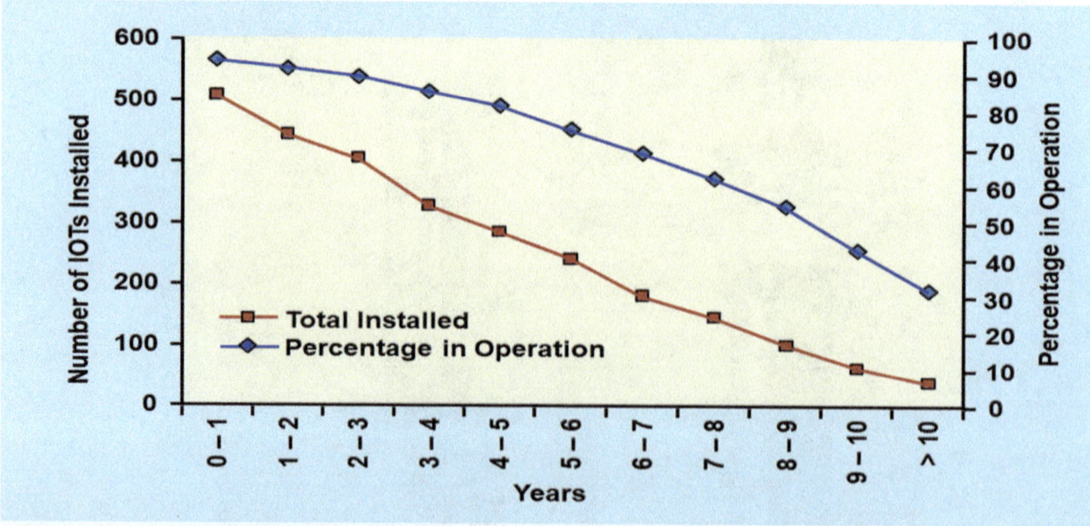

Figure 10-16 Percentage of initial group of over 500 L-3 IOTs still in operation as a function of time. (From M. Boyle, IEEE IVEC 2014.)

10.8.2 Single-Beam IOTs for Accelerators

One of the first IOTs developed for accelerator service was known as the Chalk River tube. This tube, which is shown in Figure 10-17, had an internal output cavity, operated at 67 kV and produced 250-kW CW at 267 MHz with 73% efficiency [7].

This tube was originally developed by Varian for the Chalk River Nuclear Laboratories of Atomic Energy of Canada Limited as a power source for a radio frequency quadrupole accelerating cavity [11]. The program was moved to Los Alamos National Laboratory, where the system was used as a test stand for advanced accelerator applications.

In a report of Los Alamos test results, Rees et al. [11] concluded, "Because of its high efficiency, the variation of its efficiency with output power, and its linearity, we have demonstrated the klystrode to have performance characteristics that are very appealing for accelerator RF systems." Rees et al. went on to list the favorable characteristics of the klystrode. They also noted that there was a problem with cathode overheating during high power operation. They had to "process the grid" three times during 667 hours of operation. "We are afraid that the cathode is being overheated by the RF drive power and that the result will be a reduced tube life." Figure 10-18(a) shows the power produced by the tube and Figure10-18(b) shows that the efficiency was over 70% for power outputs over 150 kW.

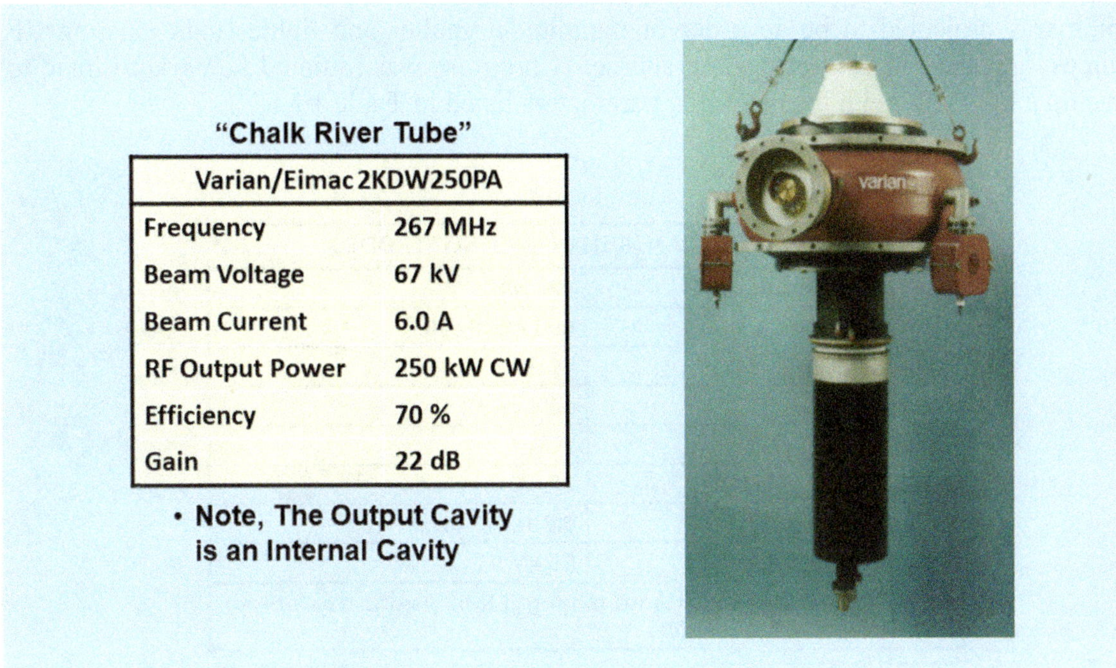

Figure 10-17 The Chalk River tube. From H. Bohlen et al., (LINAC 2004, August 16–20, 2004.)

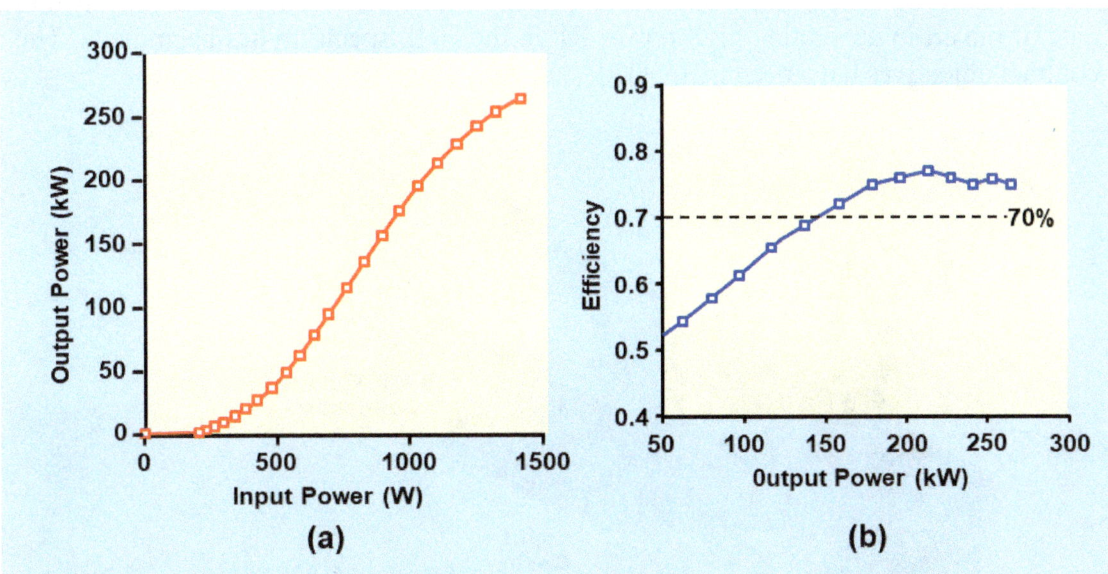

Figure 10-18 Power output and efficiency of the Chalk River tube. (From D. Rees et al., *Conference Record, IEEE Particle Accelerator Conference*, Dallas, TX, May 1995.)

In 1991, Preist and Shrader reported on another high-power klystrode development [12]. The 500-kW peak power, 50-kW average, 425-MHz tube shown in Figure 10-19 was described. The support for this program was from Los Alamos National Laboratories and was part of the Strategic Defense Initiative. In the early 1980s the need for spaceborne particle accelerators that would require megawatts of RF power was foreseen. Existing RF sources were far too heavy and too large to ever be practical in space. The Klystrode, which was projected to be an order of magnitude smaller and lighter than existing RF sources, appeared to be a promising source. A program was initiated at Varian/Eimac to develop a klystrode with the operating parameters listed in Table 10-1.

Table 10.1

EIMAC/VARIAN HIGH POWER KLYSTRODE	
Operating Frequency	425 MHz
Power Output	500 kW Peak
Pulse Length	11 ms
Duty Factor	10%
Efficiency	70%
Power Gain	20 dB
Bandwidth	5MHz (-3 dB)
Operating Voltage	85 kV
Material: Suitable for ground testing. Light weight was not an objective of this contract.	

At the conclusion of the program, Preist and Shrader stated that "A significant advance in the state of the art in generating high-power RF in the UHF spectrum has been made. The full contract objectives have been met." [12].

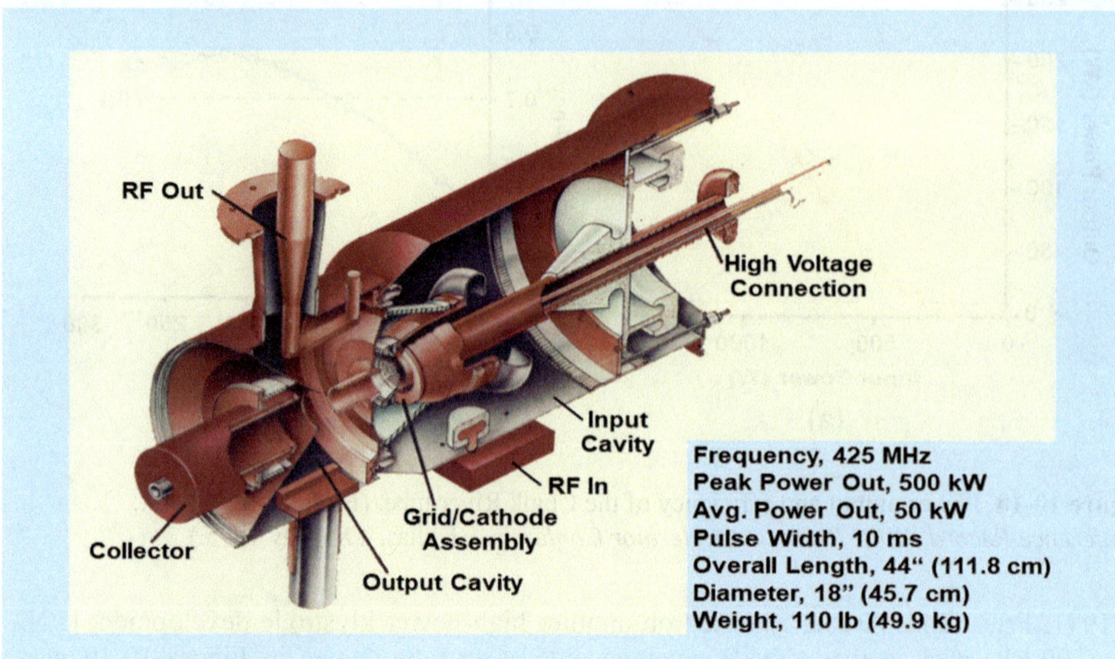

Figure 10-19 Drawing of EIMAC/Varian 500-kW, 425-MHz klystrode. (Courtesy of EIMAC/Varian.)

At this point we note that recent successful projects to produce powers substantially higher than the 500 kW reported here have used multiple beams and so the voltages are nearly a factor of two lower than the 85 kV reported here. That work is described in the last section of this chapter.

There are many facilities in the world that use single-beam IOTs for industrial and scientific services as well as accelerator applications. Most of these IOTs are based on broadcast tubes with the power limited to approximately 90 kW [13].

One of these facilities is the Diamond Light Source located in Oxfordshire, UK. This is the UK's national synchrotron science facility. Its purpose is to produce intense beams of light, the special characteristics of which are useful in many areas of scientific research. The facility became operational in 2007 and was the first synchrotron light source to use IOTs. The facility has three storage-ring amplifiers. Each amplifier combines the RF power out of four 80-kW IOTs to achieve 300-kW CW at 500 MHz.

ALBA is another synchrotron light source that uses IOTs. ALBA operates six amplifiers for the storage ring and each amplifier uses the combined output of two IOTs. Other accelerator facilities, including CERN, Brookhaven, Daresbury Laboratory, Willy Wien, and Elettra now also use IOTs [14].

Several 50- and 60- kW amplifiers are operating in three institutes in Japan [15].

Shortly after Bohlen reported that IOTs should be capable of operating in L band [6], Li described a 20-kW IOT operating at 1.3 GHz [16]. Also, CPI announced the development of the 1,300-MHz VKL-9130 IOT shown in Figure 10-20. Two versions of this tube are available. The VKL-9130A produces up to 30-kW CW at an efficiency of 65%. The VKL-9130B, for which the peak power is shown in Figure 10-21, produces up to 100 kW of pulsed power.

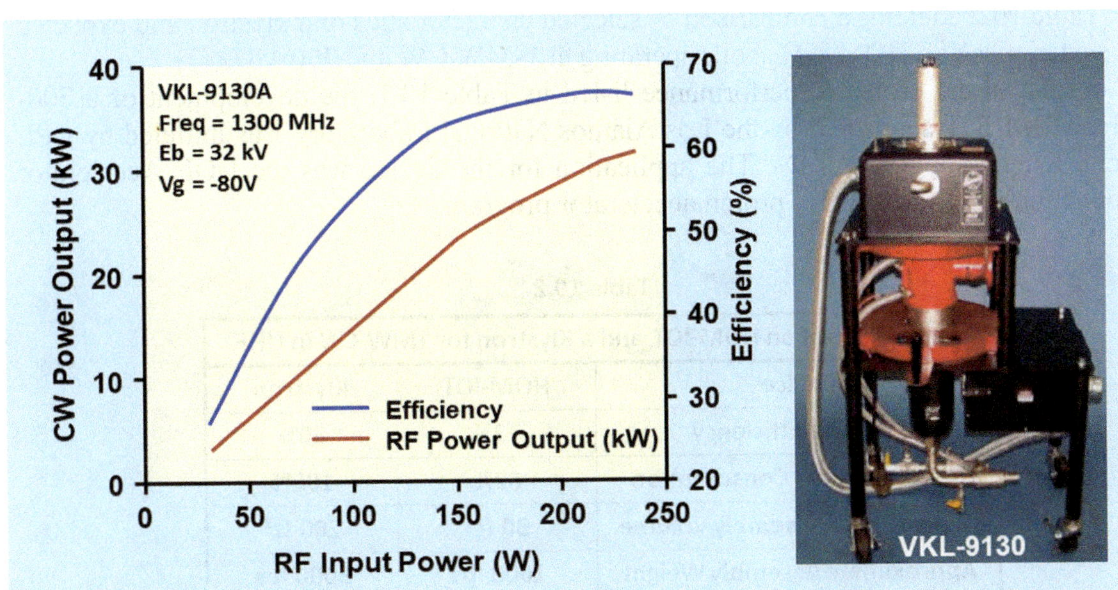

Figure 10-20 Power output and efficiency of an L-Band CW IOT. (From IEEE IVEC 2014.)

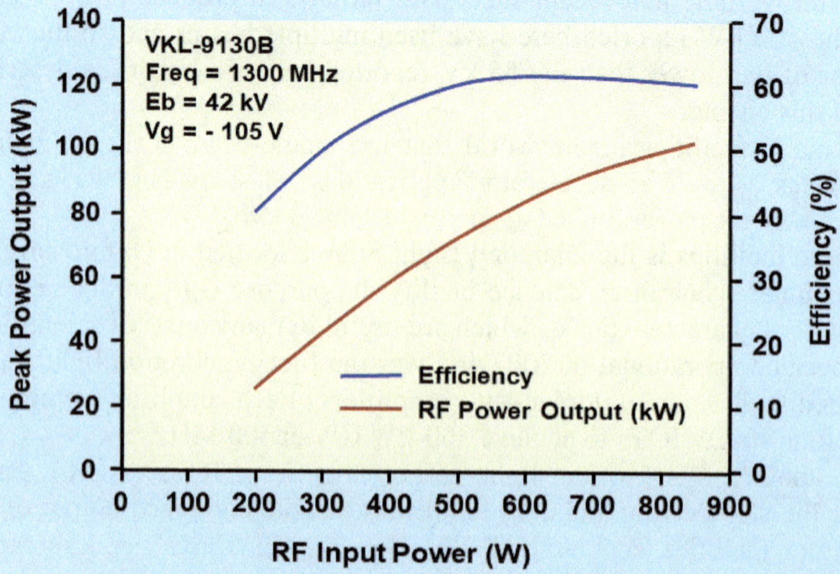

Figure 10-21 Power output and efficiency of an L-Band pulsed IOT. (From IEEE IVEC 2014.)

10.8.3 Multiple-Beam IOTs for Accelerators

Higher-order mode (HOM) IOTs use multiple beams or large electron beams with quasi-annular cross sections that interact with outer voltage maxima in TM_{0n0} cavities. The advantages are low beam voltages, low emission densities, and low energy densities.

Table 10.2 contains a comparison of selected characteristics of a klystron and expected characteristics of a HOM-IOT, both operating at 1-MW CW and 700 MHz [7].

Based on the projected performance listed in Table 10.1, the development of a 700-MHz HOM-IOT sponsored by the Los Alamos National Laboratory was initiated by CPI. The efficiency goal was 73%. The application for the device was the DOE Accelerator Production of Tritium (APT) proton accelerator program.

Table 10.2

Comparison of an HOM-IOT and a Klystron for 1MW CW in UHF		
Device	HOM-IOT	Klystron
Effective Efficiency	73%	60%
Relative Power Consumption	82%	100%
Approximate Assembly Volume	30 ft^3	200 ft^3
Approximate Assembly Weight	1000 lbs	5000 lbs
DC Voltage	45 kV	90 kV
Gain	25 dB	46 dB

From H. Bohlen, et al, LINAC 2004, August 16-20, 2004.

Unfortunately, the program was canceled before testing was complete because the entire APT proton accelerator program was canceled. However, short-pulse testing that was done in the TM 020 mode confirmed the modeling that had been carried out. The tube shown in Figure10-22 delivered more than 900 kW with a beam voltage of 31 kV. Furthermore, the results showed that the efficiency only dropped slowly as the output power was reduced [8].

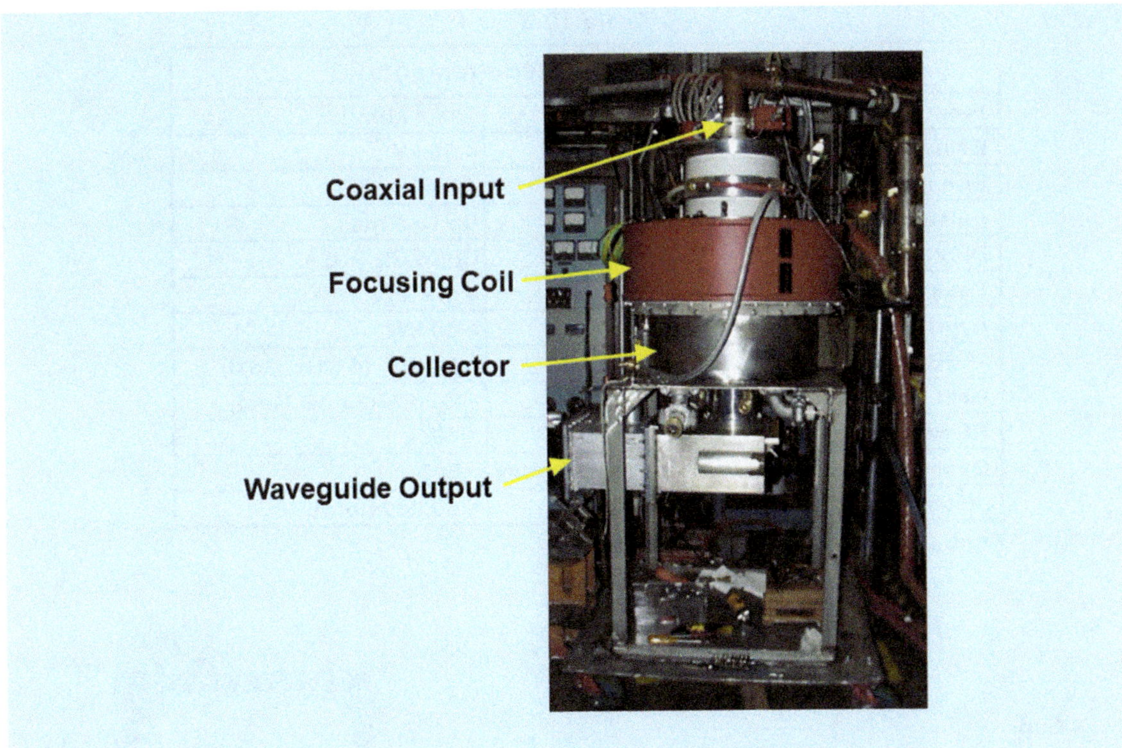

Figure 10-22 HOM-IOT for APT proton accelerator. (From H. Bohlen et al., LINAC 2004, August 16–20, 2004.)

The ESS is a new particle accelerator research facility that is being developed in Lund, Sweden. The ESS will operate at 704.4 MHz and requires RF sources capable of pulsed operation with peak power of 1.2 MW at pulse lengths up to 4 ms. ESS funded CPI/Thales and L-3 each for the development of a high-power multibeam IOT (MBIOT) to meet these requirements and the other requirements listed in Table 10.3.

The maximum power of single-beam IOTs is limited to 150 kW for long pulse operation and 50-kV maximum beam voltage. For higher powers, several devices must be operated in parallel. This development makes use of the experience and expertise at CPI and TED in development and production of high-power IOTs and other devices.

As Figure 10-23 indicates, the prototype CPI/TED MBIOT contains 10 electron beams equally spaced on a circle centered on the device axis. The beams are produced by electron guns with a proven track record for reliable operation. A solenoid provides the focusing field for the beams [17]. Figure 10-24 shows the assembled CPI/TED MBIOT.

Testing of the CPI/TED MBIOT was performed at CERN [18]. Test results for operation at a pulse length of 1 ms are shown in Figure 10-25 Power, efficiency, and gain all met or exceeded the ESS requirements. However, when conditioning was being performed at 1 MW with a 4-ms pulse length, the process was interrupted by a strong vacuum trip. It is presumed by the author that the vacuum trip was a vacuum arc. After the arc, the tube could only be operated at reduced voltages. After testing had been completed at CERN, the tube was returned to CPI for analysis and repair.

Table 10.3

ESS MB-IOT Main Requirements	
Frequency	704.4 MHz
Beam Voltage	< 50 kV
Beam Current	45 A
Pulse Length	Up to 4 ms
Duty Factor	Up to 5%
Peak Output Power	> 1.2 MW
Average Output Power	> 60 kW
-1 dB Bandwidth	> 2 MHz, (4 MHz Goal)
Gain	> 20 dB, (23 dB Goal)
RF Efficiency	> 65 %
Overall Efficiency (Including Idle Current)	> 60%
Lifetime	> 50,000 hours

From C. Yates, et al, IEEE IVEC 2016

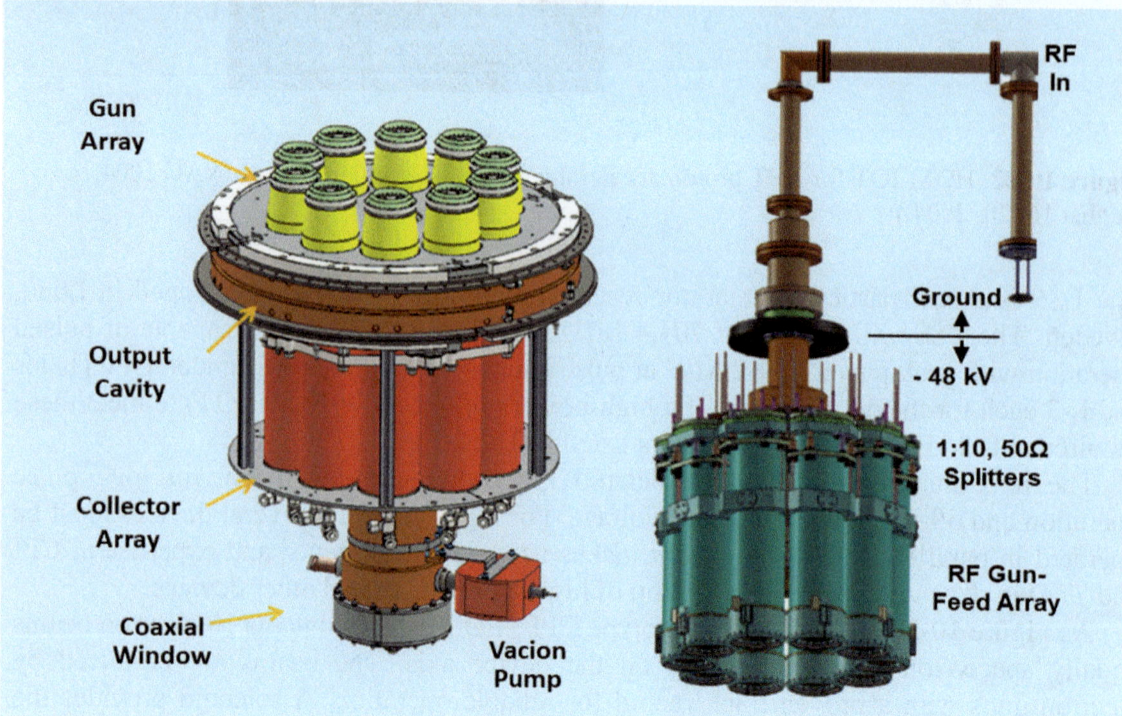

Figure 10-23 Major components of the CPI/TED MBIOT. (From A. Beunas et al., *IEEE Trans. Electron Devices*, Vol. 65, No. 6, June 2018.)

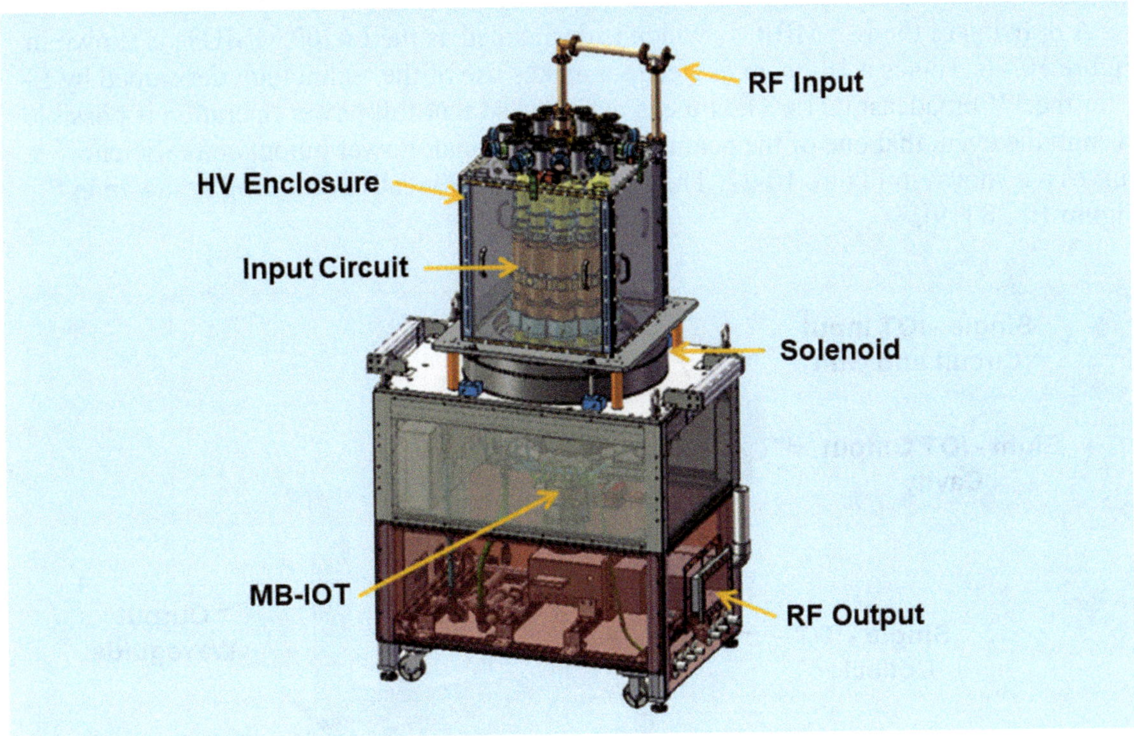

Figure 10-24 Drawing of assembled CPI/TED MBIOT. (From A. Beunas et al., *IEEE Trans. Electron Devices*, Vol. 65, No. 6, June 2018.)

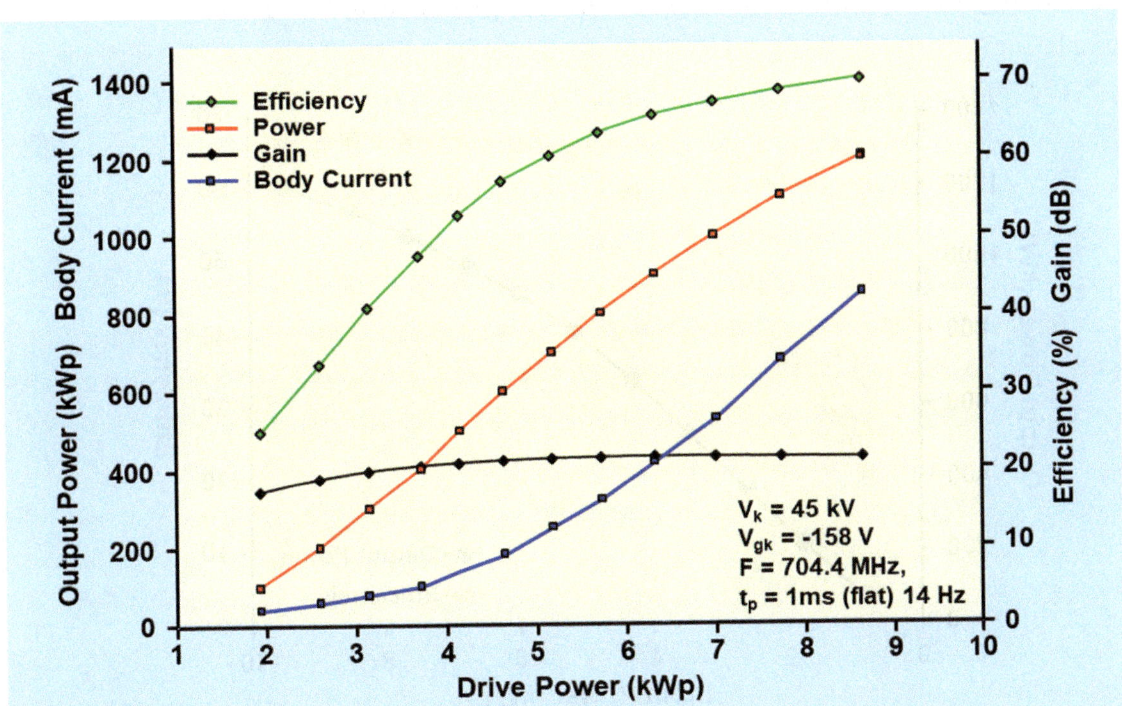

Figure 10-25 Efficiency, power output, gain, and body current for the CPI/TED MBIOT. (From A. Beunas, et al., *IEEE Trans. Electron Devices*, Vol. 65, No. 6, June 2018.)

A drawing of the L-3 MBIOT, which is designated as the L6200 MBIOT, is shown in Figure 10-26 .This is a 10-beam device that makes use of the technology developed by L-3 for the TV broadcast IOTs. The tube is designed so that full power operation is possible even in the event that one of the beams is nonoperational. Power output and efficiency curves are shown in Figure 10-27. The completed L6200 ready for testing is shown in Figure 10-28 [19].

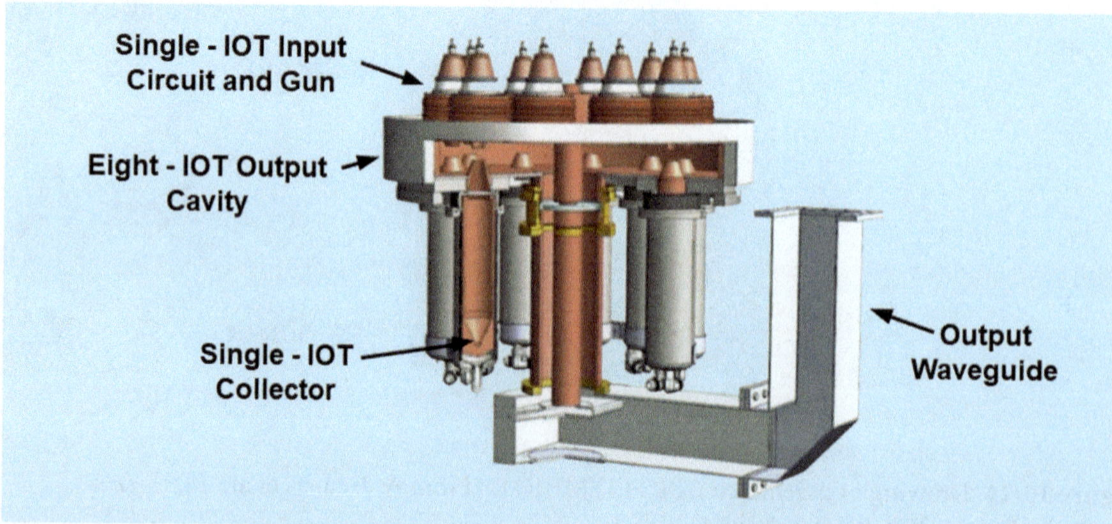

Figure 10-26 Individual IOT elements, output cavity, and output waveguide for the L-3 MBIOT. (From M. Boyle et al., *IEEE Trans. Electron Devices*, Vol. 65, No. 6, June 2018.)

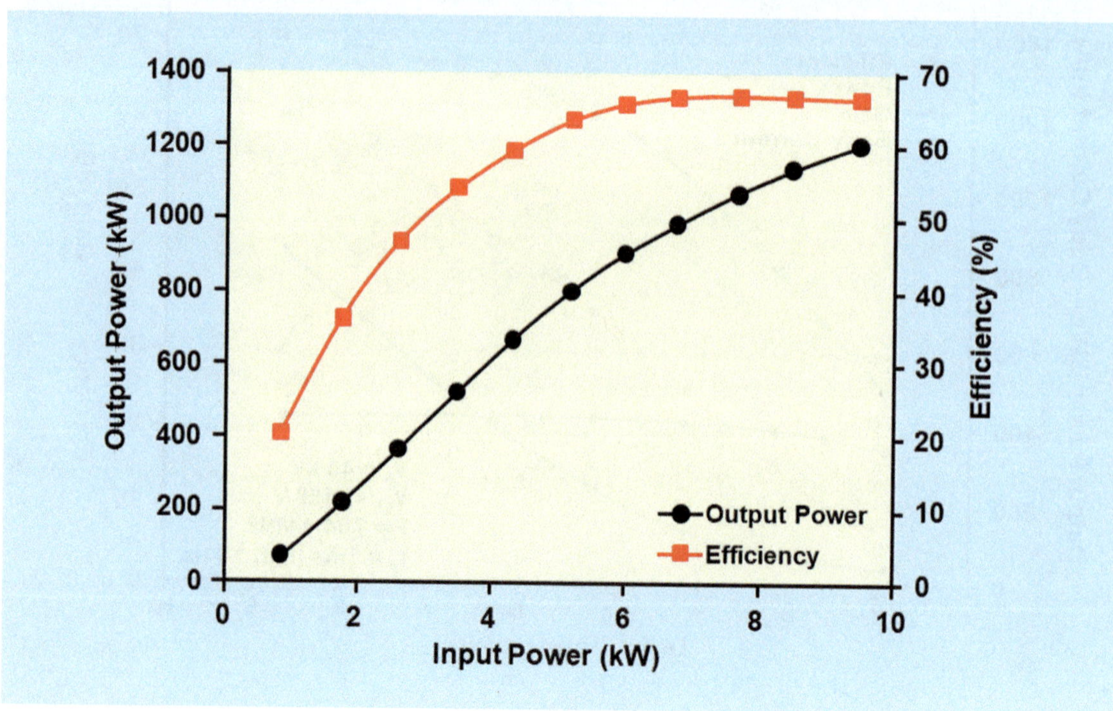

Figure 10-27 Output power and efficiency of L-3 MBIOT. (From M. Boyle et al., 18th IEEE IVEC, 2017.)

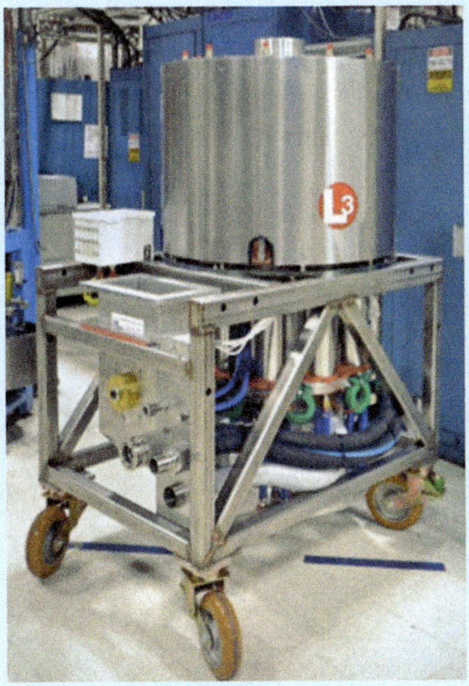

Figure 10-28 L-3 MBIOT (L6200) ready for testing. (From M. Boyle et al., 18th IEEE IVEC, 2017.)

REFERENCES

[1] A. V. Haeff, "Electron Discharge Device," U.S. Patent 2,237,878, April 8, 1941, filed Feb. 2, 1939.

[2] Wikipedia, "Inductive Output Tube."

[3] D. H. Preist and M. B. Shrader, "A high-power klystrode with potential for space application," *IEEE Trans. Electron Devices*, Vol. 38, No. 10, October 1991, pp. 2205–2211.

[4] D. H. Preist and M. B Shrader, "The klystrode—An unusual transmitting tube with potential for UHF-TV," *Proc. IEEE*, Vol. 70, No. 11, Nov. 1982, pp. 1318–1325.

[5] Michael A. Boyle, "The evolution of vacuum electronics for UHF TV broadcast," 2010 IEEE IVEC.

[6] H. Bohlen, "L-band inductive output tubes," Conference Record, 4th IEEE IVEC 2003.

[7] H. Bohlen, Y. Li, and R. Tornoe, "IOT RF power sources for pulsed and CW linacs," LINAC 2004, Lübeck, August 16–20, 2004.

[8] H. Bohlen et al., "700 MHZ HOM-IOT: Design and first test results," IEEE IVEC 2000.

[9] R. Symons, et al, "Prototype constant-efficiency amplifiers," *IEEE Trans. Broadcasting,* Vol. 47, No. 2, June 2001, pp. 147–152.

[10] M. Boyle, "IOT performance and reliability study," IEEE IVEC 2014.

[11] D. Rees et al., Characterization of a klystrode as a RF' source for high-average-power accelerators." *Conference Record, IEEE Particle Accelerator Conference*, Dallas, TX, May 1995.

[12] D. H. Preist and M. B. Shrader, "A high power klystrode with potential for space applications," *IEEE Trans. Electron Devices*, Vol. 38, No. 10, October 1991, pp. 2305–2211.

[13] Y. Li, H. Bohlen, and R. Tornoe "Development and deployment of 50 kW high efficiency IOTs for industrial and scientific services," IEEE IVEC 2003.

[14] M. Kaushik and L. M. Joshi, "Inductive output tube–a review," *Journal of Electromagnetic Waves and Applications,* Vol. 29, No, 15, August 2015.

[15] H. Bohlen, et al., "Inductive output tubes for particle accelerators," *Proceedings of EPASC 2004*, Lucerne, Switzerland.

[16] Yanxia Li et al., "IOTs step into L Band: 20 kW CW at 1.2 kW," IEEE IVEC 2004.

[17] C. Yates et al., "Development of a 1.2 MW peak, 704 MHz multi-beam inductive output tube," IEEE IVEC, 2016.

[18] A. Beunas et al., "Design, manufacturing, and performance results of a 1.2-MW peak, 704-MHz multi-beam inductive output tube," *IEEE Trans. Electron Devices*, Vol. 65, No. 6, June 2018, pp 2101–2107.

[19] M. Boyle et al., "Factory test results of the L-3 L6200 multi-beam IOT for the European Spallation Source," 18th IEEE IVEC 2017.

Chapter 11

Basic Klystrons and Their Operation

11.1 INVENTION OF THE KLYSTRON

Research leading to the development of the klystron had its start in the mid- and late 1930s. The impetus for this research was the need, emphasized by Sigurd Varian, who was an airplane pilot, for significantly improved navigation equipment. At that time, the upper frequency of operation of navigation equipment was limited by the transit time of electrons in the cathode-grid region of transmitting tubes. When the transit time approaches the period of the RF signal, gain decreases rapidly.

The first published account of a major advance that would eventually lead to the klystron (magnetron-like devices had appeared earlier) was given by A. Arsenjewa Heil and O. Heil [1]. This device, called the Heil tube, is shown in Figure 2-1.

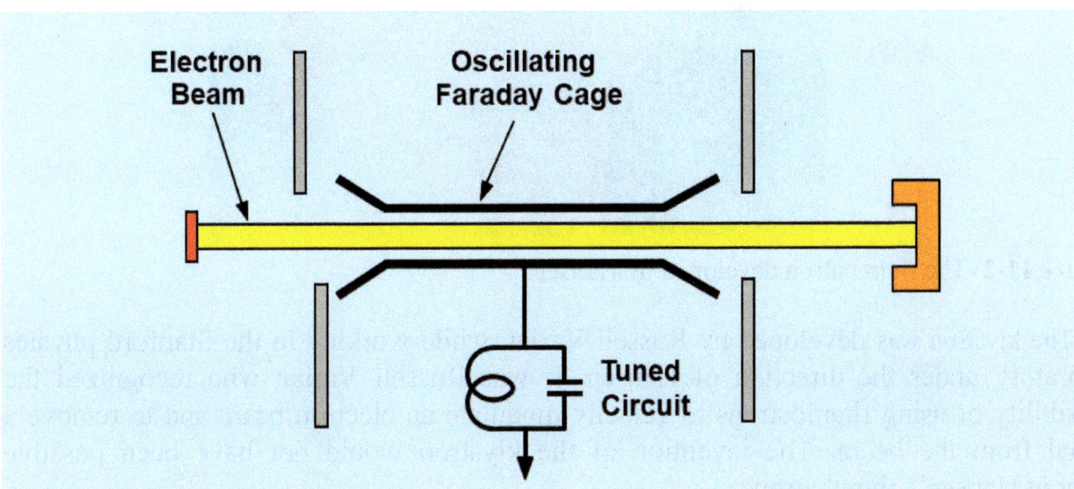

Figure 11-1 The Heil tube.

In the Heil tube, an electron beam passed through a floating drift tube, which the Heils called an oscillating Faraday cage, to a collector. The drift tube was connected to a tuned circuit. A small RF voltage across the gap at the entrance into the drift tube caused velocity modulation of the electrons. That is, the voltage across the gap accelerated some electrons and decelerated others. The faster electrons from the accelerating portion of a cycle caught up with the slower ones from the decelerating portion and so bunches of electrons were formed. When the bunches emerged from the drift tube, they induced a signal in the tuned circuit and, at the same time, enforced the velocity modulation of the beam entering the drift tube. The way in which this device operated made use of the transit time of the electrons, which had been a problem in triodes and tetrodes. The Heil tube was an experimental oscillator that apparently was not used for any practical application; however, it is credited as being the first linear-beam device to use the principle of velocity modulation of electrons.

The other technological advancement that helped make the klystron possible was the cavity resonator, called a rhumbatron, which replaced the lumped-circuit elements (capacitance and inductance) used at lower frequencies. The rhumbatron was developed by Professor William Hansen of the Physics Department at Stanford University [2]. A rhumbatron is shown in Figure 11-2.

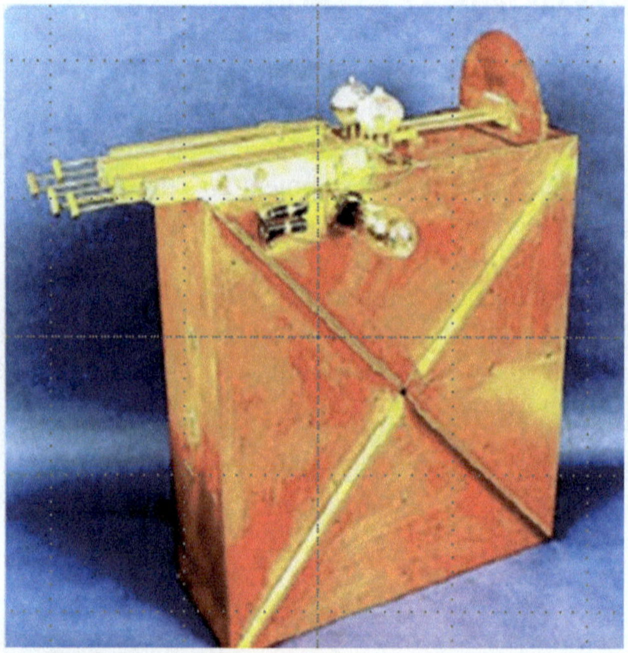

Figure 11-2 The rhumbatron developed by Hansen.

The klystron was developed by Russell Varian while working in the Stanford physics laboratory under the direction of Hansen. It was Russell Varian who recognized the possibility of using rhumbatrons to velocity modulate an electron beam and to remove a signal from the beam. The invention of the klystron would not have been possible without Hansen's rhumbatrons.

In summary, the two factors that were fundamental to the klystron invention were:

1. Velocity modulation and electron transit time were used to advantage;
2. Cavity resonators were used for interacting with an electron beam.

The invention of the klystron by the Varian brothers in 1937 was the single most important event in the development of microwave tube technology [3, 4]. This invention led directly to the invention of the cavity magnetron by Boot and Randall and to Komphner's invention of the TWT. Those inventors had been tasked to achieve improvements in klystron operation.

Figure 11-3 is an entry from Russell Varian's notebook that shows the basic two-cavity configuration of the klystron. Figure 11-4 is a photograph of the first klystron and Figure 11-5 shows the Varian brothers.

The Varians described the klystron in the *Journal of Applied Physics* in 1939 [5]. The following abstract from the Varians' paper contains a concise and very clear description of the basic operating principles of the klystron.

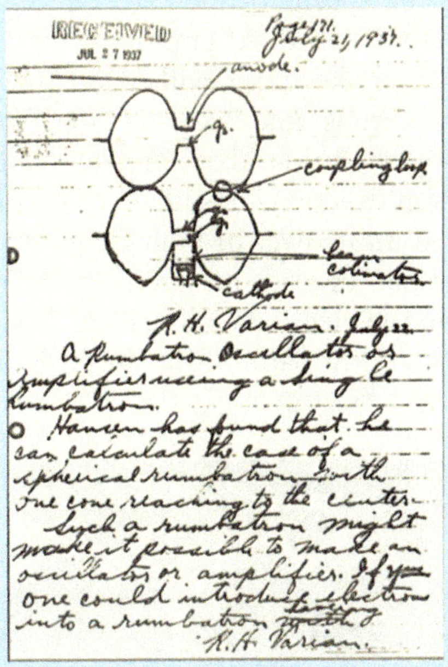

Figure 11-3 An entry from Russell Varian's notebook.

Figure 11-4 The first klystron.

"A High Frequency Oscillator and Amplifier"
Russell H. Varian and Sigurd F. Varian

Abstract

A dc stream of cathode rays of constant current and speed is sent through a pair of grids between which is an oscillating electric field, parallel to the stream and of such strength as to change the speeds of the cathode rays by appreciable but not too large fractions of their initial speed. After passing these grids the electrons with increased speeds begin to overtake those with decreased speeds ahead of them. This motion groups the electrons into bunches separated by relatively empty spaces. At any point beyond the grids, therefore, the cathode ray current can be resolved into the original dc plus a nonsinusoidal ac. A considerable fraction of its power can then be converted into power of high frequency oscillations by running the stream through a second pair of grids between which is an ac electric field such as to take energy away from the electrons in bunches. These two ac fields are best obtained by making the grids form parts of the surfaces of resonators of the type described in [this journal].

Figure 11-5 Russell and Sigurd Varian.

Figure 11-6 shows the two-cavity amplifier described by the Varians in the *Journal of Applied Physics*. An electron beam (stream of cathode rays) from the cathode at the bottom of the figure is velocity modulated by the lower resonator. Electron bunching occurs as the beam drifts toward the upper cavity. The resulting ac current in the beam induces current in the second cavity and results in RF power out of the klystron that is substantially larger than the RF input power.

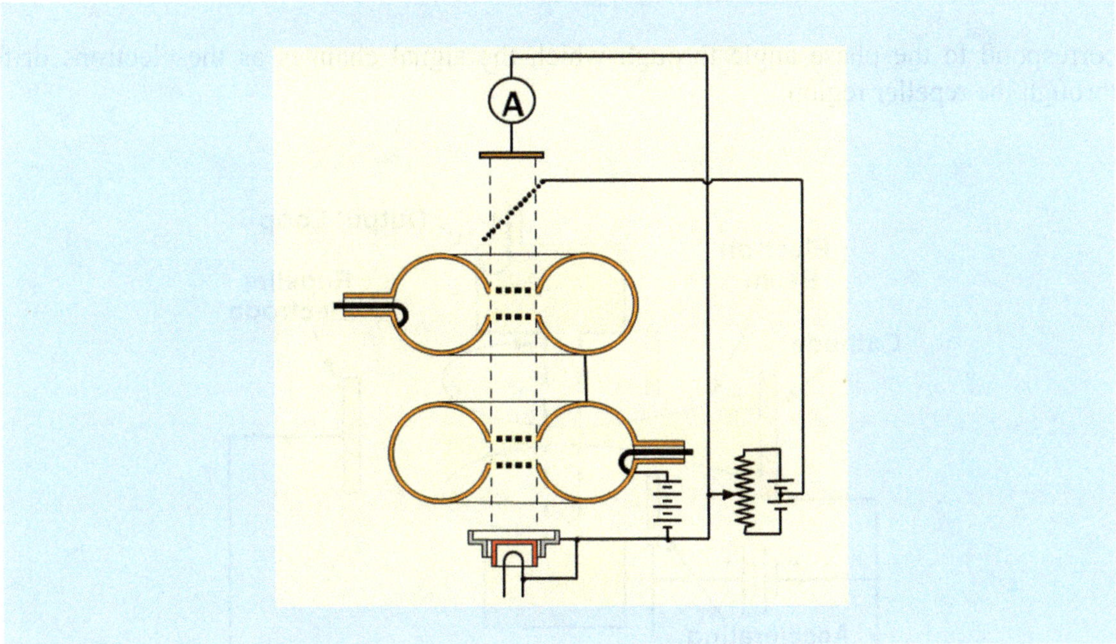

Figure 11-6 Basic klystron configuration shown by the Varian brothers. (From: R. H. Varian and S. F. Varian, *Jour. App. Phys.*, May 1939.)

The klystron was invented at Stanford but, with the outbreak of WWII, the entire group of people working on the device, including the Varians, was moved to the Sperry Gyroscope facility in Long Island, NY. The reflex klystron was developed at Sperry and was used as the local oscillator along with the Boot and Randall cavity magnetron in the development of radar. There was intense work on klystrons at Sperry but high-power klystrons were not developed until after the war. Today, the klystron remains the preferred device for producing high power at microwave frequencies for many applications.

11.1.1 Reflex Klystrons

A reflex klystron is a single-cavity oscillator. As shown in Figure 11-7, an electron beam passes through a cavity gap (normally gridded) into a region where the negative voltage on a repeller (reflector) electrode causes the electrons to reverse direction.

Assuming that a signal, such as noise induced by the electron beam, exists *across the* cavity grids, then the electron beam is velocity modulated as described in previous chapters. As the electron beam drifts into the repeller region, bunching occurs as indicated in Figure 11-8. If the bunched electrons return to the modulating gap at the correct time, they induce a current that reinforces the current in the cavity. Thus, the electron beam provides the feedback loop that causes the device to oscillate.

In Figure 11-8, the phase, α, at which the electron bunch returns to cause oscillations is 2.75 cycles after the beam initially passed through the bunching cavity. By changing the repeller voltage, the phasing of the return of the electrons to the modulating cavity can be changed. As a result, the electron bunch can be caused to return to the modulating gap 2.75, 3.75, and 4.75 cycles, and so forth after initially passing through it. As indicated in Figure 11-9, oscillations can occur in any one of several modes that correspond to the phase angle through which the signal changes as the electrons drift through the repeller region.

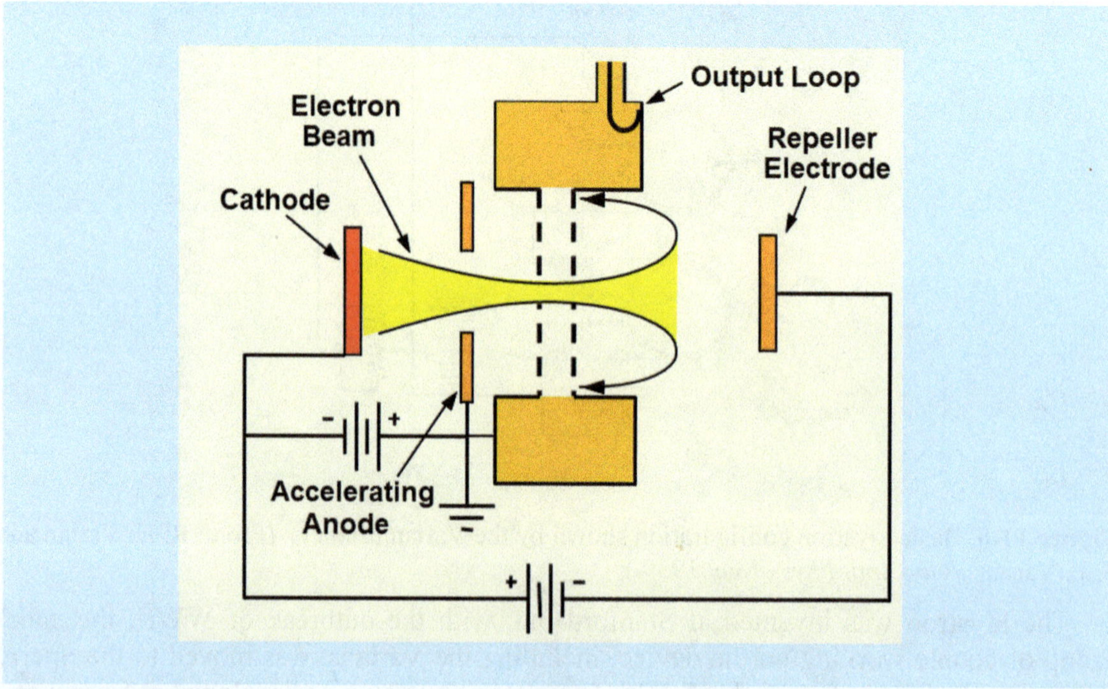

Figure 11-7 Principles of operation of a reflex klystron.

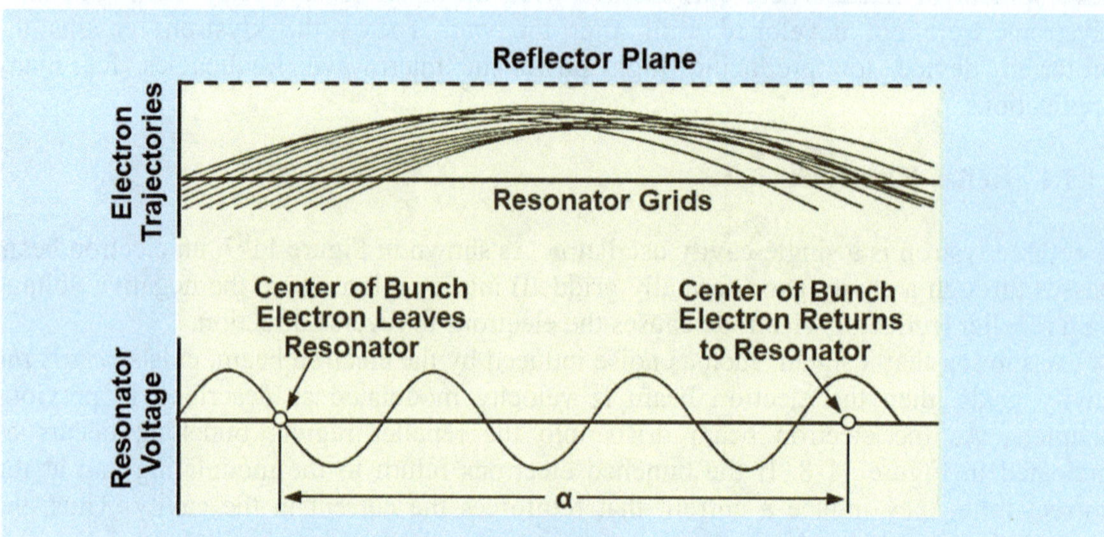

Figure 11-8 Electron bunching in a reflex klystron.

As is also indicated in Figure 11-9, the oscillation frequency in any mode of operation is a function of the time that it takes the electron bunch to return in that mode. For example, if the repeller voltage is above the mode-center value, then the electron transit time is less than the mode-center value and the frequency is above the mode-center value. As a result, the RF current induced in the cavity is at a frequency above the resonant frequency of the cavity (mode-center frequency), so the frequency of operation is above the resonant frequency. Power output is reduced because operation off resonance reduces the cavity impedance and gap voltage.

Reflex klystrons are low-power (usually less than a watt) devices that have historically been used as local oscillators. They are low-noise devices that are not difficult to phase lock. Table 11.1 contains operating parameters for the X-band, 1W, reflex klystron in Figure 11-10.

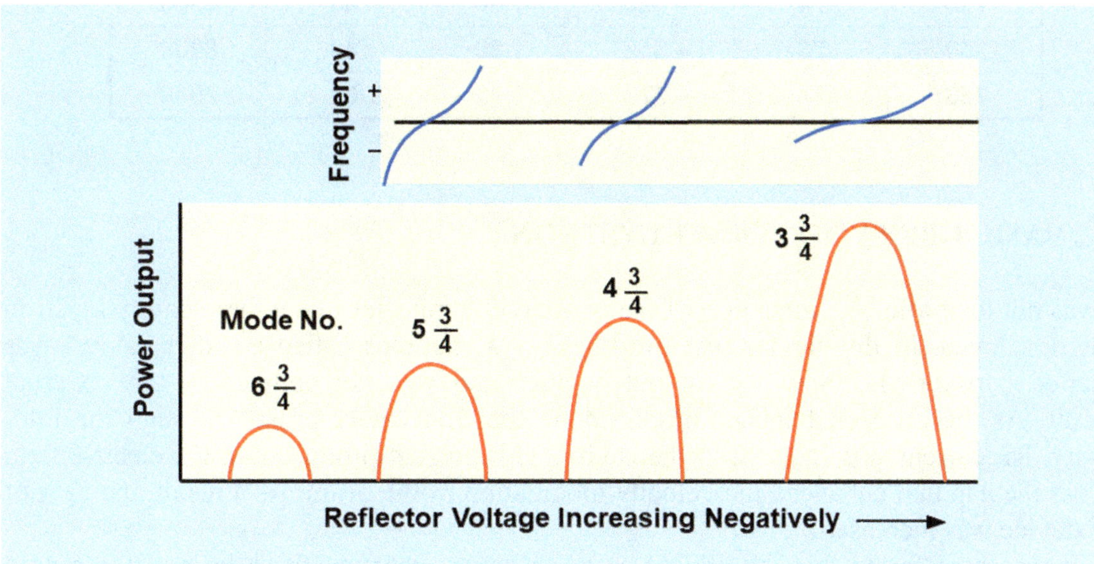

Figure 11-9 Tuning of a reflex klystron.

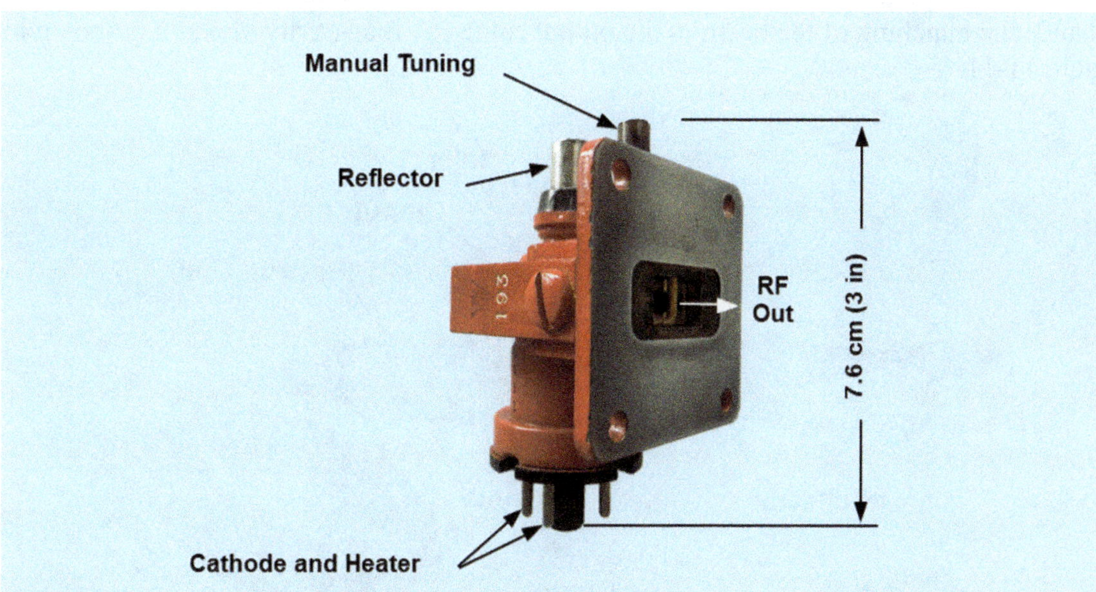

Figure 11-10 Varian VA 225A reflex klystron.

Table 11.1 Varian VA 225A Reflex Klystron					
Beam Voltage	Beam Current	Reflector Voltage	Power Out	Electronic Tuning	Frequency
Vdc	mA	Vdc	W	MHz	MHz
750	74	– 425	1.60	43	8500
750	73	– 352	1.80	54	8000
750	74	– 273	1.72	68	7500

11.2 BASIC OPERATION OF A KLYSTRON

It was not long after the invention of the two-cavity amplifier that a three-cavity klystron was developed. In this device, the third cavity (sometimes called an idler cavity) was inserted between the input and output cavities and was not coupled to any external circuit. As the electron bunches produced by the first cavity passed through the idler cavity, RF current was induced in the cavity. This current produced an RF electric field across the gap that enhanced the velocity modulation of the beam. As a result, the gain of the device was increased.

It was next found that, by adding a fourth cavity between the idler and the output cavity, the efficiency could be increased. This fourth cavity, which was called the penultimate (next to last) cavity, was tuned to a higher frequency than the center frequency of operation of the tube. The effect, as is explained in Chapter 8, was to enhance the bunching of the beam in the output cavity. A four-cavity klystron is shown in Figure 11-11.

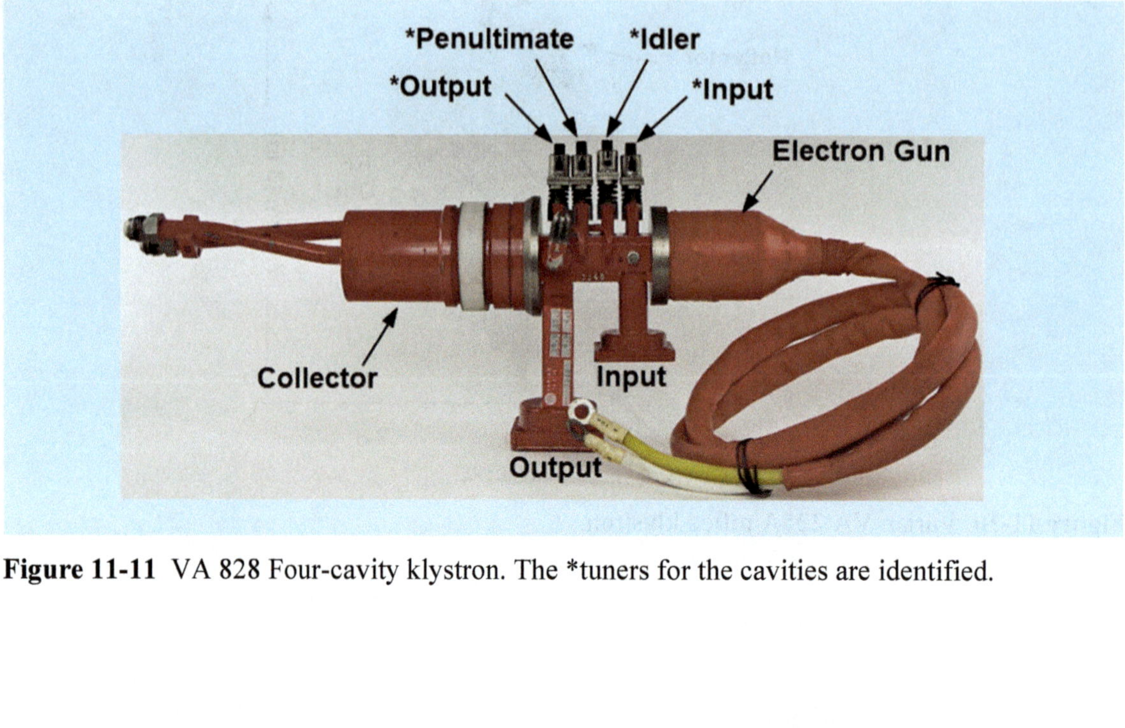

Figure 11-11 VA 828 Four-cavity klystron. The *tuners for the cavities are identified.

Two-, three-, and four-cavity klystrons are relatively narrowband devices. The possibility of increasing the bandwidth by adding idler cavities was postulated in the 1940s. However, increasing the bandwidth in this way is not as easy as it first appears to be because the cavities of a klystron are all coupled to the same electron beam and so their tuning characteristics are not independent. If not tuned properly, some of the idler cavities will demodulate the beam and produce holes (regions of reduced gain) in the frequency response. It was not until 1957 that Kreuchen et al. [6] satisfactorily analyzed the multicavity klystron and described a method for increasing bandwidth with reasonably flat power output across the band.

In addition to properly tuning the idler cavities, it is necessary to have an output structure that can cover the required bandwidth. It was the development of double-tuned and triple-tuned cavities along with the merging of TWT and klystron technologies that resulted in devices with bandwidths exceeding 11%. The configurations of some of these devices are shown in Chapter 12.

In principle, the gain of the klystron amplifier can be increased indefinitely by adding cavities along the beam. In practice, with a six-cavity amplifier, a stable gain in excess of 111 dB has been observed. Normally, the gain is limited to 50 to 60 dB. At higher levels, the shot noise of the beam is increasingly amplified, and the signal-to-noise ratio of the output is reduced. In addition, to obtain stability, there must be no feedback from output to input, either because of leakage of transmission line connections, RF modulation of collector leads, and feedback to the electron gun, or reflected electrons within the beam.

The elements of a five-cavity klystron are shown in Figure 11-12. The intermediate cavities are used to increase the gain and usually the bandwidth. Often, the spacings between cavities vary in the way reported by Mihran and discussed in Chapter 8. The intermediate cavities are often one-quarter of a plasma wavelength ($\lambda_q/4$) apart. The spacing between the penultimate cavity (next to last cavity) and the last cavity is less than $\lambda_q/4$ because of the ballistic nature of the bunching process.

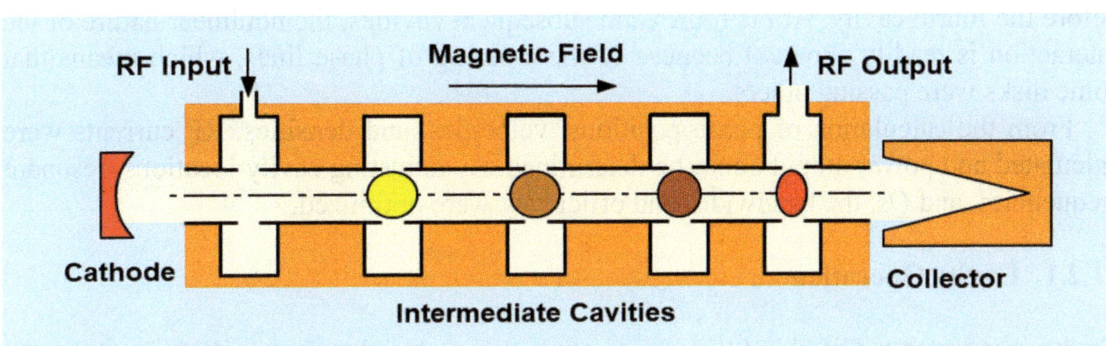

Figure 11-12 Basic elements of a five-cavity klystron.

In most modern klystrons, the spacings between intermediate cavities are not $\lambda_q/4$. The spacings are adjusted to optimize bandwidth and/or efficiency. Spacings for high-efficiency and extremely broad bandwidths are discussed in Chapter 12. Optimization is often done using computational techniques and a disk model of the electron beam. An example of a calculation for a seven-cavity klystron is shown in Figure 11-13. This figure is basically an

Applegate diagram in which dc velocities of the electrons have been removed. Each trace represents the phase of a disk of electrons relative to the undisturbed phase. Thirty electron disks per RF cycle were used in Figure 11-13 and two RF cycles are shown. The pairs of dashed vertical lines show positions of cavities. In this case the last cavity was an extended interaction cavity (to be discussed in Chapter 12) with two gaps.

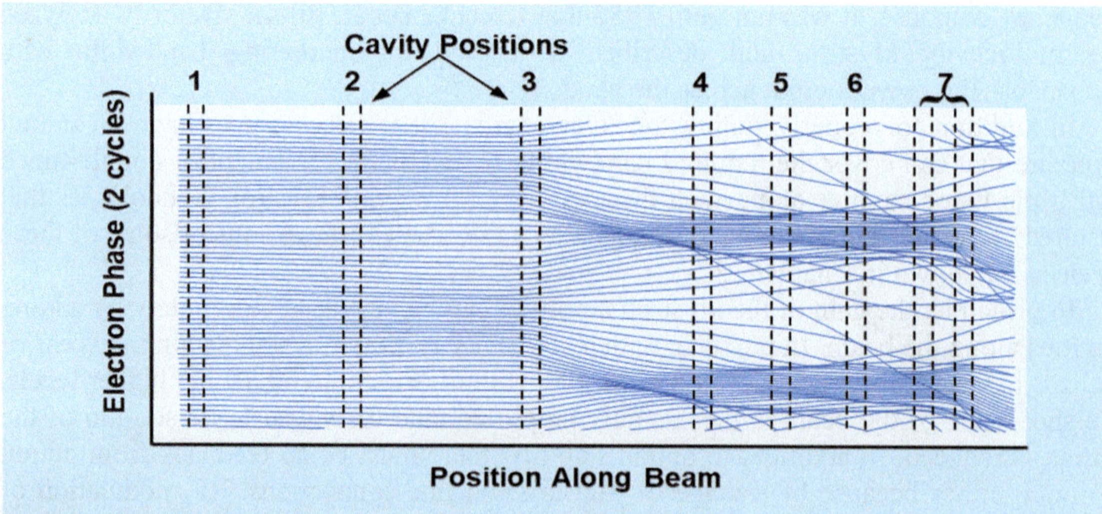

Figure 11-13 Electron phase diagram for a seven-cavity klystron.

In the regions of cavities 1 and 2, where the signal levels were very small, the change in velocities of the disks as they pass through the gaps, shown by the leading or lagging phase, was small. At the third gap, the signal was large enough to produce large velocity modulation, the magnitude of which is indicated by the slopes of the phase lines. The resulting bunching of the beam is indicated by the convergence of the phase lines just before the fourth cavity. At the fourth and subsequent cavities, the nonlinear nature of the interaction is readily apparent because of the crossing of phase lines, which means that some disks were passing others.

From the calculation of phase positions, velocities, and densities, RF currents were calculated and power output could be determined. By adjusting cavity locations, resonant frequencies, and *Q*s, the bandwidth and efficiency were optimized.

11.2.1 Cavity Operation

Cavity resonators are used in klystrons to apply a signal to the electron beam and then to remove the amplified signal from the beam. From a circuit point of view, each resonator can be considered to be a resonant parallel LC circuit that has been modified so that it will operate at microwave frequencies. This is illustrated in Figure 11-14. As frequency is increased, the multiturn-inductor, multiplate-capacitor circuit in Figure 11-14(a) is eventually replaced by the circuit in Figure 11-14(b) in which the inductor has a single half turn and the capacitor has a single pair of closely spaced parallel plates. To further increase frequency, the capacitance is reduced by increasing plate spacing and inductance is reduced by adding inductors in parallel, as indicated in Figure 11-14(c).

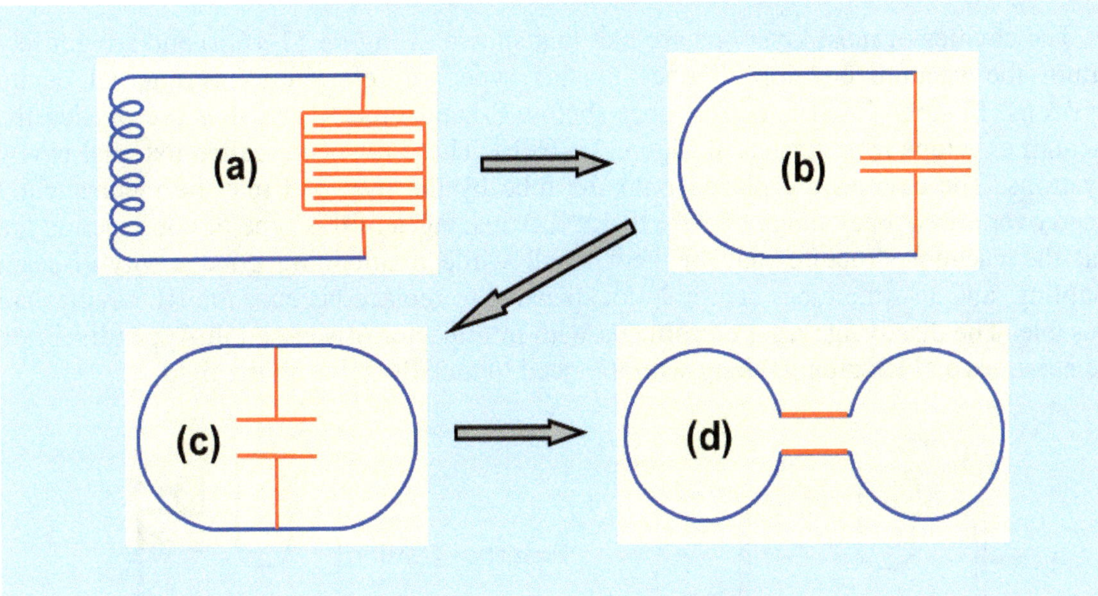

Figure 11-14 Modification of parallel resonant circuit for operation at microwave frequencies.

At high-power levels, the grids cannot be used because they would be destroyed by the electron beam. The natural evolution of Figure 11-14(d) is Figure 11-15(a). To simplify fabrication, the cavity usually takes the shape in Figure 11-15(b). This type of cavity is referred to as a doubly reentrant cavity. The ends of the tubular electron beam tunnels (referred to as the drift tube tips) are the reentrant portions of the cavity. The capacitance of the cavity results from the energy stored in the electric field between the drift tube tips.

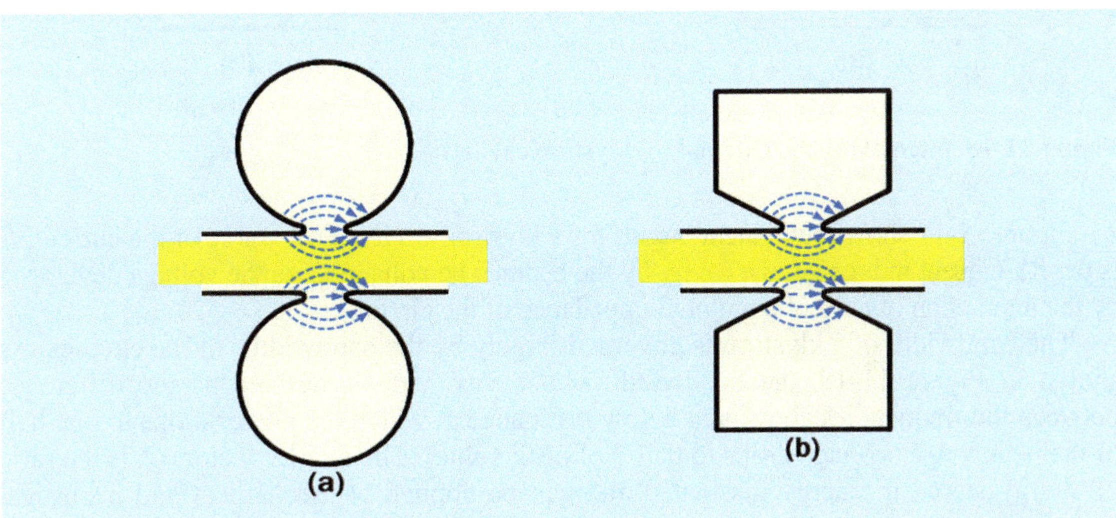

Figure 11-15 Gridless cavities for high-power klystrons.

The electric field pattern (dashed lines in Figure 11-15) is not as well defined in the gridless cavity as it is in the gridded cavity. Still, the field exists in the central region of the cavity and is available for interaction with an electron beam.

The cavities of most klystrons are like that shown in Figure 11-16(a) and are entirely within the vacuum envelope. Tubes of this type are referred to as internal cavity klystrons. In some low-frequency tubes (below S-band), the RF cavities are outside the vacuum envelope as indicated in Figure 11-16(b). These tubes are called external cavity klystrons. The cavities are placed onto the tube by the user and may be mechanically tuned over a frequency range of 10% to 15%. Advantages of this type of construction are that the vacuum structure may be kept small while maintaining good cavity-to-beam coupling, and also the tubes can be less expensive to replace because the RF cavities are reusable. The disadvantages are problems with multipactor (discussed in Appendix E) on the ceramics and RF arcing, along with the need to tune the tubes in the field.

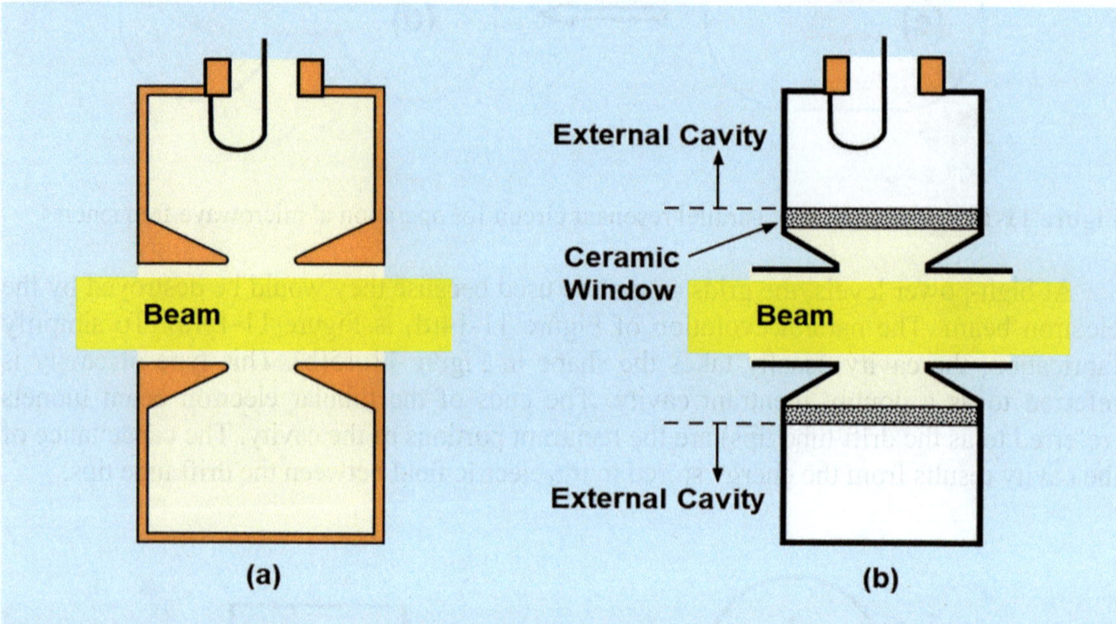

Figure 11-16 Internal (a) and external (b) klystron cavities.

Figure 11-17 shows equivalent circuit for a klystron cavity. The source of the current, I_i, is the RF current induced in the cavity by the beam. The voltage, V, is the voltage developed by the interaction of current I_i with the impedance of the circuit.

The bandwidth of a klystron is governed largely by the bandwidths of the cavities. As shown in Figure 11-18, the bandwidth of a cavity, $\Delta\omega$, is defined as the difference between the frequencies above and below resonance at which the power drops to one half of the peak value (voltage drops to 0.707 of peak value). The quality factor, Q, is the ratio of energy stored to energy dissipated and can be applied to mechanical and acoustical systems as well as electrical circuits A tuning fork that rings for an extended period after being excited has a high Q. The piece of putty that emits a dull thud when being struck by a bat has a low Q.

The Q of a system can be expressed as

$$Q_c = \frac{\omega_o}{\Delta\omega} \tag{11-1}$$

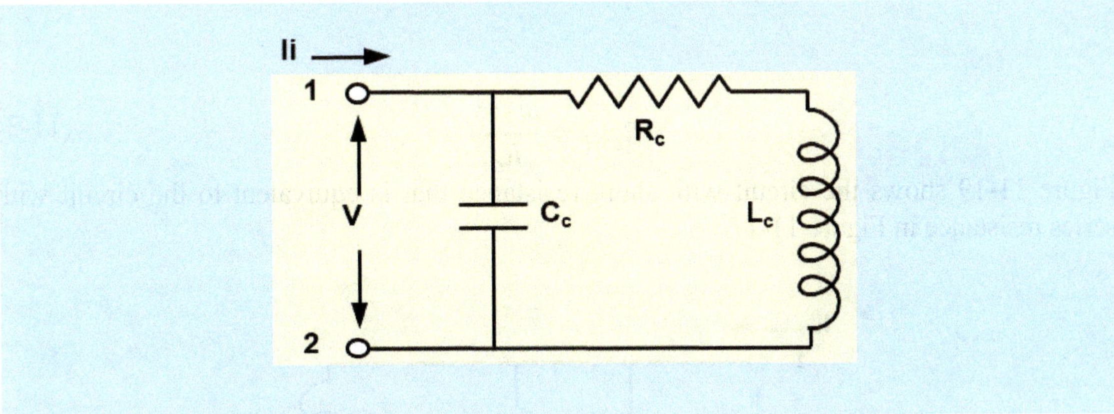

Figure 11-17 Equivalent circuit for a klystron cavity with R and L in series.

For an electrical circuit, $\Delta\omega$ is its bandwidth so a circuit with a large bandwidth has a low Q.

Using the circuit elements in Figure 11-19, the Q of the series RLC equivalent circuit for a cavity is defined as

$$Q_c = 2\pi\frac{peak\ energy\ stored\ in\ L_c}{energy\ lost\ per\ cycle} = 2\pi\frac{(I^2/2)L_c}{(I^2/2)R_c\tau} = \frac{\omega L_c}{R_c} \quad (11\text{-}2)$$

Where I is the peak current through R_c and L_c and τ is the period of the current waveform.

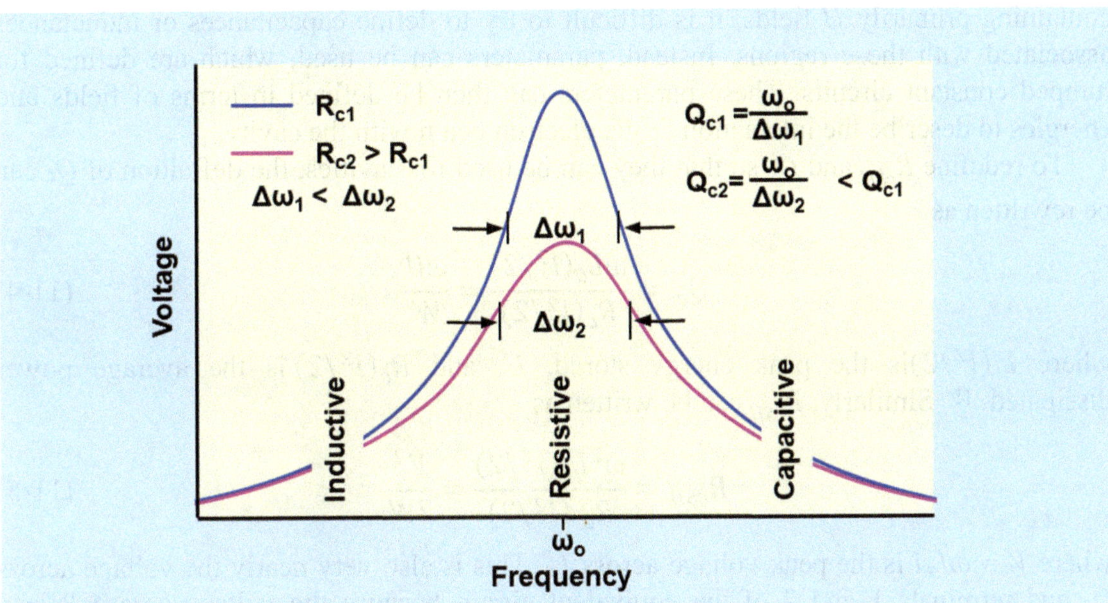

Figure 11-18 Effect of R on the Q of a resonant circuit.

The expression for cavity Q (11-2) has been derived using circuit parameters, L_c and C_c. It will be beneficial to define the loss in the cavity in terms of a shunt resistance rather than as a series resistance. The shunt resistance equivalent to the series resistance is

$$R_{cSH} = \frac{\omega^2 L_c^2}{R_c} \tag{11-3}$$

Figure 11-19 shows the circuit with shunt resistance that is equivalent to the circuit with series resistance in Figure 11-17.

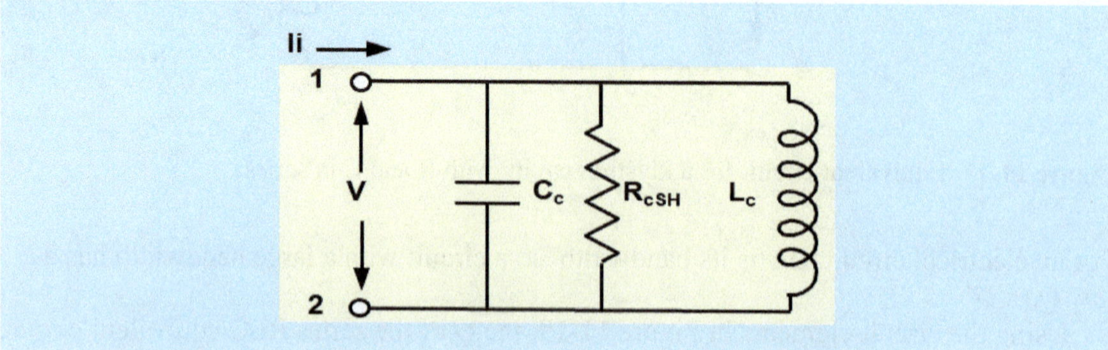

Figure 11-19 Equivalent circuit for a klystron cavity with *R* and *L* in parallel.

Next, the primary problem is to find parameters that describe the impedance characteristics of a cavity at and near resonance. Parameters are needed that will determine the voltage generated by the induced beam current as a function of frequency. Even though a klystron cavity has regions containing primarily *E* fields and those containing primarily *H* fields, it is difficult to try to define capacitances or inductances associated with these regions. Instead, parameters can be used, which are defined for lumped constant circuits. These parameters can then be defined in terms of fields and energies to describe the interaction of the electron beam with the cavity.

To redefine R_{cSH} and Q_c so that they can be used for cavities, the definition of Q_c can be rewritten as

$$Q_c = \frac{\omega L_c (I^2/2)}{R_c (I^2/2)} = \frac{\omega U}{W} \tag{11-4}$$

where $L_c(I^2/2)$is the peak energy stored, *U*, and $R_c(I^2/2)$ is the average power dissipated, *W*. Similarly, R_{cSH} can be written as

$$R_{cSH} = \frac{\omega^2 L_c^2 (I^2/2)}{R_c\ (I^2/2)} = \frac{V^2}{2W} \tag{11-5}$$

where $V = \omega L_c I$ is the peak voltage across L_c. This is also very nearly the voltage across C_c and terminals 1 and 2 of the equivalent circuit because the voltage across *R* in a resonant cavity is small compared to the voltage across the reactive elements. Finally, the ratio of R_{SH} to *Q* is

$$\frac{R_{cSH}}{Q_c} = \frac{V^2}{2\omega U} \tag{11-6}$$

Now, notice that the loss in the cavity, W, is absent in this expression. So, R_{SH}, *Q*, and ω are the three quantities that can be used to describe any cavity. Also, their definitions can be given in terms of cavity quantities. (As indicated in Figure 11-20, U is the peak energy stored and W is the power lost.) These quantities are for an unloaded cavity. With a load, loss and beam loading, the additional power dissipated must be added to W. Then, both R_{SH} and Q are reduced, but their ratio, R_{SH}/Q, is a geometrical property of the cavity and remains unchanged.

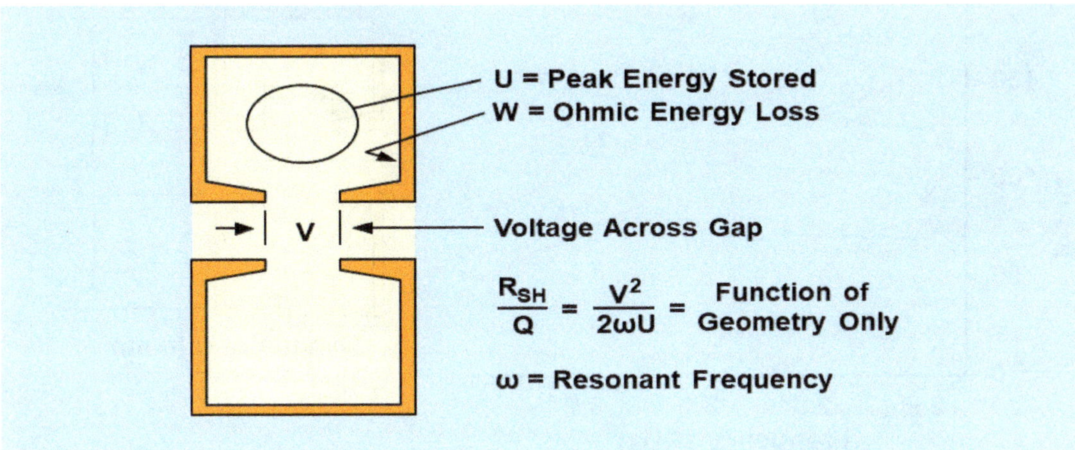

Figure 11-20 Quantities used to define cavity performance.

The value of R_{SH}/Q is a figure of merit for a cavity and is independent of Q for any particular cavity geometry. Thus, the shunt impedance, R_{SH}, may be set to the value required for matching the external load to the cavity by adjusting the loaded Q. This then fixes the Q and, since Q is inversely proportional to the bandwidth, the expected frequency response of the cavity (and the klystron) is determined. The lower the beam impedance is and the higher the R_{SH}/Q is, the wider the bandwidth is.

The R_{SH}/Q of a cavity of the type shown in Figure 11-20 (referred to as a reentrant cavity) varies with the dimensions of the cavity, as shown in Figure 11-21. It is possible to obtain values of R_{SH}/Q of over 150. Often, practical considerations limit the degree of design freedom for the cavity shape, but an R_{SH}/Q of 120 is not difficult to obtain [8].

11.2.2 Coupling to a Cavity

In the drawing of the original Varian klystron (see Chapter 2), small coupling loops attached to the ends of coaxial lines are shown for injecting a signal into a cavity or taking energy from a cavity. In modern klystrons, either loops or waveguide irises are used for coupling energy to a cavity. A typical coupling loop is shown in Figure 11-22 and cavity-to-waveguide coupling is shown in Figure 11-23.

The principle of operation of coupling loops is straightforward. To inject a signal into a cavity, an RF current, I_1, is injected into the coupling loop as is indicated in Figure 11-24. As a result, some magnetic field lines are established within and around the coupling loop and some circle through the toroidal portion of the cavity. These field lines induce a current flow, I_2 in the walls of the cavity. The resulting flow of charge produces an electric field, E, across the capacitive portion of the cavity. As the RF input current, I_1, oscillates, the magnetic field, the current, I_2, and the electric field, E, are all caused to oscillate.

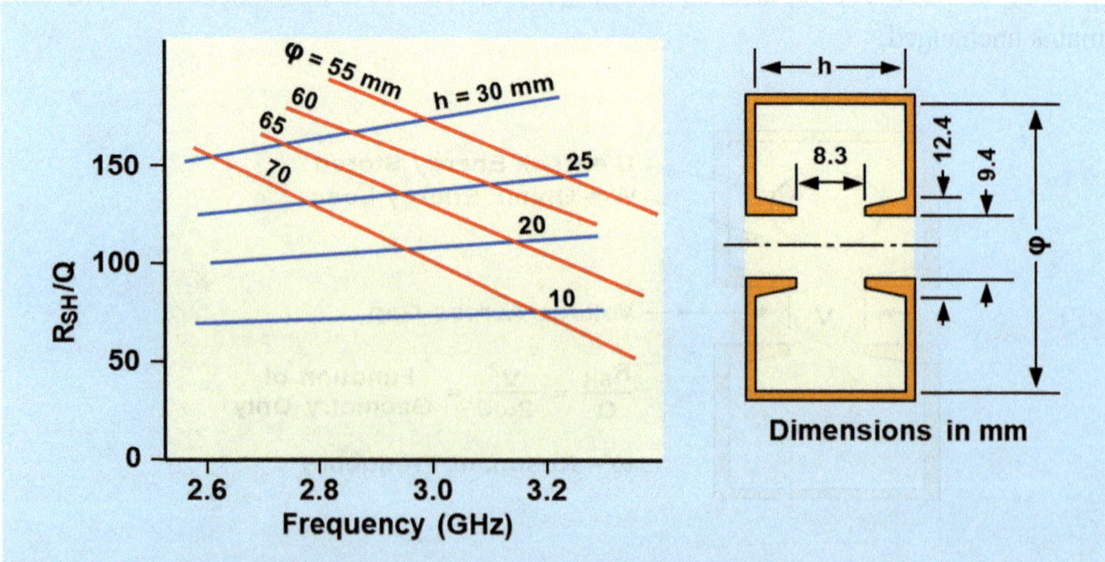

Figure 11-21 R_{SH}/Q as a function of dimensions for reentrant S-band klystron cavities. (Adapted from: G. Faillon, *Technical Digest*, IEDM. ©1985 IEEE.)

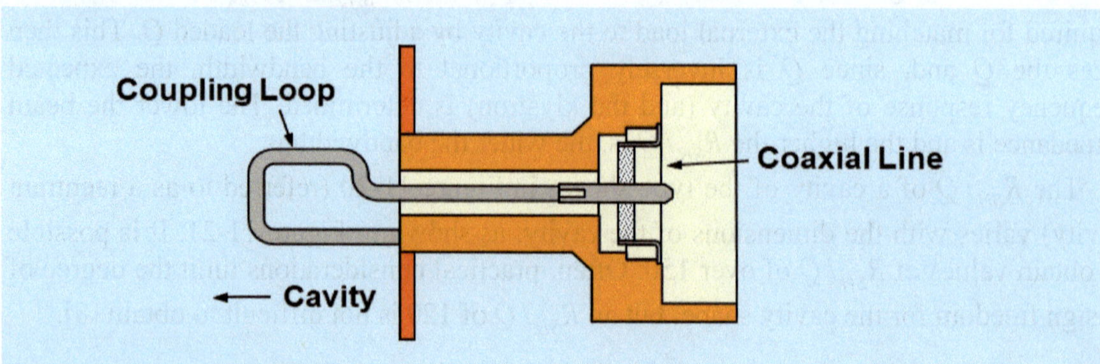

Figure 11-22 Loop for coupling energy into or out of a cavity.

When energy is extracted from a cavity by a coupling loop, the situation described in the previous paragraph is reversed. A modulated electron beam passing through the capacitive portion of the cavity induces current, I_2, in the walls of the cavity. The way in which current is induced is described in Chapter 9. Associated with the current, I_2, there is a magnetic field within the toroid. A portion of the magnetic field links the coupling loop and induces a current, I_1, in the output coaxial line and, thus, in the load. The direction of the induced current is such that it opposes the flux producing the current.

When iris coupling of a waveguide is used, the orientation of the waveguide is such that the electromagnetic field propagates directly into the cavity and establishes an electric field across the interaction gap.

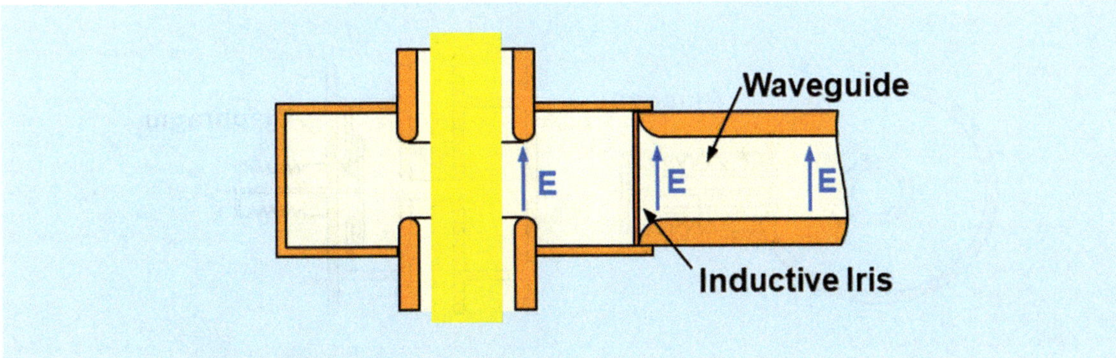

Figure 11-23 Iris coupling of waveguide to cavity.

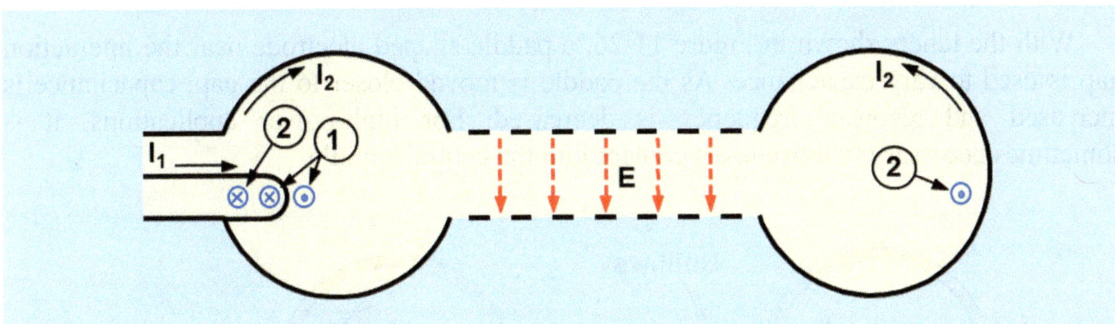

Figure 11-24 Use of a coupling loop to inject a signal into a cavity.

11.2.3 Tuners and Tuning

To optimize the bandwidth and/or the power output of a klystron, it is necessary to be able to tune the cavities. That is, it must be possible to adjust the resonant frequencies of the cavities. As was shown in the discussion associated with Figures 11-14 and 11-15, the outer donut-shaped region of a cavity is the inductive portion and the region near the interaction gap is the capacitive portion. Tuners are used to vary either inductance or capacitance.

Two inductive tuners are shown in Figure 11-25. With either the plunger shown in the left side of the figure or the diaphragm shown to the right, the volume of the cavity can be adjusted. Inductance varies directly with cavity volume and thus, as the plunger or diaphragm is pushed into the cavity, the inductance is reduced and the resonant frequency is increased. In some plunger-type tuners, chokes are used to reduce current flow in the tuner. In others, the plunger makes electrical contact with a channel surrounding it through spring-loaded fingers.

Capacitive tuning of a klystron can be achieved by physically distorting the cavities (sometimes referred to as dent tuning) or by using tuners like those shown in Figures 11-22 and 11-23. Extreme examples of tubes in which the physical distortion technique is used are the 15-ft. long B factory klystron (BFK) and a 2-inch long reflex klystron. In both of these tubes, tuning is accomplished by moving an end wall of one or more cavities slightly and thereby changing the gap length and capacitance. The BFK and the reflex klystron are described in the following chapter.

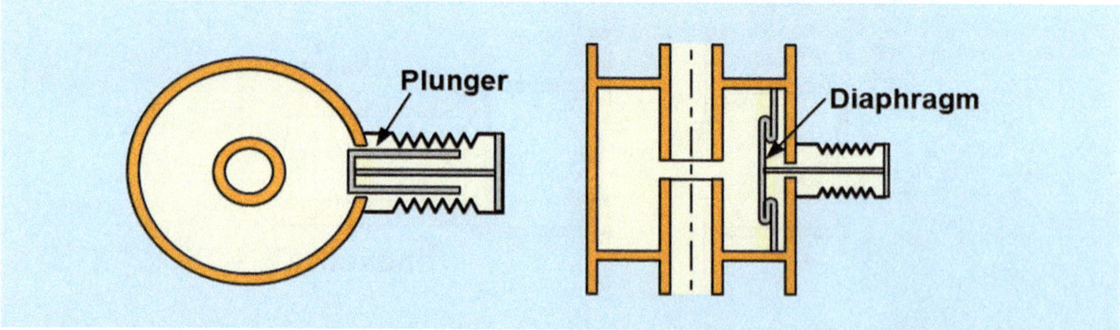

Figure 11-25 Inductive tuners. (Adapted from: *Power Klystrons Today* by M. J. Smith and G. Phillips, copyright 1995 by Research Studies Press, Ltd. [7].)

With the tuners shown in Figure 11-26, a paddle-shaped electrode near the interaction gap is used to vary capacitance. As the paddle is moved closer to the gap, capacitance is increased and resonant frequency is decreased. For high-power applications, it is sometimes necessary to introduce a coolant into the central spindle.

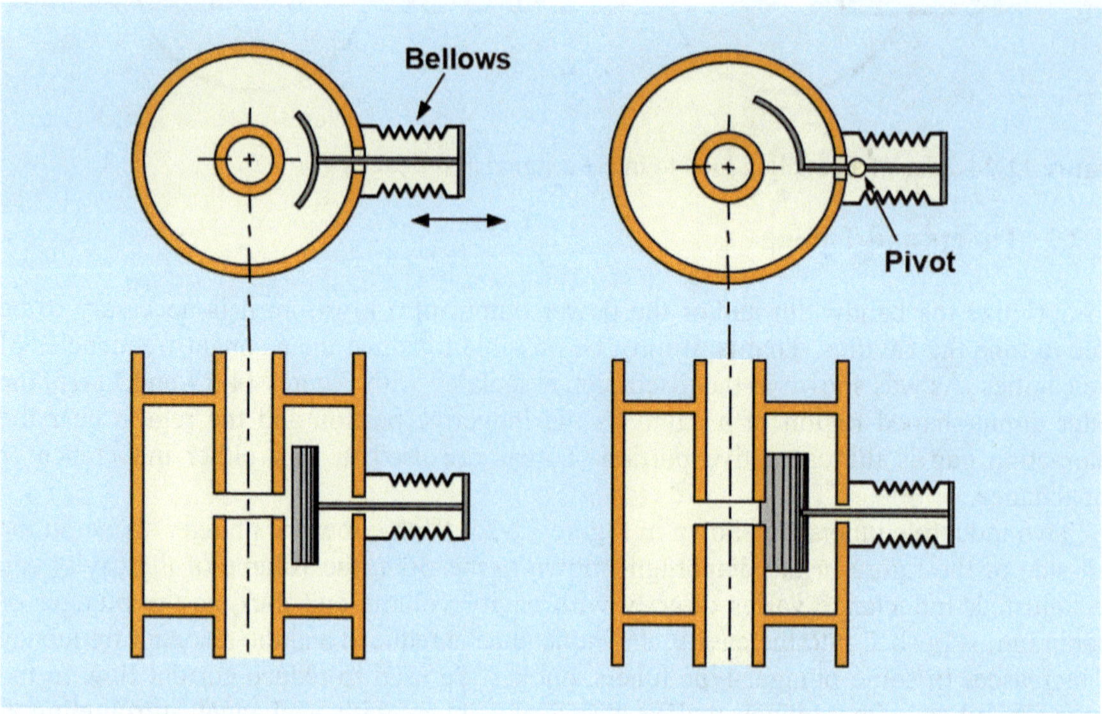

Figure 11-26 Capacitive tuners. (Adapted from: *Power Klystrons Today* by M. J. Smith and G. Phillips, copyright 1995 by Research Studies Press, Ltd. [7].)

A major application for capacitive tuning is in the VA-889F klystron, which is shown in Figure 11-27(a). This klystron is used in the AS SPS-49 radar that was first introduced in the early 1970s and is still in widespread use. The VA-889F operates at low L band and is tunable over an 11% frequency range. The tuner configuration is that shown on the right in Figure 11-26. Parts of the tuners in the VA-889F are shown in Figure 11-27(b).

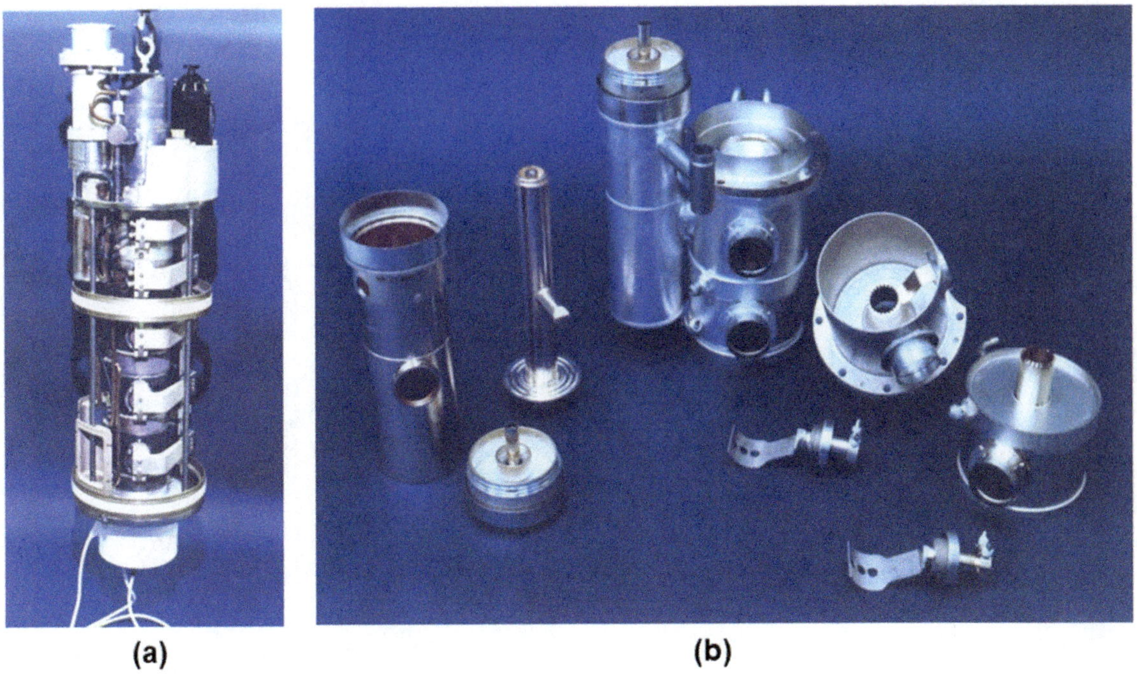

Figure 11-27 (a) The VA-889F klystron used in the AN SPS-49. (b) Cavities and tuners used in the VA-889F.

11.3 KLYSTRON DRIVER SECTIONS

A klystron consists of a driver section and an output section as indicated in Figure 11-28. The driver section functions as an RF current source for driving the output section. The cavities in the driver section are optimized to produce the required bandwidth and gain to drive the output section. The output section must produce the required power over the specified bandwidth.

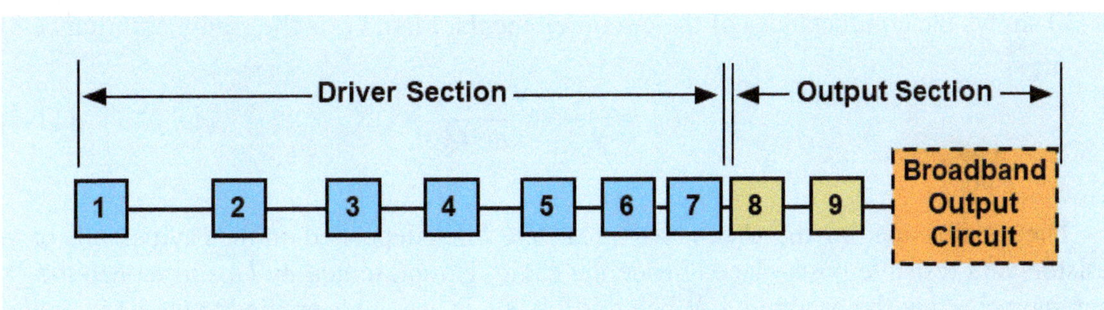

Figure 11-28 Sections of a klystron for producing gain, bandwidth, and power output.

Figure 11-29 shows the peak RF current growth in the five-cavity driver section of an eight-cavity klystron. The output section in this case consisted of two penultimate cavities and an output circuit.

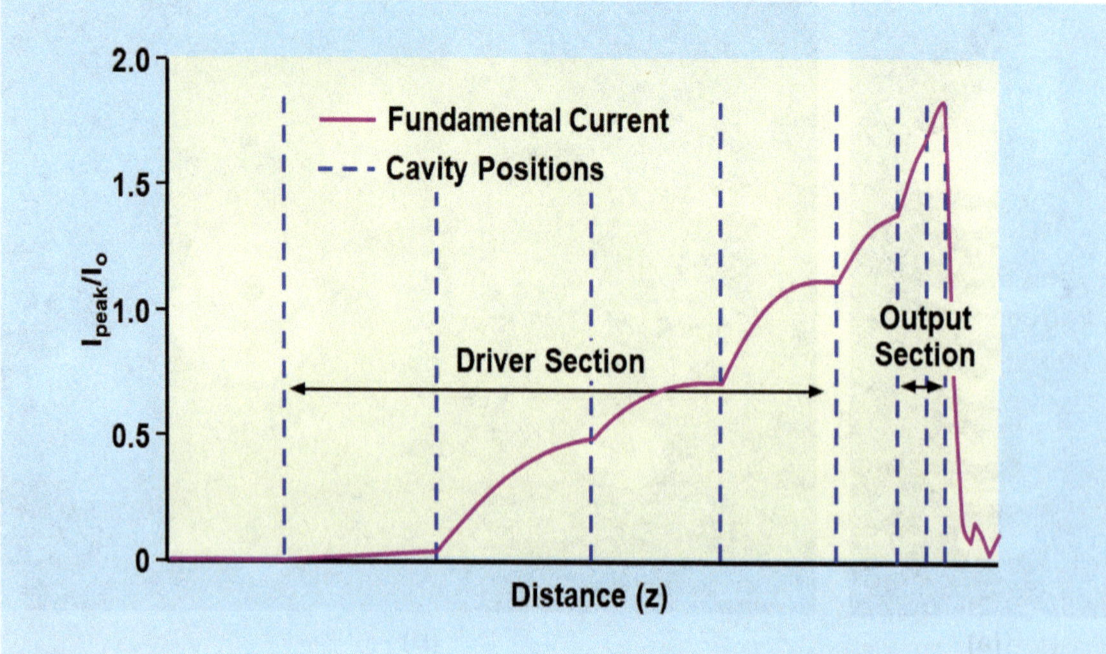

Figure 11-29 RF beam current generated by driver section.

11.3.1 Gain and Bandwidth

In this section, we will calculate the two-stage gain and bandwidth of two stages of a driver section and then comment on the procedure to follow to determine gain and bandwidth of a more complex driver section.

The formalisms of the relations that lead to the calculation of gain and bandwidth for a multicavity driver section are reasonably straightforward, but the computer codes and their use for actually doing the calculations and designs are outside the scope of this book (and many of the codes are proprietary).

It will be useful to use conductances to represent the elements of the circuit. Figure 11-30 shows the conductances of the circuit elements. First, G_c is the cavity admittance

$$G_c = \frac{1}{R_{cSH}} = \frac{R_c}{\omega^2 L_c^2} \tag{11-7}$$

The resistive loading that results from resistive films deposited on the cavity walls or to resistors and resistive posts placed inside the cavity is represented by G_{QL} to reduce the Q, thereby increasing the bandwidth. When this loading is used, it is applied to the input and/or intermediate cavities to match impedances or to reduce the Q and increase the bandwidth of the cavity. G_{QL} is related to the series loss R_{QL} by

$$G_{QL} = \frac{R_{QL}}{\omega^2 L_c^2} \tag{11-8}$$

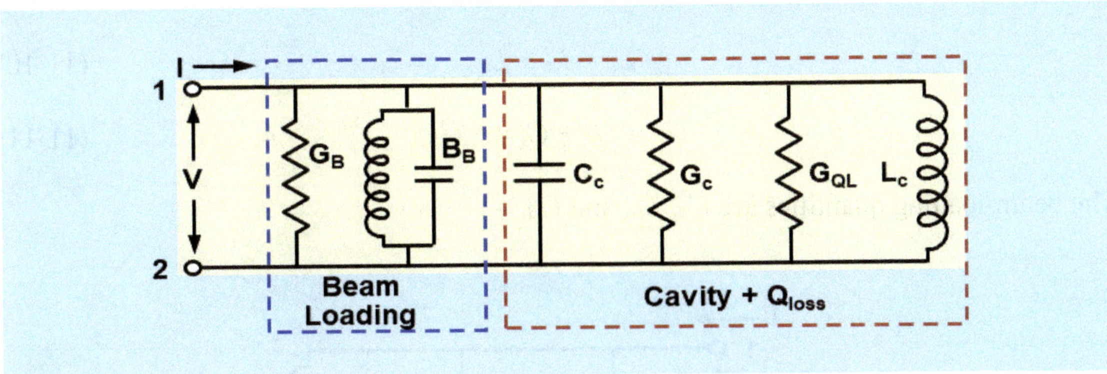

Figure 11-30 Equivalent series circuit containing loss and beam loading.

Next, consider beam loading. We showed in Chapter 8 that beam loading consists of a reactive component, B_B, as well as a resistive component, G_B. The beam-loading admittance is approximately 0.1 to 0.15 times the DC beam admittance. Beam loading in terms of its resistance, which is the inverse of beam-loading conductance, is plotted in Figure 11-31 as a function of beam voltage for $G_B/G_o = 0.15$ and microperveances of 1 and 2. Resistance varies inversely with the square root of voltage and so beam loading at high voltages is significant. Cavity Q is inversely proportional to loading, and bandwidth is inversely proportional to Q so this heavy loading is a large part of the reason why very high-power klystrons can have relatively broad bandwidths.

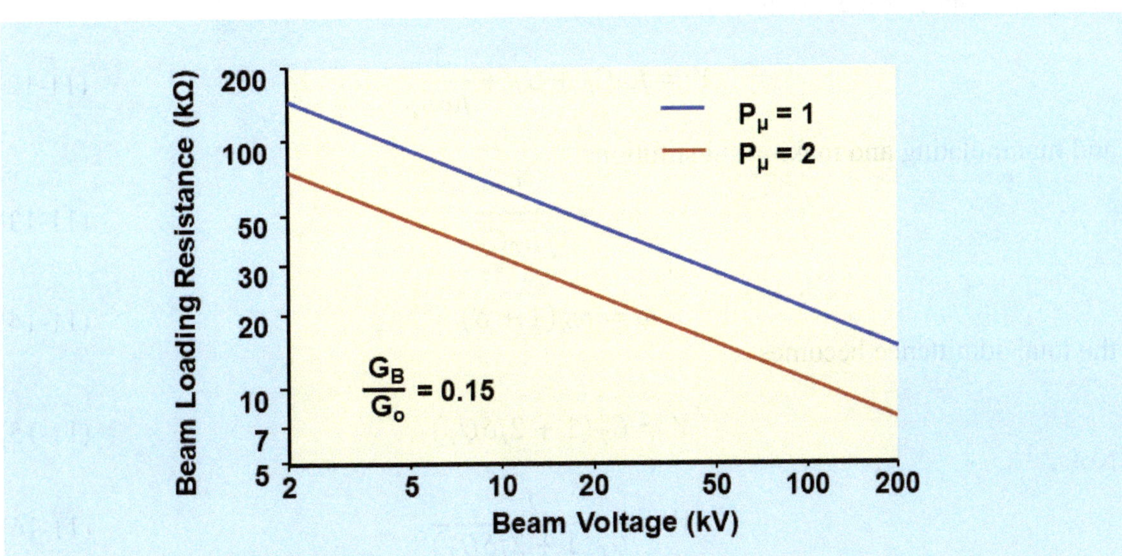

Figure 11-31 Beam-loading resistance varies inversely with the square root of voltage and becomes significant at high voltages and high values of microperveance, P_μ.

For the considerations of gain and bandwidth of the driver section, the equivalent circuit in Figure 11-32 for the cavity with beam loading and added loss to control Q_T will be considered with the following parameters:

$$C_T = C_c + C_B \tag{11-9}$$

$$L_T = L_c + L_B \tag{11-10}$$

$$G_T = G_c + G_{QL} + G_B \tag{11-11}$$

The beam-loading quantities are C_B, L_B, and G_B.

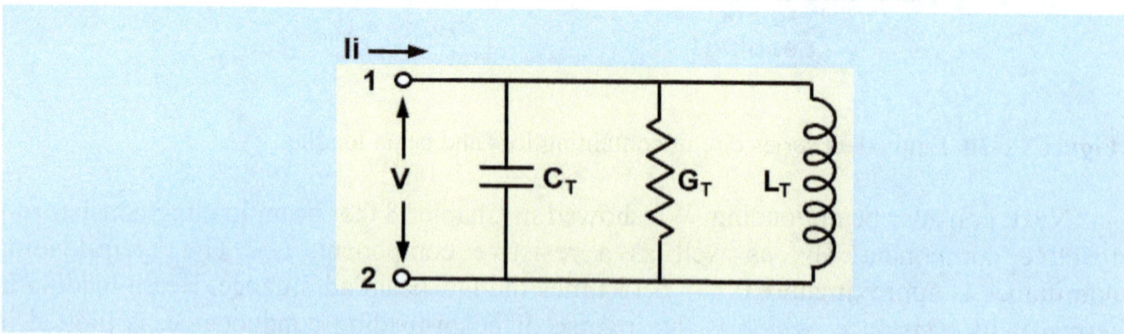

Figure 11-32 Equivalent circuit in which C_T and G_T contain cavity, beam-loading, and Q_{loss} quantities and L_T contains cavity and beam loading quantities.

The voltage produced at the cavity gap, V, can be written to take into account the effects of the total conductance, G_T, and the total susceptance C_T and L_T. Starting with the total admittance of the circuit, which is

$$Y = j\omega C_T + G_T + \frac{1}{j\omega L_T} \tag{11-12}$$

and manipulating and making substitutions

$$\omega_o = \frac{1}{\sqrt{L_T C_T}} \tag{11-13}$$

$$\omega = \omega_o(1 + \delta) \tag{11-14}$$

the total admittance becomes

$$Y = G_T(1 + 2j\delta Q_T) \tag{11-15}$$

Now,

$$V = \frac{I}{G_T(1 + 2j\delta Q_T)} \tag{11-16}$$

where I is the current induced in the cavity. For ballistic bunching, the induced current was discussed in Chapter 8 and is

$$I = M_2\, 2I_o\, J_1\left(\theta\frac{\alpha M_1}{2}\right) \tag{11-17}$$

As shown in Figure 11-33, θ is the drift angle from the modulating cavity to the cavity in which current is being induced, M_1 is the gap coupling coefficient of the modulating cavity and V_1 is the voltage at the first cavity. Here, beam loading and any additional terms due to velocity modulation on the beam as it enters the gap are being ignored. The velocity modulation term is usually small and the beam loading can easily be included as an equivalent admittance in parallel with the other elements of the equivalent circuit.

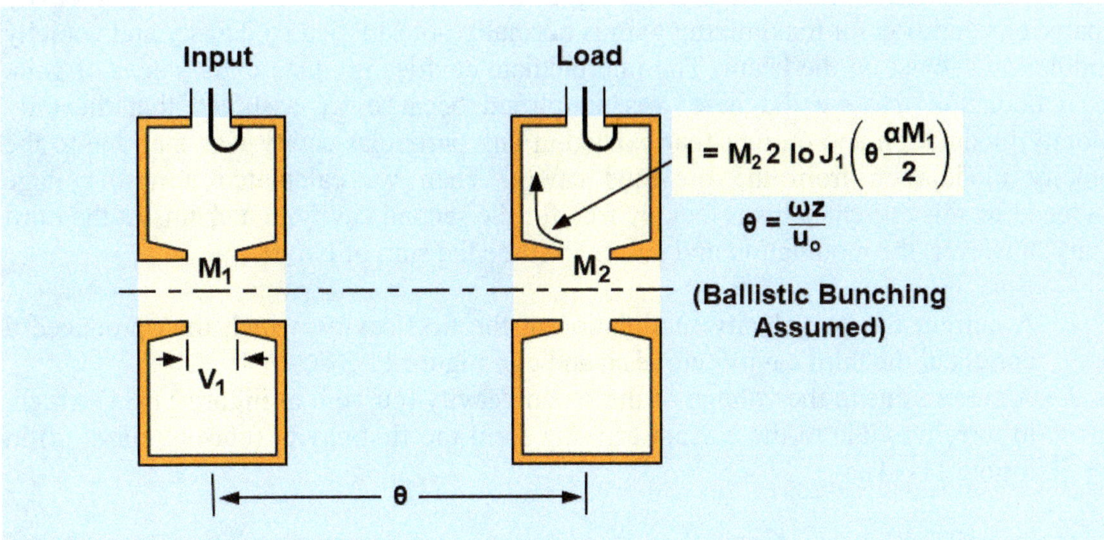

Figure 11-33 Illustration of current coupling in a two-cavity klystron.

Now, power is maximized when $G_L = G_C$; that is, when the load resistance (considered as a shunt resistance in the equivalent circuit) equals the shunt resistance of the cavity. This condition means, of course, that the total power is divided equally between internal losses and the external load. We should note here that beam loading changes these results only by adding the beam-loading conductance to the internal losses.

It is also possible to calculate voltage gain using considerations similar to those used for power. In most cases, a klystron amplifier has more than two cavities. The cavities between the input and output are intended to provide greater gain than is possible with just two cavities. The current arriving at gap n due to velocity modulation from previous gaps produces a voltage across n that is greater than the previous gap voltages. By using several such cavities, we can then produce the required maximum current at the output cavity with much smaller voltage (and power) input at the first cavity than in the simple two-cavity klystron.

The single-stage voltage gain from cavity m to cavity n can be calculated from (11-16) and (11-17). First, we note that for small values of X, $J_1(X) \approx X / 2$ so that the equation for induced current in gap n becomes

$$I_n = M_n M_m \frac{I_o}{V_o} \frac{\theta}{2} V_m \tag{11-18}$$

Now, substituting this into (11-16) and, assuming that there is no external load (as is the case in most intermediate cavities in a klystron), then the voltage gain becomes

$$\frac{V_n}{V_m} = M_n M_m \frac{I_o}{V_o} \frac{\theta}{2} \frac{R_{SH}}{(1+2j\delta Q)} \tag{11-19}$$

This relation is the single-stage gain. As expected, the gain and gain-bandwidth product depend on R_{SH}/Q.

The typical klystron has several stages. The quarter plasma wavelength cavity separation condition for maximizing gain is normally not satisfied so density and velocity modulation coexist on the beam. The intermediate cavities produce extra stages of gain. The calculation that we did here was simplified because we assumed that the only velocity modulation and current that existed at any particular cavity was that due to the velocity modulation from the previous cavity. Then we calculated a new voltage produced by this current. This is exactly true for the second cavity in a chain. At the third cavity, however, the modulation and current will be the sum of two terms:

1. A current due to velocity modulation at the first cavity, which then produced a current at the third cavity (curves a and c in Figure 11-34);
2. A current due to the voltage of the second cavity (curve b in Figure 11-34) which, in turn, was due to the current arriving from the first cavity (current curve (a) in Figure 11-34).

There will still be some persistent velocity modulation from two earlier cavities (curve d in Figure 11-34), and the resulting currents will produce voltage on the third cavity. This voltage contributes to the modulation of the beam at that cavity so that modulation continues on to the next cavity.

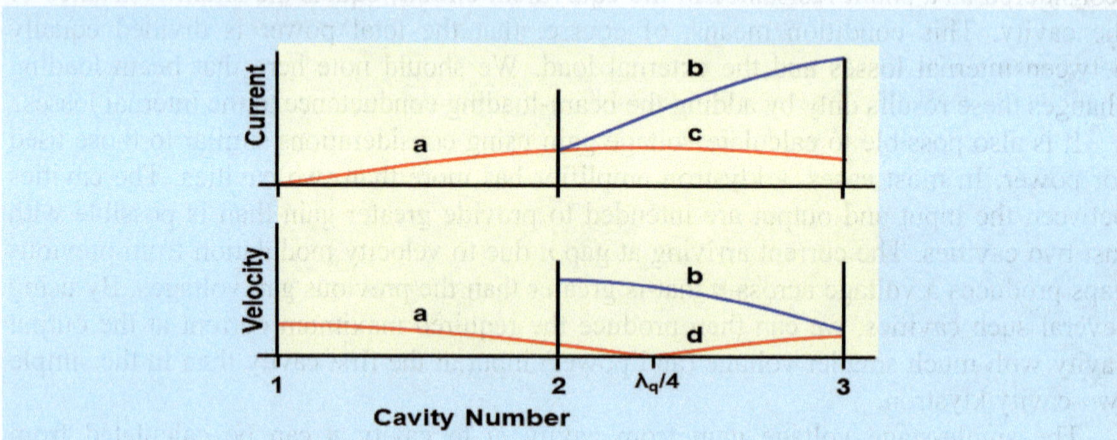

Figure 11-34 RF currents and velocities in the first three stages of the driver section. The cavity spacing is about ¾ of the quarter-plasma wavelength.

This process continues throughout the length of the interaction structure. At each cavity there are currents that depend directly on the voltage at some previous cavity through velocity modulation converted into current by bunching. In addition, there are terms that contain currents produced by voltages at intermediate cavities by currents arriving at earlier cavities. The phases with which these modulations are initiated on the beam at each gap are different, and careful account has to be kept of all the amplitudes and phases as the beam interacts with successive cavity gaps. These so-called feed-forward terms make the calculation of klystron performance very complex.

The components required for determining gain in a multicavity klystron have been illustrated by Symons and Vaughan [9] as shown in Figure 11-35. Here, gap impedances are indicated by Z_1, Z_2 ... where the subscripts indicate cavity numbers. The quantities g_{21}, g_{32}, ... are transconductances defined by, for example,

$$g_{21} = \frac{I_2}{V_1} \tag{11-20}$$

and I_2 is the current induced in cavity two by the voltage, V_1, at gap one. The total current gain for the five-cavity klystron is given in (11-21).

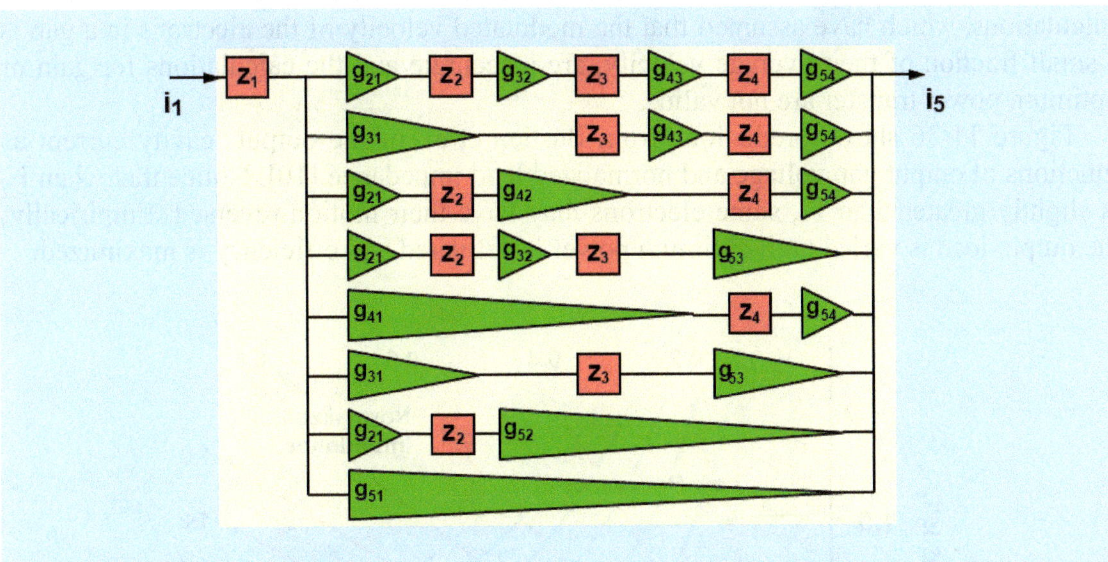

Figure 11-35 The equivalent circuit of a five-cavity klystron. (From: R. S. Symons and J. R. M. Vaughan, *IEEE Trans. Electron Devices*, October 1994. © 1994 IEEE.)

$$\begin{aligned}\frac{I_5}{I_1} = Z_1(&g_{21}\, g_{32}\, g_{43}\, g_{54}\, Z_2\, Z_3\, Z_4 \\ &+ g_{31}\, g_{43}\, g_{54}\, Z_3\, Z_4 \\ &+ g_{21}\, g_{42}\, g_{54}\, Z_2\, Z_4 \\ &+ g_{21}\, g_{32}\, g_{53}\, Z_2\, Z_3 \\ &+ g_{41}\, g_{54}\, Z_4 \\ &+ g_{31}\, g_{53}\, Z_3 \\ &+ g_{21}\, g_{52}\, Z_2 \\ &+ g_{51})\end{aligned} \tag{11-21}$$

Thus, the klystron is unlike an ordinary amplifier where the input voltage (or current) to one stage is proportional to the output voltage (or current) of the previous stage. The klystron behaves like a multistage amplifier where there are voltages from each stage fed into the stage ahead of them and also fed around into the following stages. Another factor that must be considered is that it is often desirable to tune the cavities in various ways so that the resonant frequency of different cavities is different. In addition, some cavities may be loaded to lower the Q and thereby increase the bandwidth.

11.4 KLYSTRON OUTPUT SECTION

Most of the discussions up to this point have been for gap voltages very much less than the beam voltage ($V_n < V_o$) so that small signal theory could be applied in calculating transit effects, power transfer, and so forth. However, when it is desired to maximize output power, then $V_n \approx V_o$, or more strictly, when $M_n V_n \approx V_o$. Then small-signal calculations, which have assumed that the modulated velocity of the electrons in a gap is a small fraction of their average velocity, are inaccurate and the calculations for gain or optimum power transfer are not valid.

Figure 11-36 shows predictions from Bastien et al. of the output cavity current as functions of output gap voltage and normalized load impedance [10]. Notice that when V_n is slightly greater than V_o, some electrons may have their motion reversed. Empirically, the output load is varied until optimum power is delivered and efficiency is maximized.

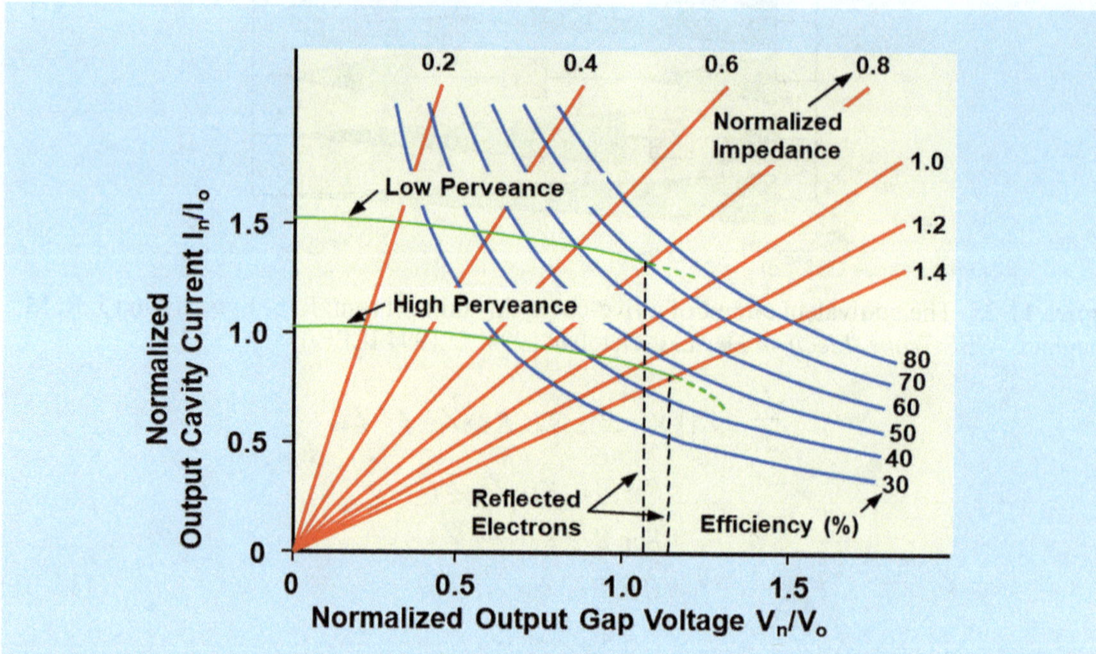

Figure 11-36 Predicted output cavity characteristics for a high-efficiency klystron. (From C. Bastien, G. Faillon, and M. Simon, 1082 IEDM. © 1982 IEEE.)

This corresponds to $G_L > G_C$ so that far more of the power is delivered to the load than to the output cavity. This additional loading will reduce the voltage across the output gap enough so that electron motion is not reversed. Instead, in the ideal case the electrons will be just slowed down to zero velocity as they pass through the gap and will have given up all of their energy to the cavity plus loads. Then as much power as possible will have been extracted from the electrons.

These conditions apply in all high-efficiency klystrons. High efficiency is obtained because most of the energy taken from the electrons is delivered to the external load rather than being dissipated in losses in the cavity. To calculate optimum power transfer under these conditions, it is necessary to go through a numerical computation of electron trajectories through the output gap and find the conditions of optimum energy loss by the electrons. Then, electrons will be passing each other and, in some cases, will be going backward. Highly efficient klystrons operate with these conditions in the output cavity.

The output section usually consists of one or two penultimate cavities and an output structure, the configuration of which depends largely on the bandwidth that is required. For narrow bandwidths and power levels that are not excessive, the output structure is usually a single reentrant cavity. In discussing the power output and gain of a klystron and their dependence on cavity parameters, it is necessary to consider the load to be connected to the cavity. The load can be coupled to the cavity in any one of many ways, for example, with coupling loop as shown in Figure 11-37 or with an aperture in the cavity wall, and the load can be placed somewhere out in an external waveguide. Also, the load can have a reactive component that appears in the equivalent circuit as additional L or C and that changes the resonant frequency. All of these require suitable parameters for their description. For our purposes, we will assume that the load impedance has been transformed and can be represented by a resistance in parallel with R_{SH} as is shown in Figure 11-38.

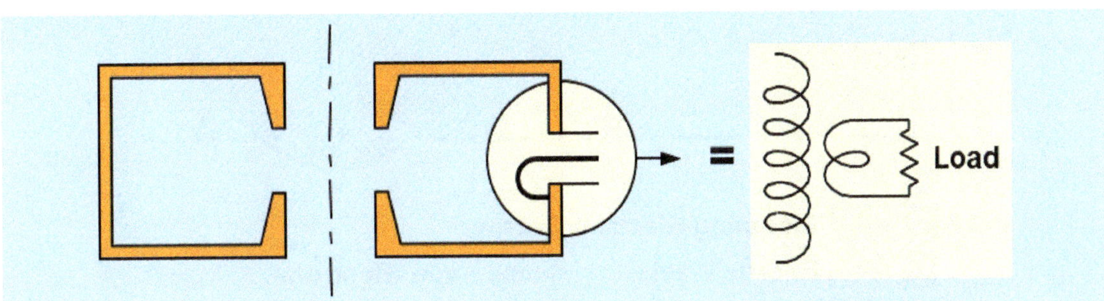

Figure 11-37 Circuit representation of coupling loop.

Most klystrons are very powerful devices and efficiency is often very important. Figure 11-39 is useful in illustrating output stage factors that are important in maximizing output cavity to be sufficient to nearly stop the electrons. Figure 11-39 illustrates the problem when the gap voltage is excessive:

1. The green cross-hatched area represents the energy of the electrons entering the klystron output cavity. The energy of these electrons is actually eV_o.
2. The red cross-hatched area represents phases of the gap voltage when energy is expended in accelerating electrons. Clearly, this degrades efficiency.
3. The blue cross-hatched area represents the phases of the gap voltage when energy is extracted from electrons as they are decelerated.
4. The black cross-hatched area represents the phases of the gap voltage when electrons are stopped and reflected by the gap because the voltage is too high.

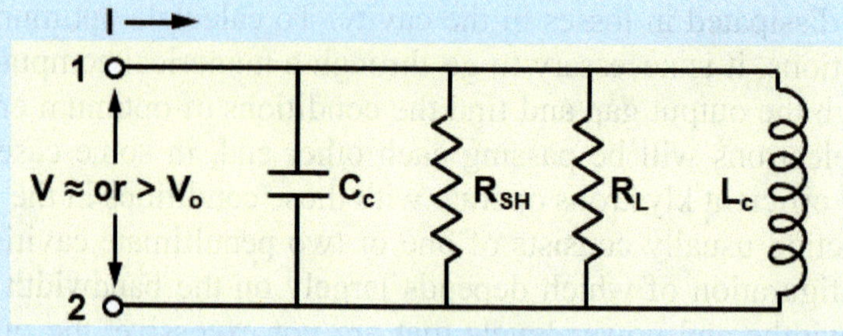

Figure 11-38 In the output section, optimum loading occurs when the circuit voltage ≈ or > dc beam voltage.

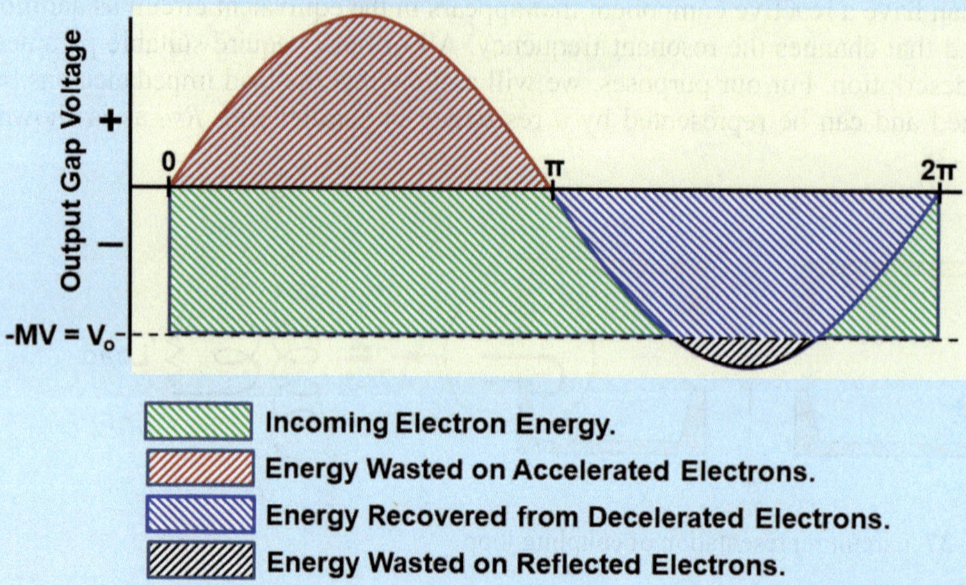

Figure 11-39 In the output section, optimum loading occurs when the circuit voltage ≈ or > dc beam voltage.

When the voltage produced across the circuit, V, is greater than the dc beam voltage, V_o, and electrons passing through the gap may be stopped and turned back, then it is necessary to increase the loading (decrease R_L). This will result in more power being delivered to R_L than to R_{SH} and an optimum adjustment of R_L can be done empirically by maximizing power to R_L. Optimum design and performance of klystrons is normally obtained under this condition. Under any circumstance, the bandwidth is still optimum when R_{SH}/Q is maximum.

REFERENCES

[1] A. Arsenjewa-Heil, and O. Heil, "Eine neue methode zur Erzeugung kurzer ungeämpfter elektromagnetisher Wellen von großer Intensität," *Zeit. für Physik*, Vol. 95, 1935, pp. 752–762.

[2] W. W. Hansen,"A type of electrical resonator," *J. of App. Phys*. Volume 9, No. 10, October 1, 1938.

[3] N. H. Pond, *The Tube Guys*, Russ Cochran, Publisher, West Plains, MO, 2008.

[4] G. Caryotakis, "High power klystrons: theory and practice at the Stanford Linear Accelerator Center," SLAC-PUB 10620, August 2004, revised January 2005.

[5] R. H. Varian, and S. F. Varian, "A high frequency oscillator and amplifier," *Jour. App. Phys.*, Vol. 10, May 1939, pp. 321–327.

[6] K. H. Kreuchen, B. A. Auld, and N. E. Dixon, "A study of the broadband frequency response of the multicavity klystron amplifier," *Jour. App. Phys.*, May 1957, pp. 529–567.

[7] M. J. Smith, and G. Phillips, *Power Klystrons Today*, Taunton, Somerset, England: Research Studies Ltd., 1995.

[8] G. Faillon, *Technical Digest,* IEDM. ©1985 IEEE.

[9] R. S. Symons, and J. R. M. Vaughan, "The linear theory of the clustered-cavity klystron," *IEEE Trans. Plasma Science*, Vol. 22, No. 5, October 1994, pp. 713–718.

[10] C. Bastien, G. Faillon, and M. Simon, "Extremely high-power klystrons for particle accelerators," *Technical Digest,* IEDM, 1982, pp. 190–194.

Chapter 12

Klystron Performance

The original klystron drawings by the Varian brothers indicated that the tubes could be either oscillators or amplifiers. In fact the first major application for klystrons was as a local oscillator in radar systems. The combination of the magnetron and the reflex klystron won WWII for the Allies. For very-low-power applications, such as the local oscillator, klystrons have been replaced by solid-state devices. However, for a wide range of applications, including those listed here, klystrons are the amplifiers of choice:

- Linear accelerators, medical, physics;
- Radar, airborne, aerial survey, missile or tracking;
- Communications, C-band and X-band earth-satellite links;
- Tropospheric scatter transmitters;
- Industrial microwave heating;
- Television (being phased out by IOT).

The powers range from a few tens of watts CW to at least a gigawatt pulsed. A klystron producing 1 MW of CW power at X band was demonstrated over half a century ago. Operating frequencies range from a few hundred megaherz to several hundred gigahertz. Bandwidths in excess of 10% are now routine and 30% has been predicted. Efficiencies as high as 80% have been demonstrated and 90% is being predicted and pursued with confidence. In this chapter, the power, bandwidth, and efficiency characteristics of klystrons are described. Most of the discussions center on conventional single-beam (sometimes referred to as pencil-beam) solenoid focused devices. However, progress with multiple-beam klystrons (MBKs) and proposed sheet-beam klystrons (SBKs) are also described.

12.1 LEGACY KLYSTRONS

The VKR 7828 inductively tuned klystron is a five-cavity version of the VA 828 klystron shown in Figure 11-11. This tube is used as an illuminator by the U.S. Navy. The VA 889F shown in Figure 11-27 is a capacitively tuned five-cavity klystron used by the U.S. Navy in the AN-SPS 49 shipboard radar shown in Figure 12-1. These tubes are excellent examples of klystrons that have served us well for several decades.

Figure 12-1 This is the AN-SPS 49 radar in which the VA 889F capacitively tuned klystron is used.

The configuration of these klystrons is basically that described by Mihran [1] and Webber [2] in the 1950s and is as shown in Figure 12-2. The first four cavities are a quarter of a plasma wavelength ($\lambda_q/4$) apart. The spacing between the fourth and fifth cavities is reduced wave behavior to about 60% of $\lambda_q/4$. These spacings are based on the space charge of the electron beam for cavities one through four and the ballistic behavior between cavities four and five. The spacings are selected as indicated in Figure 12-3 to maximize the RF current gain at each cavity that results from the modulation placed on the beam by the previous cavity.

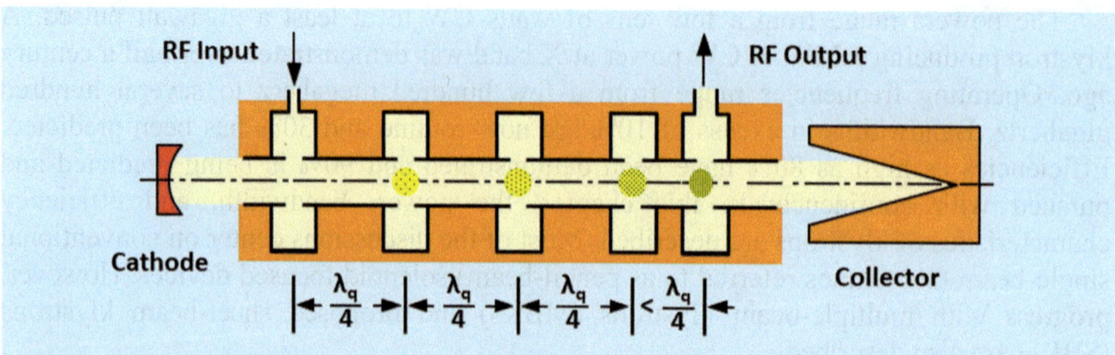

Figure 12-2 Cavity spacing used on the legacy five-cavity klystrons.

These klystrons can be synchronously tuned, efficiency tuned, or broadband tuned. In all cases, the tubes are usually tuned for maximum power output. With synchronous tuning, all cavities are tuned to the same frequency and gain is maximized.

For high-efficiency tuning, a klystron is first synchronously tuned and then the penultimate (next to last) cavity is tuned upward to maximize power. By being tuned upward, the penultimate cavity is inductive at the operating frequency and this enhances the bunching process at the output cavity.

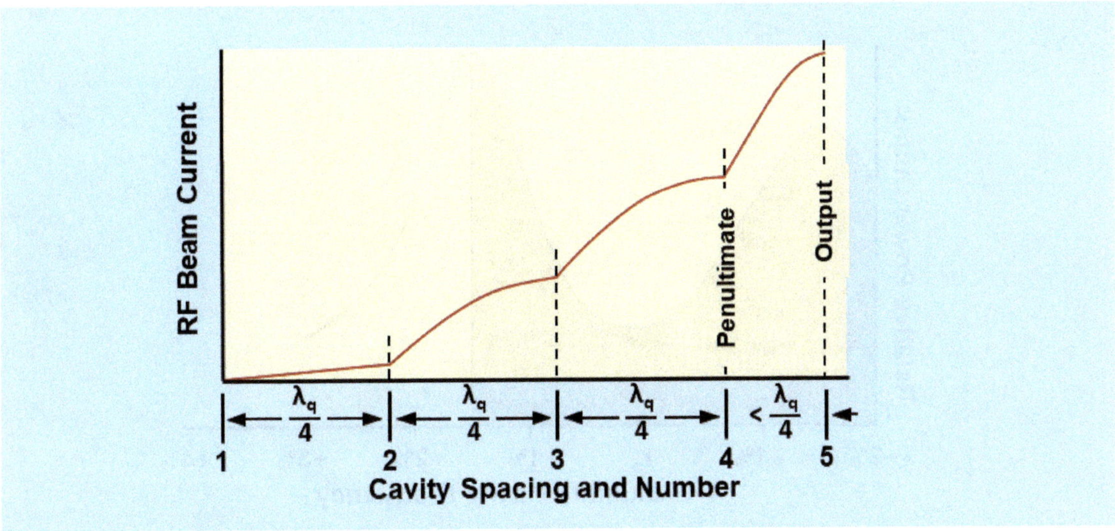

Figure 12-3 RF beam current growth in a five-cavity legacy klystron.

To broadband tune a klystron, it is first synchronously tuned. Then, as indicated in Figure 12-4, the intermediate cavities are detuned downward and upward to achieve the desired bandwidth and gain flatness. Finally, to maximize power output, the penultimate cavity is tuned to a frequency above the operating band of the tube. The bandwidth of the output cavity determines the ultimate bandwidth of the tube. For broad bandwidth, a high perveance beam is normally used to achieve low beam impedance, which, in turn, results in a low beam-loading impedance in the output cavity. Also, bandwidth can be broadened by using one of the broadband circuits discussed later in this chapter.

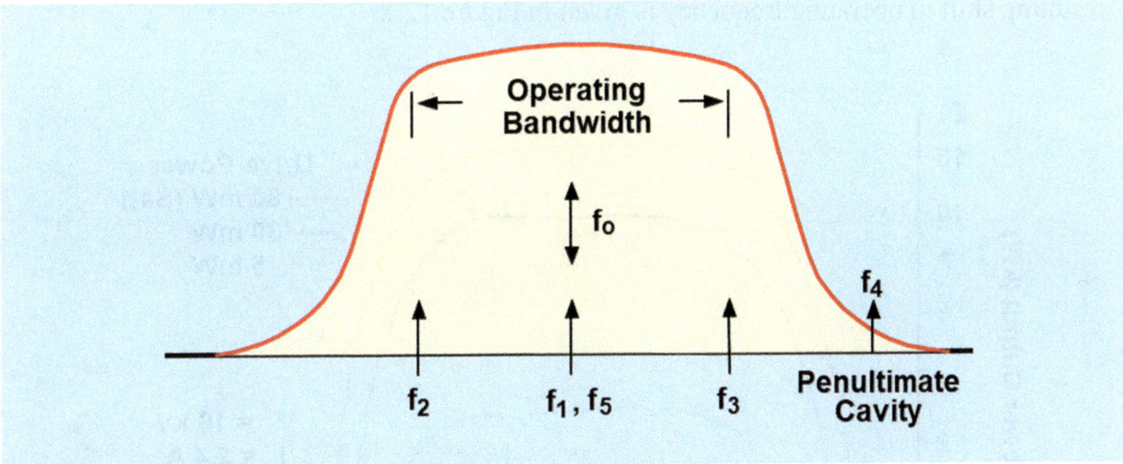

Figure 12-4 Typical cavity resonant frequencies for a broadband-tuned klystron. (From: A. Staprans et al., *Proc. IEEE*, March 1973. ©1973 IEEE.)

Tuning characteristics of a penultimate cavity are shown in Figure 12-5. With a klystron operating at or near full output power, it is important not to tune the penultimate cavity down in frequency into the shaded region indicated in Figure 12-5. This is because there is no load on the penultimate cavity, and the voltage induced could be high enough to cause breakdown across the cavity gap.

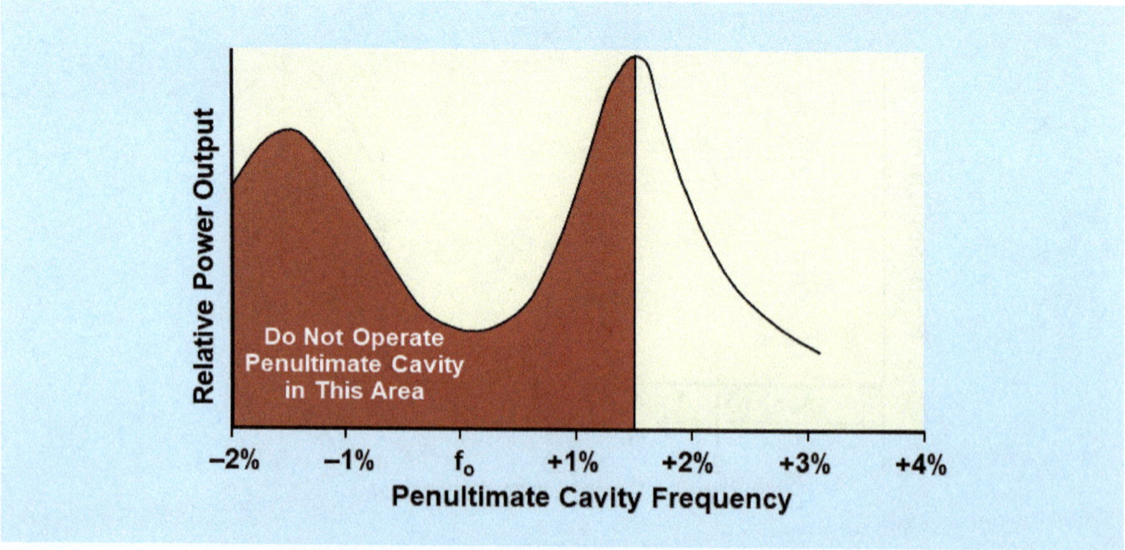

Figure 12-5 Penultimate cavity tuning characteristics. (From: *Microwave Tube Manual*, by Varian Associates, Air Force Pub. Number T.O.00-25-251, October 1979.)

Figure 12-6 shows the output characteristics for the VA-884C and VA-884D inductively tuned klystrons when they are broadband tuned. The center frequency of operation of these klystrons can be varied while maintaining the desired broadband characteristics by simultaneously tuning all cavities. This can be done mechanically with a channel tuner mechanism like that illustrated in Figure 12-7. With this mechanism, which is sometimes manually operated and sometimes motor-driven, a gear-coupled drive is used to simultaneously change the position of the tuning plunger in all cavities. An example of the resulting shift in operating frequency is given in Figure 12-8.

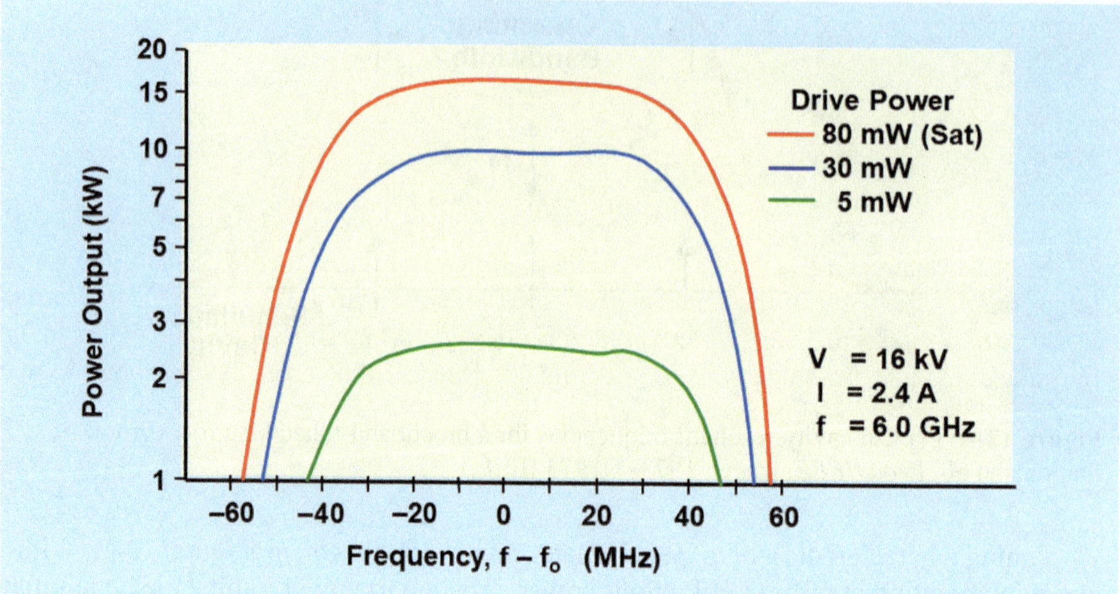

Figure 12-6 Output characteristics for the VA-884C and VA-884-D klystrons. (From: A. Staprans et al., *Proc. IEEE*, March 1973. ©1973 IEEE.)

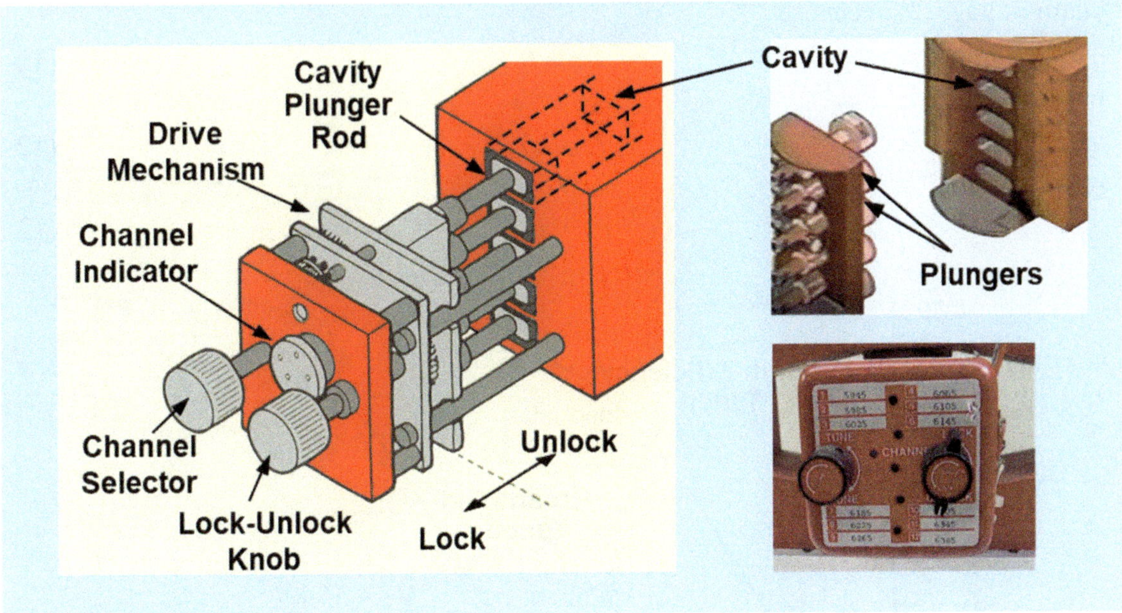

Figure 12-7 Klystron channel tuner mechanism. (From: A. Staprans et al., *Proc. IEEE*, March 1973. ©1973 IEEE.)

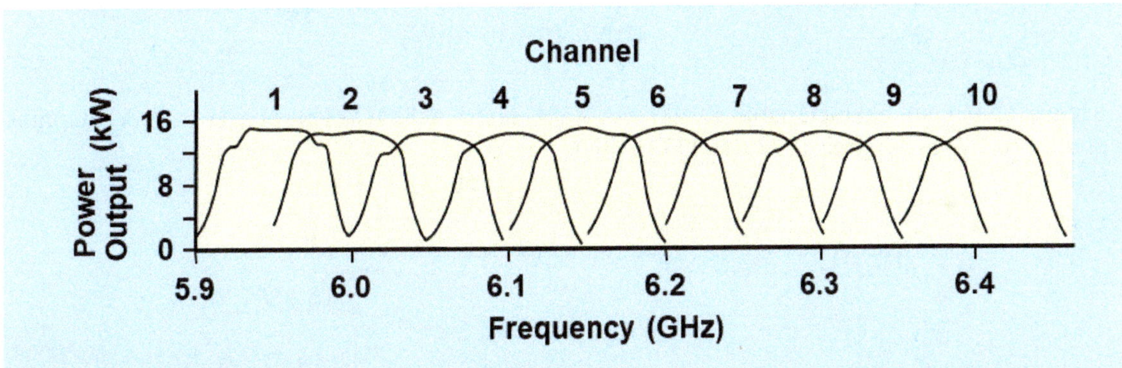

Figure 12-8 Channel-tuning characteristics of VA-884D klystron. (From: A. Staprans et al., *Proc. IEEE*, March 1973. ©1973 IEEE.)

Transfer characteristics for a klystron are shown in Figure 12-9. As is typical for all linear-beam tubes, output power usually varies linearly with drive power at low power levels. At high drive levels, there is a point where optimum bunching occurs in the output cavity and output power is maximized. Beyond that drive level, output power decreases.

When a klystron is broadband tuned, there is significant decrease in gain compared with synchronous or efficiency tuning. This is clear from Figure 12-9, which shows an increase in drive power of nearly two orders of magnitude required for broadband-tuned operation compared with efficiency-tuned operation.

The output power of a klystron (as well as for other microwave tubes) varies dramatically with beam voltage as shown in Figure 12-10. This is as expected for at least two main reasons. In the first place, beam power, P_o, varies as the 5/2 power of beam voltage, V_o, because

$$P_o = I_o V_o \tag{12-1}$$

but

$$I_o = p V_o^{3/2} \tag{12-2}$$

so

$$P_o = p V_o^{5/2} \tag{12-3}$$

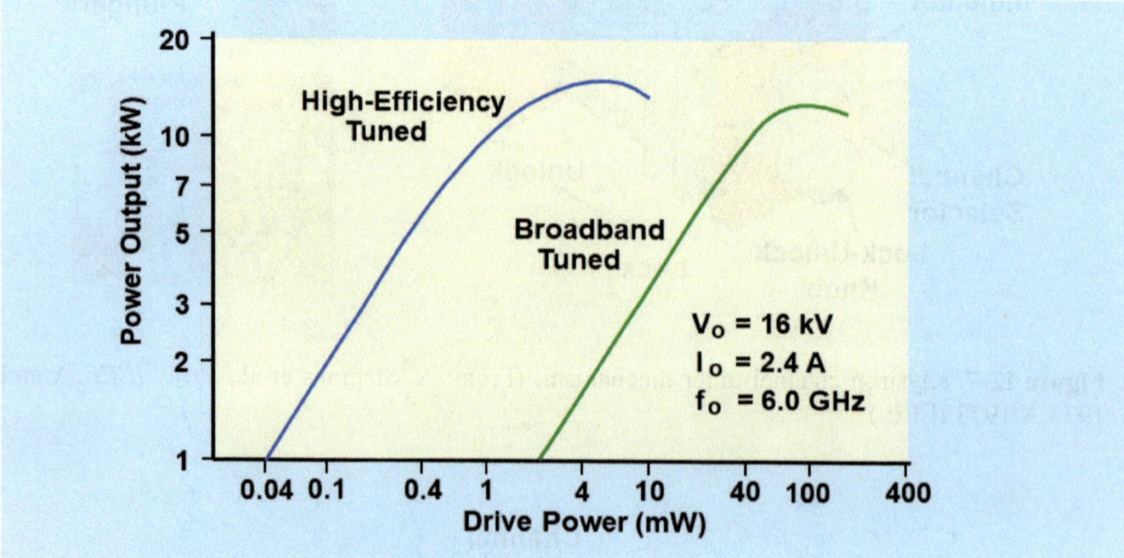

Figure 12-9 Gain characteristics for the VA-884C and VA-884D klystrons. (From: A. Staprans et al., *Proc. IEEE*, March 1973. ©1973 IEEE.)

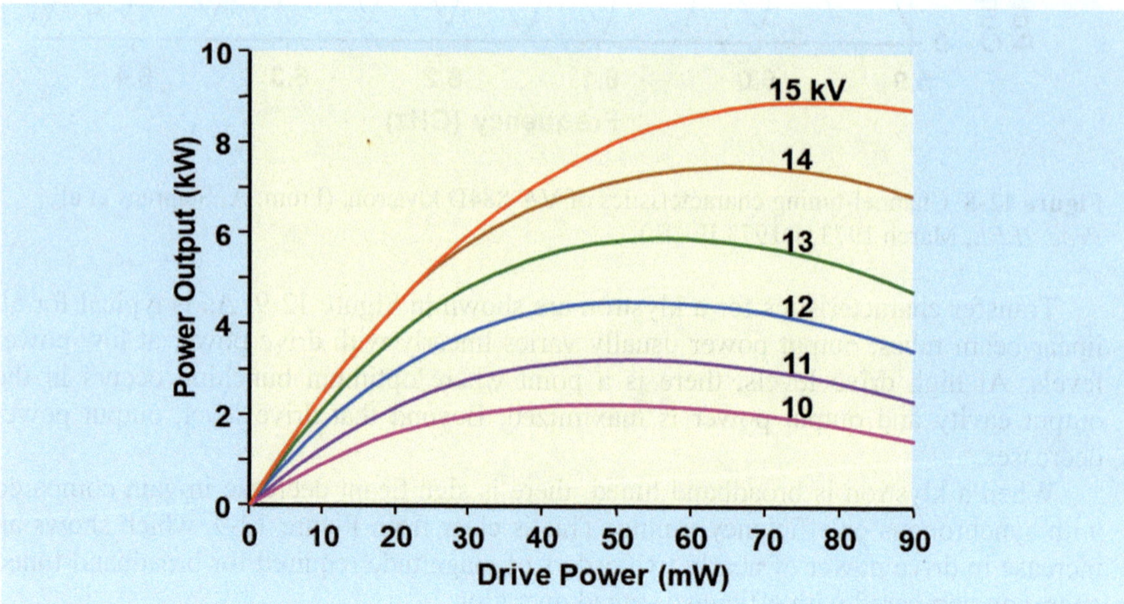

Figure 12-10 Output power as functions of drive power and beam voltage. (From: *Microwave Tube Manual*, Varian Associates, Air Force Publication Number T.O.00-25-251, October 1979.)

where I_o is the beam current and p is perveance. Thus, an increase in beam voltage by a factor of 1.5 (as shown in Figure 12-10) is accompanied by an increase in beam power of a factor of 2.8. For constant tube efficiency, RF power output would go up by the same amount. Efficiency is not constant, however, but decreases with decreasing voltage because of changes in beam bunches and diameter at voltages below the design voltage.

12.2 POWER

12.2.1 Pulsed Operation

Klystrons are capable of generating microwave power from over a megawatt CW to over a GW pulsed. Symons [3] provided a technique for estimating the power that should be obtainable from a single pencil-beam klystron. Of course, the RF power is primarily dependent on the beam power, which is the product of beam voltage and beam current, so it is necessary to determine limits on beam voltage and current in order to estimate power capability.

12.2.2 Limits on Beam Voltage

The most obvious limit on beam voltage is the onset of arcing in the electron gun. This limit depends on pulse length, as Table 12.1 indicates. Note that hold-off voltage decreases with increasing pulse length. Symons pointed out that the use of intermediate potential grading electrodes in the gun could make higher operating voltages possible.

Table 12.1
Voltage Limits Achieved in Practice

Pulse Length	Voltage	Manufacturer	Band
0.1 μs	700 kV	Varian	S
2.0 μs	475 kV	SLAC	S
6.5 μs	315 kV	SLAC	S
10 μs	300 kV	Litton	L
Continuous	200 kV	Varian	S

Source: R. S. Symons, "Scaling laws and power limits for klystrons," IEDM, 1986.

Other factors, beside gun arcing, limit the operating voltage of a klystron. For efficient operation, the voltage across the output gap must be sufficient to remove most of the energy from the electron beam as it passes through the gap. This happens when the peak voltage across the output gap is nearly equal to the beam voltage. However, arcing can place a limit on gap voltage and, consequently, on beam voltage.

There is also a limit imposed on cavity gap voltage by the RF losses in the output cavity. RF losses occur in cavities because of microwave currents flowing in the cavity walls. These losses are proportional to the square of the cavity gap voltage. Also, pulsed heating of the cavity surface can lead to cracking and fatigue failure. It is important to be able to estimate these losses for at least two reasons:

1. Adequate cooling for the cavity walls can be provided;
2. The operating power level of the cavity can be limited so that losses do not exceed the cooling that has been provided.

One approach to estimating cavity losses has been described by Symons and Jory [4]. They have defined a figure of merit, F_m, which relates the square of the peak cavity gap voltage (which is a measure of energy stored in the cavity) to the power dissipated in the cavity walls. When energy and power are expressed in terms of R_{SH}/Q and Q_o, it is seen that F_m is invariant as all cavity dimensions are scaled with wavelength. This figure of merit is

$$F_m = \frac{V^2}{P_L / A} \frac{\delta}{\lambda^3} = \left(\frac{R_{SH}}{Q} \right) \left(Q_o \frac{\delta}{\lambda} \right) \left(\frac{A}{\lambda^2} \right) \tag{12-4}$$

in which V is peak voltage, P_L/A is the loss per unit area, δ is the skin depth, λ is the operating wavelength, and Q_o is the Q of the cavity with no load attached.

Symons pointed out [4] that not only are the three quantities on the right side of this equation invariants for wavelength scaling of a particular cavity geometry, but for klystrons, they vary over extremely limited ranges. In most klystrons, $R_{SH}/Q \approx 125$, $Q_o\delta/\lambda \approx 0.8$, and $A/\lambda^2 \approx 1$, and so $F_m \approx 90$. Now, using this value of F_m and the value of δ for copper, (12-4) can be rewritten as

$$V^2 = 0.83 \times 10^6 \lambda^{5/2} \frac{P_L}{A} \tag{12-5}$$

For pulsed devices, the average power loss is the peak power loss reduced by the duty cycle, D_u. For this relation to be useful in estimating power losses, a value must be assigned to the peak voltage, V. This can easily be done in the case of a klystron output cavity operating so as to extract maximum power from the electron beam passing through it. In that case, the peak voltage across the cavity gap is approximately the beam voltage.

Figure 12-11 contains plots of average power loss for various voltages and operating wavelengths. The usual cooling capability in klystrons is limited to about 1 kW/cm^2; however, dissipations in excess of 10 kW/cm^2 have been demonstrated for use in crossed-field devices.

Kilpatrick [5] proposed a criterion for the onset of RF arcing across gaps and Peter et al. [6] proposed the modification

$$f = 1.6\ E^2\ e^{-8.5/E} \tag{12-6}$$

where f is the frequency in megahertz and E is the electric field in megavolts per meter.

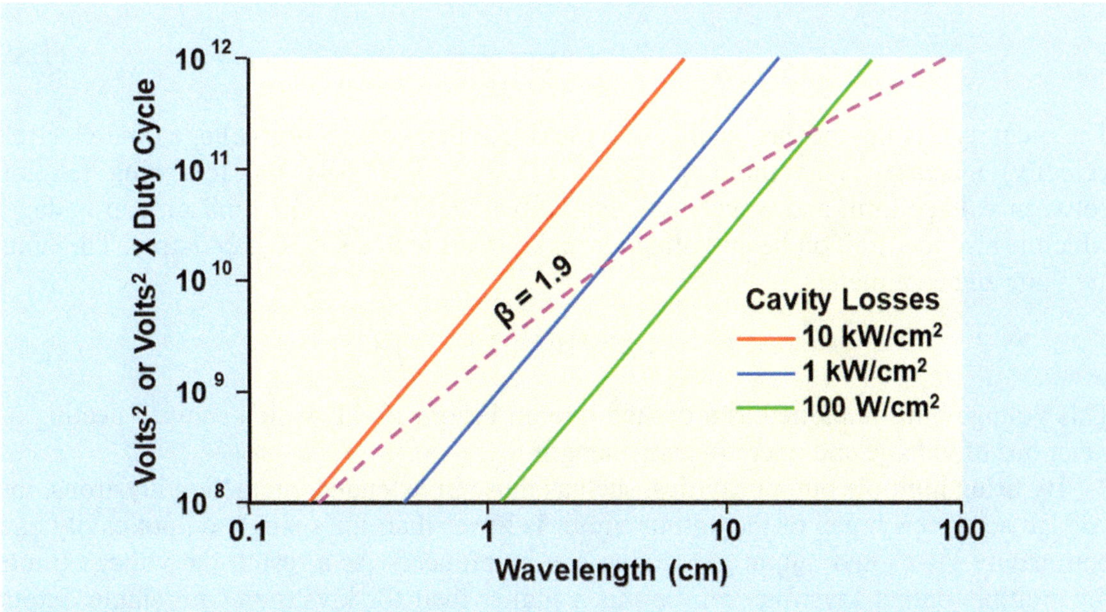

Figure 12-11 Solid lines: RF cavity losses as functions of cavity gap voltage and wavelength. Dashed line: Limits on beam voltage due to cavity losses and RF arcing. (Adapted from: R. S. Symons, 1986 IEDM. ©1986 IEEE.)

It should be noted that Loew and Wang [7] demonstrated an RF breakdown limit far above that given by (12-6). Their results were obtained for electron linac structures. However, Phillips and Sprehn [8] found that the RF breakdown in their high-power klystron work matched well with (12-6). They pointed out that a major difference between the klystron and the linac is the presence in the klystron of a nearby intense electron beam. They hypothesize that stray electrons or x-ray photons strike the cavity walls and initiate breakdown at a relatively low field gradient more like that given by (12-6).

The use of (12-6) is simplified by noting that the fields used in high-power klystrons are high enough that the exponential term is ~ unity. An approximation for the magnitude of the electric field can be found by examining the gap transit angle. First of all, the field is related to gap voltage by

$$E = \beta \frac{V}{\ell} \tag{12-7}$$

where ℓ is the gap length and β is the field enhancement factor resulting from the field concentration at the gap edges (typically 1.9 for klystron gaps).

Because the gap transit angle is nearly 1 radian in all klystrons, the gap length can be determined to be

$$\ell = \frac{\lambda}{2\pi}\left[1 - \gamma^{-2}\right]^{1/2} \tag{12-8}$$

where, with V in volts,

$$\gamma = 1 + \frac{V}{511{,}000} \tag{12-9}$$

The quantity γ is the increase in electron mass that occurs as beam voltage (and electron velocity) increase. By combining (12-6), (12-7), and (12-8), the following relation between voltage limit and wavelength can be obtained. This is the limit on gap voltage, which is also the limit on beam voltage for a klystron with a single-gap output. The units are volts and centimeters.

$$V^2 = 1.32 \times 10^{10}\left(1 - \gamma^{-2}\right)\lambda \tag{12-10}$$

This voltage limit is plotted as a dashed line on Figure 12-11, which showed heating as functions of voltage and operating wavelength.

By using multiple output cavities, such as those in extended interaction klystrons, the voltage across each gap of the output circuit is lower than for a single output cavity and both cavity losses and output gap arcing can be reduced. As a result, the voltage limits for multiple output klystrons can be much higher than for klystrons with single output cavities.

12.2.3 Limits on Beam Current

There are several limits to beam current in a linear-beam tube. The most obvious limits are evident in the relation for RF power output, which is

$$P = \eta I V = \eta P_\mu \times 10^{-6}\, V^{5/2} \tag{12-11}$$

where η is efficiency and P_μ is the microperveance of the electron gun. It is very difficult to design guns with $P_\mu > \sim 2$ because of space-charge effects on potential contours within the guns. Since microperveance is limited and beam voltage is limited, beam current is limited.

Symons [3] observed that efficiency is fairly well described by the relation

$$\eta = 0.9 - 0.2\, P_\mu \tag{12-12}$$

Others, for example Phillips and Sprehn at SLAC [8], found somewhat different relations; however, we will use Symons' relation here.

The reason for the variation of efficiency with microperveance is discussed in detail in Section 12.4 and is that as perveance is increased, it becomes more difficult to form electron bunches that are optimally configured for maximum power extraction from the electron beam. As a result, for maximum efficiency, perveance must be limited, which limits beam current.

As the frequency of operation of a klystron is increased, other factors come into play in limiting beam current. Beam current is *JA,* that is, the product of current density and the cross-sectional area of the electron beam. The area of the beam is limited by the drift tunnel radius, which scales with wavelength. Symons pointed out that the electron beams in most klystrons have a radius that is approximately one-half of a radian of electron transit angle.

The beam current density is the cathode emission density multiplied by the area convergence ratio of the gun. For impregnated cathodes and long life, a current density of 5 A/cm^2 can be obtained with little dependence on pulse length. For short pulses, Symons assumed that an oxide-coated cathode could be used. The emission density, J_c in A/cm^2, is given by the empirical relation

$$J_c^{2.61}\tau = 814 \tag{12-13}$$

where τ is the pulse length in microseconds.

If we assume a maximum area convergence ratio of 200:1, then the obtainable beam current density is 1,000 A/cm^2 for $\tau \geq 10$ μs, 2,600 A/cm^2 for $\tau = 1$ μs, and 6,300 A/cm^2 for $\tau = 0.1$ μs.

12.2.4 Estimate of Obtainable Power

Regardless of frequency, space-charge effects limit perveance and therefore beam current and power. By combining (12-6) and (12-7), we find that the dependence of RF power on perveance and on voltage is

$$P = P_\mu\left(0.9 - 0.2\,P_\mu\right)10^{-6}V^{5/2} \tag{12-14}$$

As pointed out in the previous section, as frequency increases, beam area decreases. The limit on beam current is

$$I = J\frac{\lambda^2}{16\pi}\left(1 - \gamma^2\right) \tag{12-15}$$

The limit on J is obtained from the maximum cathode current density and the area compression ratio.

Now, by combining (12-15) with (12-12) and (12-11), the following expression for power as functions of wavelength, J, V, and P_μ, can be obtained.

$$P = \left(0.9 - 0.2\ P_\mu\right)\frac{\lambda^2 J V}{16\,\pi}\left(1 - \gamma^{-2}\right) \tag{12-16}$$

We now have two limiting relations, (12-14) and (12-16), for obtainable power. Figure 12-12 shows these power limits plotted as a function of $\lambda^2 J$ with voltage as a parameter.

The procedure outlined by Symons for establishing the reasonableness of a power requirement for a klystron is as follows:

1. Find the maximum allowable beam voltage from Table 12.1 and Figure 12-11;
2. Using Figure 12-12 or (12-16) and (12-14), see if the desired power can be achieved.

As an example, consider the 5045 S-band klystron manufactured by the Stanford Linear Accelerator Center (SLAC) and shown in Figure 12-13. A total of 245 of these klystrons operate in the SLAC accelerator. During the past 30 years, over 800 of these tubes have been built at SLAC. The mean time between failures (MTBF) is over 100,000 hours [9].

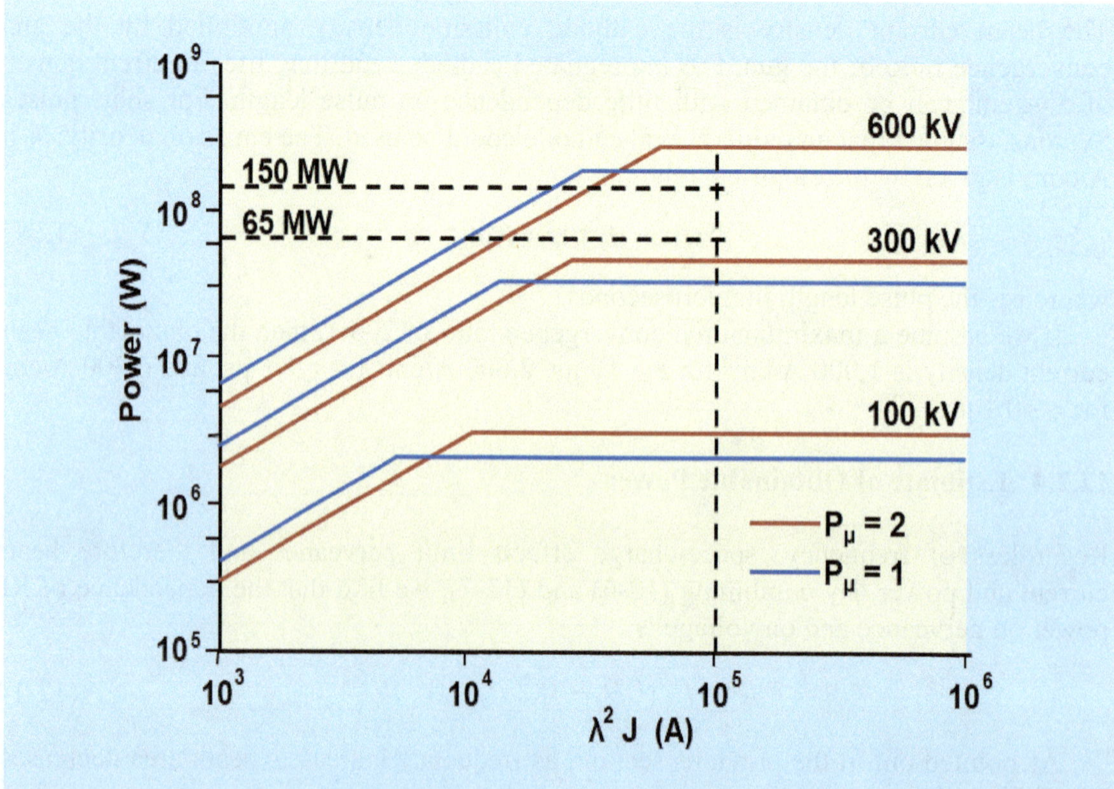

Figure 12-12 Estimate of obtainable power. (From: R. S. Symons, 1986 IEDM. © 1986 IEEE.)

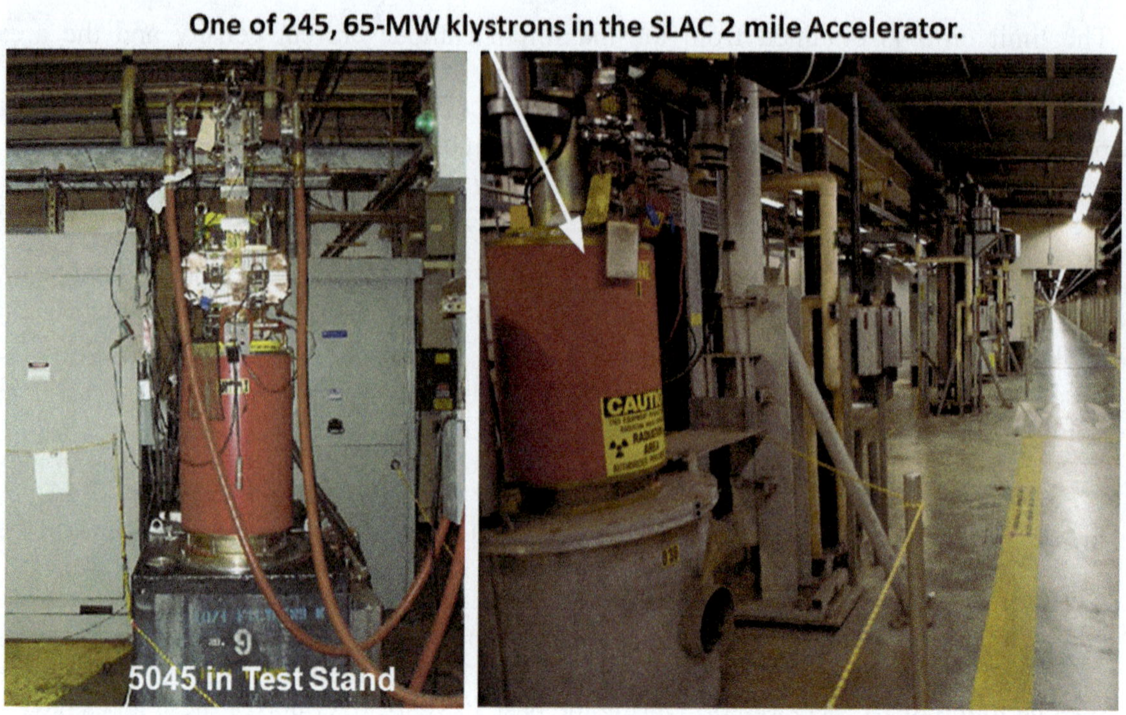

Figure 12-13 In the Stanford Linear Accelerator Center, 245 SLAC-5045, 65-MW klystrons are used to power the accelerator.

The operating parameters for the 5045 are listed in Table 12.2. Note that this tube operates at 350 kV, has a microperveance of 2, and is pulsed at 180 pps with a 3.5-μs pulse width [9]. Also, and not shown here, a scandate cathode is used. From Figure 12-10, it can be seen that it is operating close to Kilpatrick's criterion limit for RF output gap arcing of about 10^{11} square volts. The product, $\lambda^2 J$, for this device is about 10^5 and thus, from Figure 12-12, it is reasonable to expect an output power of 65 MW at a microperveance of 2.

Table 12.2
Operating Parameters for SLAC-5045, 65-MW, S-Band Klystrons

Frequency	2.856 GHz
Beam voltage	350 kV
Microperveance	2
Peak output power	65 MW
Average output power	41 kW
RF pulse width	3.5 μs
Pulse rep rate	180 Hz
Gain	50 dB
3 dB bandwidth	20 MHz
Saturated efficiency	45%
Cathode current density	8 A/cm^2

There are several possible techniques for developing klystrons with power capabilities above those predicted by Symons' procedure. As pointed out earlier in this section, by using multiple output cavities, output gap voltage and output gap arcing can be reduced so there is the possibility of using higher beam voltages.

The use of voltage grading electrodes in the electron gun could make possible higher gun voltages. True discussed the use of grading electrodes in the design of a 1-MV, 1,825-A Pierce gun [10]. Three electrodes were used to limit the voltage gradients to less than 200 kV/cm and True said that the gun should operate reliably for pulse lengths up to 5 μs.

There is considerable effort at the present time to develop multiple-beam klystrons (discussed later in this chapter). These generate extremely high powers because many beams, each operating at a microperveance near unity, can be used with the result that the klystron has an overall microperveance ~10 or higher. It is possible that a similar increase in microperveance can be achieved by using a beam in the form of a flat sheet.

The Stanford Linear Accelerator Center klystron group designed, fabricated, and tested high-power klystrons, which range from 477 MHz to 11.424 GHz and from 1 MW under continuous operation to 150 MW at a 3-μs pulsewidth. Particularly noteworthy are the 150-MW S-band tube and a 75-MW X-band tube. The operating parameters for the 150-MW S-band tube are listed in Table 12.3 [11]. The tube has operated at powers greater than 150 MW. The tube uses a two-cell output cavity to increase efficiency, a scandate cathode, and lossy drift tubes to improve stability. Figure 12-12 indicates that this tube is at the upper limit of what should be possible.

Table 12.3
Operating Parameters for 150-MW S-Band SLAC Klystron

Beam voltage	535 kV
Beam current	700A
RF pulse width at rep rate	3 μs at 60 Hz
RF output power	150 MW
Saturated gain	> 50 dB
Efficiency	≥ 40%
Operating frequency	2,998 MHz
Solenoid focusing field	2,100 gauss (3×B_B)

Source: D. Sprehn et al., SLAC, 1994

The 75-MW X-band program was established to provide klystrons for the SLAC version of the Next Linear Collider. Phillips and Sprehn described the development efforts for 75-MW solenoid-focused and PPM-focused klystrons [8]. The solenoid-focused tube operated at 75 MW and 450 kV with pulse lengths up to 1.2 μs and an efficiency of 50%. The PPM-focused klystron shown in Figure 12-14 operated at 71 MW and 463 kV with an efficiency between 53% and 63%. Assuming that the beam current density was comparable to that in the S-band tubes, these X-band tubes were operated above the region that the Symons' plots in Figure 12-12 identify as reasonable

Klystrons rated at 2.75-MW peak and 275-kW average were developed by Litton and CPI for the European Spallation Neutron Source. The Litton tube is shown in Figure 12-15. Litton observed that this average power level was equivalent to the power of a 310-horsepower Ford Excursion vehicle plus 734 60-W incandescent light bulbs. Even though these tube development efforts were successful, the tubes were not used because of a design change in the accelerator.

Several decades ago, it was thought that the way to achieve high resolution with a radar system was to use a radar operating with very-short, high-power pulses. The klystron selected for that radar was the 1 0ns, 1 GW tube developed by Varian and shown in Figure 12-16. A figure of merit for the capability of conventional microwave tubes is pf^2 where p is power and f is frequency. When expressed in GW x GHz^2, that figure of merit for this tube is 9, which still stands as a record today. Many studies were carried out to determine the feasibility of this short-pulse, high-power system. For example, it was not known if this pulse could even be propagated through the atmosphere without causing the air to break down. One of the biggest problems with the system was the enormous power modulator that was required to power the klystron. The efforts to use this system for a high-resolution radar were terminated when it was demonstrated that the desired resolution could be obtained using a lower power level and longer pulse width and sweeping the frequency during the pulse.

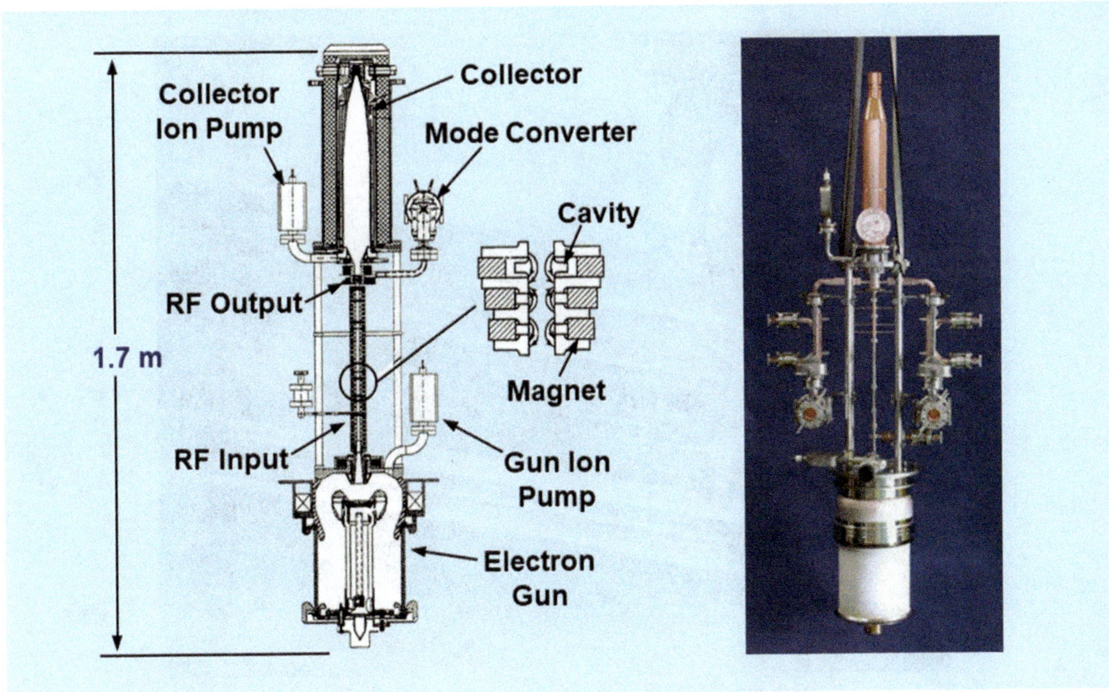

Figure 12-14 The SLAC 71-MW, X-Band, PPM-focused klystron.

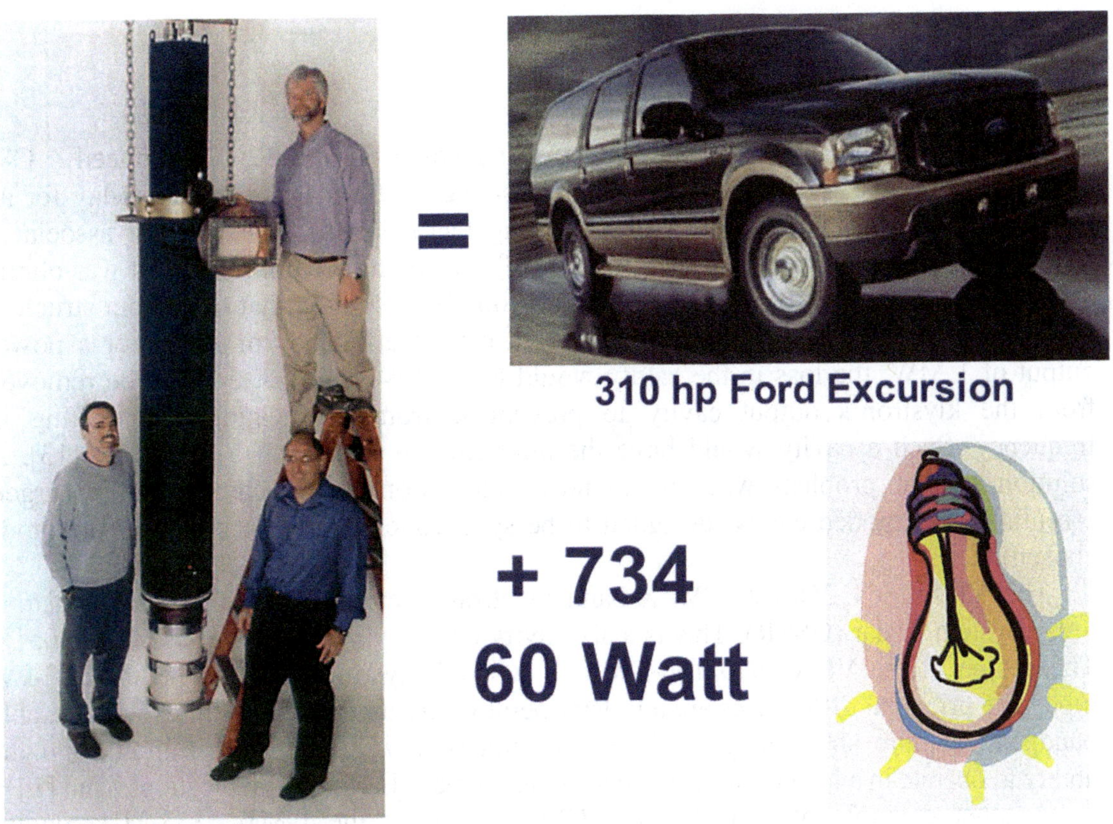

Figure 12-15 The Litton L6048 2.75MW-peak, 275-kW avg, 805-MHz klystron developed for the Spallation Neutron Source.

Figure 12-16 1-GW, 10-ns, CPI klystron.

12.2.5 CW Operation

In 1969, the Varian 3030A X-band klystron shown in Figure 12-17 produced a CW output of 1030 kW and that accomplishment stands as the state of the art today for an experimental klystron [12]. The power-limiting factor in this tube was not associated with the generation, transport, and collection of the electron beam. Instead, it was ohmic losses in the output cavity. This can be understood by assuming that the output structure consisted of a single cavity with an excellent circuit efficiency of 99%. For a power output of 1 MW, the loss in the cavity would be 10 kW. This power must be removed from the klystron's output cavity to prevent it from overheating and drifting in frequency. Such a cavity would have the approximate surface of a matchbox [13]. A solution to that problem was an extended interaction output cavity (i.e., a larger, essentially overmoded cavity, designed to be synchronous with the beam for the mode chosen).

The VKX-7864 250-kW CW X-band klystron is currently used in the Goldstone Solar System Radar (GSSR). This is a six-cavity device that operates at 51 kV, 11.05 A. The design of the VKX-7979 100-kW CW X-band klystron is similar to the 250-kW tube. The principal differences are a lower operating frequency, a two times wider usable bandpass, a more stringent gain slope variation of 0.05 dB/MHz maximum, and the ability to operate in a lower power mode to cover the near-Earth communications band [14].

Another example of a high-power CW klystron is the 1-MW, 353-MHz device manufactured by Thales [15]. The Thales tube, the TH 2089, was developed for CERN, the European Nuclear Research Center. A TH 2089 is shown in Figure 12-18. The weight of the klystron, electromagnet, and support structure is 2,400 kg and the dimensions are 4.80 × 1.85 × 1.20 meters.

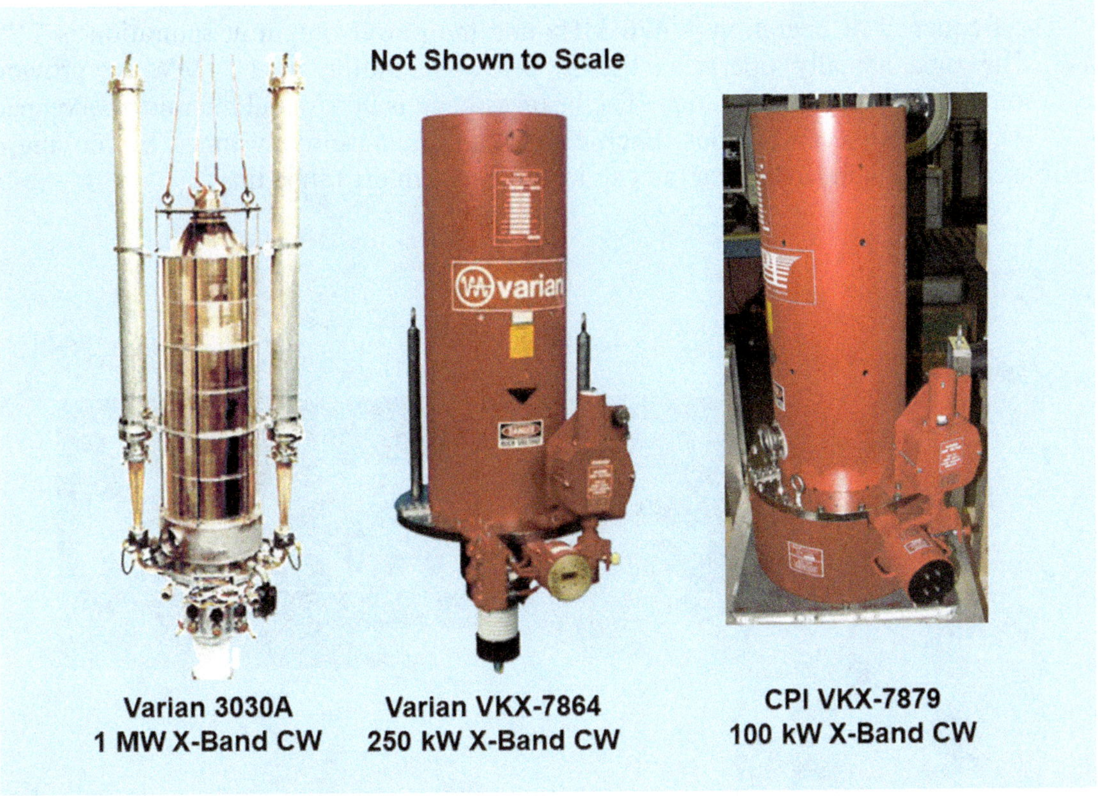

Figure 12-17 CW CPI klystrons.

Figure 12-18 The 1-MW Thales TH 2089 CW klystron.

Another CW klystron developed for accelerator service is shown in Figure 12-19. This tube is referred to as the B factory klystron (BFK) and is 15 ft. long and is operated horizontally in the SLAC PEP II collider. It is shown with its solenoid on the right in Figure 12-19. The B factory refers to the generation of B meson particles, which allows physicists an opportunity to study the subtle differences in the behavior of matter and antimatter.

The frequency of operation is 476 MHz and the power output at saturation is 1.25 MW. The tube actually operates slightly below saturation at 1.1 MW to provide headroom for amplitude modulation. The beam voltage is 90 kV and the microperveance is 1. The tube has seven cavities. Each cavity can be capacitively tuned by moving a flexible cavity wall and changing the gap length between drift tube tips.

Figure 12-19 The SLAC B factory 476-MHz, 1.25-MW CW, 15-ft. long klystron.

12.3 BANDWIDTH

Traveling-wave tubes are usually considered to be broadband amplifiers, while klystrons are thought to have narrow bandwidths. To make a correct comparison, however, output power must be specified. At power levels up to about a kilowatt, TWT bandwidths can exceed an octave because broadband helix circuits are used. When the average power level becomes high enough so the coupled-cavity circuits must be used, the bandwidth drops considerably and at the 100- to 200-kW level, the bandwidth is no wider than ~15% [16].

In the case of the klystron, the lower the cavity Q is, the wider the bandwidth is. The cavity Q, in turn, varies with beam impedance, R_o, which may be related to tube operating parameters as follows:

$$R_o = \left(\frac{\eta}{P_{RF}} \right)^{1/5} \frac{1}{p^{4/5}} \tag{12-17}$$

where η is efficiency and p is perveance. As a result, R_o and Q decrease and bandwidth increases as PRF and p increase. Within reasonable limits of perveance and efficiency (two opposing characteristics), the above relation indicates that it is two to three times easier to achieve the same bandwidth at 20 MW than it is at 200 kW. In fact, there are now klystrons available that deliver 3–10 MW and have a bandwidth of 10% or more, while at 100–200 kW, bandwidth is usually limited to 3% or 4%.

Examples of high-power klystrons capable of 10% bandwidth are the VA-812C [17] and the VA-842 that operated at 400–450 MHz. These two tubes were nearly identical in appearance but had different performance characteristics. The peak power output of the VA 812C was 8 MW and the average power was 30 kW. The peak power output of the VA-842 was 1.25 MW and the average was 75 kW. The VA-842, which is shown in Figure 12-20, was used in the Ballistic Missile Early Warning System.

The operating characteristics of the VA-812C are listed in Table 12.4 and the efficiency as a function of frequency is shown in Figure 12-21. Because these tubes operated at a low frequency, they were very large and this, in turn, made the use of very high perveance beam possible. Also, to achieve the flat broadband response, each of the four cavities in the driver section had a different gap length and a different loading. Broadbanding of the output circuit was achieved by using a triple-tuned circuit

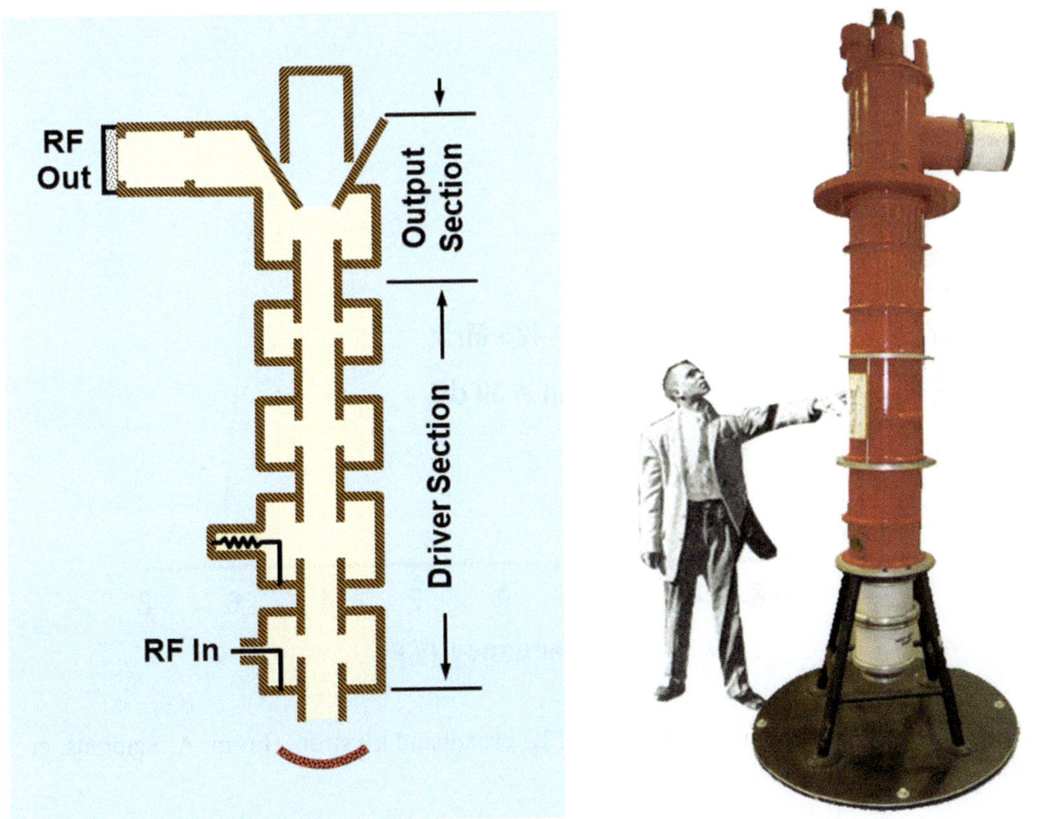

Figure 12-20 The VA-842.

At microwave frequencies in the power range from about 200 kW to a few megawatts, bandwidths are required that conventional klystrons and TWTs cannot provide. As a result, special designs have been developed that make klystrons operating at bandwidths of 10% or more possible. The driver sections and the output sections of these tubes are discussed in the following sections.

Table 12.4
VA-812C Operating Characteristics

Frequency	400-450 MHz
Peak beam voltage	145 kV
Perveance	2.5×10^{-6}
Peak power output	8 MW
Average power output	30 kW
Pulse length	6 μs
Efficiency	40%
Electronic bandwidth	12%
Gain	30 dB

Source: A. Staprans et al., *Proc. IEEE*, March 1973

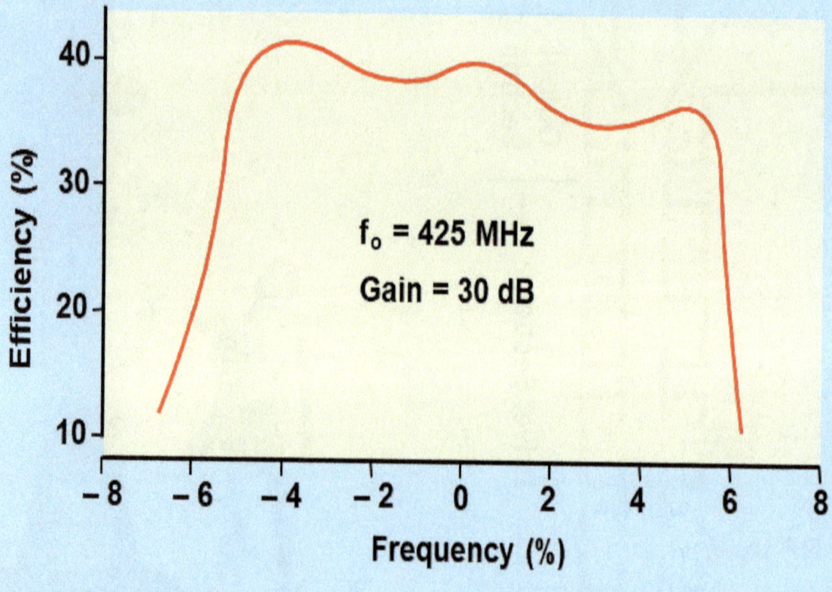

Figure 12-21 Frequency response of the VA-812C broadband klystron. (From: A. Staprans, et al., *Proc. IEEE*, March 1973. ©1973 IEEE.)

12.3.1 Driver Section

The driver section consists of all of the cavities up to the output structure. In conventional narrowband tubes, the driver section usually has only three or four cavities and the output structure is a single cavity. In broadband klystrons, the driver section may consist of eight to ten or more cavities. In fact, for some clustered-cavity klystrons, the use of over 20 cavities in the driver section has been suggested. When large numbers of cavities are used, the interactions between cavities become extremely complex because he modulation produced by each cavity is carried to all subsequent cavities by the electron beam.

In trying to achieve wide bandwidths for driver sections, it is not sufficient to simply spread the tuning frequencies of the individual cavities over the desired bandwidth. The gain dips must be taken into account, and especially those of the cavities tuned to the low end of the band. Those gain dips can be moved to the high end of the band by lengthening the associated drift tubes. Also, the cavity *Q*s may be adjusted to help control the overall shape of the composite gain curve.

An example showing the gain curves of seven cavities in a broadband driver section is shown in Figure 12-22. Here it is clear that long drift tubes have been used on the cavities at the low end of the band. The low-frequency (left-hand) gain curves are very broad and resemble, except for the broad resonances produced by low *Q*s, the curve for 90□ drift tubes in Figure 9-14. The gain dips of these low-frequency cavities are compensated for by the gain poles of the cavities tuned to the high end of the band.

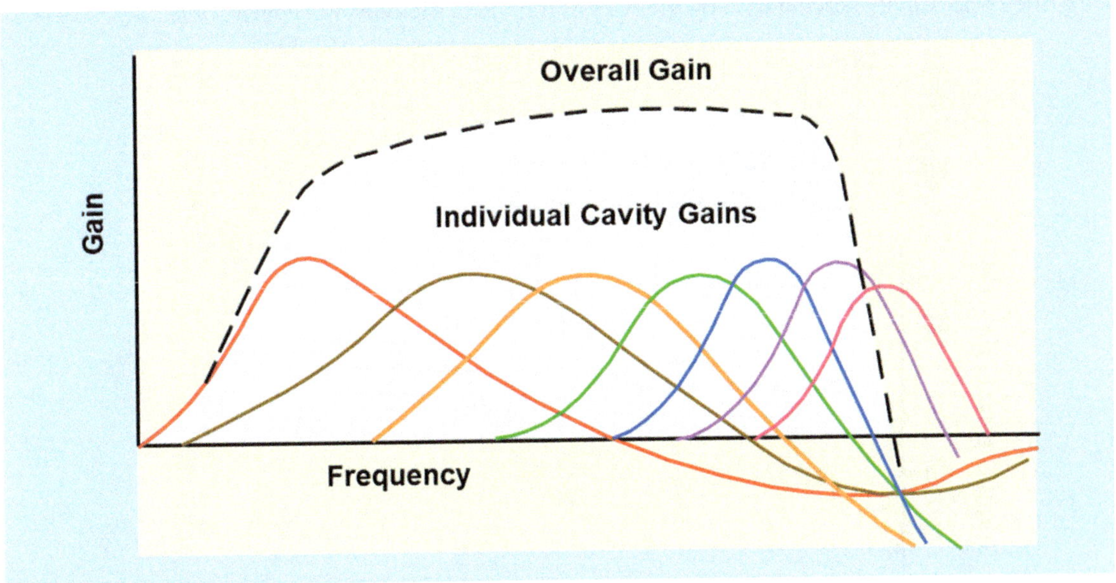

Figure 12-22 Cavity gain curves for a broadband klystron. (Adapted from: *Microwave Tube Transmitters*, L. Sivan, New York: Chapman and Hall, 1994. ©1994 L. Sivan.)

In his report on the design of broadband klystrons, Metivier [18] showed the improvement in overall gain performance of an eight-cavity klystron that could be achieved with uneven spacings of cavities in the driver section. The cavity arrangements that he described are shown in Figure 12-23. Cavity spacings are shown along with *Q*s and frequencies to which the cavities were tuned.

Small-signal gain curves corresponding to the cavity spacings in Figure 12-23 are shown in Figure 12-24. With evenly space cavities, a large gain dip occurred at the low end of the frequency band. By increasing the spacings between cavities 1 and 2 and between 2 and 3, which were tuned to the low end of the band, the gain dips produced by these cavities were moved to the high end of the band. There, these gain dips compensated for a large gain peak that had previously occurred. The overall small-signal gain ripple was reduced to 7 dB, which, with the gain compression occurring at saturation, resulted in an overall gain ripple of less than 1 dB at constant drive.

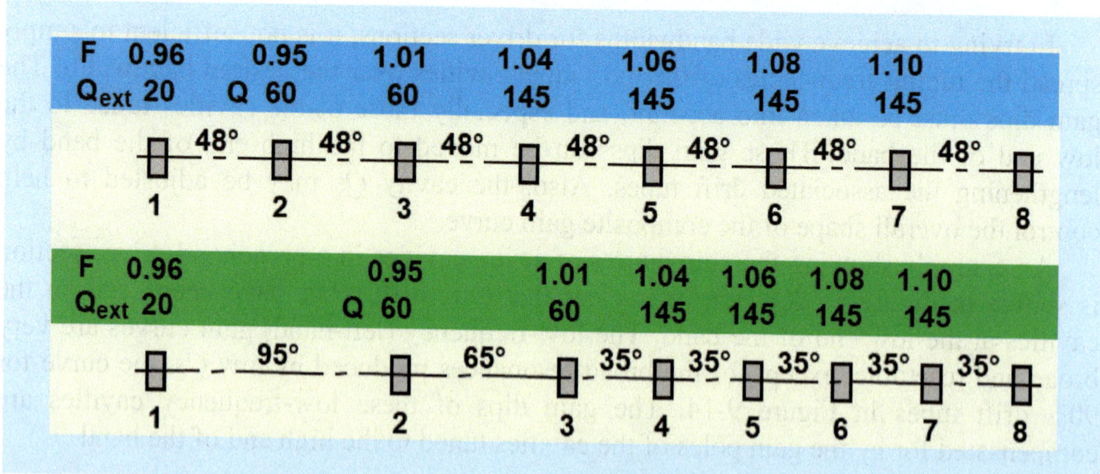

Figure 12-23 Cavity spacing used by Metivier. (From: R.L. Metivier, *Microwave Jour.*, April 1971.)

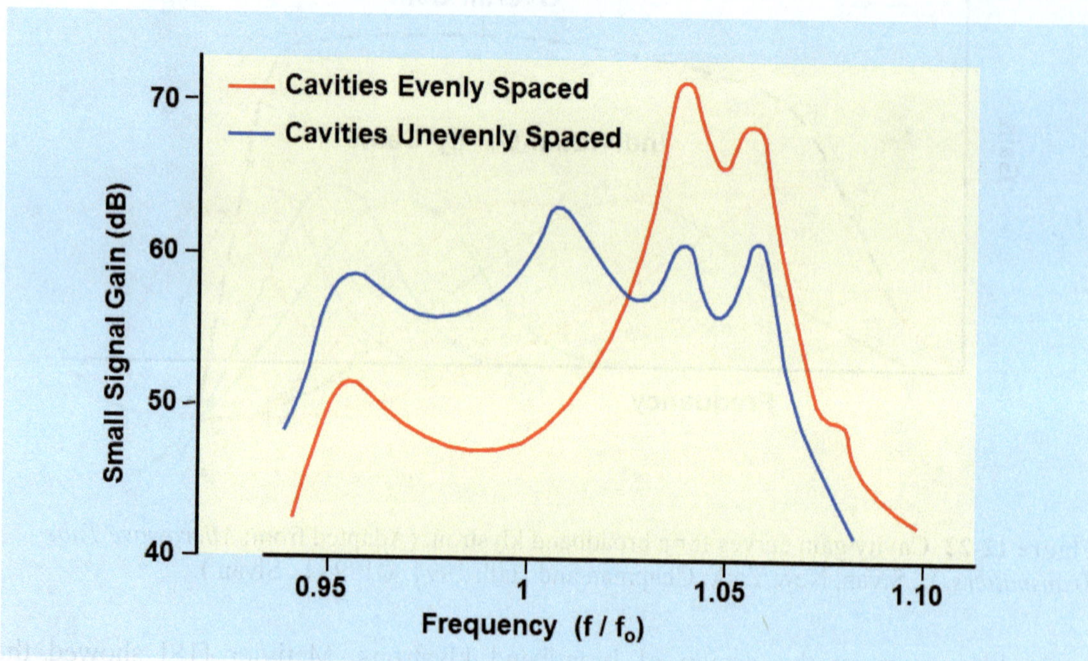

Figure 12-24 Gain of eight-cavity klystrons with cavity spacings used by Metivier. (From: R. L. Metivier, *Microwave Jour.*, April 1971.)

In his patent of a high-efficiency broadband klystron [19], Friedlander described a klystron in which "the gain-bandwidth product and amplitude response is made flatter by successive intermediate floating cavities downstream of the input cavity tuned to successively higher frequencies, and the drift lengths between them successively shorter." The frequencies to which the cavities are tuned in Friedlander's klystron (called the *tapered-drift-length klystron*) are indicated in Figure 12-25. The drift-length tapering is indicated in Figure 12-26.

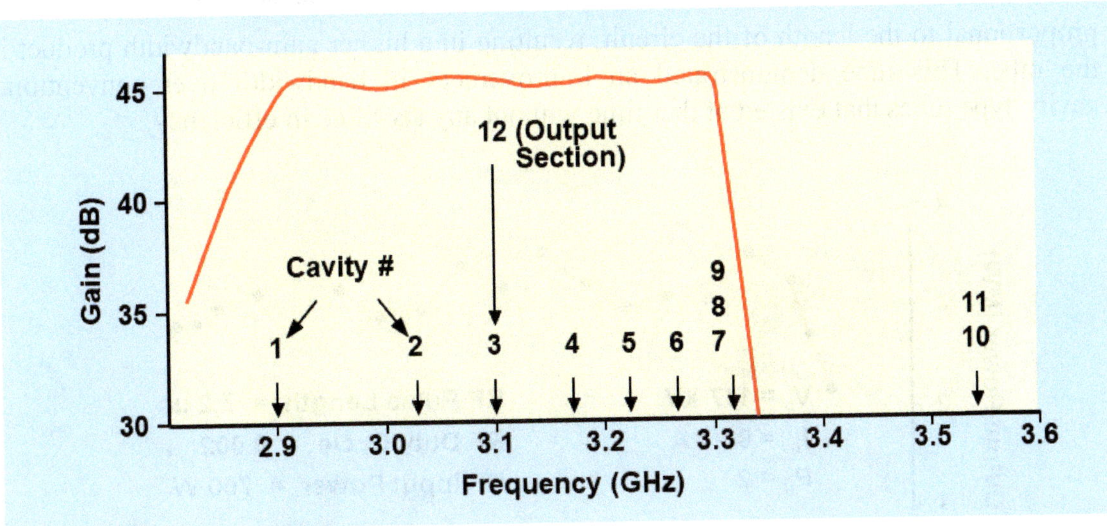

Figure 12-25 Cavity tuning of a tapered-drift-length klystron. (Adapted from: F. I. Friedlander, U.S. Patent No. 4,764,710, August 16, 1988.)

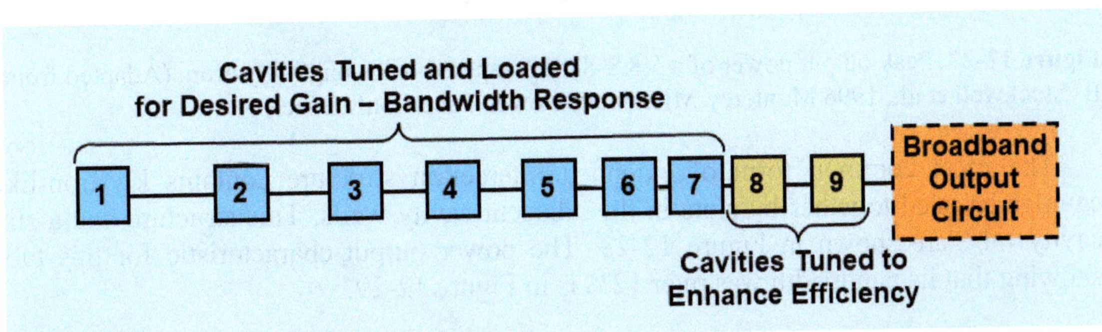

Figure 12-26 Cavity configuration in the driver section of a tapered-drift-length klystron. (Adapted from: F. I. Friedlander, 1988 Monterey Microwave Power Tube Conference.)

The power output of a tapered-drift-length klystron for a radar system is shown in Figure 12-27. This tube has nine bunched cavities and one filter-loaded output cavity (discussed in the following section) and has a bandwidth of 10%. This tube was completely designed using a variety of computer codes. The tube met specification right out of the box (when it was first operated). Minor adjustments were made to increase the operating margin [20].

Other efforts to increase bandwidth are based on the fact that if a single cavity is replaced by a pair of cavities and the gap voltage of each cavity is half the gap voltage of the single cavity, then the stored energy in the pair of cavities would be half that of the single cavity. The voltage applied to the beam would be the same in either case. Because the stored energy has been decreased, Q has been decreased and bandwidth has been increased.

In an early effort to increase the bandwidth of a klystron, Chodorow and Wessel-Berg used structures resembling sections of TWT circuits in place of the cavities normally used in klystrons [21]. These structures are referred to as extended interaction structures. The klystron built by Chodorow and Wessel-Berg contained three extended interaction structures, each of which was formed by a short section of a ring-bar circuit that acted as a half-wave-long resonant circuit. The interaction impedance is generally proportional to the length of the circuit, resulting in a higher gain-bandwidth product in the tube. This tube demonstrated an improvement in bandwidth over conventional cavity-type tubes that existed at that time without any sacrifice in efficiency.

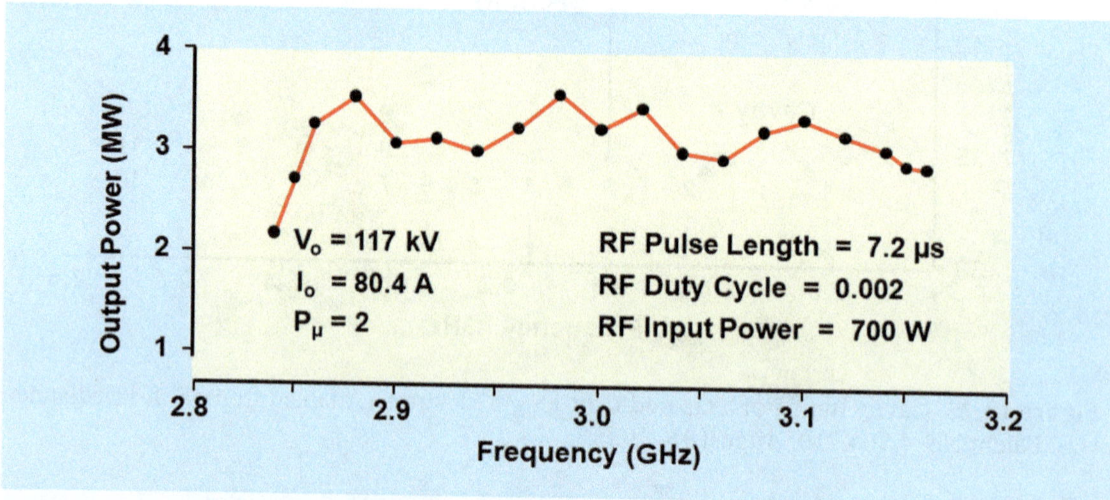

Figure 12-27 Peak output power of a VKS-8345A tapered-drift-length klystron. (Adapted from: B. Stockwell et al., 1996 Monterey Microwave Power Tube Conference.)

The most common form of extended interaction structure contains klystron-like cavities coupled together by slots in the adjacent cavity walls. This structure and a six-cavity tube are shown in Figure 12-28. The power output characteristic for this tube showing that its bandwidth was over 12% is in Figure 12-29.

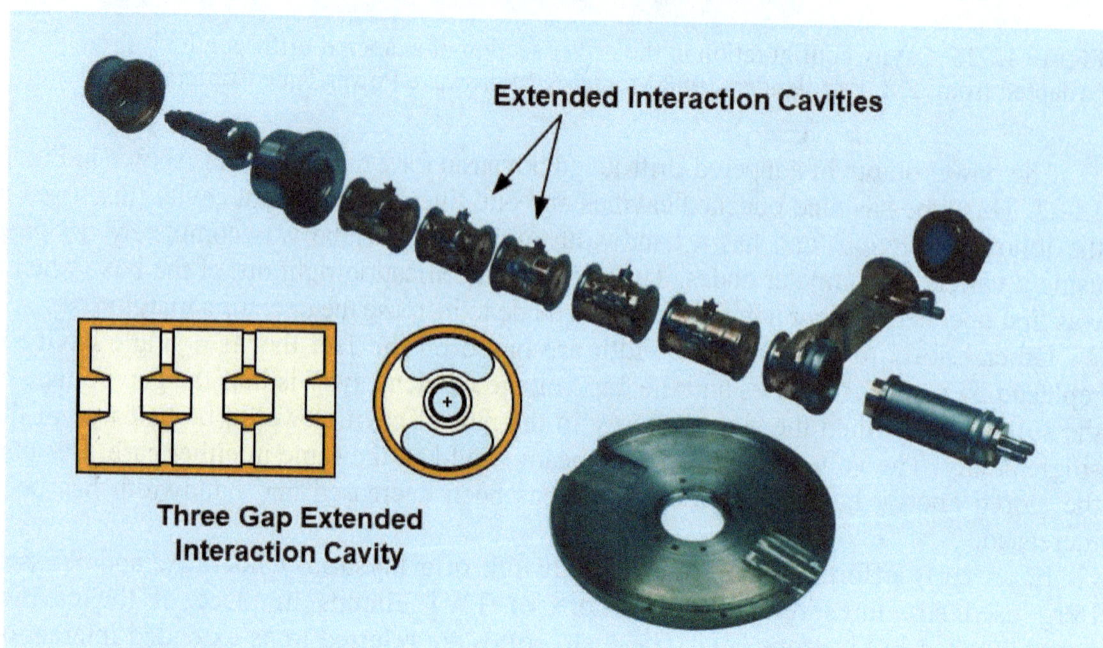

Figure 12-28 A six cavity extended interaction klystron. (Photo courtesy of Rome Air Development Center.)

Symons et al. showed, in the clustered-cavity klystron, that the coupling slots between cavities are not necessary [22]. In fact, Symons pointed out that the coupling between cavities in an extended interaction structure introduces a zero in the response between the poles associated with the individual cavities. This zero places constraints on tuning and loading that are not present when the zero is eliminated by eliminating the coupling. Figure 12-30 gives a comparison of a conventional stagger-tuned klystron with a clustered-cavity klystron.

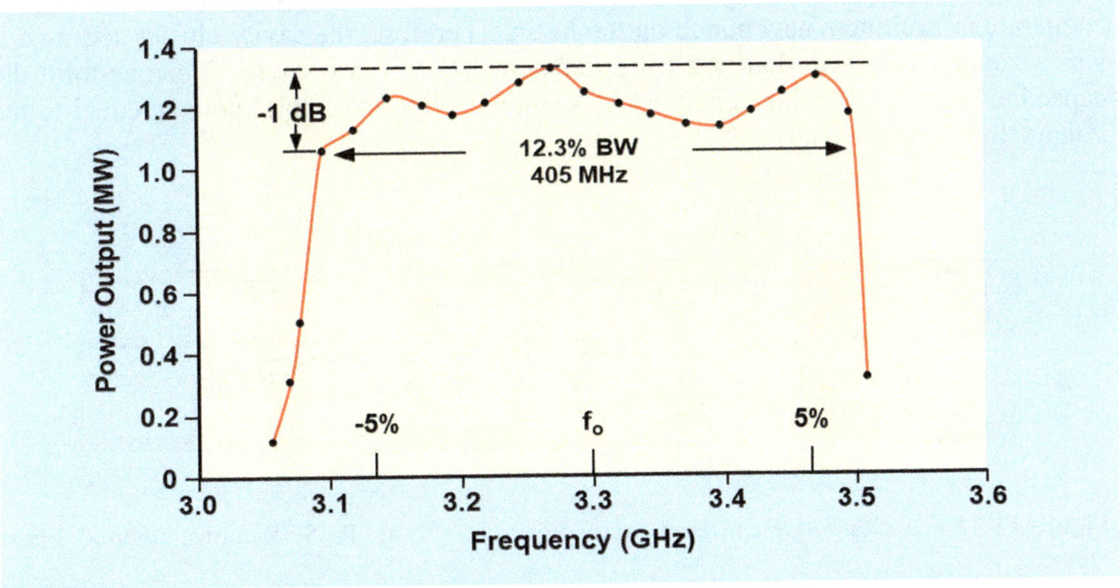

Figure 12-29 Power output characteristic for the extended interaction klystron shown in the previous figure. (Courtesy of Rome Air Development Center.)

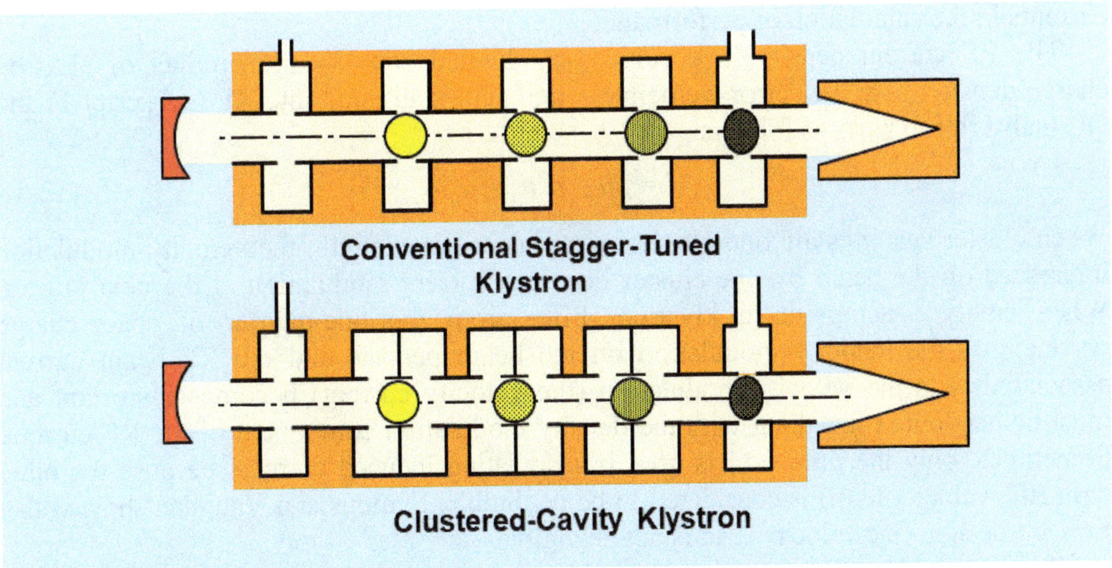

Figure 12-30 Comparison of a stagger-tuned klystron with a clustered-cavity klystron. (From: R. S. Symons and J. R. M. Vaughan, *IEEE Trans. Plasma Science*, October 1994. ©1994 IEEE.)

The cavity clusters can take on any one of a variety of configurations as indicated in Figure 12-31. In each of the clusters, the gaps of the cavities are as close together as possible so that there is essentially no gain between them. Each cavity in a cluster has resistive loading such that its Q is one-half (for a cluster of two cavities) or one-third (for a cluster of three cavities) of the Q of the single cavity it replaces. Each cavity in a cluster is driven by essentially the same current that was generated by the bunching at the previous cluster. So, not only are the gap voltages essentially equal, but also because they are excited by the same bunch, they are in exactly the right phase relationship to cooperate in exciting a new bunch on the beam. Therefore, the cavity cluster acts as if it was a single cavity with half the capacitance and Q (for two cavities) or one-third the capacitance and Q (for three cavities) of a single cavity and a total voltage equal to the voltage for a single cavity.

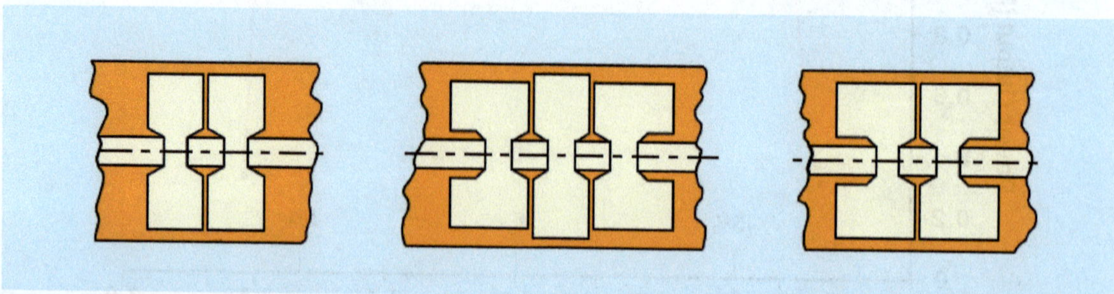

Figure 12-31 Cavities for a clustered-cavity klystron. (From: R. S. Symons, internal Litton report.)

Figure 12-32 shows the calculated small-signal gain of a prototype 4-MW cluster-cavity klystron. Notice that the bandwidth exceeds 10% and that the gain ripple is very small. In this plot, Symons and Vaughan illustrated the necessity of including velocity currents in the calculation of performance.

The RF current density in a velocity-modulated beam is the product of electron charge density, ρ, and electron velocity, u, and can be divided into RF (subscript 1) and DC (subscript 0) parts as follows:

$$J = \rho u = \rho_1 u_o + \rho_o u_1 \tag{12-18}$$

When cluster spacings are one quarter of a plasma wavelength, the velocity modulation impressed on the beam by one cluster becomes current modulation at the next cluster. When cavity spacings in a klystron differ from the one-quarter of space-charge wavelength, the velocity modulation on the beam persists and the RF beam current associated with the velocity modulation (the velocity current) becomes important and must be accounted for along with the density modulation component of the RF current. Sometimes, only the $\rho_1 u_o$ term is used in calculating induced currents because the other term (the velocity term) is considered to be negligible. Symons and Vaughan showed that for their designs, the velocity term is not negligible.

Symons and Vaughan state that it should be possible to build clustered-cavity klystrons with bandwidths from 5% at 5 kW to 30% at 50 MW [22]. The calculated small-signal gain curve for a 50-MW clustered-cavity klystron is shown in Figure12-33 [23].

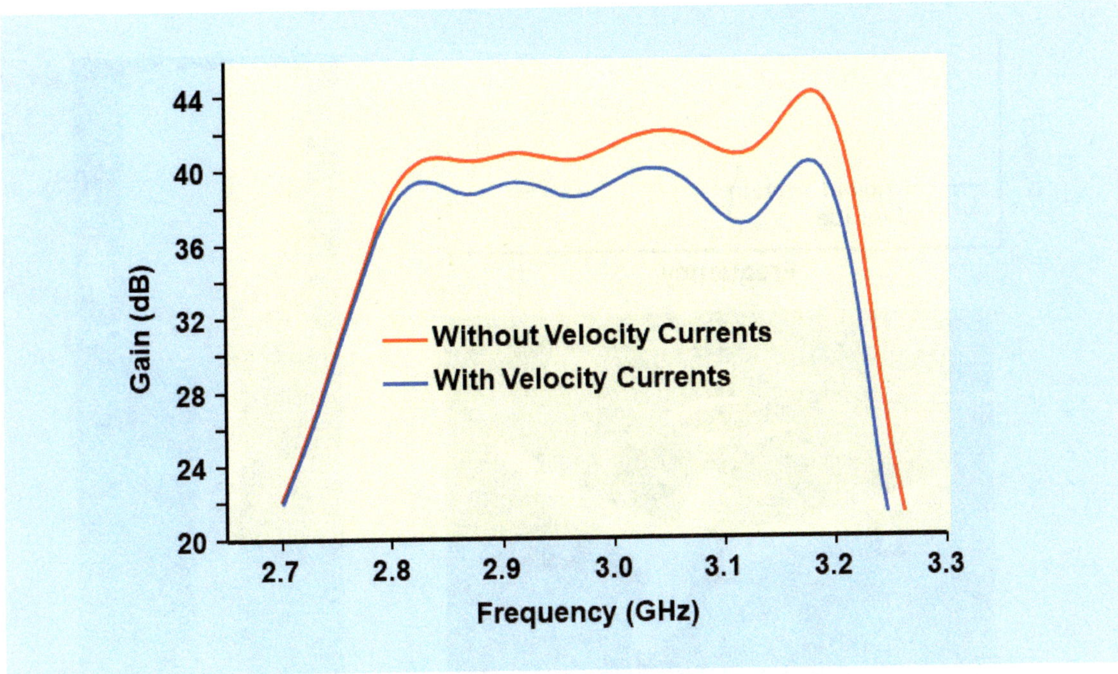

Figure 12-32 Calculated small-signal gain of a prototype clustered-cavity klystron. (From: R. S. Symons and J. R. M. Vaughan, *IEEE Trans. Plasma Science*, October 1994. ©1994 IEEE.)

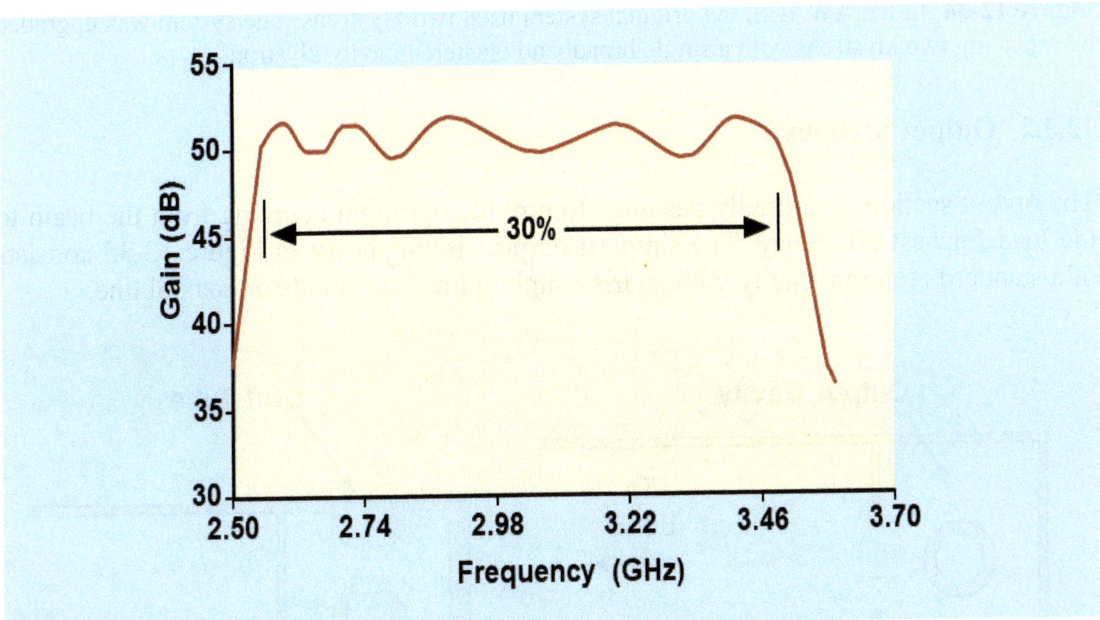

Figure 12-33 Calculated small-signal gain of a 50-MW clustered-cavity klystron with a bandwidth of 30%. (Adapted from: J. G. Siambis and R. S. Symons, *Ultra-Wideband, Short-Pulse Electromagnetics*, Plenum Press, 1993.)

A very important application for the clustered klystron is shown in Figure 12-34. A single broadband klystron was developed to replace two narrowband klystrons in the AWACS (airborne warning and control system) radar.

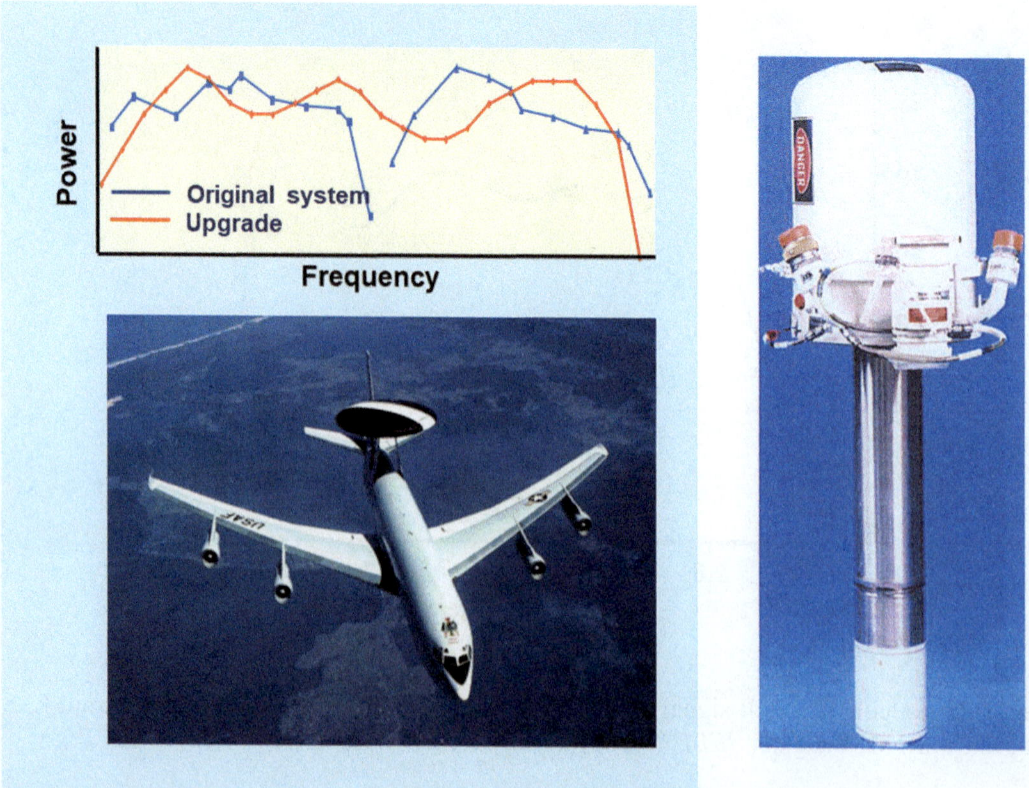

Figure 12-34 In the AWACS, the original system used two klystrons. The system was upgraded by replacing two klystrons with a single broadband clustered-cavity klystron.

12.3.2 Output Sections

The output section is normally designed to provide optimum coupling from the beam to the load for best efficiency. The simplest output circuit shown in Figure 12-35 consists of a standard reentrant cavity with an iris coupling into waveguide or coaxial line.

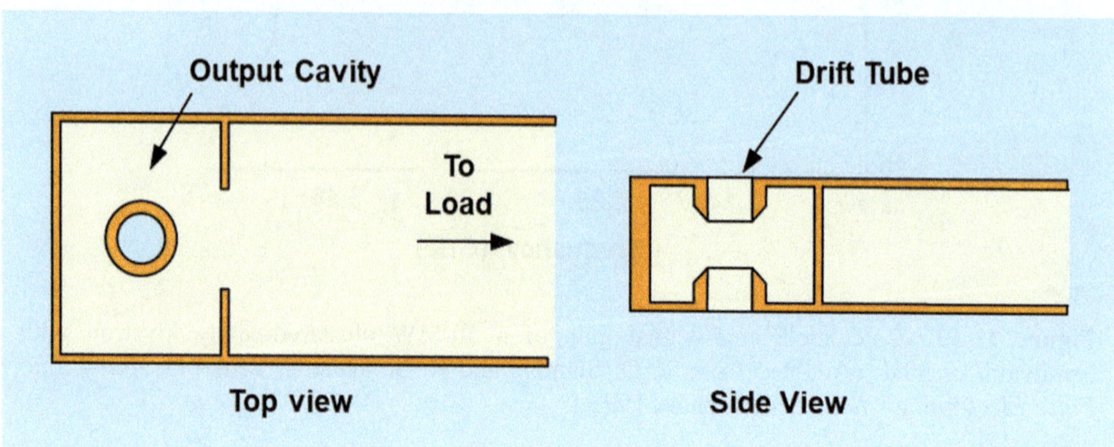

Figure 12-35 Standard reentrant cavity with an iris coupling into waveguide.

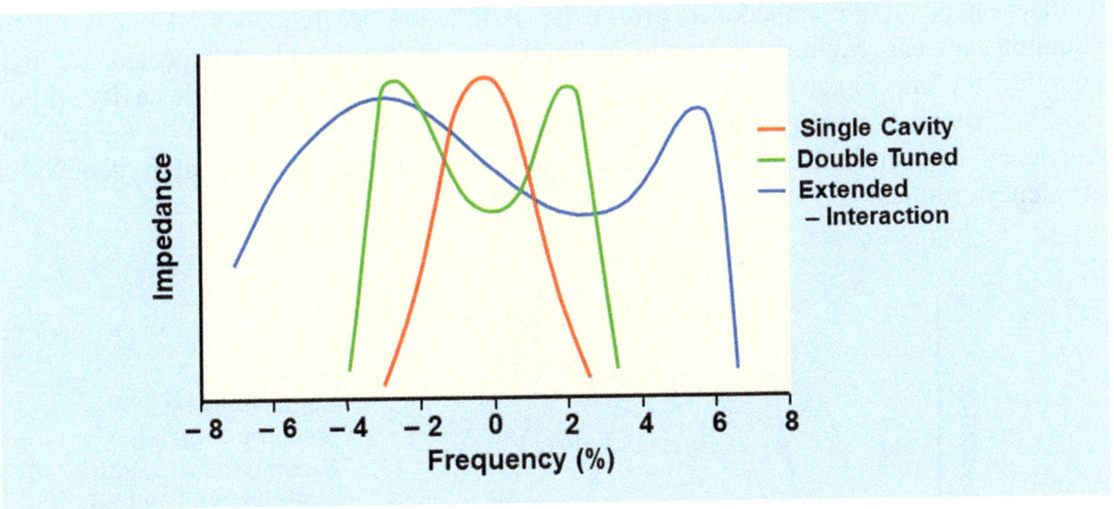

Figure 12-36 Impedance characteristics of single-tuned, double-tuned, and extended interaction circuits. (Adapted from: *Power Klystrons Today* by M. J. Smith and G. Phillips, ©1995 by Research Studies Press, Ltd.)

The double-tuned cavity consists of a conventional reentrant cavity coupled to the beam along with a filter cavity between this cavity and the load, as shown in Figure 12-37. The equivalent circuit for the double-tuned cavity is given in Figure 12-38. Both cavities are tuned to the same frequency and the combination makes it possible to load the structure so that optimum impedance occurs at the gap over a greater bandwidth than with a single cavity.

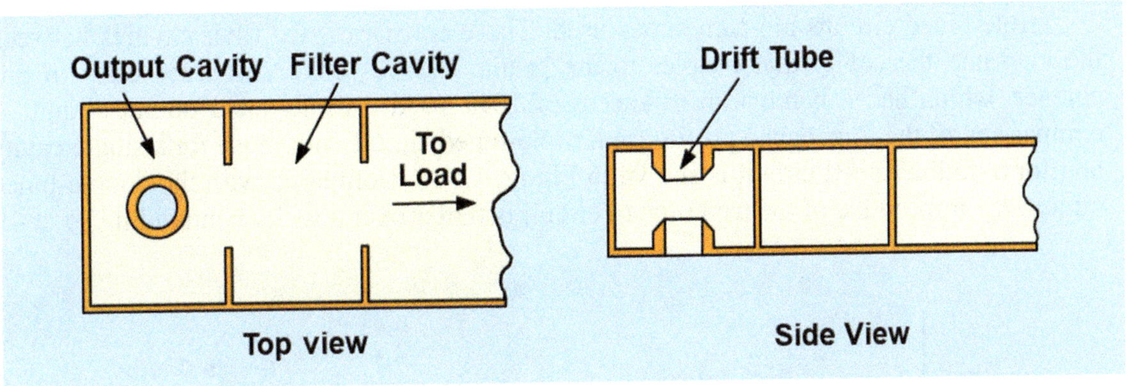

Figure 12-37 Double-tuned output circuit.

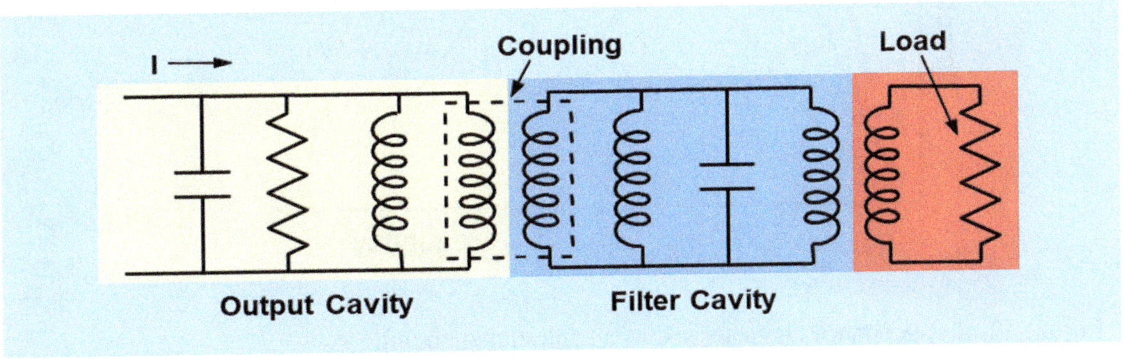

Figure 12-38 Equivalent circuit for a double-tuned output cavity.

The shape of the impedance profile (as a function of frequency) depends on the coupling between cavities, as shown in Figure 12-39. As would be expected, for light coupling, the impedance presented to the beam is similar to that for a single cavity without the filter cavity. For heavy coupling, the response has two peaks centered on the resonant frequency. In between, there is a coupling that produces the maximum bandwidth-efficiency product.

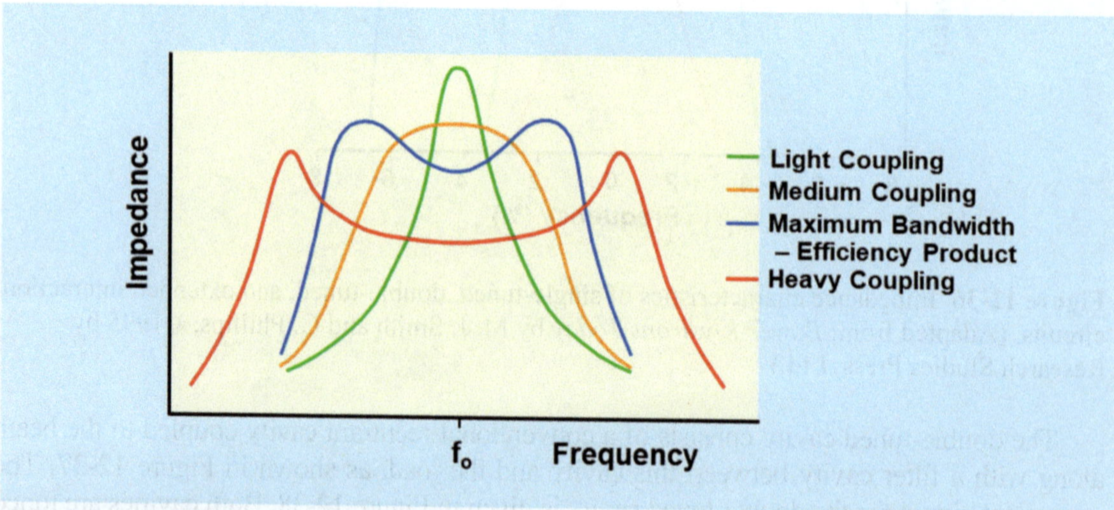

Figure 12-39 Effect of coupling on impedance characteristic of a double-tuned circuit. (Adapted from: *Power Klystrons Today* by M. J. Smith and G. Phillips, ©1995 by Research Studies Press, Ltd.)

Triple-tuned circuits are sometimes used. These consist of two filter cavities between the load and the cavity that couples to the beam. The VA-812C discussed early in this chapter, which had a bandwidth in excess of 10%, used a triple-tuned output circuit. A comparison of the impedance profile for a triple-tuned circuit with those for a single cavity and for a double-tuned circuit is shown in Figure 12-40. Compared with the double-tuned circuit, the impedance of the triple-tuned circuit is flatter over a wider bandwidth.

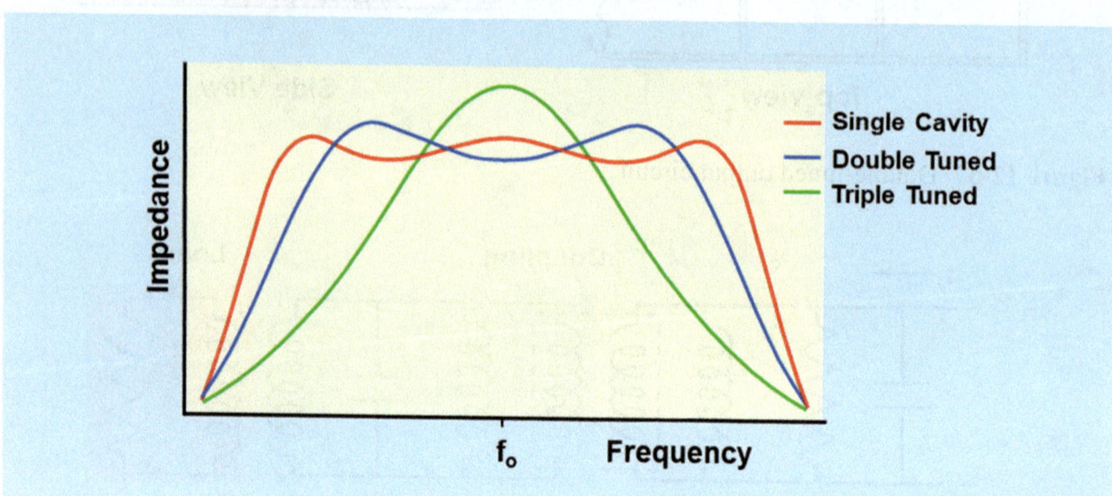

Figure 12-40 Impedance characteristic of a triple-tuned circuit.

Extended interaction circuits, which were discussed briefly in the previous section, are often used as output sections on klystrons. The impedance characteristic of an extended interaction structure is shown in Figure 12-36. An extended interaction structure consists of heavily coupled cavities as shown in Figure 12-41. In one mode of operation, the fields in all cavities are in phase and the electron transit angle between gaps is 2π radians. The other common mode of operation is similar to the fundamental backward wave mode of coupled-cavity TWT operation. That is, the fields in adjacent cavities are approximately 180° out of phase.

In comparing an extended interaction resonator with a single-gap output cavity, the bandwidth is, of course, increased. Perhaps more important is the fact that the power-handling capability is increased. This is because the total output voltage, which must be applied across a single gap in a conventional klystron, is divided among the output cavities of the extended interaction structure. The high-power capability of the extended interaction structure was perhaps best illustrated by a 1-MW average-power X-band device that used a five-gap coupled-cavity resonator like that shown in Figure 12-41. A three-gap version with output waveguide is shown in Figure 12-42.

The cloverleaf structure in Figure 12-43 has been used as the output section for a klystron. The cloverleaf structure is a fundamental forward wave circuit that was developed for use in high-power traveling wave tubes [24]. Fundamental forward-wave TWT circuits are discussed in Chapter 14. When the cloverleaf structure is used as the output section of a klystron, the tube is called a twystron. Twystrons are discussed in Chapter 15, which deals with hybrid tubes.

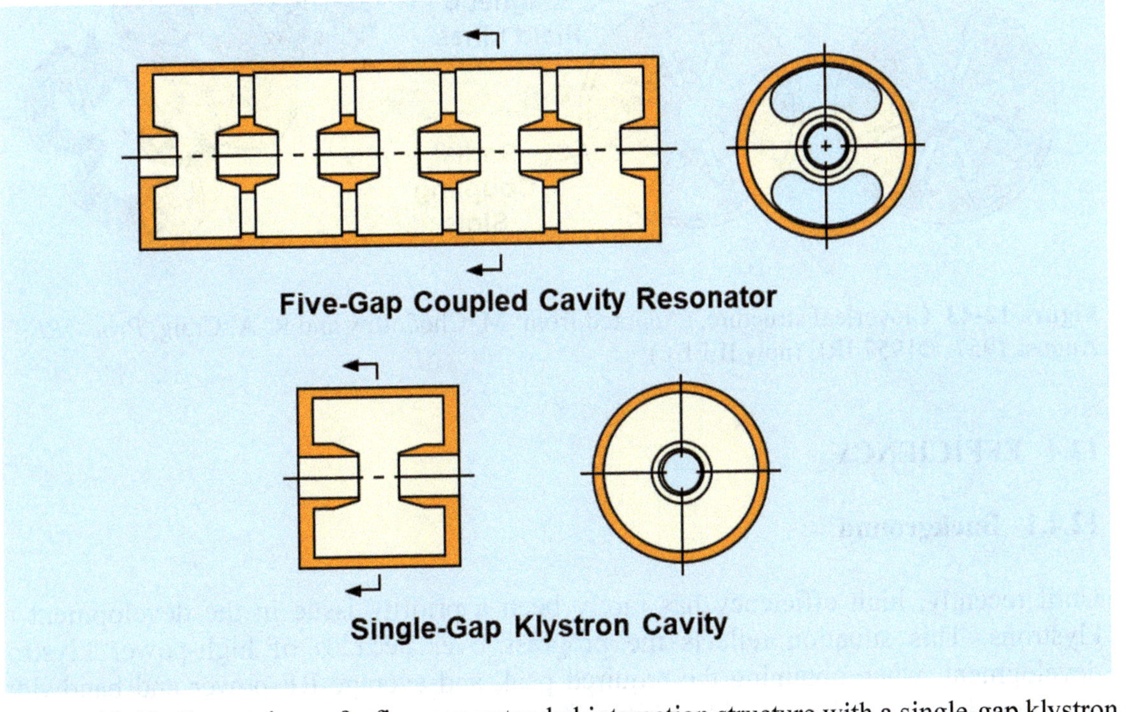

Figure 12-41 Comparison of a five-gap extended interaction structure with a single-gap klystron cavity. (From: A. Staprans et al., *Proc. IEEE*, March 1973. ©1973 IEEE.)

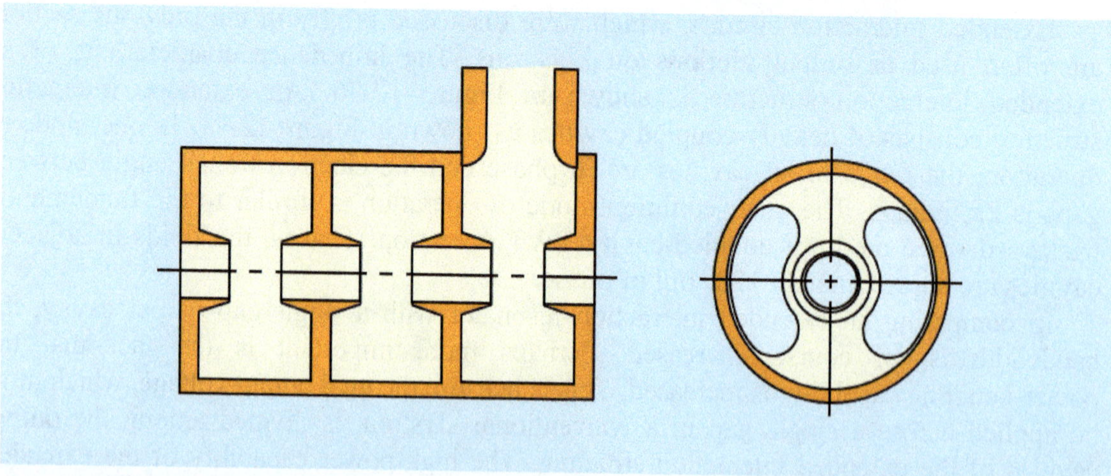

Figure 12-42 Three-gap coupled cavity output structure.

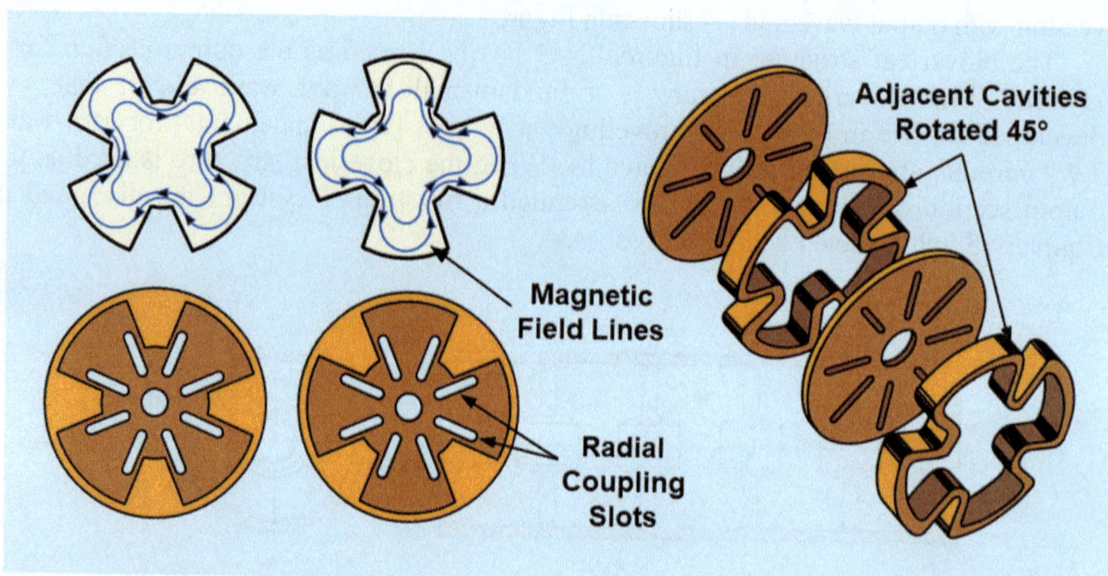

Figure 12-43 Cloverleaf structure. (Adapted from: M. Chodorow and R. A. Craig, *Proc. IRE*, August 1957. ©1957 IRE (now IEEE).)

12.4 EFFICIENCY

12.4.1 Background

Until recently, high efficiency has rarely been a priority issue in the development of klystrons. This situation reflects the progress over decades of high-power klystron development, when obtaining the required peak and average RF power and bandwidth was more important than providing high efficiency.

Most klystrons operate in the 40% to 55% range and a few are capable of as much as 65%. One exception is the klystron development in Russia in 1984, which was reported by Marrelli. This was a 15-kW, 80% efficient, CW, L-band, device.

The only exception in the United States is the work of Lien at Varian nearly half a century ago. He examined several cavity configurations as shown in Figure 12-44 and reported efficiencies up to 75% for a device with six fundamental cavities and two second harmonic cavities [25].

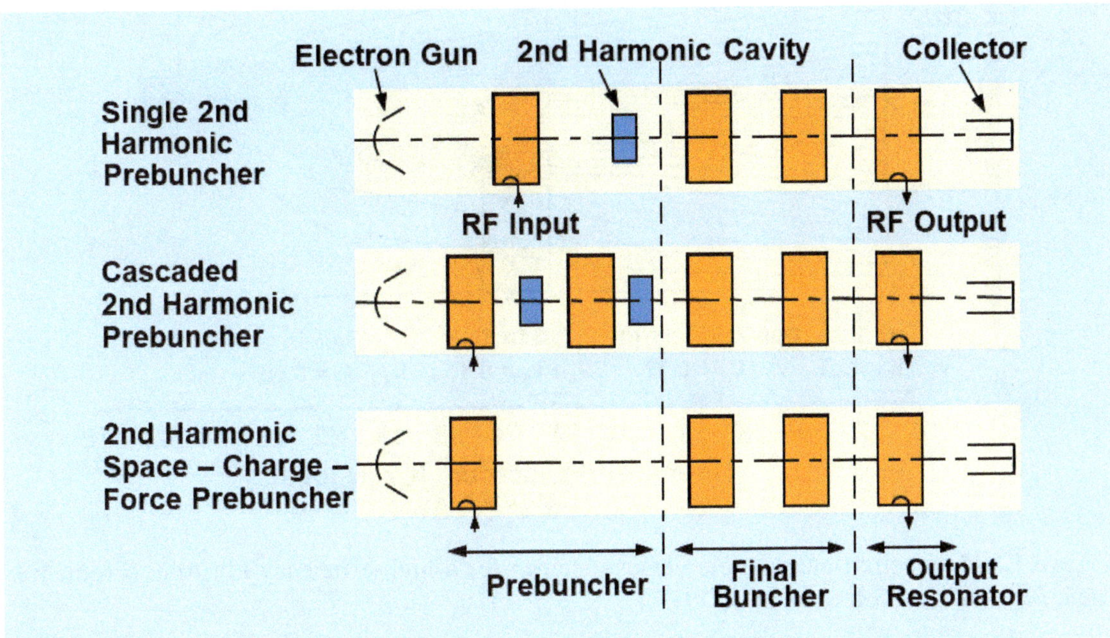

Figure 12-44 High-efficiency klystron circuits using second-harmonic bunching. (From: E.L. Lien, *Proc. 8th MOGA*, September 1970.)

This high efficiency was achieved in a 50-kW device operating at 2,450 MHz. In the same reference, Lien reported an overall efficiency of 70% for an 800-MHz klystron with a power of about 70 kW. These efficiencies were achieved without the use of depressed collectors. Some of the more significant aspects of this high-efficiency work are summarized here.

Perhaps the most important consideration in achieving the efficiencies in [25] was the use of second-harmonic bunching to increase the RF beam current. In narrowband tubes, the combinations of fundamental frequency and second harmonic cavities were used. In an alternate approach, applicable to broadband devices, an abnormally long drift length was used. In this case, the harmonic bunching was produced by second-harmonic space-charge forces within the beam rather than with a second-harmonic field.

The electron phase plot for the case of the single second harmonic prebuncher shows computed trajectories for 32 reference electrons in one RF period. This plot is an Applegate diagram in which the dc components of electron velocity have been removed. Thus, the trajectories of electrons with only a dc component of velocity have trajectories parallel to the abscissa. The trajectories of electrons with a velocity higher than the dc velocity have a positive slope and those with velocities lower than the dc velocity have a negative slope. It should be noted that the majority of the electrons are contained within the bunch at the location of the output cavity and that the interbunch region is essentially free of electrons. In addition, the velocity spread among the electrons is small.

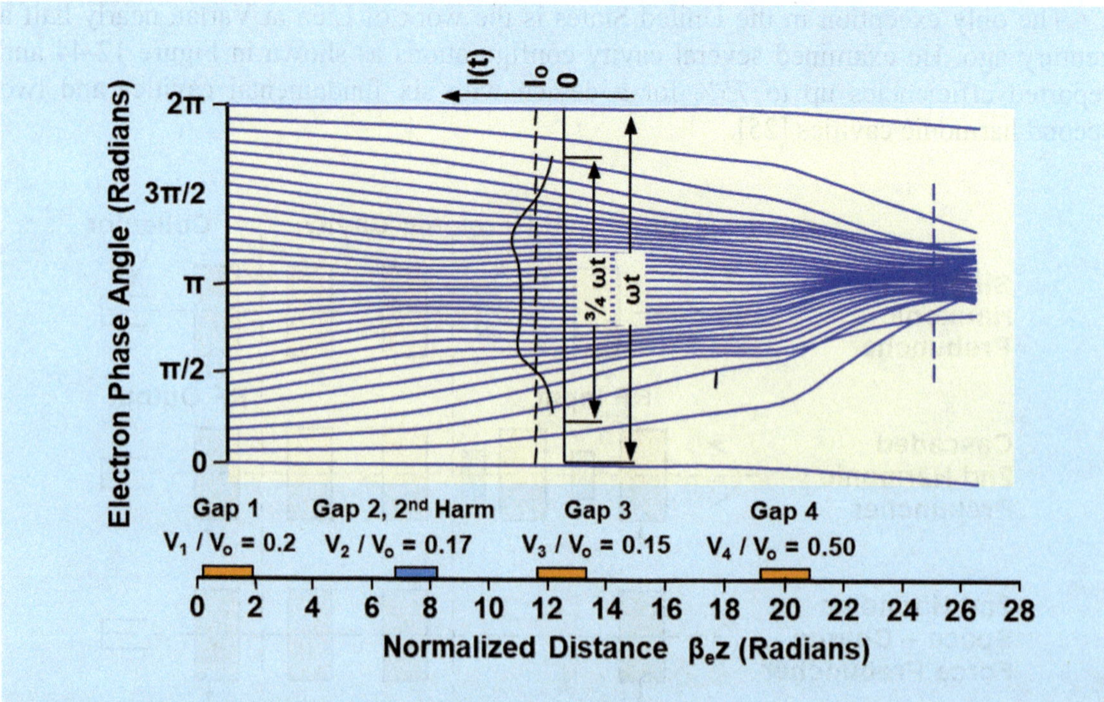

Figure 12-45 Electron phase angle versus distance for a high-efficiency klystron. (From: E.L. Lien, *Proc. 8th MOGA*, September 1970.)

The fundamental and second-harmonic frequency components of beam current are shown as functions of normalized distance along the tube in Figure 12-46. The maximum fundamental current produced by the bunches at the output resonator is 1.8 times the dc beam current. Lien noted that higher currents could be produced, but the velocity spread among electrons is increased and efficiency is not increased.

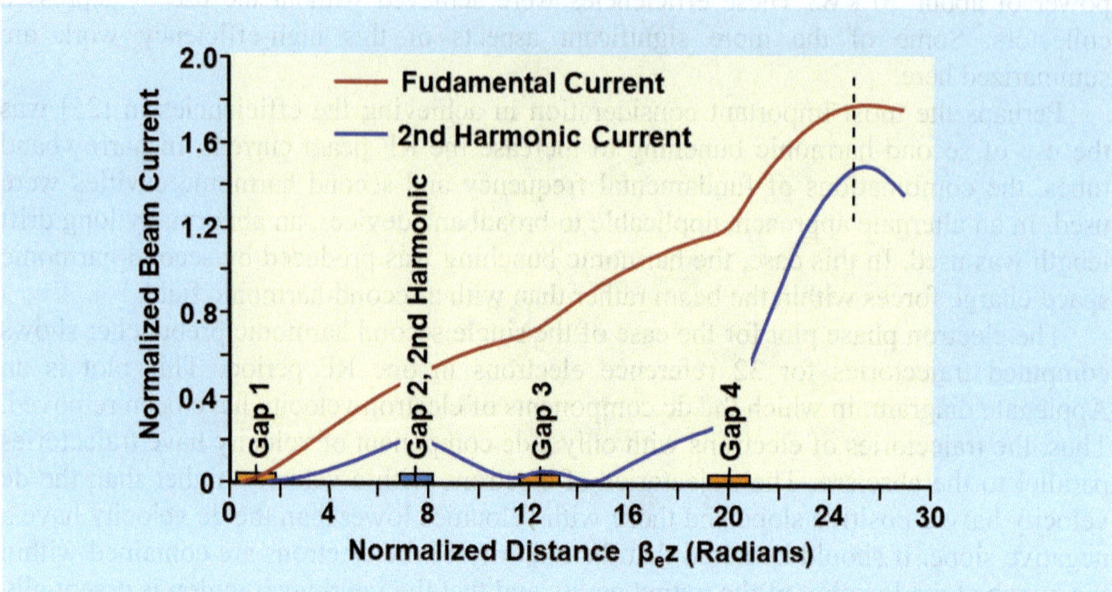

Figure 12-46 Normalized RF beam current for a high-efficiency klystron. (From: E. L. Lien, *Proc. 8th MOGA*, September 1970.)

Output power and efficiency curves for the 2,450-MHz tube, which had two second-harmonic cavities in addition to six fundamental-frequency cavities, are shown in Figure 11-47. It should be reiterated that these high efficiencies were attained without the use of depressed collectors.

In the early 1970s, NASA investigated the concept indicated in Figure 12-48 of solar power plants in space. This was one of the earliest proposed programs that would require large numbers of high-efficiency microwave sources. Microwave generators would transmit to earth energy collected by huge arrays of solar cells in space. Klystrons were considered for possible use as the microwave generators. NASA investigated the efficiency limits of klystrons. Figures 12-49, 12-50, and 12-51 are adapted from Kosmahl and Albers [26], who in 1973 published a detailed analysis of energy extraction in the output cavity of a klystron. That paper illustrated the importance of accurately modeling axial and radial fields in the interaction region.

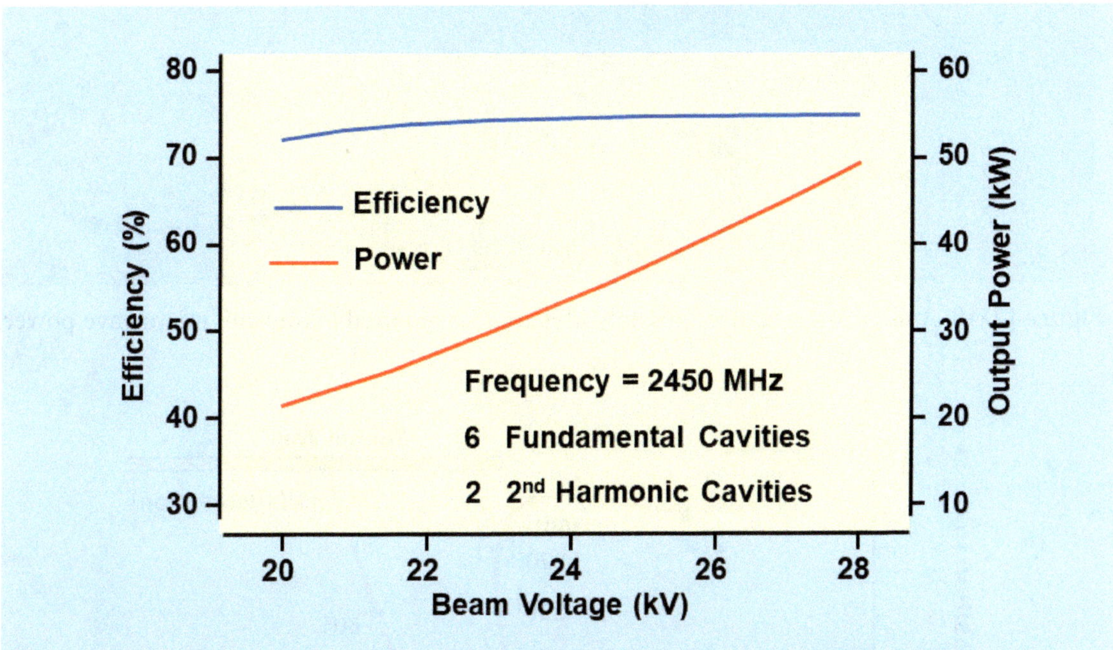

Figure 12-47 Power and efficiency for the VKS-7773, 2,450-MHz klystron. (From: E. L. Lien, *Proc. 8th MOGA*, September 1970.)

A new method for computing space-charge forces between axially and radially deformable rings was developed. Rigorous expressions for electric and magnetic fields in the tunnel and gap were used. An example of the result of the computations is shown in the sketch of the right half of the output gap in Figure 12-49. The left half of the gap, which is not shown, is a mirror image of the right half.

The trajectory of an outer ring of charge at various phases of the RF is shown in Figure 12-49. At an RF phase of 300°, the ring rapidly loses the axial component of velocity and is pushed inward. As it moves inward, its axial motion stops and reverses at 320° and then reverses again at 360°. The ring then enters the accelerating phase of the RF, moves out radially, and is intercepted at an RF phase of 83°.

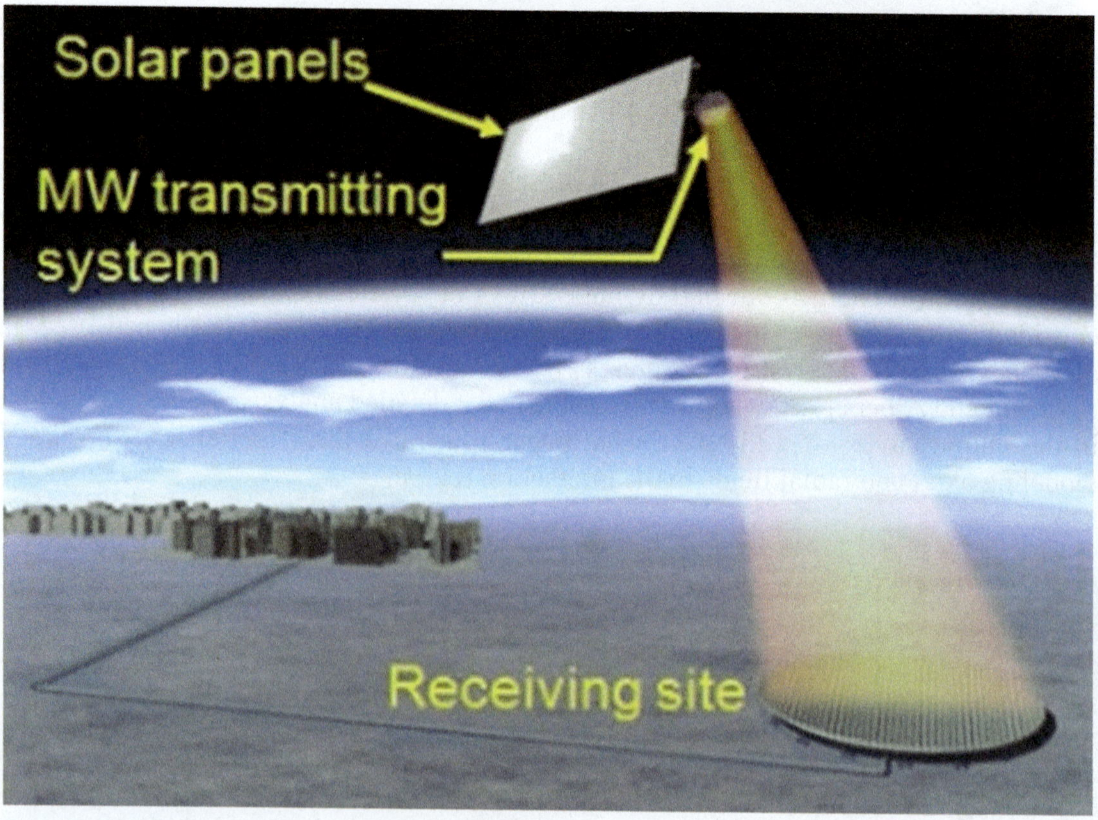

Figure 12-48 A microwave source, possibly klystrons, is required to transmit microwave power.

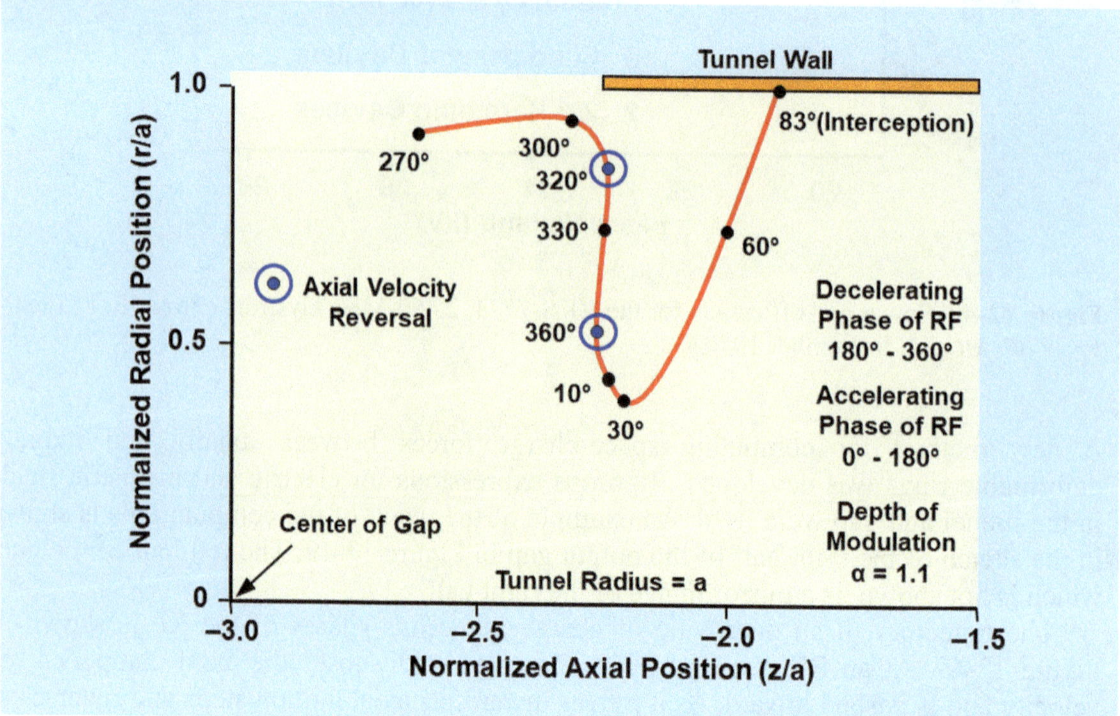

Figure 12-49 Trajectory of an outer ring of charge in the output cavity of a klystron. (Adapted from: H. G. Kosmahl and L. U. Albers, *IEEE Trans. Electron Devices*, October 1973. ©1973 IEEE.)

The plot of efficiency as a function of perveance in Figure 12-50 is nearly linear and shows efficiency decreasing with increasing perveance, which is in agreement with experimental results. An important conclusion of the study shown by Figure 12-51 is that efficiency decreases with increasing normalized tunnel radius $\beta_e a$ for b/a constant in the range 0.4 < 1.0. It is likely that efficiency continues to decrease as $\beta_e a$ is increased. Notice that an efficiency of 80% was predicted with an interception of 6%. The use of multistage depressed collectors for enhancing efficiency was not included in this study.

In 1980, Kosmahl again predicted that an efficiency of 80% or more might be achieved and included the use of multistage depressed collectors (MSDC) [27]. That technique has been very effective in raising the efficiencies of traveling-wave tubes to over 70%. The problem in using an MSDC on the typical klystron that has a reasonably high efficiency is that the energy spread of the electrons in the spent beam is, as Dr. Henry Kosmahl said, "ferocious" [27].

The curves in Figure 12-52 show the energy distribution of the electrons in the spent beam of a TV klystron and a TWT. Notice that in the klystron beam, some electrons have been stopped or nearly stopped and some have been accelerated to energy levels well above the energy with which they left the electron gun. This wide spread in energies limits the efficiency of the MSDC to 40% to 50% because an excessive number of stages would be required to sort the electrons and because the energies of electrons greater than the gun energy would be very difficult to recover.

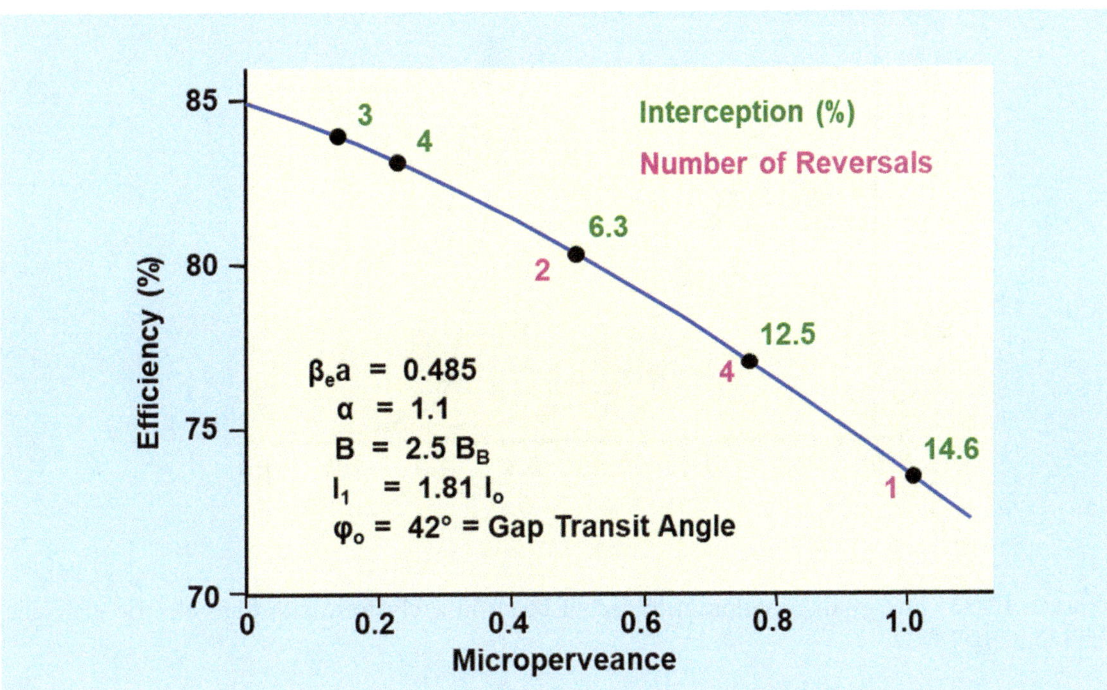

Figure 12-50 Efficiency versus perveance assuming constant bunching level. Interceptions and velocity reversals are listed at computed points.

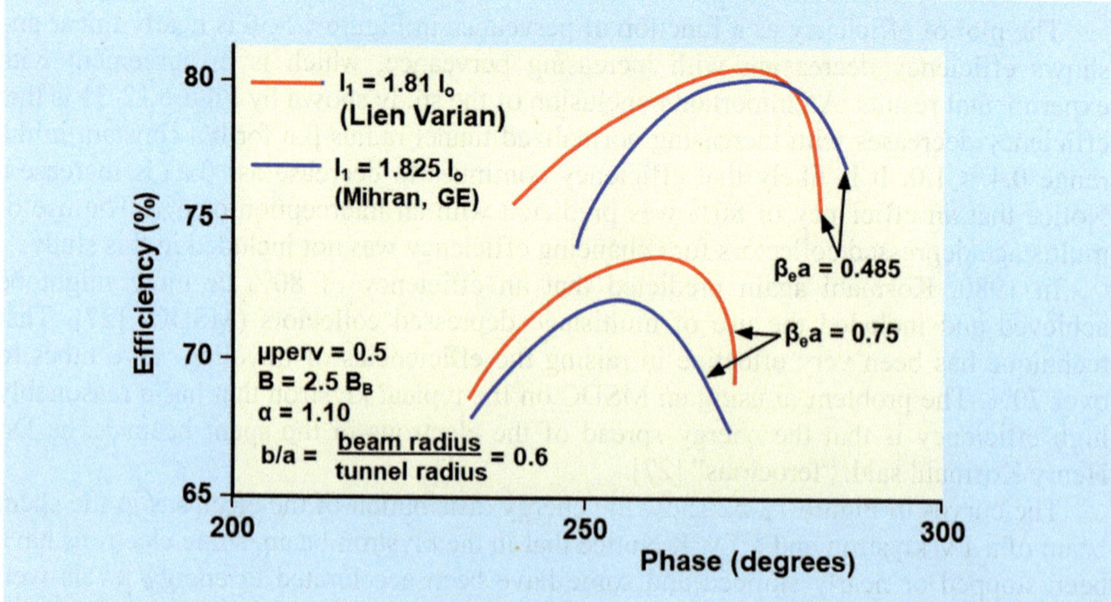

Figure 12-51 Internal conversion efficiency computed with Kosmahl and Alber's program for General Electric and Varian high-efficiency designs.

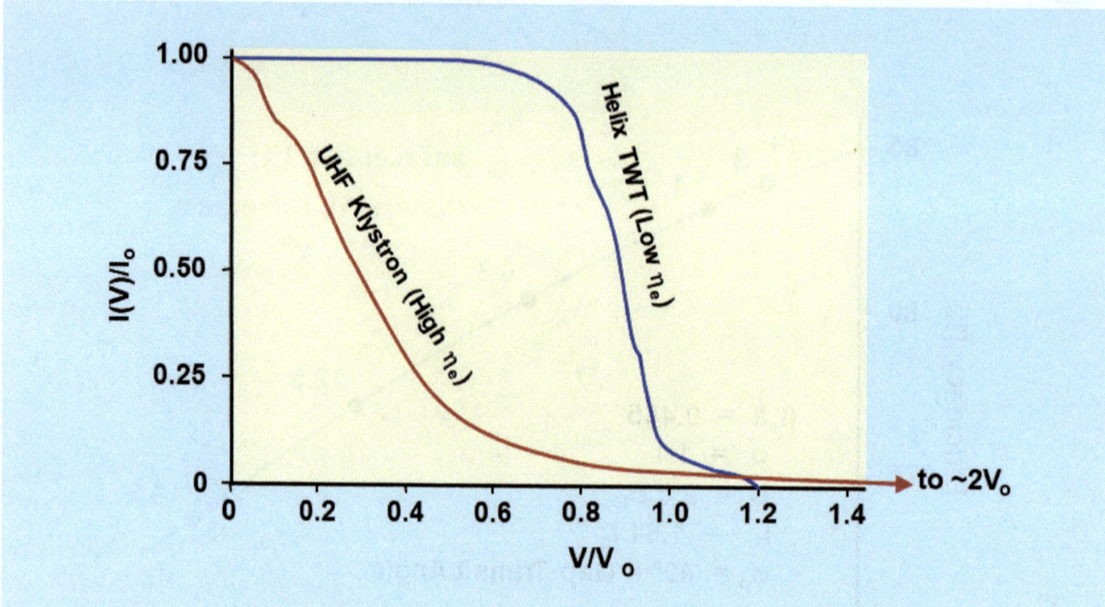

Figure 12-52 Energy distribution in the spent beam in a klystron with high electric efficiency and in helix TWT.

12.4.2 The Need for Extremely High Efficiency

Several projects that are presently in the planning stage, if implemented, will require tens to hundreds of megawatts of RF power. Some examples are [28]:

- Compact Linear Collider, CLIC, 180 MW;
- Future Circular Collider, FCC, 100 MW;
- European Spallation Source, ESS, 92 MW;
- International Linear Collider, ILC, 88 MW.

The very large cost of providing prime power to these projects is the impetus for maximizing the efficiency of these projects. A large portion of the prime power is used by the klystrons that convert the prime power to the microwave power needed to drive the particle accelerators. The efficiency of these klystrons today is about 65%. It is estimated that if that efficiency can be increased to 85%, the annual saving in the cost of power would be 9.3 M Euro ($11 M, 2017 dollars). If the efficiency could be raised to 90%, the annual saving would be 10.7 Euro ($12.7 M, 2017 dollars) [29].

With 75 years of development, klystron technology was mature and there was little room for further development. Experimental results from many devices showed that low perveance was required for high efficiency. There was no evidence, however, that efficiencies approaching 80% to 85% could be achieved. In fact, 75% was thought by many to be the practical limit. Still, the need for higher efficiencies persisted and so now there is an international effort underway to explore the possibility of achieving klystron efficiencies of 90% or higher [29].

The classical technique for maximizing efficiency was described in Section 10.1. A klystron is first synchronously tuned and then the penultimate cavity is tuned upward to maximize power. By being tuned upward, the penultimate cavity is inductive at the operating frequency and this enhances the bunching process at the output cavity.

The statement that was made by Chodorow [30] about this operation was "electrons will certainly be passing each other and, in some cases, some electrons will be going backwards while others are going forward…most highly efficient klystrons operate with these conditions in the output cavity."

To illustrate what must be done to obtain efficiencies in the 85% to 90% range, we will start by examining the electron phase diagram shown in Figure 12-53 for a high-power klystron for which efficiency could be improved. In this drawing, cavity locations are denoted by vertical lines and sinusoids. The phase angles during which electrons are accelerated are denoted by plus signs and the phases during which deceleration occurs are denoted by minus signs. The majority of the electrons passing through the output cavity are decelerated as indicated by arrow b. The electrons that pass through the cavity during an accelerating phase are denoted by arrow a.

Figure 12-54 illustrates the phases of the output voltage during which energy is wasted and /or recovered. The dashed line indicates the voltage on the electron gun that accelerates all electrons in the beam. The red line indicates the effective output-gap voltage. That is, the voltage that the electrons are exposed to, which is the gap coupling coefficient times the actual gap voltage. The limit on that voltage is $-V_0$ if the reflection of electrons by the output gap is to be avoided.

The red cross-hatched area represents the phases of the voltage during which energy is expended on electrons that are accelerated. This waste of energy detracts from efficiency. The blue cross-hatched area represents phases during which energy is recovered from decelerated electrons and efficiency is enhanced. The green cross-hatched area represents phases during which energy remains in the electrons.

To maximize efficiency, the red area must be eliminated (no electrons enter the output gap during the accelerating phase) and the green area must be minimized. Figure 12-55 shows the phases to which the electrons must be limited to obtain high efficiency. Notice that the green cross-hatched area is very small so most of the bunch energy is recovered.

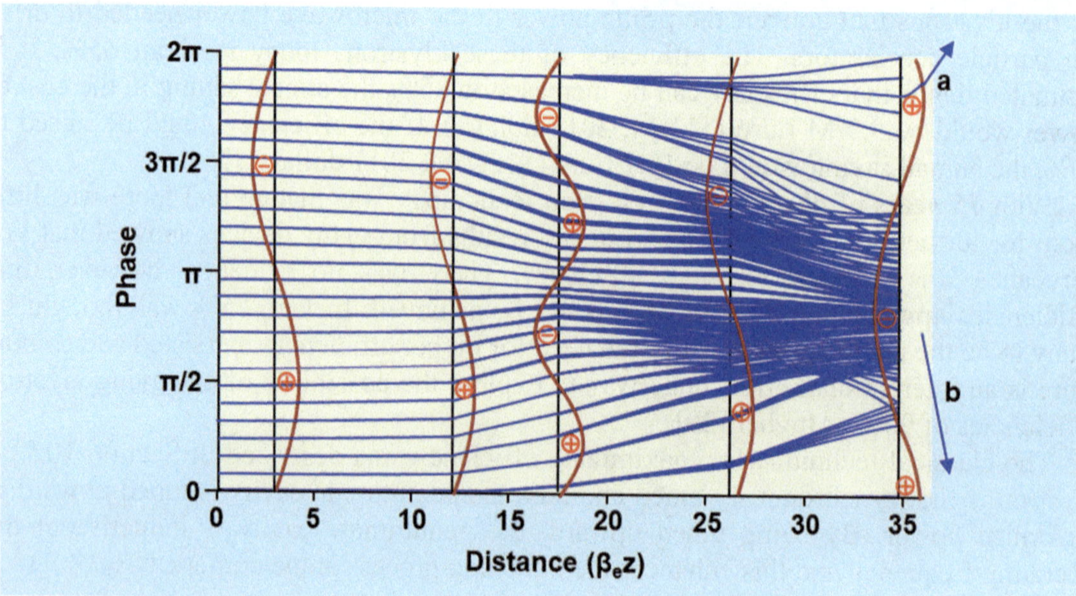

Figure 12-53 Electron phase diagram for a high-power klystron. (Courtesy of Thales.)

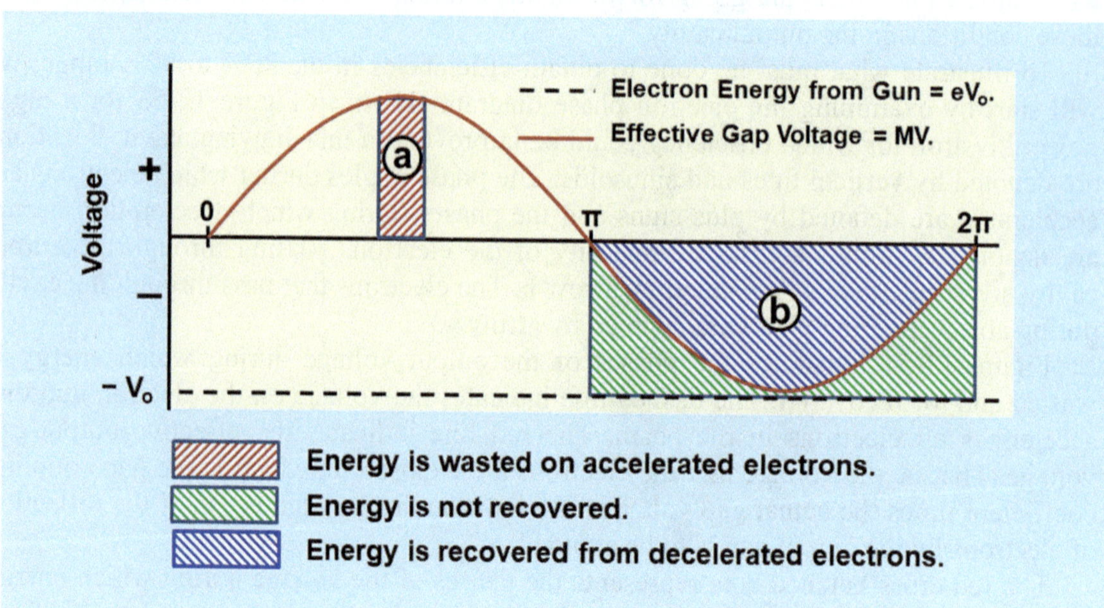

Figure 12-54 Phases of output voltage during which energy is wasted, recovered, or not recovered.

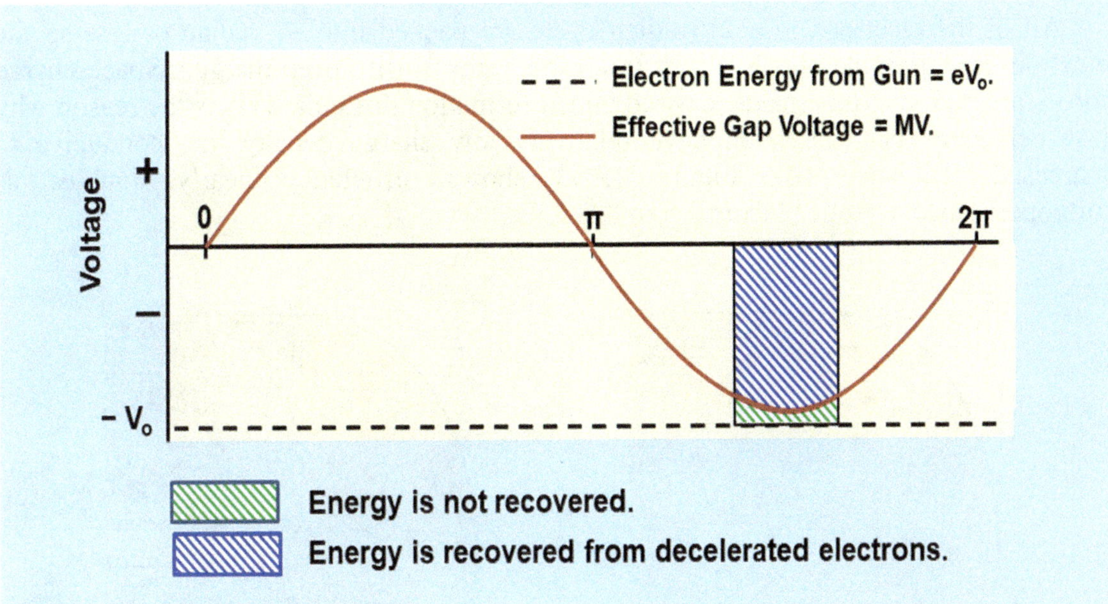

Figure 12-55 This plot shows the bunch duration and the timing of the bunch arrival at the output cavity that result in high efficiency.

Figure 12-56 is an electron phase diagram that shows bunching for a high-efficiency klystron [31]. Electrons are packed into a tight bunch that traverses the output gap when the maximum decelerating voltage occurs.

The sketch in Figure 12-57 shows the size of the bunch relative to the size of the cavity through which the bunch is passing. All dimensions in radians are scaled from a gap width of 2 cm for operation at 1 GHz and 50 kV [27]. The gap length in this case is 1 radian, which is normal for a high-power klystron. Another bunch about π radians long in the process of being compressed is also shown in an idler cavity.

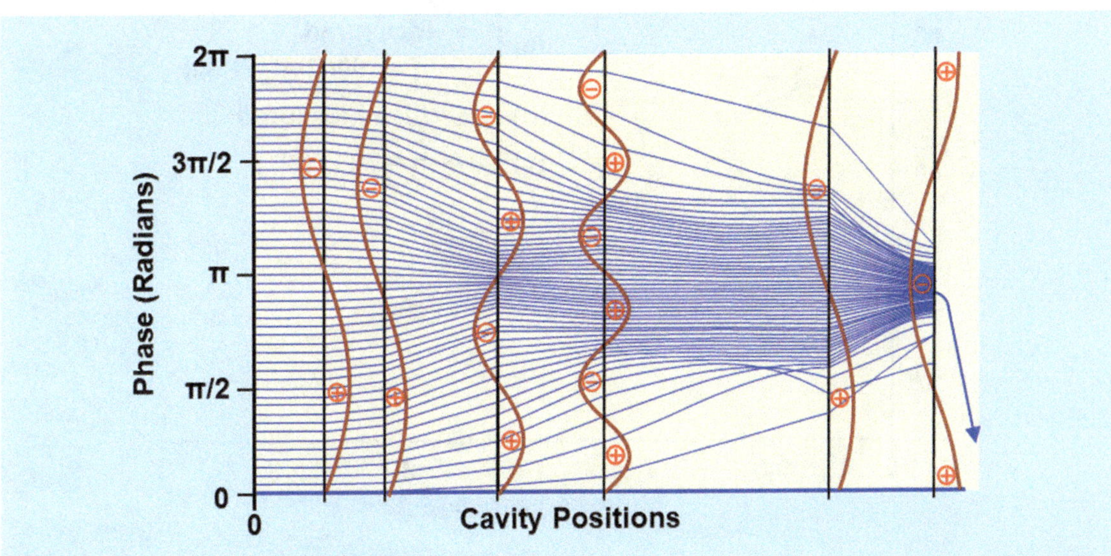

Figure 12-56 Electron phase diagram for high-efficiency operation. (Adapted from V. Hill et al., Paper no. 197, IVEC, 2016.)

All of the electrons in a 2π-radian cycle are packed into ~1 radian or less so the electron density in the bunch is ~2π times the beam density from the gun. Space-charge forces are high and this makes desired bunch formation difficult. This is the reason why low perveance beams, which have relatively low charge density, are conducive to increased efficiency. As Figure 12-58 shows, efficiency nearly doubles as microperveance is reduced from 2.5 to 0.5.

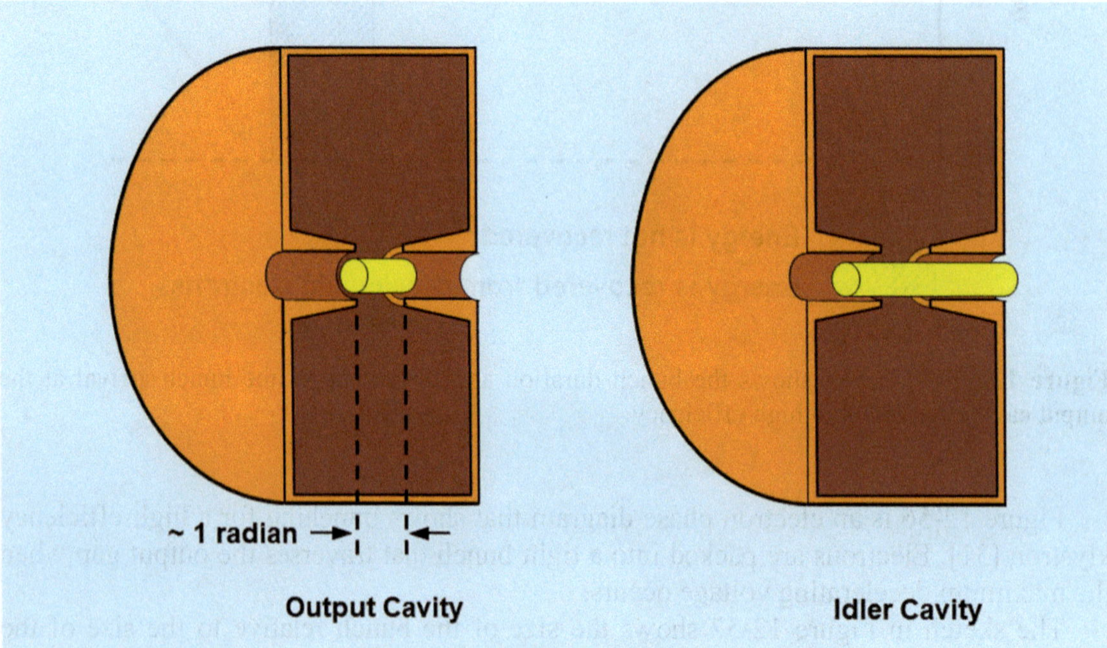

Figure 12-57 Output cavity with gap width of 1 radian and a square bunch 1 radian long.

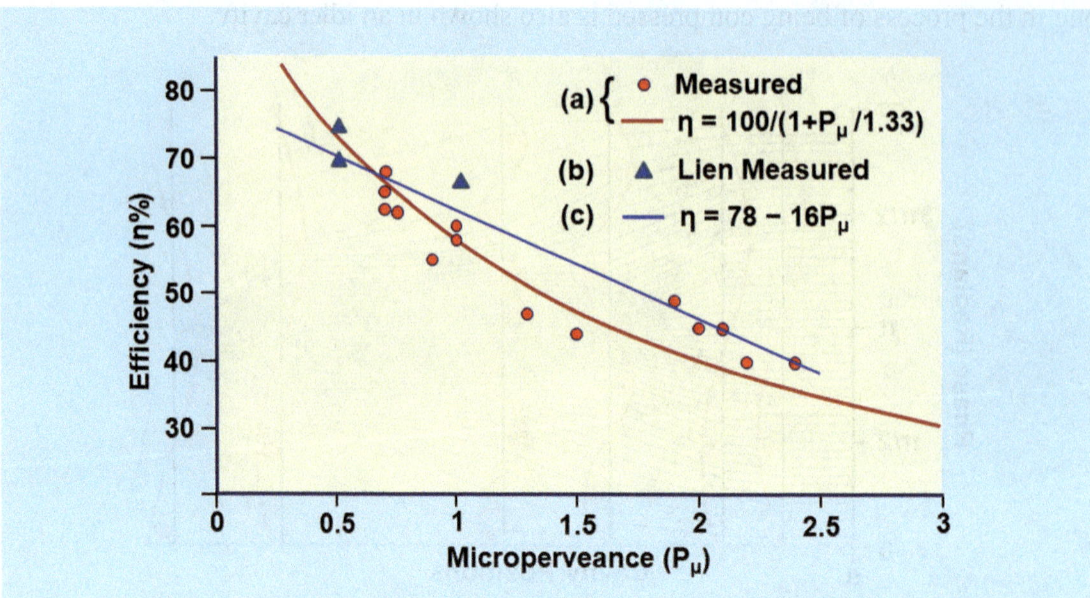

Figure 12-58 These data show that, for high efficiency, low perveance is used.

Low perveance is important. What else is important?

1. First, it is essential to avoid having stragglers that are in the accelerating phase of the output voltage.
2. Electrons should be traveling parallel to the axis so that no energy that cannot be recovered is wasted on radial velocity components (see Figure 12-49).
3. As the bunch enters the decelerating field, the electrons in the lead should be traveling slower than the average. As the bunch passes through the decelerating field of the gap, all electrons leave the decelerating field with same reduced velocity.
4. If an MSDC is going to be used to recover the remaining energy of the electrons, the electrons should all come to the same speed as they enter the decelerating field of the collector.

Several techniques are being considered for optimally bunching a beam. All of these methods serve to bring the outlying electrons, which are in the antibunch loss-producing phases, into the bunches. The sources of these electrons are clearly shown in the Applegate diagram from Chapter 8, which is repeated here as Figure 12-59. These electrons pass through the first beam-modulating cavity when the modulating voltage is low so that they don't participate strongly in the bunching process.

With the traditional approach to tuning a klystron for maximum efficiency, each cavity is placed close to the position where the RF beam current reaches a maximum. The resulting cavity spacing in the driver section is about $\lambda_q/4$. A penultimate cavity tuned inductively is used in an attempt to sweep the outlying electrons into bunches.

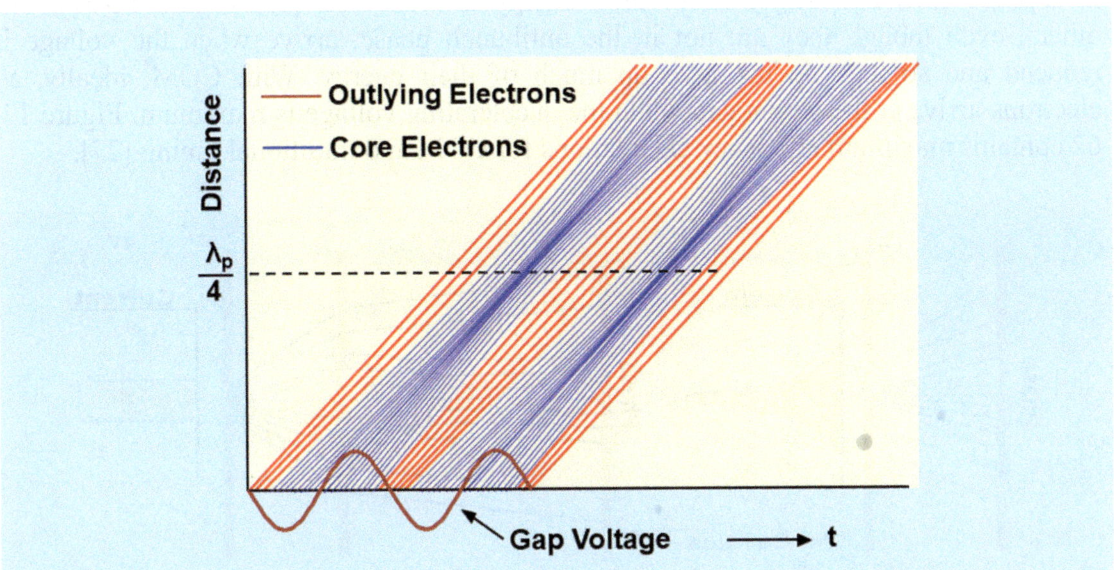

Figure 12-59 Applegate diagram showing trajectories of outlying electrons that must be brought into the bunches to achieve very high efficiencies.

12.4.3 COM and BAC Tuning

The core oscillation method (COM) for processing the beam, shown in Figure 12-60, divides the bunches into the core, in which strong space-charge forces occur and the periphery, where the space-charge forces are weak [27]. Then, the spacing between cavities is made much longer than the spacing for maximum gai0n, $\lambda_q/4$, to wait for the outlying electrons to move into the bunches. Computed efficiencies for the COM are in the 85%–90% range. The primary problem with this technique is that implementation requires extremely long tubes. This is especially true when low-perveance beams, for which the electron plasma wavelength is very long, are used.

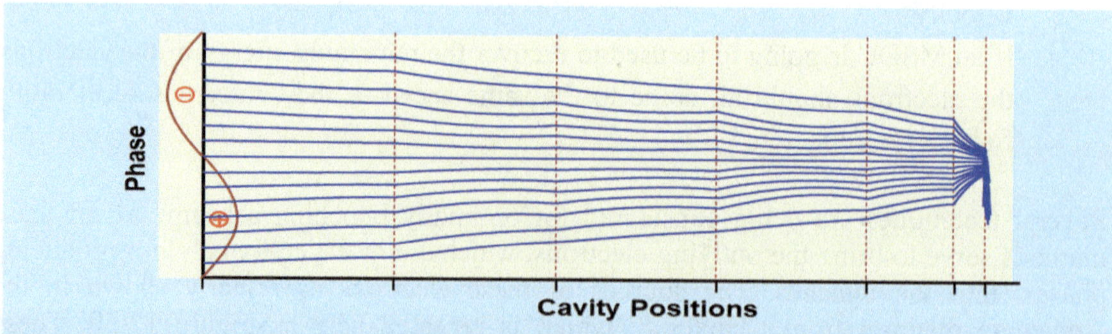

Figure 12-60 Core oscillation method for bunching a beam to achieve high-efficiency klystron operation.

While searching for the best technique for achieving high efficiency, it is essential to keep the optimum bunch parameters in mind. As the electron beam approaches the output cavity with traditional tuning, only the electrons in the center of the bunch arrive in a phase such that nearly all of their energy is recovered {see Figure 12-61). The others, even though they are not in the antibunch phase, arrive when the voltage is reduced and so they do not give up much of their energy. With COM, ideally, all electrons arrive at the output gap when the decelerating voltage is maximum. Figure 12-62 contains maximum efficiency calculations for COM and traditional tuning [27].

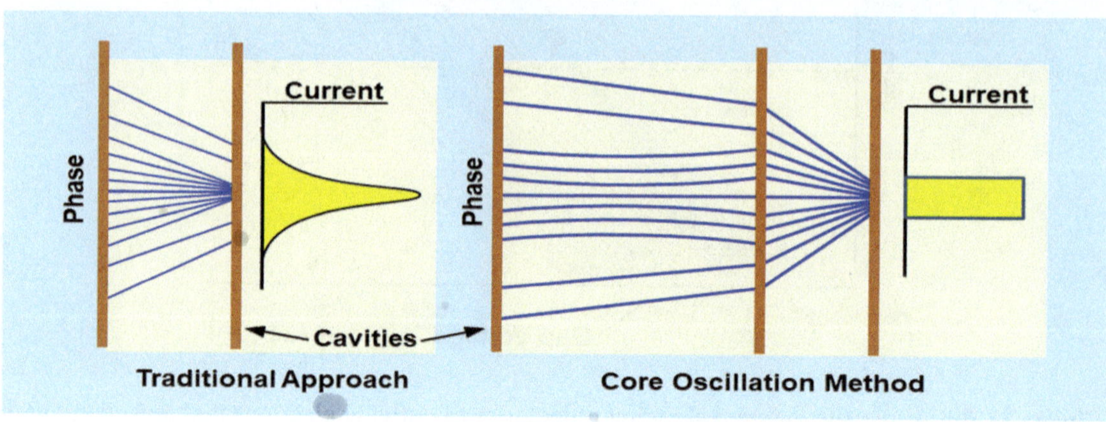

Figure 12-61 Comparison of electron trajectories at the output cavity for traditional klystron tuning with limited efficiency and COM tuning with high efficiency.

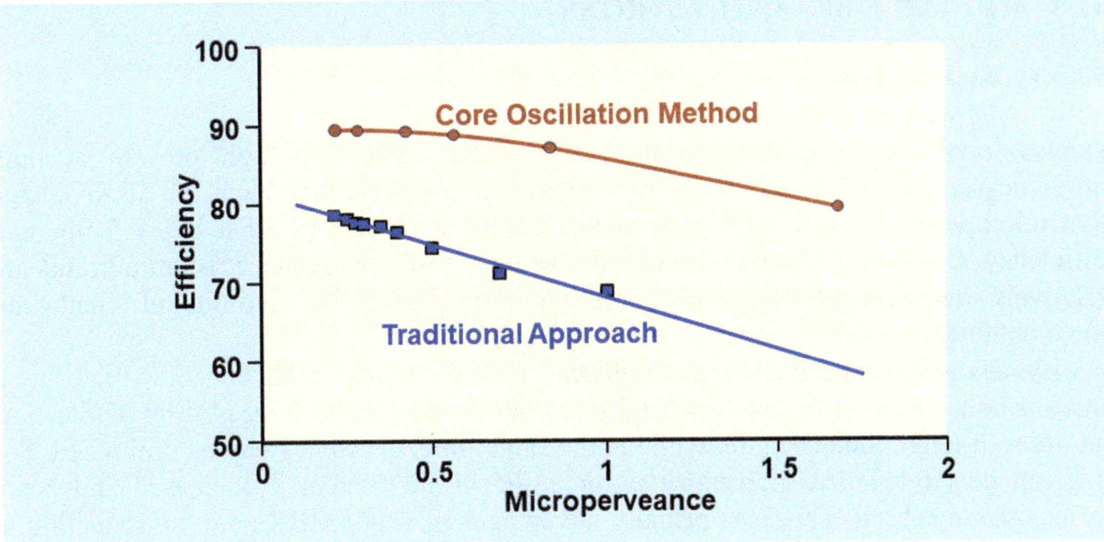

Figure 12-62 The maximum predicted efficiency calculated for klystrons in which the COM technique was used are 10% to 15% higher than for the traditional approach [27].

Another technique for processing the beam is referred to as the bunch, align, collect (BAC) method [32]. This technique is shown in Figure 12-63. The processing of the beam is divided into segments. In each segment, the beam is first bunched, then the electron trajectories are aligned. That is, the electrons are forced to drift and follow parallel paths. Finally, as the beam drifts through the collection region, space-charge forces cause the core to spread, while the outlying electrons, which are affected very little by these forces, move into the bunches. Using the BAC technique, the klystron interaction region is half as long as with the COM technique and is nearly as effective in enhancing efficiency.

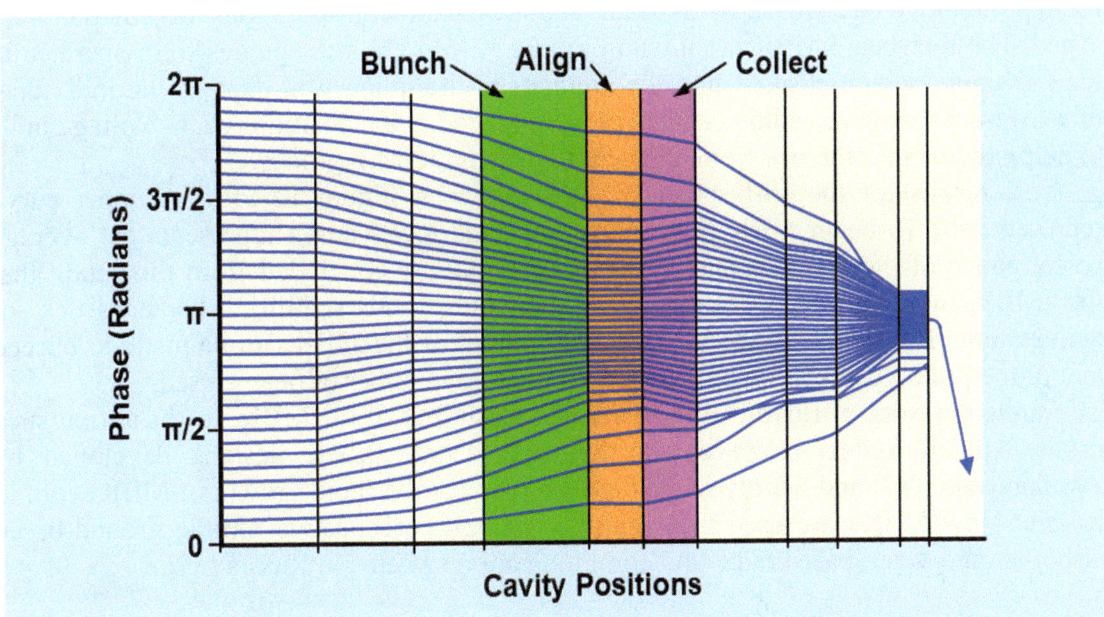

Figure 12-63 The BAC technique for bringing outlying electrons into the bunch.

12.5 MULTIPLE BEAM KLYSTRONS

12.5.1 Background

Conventional single-beam klystrons can produce extremely high powers at high efficiency and with bandwidths of 10% or more. However, they have the disadvantage that microperveance is limited to a maximum value of ~2 and must be ~0.5 for high efficiency. Because of these relatively low values of microperveance, beam currents are relatively low and very high voltages are required if high power and high efficiency are to be obtained.

By using a device containing multiple electron beams, each one of which has a microperveance of ~1 or less, the total microperveance can be increased by as much as an order of magnitude or more, while at the same time, efficiency can be optimized. For a given power level, as perveance is increased, beam current is increased and beam voltage is decreased. Thus, the primary advantages of using MBKs are the reduction in voltage that occurs and the ability to optimize efficiency.

In addition, because voltage is reduced, the length of the device is reduced. Because the individual beam currents are reduced, the magnetic field required for focusing is reduced. As a result of these two factors, for a given power level, the weight of the MBK is much less than the weight of a single-beam klystron. Other advantages resulting from the use of MBKs are that power supplies are smaller, arcing and breakdown are reduced, it may be possible to use air insulation rather than oil, and x-ray shielding can be reduced or eliminated.

Apparently, the concept of the MBK was first proposed in the U.S.S.R. (now Russia) by Kovalenko [33] and in France by Bernier [34]. Then, in the 1960s, several versions of MBKs were reported [35, 36]. Some results of the feasibility study in [35] are shown in Figures 12-64 and 12-65. For flexibility and convenience, Boyd et al. used conventional external cavity klystrons and inserted them into multiwavelength waveguides to test the basic principles of the MBK. Four-beam and 10-beam configurations of MBKs were tested. The four-beam MBK is shown in Figure 12-64. The waveguides were operated in the $\pi/2$ mode, which ensured that the voltages at the interaction gaps of the individual tubes were the same. Loading capacitors were placed at the positions of the voltage nulls to help prevent interference from adjacent modes in the waveguides.

Test results for the 10-beam MBK are shown in Figure 12-65. The upper curve represents the 10-beam MBK performance and the lower curve represents the average performance of the single-beam klystrons. The authors concluded from this study that the MBK has the gain, efficiency, bandwidth, and stability characteristics of conventional klystrons. However, the power output of the MBK can be made to exceed that of the prototype klystron by a factor equal to the number of beams used.

While the work of Boyd et al. showed the feasibility of the MBK, it was not pursued in the United States. However, in the early 1960s, MBK designs developed by Zusmanovsky [37] and Korolyov [38] led the way for the rapid growth of MBK work in Russia. Now MBKs are used in a majority of new, prospective mobile ground-based airborne- and space-based radar and communication systems in Russia [39].

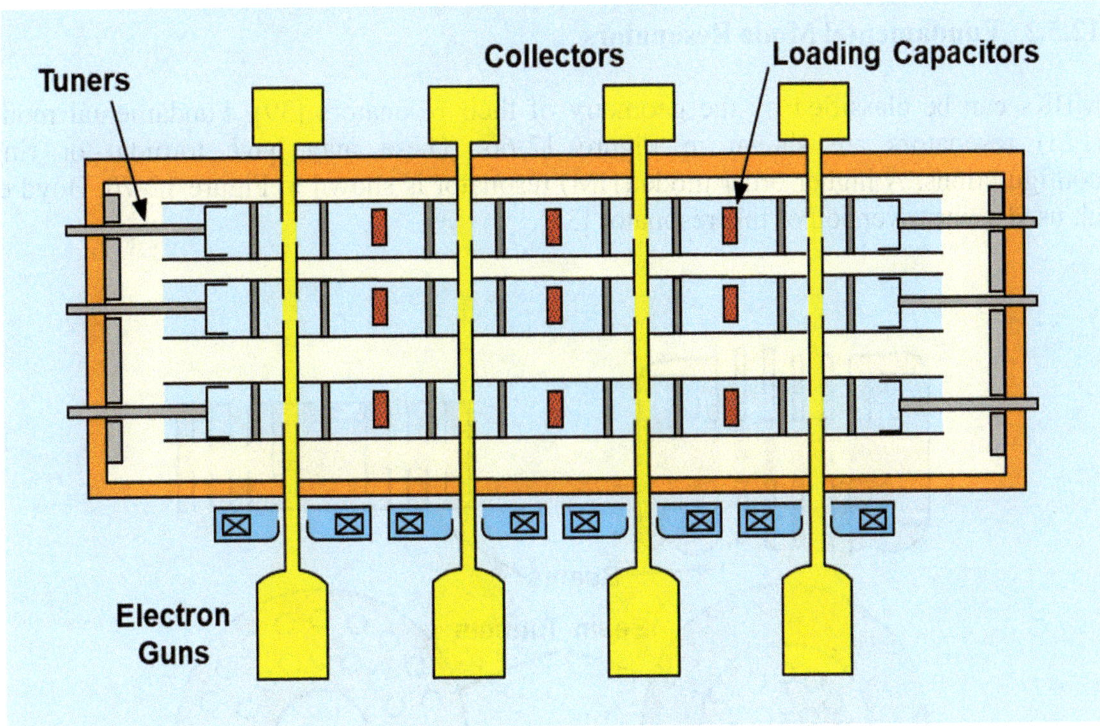

Figure 12-64 Configuration of four-beam MBK. External cavity klystrons without cavities were combined in a common RF structure. (Adapted from: Boyd et al., *IRE Trans. Electron Devices,* May 1962. ©1962 IRE (now IEEE).)

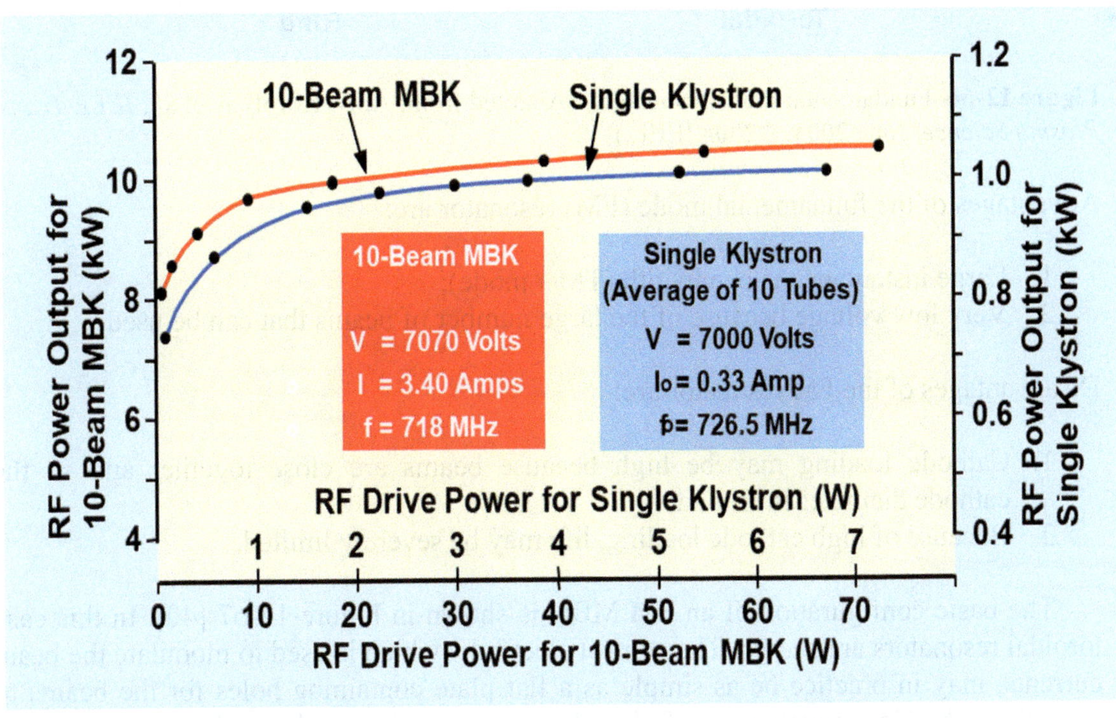

Figure 12-65 RF power characteristics of the single klystron and the 10-beam MBK. (Adapted from: Boyd et al., *IRE Trans. Electron Devices,* May 1962. ©1962 IRE (now IEEE).)

12.5.2 Fundamental Mode Resonators

MBKs can be classified by the geometry of their resonators [39]. Fundamental mode (FM) resonators are shown in Figure 12-66. These may have toroidal or ring configurations. A higher-order mode (HM) resonator is shown in Figure 12-70. Boyd et al. used a linear version of this resonator [35].

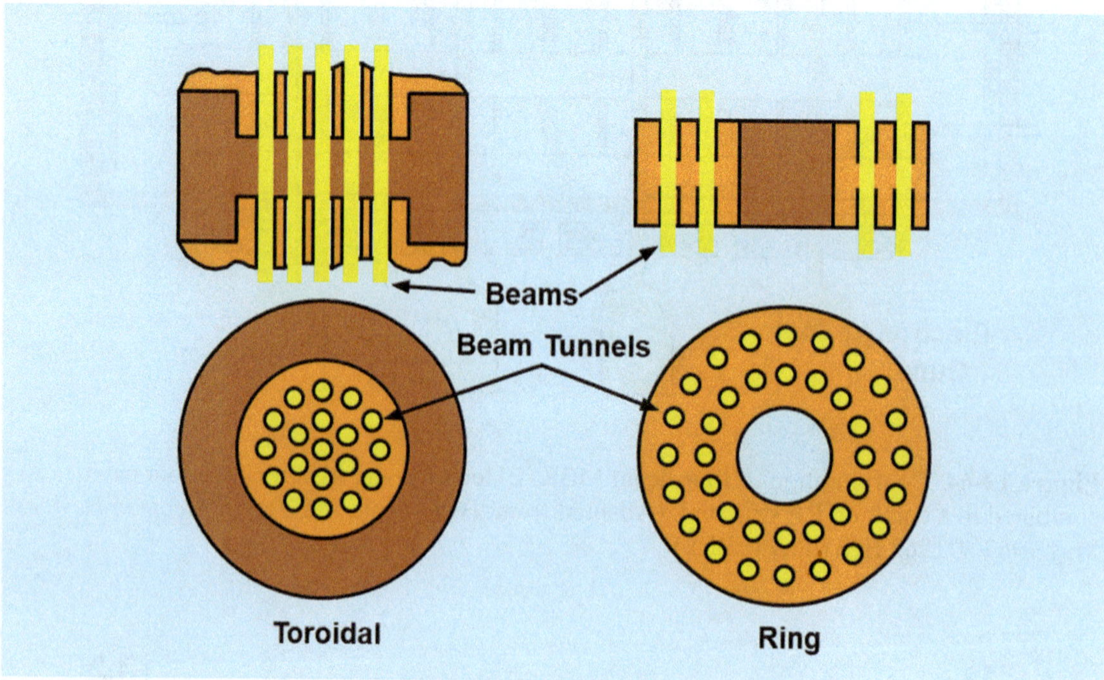

Figure 12-66 Fundamental mode resonators. (Adapted from: A. N. Korolyov et al., *IEEE Trans. Plasma Science*, June 2004. ©2004 IEEE.)

Advantages of the fundamental mode (FM) resonator are:

1. Large instantaneous bandwidth (TM_{01} mode);
2. Very low voltage because of the large number of beams that can be used.

Disadvantages of the FM resonator are:

1. Cathode loading may be high because beams are close together and so the cathode diameter is limited;
2. Because of high cathode loading, life may be severely limited.

The basic configuration of an FM MBK is shown in Figure 12-67 [40]. In this case toroidal resonators are shown. The control electrode, which is used to modulate the beam currents, may in practice be as simple as a flat plate containing holes for the beams to pass through. No convergence of the electron beams near the cathode assembly is shown, although electron guns with a relatively small area convergence ratio are sometimes used.

In [39], an excellent overview is given of the range of capabilities of Russian FM MBKs. Devices capable of RF powers from watts (in miniature MBKs) to hundreds of kilowatts are described. Emission densities as high as 30 A/cm^2 are mentioned along with lifetimes of thousands of hours.

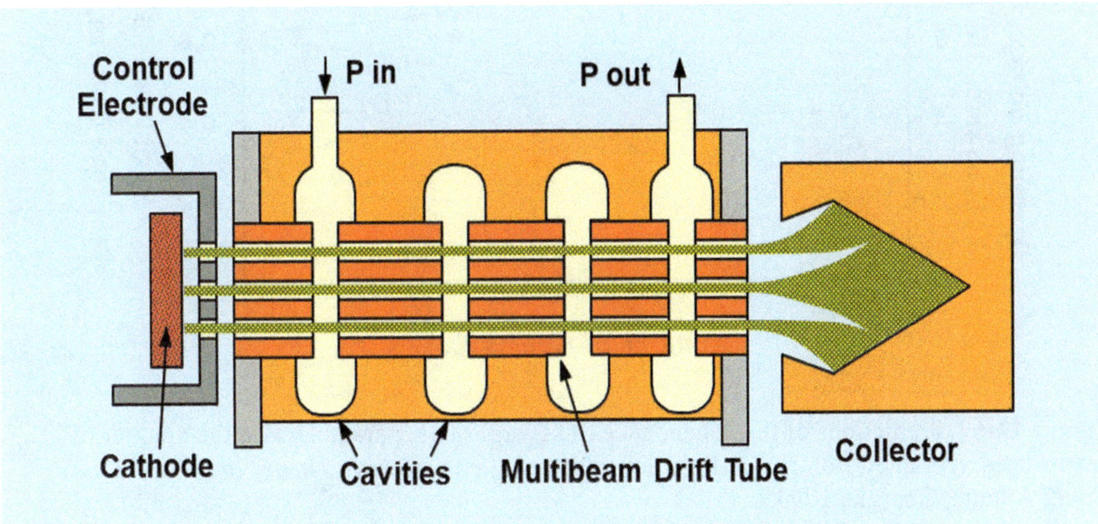

Figure 12-67 Basic configuration of a fundamental mode MBK. (Adapted from: A. S. Pobedonostsev et al., *Microwave Theory and Techniques Conf.*, Atlanta, Georgia, 1993.)

Figures 12-68 and 12-69 show predicted parameters of MBKs operating at 10 GHz with a cathode emission density of 15 A/cm^2 [40]. Notice that a 100-kW device containing 19 beams would operate at 10 kV. By comparison, a single-beam klystron that produced 100 kW would have to operate at 30 kV or more.

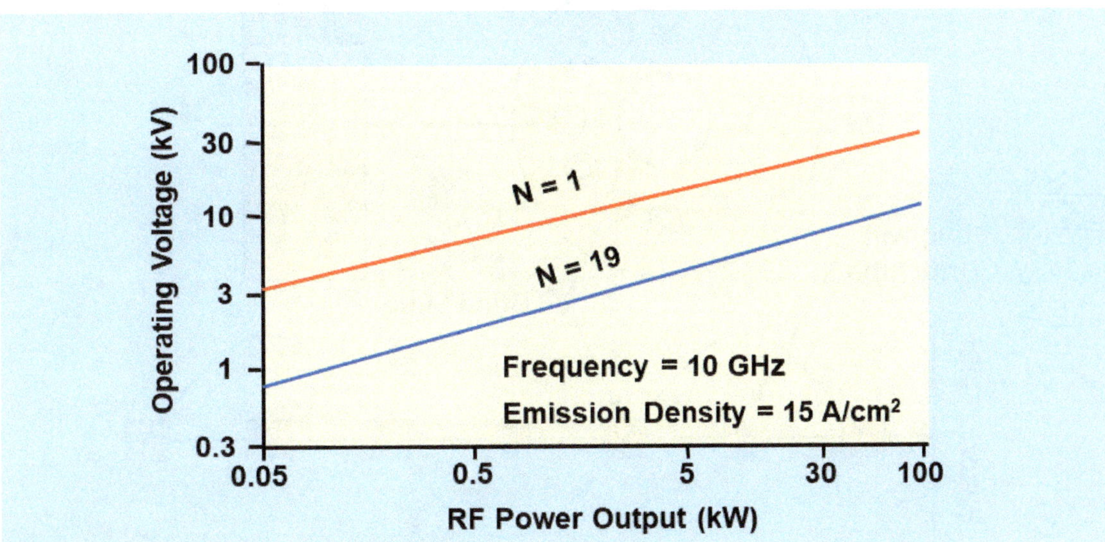

Figure 12-68 Predicted operating voltage as a function of power output for single beam and 19-beam MBKs. (Adapted from: A. S. Pobedonostsev et al., *Microwave Theory and Techniques Conf.*, Atlanta, Georgia, 1993.)

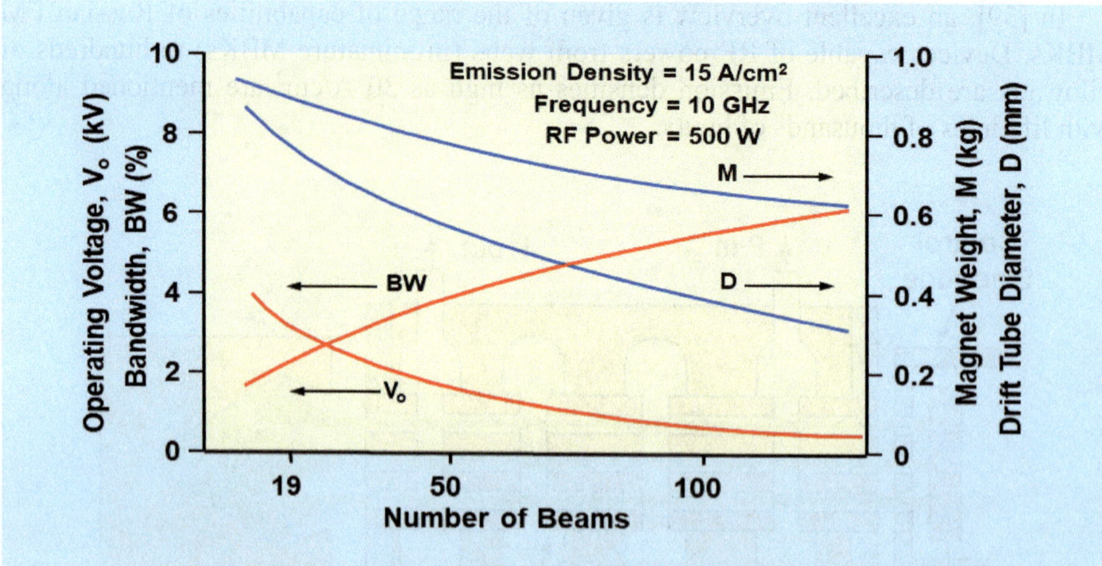

Figure 12-69 Predictions of the dependence of several MBK parameters on the number of beams used. (Adapted from: A. S. Pobedonostsev et al., *Microwave Theory and Techniques Conf.*, Atlanta, Georgia, 1993.)

12.5.3 Higher-Order Mode Resonators

The higher-order mode (HM) resonator MBK overcomes some of the limitations of the FM MBK but has other restrictions. A higher mode resonator is shown in Figure 12-70.

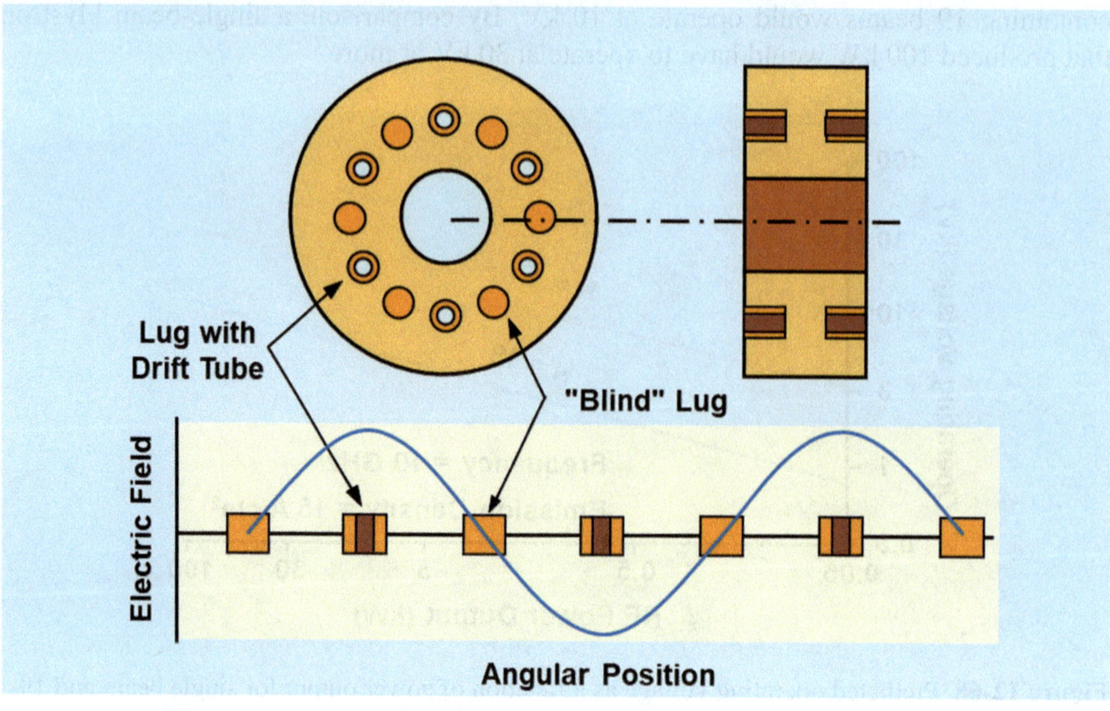

Figure 12-70 Higher-order mode resonator. (Adapted from: A. N. Korolyov et al., *IEEE Trans. Plasma Science*, June 2004. ©2004 IEEE.)

1. Cathode loading is low because beam separation can be large so cathode size can be much larger than for FM devices;
2. Operation at very high average power levels is possible because of the large separation between beams and the improved cooling that is possible.

The disadvantages of the HM resonator are:

1. Bandwidth is narrow because of the RF circuit that is used;
2. It is increasingly difficult to focus beams as the distance from the center of the magnetic field increases. This increases the difficulty in fabricating electron guns.

The difficulty in beam focusing arises because of the radial components of magnetic flux that exist near and in the electron guns as well as in the drift space along the length of the klystron. As shown in Figure 12-71, this field component causes theta-directed forces on the electron beams. The resulting effect on the beams is shown in Figure 12-72.

Bastien et al. [41] analyzed the radial flux distribution in an electromagnet without pole pieces and found that B_r is proportional to the distance from the magnet axis and to the square of the magnet radius. For this reason, cavities with drift tubes packed as closely as possible to the magnet axis should be used and the radius of the magnet should be as small as possible.

For the electromagnet to be used on an MBK, Bastien et al. used pole pieces on the ends of the magnet and also, to reduce B_r to an acceptable level, divided the electromagnet into separate coils and used pole pieces between the coils.

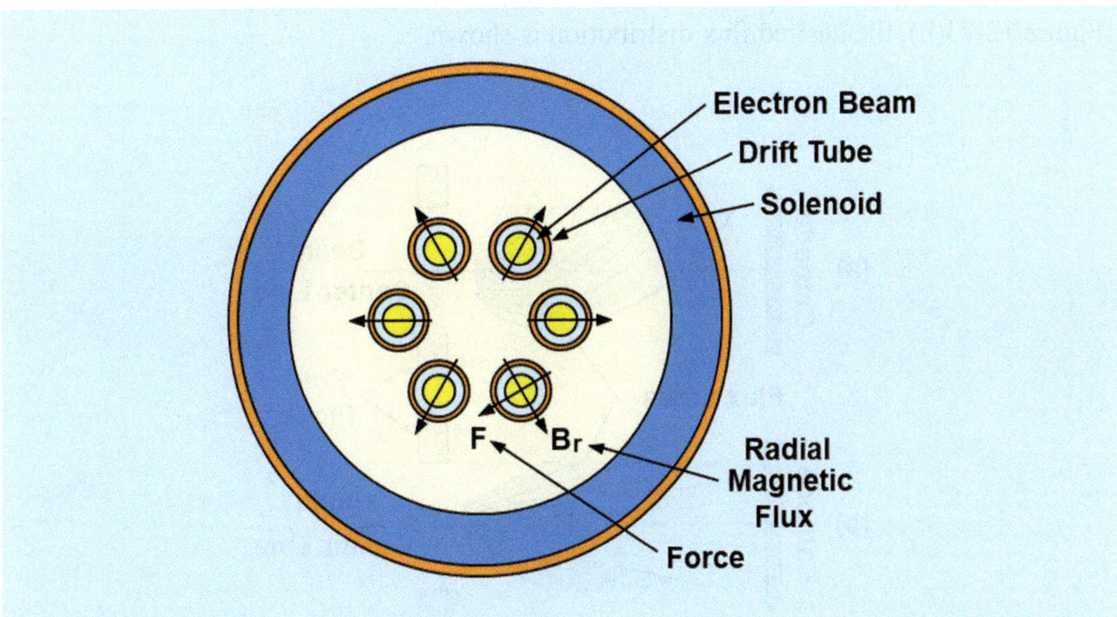

Figure 12-71 Theta-directed force resulting from the radial components of magnetic flux. (Adapted from: C. Bastien et al., Report No. SLCET-TR-87-0713-F, March 1989.)

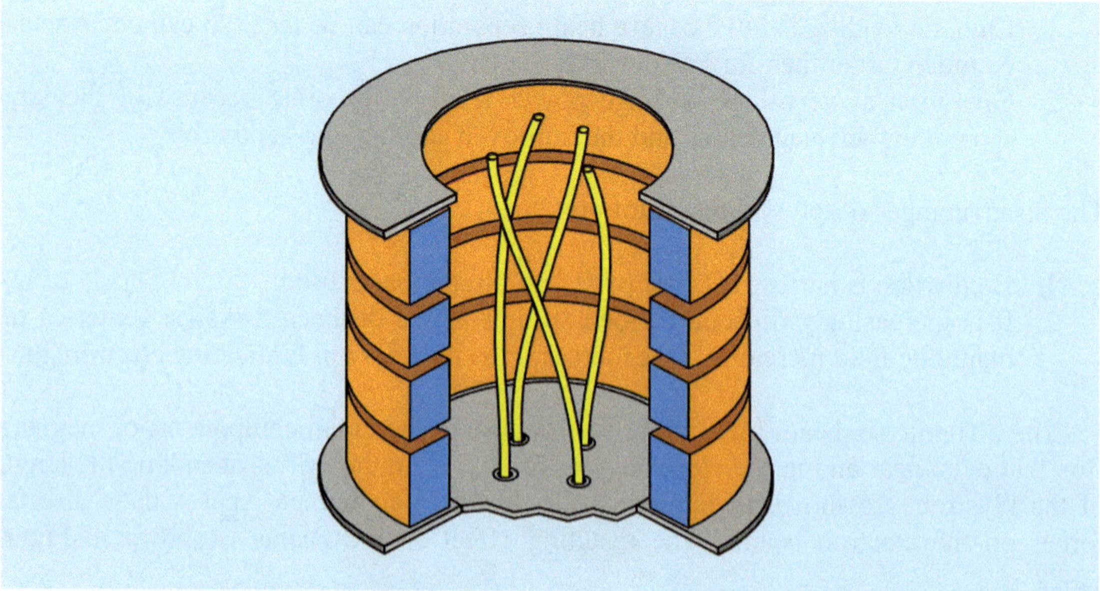

Figure 12-72 Effect on electron beams of radial magnetic flux components near and in the electron gun. (Adapted from: C. Bastien et al., Report No. SLCET-TR-87-0713-F, March 1989.)

When magnetic flux is required in the guns of an MBK to permit confined (immersed) flow focusing, the problem of radial flux components in the guns and within the apertures in the pole pieces becomes particularly severe. This problem is illustrated in Figure 12-73 [42]. In Figure 12-73(a), the field pattern that would exist in a gun of an MBK is shown if no attempt was made to eliminate the radial components of flux. In Figure 12-73(b), the desired flux distribution is shown.

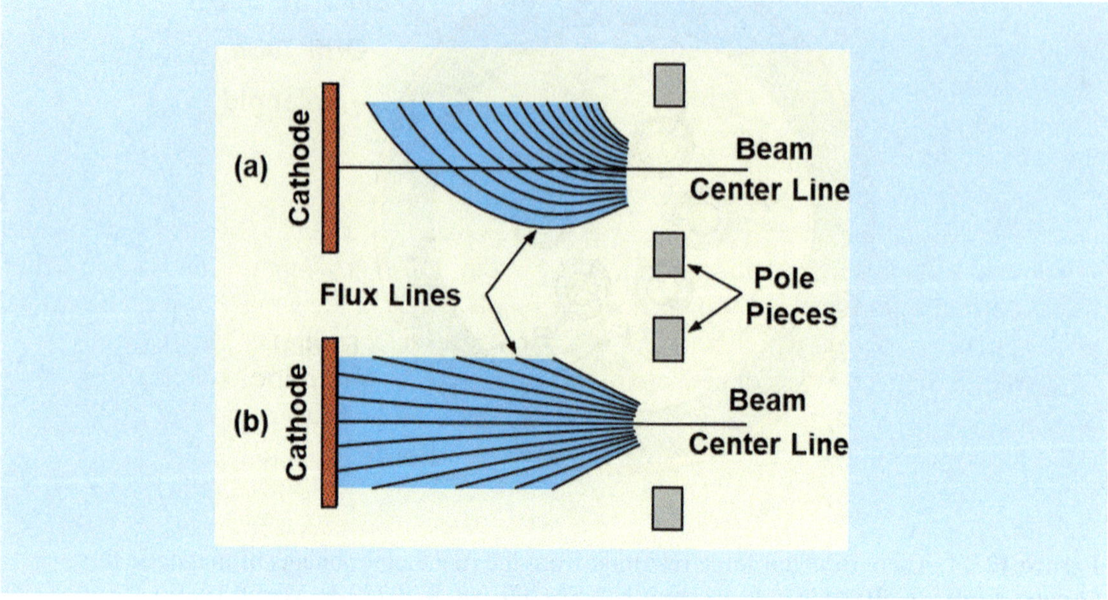

Figure 12-73 (a) Uncorrected and (b) corrected magnetic flux lines for off-axis electron guns. (Adapted from: R. L. Ives et al., *Proc. 6th Workshop on High Energy Density and High Power RF*, 2003.)

A technique for obtaining the desired flux distribution is shown in Figure 12-74. The asymmetry of the pole piece provides a magnetic reluctance configuration that eliminates B_r.

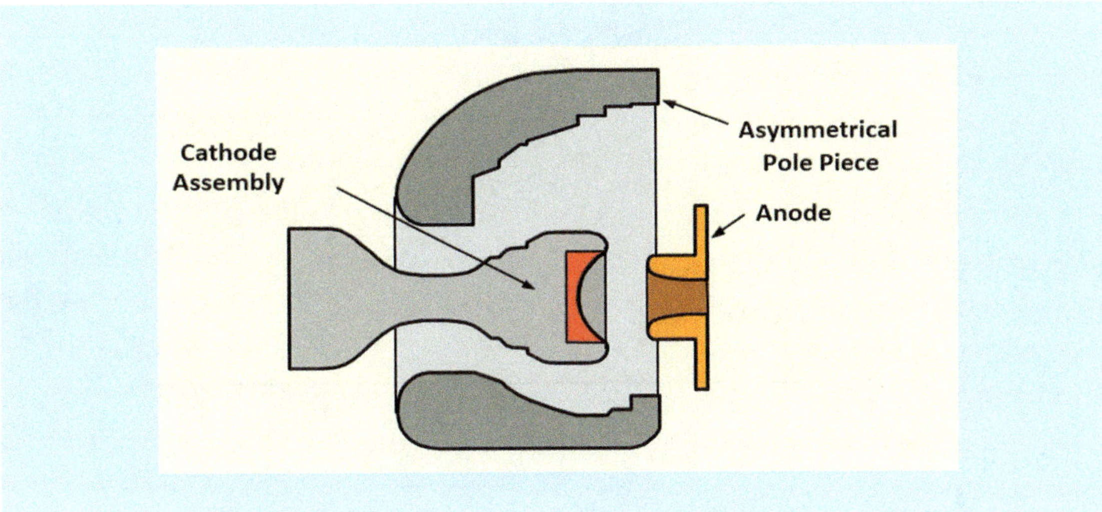

Figure 12-74 Asymmetric pole piece for correcting flux distribution in a gun. (Adapted from: R. L. Ives et al., *Proc. 6th Workshop on High Energy Density and High Power RF*, June 2003.)

An application for MBKs that will require very-high-power levels is the TESLA Linear Collider at Deutsches Elektronen-Synchrotron (DESY) [43]. The requirement to use MBKs came about when the specifications for the RF sources were established. The power requirement per tube was set at 10 MW at an efficiency of 70%. The pulse length requirement was 1.5 ms. Reaching this goal with a single-beam klystron would require a beam voltage of 241 kV. Quoting from [43]: "The construction of a klystron meeting these parameters seems to be impossible. The solution to this problem is the use of several low perveance beams in parallel at lower voltage in one vacuum vessel. This technology is utilized in the multi beam klystron."

Because the frequency of operation of TESLA is relatively low (1.3 GHz), device dimensions are large enough that either FM or HM MBKs could be considered to fulfill the RF source requirements. Figure 12-75 from Communications and Power Industries (CPI) indicated that the cathode emission density for the FM configuration would be excessive if long life was to be achieved [44]. Because the CPI goal was a design for which emission density could be limited to 2 A/cm^2 to ensure long life, the higher-order-mode configuration was chosen for its development effort. This made it possible to use cathodes of larger size, which resulted in low cathode current density loadings.

Low loading makes reduced operating temperatures and an exponentially reduced rate of barium depletion from the cathode possible. This results in an extended cathode life and an extended period of arc-free operation of the electron gun. The relative disadvantage is an increase in the complexity of the klystron, particularly in the configuration required to focus and transport the off-axis electron beams. In spite of the complexity, each beam propagates through the klystron circuit with 99.5% dc beam transmission and 97.5% transmission at fully saturated RF conditions and for full operating duty.

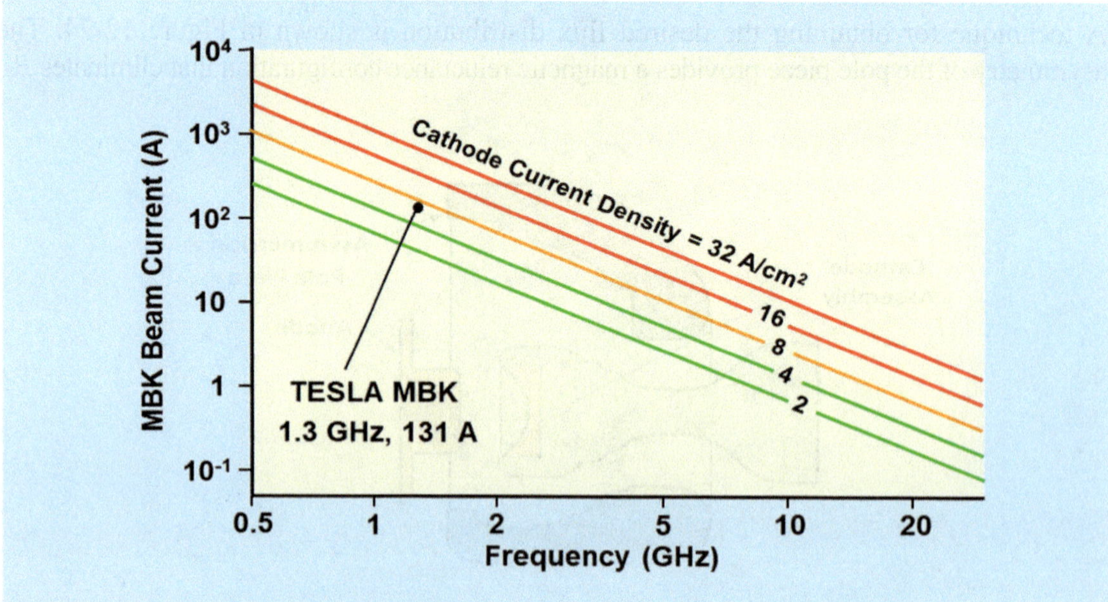

Figure 12-75 Cathode loading predicted by Communications and Power Industries (CPI) for the FM configuration of the TESLA MBK.

Two versions of the CPI TESLA klystron are shown. The vertical version without solenoid is shown on the left in Figure 12-76 and the horizontal version is on the right. Operating parameters are listed in Table 12.5.

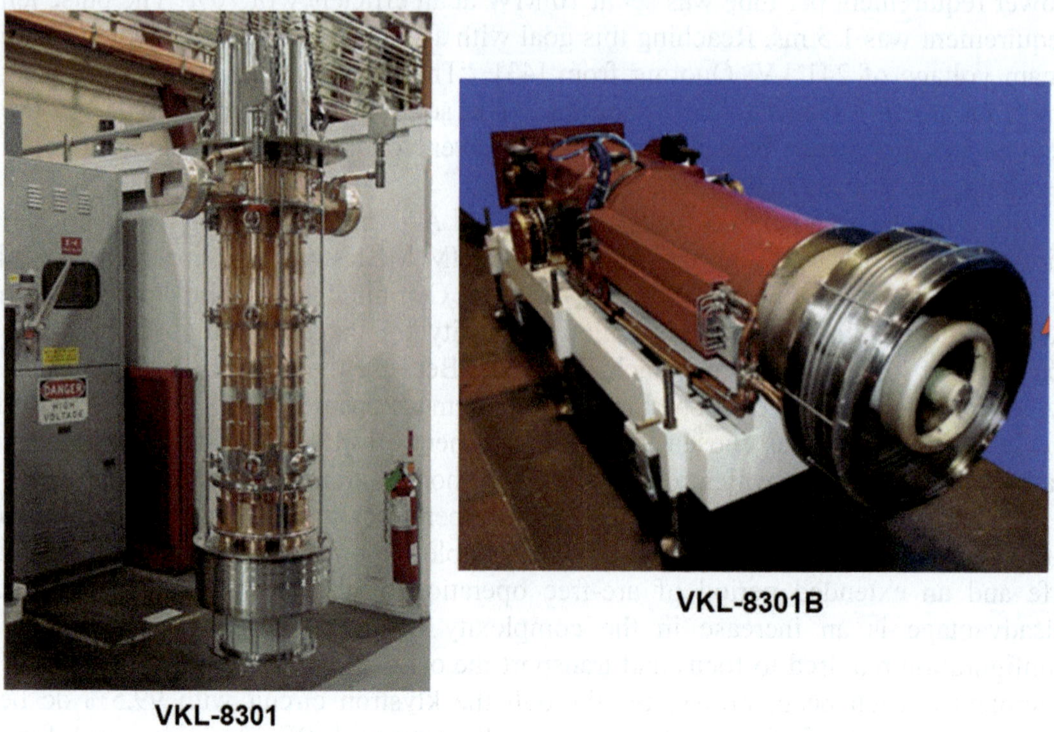

Figure 12-76 CPI 10 MW MBK for TESLA, VKL-8301 (left) and VKL-8301B (right).

Table 12.5
Operating Parameters for the VKL-8301B

Parameter	Value
Frequency	1300 MHz
Peak Power Output	10 MW
Average Output Power	150 kW
Beam Voltage	120 kV
Beam Current	141A
Efficiency	59%
Saturated Gain	48 dB
-1 dB Bandwidth	±3.5 MHz
RF Pulse Width	1.5 ms
Cathode Loading	2.2 A/cm^2
Klystron Length	101 in, 257 cm
Klystron Weight	2000 lbs, 900 kg
Electromagnet Diameter	35 in, 88 cm
Electromagnet Weight	3200 lbs, 1450 kg

Thales chose the FM configuration for their device. The cathode emission density is 5.5 A/cm^2 and the device has been operating at nominal performance level for several years (for an overview, see [44]). The power output of the Thales tube is shown in Figure 12-77 and the measured performance characteristics are listed in Table 12.6.

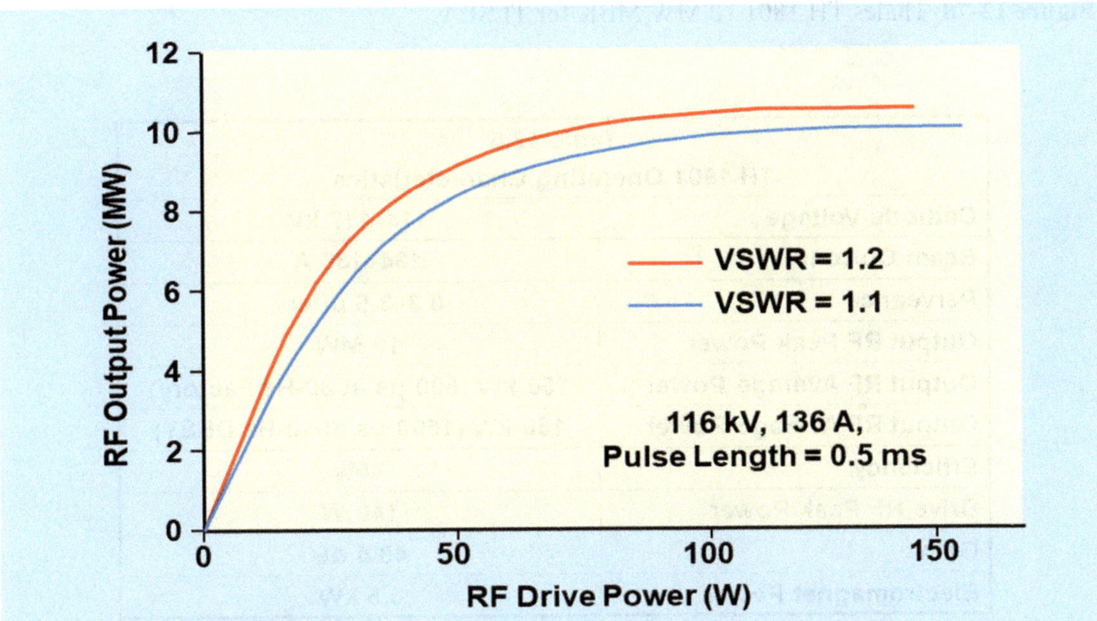

Figure 12-77 Power output characteristics of the Thales TH 1801 MBK [44].

The Thales TH 1801 without focusing solenoid is shown in Figure 12-78 along with a drawing of the tube and a photo of the electron gun. The Thales tube illustrates very well the benefits of the MBK. With a beam voltage only a little over 100 kV, a power of 10 MW is generated at an efficiency of 65%. An additional benefit that the multiple beam and low voltage parameters provide is reduced device size.

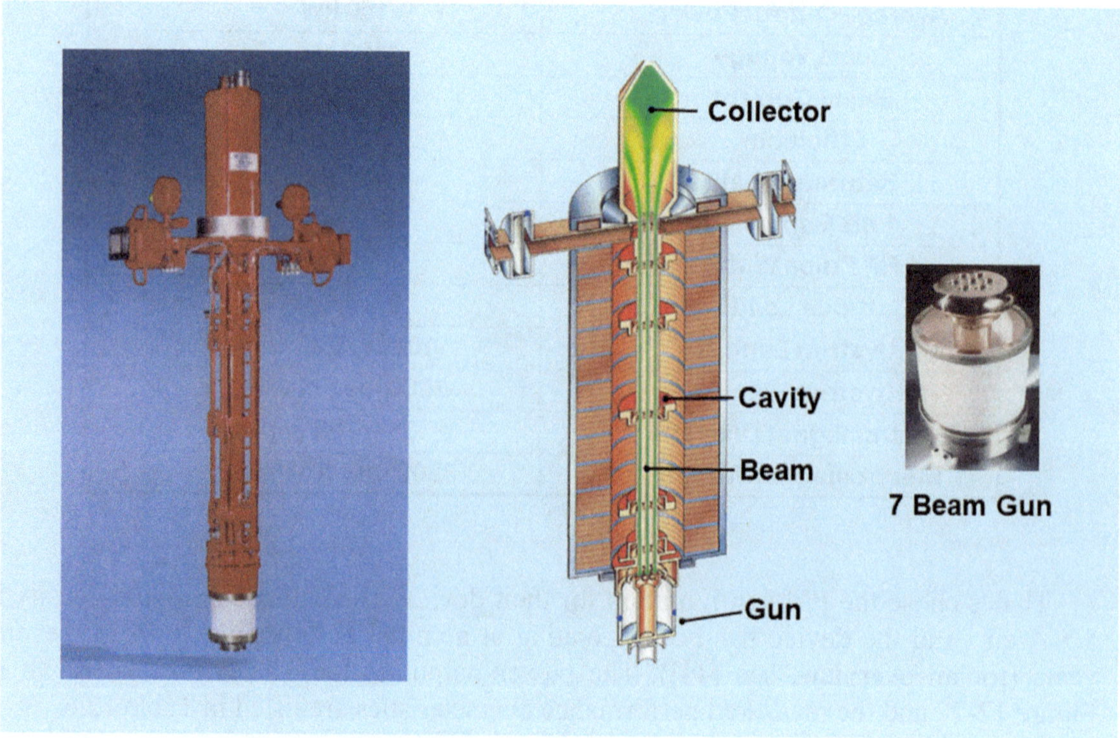

Figure 12-78 Thales TH 1801 10 MW MBK for TESLA.

Table 12.6
TH 1801 Operating Characteristics

Cathode Voltage	114–117 kV
Beam Current	134–136 A
Perveance	3.3–3.5 μPerv
Output RF Peak Power	10 MW
Output RF Average Power	150 kW (500 μs at 30-Hz Factory)
Output RF Average Power	150 kW (1500 μs at 10-Hz DESY)
Efficiency	65%
Drive RF Peak Power	140 W
Gain	48.5 dB
Electromagnet Power	5.5 kW

Toshiba has also developed an MBK for TESLA. That device, the E3736, is shown in Figure 12-79. The size of the device is substantial because of the relatively low operating frequency and the high-power requirement.

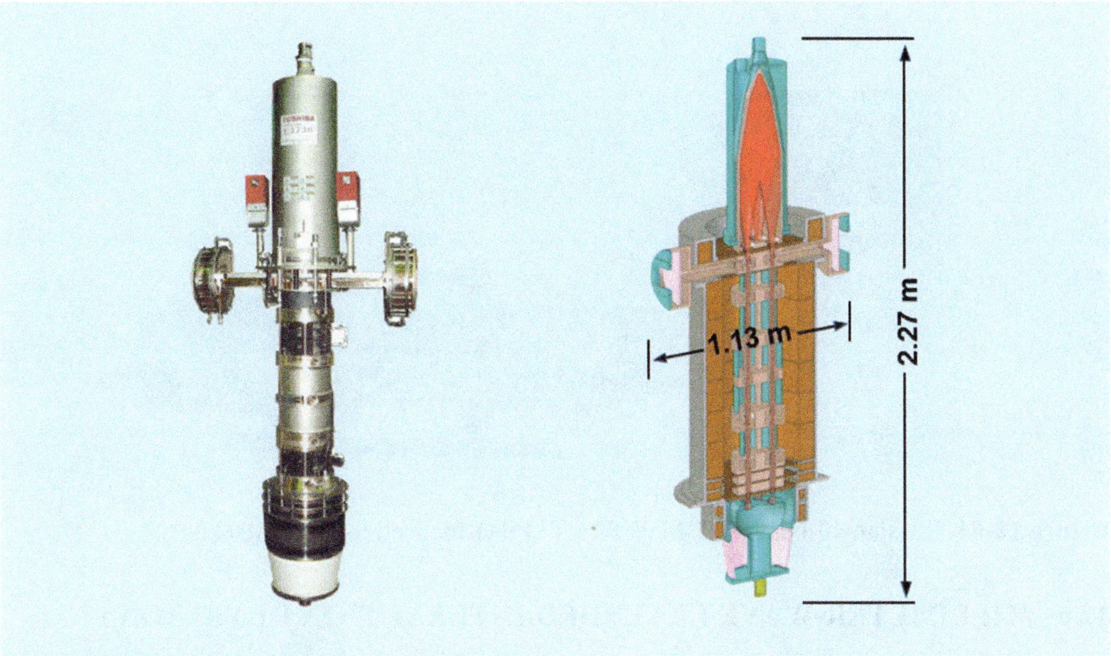

Figure 12-79 Toshiba E3736 MBK being developed for TESLA [44].

The U.S. Naval Research Laboratory is developing a 3.27-GHz MBK for beam transport and beam-wave interaction studies (Figure 12-80) [45]. This is an eight-beam four-cavity device operating in the fundamental mode. The gun is magnetically shielded and operates at 32A and 45 kV. Detailed results of calculations of electron energies, RF current, and power are given in [45].

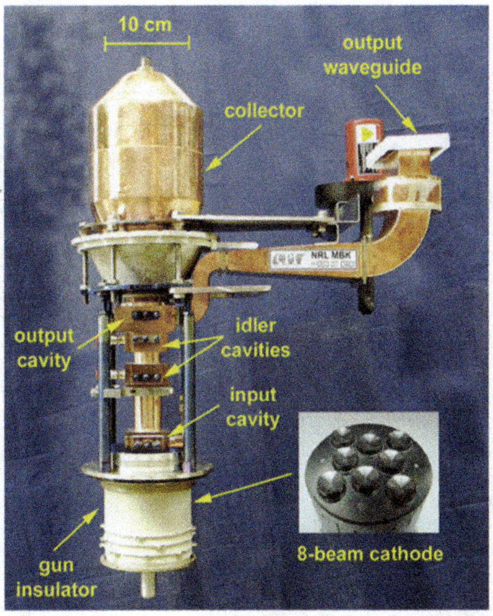

Figure 12-80 Compact. eight-beam, four-cavity NRL 600 kW S-Band MBK.

The 40-beam, PPM-focused MBK shown in Figure 12-81 has been developed by Guzilov in Russia [46]. This device employs the new BAC focusing technology described in the efficiency section (12.4) of this chapter and is designed to operate at an efficiency of 77%. The tube operates at 52 kV and produces a peak power of 7.5 MW.

Figure 12-81 Russian 40-beam, 7.5-MW, 52 kV, PPM-focused S-band MBK.

12.6 MILLIMETER-WAVE EXTENDED-INTERACTION KLYSTRONS

Since the 1970s, CPI Canada has been developing and manufacturing extended interaction klystrons (EIKs) and oscillators (EIOs) with a wide range of capabilities. The blue cross-hatched area in Figure 12-82 encompasses the power and frequency characteristics of the devices manufactured to date by CPI. The red cross-hatched areas indicate the power and frequency ranges that engineers at CPI think may be possible to fill with EIK-like devices [46].

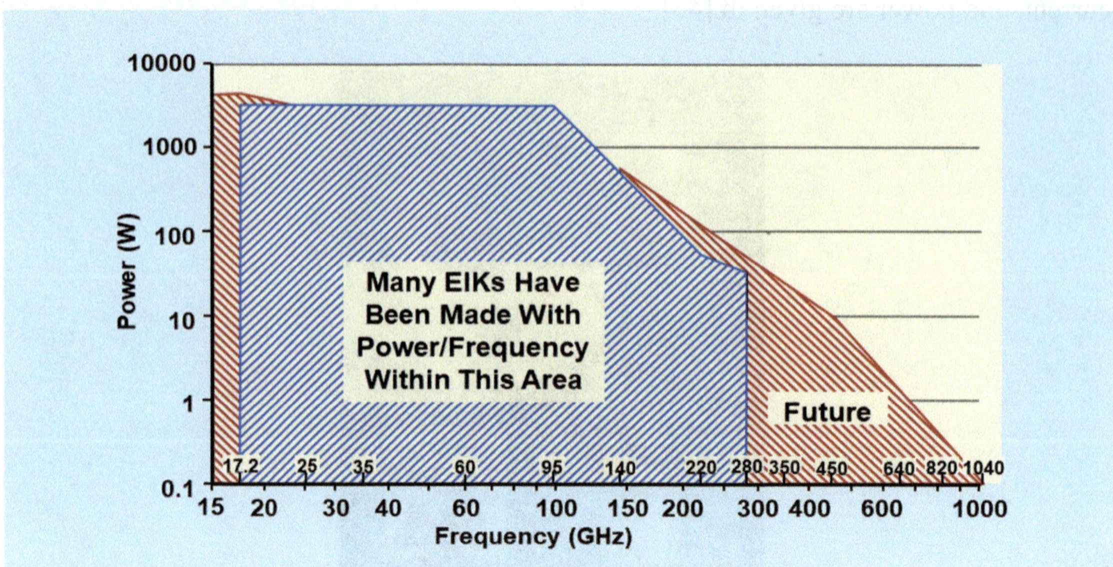

Figure 12-82 RF power capabilities of EIKs. (Adapted from D. Berry et al., *IEEE Trans. Electron Devices*, Vol. 61 No.6, June 2014.)

In EIKs, klystron technology is combined with coupled-cavity TWT technology to make amplifiers and oscillators that can operate at mm-wave frequencies. As shown in Figure 12-21, EIKs operating from 17 to 280 GHz have been developed. EIKs can be amplifiers or oscillators and can be operated CW or pulsed. Pulsed operation is accomplished with a single-aperture grid, which also serves as the focus electrode.

12.6.1 EIK Components

The major components of an EIK are shown in Figure 12-83. Very briefly, operation is like that in all linear beam tubes; that is, the electron gun (1) produces an electron beam that is focused by magnets (2) and passes through cavities (3) to the collector (4).

In extended interaction devices, as indicated in Figure 12-83, each cavity of the conventional klystron is replaced by a multigap cavity. The multigap structure in the cavity is a section of a resonant slow wave circuit, which is known as a ladder circuit. Several ladder circuits are shown in Figure 12-84. The size of the circuit scales inversely with frequency and, at frequencies high in the mm-wave range (~300 GHz), where the free space wavelength approaches 1 mm (.040 in), the average slot length is ~0.5 mm (.020 in) [50].

The dimensions of the circuit in an EIA such as the 94-GHz device described in [47] can be readily estimated. The RF fields in adjacent slots are about 180° out of phase. The beam voltage is 16 kV and the electrons travel at one-fourth of the speed of light. The electrons must reach adjacent slots every half cycle, so, at 95 GHz, the slot spacing is

$$\text{slot spacing} = 0.25\,\frac{c}{2f} = 0.395\,\text{mm} = 0.0156\,\text{inch} \tag{12-19}$$

The average slot length is about a half wavelength long or 1.6 mm (0.064 inch). The beam radius is 0.125 mm (0.005 inch) and the beam focusing flux density is 10,000 G.

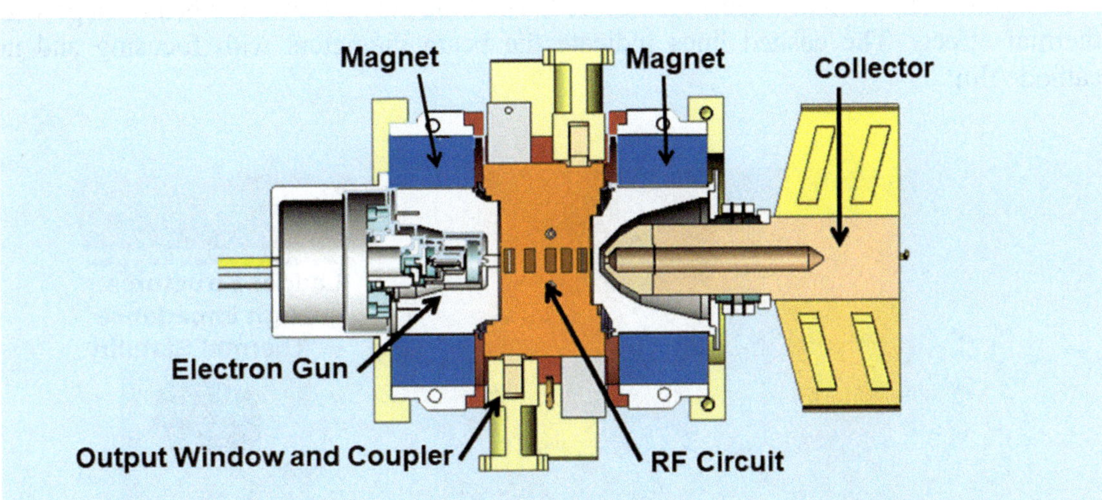

Figure 12-83 Major components of an EIK. (From D. Berry et al., *IEEE Trans. Electron Devices*, Vol. 61 No. 6, June 2014.)

There is a hole for an electron beam in the center of each vane of a ladder circuit. In the most advanced EIK designs, a 100-mA thermal beam is focused through a tunnel with a diameter of 0.005 in (125 μm). The overall length of the RF circuit is maintained below 3 cm and common sets of magnets are used with magnetic fields of approximately 0.9 T. It is also noted that this beam is suitable for terahertz (THz) applications.

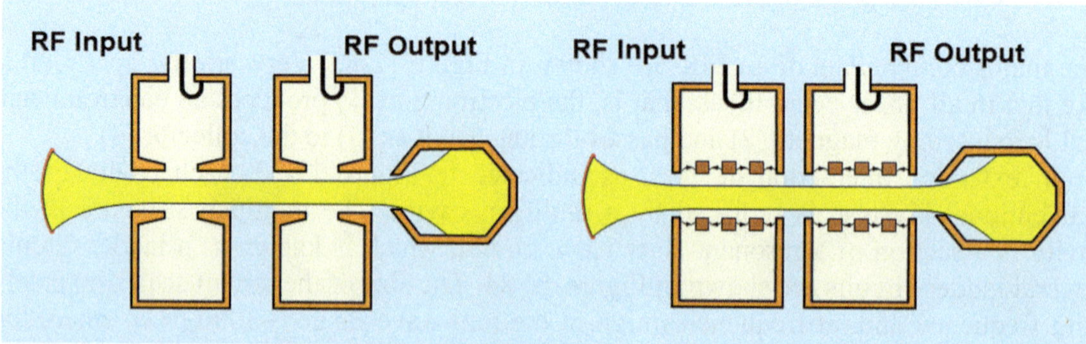

Figure 12-84 Conventional klystron circuits (left) and extended-interaction klystron ladder circuits used in mm-wave EIKs (right). (Courtesy of CPI Canada.)

To help ensure long life, an M cathode is used and the emission density is usually limited to a few A/cm^2. However, the density of the electron beam is usually required to be several hundred A/cm^2 so the gun must have an area compression ratio, A_c, of ~100:1 or more. Because the cathode temperature is ~ 1,000°C, the beam produced by the gun must have a temperature of ~100,000 °C or more and, because the beam diameter is only a fraction of a mm, the beam profile is controlled by thermal effects [48, 49].

The effect of cathode temperature on a small beam with no magnetic field is shown in Figure 12-86. The universal beam spread curve defines the beam shape leaving a Pierce gun when thermal effects are ignored. The blue curve shows the beam shape with thermal effects. The dashed lines indicate the beam diameters with focusing and no cathode flux.

Figure 12-85 Ladder circuits used in mm-wave EIKs. (Courtesy of CPI Canada.)

To limit the expansion of the thermal beam, a magnetic flux density of 0.9 T is used. Samarium cobalt magnets are used to produce this field. The magnetic circuit is designed to provide capability for deliberate localized shunting to deflect the lines of the magnetic field and steer the electrons along the beam tunnel.

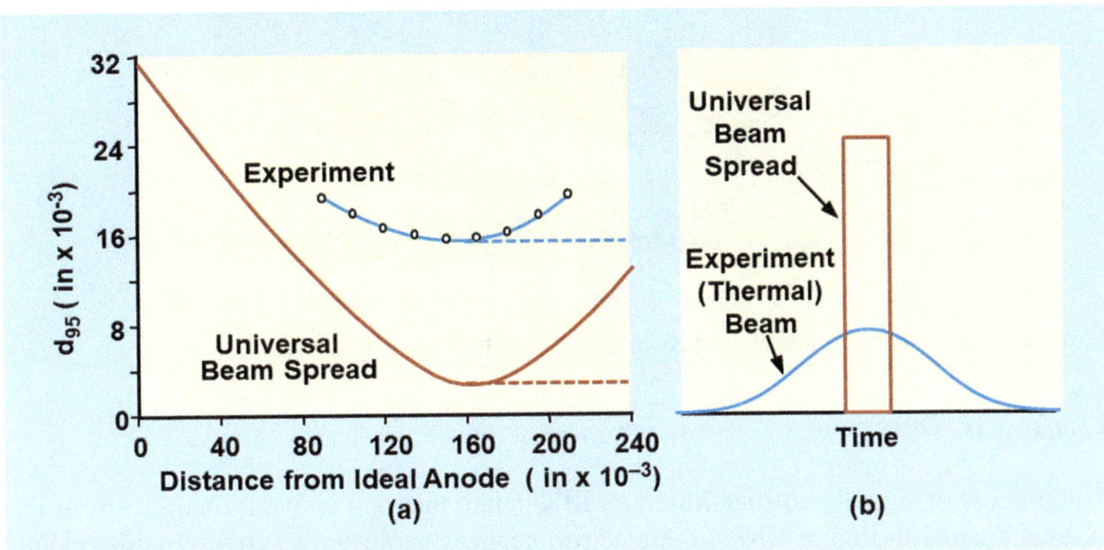

Figure 12-86 Thermal velocity effects on a very small electron beam. (a) Effect on beam diameter. (Adapted from Figure 6-21.) (b) Effect on current density.

A schematic of a two-cavity EIK is shown in Figure 12-83 and a drawing of a three-cavity device is shown in Figure 12-87. The electron beam passes through a hole in the center of the circuit as shown in Figure 12-88.

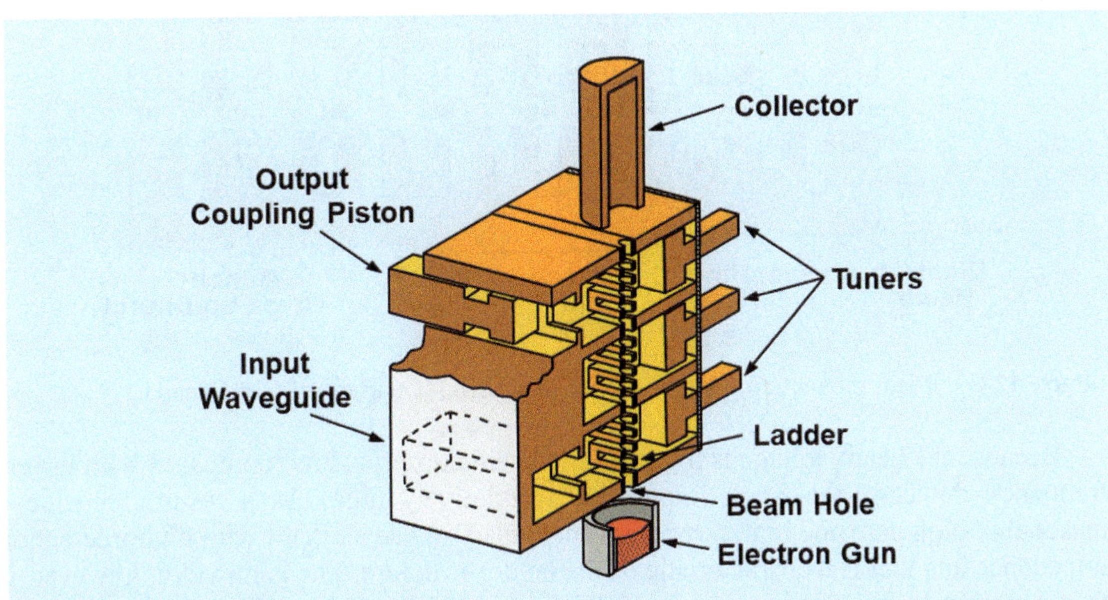

Figure 12-87 Cross-sectional view of an extended interaction amplifier. (Courtesy of CPI Canada.)

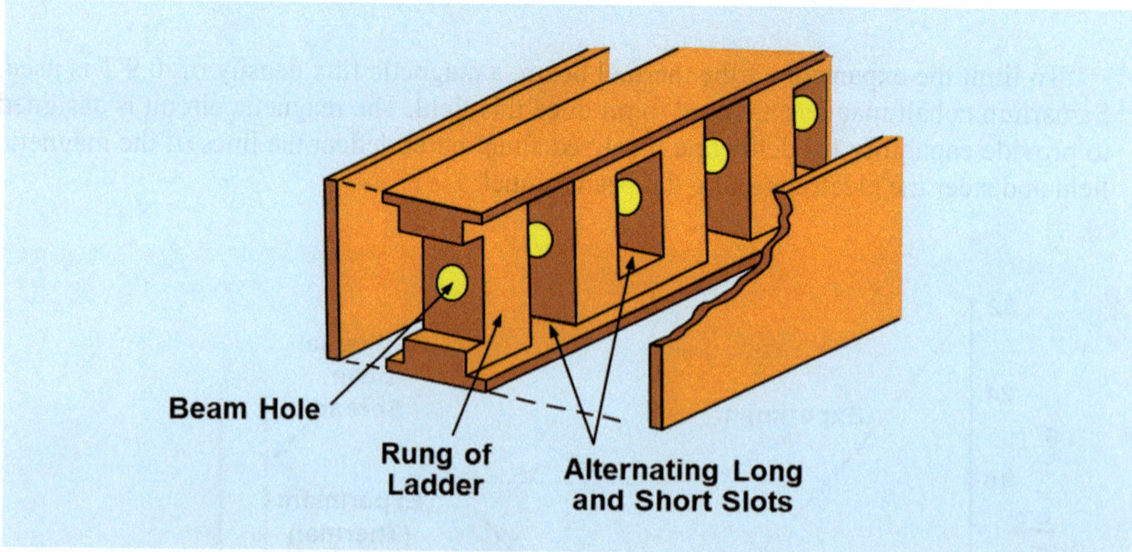

Figure 12-88 Ladder structure used in an EIK. (Courtesy of CPI Canada.)

12.6.2 EIK Operation

The process of signal amplification in an EIK is like that in a conventional klystron. That is, as indicated in Figure 12-89, the electron beam is subjected to an alternating electric field as it enters the RF structure. Some electrons are accelerated and some electrons are decelerated. The accelerated electrons catch up with the decelerated electrons to form bunches. Compared with a conventional klystron, the bunching process is enhanced in the EIK.

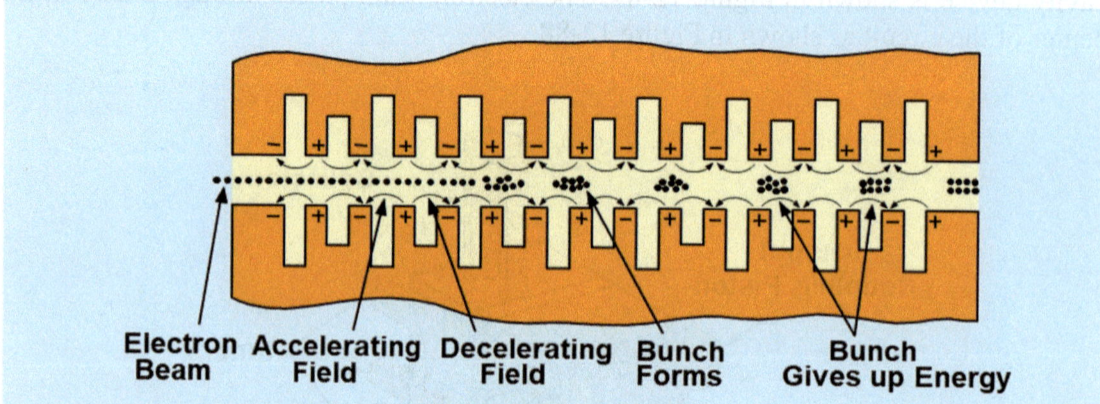

Figure 12-89 Beam interaction with the circuit in an EIK. (Courtesy of CPI Canada.)

Because the beam voltage is high and the beam current is low (compared with lower-frequency devices), the beam impedance is relatively high. As a result, to achieve reasonably high gain and high power, it is necessary to use a circuit with high interaction impedance and that is a characteristic of the ladder structure. The bandwidth, however, is fairly narrow as indicated in Figure 12-90 for a 140-GHz device. Here, at 1 dB below maximum power, the bandwidth is 0.16% [49].

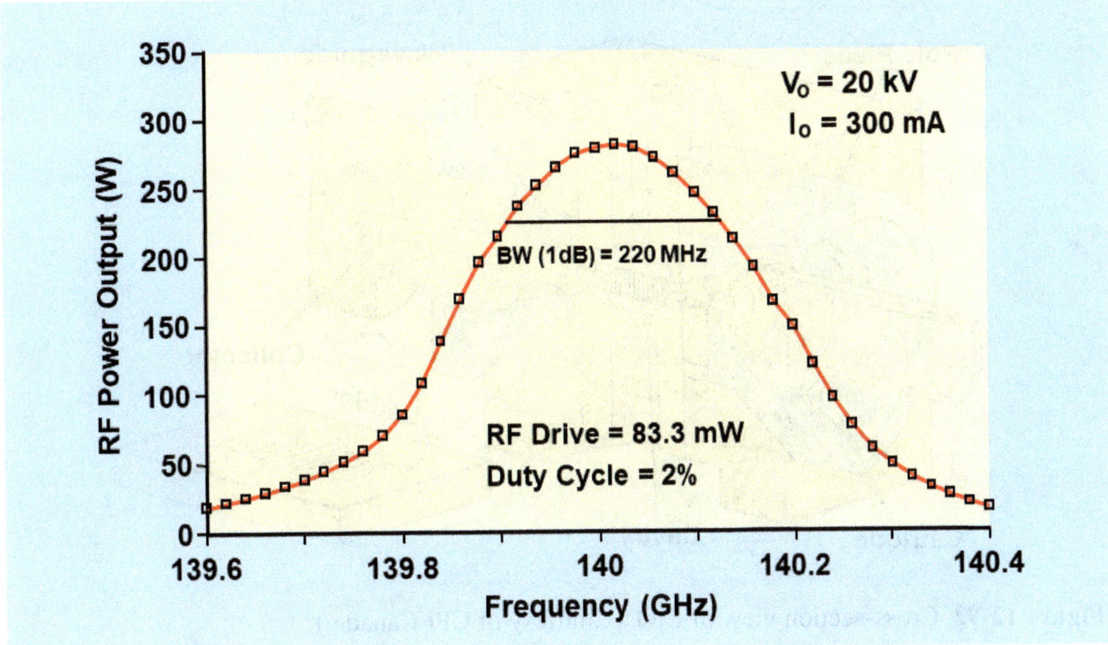

Figure 12-90 Bandwidth characteristic of a 140-GHz EIA. (Courtesy of CPI Canada.)

The EIO is a single-cavity device with interaction gaps that function like a coupled-cavity TWT with extremely strong cavity-to-cavity coupling. With high enough beam currents, oscillations are sustained. The cavity can be mechanically tuned over a bandwidth of about 4%. By varying beam voltage, up to 0.4% tuning is possible. A schematic of an EIO is shown in Figure 12-91 and a cross-section view is shown in Figure 12-92.

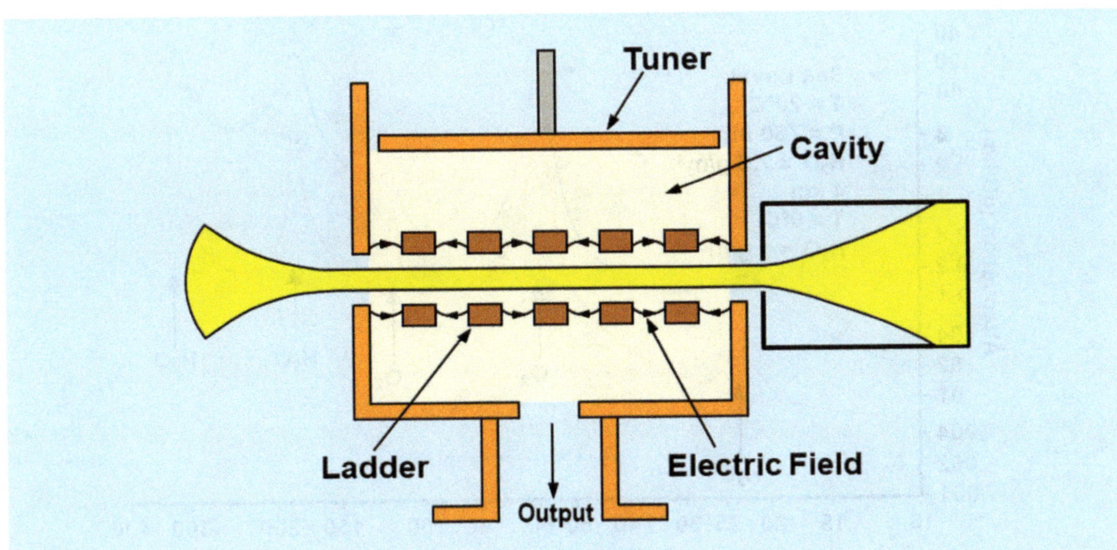

Figure 12-91 Schematic of extended interaction oscillator. (Courtesy of CPI Canada.)

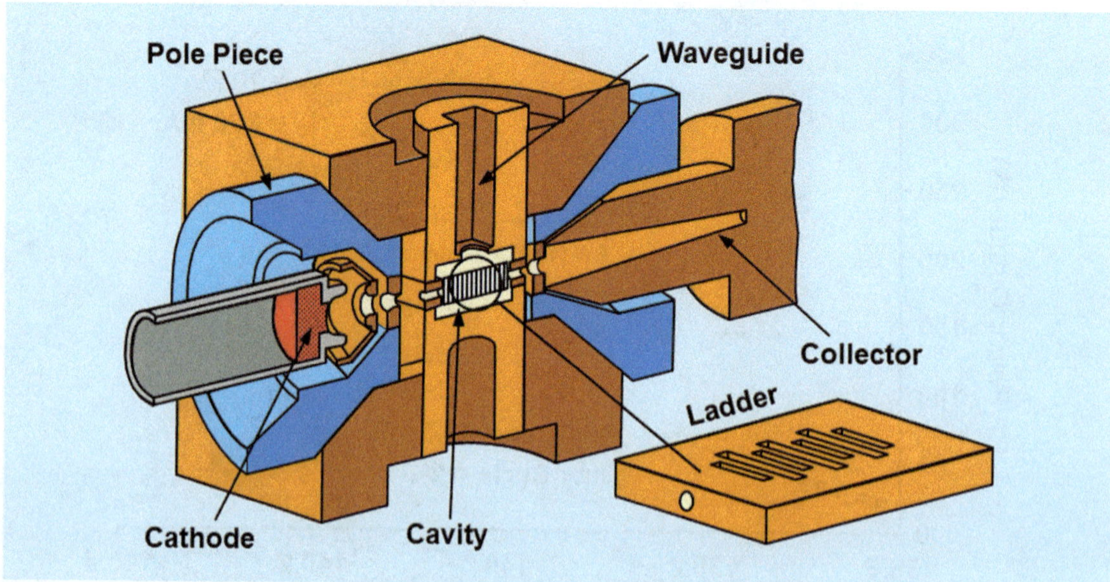

Figure 12-92 Cross-section view of EIO. (Courtesy of CPI Canada.)

12.6.3 Applications

Applications for EIKs include active denial, cloud profiling, tracking small objects in space, SatCom communications, and DNP-NMR spectrometers.

The transmission characteristics of the atmosphere shown in Figure 12-93 are useful in understanding why many of the operating frequencies of EIKs are chosen. Notice that water and oxygen produce substantial attenuation at several frequencies. Recent prestigious applications are for measuring and monitoring the weather activities on earth from satellites in space.

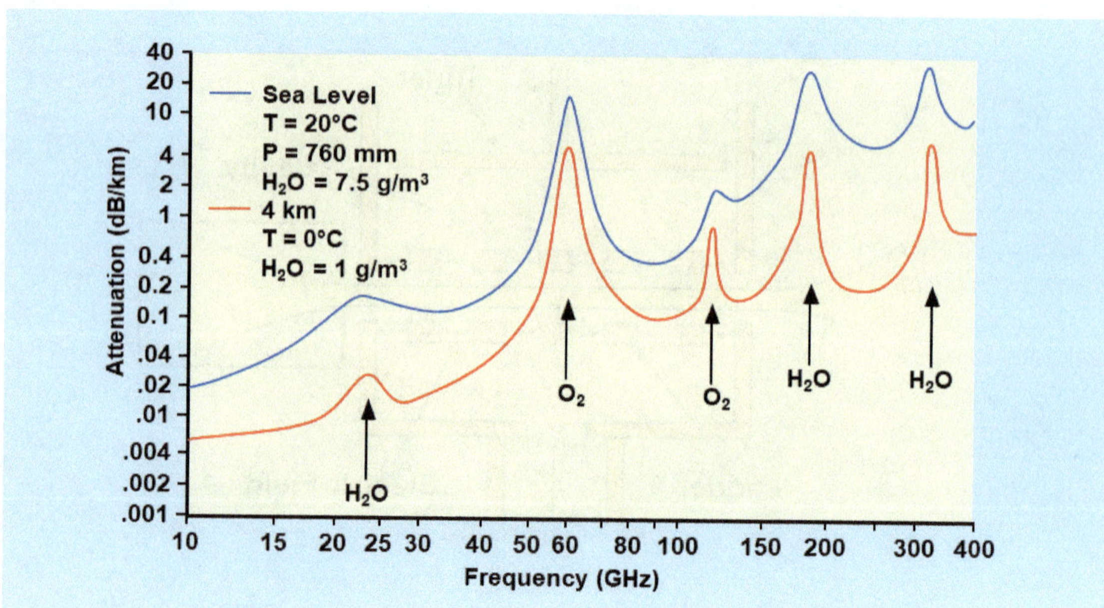

Figure 12-93 Attenuation of the atmosphere. (Federal Communications Commission, Office of Engineering and Technology, Bulletin Number 70, July1997.)

The EIK for which the RF power characteristic is shown in Figure 12-94 was developed for use in the 94-GHz cloud profiling radar on NASA's CloudSAT satellite [48]. This EIK is the first space-qualified high-power W-band amplifier. The device uses a single-stage depressed collector to help achieve an overall efficiency of 32%.

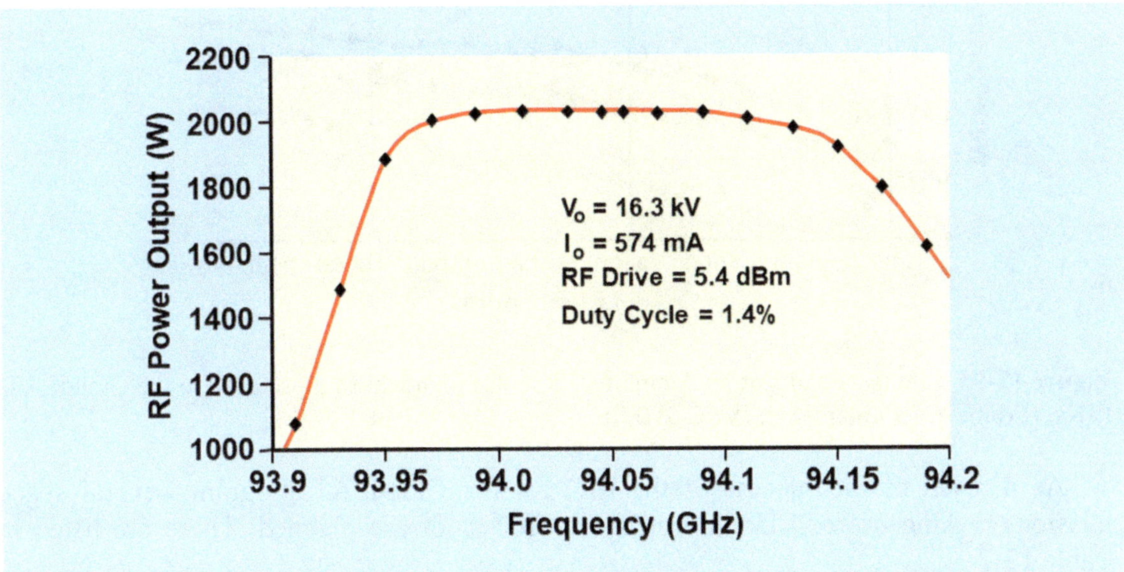

Figure 12-94 RF power characteristic of EIK used for the NASA CloudSAT Program. (Adapted from: A. Roitman et al., *IEEE Trans. Electron Devices*, May 2005. © 2005 IEEE.)

The CloudSAT mission goal was 22,500 hours at a confidence level of 99%. To achieve this mission goal, the life of the EIK and its cathode had to be at least 22,500 hours with a confidence level of 99%. The area convergence ratio of the gun was selected to be 100:1, and this resulted in a cathode emission requirement of 11.3 A/cm^2.

The requirement of 11.3 A/cm^2 for a life of 22,500 hours with a confidence level of 99% is well beyond the capability predicted by Grant and Falce for M cathodes (see Chapter 4). However, Green et al. point out that "a number of improved versions of M cathodes have been invented. Advances in deposition techniques have enabled a high degree of control to be achieved over the composition microstructure, crystal orientation and diffusion properties of the cathode coating. We can now fabricate coated dispenser cathodes with excellent high current density performance and repeatability."

Figure 12-95 shows the result of a life test performed by CPI at 10 A/cm^2 on cathodes identical to those used in the CloudSAT EIKs (operational and backup). After 24,000 hours of continuous operation, the work functions were all low enough to ensure FSCL (fully space charge limited) operation of the cathodes.

The operation of the CloudSAT EIK was flawless. Based on the flawless performance of the radar in orbit and the high data yield, the CloudSAT mission was extended from 2 years to 5 years [49] After 4 years of operation, it had demonstrated approximately 1-dB reduction in power output. In 2016, after 10 years of continuous operation, the CloudSAT satellite was retired from service. Not only was the mission a success, but the operation of the EIK cathode at 11.3A/cm^2, which was well above the life prediction of Grant and Falce, was demonstrated.

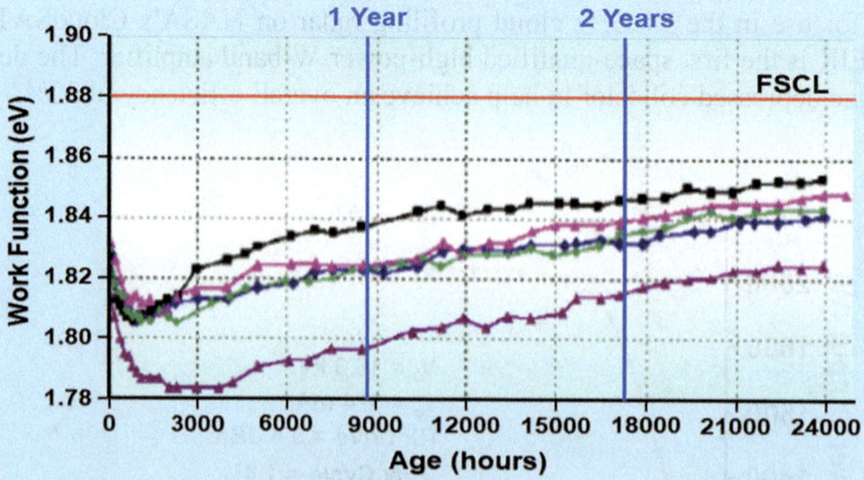

Figure 12-95 Life test results at 10 A/cm2 for cathodes identical to those used in the CloudSAT EIKs. (From: A. Roitman et al. IVEC 2007.)

As a result of the success of the EIK for the CloudSAT program, several space missions making use of EIKs are either in process or are planned. These are listed in Table 12-7.

Table 12-7
Space Missions for CPI EIKs. (Courtesy of CPI Canada)

Mission	Instrument	EIK Specification	Development Status
CloudSAT	Cloud Profiling Radar	94 GHz 1.7 kW Pulsed Amplifier	Flight Models in Operation
EarthCARE	Cloud Radar with Doppler Function	94 GHz 1.7 kW Pulsed Amplifier	Flight Models Delivered
SWOT	Ka-Band Radar Interferometer	35.75 GHz 1.5 kW Pulsed Amplifier	Demonstration Model Fabricated
CoReH2O	Dual Band/Dual Polarization SAR	X & Ku-band 2.5 kW Pulsed Amplifiers	Engineering Models Fabricated and Tested
EGPM	Precipitation Radar	35.5 GHz 1 kW Pulsed Amplifier	EM1 Fabricated and Tested
JIMO	Topo-Mapping Radar	35.5 GHz 3 kW High Duty Amplifier	Development of High Duty Beamstick Complete
ATOMS	Active Atmospheric Sounder	183 GHz 5 W CW Oscillator	EM1 Delivered to JPL
MetOpSG	Scatterometer	5.3 GHZ CW Pulsed Amplifier	Pre-Development

Flight EIKs having improved lifetime design have been delivered for the EarthCARE mission. For the same output power as the CloudSAT EIK, the cathode current density was reduced by 25% with expected life extension by 30% to 50%. A launch of EarthCARE is scheduled for June 2022.

The Surface Water and Ocean Topography (SWOT) mission is a joint mission between NASA, the Centre Nationale d'Etudes Spatiale (CNES), and the Canadian Space Agency (CSA). The mission is managed by the Jet Propulsion Laboratory (JPL) and is scheduled for launch in April 2021. A synthetic aperture Ka-band radar interferometer will be used to determine the heights of inland water bodies, major rivers and mesoscale ocean topographic features. An EIK and its high-voltage power supply will provide 1,500 watts of pulsed RF power from the radar.

The height measurements will be made by detecting subtle differences in the phase of radar returns received at two antennas separated by 10 meters. The phase error budget is such that the allocation of phase error for the EIK amplifier is very tight. This results in tight amplitude and phase pushing factors for the EIK and in tight voltage stability requirements for the power supply. A demonstration model and its power and phase characteristics are shown in Figure 12-96 along with an artist's rendition of the SWOT satellite [50]. The SWOT antennas are mounted on arms that extend out to the right and left.

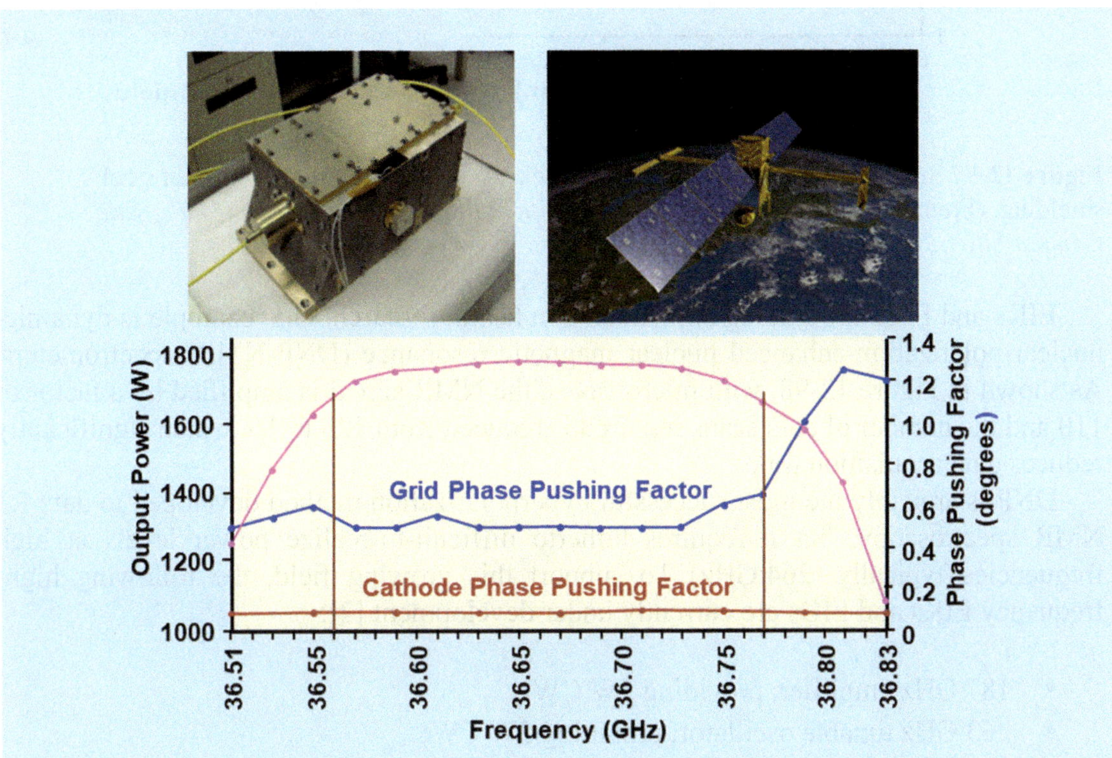

Figure 12-96 Power output and phase pushing for the demonstration model of the CPI 35.75 GHz EIK (upper left) to be used on SWOT (Surface Water and Ocean Topography) mission (upper right). The demonstration model operated at a slightly higher frequency than the SWOT EIK. (Adapted from W. Harvey et al., IVEC, 2016.)

Several components on the SWOT satellite, including RF components (ferrites, isolators, circulators) and magnetometers for attitude control, are sensitive to magnetic fields. To reduce the stray field produced by the EIK, two EIKs are used and these are oriented opposite to each other to produce some field cancellation. Orientation alone is not sufficient to adequately reduce the stray EIK fields so shielding is also required. Figure 12-97 shows the effect of shielding with 1.5-mm steel. At a distance of 24 cm from the EIK wall, the field level is dropped to a value comparable to the earth's field.

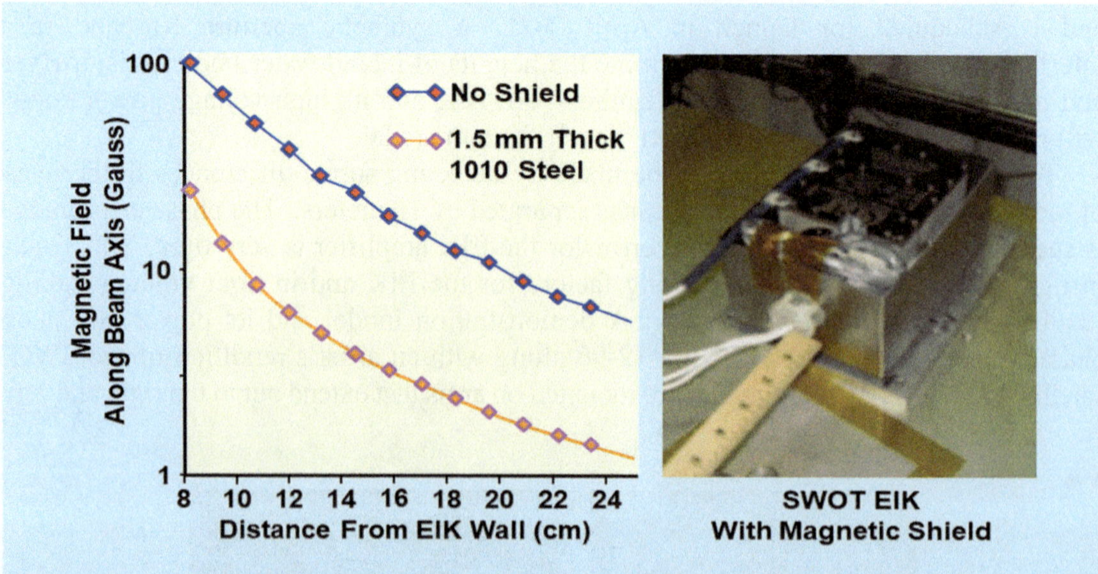

Figure 12-97 Stray magnetic field for back-to-back SWOT EIKs with and without steel shielding. (From W. Gonzales et al, *IEEE International Symposium. on Electromagnetic Compatibility*, July, 2016.)

EIKs and EIOs are finding applications in health research. One example is dynamic-nuclear-polarization-enhanced nuclear magnetic resonance (DNP-NMR) spectrometers. As shown in Figure 12-98, with microwaves, the NMR signal is amplified by a factor of 110 and the number of data scans required is reduced from 128 to 16, which significantly reduces data acquisition time.

DNP is arguably the most successful hyperpolarization method developed to date for NMR spectroscopy, but it requires hitherto difficult-to-realize power levels at high frequencies (typically ~264 GHz). To support this growing field, the following high-frequency EIKs and EIOs are currently under development [2]:

- 187 GHz amplifier, providing 5W CW;
- 263 GHz tunable oscillator, providing 5W CW;
- 264 GHz amplifier, providing 20W pulsed.

In order to be suitable for DNP, the millimeter-wave source must be highly stable in frequency and amplitude. Also, the electron spin resonance (ESR) spectrum, which is excited by the RF, may span hundreds of megahertz, so having wide bandwidth or a tunable source is also desirable. In addition, the RF source should be compact in order to reside close to the NMR magnet, thereby minimizing the significant waveguide losses that occur at millimeter and submillimeter wavelengths.

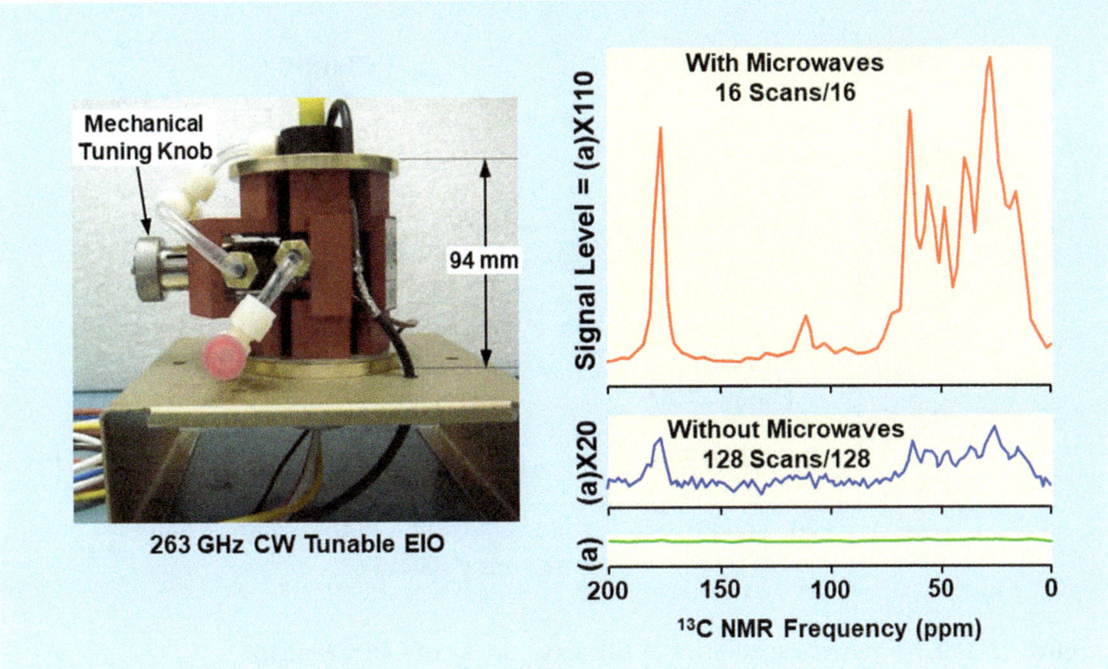

Figure 12-98 DNP/NMR signal is enhancement with a tunable 263-GHz EIO.)From R. MacHattie, 2014 COST Hyperpolarization.)

With proper design and low gas content, both the EIO and the EIA have very good phase noise performances [48]. Figure 12-99 contains plots of noise characteristics of several EIKs.

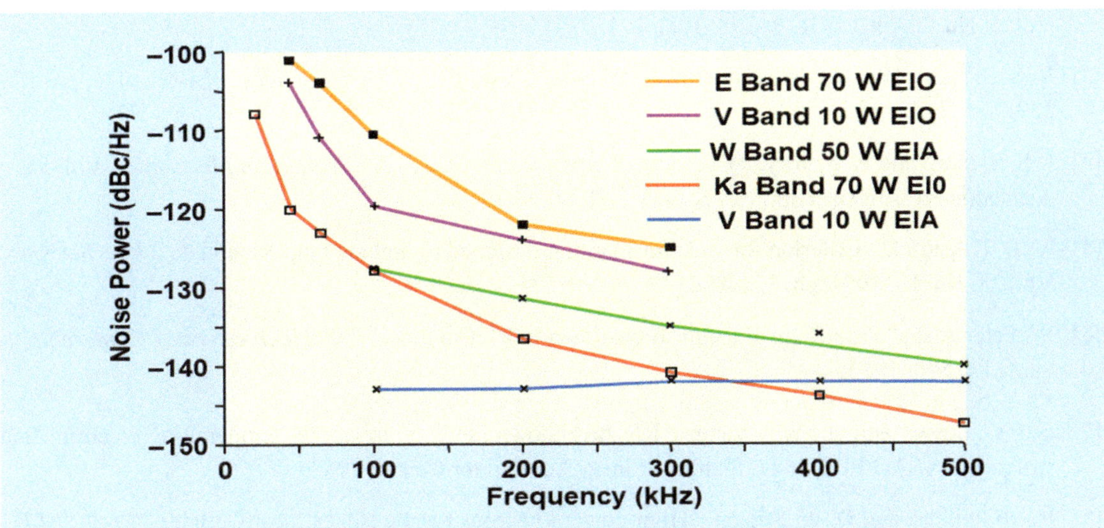

Figure 12-99 Noise performance of several EIKs. (Courtesy of CPI Canada.)

Figure 12-100 displays the RF power capabilities of many of the EIKs manufactured by CPI Canada. Power levels range from 3 kW at 30 GHz to 30W at 280 GHz. The state of the art is clearly represented by the CloudSAT device with 2 kW at 95 GHz.

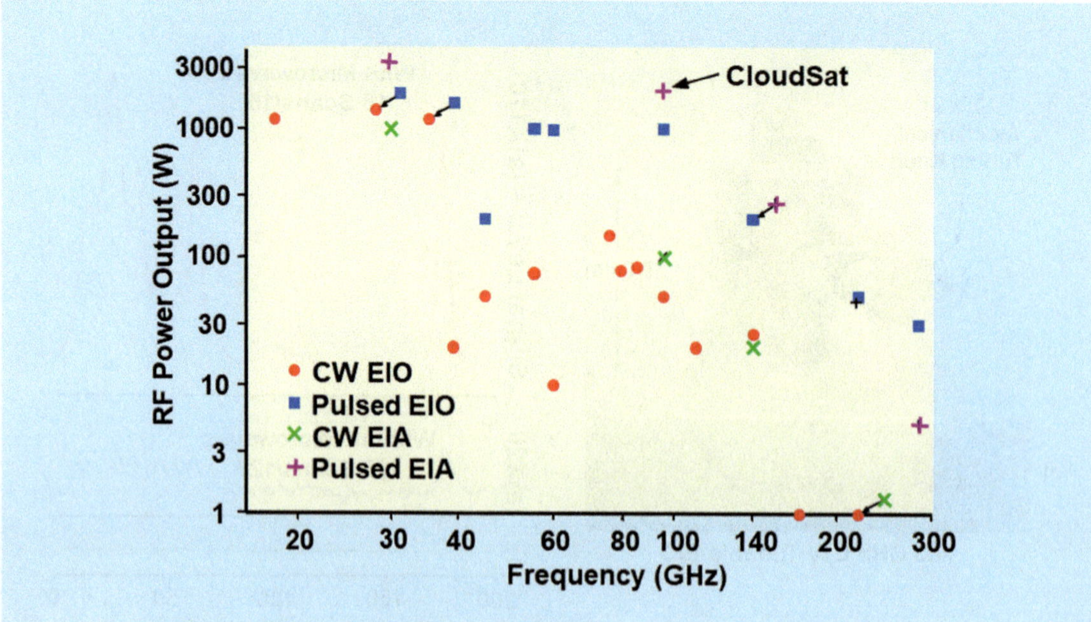

Figure 12-100 RF power capabilities of EIKs. (Courtesy of CPI Canada.)

REFERENCES

[1] T. G. Mihran, "The effect of space charge on bunching in a two-cavity klystron," *IRE Trans. Electron Devices*, Vol. ED-6, No. 1, January 1959, pp. 54–64.

[2] S. E. Webber, "Ballistic analysis of a two-cavity finite beam klystron," *IRE Trans. Electron Devices*, Vol. 5, No. 2, April 1958, pp. 98–108.

[3] R. S. Symons, "Scaling laws and power limits for klystrons," *Technical Digest,* IEDM, 1986, pp. 156–159.

[4] R.S. Symons and H. R. Jory, " Cyclotron Resonance Devices," in *Advances in Electronics,* Vol. 55, Academic Press, New York, N.Y. (1981).

[5] W. D. Kilpatrick, "Criterion for vacuum sparking designed to include both RF and dc," *Rev. Sci. Inst.*, Vol. 28, No. 10, 1957, pp. 824–826.

[6] W. Peter et al., "Criteria for Vacuum Breakdown in RF Cavities," *Particle Accelerator Conference*, Santa Fe, NM, 1983.

[7] G. A. Loew, and J. W. Wang, "RF breakdown studies in room temperature electron linac structures" SLAC PUB 4647: Stanford Linear Accelerator Center, 1988.

[8] R. M. Phillips, and D. W. Sprehn, "High-power klystrons for the Next Linear Collider," *Proc. IEEE*, Vol. 87, No. 5, May 1999, pp. 738–751.

[9] PUB-16045. Presented at the 2nd International Particle Accelerator Conference (IPAC 2011) San Sebastian, Spain, September 4 - 9, 2011.

[10] R. True, "Beam optics calculations in very high-power microwave tubes," *Technical Digest,* IEDM, 1991, pp. 403–406.

[11] Sprehn, R. M. Phillips, and G. Caryotakis, "The design and performance of 150-MW S-band klystrons," SLAC-PUB-6677, Stanford Linear Accelerator Center, September 1994.

[12] W. Luebke, and G. Caryotakis, "Development of a one megawatt CW klystron," *Microwave Jour.*, Vol. 9, No. 8, August 1966.

[13] G. Caryotakis, "high power klystrons: theory and practice at the Stanford Linear Accelerator Center," SLAC-PUB 10620 August 2004, revised January, 2005.

[14] A. Balkcum, "Design and operation of a 100 kW CW X-band klystron for spacecraft communications," IVEC 2012.

[15] Thales, "TH 2089 High power CW klystrons,"

[16] G. Faillon, "A 200 kW S band klystron with TWT bandwidth capability," *Technical Digest,* IEDM, 1985.

[17] A. Staprans, E. W. McCune, and J. A. Ruetz, "High-power linear-beam tubes," *Proc. IEEE,* Vol. 61, No. 3, March, 1973, pp. 299–330.

[18] R. L. Metivier, "Broadband klystrons for multimegawatt radars," *Microwave Jour.*, April 1971, pp. 29–32.

[19] F. I. Friedlander, "High-efficiency broad-band klystron," U.S. Patent 4,764,710, August 16, 1988.

[20] B. Stockwell, "Tapered drift length klystron," 1996 Power Microwave Tube Confrence, Monterey, CA, April 1996.

[21] M. Chodorow, and T. Wessel-Berg, "A high-efficiency klystron with distributed interaction," *IRE Trans. Electron Devices*, Vol. ED-8, No. 1, January 1961, pp. 44–55.

[22] R. S. Symons et al., "An experimental clustered-cavity klystron," *Technical Digest,* IEDM, 1987, pp. 153–156.

[22] R. S. Symons, and J. R. M. Vaughan, "The linear theory of the clustered-cavity klystron," *IEEE Trans. Plasma Science*, Vol. 22, No. 5, October 1994, pp. 713–718.

[23] J. G. Siambis and R. S. Symons, "Ultra-wideband clustered-cavity klystron," *Ultra-Wideband, Short-Pulse Electromagnetics,* Plenum Press, 1993.

[24] M. Chodorow, and R. A. Craig, "Some new circuits for high-power traveling-wave tubes," *Proc. IEEE*, Vol. 45, No. 8, August 1957, pp. 1106–1118.

[25] E. L. Lien, "High efficiency klystron amplifiers," *Proc. 8th International Conference on Microwave and Optical Generation and Amplification,* Amsterdam, the Netherlands, 1970.

[26] H. G. Kosmahl, and L. U. Albers, "Three-dimensional evaluation of energy extraction in output cavities of klystron amplifiers," *IEEE Trans. Electron Devices*, Vol. ED-20, No. 10, October 1973, pp. 883–890.

[27] H. G. Kosmahl, "Analytic investigation of efficiency and performance limits in klystron amplifiers using multidimensional computer programs; multistage depressed collectors; and thermionic cathode life studies," *Solar Power Satellite Microwave Power Transmission and Reception*, NASA CP 2141, 1980, p. 206.

[28] I. Syratchev, High efficiency klystron technology,, International Workshop on Breakdown Science and High gradient technology, June 13-16, 2017, Valencia, Spain.

[29] A. Baikov, C. Marrelli, and I. Syratchev, "Toward high-power klystrons with rf power conversion efficiency on the order of 90%," *IEEE Trans. Electron Devices*, Vol. 62, No. 10, October 2015, pp. 3406–3412.

[30] Chodorow, Lecture notes Stanford University course no. EE326B.

[31] V. Hill, C. Marrelli et al., "Particle-in-cell simulation of the third harmonic cavity f-tube klystron," IVEC 2016, IEEE

[32] I. A. Guzilov, O. Y. Maslennikov and A. V. Konnov, "A way to increase the efficiency of klystrons," Vacuum Electronics Conference (IVEC), 2013 IEEE 14th International, Paris, 2013, pp. 1-2.

[33] V. F. Kovalenko, "The electron tube of klystron type," Author's certificate, No. 72 756, class 21.13, 1940.

[34] J. Bernier, "Multiple beam device," French patent 992,853, September 15, 1944.

[35] M. R. Boyd et al., "The multiple beam klystron," *IRE Trans. Electron Devices,* Vol. ED-9, No. 3, May 1962, pp. 247–252.

[36] W. J. Pohl, "The design and demonstration of a wideband multiple-beam traveling-wave klystron," *IEEE Trans. Electron Devices*, Vol. ED-12, No. 6, June 1965, pp. 351–368.

[37] S. A. Zusmanovsky, "The multiresonator klystron," Author's certificate No. 15 556, January 5, 1955.

[38] S. V. Korolyov, "About the possibility of reducing the transit klystron weight and dimensions," *Elektron Tekh*, Vol. ser. 1, No. 9, 1968, pp. 176–184.

[39] A. N. Korolyov et al., "Multiple-beam klystron amplifiers: performance parameters and development trends," *IEEE Trans. Plasma Science*, Vol. 32, No. 3, June 2004, pp. 1109–1118.

[40] A. S. Pobedonostsev et al., "Multiple-beam microwave tubes," *Microwave Theory and Techniques Conference*, Atlanta, GA, June 1993.

[41] C. Bastien, M. Bres, and G. Faillon, "Compact multibeam klystron study," *Final Report no. SLCET-TR-87-0713-F*, March 1989.

[42] R. L. Ives et al., "Development of multiple beam guns for high-power RF sources for accelerators and colliders," *Proc. 6th Workshop on High Energy Density and High Power RF,* Berkeley Springs, WV, 2003.

[43] A. Beunas et al., "A high efficiency long pulse multi beam klystron for the TESLA linear collider," mdk2001.web.cern.ch/mdk2001/Proceedings/Session12/beunas.pdf, 2001.

[44] tesla.desy.de/new pages/Hamburg meeting 4 2004/itrp public program.htm," 2004.

[45] K. T. Nguyen et al., "High-power four-cavity S-band multiple-beam klystron design," *IEEE Trans. Plasma Science*, Vol. 32, No. 3, June 2004, pp. 1119–1135.

[46] C. Nilsen, "95 GHz 1 kW peak high performance radar transmitter," *Proc. Military Electronics 90*, London, 1990, pp. 115–120.

[47] A. Roitman, D. Berry, and B. Steer, "State-of-the-art W-band extended interaction klystron for the CloudSAT Program," *IEEE Trans. Electron Devices*, Vol. 52, No. 5, May 2005, pp. 895–898.

[48] C. Nilsen, "Low noise mm wave transmitted tube oscillators," *Microwave Jour.*, July 1988, pp. 91–97.

Chapter 13

Traveling-Wave Interaction

13.1 INVENTION OF THE HELIX TWT

There are at least three parties that claim credit for the invention of the TWT [1].

13.1.1 Haeff

The possibility of a traveling-wave interaction between an electron beam and a nearby RF circuit was recognized in 1933 by Haeff [2-4]. In patents filed in that year, Haeff described electron beam deflection tubes to be used as detectors or oscillographs that contained some of the features of modern helix traveling-wave tubes. The basic concept of Haeff's devices is shown in Figure 13-1. An electron beam (11) passing between two circuits (6, 7) to which an RF signal (22, 23) has been applied is deflected transversely. The beam is collected alternately by collectors 1 and 2. The signal output from the collectors is the detected amplitude of the RF input signal.

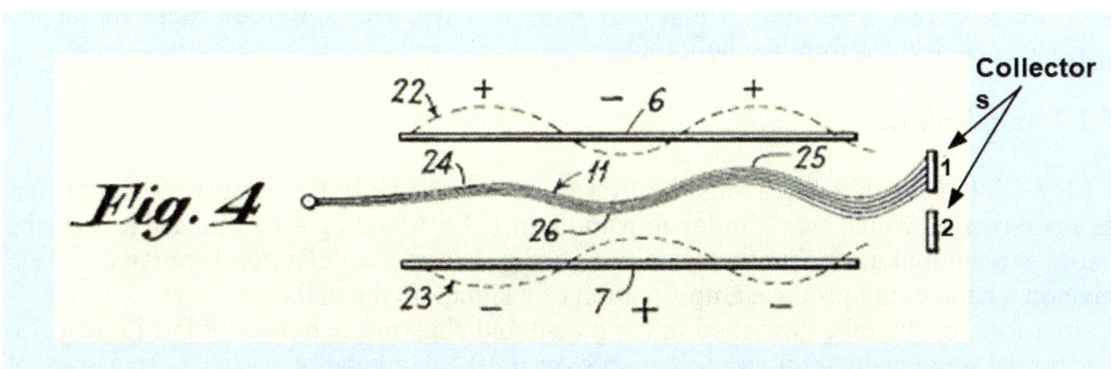

Figure 13-1 Transverse beam deflection in A. V. Haeff's traveling-wave detector: Patent filed October 1933.

In one of Haeff's devices, which is shown in Figure 13-2, a wave on a helix (126) applies an electric field transverse to the direction of motion to a nearby hollow electron beam (123) surrounding the helix. The beam and wave travel at the same velocity so the diameter of the beam is modulated by the wave. This causes the beam to expand and contract and thereby sweep back and forth across collector electrodes 124 and 125.

As a result, the signal output from electrode 124 and 125 is the detected amplitude modulation on the RF signal applied to the device at 128.

The significance of the Haeff devices is that Haeff used a slow wave circuit (helix in Figure 13-2) to slow the wave velocity and obtain synchronism between the velocities of the wave and the beam. Thus, the Haeff devices were traveling-wave tubes, however, they differed from modern traveling-wave devices in that they used a transverse interaction of the beam with a circuit rather than a longitudinal interaction.

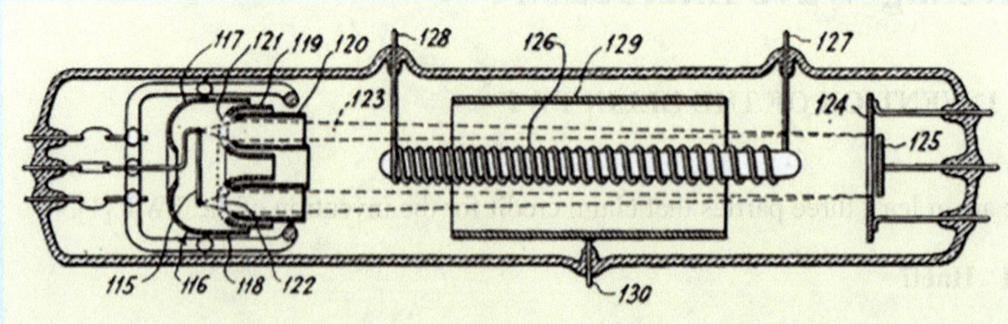

Figure 13-2 A. Haeff's traveling wave detector: Patent filed October 1933.

There is no reference in the Haeff patents to the longitudinal modulation (bunching) used in all devices that are called traveling wave tubes today. Also, as observed by Wathen [1], "Though the tubes described by Haeff at this time contain several of the features common to traveling-wave amplifiers which do, indeed, utilize the true traveling or growing wave mechanism of amplification, there is no hint in their description that these tubes operate according to that principle, or that useful amplified radio frequency energy can be derived from the helices."

13.1.2 Lindenblad

In May 1940, Lindenblad [5] was the first to describe a helix traveling-wave amplifier, the operation of which was similar to a modern helix traveling wave tube. He was the first to explain that a synchronous axial interaction between an electron beam and the RF wave on a helix could produce amplification of a signal on the helix.

In a form of the tube described in the patent and shown schematically in Figure 13-3, Lindenblad produced useful amplification over a 30-MHz band at a carrier frequency of 390 MHz. In this tube the signal was applied to the beam with a grid in the gun. In other experiments, the signal was applied to the helix.

It is interesting that Lindenblad recognized the value of tapering the pitch of the helix to maintain synchronism with the electron beam, the velocity of which is reduced as energy is extracted to produce amplification. He observed that without a gradual tapering of the helix, the decrease in the average electron velocity may cause a cumulative phase drag that becomes large enough to cause acceleration of electrons by the wave on the helix. This is undesirable because some of the RF circuit power would be returned to the beam.

Wathen reported that the Lindenblad tube was developed in connection with transmitter research in which Haeff's inductive output tubes were being used as oscillators and amplifiers. The first Lindenblad tube was apparently a modified version of one of Haeff's inductive output tubes. Lindenblad extended the glass vacuum envelope of the Haeff tube and replaced the cavity resonator with a helix. The helix was wound around the outside of the glass envelope and had a pitch such that the axial velocity of a wave on the helix was equal to the velocity of an electron beam inside the envelope. The tube was excited by placing the output of a small triode oscillator on the control grid or by coupling the input signal to the input end of the helix. This test represented the first known demonstration of the practical use of the traveling-wave principle in an amplifier characterized by the inclusion of all of the significant features of the modern helix traveling-wave amplifier.

In his patent, in addition to the helix, Lindenblad described the use of a helical waveguide as the slow wave circuit. He also stated that the helical conductors can, if desired, be placed within the envelope of the tube and surround the electron beam.

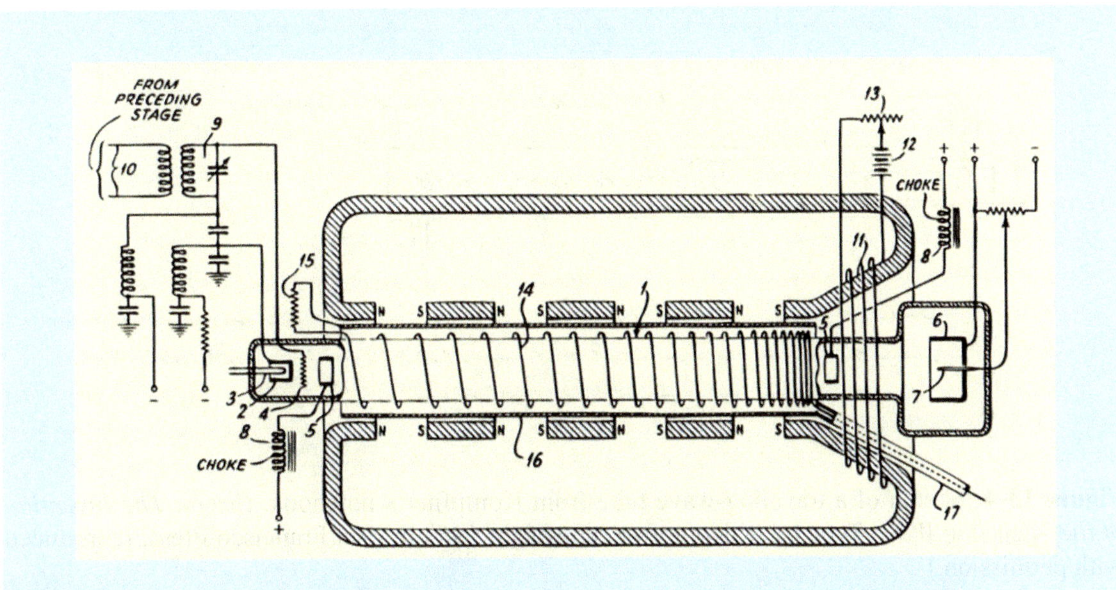

Figure 13-3 N. E. Lindenblad's traveling-wave amplifier: Patent filed May 1940.

13.1.3 Kompfner

According to Wathen, the next serious research on traveling-wave tubes appears to have begun in England at Birmingham University in 1943 and was carried on subsequent to 1944 at the Clarendon Laboratory of Oxford University. In this research, Kompfner [6–10], who was apparently unaware of Lindenblad's earlier work, reported that he realized in late 1942 (shortly after Lindenblad's patent was issued) that the basic growing-wave principle of the magnetron could be used for amplification of RF signals. His objective was to develop an amplifier capable of a sensitivity and noise figure comparable to the best of the crystal-mixer receivers available at that time. Shown in Figure 13-4 is Kompfner's first sketch of a traveling-wave tube.

Kompfner studied the possibilities of using the helix as a nonresonant slow-wave circuit by examining the fields within the helix and between the helix and the outer conductor surrounding it. He did this by observing the deflection on a fluorescent screen of an electron beam when RF power was sent through the helix. He discovered that the axial field near the helix was much more intense than that near the axis. Extensive preliminary studies were carried out on a large helix using probe measurements. The feasibility of using the helix was demonstrated.

Kompfner built his first traveling-wave tube in early 1943. He summarized the operation of the tube as follows:

> When the radio frequency power emerging from the helix with the beam switched on was compared with the radio frequency power without the beam, it was found that, at a beam voltage of 2440 volts, there was an increase of 49%, while at a beam voltage of 2200 volts, there was a decrease of 40%.

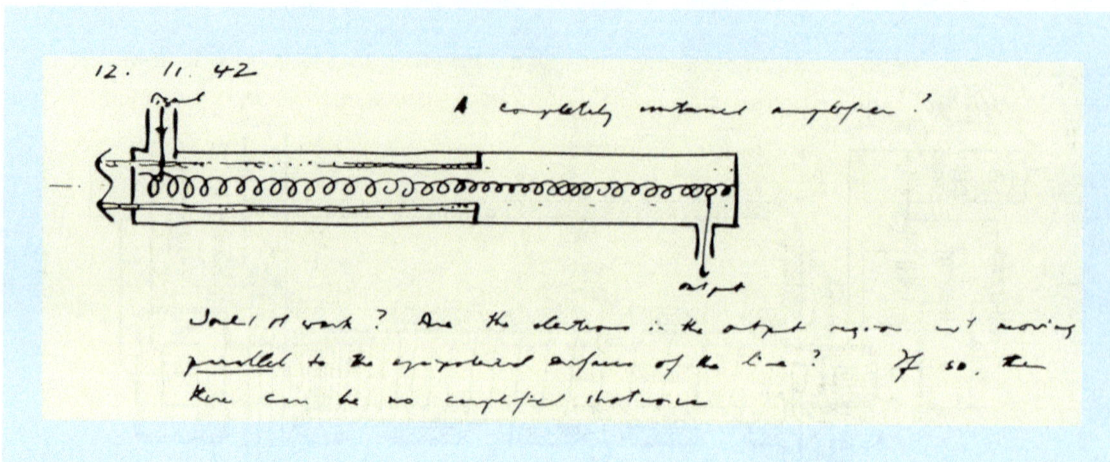

Figure 13-4 Sketch of a traveling-wave tube from Kompfner's notebook. (From: *The Invention of the Traveling Wave Tube* by R. Kompfner, copyright 1963 by San Francisco Press, reproduced with permission.)

After various measurements on density modulation as a function of beam potential, and after observations on shot noise, a tube was designed and built. The tube shown in Figure 13-5 was successfully tested in December 1943. With an electron beam current of 119 μA at 1,830V, a power amplification of 6 was achieved at a frequency of 3.3 GHz. The noise factor was 14 dB and was observed to depend on the fraction of the total current that arrived at the collector electrode. A further improvement in design increased power amplification to 14 and reduced noise factor to 11 dB. In theoretical studies completed before 1944, Kompfner developed first-order theories for power amplification and noise factor. The important result was that the wave tends to grow exponentially with distance along the transmission line.

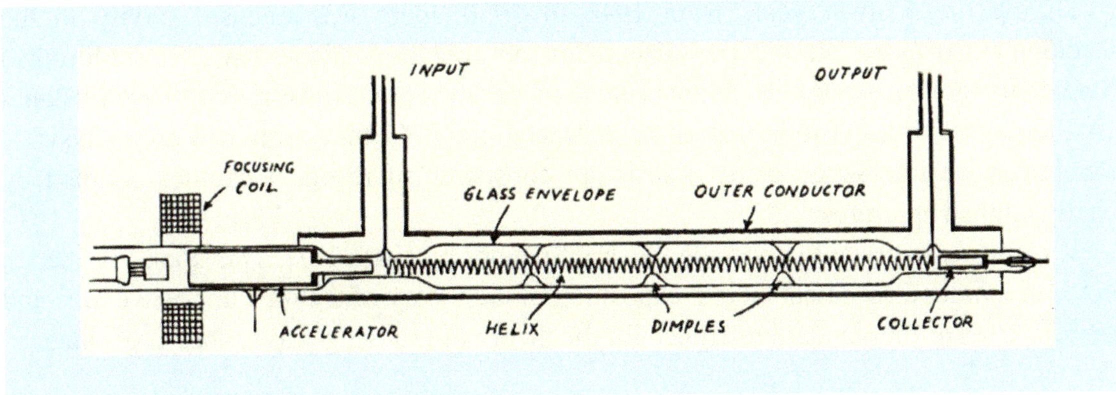

Figure 13-5 Kompfner's experimental helix amplifier. (From Rudolf Kompfner, *Proc. IRE*, February 1947. ©1947 IRE (now IEEE).)

13.1.4 Early Work in the United States

The first public announcement of the British wartime work on helix traveling-wave tubes was made at the Fourth Institute of Radio Engineer's Electron Tube Conference at Yale on June 27 and 28, 1946. At the same conference, work in the United States by J. R. Pierce and L. M. Field at Bell Telephone Laboratories was described. Subsequently [11], Pierce and Field described helix traveling-wave tubes like that shown in Figure 13-6 (figure taken from Patent 2,602,148 by J. R. Pierce [12]). One unique feature of this device, described by Pierce, was the use of longitudinal insulating rods for supporting and accurately positioning the helix. Another was the use of a system of solenoids to produce a uniform magnetic field for focusing the electron beam. Pierce also described techniques for providing loss to suppress backward-traveling waves and oscillations.

In the patent by Field [13], the suppression of oscillations through the use of thin layers of colloidal graphite on the ceramic support rods for the helix is described. Near the midpoint of the helix, the conductivity of the coating was increased to provide the proper dissipation of reflected energy with a minimum of reduction in the gain of the forward-growing wave.

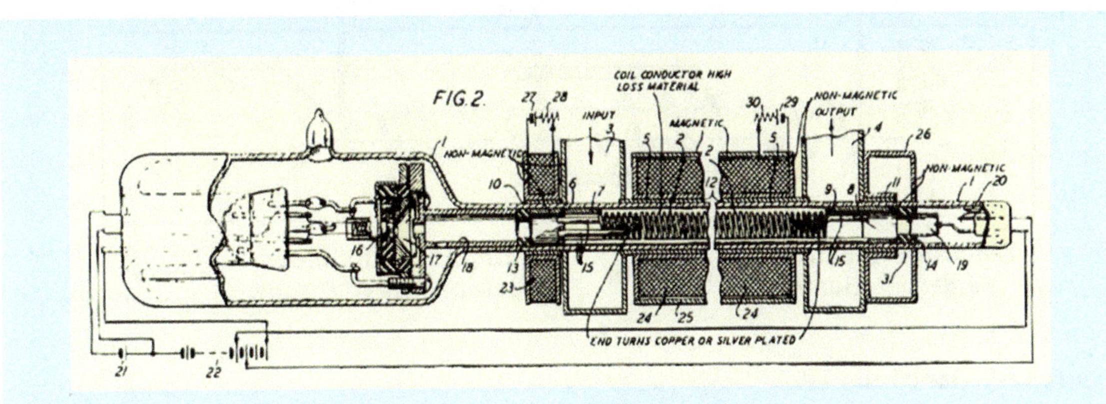

Figure 13-6 Helix amplifier shown in U. S. Patent 2,602,148 by J. R. Pierce.

During the postwar years from 1946 to 1950, there was intense activity in the development of adequate theories to explain the operation of the TWT. In addition to Pierce's theoretical work [14], noteworthy contributions [15, 16] were made by A. Blanc-LA Pierre and P. Lapostolle in France in December 1946 that contained an analysis of traveling-wave interaction using a structure consisting of resonant cavities coupled by axially aligned apertures.

It seems appropriate to conclude this brief history by citing [17]. This book (by Pierre) provided a unified coordination and exposition of the theory of the traveling-wave tube and has been used as the standard reference for the theory of traveling-wave tubes ever since.

13.2 BASIC OPERATION OF THE HELIX TWT

There are two basic categories of TWTs. The helix TWT, the main elements of which are shown in Figure 13-7, is a relatively low-power (usually tens to hundreds of watts) broadband (over 2 octaves is possible) device. Ring bar and ring loop TWTs are included in this category. The second TWT category is coupled-cavity TWTs. Fundamental forward-wave TWTs are capable of powers of megawatts, but bandwidth is limited to about 10%. TWTs with a fundamental backward wave circuits that are operated in the ford wave mode are capable of powers up to several hundred kilowatts and bandwidths up to 30% or more. TWTs with folded waveguide circuits are in this category.

Conceptually, the electron gun, electron beam, and collector are the same for all TWTs. However, there are substantial differences in the RF circuits. In spite of the differences, the operating principles are the same. For convenience, the helix is used in the following discussions.

The following discussion is a qualitative explanation of the operation of a TWT. Elements of the classical Pierce theory of TWT operation are included to support explanations. Also, many of the parameters defined by Pierce are now part of the accepted vocabulary of the TWT industry. These parameters are described in [1].

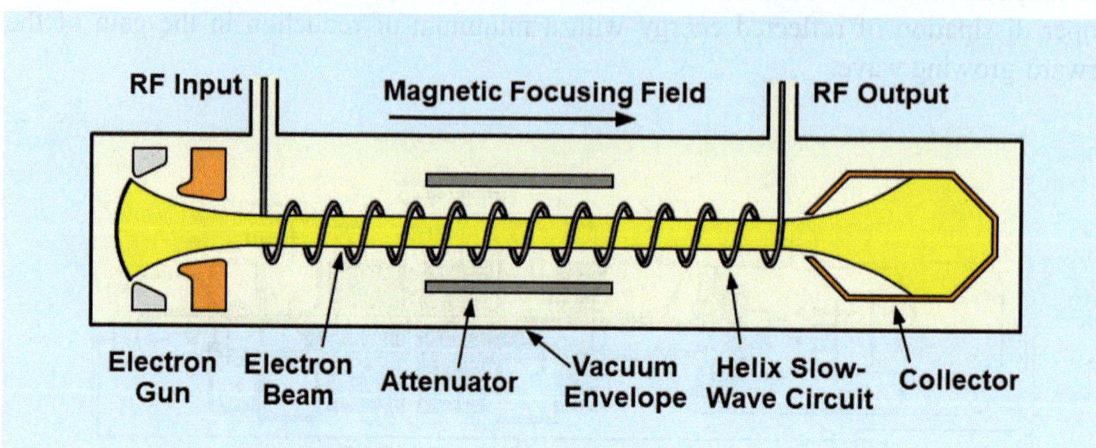

Figure 13-7 Basic helix TWT.

13.2.1 The Helix

For the helix TWT, it is helpful in explaining the RF behavior to first consider a single-wire transmission line above a ground plane as shown in Figure 13-8. Charges on the transmission line and the RF electric field pattern are shown at an instant in time. The RF magnetic field pattern is not shown and will not be considered because the RF magnetic forces on the electron beam in a microwave tube are insignificant compared with the forces produced by the electric field. If the RF generator is on the left end of the transmission line in Figure 13-8 and if a matched load is on the right end, then the charge and field patterns move to the right with constant amplitude. The velocity of propagation is that of light and is independent of frequency.

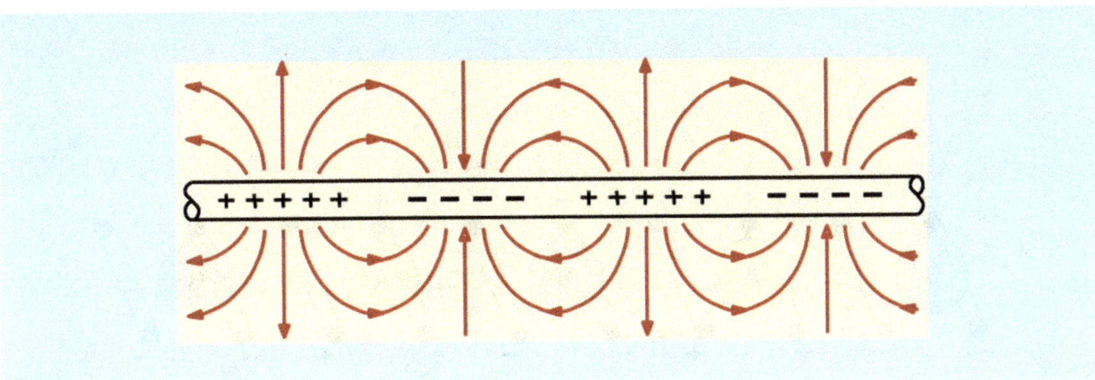

Figure 13-8 RF charge and electric field patterns for a single-wire transmission line above a ground plane.

Next, assume that the single-wire transmission line is formed into a helix as shown in Figure 13-9. An RF signal applied to the left end of the helix will travel at a velocity near that of light in a helical path along the conductor. The average velocity in the axial direction, which is called the *phase velocity* (usually denoted by v_p) of the helix, will be the velocity in the helical direction reduced by the pitch of the helix. The polarity of the signal will alternate every half-wavelength along the helical conductor. In Figure 13-9, two complete turns are required for each half-wavelength.

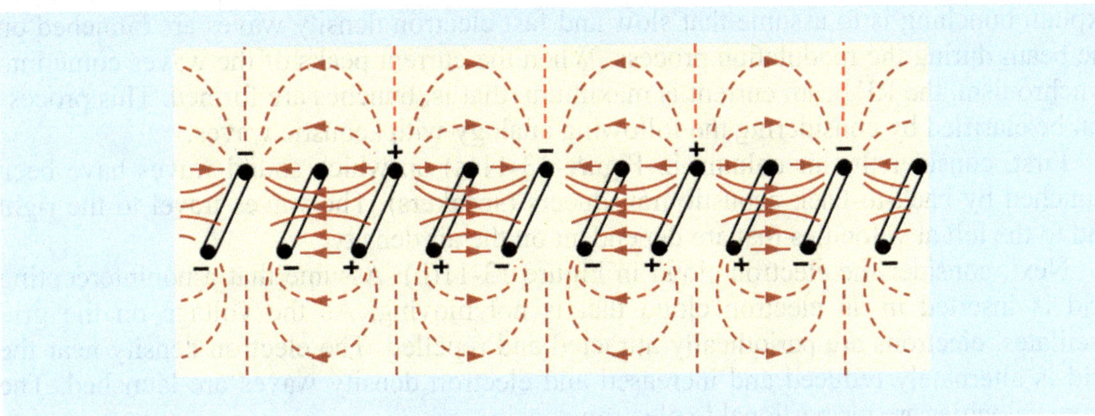

Figure 13-9 RF charge and electric field patterns for a helix.

Electric field lines extend from regions of positive charge to regions of negative charge with a configuration similar to that shown for the transmission line in Figure 13-8. Other than the reduced velocity, the primary difference between the field pattern for the helix and that for a single-wire transmission line is that there is an electric field inside the helix with large axial components.

When an electron beam is injected along the axis of the helix, the axial electric field components accelerate some electrons and decelerate others. In Figure 13-10, the forces on the electrons would be toward regions A and away from regions B. If the axial velocity of the electric field and the beam are nearly the same, then electrons will experience a continual force toward region A as the beam travels through the helix. As a result, electron bunches will start to form in regions labeled A.

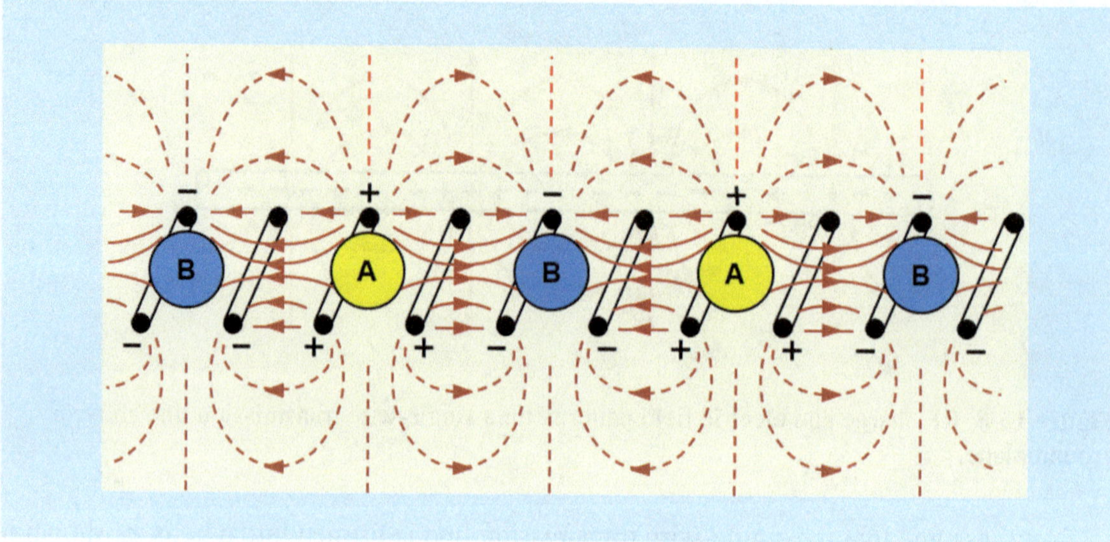

Figure 13-10 RF charge and electric field patterns for a helix.

13.2.2 Space-Charge Waves

We have said that bunching of electrons with a gap occurs because some electrons are accelerated by the electric field in the gap and some are decelerated. When the fast electrons catch up with the slow ones, bunches occur. Another more sophisticated way to explain bunching is to assume that slow and fast electron density waves are launched on the beam during the modulation process. When the current peaks of the waves come into synchronism, the RF beam current is maximum, that is, bunches are formed. This process can be clarified by considering the following analogy with acoustic waves.

First, consider the air column in Figure 13-11(a) on which sound waves have been launched by back-to-back acoustic transducers (speakers). The waves travel to the right and to the left at velocities that are dependent on the air density.

Next, consider the electron cloud in Figure 13-11(b). Assume that a nonintercepting grid is inserted in an electron cloud that is not moving. As the voltage on the grid oscillates, electrons are periodically attracted and repelled. The electron density near the grid is alternately reduced and increased and electron density waves are launched. The wave velocities are proportional to electron density.

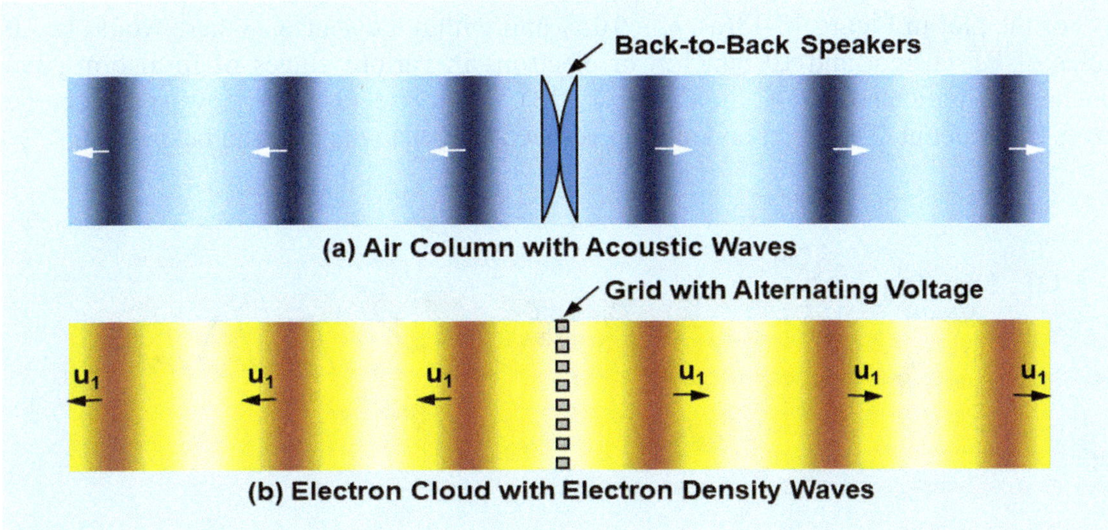

Figure 13-11 Acoustic waves and electron density waves.

Next, as indicated in Figure 13-12, let the electron column move to the right with velocity u_o which is much greater than the velocities of the density waves. The moving electron column is an electron beam. Both density waves are moving to the right with the beam. One is moving faster than the beam and is called a *fast space-charge wave*. The other is moving slower than the beam and is called a *slow space-charge wave*.

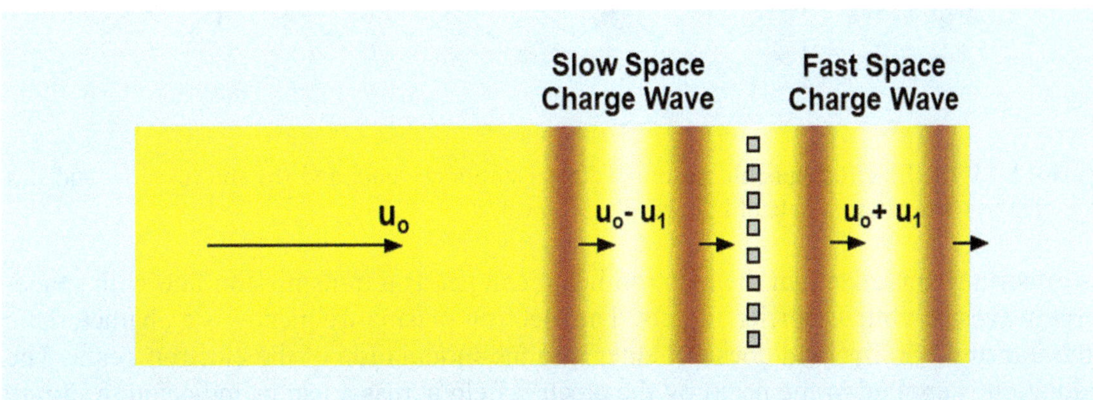

Figure 13- 12 Space-charge waves.

Let

$$u_1 = u_o\omega_p/\omega \tag{13-1}$$

where ω_p is the plasma frequency and ω is the modulation frequency.
The velocities of the space charge waves are

$$v_{fast} = u_o(1 + \omega_p/\omega) \tag{13-2}$$

and

$$v_{slow} = u_o(1 - \omega_p/\omega) \tag{13-3}$$

For the plot in Figure 13-13(a), $\omega = 10\omega_p$ and within a distance λ_p there would be 10 cycles of RF current and 10 bunches of electrons in various stages of formation (five bunches are indicated in $\lambda_p/2$ in Figure 13-13). Thus, the fast and slow space-charge waves travel about 10% faster and slower, respectively, than the electron beam.

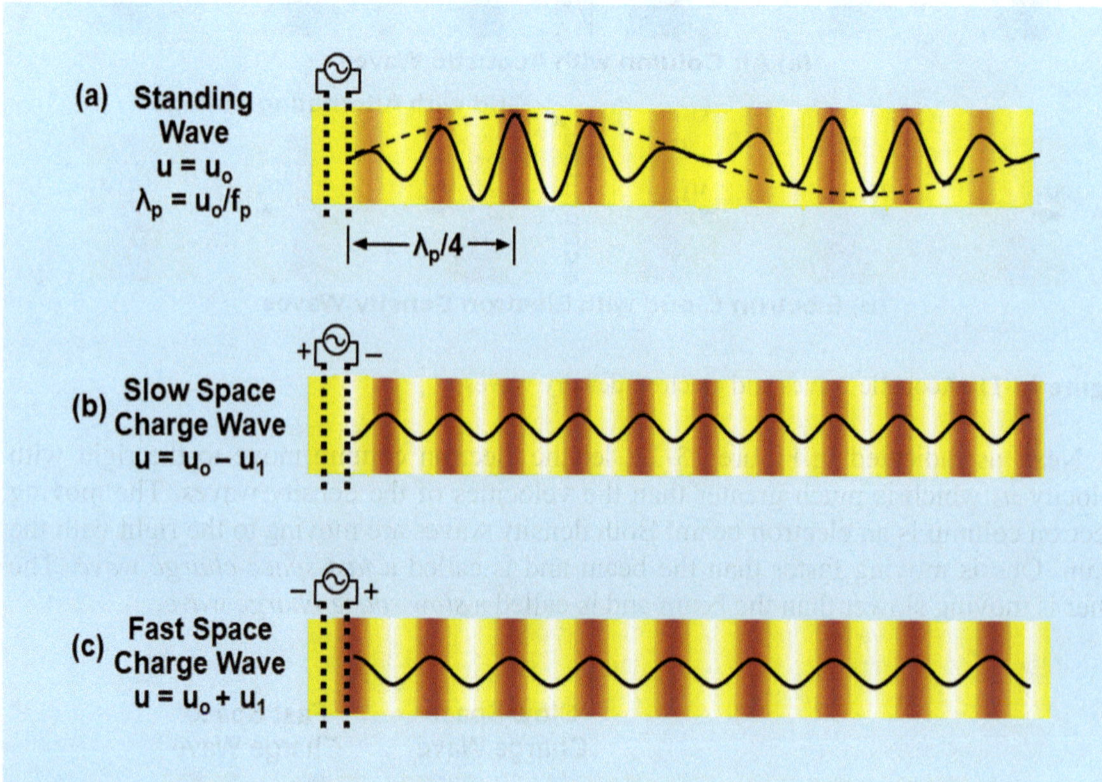

Figure 13-13 (a) Electron bunching produced by a stationary gap, and (b) and (c) slow and fast space-charge wave components of bunching waveform.

It is important to realize that the electron beam can act as a transmission line with space-charge waves serving as carrier signals. The electron plasma frequency is a characteristic of the transmission line and does not enter into the modulation of the electron beam. The modulation is applied to the beam by the electric field across a gap as indicated in Figure 13-14.

13.2.3 Plasma Frequency Reduction Factor

When the effects of a metallic structure near the beam are taken into consideration, it is found that the plasma frequency is reduced, and some field lines terminate on the metallic structure rather than a nearby bunch. The amount by which the plasma frequency is reduced is called the *plasma frequency reduction factor* and is usually denoted by *F*. This factor depends on the stiffness of the electron beam and the proximity of the metallic structure. Two examples, one for confined flow, which is a very stiff beam, and one for Brillouin flow, which is a very poorly controlled beam, are given here.

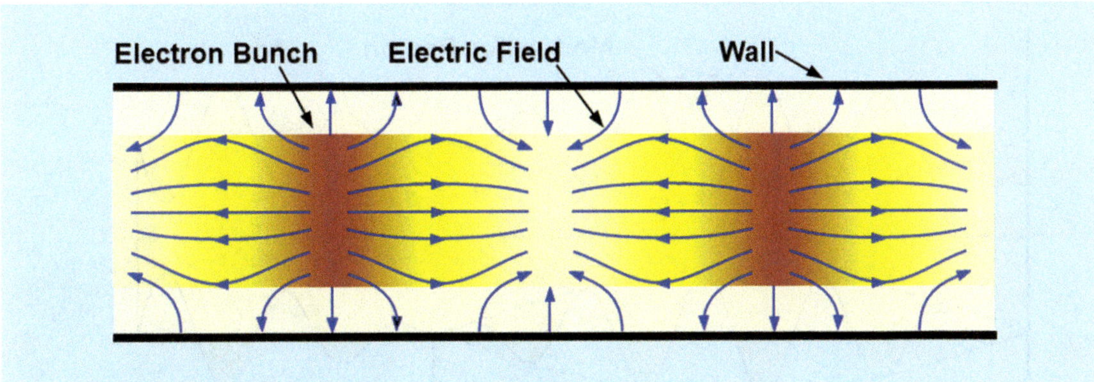

Figure 13-14 Wall effects on plasma frequency.

For a confined beam, the effect on plasma frequency is shown in Figure 13-15. Here ω_q is the reduced plasma frequency and F is the *plasma frequency reduction factor*. Note that if the normalized beam radius, $\omega b/u_o = 1$ and $n = 1.5$, the value of F is about 0.53, so the plasma frequency is one-half of the value calculated for an infinite beam.

When transverse electron velocities are present, as in a Brillouin-focused beam, the plasma frequency is reduced even more than it is in a confined beam. The reason for this is that for cavity modulation of a Brillouin beam at small signal levels, as shown in Figure 13-16, the electron bunching exists as a modulation of the radius of the beam rather than as a modulation of the beam density. To simplify the analysis, the modulated Brillouin beam is modeled as a beam with uniform charge density, having bunches of charge on the surface. Figure 13-17 shows the charge and field distributions for Brillouin and confined beams.

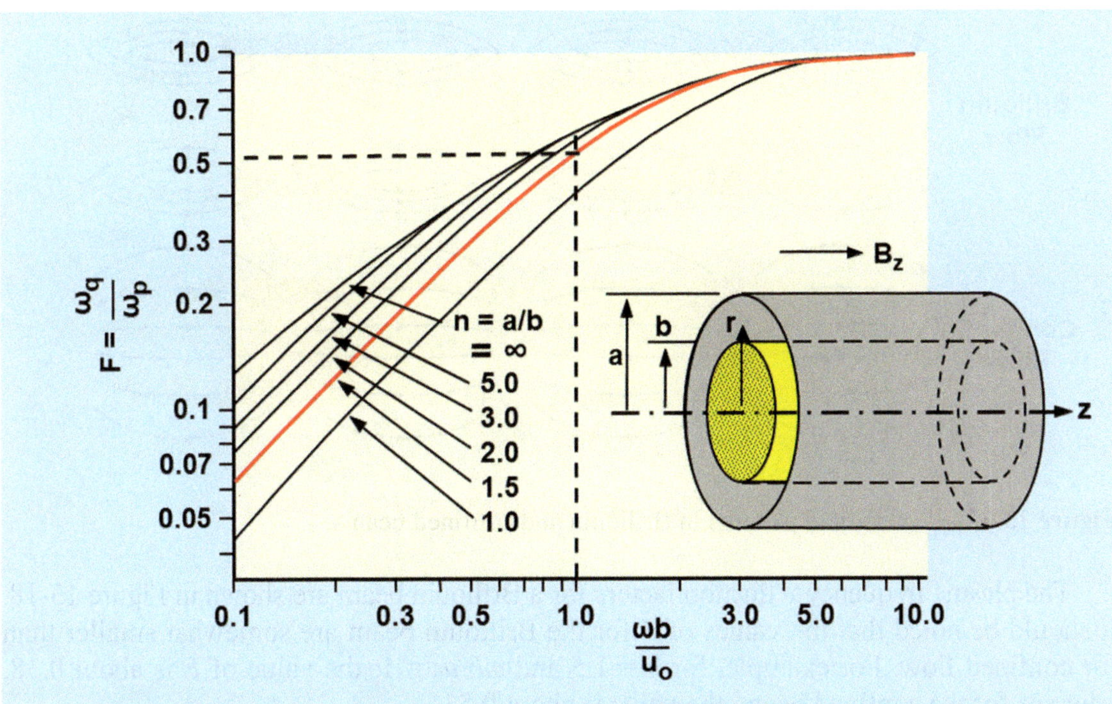

Figure 13-15 Plasma frequency reduction factor for a confined beam.

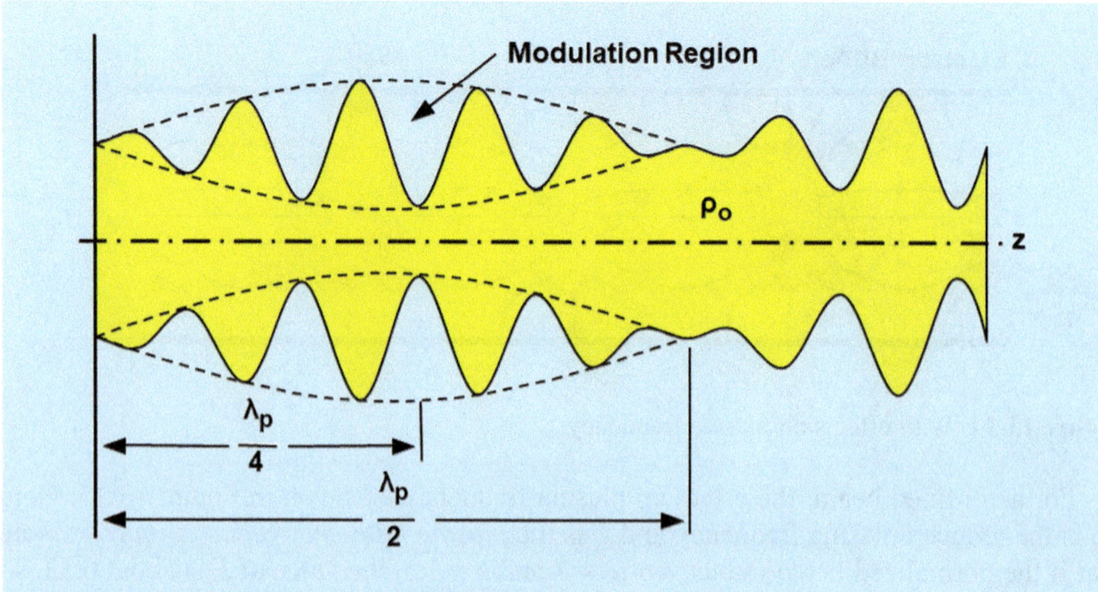

Figure 13-16 RF modulation of the beam radius for a Brillouin beam.

In Figure 13-17, note that the field lines can bow inward as well as outward from the surface bunches on the Brillouin beam. As a result, the axial component of the field and the space-charge forces are smaller than for the confined beam. This results in a lower plasma frequency and a larger plasma wavelength for the Brillouin beam than for the confined beam.

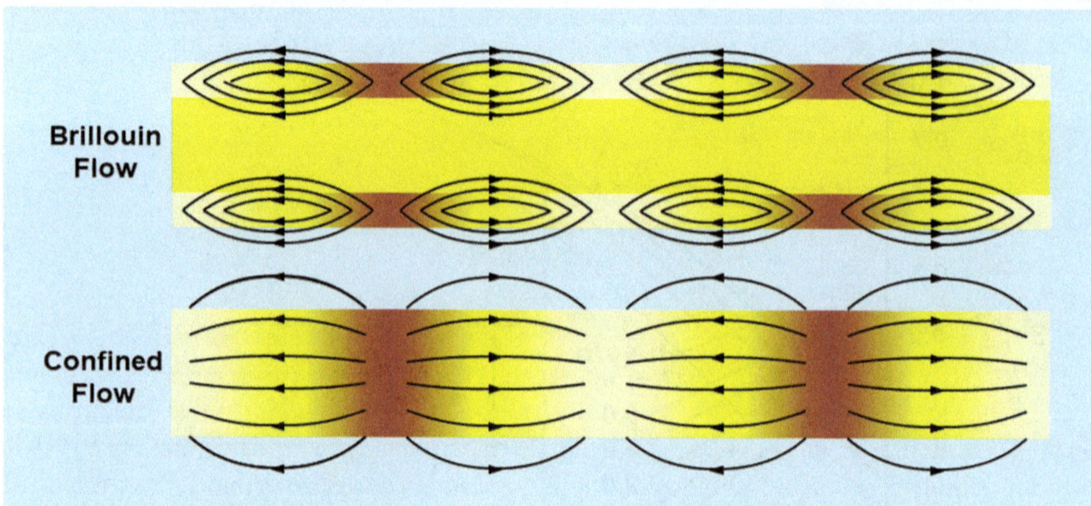

Figure 13-17 Electric field patterns in Brillouin and confined beams.

The plasma frequency reduction factors for a Brillouin beam are shown in Figure 13-18. It should be noted that the values of F for the Brillouin beam are somewhat smaller than for confined flow. For example, for $n = 1.5$ and $\omega b/u_o = 1$, the value of F is about 0.38, whereas for the confined beam, the value is about 0.53.

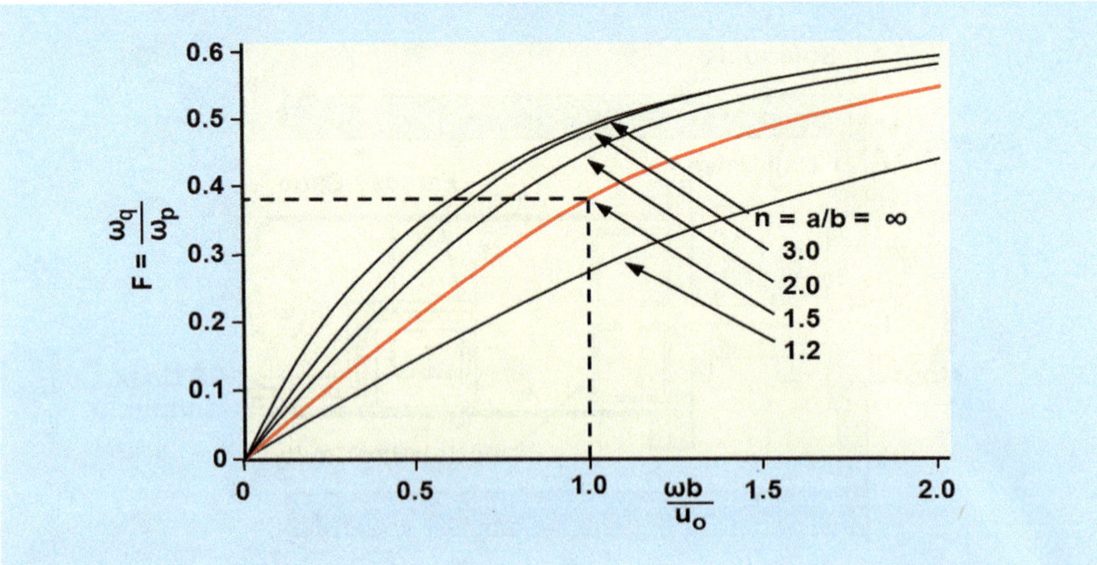

Figure 13-18 Plasma frequency reduction factor for a Brillouin beam.

13.2.4 Experimental Verification

The existence of space-charge waves on a modulated beam has been verified experimentally. Furthermore, the fact that these waves exist as a modulation of the radius rather than the density of a Brillouin beam has been demonstrated. The RF beam analyzer used for the experiments is shown in Figure 13-19 [4, 5]. This was the analyzer used for the dc beam measurements described in Chapter 6. The analyzer was modified for RF current measurements. The beam from a shielded Pierce gun was focused with a solenoid. A vacuum ball valve was used to permit changes to be made in the measurement apparatus without exposing the gun and cathode to air (which could have changed the beam characteristics). The beam was velocity-modulated at approximately 2,000 MHz. A beam-scanning mechanism with a 0.010-inch aperture was used for sampling currents at various transverse and axial positions.

Plots of the fundamental component (harmonics could also be observed) of RF beam current as functions of the radial and axial position in the beam are shown in Figure 13-20. The RF drive level was small enough so that small signal conditions existed. At the smallest drive level, the RF current was entirely on the surface of the beam as is predicted for a Brillouin beam. The predicted space-charge wave behavior of the electrons is clearly illustrated by four half-plasma wavelengths of variation of the beam current.

Electron energy measurements were made to obtain a measure of the RF velocities of electrons in the beam. The Faraday cage used for collecting the electrons passing through the 0.010-inch aperture in the beam-scanning mechanism was biased to a negative voltage (4,800V) comparable to the dc beam voltage. In addition, a 10,000-MΩ resistor was placed in series with the bias batteries. With this large resistance, only a fraction of a microampere of current to the Faraday cage caused an additional bias of several hundred to several thousands of volts. A fraction of a microampere was less than 0.1% of the normal peak Faraday cage current and consisted therefore of only those electrons having the maximum energy.

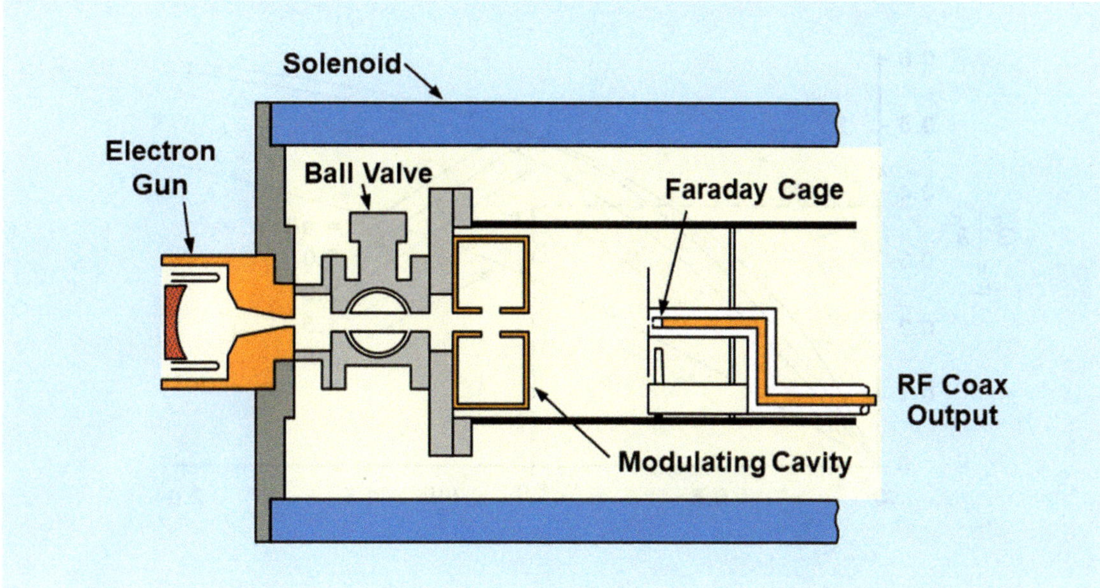

Figure 13-19 RF beam analyzer. (From A. S. Gilmour, Jr., *IEEE Trans. Electron Devices*, July 1972. ©1972 IEEE.)

Shown in Figure 13-21 are the peak energies (expressed in terms of the bias voltage on the Faraday cage) for small-signal operation. As expected, the variation of energy was a series of half sinusoids with positions of peak values occurring at distances where the RF current was zero.

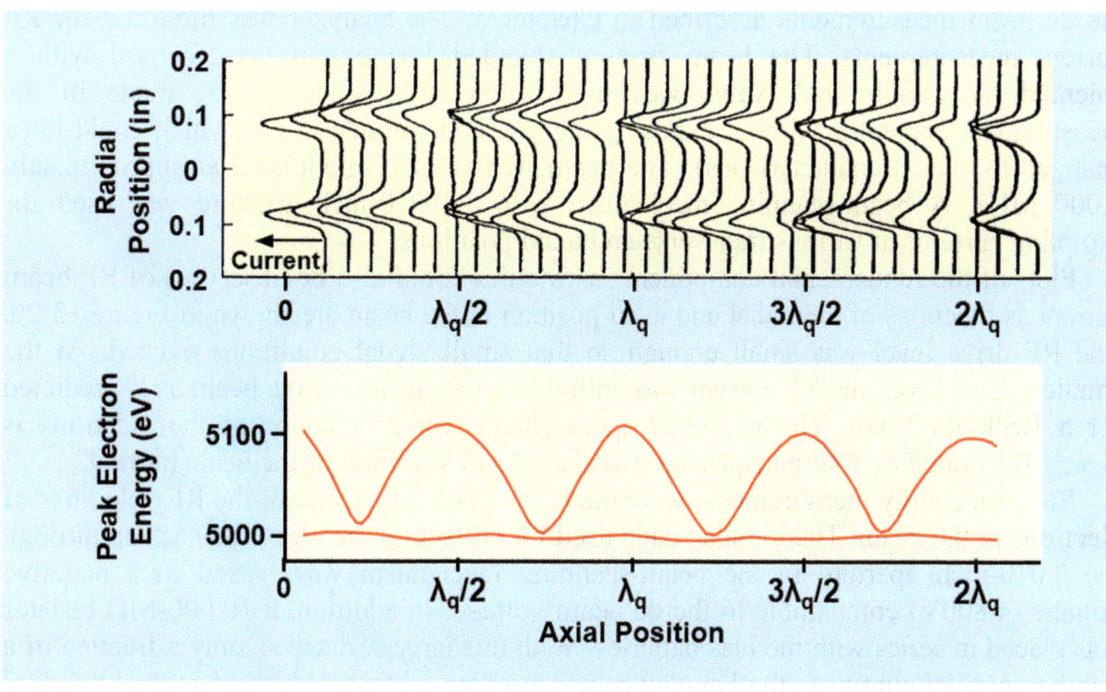

Figure 13-20 RF current distributions and peak electron energy in a velocity-modulated Brillouin electron beam.

The initial bunching action in a TWT is like that produced by a gap (fast electrons catch up with slow electrons to produce a bunch, or, alternatively, fast and slow space-charge waves are launched). As indicated in Figure 13-21, the bunches that form in the beam repel electrons on the helix causing those electrons to travel to the right and to the left. The electrons on the helix travel at the phase velocity, v_p, of the helix, which we will assume is nearly the same as the velocity of the slow space-charge wave in the beam.

Relative to a bunch, electrons on the left move left at $2v_p$. This produces a deficiency of electrons, which makes the circuit positive. Relative to the bunch, electrons on the right, which are moving at approximately the bunch velocity, are nearly standing still. As a result, electrons accumulate ahead of the bunch and make the circuit negative.

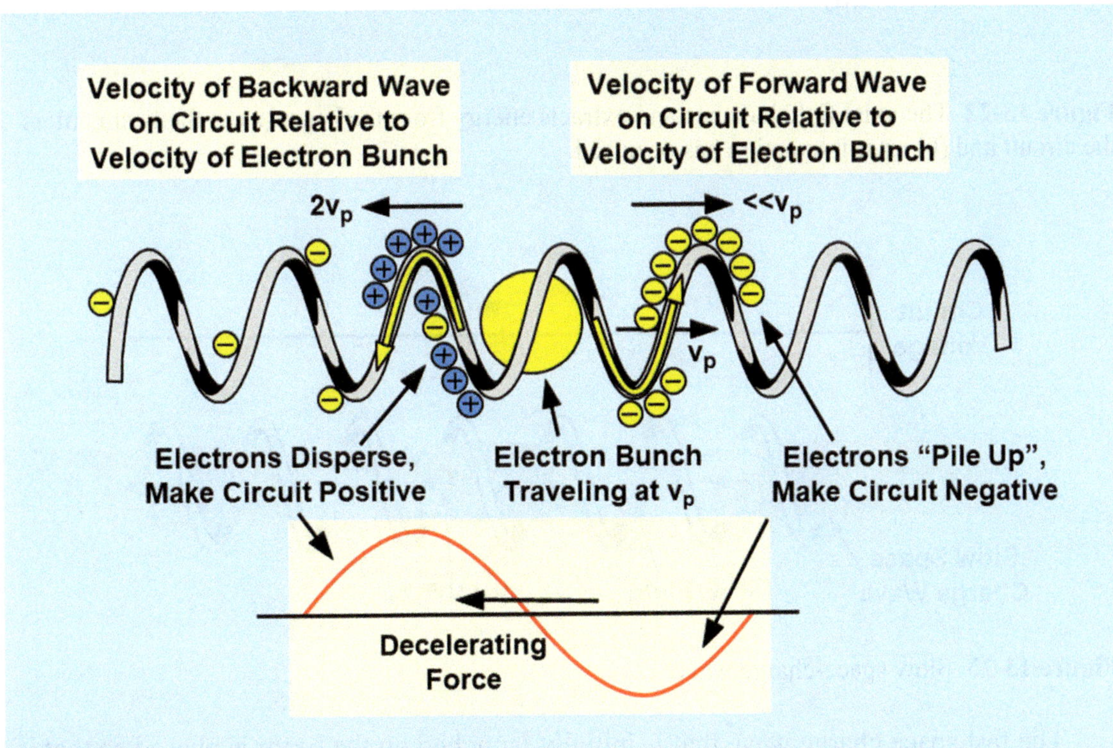

Figure 13-21 Bunches repel electrons on circuit. The interaction described here is the mechanism that produces gain in a TWT amplifier.

The increase in electron charge ahead of the bunch and decrease behind the bunch produces a voltage waveform on the helix as shown in Figures 13-22(b) and Figure 13-23 that is phase-shifted 90° to the left of the initial waveform. As the voltage waveform on the helix shifts to the left, decelerating field regions move into the regions where the bunches are located and so energy is extracted from the electrons, causing the wave on the circuit to grow. As more and more electrons are forced into the bunch, the induced waveform rapidly becomes larger than the initial waveform.

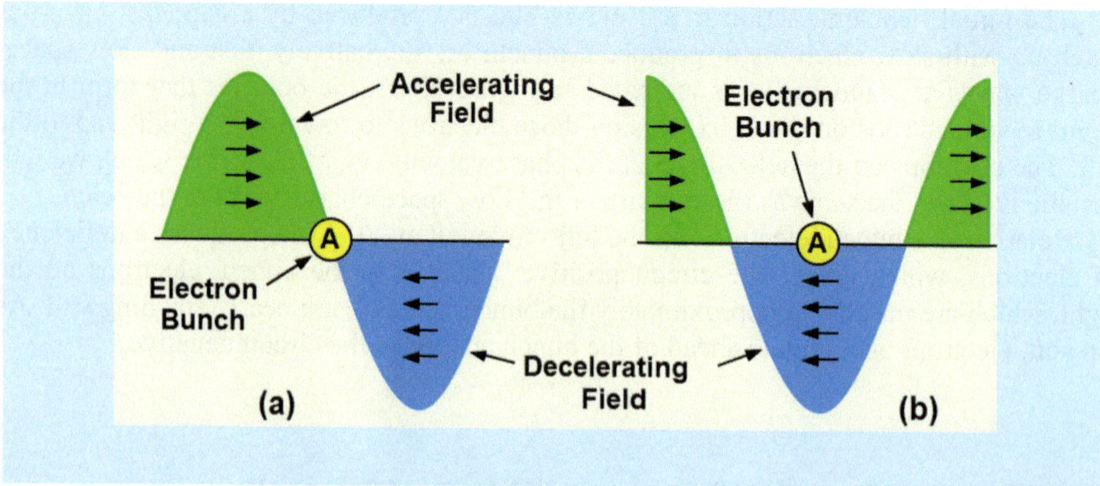

Figure 13-22 The axial field bunches and extracts energy from the beam (a) as the beam enters the circuit and (b) after interaction has occurred.

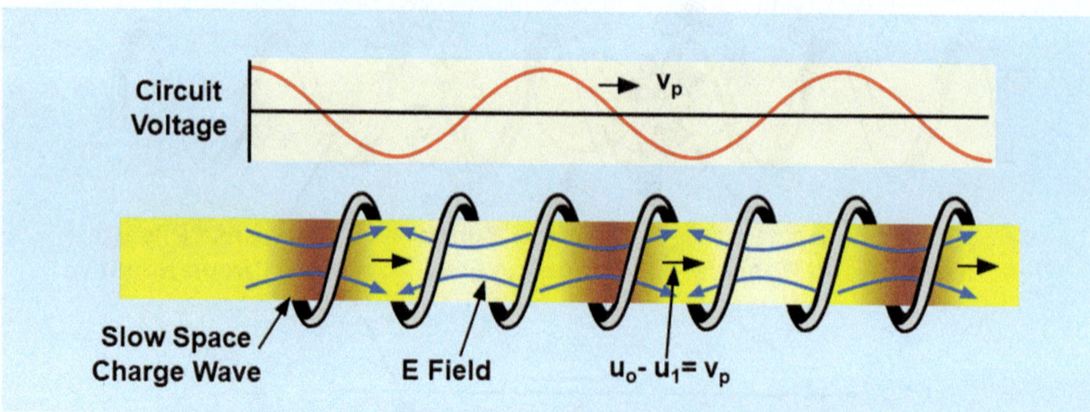

Figure 13-23 Slow space-charge wave.

The fast space charge wave that is initially launched on the beam is phased so that it gives up energy to the beam. As a result, it is rapidly attenuated.

13.3 TWT GAIN

Now that the mechanism that produces gain has been described, we can proceed to determine the factors that affect gain. The following discussion is limited to an outline of the Pierce theory

In the Pierce theory, the RF circuit of a TWT is modeled as a transmission line as shown in Figure 13-24. This model serves as a starting point, but then the analysis is generalized so that the exact form of the model is not important. The line has distributed inductance and capacitance per unit length of L_ℓ and C_ℓ. The beam with current i_z is assumed to pass very close to the circuit and, from Ramo's theorem, the current induced in the circuit is $I = i_z$.

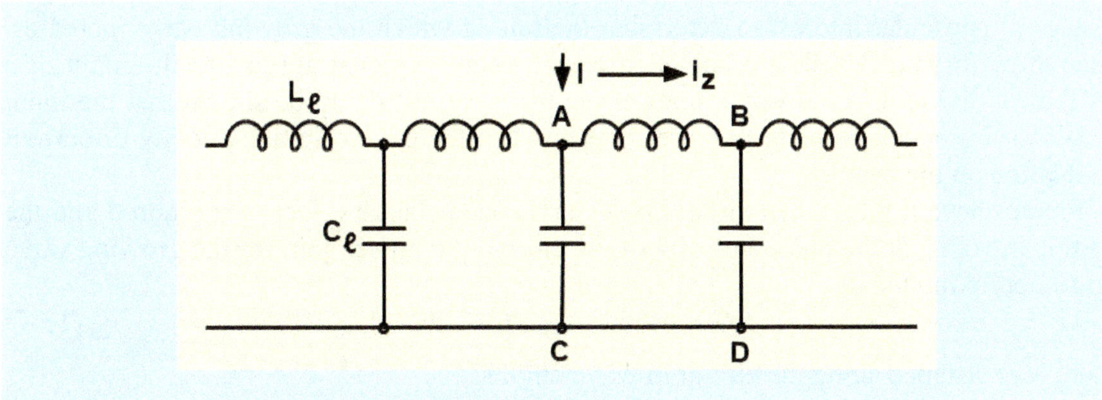

Figure 13-24 Transmission line model for a slow-wave circuit.

The relation between the RF current in the beam, i_z, and the electric field, E_z, is sometimes referred to as the electronic equation. This equation is complex and can be found from an analysis of the effect of space charge on electron bunching [17]. For our purposes here, it is sufficient to acknowledge that the RF beam current is a function of the RF electric field in the beam; that is,

$$I_z = f(E_z) \tag{13-4}$$

The electric field acting on the beam is written as a function of the current in the circuit (and the current in the beam). The resulting equation is called the circuit equation and has the format

$$\mathrm{E}_z = f(i_z) \tag{13-5}$$

The electronic equation (13-4) and the circuit equation (13-5) provide two relationships between i_z and E_z. These may be combined to eliminate i_z and E_z and obtain a relation describing the waves on the beam and circuit. This relation, sometimes called the determinantal equation, is a fourth order equation, which describes four waves on the system. These waves are shown in Figure 13-25.

Two of the waves are the fast and slow space-charge waves. The other two are the forward- and backward-circuit waves. If the phase velocity of the forward circuit wave is close to the slow space-charge wave velocity, then the interaction described in the previous section can occur and this wave will grow with distance. The fast space-charge wave will gradually give up its energy to the slow circuit wave. The backward circuit wave is only slightly affected by the beam so the velocity of this wave is basically the velocity of cold (no beam) circuit.

For reasons that will become clear later, Pierce defines a quantity C, used in the determinantal equation and called the *gain parameter*, by

$$C^3 = \frac{Z_o}{4V_o / I_o} \tag{13-6}$$

The beam impedance is V_o/I_o, so C^3 is one-quarter of the ratio of the circuit impedance to the beam impedance. Because Z_o is usually on the order of tens of ohms and V_o/I_o is on the order of thousands to tens of thousands of ohms, C^3 is a very small quantity so C is usually in the 0.01 to 0.1 range.

It is of particular interest to determine the rate at which the growing wave increases. Since three forward waves are launched on the beam, a signal applied to the input of a TWT must divide in some way among these three waves. It can be shown that the input signal divides equally among the three waves when there is no initial velocity or current modulation on the beam.

Pierce showed that, when circuit losses and space charge effects are ignored and the beam is traveling at the phase velocity of the circuit, the power gain for the growing wave can be approximated by

$$G = -9.54 + 47.3\,C\,N \ \text{dB} \tag{13-7}$$

where N is distance along the circuit in wavelengths.

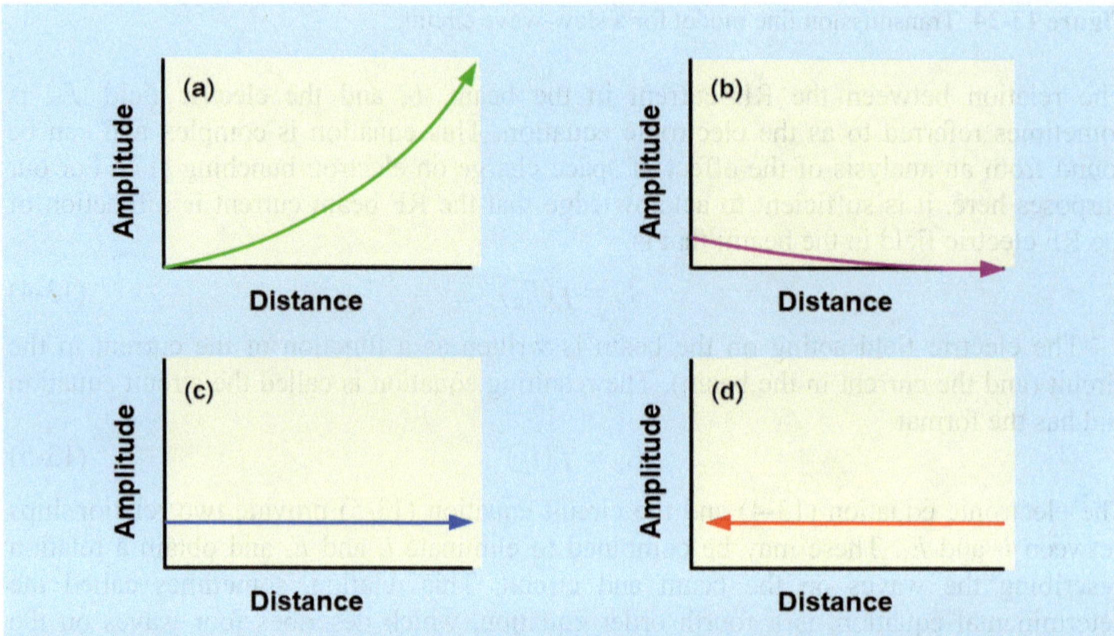

Figure 13-25 Waves in a traveling wave tube: (a) slow space-charge wave that interacts with the slow circuit wave to produce gain, (b) fast space-charge wave that decays as it gives up energy to the circuit, (c) slow circuit wave that provides energy to make the slow space-charge wave grow, and (d) fast backward-circuit wave that interacts very weakly with the electron beam.

The reason that C is called the gain parameter is now clear. Gain is directly proportional to C. Power gain as a function of CN is shown in Figure 13-26. The solid curve includes the effects of all three input waves.

Initially, the voltage does not change with distance and there is no gain. This is because the electron beam has no RF current and cannot induce a signal into the circuit until it becomes bunched. Eventually, the increasing wave predominates over the other two and the slope of the line must correspond to that for the growing wave; that is, 47.3.

The dashed line represents the increasing wave, which starts at –9.54 dB (because only one third of the input signal is used) and has a slope of 47.3. This result is reasonably good for $CN > 0.2$.

If the circuit has a phase velocity nearly equal to the slow space-charge wave velocity, the interaction of the wave on the circuit with the beam will slow beam electrons to the bunch velocity. The energy extracted from the electron beam in slowing electrons to the bunch velocity is transferred to the circuit field thereby producing amplification of that field. The interaction of the beam and circuit produces an exponential growth of the circuit voltage.

13.3.1 Effect of Loss

In addition to the losses that occur naturally in RF circuits, additional losses may be introduced to attenuate the backward wave. The backward wave normally originates from reflections either inside or outside the tube and can serve as a feedback mechanism to cause oscillations or large variations in gain (because of beating effects with the input signal).

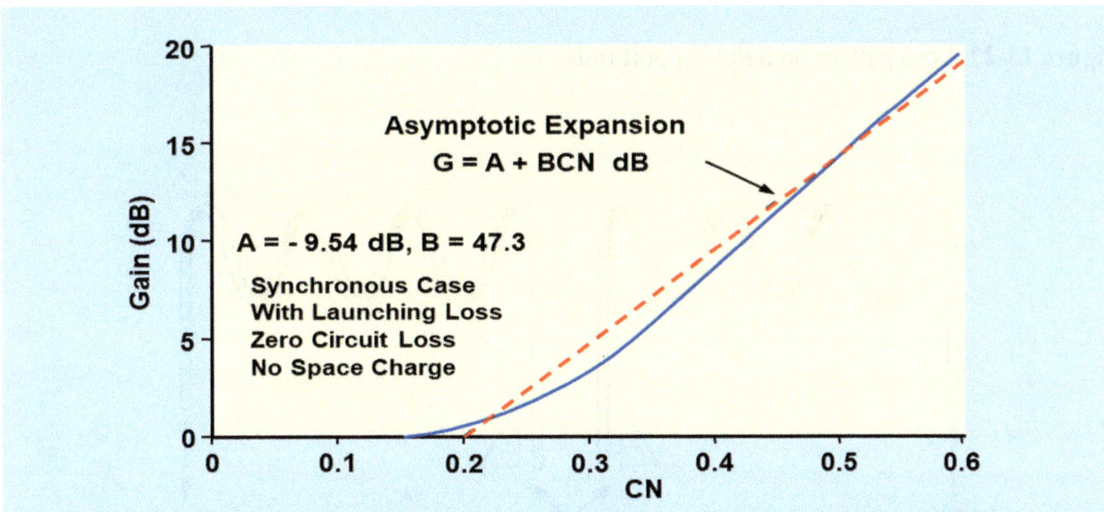

Figure 13-26 Power gain in a TWT for the synchronous case with zero loss and no space charge. (Adapted from: *Traveling Wave Tubes*, by J. R. Pierce, copyright 1950 by D. Van Nostrand Inc.)

The gain of a TWT can be $10^4 - 10^5$ or 40 – 50 dB. The attenuator shown in Figure 13-27 and/or a sever like that shown in Figure 13-28 is used to suppress the backward traveling waves. As shown in Figure 13-29, the forward-traveling wave on the circuit is also attenuated; however, the electron beam carries the signal through the attenuation region and then signal amplification resumes.

Pierce took loss into account by introducing the loss parameter, d, into the circuit propagation constant. The loss per wavelength, L, is

$$L = 54.5\, Cd \text{ dB} \tag{13-8}$$

The most significant effect of attenuation is, of course, that the gain of the growing wave is decreased. However, Pierce showed that for small losses, the reduction in gain for the growing wave is about one-third of the circuit loss.

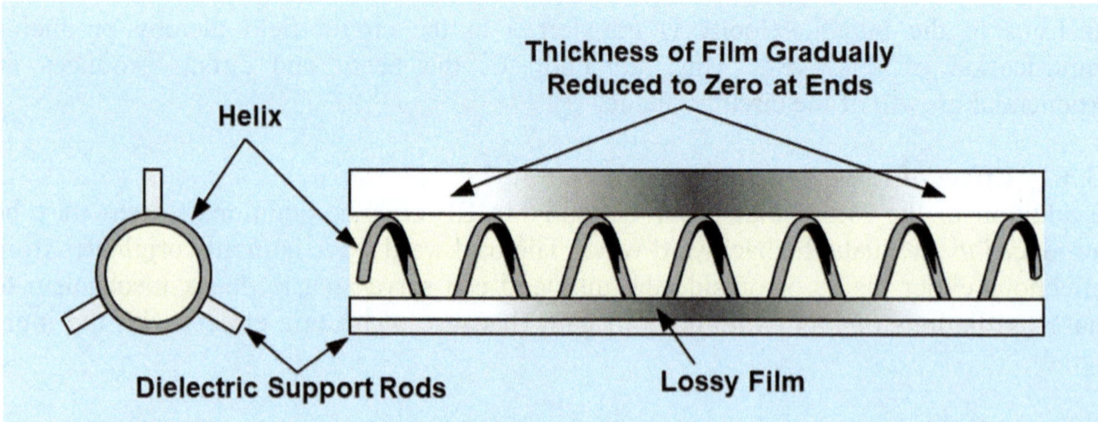

Figure 13-27 Loss pattern on helix support rods.

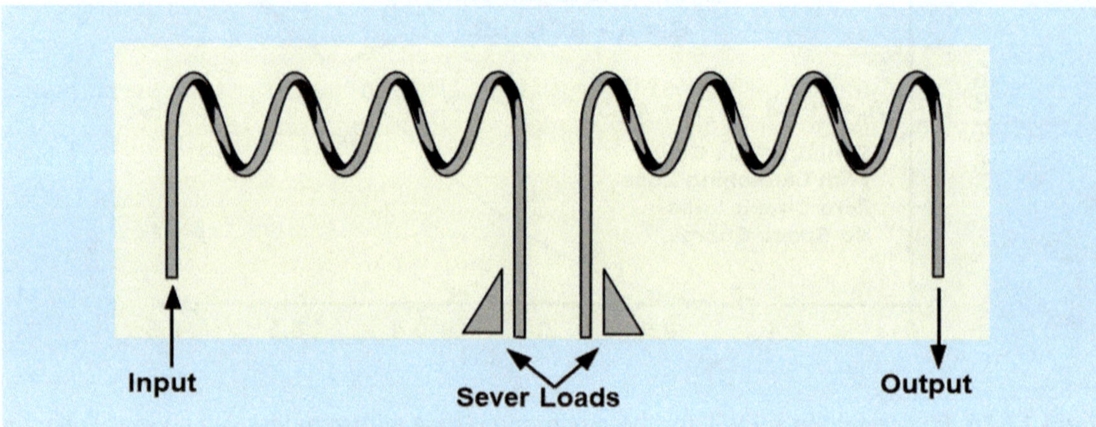

Figure 13-28 A two-section severed helix.

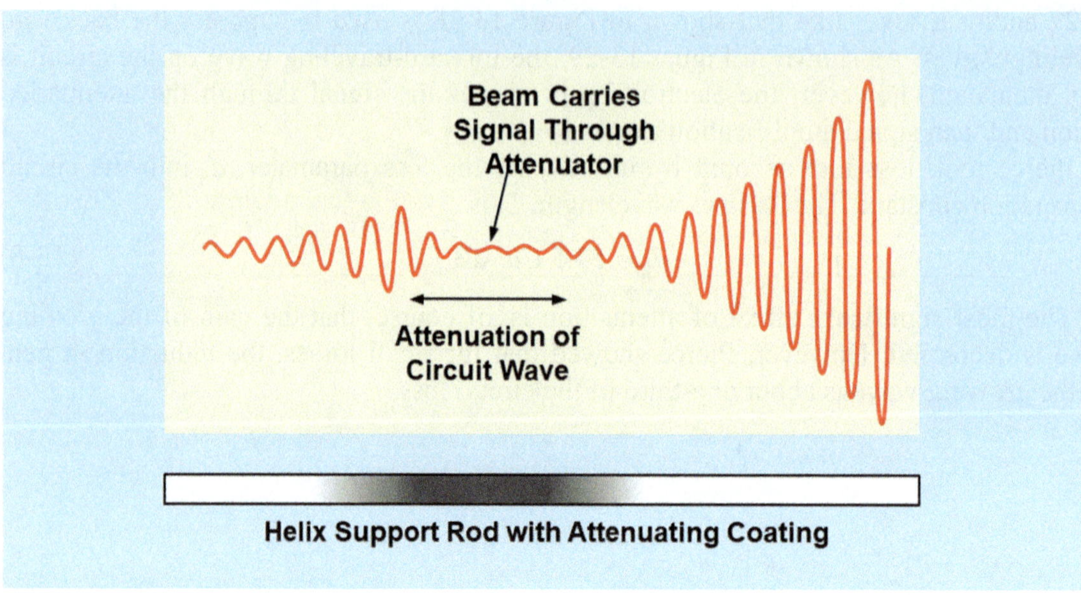

Figure 13-29 Attenuator pattern.

13.3.2 Effect of Space Charge

There are two primary effects of the density of the space charge in the electron beam. First, the velocities of the space-charge waves are directly affected. As electron density is increased, the velocity of the slow space wave decreases, and this has a direct impact on the design of the circuit because the circuit phase velocity must be close to the slow space-charge wave velocity. Second, as in a klystron, optimum bunching occurs at low perveance and, therefore, low electron density. Pierce accounted for the effects of space charge with his space charge parameter, *QC,* which can be written as

$$QC\left(=\frac{\omega_q/\omega}{2C}\right)^2 \tag{13-9}$$

Now the ratio of plasma frequency to RF frequency is a measure of space charge debunching forces. Pierce's gain parameter, *C*, is a measure of the intensity of interaction between the slow wave circuit and the electron beam. As a result, *QC* is a measure of the relative strength of the debunching forces to the bunching forces. Thus, *QC* has been adopted to describe the seriousness of space charge effects in traveling-wave tubes. Figure 13-30 displays the variations of gain and phase velocity with space charge.

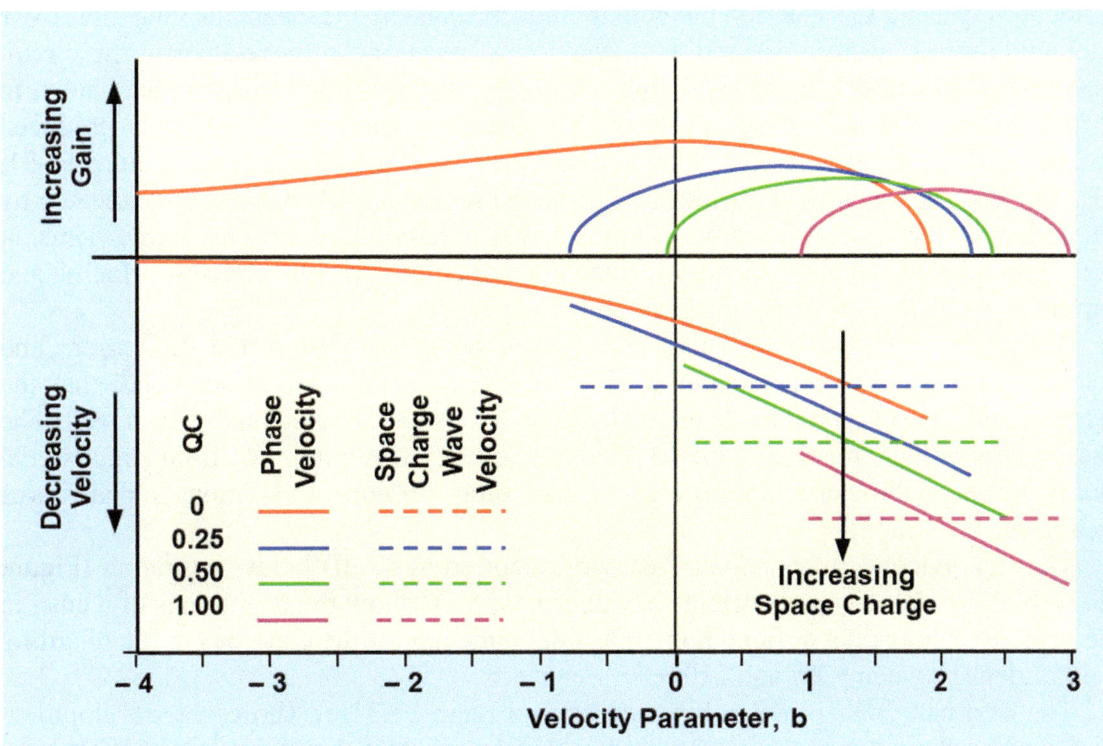

Figure 13-30 Effect of space charge on gain and velocity for a growing wave.

Another of Pierce's parameters, the velocity parameter, *b*, is used as the ordinate in Figure 13-30. The velocity parameter is useful in relating the dc electron velocity, u_o, to the cold phase velocity (the phase velocity absence in the of electrons) of the circuit, v_p.

$$u_o = (1 + C\,b)\,v_p \quad \text{or} \quad b = \frac{u_o - v_p}{C\,v_p} \tag{13-10}$$

For $b > 0$, the electrons go faster than the undisturbed wave, and for $b < 0$, the electrons go slower than the undisturbed wave. For $b = 0$, the electrons travel at the velocity of the undisturbed wave.

Several effects of space charge are apparent in Figure 13-30. The most important is that the range of beam velocities (indicated by b) over which gain occurs is severely restricted. Furthermore, the velocity at which there is maximum gain in the increasing wave is now higher than in the absence of space charge. Also, the maximum gain is reduced.

3.4 SATURATION

13.4.1 Discussion of Interactions

As the waves grow, the interaction between the beam and circuit becomes nonlinear. Perhaps the clearest paper ever published describing nonlinear phenomena in a TWT was that by Cutler [18]. A large-scale model of a TWT was constructed. Using an electron deflection system, the electron velocities and currents in the beam leaving the TWT interaction region were measured as a function of phase position relative to the circuit wave. Some of Cutler's results showing the events leading up to saturation are shown in Figure 13-31. The RF drive levels are in decibels below the level that produced saturation. The circuit voltage is shown at the top of each of the frames of Figure 13-31. RF current is indicated by the width of the shaded region and RF velocity is indicated by the deflection of the shaded region from the center horizontal line in each frame. Thus, in each frame there is a complete story of fundamental circuit voltage, electron velocity, and current as a function of phase, for a particular drive level.

In Figure 13-31(a), the drive level was well below saturation and the current and velocity patterns were representative of small-signal conditions. Cutler noted that the velocity and current modulations at small-signal levels check calculated values well. The phase of the voltage relative to the RF electron velocity was computed from small-signal theory, but everything else in Figure 13-31, including subsequent variations in phase, was measured.

The first evidence of nonlinear behavior occurred at 17 dB below saturation (Figure 13-31(b)). The velocity and current (in Cutler's words) "show the beginnings of a cusp in the velocity curve and a definite non-sinusoidal bunching of the electrons in the retarding field region (between +30° and 210°)."

The next pattern at 10 dB below saturation, Figure 13-31(c), shows the development of a spur on the velocity curve and further development of the electron bunch between +60° and +180°. Cutler observed that the velocity within the bunch did not change significantly from lower drive levels. For this to be true, since the bunch is in a decelerating portion of the circuit field, the space-charge field must compensate for the circuit field. Near 60°, the decelerating field must increase rapidly to produce the sharp deceleration of electrons and the development of the velocity spur. Furthermore, Cutler observed that the decelerating field must be much larger than the circuit field because electrons just behind the cusp are much more sharply decelerated than those preceding the cusp. He concluded that there are very sharply defined space charge fields much stronger than the circuit field.

The next two patterns of electrons from the high-velocity regions continued consolidation of the main bunch just ahead of the spur. Apparently, because the electron velocities within Figure 13-31(d, e)) show continuing development of the spur, continued movement the bunch is still nearly the same as at lower drive levels, and the space charge and its associated field in the bunch have increased enough to compensate for the circuit field so that the net field remains near zero.

At 4 dB below the saturation drive level, the velocities of a large fraction of the electrons in the spur have increased because the spur has moved well into the accelerating region. The main bunch of electrons to the right of the spur has been concentrated into the 60°–120° phase interval. Apparently, the space-charge field in advance of this region no longer compensates for the decelerating circuit field so electron velocity is decreasing.

At saturation, the tip of the spur has formed into a bunch that is nearly equal to the main bunch and is in a circuit field that is nearly equal and opposite to that in the main bunch. The few electrons (indicated by the dashed line) ahead of the main bunch are being decelerated into a second low-velocity loop. The reason for saturation is that the power being extracted from the circuit by the spur bunch is nearly the same as the power being added by the main bunch.

13.4.2 Estimates of Maximum Efficiency

Early estimates of the maximum efficiency that might be achieved in a TWT were made without the aid of Cutler's experimental results and without the aid of computer models. Slater [19] suggested that energy extraction from the beam in a TWT could be analyzed by observing the motions of electrons while traveling along at the speed of the circuit wave. We will use one of the frames from Cutler's results (Figure 13-31(d)) to illustrate Slater's suggestion. That frame is shown in Figure 13-32. The initial beam velocity is u_o. The velocity of the circuit wave is Δu_o below the beam velocity. Slater suggested that electrons eventually become trapped and oscillate in the region of the maximum decelerating circuit field. This trapping is clearly evident in Figure 13-32, where a bunch is shown to have formed ahead of the velocity cusp in the decelerating field.

If the electrons initially have an average velocity, u_o, greater than that of the circuit wave by Δu_o, and they are reflected by the decelerating field, they cannot acquire a velocity backward relative to the wave greater than Δu_o. Those electrons in the small bunch to the left of the velocity cusp in Figure 13-32 fall into this category. These have been decelerated below the initial beam velocity by $2\Delta u_o$. The maximum efficiency would occur if all of the electrons (rather than the small fraction shown in Figure 13-32) were reflected and decelerated by the velocity increment $2\Delta u_o$.

Pierce [1] followed Slater's line of reasoning and used the Pierce small-signal expression for the velocity of the growing circuit wave to determine Δu_o. Pierce noted that "Such considerations are complicated by the fact that the phase velocity of the wave in the large-signal region will not be the same as its phase velocity in the small-signal region. It is interesting, however, to see what limiting efficiencies this leads to."

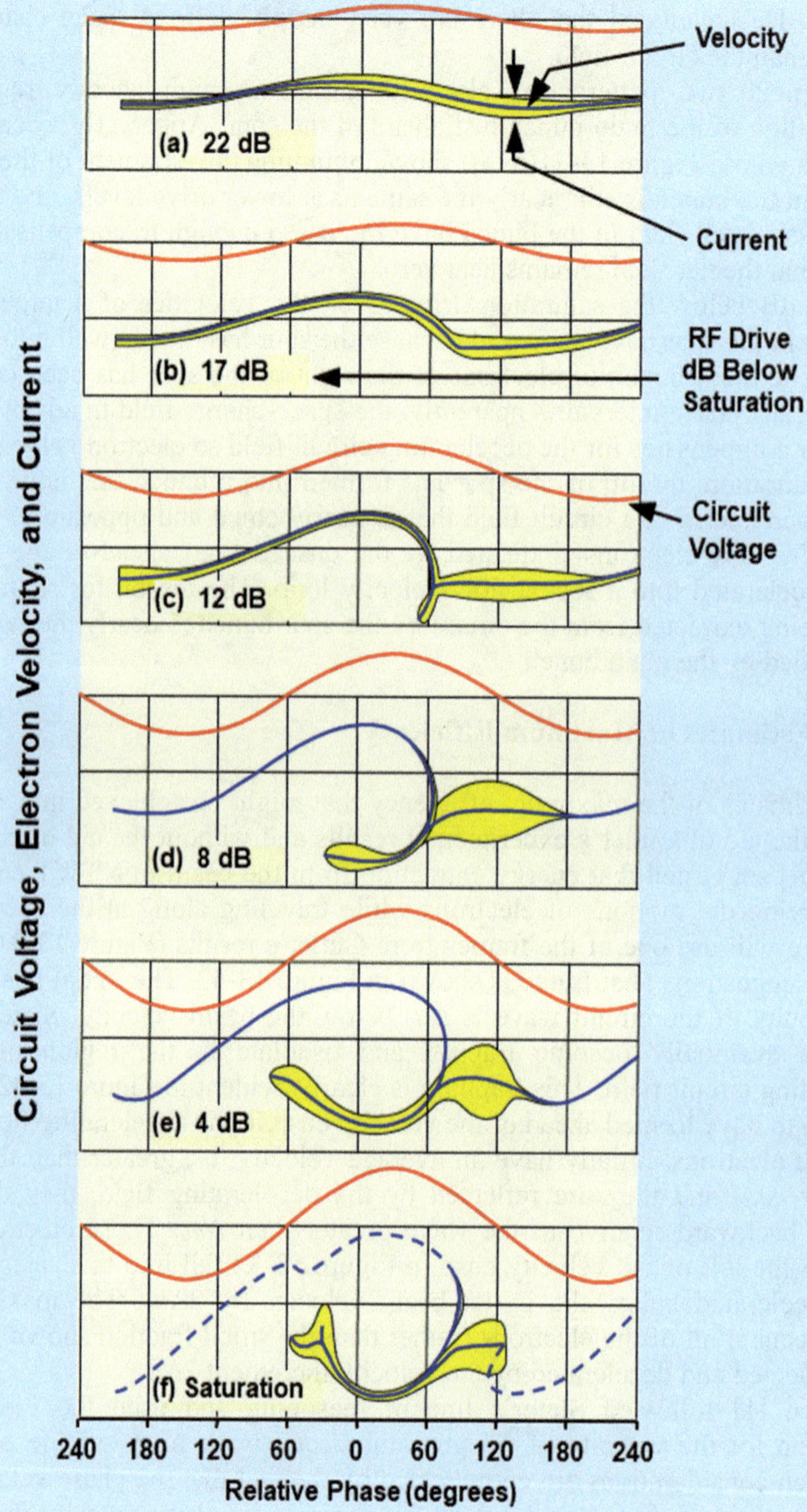

Figure 13-31 Electron bunch behavior as a function of drive level relative to saturation drive. (Adapted from: C. C. Cutler, *Bell Sys. Tech. Jour.*, July 1956.)

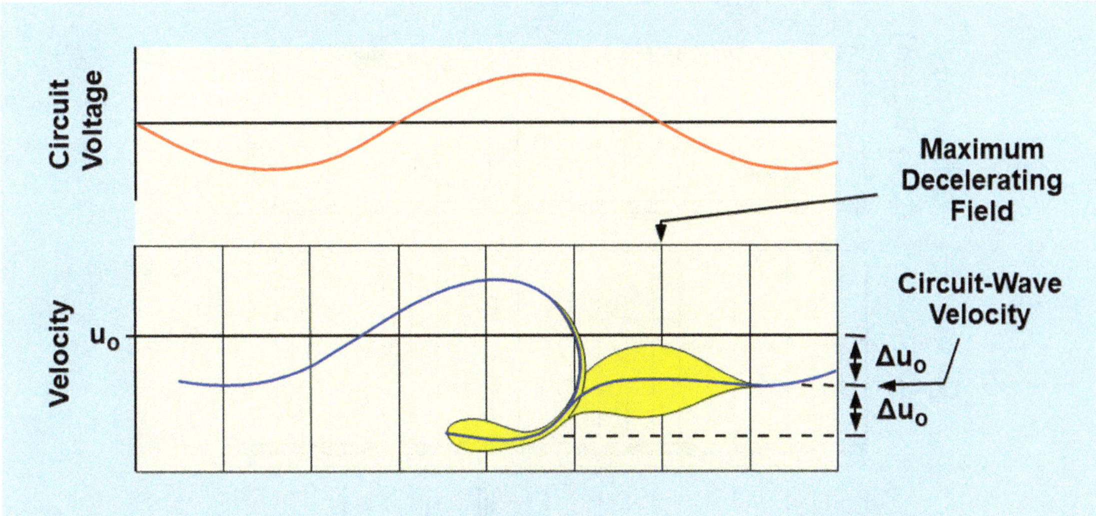

Figure 13-32 Considerations in Slater's estimate of maximum TWT efficiency.

Pierce found that, for $Cy_1 << 1$,

$$\eta \approx -4 C y_1 \tag{13-11}$$

As an example, for a value of y_1 of –1 and C of 0.05, the maximum efficiency is predicted to be about 20%, which is high but not totally unrealistic.

It should be noted in Cutler's work (Figure 13-31(f)) that the maximum efficiency (saturation) actually occurred when the electron bunch to the left of the velocity cusp was nearly equal to the main bunch. The velocities of the two bunches were nearly the same and were about the velocity of the circuit wave (Δu_o below u_o). As a result, the efficiency in Cutler's experiment is predicted to be about half that in (13-11) or

$$\eta = -2 C y_1 \tag{13-12}$$

As the interaction process continues, more electrons are slowed to the bunch velocity and space-charge forces within the bunches continue to increase. Eventually, these forces become large so that a portion of each bunch is retarded in phase enough so that it leaves the decelerating field region and enters an accelerating field (see electron density plots in the right third of Figure 13-33). Electrons in the acceleration portion of each bunch extract energy from the circuit field. Eventually, as the bunches continue to fall back in phase, energy extracted from the circuit wave becomes equal to the energy supplied and the wave on the circuit stops growing. At this point, saturation is said to occur and the signal amplitude is maximum.

The power transfer characteristic for a klystron or TWT is usually like that shown in Figure 13-34. At small-signal levels, operation can be extremely linear. As saturation is approached, operation becomes nonlinear. For some applications, such as communications, linearization techniques must be used and/or the tube operating range must be limited.

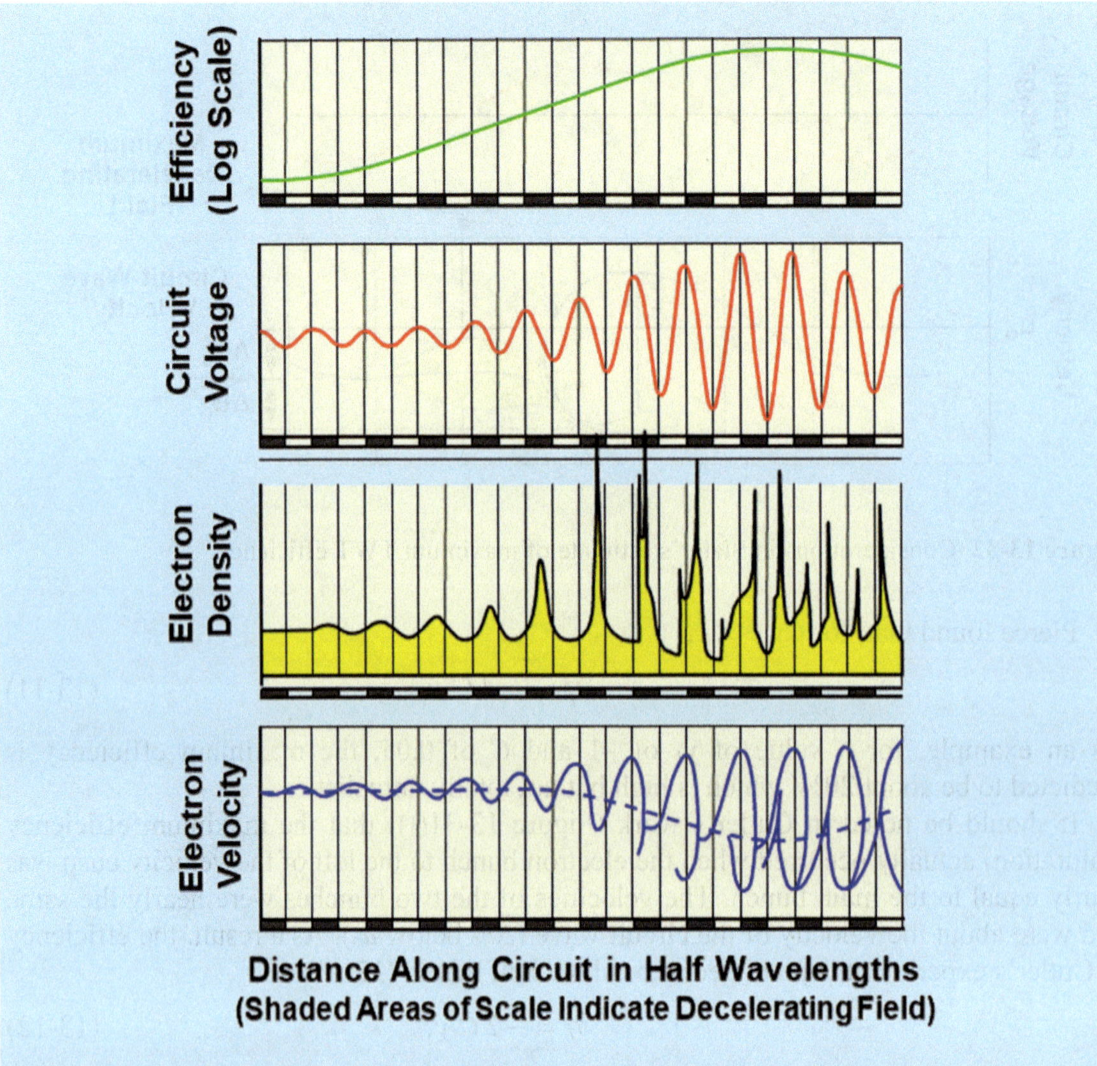

Figure 13-33 Characteristics of interactions in a TWT determined by computer analysis. (Adapted from: P. Hess, Ph.D. dissertation, University of California at Berkeley, 1960.)

A major reason for saturation of the power output in TWTs is the reduction in average beam velocity caused by the transfer of electron energy to the circuit fields. An example of this reduction in average beam velocity is shown in Hess's results in Figure 13-33. The dashed line in the electron velocity plot shows average beam velocity. As we have discussed previously, partly as a result of the reduced velocity near saturation, electrons fall back out of phase with the circuit field, move into accelerating regions, and extract energy from the circuit.

This technique of modifying the circuit to enhance efficiency (and often other TWT characteristics) is known as velocity tapering. Early attempts to use velocity tapering produced little or no improvement in efficiency over untapered circuits. The advent of computer analyses (see [22], for example) has resulted in the design of tapers that produce appreciable improvements in efficiency. In [23], Meeker and Rowe reported an efficiency improvement from 18% to 35% through the use of a variable-pitch helix.

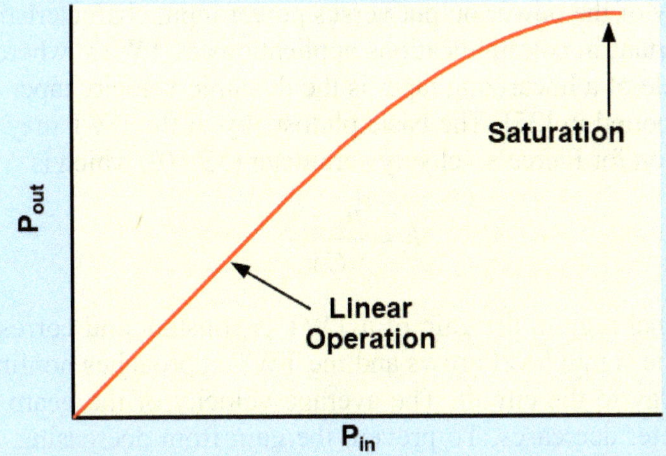

Figure 13-34 Klystron and TWT transfer curve.

Another example of a computer-designed taper reports an RF efficiency of 22.6% and a peak power of 1,000W [24]. Previously, the interaction structure for this particular tube had a piecewise linear taper that consisted of sequences of cavities of equal length and cavity sequences of linearly tapering length. The RF efficiency of that tube was 9.6% and the peak power output was 420W. In the phase-adjusted taper, the phase of the electron bunches with respect to the phase of the circuit field changes gradually from a value (about 0°) conducive to the formation of strong bunches at the input end of the taper to a value (about 90°) conducive to strong transfer of beam power to the circuit at the output end of the taper. A comparison of the efficiencies of the TWTs with the piecewise linear taper and with the phase-adjusted taper is shown in Figure 13-35.

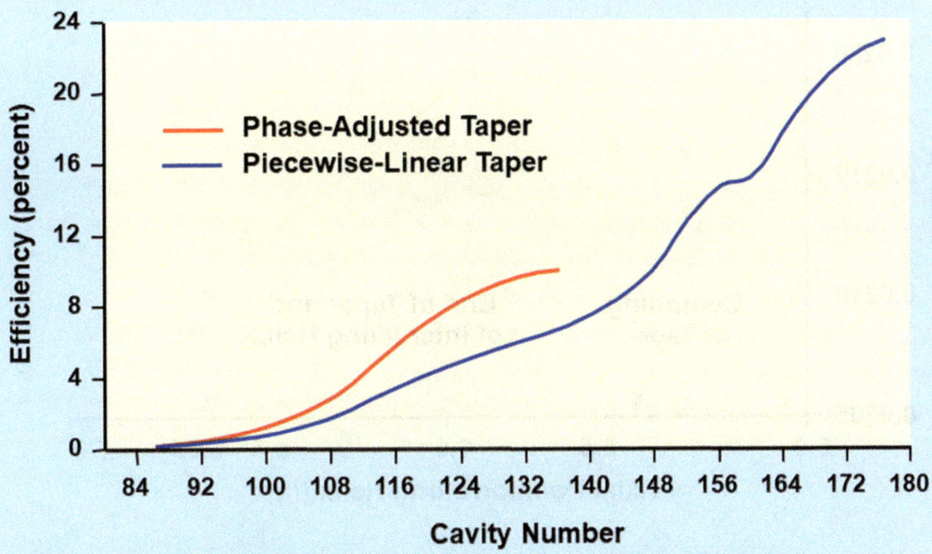

Figure 13-35 Computed RF efficiencies for a 20-GHz TWT with a phase-adjusted taper and one with a piecewise linear taper. (From: *NASA Tech Briefs*, September 1992.)

In addition to enhancing efficiency, a properly designed velocity taper can be used to enhance the linearity of the power output versus power input characteristic of a TWT. This is particularly important in communications applications of TWTs, where low distortion is required. An example of a linearizing taper is the dynamic velocity taper (DVT) developed by Kosmahl and reported in [25]. The basic philosophy of the DVT may be understood by examining the relation for Pierce's velocity parameter (13-10), which is

$$b = \frac{u_o - v_q}{C v_q} \tag{13-13}$$

In the small-signal region, the gain of a TWT is constant and corresponds directly to the value of b. As the signal level grows and the TWT approaches nonlinear operation the beam gives up energy to the circuit. The average velocity of the beam decreases and so the velocity parameter decreases. To prevent the gain from decreasing, it is necessary to modify the circuit and v_q so that the velocity parameter remains constant.

Results are presented in the referenced report [25] for TWTs with and without DVTs. For one of the TWTs with the DVT (the 13-GHz, 50-W Hughes 8802 experimental TWT), the pitch variation is shown in Figure 13-36 and the resulting improvement in linearity and power output is shown in Figure 13-37.

It has been recognized since the original invention of the TWT (see the varying pitch of the helix in Figure 13-3 from Lindenblad's patent) that efficiency might be improved if the circuit is modified so that the velocity of the circuit wave decreases along with the beam velocity. Then, even though the bunches are slowed down as they lose energy to the circuit, they will remain in phase with the circuit wave and continue to deliver energy.

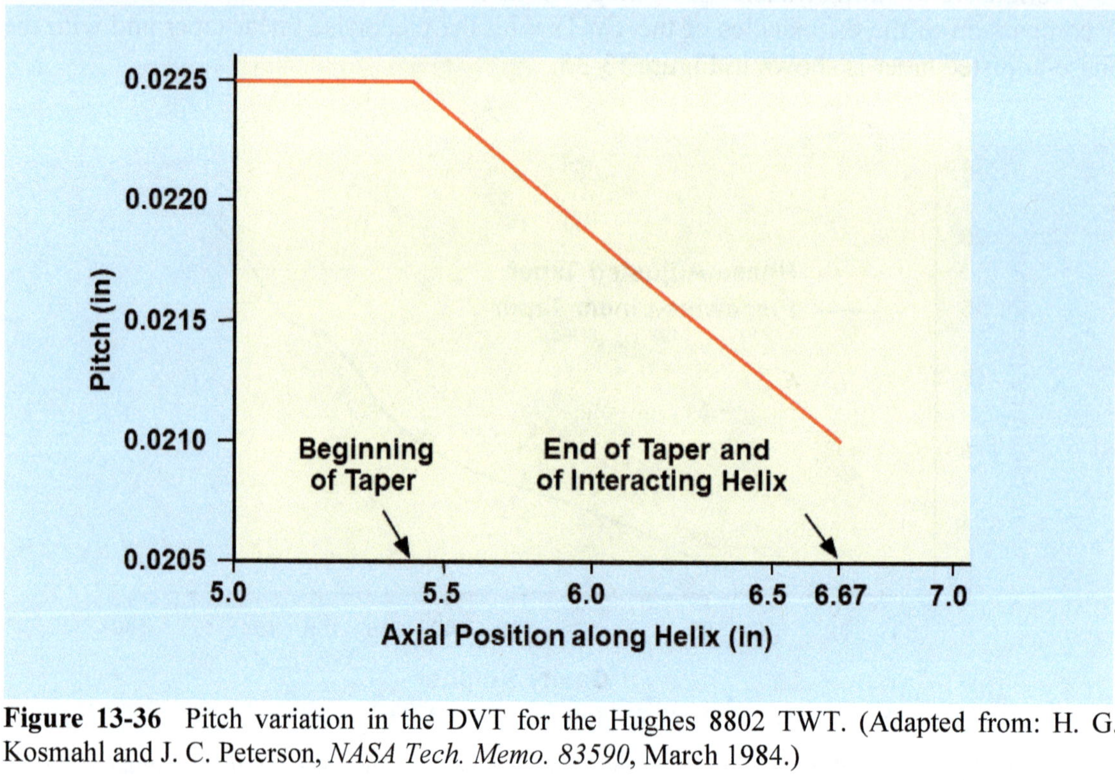

Figure 13-36 Pitch variation in the DVT for the Hughes 8802 TWT. (Adapted from: H. G. Kosmahl and J. C. Peterson, *NASA Tech. Memo. 83590*, March 1984.)

Note in Figure 13-30 that the pitch decreases linearly (this decrease is actually the first part of an exponential decrease) by about 7% over the final 18% of the length of the helix. The results of this change in pitch are impressive, as shown in Figure 13-37. The linearity near saturation is significantly improved, and the power characteristic approaches that of a hard limiter at saturation. (The discrepancy between computed and measured powers resulted from the use of cold rather than hot helix loss values in the computation.) The power output is increased about 1.5 dB and the RF efficiency (efficiency of conversion of beam power to RF output power) is about 16%, which is 5% better than without the DVT. The relative phase delay near saturation is about 30° and the AM to PM conversion does not exceed 3°/dB, which is nearly the same as that in an unmodified TWT.

Finally, Kosmahl and Peterson noted that with an RF efficiency of 16%, a well-designed multistage depressed collector should raise the overall efficiency above 50%. By replacing the beryllium oxide (BeO) support rods for the helix with diamond, losses could be reduced, the interaction impedance would be improved, and the overall efficiency would increase to about 65%.

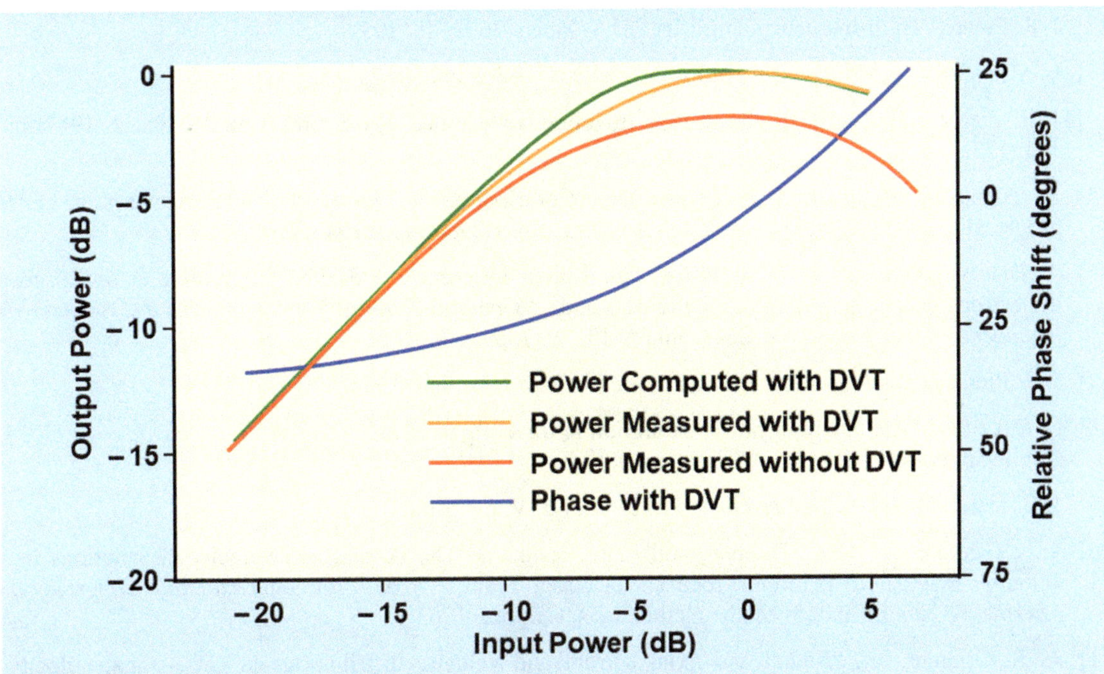

Figure 13-37 Effect of DVT on power and phase for the Hughes 8802. (From: H. G. . Kosmahl and J. C. Peterson, *NASA Tech. Memo. 83590*, March 1984.)

REFERENCES

[1] R. L. Wathen, "A history of the traveling wave tube," *Report No. 5203-5020, Sperry Gyroscope Company*, May 18, 1953.

[2] V. Haeff Andrew, "Device for and method of controlling high frequency currents," U.S. Patent 2064469, 1936.

[3] V. Haeff Andrew, "Device for and method of controlling high frequency currents," U.S. Patent 2233126, 1941.

[4] J. Copeland and A. Haeff, "Andrei Haeff and the amazing microwave amplifier," *IEEE Spectrum*, September 2015.

[5] E. Lindenblad Nils, "Electron discharge device system," U.S. Patent 2300052, 1942.

[6] R. Kompfner, "Travelling wave valve," *Wireless World,* Vol. 52, No. 11, 1946, pp. 369–372.

[7] R. Kompfner, "The traveling-wave tube as amplifier of microwaves," *Proc. IRE,* Vol. 35, No. 2, 1947, pp. 124–127.

[8] R. Kompfner, "Traveling-wave tube as amplifier of microwaves," *Wireless Engineer,* Vol. 24, No. 288, September 1947, pp. 255–266.

[9] R. Kompfner, "On the operation of the traveling wave tube at low level," *British Institution of Radio Engineers*, August–September 1950, pp. 283–289.

[10] R. Kompfner, *The Invention of the Traveling Wave Tube*, San Francisco, CA: San Francisco Press, 1963.

[11] J. R. Pierce, and L. M. Field, "Traveling-wave tubes," *Proc. IRE,* Vol. 35, No. 2, 1947, pp. 108–111.

[12] J. R. Pierce, "High-frequency amplifier," U.S. Patent 2602148, 1952.

[13] L. M. Field, "High-frequency amplifying device," U.S. Patent 2575383, 1951.

[14] J. R. Pierce, "Theory of the beam-type traveling-wave tube," *Proc. IRE,* Vol. 35, No. 2, 1947, pp. 111–123.

[15] A. Blanc-LA Pierre, and P. Lapostolle, "Contribution a l'etude des amplificateurs a ondes progressives," *Annales des Telecommunications*, No. 1, December 1946, pp. 283–302.

[16] A. Blanc-Lapierre, and P. Lapostolle, "Sur l'interaction entre une onde progressive et un faisceau d'electrons de vitesse voisine de celle de l'onde," *Comptes Rendue Rebdomadaires des Seances de l'Academie des Sciences*, No. 224, January 13, 1947, pp. 104–105.

[17] J. R. Pierce, *Traveling-Wave Tubes*, New York: Van Nostrand, 1950.

[18] C. C. Cutler, "The nature of power saturation in traveling wave tubes," *Bell Sys. Tech. Jour.,* Vol. 35, July 1956, pp. 841–876.

[19] J. C. Slater, *Microwave Electronics*, Princeton, NJ: Van Nostrand, 1950.

[20] A. S. Gilmour, Jr., G. C. Dalman, and L. F. Eastman, "The current and velocity distributions in a velocity modulated Brillouin focused electron beam," *Proc. 4th International Congress on Microwave Tubes,* Eindhoven, the Netherlands: Centrex, 1963.

[21] A. S. Gilmour, Jr., "Radial and axial current and velocity distributions in large-signal velocity-modulated electron beams," *IEEE Trans. on Electron Devices,* Vol. ED-19, No. 7, July 1972, pp. 886–890.

[22] P. Hess, "Traveling wave tube large signal theory with application to amplifiers having dc voltage tapered with distance," Ph.D. dissertation, University of California at Berkeley, 1960.

[23] J. G. Meeker and J. E. Rowe, "Phase focusing in linear-beam devices," *IRE Trans. Electron Devices,* Vol. 9, No. 3, 1962, pp. 257–266.

[24] NASA Lewis Research Center, "Coupled-cavity traveling-wave tube has phase-adjusted taper," *NASA Tech Briefs*, September 1992, pp. 26–30.

[25] H. G. Kosmahl and J. C. Peterson, "A TWT amplifier with a linear power transfer characteristic and improved efficiency," NASA Technical Memorandum 83590, prepared for the Tenth Communications Satellite Systems Conference, Orlando, FL, March 18–22, 1984.

Chapter 14

Helix TWT Performance

The TWT invented by Lindenblad used a helix as the slow-wave circuit. During the formative years of TWT development, hundreds of other propagation structures were investigated, but the helix remains as the circuit of choice when a broad bandwidth is required. The major parts of a helix TWT are shown in Figure 14-1. Major applications for helix TWTs include radar, communications, and electronic countermeasures (ECM). The purpose of this chapter is to describe the characteristics of helix TWTs that have made them the mainstay of the TWT industry.

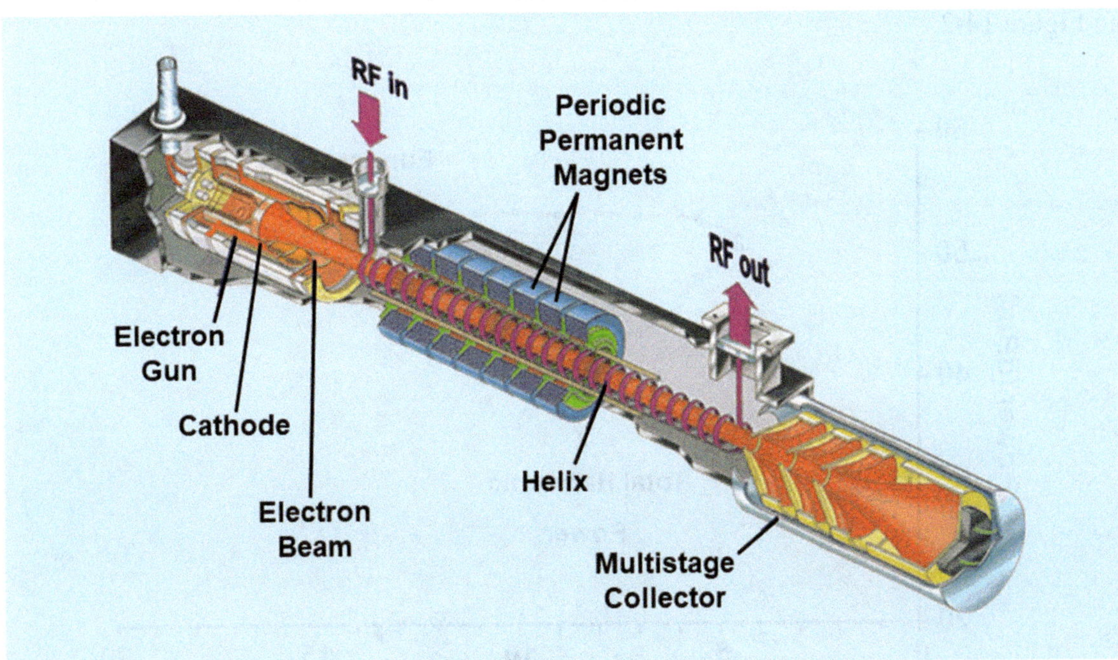

Figure 14-1 Parts of a TWT. (Courtesy of Thales.)

There is no clear dividing line between the domains of TWTs and solid-state devices because several factors (see [1] for a relevant discussion) other than average power and frequency must be considered in many applications. Some of these factors are:

1. In general, TWTs are far more efficient than solid-state devices. With the use of the appropriate collector technology, TWT efficiencies can approach or, in some cases, exceed 70%.

2. TWTs can operate at much higher temperatures than solid-state devices. This, coupled with higher efficiency, means that tubes can be smaller and lighter than solid-state devices and still eliminate waste heat.
3. Because of their extremely high reliability in space (MTBF < 100 FITs where the FIT has been adopted from the solid-state device community and equals one failure in one billion hours of operation), TWTs are the amplifiers of choice for most satellite applications. Of particular note is the performance of the X-band TWT on the Voyager I satellite, which was still in operation after 37 years in space. This TWT was the backup for the solid-state amplifier that failed early in the mission.
4. A factor of great importance in many applications is the large bandwidth for some TWTs. This can be over 2 octaves for some helix-type devices.

14.1 BANDWIDTH

The extremely broad bandwidth of the helix is the single most important property of this circuit. No other circuit has ever rivaled the helix in bandwidth and most do not even come close. Helix TWTs can now be made with a bandwidth of over 2 octaves, as shown in Figure 14-2.

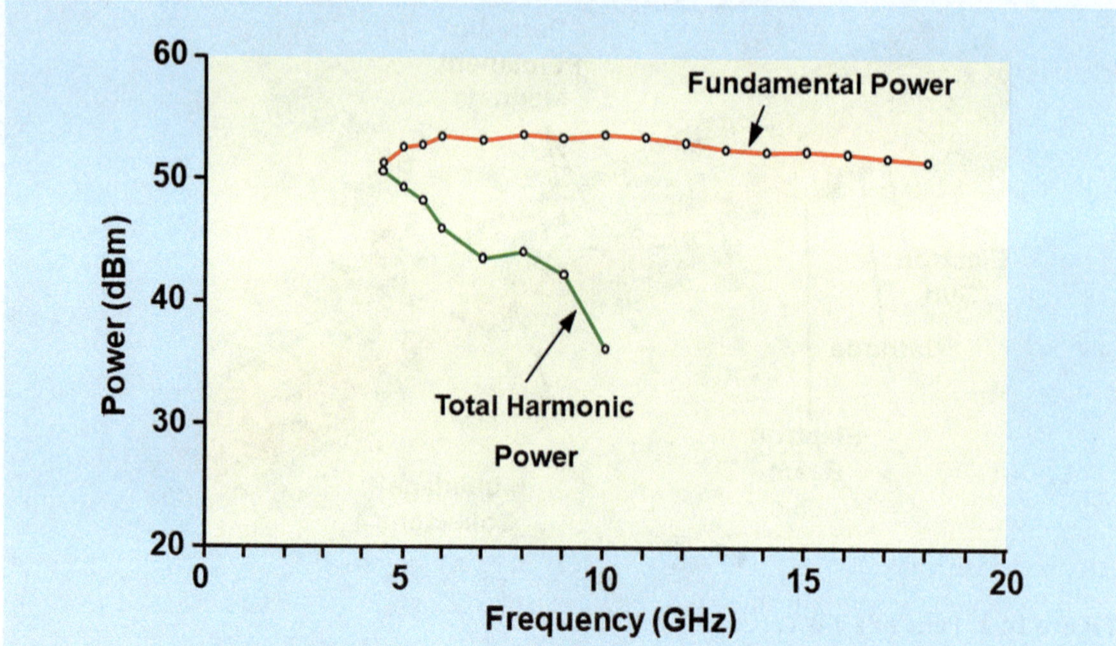

Figure 14-2 Saturated output power versus frequency for a Northrup TWT.

The broad bandwidth of the helix results from its low dispersion. As was described in the previous chapter, the helical circuit can be thought of as being derived from a single-wire transmission line for which the dispersion is zero. Shown in Figure 14-3(a) is a helix and in Figure 14-3(b) a view of the helix when it has been cut at the points marked by x and unrolled.

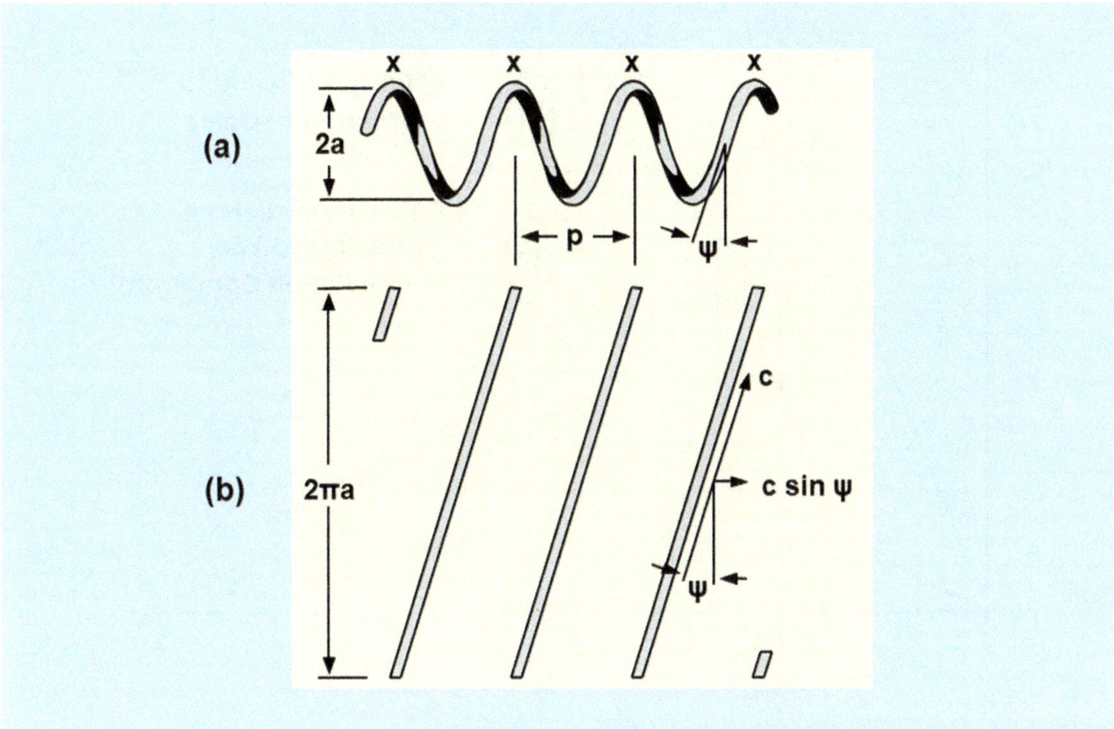

Figure 14-3 (a) Helix and (b) view of helix cut at each x and unrolled.

A wave propagating along a wire with the speed of light, *c*, advances in the axial direction with phase velocity

$$v_p \approx c \sin \psi \tag{14-1}$$

The angle ψ is the helix pitch angle given by

$$\cot \psi = \frac{2\pi a}{p} \tag{14-2}$$

where a is the radius and p is the pitch of the helix.

The ideal ω-β diagram for the helix is shown in Figure 14-4. The phase velocity, v_p, is represented by a straight line through the origin. Thus, in this case, the circuit has no dispersion. The dashed line represents the beam velocity, u_o, and has a slightly greater slope than the phase velocity line. This is so that the slow space-charge wave (the electron bunches), which has a velocity slower than the electron beam, is synchronous with the phase velocity.

In theory, a circuit with the ω-β characteristic in Figure 14-4 could be used over an almost unlimited bandwidth. In practice, bandwidth is limited by several factors. First of all, for a given circuit, as the frequency of operation is decreased, the length of the circuit, expressed in wavelengths, decreases. Since gain is proportional to circuit length in wavelengths, gain decreases.

As the frequency of operation is increased, the electric field pattern on the helix shrinks as is shown in Figure 14-5. As a result, the interaction of the field with the electron beam decreases and gain decreases. Maximum gain occurs at about four turns of the helix per wavelength (or $\pi/2$ radians per turn).

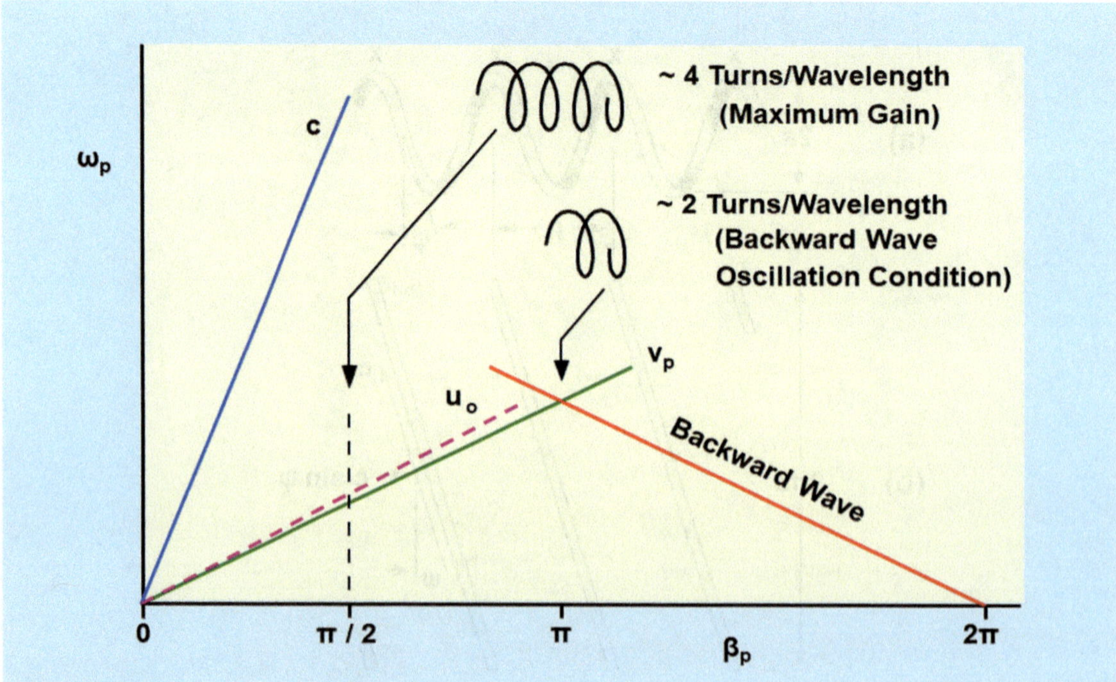

Figure 14-4 The ω-β diagram for a helix with no dispersion.

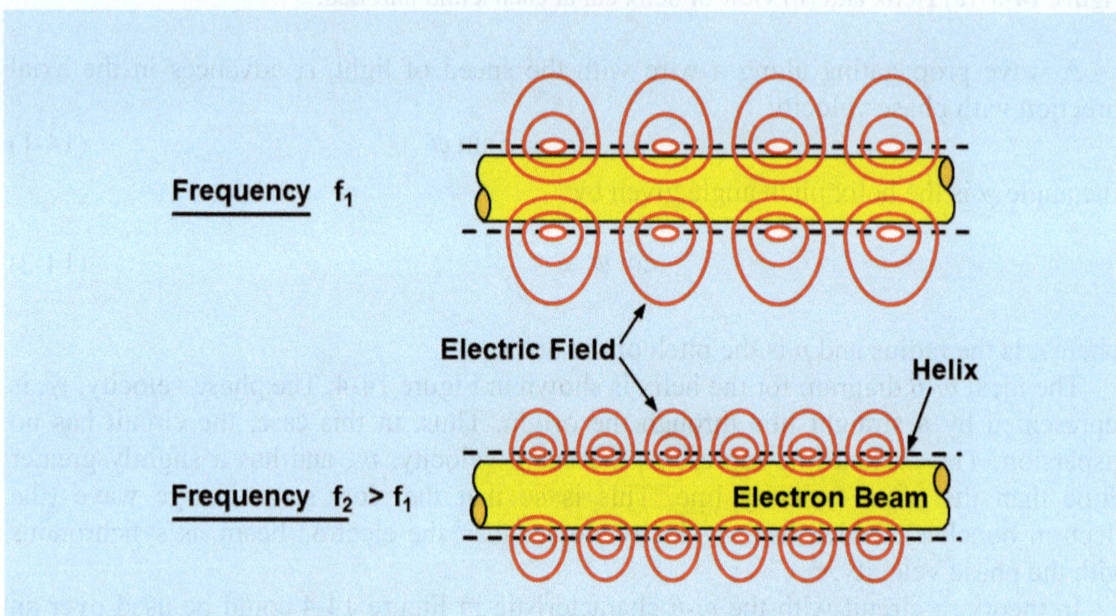

Figure 14-5 Change in the electric field that occurs as frequency is increased.

Another factor that limits bandwidth is dispersion. In reality, a helix with its supporting structure does have dispersion; this is discussed in Section 14.1.1.

A very serious factor that limits bandwidth and also the power capability of the helix is the backward-wave oscillations (BWO) that occur at $\beta_P = \pi$ radians. The reasons for these oscillations and techniques for controlling them are discussed later in this chapter.

If no techniques are used to control dispersion and oscillations, the bandwidth of the helix TWT is limited to about 1 octave at most. With dispersion control and BWO suppression, bandwidth can be increased to 2 or more octaves.

14.1.1 Dispersion

Unless corrective actions are taken, a helix has dispersion because phase velocity varies as shown in Figures 14-6 and 14-7. When compared with a coaxial line (with no dispersion), from which the helix can be considered to be derived, the helix has electric and magnetic coupling from turn to turn that is absent in the coax. This is one reason for the dispersion.

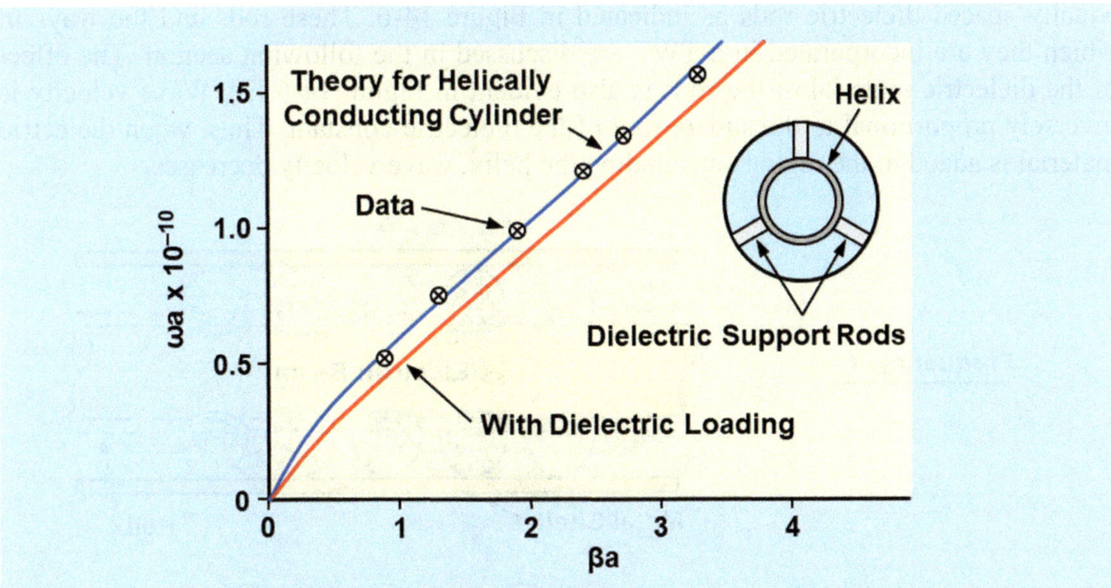

Figure 14-6 The ω-β diagram for a wire helix with a 10° pitch angle. (Adapted from: *Power Travelling Wave Tubes* by J. F. Gittins, published 1965 by American Elsevier, Inc.)

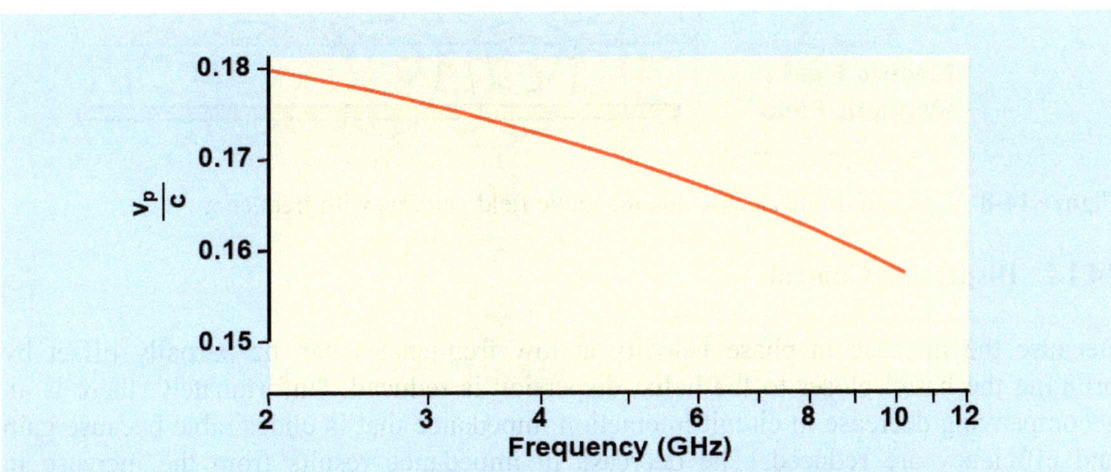

Figure 14-7 Variation of phase velocity with frequency with no dispersion control.

Another factor to consider is that at high frequencies (and a relatively small number of turns per wavelength), the electric and magnetic fields are concentrated between turns of the helix. As the frequency is reduced and the number of turns per wavelength is increased, the fields extend farther from the helix, as indicated in Figure 14-8. Therefore, a metal shell (barrel) surrounding the helix will intercept a greater portion of the fields at low frequencies than at high frequencies. (Note in Figure 14-8 that only the position of the barrel is shown. The distortion of the field patterns that the barrel produces is not shown.) As the barrel is brought closer to the helix, the effect increases, and the wave velocity on the helix decreases. The effect on velocity is greater at low frequencies than at high frequencies.

The helix is supported by an insulating structure that often takes the form of three equally spaced dielectric rods as indicated in Figure 14-6. These rods and the ways in which they are incorporated in a TWT are discussed in the following section. The effect of the dielectric material of the rods is also evident in Figure 14-6 [2]. Wave velocity is inversely proportional to the square root of the dielectric constant. Thus, when dielectric material is added in the region surrounding the helix, wave velocity decreases.

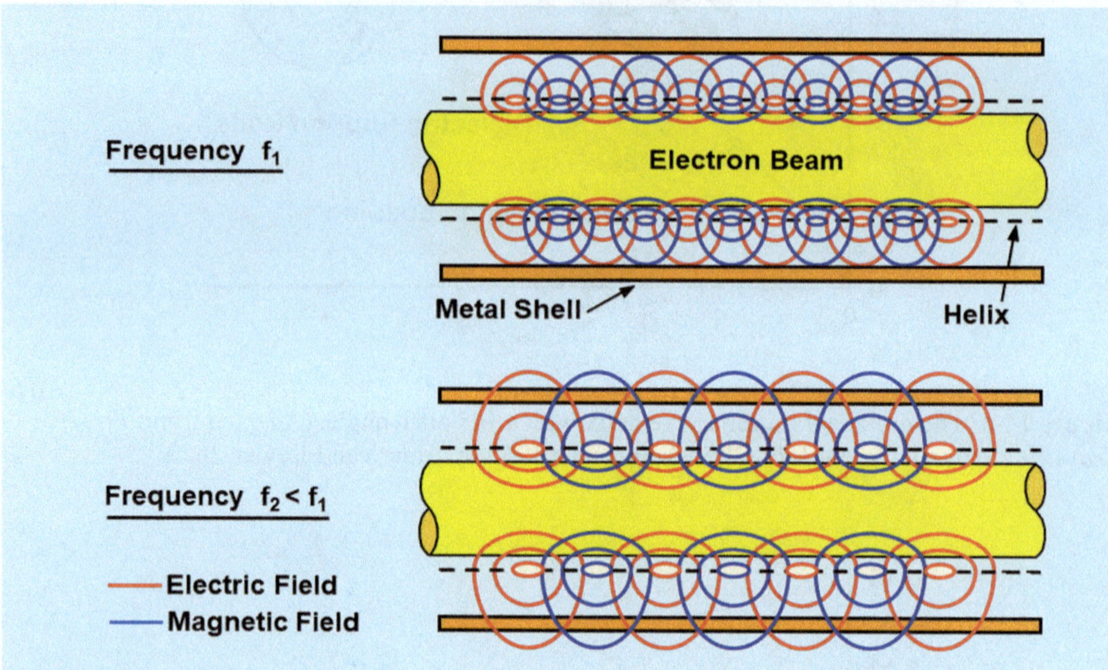

Figure 14-8 Variation of the electric and magnetic field patterns with frequency.

14.1.2 Dispersion Control

Because the increase in phase velocity at low frequencies can be partially offset by bringing the barrel closer to the helix, dispersion is reduced. Unfortunately, there is an accompanying decrease in circuit interaction impedance that is undesirable because gain and efficiency are reduced. The decrease in impedance results from the increase in capacitive loading and inductive coupling of the helix currents to the barrel and the flow of current in the circumferential direction in the barrel.

Ideally, if a metallic shell that conducted only in the axial direction could be placed around the helix, there would be no inductive coupling and no circumferential currents could flow in the shell. Still, the shell would have a greater effect on the electric field at low frequencies than at high frequencies so control of dispersion would be possible.

In practice, an axially conducting shell is approximated by one of the techniques illustrated in Figure 14-9 [3]. One of the four configurations shown (labeled unloaded) is used for comparison purposes and has only the outer metal shell, which supports the helix through dielectric support rods. In the other three configurations, either solid, vane-shaped, or T-shaped metallic elements, which are frequently referred to as anisotropic loading elements, are used to provide the desired loading.

Experiments have been performed to provide a basis for the choice of the best loading technique to use for controlling dispersion. Shown in Figure 14-10 are phase velocity and Pierce's velocity parameter as functions of frequency for three of the configurations in Figure 14-9. Tests were performed with two T-shaped elements having different spacings from the helix. The results were obtained in tubes containing these configurations, which were nearly identical in all other respects.

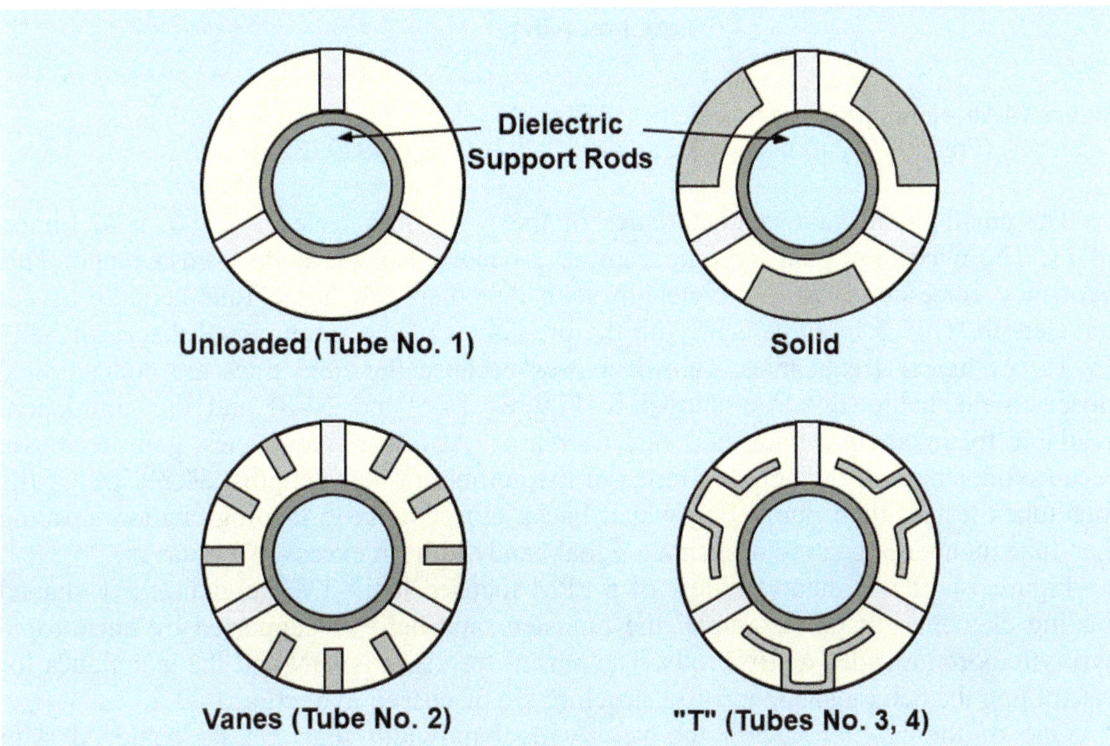

Figure 14-9 Techniques for dispersion control by approximating an axially conducting shell. (From: J. L. Putz and M. J. Cascone, 1979 IEDM. © 1979 IEEE.)

With tube number 1, phase velocity increases and *b* decreases as frequency is reduced. In tube number 2, the vanes produced a virtually flat phase-velocity characteristic over nearly the entire frequency range. In tubes number 3 and 4, the T-shaped segments produced negative dispersion characteristics. That is, the phase velocity decreased at low frequencies.

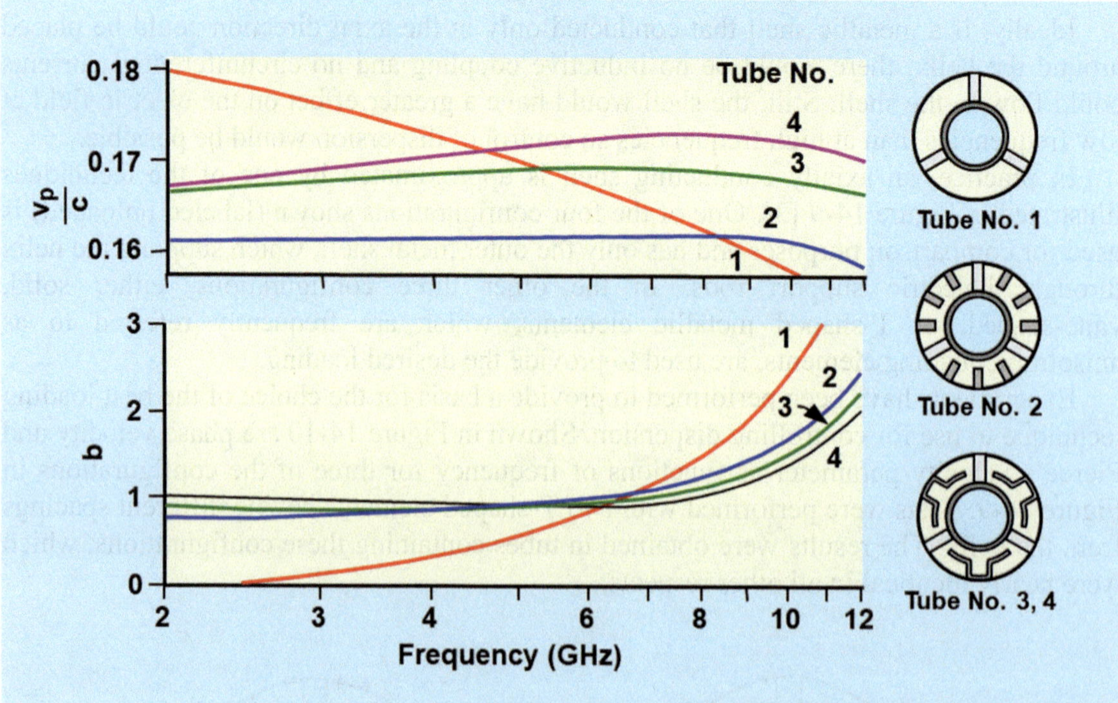

Figure 14-10 Normalized phase velocity and Pierce's velocity parameter as functions of frequency. (From: J. L. Putz and M. J. Cascone, 1979 IEDM. © 1979 IEEE.)

The small-signal gain characteristics of the four tubes tested are shown in Figure 14-11. The maximum gain for tube number 1 occurred in the 6- to 7-GHz range. This frequency corresponds to a wavelength such that there are about four helix turns per wavelength. This is in agreement with the prediction of the small-signal theory of helix TWTs. At higher frequencies, gain decreases because the field lines are concentrated closer to the helix (this was shown in Figures 14-5 and 14-8) and the component available for interaction with the beam decreases. At lower frequencies, gain decreases because the length of the helix, in terms of the number of wavelengths, decreases. Of the four tubes tested, tube number 4, which had a closer helix-to-loading-element spacing than tube number 3, had the best small-signal bandwidth (in excess of 2 octaves).

Figure 14-12 is a cutaway view of a PPM-focused helix TWT containing T-shaped loading elements. In this example, the tungsten tape helix is supported by anisotropic pyrolytic boron nitride (APBN) rods. The reason for using APBN and the techniques for assembling the helix and support-rod structure are discussed in Section 14.3.2.

One of the problems with the very wide bandwidth that can be achieved with anisotropic loading is the high harmonic power that is generated at low frequencies. This is shown in Figure 14-2 and in Figure 14-13. The harmonics result from the velocity modulation and bunching processes that are highly nonlinear. At the low end of the operating band of a broadband helix TWT, the second- and even the third-harmonic signals generated in the beam are within the bandwidth of the tube and are amplified. At the lower end of the band, harmonic power can rival fundamental power.

One way to reduce harmonic output power is with harmonic injection. An indication of the effectiveness of this technique is illustrated in Figure 14-13. When a second harmonic signal of the proper phase and amplitude is injected along with the fundamental signal, the harmonic output power drops and the fundamental power increases. This occurs when the injected second-harmonic signal is 180° out of phase with the second harmonic signal generated by the beam-circuit interaction process. With proper design of the input circuit, the cancellation process can be made reasonably noncritical.

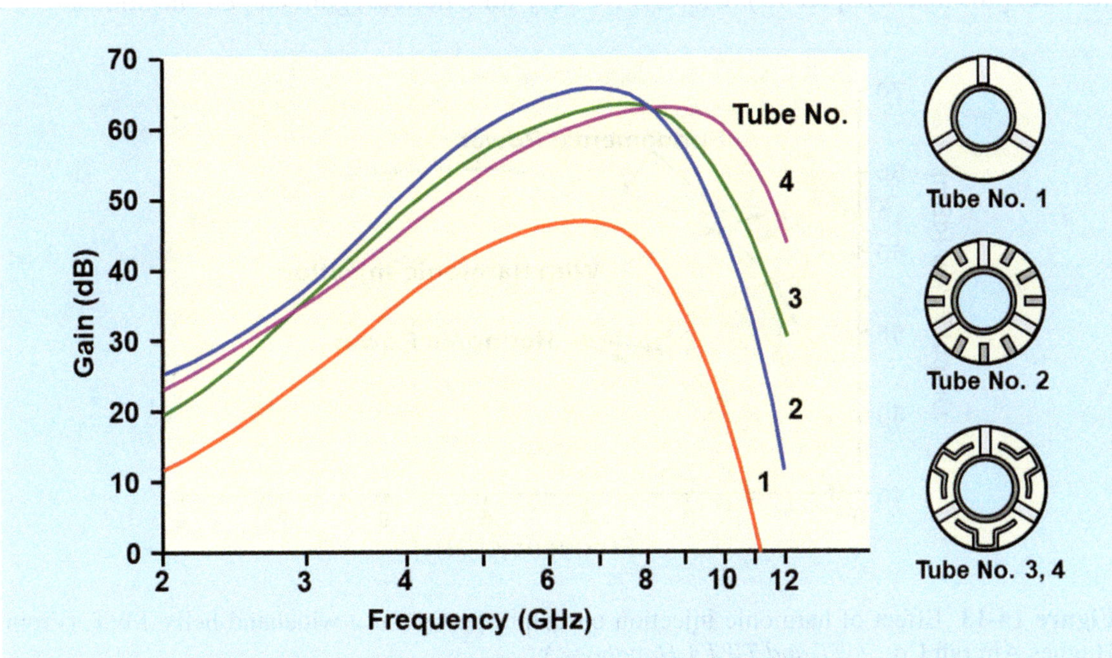

Figure 14-11 Small-signal gain for various loading configurations. (From: J. L. Putz and M. J. Cascone, 1979 IEDM. © 1979 IEEE.)

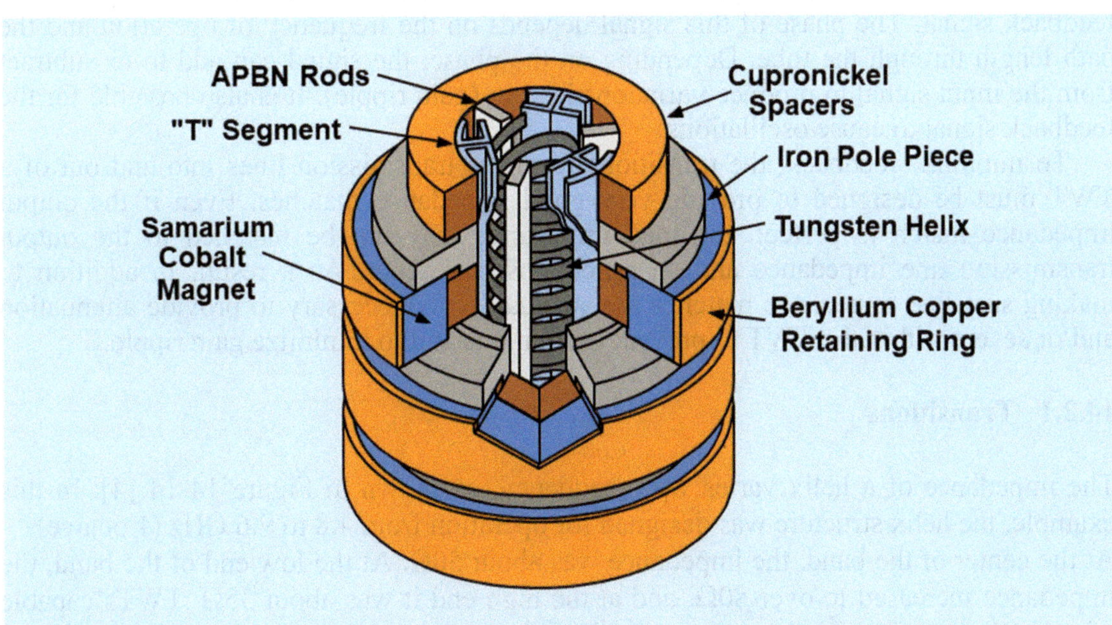

Figure 14-12 PPM-focused helix TWT with T-shaped loading elements. (Courtesy of Varian Associates.)

14.2 GAIN

At small signal levels, the gain produced by the interaction of the beam and the helix is proportional to *CN* where *C* is Pierce's gain parameter and *N* is the length of the helix in wavelengths (see Chapter 13). Clearly, for very long helices, gain can be enormous.

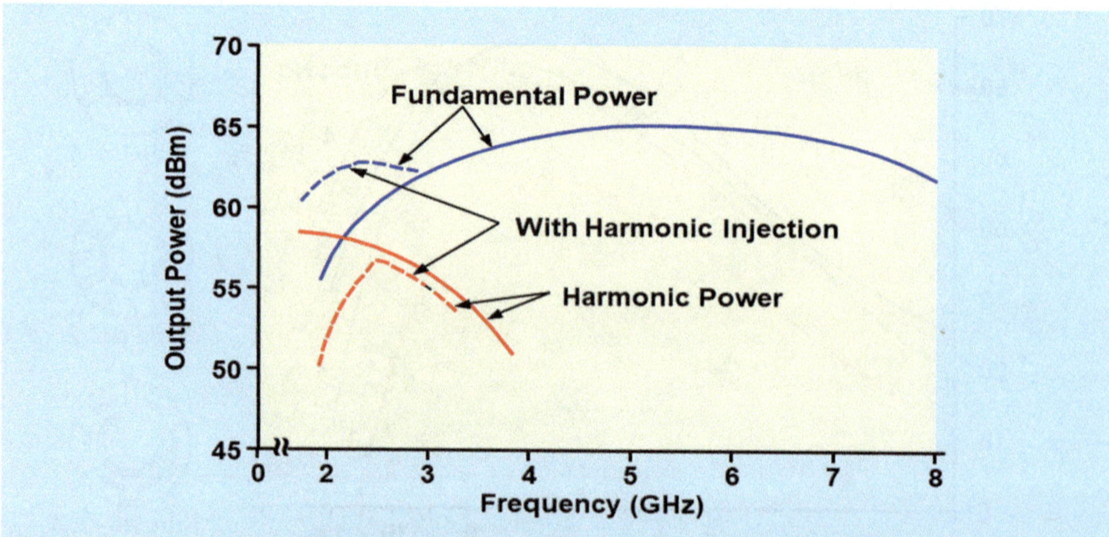

Figure 14-13 Effect of harmonic injection on output power for a wideband helix TWT. (From: Hughes Aircraft Co., *TWT and TWTA Handbook.*)

In a TWT, one of the four waves on the circuit travels in the backward direction. This wave can carry power reflected from the load or output region backward to the input. If there is a mismatch at the input, a portion of the signal will be reflected to provide a feedback signal. The phase of this signal depends on the frequency of operation and the path length through the tube. Depending on the phase, the signal can add to or subtract from the input signal to produce variations in gain (gain ripple). It is also possible for the feedback signal to cause oscillations.

To minimize feedback, the transitions from the transmission lines into and out of a TWT must be designed to provide very good impedance matches. Even if the output impedance match is perfect, the load impedance may not be matched to the output transmission line impedance and so reflections can occur. As a result, in addition to making sure that impedance matches are adequate, it is necessary to provide attenuation and/or severs within the TWT to prevent oscillations and to minimize gain ripple.

14.2.1 Transitions

The impedance of a helix varies with frequency, as shown in Figure 14-14 [4]. In this example, the helix structure was designed for operation from 4.8 to 9.6 GHz (1 octave). At the center of the band, the impedance was about 50Ω. At the low end of the band, the impedance increased to over 80Ω, and at the high end it was about 35Ω. TWTs capable of operating over more than an octave have even greater impedance variations. As a result, the impedance transitions at the input and output of these TWTs must be able to match the wide variation in helix impedance.

Designs for broadband transitions are based on the idea illustrated in Figure 14-15. Here the tapered section of a coaxial line is used to match the low impedance of the coax on the left to the high impedance of the coax on the right. An excellent impedance match is realized if the tapered transition is long compared with the wavelength of the signal.

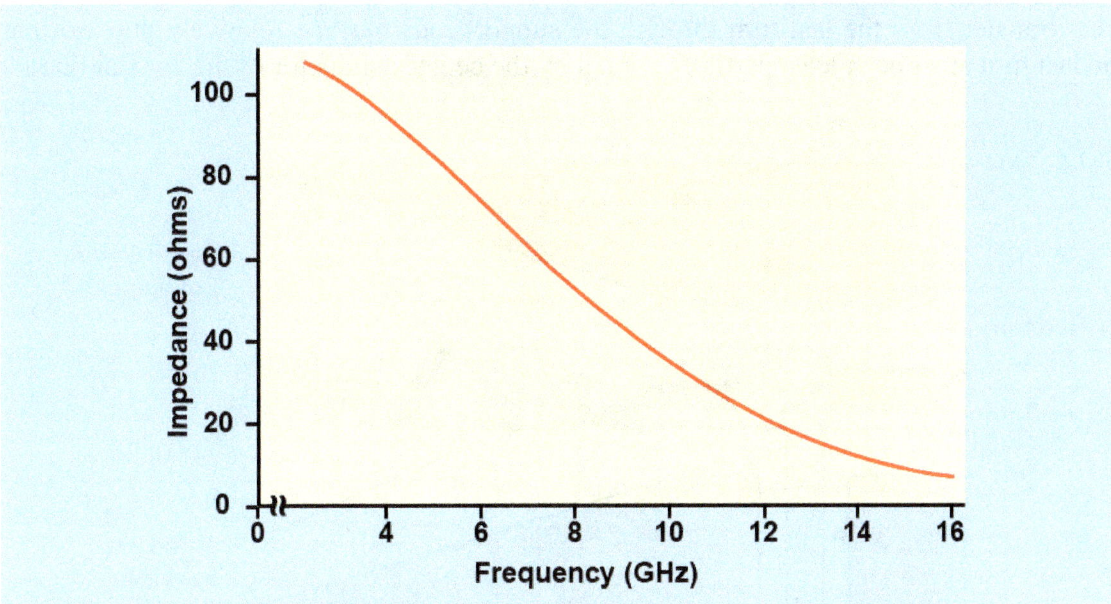

Figure 14-14 Helix impedance for an octave-bandwidth TWT. (From: A. S. Gilmour, Jr. et al., *IEEE Trans. Electron Devices*, October. 1979. © 1979 IEEE.)

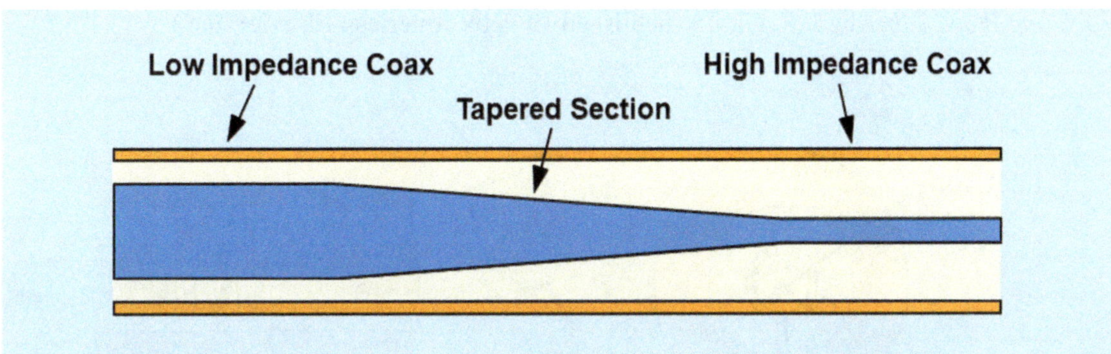

Figure 14-15 A tapered transition that matches a low impedance coaxial line to a high-impedance line.

In a TWT, a very broadband impedance match to the helix could be made with the transition shown in Figure 14-16 [2]. A gradual change in geometry is made so that the helix slowly changes into a coaxial line. As long as the overall transition is long compared to the wavelength, there would be an excellent impedance match.

Unfortunately, transitions like that shown in Figure 14-16 have the very serious disadvantage of being long. If used, they would increase the length and weight of a TWT. In practice, transitions are normally made after the signal is removed from the helix. For example, the center conductor of the coaxial line can be tapered as shown in Figure 14-17.

A problem with the configuration in Figure 14-17 is that it requires a relatively long unsupported section of the center conductor of the coaxial line. Under severe shock and vibration conditions, the weld joint at the end of the helix may fail. Also, the last turn of the helix tends to be more poorly supported than other turns when the triangulation or pressure-insertion assembly techniques to be described in the following section are used. Heat transfer from the last turn through the support rods may be relatively poor so that the last turn must be at least partially cooled by the center conductor of the coaxial line.

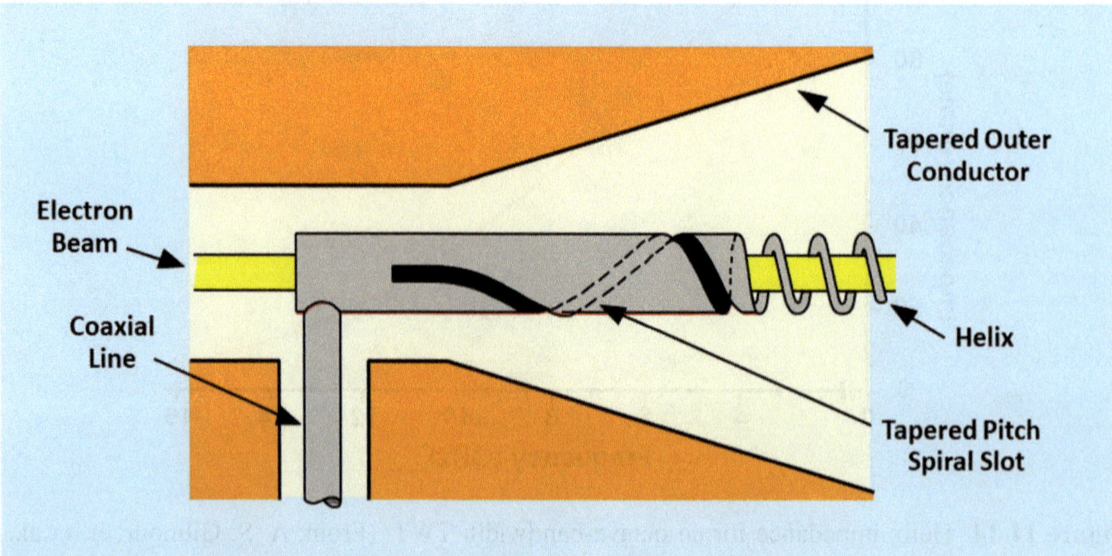

Figure 14-16 A very broadband transition from a helix to a coaxial line. (Adapted from: *Power Travelling Wave Tubes* by J. F. Gittins, published 1965 by American Elsevier, Inc.)

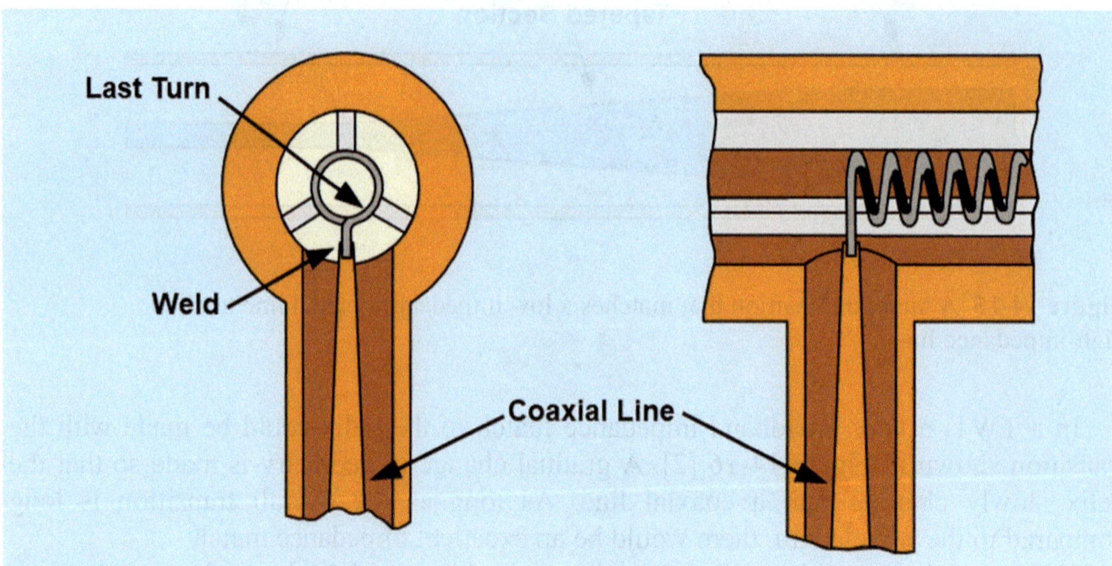

Figure 14-17 Transition using a tapered coaxial line. (Adapted from: *Power Travelling Wave Tubes* by J. F. Gittins, published 1965 by American Elsevier, Inc.)

To provide adequate cooling and help prevent failure of the weld, there should be a support for the center conductor located as close to the helix as possible. For example, in the output window of a TWT, the insulating support for the center conductor is usually only a fraction of a wavelength from the connection to the helix. The material of this support is usually beryllium oxide, which is a very good thermal conductor. The beryllium oxide is brazed to the center conductor and to the outer conductor and serves as part of the vacuum envelope.

Also, it should be pointed out that Teledyne uses a flexible strap to connect the end of the helix to the center conductor. This prevents movements of the center conductor (because of thermal expansion and contraction) from being transmitted to the helix.

14.2.2 Attenuators and Severs

Figure 14-1 illustrates a TWT having no attenuators or severs. The limiting gain of this device may be readily estimated. If the reflection coefficients at the output and input terminals of the TWT are, respectively, ρ_o and ρ_i, then oscillations may occur when

$$G - L - \rho_o - \rho_i > 0 \tag{14-3}$$

where G (gain), L (circuit loss), and the reflection coefficients are expressed in decibels.

Although the tube may be very well matched to the signal source and to the load, there will normally be reflections at input and output terminals because of the difficulty in making impedance transitions from the RF structure. As a result, it may be difficult to make ρ_o and ρ_i much greater than 10 dB (10% reflections). If the cold loss of the circuit is on the order of 6 dB, then the limiting gain is on the order of 26 dB.

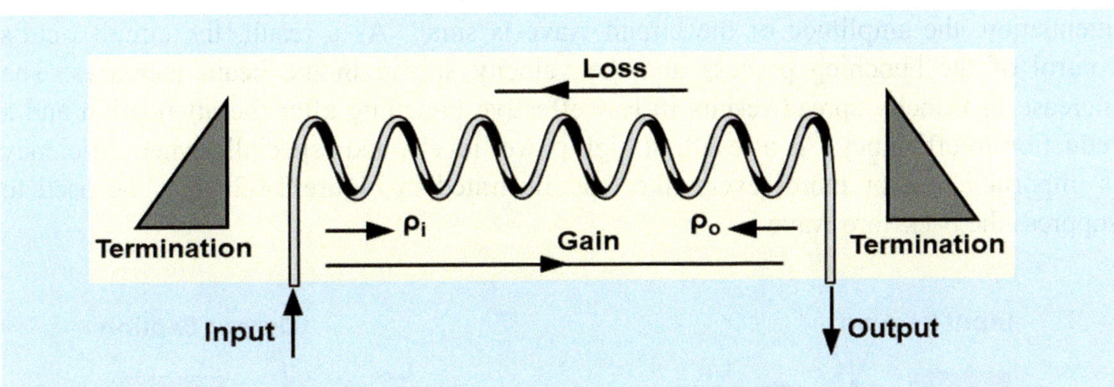

Figure 14-18 Quantities used to determine the oscillation condition. (Adapted from: *Power Travelling Wave Tubes* by J. F. Gittins, published 1965 by American Elsevier, Inc.)

To prevent large variations in gain with frequency, the actual gain cannot be as high as 26 dB and, in fact, is sometimes limited to about 20 dB. To permit increased gain, attenuators and severs are used in TWTs.

A common form of attenuation is shown in Figure 14-19. Portions of the support rods for a helix are coated with a lossy material. The thickness of the film is tapered to zero at the ends of the attenuation region to obtain a good impedance match over a wide range of frequencies.

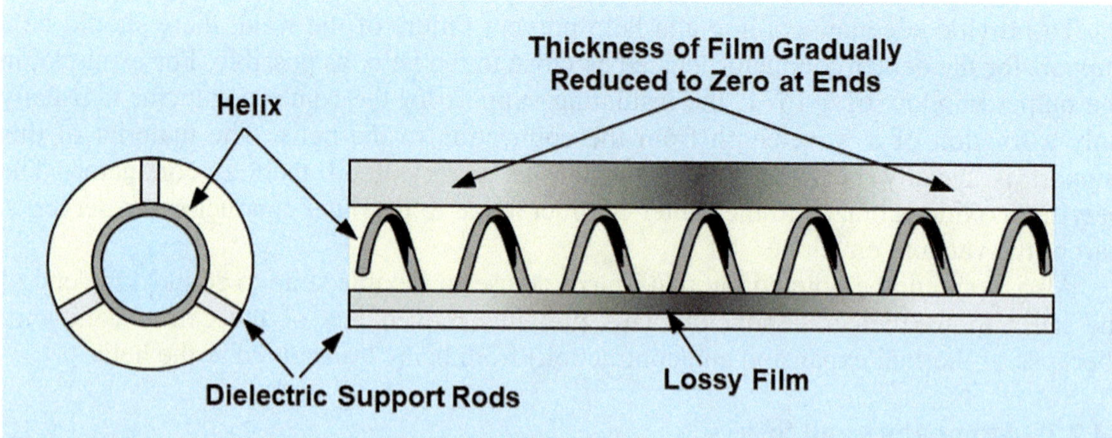

Figure 14-19 Lossy film attenuator for use with a helix.

Carbon films have often been used as the attenuator material. However, the loss has been found to change with time, probably as the result of electron bombardment [5, 6, 7]. This can lead to excess gain, instabilities and BWO. Anand and Morris reported the use of tantalum rather than carbon as a way to produce stable films that did not change attenuation with time [5].

Forward waves as well as backward waves are attenuated by the lossy film. However, we must remember that Pierce showed that the reduction in gain is only about one-third of the attenuation in the backward wave. Also, the loss in gain in the forward direction may be compensated for by an increase in length of the interaction region.

Unfortunately, if a good impedance match is to be realized, the attenuator must extend for a considerable length along the RF structure and, where there is significant attenuation, the amplitude of the circuit wave is small. As a result, the circuit field's control of the bunching process and the velocity spread in the beam increases. The increase in velocity spread results in less effective bunching after the attenuation and a reduction in efficiency. As a result, at high-power levels, and especially when efficiency is important, one or more severs like that illustrated in Figure 14-20 may be used to suppress the backward wave.

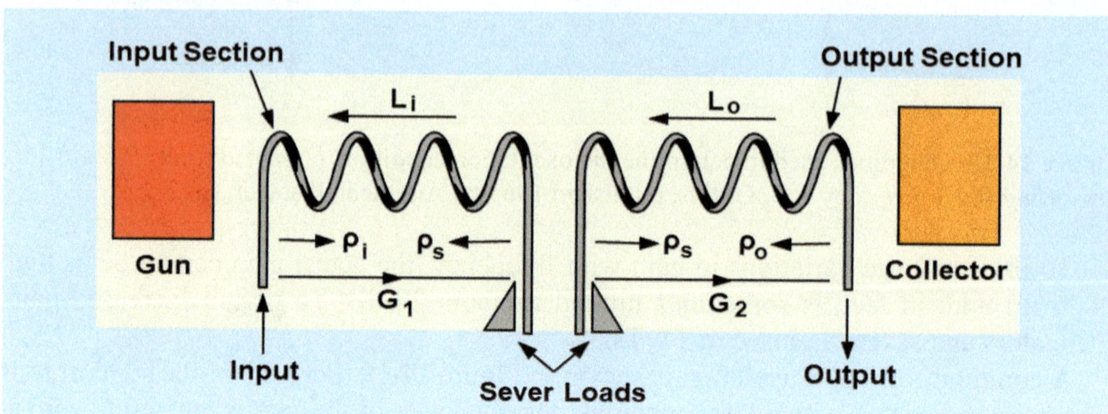

Figure 14-20 TWT containing a two-section severed helix. (Adapted from: *Power Travelling Wave Tubes* by J. F. Gittins, published 1965 by American Elsevier, Inc.)

The sever prevents reflections from the load and the output terminal from reaching the input terminal. Although the forward-growing wave on the input section of the helix is lost at the sever, current and velocity modulation remain on the electron beam to carry the signal across the sever region for further amplification in the output section.

As the beam drifts across the sever region, bunching degrades and affects the final efficiency of the tube. As a result, the sever region should be as short as possible. Also, the amount of degradation depends on the intensity of bunching in the input section of the slow-wave structure. Thus, for maximum efficiency, the gain of the input section should be relatively low and the gain of the output section should be as high as possible.

Just as there is a limit of about 20 dB to the practical gain of a tube without a sever, there is a similar limit to the gain of each section of a tube with severs. Referring again to Figure 14-20, the reflection coefficients at the output, ρ_o, at the sever, ρ_s, and at the input, ρ_i, along with the circuit attenuations L_o and L_i, must be taken into consideration in determining gain levels. Because it is usually possible to keep the reflection coefficients at the sever terminations less than those for the input and output, the gain per section of a severed tube can be higher than that for a single-section tube. On the other hand, there is some signal loss in the sever region so that an overall gain from a two-section tube of about 40 dB is practical. With additional severs, higher gains are possible.

A cross-sectional view of a broadband TWT containing an impedance transition, attenuators, and a sever is shown in Figure 14-21. The impedance transition at the output is made by varying the height of the waveguide in steps. The attenuators are made by depositing lossy material on the dielectric support rods for the helix. The sever occurs where the circuit is physically divided into two sections. Lossy terminations are used to absorb signals reflected by the sever.

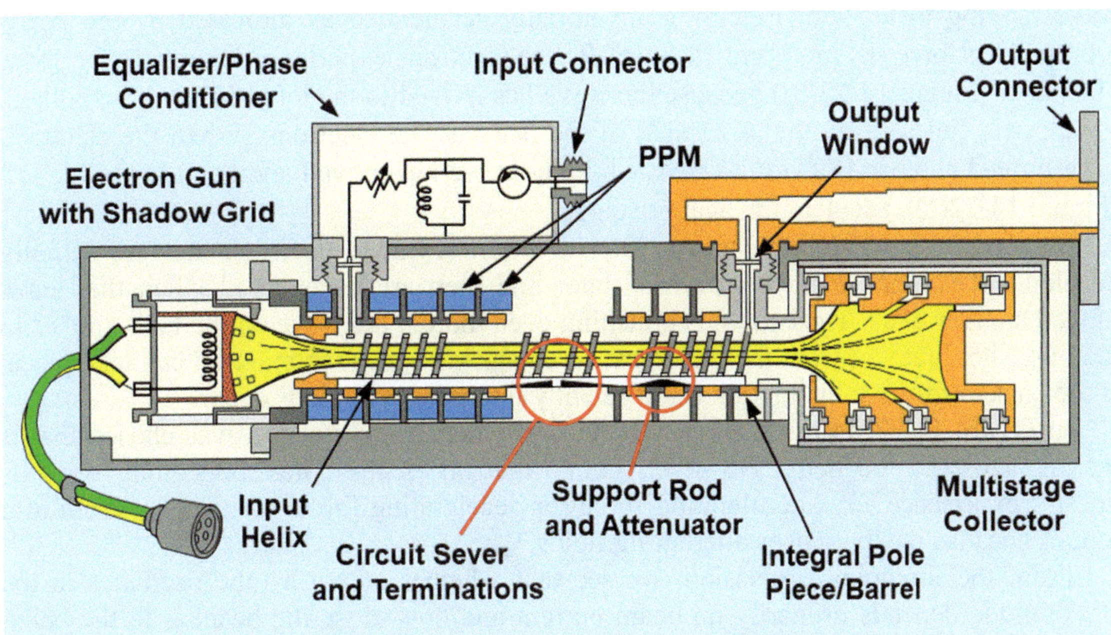

Figure 14-21 Helix TWT for electronic countermeasures use. (Courtesy of Varian Associates.)

14.3 POWER

Although a peak power of 10 kW has been demonstrated for a helix TWT [10], peak power is usually limited to a few kilowatts. At the end of this chapter, TWTs with ring-bar and ring-loop slow-wave structures are discussed. These structures are said to be derived from helices. Ring-bar and ring-loop TWTs are capable of peak powers of tens of kilowatts.

The average power of the helix TWT depends on the frequency of operation because size scales inversely with frequency. At low frequencies, average powers in the kilowatt range are possible. In the millimeter range, average power is limited to tens of watts.

In this section, the reasons for the peak and average power limitations are discussed.

14.3.1 Peak Power

A major problem that limits the peak power of helix TWTs is backward-wave oscillations (BWO). When these oscillations occur, there is a reduction in power at the desired frequency of operation and so the oscillations produce in-band power holes. In addition, the BWO signal may induce oscillations or may intermodulate with the main signal to produce spurious signals inside the operating band. Before discussing the reasons for the limits on peak power, it is important to have an understanding of backward-wave oscillations.

The BWO problem can be explained with the aid of Figure 14-22. At a frequency where the electron transit angle from turn to turn is 180°, there are two turns per wavelength or 180° per turn and a backward-wave interaction can occur. Consider electrons near the helix moving to the right (designated by the red dots) and a backward wave moving to the left. Electrons are initially accelerated as indicated by the arrow representing force (F) in Figure 14-22(a). These electrons experience essentially no force 180° later (Figure 14-22(b)) because the wave has moved to the left 180° and the nulls in the electric field are on the same side of the helix as the electrons. When the electrons have moved another 180° to the right while the wave has moved another 180° to the left (Figure 14-22(c)), acceleration again occurs.

Similarly (shown by the green dots in Figure 14-22), electrons that are initially decelerated experience no forces 180° later and then are decelerated when they have moved another 180°. Thus, there is bunching even though the wave on the helix moves in the opposite direction from the electrons, so a backward-wave interaction can take place. If the backward-wave interaction is sufficiently intense, oscillations occur.

In Figure 14-22, to illustrate the BWO phenomenon, an electron was chosen that is moving close to the helix. However, notice that an electron traveling along the axis doesn't experience a net accelerating force (or decelerating force) as it moves from turn to turn because it experiences alternating fields.

From the foregoing discussion, we see that whether or not a tube oscillates in the BWO mode depends primarily on beam current and how close the beam is to the helix. An illustration of this dependency is shown in Figure 14-23 [9].

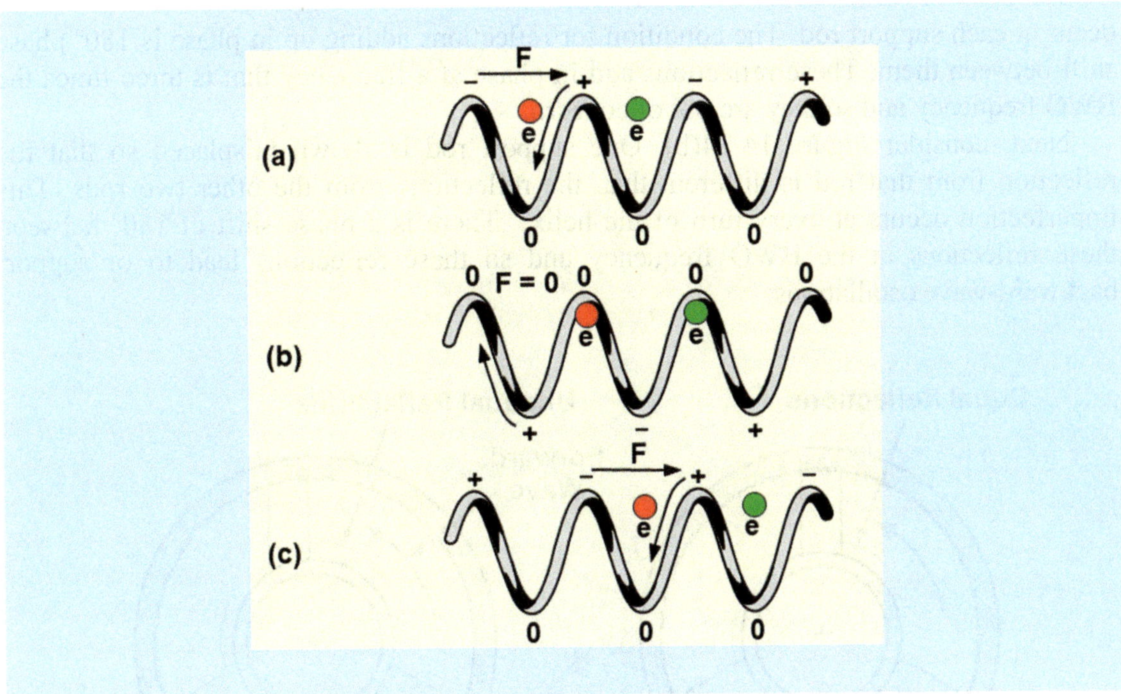

Figure 14-22 Backward-wave interaction on a helix at two turns per wavelength.

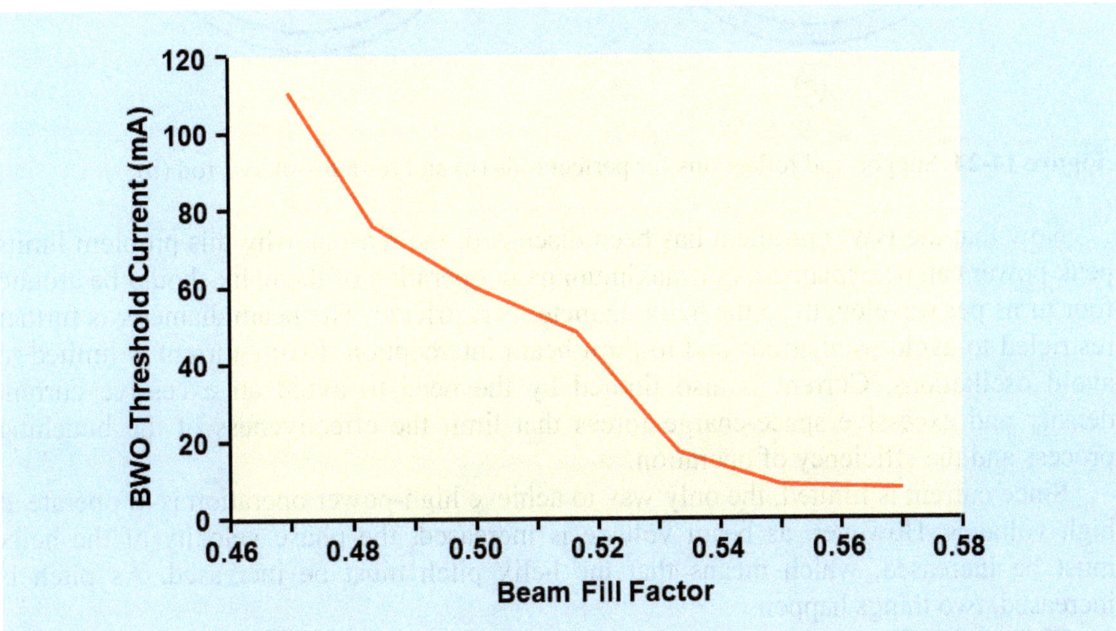

Figure 14-23 Dependence of start oscillation current on beam size. (From: P. Birtel et al., 2006 IEEE IVEC. © 2006 IEEE.)

Backward-wave oscillations can be exacerbated by defects in the circuit construction. For example, consider the situation shown in Figure 14-24. In Figure 14-24(a), the helix is assumed to be perfect and the support rods are identical, placed exactly 120° apart, and so forth. Assume the forward wave is at the BWO frequency so that there are two turns per wavelength or 180° per turn. As the wave travels around the helix, small reflections occur at each support rod. The condition for reflections adding up in phase is 180° phase shift between them. These reflections add in phase at a frequency that is three times the BWO frequency and so they are not of concern.

Next, consider Figure 14-24(b). One support rod is shown misplaced so that the reflection from that rod is different than the reflections from the other two rods. This imperfection occurs at every turn of the helix. There is a phase shift of 180° between these reflections at the BWO frequency and so these reflections lead to or support backward-wave oscillations.

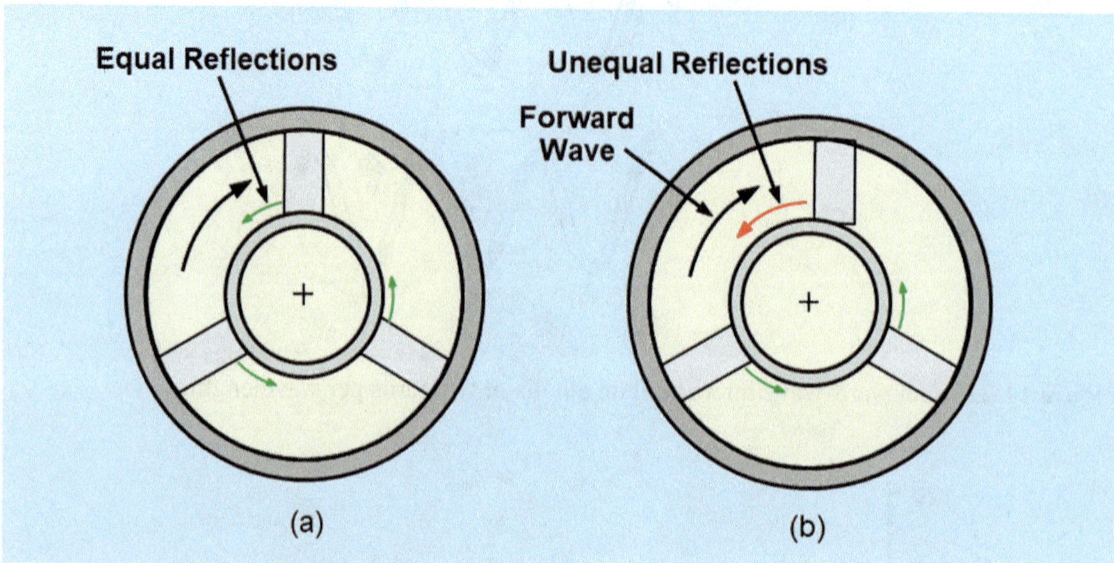

Figure 14-24 Support rod reflections for perfect rods (a) and for a misplaced rod (b).

Now that the BWO problem has been discussed, the reasons why this problem limits peak power can be explained. For maximum gain, operation of the helix should be around four turns per wavelength so the helix diameter is restricted. The beam diameter is further restricted to avoid oscillations and to limit beam interception. Beam current is limited to avoid oscillations. Current is also limited by the need to avoid an excessive current density and excessive space-charge forces that limit the effectiveness of the bunching process and the efficiency of operation.

Since current is limited, the only way to achieve high-power operation is to operate at high voltages. However, as beam voltage is increased, the phase velocity of the helix must be increased, which means that the helix pitch must be increased. As pitch is increased, two things happen.

First, to maintain about four turns per wavelength, the helix diameter must be decreased, so for a given beam size, the possibility of oscillations is increased. This forces a reduction in beam size and current and a reduction in power capability. Second, as pitch is increased and the electric fields interacting with the beam are decreased, the tube becomes more prone to oscillations.

Because the BWO frequency is such that the beam has a 180° transit angle from turn to turn, the BWO frequency is tunable with beam voltage. In fact, tubes designed for operation in the BWO mode can be voltage-tuned over a reasonably broad frequency range. Helix BWOs are low-power devices typically generating tens of milliwatts. Operating frequencies extend well into the millimeter frequency range. These devices have, for the most part, been replaced by solid-state signal sources.

The most common technique for suppressing backward-wave oscillations is to introduce a change in the slow-wave circuit that changes the phase velocity. This can be done by changing the diameter of the helix, the number of helix support rods, or, as shown in Figure 14-25, the pitch of the helix [10]. In this case, the relatively small pitch of the right-hand side of the helix is such that energy is transferred from the beam to the backward wave and oscillations occur. As the backward wave travels to the left, it encounters a helix section with a larger pitch. Thus, the wave travels faster and, if the pitch is correct, the wave travels at the velocity of the fast space-charge wave. As a result, energy is transferred from the backward wave back into the beam and the backward wave is attenuated.

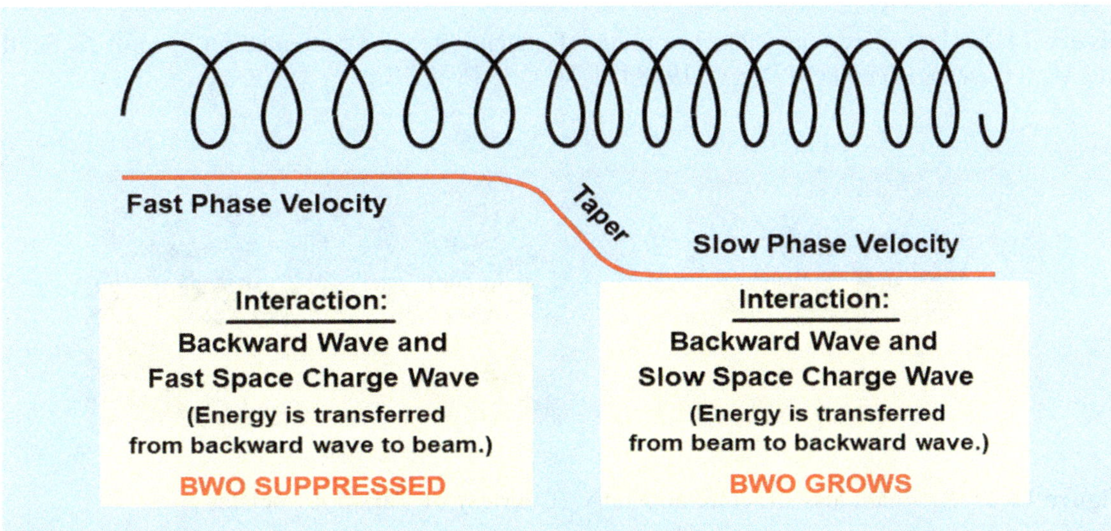

Figure 14-25 Illustration of the use of a pitch change to suppress backward-wave oscillations. (From: A. Scott and M. J. Cascone, *Technical Digest*, 1978 IEDM. © 1978 IEEE.)

Another method for suppressing BWO is the introduction of frequency-sensitive attenuation into the helix circuit. A technique developed by Varian Associates (now CPI) for producing frequency-sensitive loss is illustrated in Figures 14-26 and 14-27 [10]. Varian refers to this loss as resonant loss. Meander-line circuits are placed on the support rods in the regions between turns of the helix. The circuits are fabricated using standard microwave integrated circuit technology. The total length of each line is equal to half the wavelength at the BWO frequency so that the lowest resonant frequency is the BWO frequency.

The amount of loss that can be introduced with the resonant loss technique can be several tens of decibels. The power output for a 10-kW helix tube designed for operation in the 3.0- to 6.0-GHz range is shown in Figure 14-28 [8]. The BWO frequency of this tube was just above the top end of the band at about 8.0 GHz. Note the very sharp drop in power at the high end of the band where resonant loss becomes effective.

A problem that has been identified in using resonant loss is that a harmonic of the signal being amplified by the TWT may be at the frequency for which the resonant loss circuits have been designed. As a result, the resonant loss structures can be destroyed.

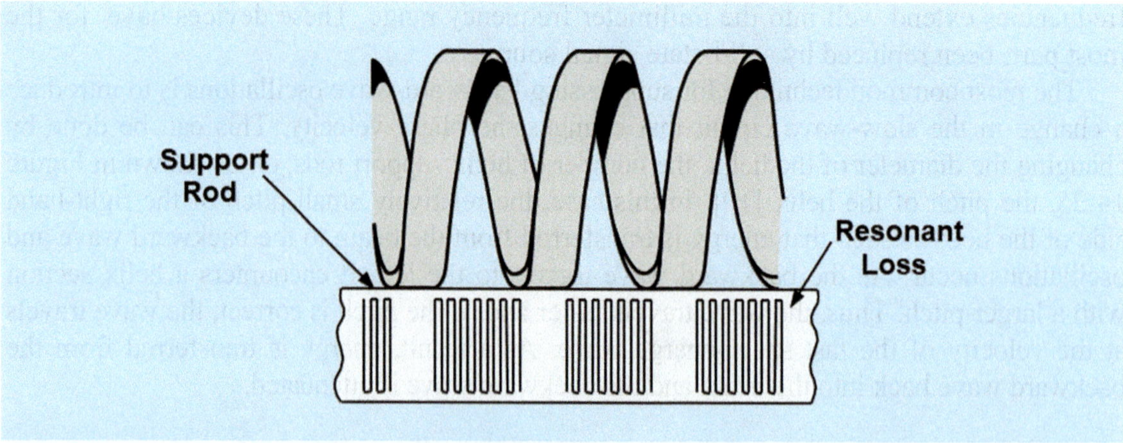

Figure 14-26 Use of resonant loss to produce frequency sensitive attenuation. (From: A. Scott and M. J. Cascone, *Technical Digest*, 1978 IEDM. © 1978 IEEE.)

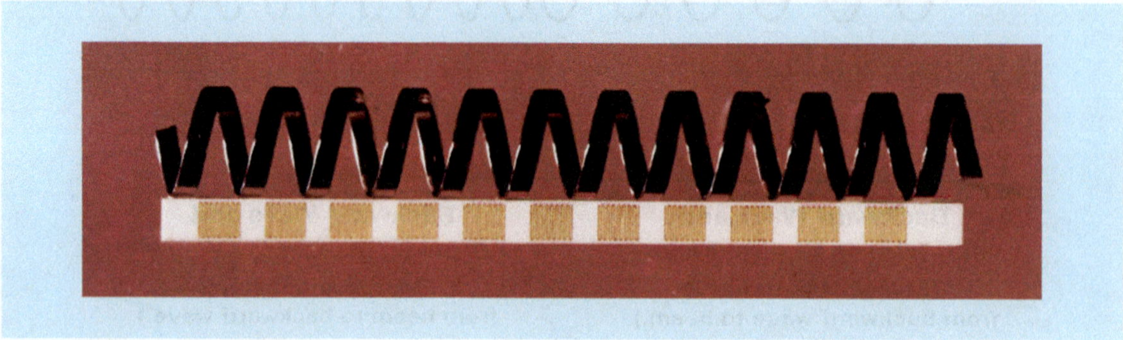

Figure 14-27 Resonant loss structure and helix. (Courtesy of Varian Associates.)

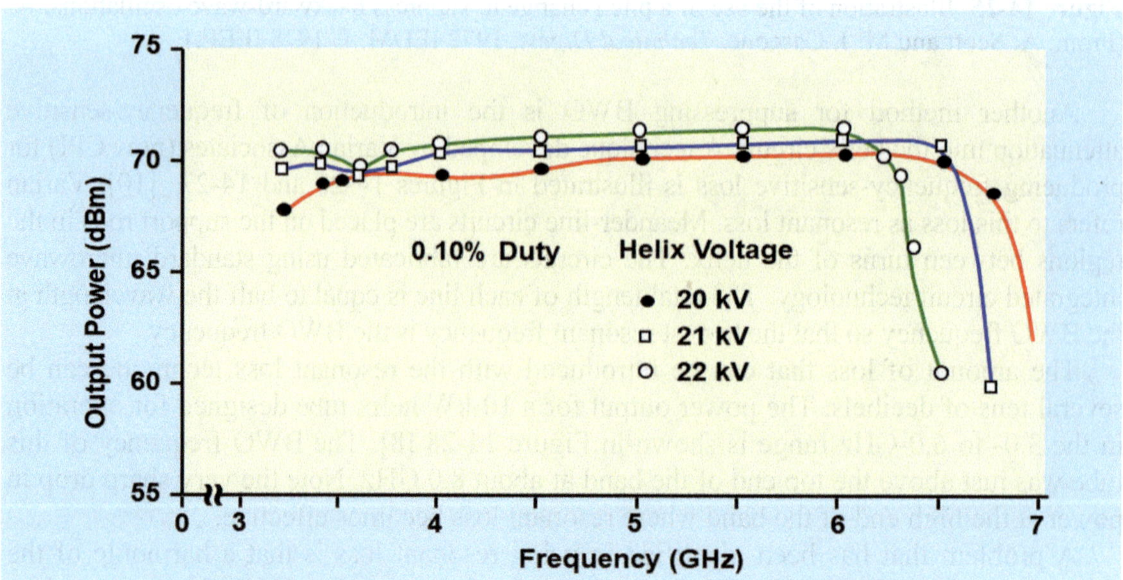

Figure 14-28 Saturated power output of 10-kW helix TWT with resonant loss. (From: Alan R. Jung, *Technical Digest*, 1978 IEDM. © 1978 IEEE.)

In Figure 14-29, estimates are shown of the peak output power that may be attainable using BWO suppression techniques. Note that peak power levels in excess of 100 kW are predicted to be possible in PPM-focused tubes using resonant loss.

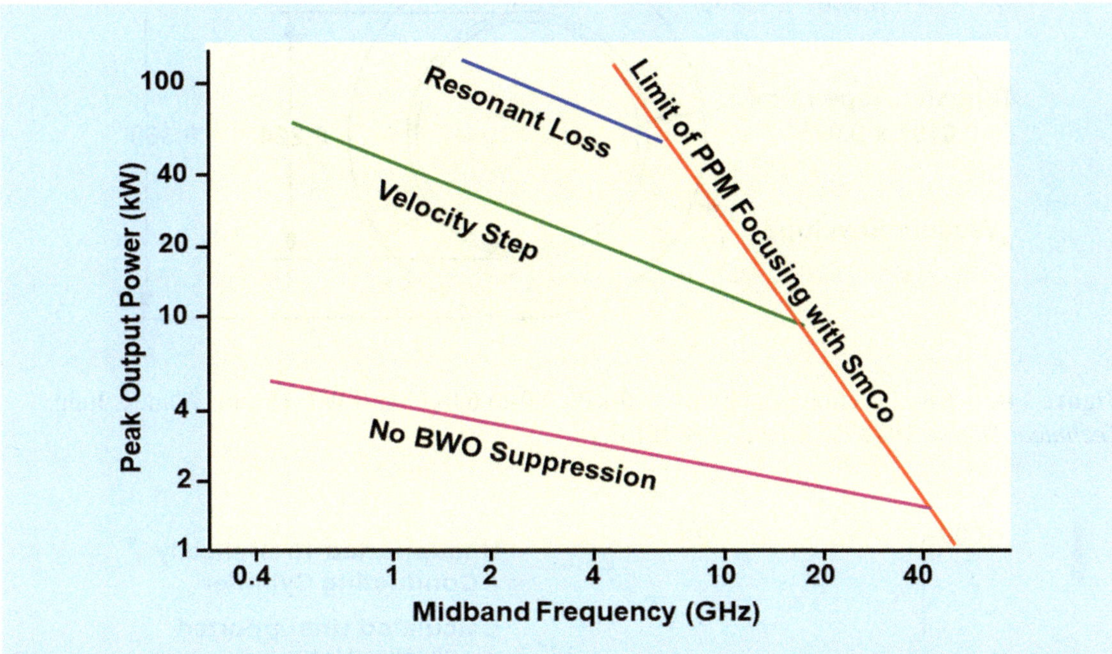

Figure 14-29 Obtainable peak power for helix TWTs with various oscillation-suppression techniques. (From: Alan R. Jung, *Technical Digest*, 1978 IEDM. © 1978 IEEE.)

14.3.2 Average Power

The helix of a TWT is a relatively delicate structure. As an example, a cross section is shown in Figure 14-30 for the helix of the 10-kW TWT described above, which was designed to operate from 3.0 to 6.0 GHz. The helix is made of 0.010-inch by 0.075-inch tungsten tape. If the TWT is to be rugged and capable of operating at high average-power levels, this helix must be very well supported.

The use of metallic supports is out of the question because of their effect on the RF characteristics of the circuit. A dielectric cylinder completely surrounding the helix could conceivably provide excellent support and cooling. Unfortunately, the dielectric constants of suitable ceramic materials are high. As a result, their use in a cylinder would severely reduce the circuit interaction impedance of the helix. In fact, to minimize the loading effect of a ceramic support structure, it is necessary to minimize the amount of ceramic material used. As a result, thin, rectangular support rods like those shown in Figure 14-30 are often used. Even these load the helix to some extent, as is indicated by the data plotted in Figure 14-31.

Two dielectric materials are widely used for supporting the helices of TWTs: BeO and APBN. In Figure 14-32, these two materials have relatively high thermal conductivity. Diamond has an outstanding thermal conductivity. CVD (chemical vapor deposited) diamond is comparable to type llA diamond. There has been interest in using diamond support rods in helix TWTs for many years. However, there have been no reports that diamond is being used in helix tubes.

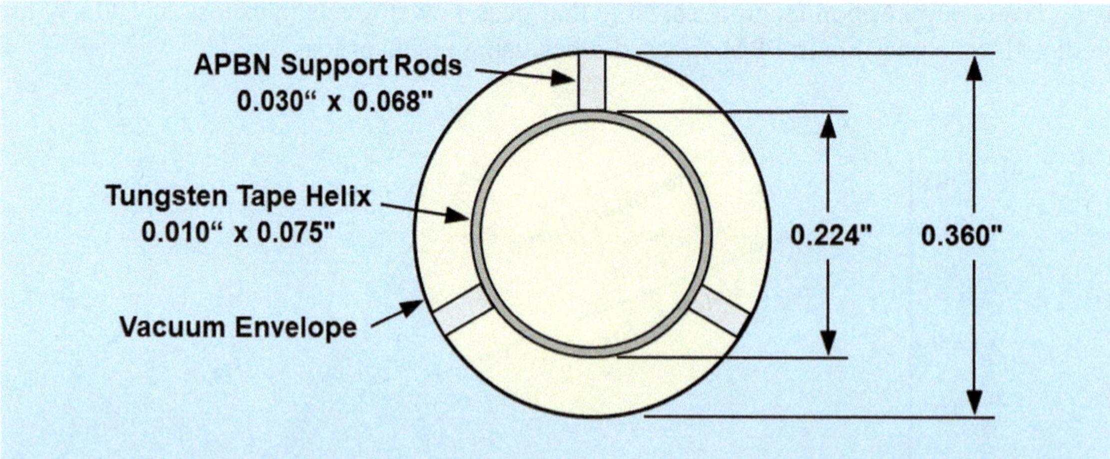

Figure 14-30 Cross section of helix of a 10-kW, 3.0-to 6.0- GHz TWT. (From: Alan R. Jung, *Technical Digest*, 1978 IEDM. © 1978 IEEE.)

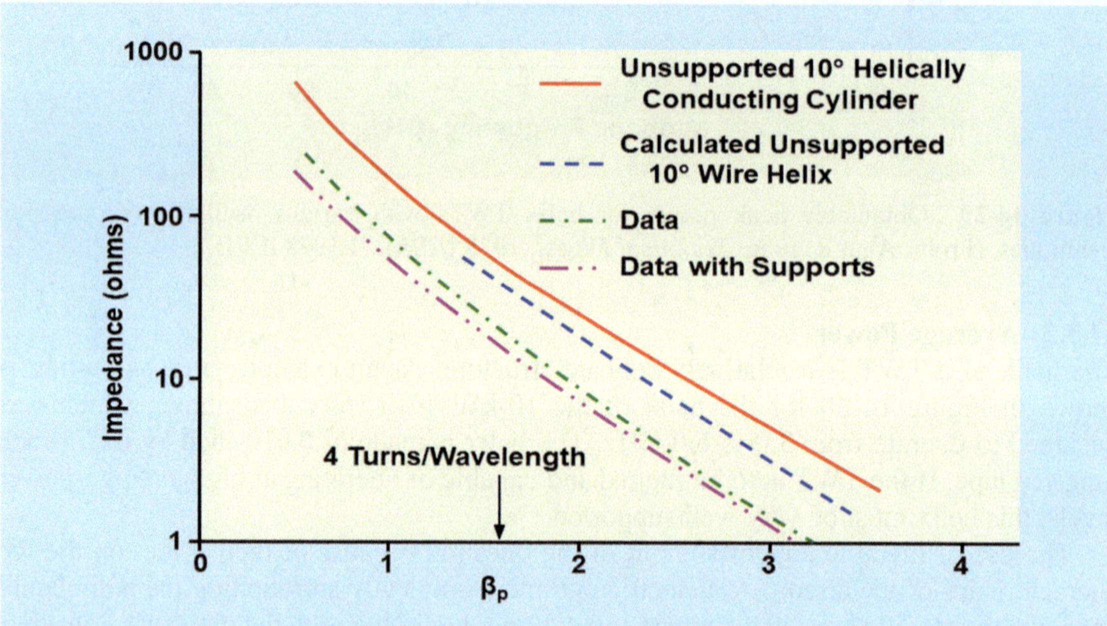

Figure 14-31 Coupled-interaction impedance of a helix with and without supports. (Adapted from: *Power Travelling Wave Tubes* by J. F. Gittins, published 1965 by American Elsevier, Inc.)

With the use of high thermal conductivity ceramics, the temperature drop across the support rods is not a severe problem in high-power helix TWTs. Instead, the major problems are the thermal resistances of the interfaces between the helix and the support rods and between the support rods and the barrel. In Figure 14-33, the thermal drops across these interfaces account for 70% of the total temperature difference between the hottest point on the helix and the barrel. The temperature drops across the interfaces are extremely dependent on the pressure applied to the interface as shown in Figure 14-34. The surface finishes of the materials in contact are also important.

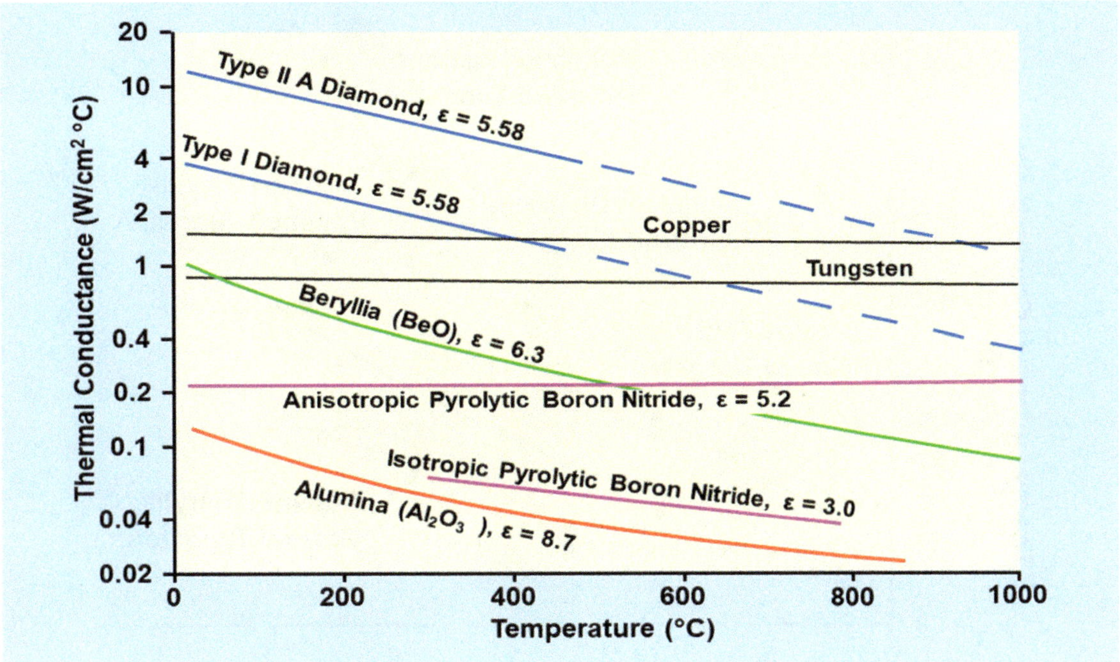

Figure 14-32 Thermal conductivities of some metals and ceramics. (From: A. E. Jacquez and A. Scott, *Final Report,* AFWAL-TR-81-1176, October 1981.)

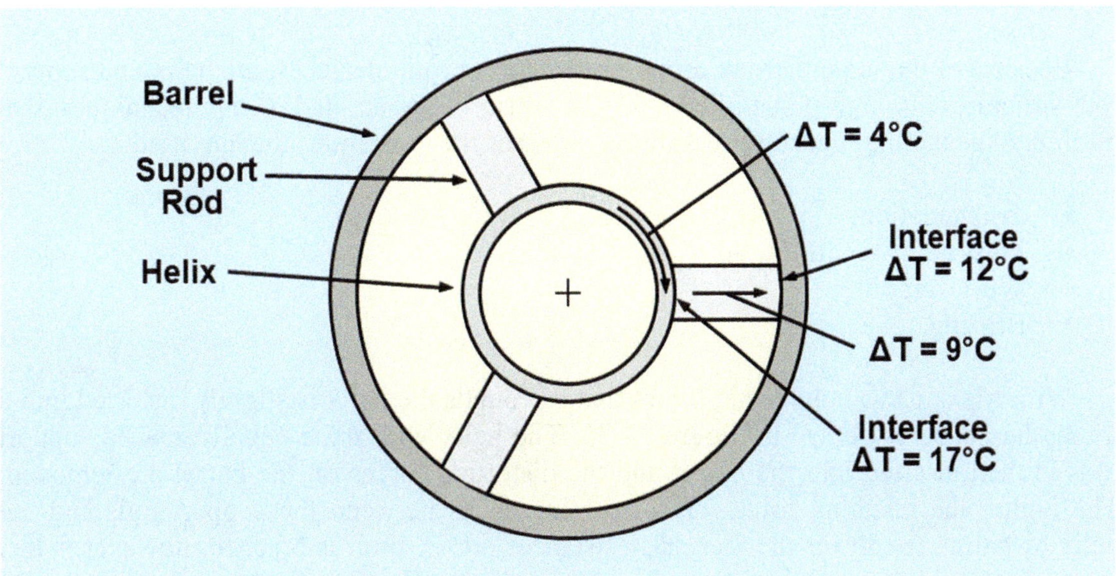

Figure 14-33 Temperature drops calculated for a tungsten helix and APBN rods dissipating 1 watt. (From: A. Scott and M. J. Cascone, *Technical Digest*, 1978 IEDM. © 1978 IEEE.)

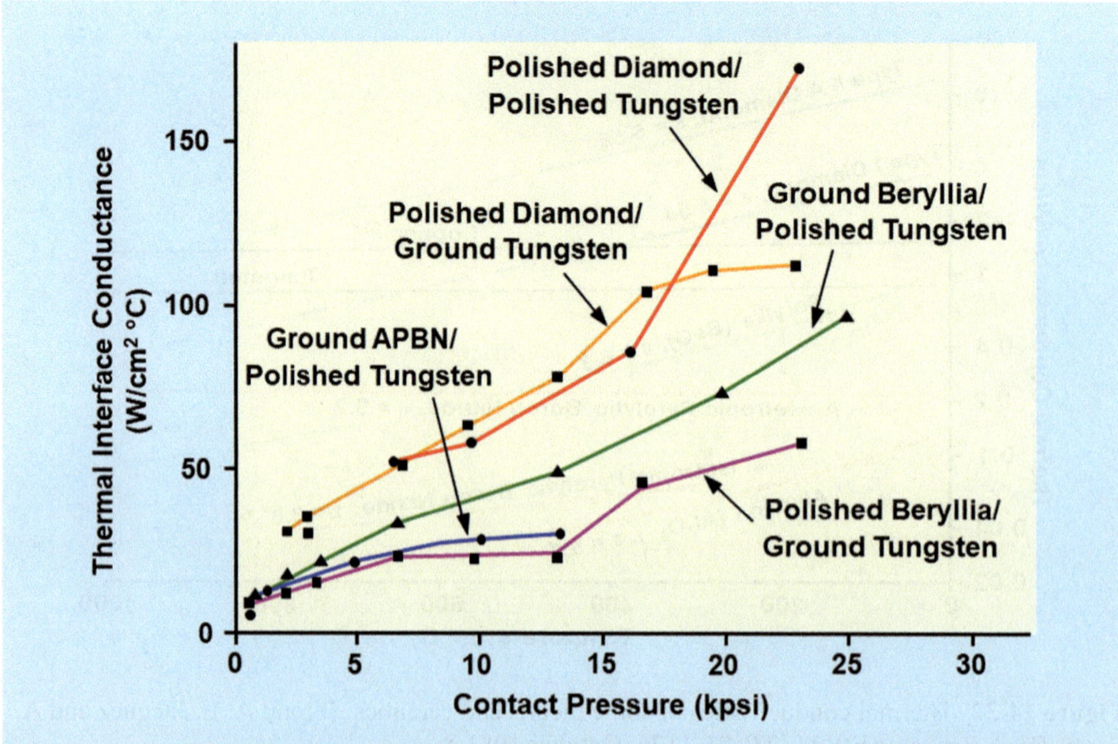

Figure 14-34 Thermal interface conductivities for several dielectrics in contact with tungsten. (From: A. E. Jacquez and A. Scott, *Final Report,* AFWAL-TR-81-1176, October 1981.)

Because of the large thermal drops at the support rod interfaces, the helix and support rod structures in high-power helix TWTs must be assembled using techniques that minimize the thermal interface resistances. Four of the techniques that are used are

- Triangulation;
- Pressure or hot insertion;
- Wire wrapping;
- Brazing.

With triangulation, the metal barrel that surrounds the helix is slightly distorted into a triangular shape as shown in Figure 14-35. The helix with three equally spaced support rods is then inserted into the barrel and the distorting forces on the barrel are removed. The higher the resulting forces are at the interfaces between the support rods and the helix or barrel, the lower the thermal resistances are. A limit is reached, however, when the forces are high enough to distort the helix excessively. For maximum helix strength, the material normally used is tungsten or molybdenum.

It should be pointed out that with the triangulation technique the barrel is not a solid brazed structure (like that shown in Figure 14-12). Instead, it is a relatively thin metallic tube over which the magnetic pole pieces are slid after the helix has been inserted. With this technique, there is a relatively poor thermal interface between the barrel and the pole pieces. This problem is alleviated by making the pole pieces fit as tightly as possible on the barrel. Then, as the helix and barrel expand because of heating, the contact between the barrel and pole pieces improves so heat transfer improves.

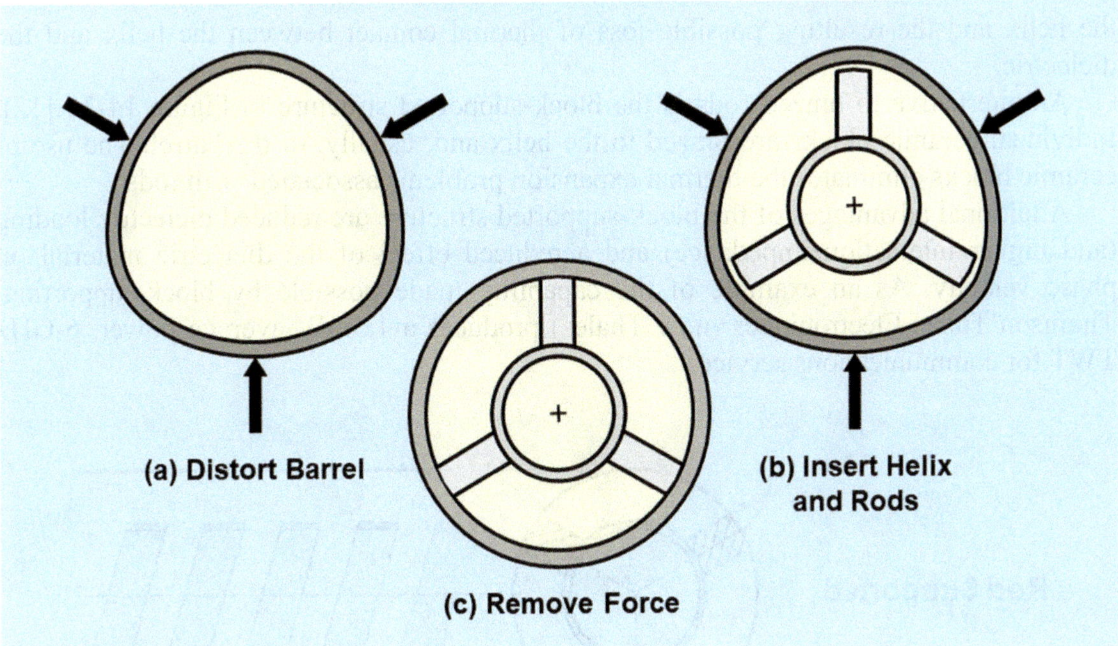

Figure 14-35 Triangulation procedure.

With the insertion technique, the helix and rod assembly is inserted into a barrel structure like that shown in Figure 14-12. In some cases, the barrel is slightly expanded by heating before the helix and rod assembly is inserted. This is referred to as hot insertion. In other cases, the helix and rod assembly are machined with tolerances for an interference fit and then inserted under high pressure into the cold barrel. The temperatures given in Figure 14-33 are for a pressure-inserted structure. As with the triangulation technique, the helix materials used are tungsten or molybdenum.

With wire wrapping, a tungsten helix that may be copper plated is held in compression by support rods (usually four) that are wrapped with wire under high tension. This ensures that uniform pressure is applied to all portions of the helix. This technique simplifies fabrication and assembly of high-frequency tubes [11].

The most effective way of reducing the thermal interface resistances is by brazing. While brazing sounds attractive, it can be difficult to implement. If rods that have been metallized on the surfaces to be brazed are used, there are difficulties with thermal expansion problems. Also, the support rods must be brazed to every turn of the helix. There may be several tens of contact points of one support rod with a helix and several support rods (usually three). Every contact point of every rod must be brazed for proper protection of the helix. After brazing, the metallization of the support rods must be removed in the regions between turns of the helix.

With the brazing technique, a wider choice of helix materials is possible than with the triangulation or pressure-insertion techniques. In addition to tungsten or molybdenum, copper can be used. While tungsten and molybdenum can be copper-plated to reduce RF losses, only the use of pure copper throughout the helix leads to the maximum possible transfer of heat and the ultimate loss reduction. Finally, it should be pointed out that in brazed structures, the helix is mechanically bonded to the dielectric support material. There is reduced probability, therefore, that helix failure will result from overheating of the helix and the resulting possible loss of thermal contact between the helix and the dielectric.

An alternative to brazed rods is the block-supported structure in Figure 14-36 [12]. Individual ceramic blocks are brazed to the helix and, usually, to the barrel. The use of ceramic blocks eliminates the thermal expansion problems associated with rods.

Additional advantages of the block-supported structure are reduced dielectric loading (and higher interaction impedance) and a reduced effect of the dielectric material on phase velocity. As an example of the capability made possible by block-supporting, Thomson Tubes Electroniques (now Thales) produces a 1.5-kW average power, 6-GHz TWT for communications service.

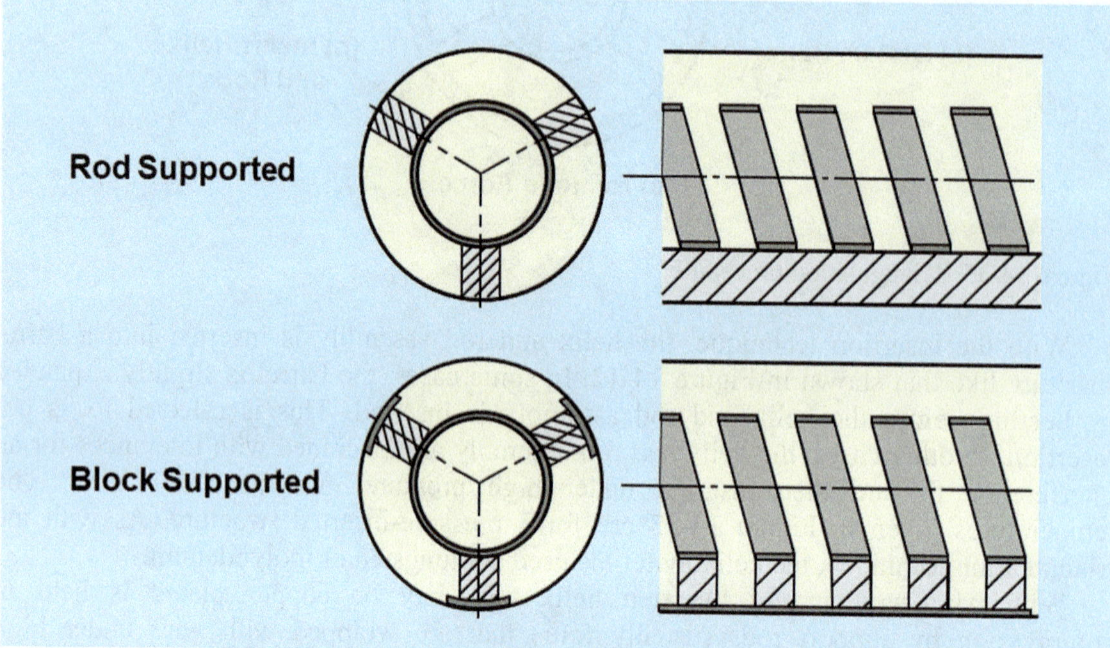

Figure 14-36 Comparison of rod-supported and block-supported tape-helix structures. (From: Yukio Hiramatsu, *Technical Digest*, 1979 IEDM. © 1979 IEEE.)

Comparisons of the advantages of various techniques are often misleading because test conditions differ. Still, because of the continuing emphasis on higher powers at higher frequencies, it is important to compare the techniques. A comparison of the triangulation, pressure-insertion, and brazing techniques is given in Figure 14-37 [13]. The base-plate temperature for the test fixture in these tests was 100°C. Thus, the temperature rise of the brazed helix (which was copper) was about two thirds of that for the insertion (shrunk) technique and about one half of that for the triangulation technique.

It is important to note that the results in Figure 14-37 were obtained by passing a dc current through the helices. RF losses and at least a small amount of additional heating are therefore not included. The additional heating may be larger in the metallized, brazed helix than in the other two.

In general, TWT dimensions scale with wavelength. At millimeter-wave frequencies, dimensions become extremely small and miniaturized helices and support rods are used. For example, rectangular support rods having cross-sectional dimensions of 0.010 inch by 0.020 inch or smaller are common. As helix assemblies become smaller, heat dissipation problems become more severe, and the thermal management techniques described in the previous sections become even more important than they are at lower frequencies.

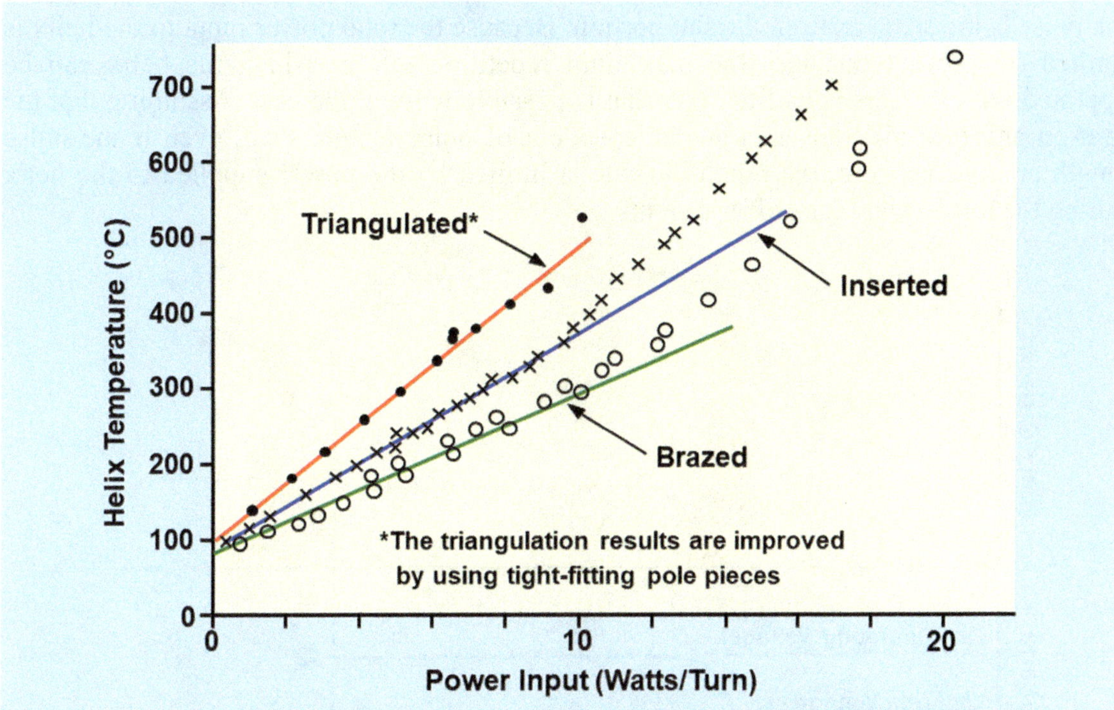

Figure 14-37 Comparison of thermal characteristics of triangulated, pressure-inserted (shrunk), and brazed helices. (From: T. Mosser et al., 1980 Microwave Power Tube Conference.)

The brazing technique, which offers excellent thermal capabilities, is very difficult to implement in the millimeter-wave range. As a result, at present, pressure insertion is widely used and average power levels are limited to about 100 W.

The use of diamond may make significant improvements in average power capability possible. Type II-A diamond has a thermal conductivity several times that of copper. With diamond supports, it is predicted that average power levels as high as 200W may be achievable at frequencies as high as 50 GHz [14].

14.3.3 Repetition Rate Limitations

When the grid of a pulsed TWT is switched from the bias voltage to the on voltage with finite rise time, there is a surge of helix current above the normal value. There is a similar surge when the grid is switched back to the bias level. The extra helix dissipation caused by these surges limits the maximum repetition rate at high duty cycles. This is of concern because, in some applications, it is necessary to operate a pulsed TWT at a duty cycle much higher than normal for relatively short bursts while maintaining the long-term average duty within the usual limitations.

Consider the hypothetical case in Figure 14-38 for example. The time required to switch the grid from completely off (bias level) to the operating voltage is 20 ns. The nominal pulse risetime from 10% to 90% is 10 ns. The duration of the flat portion of the pulse is 40 ns.

The total current in the ears of the current pulse is approximately the same as in the flat portion of the pulse. Assuming that all of the electrons in the grid current have been accelerated to the full beam voltage, then the intercepted power resulting from the ears of the pulse is the same as from the flat portion. Because the total power input to the helix is limited (to prevent damage) the maximum repetition rate at which the helix can be operated with the ears is half the rate that is possible without the ears. Assuming that the turn-on and turn-off transients are independent of pulse length, then, even if the pulse length approaches zero, the repetition rate is limited by the power supplied to the helix during the turn-on and turn-off transients.

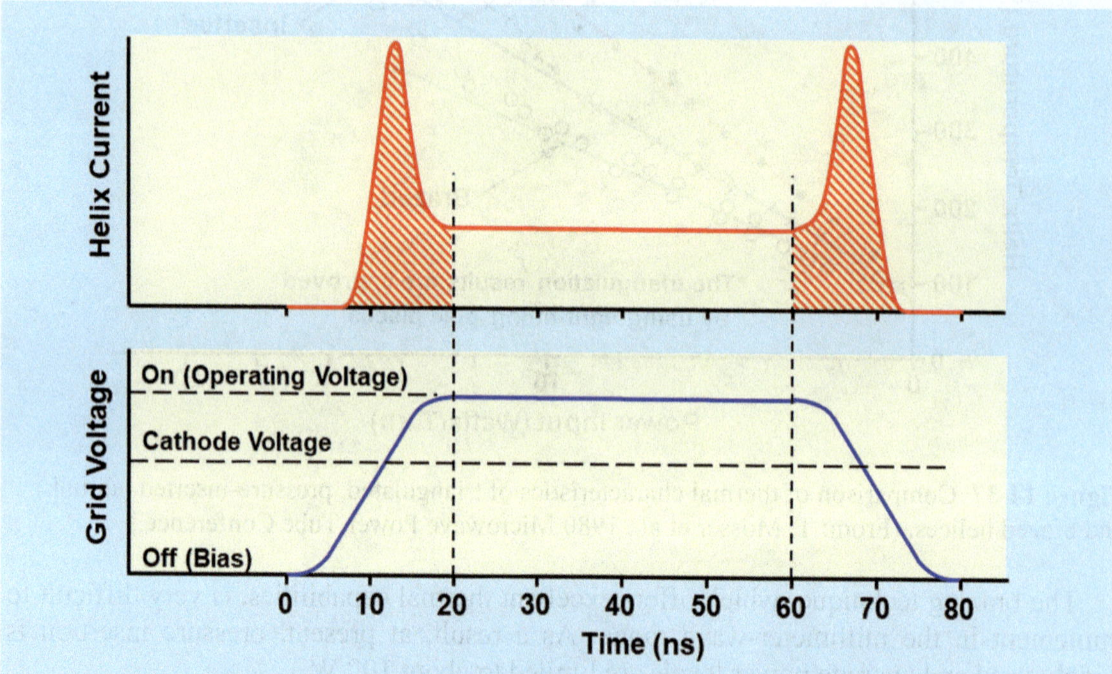

Figure 14-38 Grid pulse with turn-on and turn-off transients.

14.4 EFFICIENCY

In applications for TWTs where prime power is at a premium, efficient operation is essential. The efficiency of TWTs for space operation has increased steadily since their first use on satellites and, as is shown in Figure 14-39 for a Ku-band device, has passed the 70% mark [15]. Figure 14-40 shows 65% efficiency for an L-band device [16]. These are remarkable achievements because the basic electronic efficiency (the efficiency of conversion of beam power to RF power on the helix) for a typical helix TWT is on the order of 20% at best.

The impact of the high-efficiency devices in Figure 14-39 on the spacecraft power system is compared with solid-state power amplifiers (SSPAs) in Figure 14-41. For the same RF power output, the dc input required by the SSPA is nearly three times that of the TWTA (the TWT plus its power conditioning system). When radiation-cooled TWTs are used, the waste heat that must be dissipated by the spacecraft is over an order of magnitude below that for SSPAs. When these power considerations are combined with the demonstrated superior reliability of TWTs compared with SSPAs [17], it is clear why TWTs are the amplifiers of choice for many spacecraft applications.

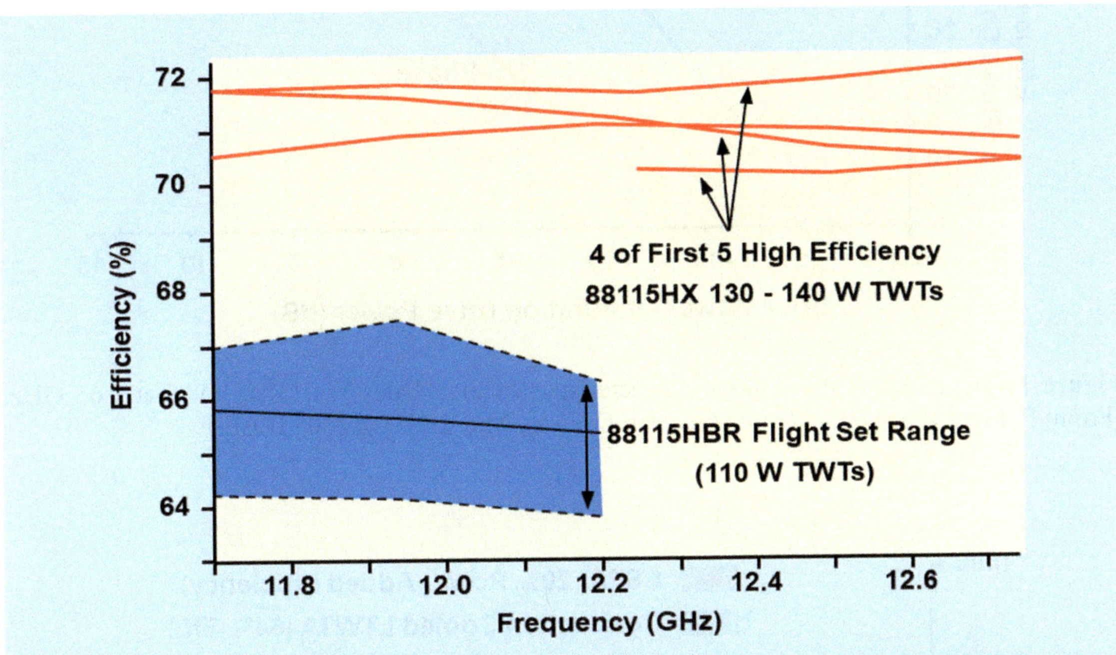

Figure 14-39 Overall efficiencies of the initial set of production Ku-band TWTs compared to the previous model. (From: W. Menninger et al., *IEEE Trans. Electron Devices*, May 2005. © 2005 IEEE.)

The achievement of 70% efficiency in a helix TWT requires great attention to detail in the design and fabrication of all aspects of the electron optics system and the RF structure. Most important, perhaps, is the use of a multistage depressed collector for recovering the power remaining in the electron beam after it has interacted with the RF structure. These collectors are discussed in Chapter 16.

The design of the RF circuit must be optimized for several factors at the same time. These include electronic efficiency, circuit efficiency, collector efficiency (the energy distribution of the electrons entering the collector must be controlled), and linearity.

Circuit efficiency is the efficiency with which the helix delivers the RF power generated in the TWT to the output port. Circuit efficiency is important because power lost to heating the circuit is no longer available in the electron beam for recovery by the collector. Collector efficiency is the efficiency with which the collector recovers power from the electron beam leaving the interaction region of the TWT. Figure 14-42 shows that an increase in circuit efficiency of 10% will result in an increase in overall efficiency of ~5% when an efficient collector is used.

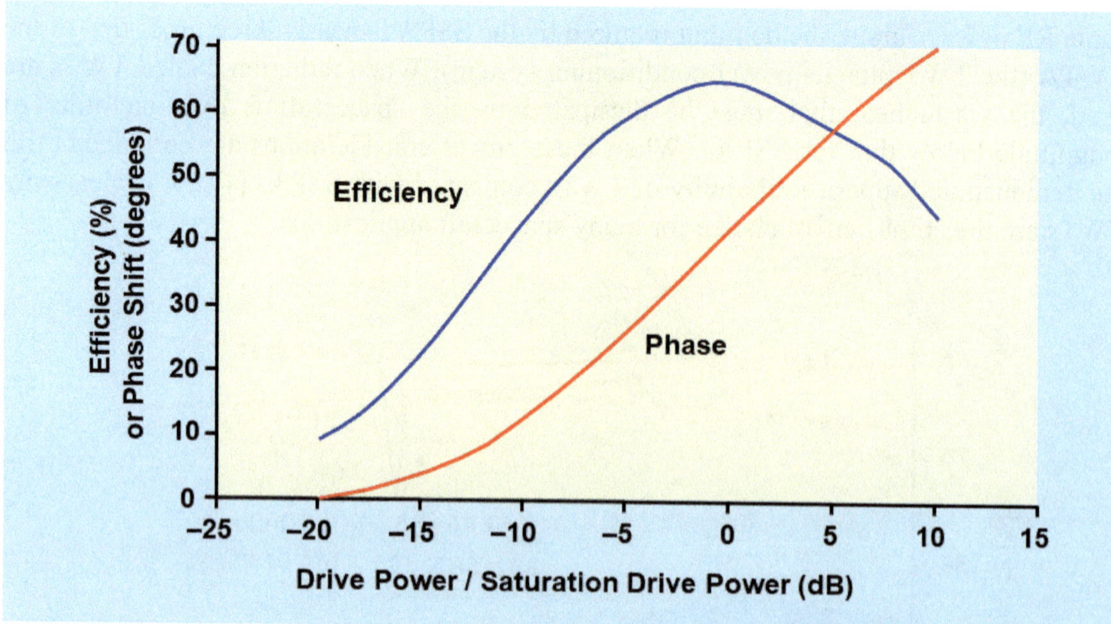

Figure 14-40 Efficiency and phase characteristics of the Thales Worldstar TWT at 1.65 GHz. (From: P. Ehret et al., *IEEE Trans. Electron Devices*, May 2005. © 2005 IEEE.)

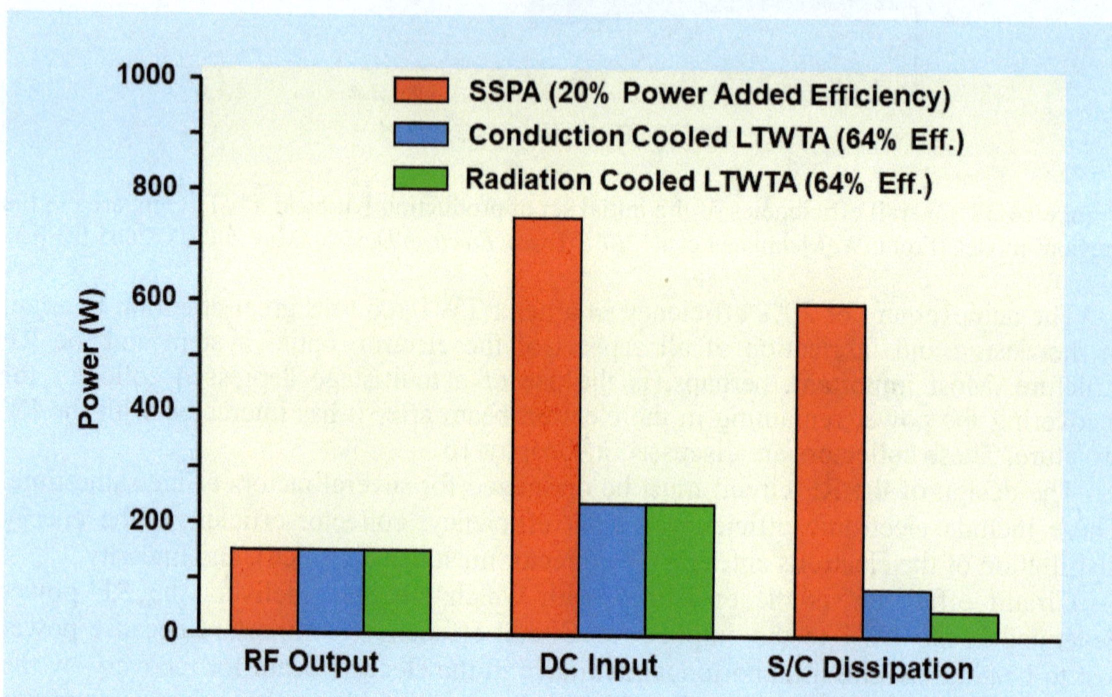

Figure 14-41 Amplifier output powers, dc input powers, and waste heat that a spacecraft must dissipate. (From: W. Menninger et al., *IEEE Trans. Electron Devices*, May 2005. © 2005 IEEE.)

Circuit efficiency is dependent primarily on the effective resistance of the circuit material. At high frequencies, dielectric losses are also important. Comparisons of measured and theoretical loss for a copper-plated helix are given in Figure 14-43.

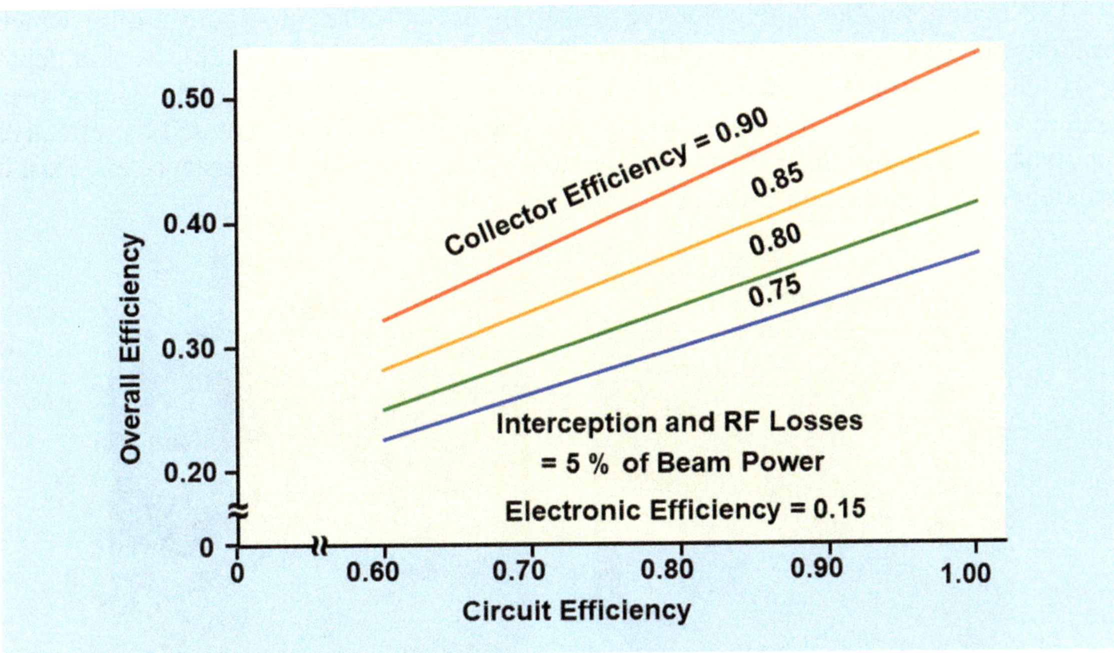

Figure 14-42 Effect of circuit efficiency on overall efficiency. (Adapted from: A. S. Gilmour, Jr. et al., *IEEE Trans. Electron Devices*, October 1979. © 1979 IEEE.)

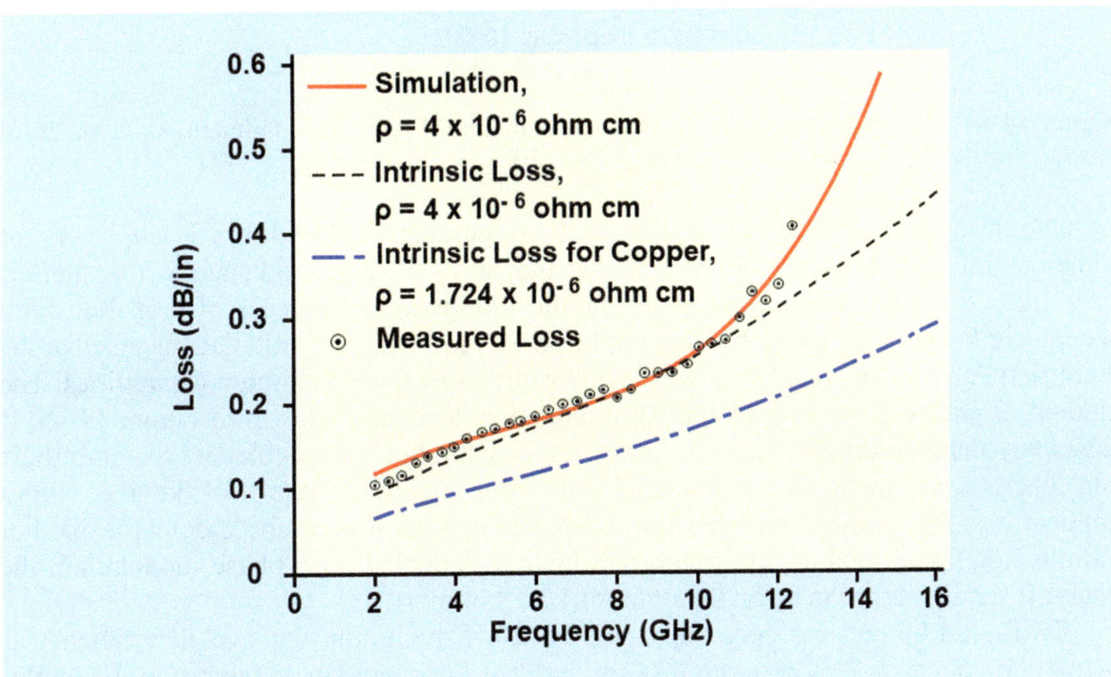

Figure 14-43 Measured and theoretical loss for a copper-plated helix. (From: A. S. Gilmour, Jr. et al., *IEEE Trans. Electron Devices*, October 1979. © 1979 IEEE.)

The excellent fit of the theoretical curves to data when a resistivity of 4×10^{-6} ohm cm is assumed leads to the conclusion that 4×10^{-6} ohm cm was the effective resistivity of the copper surface of the helix. This value is more than twice the bulk resistivity of copper.

The reason for the high effective resistivity is indicated in Figure 14-44, which contains photomicrographs of the surface of the copper-plated helix [18]. The skin depth at 10 GHz was small compared with the sizes of the irregularities on the orange-peel textured copper surface. As a result, the path length for the RF current and the effective resistance were much larger than for a smooth surface. To minimize resistance and loss, it is essential to properly polish the helix surface.

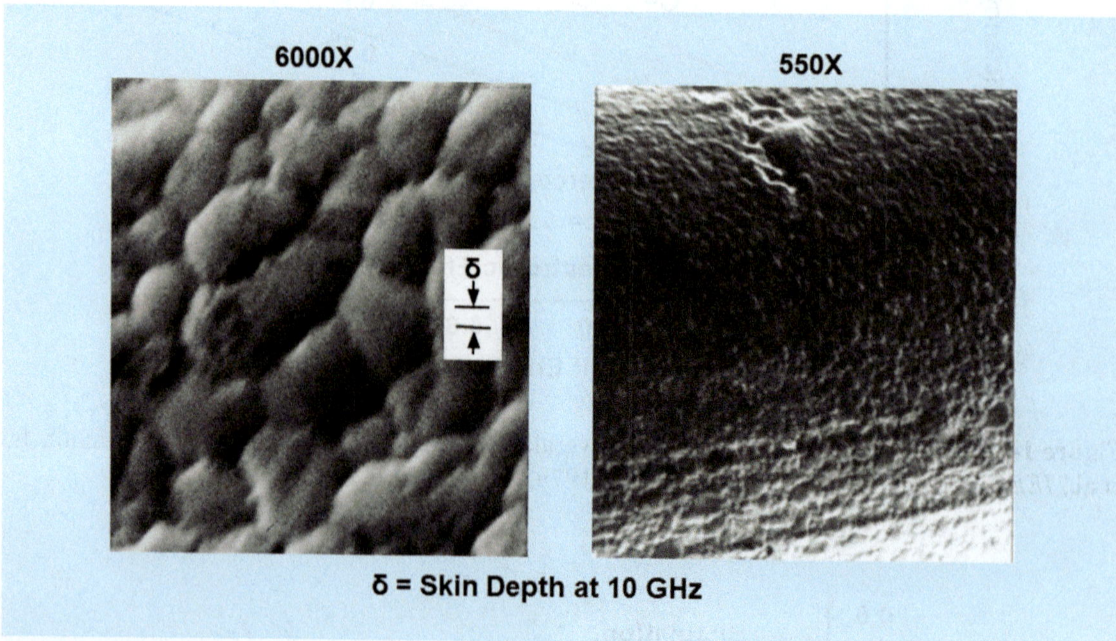

Figure 14-44 Photomicrographs of a copper-plated helix. (From: A. S. Gilmour, Jr. et al., *IEEE Trans. Electron Devices*, October 1979. © 1979 IEEE.)

One of the very unfortunate aspects of the requirements placed on RF amplifiers for some communications applications, whether they are TWTs or solid state, is that there is very little tolerance for nonlinearities in the amplification process. To operate with acceptable linearity, operation of the amplifiers has to be backed off. This means that the amplifiers must be operated at power levels well below their maximum capabilities. The amount of back-off depends on the RF modulation scheme as shown in Figure 14-45. If frequency modulation (FM) is used, operation at saturation is possible and the amplifiers can operate at maximum efficiency. For some of the phase-shift keying (PSK) applications, the back-off required for acceptable operation is on the order of 4 dB. For various amplitude modulation schemes, which are coupled with phase modulation, the back-off requirement can be on the order of 12 dB or more.

TWTs can be and are operated in the back-off mode, but the overall efficiency is below the values of 70% or more that are realized for operation at saturation. Examples of the degradation in efficiency that can occur at reduced power levels are shown in Figure 14-46. (Here efficiency was not necessarily optimized for operation below saturation.)

At 10 or more decibels below saturation, power usage that would normally be considered to be nearly insignificant becomes important. An example is the power for heating the cathode. This could be eliminated with a cold cathode. Losses due to secondary emission effects in the collector would have to be reduced. Beam interception would have to be virtually eliminated. If these losses, as well as power conditioning losses could be all but eliminated, very efficient operation at 10 or more decibels below saturation could be achieved.

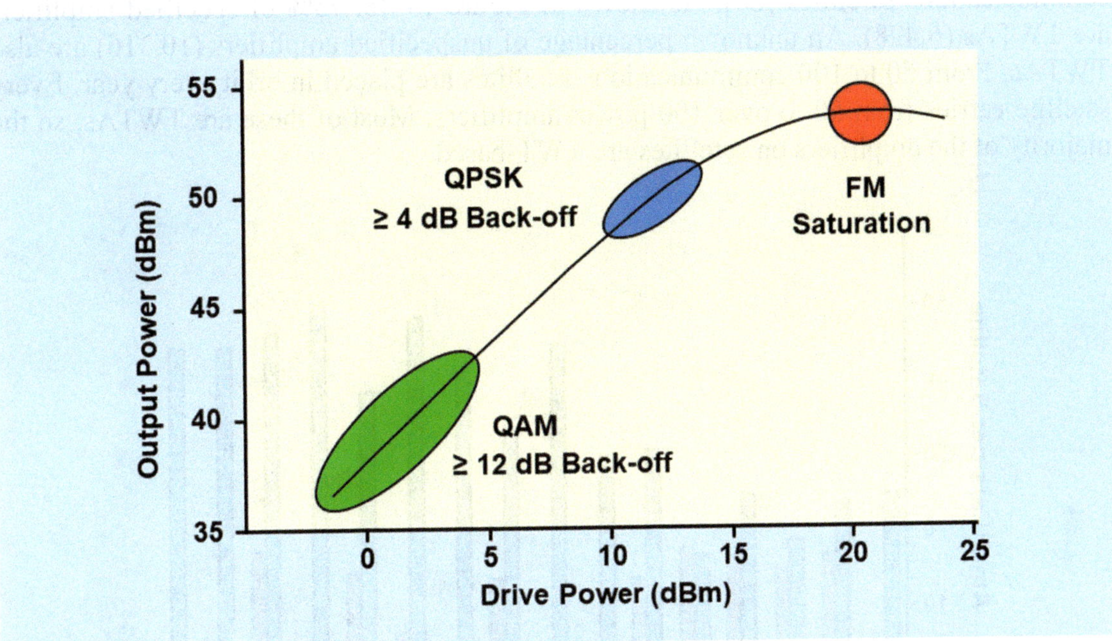

Figure 14-45 Reduction in TWT output power required for acceptable operation with various modulation techniques. (From: K. Tsutaki et al., *IEEE Trans. Electron Devices*, May 2005. © 2005 IEEE.)

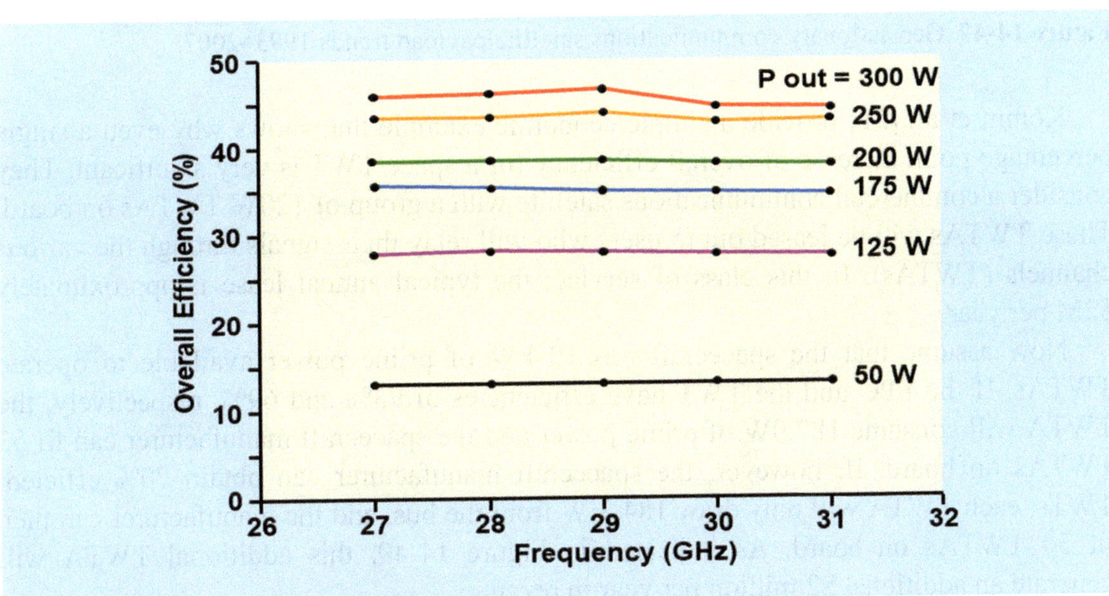

Figure 14-46 Reduction in TWT efficiency that results from operation below saturation. (From: A. Gallien et al., 2006 IVEC. © 2006 IEEE.)

14.5 SATELLITE COMMUNICATIONS

As shown in Figure 14-47, satellite requirements for TWTs have grown very rapidly in recent years [19]. At present, there are over 18,000 power amplifiers on over 500 communications satellites [20]. As shown in Figure 14-48, 75% of specified amplifiers are TWTAs (6,428). An unknown percentage of unspecified amplifiers (10,316) are also TWTAs. From 50 to 100 communications satellites are placed in orbit every year. Every satellite carries from 50 to over 100 power amplifiers. Most of these are TWTAs, so the majority of the amplifiers on satellites are TWT-based.

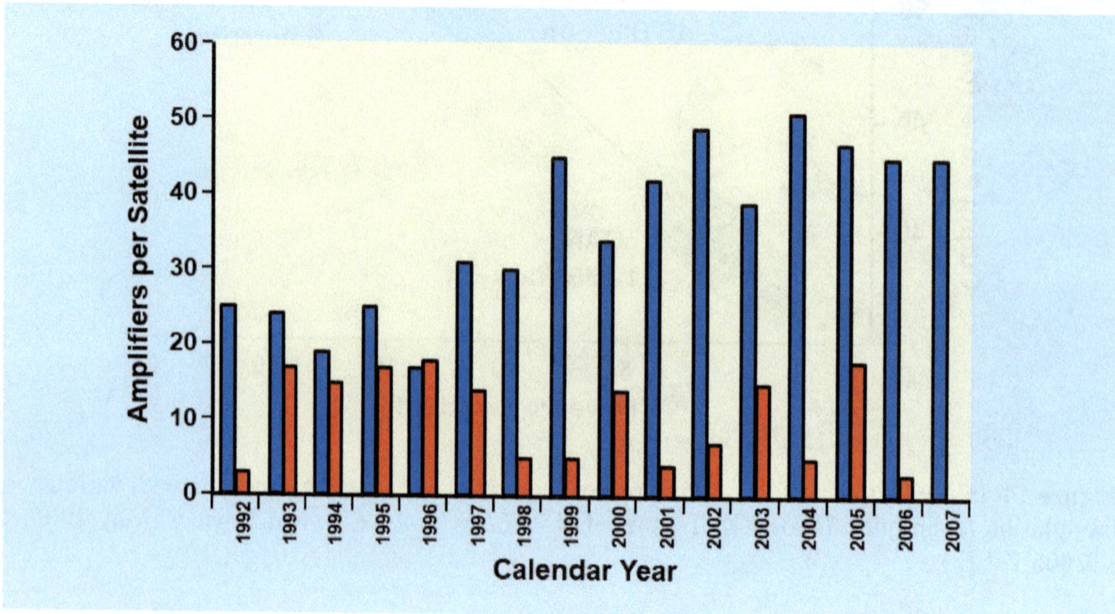

Figure 14-47 Geostationary communications satellite payload trends 1993–2007.

Komm et al. [21] provide a simple economic example that shows why even a single percentage point increase of overall efficiency for a space TWT is very significant. They consider a commercial communications satellite with a group of 120W TWTAs on board. These TWTAs will be leased out to users who will relay their signals through the various channels (TWTAs). In this class of service, the typical annual lease is approximately $2M per year.

Now assume that the spacecraft has 10 kW of prime power available to operate TWTAs. If the EPC and the TWT have efficiencies of 93% and 69%, respectively, the TWTA will consume 187.0W of prime power and the spacecraft manufacturer can fit 53 TWTAs on board. If, however, the spacecraft manufacturer can obtain 70% efficient TWTs, each TWTA will only draw 184.3 W from the bus, and the manufacturer can then fit 54 TWTAs on board. As indicated in Figure 14-49, this additional TWTA will generate an additional $2 million per year in revenue.

For space-based applications, the operational lifetimes now required are over 150,000 hours (18 years). At $2 million per year, this additional TWTA would produce $36 million in additional revenue.

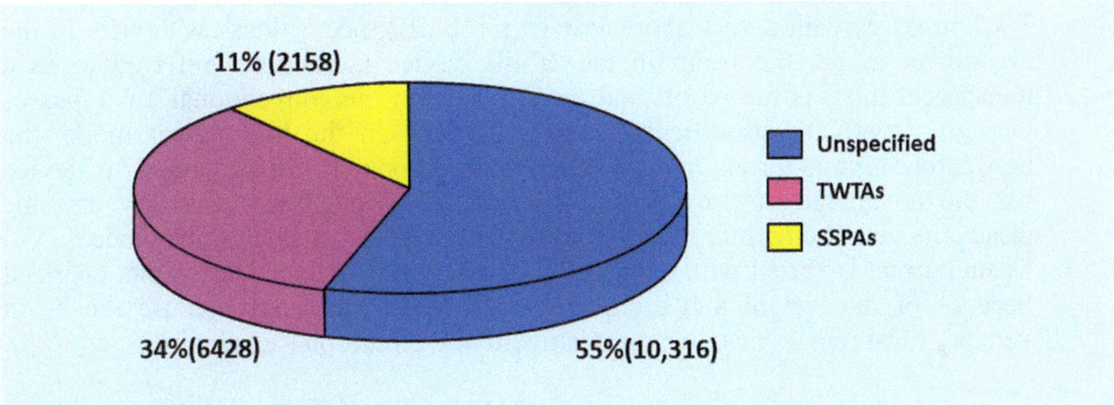

Figure 14-48 Communications satellite power amplifiers 1983–2016.

Figure 14-49 The value of TWTA efficiency was detailed by Komm et al. several years ago.

14.6 DUAL-MODE OPERATION

The purpose of a dual-mode TWT is to provide two different RF power levels (separated by as much as 10 dB) efficiently in a single device. This is important in airborne electronic countermeasures systems, for example, where volume and weight must usually be minimized. In these systems, the use of two tubes, each of which is designed for operation at one of the power levels, often results in excessive system weight and volume. There are basically three ways in which a single tube can be efficiently operated at two power levels:

1. Beam current and voltage are held constant and RF drive level is changed. Efficiency is obtained by using a multistage depressed collector to recover spent-beam power. This technique is discussed further in Chapter 16.
2. A complex tube, basically consisting of two tubes in one envelope, is used. This device is shown in Figures 14-50 and 14-51. The tube consists of a conventional

TWT in series with a very short, low-gain (10 dB), nearly lossless device. In the low-power mode, the beam in the 10-dB device (sometimes referred to as a transparent tube) is turned off and the signal from the conventional TWT passes through it with no amplification and little loss. In the high-power mode, the transparent tube is turned on to provide an additional 10 dB of gain. This device has the advantage of providing a full 10 dB of pulse-up capability and the disadvantage of being more complex and heavier than the single-tube device.

3. Beam current is varied with a gridded electron gun. (Beam voltage is not changed because of the weight and complexity of a modulator and because the beam velocity must remain near synchronism with the circuit phase velocity.)

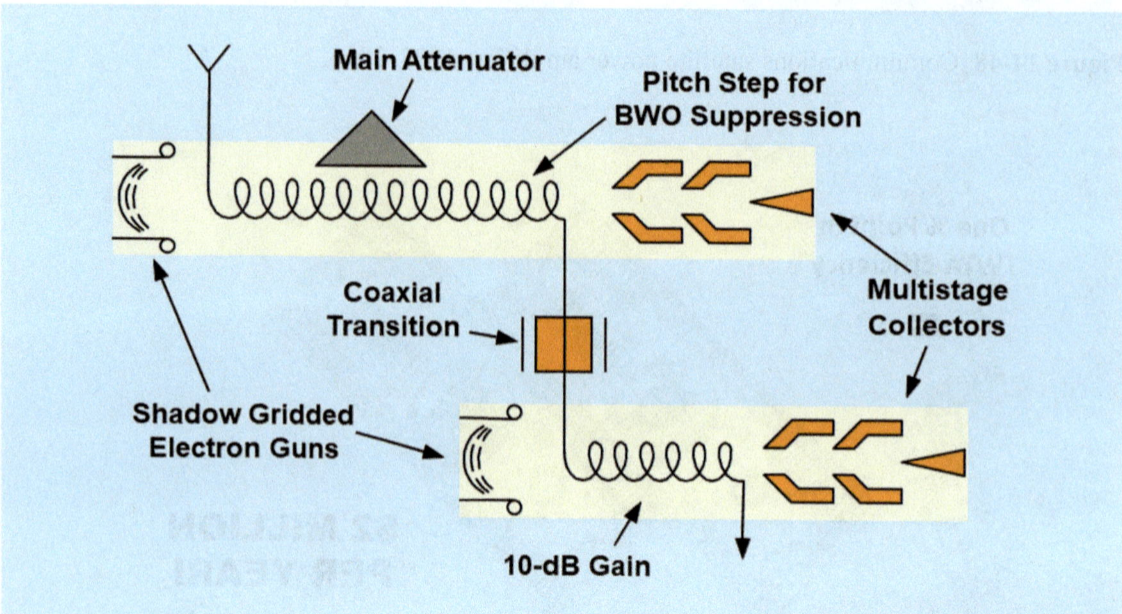

Figure 14-50 Schematic drawing of a dual-beam, dual-mode TWT. (Courtesy of Varian Associates.)

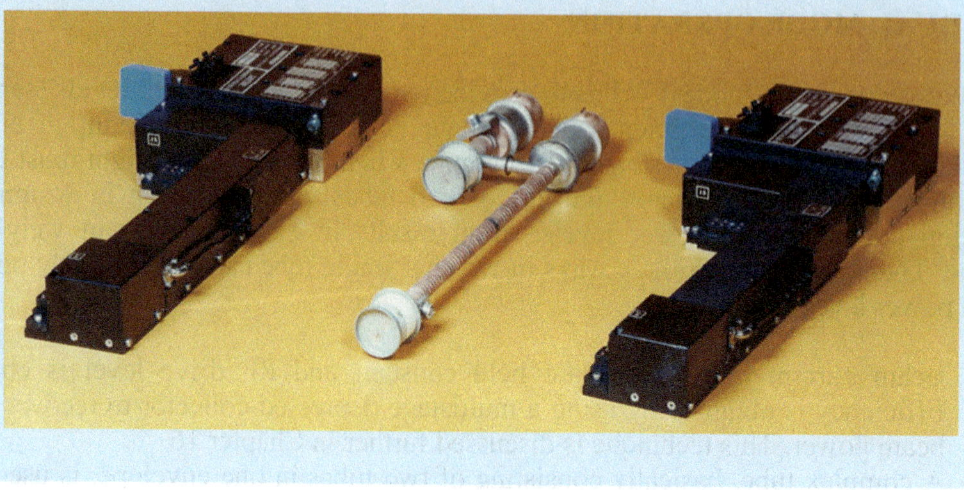

Figure 14-51 Dual-beam, dual-mode TWTs. (Courtesy of Northrop Corp.)

The third technique (changing beam current) for achieving dual-mode operation is difficult to implement when the power separation between modes must be more than a few decibels. One difficulty encountered is that of focusing both the low-power and the high-power beams with the same magnet structure.

Assuming constant efficiency, beam power and beam current must vary in direct proportion to RF power. If beam diameter is held constant, then beam charge density varies directly with beam current. Finally, the radial electric field in the beam is proportional to the charge density and, therefore, to the power level. If the PPM structure is designed for the low-power mode, the magnetic field will be inadequate to offset the electric field in the high-power mode. Of course, the opposite situation exists if the PPM structure is designed for the high-power mode. This problem may be alleviated by changing the beam diameter when the beam current is changed.

The other difficulty encountered when the beam current is changed is that the beam impedance changes and this changes Pierce's gain parameter, C. As a result, the RF characteristics of the tube change with beam current, making it difficult to obtain the desired performance at two power levels that differ by a large amount.

14.7 MICROWAVE POWER MODULES

Until three decades ago, there was intense competition between the solid-state and the traveling-wave tube communities to fill many microwave amplifier requirements. Now, the low-noise characteristic of solid-state devices is being combined with the efficient power-generation capability of TWTs to provide compact amplifiers called microwave power modules (MPMs) [22]. As indicated in Table 14.1, the capabilities of MPMs far surpass the capabilities of individual SSPAs or TWTAs. As a result, MPMs are replacing SSPAs and TWTAs in existing applications for microwave amplifiers and new markets are emerging that could not be filled by either SSPAs or TWTAs. An example of an MPM is shown in Figure 14-52. The weight and volume of the MPM are a factor of 5 below the weight and volume of the TWTA. The same benefits are realized when comparing the MPM with a SSPA.

Table 14.1
MPM Advantages Compared to Traditional TWTAs and SSPAs
(100 Watt Ku Band Example)

	MPMs vs. TWTAs	MPMs vs. SSPAs
Size*	5:1 Reduction	8:1 Reduction
Weight*	5:1 Reduction	8:1 Reduction
Noise	100:1Reduction	-2 dB degradation
Efficiency	50% Improvement	3:1 Improvement

*Includes power supply and cooling.

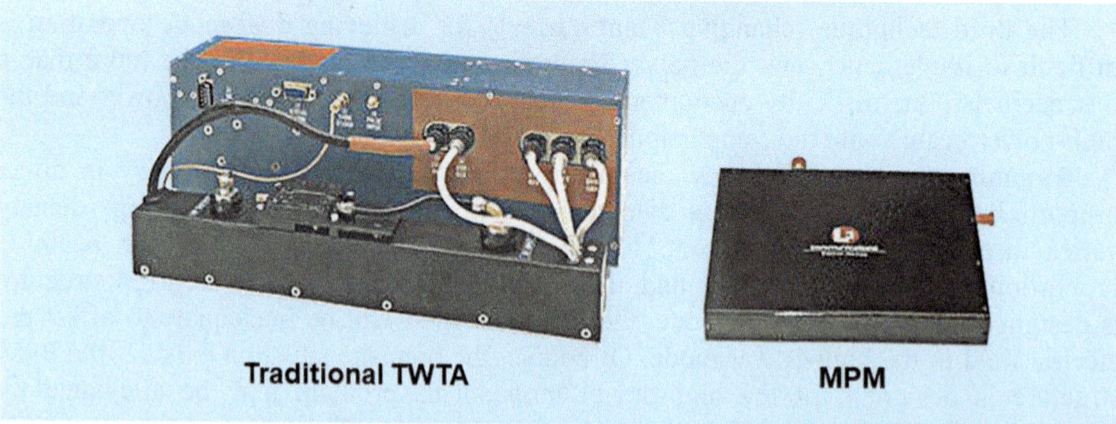

Figure 14-52 The MPM size advantage.

The gain of the MPM is divided approximately equally between the SSA and the TWT. As shown in Figure 14-53, the SSA and TWT are combined with a power conditioner and controller in a single package that is highly miniaturized and can be designed to operate in the microwave or millimeter-wave frequency bands.

As indicated in Figure 14-53, in addition to low noise, the SSA has functionality; that is, it has strong signal processing capabilities. In addition to high power, the TWT is capable of operating efficiently over a wide band of frequencies. The power conditioner delivers low voltage to the SSA and high voltage to the TWT with high efficiency. Special transformers have been developed that are very compact and have form factors that make it possible for them to fit into the very compact MPM packages. The micro-controller controls the operation of the module and reports on the module status, including the various voltages, currents, and temperatures. By integrating these components into a single package, the MPM offers the performance advantages relative to typical TWTAs and SSPAs shown in Table 14.1 [23].

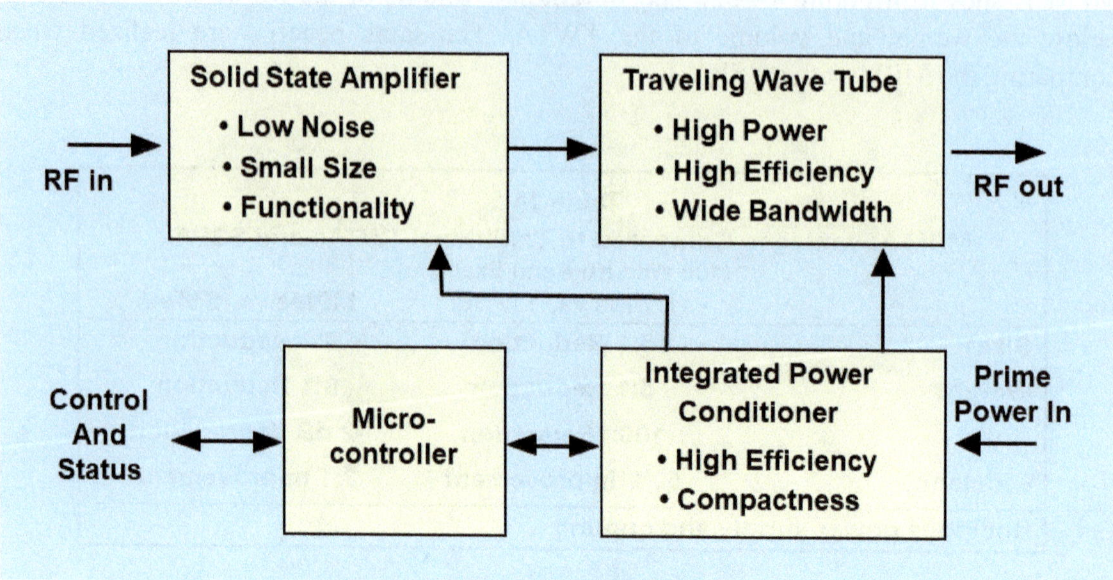

Figure 14-53 Basic components of a MPM.

The concept of the microwave power module first surfaced in 1988 when the broadband medium-power phased-array module was being developed [22]. This MPM was originally studied by the 1989 Microwave Power Module Panel. This panel was chartered to explore the potential of an integrated vacuum-electronic (TWT)/MIMIC (SSA) device that could be broadly applied in next-generation microwave systems, including EW, radar, and communications. This panel brought government and industry experts from the systems and device areas together to define the optimum power module.

The recommendations of the panel led to a triservice initiative in 1991 for the development of MPMs for electronic countermeasures and radar. The original specifications that were set as a goal for the program are listed in Table 14.2. Notice, in particular, that the power levels were ~100W at noise figures of 10 dB or lower.

Table 14.2
Tri-Service MPM Goal Specifications

Parameter	ECM	Radar
Frequency	6 – 18 GHz	7 – 11 GHz
RF Power Output	50 – 100 W	> 100 W
Duty Cycle	100%	50%
Saturated Gain	> 50 dB	> 50 dB
Efficiency	> 33%	> 40%
Noise Power	–105 dBm/Hz	–157 dBc/Hz at 1 kHz
Noise Figure	10 dB	< 10 dB
Spurious	< –10 dBc	< –60 dBc
Size	4 × 6 × 0.32 inches	2 × 6 × 1 inches

Phase 1 of the development effort was funded by the Tri-Service Vacuum Electronics Program established by the Naval Research Laboratory and the Vacuum Electronics Initiative established by the Defense Advanced Research Projects Agency (DARPA) [23]. Five industry teams participated in this development, which was initiated in 1991. Each team developed its version of the MPM. Various levels of specification compliance were achieved in 1992 and 1993. Each team demonstrated significant advancements over the existing state of the art.

Northrop came very close to meeting all of the requirements of Table 14.1 with its wideband MPM dubbed the ultraband MPM [24]. This device produced 125W (minimum) from 4.5–18-GHz. The dimensions were 8.3 inches by 4.8 inches by 1.0 inch and the weight was 3.6 lbs. The power and gain characteristics of the TWT used in this MPM are shown in Figure 14-54. The gain variation of this device is compensated for by the SSA and so the power output was nearly constant power over the entire frequency range. As Figure 14-55 shows, the efficiency of the ultraband MPM was as high as 45%.

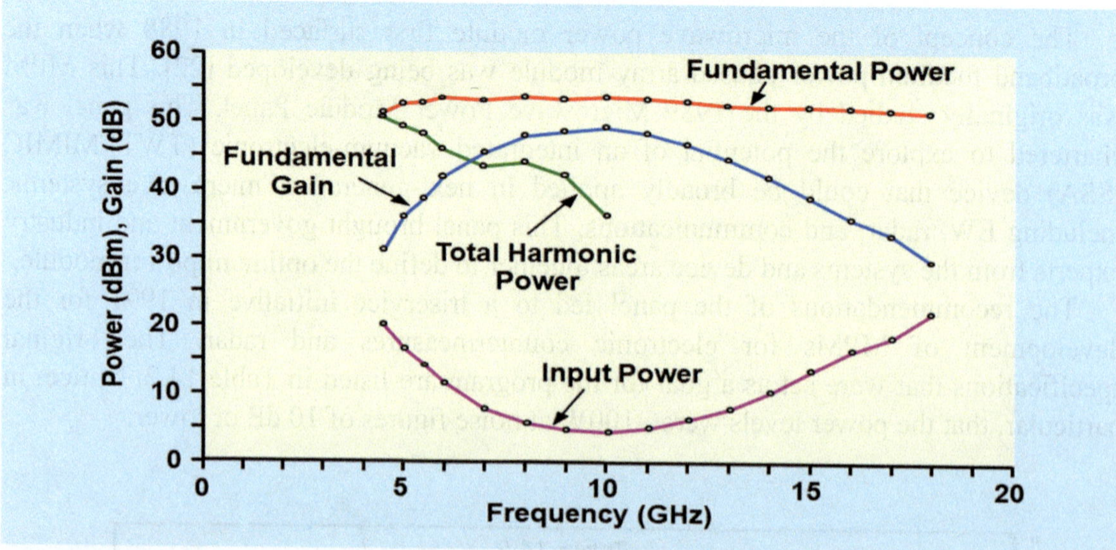

Figure 14-54 Power and gain characteristics of the TWT used in the Northrop ultraband MPM. (From: C. R. Smith et al., *Proc. IEEE*, May 1999. © 1999 IEEE.)

This initial success was followed by refinements to the MPM subassemblies and then a demonstration of a linear subarray of MPMs in 1994 [22]. Also, a major application for MPMs is and has been in unmanned airborne vehicles (UAVs). In fact, MPMs have proven to be an enabling technology for UAVs. The 1997 Annual United States Defense Report notes the following:

> Because of the UAV size, weight, and power constraints, installing both surveillance radar and a satellite communications link in the Predator would not have been feasible without advanced Microwave Power Module technology developed by the Navy. This technology provides a factor of 30 increase in power density and a factor of 10 reduction in volume. Because it operates over a broad frequency range, it offers new opportunities for integrating communications, radar, and electronic combat systems.

The frequency and power capabilities of MPMs are shown in Figure 14-56. This figure also shows the frequency/power range where solid-state devices are generally used, and the same for vacuum devices. MPMs fill much of the gap where neither technology is clearly superior.

Through the Tri-Service Vacuum Electronics Program, other Department of Defense programs, and investment by MPM manufacturers, the MPM concept was expanded to other frequency ranges and power levels. This includes development efforts toward MPMs for the following frequency bands:

- 2–8 GHz;
- High efficiency C band;
- 6-18 GHz (original MPM development);
- 4.5 -18 GHz;

- 2–18 GHz;
- 18–40 GHz (MMPMs);
- 40–46 GHz (MMPMs).

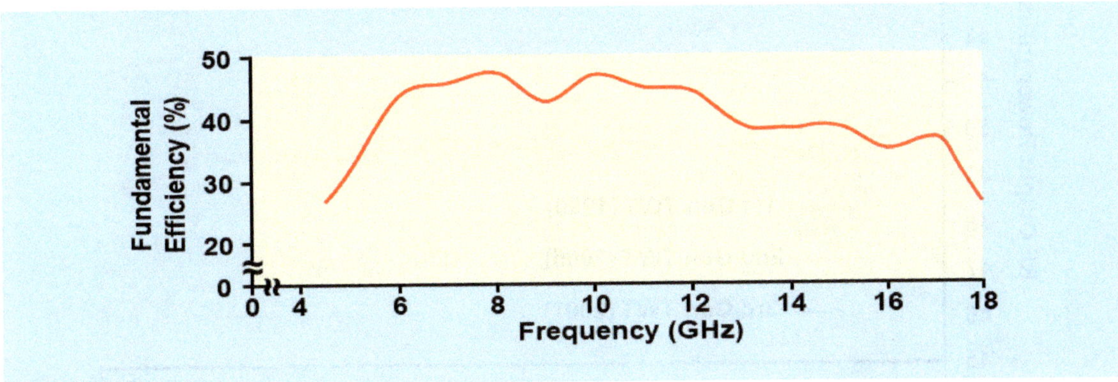

Figure 14-55 Efficiency of the Northrop ultraband MPM. (From: C. R. Smith et al., *Proc. IEEE*, May 1999. © 1999 IEEE.)

The RF performance of the MPM is dominated largely by the performance of the miniature helix TWT. These mini-TWTs are designed to be ultracompact while operating at high efficiency, with low levels of RF spurious and harmonic outputs. Figure 14-57 shows the output power evolution of IJ-band miniTWTs at L-3 Communications over the past 18 years [25]. Third-generation IJ-band MPMs utilizing the latest TWTs are capable of providing upward of 250W of power at the center of the band.

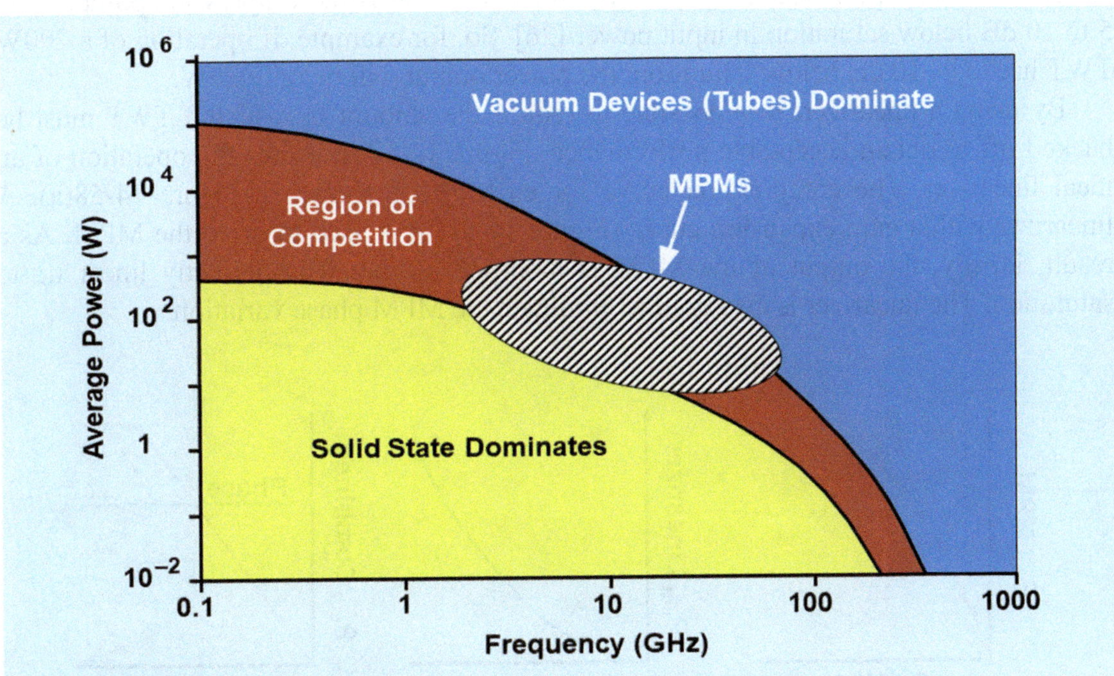

Figure 14-56 The domain of MPMs.

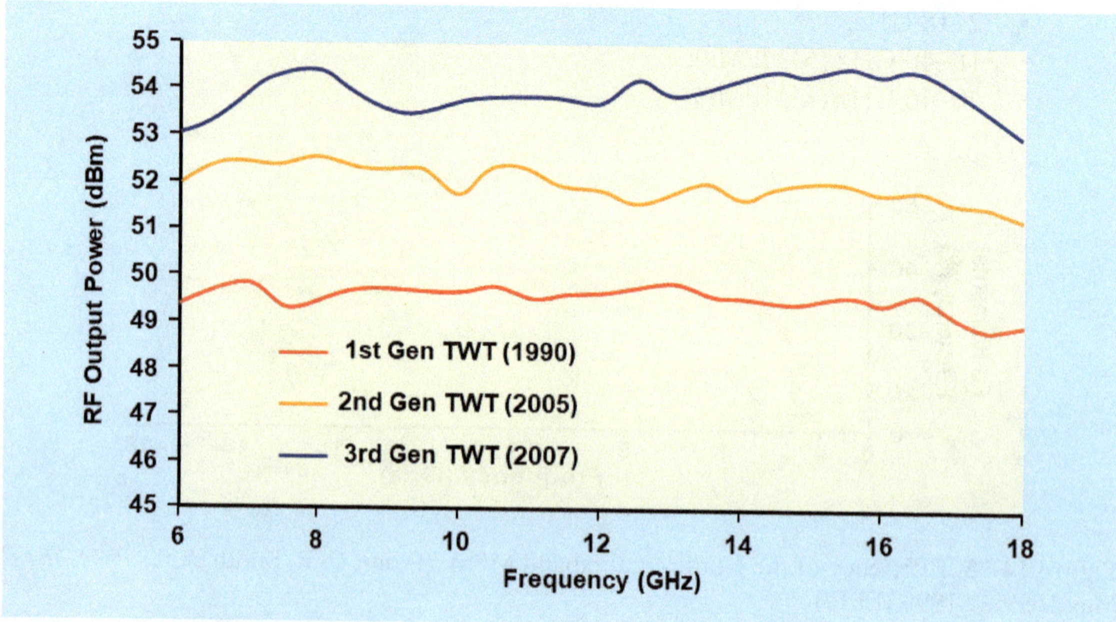

Figure 14-57 Evolution of MPM capability through TWT refinements.

MPMs are finding increased usage for communications applications. However, because of the gain compression that occurs in TWTs as they are driven toward saturation, they have to be backed off (operated below saturation). The separation of the 1-dB gain compression point from the maximum power (saturation) point is a good indicator of linear performance. Typically, for TWTs, the 1-dB compression point occurs 5 to 10 dB below saturation in input power [26]. So, for example, if operation of a 200W TWT has to be 10 dB below saturation, the power output will be only 20W.

By using a linearizer, it is possible to reduce the amount by which a TWT must be backed off to obtain acceptable performance. Figure 14-58 indicates the operation of an ideal linearizer. The compression curve for an MPM is shown in Figure 14-58(a). A linearizer with a gain expansion curve (Figure 14-58(b) is used to drive the MPM. As a result, ideally, the output of the MPM is modified so that it is perfectly linear up to saturation. The linearizer is also used to eliminate the MPM phase variation.

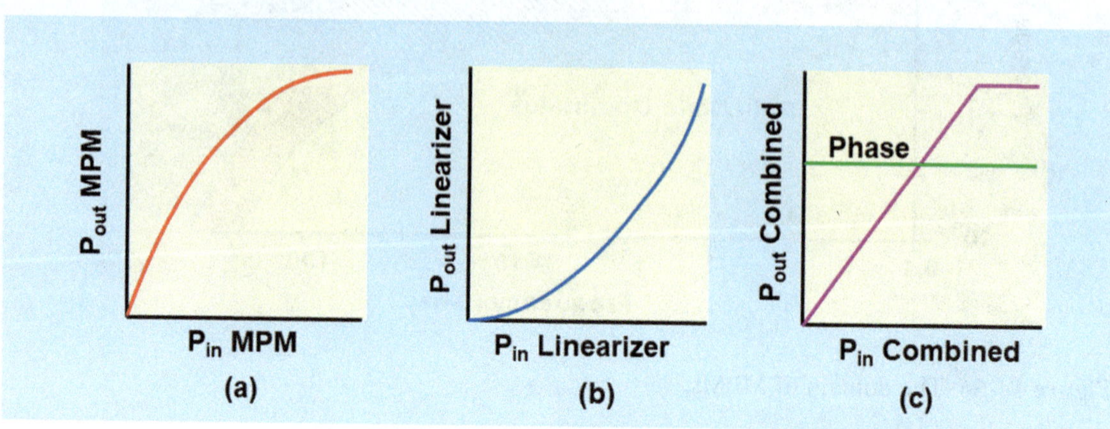

Figure 14-58 Idealized linearizer: Power out of MPM is linear up to saturation.

Figure 14-59 shows the output power, gain, and phase as a function input power for a test MPM and a linearized MPM (L-MPM) at C-band [26]. It can be seen that the 1-dB compression point was moved from about 5 dB from saturation to within 2 dB. This was equivalent to doubling the useful power output of the MPM.

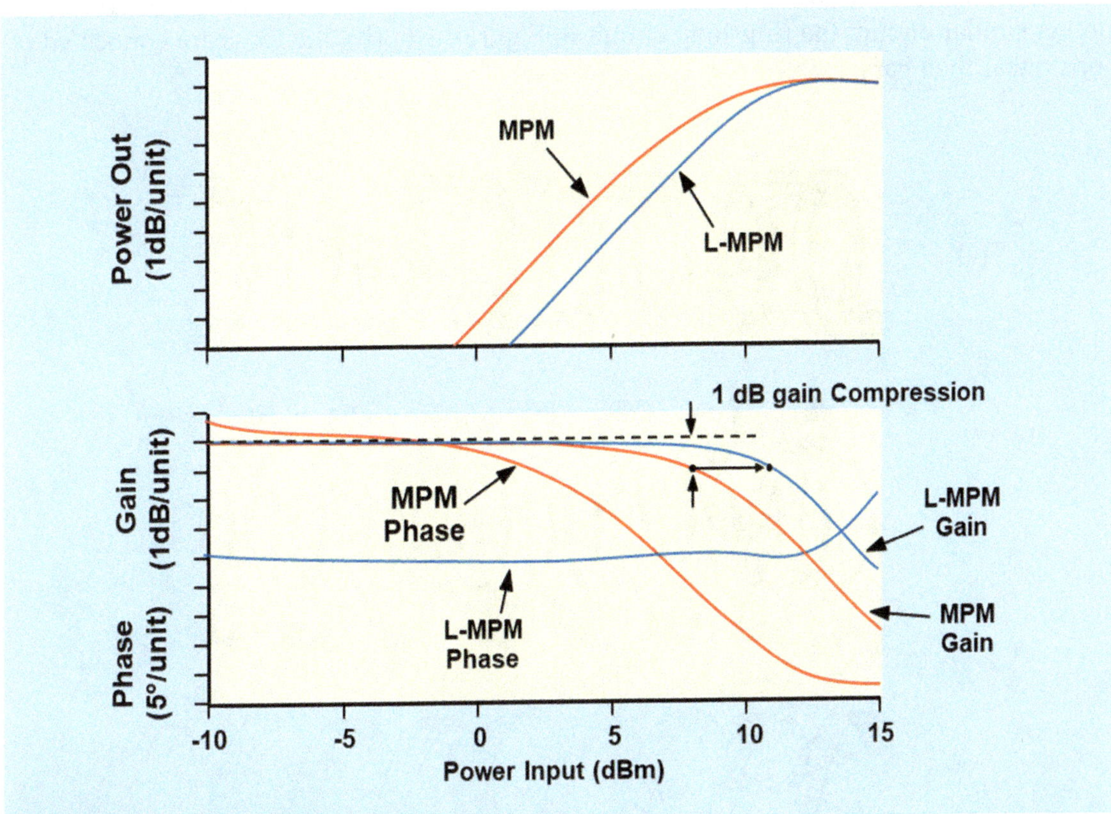

Figure 14-59 MPM power, gain, and phase with (L-MPM) and without (MPM) linearization. The 1-dB compression point moves 3 dB toward saturation.

The change in phase angle from small signal to saturation is another important indicator of linearity. Figure 14-59 also shows that the phase change between small signal and saturation was reduced from more than 45° to less than 1° by linearization.

Over the past decade, the MPM has undergone significant transformations, adapting to the needs of a wide variety of emerging radar and communications systems. In addition, the success of UAVs and mobile SATCOM terminals in various military theaters around the world has cultivated a renewed interest in MPM technology due to the size, weight, and efficiency advantages they offer when compared to traditional SSPAs and TWTAs.

As pointed out in [22], the Tri-Service investment has brought forth a new miniaturized RF amplifier capability for the RF electronics industry. Now, in addition to the ECM and radar applications, MPMs are being developed for commercial communications use such as satellite uplinks, line-of-site terrestrial links, and local multipoint distribution service.

14.8 RING-BAR and RING-LOOP TWTS

The main reason for the use of ring-bar or ring-loop circuits is to make possible operation at higher power levels than is possible with helix TWTs while avoiding backward-wave oscillations. Basically, the ring-bar circuit, which is shown in Figure 14-60(a), is derived from the contra-wound helix shown in Figure 14-60(c) and is far easier to fabricate. In another similar circuit, the ring-loop circuit in Figure 14-60(b), the rings are connected by loops rather than bars.

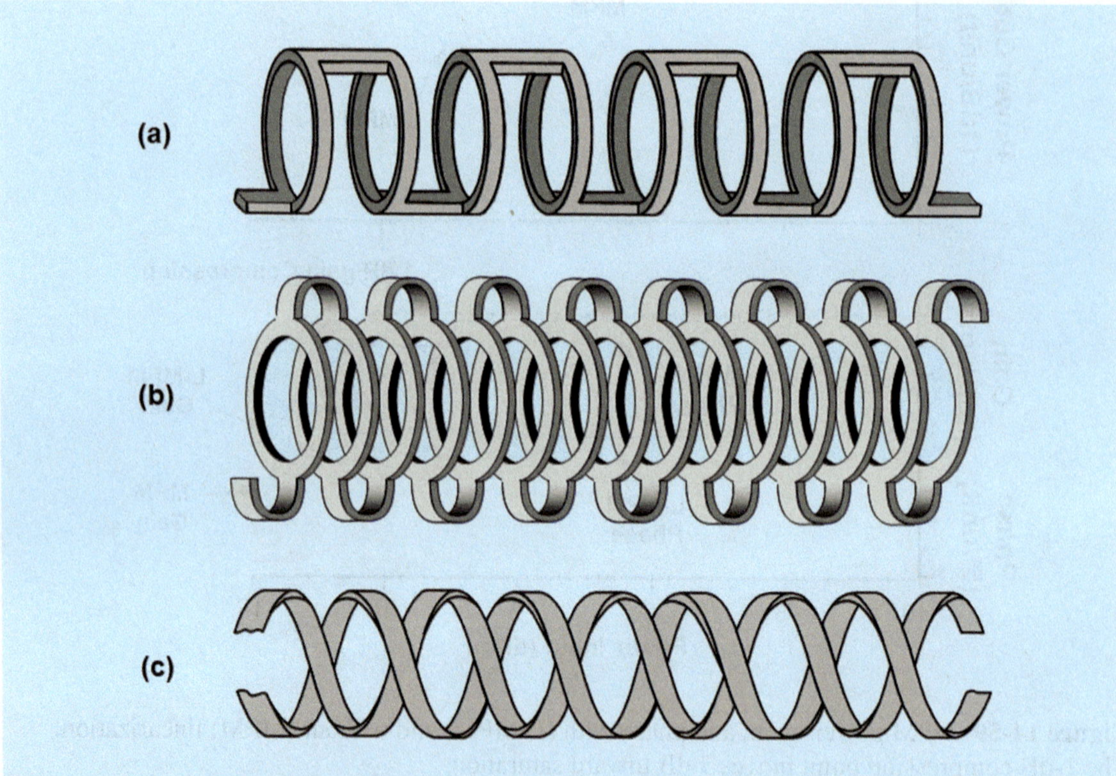

Figure 14-60 (a) Ring-bar, (b) ring-loop, and (c) contra-wound helix circuits.

These circuits, described by Chodorow and Chu [27] and by Birdsall and Everhart [28], suppress backward-wave oscillations and can be operated at higher voltages than a single helix. Also, because BWO is suppressed, it is possible to fill more of the helix tunnel with the electron beam so a larger beam with more current can be used. The penalty for these advantages is that the bandwidth is less than that of a single helix.

To understand the operation of these circuits and how they suppress backward-wave interactions, it is helpful to examine the contra-wound helix shown in Figure 14-61 and to compare its characteristics with those of a single helix. The pitch of the helices making up the contra-wound helix is the same as that of the single helix. For the moment, interactions between the two helices of the contra-wound and single helices are assumed to be the same. As a result, for a given frequency, the voltage distributions indicated in Figure 14-61 for the single helix and the contra-wound helix are the same. For the BWO discussion, these are those associated with two turns per wavelength.

The interaction of electrons with a backward wave on a single helix was discussed earlier in this chapter. To review, electrons that are initially accelerated (red dot, top left, Figure 14-61) experience essentially no forces 180° later (middle left) and then are accelerated again 360° later (bottom left). Similarly, and not shown in Figure 14-61, electrons that are initially decelerated experience no forces 180° later and then are decelerated 360° later. Thus, there is bunching and gain, even though the wave on the helix moves in the opposite direction from the electrons.

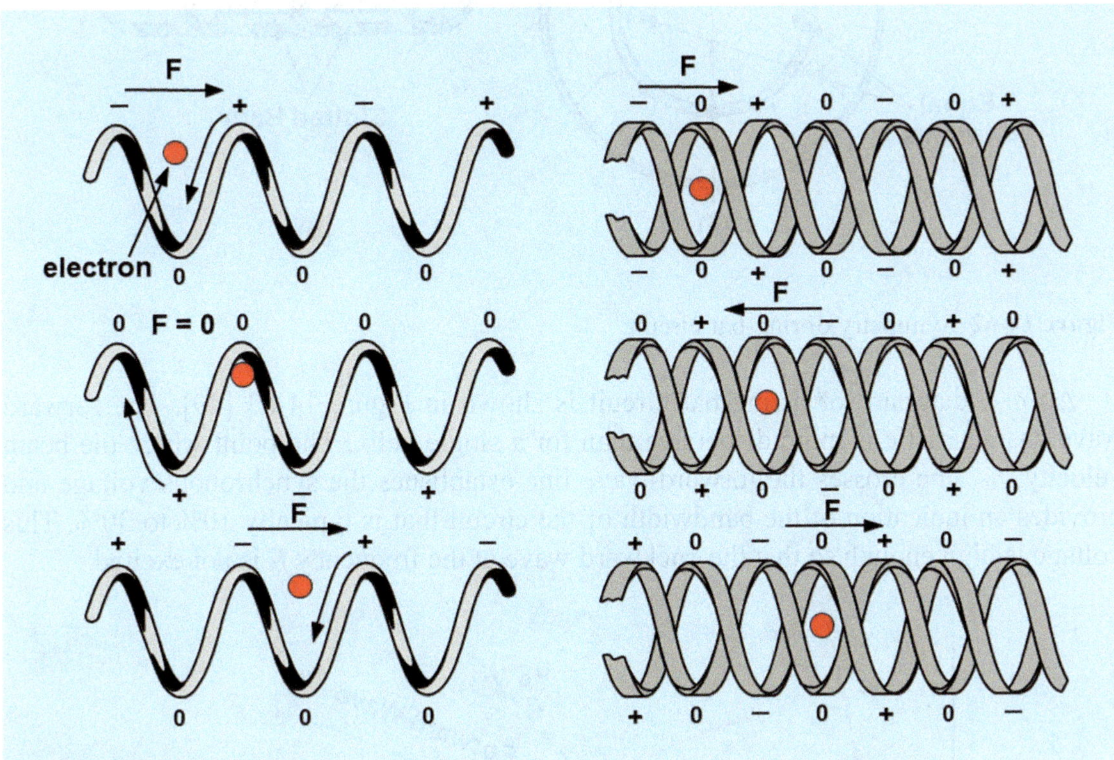

Figure 14-61 Backward-wave interaction on a single and a bifilar helix for two turns per wavelength.

For the contra-wound helix shown in Figure 14-61, it is assumed that the electrons and waves have the same velocities as for the single helix. In this case, however, electrons that are initially accelerated (top right) are decelerated 180° later (center right) and then accelerated again 360° later (bottom right). Because on each cycle the accelerating and decelerating forces tend to cancel each other, there is no net gain that can lead to backward-wave oscillations. As a result, higher operating voltages, currents, and RF power levels are possible with the contra-wound helix than with the single helix (with no suppression techniques being used). The same arguments apply to the ring-bar and ring-loop circuits, so these circuits can operate without BWO at much higher voltages than can the single helix.

For maximum effectiveness in preventing BWO, the two helices that make up the contra-wound helix must be identical. Similarly, with the ring-bar and ring-loop structures, it is important to keep the structures symmetrical about a plane passing through the center of the circuit as shown in Figure 14-62.

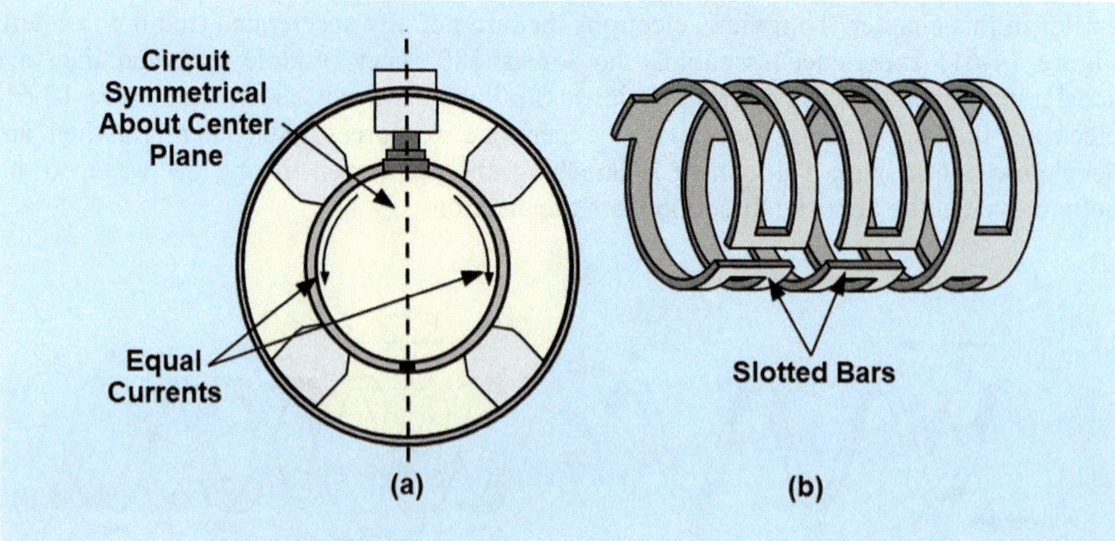

Figure 14-62 Symmetry of ring-bar circuit.

An ω-β diagram for a ring-bar circuit is shown in Figure 14-63 [29]. The forward wave characteristic is more dispersive than for a single helix. The point where the beam velocity, u_o, line crosses the forward-wave line establishes the synchronous voltage and provides an indication of the bandwidth of the circuit that is typically 10% to 30%. This voltage is high enough so that the backward wave at the frequency f_π is not excited.

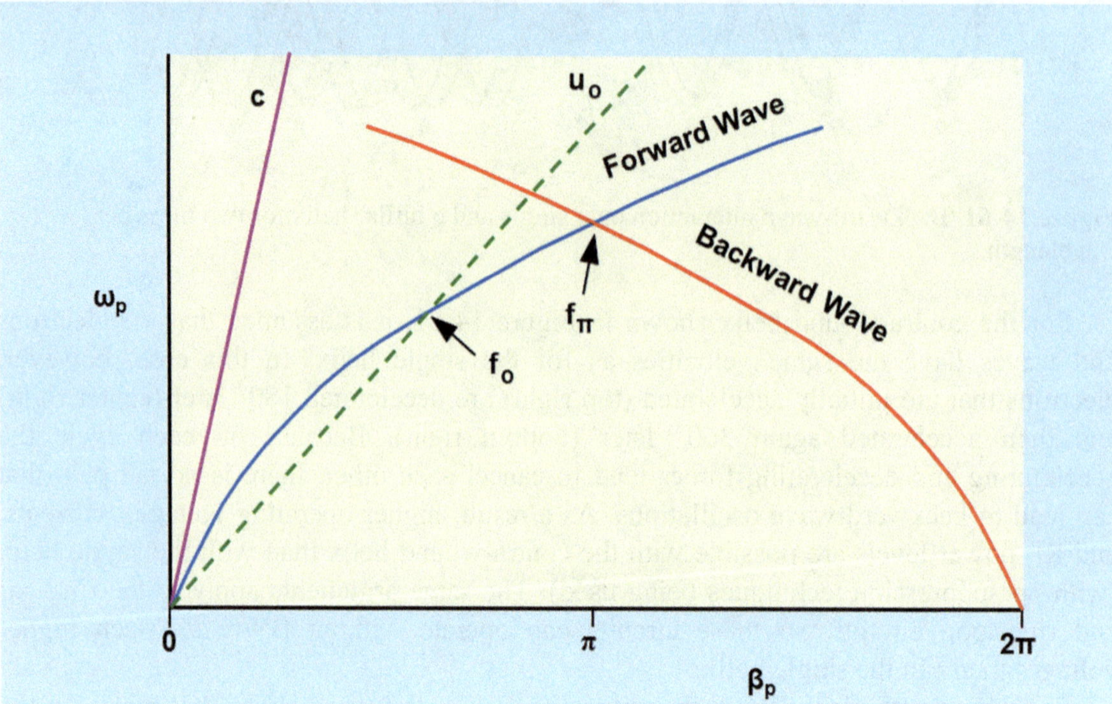

Figure 14-63 The ω-β diagram for a ring-bar structure. (Adapted from: R. LeBorgne et al., 1990 IEDM. © 1990 IEEE.)

The designs of the ring-bar and ring-loop circuits are more complex than a helix because the phase velocity is controlled by the width and length of the bar or the loop. Reference [30] describes progress in designing these circuits.

Figure 14-64 shows the variation of phase velocity with frequency for a millimeter-wave tube [29]. At the center of the operating band, the phase velocity is approximately 25% of the speed of light and the operating voltage is about 18 kV.

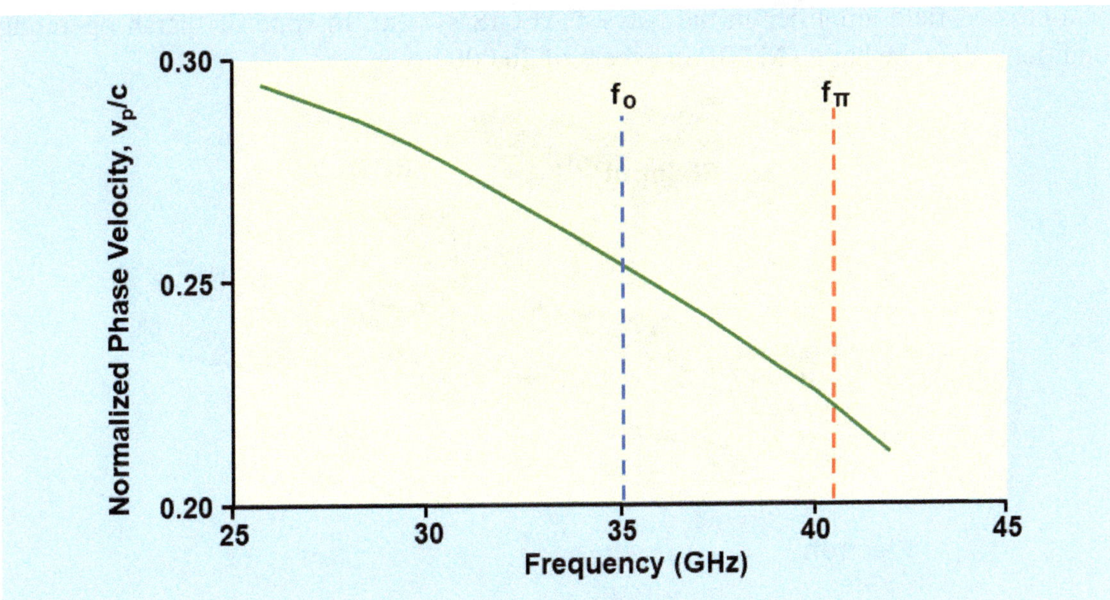

Figure 14-64 Phase velocity of the ring-bar circuit in a Ka-band TWT. (Adapted from: R. LeBorgne et al., 1990 IEDM. © 1990 IEEE.)

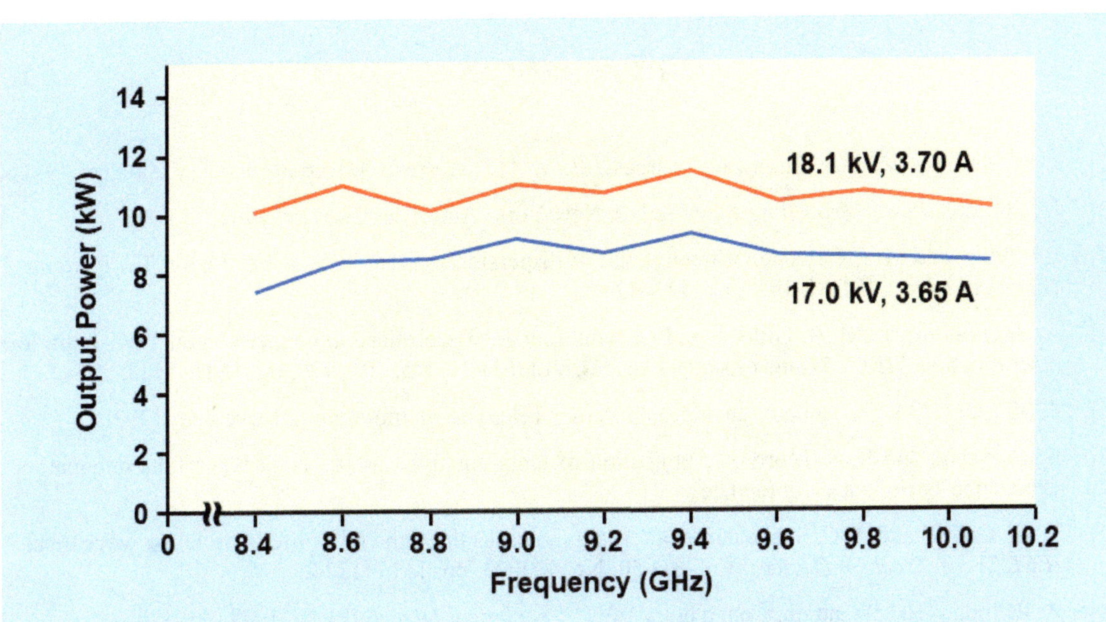

Figure 14-65 Output power characteristic for a ring-bar TWT. (From: Hughes Aircraft Co., *TWT and TWTA Handbook.*)

Figure 14-65 contains a plot of the power and bandwidth capabilities of an X-band ring-bar TWT. Note that the power is very constant over a bandwidth of nearly 20%. Ring-bar TWTs are capable of very high powers at lower frequencies. Examples are the QKW1617(A) and the QKW1818, which were made by Raytheon and had power capabilities of 160-kW peak and 10-kW average [31].

The tube shown in Figure 14-66 is manufactured by Teledyne and is used as a driver for a crossed-field amplifier in the Navy's AEGIS system. In spite of harsh operating conditions, the tube has an MTBF in excess of 20,000 hrs.

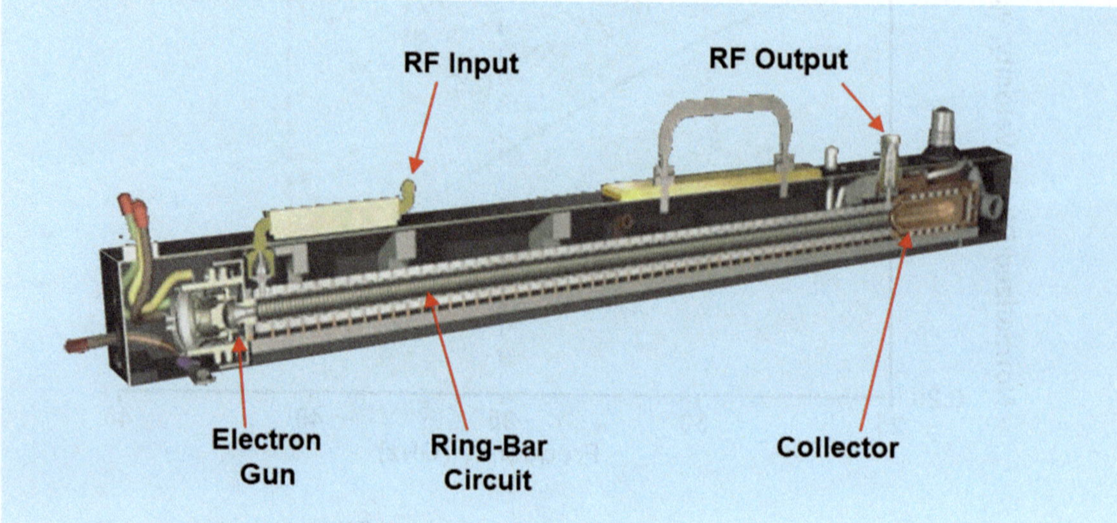

Figure 14-66 Ring-bar TWT. (Courtesy of Teledyne MEC.)

REFERENCES

[1] R. S. Symons, "Modern microwave sources," *IEEE AESS Systems Magazine*, January 2002, pp.19–26.

[2] J. F. Gittins, *Power Travelling Wave Tubes*, New York: American Elsevier, 1965.

[3] J. L. Putz, and M. J. Cascone, "Effective use of dispersion shaping in broadband helix TWT circuits," *Technical Digest*, IEDM, 1979, pp. 422–424.

[4] A. S. Gilmour, Jr., M. R. Gillette, and C. Jenn-Tsung, "Theoretical and experimental TWT helix loss determination," *IEEE Trans. Electron Devices*, Vol. ED-26, No. 10, 1979, pp. 1581–1588.

[5] P. M. Lally, "TWT attenuator chamges," Internal Teledyne Memorandum, November 1993.

[6] R. P. Anand and R. S. Morris, "Application of tantalum films as traveling wave tube attenuators," 1968 Tube Techniques Conference.

[7] D. M. Goebel and A. C. Schneider, et al. "Gain increases through end of life in traveling wave tubes." *IEEE Trans. Electron Devices*, Vol. ED-50, No. 4, 2003, pp. 1117–1124.

[8] A. R. Jung, "10 kW and up, from a helix TWT?" *Technical Digest,* IEDM, 1978, pp. 530–553.

[9] P. Birtel et al., "Simulation of the BWO threshold current in a helix TWT," *Trans. IEEE IVEC*, 2006, pp. 403–404.

[10] A. Scott, and M. J. Cascone, "What's new in helix TWTs?" *Technical Digest,* IEDM, 1978, pp. 526–529.

[11] R. Harper, and M. P. Puri, "Heat transfer and power capabilities of EHF helix TWTs," *Technical Digest,* IEDM 1986, pp. 498–500.

[12] Y. Hiramatsu, "Study of block supported helix circuits for high-power TWTs," *Technical Digest,* IEDM, pp. 416–417, 1979.

[13] T. Mosser, W. Pinger, and D. Zavidil, "High-power brazed helix fabrication," *Microwave Power Tube Conference*, 1980.

[14] O. Sauseng, A. E. Manoly, and A. Hall, "Thermal properties and power capability of helix structures for millimeter waves," *Technical Digest, International Electron Devices Meeting*, 1978, pp. 534–537.

[15] W. L. Menninger et al., "70% efficient Ku-band and C-band TWTs for satellite downlinks," *IEEE Trans. Electron Devices*, Vol. 52, No. 5, pp. 673–678, 2005.

[16] P. Ehret et al., "L-band TWTAs for navigation satellites," *IEEE Trans.on Electron Devices*, Vol. 52, No. 5, May 2005, pp. 679–684.

[17] J. M. Weekley, and B. J. Mangus, "TWTA versus SSPA: a comparison of on-orbit reliability data," *IEEE Trans. Electron Devices*, Vol. 52, No. 5, May 2005, pp. 650–652.

[18] A. S. Gilmour, Jr., "SEM analysis of TWT helix samples," *IEEE Trans.on Electron Devices*, Vol. ED-24, No. 6, June 1977, pp. 774–775.

[19] Kevin P. Mallon, TWTAs for Satellite Communications: Past, Present and Future, IVEC 2008.

[20] W. Q. Lohmeyer, et al. Communication satellite power amplifiers, current and future SSPA and TWTA technologies, *International Journal of Satellite Communications and Networking,* Vol. 34, No. 2, March/April 2016, pp. 95–113.

[21] D.S. Komm, R.T. Benton et al, "Advances in Space TWT Efficiencies," *IEEE Trans. Electron Devices,* Vol. 48, No. 1, January 2001. Pp. 174-176.

[22] R. H. Abrams Jr and R. K. Parker, "Introduction to the MPM: what it is and where it might fit,".*Microwave Symposium Digest, 1993., IEEE MTT-S International*, pp. 107–110.

[23] T. Ninnis, "Microwave power modules for UAVs," L3 Communications Electron Devices Division.

[24] C. R. Smith, C. M. Armstrong, and J. Duthie, "The microwave power module: a versatile RF building block for high-power transmitters," *Proc. IEEE*, Vol. 87, No. 5, May 1999, pp. 717–737.

[25] R. Duggal, A. Donald, and T. Schoemehl, "Technological evolution of the microwave power module (MPM)," Conference Record, IVEC 2009.

[26] A. Katz, R. Gray, et al., "Performance of microwave and millimeter wave power modules (MPMs) with linearization," MILCOM 2005.

[27] M. Chodorow, and E. L. Chu, "Cross-wound helices for traveling wave tubes," *Jour. App. Phy.*, Vol. 26, No. 1, January 1955, pp. 33–43.

[28] C. K. Birdsall and T. E. Everhart, "Modified contra-wound helix circuits for high-power traveling-wave tubes," *IRE Trans. Electron Devices,* Vol. ED-3, 1956, pp. 190–204.

[29] R. H. LeBorgne et al., "Development of an 800 W, Ka-band, ring-bar TWT," *Technical Digest,* IEDM, 1990.

[30] D. R. Dyson, M. J. Clark, and V. H. Smith, "Numerical analysis of ring-loop and ring-bar slow wave structures for traveling wave tubes," *Trans. IEEE International Vacuum Electronics Conf.*, 2003, pp. 46–47.

[31] M. Skolnik, G. Linde, and K. Meads, "Senrad: an advanced wideband air-surveillance radar," *IEEE Trans. Aerospace and Electronic Systems*, Vol. 37, No. 4, October 2001, pp. 1163–1175.

Chapter 15

Coupled-Cavity TWTs

In coupled-cavity TWTs [1, 2], several tens of klystron-like cavities are used for the slow-wave structure. Except for sever regions, these cavities are electromagnetically coupled so that an RF signal can propagate along the structure. Coupling techniques take on widely differing forms and are discussed in this chapter. If an electron beam is passed through the structure and if it has a velocity near the phase velocity of the circuit, then amplification can occur.

With appropriate choices of cavity dimensions and cavity-coupling techniques, electron beam velocity can be high, and operation at high peak-power levels is possible. Also, coupled-cavity structures are mostly metallic (copper and iron) and have geometries such that thermal resistances are very low. As a result, high average powers are possible.

15.1 BASIC OPERATING PRINCIPLES

Figures 15-1 and 15-2 are useful in describing the interaction of the beam with the circuit in a coupled-cavity TWT. The input circuit supplies a signal to the first cavity through a magnetic coupling loop or an impedance-matching section from a waveguide. As with the helix TWT, there are launching losses so only a portion of the input signal is used to initiate beam modulation by establishing voltage V_1 across the gap in the first cavity. Another part of the input signal propagates from the first cavity to the second (from A to B through one of the coupling structures to be described in the following sections).

15.1.1 Beam-Circuit Interaction

The voltage across the first gap modulates the electron beam velocity so that bunches develop as the beam moves toward the second cavity. The bunches induce a signal into the second cavity, which splits into two equal wavelets, one traveling to the right and the other to the left.

As indicated in Figure 15-2 and described in connection with bunching in a helix TWT, the bunches that form in the beam repel electrons on the circuit causing those electrons to travel to the right and to the left. The electrons on the circuit travel at the phase velocity, v_p, of the coupled-cavity circuit, which is nearly the same as the velocity of the electron bunches in the beam. Relative to a bunch, electrons on the circuit move left at $2v_p$. This produces a deficiency of electrons, which makes the circuit positive. The reverse traveling wave is eliminated by loss or severs not shown in Figure 15-1.

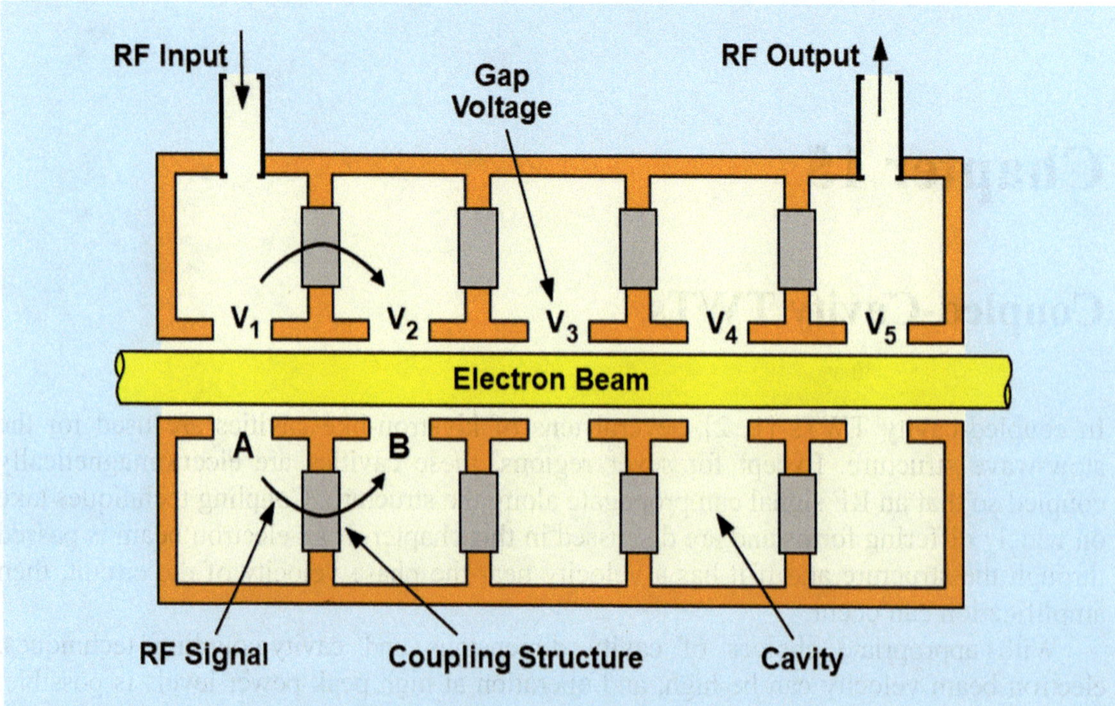

Figure 15-1 Basic coupled-cavity circuit.

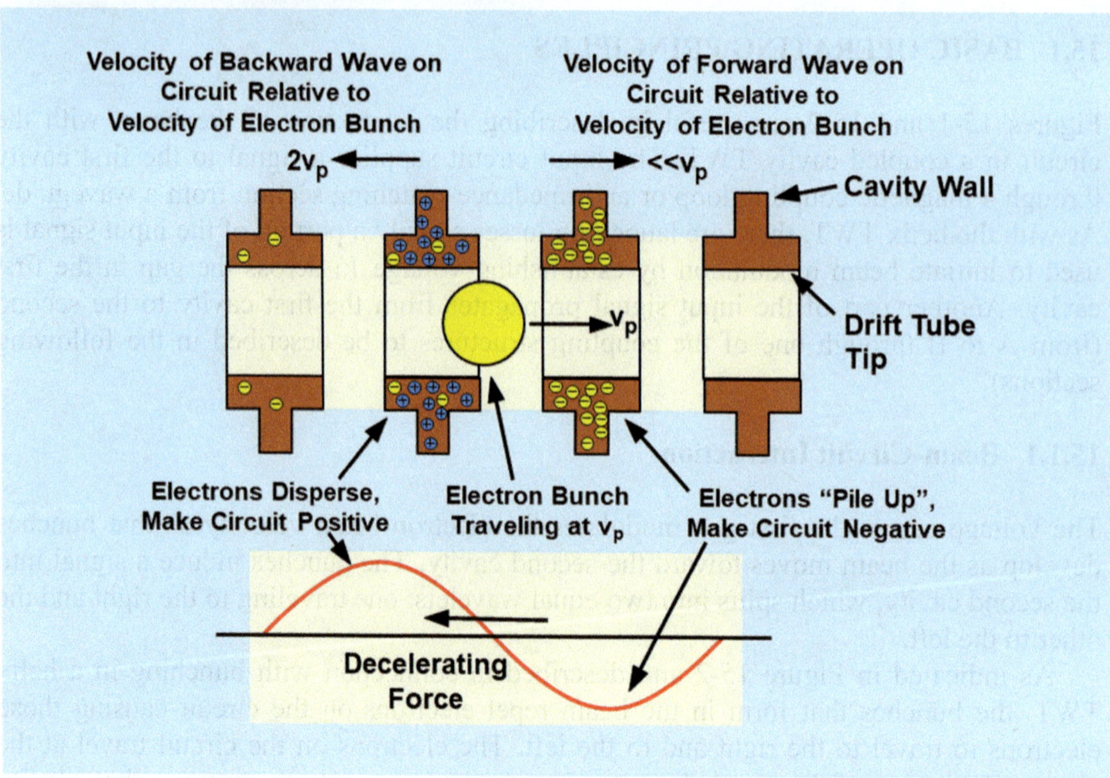

Figure 15-2 Basic beam-circuit interaction process in a coupled-cavity TWT. Only the drift tube tips and small portions of the cavity walls are shown.

Relative to the bunch, electrons on the right, which are moving at approximately the bunch velocity, are nearly standing still. As a result, electrons accumulate ahead of the bunch and make the circuit negative.

The increase in electron charge ahead of the bunch and decrease behind the bunch produce a voltage waveform on the circuit such that decelerating field regions move into the regions where the bunches are located. As more and more electrons are forced into the bunch, the induced waveform grows so the RF power on the circuit increases.

The voltage induced in the second cavity by the beam combines with the voltage fed forward from the first gap through the coupling structure to produce voltage V_2 across the second gap. This voltage enhances the velocity modulation of the beam, thereby increasing the ac current in the beam. This process is repeated at subsequent gaps with each gap voltage being the combination of the voltage coupled from the previous gap and the sum of all the components induced by the ac beam current.

15.1.2 Cavity-to-Cavity Coupling

There are two fundamental techniques for coupling a signal from cavity to cavity. These are based on whether the electric field is coupled or if the magnetic field is coupled. There are large differences between the operation of tubes with these two coupling techniques. Figure 15-3 shows how cavity-to-cavity coupling with electric and magnetic field works.

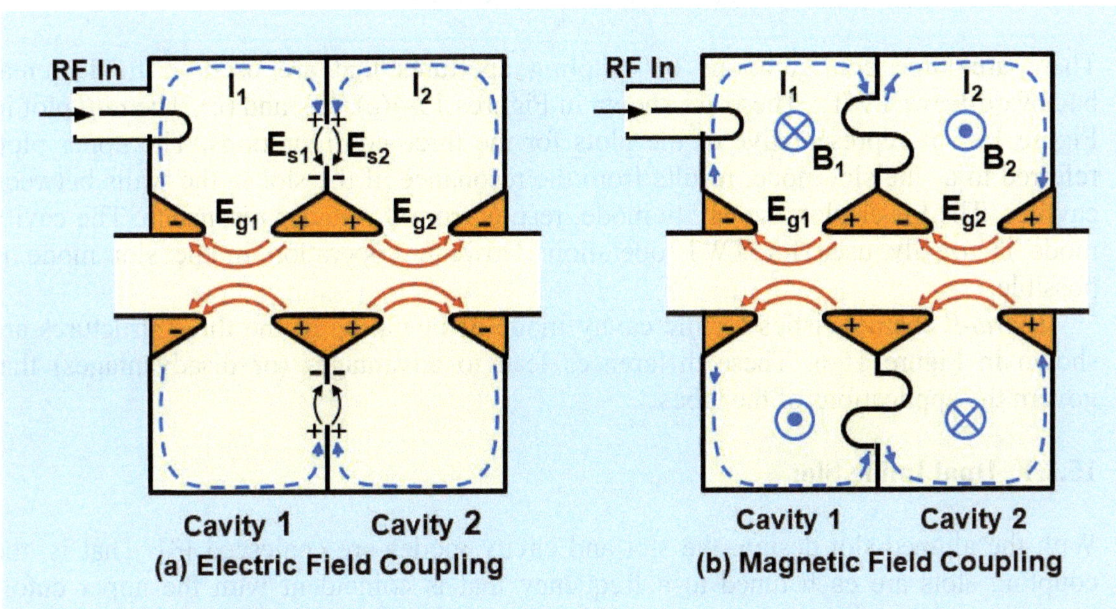

Figure 15-3 Cavity-to-cavity coupling.

As Figure 15-3(a) shows, electric field coupling is accomplished with slots in the plate between cavities. Assume, at first, that there is no signal in the cavities so all currents and electric fields are zero. When an RF signal is injected into cavity 1 (with a magnetic coupling loop for example), current I_1 flows and this produces electric field E_{s1} across the slot. The slot field in cavity 2 is the same as that in cavity 1. To produce this field, I_2 must flow in cavity 2. The charge separations that the current flows produce establish electric fields E_{g1} and E_{g2} across the cavity gaps. These are the fields that interact with the electron beam. Note that the field across gap 2 is opposite to that across gap 1. For a bunching action that is initiated in gap 1 to be reinforced in gap 2, either the wave on the circuit must travel in a direction opposite to that of the electron beam or the beam velocity must be such that the gap 2 voltage reverses phase before the bunch arrives. This is basically a backward-wave circuit and is referred to as a fundamental backward-wave circuit.

As shown in Figure 15-3(b), with coupling of the magnetic field from cavity to cavity, it is possible to make the electric field across the gap, E_{g2}, be in phase with the field E_{g1}. The dual-coupling loops shown in Figure 15-3(b) reverse the magnetic field at the coupling aperture and produce a 180° phase shift in the electric field. As a result, the field across gap 2 is in phase with the field across gap 1. For a bunching action that is initiated in gap 1 to be reinforced in gap 2, the beam velocity must be high enough that the bunching initiated in gap 1 is reinforced in gap 2. This is referred to as a fundamental forward-wave circuit.

These electric and magnetic field coupling techniques are described in greater detail in the following sections where fundamental backward-wave and fundamental forward-wave circuits are discussed.

15.2 FUNDAMENTAL BACKWARD-WAVE OPERATION

There are three configurations of coupling apertures that are used in fundamental backward-wave TWTs. These are shown in Figures 15-4(a), (b), and (c). The ω-β plot in Figure 15-5 is representative of the plots for the three configurations. The upper plot, referred to as the slot mode, results from the resonance of the slot in the walls between cavities. The lower plot, the cavity mode, results from the cavity resonance. The cavity mode is usually used for TWT operation. However, operation in the slot mode is possible.

The ω-β characteristics for the cavity mode of operation of the three structures are shown in Figure 15-6. These differences lead to advantages (or disadvantages) that govern the applications of the tubes.

15.2.1 Dual Inline Slot

With the aligned-slot design, the slot and cavity modes are coalesced [3]. That is, the coupling slots are each tuned to a frequency that is coincident with the upper cutoff frequency of the cavity mode. As a result, the two modes merge to form a composite coalesced mode [4].

When the slots are aligned, all of the circulating cavity current that couples to the previous cavity also couples to the following cavity. As a result, the aligned-slot structure has properties that are quite different than those for the staggered-slot structures to be discussed next. The result is that, as shown in Figure 15-6, the ω-β characteristic is much flatter than for the other circuit configurations.

Because of the flat shape, bandwidth is narrow. However, a larger value of βL can be used and so the cavities can be longer than for the staggered-slot circuits. This means that the structure can be used at a higher beam voltage to generate larger peak and average powers than the other circuits.

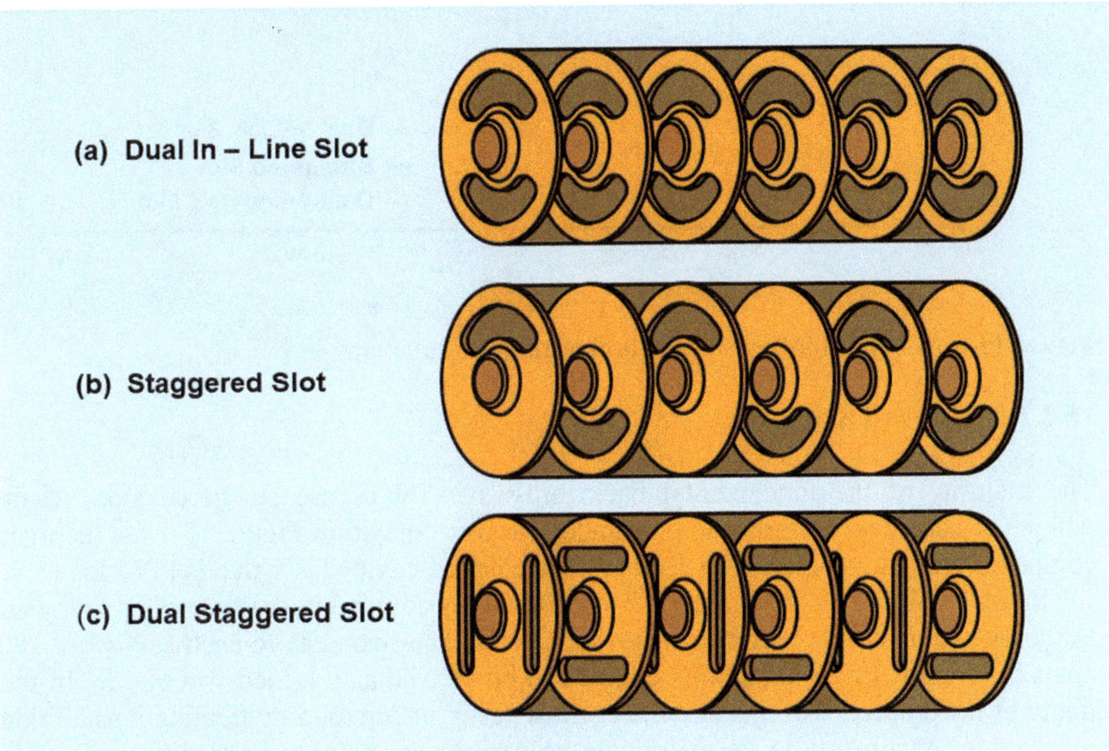

Figure 15-4 Fundamental backward-wave structures.

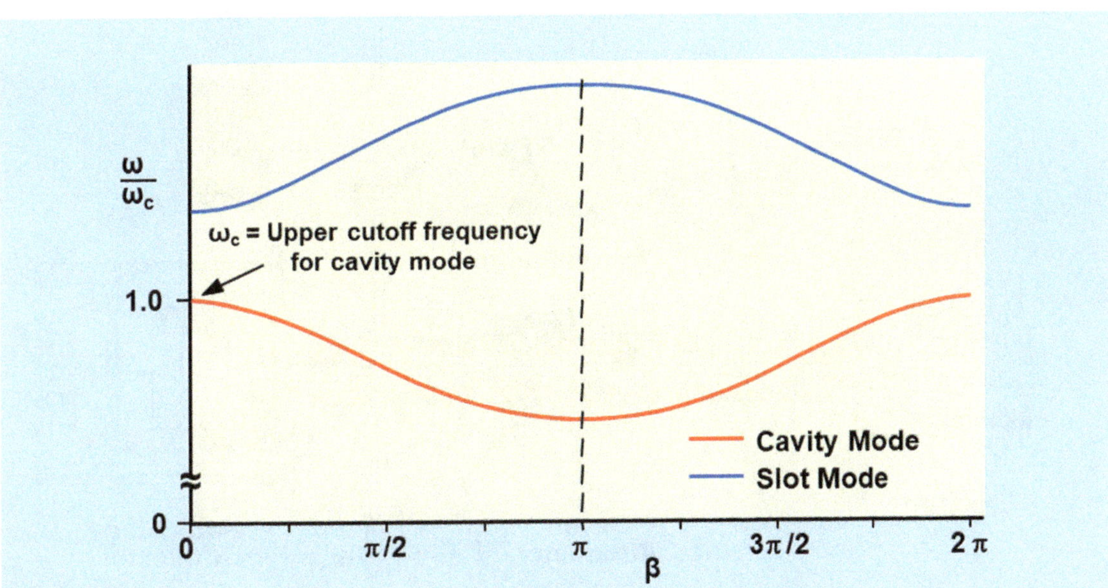

Figure 15-5 Typical ω-β plots for a fundamental backward-wave structure.

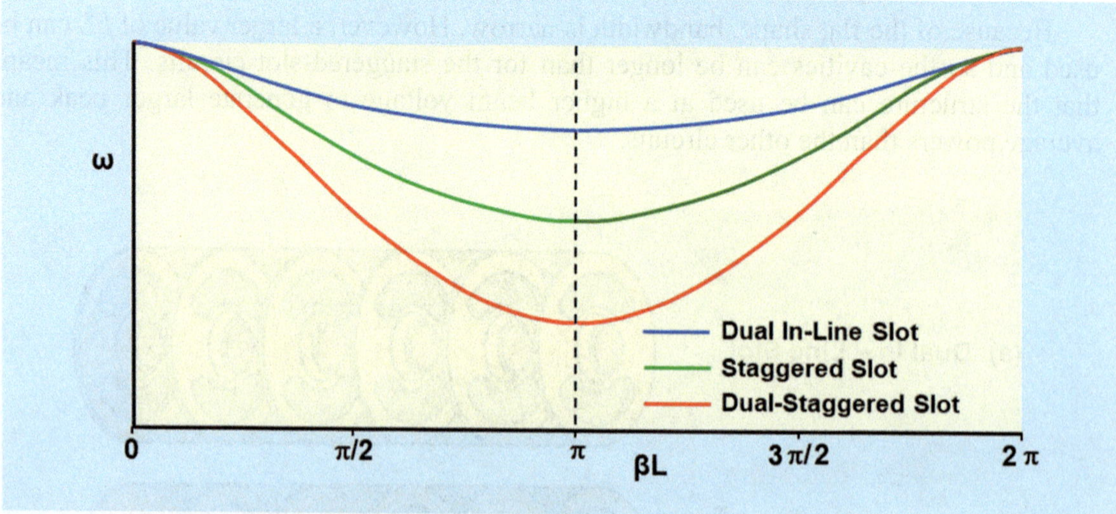

Figure 15-6 *ω-β* plots for fundamental backward-wave structures.

15.2.2 Staggered Slot

The majority of the fundamental backward-wave TWTs use staggered slot circuits, which are sometimes referred to as Hughes circuits. Shown in Figure 15-7 are the major components of a fundamental backward-wave coupled-cavity TWT that is PPM-focused.

In the TWT shown in Figure 15-7, each cavity is coupled to the following cavity and to the preceding cavity with a kidney-shaped slot. The two slots in each cavity are 180° apart in rotational position so this circuit can be viewed as a folded waveguide. In fact, many of the properties of this circuit are most easily understood by treating it as a folded waveguide. As a result, in the following discussions of *ω-β* characteristics, vane-loaded waveguides and folded waveguides are considered first.

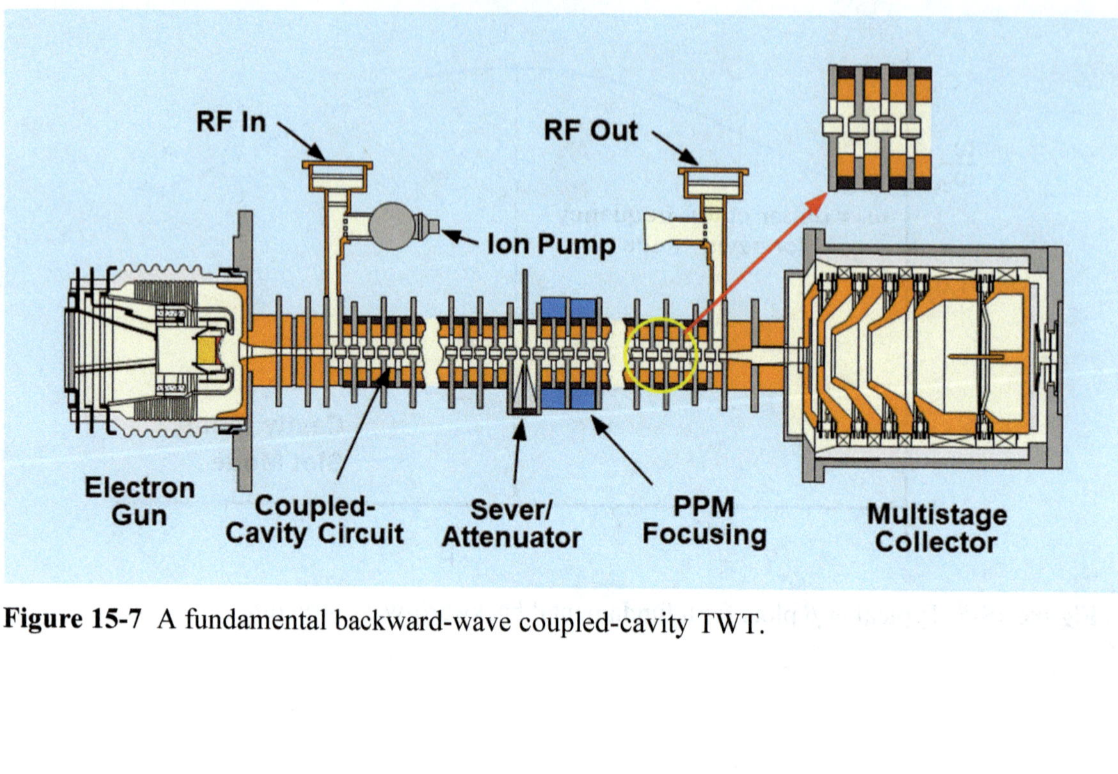

Figure 15-7 A fundamental backward-wave coupled-cavity TWT.

To examine the interaction of waves in the guide with an electron beam, consider a single electron following the serpentine path shown in Figure 15-8. Assume first that the electron travels from A to B in the same time that the wave crest in the guide travels from A to B. With respect to the direction of motion of the electron, a 180° phase reversal takes place in the electric field. Thus, as seen by the electron, the ω-β diagram is shifted by π radians as shown in Figure 15-9.

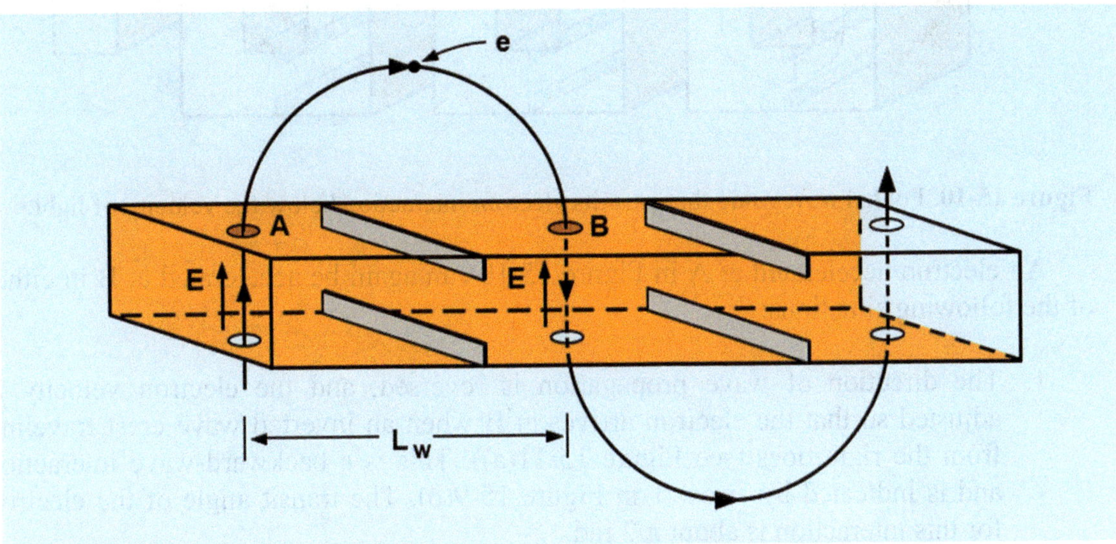

Figure 15-8 Serpentine electron path through a waveguide with vanes.

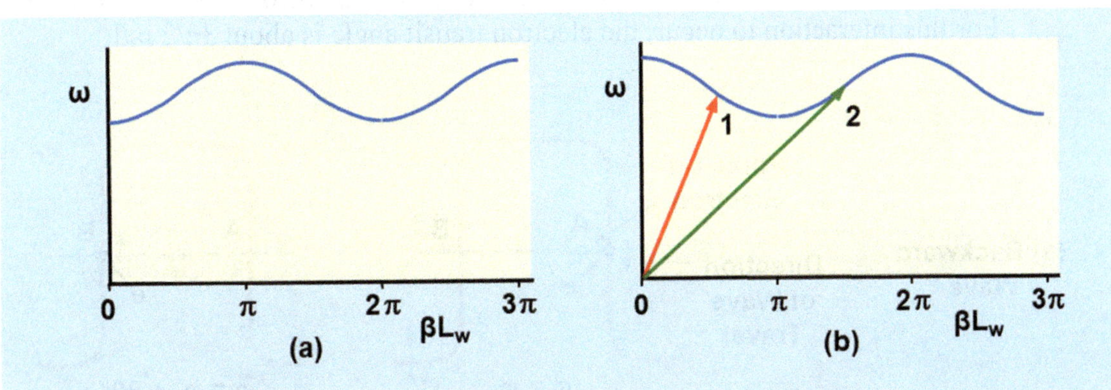

Figure 15-9 The ω-β diagram for a vane loaded waveguide (a), and as seen by an electron following the serpentine path in Figure 15-8(b).

An obvious problem with the example given here is that the electron would have to travel faster than the speed of light to move from A to B in the time that the wave does. This problem can be alleviated by folding the waveguide as shown in Figure 15-10. Assuming that no additional impedance changes occur, the ω-β diagram for the wave is not changed. From the perspective of the electron, the phase velocity of the wave and the electron velocity from A to B along path L is important, and so βL_w must be replaced by βL on the abscissa of Figure 15-9 to consider electron interaction.

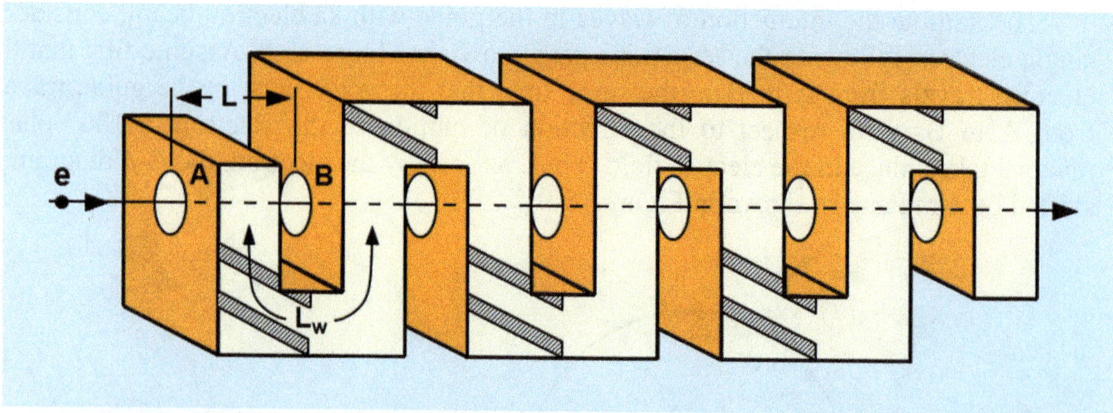

Figure 15-10 Folded waveguide that permits electron interaction below the velocity of light.

An electron accelerated at A in Figure 15-10 can again be accelerated at B in either of the following situations:

1. The direction of wave propagation is reversed, and the electron velocity is adjusted so that the electron arrives at B when an inverted wave crest traveling from the right does (see Figure 15-11(a)). This is a backward-wave interaction and is indicated by arrow 1 in Figure 15-9(b). The transit angle of the electron for this interaction is about $\pi/2$ rad.
2. As indicated in Figure 15-11(b), the electron velocity is reduced so that the next inverted wave crest traveling to the right arrives at B when the electron does. This is a forward-wave interaction and is indicated by arrow 2 in Figure 15-9(b). For this interaction to occur, the electron transit angle is about $3\pi/2$ rad.

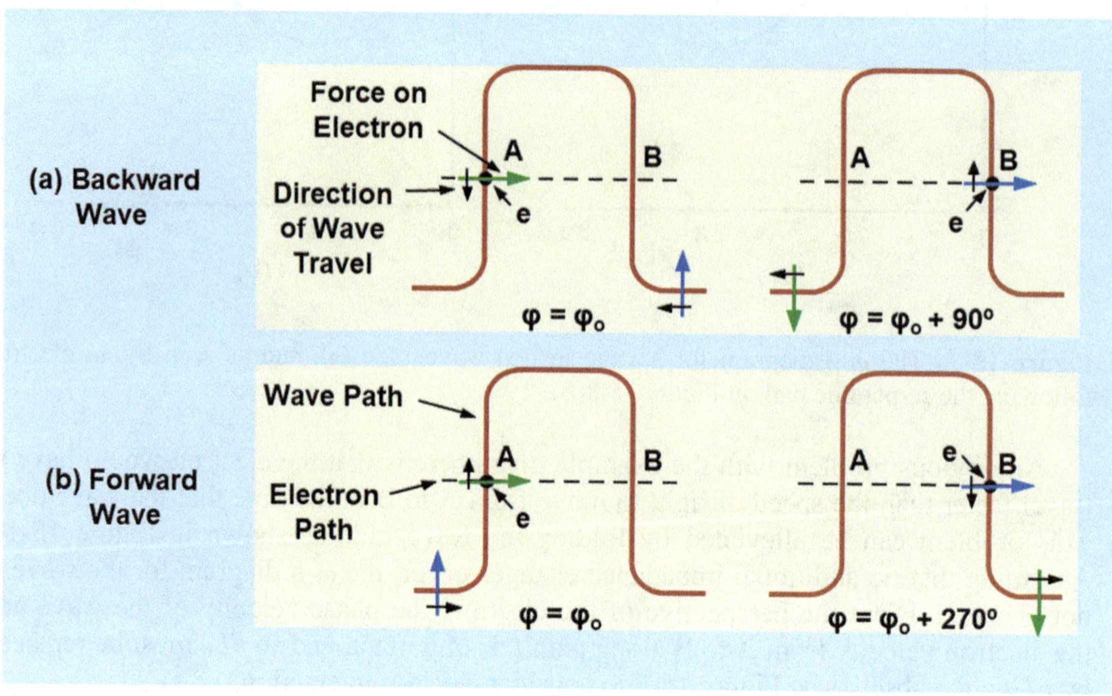

Figure 15-11 Wave propagation that reinforces acceleration of an electron.

As was shown in Figure 15-4 and 15-12, coupling between cavities in the staggered slot circuit is through kidney-shaped slots rather than through the gaps between the waveguide vanes shown in Figure 15-10. One of the variables in these slots is the angle, θ, of the coupling slot. Changes in this angle change the percentage of the cavity current involved in coupling. These changes also affect the slot inductance and the slot resonant frequency. The effect of these changes on the dispersion characteristics is shown in Figure 15-13. These changes, in turn, affect the cold bandwidth (bandwidth of circuit with no electron beam) as shown in Figure 15-14. Changes in the coupling slot angle also change the relative positions of the slot and cavity passbands as shown in Figure 15-15. Note that for large coupling angles, the slot passband can become the lower passband.

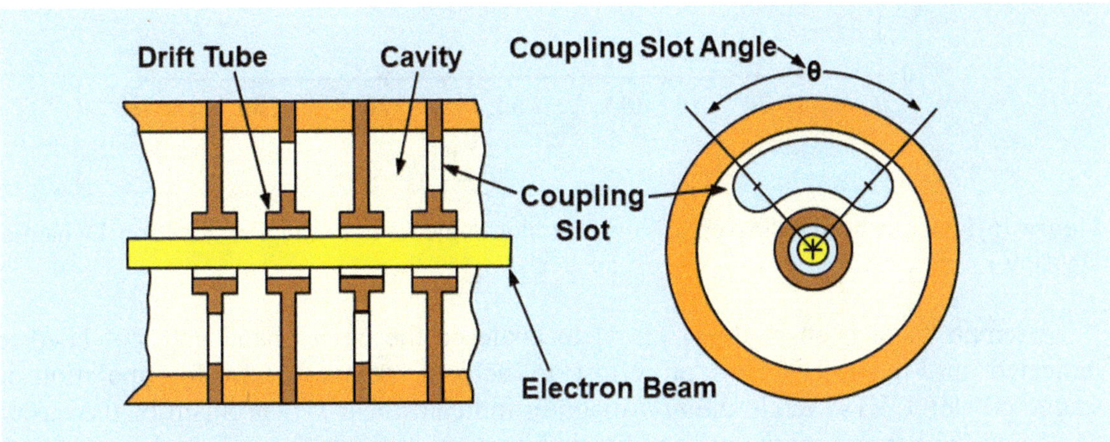

Figure 15-12 A coupling slot between cavities.

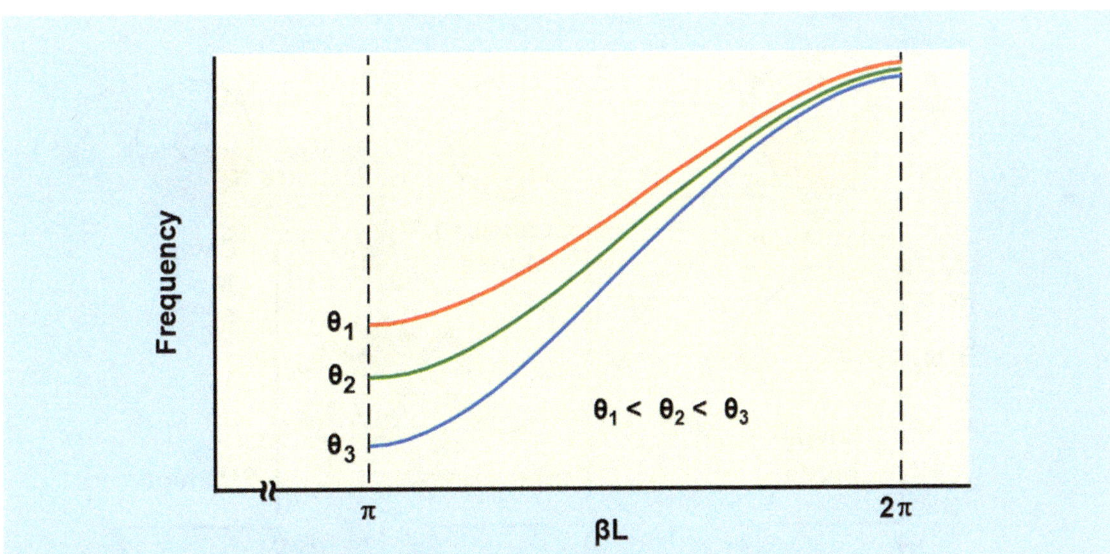

Figure 15-13 Effect of coupling angle on the ω-β diagram. (From: Hughes Electron Dynamics Division.)

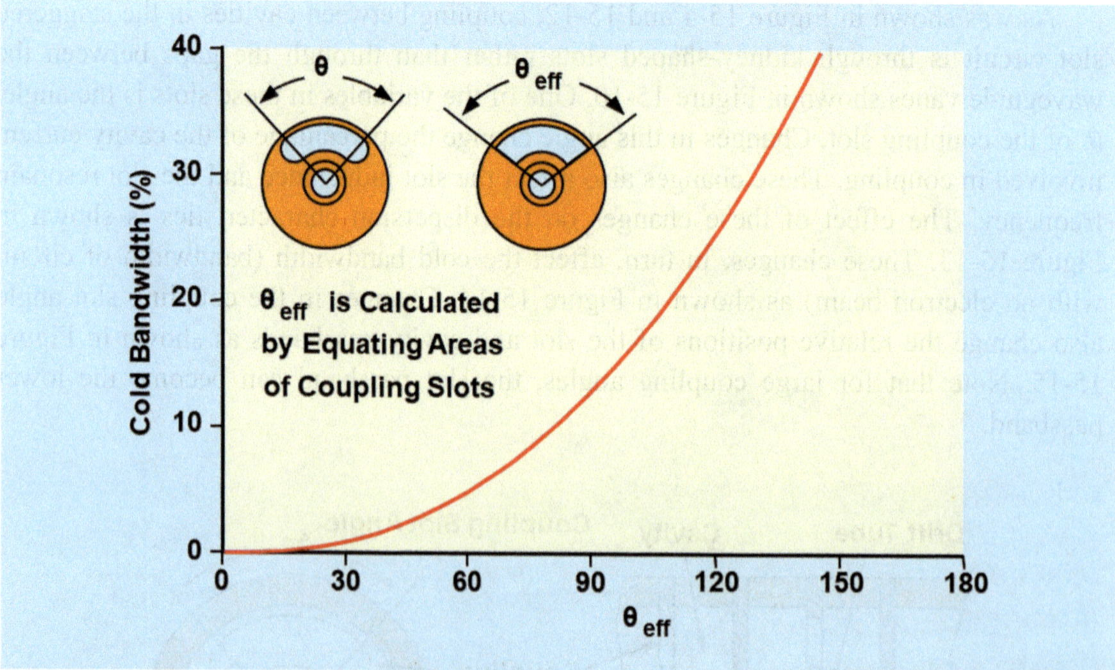

Figure 15-14 Cold bandwidth versus coupling slot angle. (From: Hughes Electron Dynamics Division.)

Attempts have been made in [3, 4] to coalesce the cavity band and slot band as indicated in Figure 15-15 in an effort to achieve octave-bandwidth operation in staggered-slot TWTs. While the ω-β diagram indicates near synchronism of the circuit wave and the beam over the desired bandwidth, two difficulties are in-band backward-wave instability and a large gain variation with frequency.

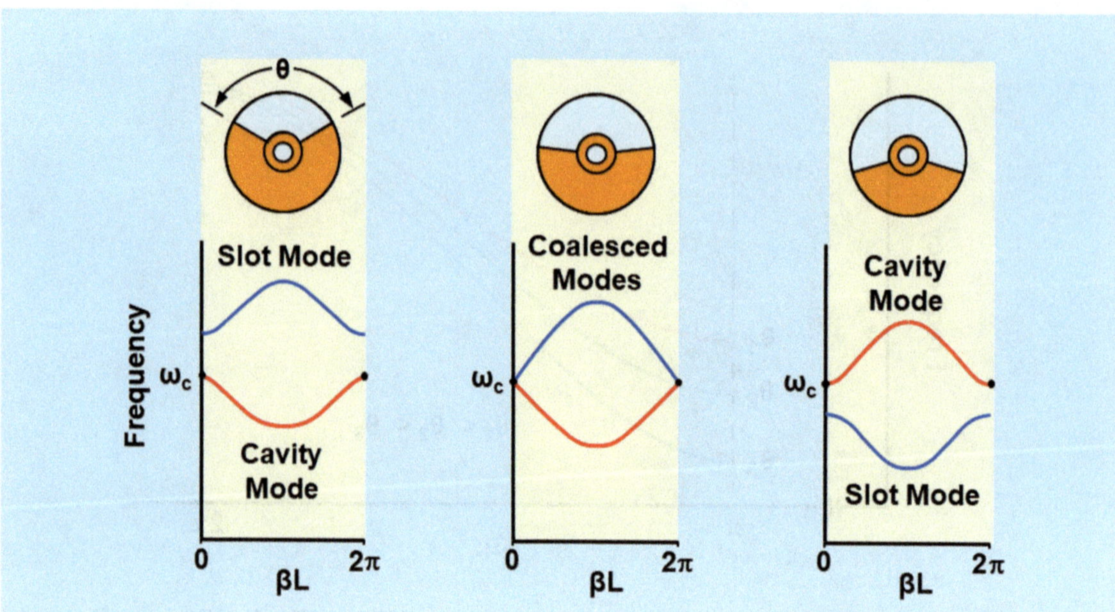

Figure 15-15 Change in passbands with increasing coupling slot angle. (From: J. R. Frey and I. Tammaru, 1981 IEDM. © 1981 IEEE.)

Normally, the cavity passband is used for interaction with the electron beam. However, it is possible to use the slot passband, and tubes designed to operate in the slot mode have existed for many years [5]. To make a tube operate in the slot mode, the frequency of the cavity passband must be increased and the frequency of the slot mode must be decreased.

The cavity passband can be increased by reducing the size of the cavity, which reduces the cavity inductance. As is shown in Figure 15-15, the slot band can be moved down by increasing the angle of the coupling slot. This has the effect of increasing the slot inductance.

These are some of the advantages to slot mode operation:

1. An important property of the slot passband is that the interaction impedance at the upper cutoff frequency of the lowest passband is zero. This results from the vanishing axial electric field component on the axis, as indicated in Figure 15-16. As a result, no special techniques for suppressing oscillations are required at this frequency.
2. Because the cavities are smaller, as Figure 15-17 shows, the PPM-focusing structure can have an inner diameter that is large enough to surround the cavities. This eliminates the need to use the cavity walls as magnetic pole pieces so the cavity walls can be copper and heat transfer is improved.
3. Construction is simplified and cost is reduced.

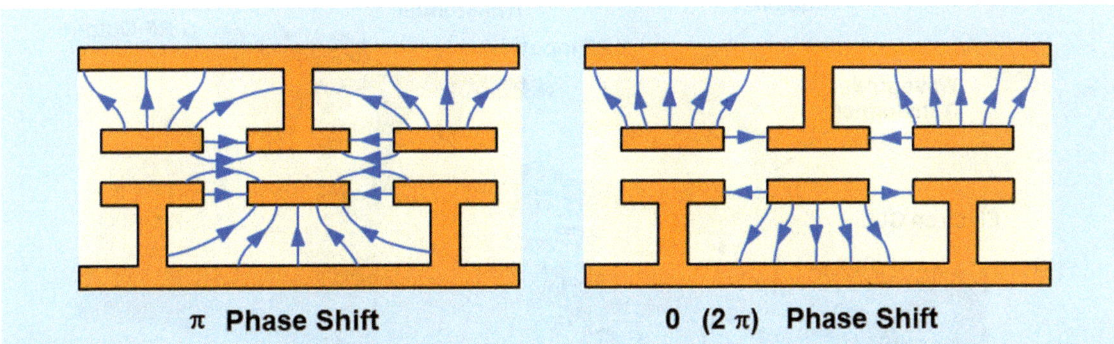

Figure 15-16 Electric field patterns for the slot-band cutoff frequencies. (From: J. R. Frey and I. Tammaru, 1981 IEDM. © 1981 IEEE.)

Sectioned staggered-slot TWTs are shown in Figures 15-18 and 15-20. The tube in Figure 15-18 is pulse operated and is used in an X-band airborne radar where there are severe restrictions on weight and size. This tube is focused with an integral solenoid. By winding the solenoid directly on the tube, weight, size, and power requirements are minimized.

Three sections separated by severs are used to obtain the required gain. Lossy terminations are used to isolate the sections and loss buttons are used to attenuate reflections. The reentrant loss pellets are shown in Figure 15-19. These are used to reduce small-signal gain ripple. According to Hughes, the buttons absorb about three quarters of the reflected power; however, they complicate matching but still they are widely used [6].

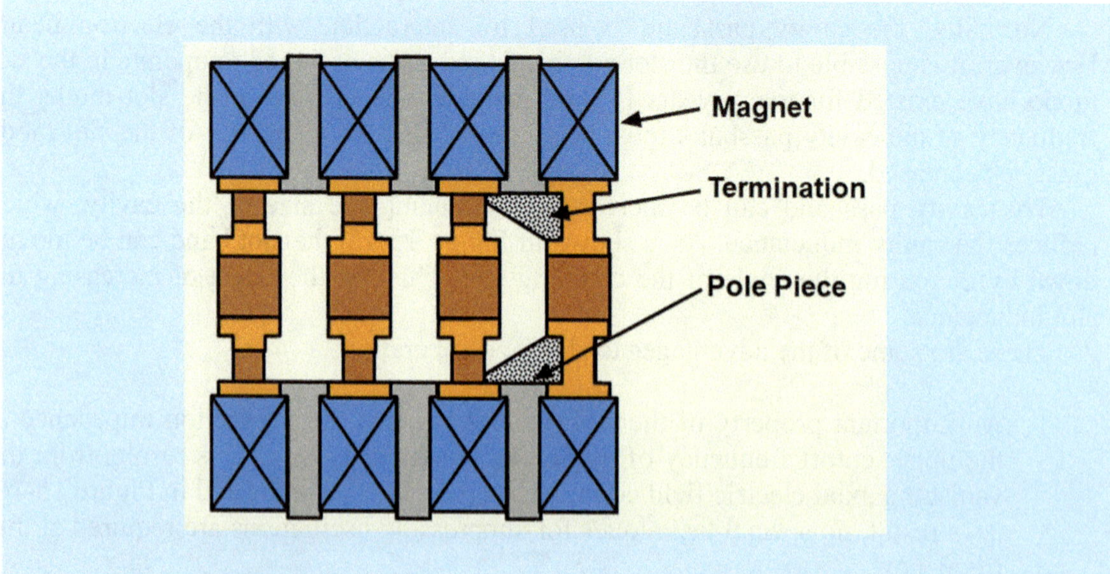

Figure 15-17 Magnet and pole piece assembly for slot-mode TWT. (From: J. R. Frey and I. Tammaru, 1981 IEDM. © 1981 IEEE.)

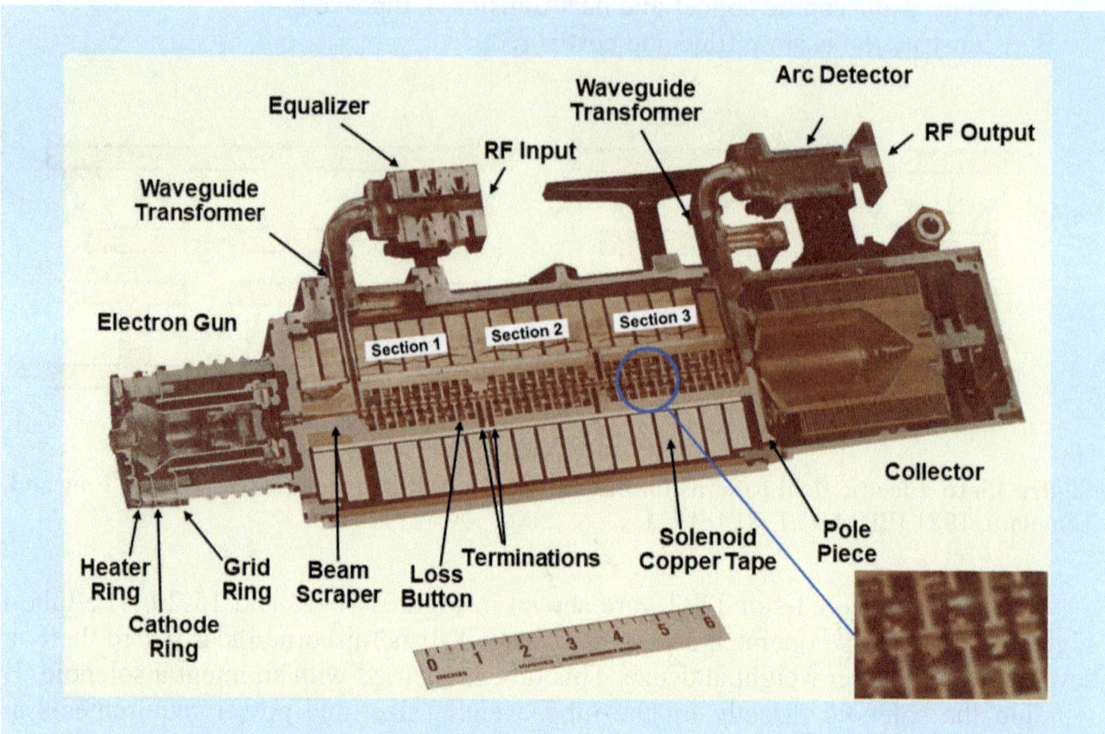

Figure 15-18 X-band, staggered slot, fundamental backward-wave TWT. (Courtesy of Hughes EDD).

Notice that there is an arc detector incorporated in the output waveguide assembly. This is needed to protect the tube from arcs in the output waveguide, which can reflect enough power to cause damage. The equalizer that is an integral part of the input waveguide assembly is used to shape the power and phase profiles of the tube output.

The tube in Figure 15-20 is also focused by a solenoid but this solenoid is not an integral part of the tube. Confined flow focusing and a depressed collector are used. This tube operates at 10-kW CW at X-band and is used in a missile tracking system.

An example of the capability of a TWT like that in Figure 15-20 is shown in Figure 15-21. The solid curves show the power output at various drive levels (referenced to the 0-dB saturation drive level of 40 mW). The overall efficiency is 20% to 30%.

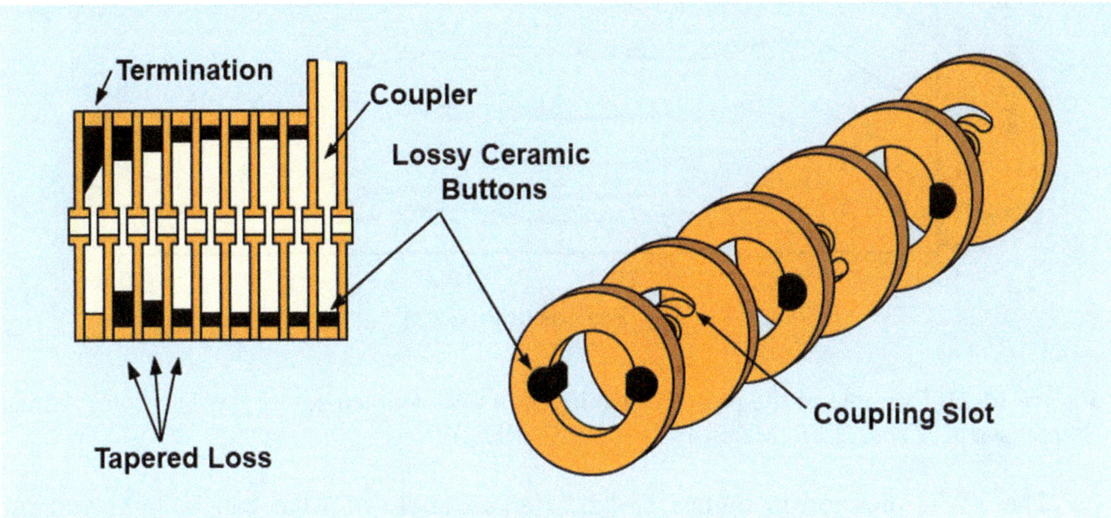

Figure 15-19 Termination and lossy ceramic buttons. (From: Hughes Aircraft Co., *TWT and TWTA Handbook.*)

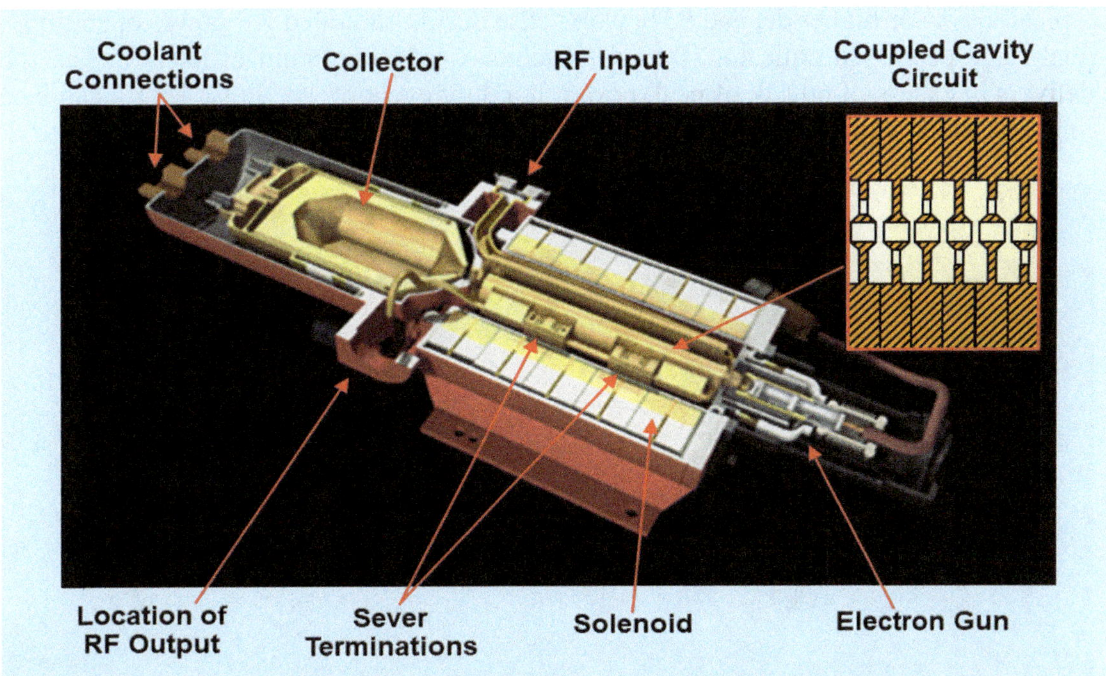

Figure 15-20 X-band, CW, staggered-slot fundamental backward-wave TWT.

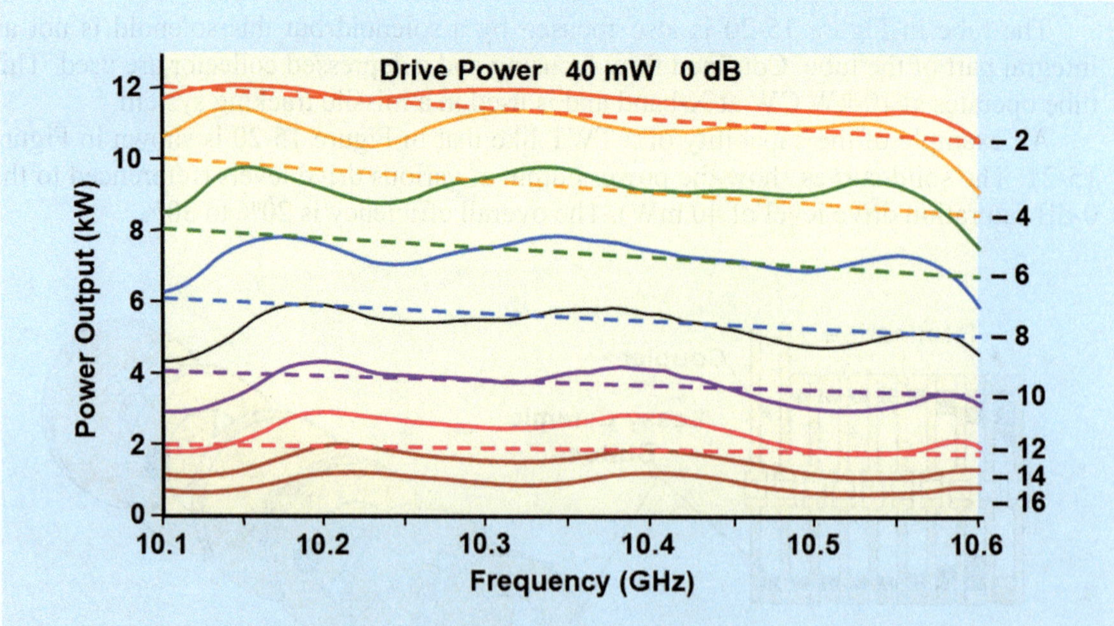

Figure 15-21 Example of the power capability of a CW coupled-cavity TWT. (From: Armand Staprans et al., *Proc. IEEE*, March 1973. © 1973 IEEE.)

The TWT, pictured in Figure 15-22, is an example of what can be achieved in a PPM-focused tube with a fundamental backward-wave circuit. A 34.5-kV, 7.4A Pierce shadow-gridded gun is used. Beam focusing is accomplished using a double period structure comprised of cloverleaf pole pieces and four pill magnets per double period. The gun is magnetically shielded so flux at the cathode is small. Beam transmission is typically 97% or higher dc, and 93% worst case during saturated RF power operation. A dual-stage depressed collector is used to achieve 45% minimum efficiency. The tube delivers in excess of 60 kW of peak power at a duty cycle of 4%. Pulse length can be as long as 150 μs with little or no degradation in performance.

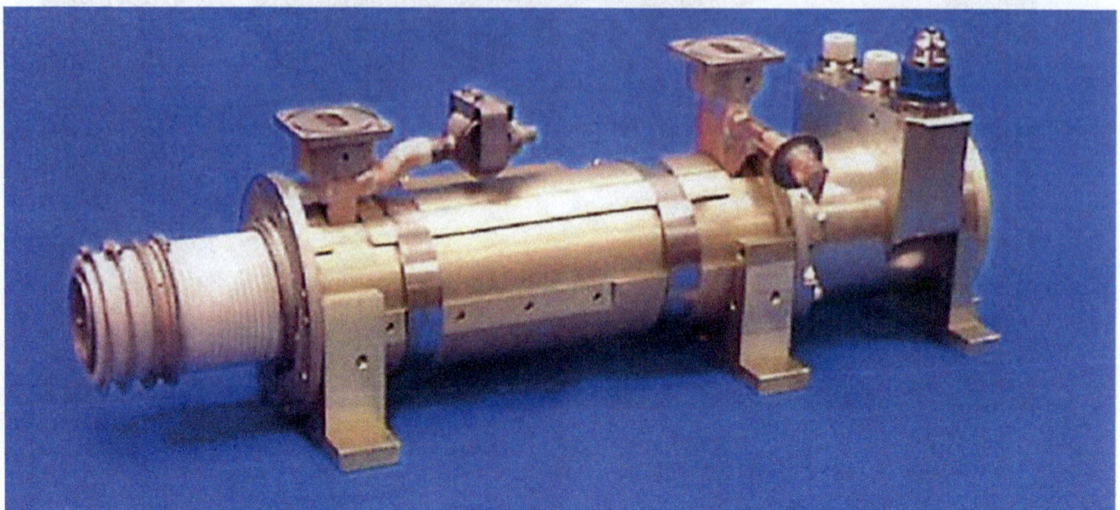

Figure 15-22 L-3 Communications 60-kW X-band coupled-cavity TWT. (From R. B. True et.al., IVEC 2001, Noordwijk, The Netherlands.)

Cavity dimensions in coupled-cavity TWTs scale directly with the wavelength of the signal being amplified. Assume that a TWT operates at a frequency near the resonant frequency of the cavities. An easy way to determine the approximate size of the cavities is indicated in Figure 15-23. Think of a cavity as a radial transmission line with a signal source at the center. Assume that the transmission line is short-circuited at a radial distance of R from the signal source. A signal propagating radially outward from the signal source will be reflected back to the signal source by the short circuit. Resonance occurs when the reflected signal arrives back at the signal source in phase with the signal generated by the signal source.

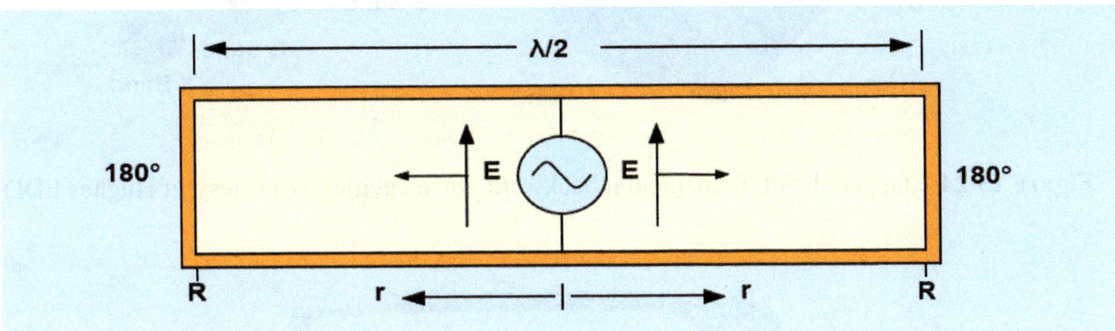

Figure 15-23 At resonance, the cavity diameter is ~ λ/2.

There is a 180° phase shift in the signal when it is reflected from the short circuit. This happens because the impinging signal causes current to flow in the short circuit and that current radiates the reflected signal. The total electric field at the short circuit has to be zero. Now, with 180° phase shift at the short, resonance will occur if there is a 90° phase shift as the signal propagates from the source to the short and another 90° shift as the signal propagates back to the source. Thus, the radius of the cavity must be λ/4 and the diameter must be λ/2 where λ is the wavelength of the signal being generated by the signal source.

Figures 15-24 and 15-25 illustrate the variation of cavity size with frequency. When coupled-cavity technology is extended to millimeter-wave frequencies, cavity dimensions become very small. The TWT circuit based on conventional cavity technology requires several dozen copper disks per section. Each disk contains a coupling slot and ferrules, which form the cavities. The disks are lathe turned, preferably by diamond-point machining. The kidney-shaped slots are cut by electric discharge milling (EDM). These disks are stacked and heated under pressure to braze or sinter them together.

15.2.3 Millimeter-Wave Construction Techniques

At millimeter-wave frequencies, the problem with this conventional construction technique is that it requires the parts to be machined to extremely tight tolerances. One reason is that errors in cavity lengths may add cumulatively when the cavities are stacked. To maintain synchronism between the beam and circuit wave, extremely tight tolerances on the height of each cavity are necessary. A second reason is that the cavity resonant frequency is critically dependent on the cavity gap width. Machining tolerances on this dimension may be 70 millionths of an inch or smaller.

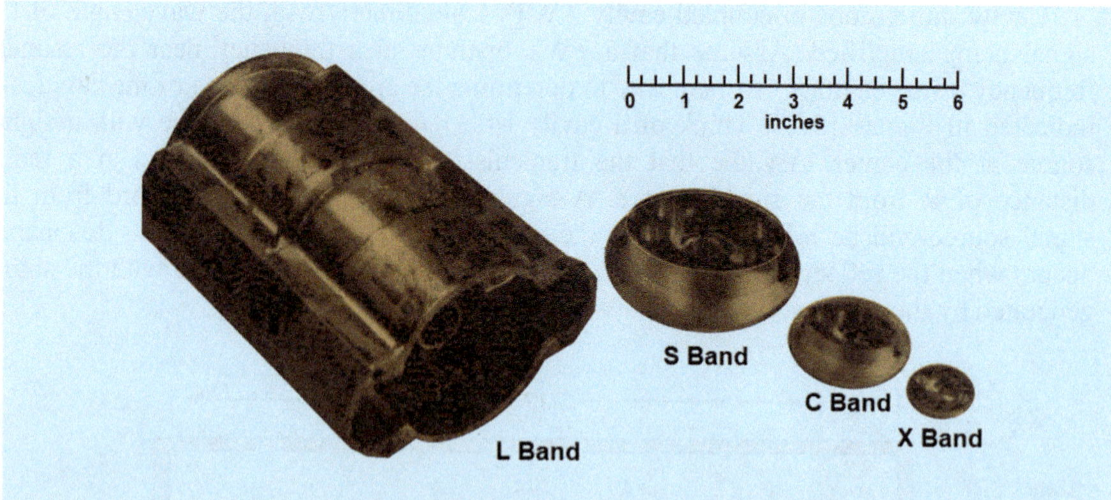

Figure 15-24 Staggered-slot, fundamental backward-wave cavities. (Courtesy of Hughes EDD).

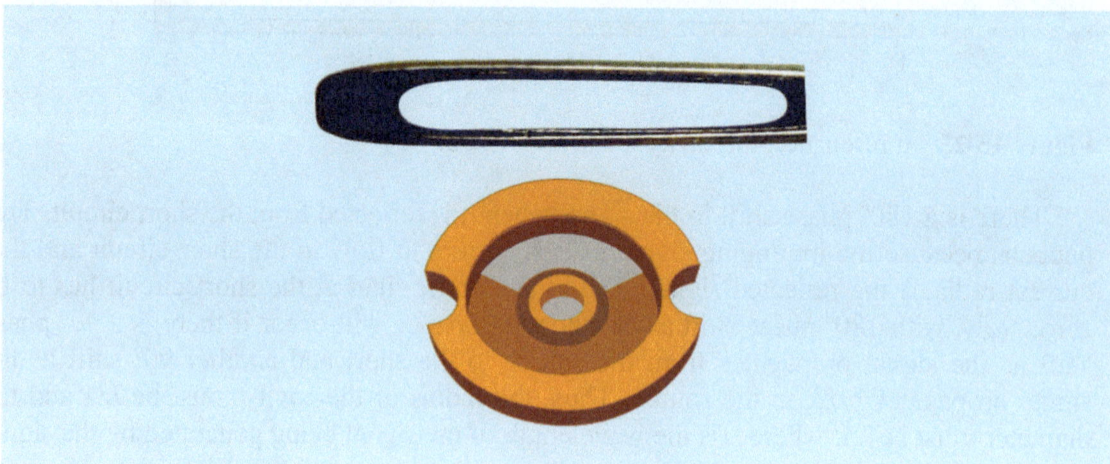

Figure 15-25 Millimeter-wave cavity. (Courtesy of Hughes EDD.)

James has shown in his 94-GHz tubes that the problem of using a large number of parts machined to extremely tight tolerances can be circumvented by using the ladder-circuit designs described in [7]. With these designs, tolerances are relaxed to 200 millionths of an inch. The cost, in turn, is reduced by a factor of about 2.

In [7] three versions of ladder circuits are described:

1. In the aligned-slot structure shown in Figure 15-26, the ladder circuit is fabricated by machining notches in a slab of copper to form a ladder or comb structure. The beam tunnel is formed by cutting half-round grooves in the tips of the comb teeth. Two identical structures are then brazed together. The coupling slots from cavity to cavity are formed by machining channels in the cover plates. The circuit is completed by brazing cover plates to the faces of the ladder assembly. Because the ferrules, which form the drift-tube tips in the cavities, are omitted with the ladder-assembly technique, gain per cavity is reduced and so the interaction circuit length is increased. Still, the advantage of needing only a few parts per section rather than several dozen is considerable.

2. In the staggered-slot structure shown in Figure 15-27, slots are machined into the edge of every other ladder rung, with the skipped rungs notched on the other side. The cover plates are flat slabs of copper.
3. In the double-staggered ladder, two coupling slots spaced 180° apart are used in each cavity wall. The pair of coupling slots is rotated 90° from cavity to cavity.

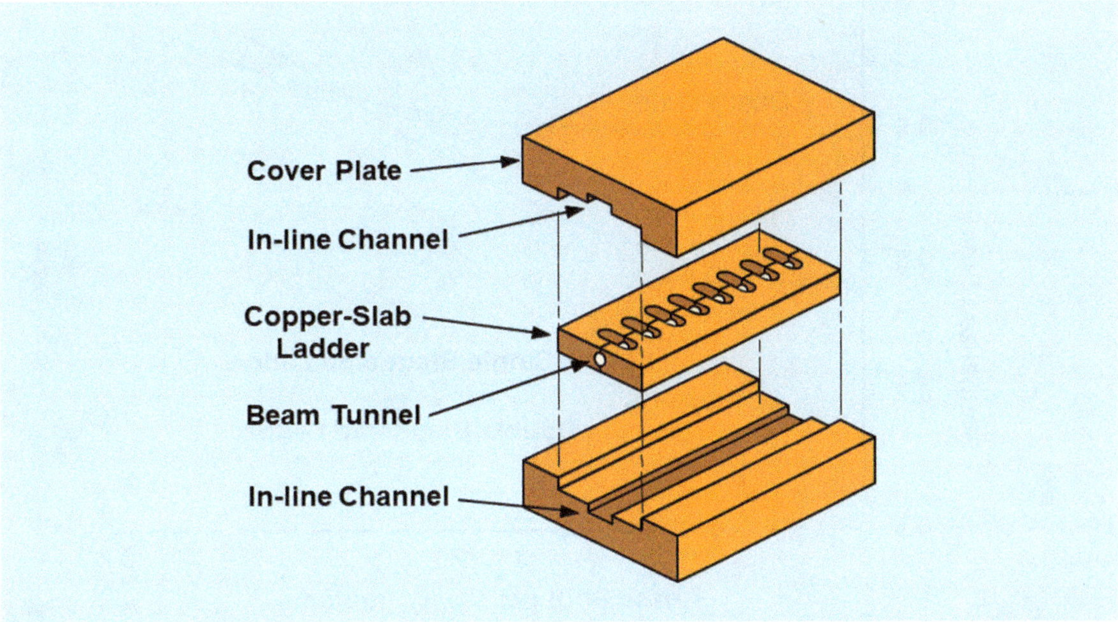

Figure 15-26 Ladder-core structure with dual in-line coupling channels. (Adapted from: B. G. James, *MSN & CT,* September 1986.)

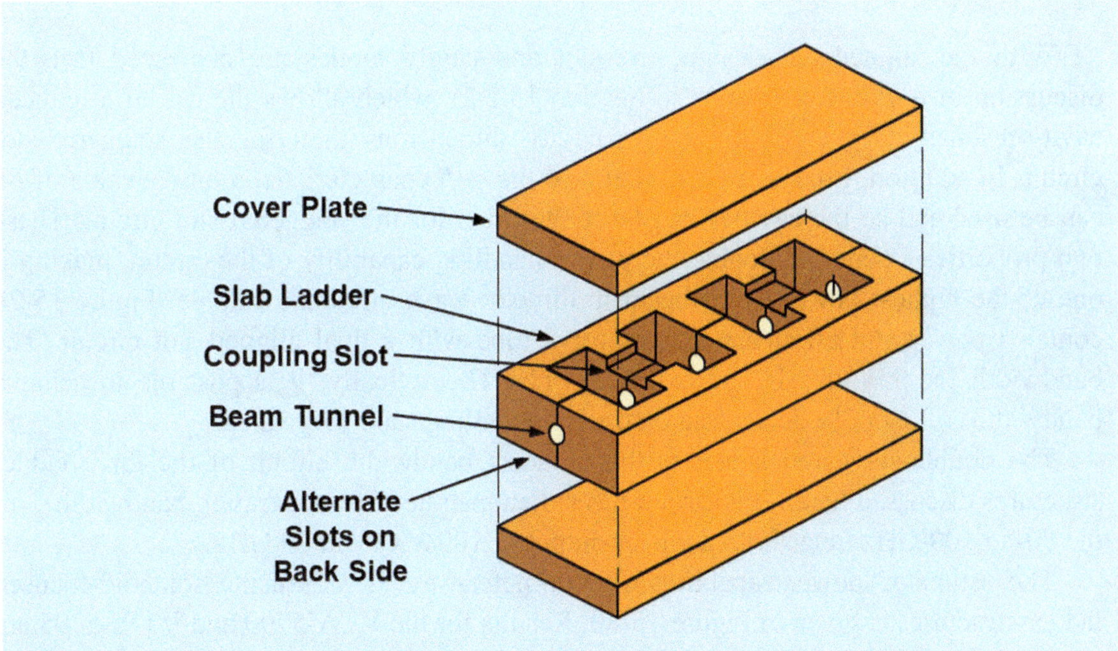

Figure 15-27 Ladder-core structure with staggered coupling slots. (Adapted from: B. G. James, *MSN & CT*, September 1986.)

As shown in Figure 15-28, there are significant differences between the ω-β characteristics for these structures. These differences lead to advantages (or disadvantages) that govern the applications of the tubes.

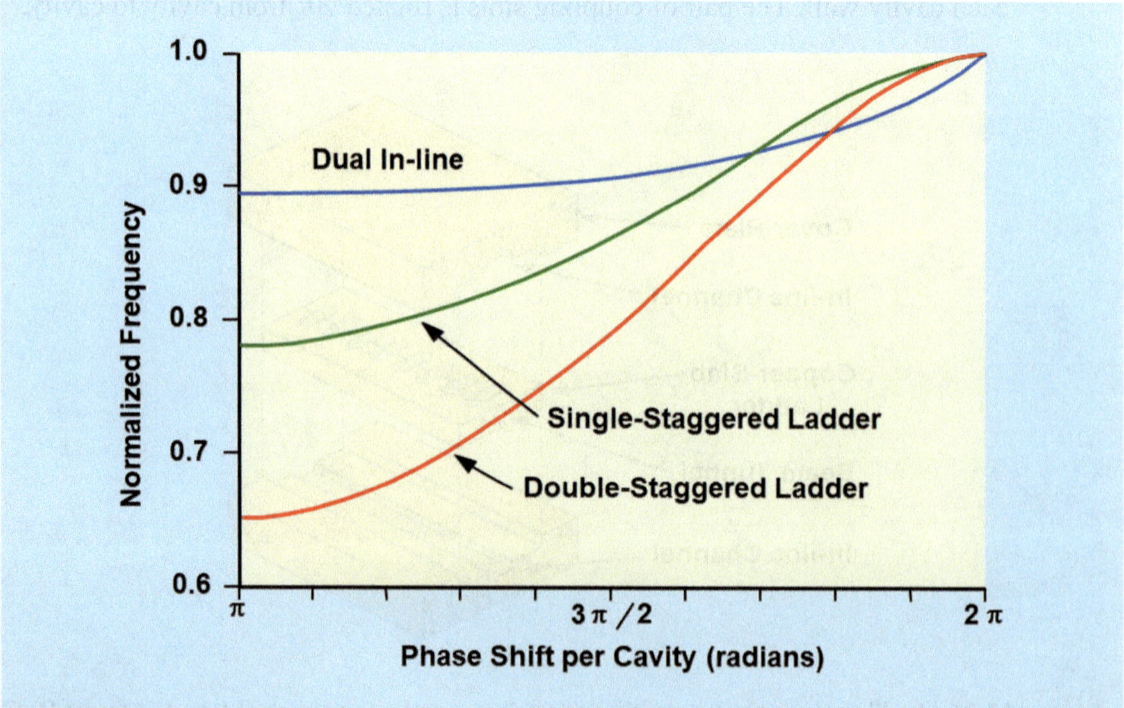

Figure 15-28 *ω-β* characteristics for ladder structures. (Adapted from: B. G. James, *MSN & CT*, September 1986.)

With the aligned-slot design, the slot and cavity modes are coalesced (see the discussion of *ω-β* characteristics in Section 15.2.2), which allows the use of a reduced cavity-resonant frequency and larger cavity dimensions than for the staggered-slot circuit. In addition, because of the nature of the *ω-β* characteristic, a larger value of βL can be used and so the cavities can be longer than for the staggered-slot circuit. These two properties essentially double the power-handling capability of the circuit, making it one of the highest power linear-beam millimeter-wave circuits available. Figure 15-29 contains power and efficiency curves for a tube with a dual aligned-slot circuit. The bandwidth for this tube is approximately 1%. Theoretically, it is possible to achieve bandwidths of over 3% depending on operating voltage and perveance.

The double-staggered ladder is the broadest bandwidth circuit of the three ladder structures discussed here. This circuit has a demonstrated 20% operating bandwidth over the 80- to 100-GHz range with a power output of 100W CW at 90 GHz.

The estimated power capability of millimeter-wave TWTs using solenoid-focused ladder structures is given in Figure 15-30. Results for the VTA-5700 and VTW-5795 are shown. When PPM-focusing is used, the estimated power capability for ladder structures is about an order of magnitude lower than shown in Figure 15-30.

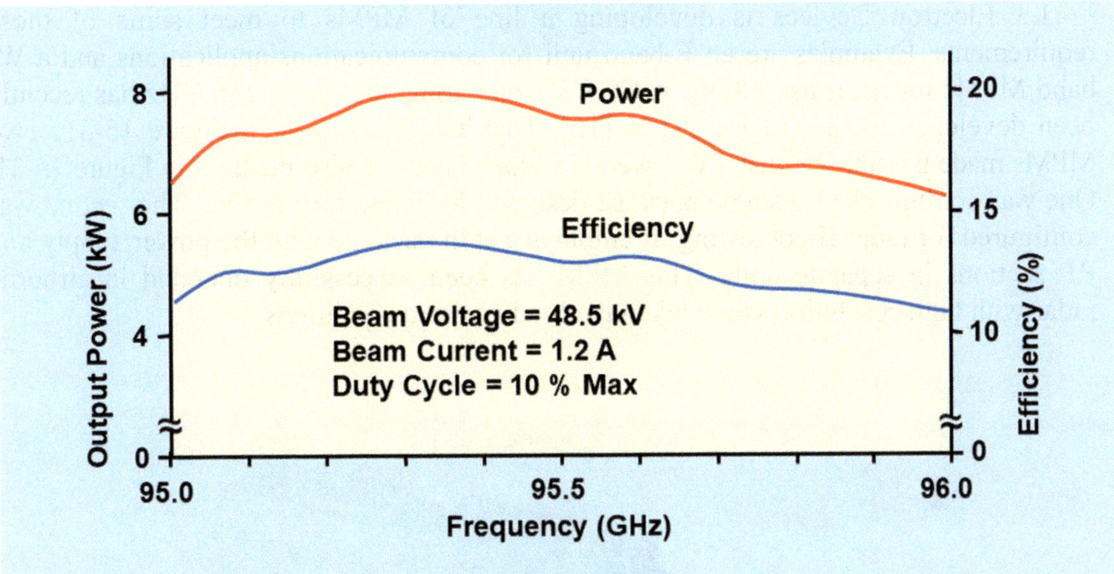

Figure 15-29 Power output for a tube with a dual aligned-slot ladder assembly. (Adapted from: B. G. James, *MSN & CT*, September 1986.)

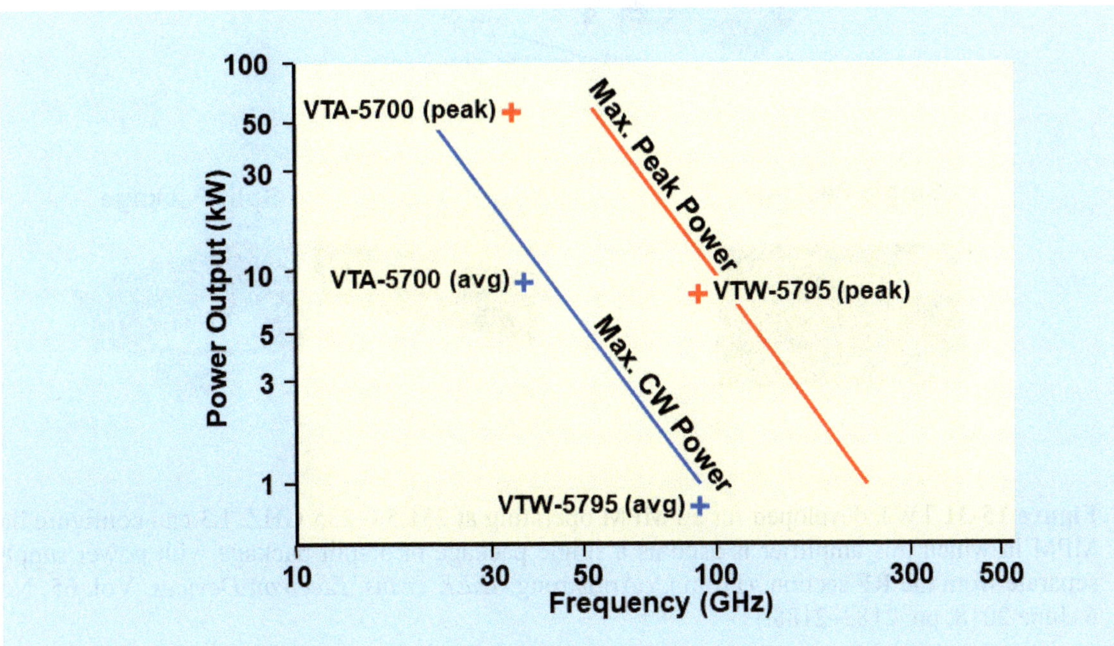

Figure 15-30 Practical power limits for millimeter-wave TWTs with solenoid-focused ladder structures. (Adapted from: B. G. James, *MSN & CT*, September 1986.)

There are growing requirements for amplifiers that operate at high millimeter wave frequencies. For example, the short wavelength of the radiation enables the development of high-resolution video frame rate synthetic-aperture radar for real time imaging under obstructed vision conditions. Bandwidth and/or power are the major limitations of existing amplifiers.

L3 Electron Devices is developing a line of MPMs to meet some of those requirements. Examples are an E-band unit for communications applications and a W-band MPMs for radar use [8, 9]. A CCTWT operating at 231.5 – 235 GHz has recently been developed for a G-band MPM [10]. That TWT is shown In Figure 15-31. Two MPMs made possible by that TWT were developed and are also pictured in Figure 15-31. One was configured as a single package designed for laboratory testing. The second was configured for radar flight testing. It employs a split package with the power supply and RF sections in separate units. This MPM has been successfully operated in airborne radar with high-resolution video taken in cloud-covered conditions.

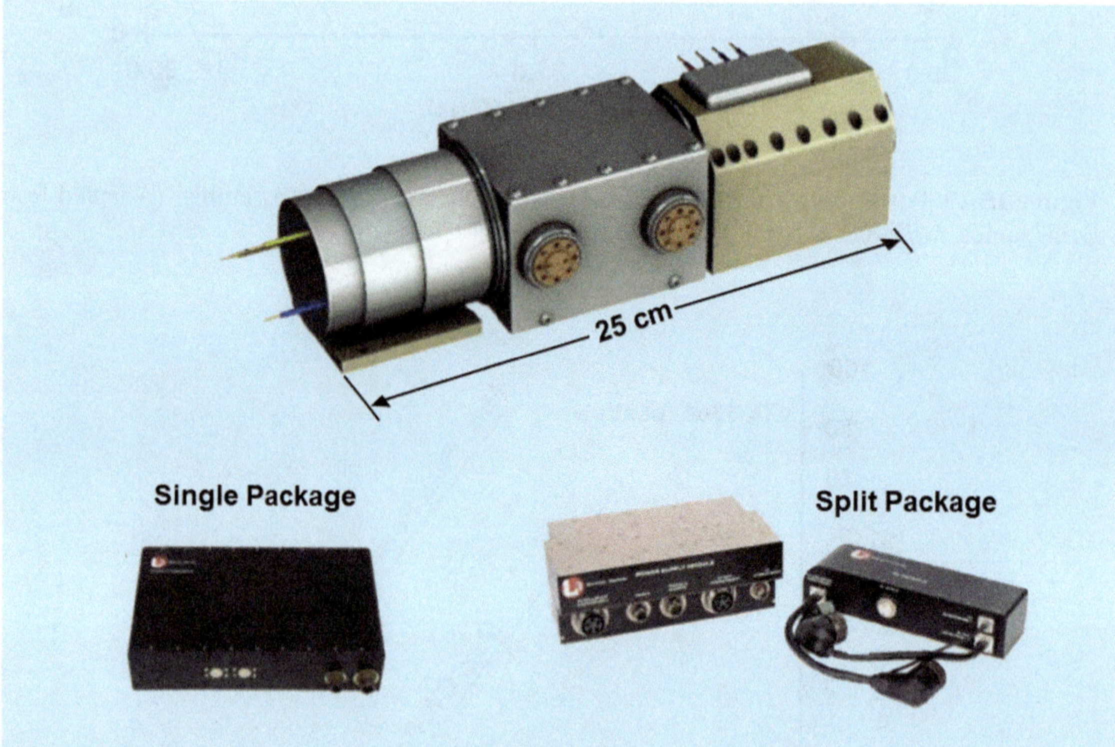

Figure 15-31 TWT developed for an MPM operating at 231.5 – 235 GHZ. L3 can configure the MPM in which this amplifier is used as a single package or a split package with power supply separate from the RF section. (From C. Armstrong, *IEEE Trans. Electron Devices,* Vol. 65, No. 6, June 2018, pp. 2183–2188.)

Some of the characteristics of the TWT used in the G-band MPM are as follows. A modulating focus electrode is used for beam control and an adjustable mod-anode is used for current adjustment. The cathode is an M-type operating at an average emission density of 6.3 A/cm^2. The nominal beam voltage is 20 kV and the dc beam current is 100 mA.

Beam focusing is accomplished with a conventional PPM stack. The dc body current is 4.5 mA and the worst-case body current with RF is 7 mA. A four stage depressed collector is used for collecting the beam.

The RF circuit is a serpentine waveguide that has a broad bandwidth and is made from copper for high power operation. The circuit is severed to suppress backward-traveling waves. The RF windows are made with CVD diamond for low loss and wide bandwidth.

The beam transmission for the TWT used in the laboratory unit was around 85% and so the laboratory model had to be operated at reduced current. The resulting efficiency of the laboratory MPM was only about 3%. The focusing of the beam in the flight test model was improved and the beam transmission was ~95%. As a result, this TWT could be operated at the full design beam current. A peak output power of 45 dBm (32 W) and an MPM efficiency of 9% were obtained for operation at 50% duty. As seen in Figure 15-32, the output power was smooth across the full 3.5-GHz operating frequency band.

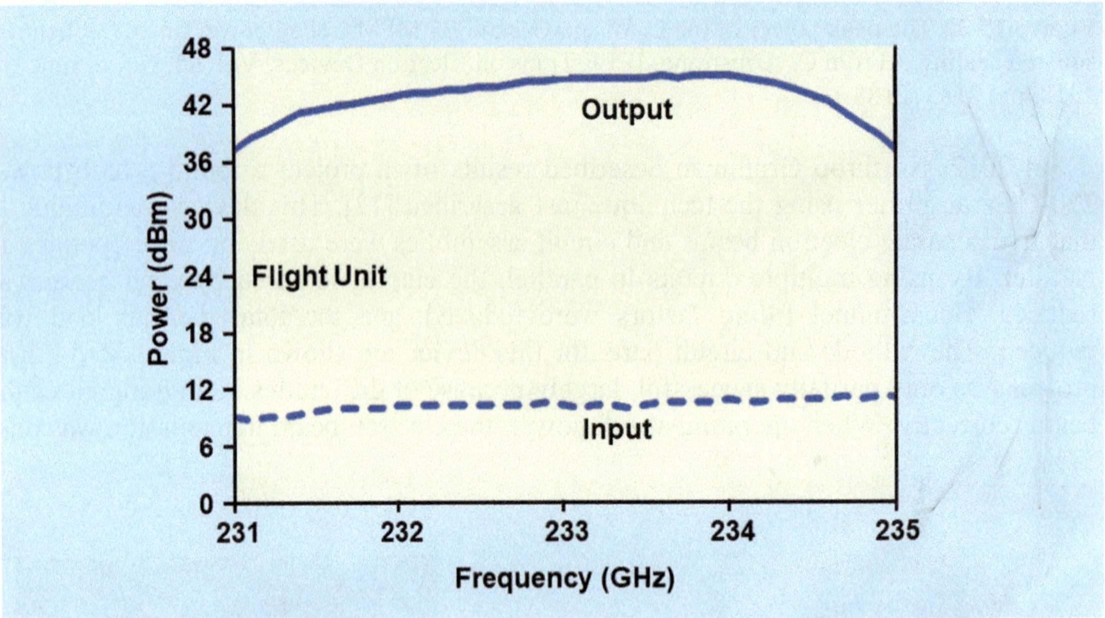

Figure 15-32 Power output for L3 G-band MPM configured for flight tests. (From C. Armstrong, *IEEE Trans. Electron Devices,* Vol 65 , No. 6, June 2018, pp 2183-2188.)

Armstrong observed that the peak power of the E, W, and G band MPMs developed by L3 obeys a power times frequency-squared scaling (Figure 15-33). "While this scaling is expected for optimized devices operating at lower frequency, it is encouraging to observe that it still applies for device operation approaching the submillimeter wavelengths."

Northrop Grumman has been developing millimeter amplifiers based on folded waveguide interaction circuits [11]. For frequencies up to 300 GHz, the required dimensions and tolerances can be realized by using high-speed milling. The circuits are fabricated in halves that are aligned and brazed together.

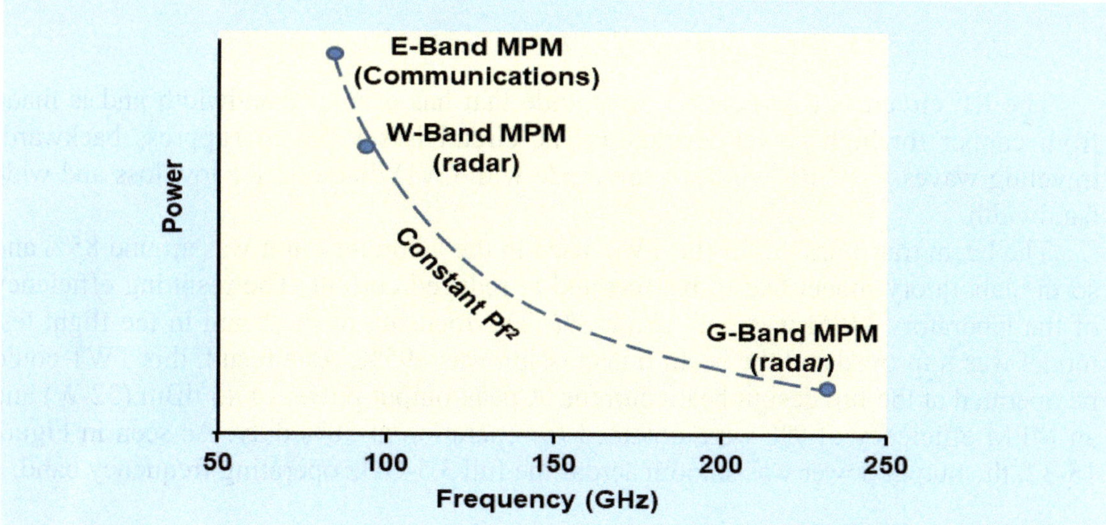

Figure 15-33 The peak power of the E, W and G-band L3 MPMs obey power times frequency-squared scaling . (From C. Armstrong, IEEE Trans on Electron Devices, Vol. 65, No. 6, June 2018, pp. 2183–2188.)

In 2012, Northrop Grumman described results of a project to build a high power 220-GHz amplifier using the technique just described [12]. This device was unique in that five separate electron beams and circuit assemblies were used and were operated in parallel. By using multiple circuits in parallel, the current required by each beam was reduced. Beam-tunnel filling factors were reduced, and the total thermal load was reduced. The cathode and circuit parts for this device are shown in Figure 15-34. The project was only partially successful, largely because of difficulties in aligning all of the beams correctly. When operating at full power, the electron beam transmission was only 57%.

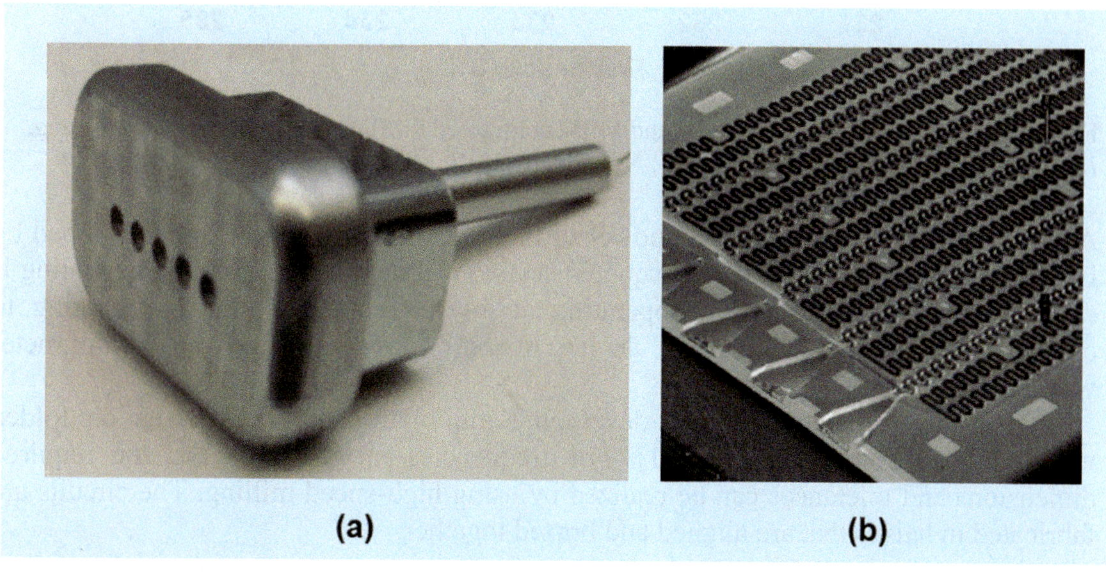

Figure 15-34 (a) Five-beam M cathode. (b) Micro-fabricated circuits. (From K. E. Kreischer et al., IEEE, IVEK.)

In 2013, Northrop Grumman described results of a program for a 233-GHz power amplifier that uses a single-folded waveguide circuit and electron beam focused by a compact permanent magnet [9]. Figure 15-35 shows a cutaway schematic of the amplifier. The beam is generated by a Pierce electron gun operating at 20 kV. A focus electrode is used and can be biased for edge emission control and modulation of the electron beam. The nominal beam current is 110 mA. A standard M-type cathode is operated at higher temperature than normal to obtain the required emission density of 17 A/cm^2.

A 6.8-kG magnet is used to focus the beam. The current density in the beam is approximately 350 A/cm^2, so alignment and focusing are critical and provisions are made for optimizing field alignment. After interacting with the circuit, the electron beam enters and is collected by the single-stage depressed collector. Beam transmission ratios of 95% to 98% are typical.

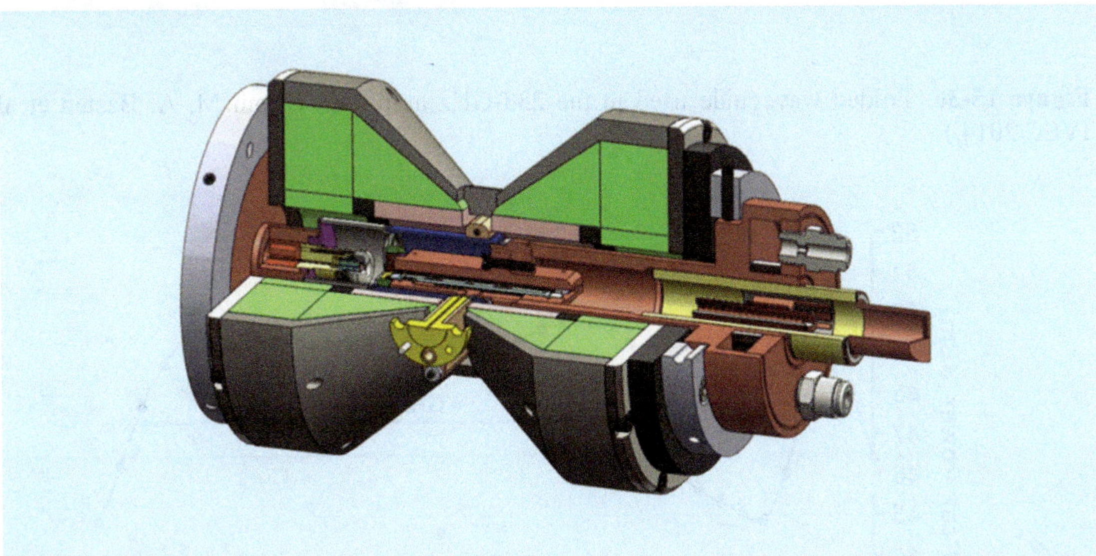

Figure 15-35 Cutaway view of 233-GHz amplifier. The green parts are the permanent magnets for focusing the electron beam. (From M. A. Basten et al., IEEE IVEC 2014.)

As Figure 15-36 shows, the folded waveguide circuit and beam tunnel along with an impedance transition and output waveguide are machined into OFHC copper halves using precision milling. The transition is used to transform the impedance of the folded waveguide interaction circuit to the impedance of the output waveguide. The amplifier is coupled with standard rectangular WR-4 waveguides to the amplifier input and output ports through the use of broadband CVD diamond windows. Two of these copper halves are joined with high accuracy with precision alignment pins and brazed together. The final assembly is joined to an electron gun and a collector.

The power output of the finished amplifier is shown in Figure 15-37 and is over 79 W (49 dBm) from 232.6 to 234.6 GHz. It is over 50 W in the band shown in Figure 15-37.

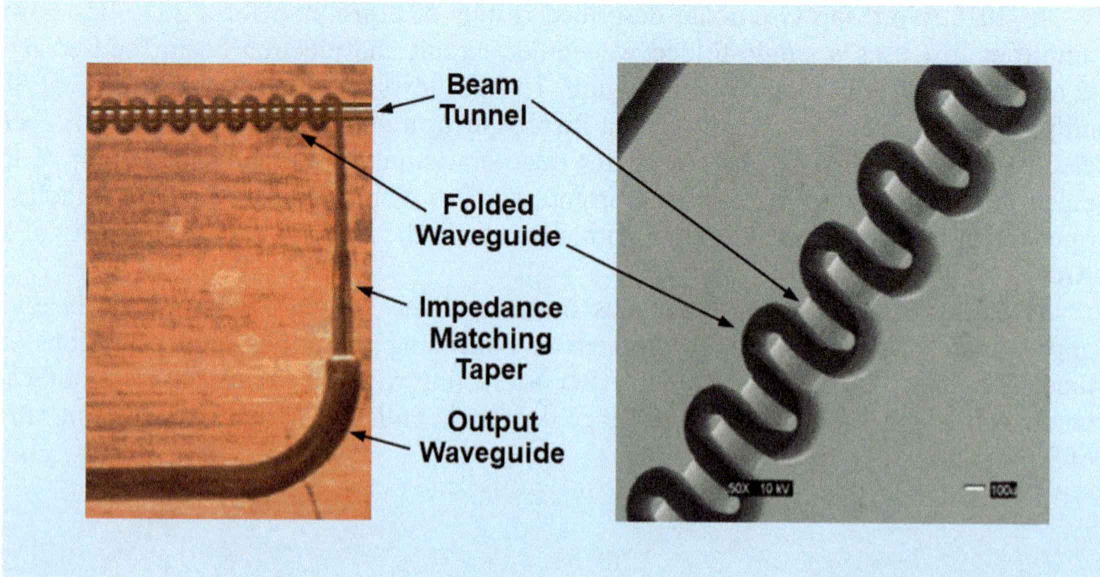

Figure 15-36 Folded waveguide used in the 233-GHz amplifier. (From M. A. Basten et al., IVEC 2014.)

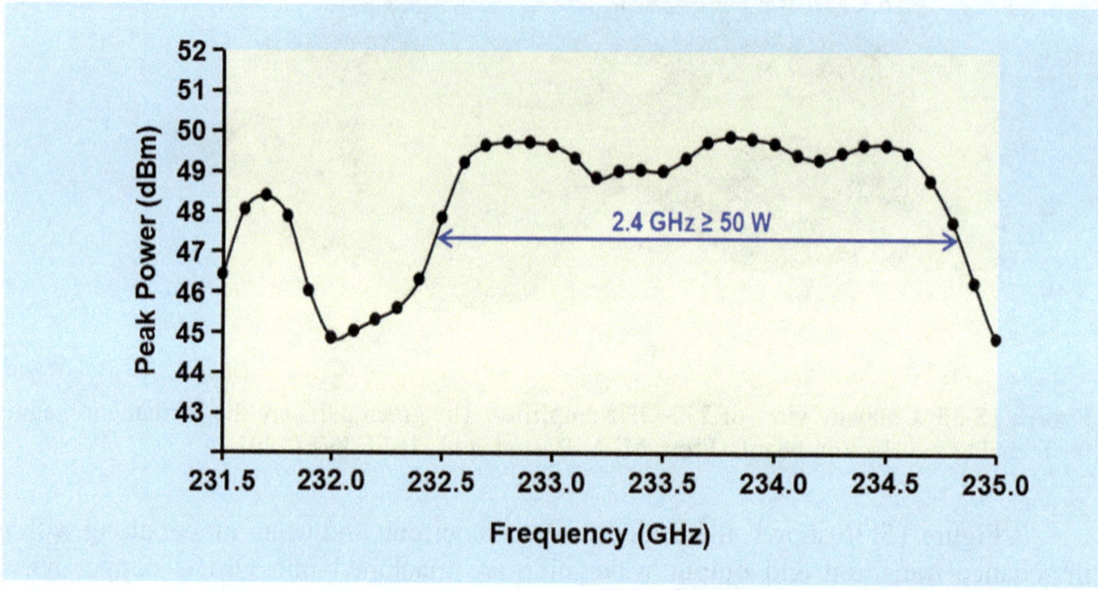

Figure 15-37 Power output from the 233-GHz amplifier. (From M. A. Basten et al., IEEE IVEC 2014.)

As frequency is increased, tolerance control becomes more and more difficult. The U.S. Naval Research Laboratory (NRL) is applying the use of ultraviolet photolithography and electroforming (UV-LIGA) to fabricate slow wave circuits. Joye et al. state that photolithographic methods using photomasks are accurate to well below the micrometer scale, and precisions of better than 100 nm are not uncommon. NRL has developed a TWT with a high-aspect-ratio (8:1) folded waveguide circuit that is capable of 60 W at 214.5 GHz [13]. The tube also achieved an instantaneous bandwidth of 15 GHz in G band in the small-signal regime. Figure 15-38 is a summary of the major steps used by NRL in fabricating a folded waveguide slow wave circuit using UV-LIGA.

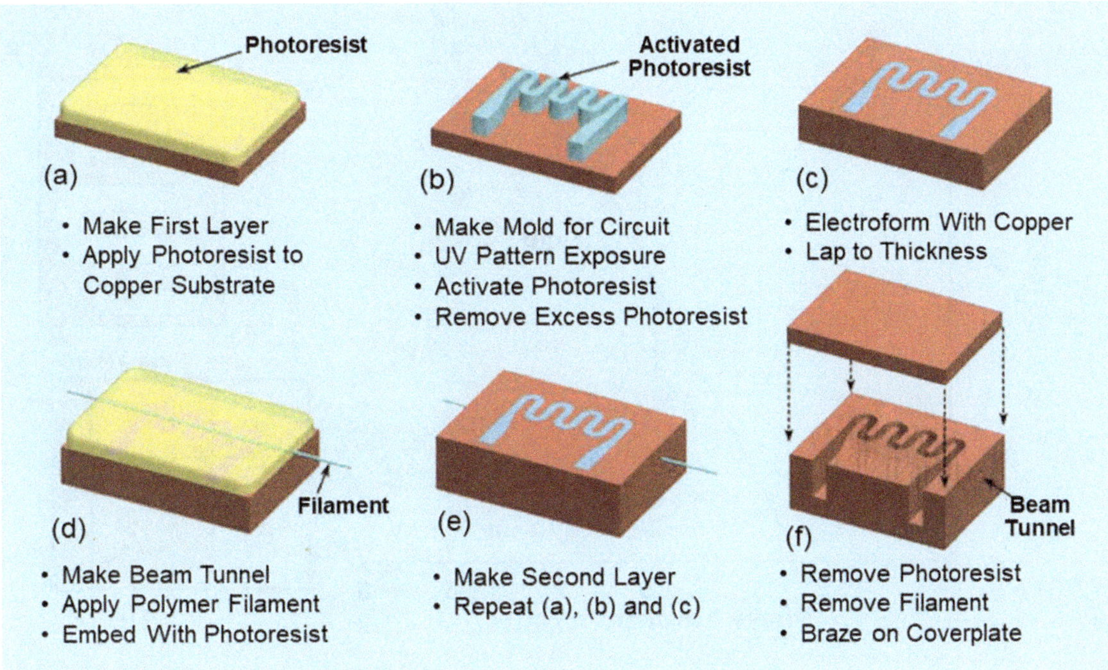

Figure 15-38 Process used by NRL to fabricate a UV-LIGA slow-wave circuit. (From: Colin Joye et al., *IEEE Trans. Electron Devices*, Vol. 61, No. 6, June 2014.)

The slow wave circuit is made in two layers. The steps in making the first layer are:

(a) Apply photoresist to a copper plate, which is one side of the circuit;
(b) Expose photoresist to UV through a photomask to make mold of the circuit;
(c) Electroform copper around the mold and lap the surface to the desired dimension.

The steps in making the second layer are:

(d) Install a polymer filament for use as the mold for the beam hole. After carefully positioning the filament. apply photoresist to embed the filament and make the second layer of the circuit. See [14] for information concerning the filament selection and use.
(e) Repeat (b) and (c).
(f) Remove the photoresist and the filament. Braze on a flat cover to complete the circuit.

The operation of a folded waveguide circuit has been discussed in detail early in this chapter. The circuit is a fundamental backward-wave circuit that is operated in a forward-wave mode. Figure 15-39 displays the dimensions and the ω-β diagram for the NRL circuit. The red lines are calculated and the green dots are measured results. Notice that the height of the waveguide is only 100 μm (~0.004"). The folded waveguide circuit was designed around parts from a G-band EIK from CPI, which limited the results that could be obtained. A 0.66 T magnetic field was used to focus the high-density 500 A/cm^2 electron beam. An example of test results is given in Figure 15-40.

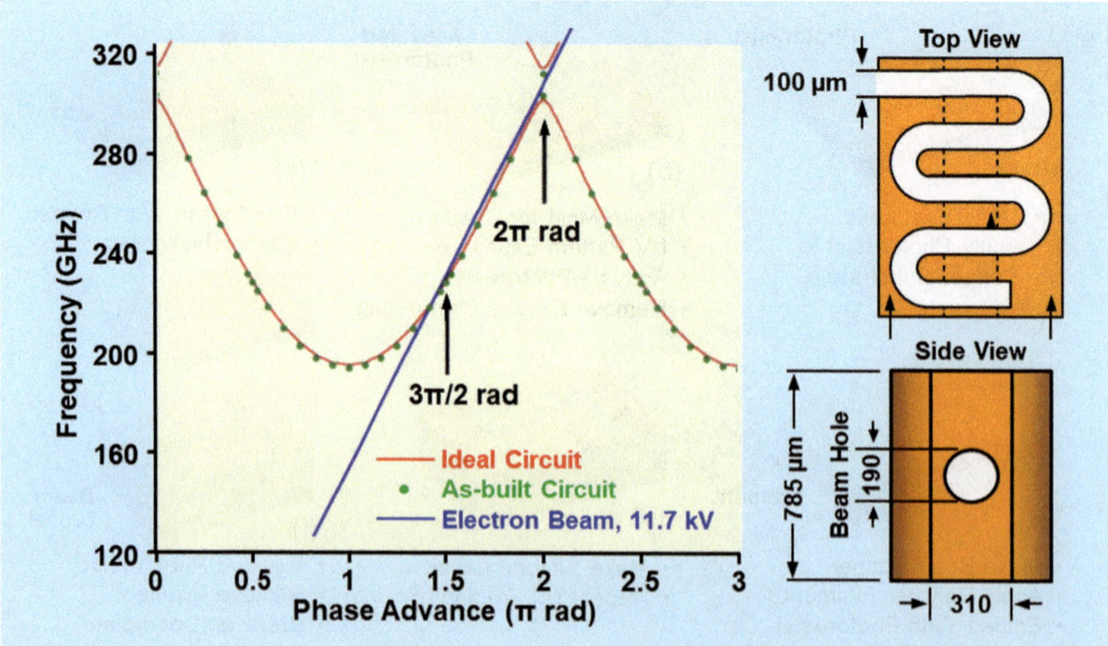

Figure 15-39 ω-β diagram for folded waveguide circuit and drawing of part of circuit.. (From: Colin Joye, et al, *IEEE Trans. Electron Devices*, Vol. 61, No. 6, June 2014.)

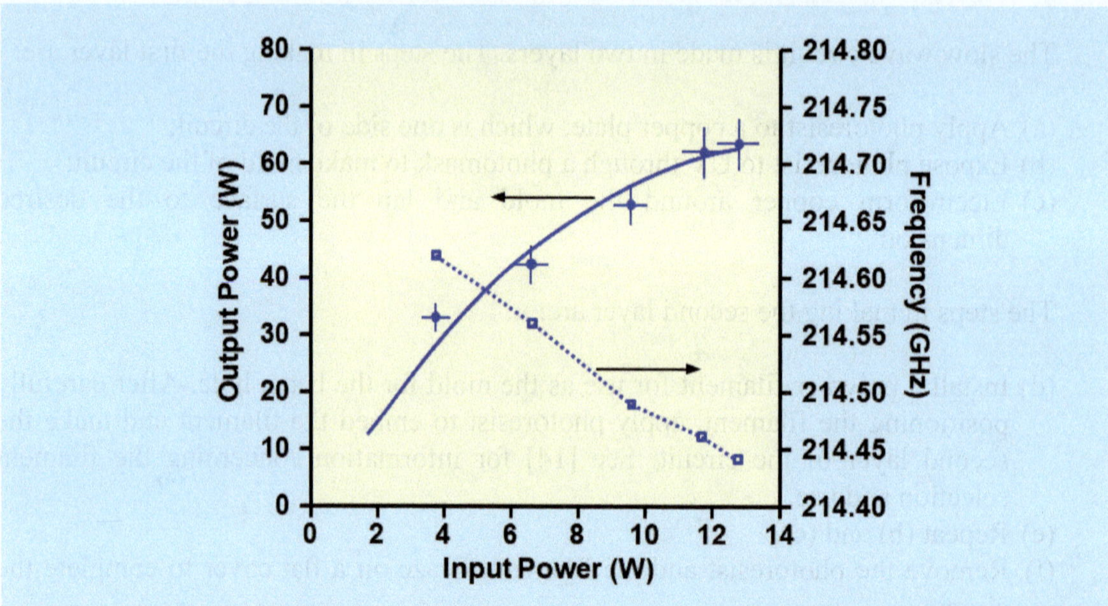

Figure 15-40 Output power as a function of drive power for UV-LIGA circuit. The frequency of the extended interaction oscillator input power source varied slowly with load. (From: Colin Joye et al., *IEEE Trans. Electron Devices*, Vol. 61, No. 6, June 2014.)

"This microfabrication technique and demonstration paves the way for a new era of vacuum electron devices that could extend into the 1–2 THz range with advances in high-current-density electron guns."

15.2.4 Dual-Staggered Slot

The dual-staggered slot structure is another fundamental backward-wave circuit. An S-band version of a tube that uses this circuit is shown in Figure 15-41. The tube in this picture has been processed and tested and is ready for final packaging. The tube has four sections that are separated by severs. Each section has 6 to 8 cavities. The parts of one of the cavities are shown in Figure 15-42. Included in this drawing are the focusing magnets and the pill-shaped loss buttons.

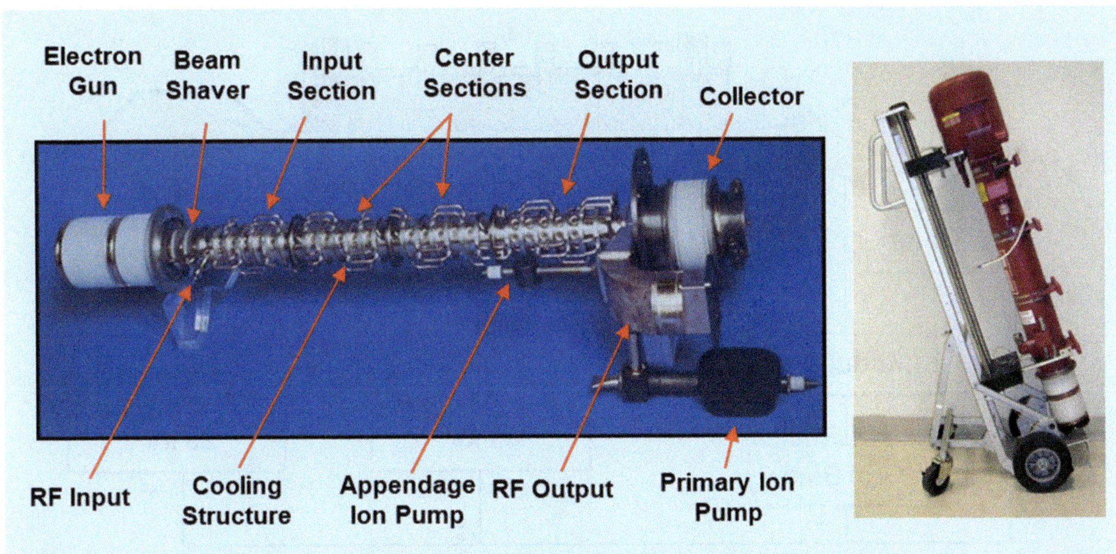

Figure 15-41 S-band double staggered slot, fundamental backward-wave TWT. (Courtesy of CPI.)

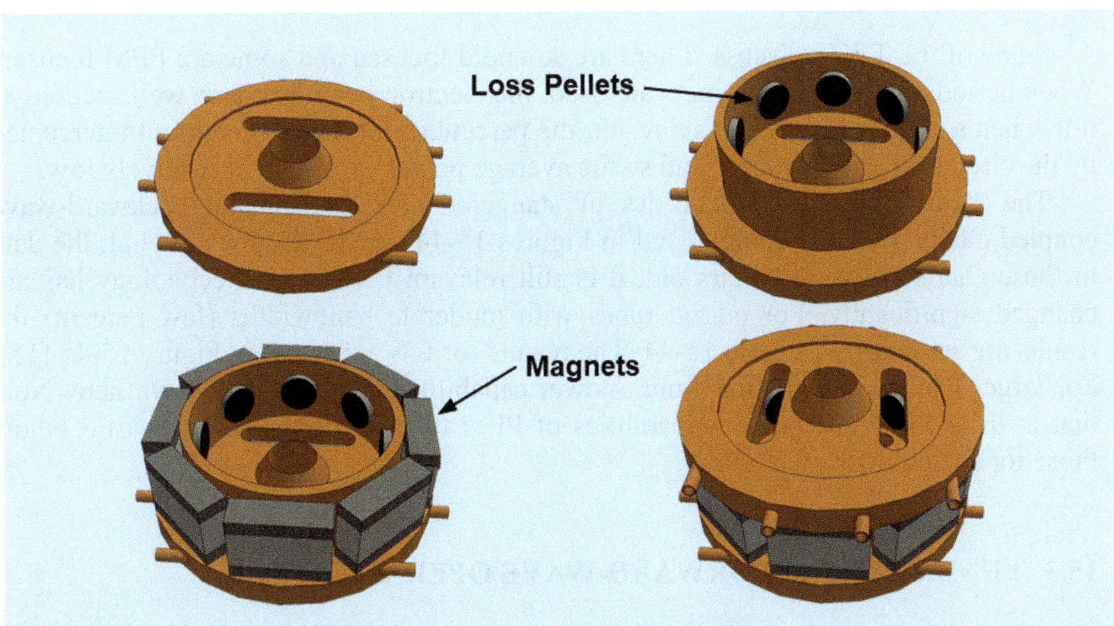

Figure 15-42 Parts including magnets and loss pellets of a dual staggered slot cavity.

A shadow-gridded gun is used. The beam current is about 9 A and the operating voltage is 45 kV. PPM focusing is used. The tube is pulse operated with a power output of 180 kW.

The diagram for the power circuit for the TWT in Figure 15-41 is in Figure 15-43. Note that a depressed collector is used and that the gun has a modulating anode. The 20-kΩ resistor in series with the modulating anode limits the current and prevents damage to the gun when an arc occurs.

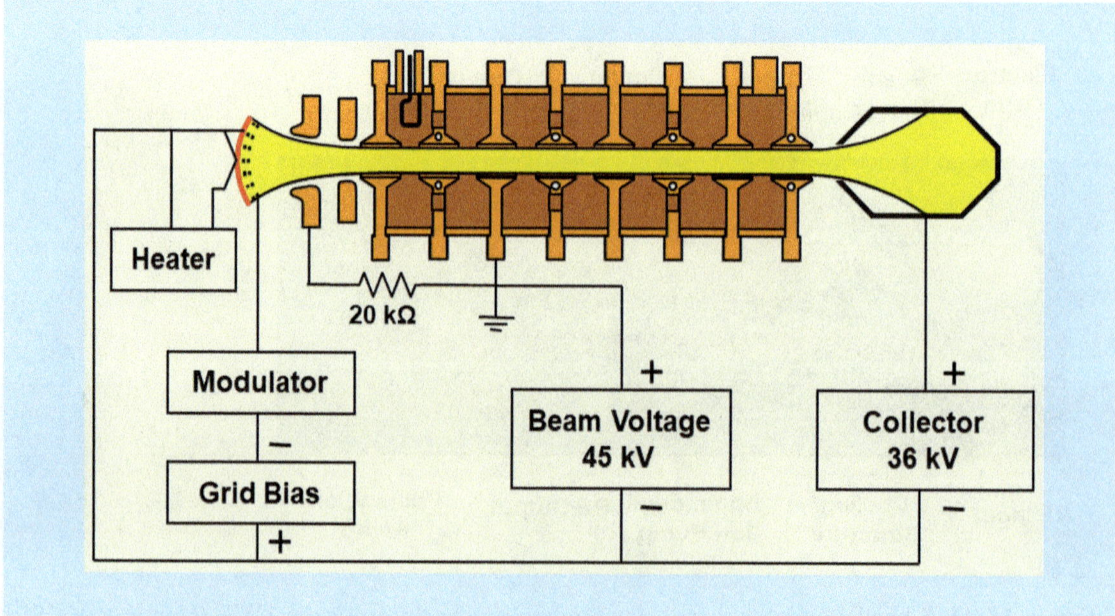

Figure 15-43 SDR TWT power supplies.

Some of the TWTs discussed here are solenoid focused and some are PPM focused. When periodic permanent magnets are used, the electron beam is not as well focused as it is when a solenoid is used. As a result, the percentage of the beam current intercepted by the circuit is relatively high, and so the average power capability is relatively low.

The status of power capabilities of staggered slot fundamental backward-wave coupled-cavity TWTs is summarized in Figures 15-44 and 15-45. Even though the data in these charts is over 30 years old, it is still relevant because the technology has not changed significantly. For pulsed tubes with moderate bandwidths (few percent) the results are summarized Figure 15-44. The results for CW tubes are in Figure 15-45 [15]. For larger bandwidths, the maximum power capability is lower than shown here. Note that in these plots, the power capabilities of PPM tubes are about a factor of 3 below those for solenoid-focused tubes.

15.3 FUNDAMENTAL FORWARD-WAVE OPERATION

Another class of coupled-cavity TWTs uses structures referred to as fundamental forward-wave circuits. The operation of these circuits may be understood with the aid of Figures

15-46 and 15-47. In these circuits, there is a mechanism for shifting the phase of the RF signal by 180°. The 180° twist shown in Figure 15-46 is an obvious way to produce the desired phase shift but would be virtually impossible to implement in a folded waveguide structure.

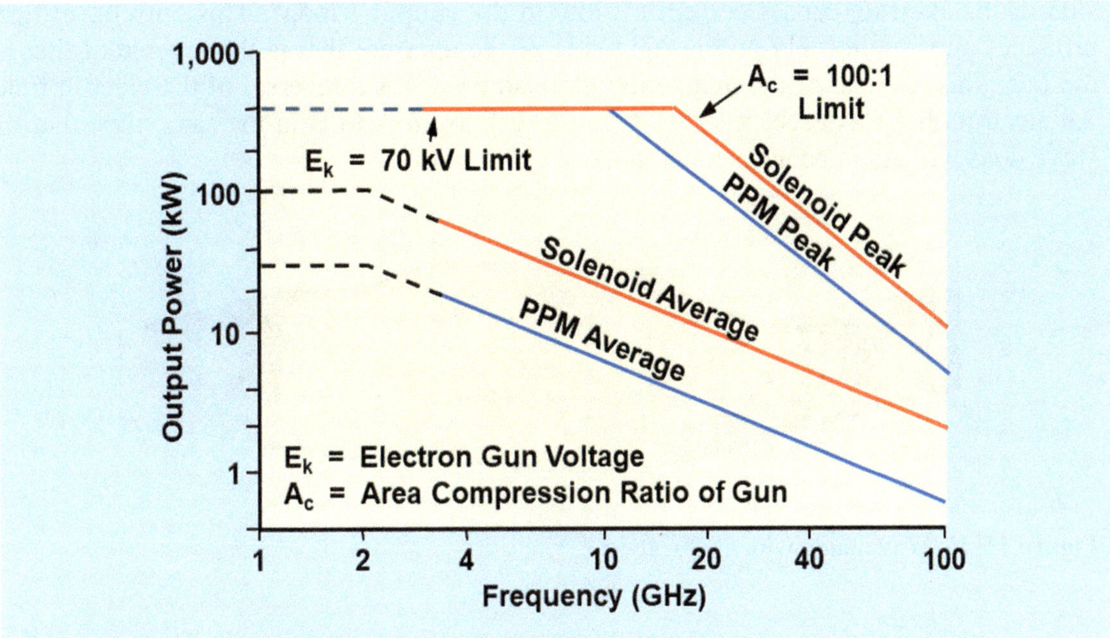

Figure 15-44 Output power for pulsed coupled-cavity TWTs with moderate bandwidths. (From: James W. Hansen, *Microwave Jour.*, 1989 State of the Art Reference.)

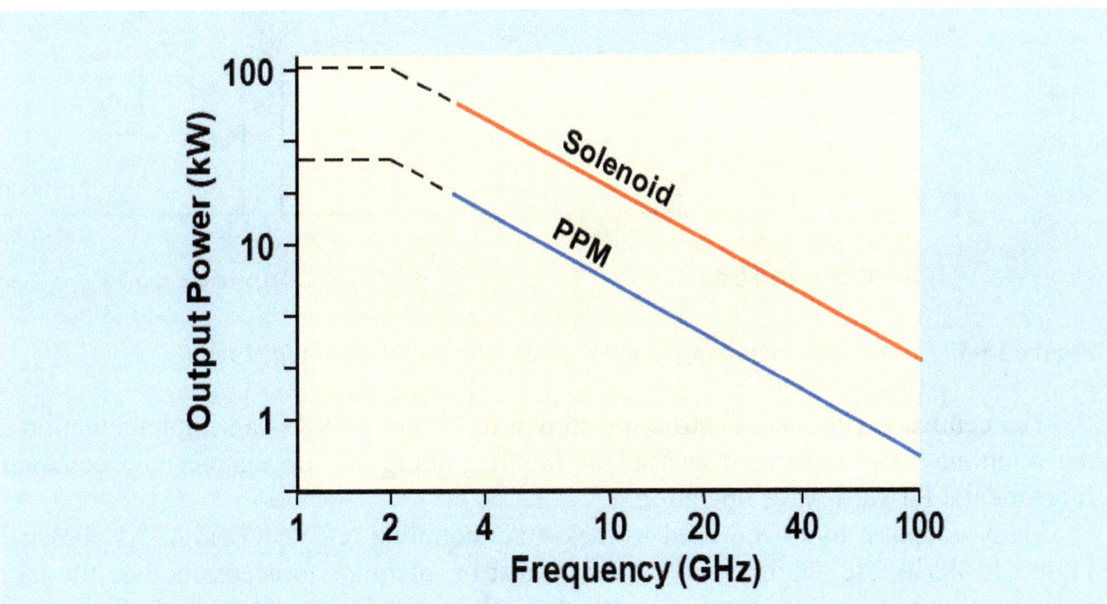

Figure 15-45 Output power for CW coupled-cavity TWTs with moderate bandwidths. (From: James W. Hansen, *Microwave Jour.*, 1989 State of the Art Reference.)

The dual-coupling loop shown in Figure 15-48 reverses the magnetic field at the coupling aperture and as a result, produces a 180° phase shift in the electric field. This action may be understood by first considering the magnetic field at the aperture in the folded guide with vanes. As indicated in Figure 15-48(a), the direction of the *H* field is the same on the two sides of the coupling aperture. In Figure 15-48(b), the magnetic field on the left side of the aperture causes current to flow in the coupling loops. This current, in turn, produces a magnetic field on the right side of the aperture that is the reverse of that on the left. This reversal of the magnetic field also produces a reversal of the electric field. An electron that was accelerated at A and travels from A to B in the same time that the wave does will again be accelerated at B.

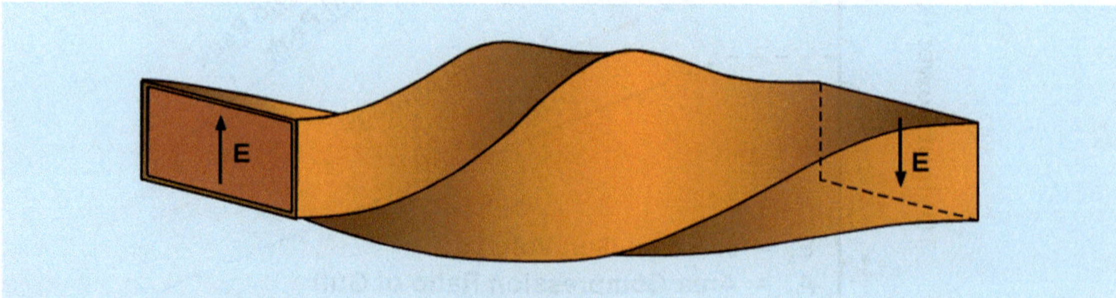

Figure 15-46 Waveguide with a 180° twist.

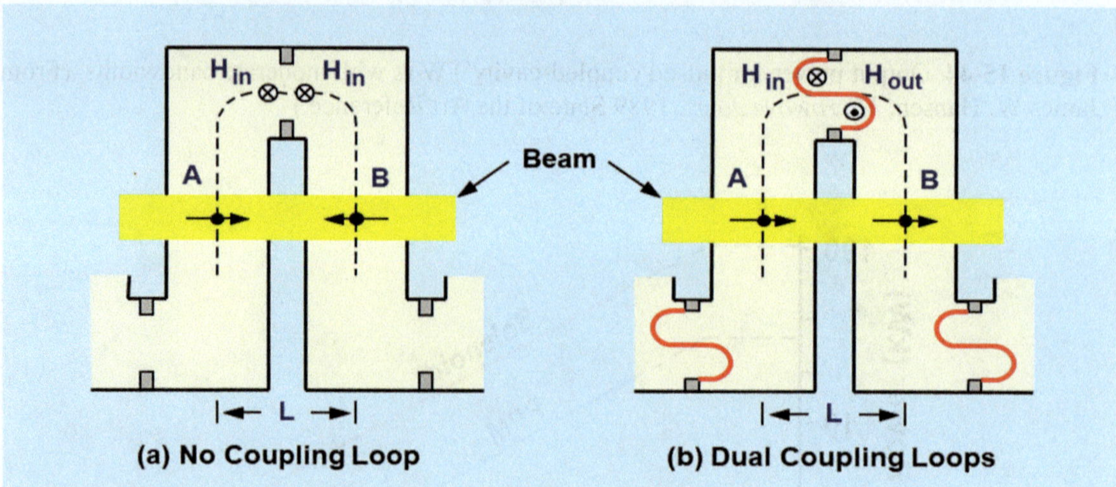

Figure 15-47 Use of a dual-coupling loop to reverse the direction of the *H* field.

The centipede (interlaced) structure shown in Figure 15-48 is an implementation of the coupling loop scheme discussed at the beginning of the chapter for achieving fundamental forward-wave operation.

Many coupling loops are used to maximize coupling and bandwidth. As shown in Figure 15-48(b), the height of each cavity must be adequate to accommodate the large coupling loops required to maximize coupling. Since the beam transit angle from cavity to cavity is only about 90°, the electron beam velocity must be high, so that the centipede structure is suitable for high-power operation.

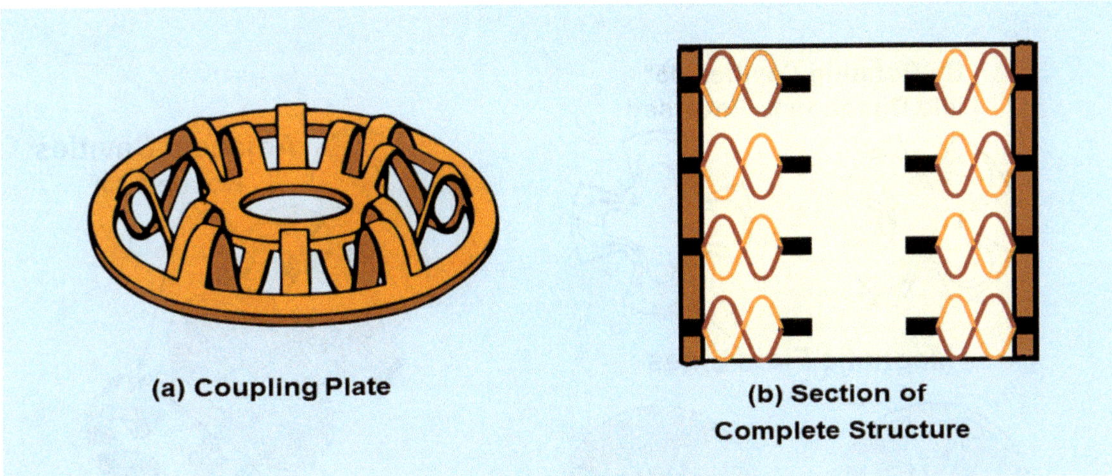

Figure 15-48 Centipede (interlaced) structure. (Adapted from: *Power Travelling Wave Tubes* by J. F. Gittins, published 1965 by American Elsevier, Inc.)

Another technique for producing a 180° phase shift in the electric field from cavity to cavity is illustrated in Figures 15-49 and 15-50. Instead of being circular cylinders, the cavity walls are formed into a four-leaf clover shape [16]. The RF magnetic flux lines are forced to take on the four-leaf clover shape. Radial slots are cut in the cavity walls to couple the magnetic field from cavity to cavity. By rotating adjacent cavities 45° the direction of the azimuthal magnetic field reverses and the polarity of the electric field reverses. As was the situation for loop coupling, an electron that was accelerated in one cavity and travels from cavity to cavity in the same time that the wave does will again be accelerated in the next cavity.

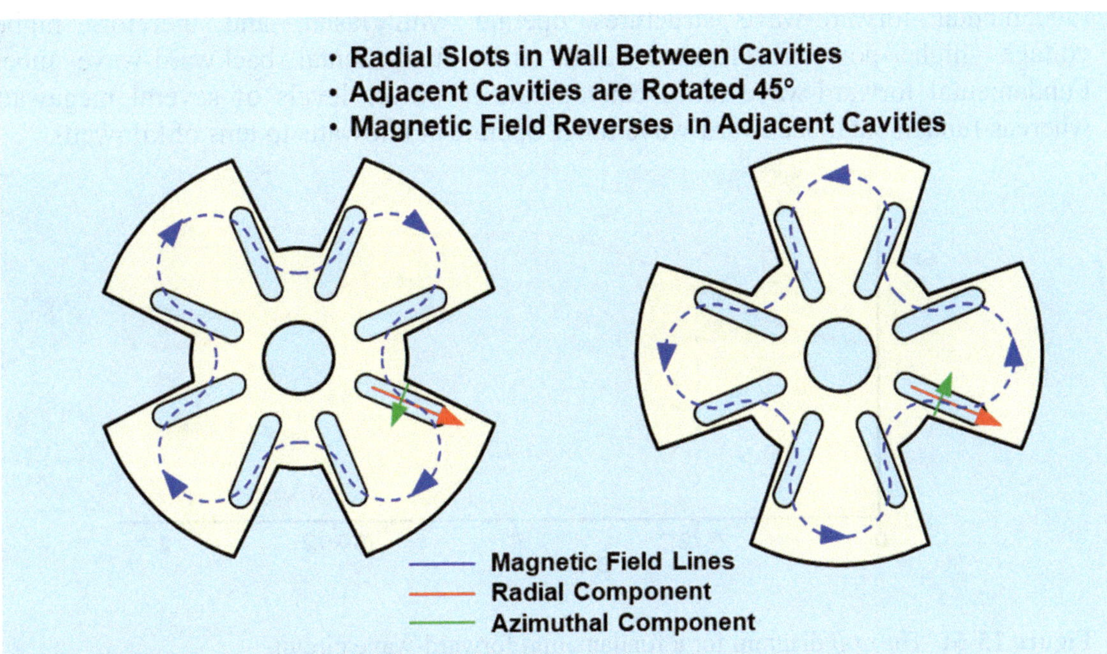

Figure 15-49 Use of cloverleaf-shaped cavities to reverse the electric field and make fundamental forward-wave interaction possible.

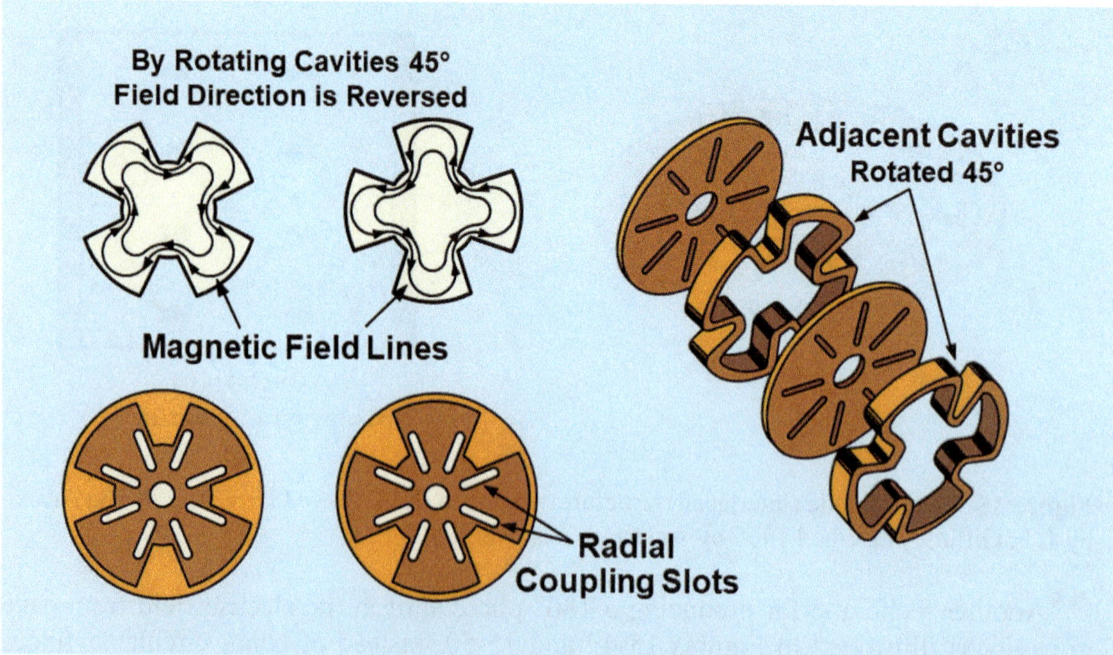

Figure 15-50 Cloverleaf structure. (Adapted from: M. Chodorow and R. A. Craig, *Proc. IRE*, August 1957. © 1957 IRE (now IEEE).)

The ω-β diagram for the loop coupled and the cloverleaf structures is shown in Figure 15-51. Interaction in this fundamental forward-wave structure occurs with electrons at a transit angle near $\pi/2$ rad. This transit angle is much smaller than that for the fundamental backward-wave structure (~$3\pi/2$ rad) so it is clear why tubes with fundamental forward-wave structures operate with faster—and therefore higher voltage—higher-powered electron beams than fundamental backward-wave tubes. Fundamental forward-wave tubes can operate at power levels of several megawatts whereas fundamental backward-wave tubes operate at kilowatts to tens of kilowatts.

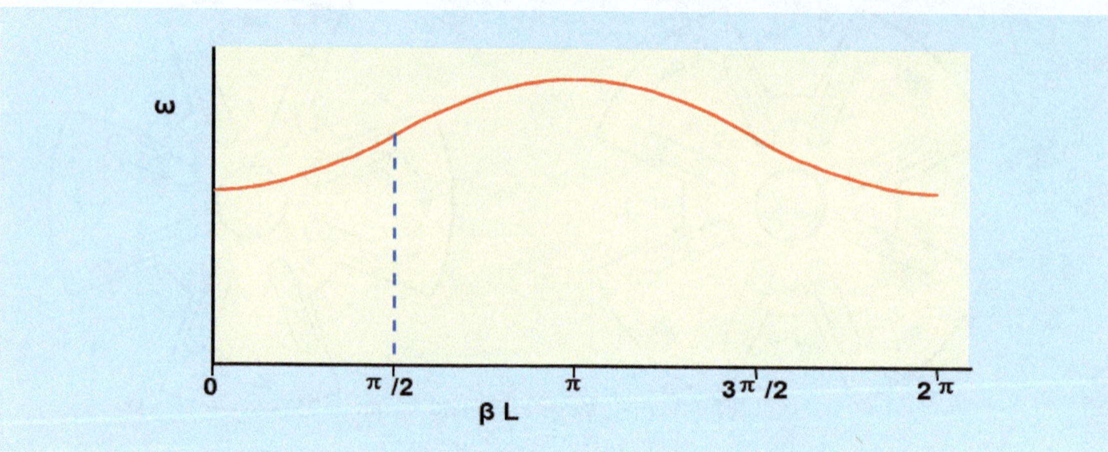

Figure 15-51 The ω-β diagram for a fundamental forward-wave circuit.

One of the techniques for terminating the centipede circuit has been to use a Kanthal loss coating as shown in Figure 15-52. This coating is made from FeCrAl, which is a high-temperature resistive alloy.

Figure 15-52 Kanthal coated centipede (interlaced) structure. (Courtesy of CPI.)

The cloverleaf structure has been used in high-power TWTs. An example, which was developed many years ago [17] and which is shown in Figure 15-53, is the Varian VA-125. This tube was made in two models – the VA-125A covered the band from 2.60 to 2.95 GHz and the VA-125B covered the band from 2.90 – 3.25 GHz. A modified Pierce gun and immersed flow focusing were used. The RF circuit had two sections. The output end of the input section was terminated to absorb the energy on the input circuit. The beam carried the signal across the termination space and reintroduced the signal on the output section. The output circuit was terminated at the input end to minimize the effect on reflections from the output end. One of the termination structures used is shown in Figure 15-55. Figure 15-54 shows the transition from the output end to waveguide.

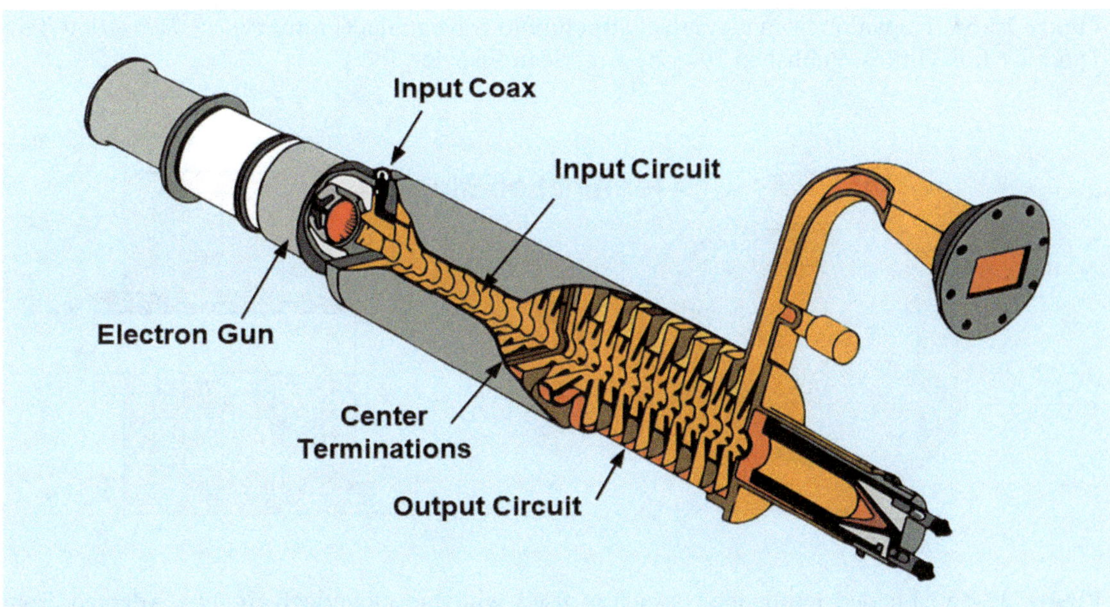

Figure 15-53 The use of a cloverleaf structure in the Varian VA-125 high-power TWT. The transition from the circuit to the output waveguide is shown in Figure 15- 44. (Adapted from: *Power Travelling Wave Tubes* by J. F. Gittins. © 1965 American Elsevier, Inc.)

The saturated power output for the VA-125A is given in Figure 15-56. These results were obtained with the drive power adjusted to maximize output. Note that output powers in excess of 4 MW were obtained. The data shown in Figure 15-57 for the VA-125B were obtained with drive power held constant at 1,200 W. This data shows that, for constant drive operation, a decrease in power output must be accepted at some frequencies.

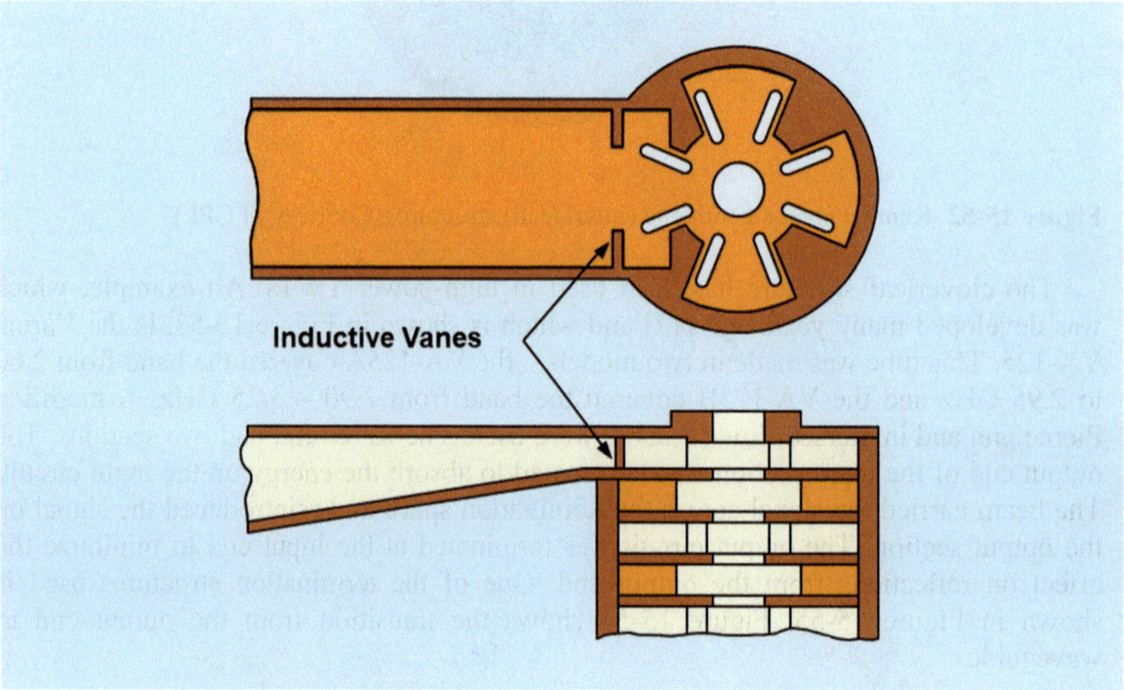

Figure 15-54 Transition from cloverleaf structure to waveguide. (From: *Power Travelling Wave Tubes* by J. F. Gittins. Published 1965 by American Elsevier, Inc.)

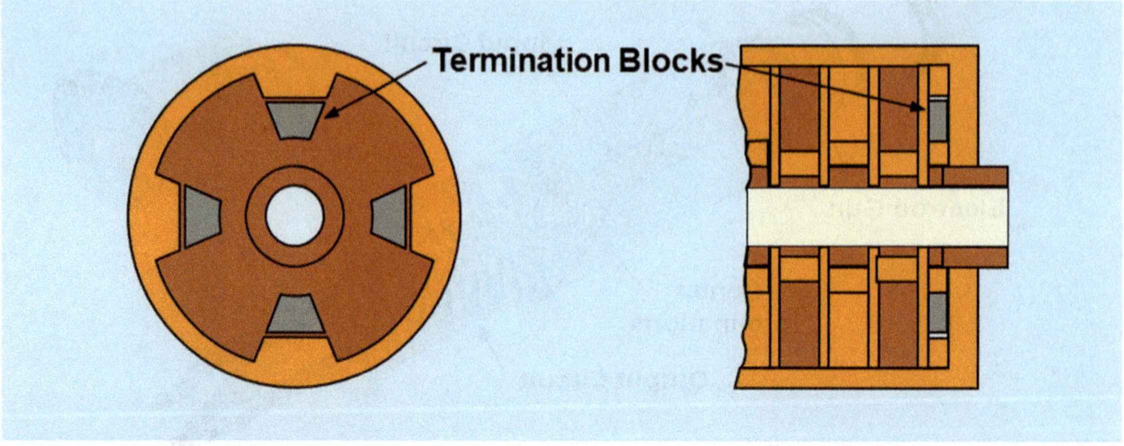

Figure 15-55 Loaded termination structure used with the cloverleaf circuit. (Adapted from: *Power Travelling Wave Tubes* by J. F. Gittins. © 1965 American Elsevier, Inc.)

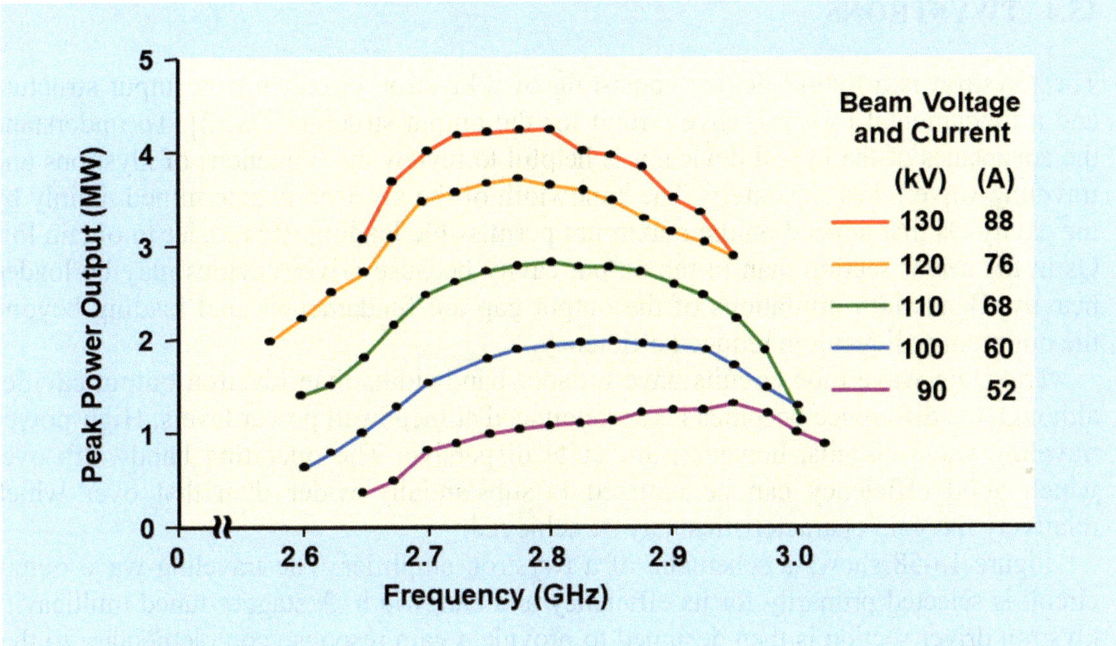

Figure 15-56 The power output for a VA-125A TWT. (From: Armand Staprans et al., *Proc. IEEE*, March 1973. © 1973 IEEE.)

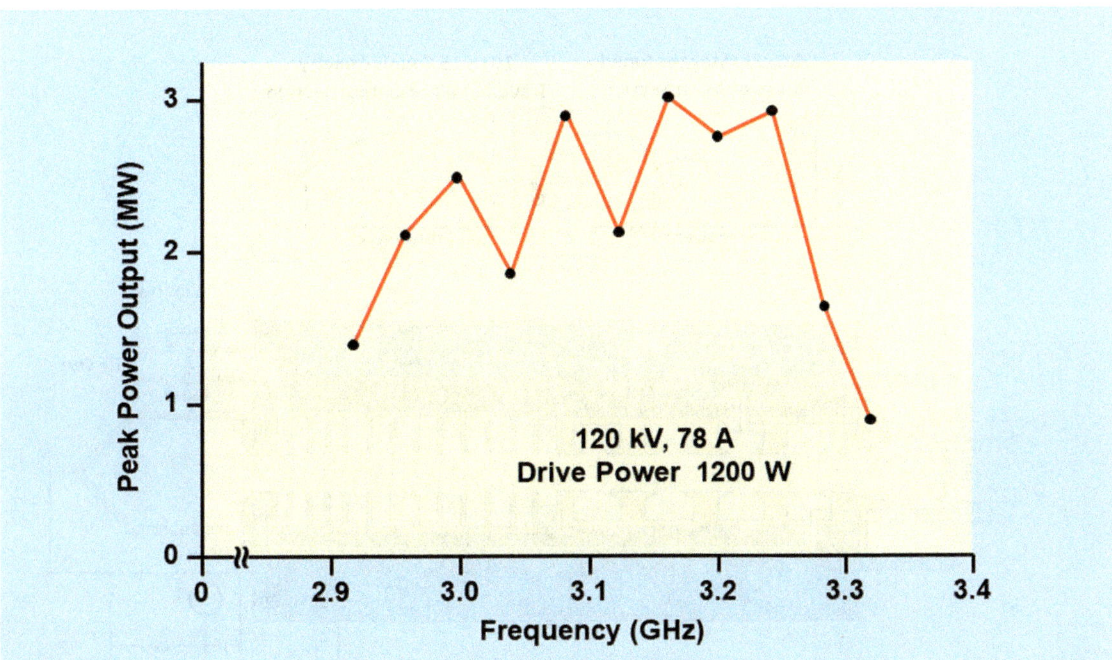

Figure 15-57 Power output for VA-125B with constant drive power. (From: J. A. Ruetz and W. H. Yocum, *IRE Trans. on Military Electronics*, April 1961. © 1961 IRE (now IEEE).)

15.4 TWYSTRONS

The twystron is a hybrid device consisting of a klystron circuit for the input structure and a fundamental forward-wave circuit for the output structure [21, 2]. To understand the advantages of the hybrid device, it is helpful to review the limitations of klystrons and traveling-wave tubes separately. The bandwidth of the klystron is determined mainly by the cavity Qs that depend on the maximum permissible loading. It is easier to obtain low Qs in the driver section than in the output cavity because driver cavities may be loaded heavily. Bandwidth limitations of the output gap are fundamental, and loading beyond the optimum will result in reduced efficiency.

Traveling-wave tube circuits have broader bandwidths than klystron output cavities although the difference becomes less pronounced at megawatt power levels. High-power traveling wave circuits, however, are quite dispersive. The operating bandwidth over which good efficiency can be realized is substantially wider than that over which relatively flat gain characteristics may be achieved.

Figure 15-58 shows a schematic of a twystron amplifier. The traveling-wave output circuit is selected primarily for its efficiency and bandwidth. A stagger-tuned multicavity klystron driver section is then designed to provide a gain response complementary to that of the input. The end result of the combination is a relatively flat gain response over the available efficiency bandwidth of the output circuit.

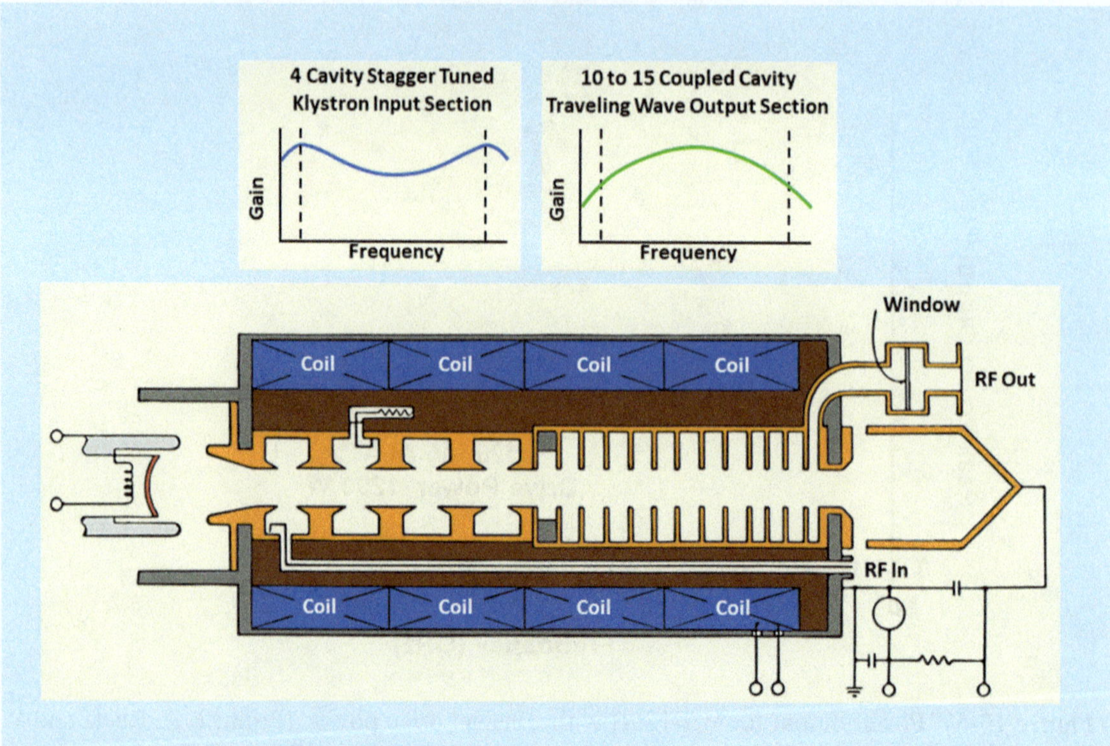

Figure 15-58 Klystron and traveling-wave parts of a twystron amplifier. (From: *Linear Beam Tubes,* Internal Varian Report by Armand Staprans.)

To help increase the bandwidth of a twystron, a high-perveance beam is used to provide strong beam loading. With this circuit and beam combination, the low gain of the TWT section at band edges is offset by the relatively high gain of the stagger-tuned-klystron driver section. The overall result, as Figure 15-59 indicates, can be a very flat gain over a bandwidth of up to 15%. The resulting power output as a function of frequency is also very flat, as shown in Figure 15-61. Twystrons with peak-power outputs up to at least 10 MW have been built. Average powers over 30 kW are available and can be increased about an order of magnitude before circuit losses and interception present fundamental limitations.

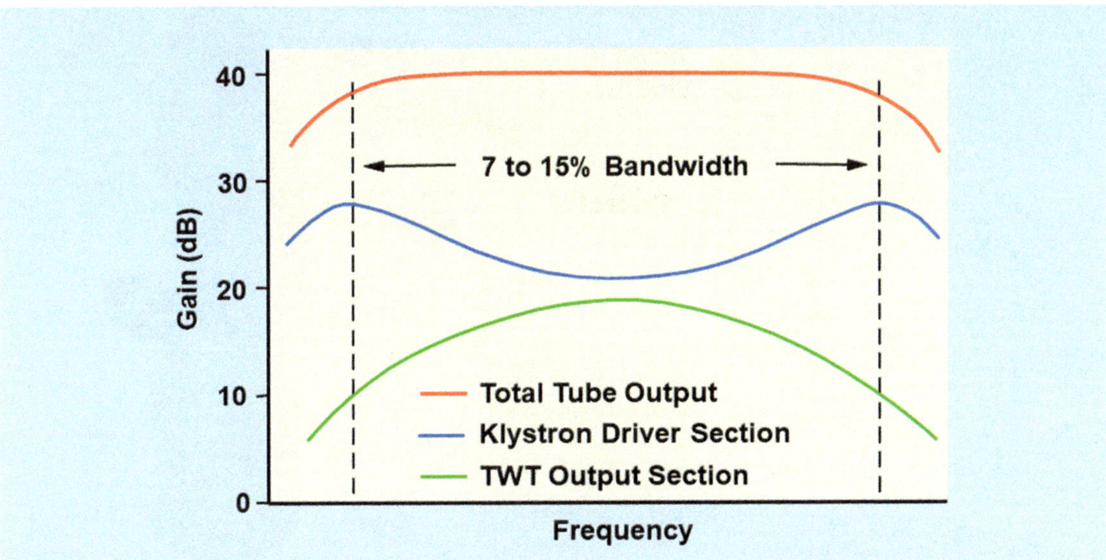

Figure 15-59 The gain of a twystron amplifier. (From: *Microwave Tube Manual* by Varian, Air Force Publication No. T.O.00-25-251, 1979.)

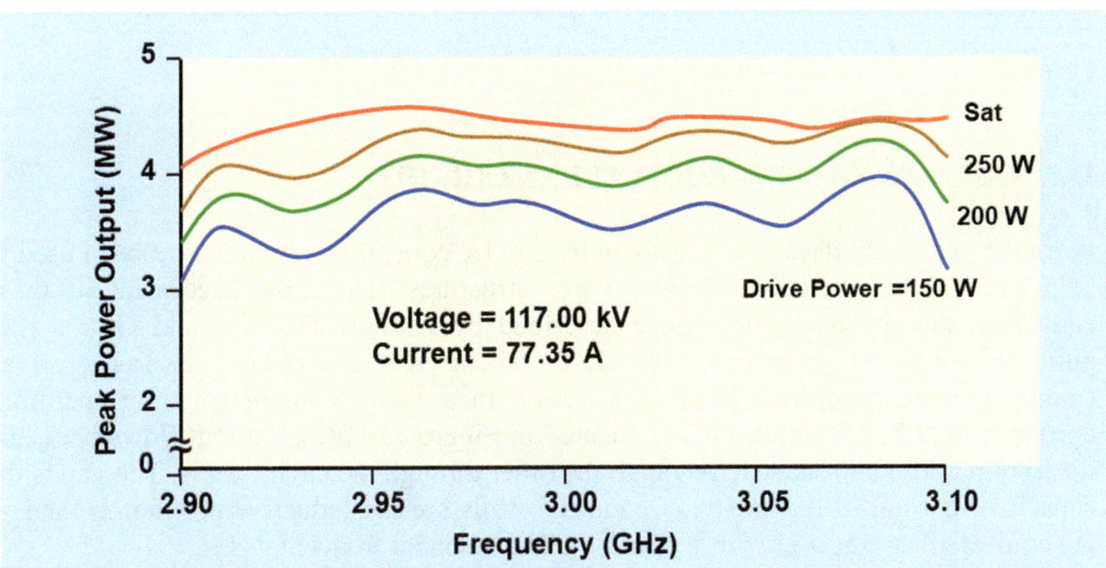

Figure 15-60 The power output for twystron amplifier VA-146E.

The VA-144 twystron shown in Figure 15-61 was developed for the ballistic missile early warning system. Two of these tubes operating in parallel were used in the missile site radar (MSR) in North Dakota in the 1970s–1990s. The VA-144 operated at 150 kV and used a modulating anode for beam control. The power capability of this tube is not known, however, it was high enough to "melt the telephones and the insulation on cables" in the anechoic test facility at the MSR. The reported cost of these tubes was $750,000 each in 1970 dollars.

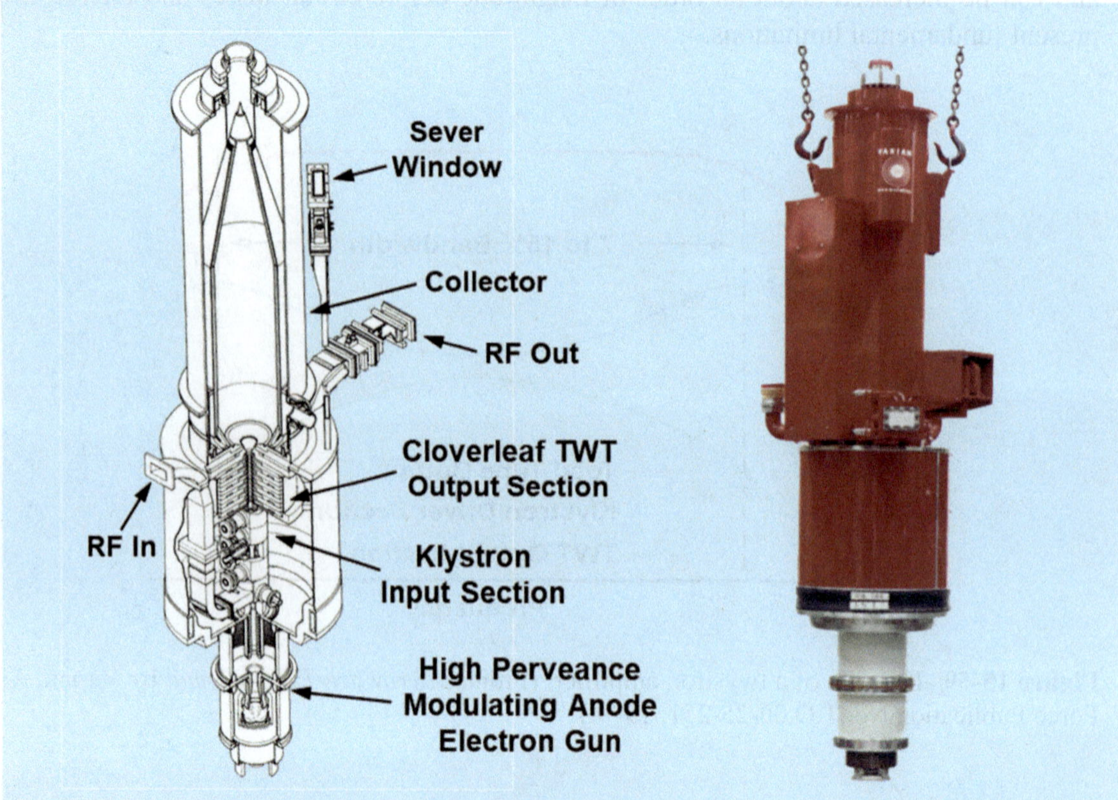

Figure 15-61 Varian VA-144 twystron.

15.5 CURNOW-GITTINS EQUIVALENT CIRCUIT

From the very early days of coupled-cavity TWTs, equivalent circuits have been used to help explain the operation of slow wave structures. The most successful of these equivalent circuits have been those described by Curnow [18, 19] and Gittins [20] published over 50 years ago. The development of these circuits is based on an examination of the current paths in a cavity. In a cavity without coupling apertures, current is distributed uniformly as indicated in Figure 15-62(a). Current flows back and forth from one side of the cavity gap to the other through the cavity walls. The gap is the capacitive portion of the cavity, C_c, and the walls are the inductive portion, L_c, and so the equivalent circuit is in Figure 15-62(b). The resonant frequency is f_c.

When coupling slots are placed in the cavity walls, some of the current paths are intercepted (Figure 15-63). In general, there can be four types of current paths:

1. I_1 between drift tube tips;
2. I_2 between the left coupling slot and the right drift tube tip;
3. I_3 between the left drift tube tip and the right coupling slot;
4. I_4 between coupling slots.

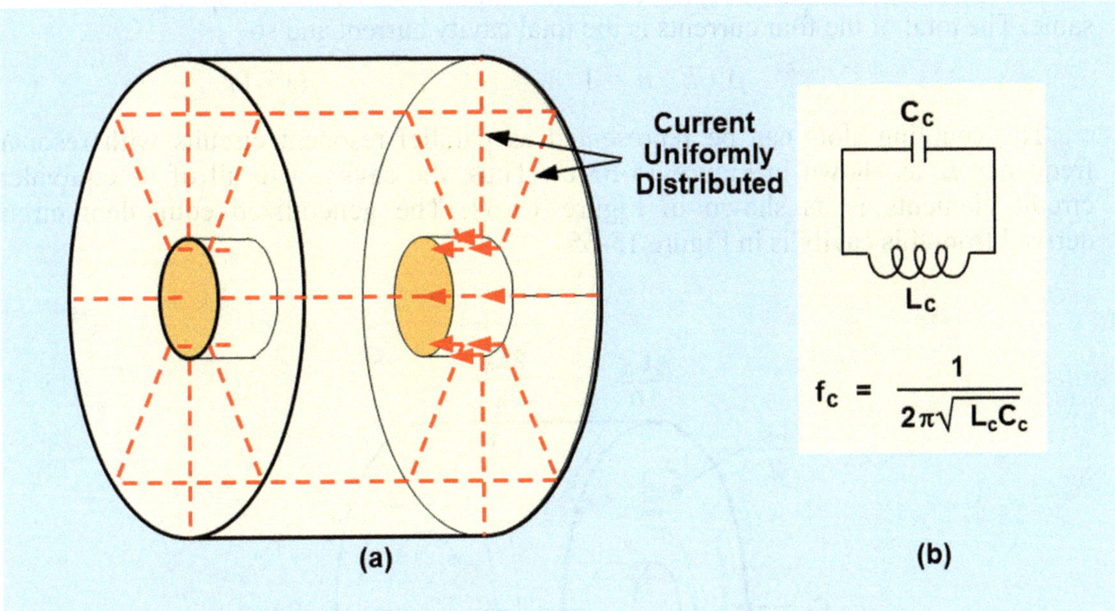

Figure 15-62 Current paths and the equivalent circuit for a cavity without coupling slots.

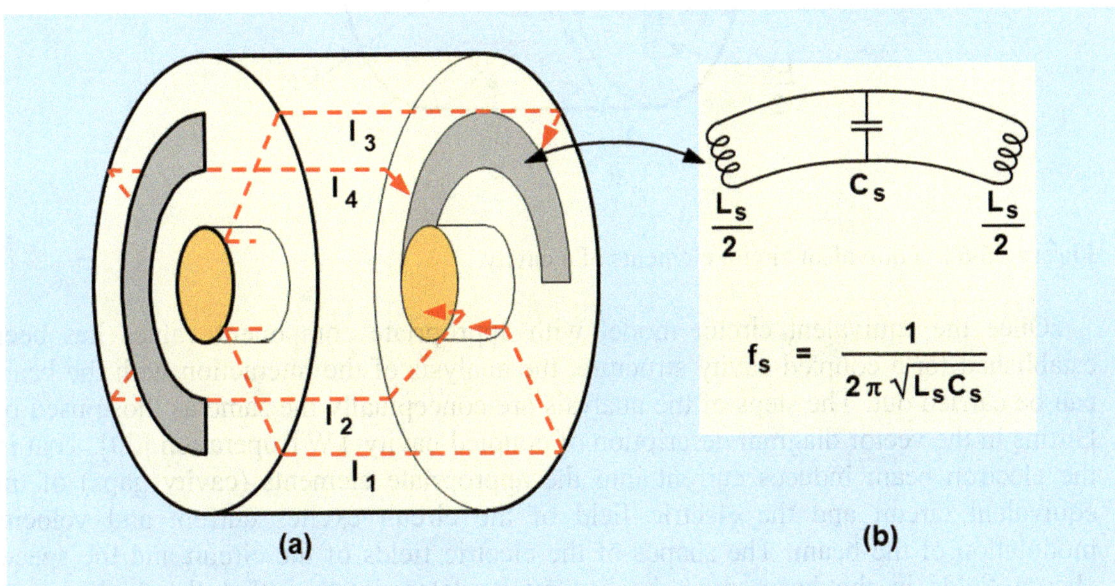

Figure 15-63 (a) Cavity currents used to define the inductance for the equivalent circuit, and (b) the equivalent circuit for a slot.

The equivalent circuit to represent this distribution of currents can be constructed by dividing the cavity inductance, L_c, into four parts:

1. L_c/p, where p is the fraction of the current not involved in coupling;
2. $2L_c/k$, where k is the fraction involved in coupling to the following cavity;
3. $2L_c/k$, where k is the fraction involved in coupling to the preceding cavity;
4. L_c/n, where n is the fraction involved in coupling both ways.

Here, it has been assumed, as is usually the case, that the two coupling slots are the same. The total of the four currents is the total cavity current and so

$$p + k + n = 1 \qquad (15\text{-}1)$$

The coupling slots can be represented as parallel resonant circuits with resonant frequency f_s, as shown in Figure 15-63(b). Thus, the cavity with all of its equivalent circuit elements is as shown in Figure 15-64. The generalized equivalent circuit derived from this cavity is in Figure 15-65.

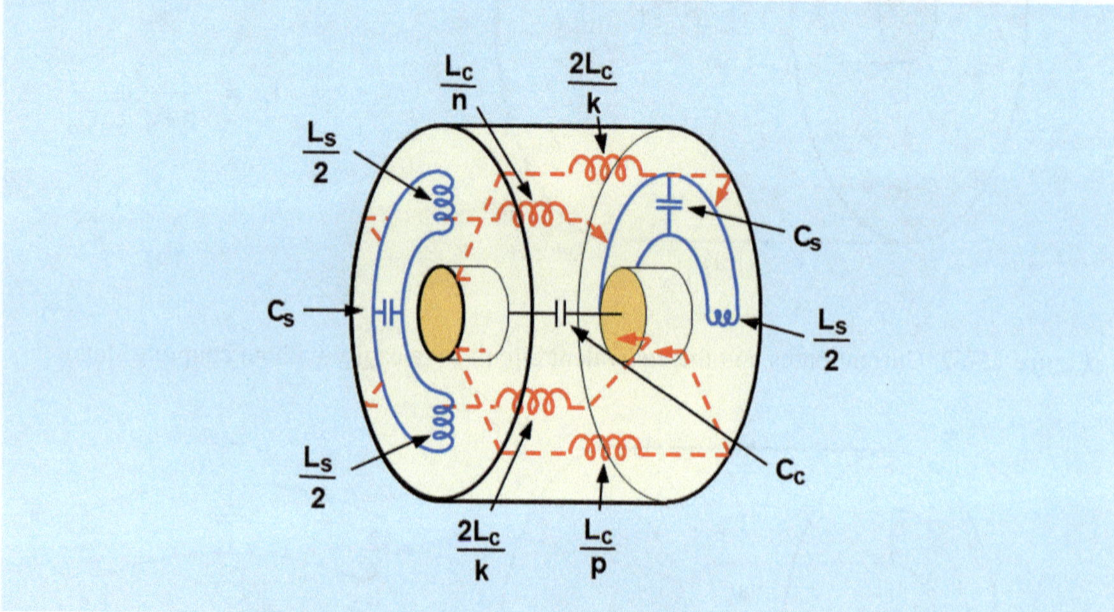

Figure 15-64 Equivalent circuit elements of a cavity.

Once the equivalent circuit model with appropriate component values has been established for a coupled-cavity structure, the analysis of the interaction with the beam can be carried out. The steps of the analysis are conceptually the same as those used by Gittins in the vector diagram description of coupled-cavity TWT operation [20]. That is, the electron beam induces current into the appropriate elements (cavity gaps) of the equivalent circuit and the electric field of the circuit excites current and velocity modulation of the beam. The shapes of the electric fields of the circuit and the space-charge fields in the beam must be accurately determined so that the analysis can properly predict tube performance.

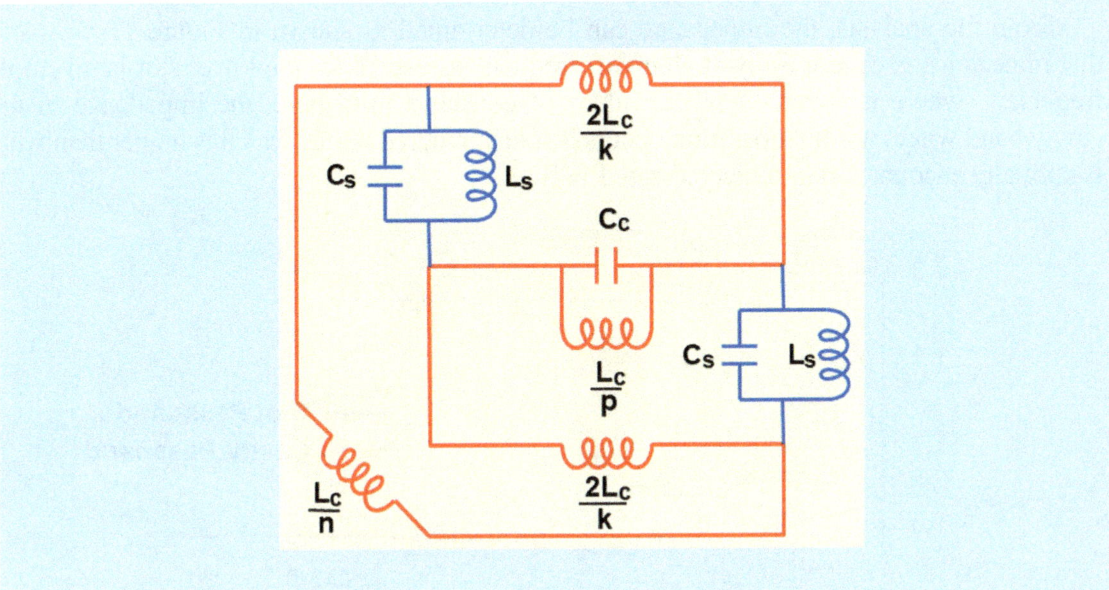

Figure 15-65 The equivalent circuit for a cavity.

Figure 15-66 is a typical ω-β diagram that can be constructed from the results of Curnow-Gittins analysis. The solid curves are the ω-β characteristics derived for one midcavity-to-midcavity cell of the equivalent circuit for the entire coupled-cavity structure. Because only one cell is used, results for other portions of the ω-β characteristic, such as those shown by the dashed lines in Figure 15-66, are not obtained. However, these can be inferred from the discussion of ω-β characteristics contained in Chapter 13. Thus, the curves from π to 2π rad are simply mirror images of those from 0 to π rad. For the staggered-slot circuit, the lower dashed curve, which indicates forward-wave interaction, is the ω-β characteristic that is needed.

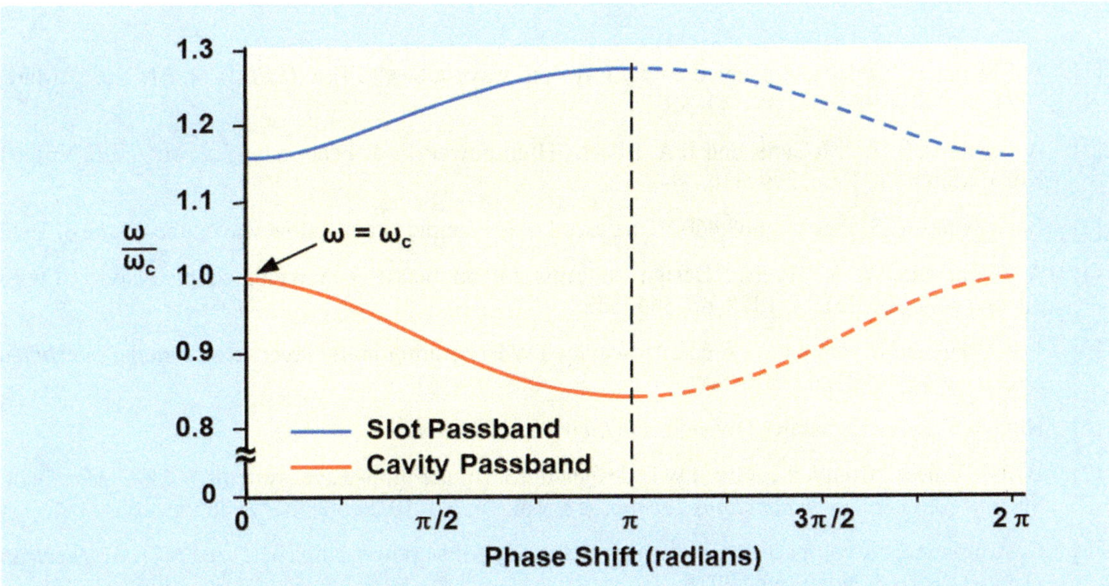

Figure 15-66 The ω-β diagram for the equivalent circuit for the staggered-slot structure.

From the analysis, the impedance can be determined as shown in Figure 15-67. Note that impedance goes to infinity at all cutoff frequencies except for the lower slot-band cutoff frequency, where it is zero. For the values of constants used here, the impedance in the cavity band where normal operation occurs is a few hundred ohms, which is higher than what is normally encountered in staggered-slot TWTs.

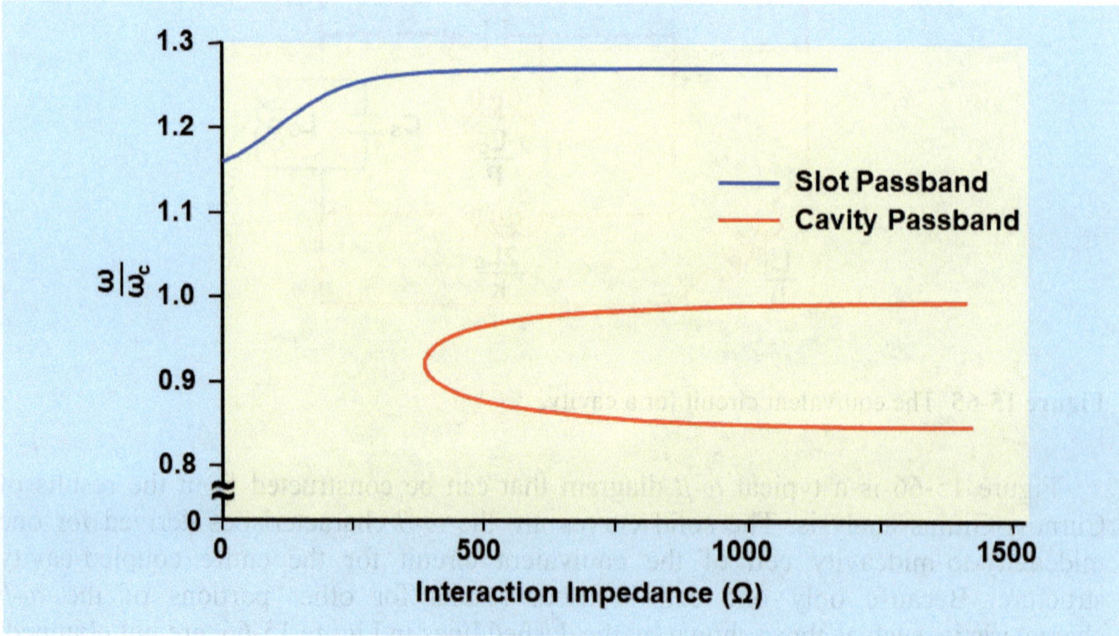

Figure 15-67 Interaction impedance derived from the circuit model for the staggered-slot structure.

REFERENCES

[1] J. T. Mendel, "Helix and coupled-cavity traveling wave tubes," *Proc. IEEE*, Vol. 61, No. 3, March 1973, pp. 280–298.

[2] A. Staprans, E. W. McCune, and J. A. Ruetz, "High-power linear-beam tubes," *Proc. IEEE*, Vol. 61, No. 3, March 1973, pp. 299–330.

[3] B. G. James, U.S. Patent 3,668,460, "Coalesced mode coupled cavity slow wave tube," June 6, 1972.

[4] A. Karp, and W. R. Ayers, "Design concepts for an octave bandwidth coupled-cavity TWT," *Technical Digest*, IEDM, 1978, pp. 546–549.

[5] J. R. Frey, and I. Tammaru, "A coupled-cavity TWT operating in the inverted slot mode," *Technical Digest*, IEDM, 1981, pp. 504–506

[6] Hughes Electron Dynamics Division, *TWT/TWTA Handbook.*

[7] B. G. James, "Coupled-cavity TWT designed for future mm-wave systems," *MSN Microwave Systems News & Communications Technology*, Vol. 16, No. 10, September 1986, pp. 105–116.

[8] C. Armstrong, "A compact extremely high frequency MPM power amplifier," *IEEE Trans. Electron Devices*, Vol. 65, No. 6, June 2018.

[9] R. Kowalczyk et al., "High efficiency E-band MPM for communications applications," in Proc. 17th IEEE Int. Vac. Electron. Conf., Apr. 2016, pp. 1–2.

[10] R. Kowalczyk et al., "A 100 watt W-band MPM TWT," in Proc. 14th IEEE Int. Vac. Electron. Conf., May 2013, pp. 1–2.

[11] M. A. Basten, et al, "233 GHz high power amplifier development at Northrop Grumman," IEEE IVEC.

[12] K. E. Kreischer, et al, "220 GHz Power Amplifier Testing at Northrop Grumman," IEEE, IVEC 2014.

[13] C. D. Joye, et al, "Demonstration of a high power, wideband 220-GHz traveling wave amplifier fabricated by UV-LIGA," *IEEE Trans.Electron Devices*, Vol. 61, No. 6, June 2014, pp 1672–1678.

[14] C. D. Joye, J. P. Calame, K. T. Nguyen, and M. Garven, "Microfabrication of fine electron beam tunnels using UV-LIGA and embedded polymer monofilaments for vacuum electron devices," *J. Micromech. Microeng.*, Vol. 22, No. 1, p. 015010, 2012.

[15] J. W. Hansen, "State of the art reference, US TWTs from 1 to 100 GHz," *Microwave Jour.*, 1989, pp. 179–193.

[16] M. Chodorow, and R. A. Craig, "Some new circuits for high-power traveling-wave tubes," *Proc. IEEE*, Vol. 45, No. 8, August 1957, pp. 1106–1118.

[17] J. A. Ruetz and W. H. Yocum, "High-power traveling-wave tubes for radar systems," *IRE Trans. Military Electronics*, April 1961, pp. 39-45.

[18] H. J. Curnow, "A general equivalent circuit for coupled-cavity slow wave structures," *IEEE Trans. Microwave Theory Tech.*, Vol. MTT-13, No. 9, September 1965, pp. 671–675.

[19] H. J. Curnow, "A new equivalent circuit for coupled-cavity structures," *Proc. 5th International Congress on Microwave Tubes*, (conference held in Paris, France), New York and London: Academic Press, September 14–18, 1964.

[20] J. F. Gittins, Power *Travelling Wave Tubes*, New York: Elsevier, 1965.

[21] A Staprans, "Linear beam tubes," Internal Varian report.

[22] A. D. Larue and R. R. Rubert, "Multi-megawatt hybrid TWTs at S band and C band," paper presented at the IEEE Electron Devices Conference, Washington, D. C. October 1964.

Chapter 16

Collectors for Linear-Beam Tubes

As indicated in Figure 16-1, there are two major considerations in the design of collectors for linear-beam tubes. These are

1. Power dissipation;
2. Power recovery.

When power dissipation is the dominant consideration, the collector is a single electrode at the voltage of the body of the tube (usually ground) and is designed to dissipate as heat all of the power in the spent beam (that is, the beam leaving the RF interaction structure of the tube). These collectors are usually used on very large devices in applications where power consumption is not a critical consideration.

When power recovery is most important, the collector may have a single electrode or it may have two or three or more electrodes. To recover power, the voltages on the electrodes must be negative with respect to the body of the tube so that the beam is slowed down before it is collected. The energy recovered in slowing the beam is returned to the power supply and the heat that must be dissipated by the collector is greatly reduced. These collectors are used where overall efficiency is an important consideration.

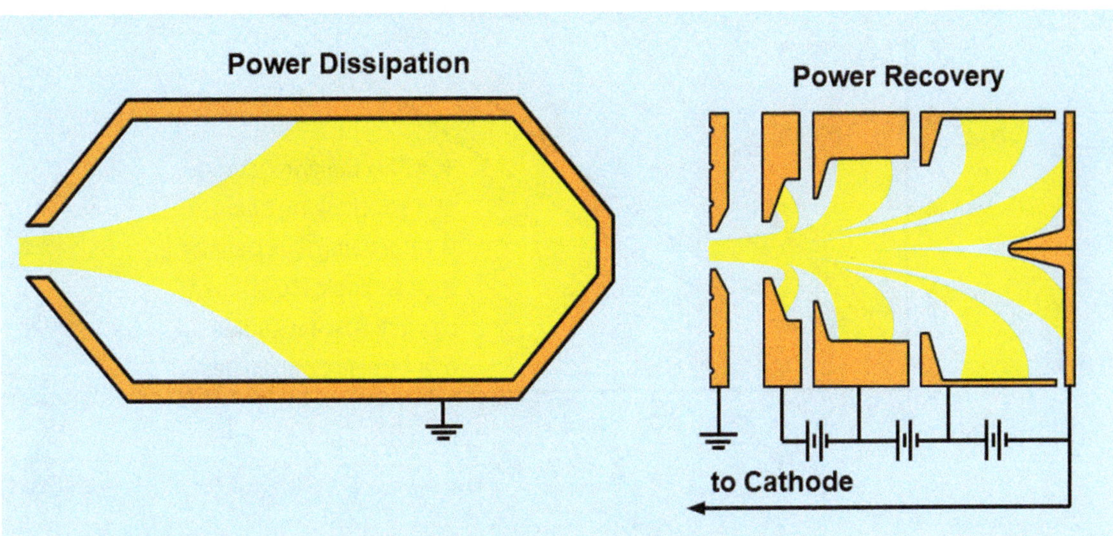

Figure 16-1 Collectors for power dissipation and for power recovery.

16.1 POWER DISSIPATION

In a linear-beam tube, in the absence of RF power, the entire beam energy must be dissipated in the collector [1]. At first glance this does not appear to be a difficult problem, since, in principle, the collector may be made as large as desired. However, it is usually desirable to keep the collector as small as intercepted power densities allow. This requires a detailed understanding of the electron trajectories inside the collector.

In most collectors, the magnetic field is small. The magnetic field is terminated in a pole piece between the end of the interaction structure and the collector. Ideally, for all focusing schemes where the cathode is also in a field-free region, such as for Brillouin flow, the beam enters the collector with only an axial component of velocity. Thus, spreading of the beam in the collector results only from space-charge forces. If magnetic flux threads the cathode, then it also threads the collector. Then there is rotation of the beam in the collector, which is directly proportional to the amount of flux through the cathode. This rotation spins the electrons out radially and so it contributes to beam spreading in the collector. The effect of the magnetic field may be shown by universal beam-spread curves in the presence of the magnetic field threading the cathode, as shown in Figure 16-2. Beam spreading is increased substantially for confined-flow beams, which may use a focusing field of about three times the Brillouin value. For constant charge density in the beam, the spreading of the beam is laminar and, in principle, power density is known everywhere.

In the ideal case, to minimize the size of a collector, the objective is to design the inside shape so that the intercepted power density is constant at the maximum tolerable value. The shape of the resulting collector (Figure 16-3(a)) is impractical because it would be difficult to fabricate and because small changes in the beam pattern would vary power dissipation at the walls and cause overheating in some regions.

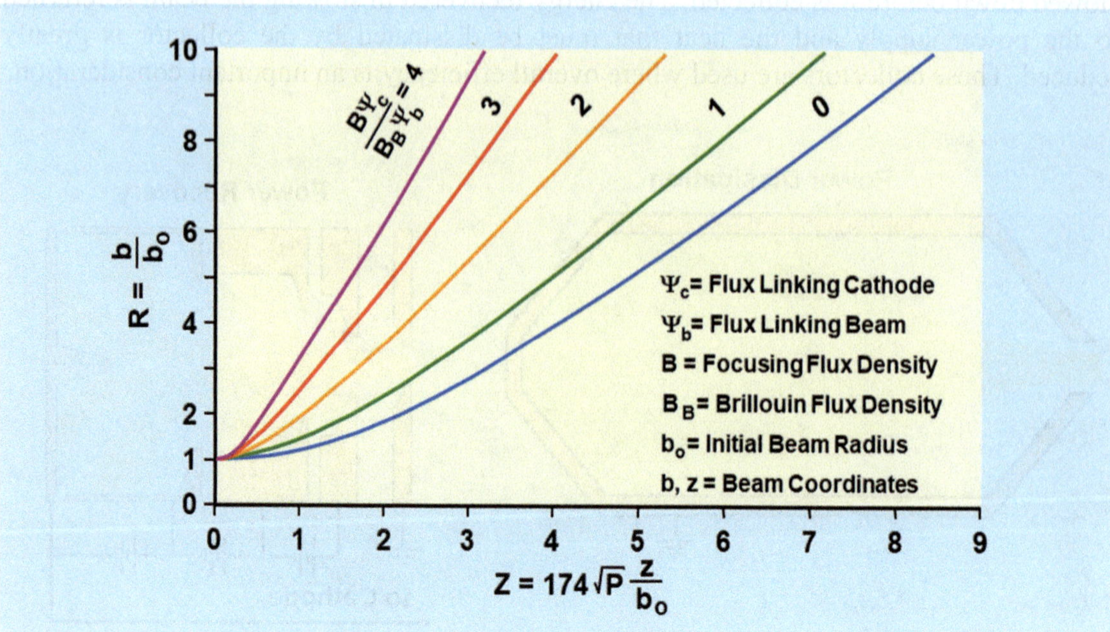

Figure 16-2 Beam spread curves for magnetically confined beams in a field-free collector. (Adapted from: *Linear-Beam Tubes* by A. Staprans, Internal Varian Report, September, 1968.)

A collector is generally designed approximately as shown in Figure 16-3(b). As shown in Figure 16-3(c), two critical areas of high dissipation exist:

1. Where the beam first impinges on the collector wall (dissipation is maximum);
2. At the end of the collector.

The errors in the idealized analysis of trajectories inside a collector are mostly such that they make the beam less divergent than occurs in an actual tube. Causes may be

1. Leakage of the magnetic field from the focusing magnet inside the collector;
2. A gradual instead of the idealized sharp cutoff of the magnetic focusing field at the entrance to the collector;
3. The presence of positive ions that neutralize the electron space-charge forces.

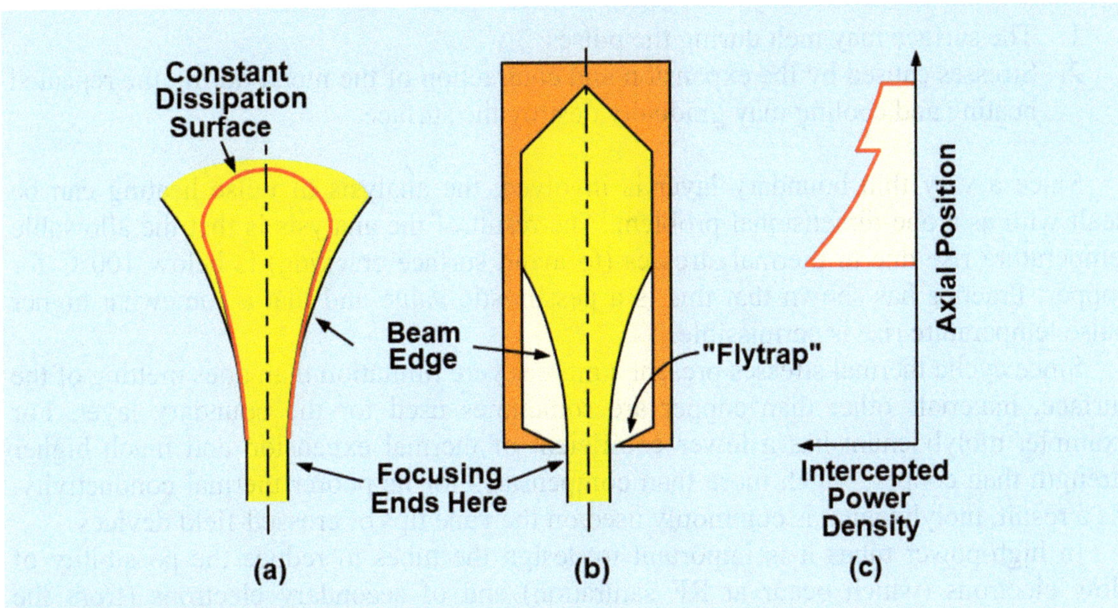

Figure 16-3 (a) Collector for constant dissipation, (b) practical collector, and (c) intercepted power density. (Adapted from: *Linear-Beam Tubes* by A. Staprans, Internal Varian Report, September 1968.)

In practice, the cylindrical wall of the collector is designed according to the best estimate of actual beam trajectories. However, the collector is made longer than the calculation would indicate, and a conical end is often used to increase the intercepting surface area. Typically, the ratio of average to maximum (where the beam is first intercepted) wall dissipation will vary between 0.25 and 0.5.

The maximum allowable intercepted power density is affected by

1. Collector wall material and thickness;
2. Method of heat transfer from the wall to the cooling medium;
3. Whether the intercepted power is continuous or pulsed.

Collectors for high-power tubes are usually made of copper because of its high thermal conductivity. If the collector wall is relatively thin, less than 1 cm, the temperature rise in the wall from the average beam power is less important than the mechanism of heat transfer to the cooling medium. For forced-flow water cooling with water channels of ordinary design, the average dissipation is limited to a working value of about 1 kW/cm^2. Dissipations up to 18 kW/cm^2 have been realized over small areas in some crossed-field devices by the use of laminated wall materials, very small water channels, and extremely high water pressures [2]. It has not been necessary to resort to these measures in linear-beam tubes.

The limits of pulse heating are less well understood than average heating. Since pulse lengths do not usually exceed a few milliseconds, the problem is essentially a surface phenomenon. During a pulse, the temperature rises to a high value in a very thin surface layer. Two problems may arise:

1. The surface may melt during the pulses;
2. Stresses caused by the expansion and contraction of the metal during the repeated heating and cooling may gradually destroy the surface.

Since a very thin boundary layer is involved, the analysis of pulse heating can be dealt with as a one-dimensional problem. The result of the analysis is that the allowable temperature rise due to thermal stresses (to avoid surface cracking) is below 100°C for copper. Practice has shown that this is a pessimistic value and that a somewhat higher pulse temperature rise is permissible.

Since cyclic thermal stresses present a more severe limitation than does melting of the surface, materials other than copper are sometimes used for the boundary layer. For example, molybdenum has a lower coefficient of thermal expansion and much higher strength than copper, which more than compensates for its poorer thermal conductivity. As a result, molybdenum is commonly used on the vane tips of crossed-field devices.

In high-power tubes it is important to design the tubes to reduce the possibility of slow electrons (which occur at RF saturation) and of secondary electrons (from the collector) striking the tube body. Therefore, the entrance to the collector is made as small as possible without actual interception of primary beam electrons. The entrance section of the collector, often of a conical shape, is sometimes referred to as the fly trap.

16.2 POWER RECOVERY

A primary consideration in selecting microwave tubes for many applications is the overall efficiency in converting dc input power to RF output power at the desired frequency. One very important part of overall efficiency is electronic efficiency, which is the efficiency of conversion of power in the electron beam to RF power. Electronic efficiency has received considerable attention since the invention of microwave tubes. In this section, it will be assumed that the electronic efficiency has been established in the design of the RF characteristics of a tube. Then, with electronic efficiency fixed, the enhancement of the overall efficiency with the collector will be discussed.

16.2.1 Power Flow

The various elements of power flow in a linear-beam tube are illustrated in Figure 16-4. Power is supplied for the heater, the electron beam, and the solenoid. The RF input power is normally at least 30 dB below the RF output power and is assumed to be negligible.

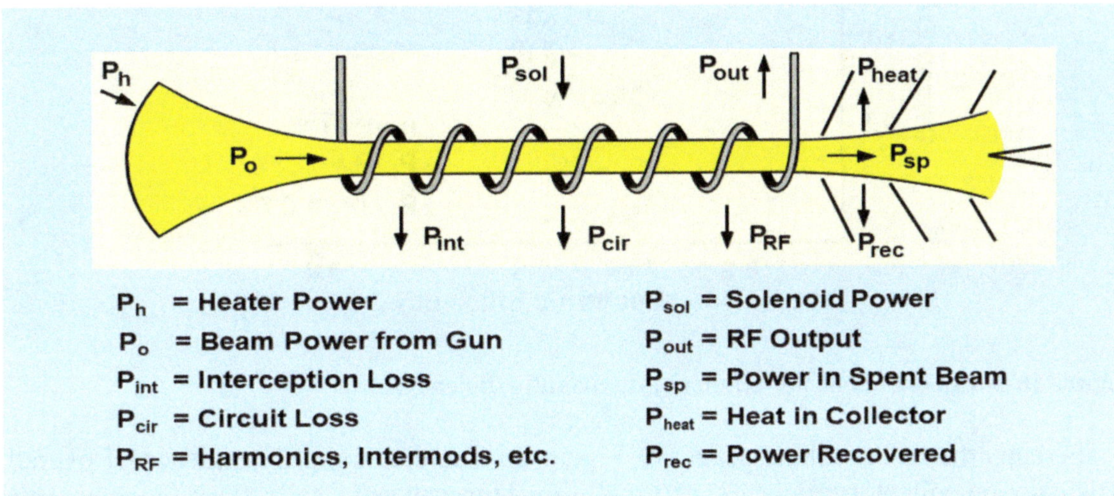

Figure 16-4 Power flow in a linear-beam tube.

Power is lost when electrons are intercepted by the circuit and by RF losses within the tube. In addition, some of the RF power that is generated is lost because it is not at the desired frequency (harmonics, intermodulation products, and so forth).

Finally, after the RF output is removed from the tube, the beam leaving the RF circuit (the spent beam) contains power that is delivered to the collector. Some of this power is converted to heat by electrons striking the collector. If the collector is designed properly, the electrons can be slowed down before they strike the collector surfaces, and the collector can thereby recover power from the spent beam [3].

Using the various elements of power identified in Figure 16-4, it is straightforward to derive power flow and efficiency relationships. It is convenient to define four efficiencies that are optimized in maximizing the effectiveness of the collector in recovering power:

- The electronic efficiency, η_e, is the basic efficiency of converting dc beam power into RF power at the desired frequency;
- The circuit efficiency, η_{cir}, is the efficiency of the circuit in delivering the power that is generated by the beam-circuit interaction to the output port of the TWT;
- The collector efficiency, η_{coll}, is the efficiency of the collector in recovering the power remaining in the spent beam;
- The overall efficiency, η_{ov}, is the efficiency of conversion of the total input power to the desired RF output power.

The following plot illustrates the effect of collector efficiency on overall efficiency for a tube with very low losses and a high circuit efficiency.

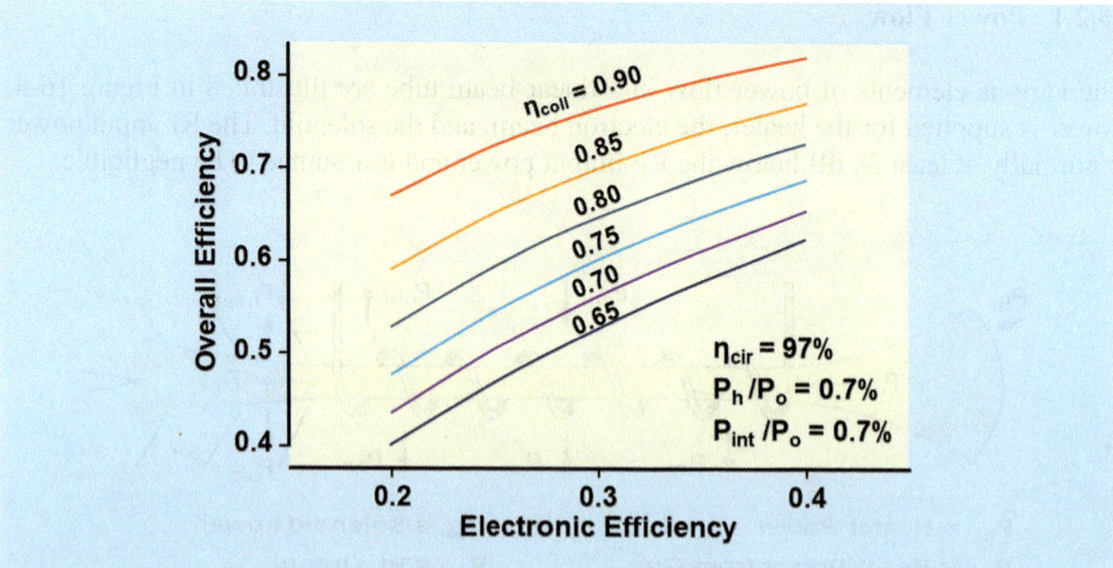

Figure 16-5 Effect of collector efficiency on overall efficiency.

Perhaps the most striking aspect of Figure 16-5 is the strong dependence of overall efficiency on collector efficiency, especially for high values of collector efficiency. This leads directly to the next section, which deals with the design of efficient collectors.

The importance of circuit efficiency is also evident in Figure 16-5. Note that a 10% increase in η_{cir} increases η_{ov} about five percentage points. This leads to considerations of circuit efficiency, which were discussed in Chapter 14.

16.2.2 Power Recovery with a Depressed Collector

A collector for a linear-beam tube is shown in Figure 16-6. As the beam passes through the pole piece, the magnetic focusing field is removed, and so space-charge forces cause the beam to expand as it enters the collector.

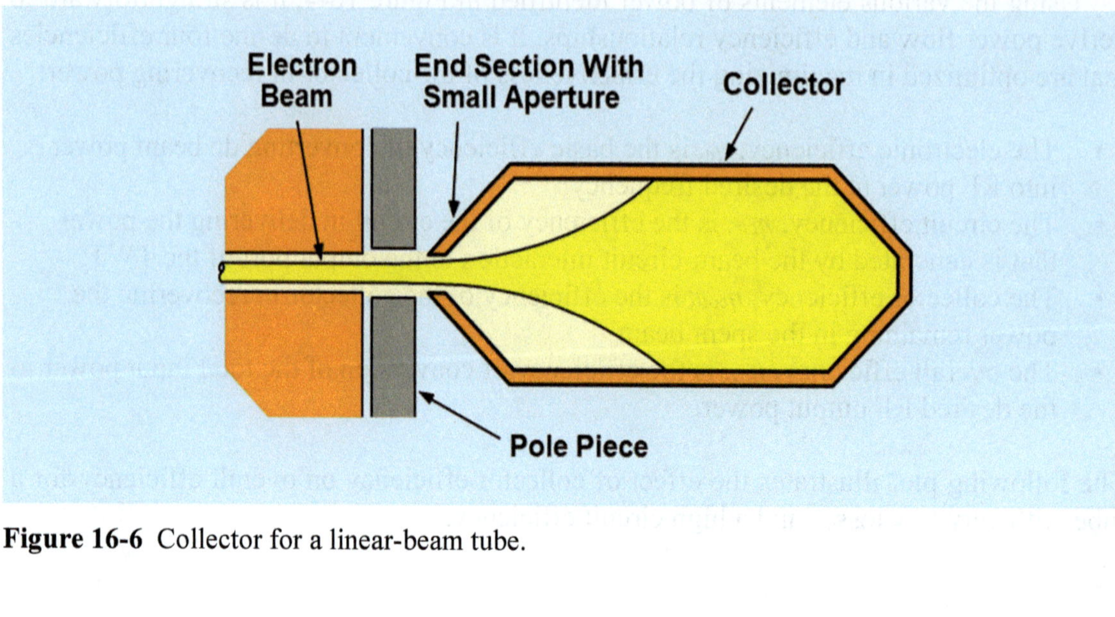

Figure 16-6 Collector for a linear-beam tube.

With the collector at the same potential as the body of the tube, electrons strike the collector at a relatively high velocity. Electron energy is converted to heat at the collector surface. With a collector like that shown in Figure 16-6, a relatively large percentage (normally 50% or more depending on electronic efficiency) of the power input to the tube is dissipated in heat in the collector.

By reducing (depressing) the voltage on the collector below the body potential, the velocities of electrons striking the collector and the heat generated are reduced. As a result, the collector recovers some of the power in the spent electron beam.

The concept of power recovery by a collector is sometimes confusing. The following example should help in understanding power recovery. Assume that a TWT with a 10-kV, 1-A electron gun produces an RF power of 2 kW as is indicated in Figure 16-7. Assume that, other than interception, there are no losses, heater power is negligible, and PPM focusing is used. With the collector at the same potential as the body and the helix (ground potential) as shown in Figure 16-7(a), the overall efficiency is

$$\eta_{ov} = \frac{2\,\text{kW}}{10\,\text{kV} \times 1\text{A}} = 20\% \tag{16-1}$$

All of the current intercepted by the circuit and collected by the collector returns through the cathode power supply. It should also be noted that to regulate the RF output, the full 10 kW of input power must be regulated.

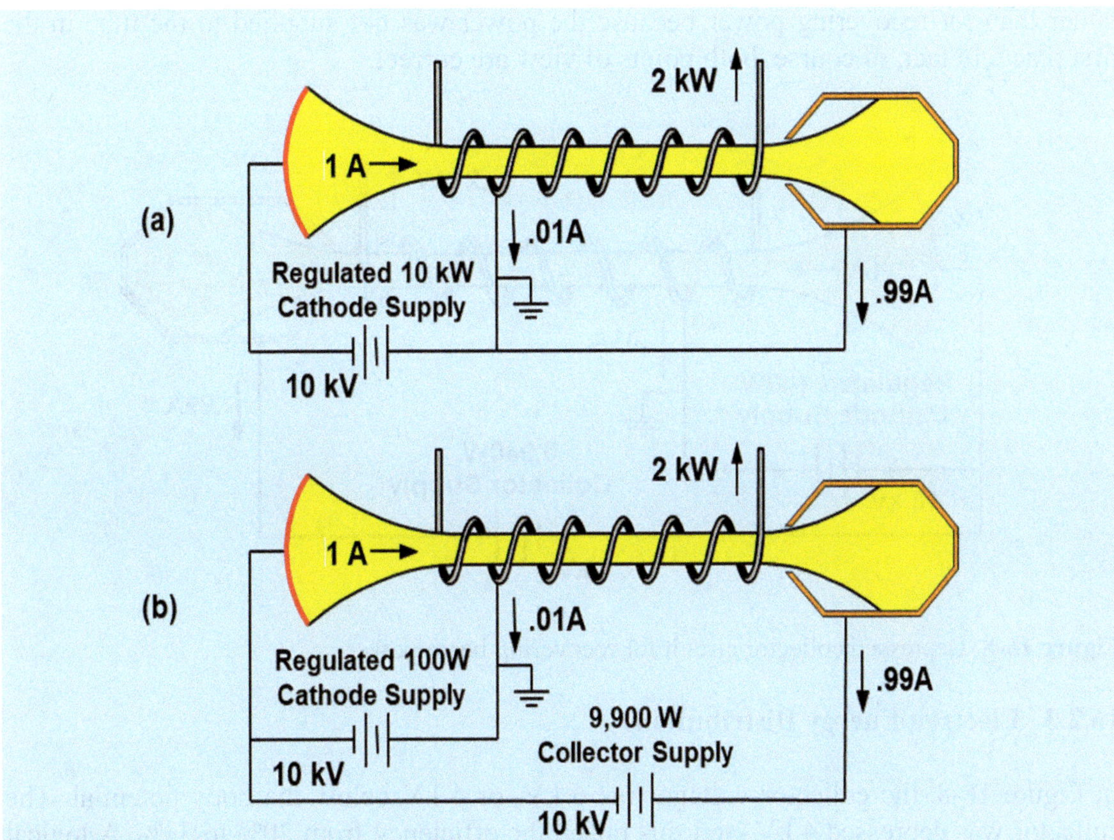

Figure 16-7 Power supply connections for collectors at ground potential: (a) common power supply, and (b) separate power supplies.

In Figure 16-7(b), a second 10-kV power supply has been provided for the collector. The total power supplied to the tube is the same as in Figure 16-7(a) and so the efficiency is still 20%. However, to regulate the RF output power, only the cathode supply must be regulated and the power provided by that supply is now only 100W. The regulation requirement for the collector supply of 9,900W is normally far less severe than for the cathode supply. As a result, while there has been no improvement in efficiency in Figure 16-7(b), a substantial reduction in power supply complexity may have been achieved.

In Figure 16-8, the voltage of the collector supply has been reduced to 6,000 V. The current provided by this supply is still the full collector current of 0.99 A; however, the power provided is only 5,940 W. The RF output power is still 2 kW and so the overall efficiency is now

$$\eta_{ov} = \frac{2{,}000\text{W}}{100\text{W} + 5{,}940\text{W}} = 33\% \tag{16-2}$$

This is a substantial improvement over the 20% obtained in Figure 16-7.

The improvement in efficiency in Figure 16-8 has been achieved by reducing (depressing) the voltage of the collector below the body and helix voltage and, thereby, slowing down the electrons before collecting them. As a result, we say that the collector has recovered power from the electron beam. It should be noted here that the power supply engineer considers collector depression to be a technique for reducing power input rather than for recovering power because the power was not supplied to the tube in the first place. In fact, of course, both points of view are correct.

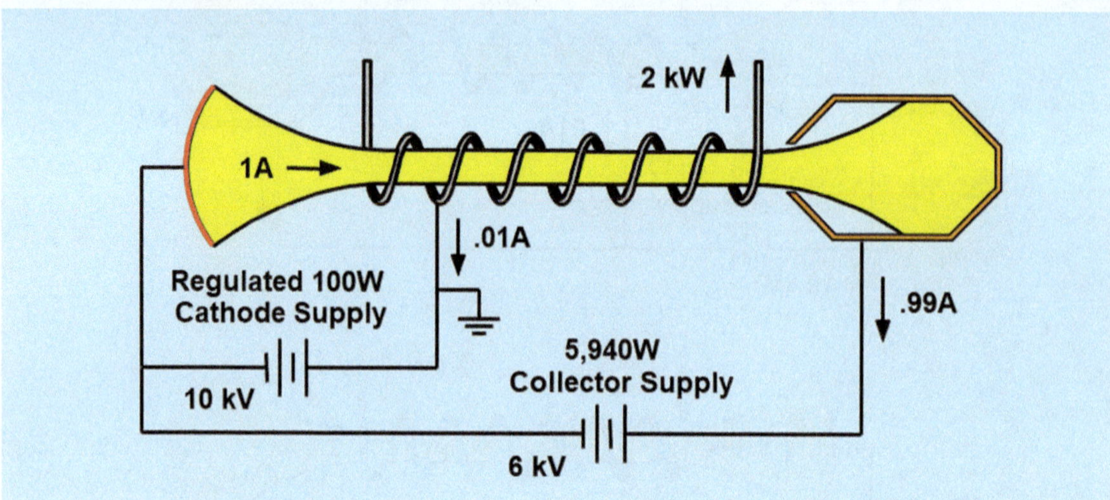

Figure 16-8 Depressed collector circuit for recovering beam power.

16.2.3 Electron Energy Distribution

In Figure 16-8, the collector voltage was 6 kV, or 4 kV below the body potential. The collector was depressed 4 kV, and this raised the efficiency from 20% to 33%. A logical question to ask is: Why not depress the collector further and improve the efficiency more? The problem is that the voltage of a depressed collector supply like that shown in Figure 16-8 is highly dependent on the distribution of electron velocities in the beam. The slower electrons will be reflected by the collector if the potential is depressed too far. With increasing depression of the collector voltage, as shown in Figure 16-9, collector current will decrease and body current will increase. Excessive body current will damage the RF structure. Also, reflected electrons traveling back through the tube induce noise in the RF circuit and can provide a signal feedback path to the TWT input.

Single-element collectors, like that in Figure 16-6, recover roughly 30% to 40% of the power in a spent beam (dependent on tube type). To recover a large fraction (over 80%) and, at the same time, prevent electrons from being reflected, the electric fields in a collector must be shaped so that electrons are nearly stopped just as they strike the collector electrodes. In addition, the fields must be shaped to prevent secondary electrons from being accelerated (thus wasting energy) or leaving the collector.

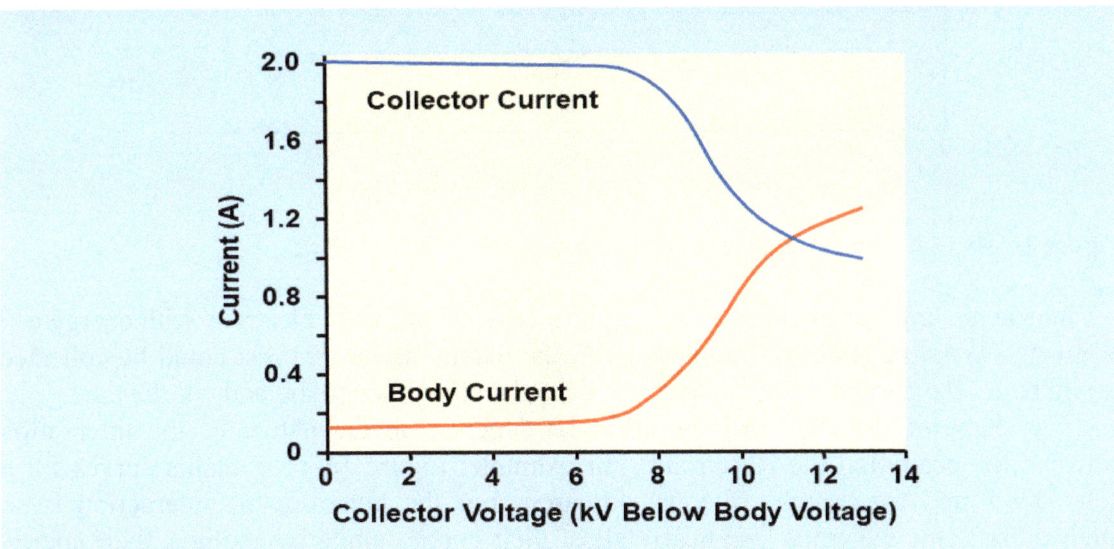

Figure 16-9 Effect of collector voltage on body current.

To better understand the problem of building a highly efficient collector, it is necessary to examine the energy distribution of the electrons in the beam entering the collector. In Figure 16-10, the current to the collector is plotted as a function of the amount by which the collector voltage is depressed, V_{coll}. It is assumed that the body current is zero as long as the collector is not depressed too far. Also, the secondary electron current is ignored. When $V_{coll} = 0$, all of the cathode current reaches the collector and so $I_{coll} = I_o$. With no RF drive, no energy is extracted from the beam and so all of the electrons entering the collector have the same energy that they had when leaving the electron gun. As a result, as the collector voltage is depressed, all of the electrons can reach the collector until $V_{coll} = V_o$. As V_{coll} becomes larger than V_o, suddenly none of the electrons can reach the collector and so I_{coll} drops to zero.

When the RF drive is applied to the tube, energy in varying amounts is extracted from (or in some cases, added to) electrons during the interaction of the RF wave on the circuit with the beam. In a helix TWT, some electrons lose as much as 30% to 40% of their initial energy (when they left the gun) and some may gain as much as 30%.

With the RF drive, as the collector voltage is depressed from the body potential, all of the cathode current reaches the collector until the collector voltage reaches the point where the slowest electrons can no longer reach the collector surface. This voltage is indicated by V_{knee} in Figure 16-10. As the collector voltage is depressed below V_{knee}, more and more electrons are reflected and so I_{coll} decreases. At some voltage V_{max}, which is actually more negative that the cathode potential, V_o, the collector current drops to zero. The small tail of the curve at voltages between V_o and V_{max} is caused by those electrons that were accelerated rather than decelerated as the beam interacted with the RF wave.

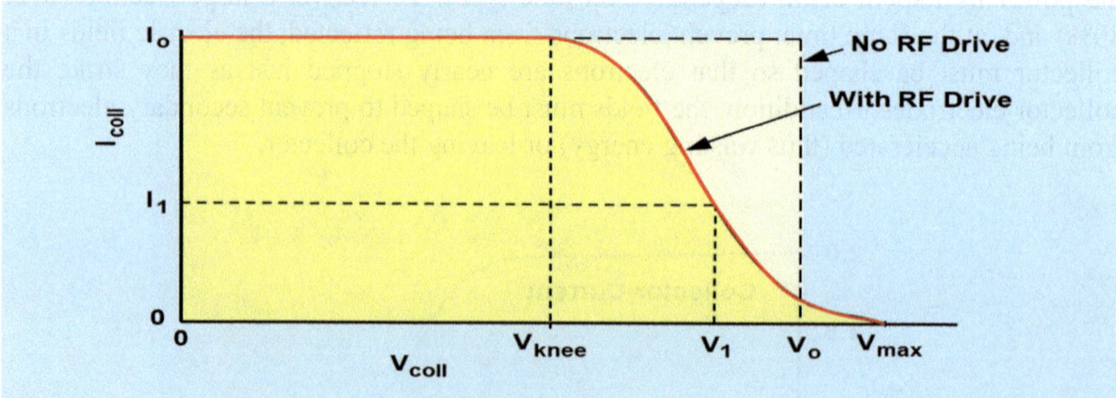

Figure 16-10 Collector current versus collector voltage.

In general, any current, such as I_1 in Figure 16-10, consists of electrons with energy eV_1 or greater. With the collector at potential V_1, the maximum current that could be collected would be I_1. The current, $I_o - I_1$, would be reflected and returned to the body of the tube.

The shape of the curve in Figure 16-10 depends on the nature of the interaction between the beam and the RF circuit. For example, Figure 16-11 contains curves for a helix TWT and for a high-efficiency klystron. For the klystron, the interaction is so intense that some electrons lose nearly all of their energy while others have their energy nearly doubled so a collector that was depressed even a small amount would reflect some current.

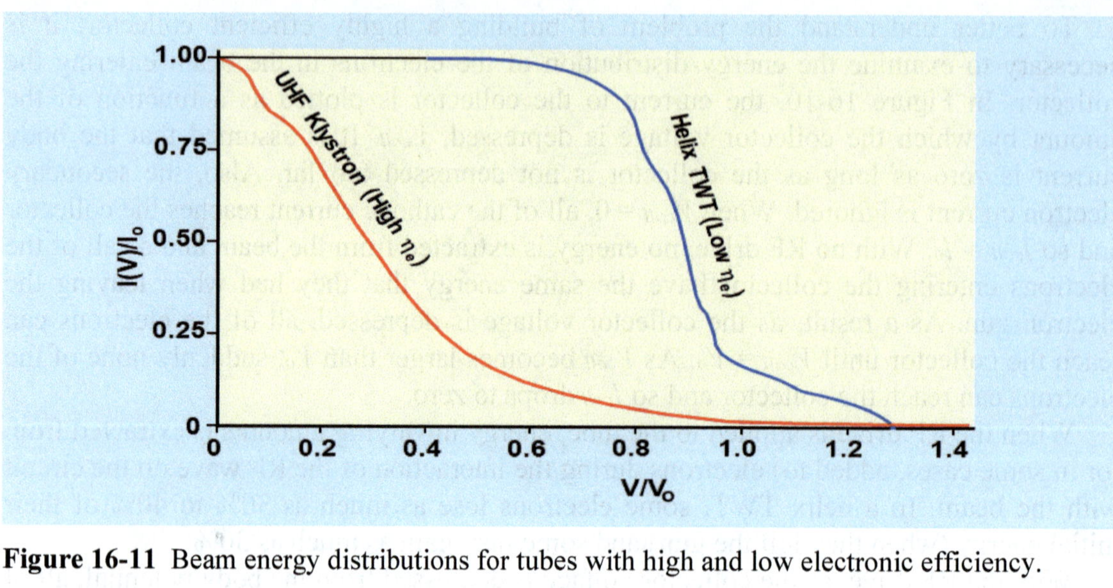

Figure 16-11 Beam energy distributions for tubes with high and low electronic efficiency.

16.2.4 Spent Beam Power

In Figure 16-12, current is plotted versus voltage and so the area under a curve represents power. For example, the area under the No RF Drive plot is simply $I_o \times V_o$, which is the dc beam power. When the RF drive is applied, as shown by the shaded area in Figure 16-12, the difference between the no-drive rectangle and the curve with drive is nearly the beam power that is converted to RF power. Some power is lost to the V_{max}–V_o tail and there are RF losses, so the RF power out of the tube is somewhat smaller than is shown by the shaded area.

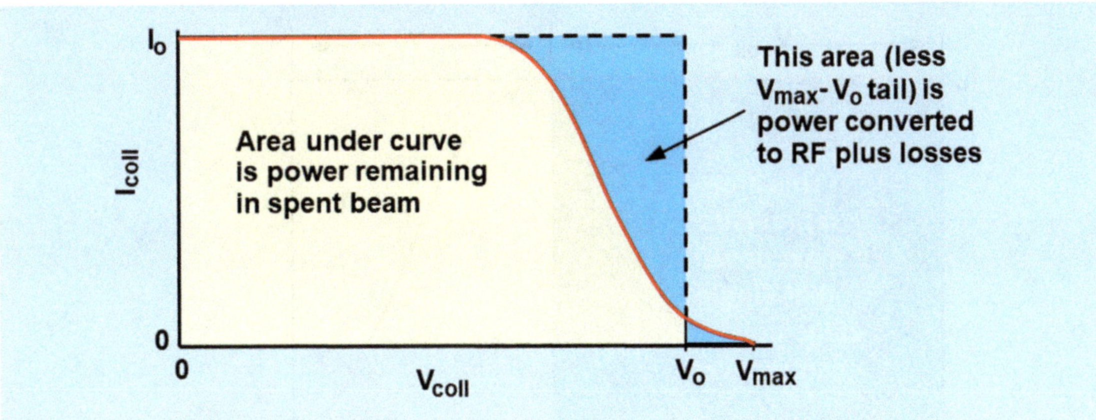

Figure 16-12 Spent beam power.

In Figure 16-12, the area under the solid curve is the power remaining in the spent beam. This is the maximum power that could possibly be recovered by an ideal collector. To recover part of that power, the power supply configuration must be as shown in Figure 16-13 (and as was shown in Figure 16-7). The collector supply voltage, V_c, is the cathode voltage, V_o, less the amount by which the collector voltage is depressed, V_{coll}. The maximum power that could be recovered by a single-electrode collector is shown by the shaded area in Figure 16-14. (The collector with a single electrode is often referred to as a single-stage collector. Collectors with several electrodes, to be discussed in following sections, are called multistage collectors.)

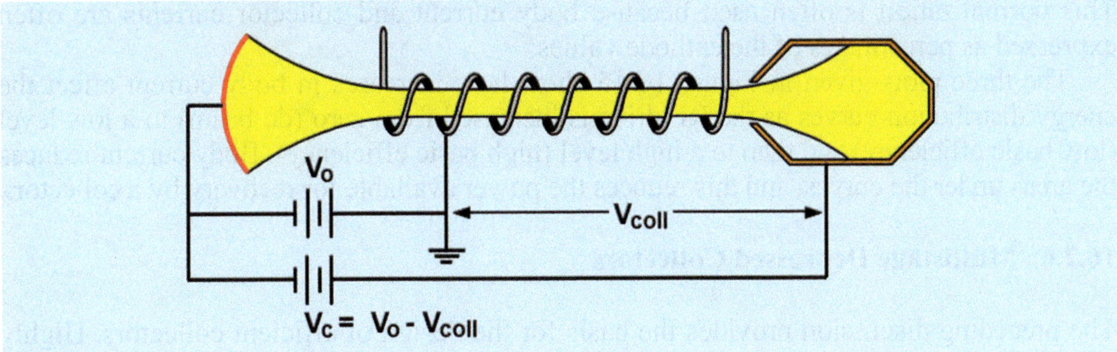

Figure 16-13 Power supply configuration for single-stage depressed collector operation.

When V_c is smaller than what is shown in Figure 16-14, electrons are reflected by the collector because the amount by which the collector is depressed, V_{coll}, is greater than the knee voltage. When V_c is greater than what is shown in Figure 16-14, all electrons are collected. In practice, it is not possible to operate a single-stage collector as close to the knee of the curve as is indicated in Figure 16-14. One reason is that V_c is not normally highly regulated and so V_c must be high enough to avoid electron reflections under all operating conditions. Another reason is that the knee of the curve is normally not as sharply defined as is shown in Figures 16-11, 16-12, and 16-14 and so excessive electron reflection (and body current) occurs with collector voltages well above that indicated in Figure 16-14.

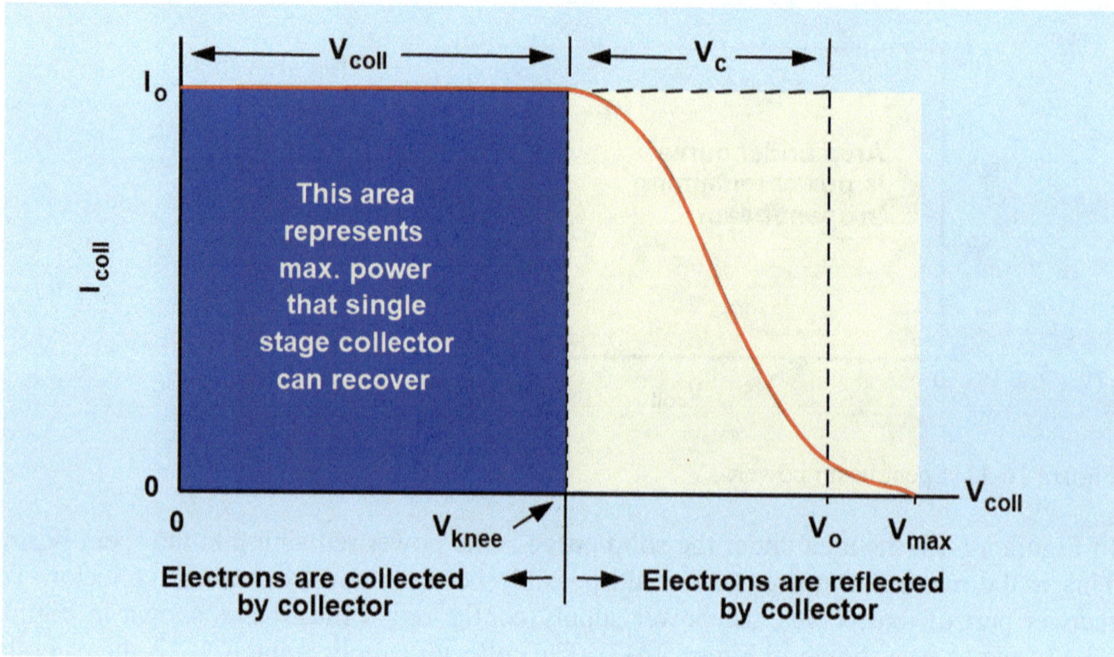

Figure 16-14 Maximum power recovered by a single-stage depressed collector.

16.2.5 Effect of Body Current

The effect of body current on the energy distribution curves is shown in Figure 16-15 [4]. Here, currents and voltages have been normalized with respect to the cathode values. This normalization is often used because body current and collector currents are often expressed as percentages of the cathode values.

The three plots given in Figure 16-15 show how increases in body current affect the energy distribution curves as the RF drive is increased from zero (dc beam) to a low level (low basic efficiency) and then to a high level (high basic efficiency). Body current reduces the areas under the curves, and this reduces the power available for recovery by a collector.

16.2.6 Multistage Depressed Collectors

The preceding discussion provides the basis for the design of efficient collectors. Highly efficient collectors use several electrodes at different potentials to selectively collect electrons at low energies. These collectors are multistage depressed collectors (MDCs).

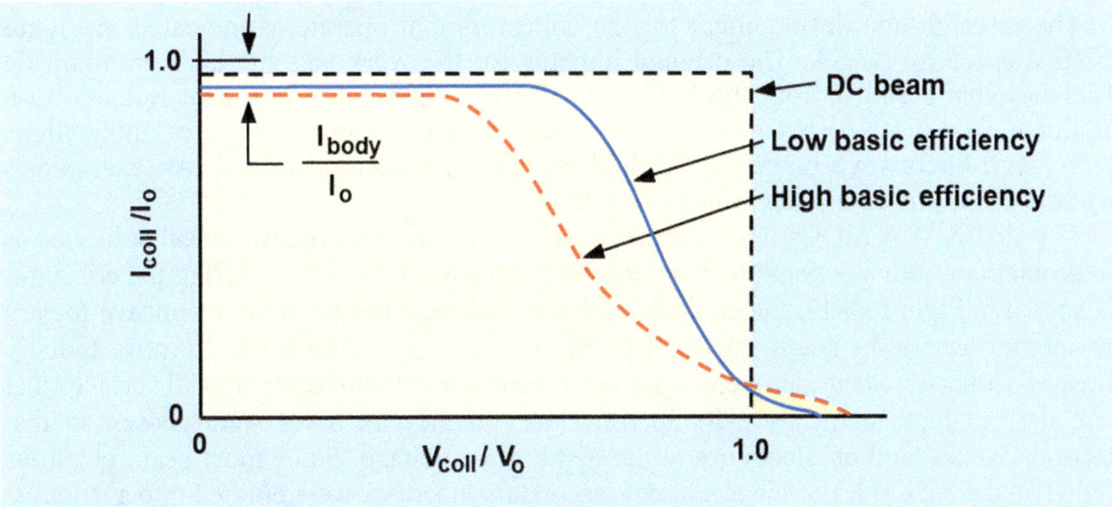

Figure 16-15 Effect of body current on energy distribution plots.

In the previous section, energy distribution curves were shown for the electron beam entering the collector. It was pointed out that the area under a curve represents the power available in the spent beam for recovery by the collector. To recover a large fraction of this power, the collector must be designed to sort the electrons in the spent beam into various energy classes. Then electrons in each energy class must be collected on an electrode at a voltage that recovers as much of that energy as possible. Figure 16-16 illustrates the recovery of a substantial amount of spent beam power by a four-stage collector. Here, as in previous discussions, the effects of secondary electrons are not yet being considered.

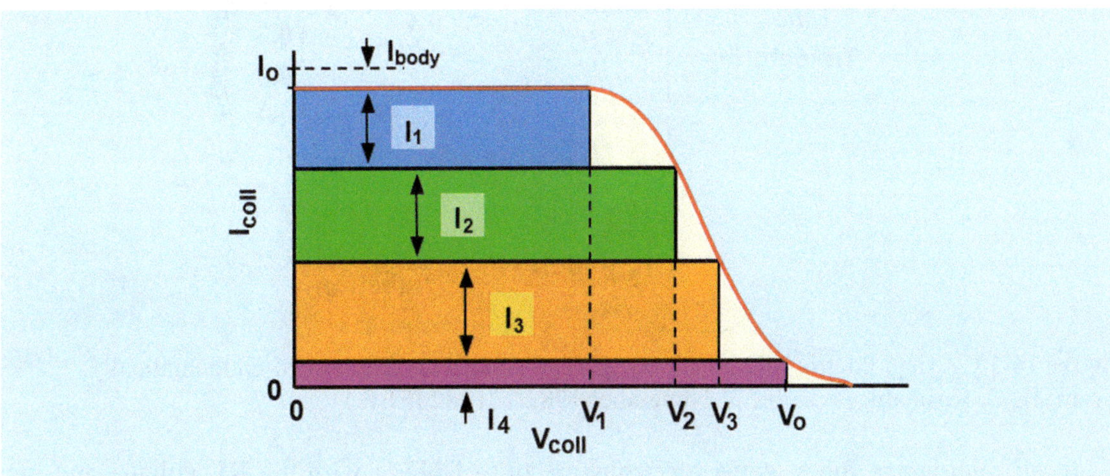

Figure 16-16 Power recovered by a four-stage depressed collector.

A collector electrode depressed to the cathode potential, V_o, can collect that current, I_4, consisting of electrons with energies in the range eV_{max}–eV_o. An electrode depressed to potential, V_3, can collect current, I_3, but not I_4 because that current goes on to the electrode at V_o. The total power recovered by the four-stage collector is represented by the shaded area in Figure 16-16 and is

$$P_{rec} = I_1 V_1 + I_2 V_2 + I_3 V_3 + I_4 V_o \quad (16\text{-}3)$$

The research and development for the collectors that operate as indicated in Figure 16-16 was led by NASA. The original impetus for the work was the high premium on efficiency that resulted from the high cost of power generation and heat reflection on satellites. Now the NASA collector technology is used in a broad range of applications where high microwave power is required and prime power is limited. An example is airborne electronic countermeasures systems [5].

The first NASA MDCs were derived by Kosmahl [5] from mathematical solutions to the problem of both slowing down and dispersing the electron beam entering the collector. As shown in Figure 16-17, the equipotential surfaces were designed to be concave toward the aperture where the beam entered the collector. This forced electrons to move radially outward. Collector electrodes were designed to coincide with the equipotential surfaces that were derived mathematically. The aperture sizes in the electrodes were chosen so that electrons would land on electrodes at the appropriate voltage. Since most beam electrons landed on the back sides of the electrodes, secondary electrons were emitted into a negative potential region and, therefore, were immediately returned to the electrode surfaces.

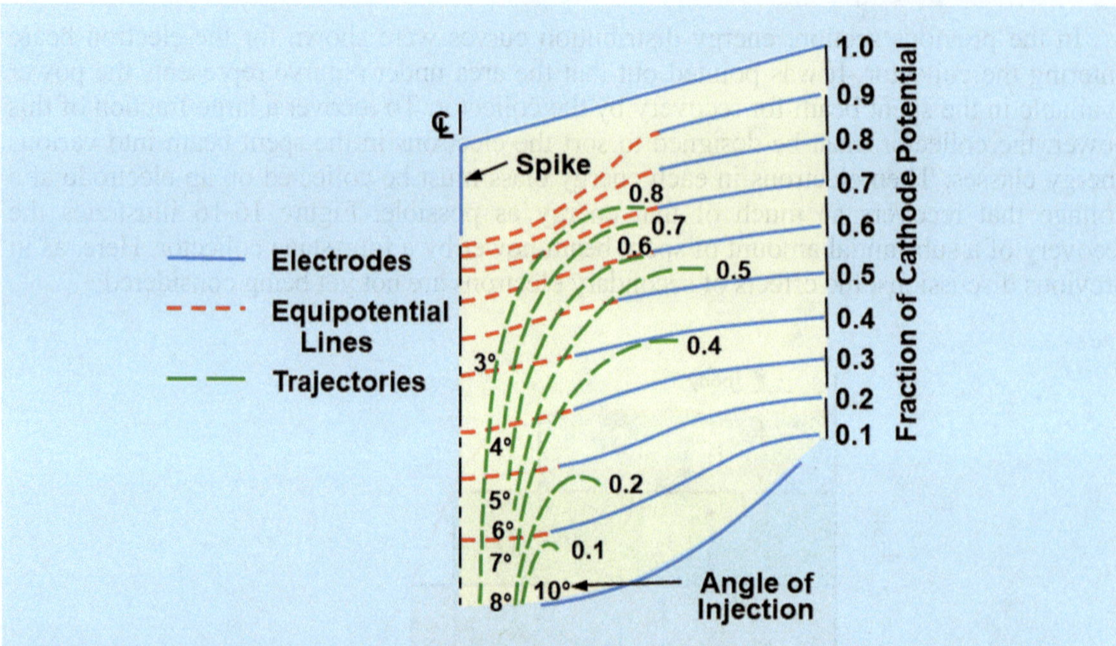

Figure 16-17 Axisymmetric MDC with equipotential surfaces determined mathematically. (From: H. G. Kosmahl, *Proc. IEEE*, November 1982. ©1982 IEEE.)

In all microwave tubes, some electrons are out of phase with the RF voltage and are accelerated, rather than decelerated, by the interaction process. These electrons leave the interaction region with energies greater than the original beam energy. In the NASA collector in Figure 16-17, the sharp spike at the back of the collector helps shape the fields to disperse high-energy electrons. In addition, the surface area presented to high-energy electrons is minimized. Thus, the generation of reflected electrons is minimized.

For an MDC to operate efficiently, the electron beam entering it must not diverge rapidly. Rapid divergence leads to large radial components of electron velocities and to energy components that cannot be recovered by the collector.

To reduce the divergence of a beam, it is sometimes necessary to expand its diameter so that space charge forces are reduced. A magnetic refocusing system, one version of which is shown in Figure 16-18, is used for this purpose. An additional benefit of the refocusing system is an improvement in the laminarity of the beam entering the collector.

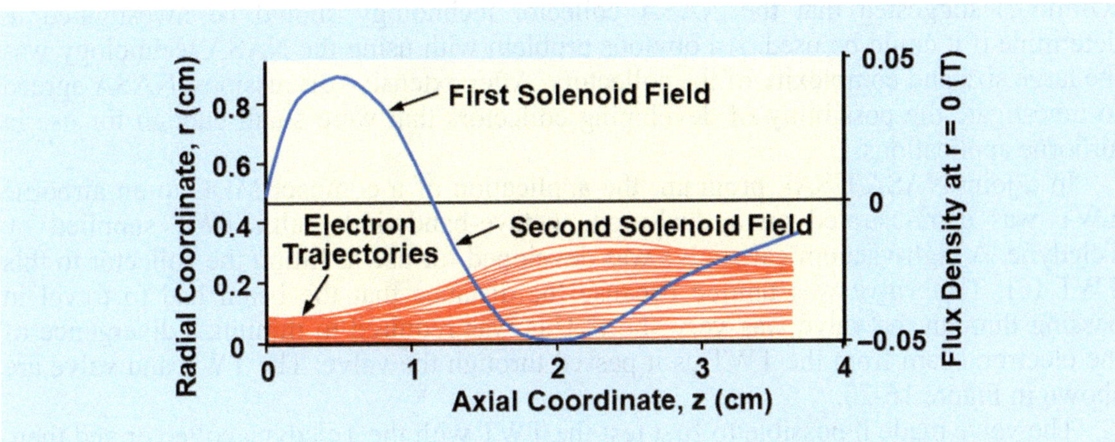

Figure 16-18 Magnetic refocusing system for reducing beam divergence and improving laminarity. (Adapted from: H. G. Kosmahl, *Proc. IEEE*, November 1982. ©1982 IEEE.)

The efficiencies of collectors patterned after the design in Figure 16-17 are very high (> 85%) and the current reflected from the collector is very small. Complexity and size are disadvantages. Figure 16-19 shows a 10 stage collector that was one of the first MDCs developed for use on a satellite. The spike in the back of the collector and the refocusing section between the collector and RF circuit are clearly shown.

Figure 16-19 One of the first multistage depressed collectors developed for use on a satellite. (Courtesy of the NASA Lewis Research Center.)

16.2.7 Individual Lens Collectors

About the time that the original NASA MDCs were being developed, there was an urgent need to increase the efficiency of TWTs used on airborne ECM systems. The author (Gilmour) suggested that the NASA collector technology should be investigated to determine if it could be used. An obvious problem with using the NASA technology was the large size and complexity of the collectors. After extensive discussions, NASA agreed to investigate the possibility of developing collectors that were small enough for use in airborne applications.

In a joint NASA/USAF program, the application of a compact MDC to an airborne TWT was demonstrated on a dual-mode octave-bandwidth helix TWT supplied by Teledyne. A high-vacuum gate valve was developed for use in fitting the collector to this TWT [6]. This valve was unique because the distance that the beam had to travel in passing through the valve was very short. This was required to minimize divergence of the electron beam from the TWT as it passed through the valve. The TWT and valve are shown in Figure 16-20.

The valve made it possible to first test the TWT with the Teledyne collector and then, with the valve closed to maintain high vacuum in the TWT, remove the Teledyne collector and attach the NASA collector. After evacuating the collector, the valve was opened and tests with the MDC could be performed.

Improvements in overall efficiency that were demonstrated are shown in Figure 16-21. An overall efficiency improvement by a factor of three or more was demonstrated across a bandwidth of an octave. The collector efficiency was nearly constant at about 85%, as shown in Figure 16-22.

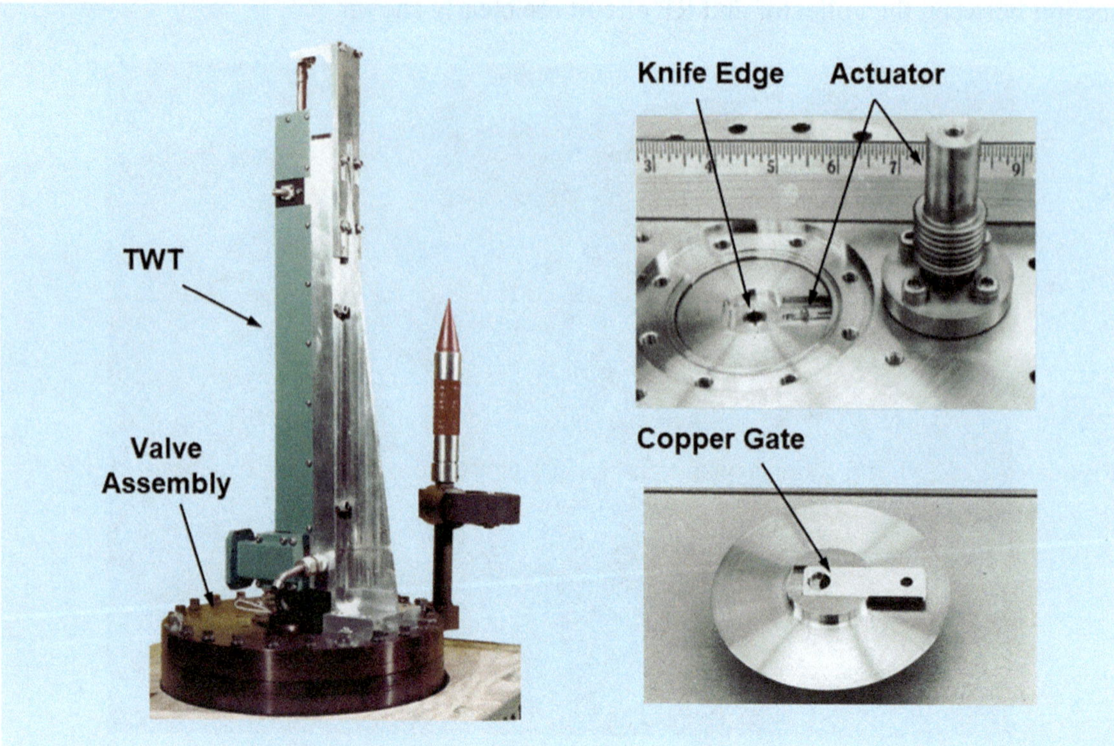

Figure 16-20 Teledyne TWT and Gilmour gate valve.

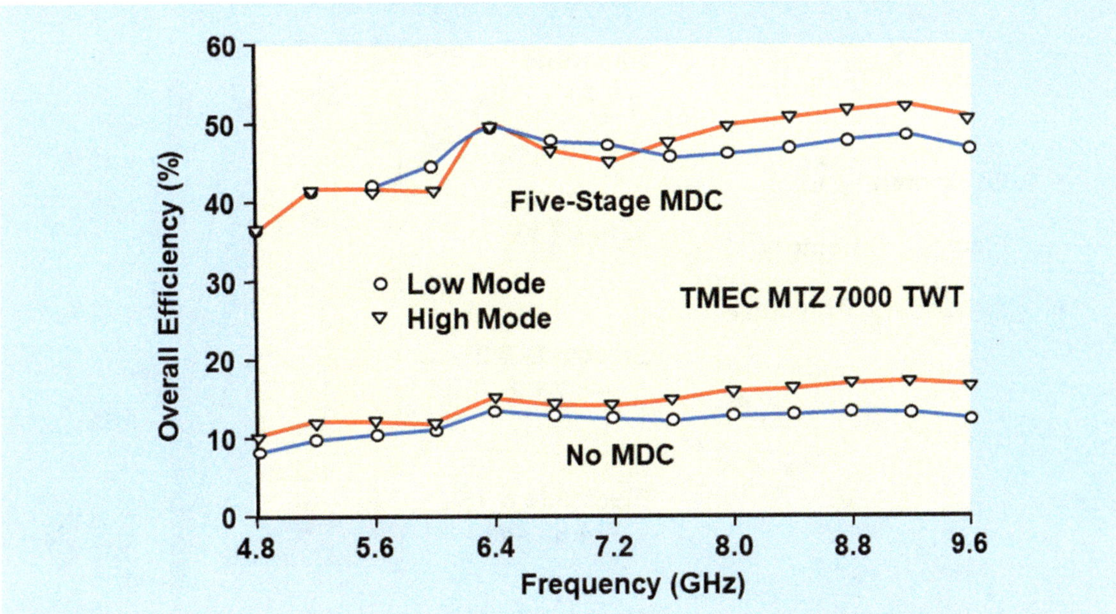

Figure 16-21 Overall efficiency of a dual-mode ECM helix TWT with and without an MDC. (Adapted from: H. G. Kosmahl, *Proc. IEEE*, November 1982. ©1982 IEEE.)

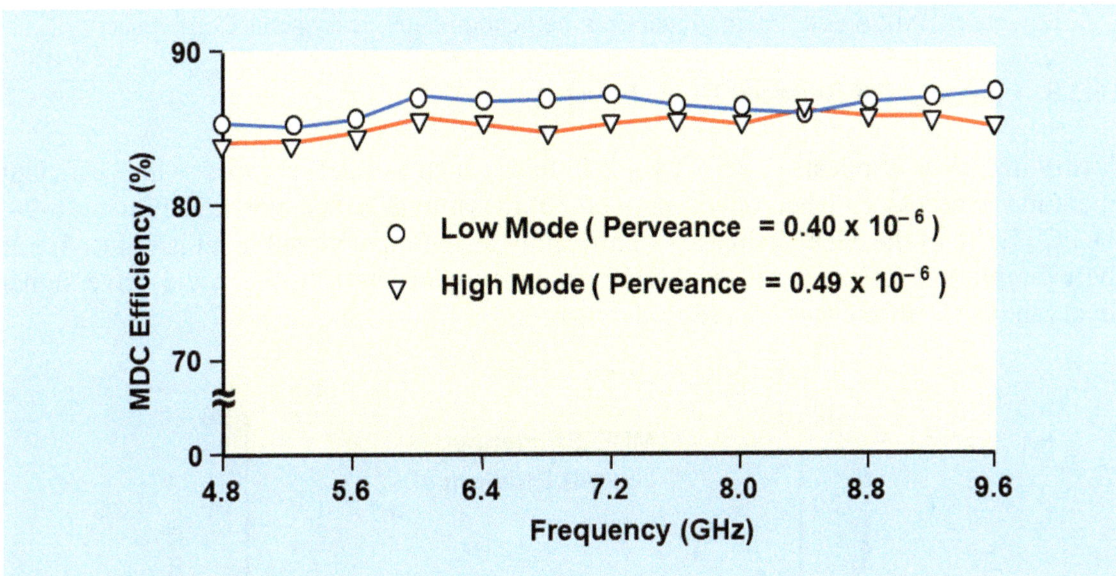

Figure 16-22 Collector efficiency at saturation. (Adapted from: H. G. Kosmahl, *Proc. IEEE*, November 1982. ©1982 IEEE.)

The new compact collectors, one of which is shown in Figure 16-23, were called individual lens collectors by NASA. In these collectors, each electrode and its associated fields were optimized to produce maximum collector efficiency while minimizing the current resulting from reflected electrons. These collectors are now widely used on helix and coupled-cavity TWTs.

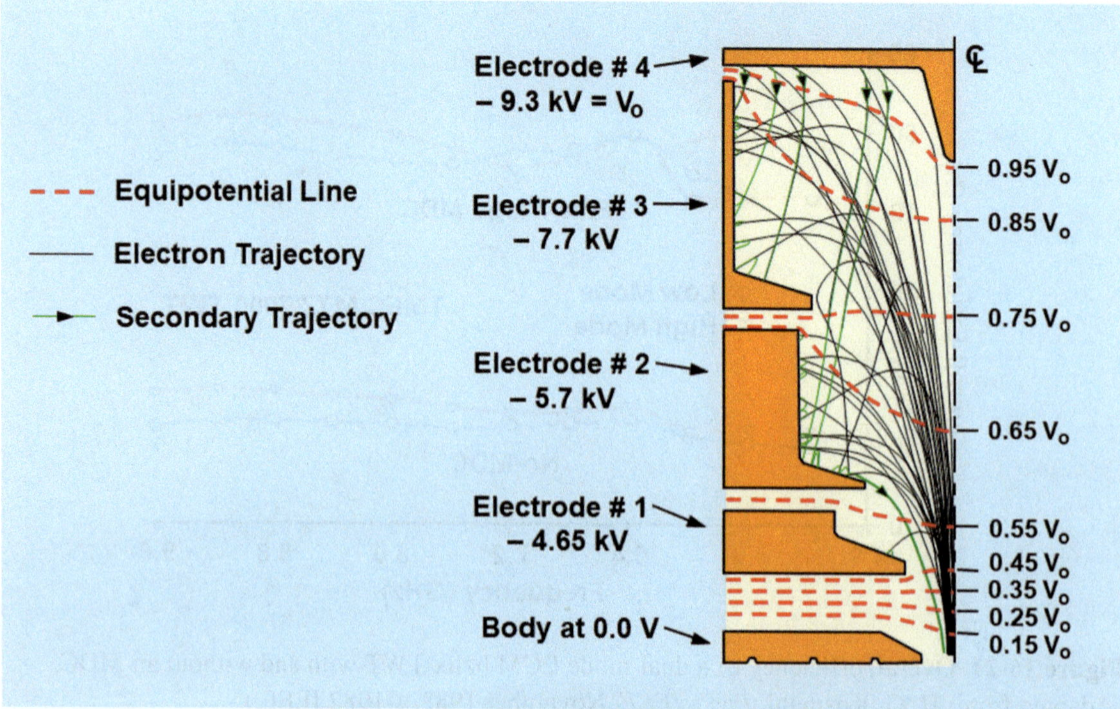

Figure 16-23 Equipotential surfaces and electron trajectories in individual lens collector. (From: J. A. Dayton, Jr., AIAA 12th International Communication Satellite Systems Conference.)

16.2.8 Operation at Reduced Drive Levels

A very important application for MDCs is in tubes, such as satellite tubes, where efficient operation is necessary when operating well below saturation. As shown in Figure 16-24, the efficiency of the MDC increases significantly as the input signal level is reduced. For the example in Figure 16-24, the MDC efficiency exceeds 97% when the drive signal level (and basic efficiency) are reduced to zero.

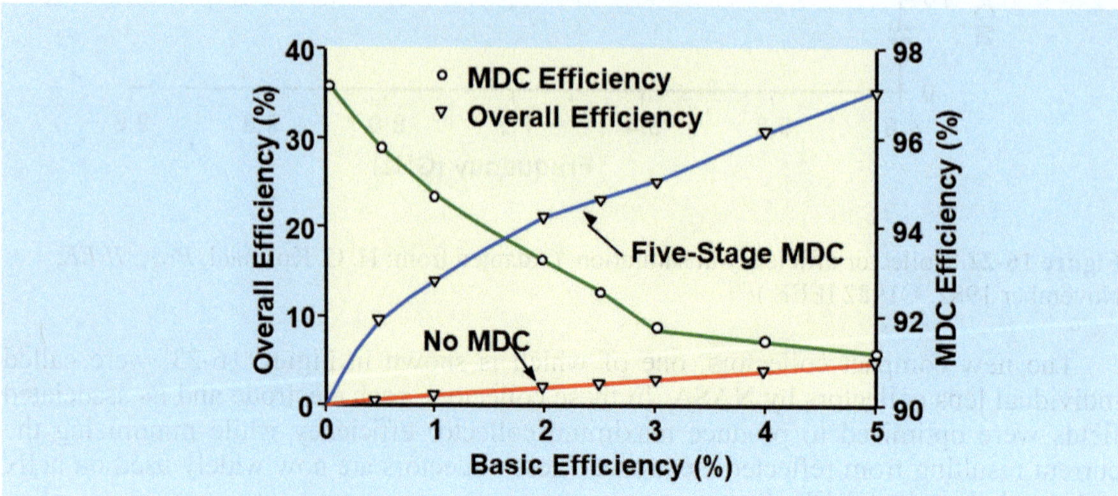

Figure 16-24 MDC efficiency in the small-signal region. (Adapted from: H. G. Kosmahl, *Proc. IEEE*, November 1982. ©1982 IEEE.)

16.2.9 Efficiency vs. Number of Depressed Stages

In Figure 16-17, the number of stages in the MDC was 10 and, in Figure 16-23, it was four. The optimum number is determined by considering the increase in tube efficiency that can be achieved with a large number of electrodes and trading this off against the increased complexity of the power supply and of tube construction.

As Figure 16-25 shows, the increase in efficiency achieved with more than about three or four stages is small and thus, for most applications, no more than three or four stages are used. The efficiencies in Figure 16-25 were computed and are based on a knowledge of the spent beam characteristics of a helix TWT.

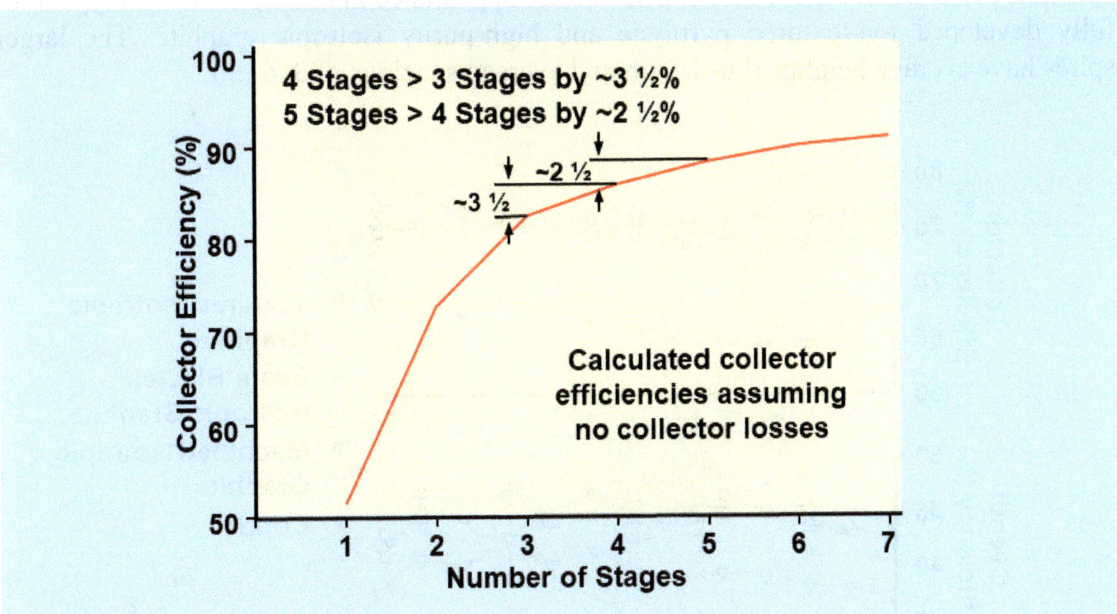

Figure 16-25 Dependence of collector efficiency on the number of collector stages.

16.2.10 Secondary Electrons in Depressed Collectors

Excessive secondary electron emission in depressed collectors can cause two problems:

1. Escape of secondaries from the collector, which may cause excessive noise, signal distortions, and heating of the circuit;
2. Current flow between collector electrodes and excessive heating of electrodes, which, in turn, results in the reduction of collector efficiency.

The second problem arises when secondary electrons released from the front surface of a collector electrode are collected by a more positive electrode. An example was shown in Figure 16-23 where trajectories for five secondary electrons released from electrode #4 are plotted. Three of these electrons strike electrode #3 after being accelerated through 1.6 kV (9.3–7.7 kV). The other two strike electrode #2 after being accelerated through 3.6 kV.

The effect of secondary electrons on efficiency is shown in Figure 16-26 [7]. Tests were performed with a four-stage MDC having various electrode surface treatments. Using electrodes with textured isotropic graphite surfaces rather than untreated copper, collector efficiency increased about 10%. The overall efficiency increased over 5%.

The technique of texturing materials for collector surfaces has been developed by the NASA Lewis Research Center [8]. The texturing is accomplished by bombarding the surface with argon ions. For the textured graphite surfaces described in [8], the ion accelerating potential was 1,000–1,500V dc, the surface current density of the sample was 3–5 mA/cm^2, and the texturing period was 1–4 hours. A scanning electron microscope photomicrograph of a fully developed textured carbon surface on a copper substrate is shown in Figure 16-27. The surface appearance in this figure is also typical of fully developed ion-textured pyrolytic and high-purity isotropic graphite. The larger spires have average heights of 6–12 μm and average spacings of 3–6 μm.

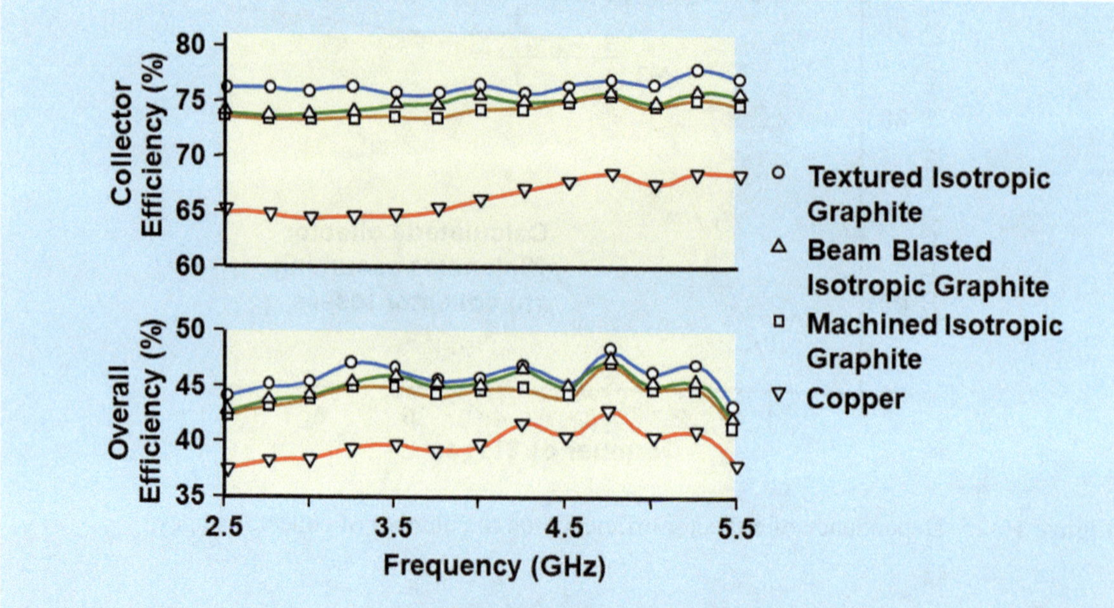

Figure 16-26 Efficiencies for various surfaces in a four-stage MDC on a TWT with η_e = 16%. (From: P. Ramins and B. T. Ebihara, *IEEE Trans. Electron Devices*, November 1986. ©1986 IEEE.)

The secondary emission characteristics of textured carbon on a copper substrate are shown in Figure 16-28. The true secondary emission coefficient for normal incidence of the primary beam is from 0.2 to about 0.25. This is significantly below the value for soot, which has, in the past, been the lowest yield surface available. Soot has not been used in microwave tubes because of its poor physical characteristics. The reflected primary yield index, π, for normal incidence ranges from about 0.5 to about 1.5. The basis for this index is the primary yield for soot with a primary electron energy of 1 keV and normal incidence. For these conditions, the value of π is unity.

In addition to texturing various carbon surfaces, NASA has developed techniques for texturing copper [9]. With one technique, sputtered molybdenum is used as a texture-inducing masking film. The true secondary electron emission ratios for these surfaces range from 0.3 to 0.5 and are a factor of 3 below ratios for untreated copper surfaces. The reflected primary yield index is a factor of about 3 to 7 below that for an untreated copper surface. The results reported in [9] were for three samples of ion-textured copper. One conclusion of the NASA work is "with further development, the molybdenum-masked, ion-textured OFHC copper surface will be a promising material for high efficiency MDC electrodes."

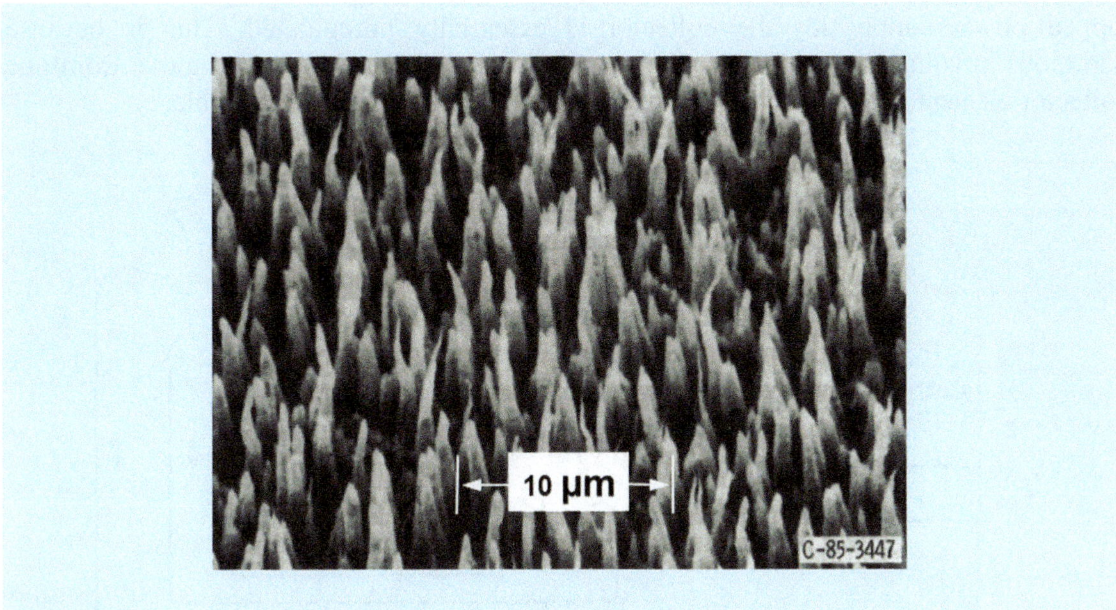

Figure 16-27 SEM photomicrograph of textured carbon on a copper substrate. (From: A. N. Curren, *IEEE Trans. Electron Devices*, November 1986. ©1986 IEEE.)

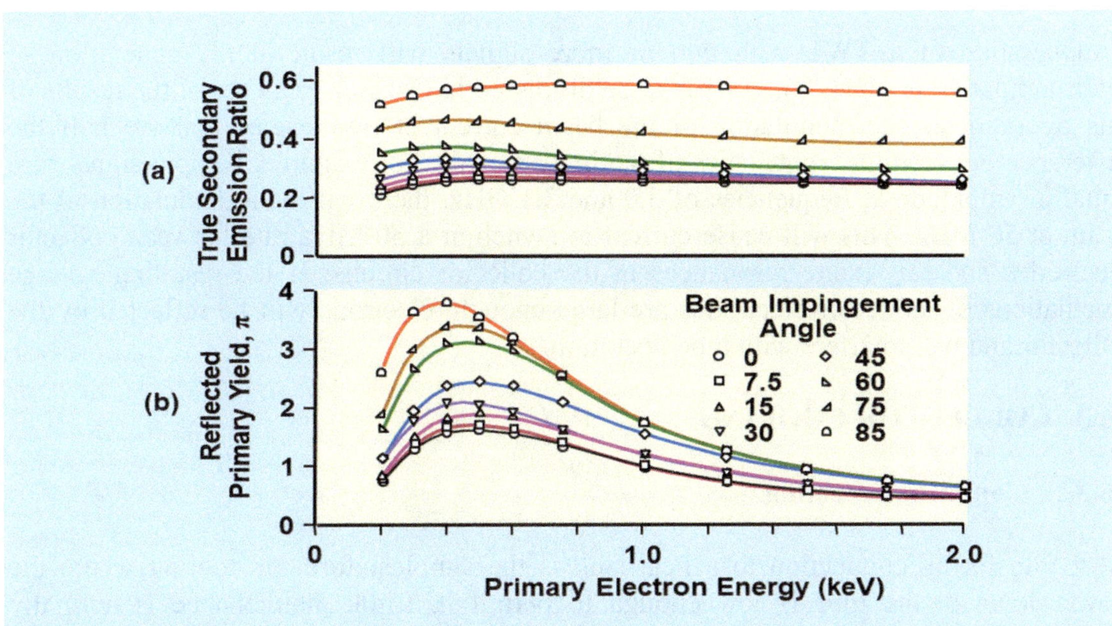

Figure 16-28 Secondary emission characteristics of textured carbon on a copper substrate. (From: A. N. Curren, *IEEE Trans. Electron Devices*, November 1986. ©1986 IEEE.)

16.2.11 Power Supply Considerations

A power supply configuration for a MDC is shown in Figure 16-29. As was the case for the single-element depressed collector, the cathode-to-anode (and helix) supply must be regulated, but it must supply only the power resulting from the interception current. The tapped power supply for the collector is essentially unregulated. This is because variations in voltages to collector elements may cause some electrons to move from one collector element to another, but the overall effect on efficiency is negligible.

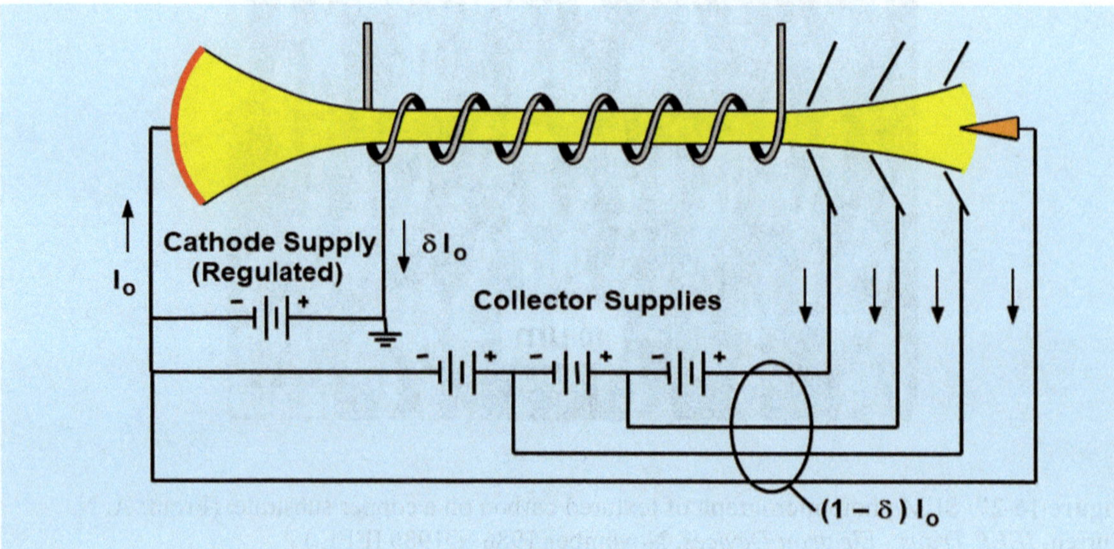

Figure 16-29 Power supply connection with a multistage depressed collector.

Operation of a TWT with two or more signals will result in the generation of intermodulation products (intermods to be discussed in Chapter 17). One of the results of this operation is the modulation of the beam current at frequencies that are half the difference between the frequencies of the input signals. For example, if input signals are equal in amplitude at frequencies of 3.0 and 3.1 GHz, there will be a modulation of the beam at 50 MHz. This will cause current to switch at a 50 MHz rate between collector electrodes and can excite resonances in the collector circuits. If the resulting voltage fluctuations of collector electrodes are large enough, electrons will be reflected by the collector and will interfere with tube operation.

16.3 COLLECTOR COOLING

16.3.1 Conduction Cooling

Heat transfer by conduction to a heat sink is the simplest form of cooling where the power level of the tube is low enough to permit it. Little maintenance is normally required. However, it is necessary to make sure that tubes are tightly bolted to the heat sink and that no gaps exist between surfaces.

16.3.2 Convection Cooling

Devices that generate up to a few watts of average power may require no cooling other than natural air convection. To take the best advantage of this type of cooling, the surface exposed to the air is often many times larger than the actual inside area of the collector. About 1 watt/cm^2 of surface exposed to open air can be handled by this approach.

16.3.3 Forced-Air Cooling

Although forced-air cooling is capable of dissipating power densities nearly as high as water cooling can, its use is limited because of practical considerations. A typical air-cooled collector, which has been used to dissipate a power density of 200 watts/cm^2, is shown in Figure 16-30. The cooling fins are brazed to the body of the collector to provide a large area multiplication factor (larger than 10) from the power-intercepting area to the air-exposed surface. One of the limitations of air cooling is blower power. To minimize the input to the air pump, it is necessary to have a large number of short air passages in parallel. An optimum cooling fin arrangement, shown in Figure 16-30, requires a blower power of from 5% to 10% of the average collector power. This is a much higher level than for water cooling. Another limitation of air cooling is that it is not easily applicable to other portions of the tube such as the interaction structure. Because of these disadvantages, forced-air cooling is generally used only for tubes up to the few kilowatts level. At higher-power levels the circulating water system is preferred.

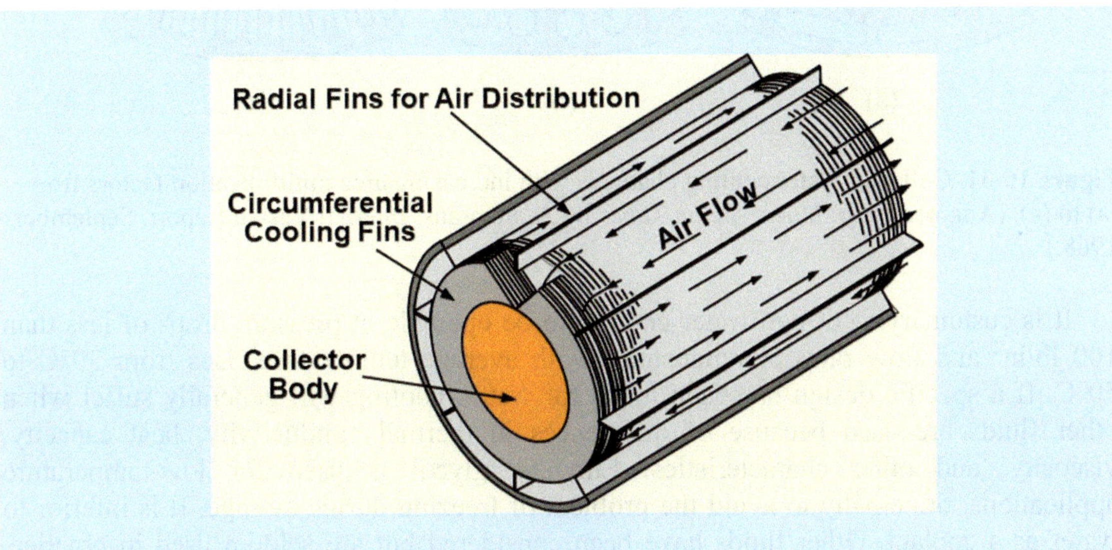

Figure 16-30 Flow patterns in an air-cooled collector. (Adapted from: *Linear-Beam Tubes* by A. Staprans, Internal Varian Report, September 1968.)

16.3.4 Forced-Flow Liquid Cooling

Forced-flow liquid cooling is by far the most often used method for cooling collectors for medium- and high-power tubes. An effective method is simply to place a water jacket around the collector and remove heat by conduction from the collector body to the water. The amount of heat that can be transferred to the water is directly proportional to the wetted surface area and the heat transfer coefficient. Both are optimized when many narrow channels with a large perimeter, but small cross-sectional areas, are used together with high velocity flow.

To avoid excessively high temperature rises in the collector walls, water channels must be located near the intercepting surface. As a result, the wetted surface area is generally limited to a factor of about three or four times the interior collector area. Higher factors could be achieved, but this requires machining of channels of a more complicated cross-section than is convenient with ordinary techniques. Some collector wall sections showing water channel shapes are shown in Figure 16-31.

The heat transfer coefficient used is usually computed on the basis of allowing no boiling of water at the surface. The heat transfer coefficient will characteristically be on the order of 1 cal/cm^2/sec/°C temperature rise in the surface film of water. Much higher heat transfer coefficients can be achieved if the water is allowed to boil, as is frequently the case with high-power collectors. Care must be taken to insure that operation occurs in the nucleate boiling region, which has a high heat transfer capability and does not enter into the film boiling region, for which the surface temperature must be very high and might cause melting of the collector material.

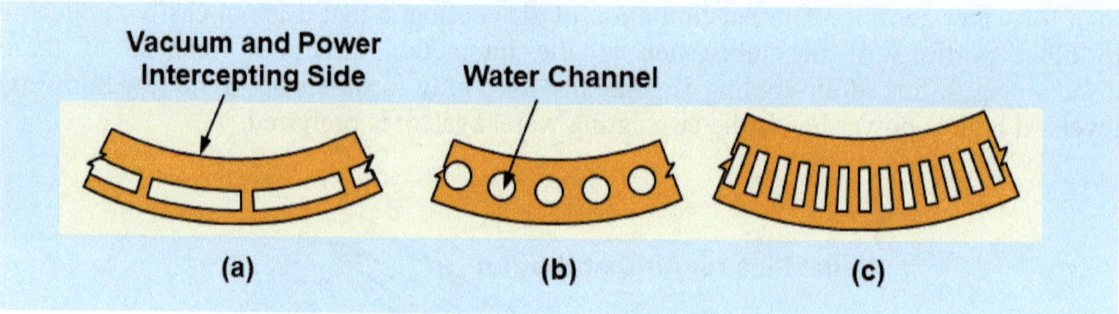

Figure 16-31 Collector water cooling channels with increasing area multiplication factors from (a) to (c). (Adapted from: *Linear-Beam Tubes* by A. Staprans, Internal Varian Report, September 1968.)

It is customary to design water cooling to be operable at pressure drops of less than 100 lb/in^2 and flow rates commensurate with average temperature rises from 30°C to 50°C. If a specific design has been made for water, cooling will generally suffer when other fluids are used because of differences in thermal conductivity, heat capacity, viscosity, and other characteristics. Ethylene glycol is used for low-temperature applications, principally to avoid the problem of freezing during storage. It is inferior to water as a coolant. Other fluids have been considered but are seldom used in practice, even though some have advantages with respect to water. Cost and availability are the most frequent deterrents to the use of most of the special fluids.

16.3.5 Vapor Phase Cooling

This method takes advantage of the latent heat of evaporation of water by allowing the water to boil at the collector surface. The steam is then condensed in a heat exchanger and returned to the collector, forming a closed system. To achieve reasonable dissipation densities, operation must be either in or near the transition between film boiling and nucleate boiling. Since this would ordinarily represent a runaway region, it can be used only when boiling equilibrium is maintained.

Beurtheret [10] pioneered this type of cooling for vacuum tubes in the Vapotron-type collectors. To understand this, the curve shown in Figure 16-32 is helpful. Heat transfer increases very rapidly with the surface temperature until the discontinuous region is reached where heat transfer actually drops as nucleate boiling changes into film boiling. At this point the surface temperature will undergo a temperature increase of about 1,000°C for the same heat flux, which is prohibitive because of burnout.

The vapor-phase-cooled collector (a section is shown in Figure 16-33) has an array of protruding bars, which, during operation, develop a thermal gradient along them such that the outer end does not boil at all. The base of the protrusions is always low in the film boiling region, and all other regions of boiling exist continuously between the two extremes. Maximum heat transfer occurs near the middle of the bar. Any small change of heat transfer in the system will shift the point of maximum heat transfer along the bar, thus readily absorbing the fluctuation without a discontinuous jump in the temperature. The detailed shape of the collector protrusion is crucial in the functioning of the system.

Vapor-phase-cooled collectors are used in high-power tubes where dissipation densities are on the order of 1 to 200 watts/cm^2 and where it is preferable to avoid a forced-flow system with water pumps. Examples are CW klystrons designed for UHF television transmitters in the 30- to 50-kW range.

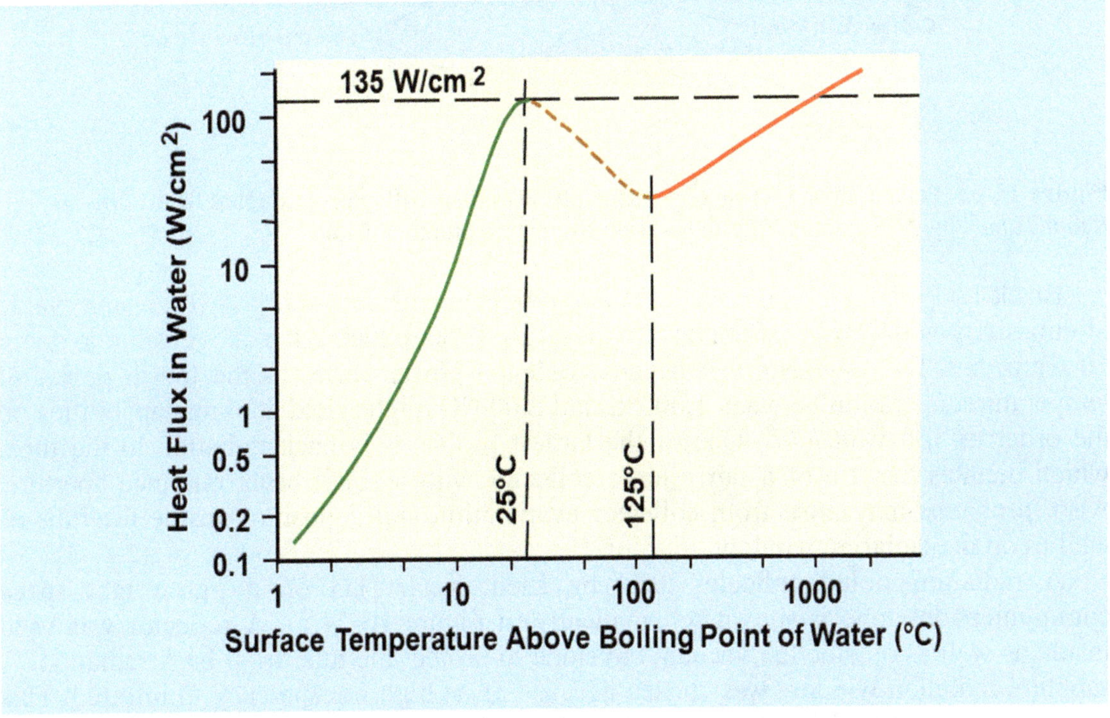

Figure 16-32 Heat transfer by boiling. (Adapted from: *Linear-Beam Tubes* by A. Staprans, Internal Varian Report, September 1968.)

16.3.6 Radiation Cooling

A primary application for radiation cooling is on TWTs used in space. There, it is possible to allow the collector to run at a high enough temperature so that heat is lost by radiation directly into space. This approach, though attractive because of its simplicity, poses certain problems. The collector temperature must be in the neighborhood of 1,000°C to provide a reasonable dissipation capability. In order to minimize temperature rise on the inside as well as to suppress lateral flow of heat to the tube body, a thin membrane-type collector is necessary. This raises problems with mechanical strength as well as gas diffusion through the collector. The latter, at least in principle, may be avoided if the tube is intended for use in a space environment. However, some tube testing at atmospheric pressure is always necessary. Thus, it is preferable to enclose the collector in a vacuum envelope essentially transparent to radiated heat. Some of the possible collector materials are the refractory metals, such as tungsten, molybdenum, and tantalum.

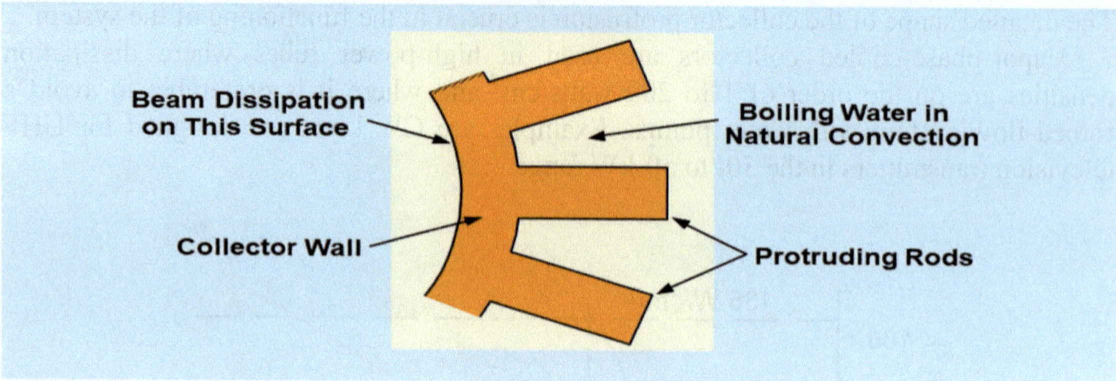

Figure 16-33 Partial cross section of a vapor-phase cooled collector. (Adapted from: *Linear-Beam Tubes* by A. Staprans, Internal Varian Report, September 1968.)

Black-body radiation at 1,000°C is approximately 12 watts/cm^2. A reasonable value of emissivity with proper carbonization might be 80%, reducing the above value to about 10 watts/cm^2. So it appears that because radiated power varies as the fourth power of temperature, operation between 1,000°C and 2,000°C might yield cooling capabilities on the order of 100 watts/cm^2. One of the largest problems is back radiation to the tube, which dictates the use of a fairly large collector with a small beam entrance aperture. Also, problems may arise from collector evaporation and deposition inside the tube as well as on the radiation window.

A radiation-cooled collector used by Lien et al. [11] on a lightweight space communications tube is shown schematically in Figure 16-34(a). A reflector was used inside as well as outside the vacuum envelope to protect the tube from back radiation. A sapphire radiation window was chosen because of its high transparency to infrared. This collector was operated at a power level of about 300 W at a temperature of 1,500°C. The collector material was tungsten. The measured radiation-cooling efficiency was about 75%, indicating that 25% of the beam power striking the collector was returned to the tube body. Figure 16-34(b) shows a radiation cooled collector being used on TWTs for satellite applications.

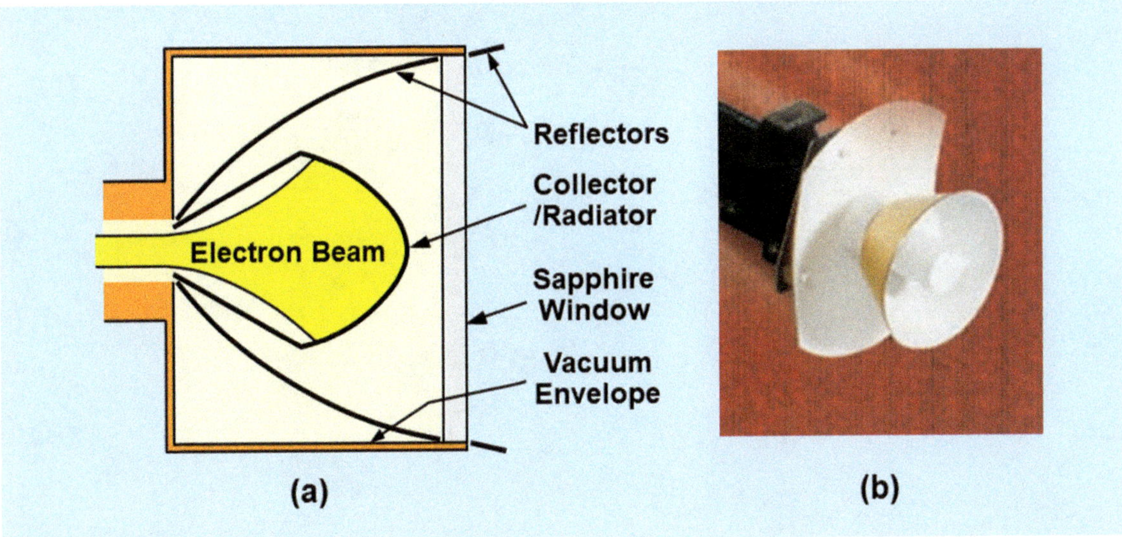

Figure 16-34 (a) Schematic layout of radiation-cooled collector. (Adapted from: *Linear-Beam Tubes*, by A. Staprans, Internal Varian Report, September 1968.) (b) Radiation-cooled TWTA advertised by Bosch.

REFERENCES

[1] A. Staprans, *Linear-Beam Tubes*, Internal Varian Report, September 1968.

[2] J. F. Skowron, "Ultra high-power amplitron," *Final Report*, RADC-TDR 64-389, Vol. 1, May 1965.

[3] H. G. Kosmahl, "Modern multistage depressed collectors—a review," *Proc. IEEE*, Vol. 70, No. 11, Nov 1982, pp. 1325–1334.

[4] I. Tammaru, "Collector lecture notes," Hughes Electron Dynamics Division.

[5] H. G. Kosmahl, and P. Ramins, "Small size 81 to 83.5 percent efficient 2 and 4 stage depressed collectors for octave bandwidth high performance TWTs," *IEEE Trans. On Electron Devices,* Vol. 24, No. 1, January 1977, pp. 36–44.

[6] A. S. Gilmour, Jr., "Bakeable UHV gate valve for microwave tube experimentation," *J. Vac. Sci. Technol.*, Vol. 13, No. 6, Nov./Dec. 1976.

[7] P. Ramins, and B. T. Ebihara, "Improvements in MDC and TWT overall efficiency through the application of carbon electrode surfaces," *IEEE Trans. On Electron Devices*, Vol. ED-33, No. 11, November 1986, pp. 1915–1924.

[8] A. N. Curren, "Carbon and carbon-coated electrodes for multistage depressed collectors for electron-beam devices—A technology review," *IEEE Trans. Electron Devices*, Vol. ED-33, No. 11, November 1986, pp. 1902–1914.

[9] A. N. Curren, K. A. Jensen, and R. F. Roman, "Secondary electron emission characteristics of molybdenum-masked, ion-textured OFHC copper," *NASA Technical Paper 2967*, January 1990.

[10] C. Beurtheret, "The Vapotron technique," *Revue Technique* C.F.T.H, No. 24, December 1956, pp. 53–82.

[11] E. L. Lien, A. Mizuhara, and D. I. Boilard, "Electrostatically-focused extended-interaction S-band klystron amplifier," *Proc. 6th International Conference on Microwave and Optical Generation and Amplification,* Cambridge, England, September 1966.

Chapter 17

Distortion in Linear-Beam Tubes

This chapter addresses the problems of distortion and noise in amplifiers, which have reasonably constant gain over a wide dynamic range. Crossed-field amplifiers have no linear range of operation and are not considered. At small-signal levels, the power output versus power input curve for a linear-beam amplifier is reasonably linear and signal distortion is small. As a device is driven toward saturation, the transfer characteristic becomes more and more nonlinear. This nonlinearity results in distortions of amplitude-modulated (AM) or phase-modulated (PM) signals for several reasons:

- AM/AM conversion;
- AM/PM conversion;
- Harmonic generation;
- Intermodulation distortion;
- Signal capturing.

In addition, other signal distortions occur that are not related to saturation effects. Noise is one example; the following are also often of concern:

- Variations of gain and power output with frequency;
- Phase nonlinearities or time delay distortion;
- Power supply and load effects (referred to as pushing and pulling).

17.1 DISTORTION RESULTING FROM SATURATION EFFECTS

17.1.1 AM/AM Conversion

The curve in Figure 17-1 shows the RF power output versus input for a TWT. The distortion of a signal is also indicated. A sinusoidal input signal to the TWT is assumed and the resulting output signal is shown. The shape of the output signal was determined graphically, as indicated by the dashed lines, by plotting output power levels for various input signal levels as a function of time. The distortion resulting from gain compression and saturation is apparent in the flat-topped shape of the output curve.

AM/AM conversion is a measure of the change in RF power output that results from a change in RF power input and is usually defined by the slope of the curve of RF power output versus RF power input expressed in dB/dB. Thus,

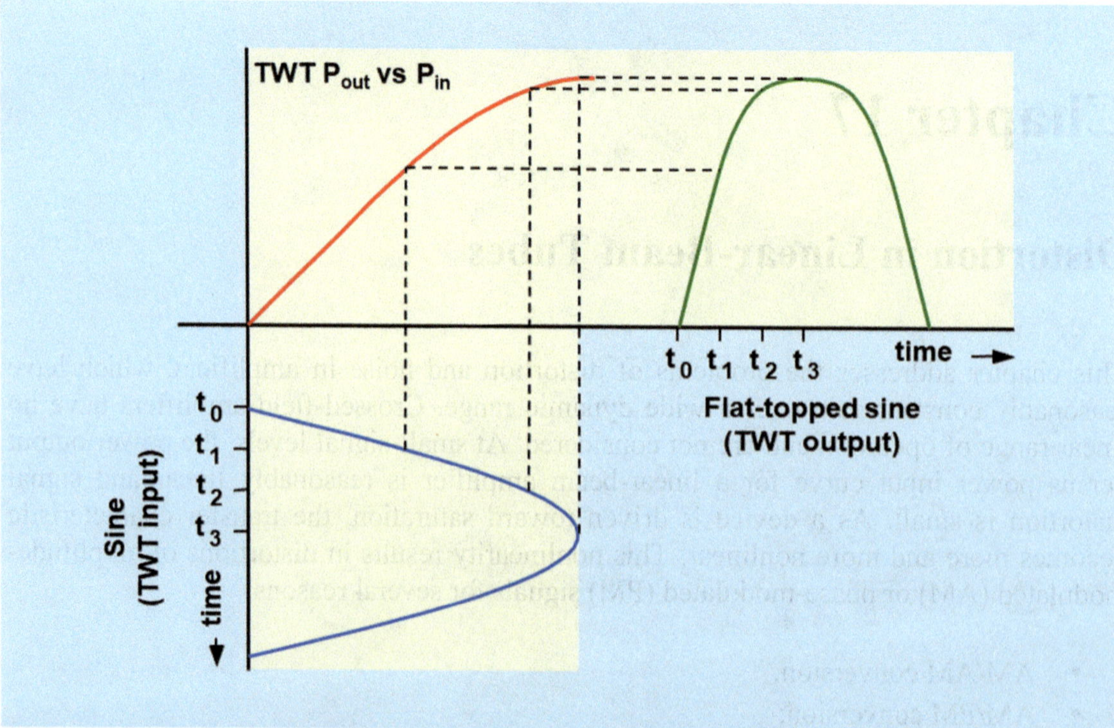

Figure 17-1 AM/AM distortion.

$$\text{AM/ AM conversion} = \frac{\Delta \text{ RF power output (in dB)}}{1-\text{dB change in input power}} \tag{17-1}$$

AM/AM conversion and gain are shown as functions of RF power input in Figure 17-2. Gain is defined as the ratio of RF power output to RF power input. Two values of gain are usually of importance. These are the small signal gain and the saturated gain. Saturated gain is usually 5 to 6 dB less than small signal gain.

In the linear (small signal) region, the AM/AM conversion is very nearly 1 dB/dB. At saturation, where no change in output occurs as the input is varied, the AM/AM conversion is 0 dB/dB.

AM/AM conversion is important in communication applications of TWTs. The specification of AM/AM conversion depends on the type of modulation used. For example, if amplitude modulation is employed and if amplitude distortion must be avoided, then it may be necessary to operate a TWT at a maximum power level well below saturation. For this operating condition, we say that the power level must be backed off from saturation.

The amount of back-off depends on the amplitude distortion that can be tolerated. Most amplitude modulation communication systems can tolerate no less than 0.9 dB/dB of AM/AM conversion [1]. If no provisions are made for linearizing a TWT, it may be necessary to back off as much as 6–8 dB to meet this AM/AM conversion requirement. Through the use of a pitch taper in a TWT [2, 3] and an external linearizing circuit, distortion is significantly reduced and it is possible to operate much closer to the saturation power level.

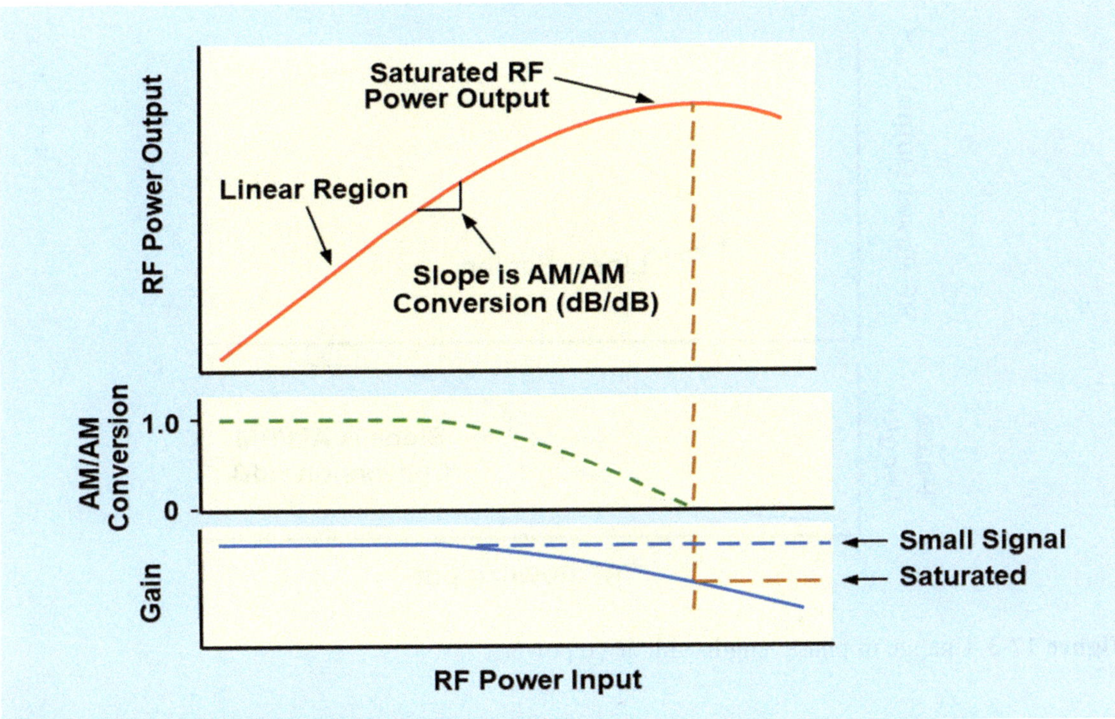

Figure 17-2 Gain and AM/AM conversion versus RF power input. (Adapted from: *L3 Communications TWT and TWTA Handbook*. 2007.)

17.1.2 AM/PM Conversion

AM/PM conversion is a measure of the change in tube phase length that results from a change in RF drive level. This occurs because, as drive level is increased, more power is extracted from the electron beam and the velocity of the beam is reduced. As beam velocity decreases, the velocity of the RF wave on the circuit is reduced and this increases the phase length of the tube. It should be noted here that the cathode voltage supply may aggravate the phase delay. As drive increases, body current usually increases, which in turn increases the load on the cathode supply, causing a reduction in voltage and an increase in phase length.

A curve showing the variation of phase length with RF input power is given in Figure 17-3. In the linear region of the RF power output curve, phase shift is relatively insensitive to drive level. As drive is increased toward saturation, the phase length increases rapidly and then becomes constant again as the TWT saturates. The slope of the phase length curve in degrees/dB and shown in Figure 17-4 is the AM/PM conversion.

The maximum value of AM/PM conversion is usually the value that is specified. This occurs at 3 to 10 dB below the input power that produces saturation. The type of modulation employed determines the need for specifying AM/PM conversion. For example, in amplitude-modulated systems, AM/PM conversion can result in amplitude distortions. In multiple carrier systems in which both AM and PM signals are present, AM/PM conversion can cause distortion or interference with the phase- or frequency-modulated signals. Finally, amplitude variations in phase- or frequency-modulated systems can cause distortions or interference.

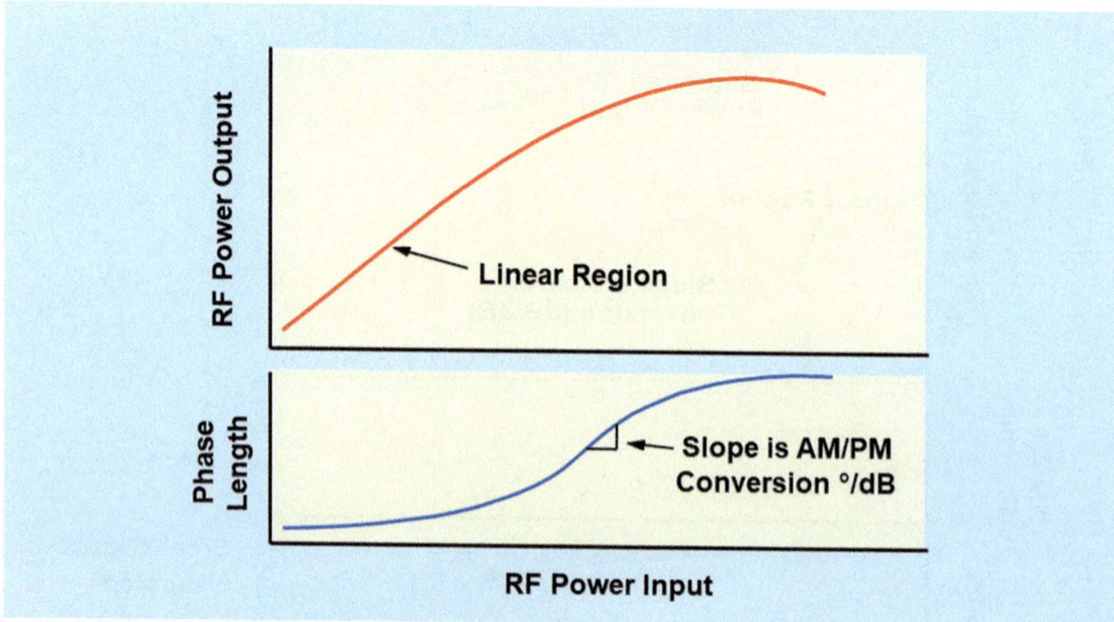

Figure 17-3 Change of phase length with drive power.

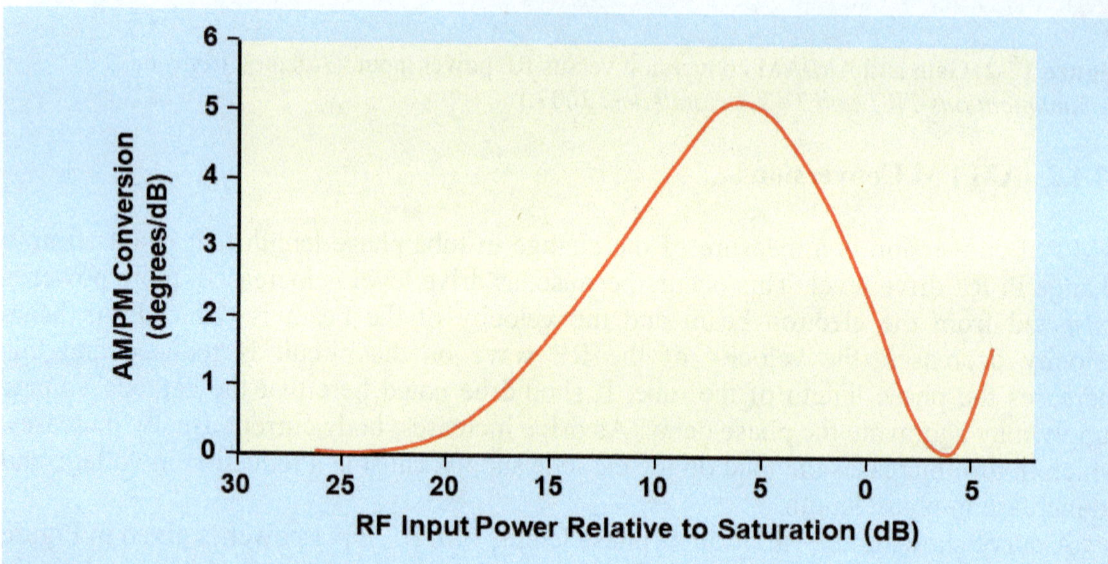

Figure 17-4 Example of variation of AM/PM conversion with RF drive. (Adapted from: *L3 Communications TWT and TWTA Handbook*. 2007.)

17.1.3 Harmonic Generation

Near saturation, the electron bunches in a TWT become sharply defined, and so the RF current is rich in harmonics. In addition, as saturation is approached and the output signal becomes distorted, harmonics are generated. This is apparent from Figure 17-5 where a sinusoidal input signal has become flat-topped because of saturation. The fundamental and third and fifth harmonic components of the distorted output signal are shown in Figure 17-5. It will be shown in the next section that these harmonics are one of the sources of the intermodulation distortion produced by an amplifier.

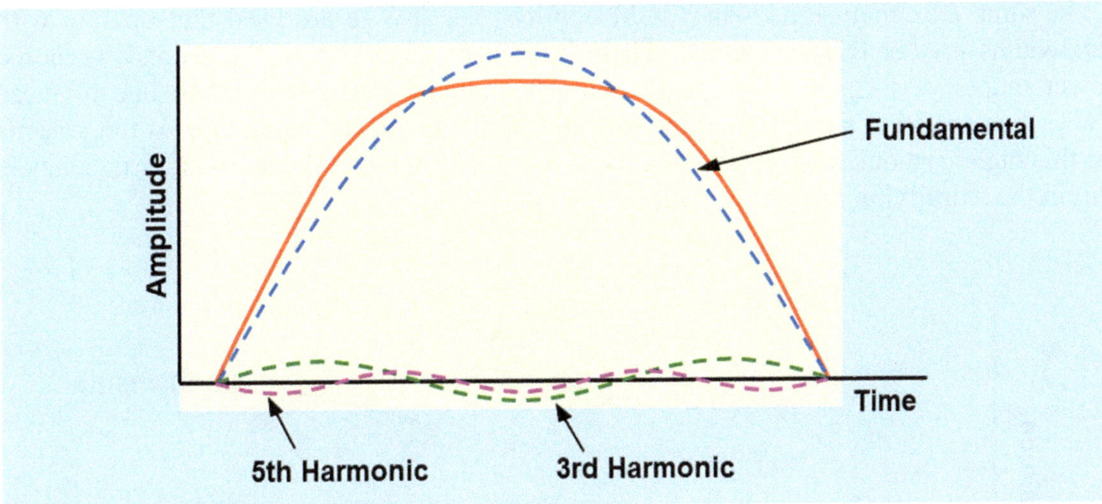

Figure 17-5 Harmonics in a flat-topped sine wave.

Harmonics are also produced as the result of AM/PM conversion. This is illustrated in Figures 17-6 and 17-7. As a TWT is driven into saturation and the beam slows down, the phase of the output signal is delayed compared with the phase at small drive levels. As a result, the output curve for a sinusoidal input becomes a tilted sine wave. This tilted wave contains harmonics as shown in Figure 17-7. The third and fifth harmonics are phase-shifted relative to those produced by amplitude distortion during saturation. These harmonics also contribute to intermodulation distortion.

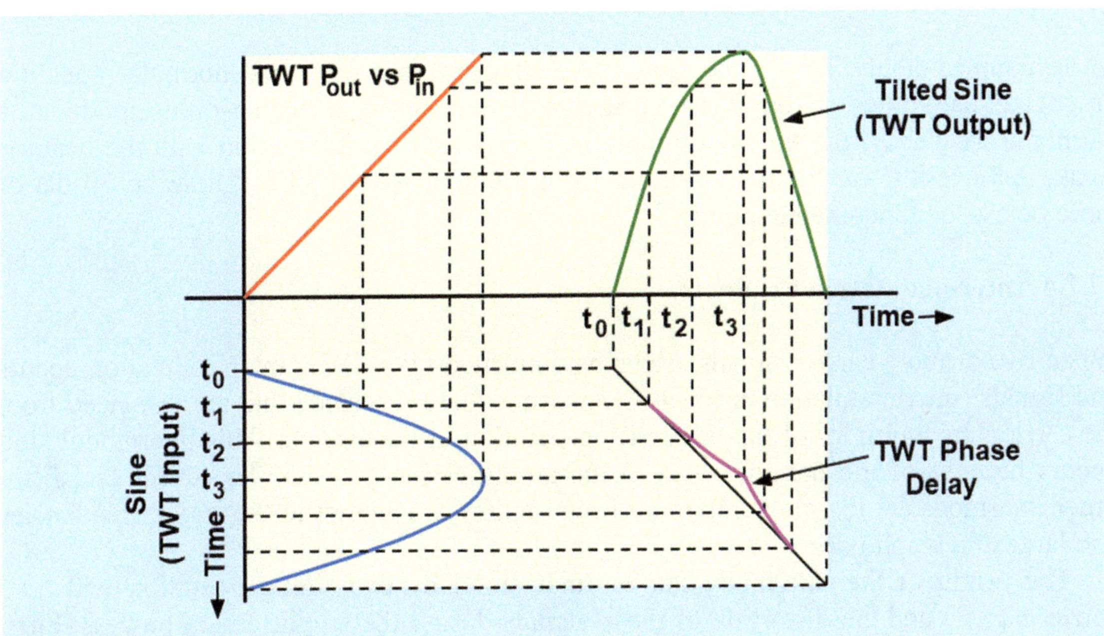

Figure 17-6 Distortion resulting from AM/PM conversion.

At saturation, helix TWTs used for communication applications usually produce second harmonic power that is 8 to 10 dB below the fundamental power. Third harmonic power at saturation is usually 23 to 18 dB below the fundamental power [2].

In some extremely broadband ECM applications, TWTs are used that operate with bandwidths greater than an octave. There have been cases where the second harmonic power output was equal to or greater than the fundamental power when one of these TWTs was driven to saturation at the low end of the frequency band. One of the reasons for this high harmonic output is that at the low end of the band, the harmonic frequency is within the amplifying band of the TWT.

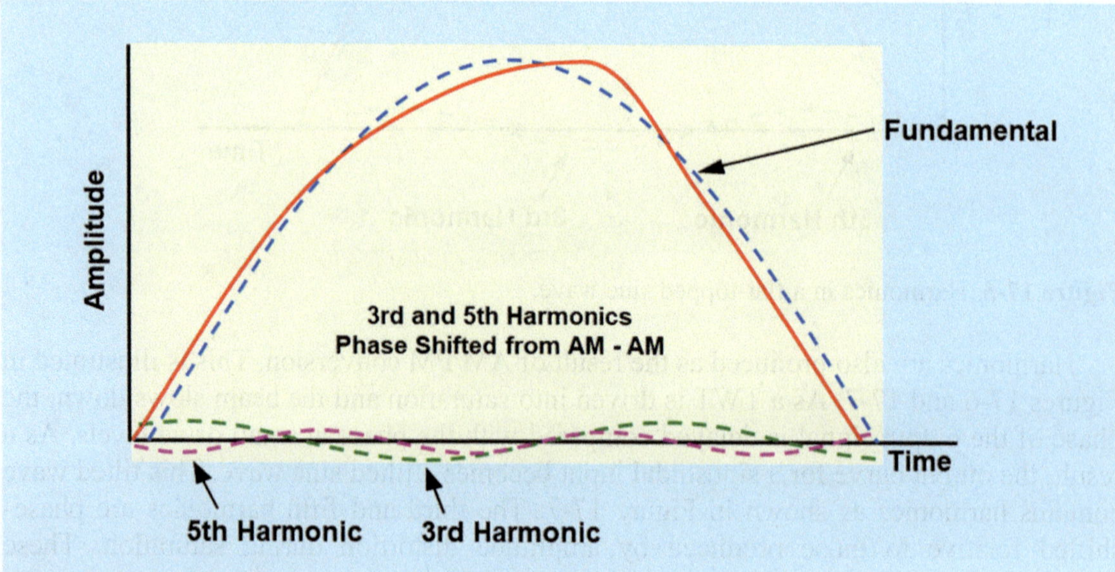

Figure 17-7 Harmonics generated by AM/PM conversion.

In coupled-cavity TWTs, the slow-wave structure is a filter that normally does not propagate harmonics. Even if the structure does contain a higher-order mode at a harmonic frequency, the field configuration is such that the interaction with the beam is weak. As a result, the harmonic output from a coupled-cavity TWT may be 30 dB or more below the fundamental power.

17.1.4 Intermodulation Products

When two or more carrier signals are being amplified by a TWT, intermodulation occurs and signals (intermodulation products or intermods) are produced that are displaced from the carriers by multiples of the frequency separation of the carriers. This intermodulation occurs because of the nonlinearities in the amplification process. The third- and fifth-order intermods are the most important because they are closest to the signal frequencies and largest in amplitude.

The origin of the intermods can be understood by examining operation with two carrier signals and the distortion of those signals during the amplification process. First, consider two carrier signals of equal amplitude as follows:

$$V_1 = V \cos \omega_1 t \tag{17-2}$$

and

$$V_2 = V \cos \omega_2 t \tag{17-3}$$

When these two signals are added together, they combine as follows:

$$V_1 + V_2 = V\left(\cos \omega_1 t + \cos \omega_2 t\right) \tag{17-4}$$

or

$$V_1 + V_2 = 2V \cos\left(\frac{\omega_2 - \omega_1}{2}\right)t \cos\left(\frac{\omega_2 + \omega_1}{2}\right)t \tag{17-5}$$

This is a signal having an average carrier frequency of $(\omega_2 + \omega_1)/2$ and amplitude modulation at frequency $(\omega_2 - \omega_1)/2$ as shown in Figure 17-8.

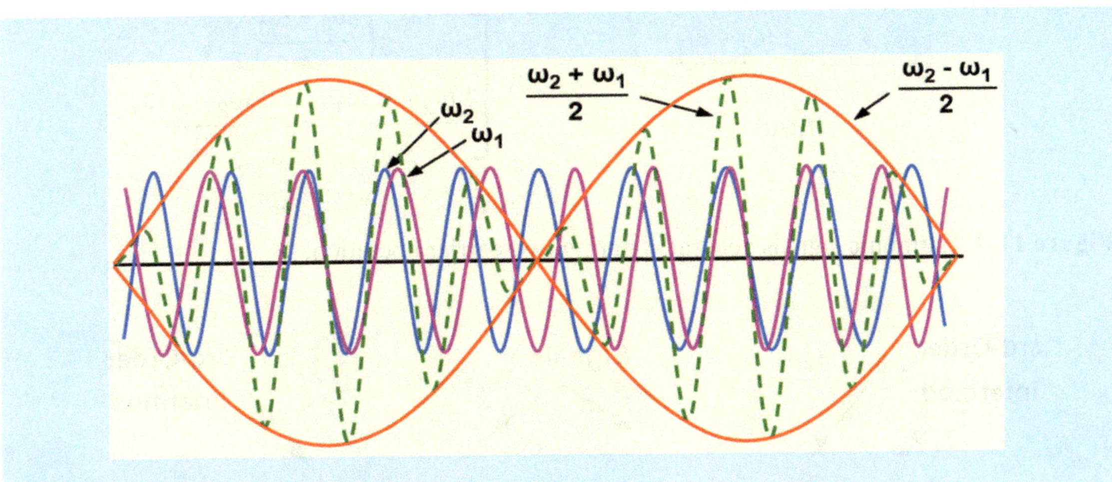

Figure 17-8 Two-carrier operation.

When the signal in Figure 17-8 is amplified by a TWT (or any other amplifier) in which saturation or AM/PM conversion occurs, the signal becomes distorted as shown in Figure 17-9. Harmonics of the modulation frequency and of the carrier frequency are generated. The total third-harmonic signal, including the high-frequency variation, may be written as

$$V_{3rd} = 2 \cos 3\left(\frac{\omega_2 - \omega_1}{2}\right)t \cos\left(\frac{\omega_2 + \omega_1}{2}\right)t \tag{17-6}$$

or as

$$V_{3rd} = \cos\left(2\omega_2 - \omega_1\right)t + \cos\left(2\omega_1 - \omega_2\right)t \tag{17-7}$$

These two cosine terms represent the third-order intermods, which are at frequencies close to the carrier frequencies as shown in Figure 17-10.

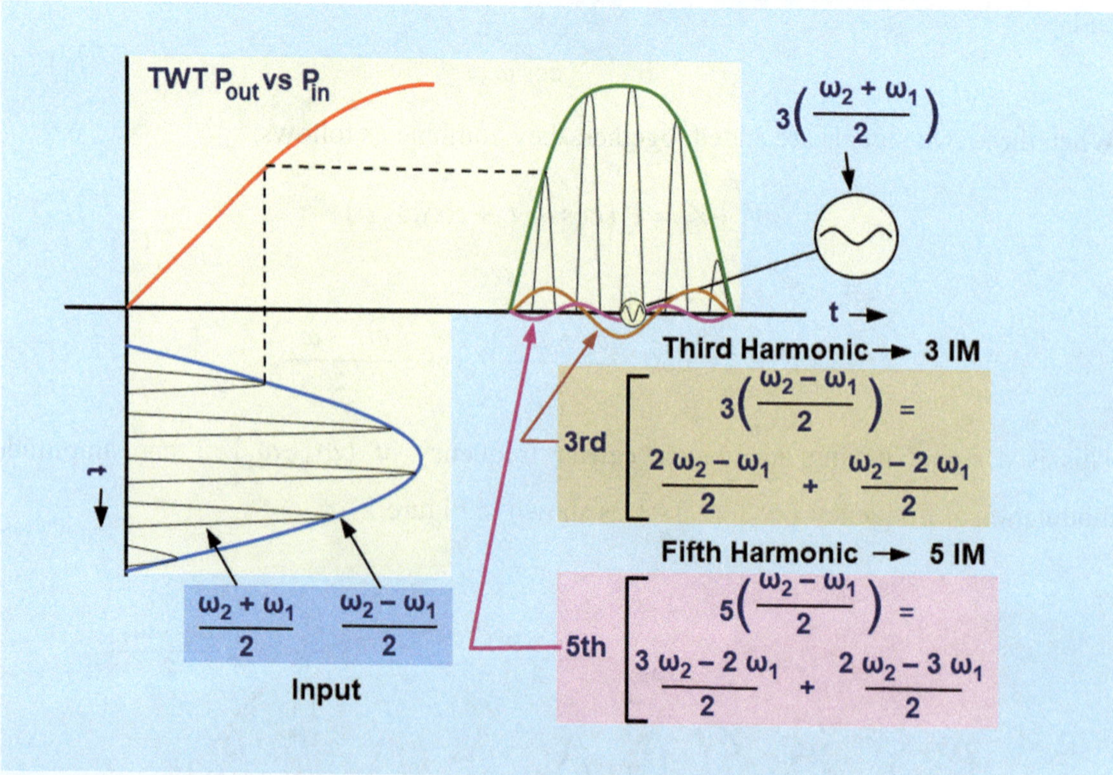

Figure 17-9 Harmonic signals generated during two-carrier operation.

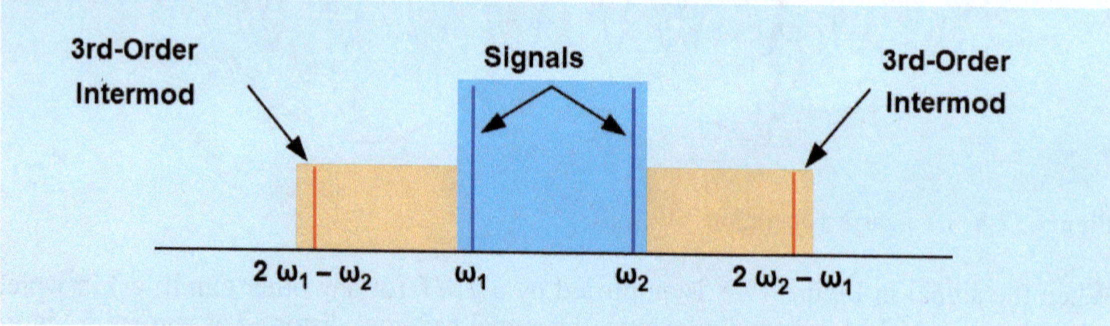

Figure 17-10 Third-order intermodulation products.

The total fifth harmonic signal including the high-frequency variation is

$$V_{5th} = 2\cos 5\left(\frac{\omega_2-\omega_1}{2}\right)t\ \cos\left(\frac{\omega_2+\omega_1}{2}\right)t \tag{17-8}$$

or

$$V_{5th} = \cos\left(3\,\omega_2 - 2\,\omega_1\right)t + \cos\left(3\,\omega_1 - 2\,\omega_2\right)t \tag{17-9}$$

These two cosine terms represent the fifth-order intermods as shown in Figure 17-11.

When a TWT approaches saturation and gain decreases, *gain compression* occurs. It is also possible, at input levels below saturation, for *gain expansion* to occur. Gain expansion (exaggerated for clarity) is compared with gain compression in Figure 17-12.

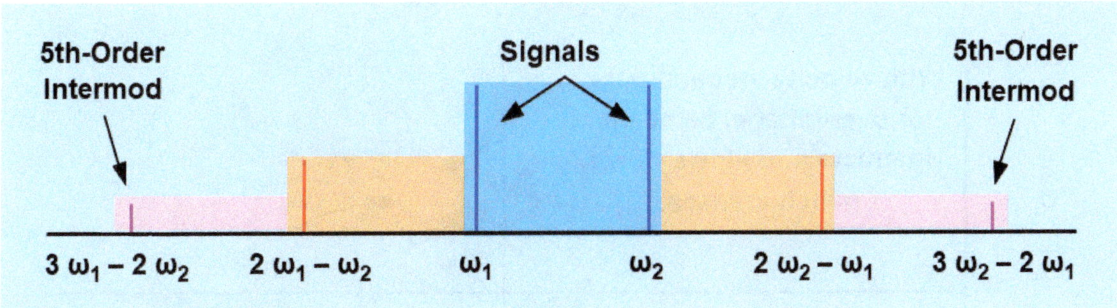

Figure 17-11 Fifth-order intermodulation products.

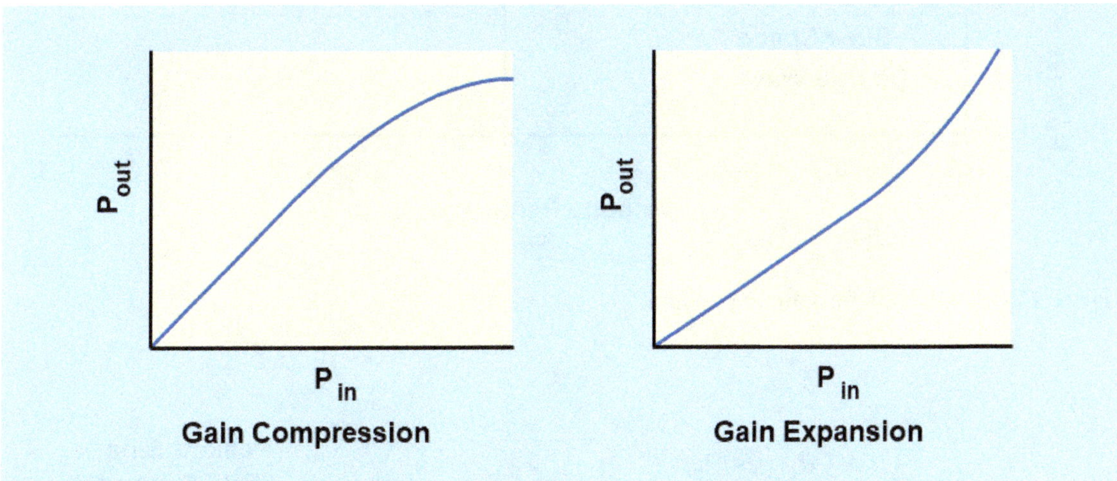

Figure 17-12 Comparison of gain compression and gain expansion.

Two causes for gain expansion (indicated in Figure 17-13) are the use of velocity steps or tapers and the use of a beam voltage greater than that which produces maximum gain (overvoltaging). If the beam is overvoltaged (the velocity parameter is too high), then the gain at small signal levels is less than optimum. As signal level is increased and more energy is extracted from the beam, then, as the beam slows down (and the velocity parameter decreases), gain increases.

Similarly, when a velocity step or taper is used, then, at small-signal levels where little energy is extracted from the beam, the beam will be going too fast for optimum interaction in the step or taper region. As signal level is increased, the beam will slow down in the step or taper region and gain will increase.

The way in which gain expansion generates third-order intermods is indicated in Figure 17-14. For a sinusoidal input, the TWT output is distorted into a peaked shape. The third harmonic that is generated is 180° out of phase with that generated during gain compression and is the source of the third-order intermod. The effect of this phase difference is evident in Figure 17-15, which shows the third-order intermod for a TWT with a velocity taper and overvoltage [4]. As the drive power is increased so that operation moves from the gain expansion region to gain compression, the rate at which the third-order intermod increases varies significantly. This is in contrast to the linear increase in intermod level with increasing drive power that is usually experienced when only gain compression is present.

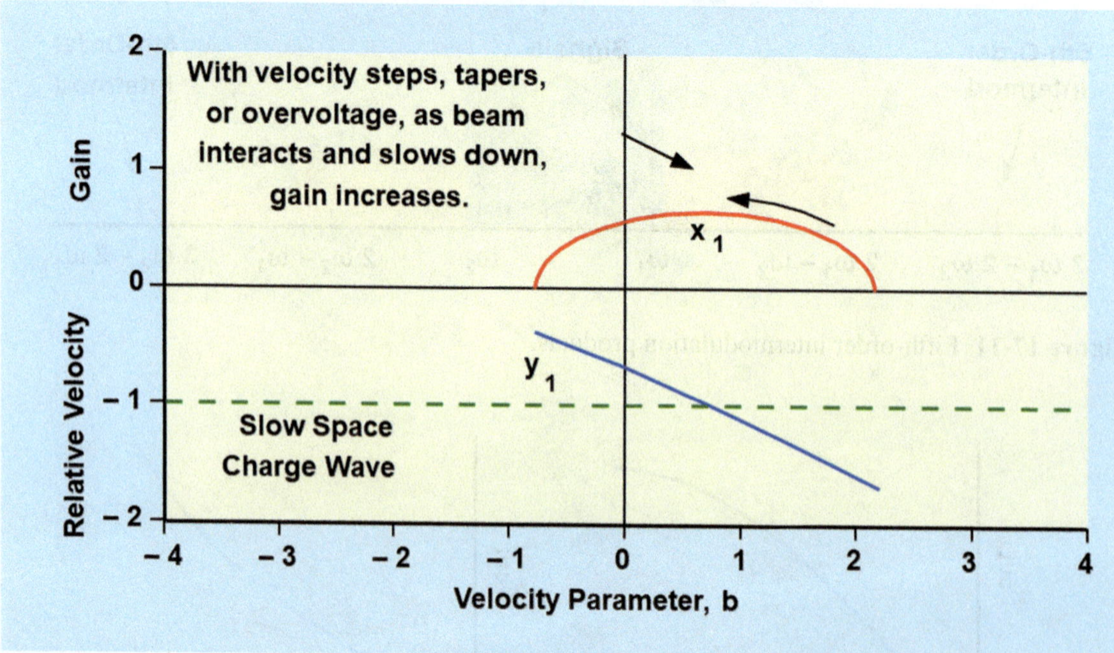

Figure 17-13 Reasons for gain expansion.

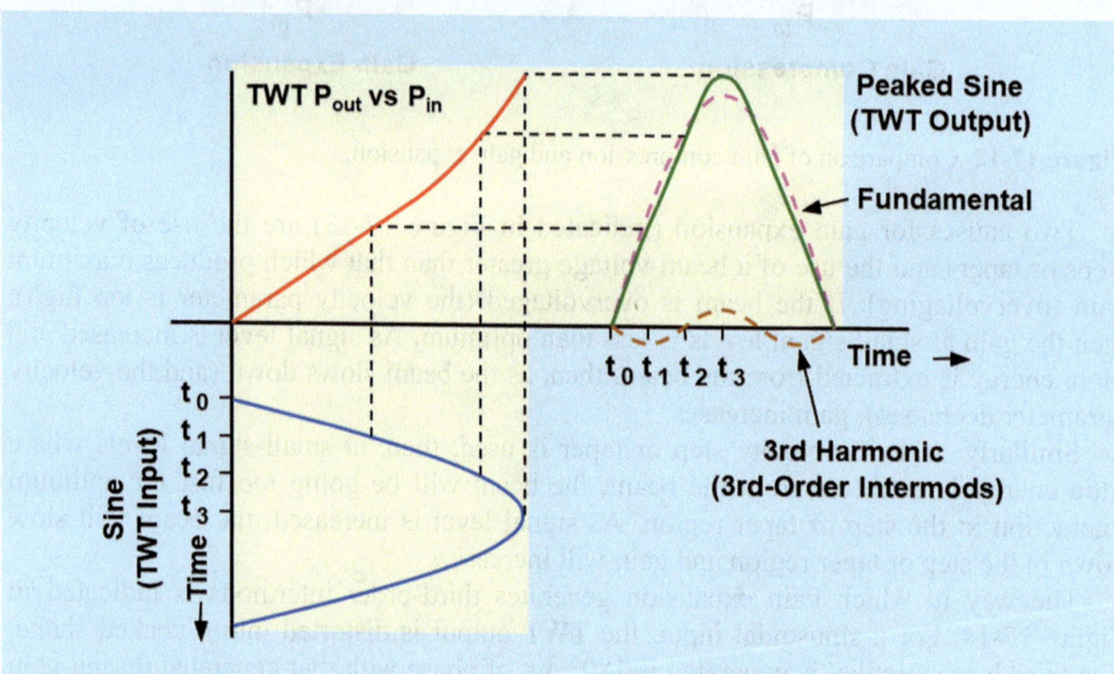

Figure 17-14 Effect of gain expansion on third-order intermodulation product.

An example of the variation of intermod levels with drive when only gain compression is present is shown in Figure 17-16. In this case, when the tube is driven to saturation, third-order intermodulation products are each 8 to 10 dB below the carrier signals (two carriers of equal amplitude are assumed). By reducing the amplitudes of the drive signals so that the TWT operates well below saturated power output, the amplitudes of the intermodulation products drops linearly and rapidly with drive power level. In the example in Figure 17-16, with the TWT operated about 10 dB below saturation, the third-order intermodulation products drop to about 40 dB below the signal output level.

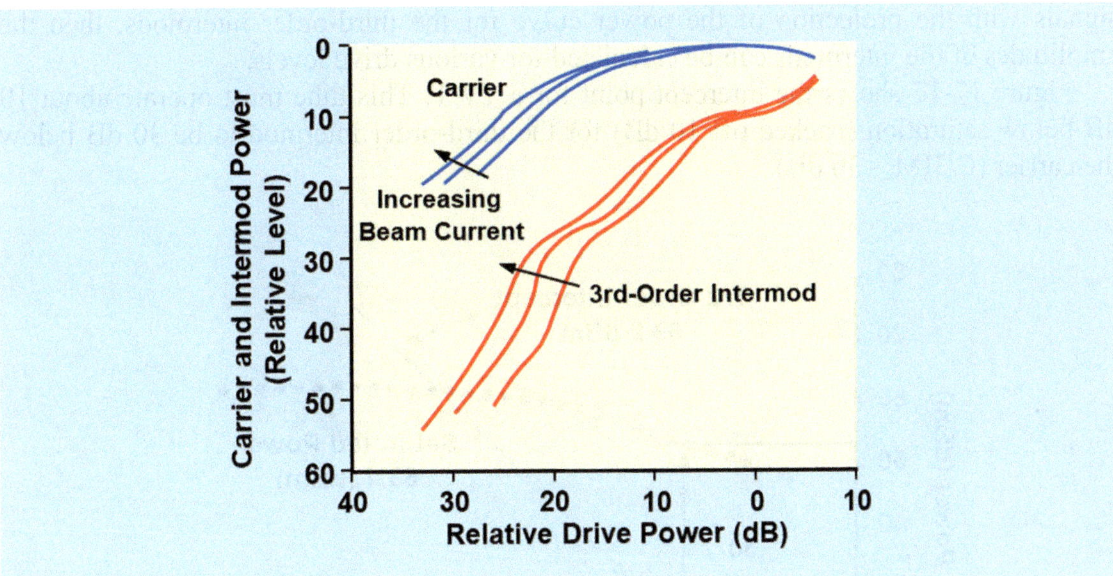

Figure 17-15 Third-order intermods for an overvoltaged TWT with a velocity taper.

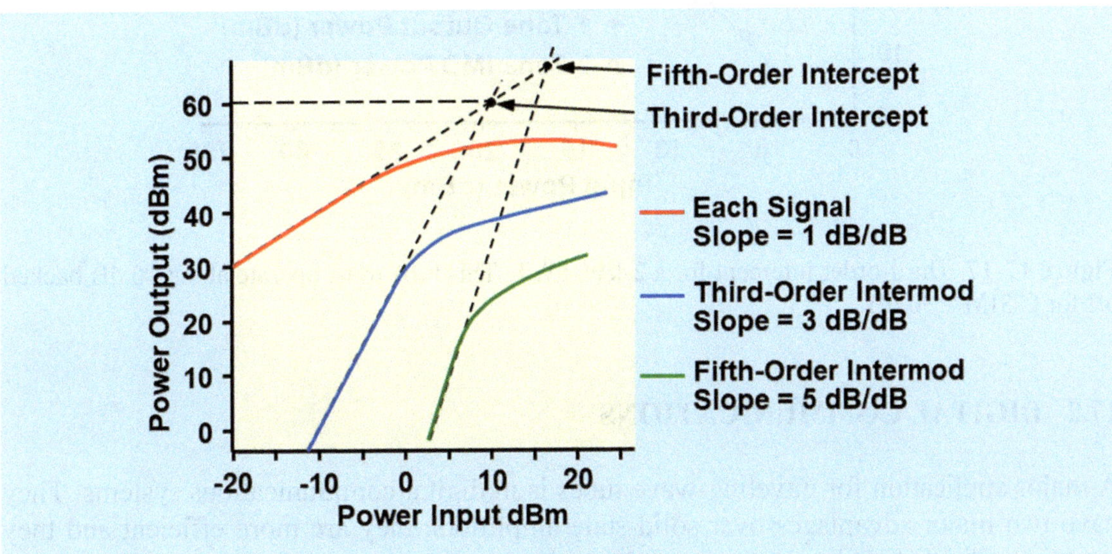

Figure 17-16 Intercept point method for specifying intermodulation distortion.

The specification of intermodulation products is not straightforward. Usually, the amplitudes of the products are specified as the maximum allowable relative to the carrier signal levels when these signals have equal amplitudes. The amplitude of the carrier signals is also usually specified relative to the saturated output power when the TWT is amplifying two signals of equal amplitude.

The method for specifying intermods illustrated in Figure 17-16 is called the *intercept point* method. This is based on an understanding of the rate at which the levels of the intermodulation products vary as drive level is varied. If one point is known, such as the intercept of the projection from the linear regions of the power curve for the carrier signals with the projection of the power curve for the third-order intermods, then the amplitudes of the intermods can be calculated for various drive levels.

Figure 17-17 shows the intercept point for a TWT. This tube must operate about 10 dB below saturation (backed off 10 dB) for the third-order intermod to be 30 dB below the carrier (C/3IM = 30 dB).

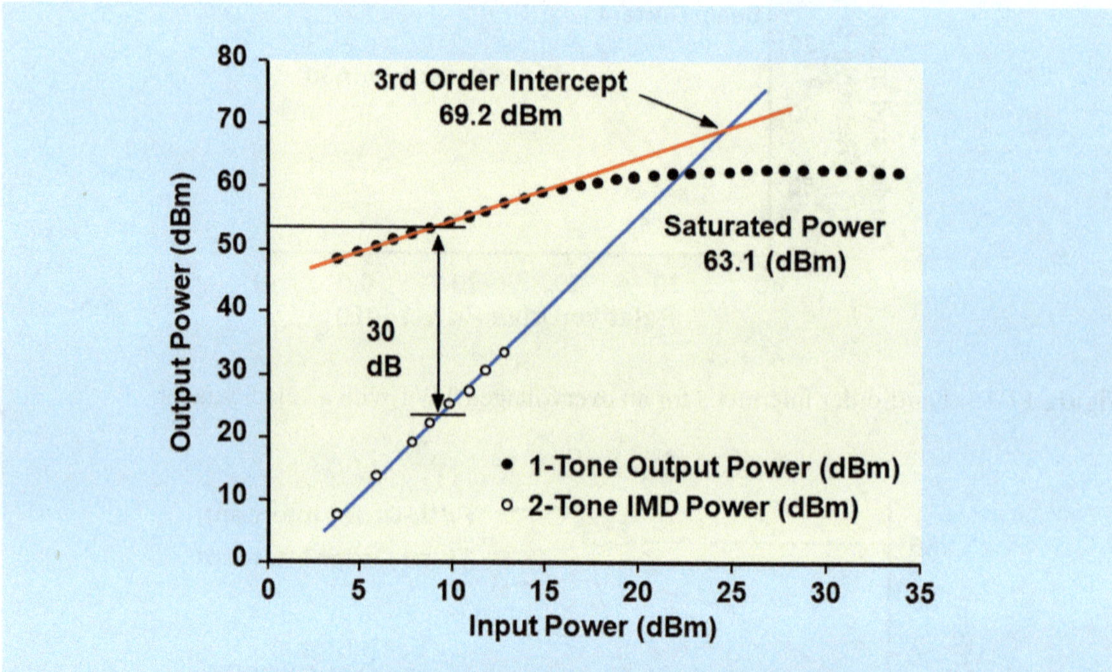

Figure 17-17 Third-order intercept for a 2-kW TWT. This tube must operate about 10 dB backed off for C/3IM = 30 dB.

17.2 DIGITAL COMMUNICATIONS

A major application for traveling-wave tubes is in digital communications systems. They have two major advantages over solid-state amplifiers: they are more efficient and they are more reliable [5, 6].

To understand the demands placed on TWTs, it is necessary to understand some of the basic aspects of digital communications. First of all, information is transmitted as a series of ones and zeros. One or more of the properties of the transmitted signal may be varied (modulated) as shown in Figure 17-18 to produce the ones and zeros.

With frequency-shift keying (FSK) the frequency of a signal is switched between two values. One frequency represents a zero and the other represents a one. With amplitude-shift keying (ASK) the amplitude of the signal is switched between two values, and with PSK, phase is switched.

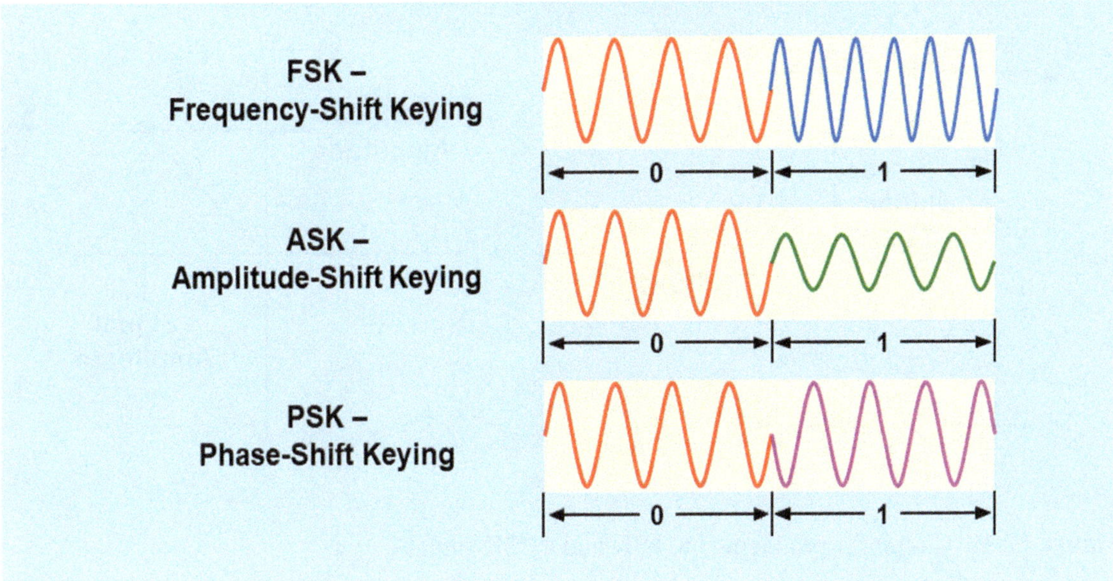

Figure 17-18 Signal properties that can be varied to transmit ones and zeros.

In the following sections, two modulation schemes, quadrature phase-shift keying (QPSK), which uses PSK, and 16QAM, which uses PSK and ASK, will be discussed. The way in which PSK and ASK are implemented is as follows. As shown in Figure 17-19, a carrier signal generated by a local oscillator is split into two paths. One path contains a phase shifter that delays the signal 90° relative to the other path.

The signal containing the information to be transmitted is split into two independent components, *I* (in phase) and *Q* (quadrature). The *I* signal is used to modulate one of the carriers from the local oscillator and the *Q* signal is used to modulate the other. The two modulated carriers are then added together. The result is a digitally modulated signal, the amplitude and phase of which are determined by the amplitudes of the *I* and *Q* modulating signals.

17.2.1 QPSK and 16QAM

Data plots, referred to as constellation plots, for QPSK and 16QAM modulation techniques are shown in Figure 17-20. With QPSK, there are four ideal positions (states) for the data points on the constellation plot. As shown in Figure 17-21, when QPSK is used, one signal serves as the phase reference and the other is phase shifted in increments that are multiples of 90°. Each phase state corresponds to information that is to be transmitted. For all practical purposes, from a TWT engineer's point of view, phase shifts occur at random. Also, phase shifts do not occur instantaneously, but instead take place over several cycles as shown in Figure 17-21.

When 16QAM is used, there are 16 states on the constellation diagram. In this case, there are two amplitude states at a phase of 45° (and at 135°, 225°, and 315°). There are also phase states at 15° and 75° into each quadrant of the constellation plot. The relative phases and amplitudes of the states in the first quadrant are shown in Figure 17-22.

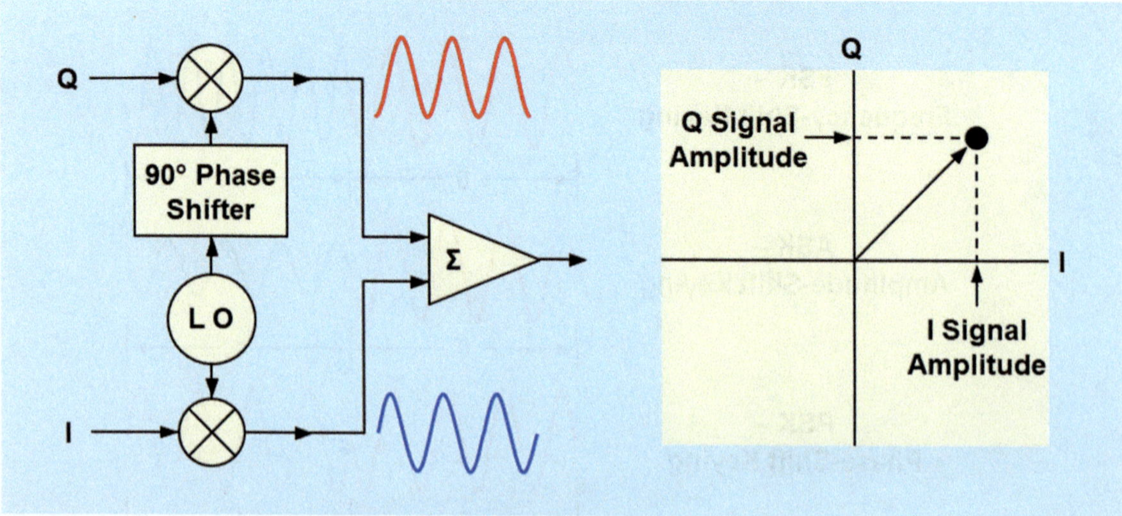

Figure 17-19 Circuit for producing an ASK and a PSK signal.

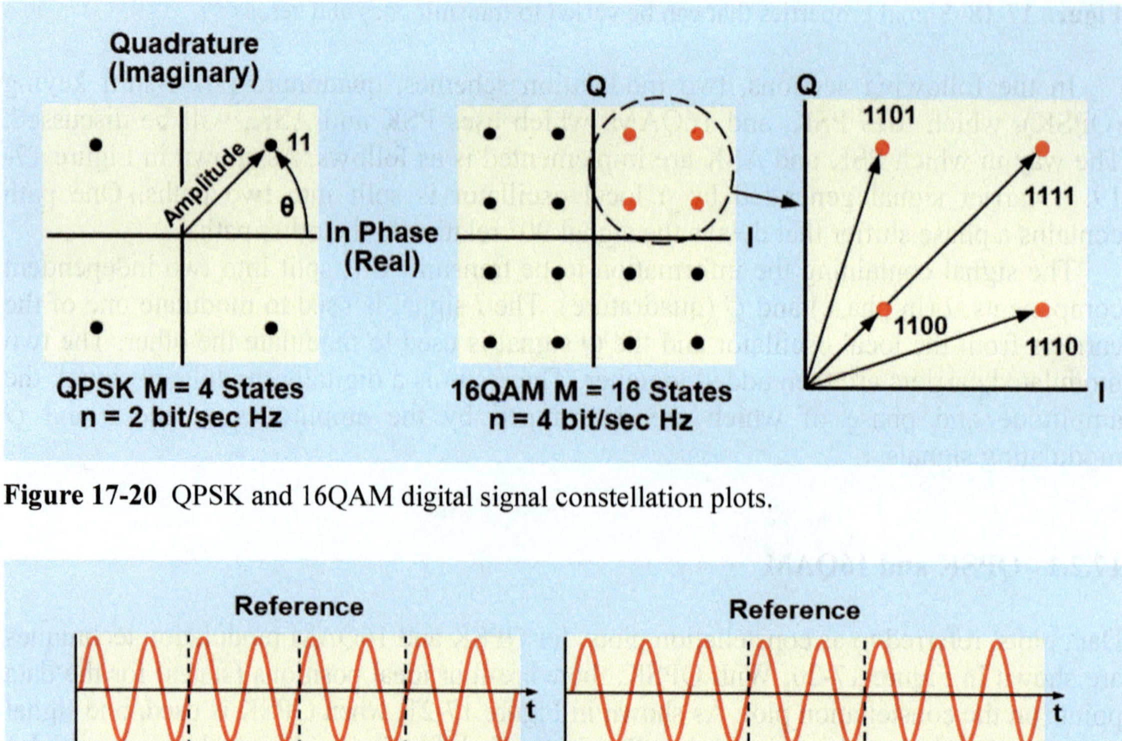

Figure 17-20 QPSK and 16QAM digital signal constellation plots.

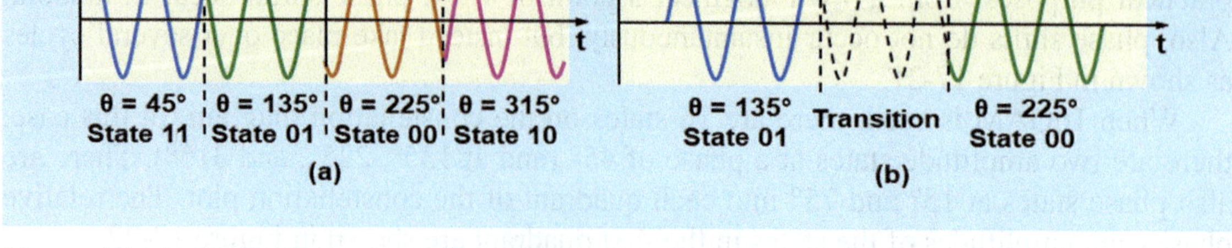

Figure 17-21 QPSK timing (a) and transition between states in QPSK modulation (b).

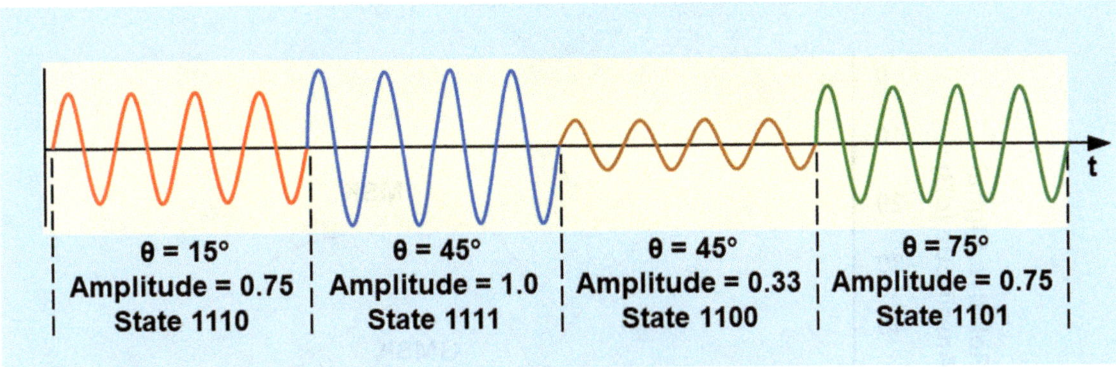

Figure 17-22 First quadrant of the constellation for 16QAM modulation.

Two refinements to FSK are minimum-shift keying (MSK) and Gaussian minimum shift keying (GMSK). These are continuous-phase, frequency-shift keying schemes. As indicated in Figure 17-23, frequency changes occur at the carrier signal zero crossings. The frequency difference between 1 and 0 states is half of the data rate. GMSK adds a Gaussian filter with no ringing, a sharp cutoff, and a narrow bandwidth to reduce sidebands. The improvement in spectral efficiency over MSK is illustrated in Figure 17-24. GMSK signals can be amplified with non-linear amplifiers and remain undistorted. This is because no part of the signal is carried as amplitude variations. Nonlinear amplifiers such as amplifiers driven to saturation are more efficient than linear amplifiers. This is important for small battery-operated transmitters like cell phones.

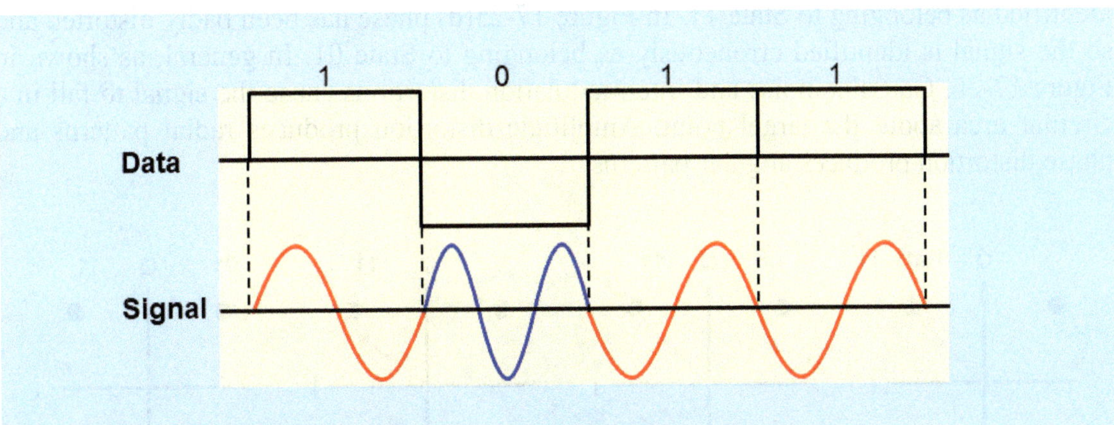

Figure 17-23 With MSK, frequency changes occur at carrier signal zero crossings.

17.2.2 Data Characteristics

Digital data are transmitted in bits that have a long enough duration to make it possible to identify the frequency, phase, and amplitude. The rates at which bits are generated and transmitted are critical characteristics of a digital communications system.

A figure of merit for a digital transmission is E_b/N_o where E_b is the energy of the bit and N_o is the noise power density. E_b/N_o is the digital system's equivalent of the signal-to-noise ratio (SNR). The digital data rate, R, for a system is the rate at which bits are transmitted. The average signal power is E_bR.

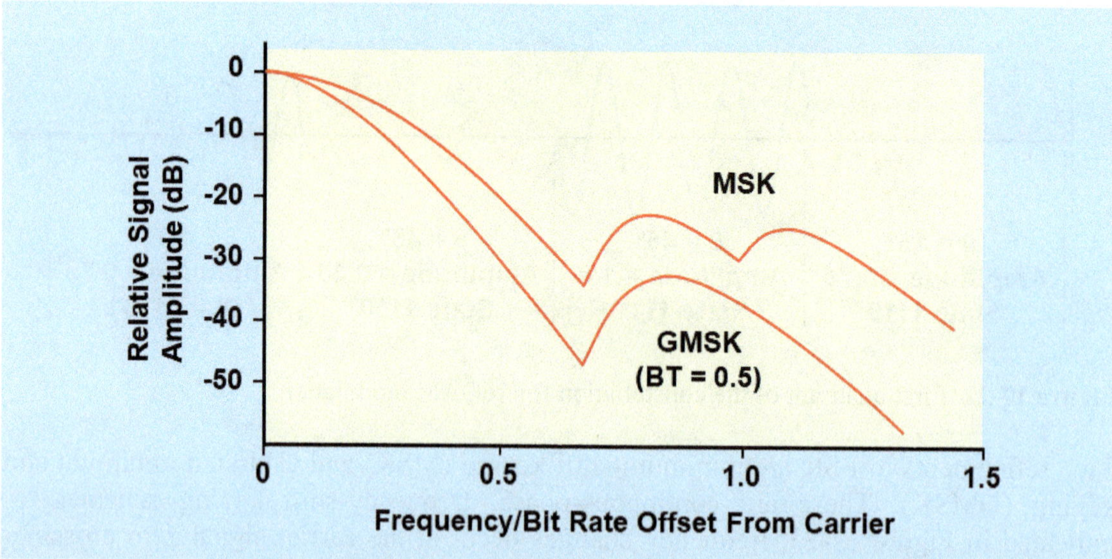

Figure 17-24 The spectral efficiency of GMSK is significantly better than MSK.

Because of noise and amplifier characteristics, impairment of bits occurs. This is illustrated for QPSK in Figures 17-25 and 17-26. A bit with no impairment is shown in Figure 17-25(a). The amplitude and phase of the signal are exactly those required to place the bit on the target spot (State 11) on the constellation diagram. In Figure 17-25(b), the amplitude is impaired and in Figure 17-25(c) the phase is impaired but the bits are still identified as belonging to State 11. In Figure 17-25(d) phase has been badly distorted and so the signal is identified erroneously as belonging to State 01. In general, as shown in Figure 17-26, Gaussian noise and intermodulation distortions cause the signal to fall in a circular area about the target point. Amplitude distortion produces radial patterns and phase distortion produces angular patterns.

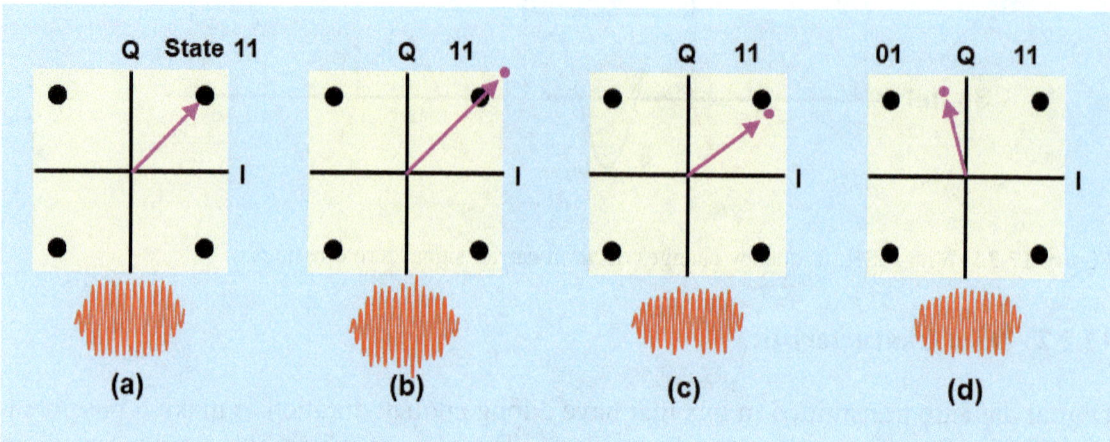

Figure 17-25 Locations of data points on the constellation diagram for various impairments.

The effect of a weak signal (low SNR) on the 16QAM constellation is shown in Figure 17-27. No impairments are shown when E_b/N_o = 11 dB. There are a few impairments (indicated by the large dots) when E_b/N_o = 6 dB and many impairments when E_b/N_o = 2 dB. These figures illustrate the importance of maintaining a high SNR.

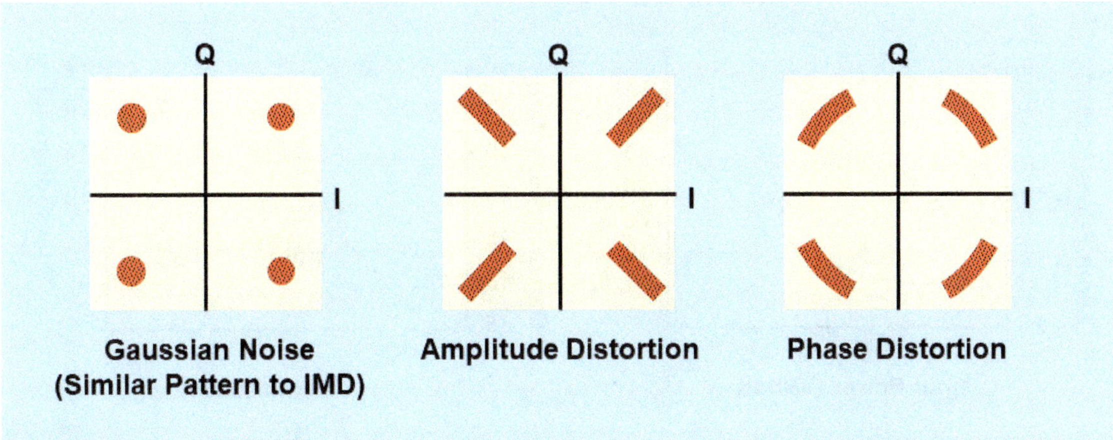

Figure 17-26 Impairment patterns produced by various distortions. (Adapted from: D. M. Goebel and W. L. Menninger, IVEC 2002. © 2002 IEEE.)

17.2.3 Amplifier Design to Reduce Distortion

From the foregoing discussion it is seen that the phase characteristics of the bits are extremely important. In QPSK, a phase error of > 45° will cause a signal to be interpreted in the wrong state. In 16QAM, phase is even more critical. Since phase states are only 30° apart, a phase error of > 15° will place the signal in the wrong state. Thus, it is clear that phase linearity is a very important characteristic of an amplifier.

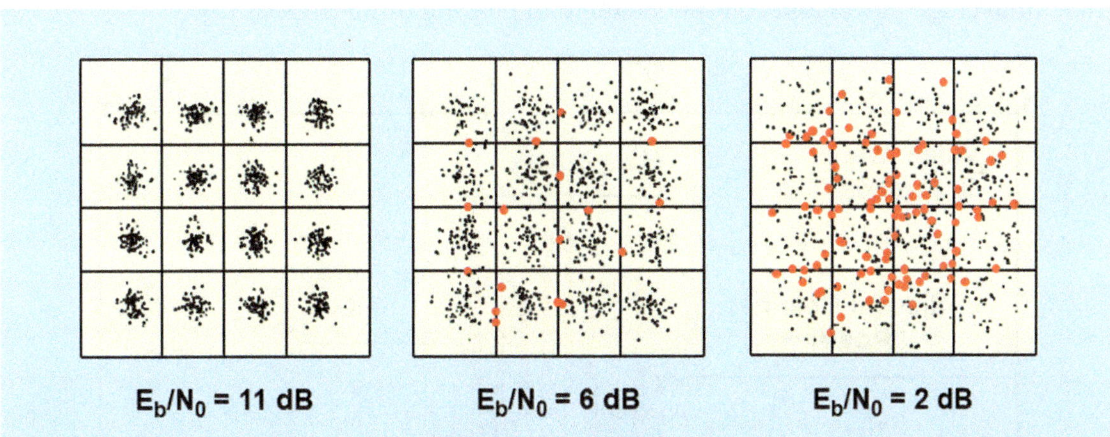

Figure 17-27 Constellation plots for 16QAM with various degrees of impairment. (Adapted from: D. M. Goebel and W. L. Menninger, IVEC 2002. © 2002 IEEE.)

Power and phase characteristics are shown for a conventional TWT in Figure 17-28(a) and for a TWT amplifier designed for phase linearity in Figure 17-28(b). The conventional design produced a phase shift of ~40° at saturation. For the TWT designed to minimize AM/PM, phase shift was < 10° near saturation. The design resulted in slight gain expansion and a nonuniform AM/AM curve [7].

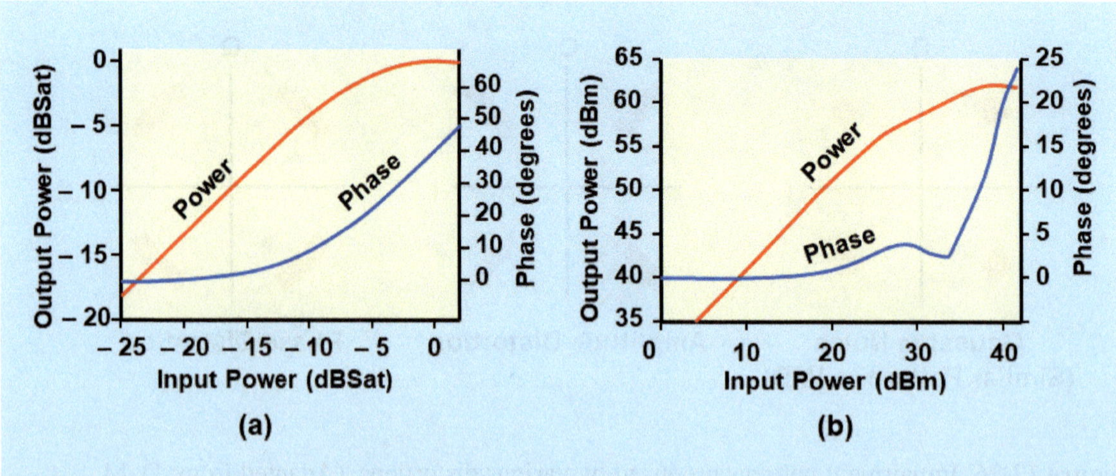

Figure 17-28 Comparison of the phase characteristics of a conventional TWT (a) and a TWT designed for phase linearity (b). (Adapted from: D. M. Goebel and W. L. Menninger, IVEC 2002. © 2002 IEEE.).

Even though TWTs are more efficient than solid-state amplifiers, their efficiency is far less than the maximum that they are capable of at saturation. This is because of the requirements for low intermodulation distortion. These requirements are in addition to the phase linearity requirements discussed earlier. Figure 17-29 shows the third- and fifth-order intermods produced by the eight-tone operation of a Boeing EDD 422H TWT at 150W of average power and continuous random phasing of the signals [8].

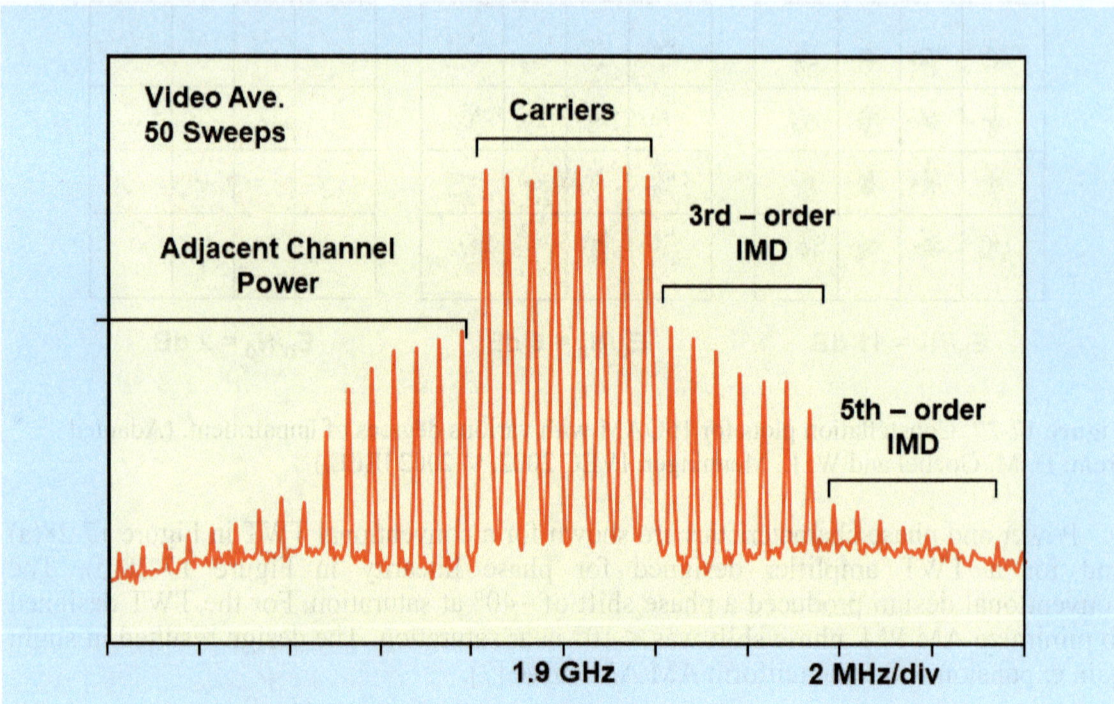

Figure 17-29 Sidebands generated by intermods of carrier frequencies. (Adapted from: D. M. Goebel and W. L. Menninger, IVEC 2002. © 2002 IEEE.)

The reductions in intermods that result from operating backed off are shown in Figures 17-30 and 17-31. When backed off operation is 9 dB below saturation, C/3IM is about 37 dBc. To achieve this level of performance, the helix design was optimized for intermod reduction. As shown in Figure 17-30, a tube with a standard helix had an intermod level about 7 dB higher than the tube with the optimized helix [8].

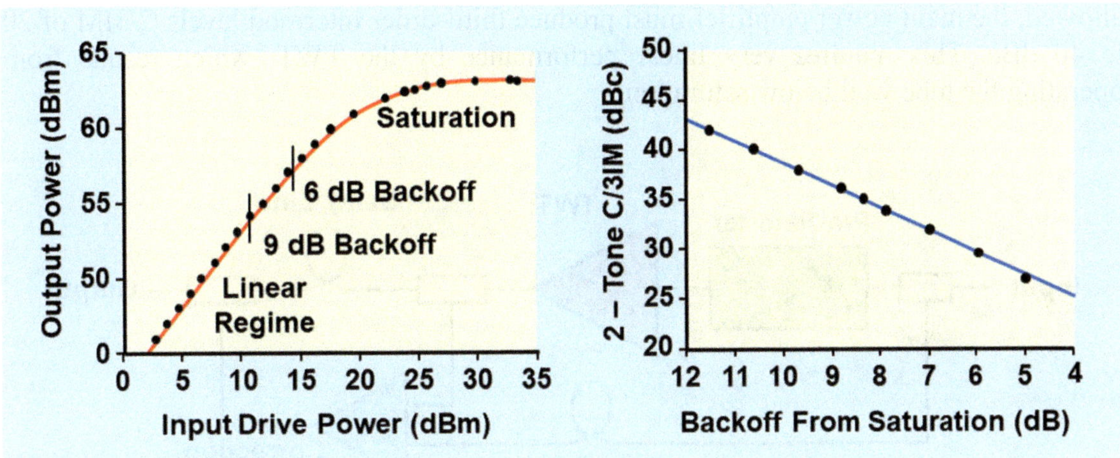

Figure 17-30 Intermod reduction that results from operating backed off. (Adapted from: D. M. Goebel and W. L. Menninger, IVEC 2002. © 2002 IEEE.)

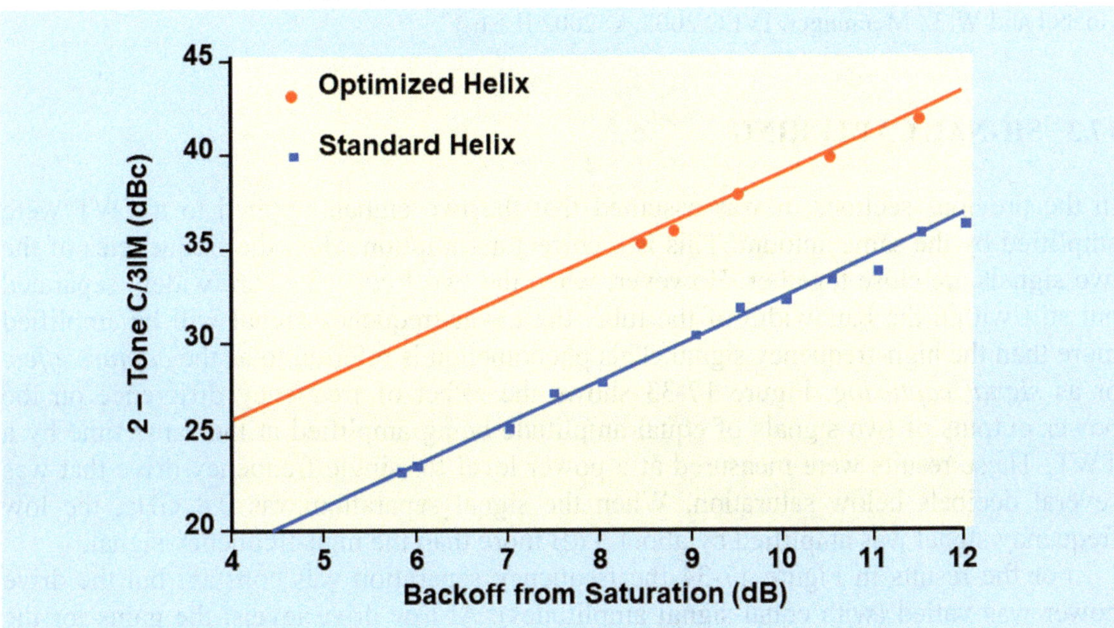

Figure 17-31 Reduction in intermod level by optimizing the helix design. (From: D. M. Goebel, et al., *IEEE Trans. Electron Devices*, 2001. © 2001 IEEE.)

To improve amplifier linearity and reduce intermodulation products, the technique [8] of predistorting the input signal to the amplifier is used (see Figure 17-32). Predistortion offsets the gain compression and phase shifts that cause intermod production. In addition, feedforward circuits [4] are often used to cancel the intermods from the TWT. The input signal and a small part of the amplifier output signal are fed into a correction amplifier, which amplifies the nonlinear signals from the TWT and mixes them out of phase with the TWT output to cancel the intermodulation products.

Predistortion can reduce the intermod level by 10 to 20 dB, and the feedforward circuit can reduce it by about 30 dB. Depending on the regulatory levels of intermods allowed, the main power amplifier must produce third-order intermod levels C/3IM of 20 to 40 dBc. This requires very linear performance by the TWT, which results from operating the tube well below saturation.

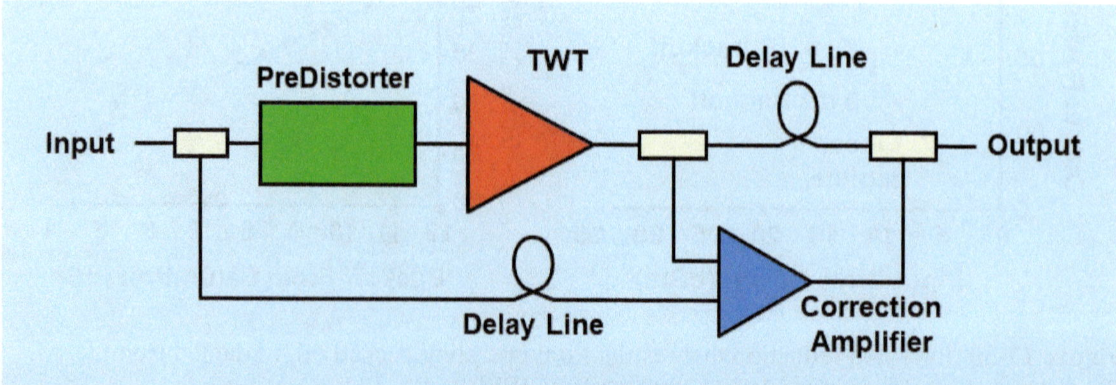

Figure 17-32 Linearizer circuit for minimizing intermodulation products. (Adapted from: D. M. Goebel and W. L. Menninger, IVEC 2002. © 2002 IEEE.)

17.3 SIGNAL CAPTURING

In the previous sections, it was assumed that the two signals applied to a TWT were amplified by the same amount. This is a correct assumption when the frequencies of the two signals are close together. However, when the two frequencies are widely separated but still within the bandwidth of the tube, the lower-frequency signal will be amplified more than the high-frequency signal. This phenomenon is referred to as the *capture effect* or as *signal capturing*. Figure 17-33 shows the effect of frequency difference on the power outputs of two signals of equal amplitude being amplified at the same time by a TWT. These results were measured at a power level for single frequency drive that was several decibels below saturation. When the signal separation was 0.8 GHz, the low frequency signal was amplified by about 3 dB more than the high-frequency signal.

For the results in Figure 17-34 the frequency separation was constant but the drive power was varied (with equal signal amplitudes). At low drive levels, the gains for the two signals were equal. As saturation was approached, the gain for the low-frequency signal became larger than for the high-frequency signal and was about 10 dB higher at saturation.

Signal capturing is of little or no importance in applications of TWTs where the separation in frequencies of the signals is small. In electronic countermeasures applications, however, very broadband TWTs are sometimes used and the frequency separation of the signals can be large. In these cases, signal capturing does occur.

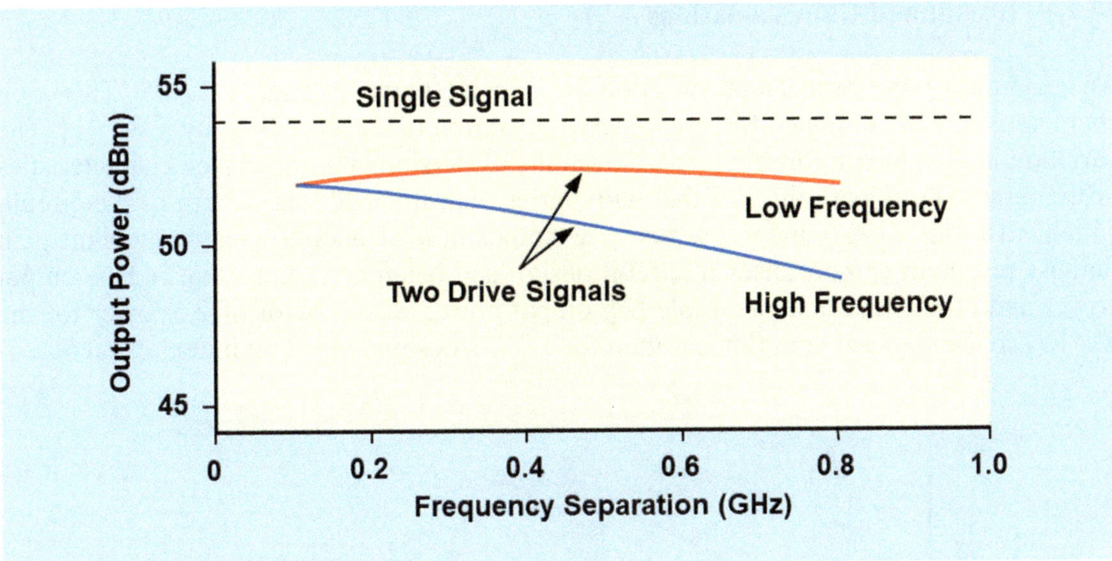

Figure 17-33 Effect of separation of signal frequencies on capturing. (From: R. True, AFTER Program Special Topics Course, November 1984.)

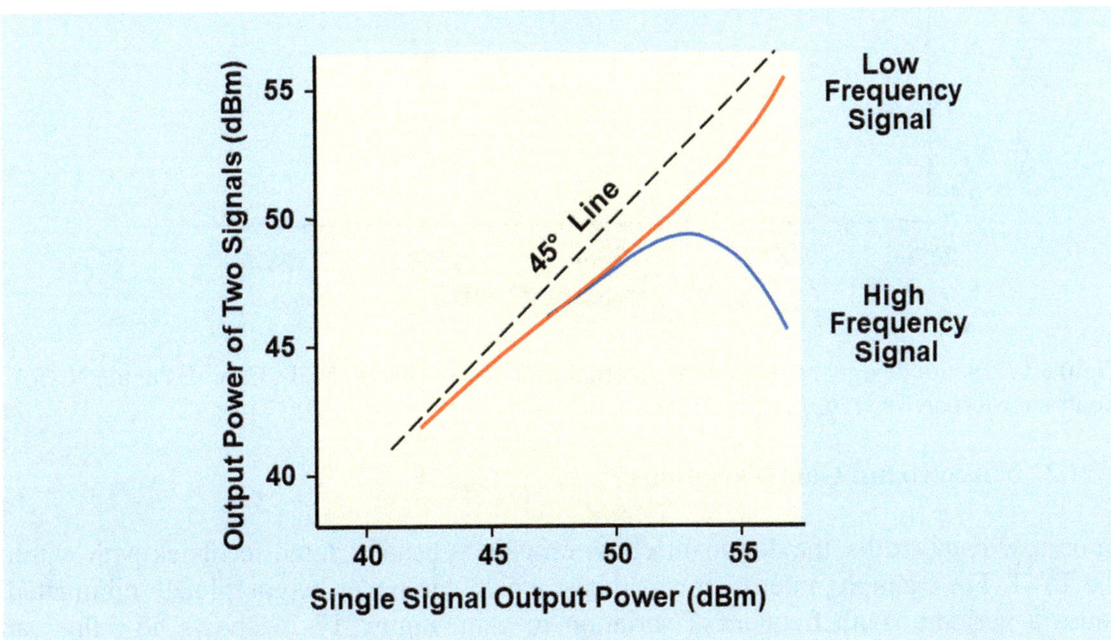

Figure 17-34 Saturation effect on signal capturing. (From R. True, AFTER Program Special Topics Course, November 1984.)

17.4 VARIATIONS WITH FREQUENCY

When the gain versus frequency characteristics of a TWT are examined, two distinct types of gain variation are usually observed. One is a broadband variation for which the gain varies slowly as a function of frequency. The other is a relatively rapid variation of gain in narrow bandwidths.

17.4.1 Broadband Gain Variations

An example of the broadband variation in gain is shown in Figure 17-35. This gain characteristic was computed for small-signal operation of a coupled-cavity TWT [1]. The variation in gain with frequency results from the dispersion and impedance characteristics of the circuit along with the fact that gain varies with the electrical length of the circuit, which, in turn, varies with frequency. The imposition of a tight specification on gain flatness results in compromises in circuit design and beam conditions that reduce output power and efficiency. In an example [1], output power was a factor of 2.5 lower for the TWT with the required gain flatness than for a TWT designed for maximum efficiency.

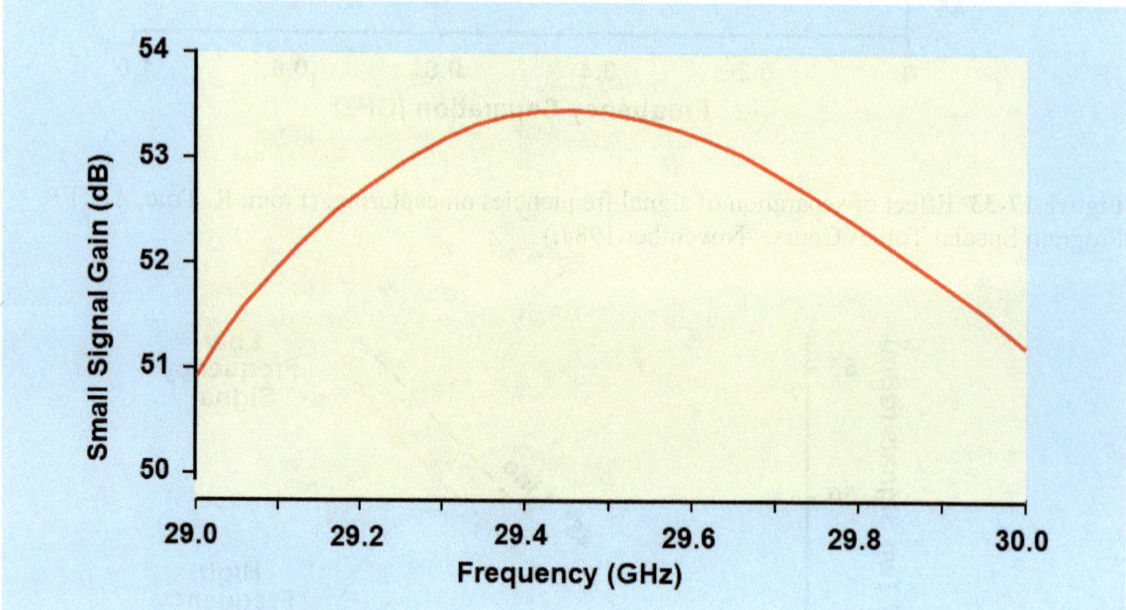

Figure 17-35 Small signal gain with no internal reflections. (From: M. L. Cabrales et al., NASA Contractor Report 185146, January 1990.)

17.4.2 Narrowband Gain Variations

In narrow bandwidths, the dominant effect on gain is usually some feedback path within the TWT. For example, internal mismatches, which can never be completely eliminated, cause a periodic (with frequency) variation in gain. Figure 17-36 shows how this can occur. A mismatch at the output end of the slow-wave structure causes the reflection of a small portion of the signal. This portion is then reflected by a mismatch at the input end of the structure. It then combines, in various phase relationships, with the desired growing wave. The result is that a gain ripple is produced as is shown in Figure 17-37. This particular curve was calculated for the millimeter-wave coupled-cavity TWT mentioned in Section 17.1.4 [4]. The frequency difference between gain peaks depends on the frequency of operation and the length of the RF circuit (which produce a closed-loop phase change of 360°) and is usually in the 100–300-MHz range.

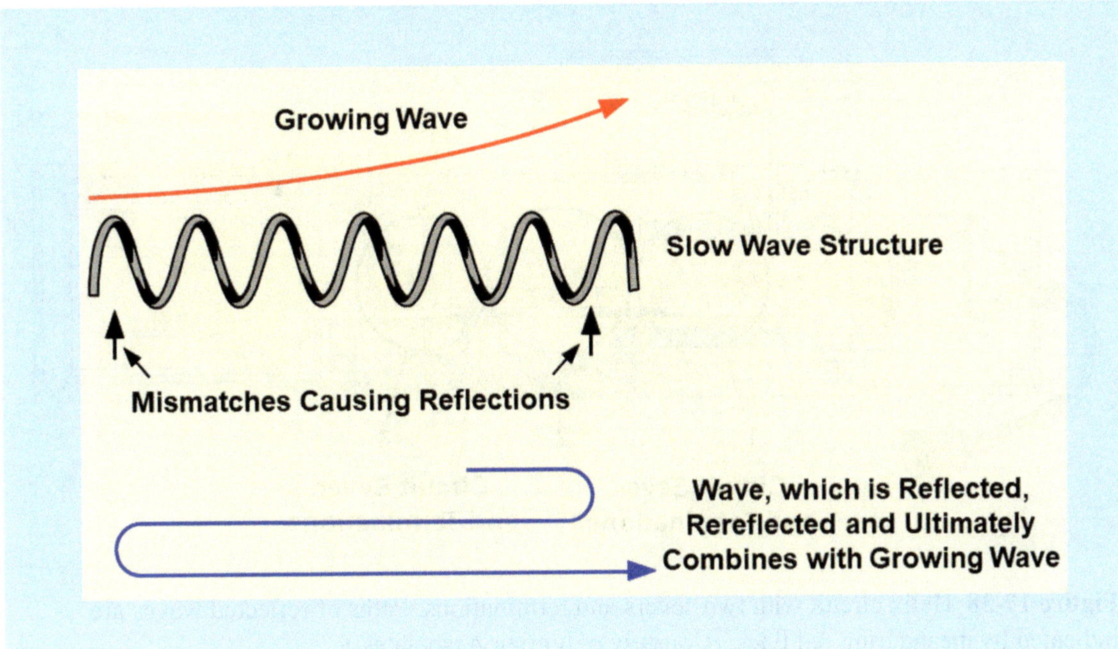

Figure 17-36 TWT section with two internal mismatches.

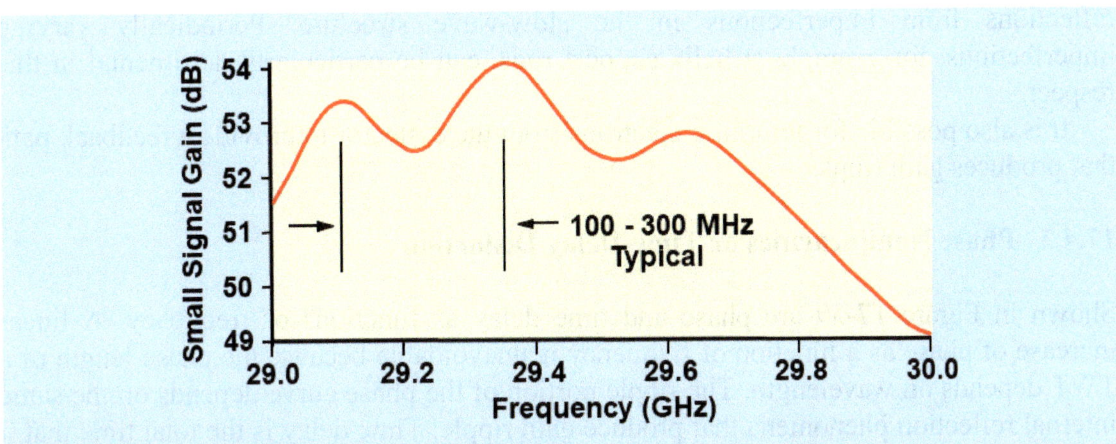

Figure 17-37 Gain ripple with 7% voltage reflections. (From: M. L. Cabrales et al., "NASA Contractor Report 185146," January 1990.)

A more complex example of internal reflections in a TWT is shown in Figure 17-38. Reflections can occur:

1. At the attenuator at the end of the input section of the TWT. These reflections are then rereflected at the input port to the TWT.
2. Within the center section of a TWT with two severs.
3. At the output port. These reflections are then rereflected at the sever or attenuator at the input end of the output section. This feedback path is usually the dominant path because gain is high and reflections from the output port vary with load characteristics. An isolator at the output helps control load reflections.

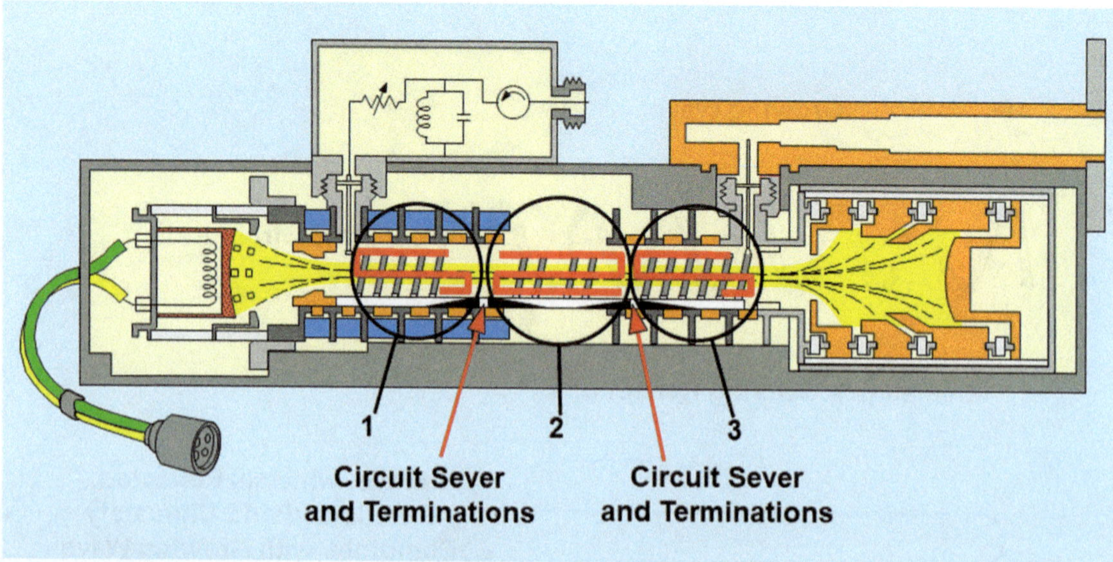

Figure 17-38 Helix circuit with two severs and terminations. Paths of reflected waves are indicated by meandering red lines. (Courtesy of Varian Associates.)

In addition to reflections caused by severs and terminations, there can also be reflections from imperfections in the slow-wave structure. Periodically varying imperfections, for example at helix support rods, can be particularly detrimental in this respect.

It is also possible for returning electrons from the collector to provide a feedback path that produces gain ripple.

17.4.3 Phase Nonlinearities or Time-Delay Distortion

Shown in Figure 17-39 are phase and time delay as functions of frequency. A linear increase of phase as a function of frequency is unavoidable because the phase length of a TWT depends on wavelength. The ripple portion of the phase curve depends on the same internal reflection phenomena that produce gain ripple. Time delay is the total time that it takes for a signal to pass through a TWT and is the derivative of phase delay. Thus, the same mechanisms that cause phase nonlinearities are responsible for time-delay distortion.

The frequency band of interest is indicated in Figure 17-39 to illustrate the fact that, for most TWTs, the narrowband effects dominate the broadband effects. This is true because most TWTs operate over only a small portion of the passband of the circuit and so band edge effects are far from the frequency band of interest.

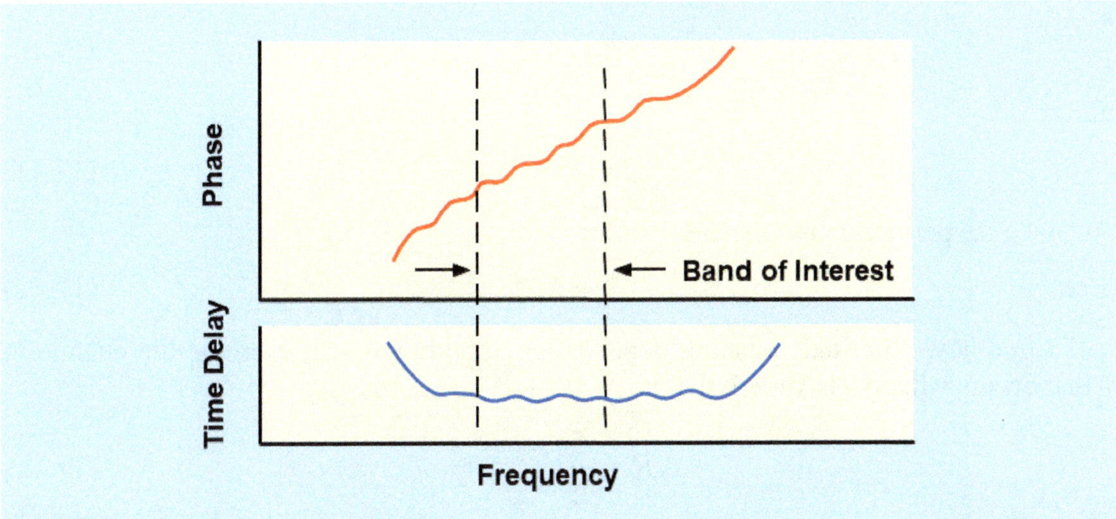

Figure 17-39 Phase nonlinearity and time delay distortion. (Adapted from: *L3 Communications TWT and TWTA Handbook*. 2007.)

17.5 PUSHING AND PULLING

When an operating parameter such as the beam voltage varies, the RF performance of a tube is affected. The pushing figure is a measure of the change in a performance characteristic that occurs as a result of a change in an operating parameter. For example, the cathode- phase pushing figure is the change in phase of the output that results from a change in cathode voltage.

There is also a pulling figure and this has to do with the effect of the load on the tube. The load impedance has an effect on the power output and on the stability of the tube.

17.5.1 Amplitude Pushing

Amplitude-pushing figures are relatively easy to determine. To measure amplitude pushing figures, the change in power output produced by a change in one of the tube operating parameters is measured. Typical specifications for the maximum values for the amplitude pushing figures for a tube with a gridded gun are as follows:

- *Cathode pushing*: 0.5 dB per 1% change in cathode voltage, V_a;
- *Grid pushing*: 0.5 dB per 1% change in grid voltage, V_g;
- *RF drive pushing*: 0.5 dB per 1-dB change in RF input power.

These pushing figures are all specified for operation at or very close to the nameplate operating conditions.

17.5.1.1 Amplitude Pushing with Gridless Guns

For constant efficiency, the RF power output, P_o, is proportional to the product of cathode current and voltage; that is

$$P_o \propto I V_a \tag{17-10}$$

but

$$I = p\, V_a^{3/2} \tag{17-11}$$

and, since the perveance is constant,

$$P_o \propto V_a^{5/2} \tag{17-12}$$

To find how amplitude pushing depends on cathode voltage, examine the change in P_o that occurs when V_a is varied; that is,

$$\Delta P_o \propto \frac{5}{2} V_a^{3/2}\, \Delta V_a \tag{17-13}$$

Now, dividing this equation by (17-12),

$$\frac{\Delta P_o}{P_o} = \frac{5}{2}\,\frac{\Delta V_a}{V_a} \tag{17-14}$$

To express this relation in terms of decibel change in power output per 1% change in cathode voltage, remember that

$$\text{dB} = 10 \log_{10} \frac{P_o + \Delta P_o}{P_o} = 10 \log_{10} \left(1 + \frac{\Delta P_o}{P_o} \right) \tag{17-15}$$

but

$$\log_{10} = 0.434 \ln \tag{17-16}$$

and

$$\ln \left(1 + \frac{\Delta P_o}{P_o} \right) \approx \frac{\Delta P_o}{P_o} \tag{17-17}$$

so

$$\text{dB} = 4.34 \frac{\Delta P_o}{P_o} = 10.85 \frac{\Delta V_a}{V_a} \tag{17-18}$$

As a result, for $\Delta V_a/V_a = 1\%$, the theoretical cathode pushing figure for a gridless gun is

$$\text{cathode pushing} = 0.11 \text{ dB } / \text{ } 1\% \text{ change in } V_a \tag{17-19}$$

The amplitude pushing that results from a change in RF drive level depends on the power transfer characteristic of the tube. If the tube is operating saturated, then, of course, RF drive pushing is zero. However, if the tube is operating below or above saturation, then RF drive pushing must be determined experimentally.

17.5.1.2 Amplitude Pushing with Gridded Guns

The calculation of the dependence of the amplitude-pushing figure on the voltages in a gridded gun is not as straightforward as for the gridless gun. Of course, for constant efficiency, the RF power output is still proportional to the product of cathode current and voltage, but the current depends on grid voltage as well as cathode voltage. The way in which current varies with these voltages depends on the construction of the grid.

For example, if the wires of the grid are very close together, then the grid-cathode region behaves as a space-charge-limited diode and is shielded from the effects of the anode. As a result, current depends on the 3/2 power of grid voltage and is independent of anode voltage. The anode penetrates to the surface of the cathode and so the cathode voltage controls the current. The grid voltage has little effect on the current.

With the grid wire spacing that is used in practice, it is found that the current is almost linearly dependent on grid voltage and that it varies as the square root of cathode voltage, that is

$$I \propto V_g V_a^{1/2} \tag{17-20}$$

This means that

$$P_o \propto V_g V_a^{3/2} \tag{17-21}$$

To find how amplitude pushing depends on cathode voltage, examine the change in P_o that occurs when V_a is varied (with V_g held constant); that is,

$$\Delta P_o \left(V_g \text{ fixed}\right) \propto V_g \frac{3}{2} V_a^{1/2} \Delta V_a \tag{17-22}$$

Now, dividing this equation by (17-21),

$$\frac{\Delta P_o}{P_o} = \frac{3}{2} \frac{\Delta V_a}{V_a} \tag{17-23}$$

In this case, in terms of decibel change in power output per 1% change in V_a, this becomes

$$\text{dB} = 4.34 \frac{\Delta P_o}{P_o} = 6.51 \frac{\Delta V_a}{V_a} \tag{17-24}$$

For $\Delta V_a / V_a = 1\%$

$$\text{cathode pushing} = 0.065 \text{ dB } / \text{ } 1\% \text{ change in } V_a \tag{17-25}$$

and so the expected cathode pushing figure is well below the specification of 0.5 dB/1% given at the beginning of this section for a gridded gun.

To find how amplitude pushing depends on grid voltage, examine the change in P_o that occurs when V_g is varied (with V_a held constant); that is,

$$\Delta P_o\,(\,V_a\text{ fixed}) \propto \Delta V_g\, V_a^{3/2} \tag{17-26}$$

Now, dividing this equation by (17-21),

$$\frac{\Delta P_o}{P_o} = \frac{\Delta V_g}{V_g} \tag{17-27}$$

and so the pushing figure relating the change in power output in decibels to a 1% change in grid voltage is

$$\text{grid pushing} = 0.0434\ \text{dB}\ /\ 1\%\ \text{change in}\ V_g \tag{17-28}$$

and this is over an order of magnitude below the specified value of 0.5 dB/1%.

17.5.2 Phase Pushing

To measure phase-pushing figures, it is necessary to make accurate measurements of relatively small changes in phase of the RF output when an operating parameter is changed. Typical specifications for the maximum values for the phase pushing figures are:

- *Cathode pushing*: 55° per 1% change in V_a;
- *Grid pushing*: 12° per 1% change in V_g;
- *Collector pushing*: 0.002° per 1 V change in collector voltage;
- *Filament pushing*: 10° per 1% change in filament voltage;
- *RF drive pushing*: 10° per 1 dB change in RF input power.

These pushing figures are all specified for operation at or very close to the nameplate operating conditions.

The calculation of phase pushing figures is not as straightforward as the calculation of amplitude pushing figures. For example, consider cathode pushing. The total phase length of the tube, φ, is proportional to the time that it takes for the RF signal to travel through the tube. That time is the length of the RF section divided by the phase velocity, v_p, of the signal on the circuit. As a result,

$$\phi \propto \frac{1}{v_p} \tag{17-29}$$

and the relation between a change in phase velocity and the resulting change in phase is

$$\frac{\Delta\phi}{\phi} = -\frac{\Delta v_p}{v_p} \tag{17-30}$$

For small-signal levels, Pierce's theory gives the relation between a change in beam velocity and the resulting change in velocity of propagation of the signal on the circuit. The result is shown in Figure 17-40. Remember, b is Pierce's velocity parameter", which relates the beam velocity to the cold circuit phase velocity. From this figure, it can be shown that the slope of the velocity lines is nearly 1/2. The reason for this is that the velocity of the slow space-charge wave (the bunch velocity) is controlled by the cold circuit phase velocity, which is constant, as well as by the beam velocity.

Since the slope is 1/2, the change in signal velocity is 50% of the change in beam velocity. As a result,

$$\frac{\Delta\phi}{\phi} = -\frac{1}{2}\frac{\Delta u_o}{u_o} \tag{17-31}$$

Finally, $u_o \propto V_a^{1/2}$ and so

$$\frac{\Delta\phi}{\phi} = -\frac{1}{4}\frac{\Delta V_a}{V_a} \tag{17-32}$$

(Changes in velocity resulting from changes in beam density and beam size are being ignored. These are only a few percent of the cathode-induced changes.) As a result, the relation for the cathode phase-pushing figure is

$$\Delta\phi = -\frac{1}{4}\frac{\Delta V_a}{V_a}\phi = -0.0025\phi \quad \text{degrees/1\% change in } V_a \tag{17-33}$$

For traveling-wave tubes, this phase length is usually ~ 5,000–10,000° and so the cathode phase pushing figure is usually

$$\Delta\phi = -0.0025 \times 5{,}000° \text{ to } 10{,}000° = 12.5° \text{ to } 25°/1\% \text{ change in } V_a \tag{17-34}$$

which is well below the maximum specified value of 55°/1%. It must be remembered that this result is based on Pierce's small-signal theory. In practice, measured cathode phase-pushing figures are a factor of about two above this result.

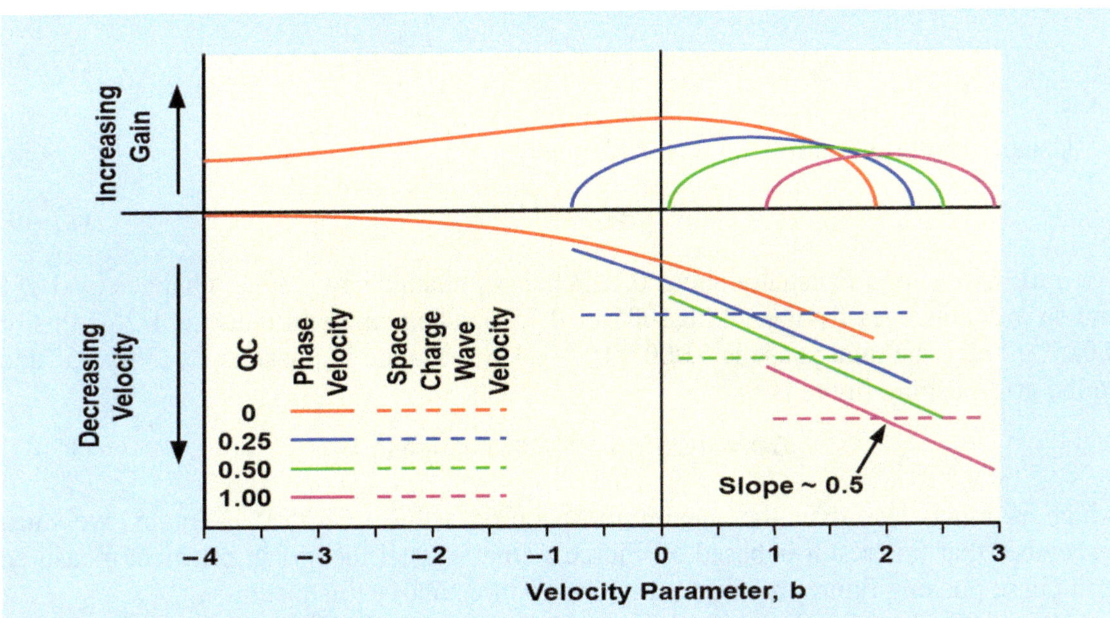

Figure 17-40 Effect of space charge on velocity and gain of the growing wave.

Grid phase pushing occurs because, as grid voltage changes, beam current changes and this, in turn, changes the electron density in the beam. (Here, beam diameter will be assumed not to change.) As electron density changes, plasma frequency changes and so the velocity of the slow space-charge wave, v_{sscw}, changes. As was the case for cathode pushing (and from Figure 17-28), the resulting change in the velocity of the signal on the circuit is one half of the change in v_{sscw}. The velocity of the slow space-charge wave is

$$v_{sscw} = \frac{u_o}{1 + \frac{\omega_q}{\omega}} \approx u_o\left(1 - \frac{\omega_q}{\omega}\right) \tag{17-35}$$

This approximation is valid because ω_q/ω is usually about 0.1.

When ω_q changes, the resulting change in v_{sscw} is

$$\Delta v_{sscw} = -u_o \Delta\left(\frac{\omega_q}{\omega}\right) \tag{17-36}$$

and so

$$\frac{\Delta v_{sscw}}{v_{sscw}} = -\frac{\Delta(\omega_q/\omega)}{1 - \omega_q/\omega} \approx -\Delta\left(\frac{\omega_q}{\omega}\right) \tag{17-37}$$

This means that

$$\frac{\Delta\phi}{\phi} = -\frac{\Delta v_p}{v_p} \approx \frac{1}{2}\Delta\left(\frac{\omega_q}{\omega}\right) \tag{17-38}$$

plasma frequency is

$$\omega_q = F\,\omega_p = F\left(\eta\frac{\rho_o}{\varepsilon_o}\right)^{1/2} \propto \left(\frac{I}{u_o}\right)^{1/2} \propto \sqrt{I} \tag{17-39}$$

Usually, for grid-controlled guns, $I \propto V_g$ and so

$$\omega_q \propto \sqrt{V_g} \tag{17-40}$$

The ratio of ω_q to ω is usually about 0.1. When V_g changes by 1%, ω_q changes by 1/2% and so ω_q/ω changes from 0.1 to 0.1005 or 0.05%. The change in phase is 1/2 of this or 0.025%. Since the total phase is 5,000°–10,000°, the change in phase is 1.25° to 2.5° and so the grid-pushing figure is

$$\Delta\phi = 1.25° \text{ to } 2.5°/1\% \text{ change in } V_g \tag{17-41}$$

which is much less than the maximum specified value of 12°/1%. Again, we must remember that this result is based on Pierce's small-signal theory. In practice, measured grid phase pushing figures are a factor of two to three above this result.

Next, consider phase pushing by the filament voltage. Fluctuations or changes in filament voltage (and current) can have two effects on tube operation:

1. For slow variations in voltage, the cathode temperature changes. If the position of the cathode changes because of thermal expansion or contraction of the cathode support structure, then the cathode current changes and this, in turn, causes phase pushing. Also, if the cathode is not completely space-charge-limited, emission will change with temperature and this will affect phase.

2. Fluctuations in filament current cause variations in the magnetic field generated by the filament. If the filament magnetic field links through the cathode surface, then the focusing of electrons in the gun and beam size are affected. As beam size fluctuates, the potential depression in the beam and the beam velocity vary, affecting the phase length of the tube.

It would appear that the collector should not have an effect on phase because the beam has completed its interaction with the circuit before entering the collector. However, fluctuations in the collector voltage can cause variations in the current reflected by the collector. Even a small current traveling backward through a tube can carry a signal from output to input. As this signal combines with the input signal in various phase relations, the amplitude and phase of the resulting signal that is launched on the circuit vary. As a result, there is an effect on the phase and amplitude of the RF output.

The primary reason why the RF drive level affects phase is that as drive is increased, more and more of the energy in the electron beam is converted to RF power output. As a result, the beam loses energy and its velocity decreases. As the velocity decreases, phase length increases.

17.5.3 Pulling

RF power generated within a microwave tube is transmitted to a load as shown in Figure 17-41. There may be (and, to some extent, always are) impedance mismatches at several places in the output transmission line where RF power is reflected back toward or away from the tube. Reflected power that enters the tube affects tube operation. This effect of the load on tube operation is referred to as pulling.

The pulling effects that may be produced by reflected power are variations in the following:

1. Amplitude of the RF output power;
2. Phase of the RF output power;
3. For an oscillator, frequency of the RF output power.

Load reflections may also cause an amplifier to produce unwanted signals.

There is no straightforward way to predict the pulling effects of a load. Instead these effects must be measured by using loads that produce known voltage standing-wave ratios with known phases.

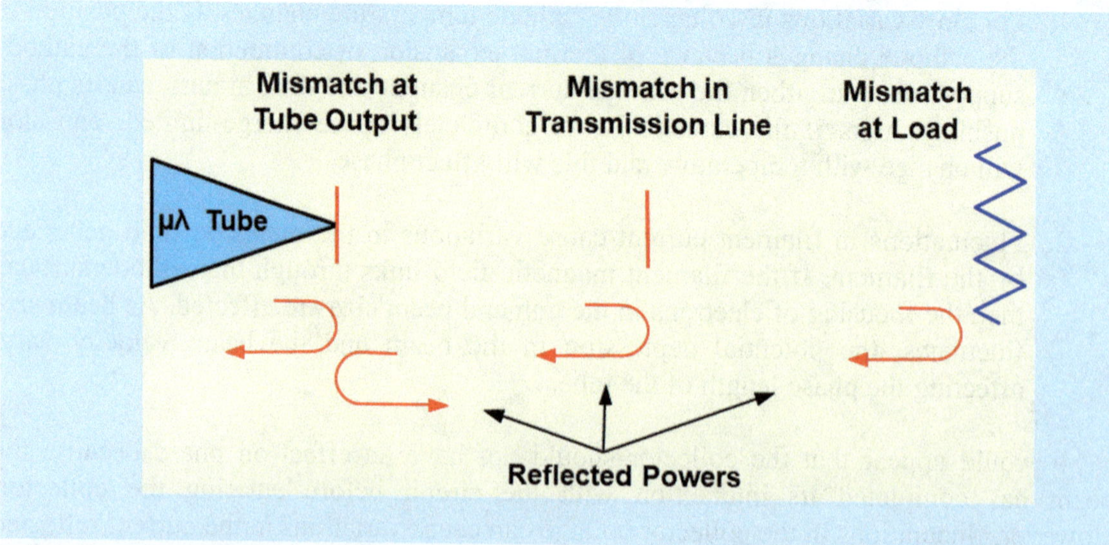

Figure 17-41 Impedance mismatches in the output transmission line of a microwave tube.

REFERENCES

[1] J. W. Hansen et al., "Systems aspects of communications TWTAs or how to deal with the tube manufacturer to your best advantage," Hughes Aircraft Company Electron Dynamics Division Applications Note, August 1982.

[2] H. G. Kosmahl and J. C. Peterson, "A TWT amplifier with a linear power transfer characteristic and improved efficiency, NASA Technical Memorandum 83590," *Tenth Communications Satellite Systems Conference*, Orlando, FL, 1984.

[3] J. W. Hansen, "Linearized TWTAs for low-distortion communications applications," *Microwave Power Tube Conference*, Monterey, CA, 1992.

[4] M. L. Cabrales et al., "Study of methods for the reduction of distortion in high-power traveling-wave tubes," NASA Contractor Report 185146, January 1990.

[5] W. L. Menninger et al., "70% efficient Ku-band and C-band TWTs for satellite downlinks," *IEEE Trans. Electron Devices,* Vol. 52, No. 5, 2005, pp. 673–678.

[6] R. Strauss, "Reliability of SSPAs and TWTAs," *IEEE Trans. Electron Devices,* Vol. 41, No. 4, 1994, pp. 625–626.

[7] D. M. Goebel and W. L. Menninger, "Design and performance of vacuum electron devices for digital communications," *Third IEEE IVEC*, 2002.

[8] D. M. Goebel et al., "Development of linear traveling wave tubes for telecommunications applications," *IEEE Trans. Electron Devices,* Vol. 48, No. 1, 2001, pp. 74–81.

Chapter 18

Noise in Linear-Beam Tubes

There are many sources of noise in a microwave tube. Some are unavoidable, but most can be minimized or eliminated through proper design and construction. This chapter provides discussions of important noise sources and techniques that can be used to minimize their effects. Noise phenomena in crossed-field devices are still not understood. The chapter starts with a description of thermal agitation noise, which is the noise basis against which all other noise is compared, and a review of the definitions of noise figure. Then, discussions center on noise in linear-beam tubes where noise phenomena are well understood.

18.1 THERMAL AGITATION NOISE

Thermal agitation noise results from the minute voltage generated by the random motion of free electrons in conductors. The voltage produced by thermal agitation has energy distributed throughout the entire frequency spectrum. In the frequency band $f_2 - f_1$, the effective voltage produced, e, is given by

$$e^2 = 4\,k\,T \int_{f_1}^{f_2} R\,df \tag{18-1}$$

where k = Boltzmann's constant, T = absolute temperature, and R = resistive component of impedance. For R independent of frequency and, letting B be bandwidth,

$$e^2 = 4\,k\,T\,R\left(f_2 - f_1\right) = 4\,k\,T\,R\,B \tag{18-2}$$

Thermal agitation noise sets the ultimate limit to the smallest signal that can be amplified without being lost in a background of noise. For example, for R = 1/2 MΩ, T = 300K, and B = 5 kHz, the effective noise voltage, e, is 6.4 μV. Signal amplitudes on the order of this voltage level (6.4 μV) or smaller would be distorted or lost in the noise.

The thermal power, N_g, generated in a resistance is

$$N_g = \frac{e^2}{R} = 4\,k\,T\,B \tag{18-3}$$

If R is connected to a matched load resistance, R_L, maximum noise transfer occurs. Then the noise voltage, e, divides equally between R and R_L and so the noise power, N_d, delivered to R_L is

$$N_d = \frac{\left(\frac{e}{2}\right)^2}{R} = k\,T\,B \tag{18-4}$$

This is the theoretical minimum noise level. At room temperature, its value is

$$N_d \;(\text{room temp}) = -144\,\text{dBW/ MHz} = \text{-174 dBm/Hz} \tag{18-5}$$

That is, for a 1-MHz bandwidth, N_d is 144 dB below 1W (i.e., 4×10^{-15} W).

18.2 DEFINITIONS OF NOISE FIGURE

For an amplifier or receiver that

1. Is noise free;
2. Has gain G;
3. Is perfectly matched to its input termination;

the output noise power is $kTBG$. The noise figure, F, for a real amplifier or receiver that generates noise is

$$F = \frac{\text{actual output noise power at 290K}}{\text{theoretical minimum noise power at 290K}} \tag{18-6}$$

or

$$F = \frac{N_1}{k\,T_o\,B\,G} \tag{18-7}$$

where N_1 is the measured output noise power at temperature T_o = 290K. The amount of noise added, N_a, by an amplifier is

$$N_a = N_1 - k\,T_o\,B\,G = (F-1)\,k\,T_o\,B\,G \tag{18-8}$$

For a noise-free amplifier, $N_a = 0$ and $F = 1$.

Noise figure can be expressed in decibels as follows

$$F_{dB} = 10\log_{10} F \tag{18-9}$$

For a noise-free amplifier, $F_{dB} = 0$.

The equivalent noise temperature, T_e, of an amplifier is defined using

$$N_a = k\,T_e\,B\,G \tag{18-10}$$

and is

$$T_e = \frac{N_a}{k\,B\,G} = (F-1)\,T_o \tag{18-11}$$

18.3 OVERVIEW OF NOISE PHENOMENA

There are three unavoidable sources of noise in a microwave tube:

- Shot (Schottky) noise;
- Velocity noise;
- Thermal agitation noise.

Shot noise results from the discrete nature of the electron and the fact that electron emission is a random process. Velocity noise occurs because of the wide distribution (Maxwellian) in velocities of electrons emitted by the hot cathode. This noise is often referred to as Rack velocity noise, named after A. J. Rack, who first described a technique for analyzing the effects of this noise. Thermal agitation noise is present because the RF circuit, which has loss, is normally at room temperature or above.

In addition to these three unavoidable noise sources, there are many other sources of noise that depend largely on some aspect of design and construction. Most of these can be made insignificant through the use of careful design and construction procedures. Some of these noise sources result from:

- Partition effect;
- Flicker effect;
- Collision ionization;
- Secondary and reflected electrons;
- Lens effects;
- Noise growth;
- Poor insulation;
- Varying insulator charges;
- Microphonics resulting from vibration;
- Multipactor;
- Power-supply induced noise, such as hum.

The most significant noises in a linear-beam tube are usually those produced by the cathode (shot and velocity). These noises are modified in significant ways as the electron beam travels through the electron gun and into the RF region of the tube.

An extremely important characteristic of noise impressed on an electron beam is that noise propagates like any other RF signal on the beam. That is, noise travels as fast and slow space-charge waves. The two sources that produce these waves are the shot effect current and the velocity noise. In the general case, the two sources are assumed to be uncorrelated, and so each source produces fast and slow waves. As these waves propagate along the beam, noise current and velocity standing waves are independently produced.

With no correlation between current and velocity noise, the minimum noise figure for a TWT is predicted to be 6.5 dB. With correlation (produced by the low-velocity drift region in the gun) noise figures below 3 dB have been achieved. It has also been found that the use of a large axial magnetic field in the acceleration region of the gun and in the input section of the RF structure further inhibits noise.

Some of the discussions in this chapter relate primarily to low-power, low-noise TWTs that are no longer used in modern systems. However, the noise phenomena in these TWTs exist, to some extent, in all TWTs so it is of value to discuss them.

18.4 NOISE IN ELECTRON GUNS

For a discussion of noise properties, it is convenient to divide the electron gun into four regions as shown in Figure 18-1. The unavoidable sources of noise at the cathode surface (Region I) are velocity noise as described by Rack and shot noise. There may also be flicker noise but this is a low-frequency noise that is negligible at microwave frequencies. Flicker noise will be discussed at the end of this section.

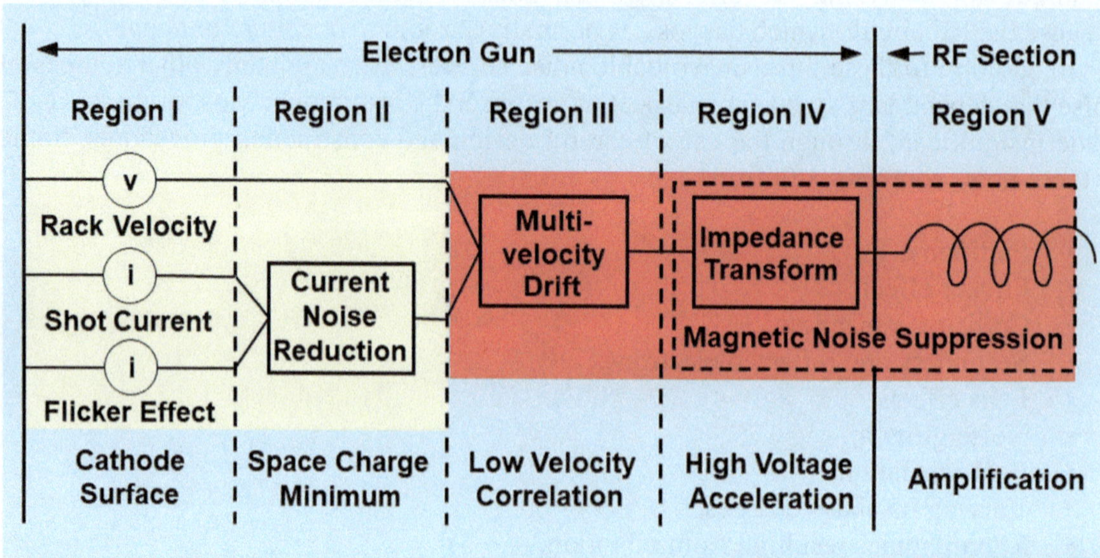

Figure 18-1 Noise regions in an electron gun.

Because of the initial emission velocities of the electrons, the potential of the space-charge cloud (Region II) near the cathode surface is slightly below cathode potential. Low-velocity electrons are returned to the cathode by this potential minimum. The amount by which the potential is depressed fluctuates as the cathode current fluctuates (shot effect). Thus, the number of electrons returned to the cathode fluctuates with time and this, in turn, changes (reduces at low microwave frequencies) the shot noise current passing the potential minimum. Additional noise may be introduced at the potential minimum by ions and reflected electrons from the collector. It turns out that the velocity distribution of the electrons is unaffected by the potential minimum.

Region III exists in low-noise TWTs through careful design of the cathode configuration and the electrodes near the cathode. In the gun of a high-power TWT this region, if it exists at all, is so short that it has a negligible effect on noise performance.

In low-noise TWTs, electrons drift at low velocities (energy <1 eV) for an extended distance in Region III. Noise current and velocity become correlated to some degree, and this reduces noise figure by about 3 dB. The reason for this correlation is not clear.

When the electron beam emerges from Region III and is accelerated in Region IV, space-charge waves are launched and a standing wave pattern is produced. As is the case in an RF transmission line, a sudden change in impedance produces an increase in the standing wave ratio. In a low-noise TWT, an increase in the standing wave ratio results in an increase in noise figure. To avoid this problem, several electrodes are used in Region IV to control the rate of voltage rise (and the rate of impedance change). In high-power microwave tubes, the function of the electron gun is to produce a well-focused high-perveance beam and no attempt is made to control the rate of impedance change.

As the electron beam emerges from the electron gun into Region V, care is taken in low-noise tubes to position the entrance of the RF structure at the noise minimum. A strong magnetic field in Regions IV and V has been found to reduce noise. The phenomena occurring in Regions I–IV are described in Sections 18.5 to 18.7.

18.5 NOISE GENERATION AT THE CATHODE

18.5.1 Shot Noise

Shot noise [1] results from the random emission of electrons from the cathode. The randomly emitted electrons are a component of the noise current in the electron beam. This noise current is coupled to nearby RF circuits as discussed in Chapter 8 and illustrated in Figure 18-2. The circuit current can be found from a Fourier analysis of the induced current pulses. The mean square of the current fluctuations, i^2, depends only on emitted current, I_o, and bandwidth, B, as follows:

$$i^2 = 2\,e\,I_o B \tag{18-12}$$

If this current is passed through a resistance, R, the noise power generated is

$$i^2 R = 2\,e\,I_o B\,R \tag{18-13}$$

For example, assuming $R = 100\Omega$, $I_o = 1\text{A}$, and $B = 1$ MHz, the noise power generated is

$$i^2 R = 3.2\times10^{-11}\ \text{W/ MHz} = -105\,\text{dB/ MHz} \tag{18-14}$$

The resulting noise figure is

$$F = 39\ \text{dB} \tag{18-15}$$

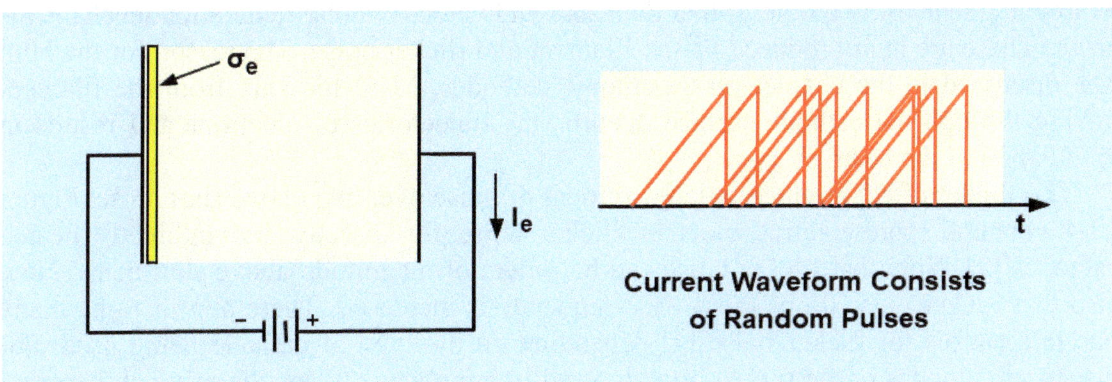

Figure 18-2 Random emission of electrons from a cathode produces shot noise.

18.5.2 Velocity Noise

Velocity noise arises from the distribution in velocities of the electrons emitted from a hot cathode. The distribution is Maxwellian and, at 888°C, is as shown in Figure 18-3.

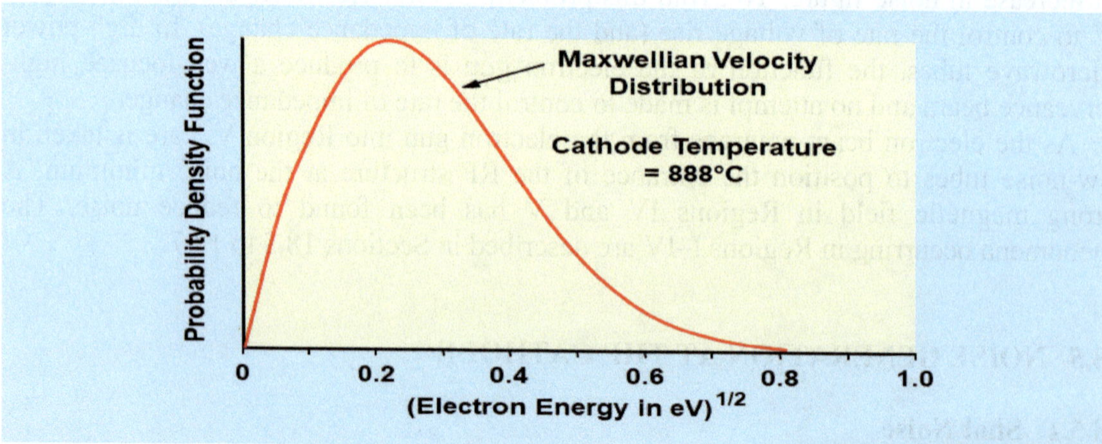

Figure 18-3 Velocity distribution of electrons from a cathode at a temperature of 888°C.

The multivelocity flow of electrons from the cathode that results from the Maxwellian velocity distribution is difficult, if not impossible, to treat analytically. Rack [2] showed that at low frequencies, this multivelocity flow can be replaced by a single-velocity flow with the same mean square fluctuation in velocity, which is

$$v^2 = (4 - \pi)\,\eta \frac{k\,T_c}{I_o} B \tag{18-16}$$

where T_c is the temperature of the cathode in Kelvins.

We will discuss in Section 18.8.1 how Peterson and Pierce used the Rack velocity expression to predict noise behavior at microwave frequencies. Cutler and Quate found surprisingly good experimental agreement with the Pierce prediction.

18.5.3 Other Noise Generation Mechanisms

At low frequencies, two other (other than shot and velocity) noise generation mechanisms occur. These are hum produced by the filament and flicker noise. The reason for the hum was discussed in the chapter on thermionic cathodes. Magnetic flux from the filament linking through the cathode surface disturbs the trajectories of electrons and results in modulation of the beam shape.

The flicker effect produces a large amount of noise over and above shot noise. Figure 18-4 contains representative plots of flicker noise for directly and indirectly heated cathodes [3]. Note that flicker noise can be orders of magnitude above shot noise. Note also that flicker noise drops rapidly as frequency is increased. There appear to be many possible causes for flicker noise [4] depending on the type of cathode being used. For dispenser cathodes of the types normally used in microwave tubes, the original theory of flicker noise proposed by Schottky may apply.

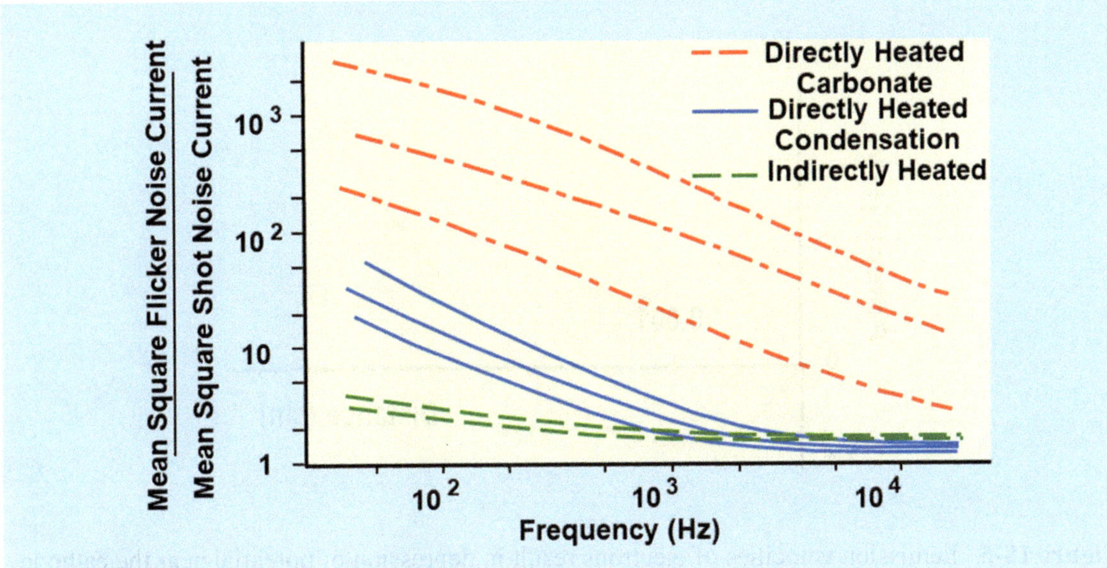

Figure 18-4 Flicker noise for some directly and indirectly heated cathodes. (Adapted from: *The Oxide-Coated Cathode* by G. Herrman and P. S. Wagener, Chapman and Hall, 1951.)

Schottky attributed the flicker effect to foreign atoms arriving and departing from the cathode surface at random. For temperature-limited cathodes, he predicted that the mean square noise current should be proportional to the square of emission current density and, at high frequencies, inversely proportional to the square of frequency. Flicker noise was predicted to be even more strongly suppressed by space charge than shot noise.

A dispenser cathode consists of a monoatomic layer of barium on oxygen on the substrate metal (tungsten in a B cathode, osmium in an M cathode, and so forth). During operation of the cathode, there is continual evaporation of atoms from the surface and migration of additional atoms from the tungsten substrate to the surface. The evaporation of atoms from the surface is a random thermal process that produces fluctuations in work function and, thereby, in emission current. These fluctuations are thought to be the primary source of flicker noise in a dispenser cathode.

18.6 THE SPACE-CHARGE MINIMUM REGION

The electron emission velocities were ignored when we examined space-charge-limited flow. Emission velocities are included in the analysis of noise behavior. Because of these velocities, there is a small depression in the potential near the cathode surface. This depression, which is indicated in Figure 18-5, reflects low energy (< ~0.2 eV) electrons and returns them to the cathode. At the potential minimum, the energies of all other electrons are reduced by ~0.2 eV.

Figure 18-6 shows plots of electron positions as functions of time [5]. Each plot also shows the distance from the cathode of an electron at time t and the time at which the electron must have left the cathode surface (depending on its initial velocity) to arrive at that position. Note that the time required for electrons to travel to the potential minimum (located at 0.0075 mm) is only ~0.02 ns.

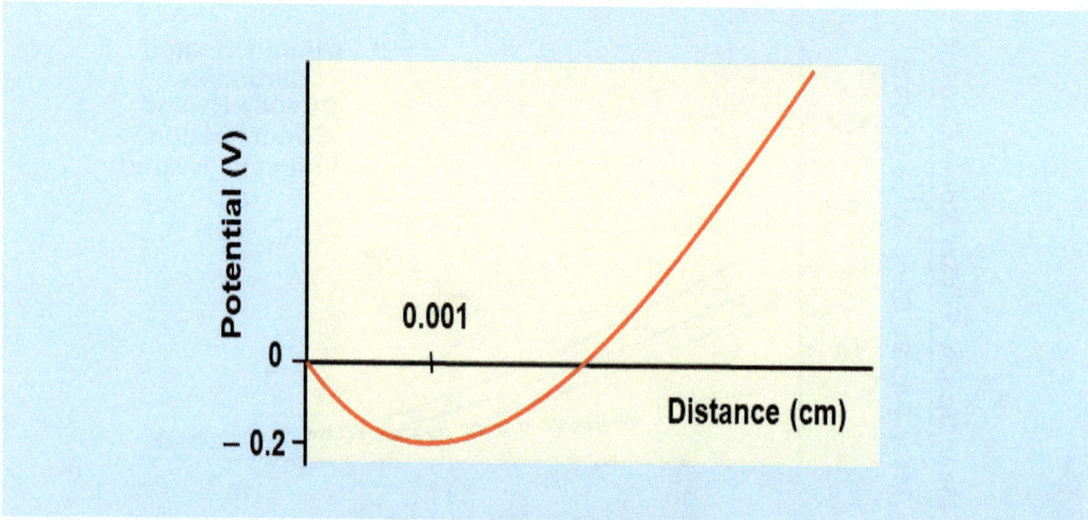

Figure 18-5 Eemission velocities of electrons result in depression of potential near the cathode.

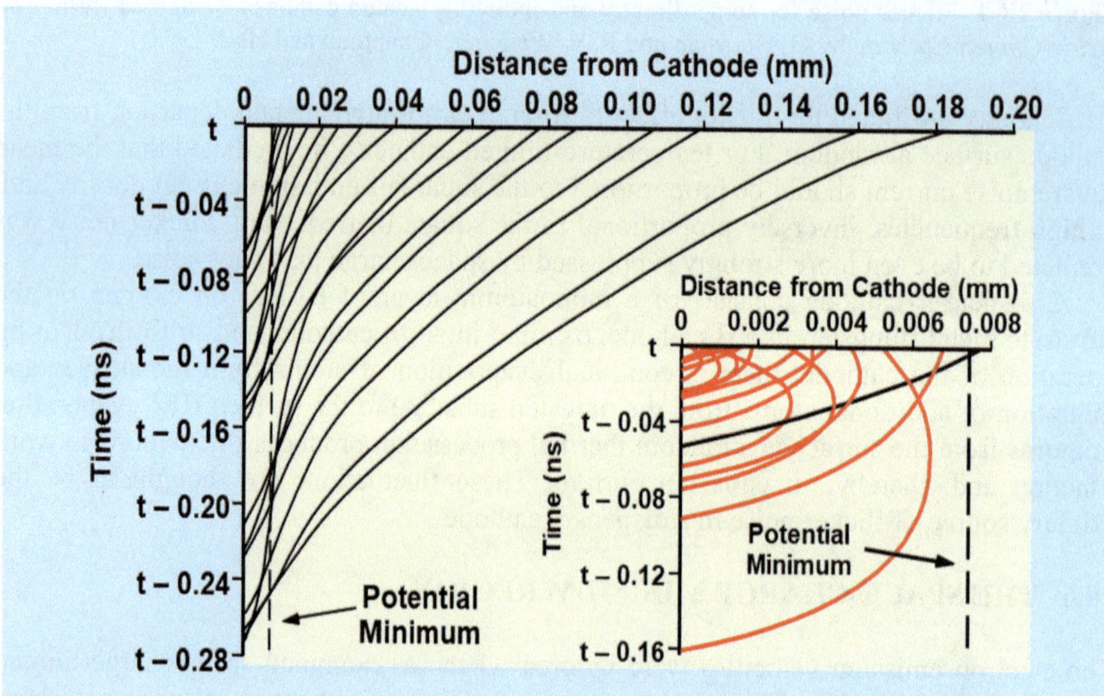

Figure 18-6 Electron velocity plots resulting from the Maxwellian velocity distribution in Figure 18-3. (From: P. K. Tien, and J. Moshman, *Jour. App. Phys.*, September 1956.)

18.6.1 Rack Noise Invariance

Because the initial velocity distribution is Maxwellian and because the velocities of all electrons that reach the potential minimum are reduced by the same amount, the velocity distribution at the potential minimum is Maxwellian. Figure 18-7 shows velocity distributions at the cathode surface and at the potential minimum.

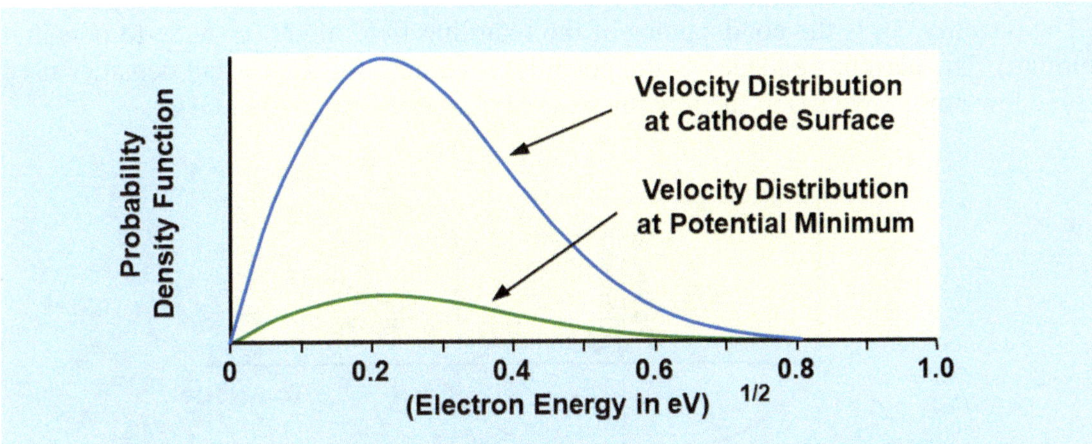

Figure 18-7 Electron velocity distributions at the cathode surface and at the potential minimum.

18.6.2 Shot Noise Reduction

Because emission is random (shot effect), the potential minimum produced by the electrons fluctuates with time. The increase in depression produced by a burst of electrons is shown in Figure 18-8. The additional depression increases the number of reflected electrons, thereby reducing the number of electrons in the burst that pass the minimum. As a result, the potential minimum reduces the shot noise current. The amount of reduction depends on frequency and approaches 100% at low frequencies. Fluctuations in the potential minimum do not change the average velocity or the velocity distribution of the electrons.

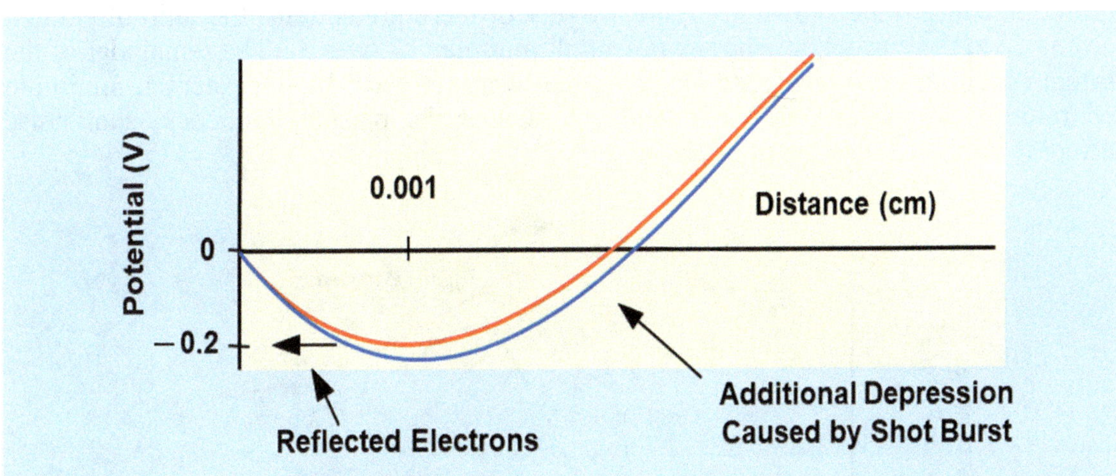

Figure 18-8 Perturbation of potential minimum by shot burst.

At frequencies approaching the electron plasma frequency, a resonance condition exists that enhances shot noise. One explanation for this resonance may be understood with the aid of Figure 18-9 [6]. The equivalent circuit for the cathode-to-potential-minimum region may be thought of as the capacitance of this region in parallel with an inductance that represents the inertial effect of the space-charge cloud at the minimum.

The quantity, G, is the conductance of the retarding field diode (cathode to potential minimum). The plasma frequency at the potential minimum for the current densities used in some low noise TWTs is in the low microwave frequency range (3–4 GHz).

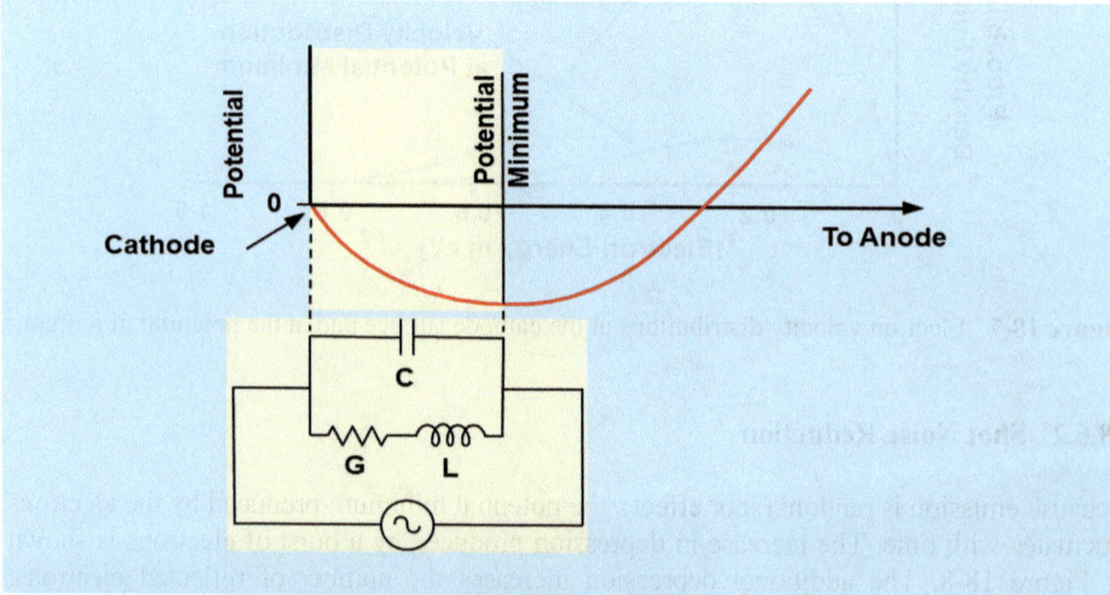

Figure 18-9 Circuit analog for potential minimum region. (From: R. W. Peter, *Noise in Electron Devices*, published by MIT and John Wiley & Sons, 1959.)

Predictions of the shot noise reduction factor Γ^2 (the amount by which the shot current is reduced) are shown in Figure 18-10. For these predictions, the ratio of emitted current, I_s, to the current passing the potential minimum, I_o, was 5. (The remainder of the current consisted of low-velocity electrons that were reflected by the potential minimum and returned to the cathode.) For frequencies near the plasma frequency, shot noise current is increased by a factor of about 2.

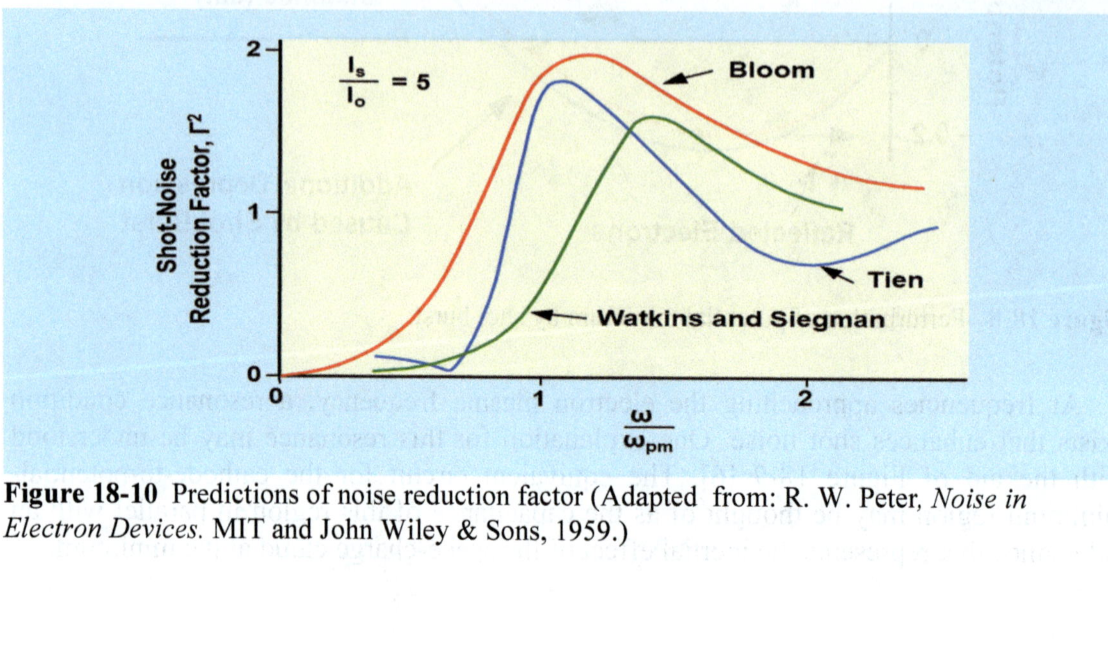

Figure 18-10 Predictions of noise reduction factor (Adapted from: R. W. Peter, *Noise in Electron Devices*. MIT and John Wiley & Sons, 1959.)

18.6.3 Other Noise Effects

Charged particles other than electrons from the cathode will also affect the potential minimum. For example, positive ions generated in or near the electron gun will travel toward the cathode. As they pass through the potential minimum they cause the voltage to rise, so additional current from the cathode can pass the minimum region. Since the ion production process is random, the fluctuations of the potential minimum that they produce are random and the resulting current fluctuations are random.

Another source of charge that can affect the potential minimum is high-velocity secondary electrons, primarily from the collector. These electrons are guided through the tube in the reverse direction by the magnetic focusing field. If their interaction with the RF fields is such that they gain (or at least do not lose) energy, they will be able to reach the potential minimum region. The secondary emission process is a noisy process and so secondary electrons in the potential minimum region produce random fluctuations of the potential minimum, which, in turn, produce an additional component of noise modulation on the emission current passing the potential minimum.

Finally, it is noted that there is evidence that noiselike fluctuations of the space-charge cloud in the potential minimum region may occur. These fluctuations occur on a large scale in the vircator (virtual cathode oscillator). In this device a dense beam of electrons is used to produce a fluctuating potential minimum. Reasonably efficient conversion of beam energy to generally noisy RF power takes place.

18.6.4 Noise Space-Charge Waves

The propagation of noise as fast and slow space-charge waves was predicted by Pierce [8] and by Peterson [9]. They carried out analyses of the noise resulting in an electron beam after acceleration of the electrons in the gun. They used the equations of Llewellyn and Peterson [10] for a space-charge limited diode. We will not reproduce the Llewellyn-Peterson equations here, but will point out that two of the initial conditions for these equations are the noise current and velocity at the potential minimum. Because the dc velocity of the electrons is zero at the potential minimum, the equations predict that the effect of shot noise current vanishes. This is equivalent to using a shot-noise reduction factor of 0, which, in turn, is equivalent to assuming that fluctuations of the potential minimum produced by the shot current completely eliminate shot noise current beyond the potential minimum. Thus, only Rack velocity noise was used to predict noise in an electron beam.

During acceleration of the electrons in the gun, some of the Rack velocity fluctuations result in current noise, and so both current and velocity noise exist in the electron beam. Pierce used the noise current and velocity at the exit from the gun as input conditions to a TWT and used his TWT theory to predict fast and slow noise space-charge waves on an electron beam.

The theories of Peterson and Pierce (and of Rack) met with considerable skepticism. However, the December 1950 issue of *Physical Review* [11] published C. C. Cutler and C. F. Quate's paper "Experimental Verification of Space-charge and Transit Time Reduction of Noise in Electron Beams." The noise-measuring apparatus used by Cutler and Quate is shown in Figure 18-11. A cavity was used to detect a portion of the noise current in the electron beam. A shutter could be moved across the beam to investigate partition current (to be discussed later in this chapter).

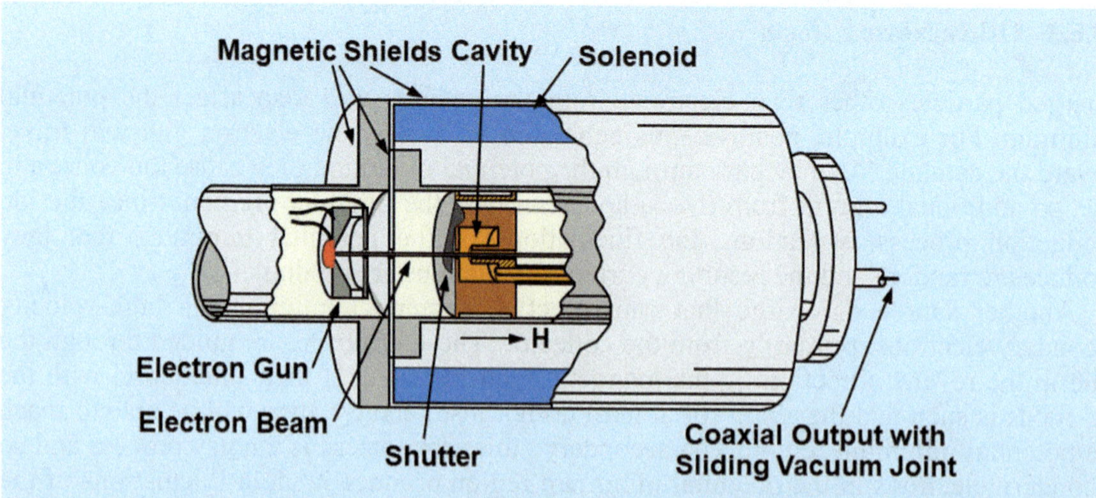

Figure 18-11 Cutler and Quate's noise measuring apparatus. (From: C. C. Cutler and C. F. Quate, *Phys. Rev.*, December 1950.)

The key result of the Cutler-Quate experiment is shown in Figure 18-12. Measured noise power was compared with the predictions of Pierce. The existence of noise space-charge waves was clearly verified. In addition, the variation of the position of the minimum with voltage and with the transit angle of the electrons in the gun agreed with the theoretical prediction. The reason that the position of the minimum differs from the predicted minimum by 0.4 in (10.2 mm) was thought to be that calculations were based on an infinite plane diode configuration rather than a cylindrical beam configuration in the gun.

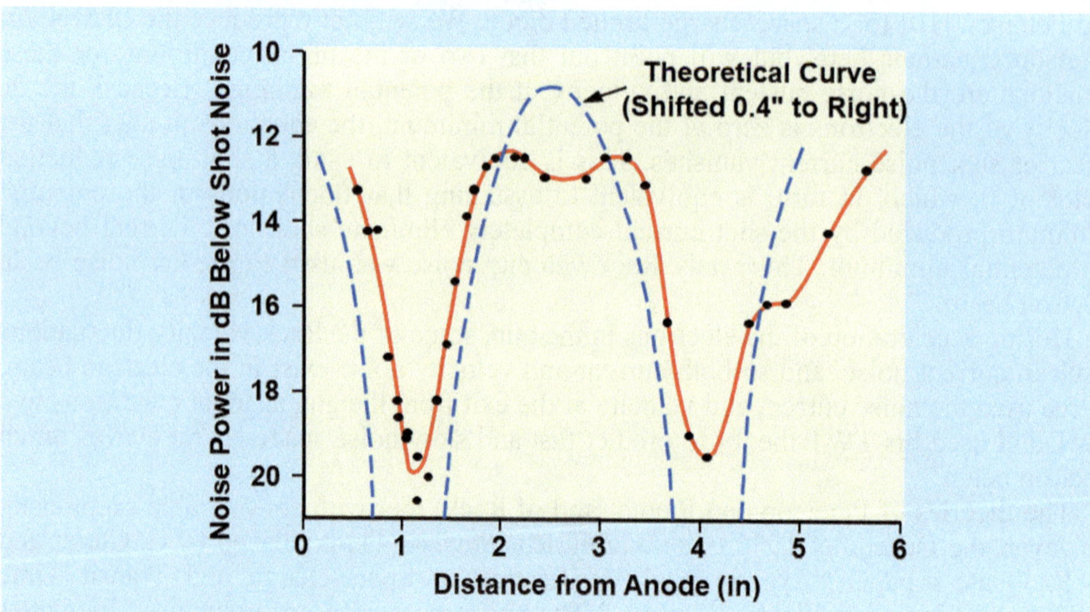

Figure 18-12 Noise power versus distance from the anode. (From: C. C. Cutler and C. F. Quate, *Phys. Rev.*, December 1950.)

The absolute values of the noise levels in Figure 18-12 were determined by comparing them with the temperature-limited shot noise. This was done by reducing the cathode temperature and measuring the noise power as a function of beam current as shown in Figure 18-13. The slope of the curve at low beam currents shows the shot noise power to be proportional to current as expected. At larger beam currents, the noise power was reduced by space charge. The greatest total reduction of noise power occurred at the first minimum and was 20 dB below the shot noise level.

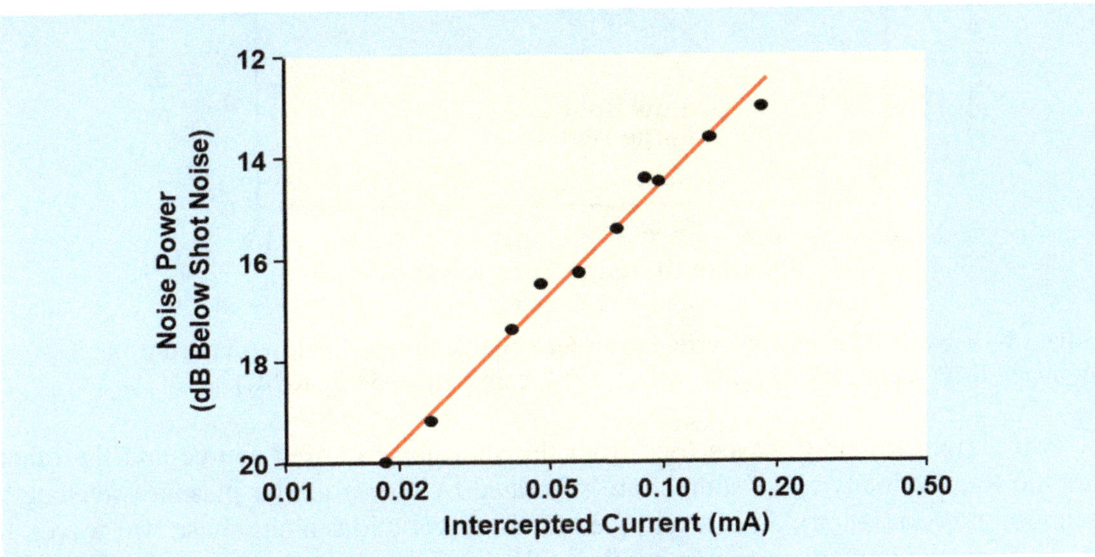

Figure 18-13 Noise power versus beam current for temperature limited emission. (From: C. C. Cutler and C. F. Quate, *Phys. Rev.* December 1950.)

18.7 RF SECTION NOISE PHENOMENA

The electron beam acts as a transmission line to carry space-charge waves launched by shot current and Rack velocity noise generators in the potential minimum region. From the shot current source, fast and slow current waves emerge and produce a current standing wave. Associated with this current wave is a velocity standing wave that is displaced one quarter of a plasma wavelength from the current wave.

18.7.1 Impedance Transformation

The Rack velocity noise generator produces fast and slow velocity waves that form a velocity standing wave. Associated with this velocity wave is a current standing wave that is displaced one quarter of a plasma wavelength from the velocity wave. These standing waves produced by the velocity generator are shown for a space-charge limited planar diode in Figure 18-14 [12]. Velocity and current waves for two space-charge conditions are shown. For a high space charge, a velocity minimum and current maximum occur near the potential minimum (where the relative distance through the diode is zero). Subsequently, as the electrons are accelerated through the diode, the noise velocity wave grows and the noise current wave decreases. For the low space-charge condition, the velocity minimum and noise current maximum occur near the anode.

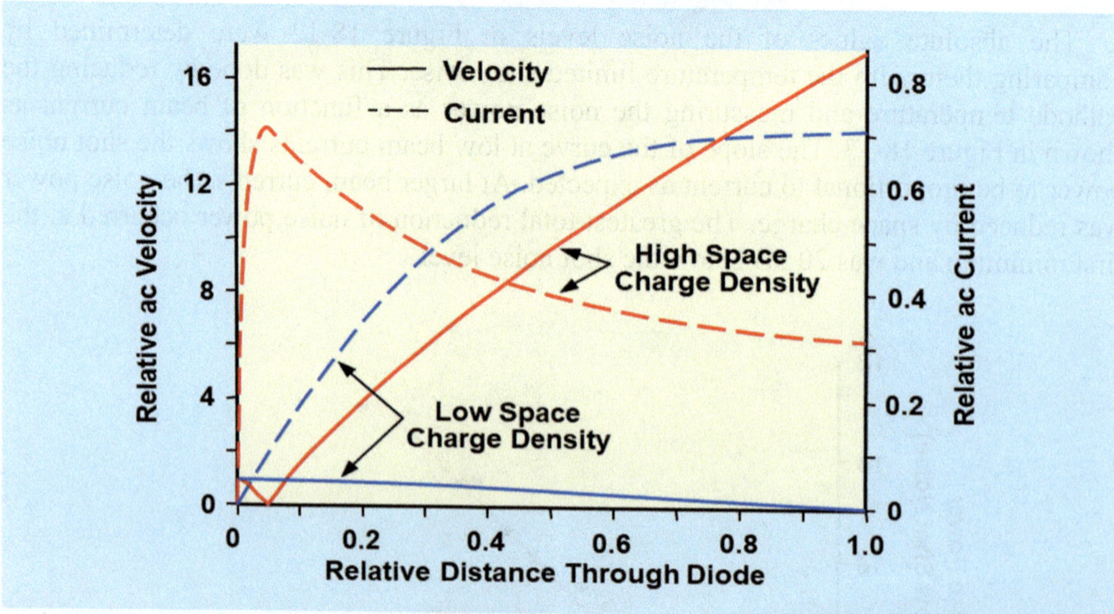

Figure 18-14 Noise velocity and current in a space-charge limited diode. (Adapted from: S. Bloom and R. W. Peter, *RCA Review*, March 1954, copyright 1954 by RCA.)

The two current standing waves (one from the shot noise current source and the other from the Rack velocity noise source) are separated by a one-quarter plasma wavelength (assuming no correlation). As a result, the current distributions along these two waves at the exit from the gun are as shown in Figure 18-15. Each of these waves contributes to the noise figure of a microwave tube.

The amplitudes of the standing waves in the electron gun are governed by the impedance of the electron beam. This impedance is very low at the potential minimum where the electron velocity is low. As the beam is accelerated and the velocity increases, the impedance increases (see Figure 18-15). The impedance at the input to the helix may be over two orders of magnitude greater than the impedance at the potential minimum.

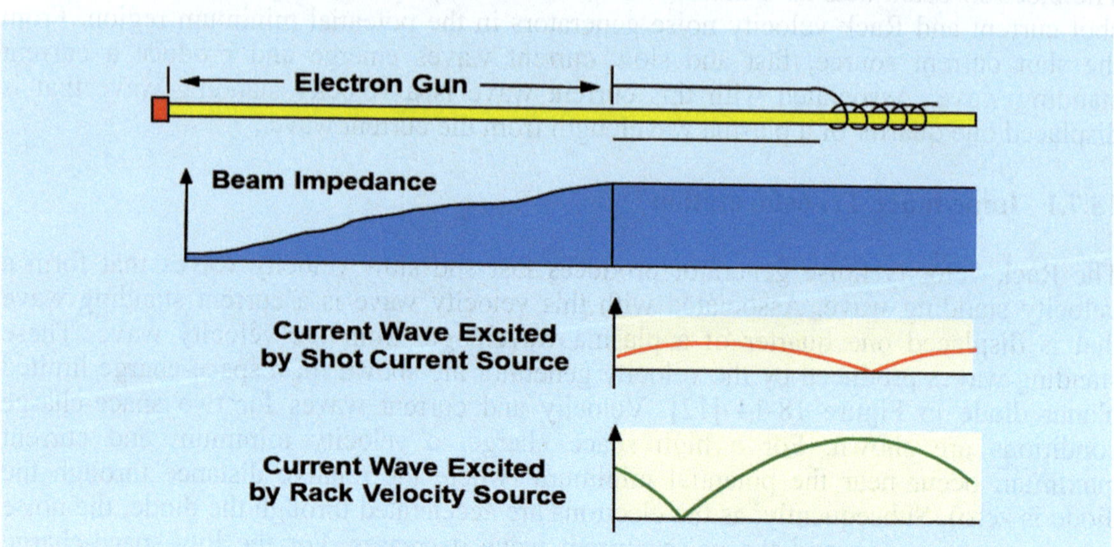

Figure 18-15 Beam impedance in gun and current waves that emerge from gun.

The transverse and axial velocities of electrons emitted from a thermionic cathode are assumed to be the same. In a low-noise gun, as electrons are accelerated, the axial velocity fluctuations decrease (see the low space-charge curve in Figure 18-16, for example). However, the transverse components of velocity are not appreciably affected. Thus, after acceleration, the transverse velocity components are much larger than the axial components.

18.7.2 Lens Effects

As electrons pass through a lens, their transverse and axial velocities are altered. If, as the electrons enter the lens, the fluctuations in their transverse velocities are larger than those in the axial velocities, then the axial velocity fluctuations are increased by the lens. This was proven theoretically and experimentally by Knechtli [13].

The situation illustrated in Figure 18-16 will help in providing a qualitative understanding of the effect of a lens on transverse velocities. The trajectories of three electrons with different transverse velocities are shown. To simplify the illustration, it is assumed that the electrons have no axial velocity fluctuations when entering the lens and that they all pass some plane, z_o, at the same time.

The electrons all pass through the lens at the same time and the trajectory of each is deflected through an angle, φ. At some later time, the electrons arrive at positions 1, 2, and 3, which are separated in the axial direction. For this to happen, the electrons must have different axial velocities. Thus, in passing through the lens, transverse velocity components have been converted to axial velocity components.

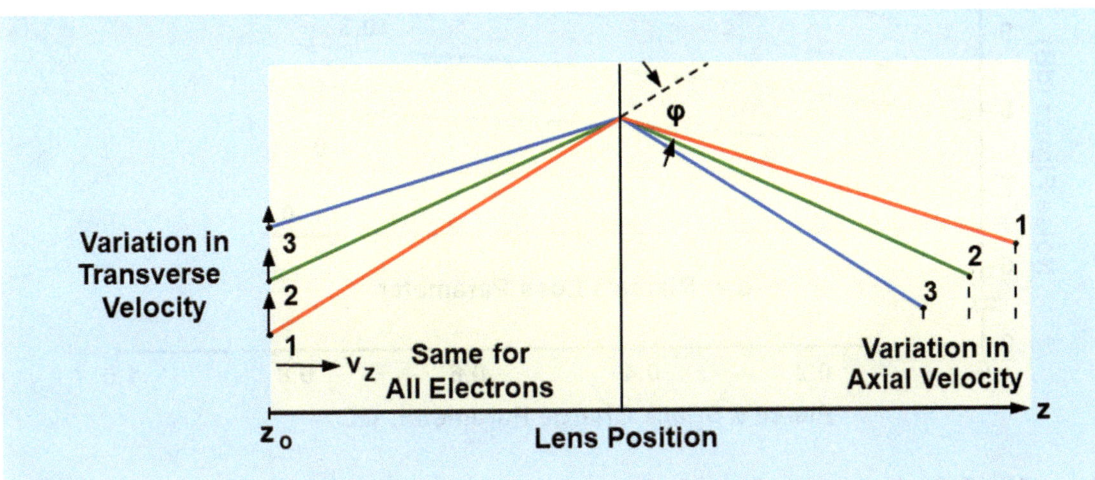

Figure 18-16 Conversion of transverse electron velocities to axial velocities by a lens.

18.7.3 Circuit Loss

There are two effects of circuit loss on the noise figure. The smaller of the two is the thermal agitation noise that is generated by the resistance of the helix. The larger effect occurs because the noise figure depends on the gain of a TWT and gain, in turn, is affected by circuit loss. The effect of circuit loss on the noise figure amounts to, at most, a few decibels as is shown in [14] and is important in low-noise TWTs. As an example of the effect of helix material on the loss and noise figure, assume that the Pierce loss parameter, d, is 0.35 and so, from Figure 18-17, the contribution to the noise figure is about 2.3 dB. Compared with tungsten, if copper with an excellent surface finish was used for the helix, the value of d would drop to about 0.2 and the improvement in the noise figure would be 0.9 dB.

18.7.4 Partition Noise

Partition noise arises because transverse components of thermal velocity cause fluctuations in the current intercepted by electrodes other than the collector. Transverse components of velocity may induce electron motion in a helix and produce a noise analogous to partition current even though no electrons actually strike the helix. A similar effect can occur in the grid cathode region, where space-charge fluctuations induce grid signals. The noise added to the beam by partition current equals the shot noise contained in an equal amount of temperature-limited current [8, 15]. Also, the partition current adds a fully random velocity component to the beam [15].

In the experiments of Cutler and Quate [11], a shutter was used (see Figure 18-11) to intercept a small fraction of the beam current and, thus, control the partition current. The resulting noise data given in Figure 18-13 show a direct correlation between noise power and partition current.

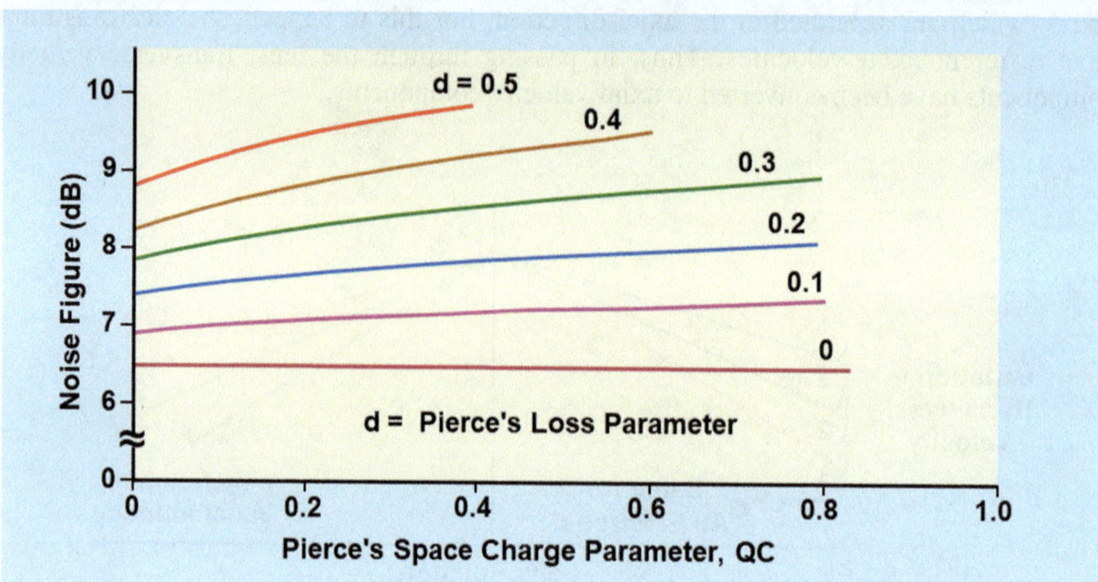

Figure 18-17 Predicted noise figure for low-noise TWTs including the effect of circuit loss.

18.7.5 Secondary Electron Interactions

The electron beam striking the collector generates a large number of slow secondary electrons and a small percentage of fast secondaries (often referred to as reflected electrons) that have energies nearly equal to the incident electron energy. The secondary electron current contains noise of the primary beam plus shot noise from the secondary emission process. If a portion of the secondary current escapes from the collector, it can introduce noise on the RF circuit through interception or by interacting with reverse traveling waves on the RF circuit. This noise is amplified and, at the input of the RF circuit, may be partially reflected and thus contribute to the total noise output.

In multistage depressed collectors, great care is taken to minimize the generation of secondary electrons and to prevent their escape from the collector. Still, some secondaries are generated and those that do leave the collector have a wide range of energies. This is because those electrons were generated from electrodes at depressed potentials. Thus, those electrons are accelerated as they travel toward the exit from the collector (and into the interaction region). The problem of noise generation by secondaries from collectors is greatly reduced with properly designed multistage depressed collectors.

18.8 OTHER NOISE SOURCES

There are several other sources of noise in a microwave tube. Some of these are the following:

1. One or more grids are often used in an electron gun to control current. Some interception of emitted current occurs, and this introduces partition noise. In addition, field distortions occur near the grid, and these introduce transverse velocity components to electron trajectories. The overall effect of the grid is to raise the apparent temperature of the cathode. In one report in which a computer code was used to simulate a shadow-gridded gun with a fictitious cathode temperature, the best correlation between computed and measured beam interception was obtained with an apparent cathode temperature of 10,750K.
2. Hum results from inadequate filtering of power supply voltages and can result in RF modulation at power frequencies and their harmonics.
3. Faulty insulation can result in breakdown, corona, or partial discharges that produce fluctuations in voltages and modulation of the RF signal.
4. Multipactor is a form of RF vacuum breakdown that occurs across vacuum gaps or on insulator surfaces. Noiselike impedance fluctuations occur that modulate the RF signal.
5. Time-varying charge accumulations on insulator surfaces, such as the support rods for the helix, can modulate the RF signal.
6. Movement of the cathode, grid, helix, and so forth as the result of shock and vibration produces modulation of the RF signal.
7. "Gas noise" is generated by ions that produce fluctuations of the potential minimum at the cathode surface and, thereby, produce noise. In addition, ions can produce oscillations, beam compression and reduced power output, RF phase shift and phase noise and reduced cathode activity, which, in turn, results in patchy emission and increased flicker and shot noise.

REFERENCES

[1] W. Schottky, "Spontaneous current fluctuations in electron streams," *Ann. Physik*, Vol. 57, December 1918, pp. 541–567.

[2] A. J. Rack, "Effect of space charge and transit time on the shot noise in diodes," *Bell Sys. Tech. Jour.*, Vol. 17, October 1938, pp. 592–619.

[3] G. Herrman and P. S. Wagener, *The Oxide-Coated Cathode*, London: Chapman and Hall, 1951.

[4] A. van der Ziel, "Low-frequency noise in vacuum tubes (flicker effect)," *Noise in Electron Devices*, New York: The Technology Press of MIT, and John Wiley & Sons, 1959.

[5] P. K. Tien and J. Moshman, "Monte Carlo calculation of noise near the potential minimum of a high-frequency diode," *Jour. App. Phys.*, Vol. 27, No. 9, September 1956, pp. 1067–1078.

[6] R. W. Peter, "Low-noise traveling-wave tubes," *Noise in Electron Devices*, New York: published jointly by The Technology Press of MIT and John Wiley & Sons, 1959.

[7] M. R. Currie and D. C. Forster, "New mechanism of noise reduction in electron beams," *Jour. App. Phys.*, Vol. 30, No. 1, January 1959, pp. 94–103.

[8] J. R. Pierce, *Traveling Wave Tubes*, Princeton, NJ: Van Nostrand, 1950.

[9] L. C. Peterson, "Space-charge and transit-time effects on signal and noise in microwave tetrodes," *Proc. IRE*, November 1947, pp. 1264–1272.

[10] F. B. Llewellyn and L. C. Peterson, "Vacuum-tube networks," *Proc. IRE*, March 1944, pp. 144–166.

[11] C. C. Cutler and C. F. Quate, "Experimental verification of space charge and transit time reduction of noise in electron beams," *Phys. Rev.*, Vol. 80, No. 5, December 1950, pp. 875–878.

[12] S. Bloom and R. W. Peter, "Transmission-line analog of a modulated electron beam," *RCA Review*, March 1954, pp. 95–112.

[13] R. C. Knechtli, "Effect of electron lenses on beam noise," *IRE Trans. Electron Devices*, April 1958, pp. 84–88.

[14] S. Bloom, "The effect of initial noise current and velocity correlation on the noise figure of traveling-wave tubes," *RCA Review*, Vol. 16, No. 2, June 1955, pp. 179–196.

[15] W. R. Beam, "Interception noise in electron beams at microwave frequencies," *RCA Review*, Vol. 16, No. 4, December 1955, pp. 551–579.

[16] T. G. Mihran et al., "The growth of peak velocity, noise, and signal along O-type beams from shielded guns," *Conference on Electron Device Research*, Cornell University, 1964.

[17] L. D. Smullin and C. Fried, "Microwave noise measurements on electron beams," *IRE Trans. Electron Devices*, Vol. ED-1, No. 4, 1954, pp. 168–183.

[18] J. M. Hammer and E. E. Thomas, "Traveling-wave tube noise figures of 1.0 dB at S-band," *Proc. IEEE*, Vol. 52, No. 2, February 1964, p. 207.

[19] J. M. Hammer and C. P. Wen, "Effect of high magnetic field on electron-beam noise," *RCA Review*, December 1964, pp. 785–789.

[20] F. N. H. Robinson, "Space-charge smoothing of microwave shot noise in electron beams," *Phil. Mag.*, Vol. 43, 1952, p. 51.

[21] S. Bloom and R. W. Peter, "A minimum noise factor for the traveling-wave tube," *RCA Review*, Vol. 15, 1954, p. 252.

[22] J. R. Pierce and W. E. Danielson, "Minimum noise figure of traveling-wave tubes with uniform helices," *Jour. App. Phys.*, Vol. 25, 1954, p. 1163.

Chapter 19

Magnetrons

Magnetrons, as well as crossed-field amplifiers (which are described in Chapter 20) are crossed-field tubes that are capable of generating very large amounts of microwave power.

In crossed-field tubes, the dc electric field is perpendicular to the magnetic field as indicated in Figure 19-1(a). The general motion of the electron beam is perpendicular to both fields. This is in contrast to linear-beam tubes (such as klystrons and TWTs, Figure 19-1(b)) where the general direction of beam motion is aligned with the fields. Crossed-field tubes are referred to as M-type tubes after the French TPOM, which in English means tubes for propagation of waves in a magnetic field.

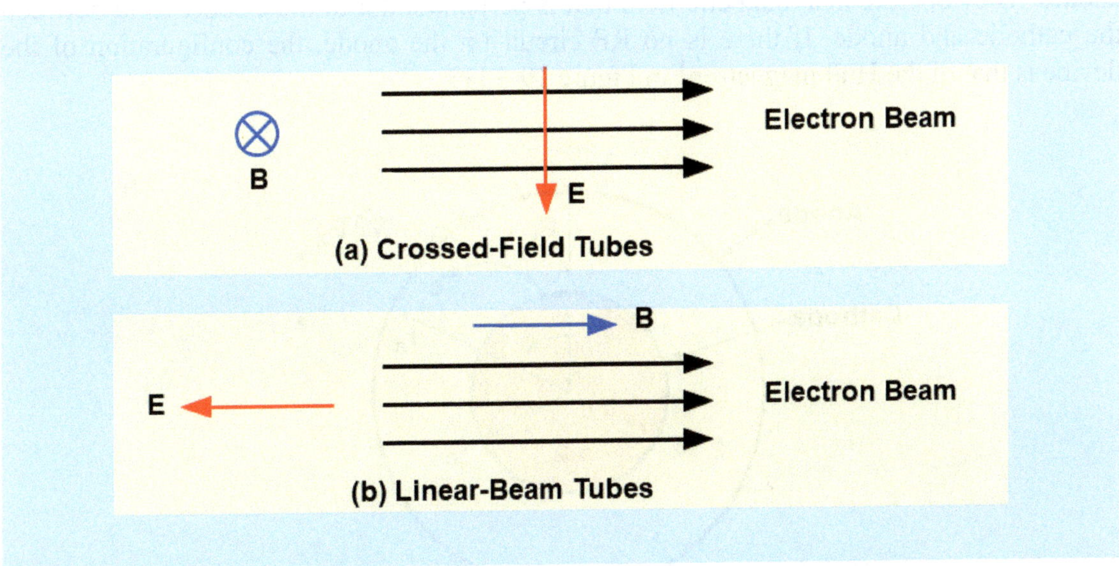

Figure 19-1 Comparison of field configurations in (a) crossed-field and (b) linear-beam tubes.

Many types of crossed-field tubes have been developed and can be categorized as indicated in Figure 19-2. With the exception of injected-beam crossed-field amplifiers, these devices are in use in deployed systems. Injected-beam crossed-field amplifiers, while highly developed, have not found application in the United States and will not be discussed.

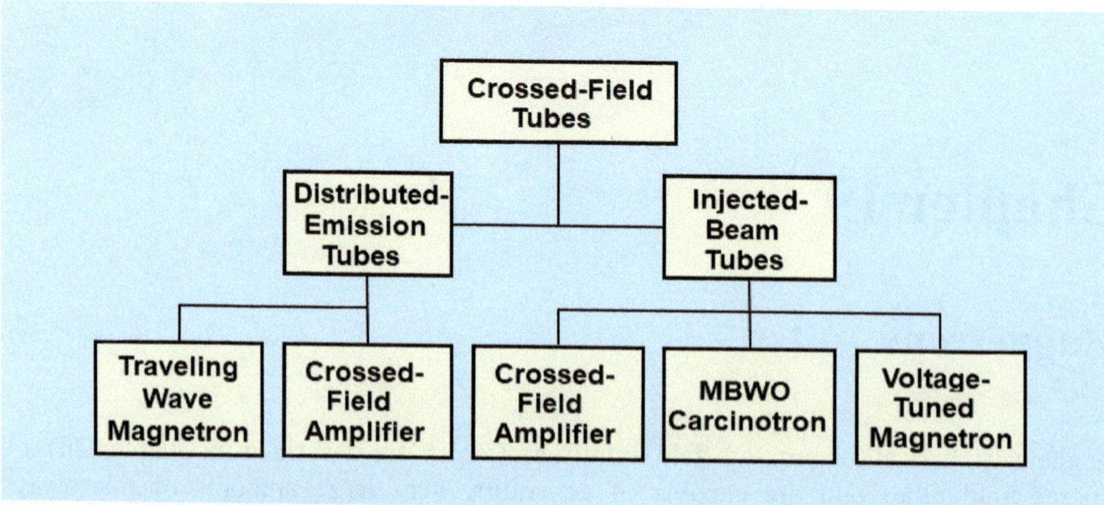

Figure 19-2 Family tree of crossed-field tubes. (From: *Microwave Tube Manual* by Varian Associates, Air Force Publication Number T.O.00-25-251, October 1979.)

19.1 ELECTRON FLOW WITH NO RF FIELDS

The magnetrons considered in this book are all cylindrical with concentric cathodes and anodes. They operate in a magnetic field that is perpendicular to the electric field between the cathode and anode. If there is no RF circuit on the anode, the configuration of the device is that of the Hull magnetron[1] in Figure 19-3 [1].

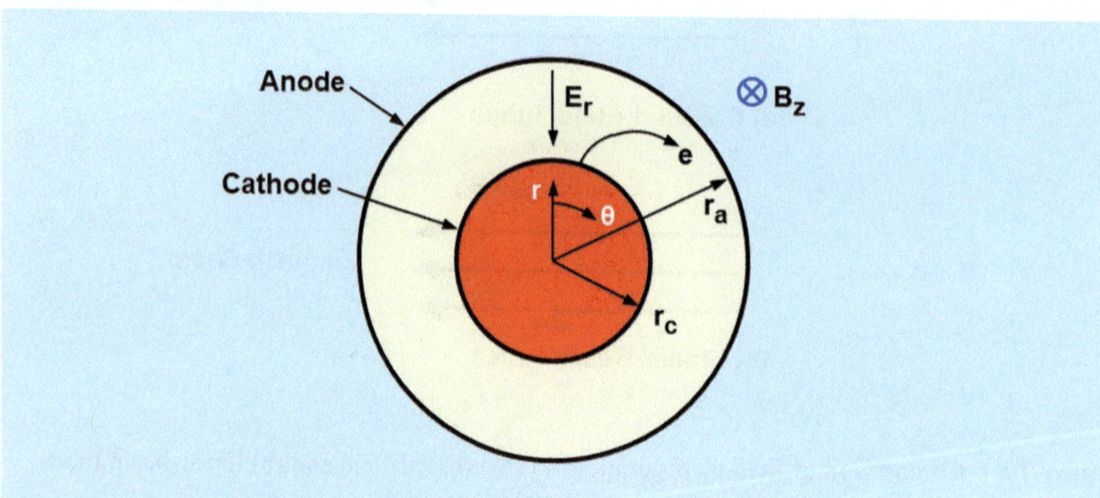

Figure 19-3 Configuration of a Hull magnetron.

[1] The first magnetron was invented by A. W. Hull and reported in 1921 [1]. This device was called a magnetron by Hull but it was not an RF device. It did not generate or amplify radio frequency signals. Hull called it "an inertialess, high-resistance valve, operated by magnetic field", that is, it was a switch tube that could be used to control current.

As electrons leave the cathode and move toward the anode, they are deflected in the θ direction. In cylindrical coordinates, if r is radially outward from the axis and θ is clockwise about the axis, then z is into the plane of r and θ.

Thus, the voltage that causes the electron to just graze the anode depends on the dimensions r_a and r_c and on the flux density but not on the way in which the voltage varies between the cathode and anode. The electron trajectory is shown in Figure 19-4(a). The anode voltage at which this occurs is called the Hull cutoff voltage and is denoted here by V_H. When $V_a > V_H$, the electron strikes the anode as indicated in Figure 19-4(b). When $V_a < V_H$, the electron returns to the cathode (Figure 19-4(c))

If the anode voltage is held constant at V_H, then, as is indicated for the ideal case in Figure 19-5, there is a critical magnetic field, *Bc*, above which no current can flow. As the magnetic field is decreased below *Bc*, all available current flows to the anode. In practice (Figure 19-6 for example), there are factors that cause the transition from full current to zero current to occur over a small change in magnetic field. Space charge is one of these factors.

The critical combination of electric and magnetic fields at which electrons just touch the anode is known as the Hull cutoff condition and is shown in Figure 19-7.

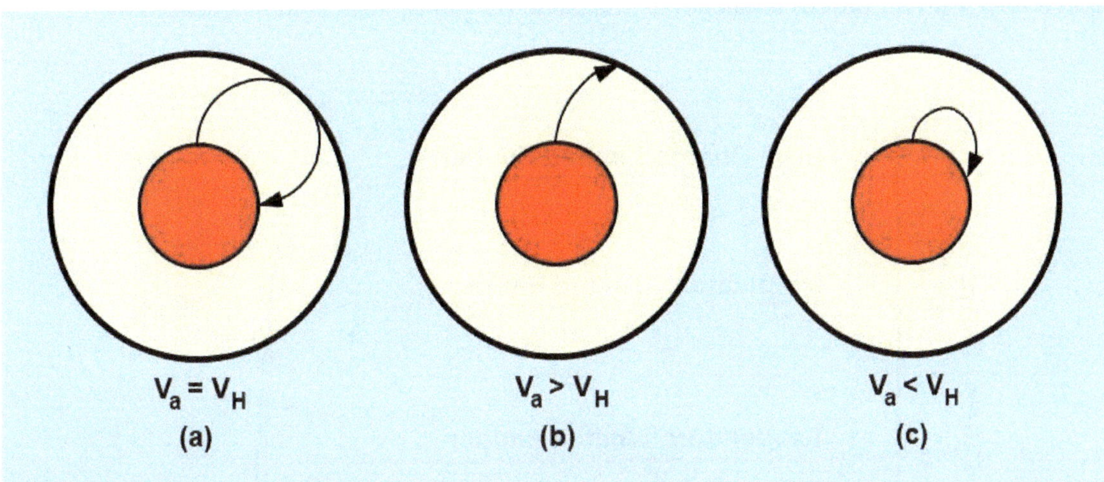

Figure 19-4 Electron trajectories for various anode potentials.

19.2 TYPES OF MAGNETRONS

There are three basic types of magnetrons:

1. Cyclotron-frequency magnetrons;
2. Negative-resistance magnetrons;
3. Traveling-wave magnetrons.

Cyclotron-frequency and negative-resistance magnetrons are no longer used. However, their basic principles of operation will be described to help form a foundation for the understanding of traveling-wave magnetrons, which are now in use.

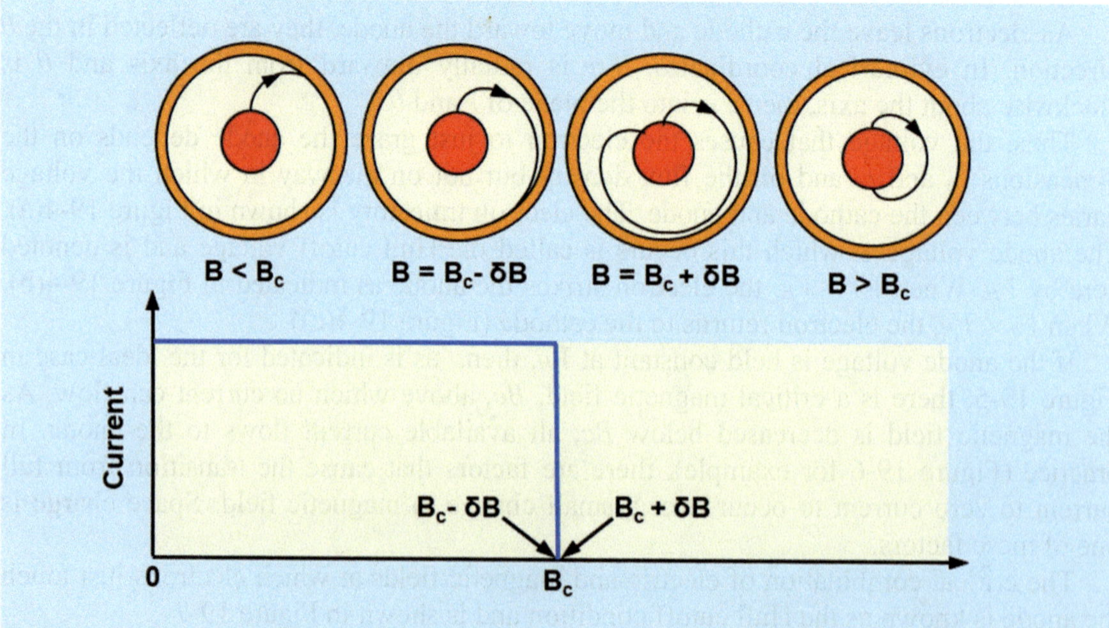

Figure 19-5 Current flow as a function of magnetic field in an ideal Hull magnetron.

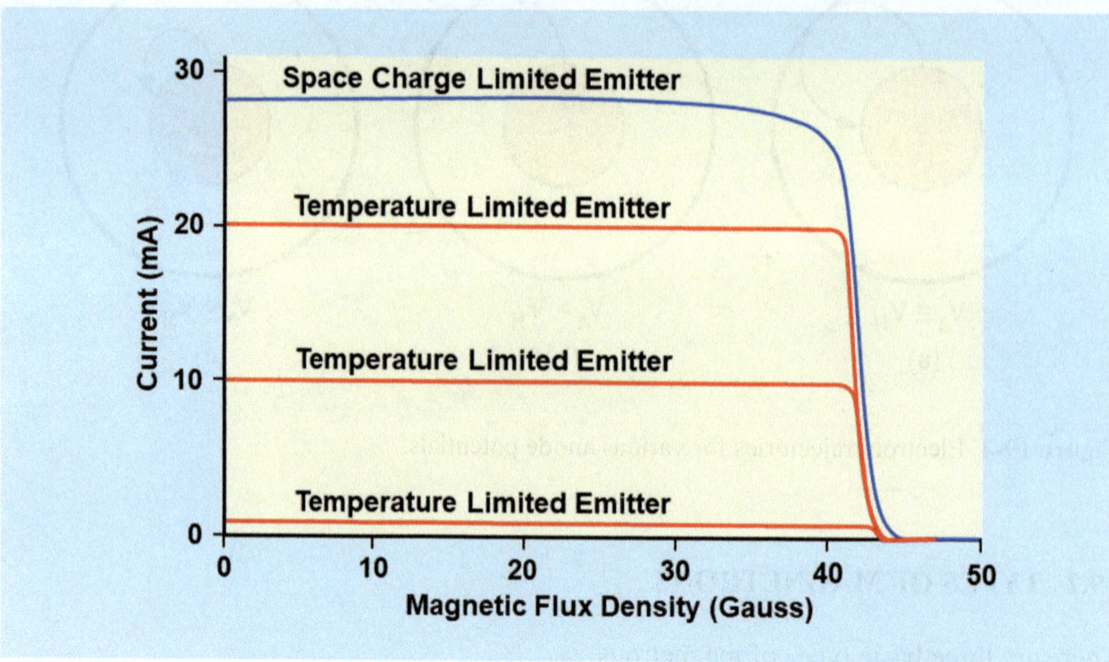

Figure 19-6 Current flow as a function of magnetic field in a real Hull magnetron [1].

The first magnetrons that generated high-frequency RF signals were cyclotron-frequency magnetrons and negative-resistance magnetrons. These were investigated independently in 1924 by Erich Habann [3] in Jena, Germany, and Napsal August Zázek [4] in Prague, Czech Republic. These magnetrons operated very close to the Hull cutoff condition and were very inefficient.

Traveling-wave magnetrons, which were developed more recently, operate at voltages and magnetic fields that are well below the Hull cutoff parabola and are very efficient. If there was no RF interaction, no current would flow in these devices. However, with RF, current does flow and the high efficiency of these devices results from the very efficient conversion of potential energy to RF energy that occurs as the electrons move from the cathode to the anode.

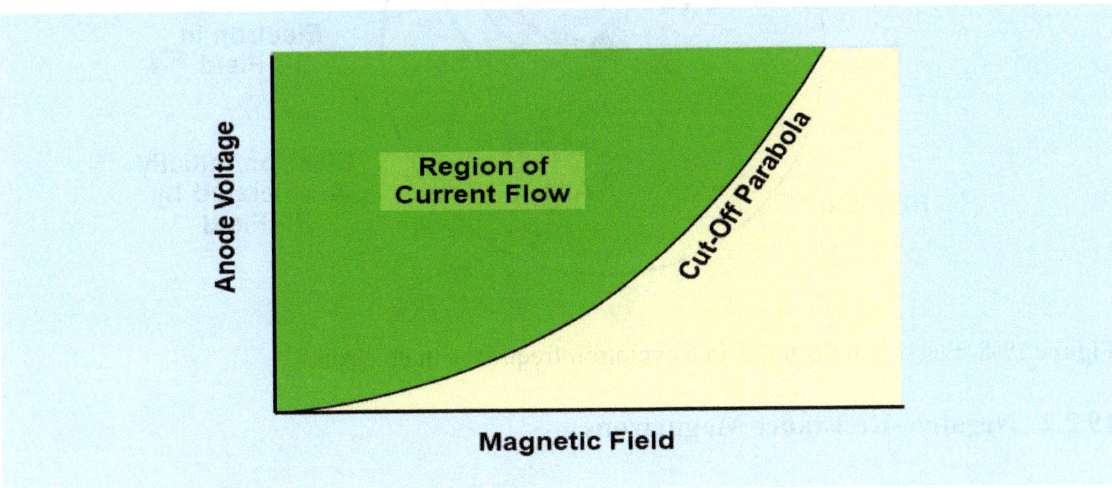

Figure 19-7 The Hull cutoff condition [2].

19.2.1 Cyclotron-Frequency Magnetrons

The first magnetron was the cyclotron-frequency magnetron. Electron trajectories in the cyclotron frequency magnetron are shown in Figure 19-8. This and other early magnetrons used cylindrical anodes with no RF structure. Efficiencies and power capabilities were very low.

Cyclotron-frequency magnetrons made use of interaction between orbiting electrons in a magnetic field and a resonant circuit tuned to the cyclotron frequency. In the simplest form of these magnetrons, the cathode and anode were coaxial cylinders and were operated in a magnetic field parallel to the electrode axis. The resonant circuit was a quarter-wavelength parallel-line resonator connected between the cathode and anode as indicated in Figure 19-8.

To produce oscillations, the voltage and magnetic field were adjusted close to the cutoff condition so that electron trajectories were like that for the dc-field case. Electrons emitted when the RF electric field aided the dc electric field landed on the anode, and so these, which extracted energy from the RF field, were removed from interaction. Electrons emitted when the RF field opposed the dc field gave up energy to the RF field and missed the anode. If the frequency of rotation of the electrons (cyclotron frequency) was close to the RF frequency, then electrons that initially gave up energy (in Region 1) arrived in Region 2 as the field reversed and again gave up energy to the field. So these electrons contributed energy to the RF oscillation. The process continued as long as the phase relationship with the field persisted or until the electrons were removed from interaction. Cyclotron frequency magnetrons operated at frequencies up to the millimeter-wave range, but had very low power output and efficiency.

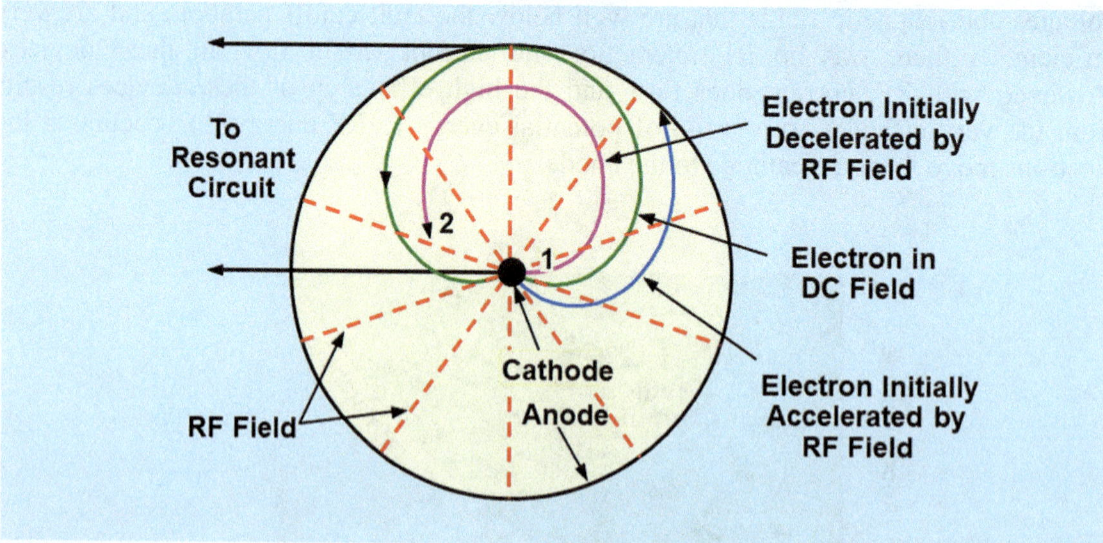

Figure 19-8 Electron trajectories in a cyclotron frequency magnetron.

19.2.2 Negative-Resistance Magnetrons

Negative-resistance magnetrons were split-anode devices as shown in Figure 19-9. A magnetic field was applied parallel to the axis and a dc voltage was applied between the cathode and the anode segments. The magnetic field was sufficient to keep electrons from reaching the anode when both segments were at the same potential as shown in Figure 19-9(a). In the absence of an RF signal, electrons would leave the cathode, move toward the anode, and then be turned back to the cathode by magnetic deflecting forces.

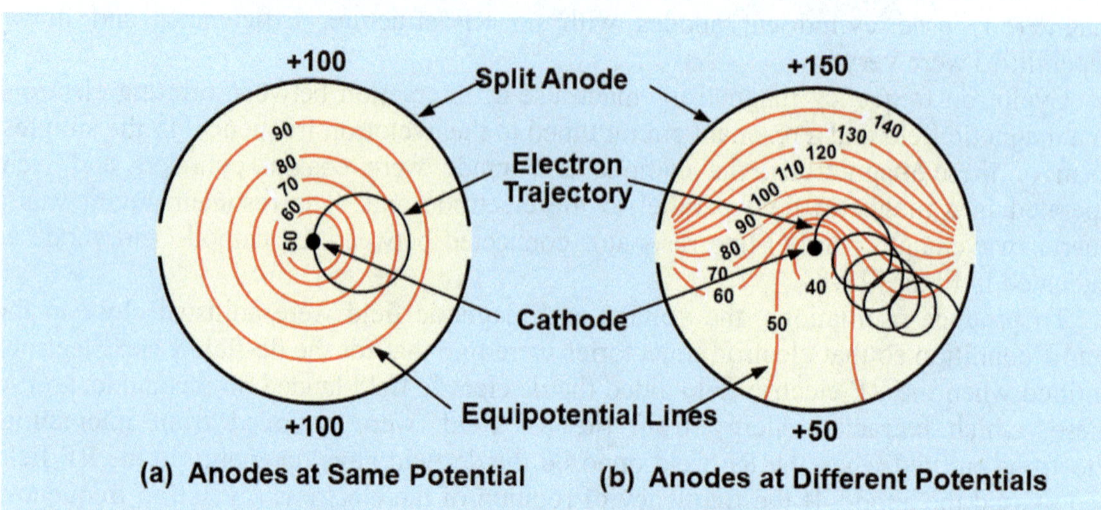

Figure 19-9 Electron trajectories in a negative-resistance magnetron with a split anode.

When an RF voltage was present and the anode segments were at different potentials, much more complicated trajectories were formed as are illustrated in Figure 19-9(b). Electrons moving in a crossed-field region tend to move along equipotential lines. As indicated in Figure 19-9(b), as one electrode segment is made negative with respect to the other, more electrons are collected on that electrode. When the voltages on the anode segments are reversed, more electrons are collected on the other electrode so a negative resistance effect occurs. When a resonant circuit was connected between the two anode segments, it was possible to obtain sustained oscillations. Negative-resistance magnetrons generally operated below 500 MHz, were low-power devices, and had low efficiency.

19.2.3 Traveling-Wave Magnetrons

Early traveling-wave magnetrons were split-anode devices similar to negative resistance magnetrons. An example is shown in Figures 19-10 and 19-11. The primary difference between the two was the magnitude of the magnetic field. In the negative resistance magnetron, the magnetic field was near the Hull cutoff value and the frequency of operation was low enough so that an electron moved in a single cycloid-like path all the way from the cathode to the anode, while the voltage on the anode segments remained nearly constant (as indicated in Figure 19-9(b)). In traveling-wave magnetrons, the magnetic field is well above the Hull cutoff field. The frequency with which the alternating voltage on the anode segments changes can be made near the rate of rotation of the electrons about the cathode.

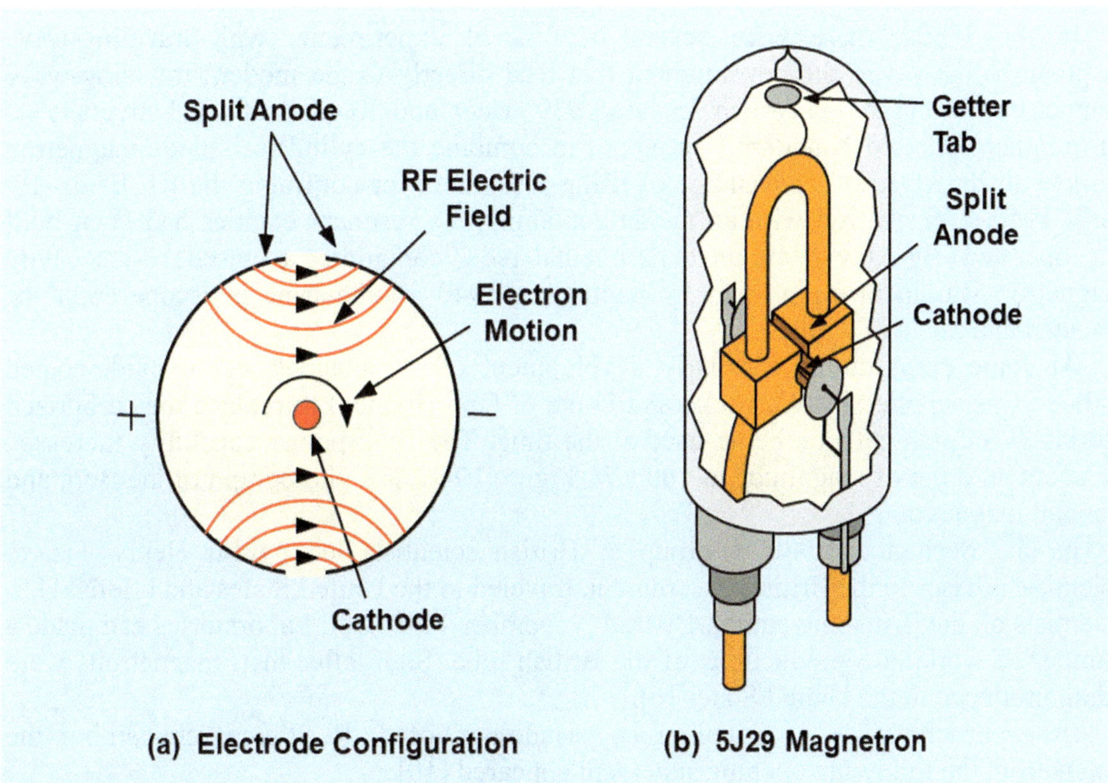

Figure 19-10 Configuration of a split-anode magnetron.

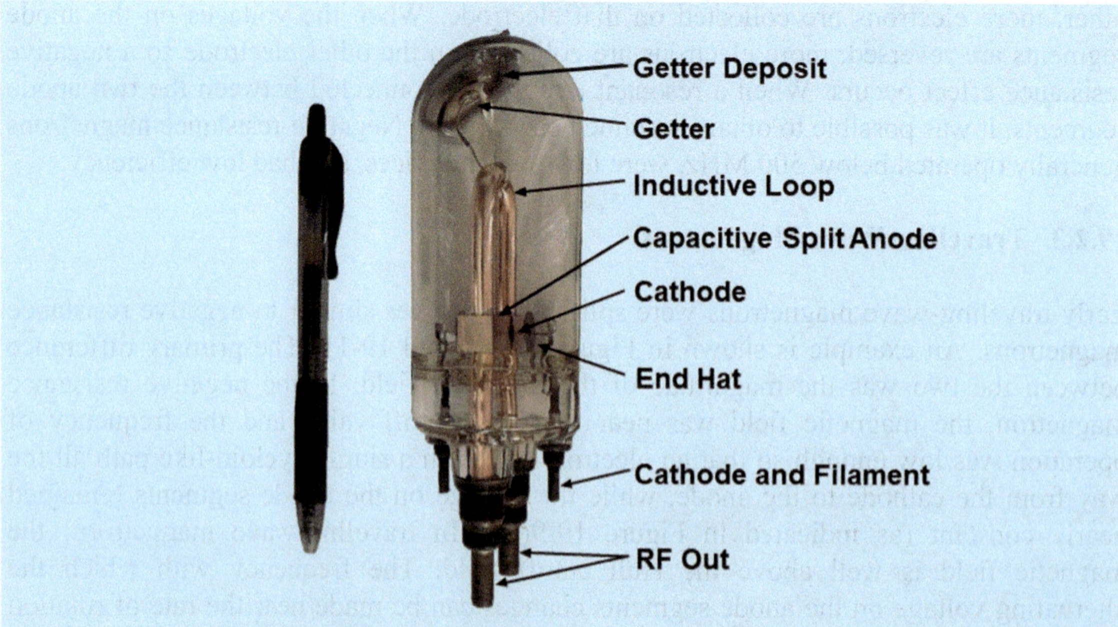

Figure 19-11 A 5J29 split-anode magnetron.

In the 1930s, there were several reports of experiments with traveling-wave magnetrons; however, the development that lead directly to the modern traveling-wave magnetron was that of the British. In late 1939, Boot and Randall at the University of Birmingham (United Kingdom.) set about to combine the cylindrical-anode-magnetron work with the klystron's advantage of using resonators for confining the RF fields. By early 1940, a magnetron with an anode containing six resonant cavities had been built and operated. By May 1940, an experimental radar containing a pulsed 10-cm cavity magnetron was in operation and by September 1940 a submarine periscope could be detected at a range of 7 miles.

At some point during the early development of the magnetron, an oxide-coated cathode was supplied by Warnecke and Ponte of CSF (France) to replace the carburized thoriated-tungsten cathode being used at the time. The peak power capability increased by about an order of magnitude to 100 kW. Figure 19-12 is a photograph of the Boot and Randall magnetron [5].

In late September 1940, a group of British scientists led by Sir Henry Tizard, scientific advisor to the British government, traveled to the United States and briefed U.S. scientists on the British invention. By mid-November 1940, Bell Laboratories had made a number of working reproductions of the British tube. Soon after that, magnetrons were being produced in the United States [6].

In an article by J. J. Coupling (a pseudonym of J. R. Pierce) concerning the magnetron, the following opening statement appeared [10]:

> "The two most important weapons of the recent war were the atomic bomb, and radar . . . and it was Maggie, the magnetron, that made Allied radar incomparably superior to the Nazi's boat."

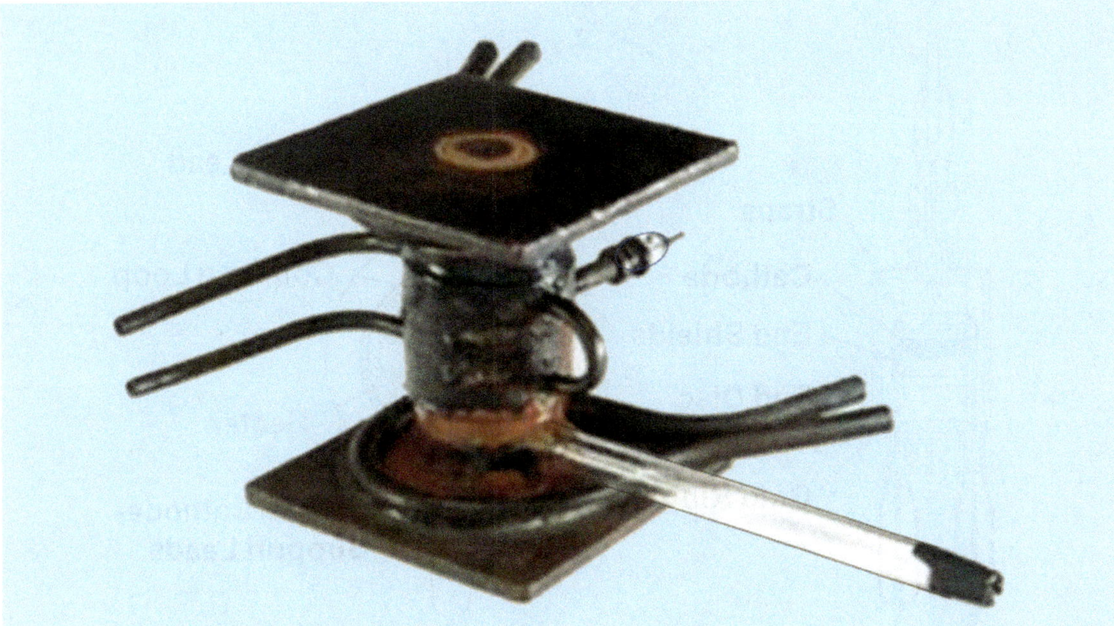

Figure 19-12 The Boot and Randall magnetron.

Modern magnetrons are all traveling-wave magnetrons. The primary differences between these devices and the 5J29 split-anode magnetron shown in Figures 19-10(b) and 19-11 are in the RF circuit. In the 5J29, the part of the resonant circuit internal to the tube consisted of an inductive loop (to which the getter is attached in Figures 19-10(b)) and the capacitance between the anode segments. The RF circuit in all modern magnetrons consists of several resonant cavities similar to those shown in Figure 19-13, which is an early version of a pulsed cavity magnetron. In this tube, the anode contains eight resonators and is concentric with an oxide-coated cathode. A permanent magnet (not shown) was used for applying a magnetic field parallel to the cathode surface.

19.3 OPERATION OF THE TRAVELING-WAVE MAGNETRON

When voltage is first applied to a traveling-wave magnetron, there are no RF fields and the electrons form a space-charge cloud as shown in Figure 19-14 that rotates about the cathode. This cloud is often referred to as a space-charge hub.

Because of fluctuations in the emission process and noise-amplifying effects in the space-charge hub, electron velocities and the charge density in the rotating space-charge cloud fluctuate randomly with time and produce a noise component to the rotating current in the space-charge hub. The frequency of the fluctuations varies over such a wide range that the noise is referred to as white noise, which implies fluctuations over an infinite frequency range. While the frequency range of fluctuations is not infinite, it is rich in components in the microwave range, and so there are components present to initiate any of the frequencies that may be supported by the cavity structure of a magnetron.

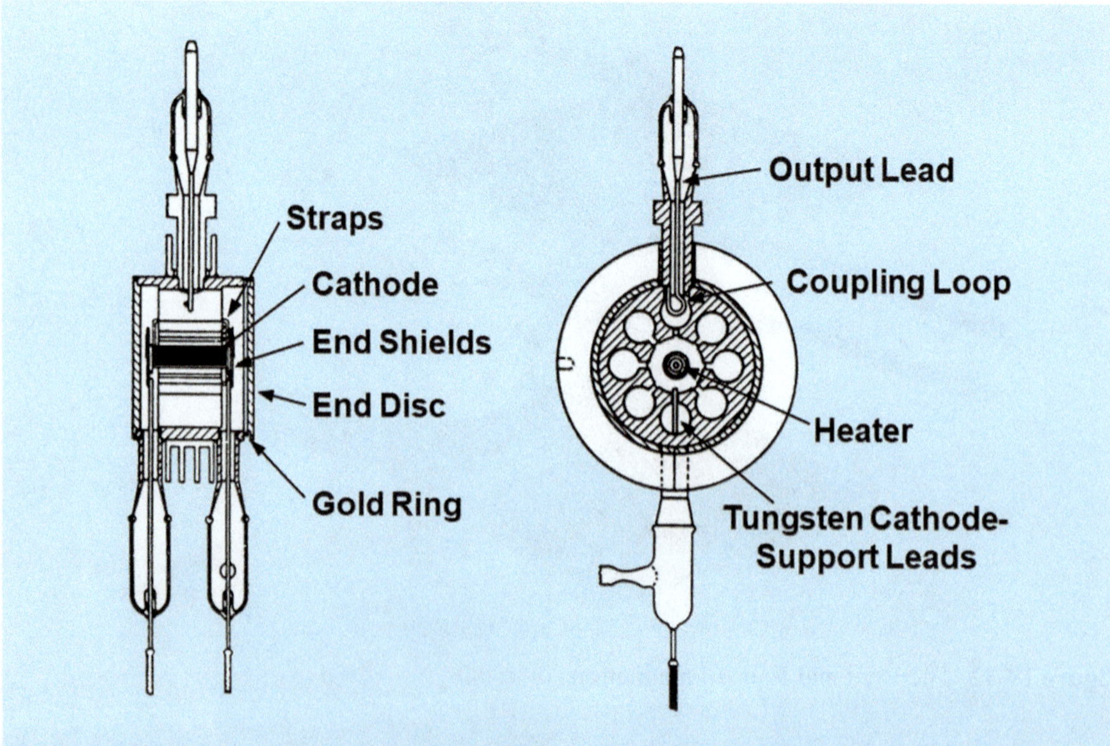

Figure 19-13 Early sealed-off pulsed cavity magnetron. (From: H. A. H. Boot and J. T. Randall, *IEEE Trans. Electron Devices*, July 1976. © 1976 IEEE.)

This noise current in the hub induces noise current in the cavities. When the component of this noise current that is at the resonant frequency of the cavities passes through a high impedance of the cavities at one of the resonant frequencies, RF voltages are produced across the cavity gaps and these amplify the RF current in the hub. RF current and power build up very rapidly (in nanoseconds) and saturate when the power supplied to the magnetron equals the power to the load plus losses.

In the following sections, the formation of the hub, the buildup of RF current, and the characteristics of the cavity resonator are discussed.

Part of the behavior of the electron hub can be explained by considering the motion of a single electron when no RF fields are present.

Referring to Figure 19-15, it is seen that there are two electron motions to consider. With a radial electric field, E, and an axial magnetic field, B, two forces act on an electron. The force from the electric field, $-eE$, is toward the anode. The force exerted by the magnetic field, $-e(u \times B)$, is perpendicular to the motion of the electron. The resulting path of the electron is approximately represented by combining a slow rotation around the cathode at radius R_o and a faster circular motion with a smaller radius r_o. The result of these two motions corresponds roughly to the motion of a point on the circumference of a wheel as it rolls around a circle somewhat smaller than the cathode in such a way that its center moves in a circle of radius R_o.

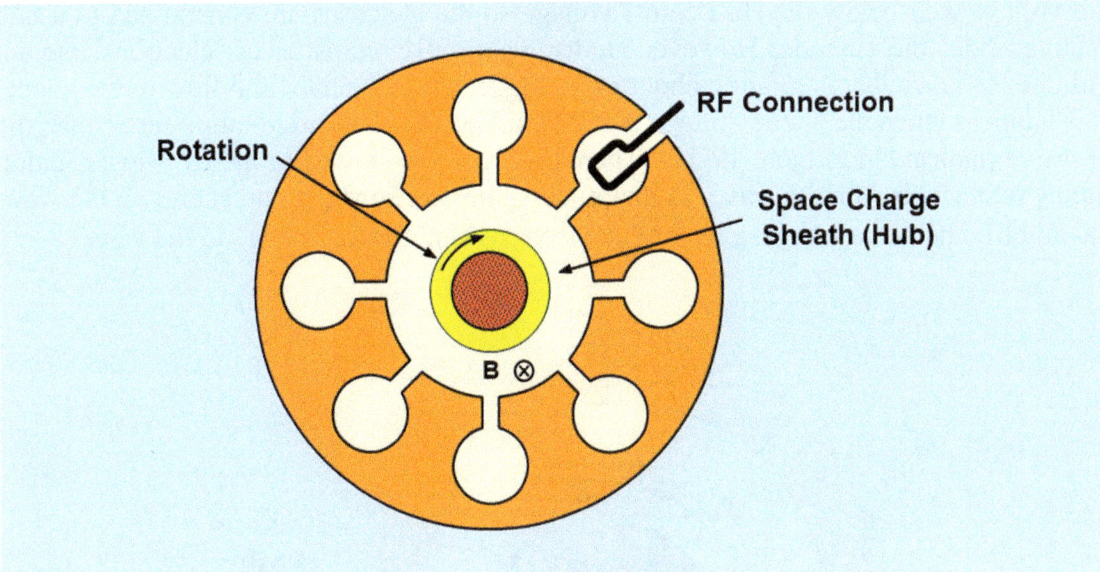

Figure 19-14 Space-charge cloud that is formed when RF is absent.

19.3.1 Hub Formation

The speed of the slow R_o rotation is given approximately by the ratio E/B. The fast r_o rotation corresponds to the cyclotron frequency so it is determined only by B. Although the angular velocity of this r_o rotation is constant, the magnitude of r_o depends on the initial kinetic energy of the electron so it may vary for different electrons. With no RF, the maximum distance that any electron can move toward the anode ($R_o + r_o$) is fixed by the ratio E/B and, in practice, is about half of the distance from the cathode to the anode.

This description of the path of a single electron is only accurate for small r_o when RF fields and space charge are negligible.

Because the electron flow is in crossed fields, the velocities with which the electron streams move are given by E/B (E_y for the linear format and E_r for the circular format). Because of space charge, the electric field near the cathode is lower than in the absence of space charge and so, as Figure 19-16 indicates, the velocities of the electron streams are reduced. This results in a condition called slipping flow.

The velocities of the electron streams are of interest because, for interaction with the RF circuit in a magnetron, there must be near synchronism between the electron velocity and the phase velocity of the circuit.

Although neither laminar flow nor a Brillouin hub can be obtained in practice, they are often considered to be present to simplify the problem of analysis. In these procedures, the top layer of space charge is assumed to be moving in near synchronism with an RF traveling wave on the anode circuit.

19.3.2 The Hartree Voltage

The Hull cutoff condition is derived when no RF fields are present. This condition defines the anode voltage as a function of the magnetic field that prevents the flow of electrons to the anode. In the previous sections, we have said that a magnetron operates at a voltage well below the Hull cutoff voltage so the electrons are constrained to a hub rotating about the cathode. However, under the right circumstances, electrons interact with the RF wave that is rotating about the cathode, leave the hub, and flow to the anode. This happens when the anode voltage is such that the electrons are rotating faster than the RF wave (indicated in Figure 19-17). Then, electrons are slowed to the RF wave rotation rate by interaction with the wave. In the process, the electrons give up energy to the wave and amplify the wave. As long as $\dot{\theta} > \dot{\varphi}$, electrons can give up energy to the wave.

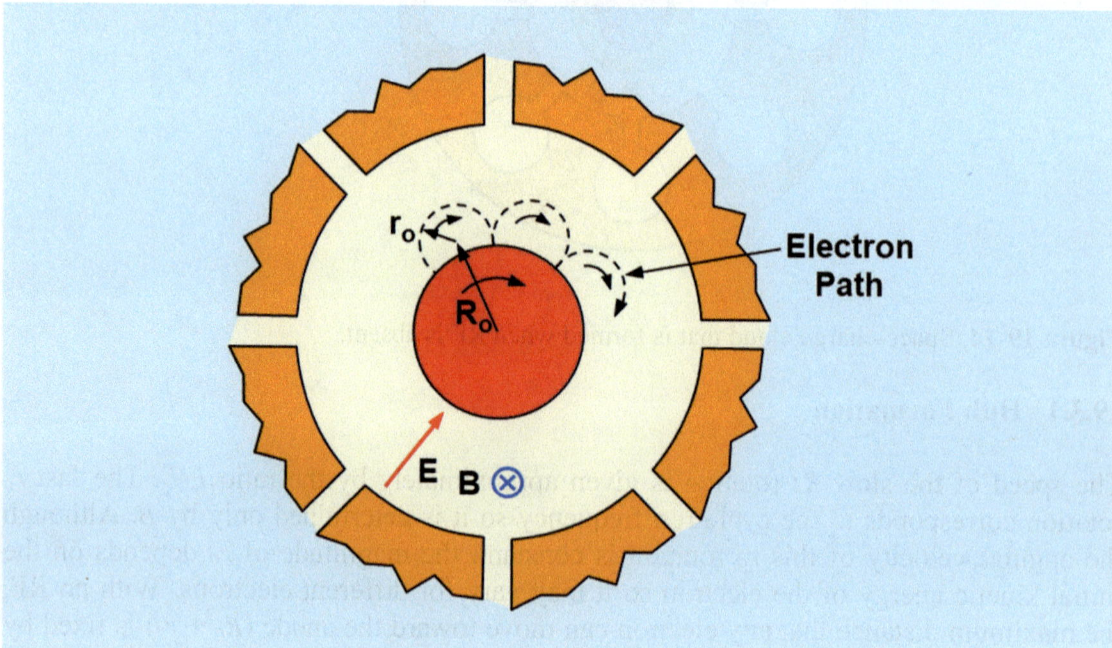

Figure 19-15 Path of a single electron when space-charge effects and RF fields are absent.

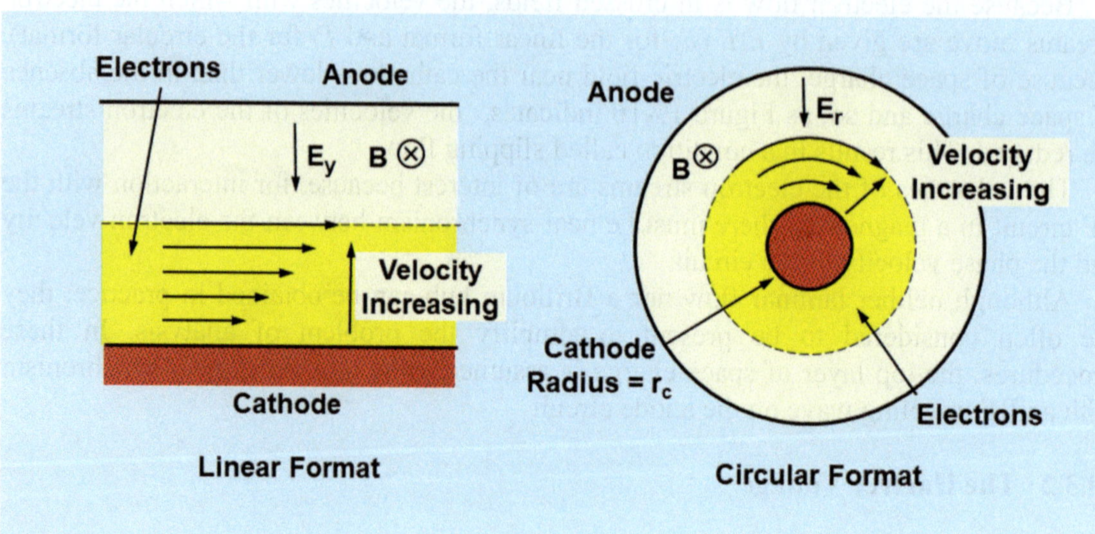

Figure 19-16 Theoretical electron flow in crossed fields in linear and in circular formats.

If the anode voltage is reduced, the rate at which the electrons tend to rotate is reduced so they give up less energy to the rotating wave. There is a critical voltage at which the rate of electron rotation equals the rate of rotation of the wave. At that voltage, the electrons can no longer give up energy to the wave and so the magnetron stops operating. Below that voltage, the electrons rotate more slowly than the RF wave and so the magnetron cannot operate. The critical voltage at which energy transfer to the wave ceases and the magnetron stops operating is called the *Hartree voltage*. The Hartree voltage varies linearly with the magnetic field and so the Hartree voltage and the Hull cutoff voltage can be plotted as shown in Figure 19-18. The significant voltages for a magnetron are between the Hartree and Hull cutoff voltages.

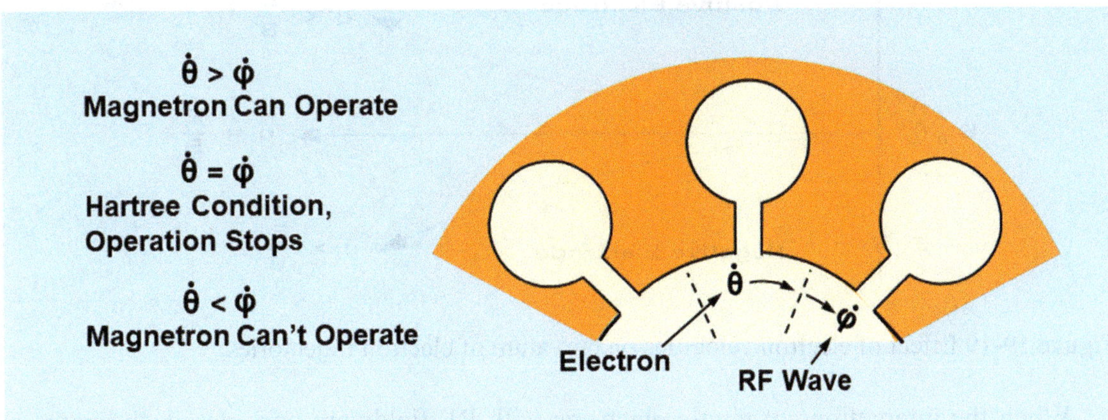

Figure 19-17 Electron and wave rotation in a magnetron.

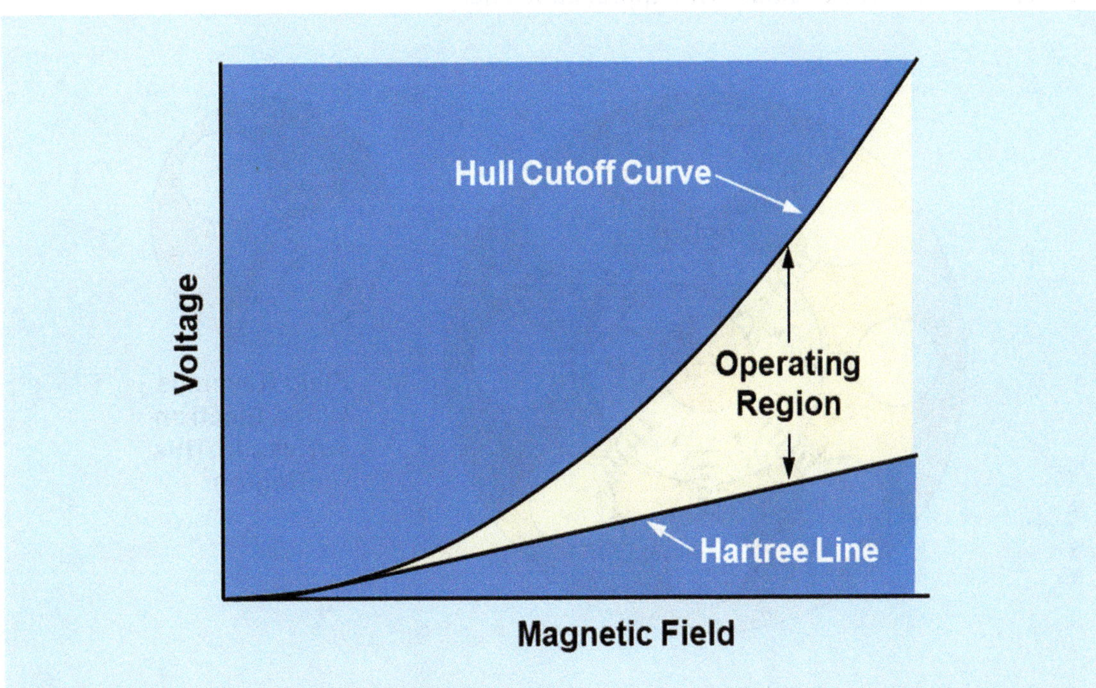

Figure 19-18 Operating region of a magnetron.

19.3.3 Spoke Formation

In considering the interactions of electrons with RF electric fields, it is helpful to review one aspect of the electron trajectories in a dc electric field, E. That is, as is shown in Figure 19-19, electrons with velocity E/B move in a straight line. Those electrons with $u > E/B$ are deflected downward in the direction of the dc electric field toward the negative electrode. Those electrons with $u < E/B$ are deflected upward in the direction opposite to the dc electric field toward the positive electrode.

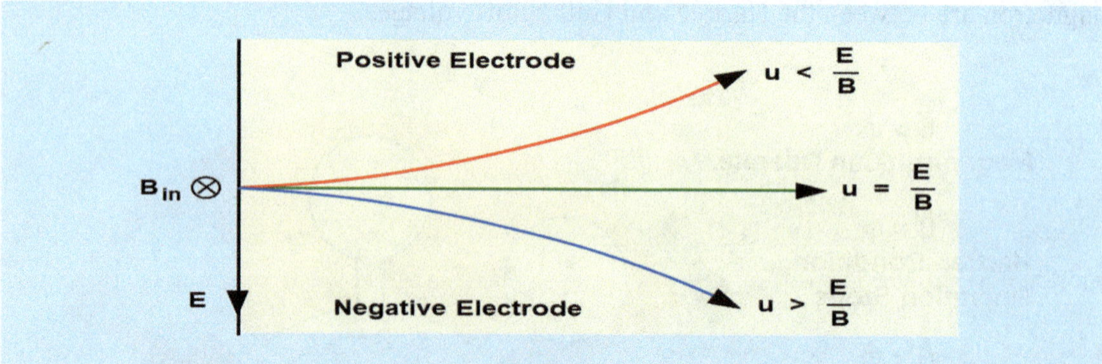

Figure 19-19 Effect of electron velocities on curvature of electron trajectories.

When the interactions of single electrons with RF fields are considered, the paths of the electrons can be traced in a relatively straightforward manner. For example, consider Figure 19-20. An electron leaving the cathode at point **A** enters an accelerating field and is driven back into the cathode with appreciable energy.

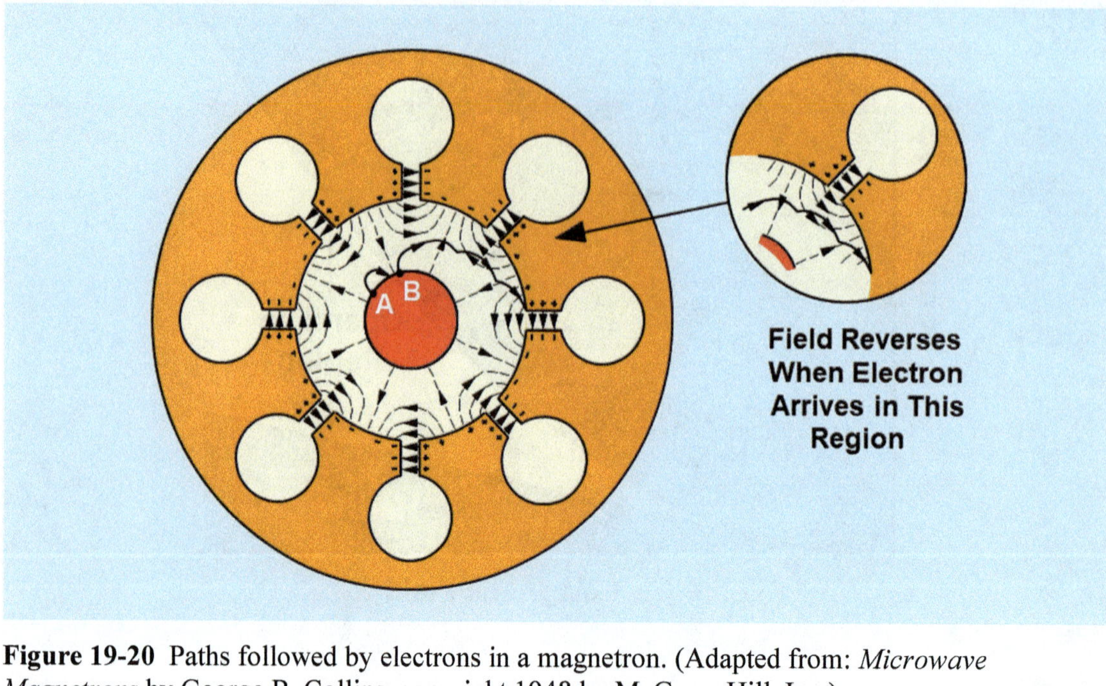

Figure 19-20 Paths followed by electrons in a magnetron. (Adapted from: *Microwave Magnetrons* by George B. Collins, copyright 1948 by McGraw-Hill, Inc.)

An electron leaving the cathode at **B** enters a decelerating field and moves toward the anode. With the correct frequency of operation, this electron will remain in synchronism with the decelerating field as it passes successive anode segments and will eventually strike the anode. In moving from the cathode to the anode, this electron acquires dc energy, which it gives up to the RF field.

Next, consider the interaction of a single stream of electrons with an RF field. It will be shown in the following section that the electric field pattern on the anode of a magnetron is a standing wave that can be resolved into two waves of equal magnitude traveling in opposite directions. Figure 19-21 shows one wavelength of the traveling electric field pattern and an electron stream with *E/B* adjusted so that it is moving in synchronism with the traveling wave. The directions of the forces on the electrons (opposite to the field directions) are indicated at the bottom of the figure and the resulting deflections of the electrons are indicated by small arrows along the electron stream.

Electrons near **A** experience decelerating forces so they move toward the anode. As they move toward the anode, the RF fields become stronger and the electrons are pushed progressively into spoke-like projections that extend to the anode. Electrons near **B** are accelerated by the RF field so they move toward the cathode. This produces depressions in the electron stream that extend to the cathode.

As with individual electrons, the interaction of a single stream of electrons with an RF field is straightforward and individual electron trajectories can be determined.

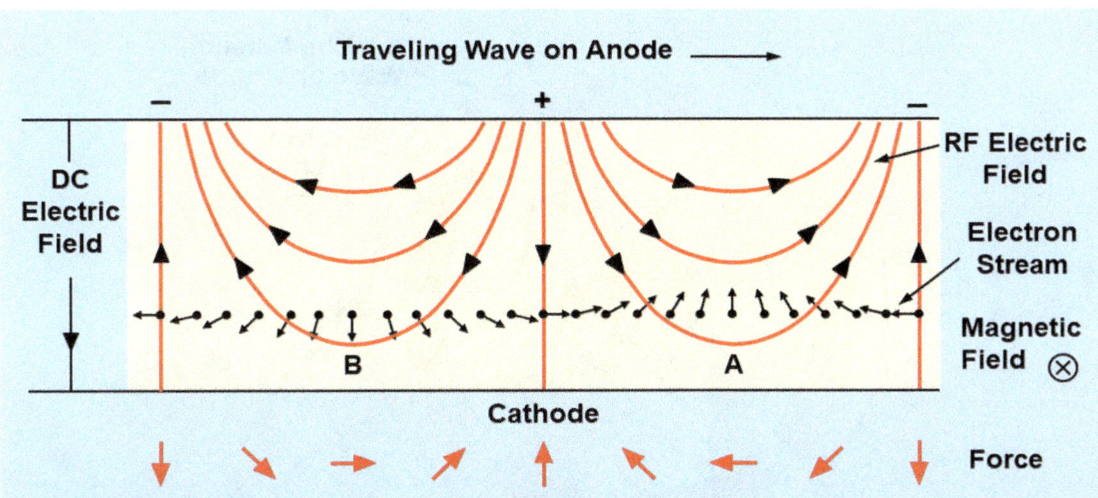

Figure 19-21 Electron deflections produced by an electric field traveling in synchronism with an electron stream.

The space-charge cloud surrounding the cathode greatly complicates the analysis of electron trajectories. However, qualitatively, the behavior of the electrons is very similar to that which has just been described. The rotating space-charge cloud is distorted into the spoke-like pattern shown in Figure 19-22.

Electrons in decelerating field regions move through the spokes from the cathode to the anode. During synchronous interaction, electrons at the surface of the hub and those transported to the anode move with a velocity approximately equal to that of the traveling wave on the anode.

The spoke-like cloud rotates with an angular velocity such that it remains in synchronism with the alternating RF voltage on the anode segments. As the spokes pass the slots between anode segments, the RF fields provide a focusing action that tends to further bunch the electrons. As indicated in Figure 19-23, an electron at position **A** on the leading edge of a spoke is in a region where the radial component of the RF field is in opposition to the dc field. As a result, from $u = E/B$, its angular velocity will be reduced so that it will fall back into the spoke. An electron at point **B** will experience an increased field and will be accelerated into the spoke. This same focusing action helps maintain synchronism between the rotating spokes and the alternating field so that the operating conditions of the magnetron (magnetic field, voltage, and so forth) are not as critical as they would appear to be at first.

19.3.4 Conversion of Potential Energy to RF Energy

The potential energy of an electron at rest on the cathode of a diode with voltage V_o on the anode is eV_o. In the absence of a magnetic field, if that electron is released from the cathode, it will be accelerated to the anode where its kinetic energy will be $_eV_o$. That energy will be dissipated as heat when the electron strikes the anode. In a magnetron, most of the potential energy of the electron is converted to RF energy rather than heat.

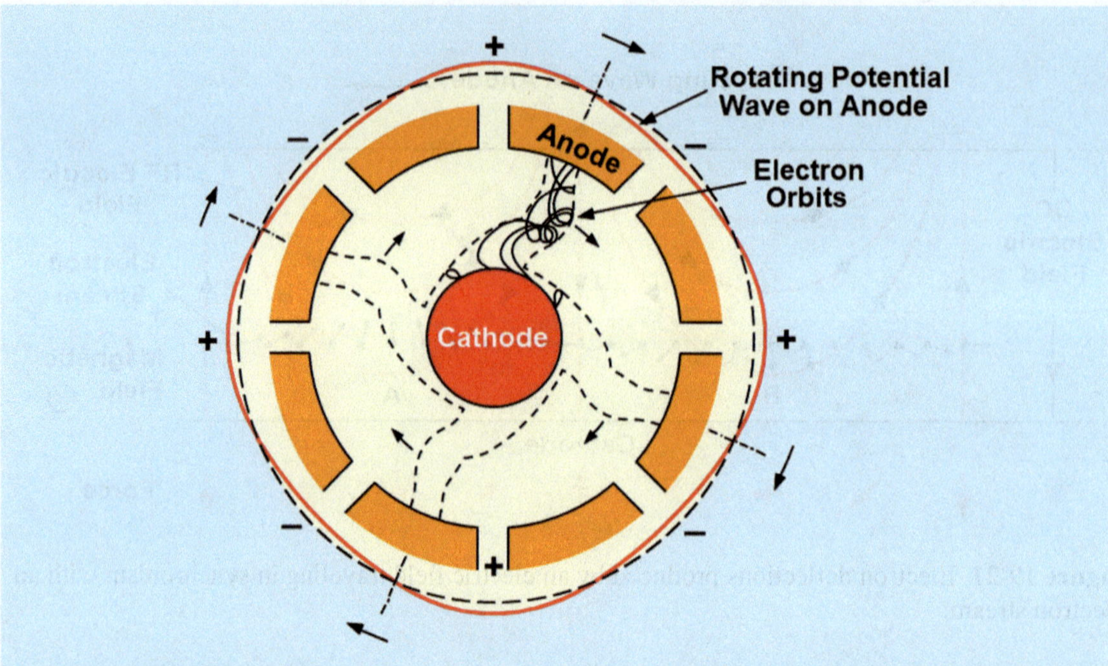

Figure 19-22 Spoke-like electron cloud in eight-cavity magnetron. (From: H. A. H. Boot and J. T. Randall, *IEEE Trans. Electron Devices*, July 1976. © 1976 IEEE.)

To understand how this energy conversion takes place, consider Figure 19-17. Only the average motion of an electron is shown. The cycloidal orbits only confuse the situation. In a), the electric field has been adjusted so the $u_x = E/B$ and the electron is traveling in a straight line. Assume that the velocity is greater than the phase velocity of the RF circuit.

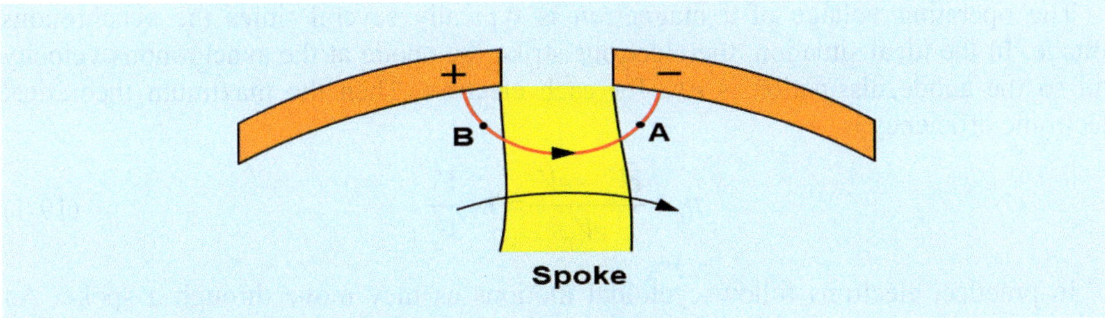

Figure 19-23 Focusing action of RF fields on space charge.

In (b), a wave on the anode traveling at phase velocity v_b has been added. The direction of the RF electric field is such that it reduces the velocity of the electron to the phase velocity. Electron energy is transferred to the RF E field. Because u_x is reduced, $u_x xB$ is reduced and the DC electric field forces the electron to move toward the anode with velocity u_y.

$u_y xB$ tends to cause the electron to move in the x direction but E_{RF} prevents that motion. In so doing, energy is again transferred to E_{RF}. Notice that u_x is forced to remain at the wave velocity, v_p. As this process continues, the electron is transported to the anode but its velocity remains constant at v_p. So, the electron acquires potential energy in moving toward the anode but that energy is transferred to E_{RF} and the electron velocity doesn't change.

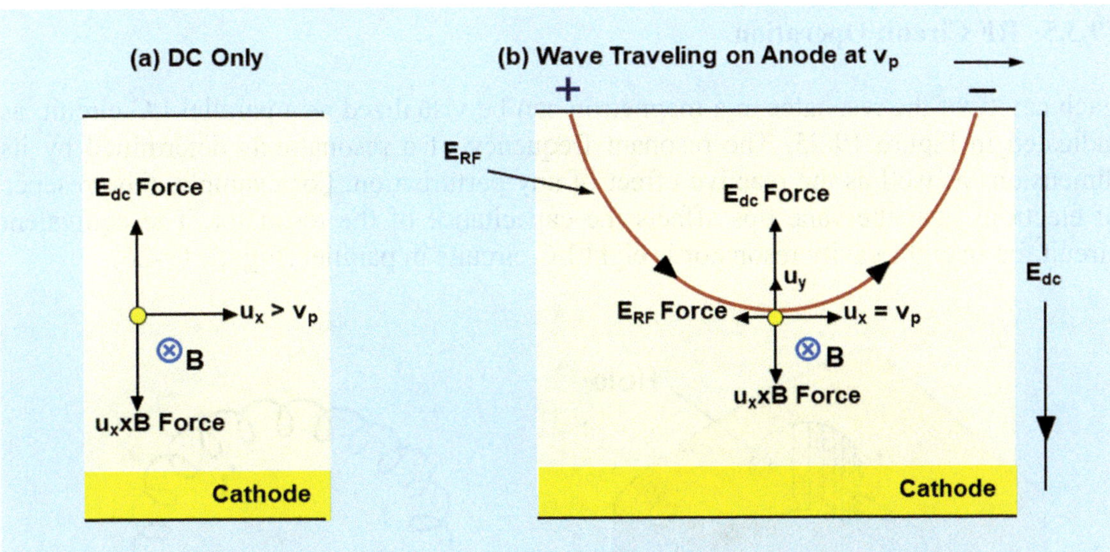

Figure 19-24 Origin of force that causes electrons to move toward anode.

The energy extracted from the power supply by an electron is the sum of its potential energy and its kinetic energy. Since the kinetic energy remains nearly constant, it is the potential energy that is converted to microwave power. This is different from a linear-beam tube where the kinetic energy of the beam is converted to microwave power and the beam slows down as a result of the energy exchange.

The operating voltage of a magnetron is typically several times the synchronous voltage. In the ideal situation, the electrons strike the anode at the synchronous velocity and so the anode dissipation is eV_o for each electron. Then the maximum theoretical electronic efficiency is

$$\eta_e = \frac{eV_a - eV_o}{eV_a} = 1 - \frac{V_o}{V_a} \tag{19-1}$$

In practice, electrons follow cycloidal motions as they move through a spoke. An electron moving at the top of a cycloid is moving at twice the synchronous velocity. An electron striking the anode with that velocity dissipates four times eV_o so the electronic efficiency on average is

$$\eta_e = 1 - \frac{kV_o}{V_a} \tag{19-2}$$

where $1 \leq k \leq 4$.

Electrons in accelerating field regions are forced back to the cathode surface. When they strike the cathode, they have sufficient energy to produce secondary emission. In a typical magnetron, the secondary electron emission is far greater than the primary emission. Secondary yield is great enough to replenish the hub and make up for the electrons that go to the anode. The back bombardment power delivered to the cathode results in heating the cathode surface. In a typical crossed-field tube, approximately 5% to 10% of the dc input power is dissipated in heating the cathode.

19.3.5 RF Circuit Operation

Each cavity of the resonator in a magnetron can be visualized as a parallel LC circuit, as indicated in Figure 19-25. The resonant frequency of a resonator is determined by its dimensions as well as the reactive effect of any perturbation. For example, the presence of electrons near the vane tips affects the capacitance of the resonator. The equivalent circuit for an eight-cavity resonator is eight LC circuits in parallel (Figure 19-26).

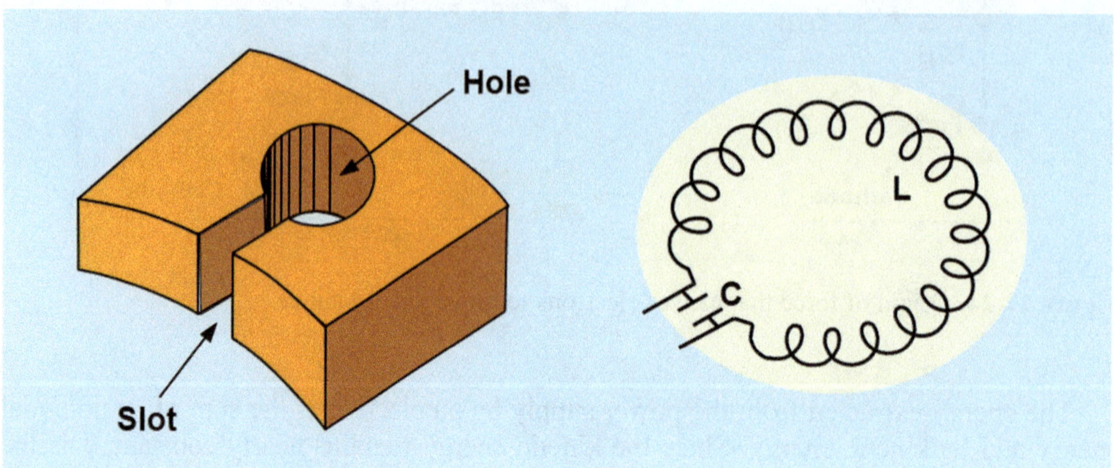

Figure 19-25 Equivalent circuit of hole-and-slot cavity.

Figure 19-26 Equivalent circuit for an eight-cavity resonator.

The magnetic and electric field patterns for each cavity are indicated in Figure 19-27. In the desired mode of operation, some of the magnetic flux lines of each cavity link adjacent cavities (two lines are shown). The fields in adjacent cavities are 180° out of phase. This is referred to as the π mode of operation. The resulting field configurations in an eight-cavity structure are shown in Figure 19-28.

Now, the reason why these magnetrons are referred to as traveling-wave magnetrons can be explained. In Figure 19-29 the cavities of an eight-cavity structure are shown in a linear configuration. The electric field pattern is a standing wave with the direction of the electric field lines across the cavity gaps oscillating back and forth as the RF voltage on the anode segments oscillates.

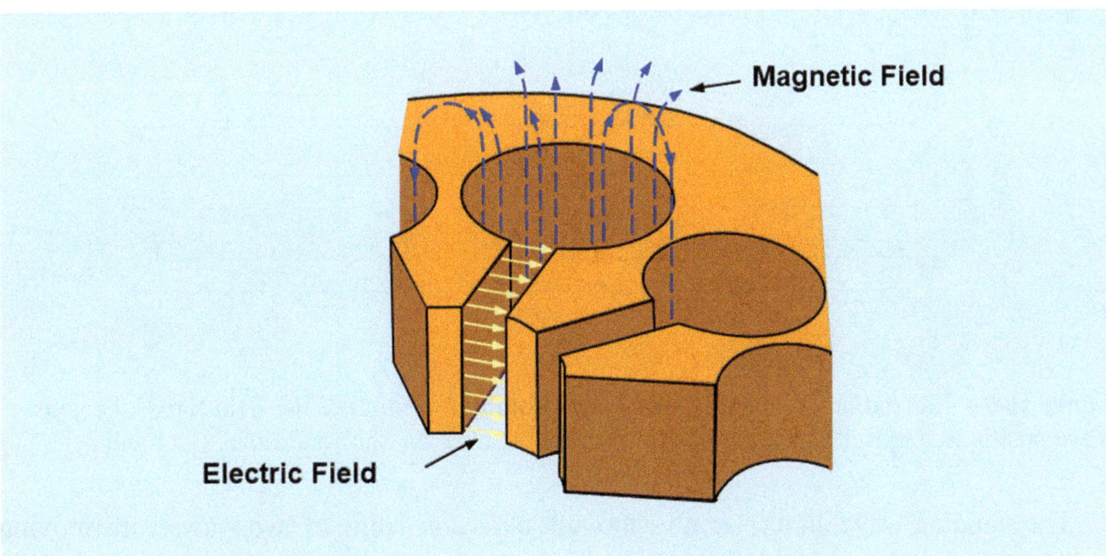

Figure 19-27 Magnetic and electric field patterns in a hole-and-slot cavity. (Adapted from: *Microwave Magnetrons* by George B. Collins. © 1948 McGraw-Hill, Inc.)

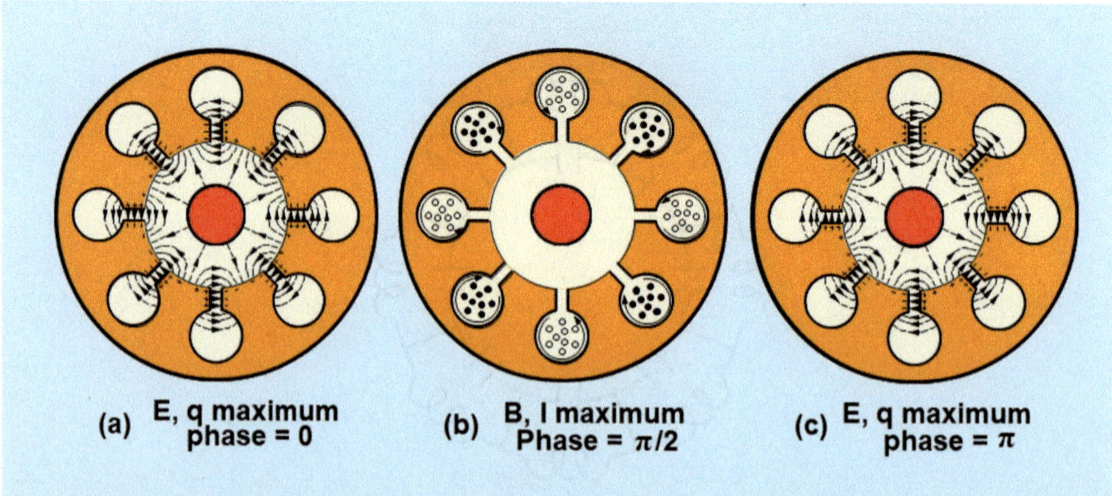

Figure 19-28 Charges, fields, and currents for the π-mode in an eight-cavity resonator. (Adapted from: *Microwave Magnetrons* by George B. Collins. © 1948 McGraw-Hill, Inc.)

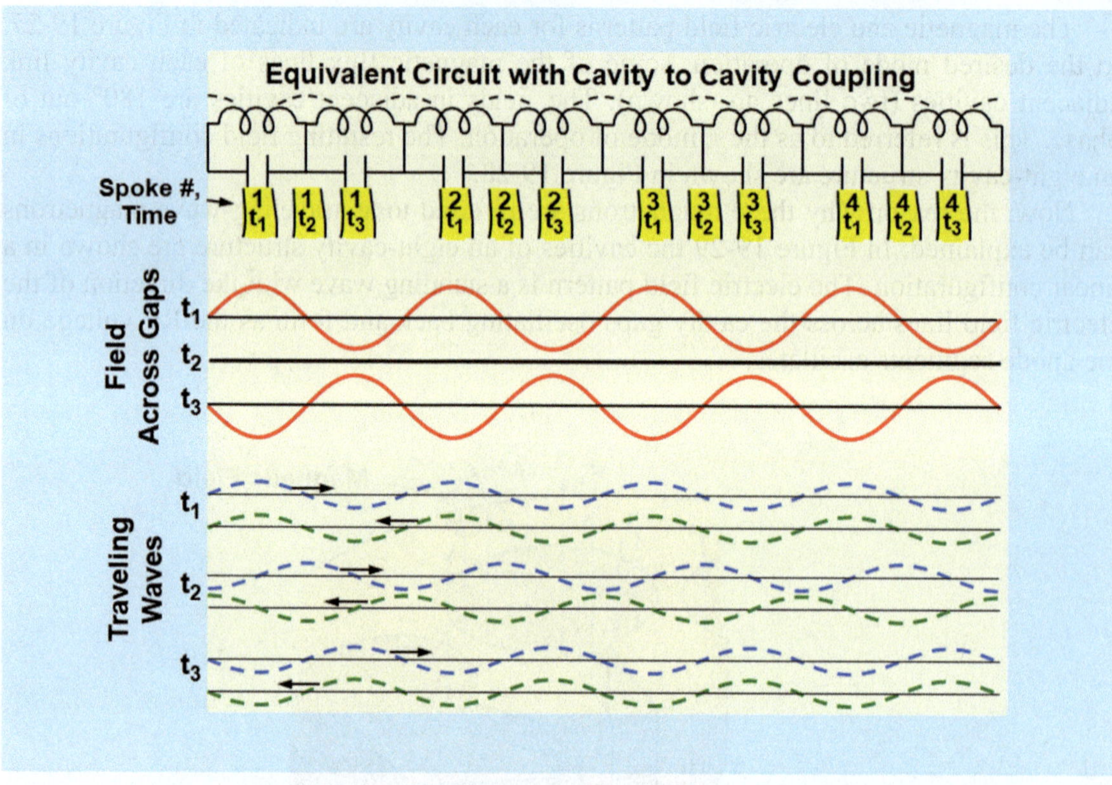

Figure 19-29 Illustration of how traveling waves combine to produce the field across the gaps. Wave positions, gap fields, and the positions of four spokes are shown at times t_1, t_2, and t_3.

The standing wave, in turn, can be thought of as consisting of two waves, one moving to the right (rotating clockwise in the magnetron) and the other moving to the left (rotating counterclockwise). At time, t_1, the left-moving and right-moving waves combine to maximize the electric field across the gaps of the first and every other cavity. At this time, forces that compress the spokes are maximum.

At time t_2, the phases of the wave moving to the right and the wave moving to the left are such that the fields of the waves are opposite and so they cancel. As the electron spokes move by the gaps between cavities, the electric field is zero.

At time t_3, the fields from the two waves again add, but this time the field is maximum at the second, fourth, sixth, and eighth gaps. Again, the forces that compress the spokes are maximum. Thus, it is clear that to optimize the interaction of the wave on the anode with the movement of the spokes, the velocity of the wave on the anode must be near the spoke velocity. Then energy exchange from the electrons to the RF field is maximum.

19.4 MODING

Modes and field patterns other than those associated with the π mode can exist and result in magnetron operation at frequencies other than the desired frequency. These other modes and the steps taken in attempts to prevent their occurrence are discussed here.

In his science-fiction article on magnetrons [10], Pierce leads into a discussion of moding by commenting on the various problems faced by early magnetron engineers and the humorous names that they applied to the peculiarities of performance encountered:

> ...The Civil War Effect, or North-South Effect had to do with differences in performance when the direction of the magnetic field was reversed. The differences proved to be due to asymmetries in the tubes, and disappeared when the tubes were made more accurately. The Blowing Out of Brains Effect had simply to do with the melting of a copper end cap which covered the end of the anode block of a small magnetron to form part of the vacuum envelope. The Gnawing Away of the Anode Effect was similar. The Gardarens Swine Effect has to do with the particularly precipitous descent of certain contours on a performance chart, and the Valley of the Shadow was a region of low efficiency lying close to a high-efficiency region known as the Heights of Abraham.
>
> Undoubtedly the worst effect was one perhaps too serious for a fancy name. It was called, simply, moding. An engineer in using an early magnetron would be adjusting the load, that is, the impedance into which the magnetron fed power, in order to get more power. Suddenly, the power, which had been rising, would drop, and he would find the tube to be oscillating feebly at some new and unwanted frequency a little removed from the old. Or, he would be gradually raising the voltage to get more power, and the same thing would happen. The tube was said to be operating in another mode of oscillation....

Moding occurs because the RF circuit of a magnetron consists of many resonators coupled together. An analogous situation exists when several identical pendulums are suspended from a common support that is not rigidly held in place. For example, in Figure 19-30, three pendulums are shown along with the three forms of motion (modes of oscillation) that they can have. The frequencies at which the pendulums swing back and forth are different for each mode. Similarly, a magnetron may have several modes of oscillation that depend on the coupling between cavities.

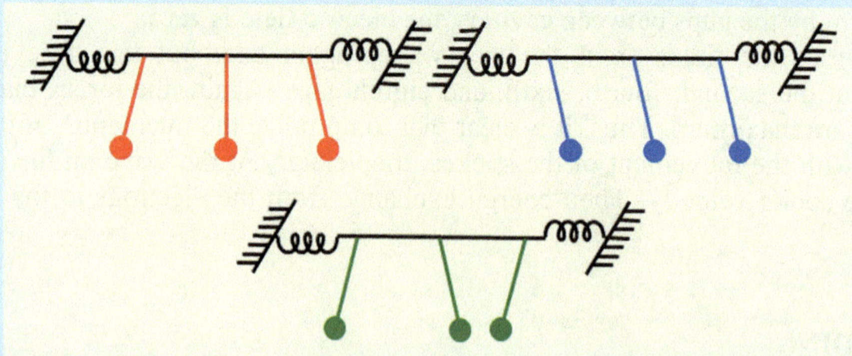

Figure 19-30 Modes of oscillation of three pendulums supported by a movable bar.

Field distributions for the modes in an eight-cavity magnetron are illustrated in Figure 19-31. The modes are designated by a mode number n, which is the number of times the RF field pattern is repeated in going around the anode once. The largest value of n is $N/2$, where N is the number of cavities. This mode is called the π mode and is always the desired mode of operation.

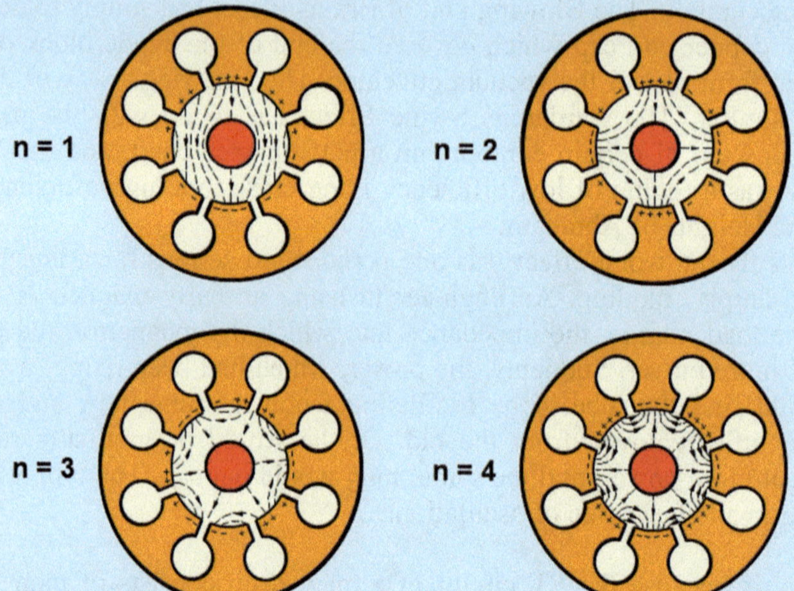

Figure 19-31 Field distributions for the modes in an eight-cavity magnetron. (From: H.A.H. Boot and J. T. Randall, *IEEE Trans. Electron Devices*, July 1976. © 1976 IEEE.)

In addition to there being several modes of oscillation for a magnetron, there is, quite literally, an infinite number of electron velocities that can excite any given mode. As an example, the situation existing in the π mode is illustrated in Figure 19-32.

The equivalent circuits of an eight-cavity magnetron are across the top of the figure. The electric field distribution along the anode is shown for various time intervals separated by one-half of a period. An electron, such as one at point **A** and time *t* is in a decelerating region and gives up energy to the RF field. An electron in the region between **A** and **B** is considered to be essentially shielded from the RF field by the anode segment.

The dashed lines represent the progress of electrons traveling in favorable phases (continually giving up energy) with the RF field. For example, electrons represented by line 1 have a transit angle from cavity to cavity of π radians and travel in synchronism with the traveling field. Electrons represented by line 3 travel at one-third of the velocity of those represented by line 1 and have a transit angle of 3π radians.

A complete cycle of oscillation of the RF circuit occurs while these electrons pass by each anode segment in traveling from one cavity to the next. Still, these electrons are synchronous with the RF field and give up energy on every third cycle of oscillation of the RF field. This situation is analogous to keeping the pendulums in Figure 19-30 moving by giving them a push on every third swing.

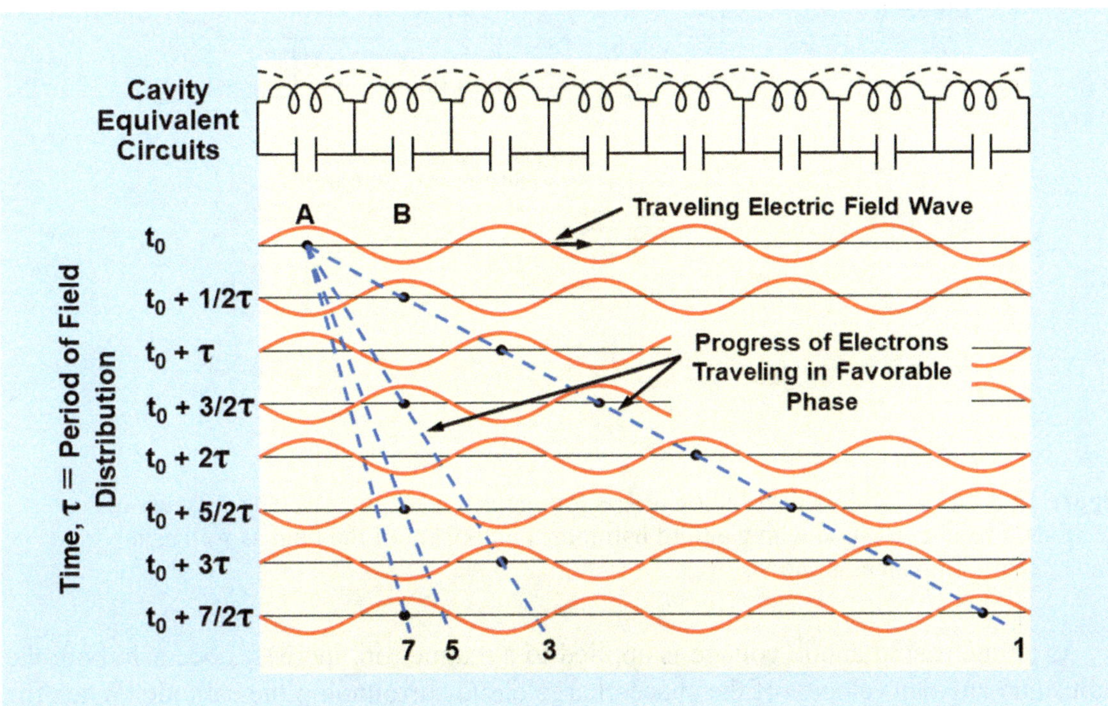

Figure 19-32 Electric field distribution in an eight-cavity magnetron operating in the π mode. (Adapted from: *Theory and Application of Microwaves* by Bronwell and Beam, © 1947 by McGraw-Hill, Inc.)

Close to the resonator gaps, the electric field is a square wave as shown in Figure 19-33. If this field is broken into its harmonic components, odd-numbered harmonics (1, 3, 5, 7, and so forth) result, and it is found that the electrons represented by line 3 are in synchronism with the third harmonic component of the field. Similarly, electrons represented by lines 5, 7 in Figure 19-32, and an infinite set not shown are synchronous with still higher-order harmonic components of the field. In principle, any of these harmonic interactions can produce oscillation in the π mode. Each of the other magnetron modes can be broken into an infinite set of harmonic components and can, therefore, in principle, be driven by electrons having an infinite range of velocities.

The resulting possible array of magnetron operating conditions is bewildering and this is, perhaps, what led Boot and Randall in their 1976 article [4] to comment: "If this point had been appreciated in 1939 it is doubtful that the magnetron would ever have been made." Fortunately, there are several selection processes, both natural and by design, which limit the range of operating conditions of a magnetron. For example, the amplitudes of the harmonic fields decrease rapidly as the harmonic number increases so that the probability of interaction at harmonics decreases. For a square wave like that in Figure 19-33, the amplitude of the third harmonic is 1/3 of the fundamental amplitude.

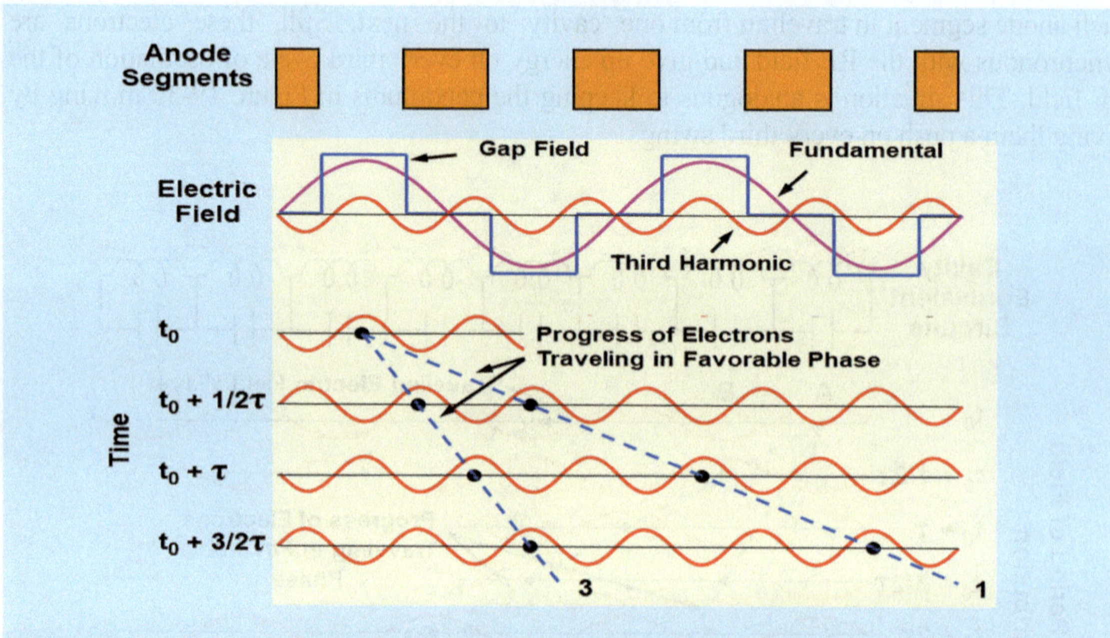

Figure 19-33 Electric field distribution across four gaps in a magnetron. The progress of electrons traveling in phase with the third harmonic component of the field as well as the fundamental component is shown.

As an increasing anode voltage is applied to a magnetron, increases occur in both the radius and angular velocity of the space-charge cloud surrounding the cathode. When the angular velocity is such that the cloud is rotating in near synchronism with one of the RF modes, or one of the harmonic components of an RF mode, noise currents on the space-charge cloud may cause oscillations to start in that mode. The voltages at which the oscillations are initiated in various modes and various harmonics are the Hartree threshold voltages. Some of these voltages are plotted as a function of magnetic flux density for an eight-cavity magnetron in Figure 19-34. The dashed line represents a harmonic of the $n = 3$ mode, which has a threshold voltage nearly identical to the $n = 4$ mode, which is the π mode. As a result, unless precautions are taken, operation may occur in the $n = 3$ mode rather than the π mode.

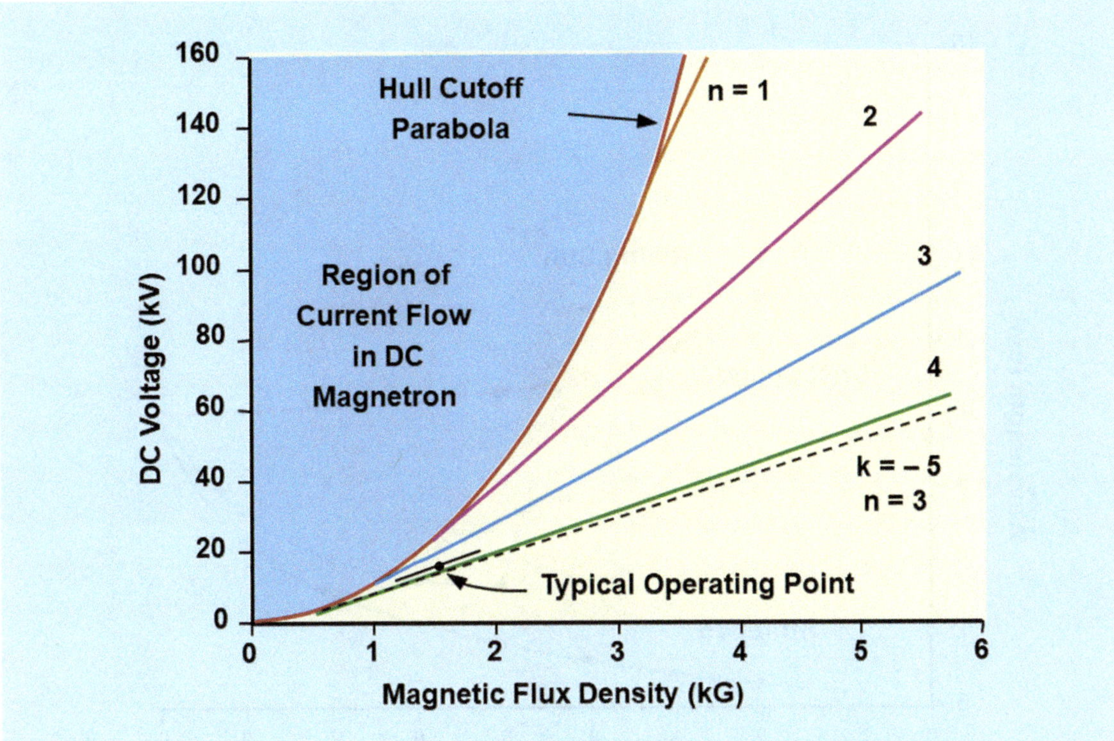

Figure 19-34 Hartree threshold voltage diagram for an eight-cavity magnetron. (From: H. A. H. Boot and J. T. Randall, *IEEE Trans. Electron Devices*, July 1976. © 1976 IEEE.)

In considering precautions that can be taken to force a magnetron to operate in the desired mode, it is important to realize that the frequencies of the various modes are very close together. This is indicated in Figure 19-35 by the unstrapped line. The separation between the $n = 3$ and $n = 4$ modes is only 1% to 2%. For structures having more than eight cavities (in practice, the number of cavities usually ranges from 8 to 18), the separation in frequency between the π mode and its nearest neighbor becomes even smaller than 1% to 2%. Because of the frequency instabilities when the mode separation is small, structural changes are made to the resonator to increase this separation.

Two techniques, the use of straps and the rising-sun structure, for increasing the mode separation are discussed at this point. The third, the use of a coaxial cavity, is discussed in Section 19.5 on coaxial magnetrons.

In the desired π mode of operation, alternate anode segments are at the same RF potential so, if alternate segments are connected together by wires (straps), no additional inductance will be introduced because the ends of the straps are at the same potentials. The straps add capacitance to the circuit, so the π-mode frequency will be altered.

In modes other than the π mode, the voltage differences between alternate anode segments is not zero, so the straps introduce inductance as well as capacitance, resulting in different frequency shifts than occur for the π mode. The effect of strapping on mode separation in 18-cavity magnetrons is shown in Figure 19-35.

In low-power tubes, mode separations of 25% to 35% are achieved with strapping. In high-power tubes, the mode separation is normally about 10% to 14%. In either case, the undesired modes are at frequencies far enough removed from the π mode so that the tube can usually be prevented from operating in these modes.

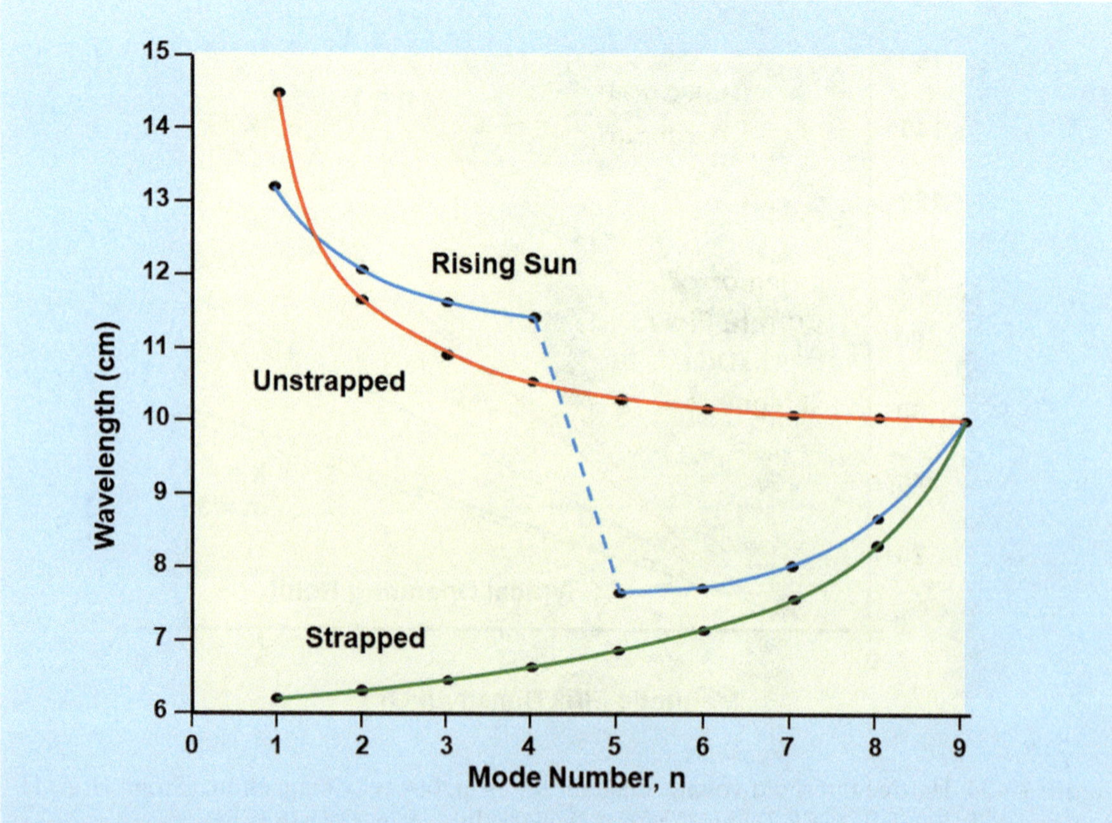

Figure 19-35 Mode spectrum for 18-cavity magnetrons. (From: G. K. Farney, *Crossed-Field Microwave Devices,* Technology Service Corporation/Crane Division Naval Surface Warfare Center, 1995.)

Shown in Figure 19-36 are examples of strapped structures and an unstrapped resonator known as a rising-sun structure. The way in which the rising-sun structure causes mode separation to occur is not as easy to understand as the strapping technique. The number of modes in the rising sun magnetron is the same as in the strapped magnetron. However, in the rising-sun magnetron, half the modes are such that the large cavities control the frequency, and so the frequencies associated with these modes are relatively low. For the higher-numbered modes (except the π mode), the field patterns are such that the small cavities control the frequency, and so the frequencies associated with these modes are relatively high. In the π mode, the frequency is midway between the resonant frequencies of the two cavity configurations. Thus, the rising-sun structure increases the frequency separation between the π mode and the two adjacent modes (one above π mode and one below π mode). By comparison, in a strapped system, all the modes are resonant at frequencies below the π mode.

The mode separation characteristic of rising-sun structures has facilitated the design and fabrication of millimeter-wavelength magnetrons where dimensions are small and it is difficult, if not impossible, to fabricate strapped resonators. However, the number of resonators is limited by the mode separation requirements and the size of the cathode is limited by the number of resonators. Cathode size also decreases as the frequency of operation is increased. This limits the current and the output power.

19.5 COAXIAL MAGNETRONS

An indication of the significance of the development of the coaxial magnetron [11] is that it is listed by Skolnik [12] along with the cavity magnetron and the high-power klystron as a significant event in the development of radar technology. The coaxial magnetron was developed as an alternative to the strapped cavity magnetron.

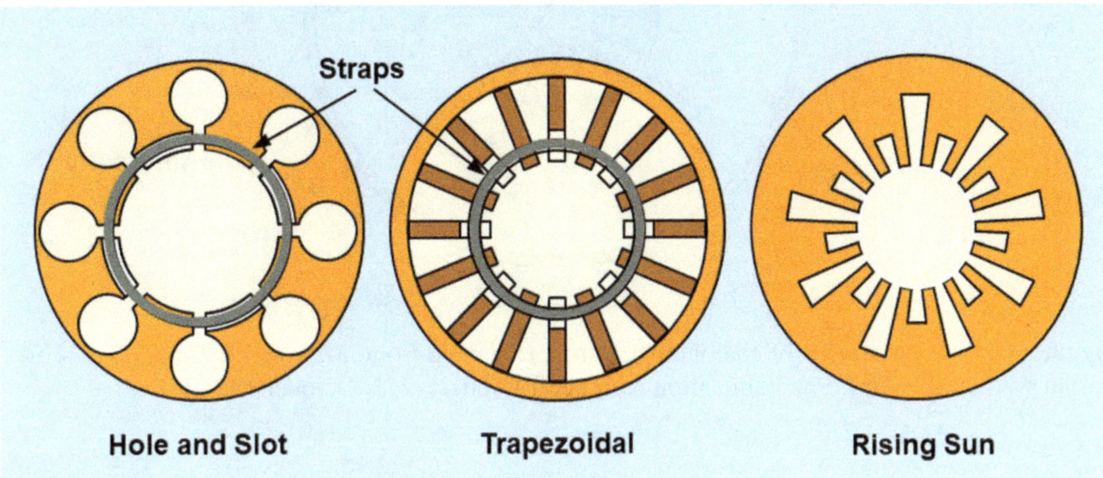

Figure 19-36 Techniques for achieving mode separation.

In the conventional magnetron, the strap-vane structure results in anode segments that are sometimes small and operate at an excessively high-power density. Close interelectrode spacing leads to occasional voltage breakdown and the cathode loading is often excessive. As a result, in early radars, the conventional magnetron was often the least reliable component of the system.

As Figure 19-37 shows, the coaxial magnetron contains an anode resonator structure very much like that in a conventional magnetron. This structure is surrounded by a single high-Q cavity operating in the TE_{011} mode. Slots in the back walls of alternate cavities of the anode resonator structure tightly couple the fields in these resonators to the surrounding cavity. In the π mode, the fields in every other cavity are in phase, and they couple in the same direction into the surrounding cavity. Thus, the surrounding coaxial cavity helps stabilize the magnetron in the desired π mode of operation.

Because straps are not required, the anode resonator for the coaxial magnetron can be larger and less complex than for the conventional strapped structure. A comparison of anode sizes for the same operating frequency and power is shown in Figure 19-38. Because the anode is larger, the cathode and interelectrode spaces can also be larger. As a result, cathode loading is lower and voltage gradients are reduced. These factors all combine in the coaxial magnetron to produce relatively conservative magnetron designs having enhanced reliability compared with conventional magnetrons.

In the coaxial magnetron, the multiple-cavity anode structure has a series of modes like those of the conventional magnetron. In addition, there are various modes of oscillation of the coaxial cavity. Fortunately, the use of the TE_{011} mode makes possible the use of attenuators to damp unwanted modes.

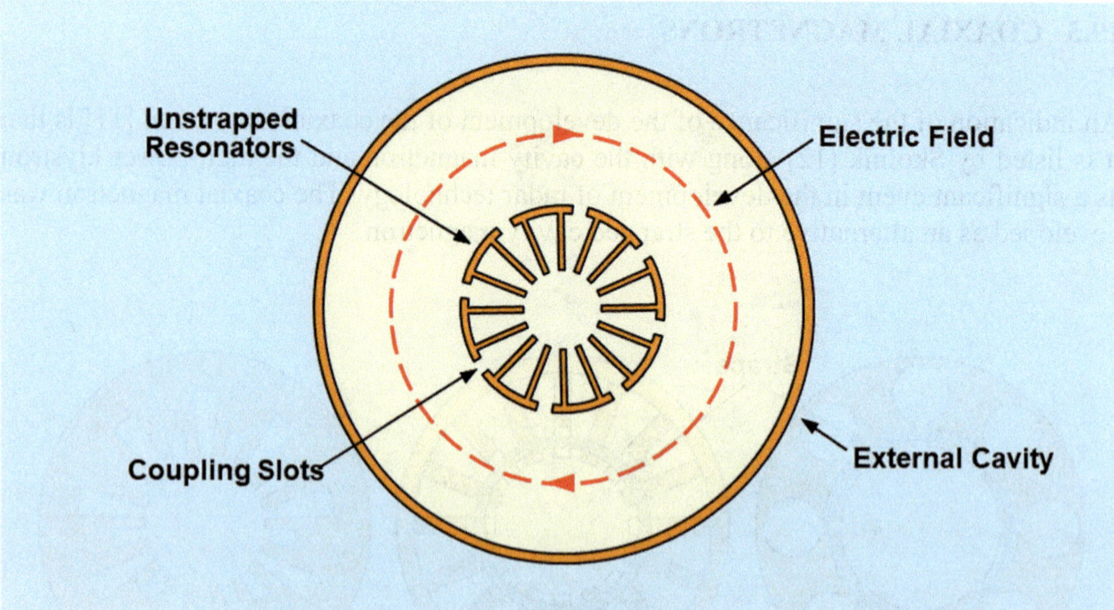

Figure 19-37 Cross section of coaxial magnetron. (Adapted from: *Microwave Tube Manual* by Varian Associates, Air Force Publication Number T.O.00-25-251, October 1979.)

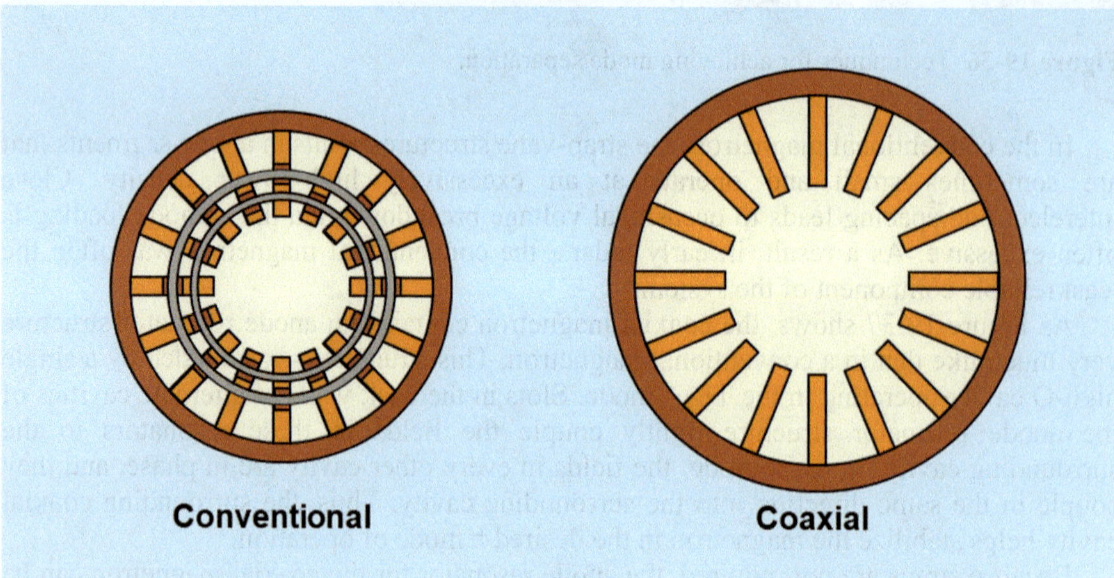

Figure 19-38 Comparison of anodes of conventional strapped magnetrons and coaxial magnetrons. (From: *Microwave Tube Manual* by Varian Associates, Air Force Publication Number T.O.00-25-251, October 1979.)

In the desired TE_{011} mode, the RF electric field exists in circular form within the cavity as indicated in Figures 19-37 and 19-39, and decreases to zero at the walls of the cavity. Current flow in the TE_{011} mode is in the walls of the cavity in circular paths about the axis of the tube. An attenuator placed as shown in Figure 19-39 at the junction of the inner cylinder to the fixed end of the cavity couples very little to TE_{011} currents, but couples heavily to other modes, thereby suppressing these unwanted modes of oscillation.

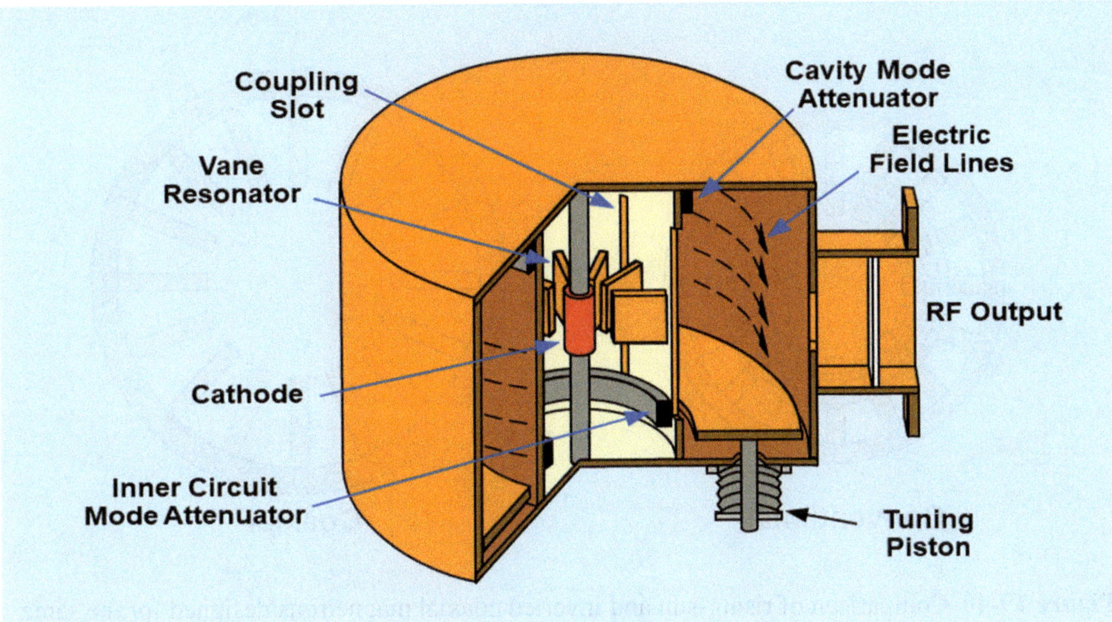

Figure 19-39 Cutaway view of coaxial magnetron. (From: *Microwave Tube Manual* by Varian Associates, Air Force Publication Number T.O.00-25-251, October 1979.)

Similarly, when the fields near the coupling slots from the resonator to the coaxial cavity are examined, it is found that the ends of the slots are free of TE_{011} current. Thus, an attenuator within the inner cylinder near the ends of the coupling slots does not couple to the TE_{011} mode currents, but it does dampen undesired modes of the vane structure.

Another advantage of the TE_{011} mode is that one of the cavity end plates can be made movable (tuning piston in Figure 19-39) without interrupting the flow of current. As a result, tuner reliability is high for coaxial magnetrons.

19.6 INVERTED MAGNETRONS

A further advantage of the coaxial-magnetron configuration is that it can be turned inside out, or inverted, as shown in Figure 19-40. In this figure, an inverted structure is compared with a conventional rising-sun magnetron designed to operate at the same frequency. In comparing the two structures, it should be noted that the cathode size in the inverted configuration is, perhaps, an order of magnitude larger than the cathode in the conventional magnetron. Also, the number of anode vanes and cavities can be large in the inverted structure because moding is controlled by the center coaxial cavity.

The inverted coaxial configuration has application in the millimeter-wave region where dimensions are extremely small. In particular, the power of the rising-sun magnetron is limited by the high cathode loading. This problem is alleviated in the inverted structure. In addition, the center TE_{011} cavity lends itself to coupling to an output waveguide operating in the circular electric mode, which has extremely low loss.

It should also be noted that experiments have been performed with inverted magnetrons at the S-band to determine if they can generate extremely high power [13].

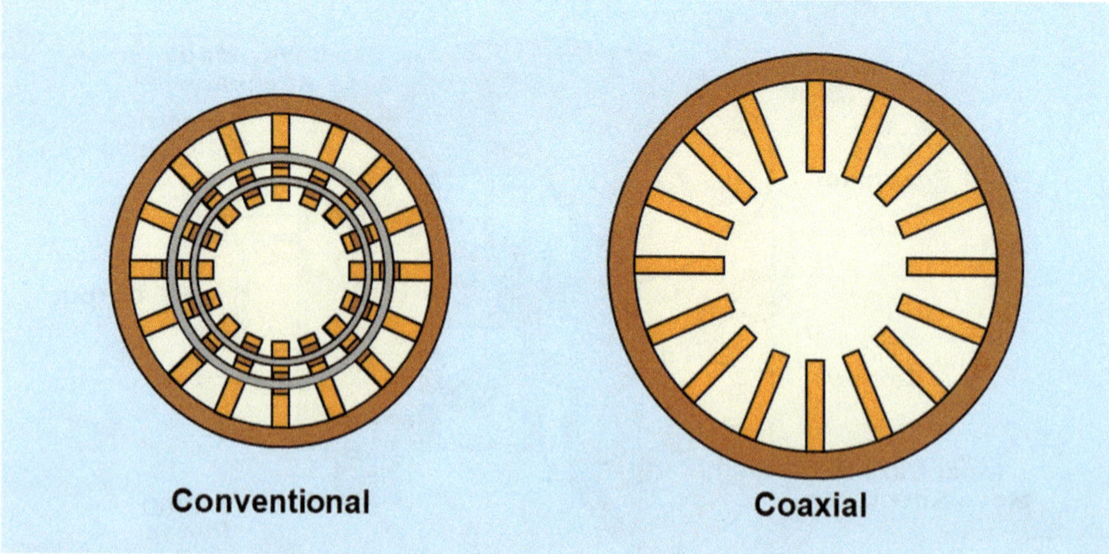

Figure 19-40 Comparison of rising-sun and inverted coaxial magnetrons designed for the same frequency. (From: *Microwave Tube Manual* by Varian Associates, Air Force Publication Number T.O.00-25-251, October 1979.)

19.7 MAGNETRON TUNING

Many techniques are used for changing the frequency (tuning) of magnetrons. In one way or another, these tuning techniques employ changes in the capacitance or the inductance of the resonant cavities. For example, as shown in Figure 19-41, the capacitance of the cavity may be varied by the movable metallic element (or dielectric element) in the vicinity of the slot portion of the cavity. The inductance of the cavity may be decreased by inserting a metallic plunger into the hole portion of the cavity. Configurations of capacitive and inductive tuning elements are shown in Figure 19-42.

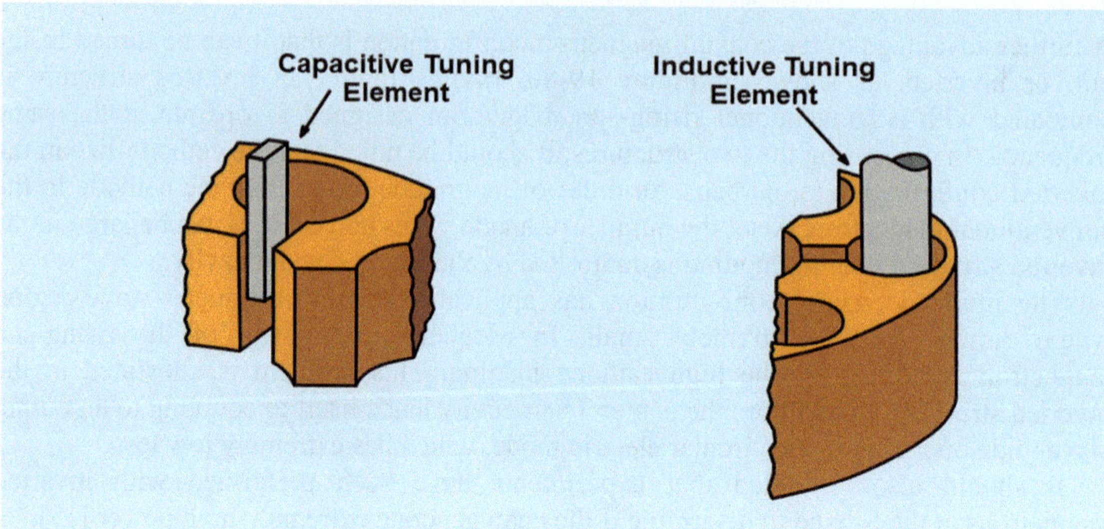

Figure 19-41 Basic tuning techniques for a hole-and-slot resonator.

Coaxial magnetrons can be tuned by changing the volume of the cavity with a movable end plate as indicated in Figure 19-39. A fine-tuning technique using a compressible ring in the wall of the coaxial cavity is also shown in Figure 19-43.

In another technique for tuning a coaxial magnetron, rotatable dielectric paddles are inserted into the high-electric-field region of the coaxial cavity. The lowest frequency occurs when the paddles are aligned with the electric field. The highest frequency occurs when the paddles are perpendicular to the electric field lines.

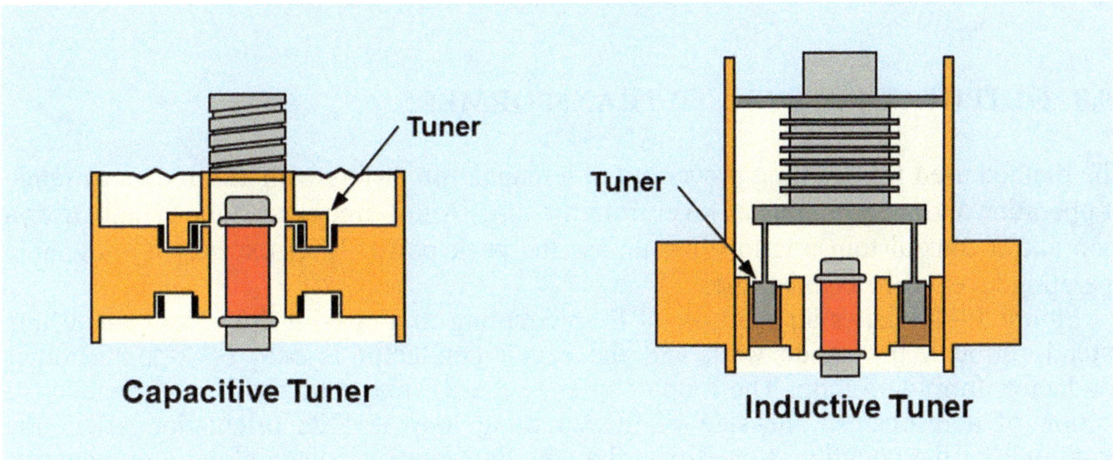

Figure 19-42 Capacitive and inductive tuning elements.

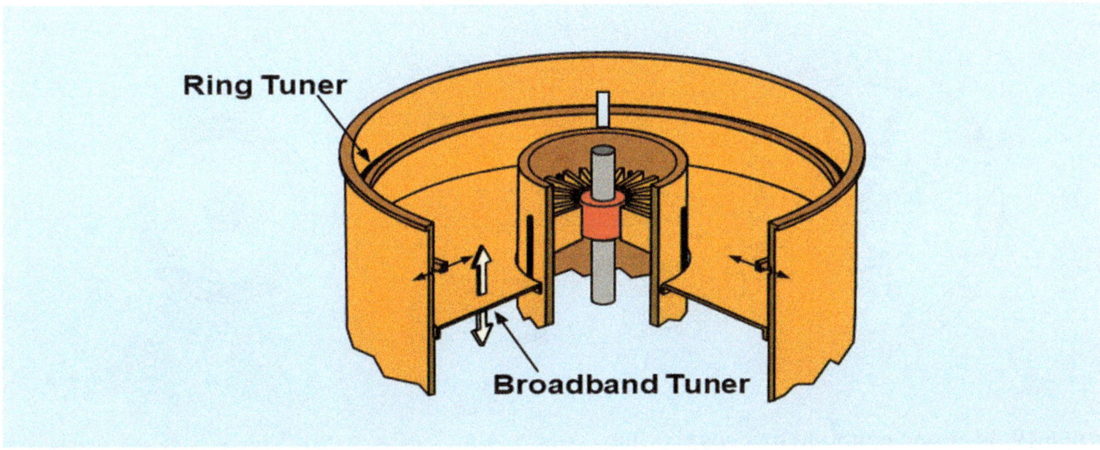

Figure 19-43 End-plate (broadband) tuning and ring tuning of a coaxial magnetron. (From: *Microwave Tube Manual* by Varian Associates, Air Force Publication Number T.O.00-25-251, October 1979.)

There are many ways for positioning tuning elements in a magnetron. Examples are the mechanical gear arrangements on early magnetrons in which the position of the tuning element was adjusted through a bellows using a gearing arrangement. Tuning speed, backlash, and drive torque are limiting factors. Early frequency-agile magnetrons were tuned by a hydraulic system that replaced the mechanical gear system. Tuning was slow and maintenance difficult, so these systems are becoming obsolete.

There are various electromagnetic techniques for positioning tuning elements. In one, the tuner operates much like the voice coil of a speaker. Current through a coil produces a magnetic field that acts with the field of a permanent magnet for axial positioning of a shaft to which the tuning element is attached. Another example is the use of an electric motor that is coupled through a gear-and-cam arrangement to position a tuning plunger. Accurate tuning can be accomplished through a resolver.

Some additional tuning techniques that make it possible to change frequency very rapidly are described in Section 19.10.2.

19.8 OUTPUT COUPLERS AND TRANSFORMERS

The method used for coupling power out of a magnetron differs with the frequency range of operation and the peak power level from the tube. A low-frequency tube usually uses a loop and a coaxial transmission line unless the peak power level exceeds the tolerable operating level of the coaxial line.

Figure 19-44 shows an example of loop coupling to a coaxial line. The coaxial line extends through the anode wall, and the center conductor is bent back to the outer conductor, forming a loop. The loop couples to the RF magnetic fields in the inductive portion of a resonator. The size of the coupling loop and its orientation affect the magnitude of the coupling. Sometimes the coupling loop is located above the inductive region of the anode in the end space region of the magnetron. This kind of coupling is called halo loop coupling.

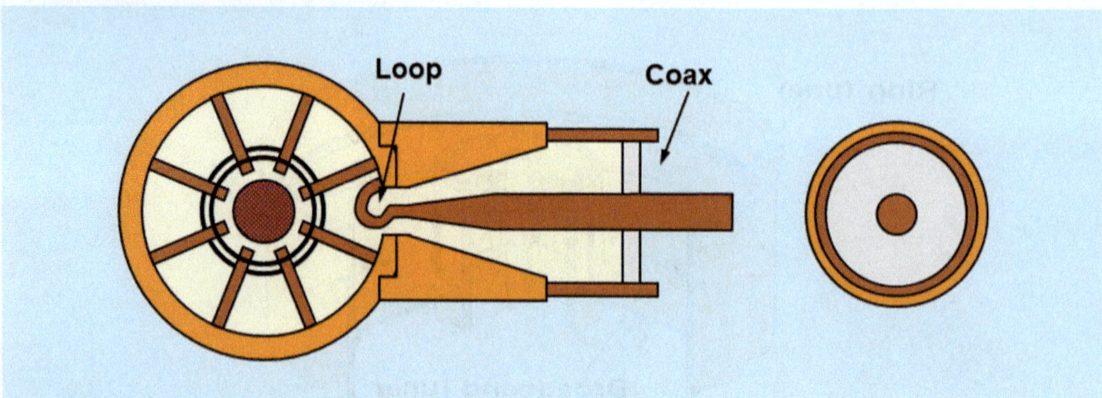

Figure 19-44 Loop coupling to a coaxial line.

At higher frequencies, iris coupling from the back of a resonator to a waveguide is used. In the example in Figure 19-45, the slot at the back of the resonator is coupled into the ridge, or capacitive region, of a ridge waveguide. The end of the waveguide is attached to a flat surface on the anode with the center at the slot opening. The cross section of the ridge waveguide is shown relative to the full-size waveguide output for the tube. The waveguide portion is called the H section for obvious reasons. The ridge waveguide section is approximately a quarter wave in length and has the appropriate impedance to optimize the coupling.

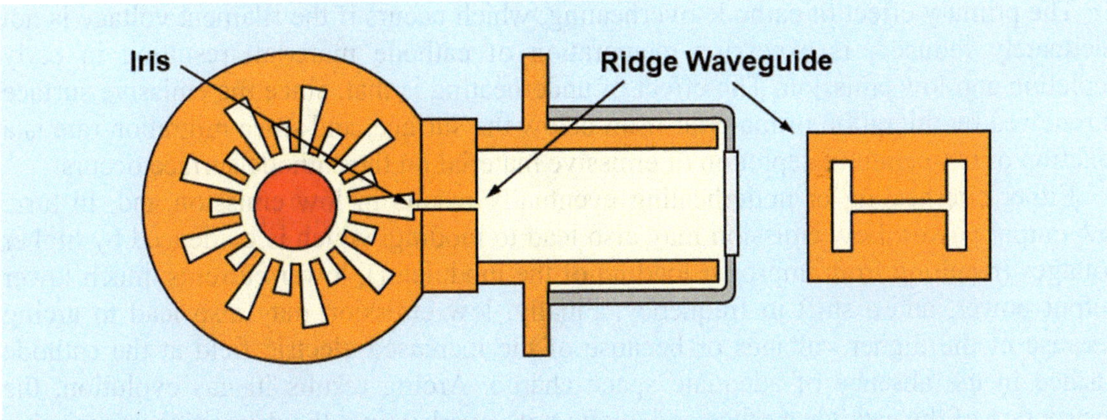

Figure 19-45 Iris coupling to a waveguide.

Although these transitions have been called matching sections, they do not provide a match in the sense of having no reflected power. In contrast, the proper impedance provides the necessary coupling to load the magnetron resonator with the correctly transformed load impedance that will allow oscillation with high circuit efficiency. The typical resonator system is over coupled to do this.

The matching impedance looking into the cavity resonator of a magnetron is very low. The exact magnitude of the impedance is difficult to calculate or to measure. It is always much smaller than that of the coaxial line or waveguide output from the magnetron. Hence, some form of impedance-matching transformer must be used. A convenient method with coaxial outputs is to use a linearly tapered center conductor to obtain both impedance and a geometry transitions. An example is shown in Figure 19-44.

19.9 CATHODE AND HEATER OPERATION

In crossed-field devices, electrons that are out of phase with the RF fields are driven back into the surface of the cathode (see Figure 19-20). These electrons produce secondary emission and heating of the cathode. The secondary emission process was discussed in the chapter on cathodes (Chapter 5). In particular, its significance was illustrated by the Philips YJ 1183 spin-tuned magnetron in which it is estimated that thermionic emission is 3 A/cm^2 and secondary emission is 23 A/cm^2 for a total cathode emission of 26 A/cm^2.

In the microwave oven magnetron, the cathode is carburized thoriated tungsten for which the secondary emission coefficient is low so thermionic emission is the main emission mechanism. Carburized thoriated tungsten is used to achieve a long tube life.

The importance of cathode heating by out-of-phase electrons cannot be overstated. Typically, 5% to 10% of the anode power input may go into cathode heating. In some magnetrons, this effect is such that the filament voltage can be removed once the tube is operating and the cathode temperature is maintained by bombarding electrons. Figure 19-46, which is a more typical case, shows how the magnetron filament voltage should be reduced as average anode current is increased if essentially constant cathode temperature is to be maintained.

The primary effect of cathode overheating, which occurs if the filament voltage is not adequately reduced, is excessive evaporation of cathode material, resulting in early depletion and low emission. The effect of underheating is that, since the emissive surface is renewed by migration of material from below the surface, and since migration rate is a function of temperature, depletion of emissive material on the cathode surface occurs.

Either overheating or underheating eventually results in low emission and, in turn, low output power. Low emission may also lead to moding, which is evidenced by higher voltages (resulting from improper loading of the modulator), lower currents, much lower output power, and a shift in frequency. Finally, low emission may also lead to arcing because of the higher voltages or because of the increased electric field at the cathode surface in the absence of adequate space charge. Arcing results in gas evolution, the destruction of the cathode surface and, along with overheating, the deposition of emissive material on critical areas.

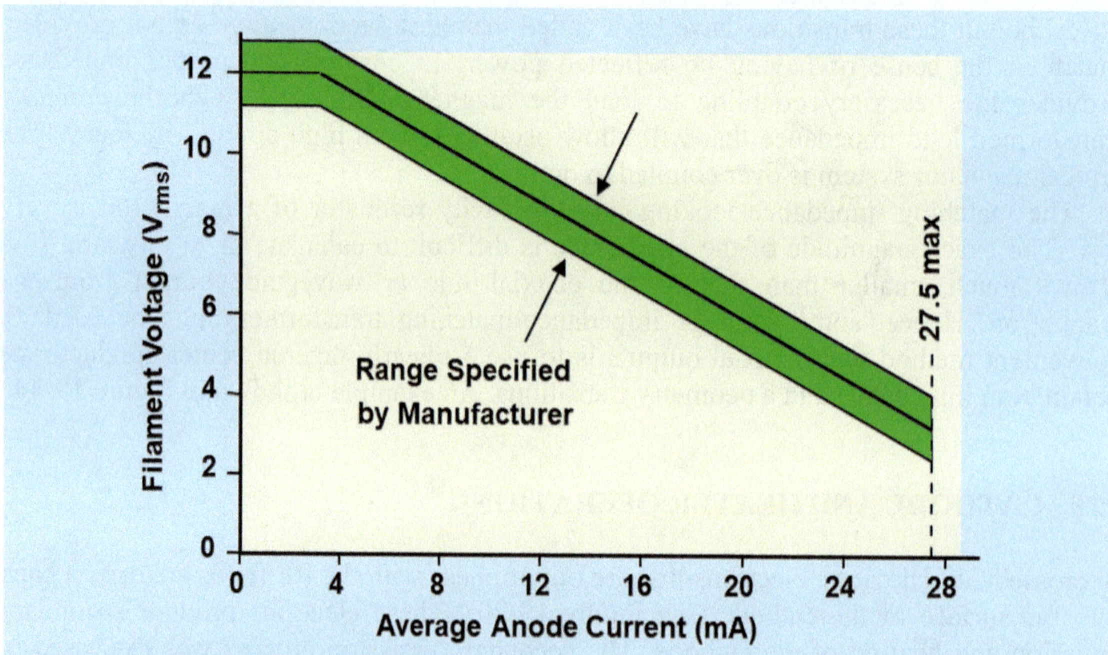

Figure 19-46 Magnetron filament voltage as a function of average anode current.

An example shown in Figure 19-47 of a particularly critical area is the end hat on the cathode in a spin-tuned magnetron. If precautions are not taken, the deposition of emissive material on the end hat leads to electron emission from it. The resulting electron current impinges on the magnetic pole piece and may heat it to the Curie temperature at which point it becomes nonmagnetic. As a result, the magnetic field collapses and the anode current may increase to the point where the anode melts.

Finally, the disastrous effect of overheating a magnetron cathode might possibly be best illustrated by the actual example in which the filament voltage schedule given in Figure 19-46 was not followed. In this case, the manufacturer of the equipment utilizing the magnetron did not properly understand the effect of cathode overheating and failed to reduce filament voltage adequately at high average anode currents. The result was a MTBF of approximately 4 hours for a magnetron costing over $30,000 (in 1980).

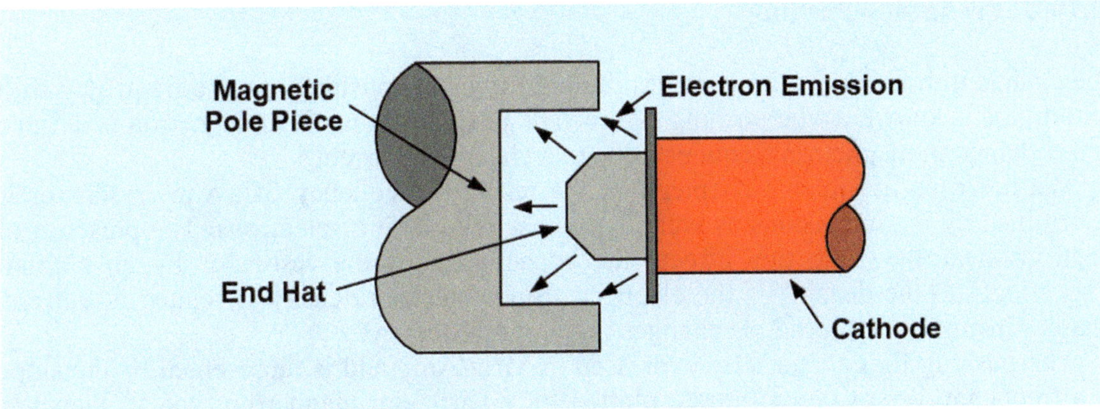

Figure 19-47 End-hat emission resulting from migration of cathode material.

19.10 PERFORMANCE

19.10.1 Voltage-Current Characteristics

The voltage-current characteristics of a magnetron are sometimes shown as in Figure 19-48. Magnetic flux density is fixed and so operation is along one of the constant Gauss lines. Normally, the optimum operating condition is indicated (26 kV and 38A). As anode voltage is varied, current changes, and so power output and, to some extent, efficiency changes. As current changes, the electronic impedance varies, which affects the resonant frequency of the resonator, and so the frequency of oscillation changes.

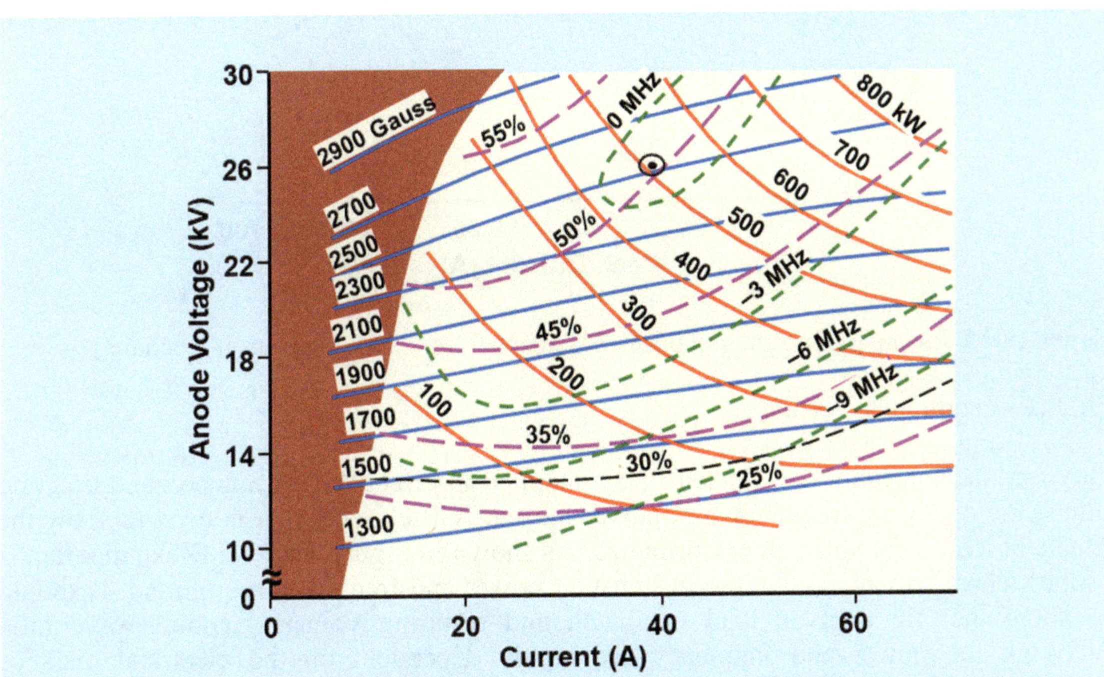

Figure 19-48 Magnetron operating characteristics. (From: *Microwave Tube Manual* by Varian Associates, Air Force Publication Number T.O.00-25-251, October 1979.)

19.10.2 Frequency Pushing

The change in frequency with current is called frequency pushing and the frequency shift per ampere is known as the pushing figure. The pushing figure of a magnetron is defined as the change in magnetron frequency due to a change in current.

We noted earlier in this chapter that the resonant frequency of a vane resonator is determined by its dimensions as well as the reactive effect of electrons. The presence of electrons near the vane tips affects the capacitance of the resonator by an amount proportional to the density of the electrons. Since electron density is related to current, changes in current will produce changes in output frequency.

The pushing figure is usually expressed in MHz/Amp and is represented by the slope of a frequency versus current curve plotted for a particular magnetron type as shown in Figure 19-49. Notice that the slope is not a constant so it is necessary to specify the range of current over which it applies. Also, since power output is proportional to current, the pushing figure at current levels well below the normal operating point is usually unimportant because the power output at these current levels is low. The primary importance of a low pushing figure is that the pushing figure will determine intrapulse FM and thereby will affect the spectral quality of the transmitting pulse.

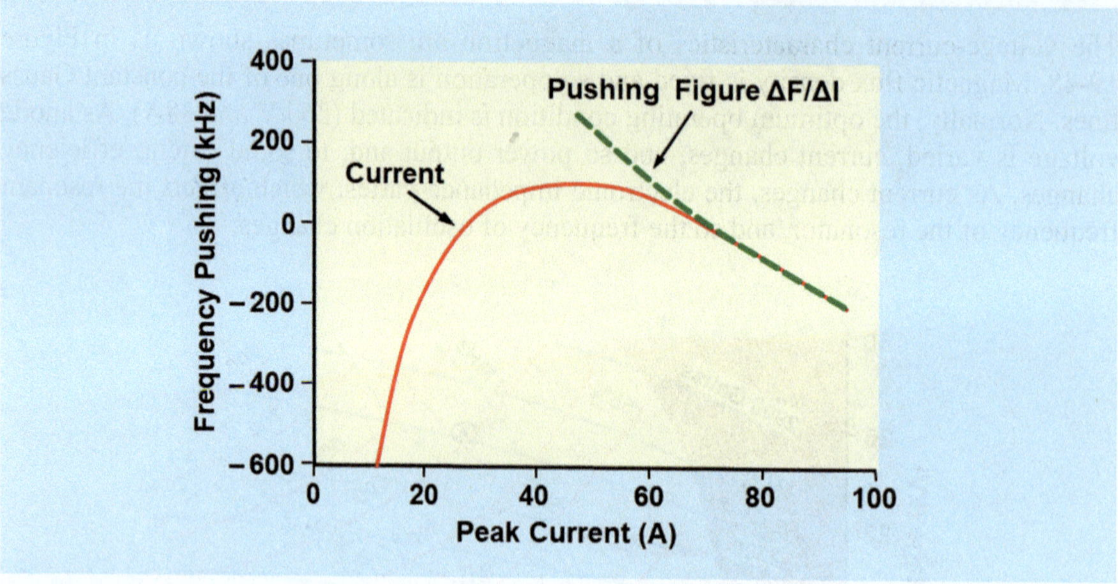

Figure 19-49 Example of magnetron pushing. (Adapted from: CPI magnetron brochure.)

19.10.3 Frequency Pulling

Varying load impedances reflected into a magnetron affect the resonance condition and with it the operating frequency and power output. A Rieke diagram is used to show the effects of load impedance on performance. As shown in Figure 19-50, a Rieke diagram is a Smith chart on which contours of constant power and frequency are plotted. It should be noted that for a given load mismatch and resulting voltage standing-wave ratio (VSWR), the power and frequency are highly dependent on the electrical distance between the tube and load. This is because a load may appear to be inductive, resistive, or capacitive depending on its distance from the tube. Thus, in Figure 19-50, for example, for a load VSWR of 1.5, the frequency may vary from –0.7 to +0.75 MHz relative to the value at the VSWR of 1.0, and the power may vary from 1.07 to 1.44 MW.

If a magnetron operates into a load, the impedance of which varies with time, then frequency and power will vary with time. As an example, a rotating antenna will usually present a varying impedance to a magnetron, so frequency and power will vary with antenna position. The use of ferrite isolators significantly reduces this variation. The resulting variation in frequency with changes in load conditions is called frequency pulling. The pulling figure is the change in output frequency that results when a fixed amplitude mismatch, located in the output waveguide of the magnetron, is moved through a distance of one half wavelength.

During the design of a magnetron, the coupling of the output waveguide to the resonator structure is selected to optimize certain performance parameters. Strong coupling increases output power and efficiency but also increases jitter and sensitivity to changes to load mismatch. Usually, the coupling is chosen to obtain the best compromise between efficiency and stability.

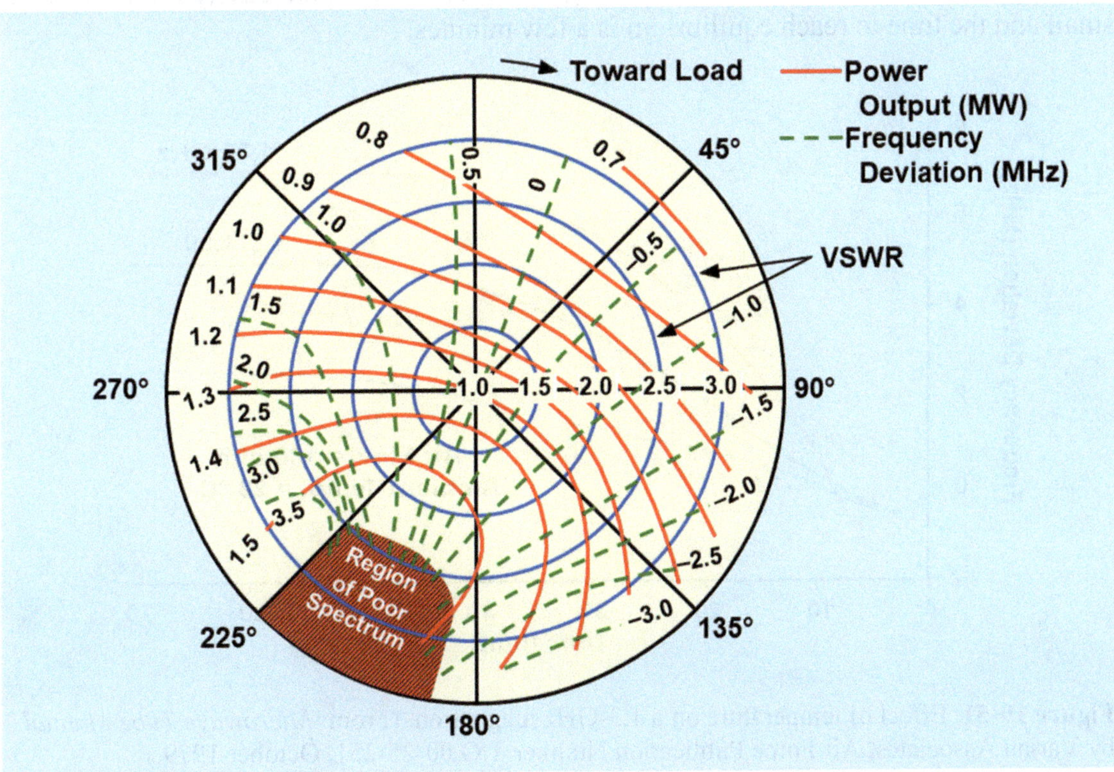

Figure 19-50 Rieke diagram for displaying the effect of load on tube operation. (From: *Microwave Tube Manual* by Varian Associates, Air Force Publication Number T.O.00-25-251, October 1979.)

The pulling figure is measured by operating the tube into a load with a specified VSWR (usually 1.5:1) and determining the frequency shift as the mismatch is moved through a distance of one half wavelength. In very high-power magnetrons a 1.3:1 VSWR is often used. When referring to the pulling figure of a magnetron the VSWR value used in the measurement should always be specified.

19.10.4 Thermal Drift

In addition to frequency pulling and frequency pushing, it should be pointed out that the operating frequency of a magnetron is subject to variations in tube temperature. When high voltage is first applied to a magnetron, the tips of the anode vanes begin to heat due to electron bombardment. Then the entire anode/cathode structure undergoes a transient change in thermal profile. During the time required for each part of the magnetron to stabilize at its normal operating temperature, the output frequency of the magnetron will drift. This is because cavity size determines frequency and cavity size is affected by temperature. The curve of output frequency versus time during the period following initial turn on is called the thermal drift curve.

Generally speaking, the maximum drift occurs during the first few minutes after turn on, and slowly approaches equilibrium. In a low-frequency magnetron, the RF circuit elements are relatively massive and the time required to reach thermal equilibrium may be as much as an hour, as indicated by Figure 19-41. At X-band, cavity dimensions are small and the time to reach equilibrium is a few minutes.

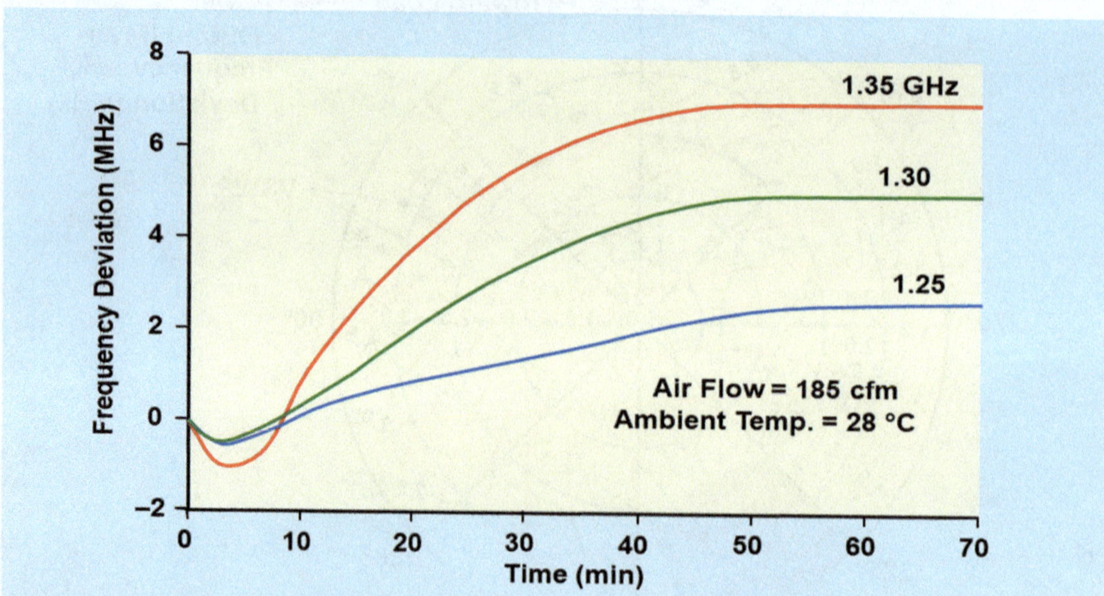

Figure 19-51 Effect of temperature on a 1.3-GHz magnetron. (From: *Microwave Tube Manual* by Varian Associates, Air Force Publication Number T.O.00-25-251, October 1979.)

Thermal drift curves for a variety of magnetron types operating at the same frequency and power can differ radically. Each type is usually designed for a particular application and small differences in internal configurations can produce large differences in the thermal drift curves. It should be noted that a thermal drift effect will also occur whenever the input power to the magnetron is changed, for example, when a change of pulse duration, or pulse repetition frequency occurs. Figure 19-52 shows thermal drift curves for a magnetron plotted as a function of duty cycle. The dashed line indicates the effect of a change in duty from 0.001 to 0.0005 after thermal equilibrium has been initially achieved.

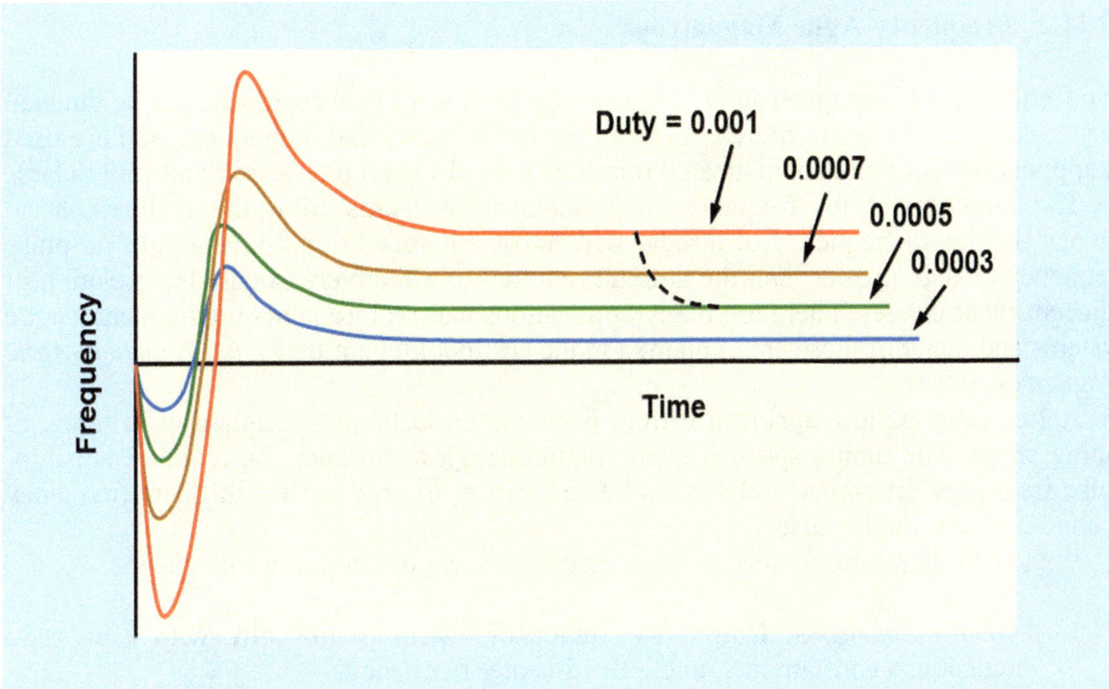

Figure 19-52 Example of thermal drift characteristic as a function of duty cycle. (Adapted from: CPI Magnetron brochure.)

In some devices, compensation techniques are used to offset the temperature effects. For example, bimetallic disk elements can be used in the cavity of a coaxial magnetron to correct for cavity expansion. Similarly, bimetallic elements can be used in the drive linkage to the tuner to compensate for temperature-induced frequency drift.

After a tube has reached thermal equilibrium and a stable operating frequency has been achieved, a change to ambient conditions, which causes a change in the magnetron temperature, will produce a change in the output frequency. The change in frequency for each degree change in temperature is the temperature coefficient for the magnetron and is usually expressed in MHz/°C. For most magnetrons the temperature coefficient is a negative (frequency decreases as temperature increases) and is essentially constant over the operating range of the magnetron.

19.11 APPLICATIONS OF MAGNETRONS

19.11.1 Conventional Magnetrons

Conventional magnetrons are considered here to be fixed frequency and mechanically tuned tubes intended for use in radar transmitters that operate primarily at one frequency. The purpose of the tuning capability is to allow operation at any selected frequency within the tuning range of the tube. Because of this, mechanical tuning by hand or by relatively slowly driven automatic means is sufficient. Most of the discussions up to this point have been related to conventional magnetrons.

19.11.2 Frequency-Agile Magnetrons

The frequency-agile magnetron [1, 14] is in another class of tubes intended to be tuned at very rapid rates. They are more complex than the conventional magnetrons and are used in applications for more sophisticated missions than, for example, search-and-find radars.

The requirement for frequency-agile magnetrons stems from the requirement to change the output frequency of a radar with sufficient speed to produce a pulse-to-pulse frequency change greater than the amount required to effectively obtain decorrelation of adjacent radar echoes. There are many applications and requirements for frequency-agile systems and many of these are explained in the CPI booklet entitled *The Frequency-Agile Magnetron Story*.

A frequency-agile magnetron differs from the basic tunable magnetron in terms of tuning speed. The tuning speed must be rapid enough to produce the required pulse-to-pulse frequency difference and yet not be so rapid as to produce a significant frequency change during a single pulse.

In the CPI literature, frequency-agile magnetrons are divided into four classes:

1. *Dither magnetrons*: Output RF frequency varies periodically with a constant excursion, a constant rate, and a fixed-center frequency.
2. *Tunable/dither magnetrons*: Output RF frequency varies periodically with a constant excursion and a constant rate. The center frequency may be slowly tuned by hand or by external servomotor drive to any point within the tunable band.
3. *Accutune magnetrons*: Output RF frequency variations are determined by the wave shape of an externally generated, low-level voltage signal. With appropriate selection of a tuning wave shape, the Accutune magnetron combines the features of dither and tunable/dither magnetrons.
4. *Accusweep magnetrons*: The output RF tuning rate and wave shape are infinitely variable within the design limits of each device. User inputs are typically any waveform from random to square wave and a +5V command.

The spin-tuned magnetron is a frequency-agile magnetron that falls under the dither magnetron category. The frequency can be varied 5% about a center value at over 1,000 tuning cycles per second. A rotary tuner mounted on a ball-bearing system in vacuum is used for very fast inductive tuning (Figure 19-53). The tuner is driven by an electric motor through the vacuum envelope by a magnetic coupling system.

The tuner is a copper cylinder that forms part of the inductive portions of the cavity resonator. As the tuner is rotated, holes in the cylinder move into and out of alignment with the cavities, and thereby inductively tune the cavities. In X-band versions of the spin-tuned magnetron, there are 16 cavities and a rotor with 16 holes (Figure 19-54). With a nominal rotor speed of 4,000 rpm, the frequency of the magnetron is changed over 1,000 complete tuning cycles per second. The frequency excursion of the magnetron is approximately 5% of the carrier frequency.

Because the ball bearings supporting the rotary tuner must operate in vacuum, bearing life would appear to be a matter of considerable concern. In the Philips versions of the spin-tuned magnetrons, bearing life is over 4,000 hours and, in most cases, exceeds cathode life.

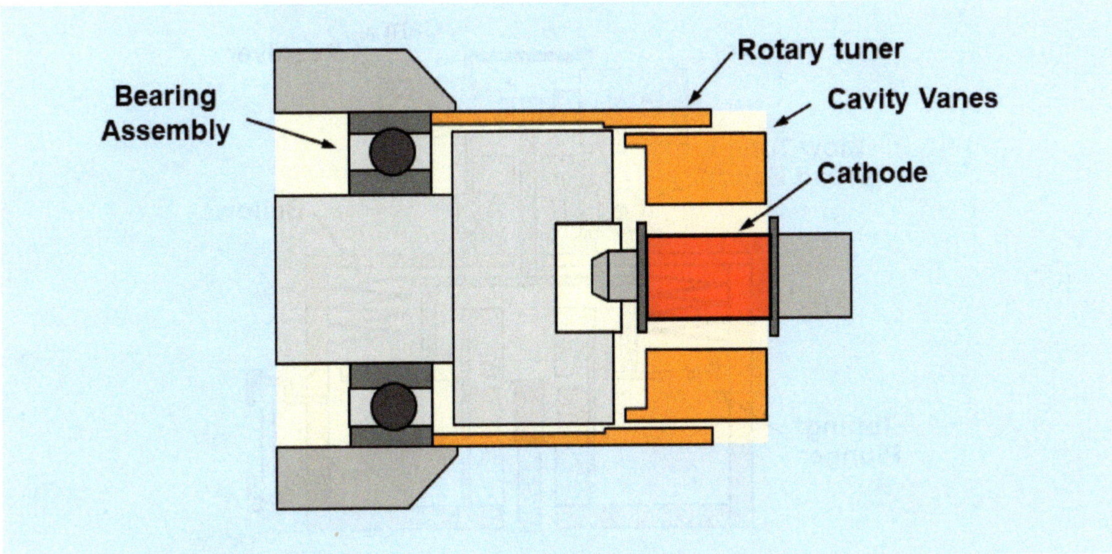

Figure 19-53 Ball bearing support assembly for the rotary tuner in a spin-tuned magnetron.

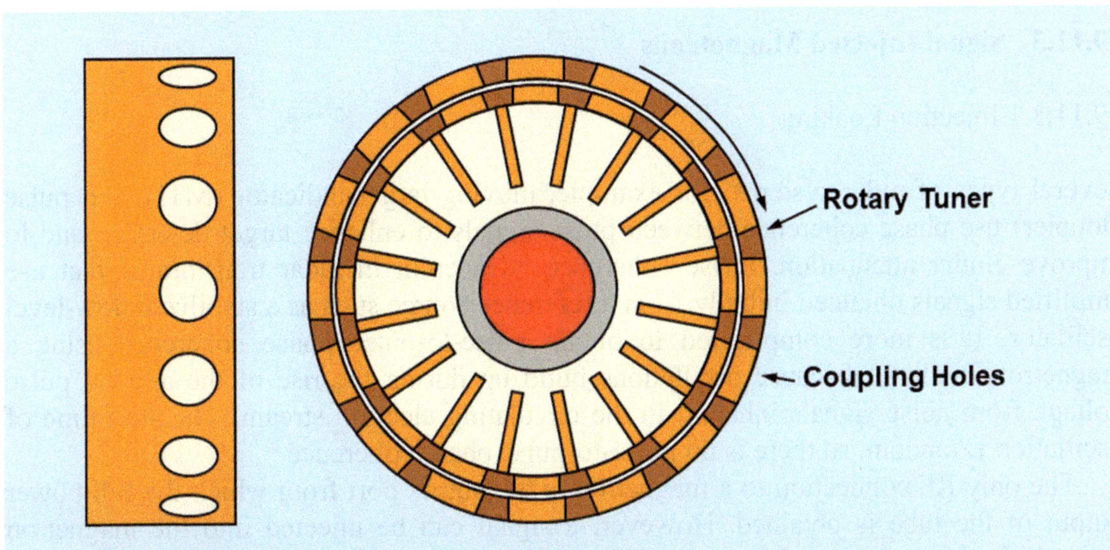

Figure 19-54 Resonator and rotary tuner for a 16-cavity spin-tuned magnetron.

The coaxial magnetron in Figure 19-55 is a tunable dither magnetron. The dither mechanism is mounted on the magnetron tuner and uses a motor with a small cam, or eccentric, ground into its shaft. As the motor shaft revolves, the eccentric imparts an up-and-down sinusoidal motion through a ball bearing to a shaft connected to the tuning plunger. Thus, the standard tuning plunger is used to dither the frequency.

Broadband tuning is accomplished in the normal manner using a geared drive. In dither-tuned tubes, the broadband tuning mechanism may move the dither mechanism as well as the tuning plunger as shown in Figure 19-55 or, in some types, the broadband tuning mechanism may move only the plunger, through a spline connection. In either case, the broadband tuner can operate independently of the narrowband tuner. The broadband tuner may be adjusted either by an external drive motor or manually.

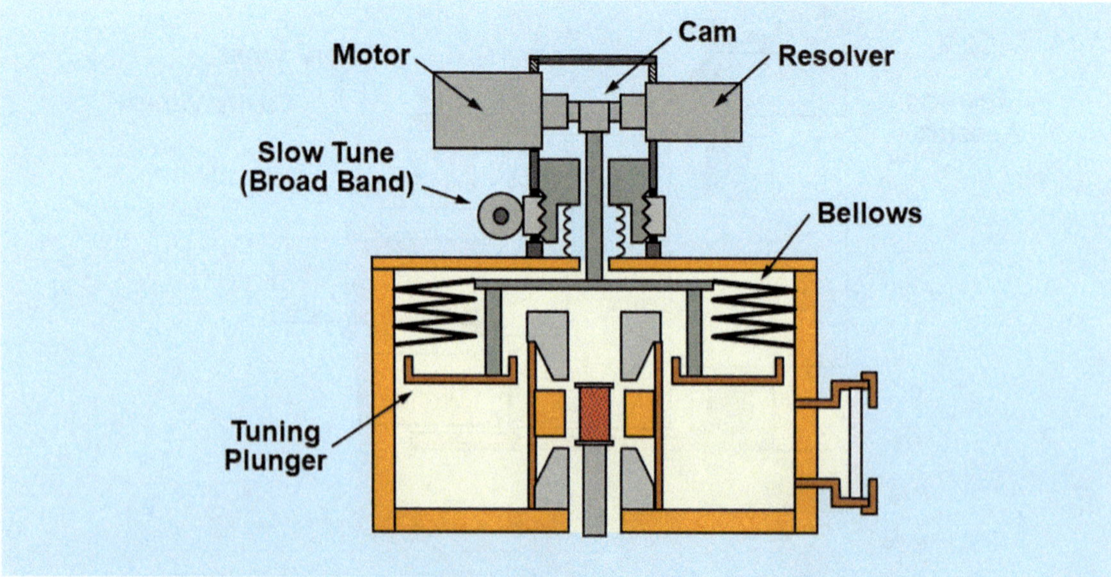

Figure 19-55 Coaxial magnetron that is dither tunable and broadband tuned.

19.11.3 Signal Injected Magnetrons

19.11.3.1 Injection Locking

Several types of radar systems (for example, moving target indicator (MTI), and pulse Doppler) use phase coherence between pulse signals to enhance target detection and to improve clutter attenuation. Phase coherence is inherent in radar transmitters that use amplified signals obtained initially from a reference source such as a stabilized, low-level oscillator. It is more complicated to obtain pulse-to-pulse phase coherence using a magnetron oscillator because oscillations build up during the rise of the applied pulse voltage from noise signals inherent in the circulating electron stream. The start time of oscillation is random, so there is no pulse-to-pulse phase coherence.

The only RF connection to a magnetron is the output port from which the full power output of the tube is obtained. However, a signal can be injected into the magnetron through the output port to help control operation of the magnetron. The signal injection can be achieved by means of a circulator as shown in Figure 19-56. The injection signal is sent into one port of the circulator, which is then redirected through a second port of the circulator into the output window of the magnetron.

The typical magnetron oscillator is heavily overcoupled for optimum efficiency so the impedance looking into the magnetron through the output port is a mismatch to the waveguide impedance. Only a portion of the injection signal is coupled into the tube. The remainder is reflected back from the magnetron and diverted to a load (load 1) through a third port of the circulator. The portion of the input signal that enters the magnetron can be used for controlling the magnetron frequency.

The magnetron power output is also directed to a load (load 2) by the circulators. Hence, the circulator provides the necessary isolation and protection of the input signal source from the magnetron power output. The maximum control occurs when the injection signal is at the same frequency as the free-running oscillation frequency of the magnetron. Injection locking can be achieved at other frequencies, but the required signal power level for proper operation increases with frequency departure from the free-running frequency.

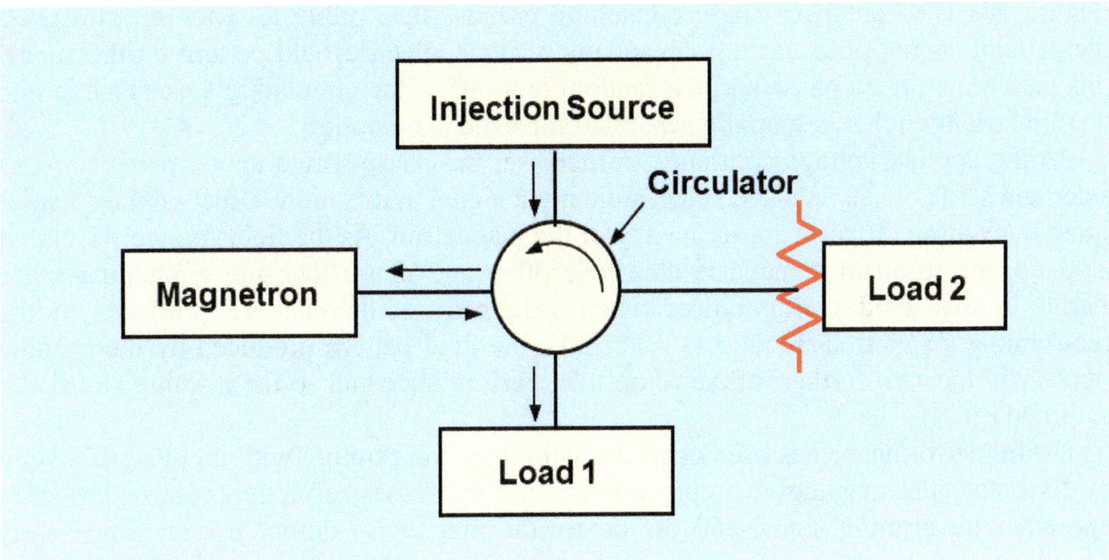

Figure 19-56 Circuit for injecting a signal into a magnetron.

Typical locking bandwidths are in the range of 1%. The locking bandwidth can be made somewhat larger by increasing the coupling to the magnetron circuit. However, this will further increase the mismatch looking into the output port and requires a larger injection signal power for proper frequency locking.

The ratio of the magnetron power output to the required input power is called the locking gain. Typical locking gains for narrow bandwidths are about 10–15 dB. This is comparable to the levels of gain available from crossed-field amplifiers but only over a significantly smaller bandwidth than the amplifier. Hence, injection-locked magnetrons are used only in those cases where satisfactory operational requirements can be achieved with the restricted bandwidths. In these cases, the lower costs of an injection-locked oscillator are preferred to the larger cost of an amplifier.

19.11.3.2 Injection Priming

Injection locking is the terminology applied to the circuit of Figure 19-56 when the magnetron output signal is fully slaved to the input signal. The magnetron, for all practical purposes, duplicates the input signal in frequency and phase throughout the entire pulse. The limited gain and bandwidth attainable with this technique has resulted in limited use of this technology. However, there is an alternate injection technique that offers many of the advantages of locking while avoiding many of the problems that have limited the use of full injection locking. This technique is called injection priming. Injection priming differs primarily from injection locking in that much smaller signals are used. Also, the priming signal is slightly lower in frequency than that of the free-running oscillator. The lower frequency is at the magnetron's start oscillation frequency. Since this frequency is lower than that of the free running magnetron, its synchronous voltage will be lower than the normal operating voltage for the tube. Therefore, the magnetron passes through the critical voltage for the start oscillation frequency in a very short time during the rise of the applied voltage pulse.

Priming has a significant effect on the magnetron during the start-oscillation period because this is when space-charge bunching is most susceptible to external influences. The priming signal produces a nonresonant, π mode, electric field pattern on the anode. This pattern begins to phase sort the random motions of the circulating space charge into a spoke formation that is spatially separated for π mode operation.

As the applied voltage continues to increase, the phase-sorted spoke pattern rotates faster and serves as an enhanced electronic input signal that is more effective than a noise signal in exciting the resonant frequency of the magnetron. As the fields on the RF circuit build up, the resonant π mode is able to capture and phase lock the circulating space charge to that field configuration. Then sustained oscillations are produced at the free-running magnetron frequency. After this, the field pattern produced by the priming source will have no further influence on tube performance and so the priming signal can be turned off.

Useful performance has been achieved with injection priming with input signals 30 to 40 dB below the magnetron output power. In some cases, effective control has been reported with priming signals 60 dB below the magnetron output power. Some early experiments were performed using low-power magnetrons, but more recent applications have been with high-power tubes.

19.11.4 Beacon Magnetrons

Beacon magnetrons are small conventional magnetrons with peak power output less than 4 kW and average power output of less than 5W. Typically, they weigh 8 ounces. The technical requirements for these magnetrons demand precise frequency control. The temperature stability factor is of great importance since it allows frequency control without additional electronics in the radar transponder. The magnetron must be tunable but must have the properties of a fixed-frequency magnetron after adjustment and locking. As a result, the techniques used for temperature compensation must work over a band of frequencies. Also, frequency stability is generally that of missile and aircraft electronic systems. This means that the tube must be stable over a temperature range that is typically –65°C to +100°C, a shock of 100G, and a vibration of 15G. Refer to Figure 19-57.

19.11.5 Microwave Oven Magnetrons

Microwave oven magnetrons supply 500–1,000W at 2,450 MHz and an efficiency of ~ 70%. They are produced in quantities of tens of millions annually at a price of about $7 each. In 1984 [6], Brown noted that there were an estimated 50 million microwave ovens in the world. If turned on at the same time, they would produce 25 GW of microwave power. In 2002 [7] Osepchuk estimated that the total number of consumer microwave ovens in the United States was well over 100 million and that there were over 250 million microwave ovens in the world. No recent estimate of these numbers has been found.

CPI VMX 1295

- **Frequency 9.7 – 9.8 GHz**
- **Peak Power Output 750 W**
- **Duty Cycle 0.001**
- **Anode Voltage 2.7 kV**
- **Anode Current 1.4 amps**
- **Pulse Width 1.0 μs**
- **Heater 5.0 volts @ 0.7 A**
- **Conduction cooled**
- **Mechanically tunable**

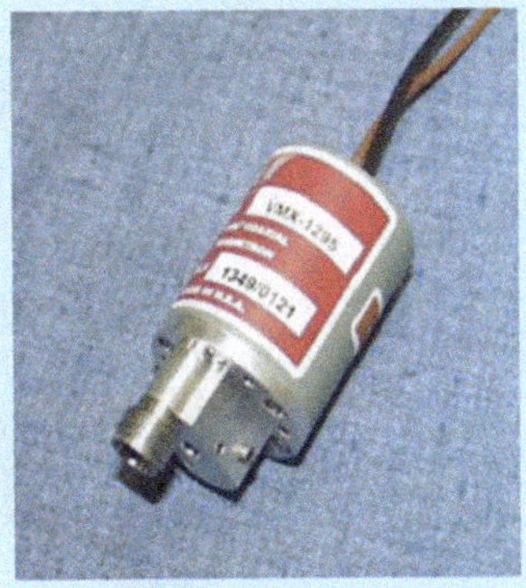

Figure 19-57 Beacon magnetron.

Microwave oven magnetrons are extremely reliable devices. There is a good chance that a microwave oven will operate for 10 years or more without requiring repairs, and at a performance level that is indistinguishable from when it was new.

The essential elements of one of these magnetrons are shown in Figure 19-58. Pole pieces carry the magnetic flux from ring-shaped magnets to the interaction region between the cathode and resonator. The resonator usually consists of 10 to 12 cavities. A 12-cavity resonator is shown in Figure 19-59. The RF output lead is connected directly to one of the resonator vanes. The cathode is directly heated carburized thoriated tungsten and comes up to operating temperature within seconds of the time that it is turned on. Even though the magnetron is very efficient, several hundred watts must be carried away by the aluminum cooling fins.

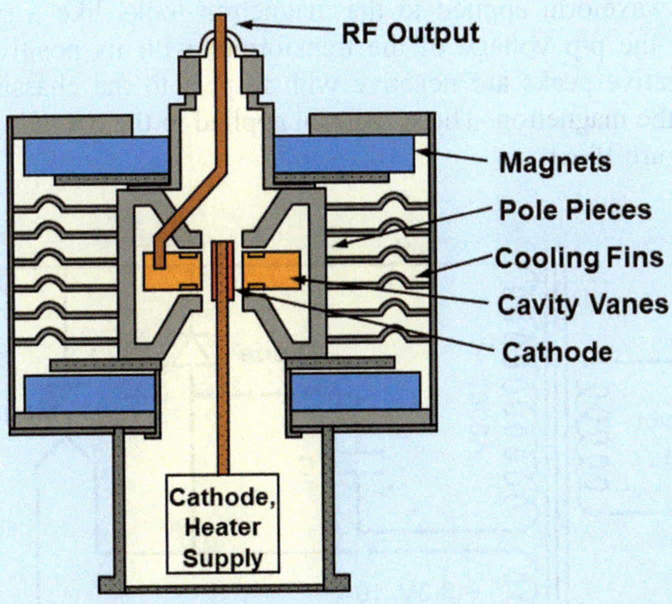

Figure 19-58 Essential elements of a microwave oven magnetron.

Figure 19-59 Resonator assembly for a 12-vane oven magnetron.

The power circuit for the magnetrons in most microwave ovens, which is shown in Figure 19-60, is a model of simplicity. The anode and chassis of the magnetron are at ground. When the transformer voltage tends to drive the cathode of the magnetron positive, the conducting diode charges the capacitor to the peak voltage of the transformer. At the peak of the negative swing of the transformer voltage, the voltage applied to the magnetron cathode is the peak voltage of the transformer plus the voltage across the capacitor. As a result, the voltage on the magnetron is nearly twice the peak output voltage of the transformer or ~5,000V. Thus, this circuit is a half-wave voltage doubler.

The voltage waveform applied to the magnetron looks like a sinusoid with a p-p voltage equal to the p-p voltage of the transformer with its positive peaks at chassis ground. The negative peaks are negative with respect to the chassis and get distorted when loaded by the magnetron. The waveform applied to the operating magnetron is like that shown in Figure 19-61.

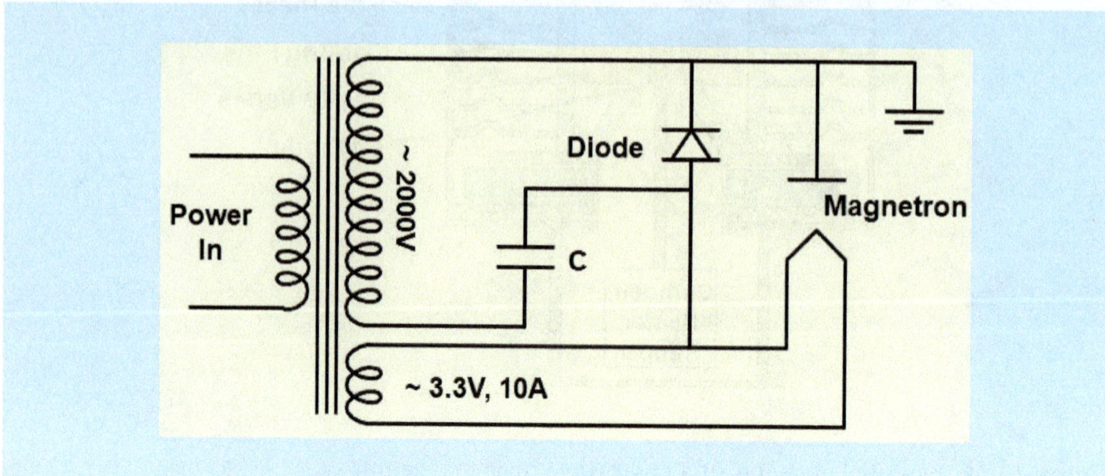

Figure 19-60 Power supply for an oven magnetron.

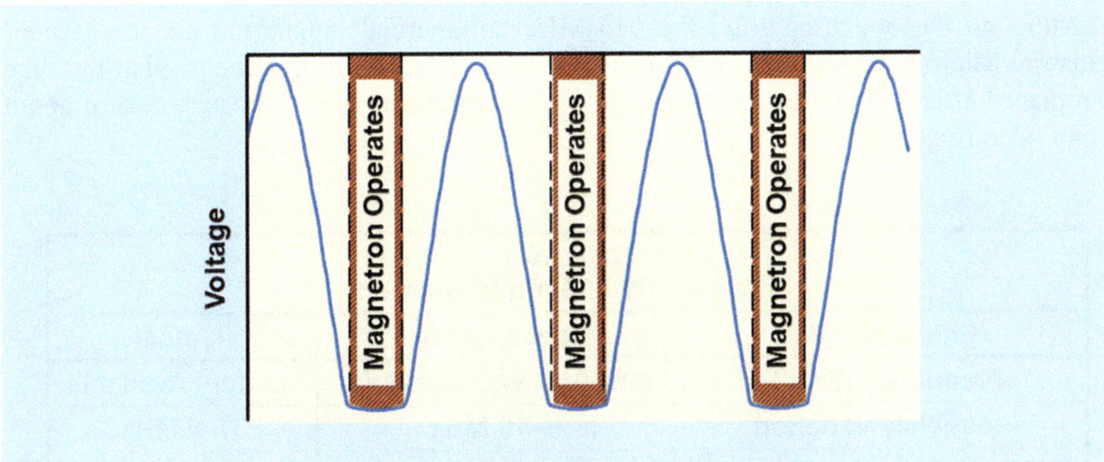

Figure 19-61 Voltage waveform applied to an oven magnetron.

19.11.6 Industrial Heating Magnetrons

Microwave heating using magnetrons has gained wide acceptance in industry because of its cleanliness, its high processing speed, high efficiency, accurate control, and flexibility. Food processing is one of the largest applications. As an example, tempering, which is a process of heating frozen meats from approximately 0°F to 27°F, can be performed. At that temperature, processing, such as slicing, grinding, or pressing can be carried out.

In addition to food processing, applications include product sterilization, wood and rubber curing, and controlled evaporation. Two frequencies, L-band and 2,450 MHz, are available for use. L-band systems can provide up to 100 kW from a single magnetron. The largest 2,450-MHz units use 30-kW magnetrons. A sketch of a commercially available 100-kW L-band magnetron, without a magnet, is shown in Figure 19-62. Note that the center portion of this device, which contains the resonator and is where the conversion from dc power to RF takes place, is less than 5 inches in diameter and less than 5 inches long.

The operating parameters for the 100-kW magnetron are listed in Table 19-1. Perhaps the most significant characteristic is efficiency, which is typically 88%.

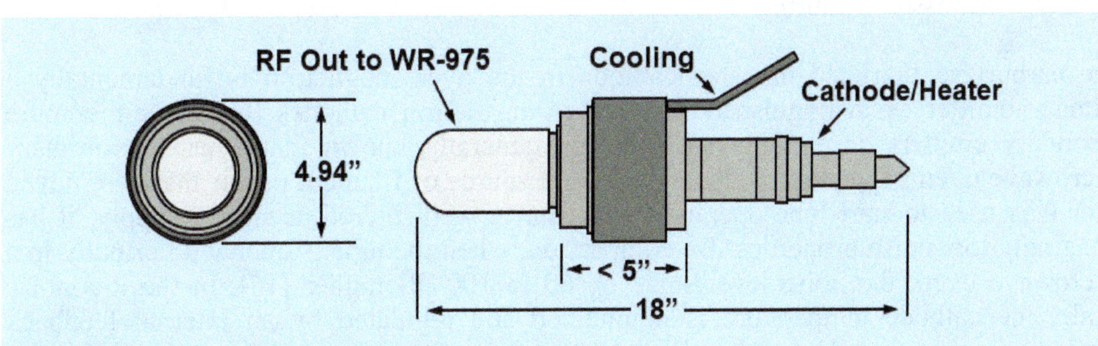

Figure 19-62 100-kW L-band industrial heating magnetron without a magnet. (Adapted from: L-3 Communications California Tube Laboratory brochure.)

Although the operating life of a 915-MHz commercial magnetron can be greater, Industrial Microwave Systems of Morrisville, North Carolina, recommends that the tube be replaced after 8,000 hours of operation. This translates to a maintenance cost of about $1 per operating hour.

Table 19.1
100-kW, CW, L-Band Magnetron

Characteristic	Specification	Typical
Frequency (fixed)	896, 915, 922, 929 MHz	Custom Available
Frequency Variation	+/- 10 MHz	+/- 5 MHz
CW Power Output, Minimum	100 kW	100 kW
Anode voltage	20 kV	19.5 kV
Anode Current	6.0 A	5.8 A
Efficiency	83%	88%
Filament Voltage (standby)	12.6 V	12.0 V
Filament Current	115 A	112 A
Load VSWR	1.1.1 Circulator Required	
RF Launch	Coaxial into WR-975	

19.11.7 Low-Noise Magnetrons

There are three regions of operation in a microwave oven magnetron. As anode current is varied, regions of high random noise, discrete spurious signals, and low noise are observed. The boundaries between these regions are remarkably reproducible and universal. Low-noise operation is discussed here.

19.11.7.1 Emission Limiting

The carburized thoriated tungsten cathode in the oven magnetron is fundamentally a primary emitter as distinguished from other magnetron cathodes that are either pure secondary emitters or partially so. It is not generally known that when the ordinary microwave oven magnetron is operated from a source of filament power that is removed after it is used to start tube operation and from a well-filtered dc power supply, it has amazingly low noise properties. By comparison, when the tube is operated normally in a microwave oven, the noise levels may be 60 to 100 dB higher [15]. In the low-noise mode, the cathode temperature is maintained and regulated by an internal feedback mechanism involving electron bombardment alone.

Experimentally, it has been found that the tube automatically operates with the cathode in an emission-limited condition. This is thought to be part of the reason for the low-noise performance because thermionically emitted electrons leave the cathode with very low velocities compared to the velocities of electrons emitted by secondary emission during normal magnetron operation. The experimental evidence that the cathode is operating temperature-limited is shown in Figure 19-63 [16]. In this figure, the observed temperature of the cathode is plotted against the anode current of the tube. This experimental relationship is seen to nearly follow the slope that is predicted by the Richardson-Dushman equation for temperature-limited emission. The two curves were matched to each other at a brightness temperature of 1,500°C. It was assumed that the magnetron current and emission current density are related by a multiplying constant.

When a magnetron is operating temperature-limited, it is observed that the signal-to-noise ratio is very high. For example, it was reported by Brown that with external microwave circuitry on the support for the cathode, spectral noise densities as low as 196 dB below the carrier were found at frequencies (other than harmonics) removed 50 MHz or more from the carrier [16].

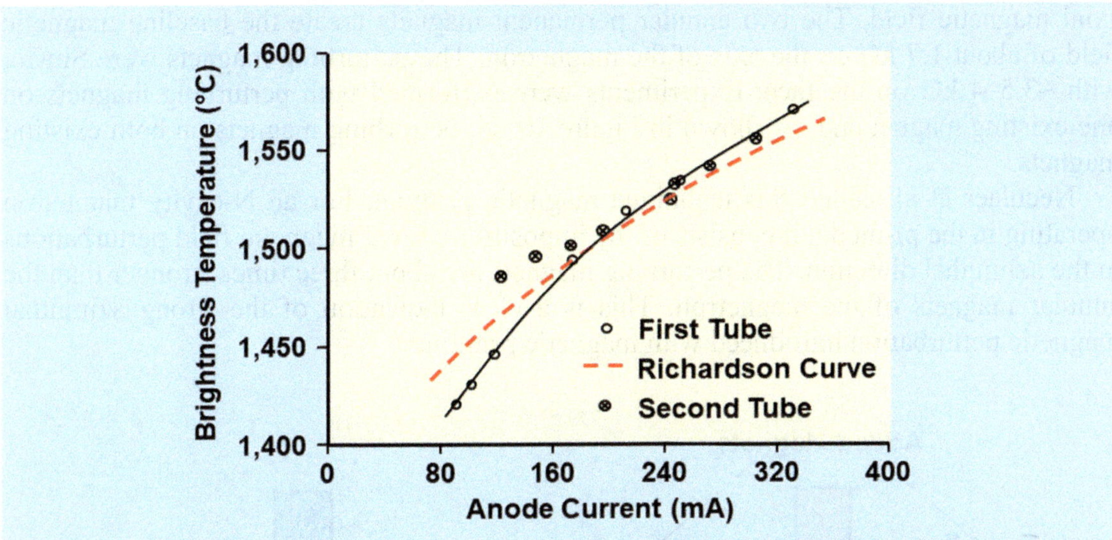

Figure 19-63 Comparison of emission limited-magnetron data with Richardson-Dushman thermionic-emission curve. (From: W. C. Brown, *IEEE Trans. Electron Devices*, November 1984. © 1984 IEEE.)

A second benefit, in addition to low noise, of operating the cathode at the lowest possible temperature, is that it is possible to design tubes for lives of decades. This is because the life of a carburized thoriated tungsten filament is highly dependent on the temperature being much greater as the temperature is reduced [15, 16]. The emission current density requirements to supply the anode current of the magnetron are fortunately low and consistent with the low temperatures needed for a very long life. Also, the internal feedback mechanism that controls cathode temperature makes it unnecessary to externally regulate cathode temperature.

One of the early applications proposed for low-noise oven magnetrons was in the 1980s for the Solar Power Satellite Program. These amplifiers are ideal for combining with slotted-waveguide radiators to form radiating modules in a low-cost, electronically steerable phased array for beamed power transmission.

Another application pointed out by Brown was that one of these tubes can be combined with external circuitry to convert it into a phase-locked amplifier with 30-dB gain without compromising its low-noise properties.

19.11.7.2 Magnetic Priming

The normal oven magnetron generates a high level of noise and spurious oscillations. Noise generated in the magnetron can leak out through filament or cathode leads and then is radiated outside from an ac power line. Solving the noise problem in magnetrons could potentially have important practical implications such as reducing unwanted interference with 2.4-GHz wireless systems and possibly providing a technique for reducing CFA noise.

Neculaes et al. [17–19] demonstrated that azimuthally varying axial magnetic fields can be used on standard oven magnetrons to produce rapid startup, low noise, and mode control with an efficiency loss on the order of 10%. The new magnetic field geometry was implemented by placing a number of small perturbing magnets on the perimeter of one of the existing annular magnets of the magnetron, providing an azimuthally varying axial magnetic field. The two annular permanent magnets create the baseline magnetic field of about 1.7 kG on the axis of the magnetron. The perturbing magnets were SmCo, with ~3.5–4 kG on the face. Experiments were performed with perturbing magnets on one existing magnet and, as shown in Figure 19-64, perturbing magnets on both existing magnets.

Neculaes et al. called this technique magnetic priming. For an N-cavity magnetron operating in the pi-mode, it consists of the imposition of N/2 magnetic field perturbations in the azimuthal direction. The perturbing magnets are about three times stronger than the annular magnets of the magnetron. This is a clear indication of the strong azimuthal magnetic perturbation introduced with magnetic priming.

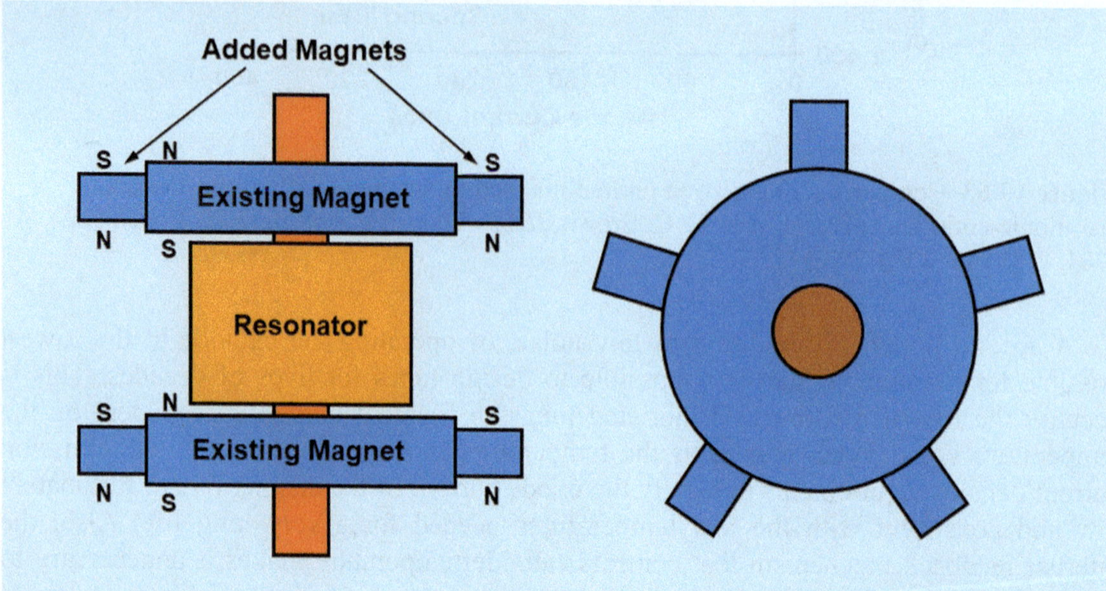

Figure 19-64 Placement of perturbing magnets on an oven magnetron. (From: Neculaes et al., *IEEE Trans. Electron Devices*, May 2005. © 2005 IEEE.)

Using azimuthally varying axial magnetic field techniques, two magnetic priming configurations have been employed; the four perturbing magnet configuration (nonoptimized) and the five perturbing magnet configuration (optimized) to stimulate the pi-mode in the 10-cavity oven magnetron, which is characterized by the existence of five electron spokes. It is clear that these magnetic priming configurations distort the axial symmetry and introduce radial and azimuthal components in the magnetic field.

The effect of magnetic priming on noise is shown in Figure 19-65. The microwave spectrum for an aged magnetron without magnetic priming is in Figure 19-54(a). The microwave signal exhibits a broadband (up to 300 MHz from the carrier) noise spectrum. The microwave noise level is some 60 dB below the carrier (20 dB above the noise floor of the spectrum analyzer). When magnetic priming is imposed, as seen in Figure 19-54(b), the microwave noise is reduced at least 20 dB to the limit of the spectrum analyzer sensitivity both close to the carrier and far from the carrier.

This magnetic priming concept has been extended for application to relativistic magnetrons. Two-dimensional and three-dimensional simulations indicate that magnetic priming could significantly decrease the startup time for these magnetrons.

19.11.8 Magnetrons for Power Beaming

Power beaming is the wireless transfer of energy. This transfer may be between ground base facilities, from ground to satellite or satellite to ground, or between satellites. Microwave beams and laser beams can be used to transport the power. When substantial amounts of power are required, magnetrons and klystrons are the microwave generators.

W. C. Brown used beamed microwave power from a magnetron to operate a miniature helicopter [8]. In 1975, the transmission of tens of kilowatts over a distance of a kilometer was demonstrated at Goldstone, California in 1975. The power source was a klystron.

When a large amount of power is to be transferred, magnetrons are preferred over klystrons because of their weight and efficiency advantages. However, magnetrons are oscillators with questionable stability. If they are to operate in parallel to produce large amounts of power that can be focused on to a distant collector, their frequency and power levels must be synchronized and, although techniques for accomplishing this have been proposed, they have not yet been demonstrated [20].

19.11.9 Relativistic Magnetrons

Relativistic magnetrons are capable of producing gigawatts of RF power for tens of nanoseconds. The operating voltages are hundreds of kilovolts and the currents are kiloamperes. Applications include testing of equipment for electromagnetic susceptibility and, possibly, the destruction of anthrax and other biological substances. A description of these magnetrons is beyond the scope of this book.

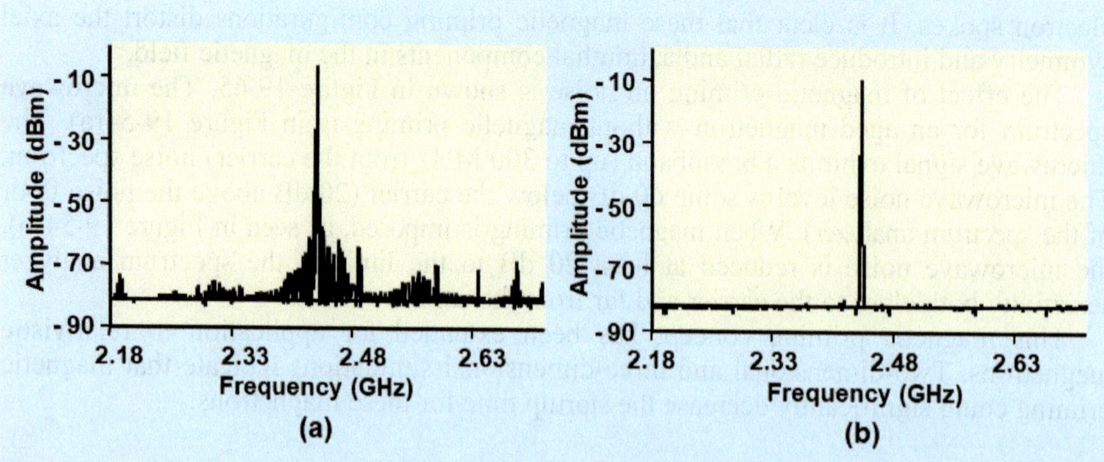

Figure 19-65 Microwave spectra (a) without and (b) with magnetic priming. (From: Neculaes et al., *IEEE Trans. Electron Devices,* May 2005. © 2005 IEEE.)

19.12 SUMMARY OF POWER CAPABILITIES

The overall publishable power characteristics of magnetrons are shown in Figure 19-66. Peak power levels well in excess of 1 MW have been available since the early days of magnetron development. In the laboratory, peak powers in the gigawatt range have been generated.

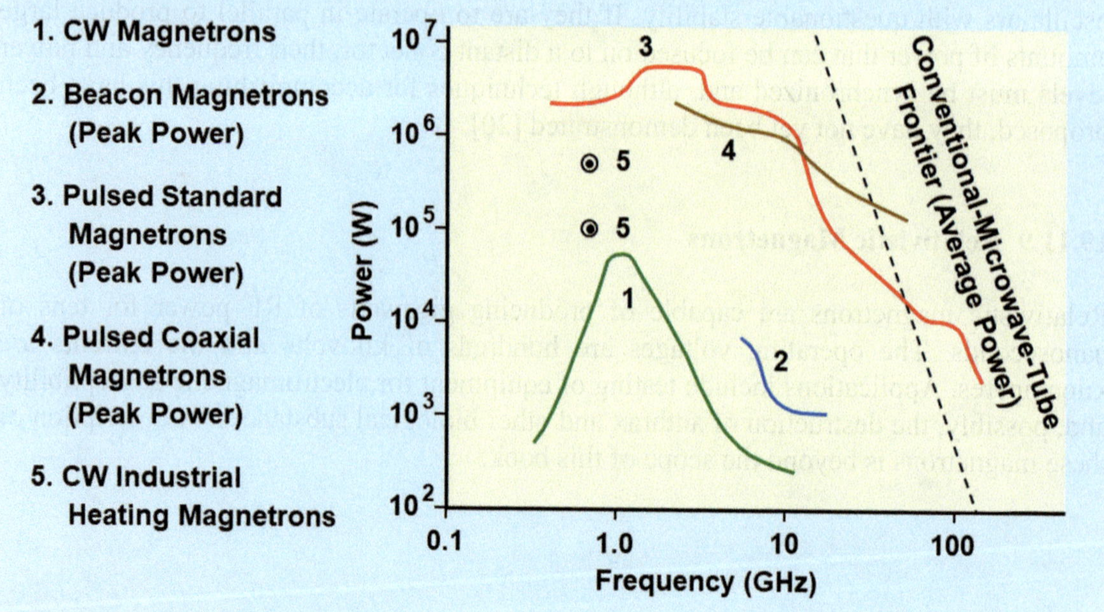

Figure 19-66 Magnetron power capabilities. (From: J. Skowron, (MPTD), Raytheon Co., Workshop on High-Power, Space-Based Microwave Systems, Los Alamos, NM, March 1985.)

Average and CW power levels are limited because the beam power must be dissipated by the RF structure. This matter is discussed further in Chapter 20 concerning crossed-field amplifiers. Still, average powers approaching the megawatt level are available. Magnetron conversion efficiency from dc to RF is normally in excess of 50% and may be as high as 90%.

Fixed-frequency and tunable magnetrons are available that weigh as little as a few ounces or as much as about 200 lbs. These weights may be reduced in weight-critical applications through the use of samarium cobalt magnets. Possible magnet configurations are shown in Chapter 20 in discussions concerning crossed-field amplifiers.

REFERENCES

[1] A. W. Hull, "The Magnetron," AIAA, Vll.,\XL, No. 9, September, 1921.

[2] A. W. Hull, "The effect of uniform magnetic fields on the motion of electrons between coaxial cylinders," *Phys. Rev.*, Vol. 18, No. 1, pp. 31-57, 1921.

[3] E. Habaan, Translated from German, "A new generator tube," *Zeitschrift fur Hochfrequenztechnik*, 24, 115–120 and 135–141, 1924.

[4] N. A. Zacek, Translated from Czech, "New method of generating undamped oscillations," *Journal for the Cultivation of Mathematics and Physics,* 53, pp. 378 – 380, 1924.

[5] H. A. H. Boot and J. T. Randall, "Historical notes on the cavity magnetron," *IEEE Trans. Electron Devices*, Vol. ED 23, No. 7, July 1976, pp. 724–729.

[6] S. Phelps, The Tizard Mission,WESTHOLME, Yardley, June 2017

[7] J. M. Osepchuk, "Microwave power applications," *IEEE Trans on Microwave Theory and Techniques,* Vol. 50, No. 3, March 2002, pp. 975–985.

[8] J. F. Hull, "A young engineer enters the microwave tube field during World War II," *IEEE Trans. Electron Devices*, Vol. ED 31, No. 11, November 1984, pp. 1611–1613.

[9] J. F. Skowron, "The continuous cathode (emitting sole) crossed-field amplifier," *Proc. IEEE*, Vol. 61, No. 3, March 1973, pp. 330–336.

[10] J. J. Coupling, a pseudonym of J. R. Pierce, *Astounding Science Fiction*, Street and Smith, 1946.

[11] Varian brochure, "Introduction to Coaxial Magnetrons."

[12] M. I. Skolnik, "The future of microwave engineers: from the technical perspective," *Microwave Jour.*, Vol. 18, 1985, pp. 24–36.

[13] W. M. Black et al., "A high-power magnetron for air breakdown studies," *Technical Digest,* IEDM, 1980.

[14] CPI brochures, "Magnetrons."

[15] W. C. Brown, "The magnetron—a low noise, long life amplifier," *Applied Microwave*, Summer 1990.

[16] W. C. Brown, *Satellite Power System (SPS) Magnetron Tube Assessment Study*, NASA Contractor Report 3383, 1981.

[17] I. Tahir, "Phase and frequency locking of magnetrons by pushing and pulling," Lancaster University.

[18] V. B. Neculaes et al., "Magnetron microwave noise reduction and magnetic priming by azimuthally varying axial magnetic fields," Fifth IEEE IVEC, 2004, pp. 158–159.

[19] V. B. Neculaes et al., "Magnetic priming effects on noise, startup, and mode competition in magnetrons," *IEEE Trans. Plasma Science*, Vol. 33, No. 1, February 2005, pp. 94–102.

[20] V. B. Neculaes et al., "Magnetic perturbation effects on noise and startup in DC-operating oven magnetrons," *IEEE Trans. On Electron Devices*, Vol. 52, No. 5, May 2005, pp. 864–871.

Chapter 20

Crossed-Field Amplifiers

20.1 INTRODUCTION

The CFA is a broadband amplifier of RF signals at microwave frequencies. It is capable of providing relatively large amounts of output power with high efficiency. The electronic bandwidth of the CFA is nominally 10.0% to 15.0%. Power levels with peaks of many megawatts and averages of tens of kilowatts are currently being obtained from production CFAs. Efficiency figures in excess of 70% are not uncommon in some crossed-field amplifiers that utilize backward-wave circuits and operate at low anode voltages.

Crossed-field amplifiers are classified as either injected-beam or distributed-emission devices. In an injected-beam CFA, an electron gun generates a ribbon-shaped electron beam that passes through the device near an RF circuit. Operation of the injected-beam CFA is similar to a TWT. Like a magnetron, the distributed-emission CFA contains a cathode adjacent to the full length of the RF structure.

20.1.1 Injected-Beam CFAs

In the late 1940s and early 1950s, there was much interest in extending the crossed-field interaction principle of the magnetron oscillator to the development of a crossed-field amplifier [1, 2]. The initial work was done largely in France on devices that were similar to traveling-wave tubes. These were injected-beam, crossed-field amplifiers with configurations like that in Figure 20-1. These devices were similar to the TWT in that they used a separate electron gun to form an electron beam. The electron beam was ribbon-shaped and was launched into the interaction region between the anode delay line (slow-wave circuit) and the negative electrode, which was called the sole by the French tube engineers.

Interaction between the RF input signal on the delay line and the electron beam produced phase sorting of the beam into spokes much like those in a magnetron. The electrons in the spokes moved toward the anode, and most were collected on it just as they are in a magnetron. Those that were not collected on the anode would exit the interaction space and be intercepted by the collector. There was no intended back bombardment of or emission from the sole electrode.

The basic elements of a backward-wave injected-beam CFA with a circular format are shown in Figure 20-2. Very-high-power pulsed devices were developed in France. In the United States, much development work was devoted to CW and low peak power injected-beam CFAs for very wideband use in airborne electronic countermeasures systems. In spite of significant development efforts, injected-beam CFAs are not currently being produced because the applications for which they might be suited are better filled by other devices. For this reason, only distributed-emission CFAs are discussed in this book and henceforth they are referred to simply as CFAs.

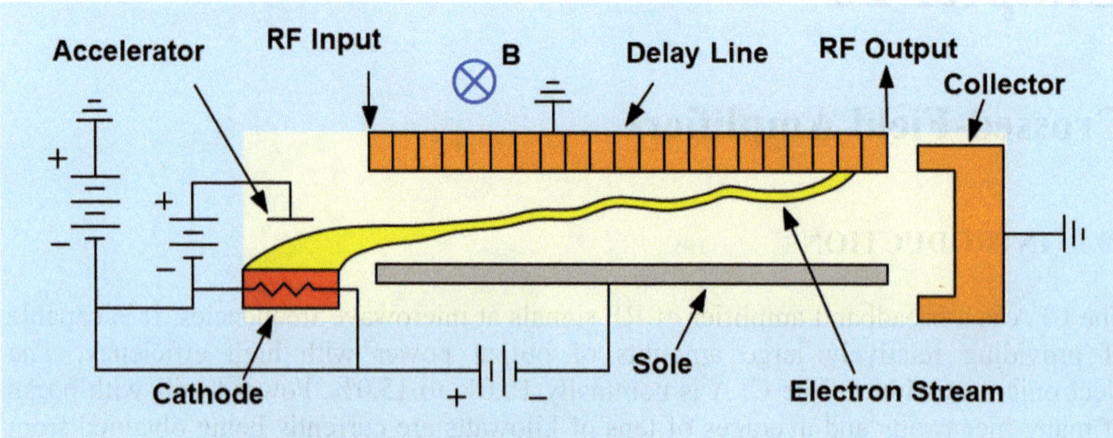

Figure 20-1 French injected-beam crossed-field amplifier.

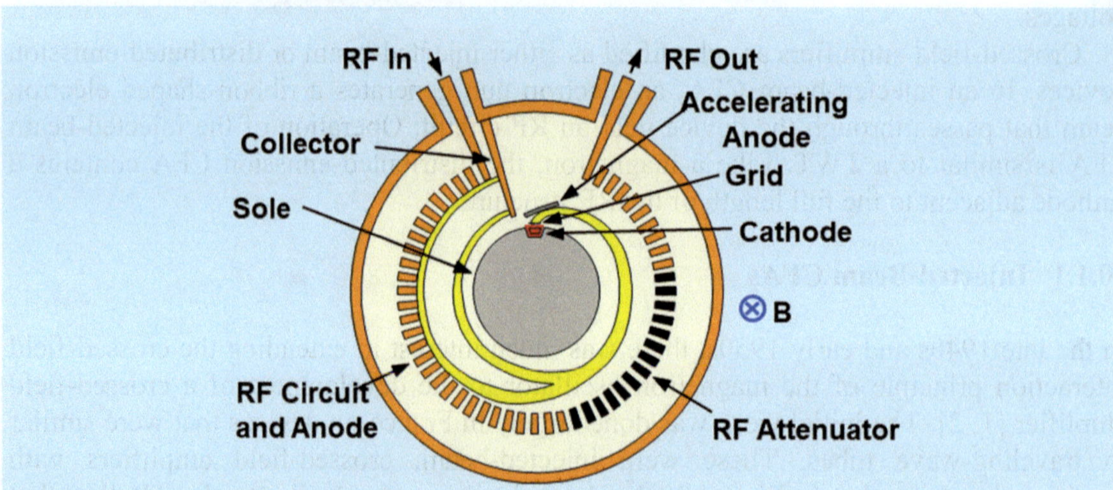

Figure 20-2 Basic injected-beam crossed-field amplifier. (From: J. Skowron, *Proc. IEEE,* March 1973. © 1973 IEEE.)

20.1.2 Distributed Emission CFAs

The first practically successful distributed emission CFA [3] was introduced by Brown in 1953 [4]. The device was similar to a magnetron, except that, as shown in Figure 20-3, the RF circuit was interrupted to provide input and output connections. The RF structure was a strapped-vane network similar to that in a magnetron. The tube had a gain of 13 dB over a frequency range of 3,780 to 3,850 MHz. The peak power output was 24 kW and the efficiency was 40%.

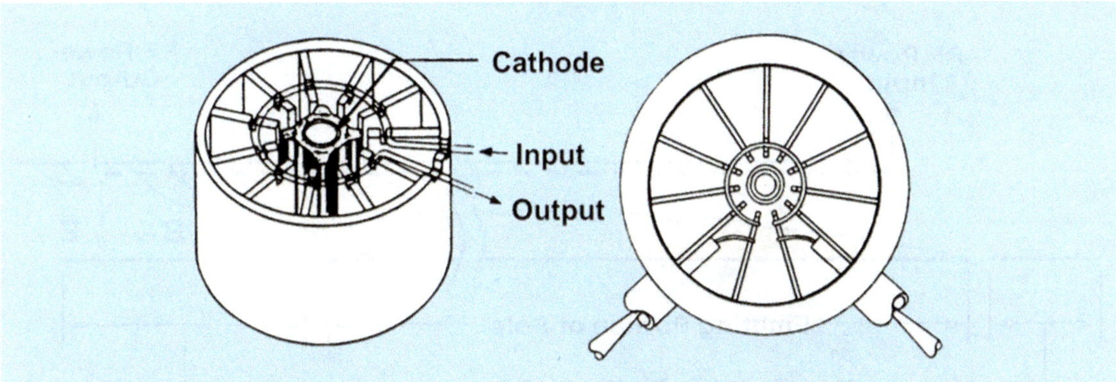

Figure 20-3 Schematic of the first device to demonstrate the principles of converting a magnetron into an amplifier. (From: W. C. Brown, *IEEE Trans. Electron Devices,* November 1984. © 1984 IEEE.)

Since the first demonstration of the distributed-emission CFA, nearly every conceivable configuration has been examined, as indicated in Figure 20-4 [5]. The circular format is the one shown in Figures 20-3 and 20-6. In the linear format device, as shown in Figure 20-5, the cathode and RF circuit are parallel [6].

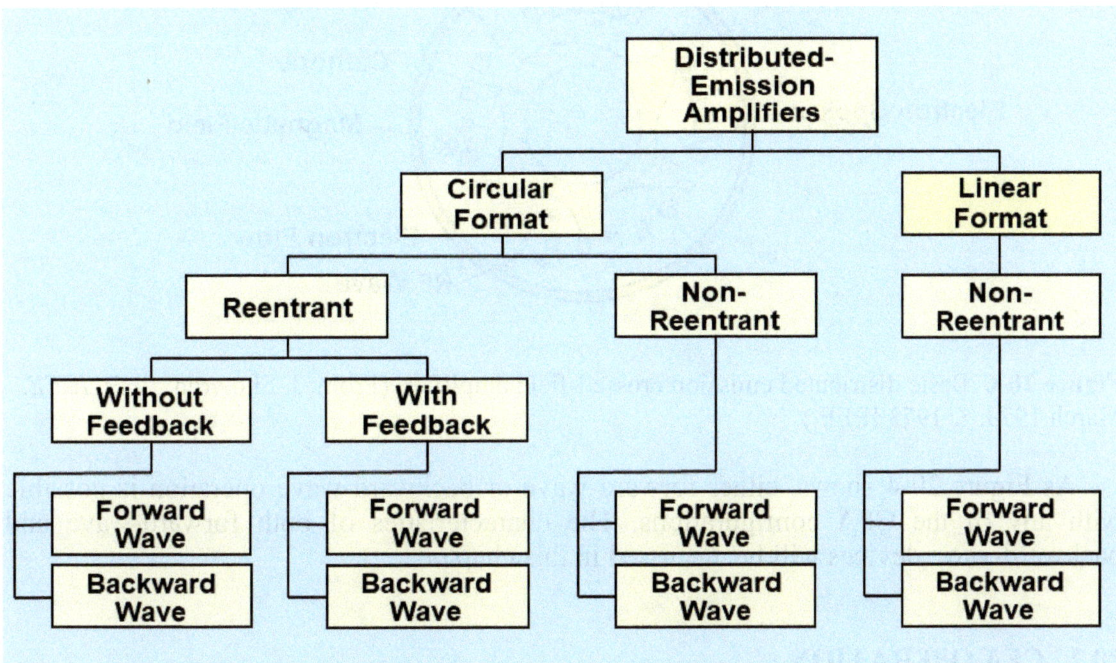

Figure 20-4 Family tree of distributed-emission CFAs. (From: *Microwave Tube Manual* by Varian Associates, Air Force Publication Number T.O.00-25-251, October 1979.)

In the reentrant circular format, the electron cloud circles around the cathode as it does in a magnetron until electrons give up sufficient energy to reach the anode. This contributes to high efficiency and, it is believed, to improved turn on and starting. In some devices, a drift region is used between the output and input ports (see Figure 20-6) so that space-charge forces cause electron bunches to disperse. This removes most of the modulation from the reentering cloud of electrons, which minimizes feedback from the output to the input.

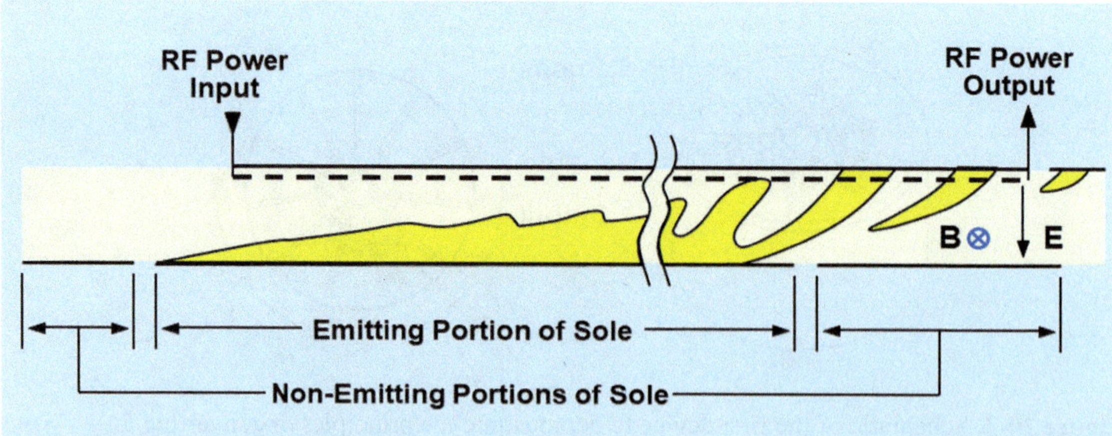

Figure 20-5 The DEMATRON, a linear format distributed-emission crossed-field amplifier. (From: G. E. Pokorny, et al., *IRE Trans. Electron Devices*, July 1962. © 1962 IRE, now IEEE.)

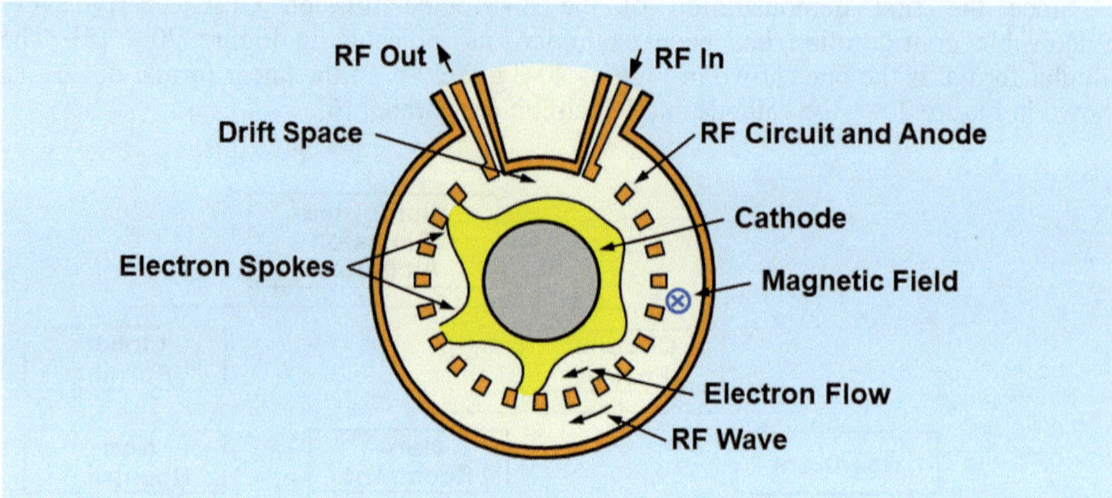

Figure 20-6 Basic distributed emission crossed-field amplifier. (From: J. Skowron, *Proc. IEEE*, March 1973. © 1973 IEEE.)

As Figure 20-4 shows, either forward-wave or backward-wave operation is possible with any of the CFA configurations. The characteristics of both forward-wave and backward-wave devices will be discussed in this chapter.

20.2 CFA OPERATION

A cross-sectional view of a CFA is given in Figure 20-7 [5]. The cathode assembly and slow wave structure are in the center of the device. Notice that even though these elements are the heart of the device and are responsible for providing the operating characteristics of the device, they make up only a very small part of the volume of the device. The magnets and magnetic circuit, the RF output structure, and the cooling apparatus make up well over 90% of the volume of the device. This section contains a qualitative description of the interaction between the electrons from the cathode and the wave on the slow-wave structure.

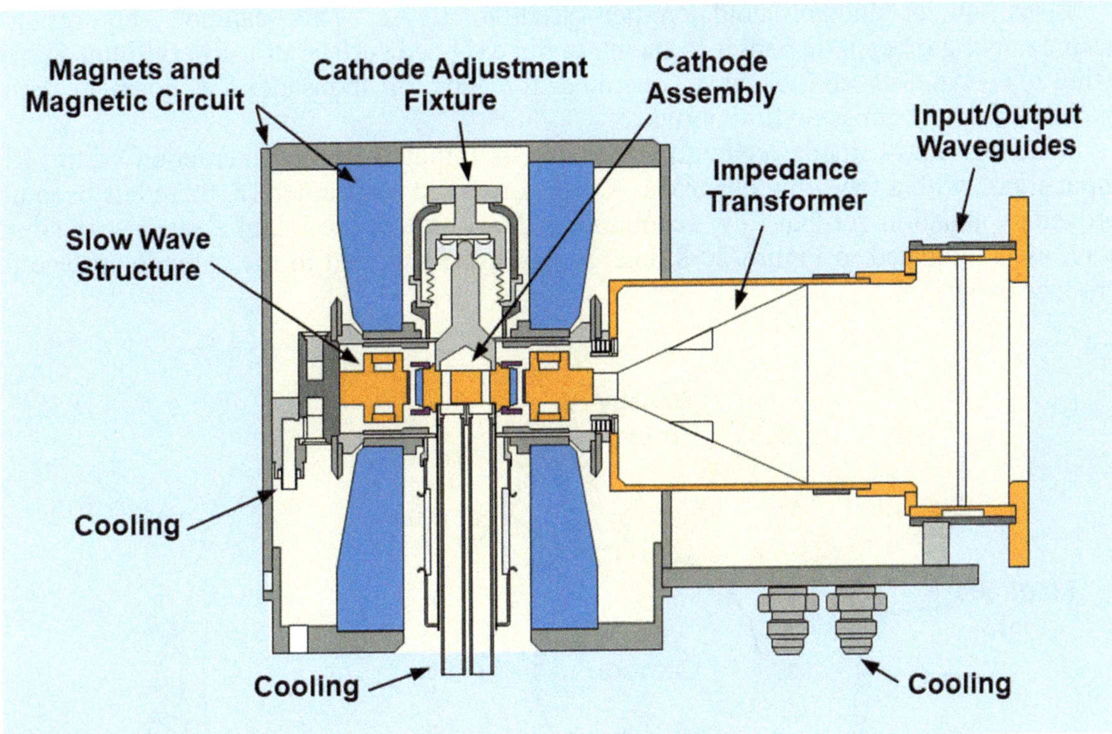

Figure 20-7 Cross section of a forward-wave distributed-emission crossed-field amplifier.

20.2.1 Electron Emission and Hub Formation

Indirectly heated thermionic cathodes and cold secondary emission cathodes are used in CFAs. The operating conditions for each type of CFA, such as pulse width, power output, and duty cycle, affect the choice of cathode. The cathode selected may be an oxide-coated type, an impregnated tungsten matrix type using barium or thorium compound impregnants, pure metal such as platinum, or variants of these types. Each has its advantages and disadvantages.

When cathodes are used that require heater power to raise their temperature for starting, a word of caution is necessary. These cathodes may become overheated when the back-heating power is added to the starting heater power. Extreme care must be taken to follow specified instructions on reducing heater power after the tube has been brought to operating power level. If this is not done, the cathode can quickly become overheated and destroyed.

The cold cathode technique is used primarily in high-power amplifiers with ≥10-kW output, although lower power levels are possible. At lower power levels, the RF electric fields are less likely to impart sufficient energy to electrons to produce the required secondary emission.

Many high-power backward-wave CFAs use liquid-cooled platinum emitters. Liquid cooling is used to dissipate the back-bombardment power. Platinum is generally used where the secondary emission ratio requirement is less than 1.80. Platinum cathodes are inert, easily mounted, and inexpensive, and can withstand high temperatures. Because the thin layer of platinum (0.010 inch) is so easy to mount on its base metal support, the platinum cathode is less expensive to build than a thermionic cathode.

Beryllium is another cold emitter used in CFAs. This cathode requires an accompanying oxygen dispenser to maintain the oxidized surface of the beryllium. Tubes using this type of emitter have operated for tens of thousands of hours without experiencing any emission limitations.

With the cold-cathode technique, emission is initiated by the interaction of the RF input signal with a few stray electrons. As was observed in Chapter 18, there has been no proven explanation yet that fully accounts for the source of the initial electrons. In some way, as is indicated in Figure 20-8, energy has to be supplied to the cathode surface to produce emission.

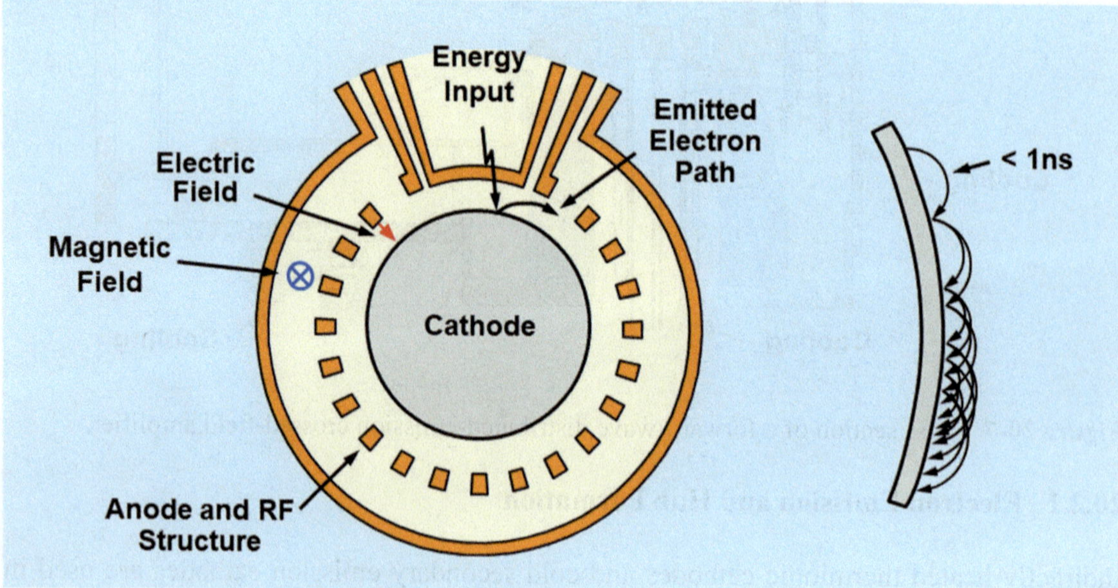

Figure 20-8 Initiation of emission in a cold cathode CFA.

Field emission and the very low level of thermionic emission that occurs at or near room temperature are possibilities. Cosmic ray ionization of internal gas molecules providing some free electrons and single-surface multipactor discharge from the cathode surface have been suggested. None have been proven conclusively. However, many hundreds of thousands of hours of reliable starting have been demonstrated on many different CFAs, and the phenomenon is now accepted as reproducible and reliable even if not fully understood.

The cathode in the CFA shown in Figure 20-7 is cold beryllium that is oxidized so that the secondary emission coefficient is ~ 2 to 3. As indicated in Figure 20-9, this cathode is liquid cooled. Because of intense electron and ion bombardment, oxygen is driven from the beryllium oxide surface and so, if corrective action is not taken, the secondary emission coefficient decreases and emission becomes inadequate. To maintain the correct state of oxidation of the cathode surface, oxygen is supplied by a heated cupric oxide source. The internal pressure of the CFA is controlled at about 10^{-6} torr with a 0.2 liter/sec ion pump. Surprisingly, the life of the cathode and of the CFA is many tens of thousands of hours.

End hats are a necessary part of a CFA or magnetron cathode. These operate at cathode potential and are designed to distort the equipotential lines near the ends of the interaction space so that inwardly directed forces are applied to the electrons. As a result, electrons are urged to move toward the center of the cathode and this prevents electron leakage from the interaction space.

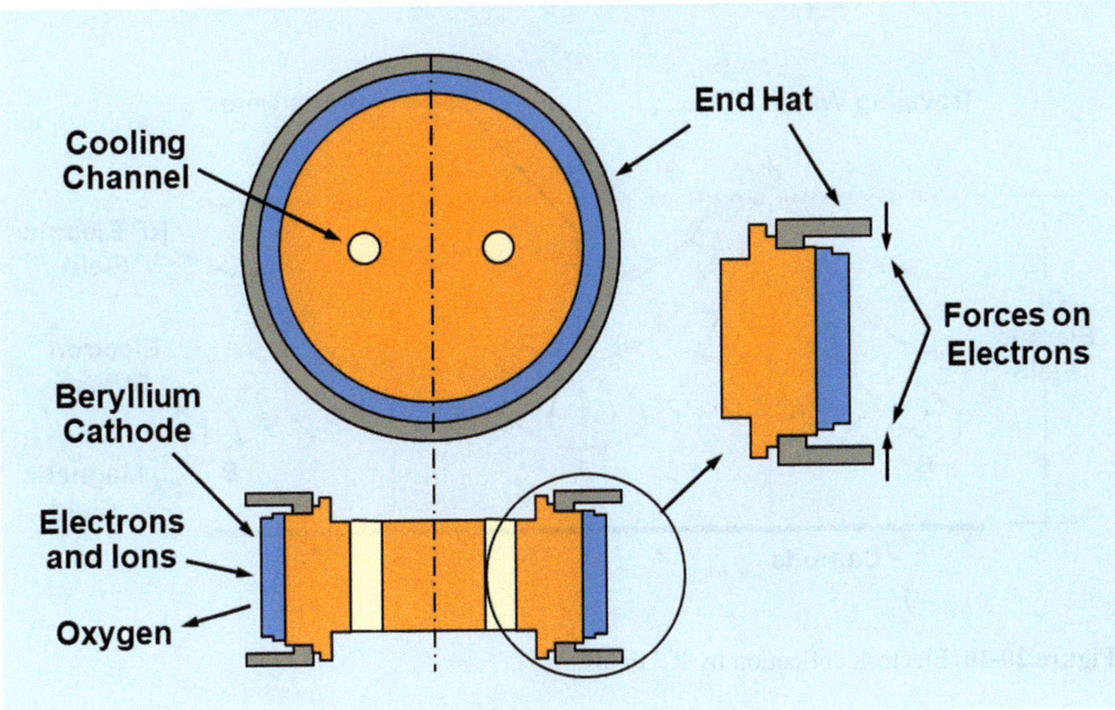

Figure 20-9 Cathode assembly for the CFA shown in Figure 20-7.

20.2.2 Spoke Formation and Growth

The basic interaction that occurs between the electron cloud near the cathode surface and the wave traveling on the slow-wave circuit is like that in a magnetron. Referring to Figure 20-10, we assume that the wave on the slow-wave circuit is traveling to the right and that the magnitudes of the dc electric field and the magnetic field have been adjusted so that the electron velocity ($u = E/B$) is the same as the wave velocity. Near A, the electric field is from left to right and so the force on the electron stream is from right to left and is tending to reduce the velocity of the electrons. As indicated in Figure 20-11, when $u < E/B$, electrons are deflected upward and the electron cloud expands toward the anode. Near B, the force is such that electrons are being accelerated and so the electrons are deflected toward the cathode.

As this process continues, the electrons in region A eventually reach the slow-wave structure (and anode) as shown in Figure 20-12. As long as the spoke and wave travel at nearly the same velocity (the phase velocity of the circuit), the wave and spoke interact to produce amplification.

It is reasonable to ask why the spoke and wave should travel in synchronism. This can be understood by examining the forces on electrons on each side of a spoke as indicated in Figure 20-12. At A, RF and dc electric fields add to accelerate electrons toward the anode. The magnetic field directs electrons toward the spoke with $v > v_p$ so electrons catch up with the spoke. At B, the RF field subtracts from the dc field and this decelerates electrons moving toward the anode. The magnetic field directs electrons in the direction of the spoke travel, but with $v < v_p$ so the spoke traveling at $v = v_p$ catches up with the electrons. As a result, the spoke is forced to travel in near synchronism with the traveling wave.

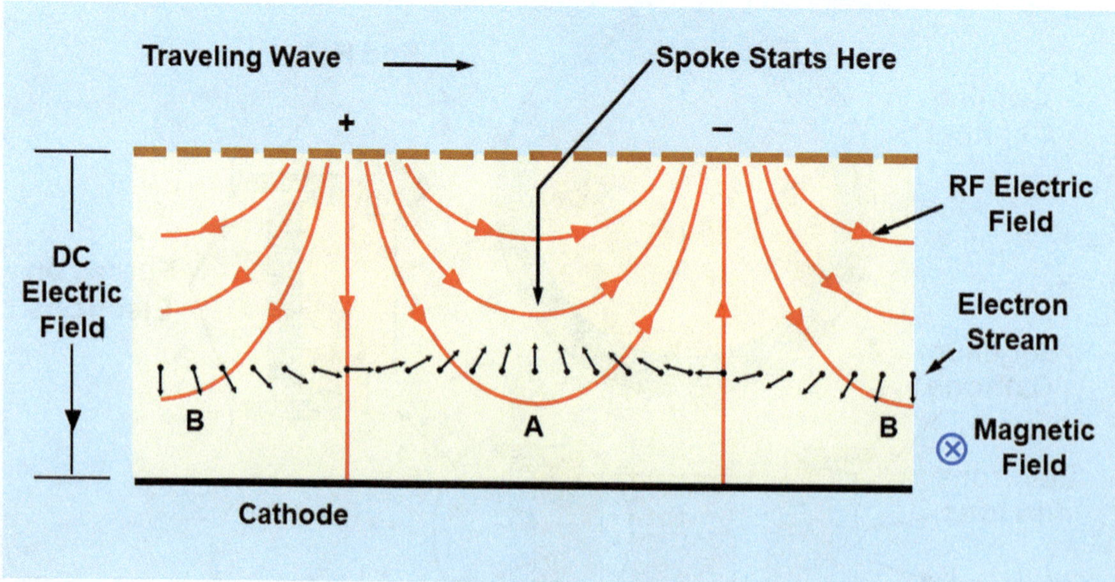

Figure 20-10 Electron deflection by RF fields.

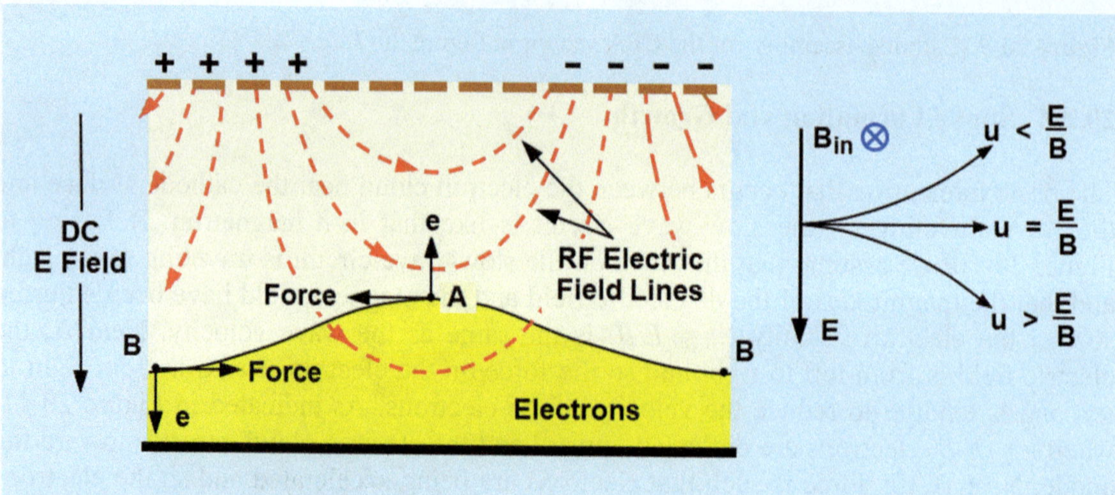

Figure 20-11 Start of spoke formation in a CFA.

As a spoke travels along, there are two mechanisms by which it causes RF current to flow in the slow-wave circuit. First, as shown in Figure 20-13, electrons in a spoke repel electrons on the circuit just as a bunch in the beam in a TWT repels electrons on the circuit. Second, and unlike a TWT, electrons move through a spoke from the cathode to the anode just as they do in a magnetron. In effect, a spoke acts as a conduit carrying RF.

Electrons on the circuit travel right and left at phase velocity, v_p. Relative to the spoke, electrons on the left move left at $2v_p$. The deficiency of electrons makes the circuit positive. Relative to the spoke, electrons on the right move at a velocity near 0. As a result, electrons accumulate and make the circuit negative. As the charges and current on the circuit increase, the electric field increases and the RF power on the circuit increases current from the cathode to the anode.

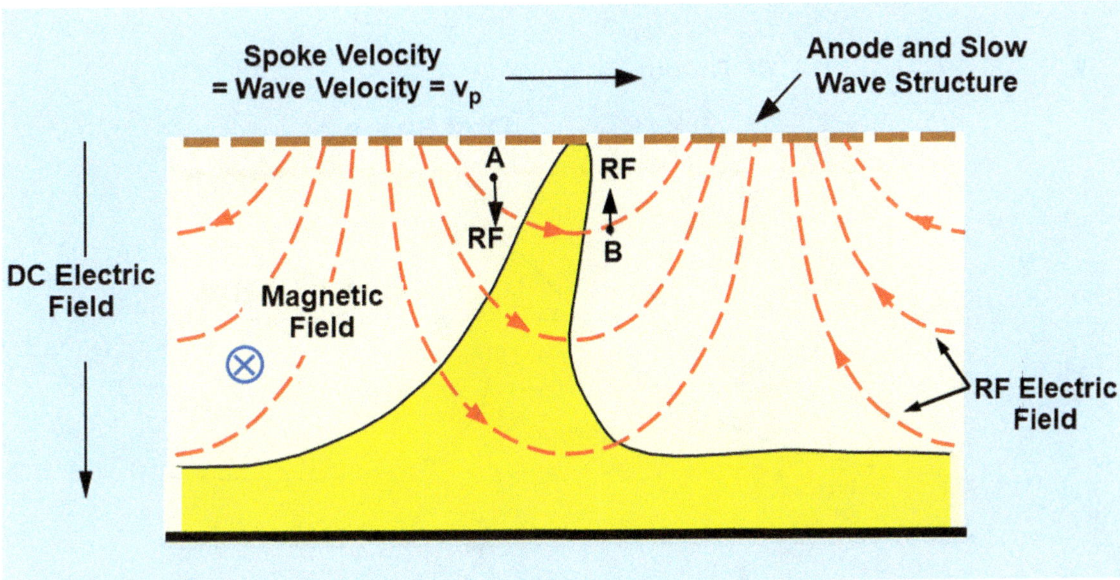

Figure 20-12 Forces that hold the spoke together.

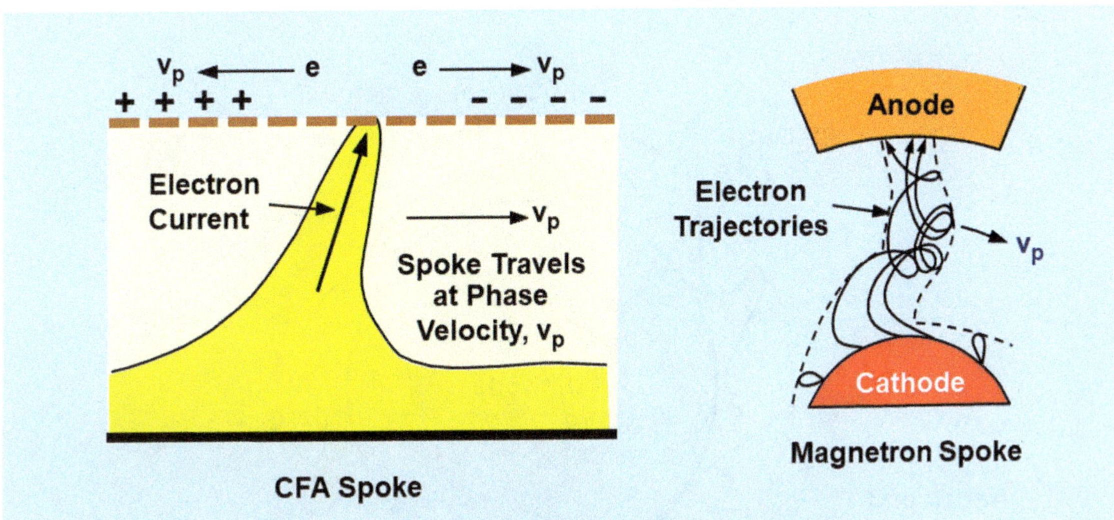

Figure 20-13 RF current flow in a CFA.

At the RF output port as shown in Figure 20-14, the circuit is terminated and RF power is removed from the tube. However, electrons remaining in the drift space continue to travel around the cathode to the RF input where they again participate in the RF interaction process. The electron stream is partially debunched in the drift space, enough so that the RF feedback from the output to input is held to an acceptable level.

The complete picture of the spoke behavior in a forward wave reentrant CFA is shown in Figure 20-15. The drift space between the output and input port provides a feedback path that helps to limit gain to about 15 dB. There are reports of CFA configurations in which part of the RF circuit is on the cathode. This helps isolate the output from the input and, reportedly, makes a gain of 30 dB or more possible. While considerable development funds went into this effort, it is not clear that any practical device resulted from it.

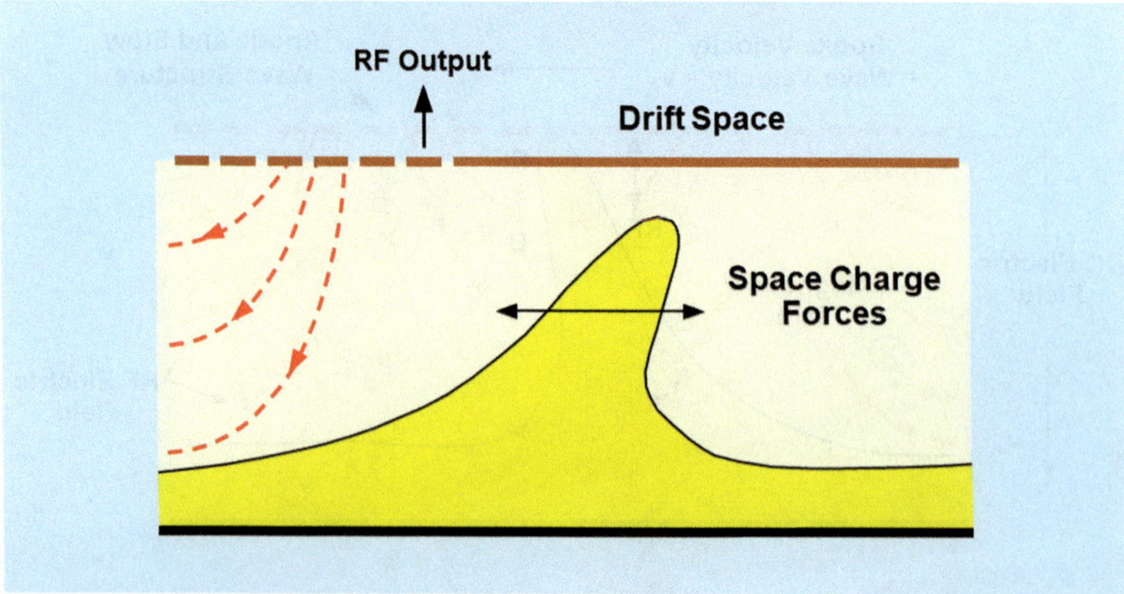

Figure 20-14 Space-charge dissipation of RF bunching in the drift space region.

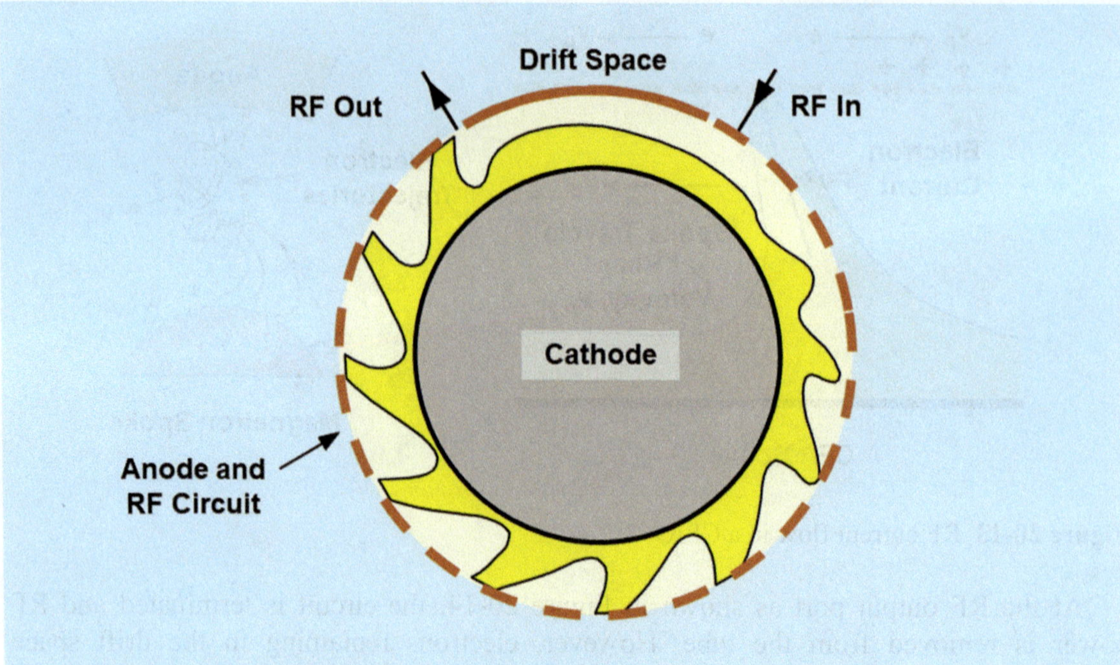

Figure 20-15 Spoke configuration in a forward-wave reentrant-beam CFA.

20.3 CFA SLOW-WAVE CIRCUITS

Crossed-field amplifiers can be forward-wave or backward-wave devices and many RF circuits have been evaluated for these devices. The selection of a slow-wave circuit is dominated by the form factor and the suitability for cooling, because the circuit serves the dual function of electron collector and slow-wave network.

The requirements for good bandwidth and flat dispersion limit the selection of circuits for forward-wave operation to some version of a helix or a meander line. Adapting these for use in a CFA will usually compromise either the bandwidth or its power-handling capability because of the necessity to deform the circuit to accommodate either conductive or liquid cooling. The helix-coupled vane and its counterpart, the stub-mounted helix, provide satisfactory compromises by retaining the good electrical features of the helix and providing cooling through the vane or stub. Ceramic-mounted circuits have provided a method for conduction cooling through the ceramic and are particularly suited to high frequencies. Figure 20-16 shows a liquid-cooled meander line [6] and a ceramic-mounted circuit.

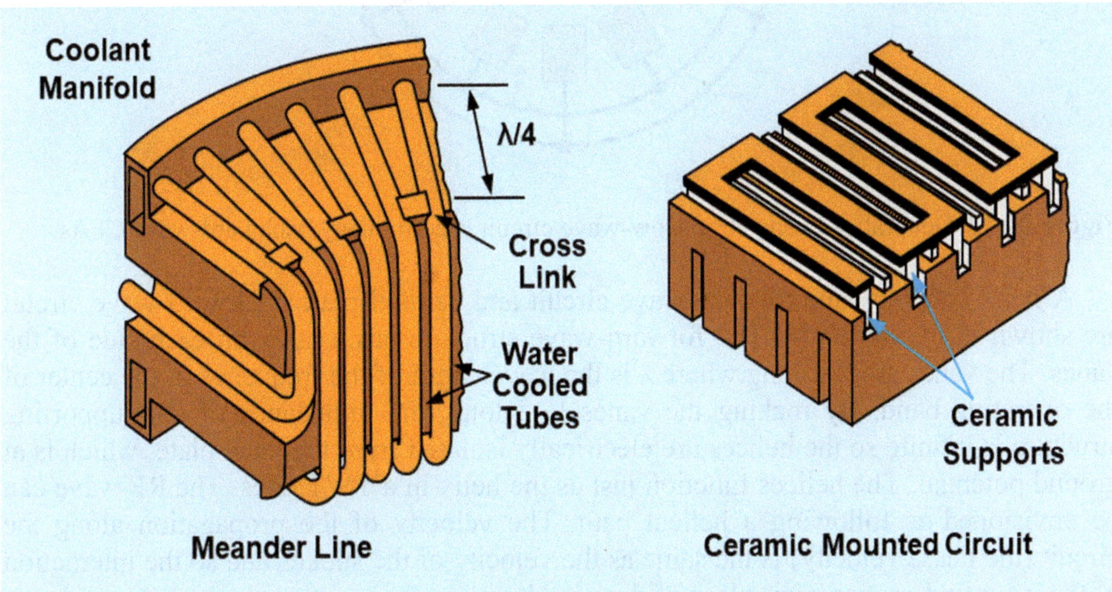

Figure 20-16 Forward-wave slow-wave circuits.

A network used by many backward-wave CFAs and which satisfies the requirements of slowing down the RF wave and providing the proper phase relationships for electron interaction is shown in a general form in Figure 20-17.

An implementation of this network is the strapped-bar line, which is similar to the strapped magnetron. The original CFA built by Brown was of this form [4]. This network is used in many backward-wave tubes that are in radar systems in the field.

The circuit as seen by the RF wave entering from the coupler is a parallel plane transmission line consisting of two straps periodically loaded by a bar or ladder line, the alternate bars or vanes being connected to the same strap. As seen by the electrons, the network consists of a series of vanes or bars exposing RF fields, which are properly phased to extract energy from the moving electrons in the tubes' interaction region.

A circuit that has been used extensively in backward-wave tubes is the parallel-plane transmission line consisting of two straps periodically loaded with bars. It is a broadband, nonresonant structure directly connected at both ends to RF input and output couplers. Because of this direct connection and its circuit characteristics, the CFA acts as a passive low loss element in both directions when no high voltage is applied.

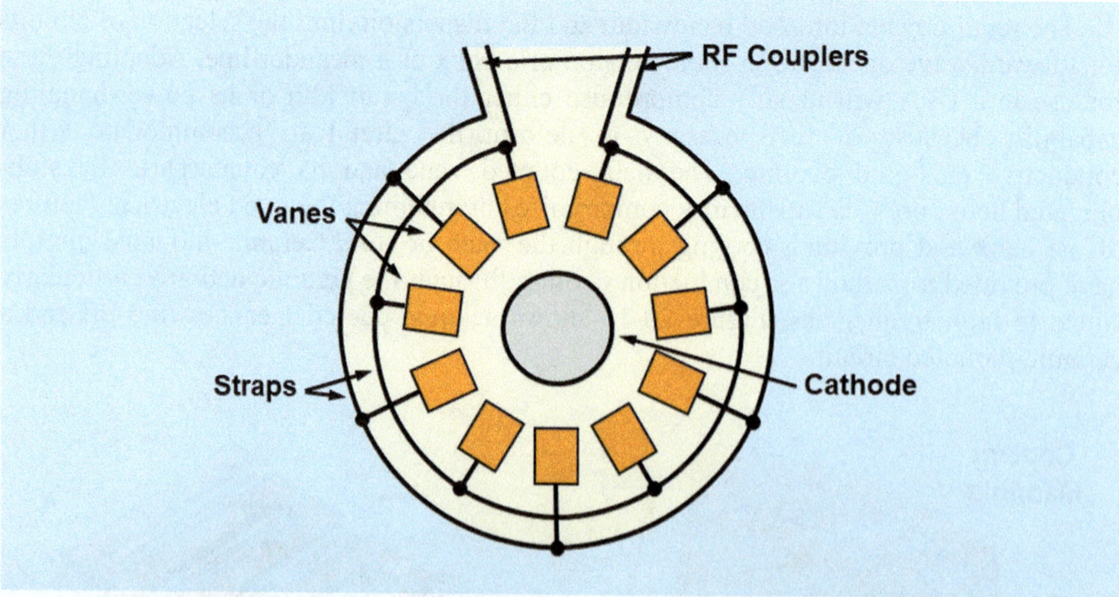

Figure 20-17 Generalized vane-strap slow-wave circuit used in many backward-wave CFAs.

A helix-coupled vane forward-wave circuit and parallel-plate backward-wave circuit are shown in Figure 20-18. The forward-wave structure has a helix on each side of the vanes. The vanes are ¼ λ long where λ is the wavelength of the frequency at the center of the operating band. By making the vanes ¼ λ long, the impedance of the supporting structure is infinite so the helices are electrically isolated from the back plate, which is at ground potential. The helices function just as the helix in a TWT does. The RF wave can be envisioned as following a helical path. The velocity of the propagation along the circuit (the phase velocity) is the same as the velocity of the spokes and so the interaction of the wave and spokes takes place as described previously.

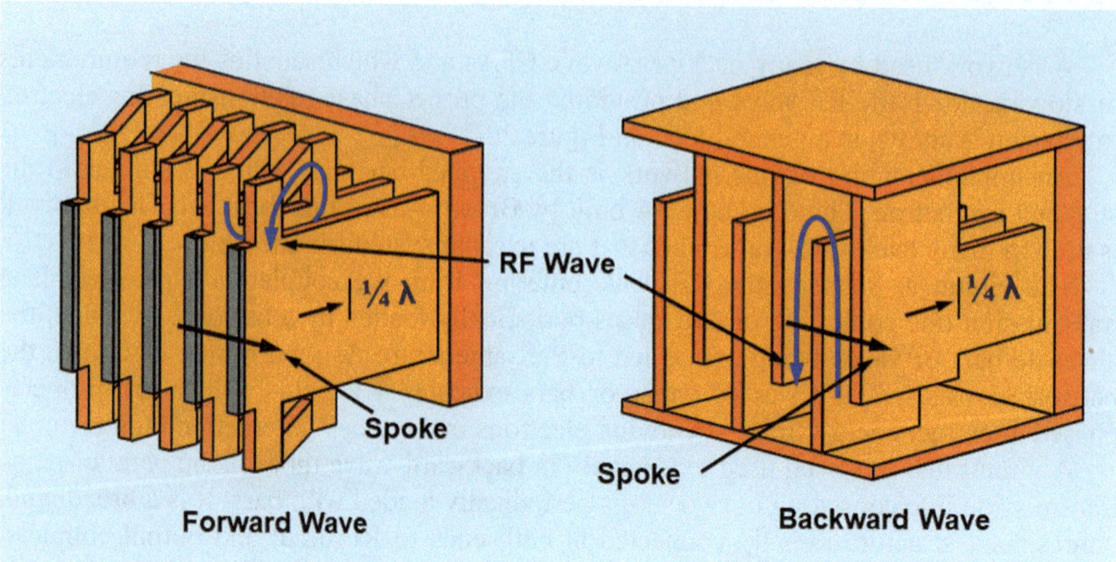

Figure 20-18 Helix-coupled vane forward-wave and parallel-plate backward-wave slow-wave circuits.

The backward-wave structure shown in Figure 20-18 is a folded parallel-plate transmission line. This line is also supported by vanes ¼ λ long. The operation of the backward-wave circuit is not as easy to understand as the forward-wave circuit is. Assume that the RF wave follows the serpentine path indicated in Figures 20-18 and 20-19. If the spoke and the wave travel in the same direction (Figure 20-19(a)), then, as the spoke and wave go from A to B, the direction of the electric field and the phase reverse (indicated by the dashed sine wave) so any modulation placed on the spoke at A is canceled at B. For backward-wave operation (Figure 20-19(b)), as a spoke goes from A to B, the wave on the circuit goes from C to B and the resulting forces are in phase with those from A so the modulation from A is reinforced.

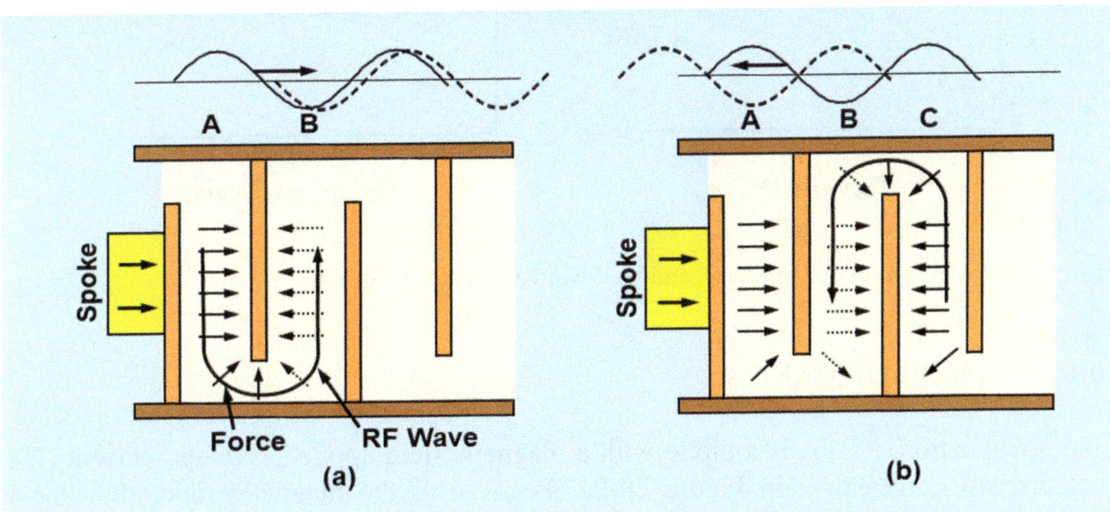

Figure 20-19 (a) Forward- and (b) backward-wave interaction with the wave on a folded transmission line.

There are significant differences between the performance characteristics of forward wave CFAs and backward-wave CFAs. The dispersion curves for these devices in Figure 20-20 show the primary differences. Recall that the slope of a straight line from the origin to any point on the dispersion curve is equal to the phase velocity of the circuit wave at that frequency; that is, $v_p = \omega/\beta$. The slope of the curve at any point is the group velocity, $v_g = \Delta\omega/\Delta\beta$, and is equal to the velocity of energy flow on the slow-wave circuit at that frequency. When both v_p and v_g have the same algebraic sign, the direction of the energy flow and the direction of the phase velocity of the circuit wave are the same. When v_p and v_g have opposite signs, then the circuit phase velocity and the velocity of the energy flow on the circuit are in opposite directions.

Notice that for the forward-wave plot in Figure 20-20 the phase velocity is nearly constant for frequencies from f_1 to f_3. An input signal at any of the frequencies within this band will be able to interact synchronously with an electron beam with a velocity nearly equal to the phase velocity so the circuit has a broad instantaneous bandwidth. Notice also that at all of the frequencies within this band, the circuit has nearly the same phase and group velocity.

The backward-wave dispersion curve shown in Figure 20-20 has a different phase velocity for each frequency as shown by f_1 and f_3 at V_1 and V_3, respectively. *Synchronous* range of frequencies at any particular beam velocity, so the amplifier has a very narrow interaction between the circuit wave and the electron beam can only occur for a narrow instantaneous bandwidth. Amplification at different frequencies requires a voltage change to the amplifier to change the electron beam velocity. Hence the backward wave tube is a voltage-tunable amplifier.

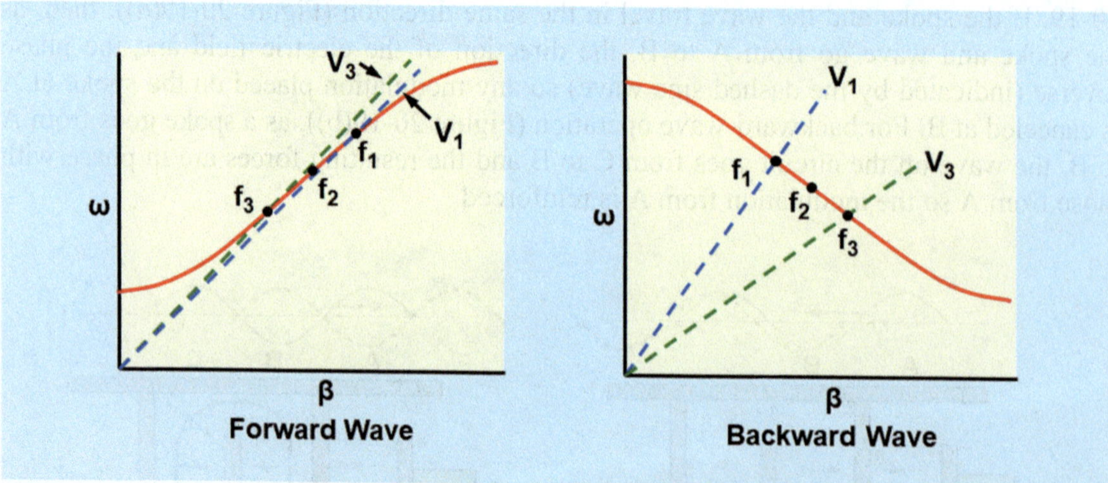

Figure 20-20 ω-β plots for forward- and backward-wave CFA circuits.

20.4 CFA PERFORMANCE

Like a magnetron, a CFA is a diode with a magnetic field applied. Voltage-current (*V-I*) characteristics are shown in Figure 20-21. Because of the magnetic field, the tube is essentially cut off until a relatively high voltage level is reached. As the magnetic field is increased, voltage must be increased to maintain the same current. Factors other than the magnetic field that affect the *V-I* characteristic are frequency (and the dispersion characteristic of the slow-wave circuits) and RF drive power.

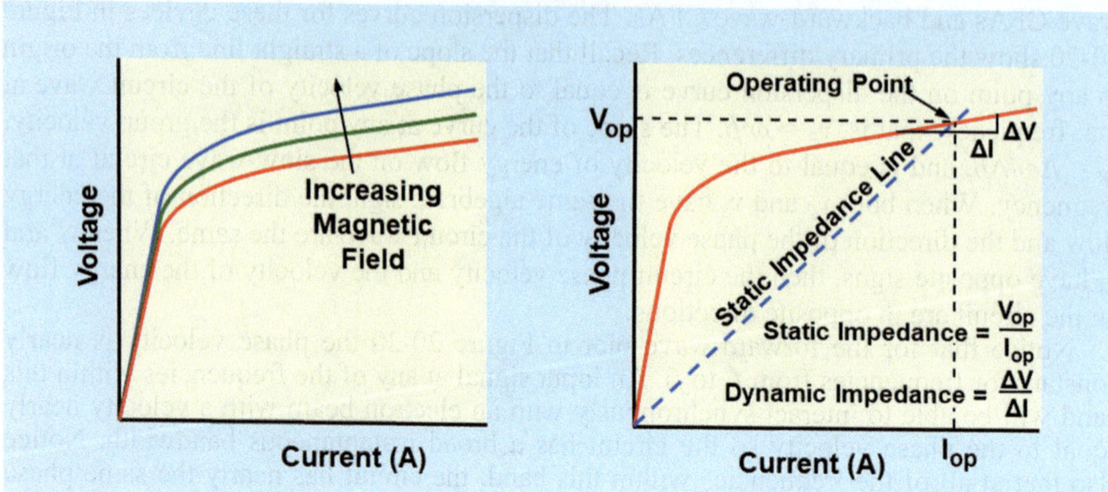

Figure 20-21 Voltage-current and impedance characteristics for a CFA.

The impedance characteristics of CFAs are of importance when considering their operation with a power supply. As Figure 20-21 shows, there is a significant difference between static impedance, which is simply the operating voltage divided by the operating current, and the dynamic impedance, which is the rate of change of voltage with current. The dynamic impedance may be an order of magnitude smaller than the static impedance. With the exception of dc-operated CFAs (to be discussed in Section 20.4.3), current control in CFAs, forward-wave or backward-wave, is accomplished by controlling (pulsing) the cathode voltage.

As indicated in Figure 20-22, the RF drive pulse is applied before high voltage is applied to the cathode. There are two primary reasons for this. First, if the amplifier uses a cold cathode and if RF drive is absent, the emission process may fail to build up. As a result, the modulator providing the cathode voltage will be essentially unloaded, so excessive voltage may be applied to the tube. This can result in internal arcing and damage to the tube.

The second reason for applying RF drive before cathode voltage is important in CFAs using cathodes heated by a heater or by back bombardment so that thermionic emission occurs. In these devices, RF drive is required to help control the space charge and to help prevent oscillations at the band edge or the generation of broadband noise.

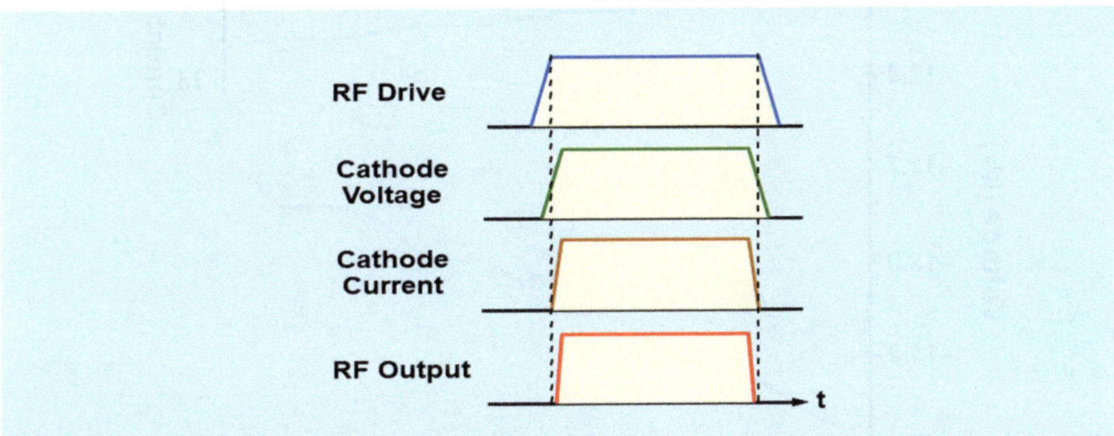

Figure 20-22 Pulse timing in a CFA. RF drive is applied first and then cathode voltage is applied.

20.4.1 Forward-Wave CFAs

The ω-β plot for the forward-wave CFA in Figure 20-20 shows that when frequency is changed from f_1 to f_3, there is a small change in phase velocity and in the corresponding optimum operating voltage. As frequency is reduced, voltage increases slowly. As a result, the *V-I* characteristic for a forward-wave amplifier is as shown in Figure 20-23.

Because voltage is nearly constant as frequency is changed, the current in a forward-wave CFA should be held nearly constant if power output is to be held constant. Figure 20-24 shows current and voltage as functions of frequency for a forward-wave CFA. For this tube type, voltage is adjusted at f_3 to obtain a specified current (22A). Then, as frequency changes, voltage changes slowly to maintain a nearly constant current.

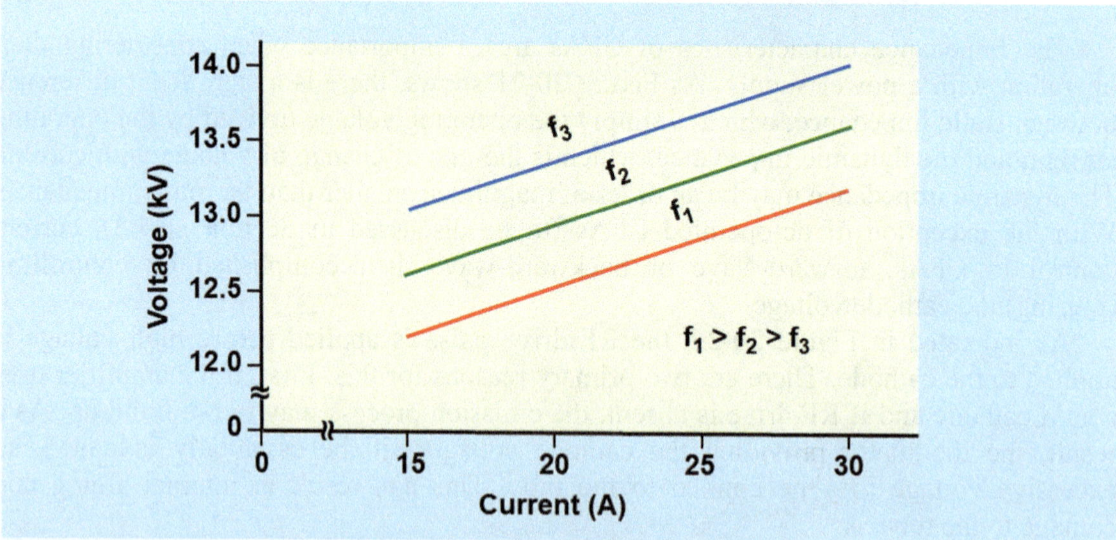

Figure 20-23 Voltage-current characteristic for a forward-wave CFA.

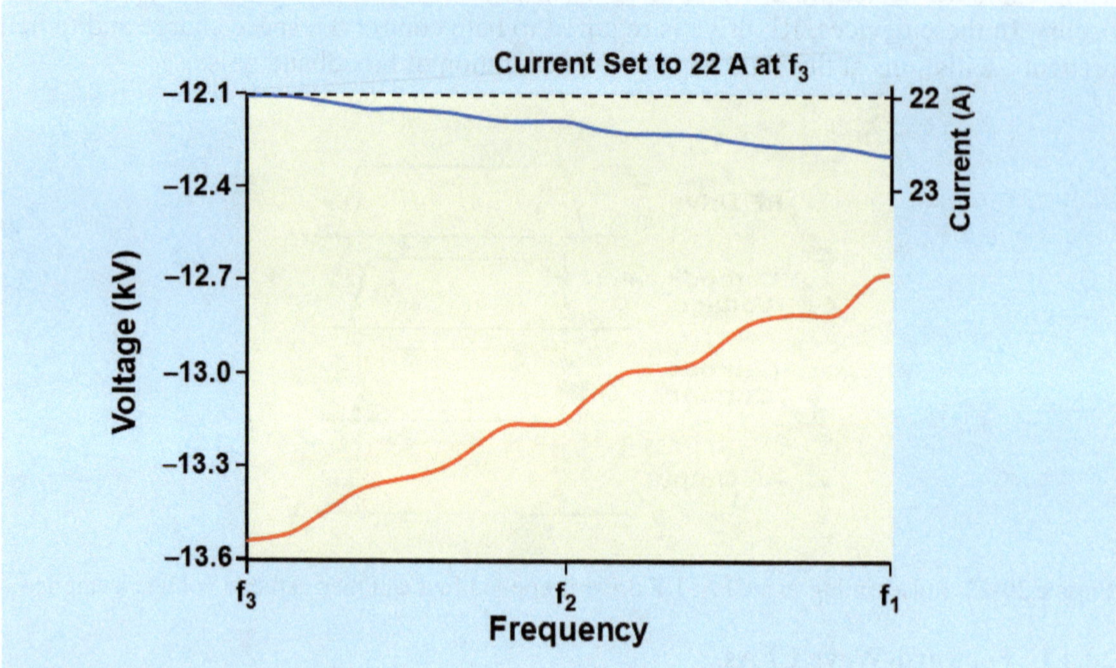

Figure 20-24 Current and voltage as functions of frequency for a forward-wave CFA.

An example of the power circuit used to control voltage as well as to turn the forward-wave CFA on and off is shown in Figure 20-25. When the CFA is turned off, the full voltage of the power supply is applied to the tetrode. As shown in Figure 20-26, which is the constant current plot for the tetrode, the voltage, which is across the tetrode when the CFA considered in this example is turned off, is 17.4 kV. To apply cathode voltage to the CFA, the control grid voltage for the tetrode is switched from –608V to ~+150V (with respect to the tetrode cathode). As the CFA frequency is varied from f_1 to f_3, the voltage across the switch tube varies automatically to apply the required voltage to the CFA while maintaining a nearly constant current.

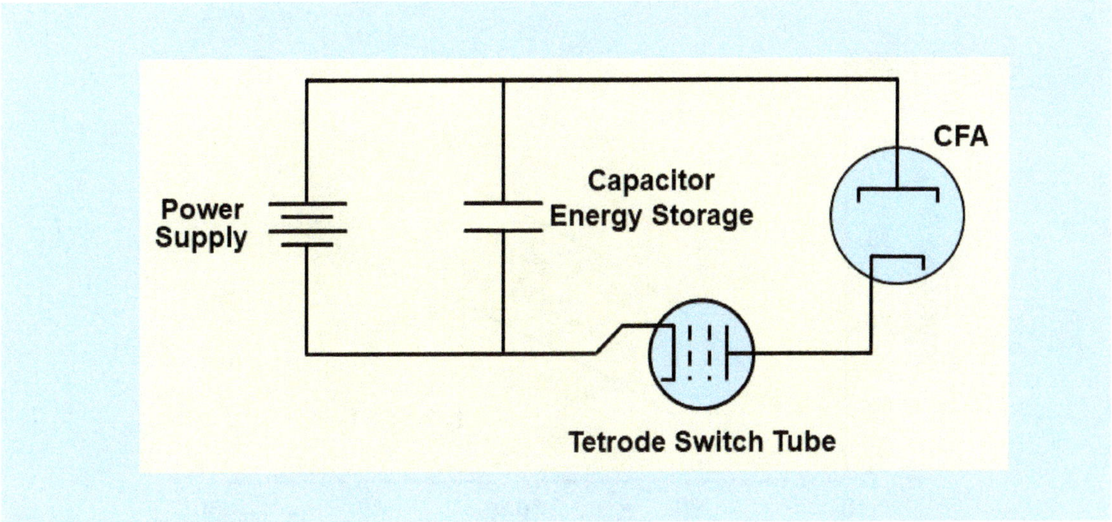

Figure 20-25 Example of the power circuit that can be used for a forward-wave CFA.

20.4.2 Backward-Wave CFAs

The voltage-current characteristic for a backward-wave CFA is shown in Figure 20-27. From the ω-β plot in Figure 20-20, it would be expected that the CFA voltage required for proper operation would decrease as frequency is decreased from f_1 to f_3 and that is what is observed in practice.

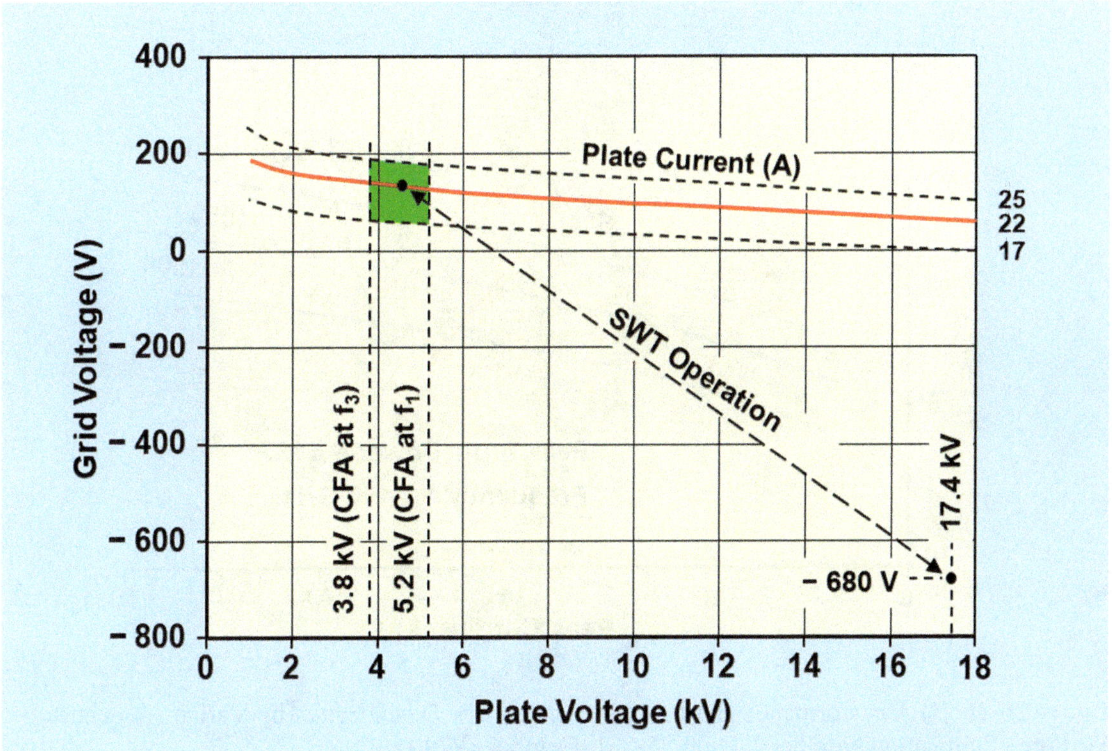

Figure 20-26 Constant current characteristic for the switch tube being considered in this example.

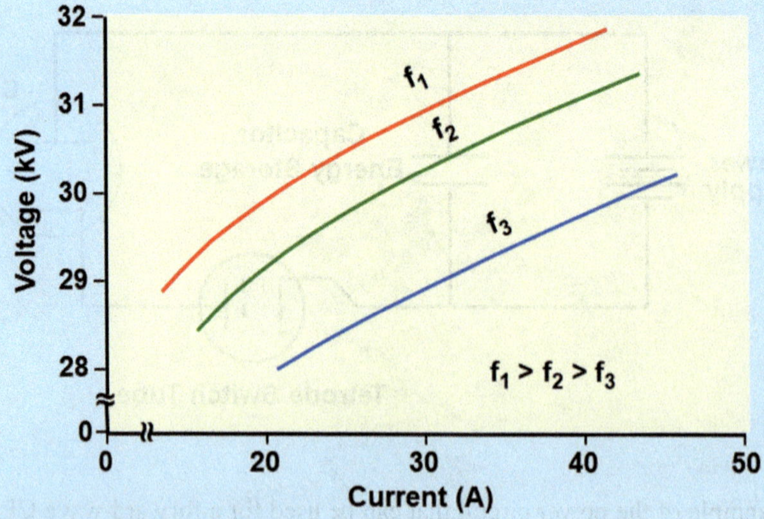

Figure 20-27 Voltage-current characteristic for a backward-wave CFA.

Another chart showing the voltage-current characteristics for a backward-wave CFA is shown in Figure 20-28. This chart shows the power output and efficiency of the device and is similar to a portion of a magnetron performance chart. The magnetic flux density is not normally a variable to the user of a CFA so the user is likely to receive a one-line version of this chart (as shown in Figure 20-29).

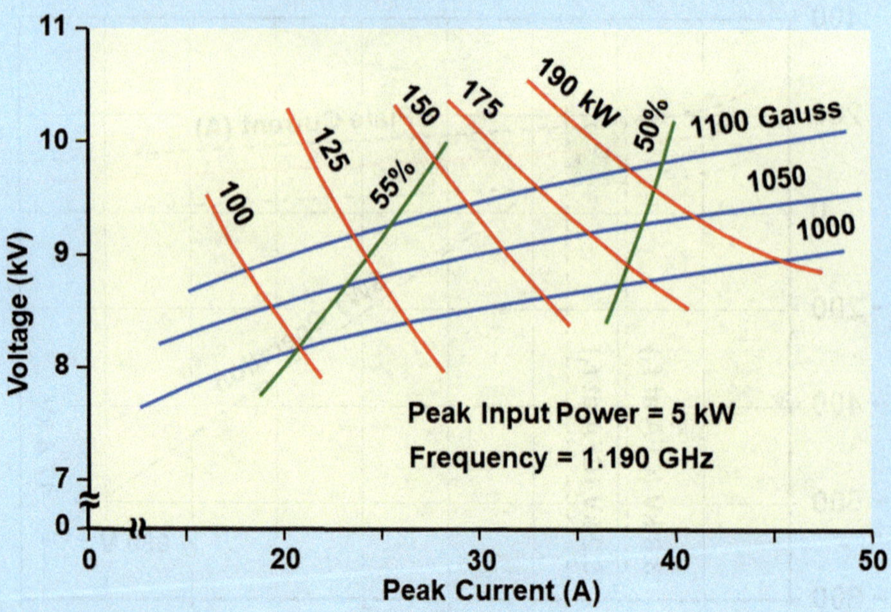

Figure 20-28 CFA performance chart. (From: *Microwave Tube Manual* by Varian Associates, Air Force Publication Number T.O.00-25-251, October 1979.)

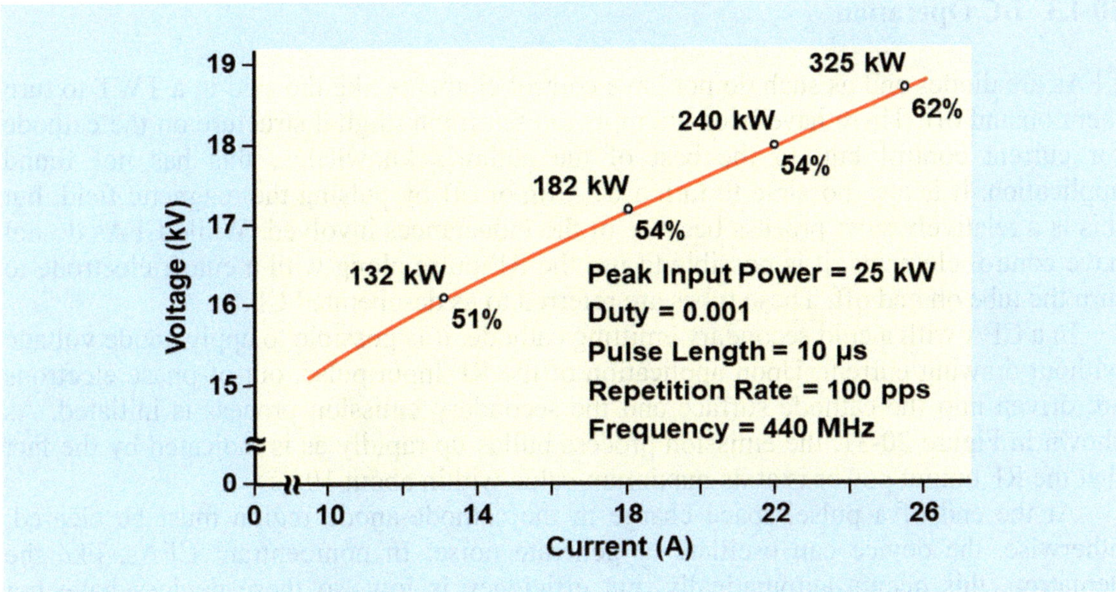

Figure 20-29 CFA one-line performance chart. (From: *Microwave Tube Manual* by Varian Associates, Air Force Publication Number T.O.00-25-251, October 1979.)

When voltage and current are dependent on frequency of operation, as in a backward-wave CFA, the performance chart will show the frequency dependence as illustrated in Figure 20-30. With this chart, the user can determine the variations in power, current, and voltage that will result from the modulation technique used in tuning the amplifier across the required frequency range.

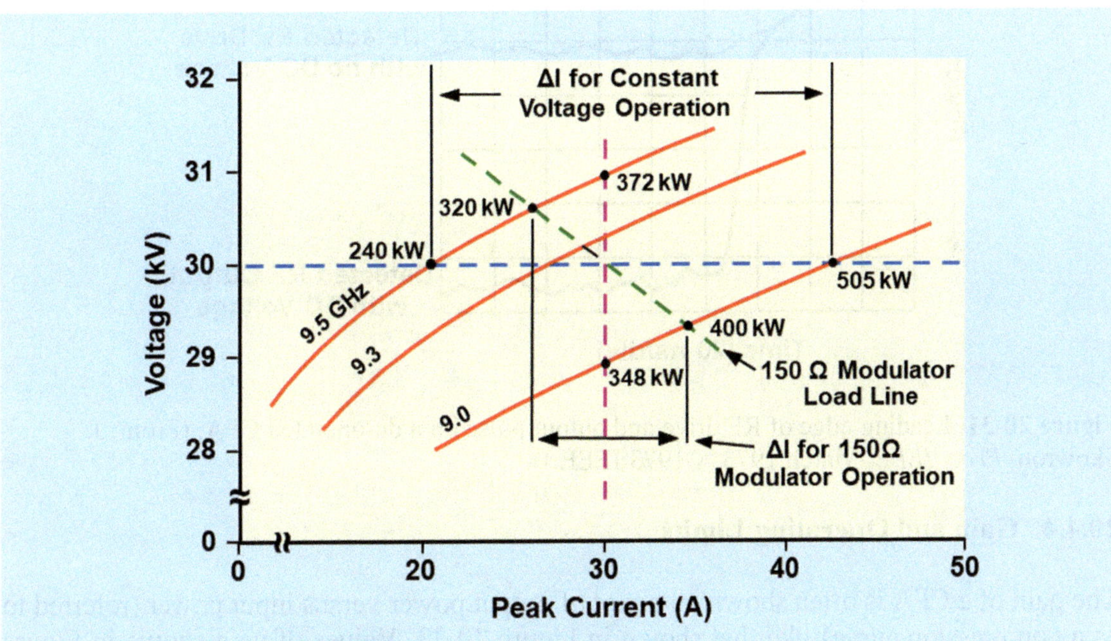

Figure 20-30 Frequency dependence of voltage and current. (From: *Microwave Tube Manual* by Varian Associates, Air Force Publication Number T.O.00-25-251, October 1979.)

20.4.3 DC Operation

CFAs are diodes and as such do not have control elements like the grid in a TWT to turn them on and off. There have been attempts to use an interdigital structure on the cathode for current control but, to the best of the author's knowledge, this has not found application. It is also possible to turn a CFA on or off by pulsing the magnetic field, but this is a relatively slow process because of the inductances involved. While CFAs do not have control elements, it is possible to use the RF pulse along with a cutoff electrode to turn the tube on and off. These tubes are referred to as dc-operated CFAs.

In a CFA with a cold secondary emitting cathode, it is possible to apply anode voltage without drawing current. Upon application of the RF input pulse, out-of-phase electrons are driven into the cathode surface and the secondary emission process is initiated. As shown in Figure 20-31, the emission process builds up rapidly as is indicated by the fact that the RF output power is at its maximum value within about 10 ns.

At the end of a pulse, space charge in the cathode-anode region must be cleared; otherwise, the device can oscillate or generate noise. In nonreentrant CFAs, like the dematron, this occurs automatically, but efficiency is low, so these devices have not found application. In reentrant CFAs, a control or cutoff electrode is pulsed positive with respect to the cathode to collect the space charge. The configuration of a CFA containing a cutoff electrode is shown in Figure 20-32.

The sequence of pulses applied to a CFA during dc operation is shown in Figure 20-32. The applied RF drive pulse produces anode current and an RF output pulse. At the end of the RF pulse, the voltage pulse to the cutoff electrode produces a short pulse of current corresponding to the charge stored in the interelectrode space.

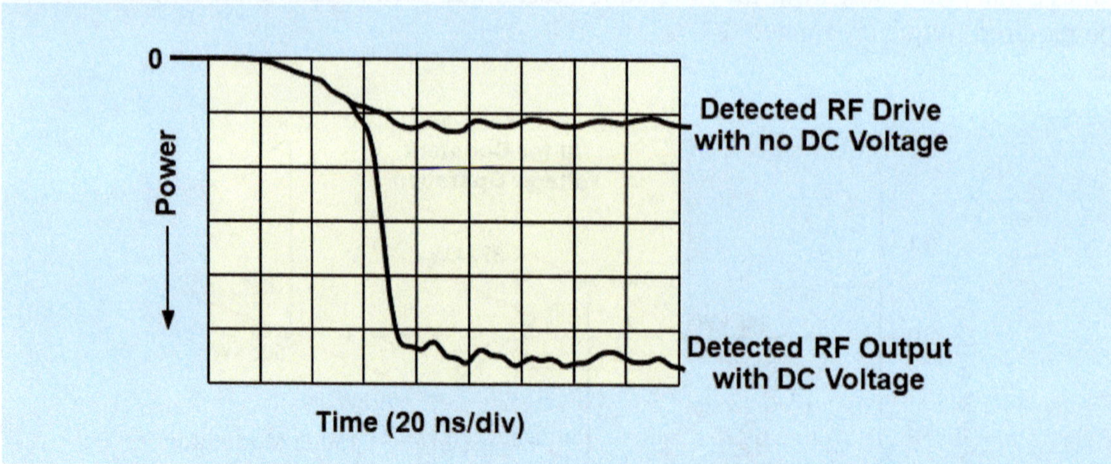

Figure 20-31 Leading edge of RF drive and output pulse on a dc-operated CFA. (From: J. Skowron, *Proc. IEEE*, March 1973. ©1973 IEEE.)

20.4.4 Gain and Operating Limits

The gain of a CFA is often shown on a plot of output power versus input power (referred to as a compression curve) like that shown in Figure 20-33. Values of gain shown in Figure 20-33 are typical and are low compared with those for linear-beam tubes. It should also be noted that the CFA does not exhibit any region in which operation is linear (constant gain).

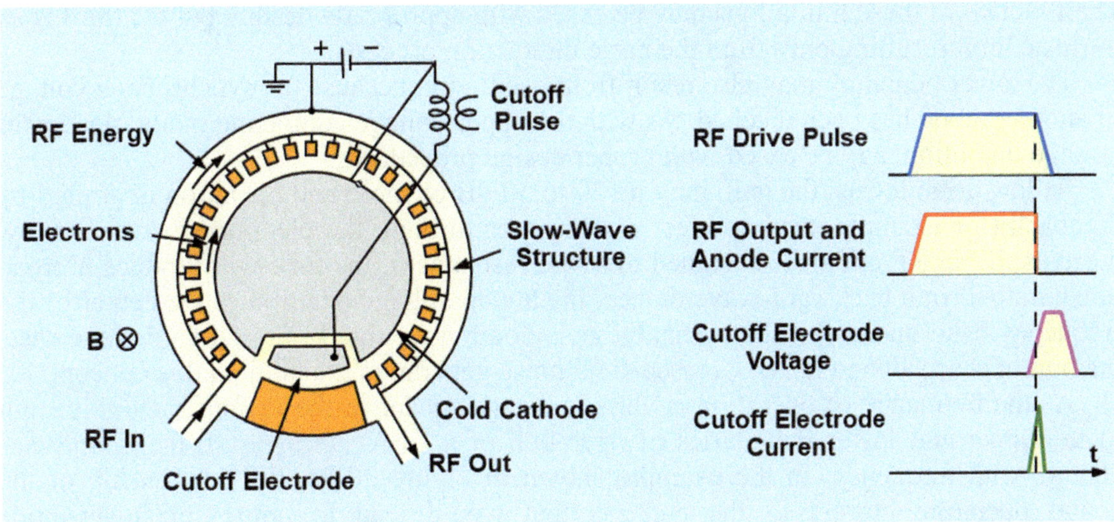

Figure 20-32 DC-operated CFA with cutoff electrode. (From: J. Skowron, *Proc. IEEE*, March 1973. © 1973 IEEE.)

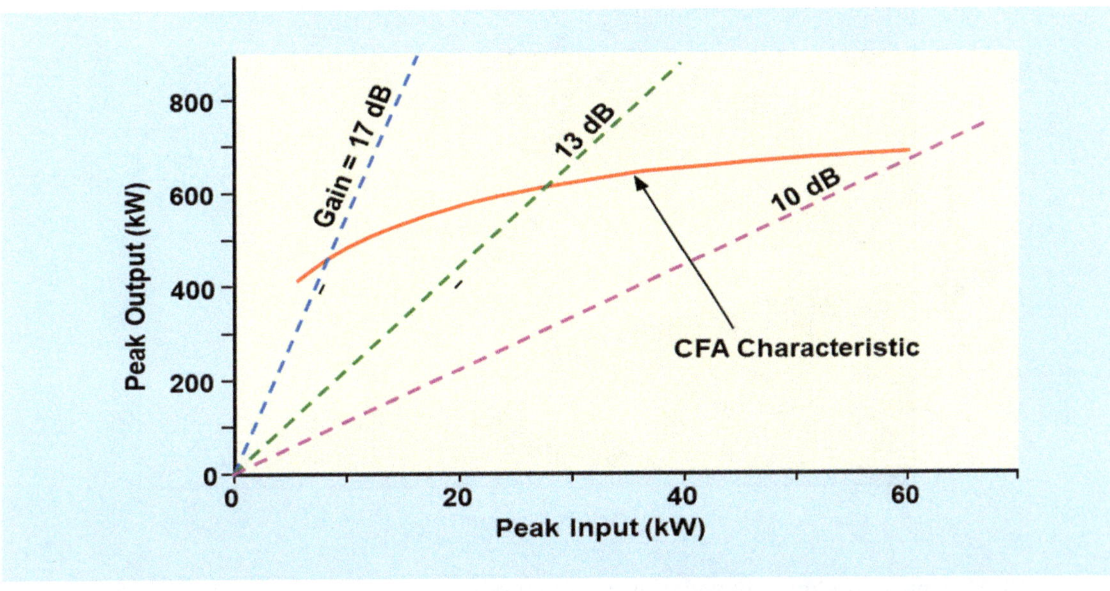

Figure 20-33 Compression curve for CFA showing lines of constant gain. (From: *Microwave Tube Manual* by Varian Associates, Air Force Publication Number T.O.00-25-251, October 1979.)

The gain-limiting factors in CFAs are varied and complex and lead to the chart in Figure 20-34, showing upper and lower boundaries of output power for given values of input power. The upper boundary can be caused by any one of the following factors:

1. A limitation on cathode current;
2. The onset of oscillation because the synchronous voltage of a competing mode has been reached;
3. Sufficiently large output powers so that the drive level is insufficient to retain lock in the main amplifying mode.

The first two of these limitations may be eased with appropriate design, but the third is an intrinsic limit resulting only from the basic interaction process.

The lower boundary may also result from oscillation because the synchronous voltage of another mode has been reached. As with the upper voltage competing mode, this lower voltage limitation may be eased with proper design procedures.

At low drive levels, the gain may be 20 to 30 dB or more and operation is limited by feedback (for example, by imperfect matching or through the electron stream). At low currents, if the RF circuit is designed to avoid resonances, the tube will produce a broad noisy output. In a backward-wave device, the anode voltage determines the center of the frequency band and so a voltage-tunable noisy output results. In the forward-wave case, the output is broadband noise. Crossed-field noise generators are built on this concept.

As the frequency of operation is varied, changes occur in the various factors leading to the upper and lower boundaries of operation of a CFA. As a result, the boundaries change with frequency. In the example shown in Figure 20-35, the relationship of the actual operating current to the currents that would lead to upper- or lower-mode instabilities is shown. In this case, a limiting factor in bandwidth is the intersection of the operating current line with the upper-mode characteristic.

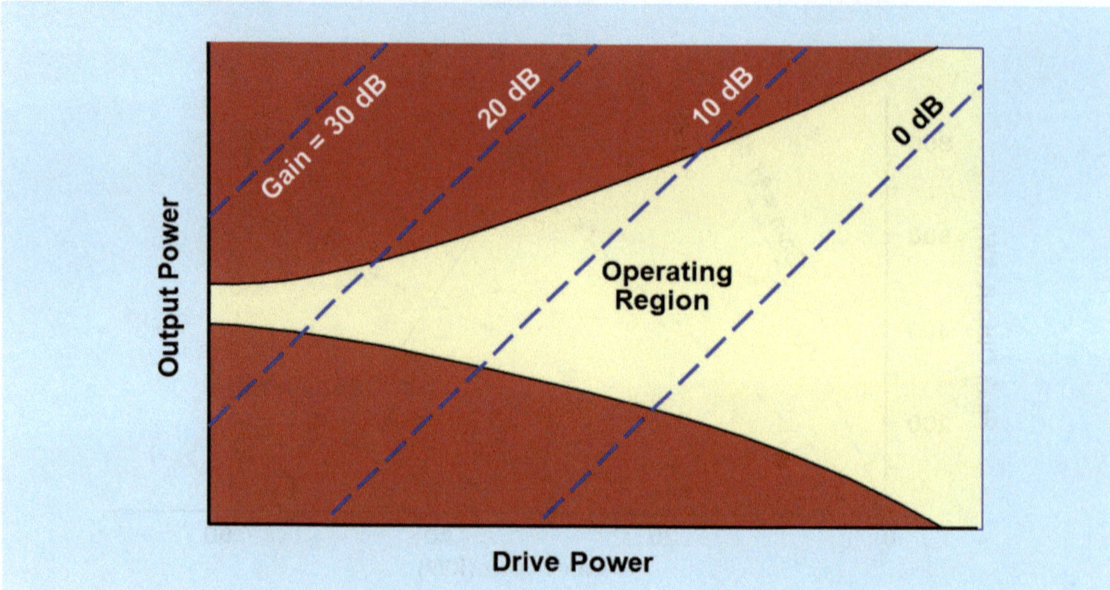

Figure 20-34 Upper and lower boundaries of output power of a CFA. (From: J. Skowron, *Proc. IEEE*, March 1973. ©1973 IEEE.)

In some CFAs it is possible to improve operating characteristics by introducing an asymmetry into the interaction region. As an example, for the forward-wave CFA discussed earlier in this chapter, with the cathode centered, performance at the high end of the band is marginal. The problem is that the spokes are out of phase with the RF field. The spoke phase can be controlled with voltage and current and current density can be controlled with cathode position. Thus, by pushing the cathode off-center, the spoke phase can be corrected. The result is that the high-frequency performance is improved and, it turns out, intrapulse noise is reduced. Figure 20-36 shows the cathode positioning device in the CFA in Figure 20-7 [5].

In practice, mode interference is often the limiting factor in determining the useful range of CFA operation. Mode interference is an oscillation that takes time to build up. In pulsed operation, the rate of cathode voltage rise has a significant effect on mode interference. Care must be taken in modulator design, however, because a fast rate of rise, which prevents a lower mode from starting, may initially result in insufficient tube current to load the modulator. As a result, voltage can swing through the values appropriate for operation in the amplifying mode and cause oscillation in an upper mode.

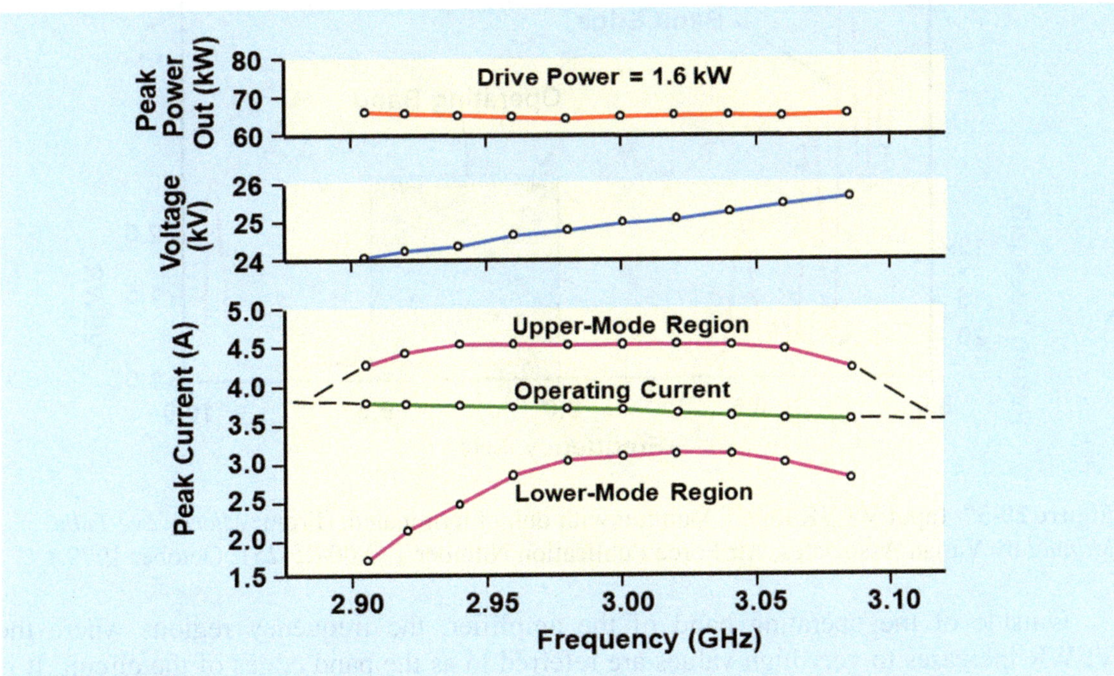

Figure 20-35 Relation of operating current to currents that lead to upper- or lower-mode instabilities in a backward-wave CFA. (From: J. Skowron, *Proc. IEEE*, March 1973. ©1973 IEEE.)

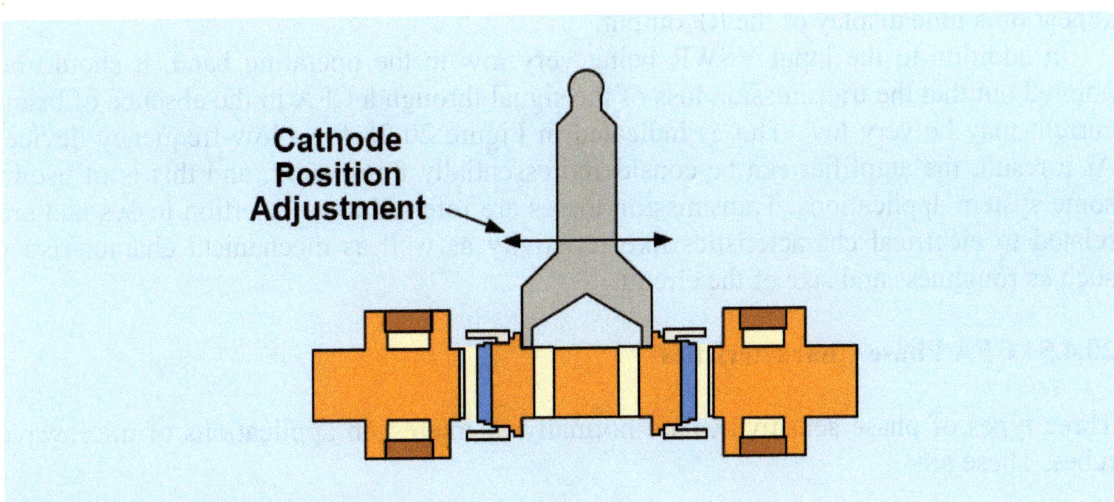

Figure 20-36 Cathode positioning device in a forward-wave CFA.

In connection with a discussion of mode interference, it is instructive to examine the impedance characteristics of a CFA. The input VSWR as a function of frequency is typically as shown in Figure 20-37. Within the operating band, the VSWR is low, typically between 1.2:1 and 1.5:1. The corresponding return loss (magnitude of reflected power) is between 20 and 14 dB below the input power.

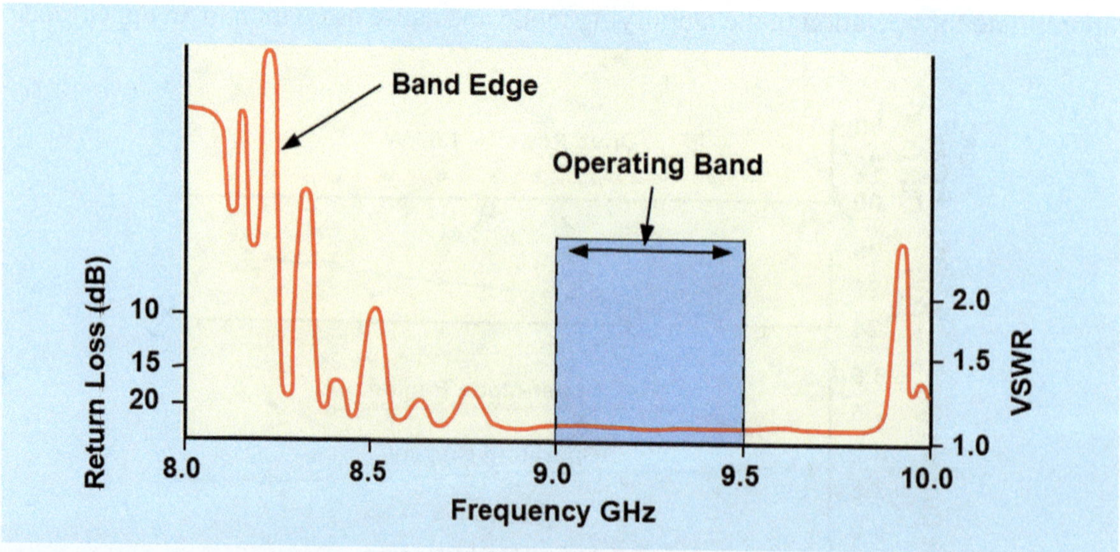

Figure 20-37 Input VSWR for CFA circuit with output terminated. (From: *Microwave Tube Manual* by Varian Associates, Air Force Publication Number T.O.00-25-251, October 1979.)

Outside of the operating band of the amplifier, the frequency regions where the VSWR increases to very high values are referred to as the band edges of the circuit. It is at one of these band edges (the low-frequency edge in a backward-wave amplifier) that oscillations may occur during the rise and fall of the voltage pulse. The resulting spurious outputs of the tube are frequently referred to as rabbit ears because of the way they appear on a time display of the RF output.

In addition to the input VSWR being very low in the operating band, it should be pointed out that the transmission loss of the signal through a CFA in the absence of beam current may be very low. This is indicated in Figure 20-38 for a low-frequency device. As a result, the amplifier can be considered essentially transparent, and this is of use in some system applications. Transmission losses are referred to as insertion losses and are related to electrical characteristics like resistivity as well as mechanical characteristics such as roughness and size of the circuit.

20.4.5 CFA Phase Characteristics

Three types of phase sensitivities are normally of interest in applications of microwave tubes. These are:

- Phase sensitivity to voltage or current;
- Phase sensitivity to RF drive;
- Phase linearity with frequency.

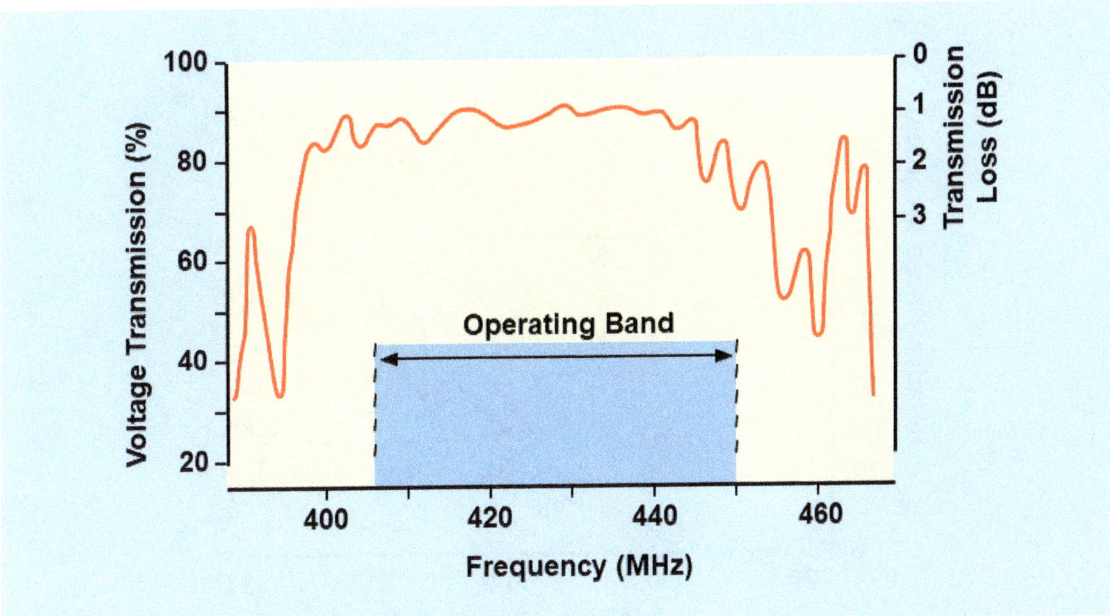

Figure 20-38 Transmission characteristic of a CFA. (From: *Microwave Tube Manual* by Varian Associates, Air Force Publication Number T.O.00-25-251, October 1979.)

In a linear-beam tube, phase shift through a tube is proportional to the transit time of the beam, which is inversely proportional to beam velocity. Beam velocity, in turn, varies with the square root of beam voltage. For a TWT, phase changes about 30° for a 1% change in beam voltage. Current varies with voltage to the three-halves power, so the phase changes about 45° for a 1% change in current.

In a CFA, the rotational rate of the spokes in the space-charge cloud is locked to the velocity of the signal on the RF circuit. The speed of the spokes tends to be controlled by the propagation characteristics of the circuit and the signal level in the circuit. Also, the electrical length of the RF circuit is short. As a result, phase sensitivity to changes in the anode voltage would be expected to be small.

In a linear-beam tube, because of the three-halves power relationship between voltage and current, the phase shift for a 1% change in voltage or current is of the same order of magnitude. In a crossed-field device, as can be seen in the voltage-current plots, a small change in voltage is accompanied by a large change in current. Thus, the phase shift for a 1% change in voltage is much larger than the phase shift for a 1% change in current.

The phase variations with voltage for three RF drive levels in an X-band CFA are shown in Figure 20-39. The phase changes indicated are for a 1% change in anode voltage. Thus, at low drive levels and high gain, the phase shift (15°/1% ΔV) is much larger than at high drive levels and low gain (3.8°/1% ΔV).

In Figure 20-40, the phase shift per 1% change in current is shown for a low-power CFA with a thermionic cathode and for a medium-power, cold-cathode device. In both cases, the phase shift at all drive levels is small. While the comparison is not strictly valid because of the large difference in devices, the phase sensitivity to voltage in Figure 20-39 is much larger than the phase sensitivity to current in Figure 20-40, as would be expected. In all cases, the phase sensitivity is much smaller than for linear-beam devices.

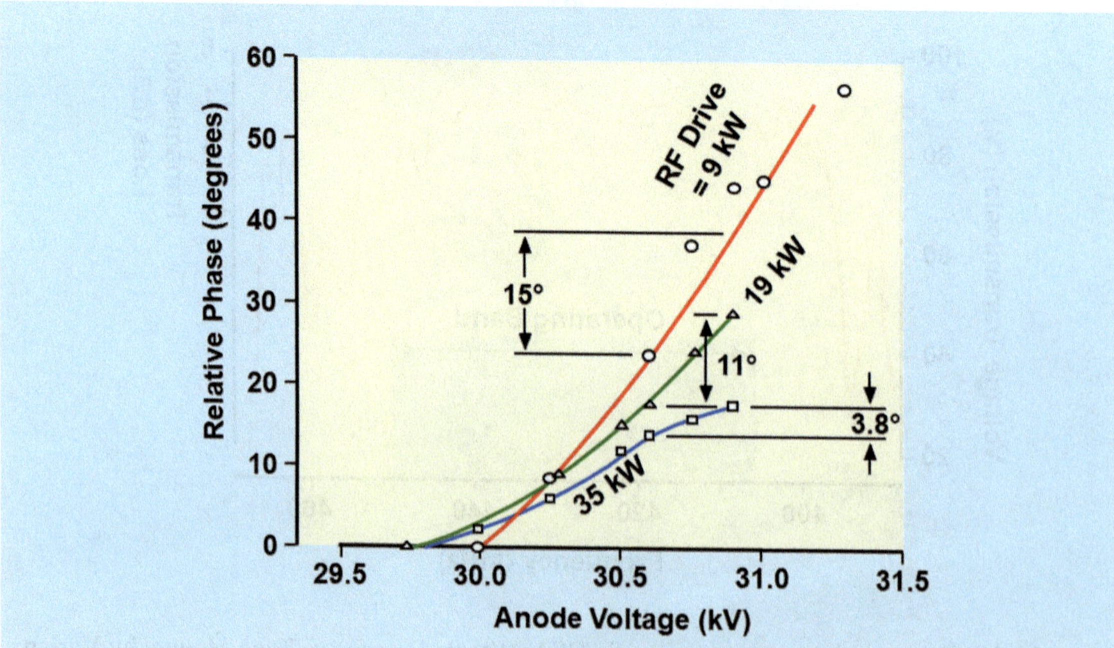

Figure 20-39 Phase variation with voltage for three RF drive levels for an X-band CFA. (From: *Microwave Tube Manual* by Varian Associates, Air Force Publication Number T.O.00-25-251, October 1979.)

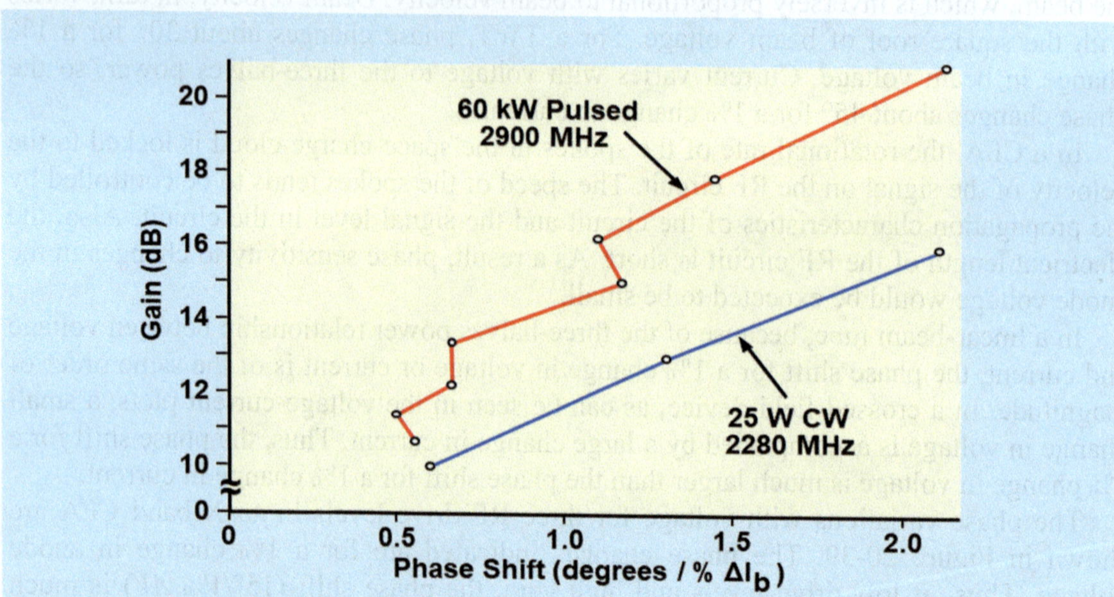

Figure 20-40 Effect of gain on phase sensitivity for low- and medium-power backward-wave CFAs. (From: J. Skowron, *Proc. IEEE*, March 1973. ©1973 IEEE.)

The phase linearity of a CFA with frequency is largely determined by the circuit reflections and the gain of the device. The effect of circuit reflection depends on electrical length and gain. Because the CFA is electrically short (compared with linear-beam devices) and gain is low, phase deviations from linearity of less than ±5° can be achieved.

The favorable phase characteristics of CFAs lead to their use in systems where arrays of devices are used and phasing is important. This is indicated in Figure 20-41. Also indicated are other applications where the transparent characteristic is used (for interpulse feedthrough) or where tubes are readily connected in parallel to upgrade power capability.

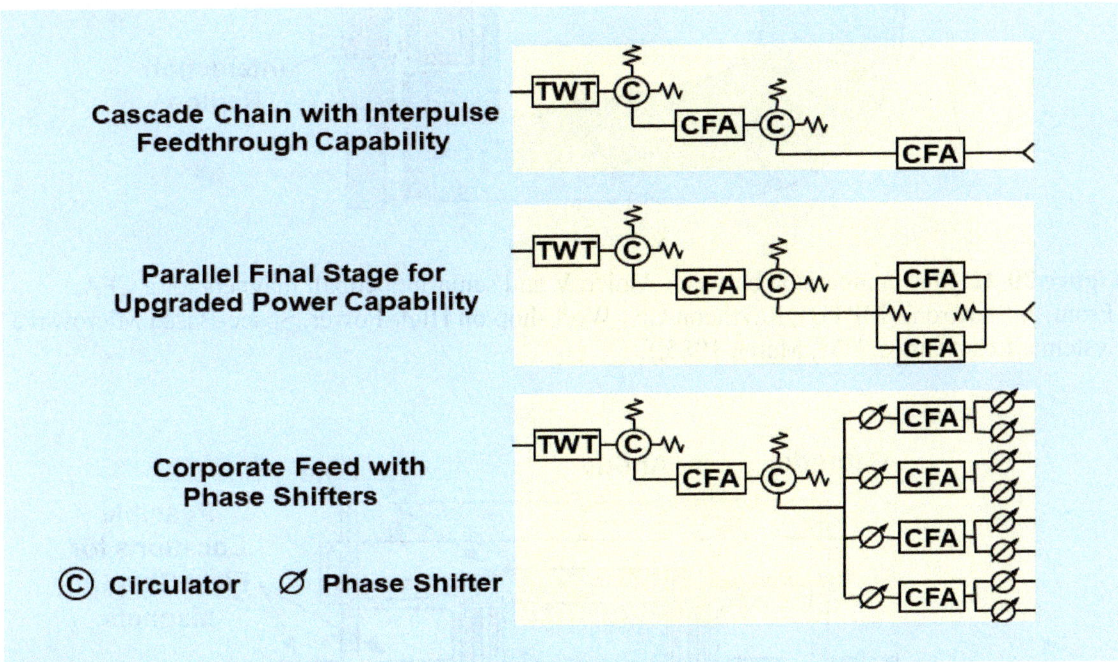

Figure 20-41 Configurations for applications of CFAs. (From: J. Skowron, *Proc. IEEE*, March 1973. © 1973 IEEE.)

20.4.6 Weight and Size Considerations

The use of samarium cobalt can significantly reduce magnet and CFA weight. For example, a size comparison between Alnico V and samarium cobalt magnets is illustrated in Figure 20-42. Still another possibility in cold-cathode CFAs is the placement of the magnet inside the cathode as illustrated in Figure 20-43. The use of samarium cobalt in this manner has been demonstrated in low-frequency CFAs.

20.5 POWER CAPABILITIES

In practice, CFAs are capable of impressive performance. For example, Figure 20-44 shows an efficiency of approximately 50% over a 10% bandwidth for a pulsed C-band amplifier having a power output of about 1 MW.

An important feature of CFAs is the relatively low voltage at which they operate. The voltage of less than 40 kV for the megawatt device in Figure 20-44 is a factor of 2 to 3 below the voltage required by a single-beam linear-beam device at the same power level. (New multiple-beam linear-beam devices operate at voltages comparable to CFAs.)

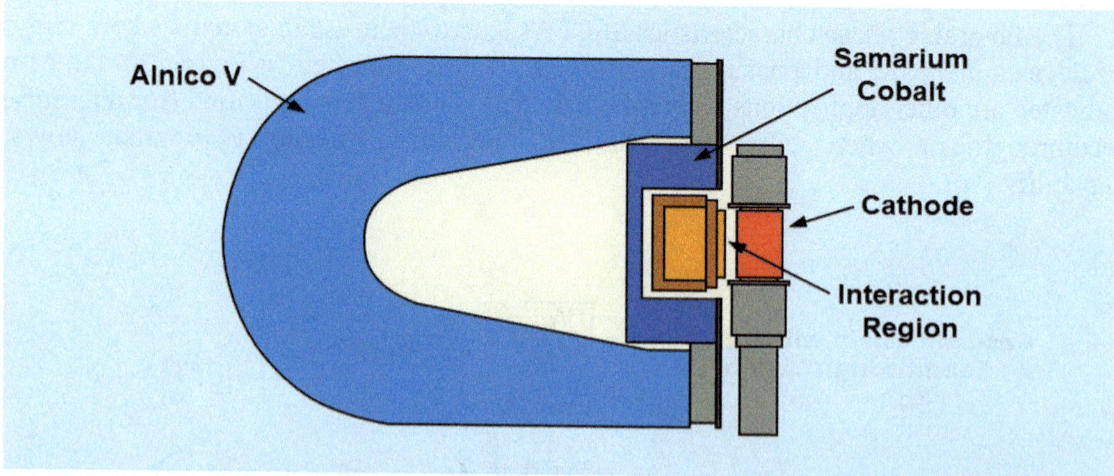

Figure 20-42 Size comparison between Alnico V and samarium cobalt magnets for a CFA. (From: J. Skowron, (MPTD), Raytheon Co., Workshop on High-Power, Space-Based Microwave Systems, Los Alamos, NM, March 1985.)

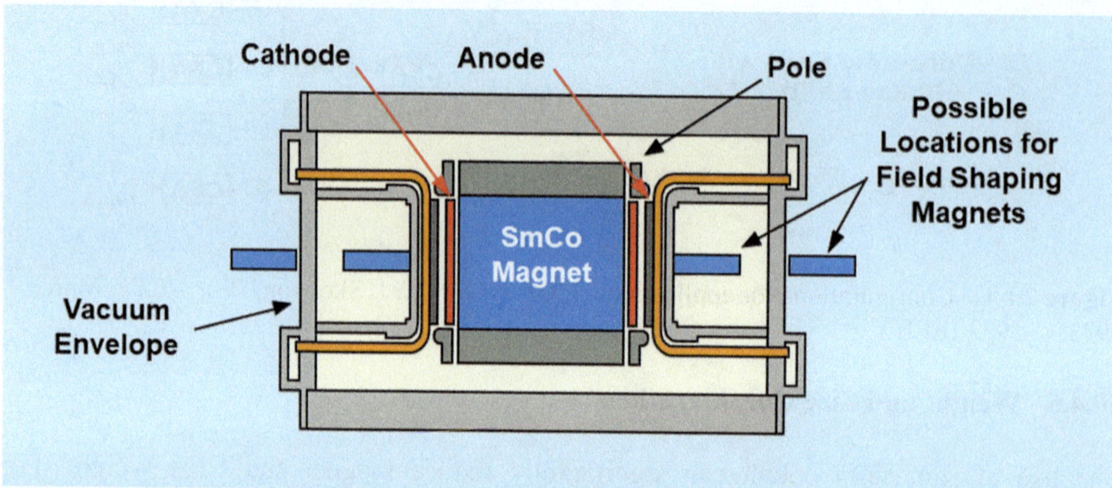

Figure 20-43 Placement of samarium cobalt magnet inside the cathode in cold-cathode CFA. (From: J. Skowron, *Proc. IEEE*, March 1973. ©1973 IEEE.)

The performance characteristics of an extremely high-power (super power) CW CFA operating at 3 GHz with a bandwidth of over 5% are shown in Figure 20-45. Output power was as high as 425 kW and efficiency exceeded 70%. The performance characteristics of a pulsed super-power CFA are shown in Figure 20-46. These high-power tubes were developed for use in very high-power radar systems, which were replaced by the development of active phased array technology using large numbers of lower-powered tubes.

The overall state of the art of publishable power capabilities of CFAs at the present time is comparable to that in 1985 and is shown in Figure 20-47. Note that megawatt power levels are available in pulsed devices at frequencies in excess of 10 GHz.

20.6 THERMAL CONSIDERATIONS

The electron beam in an efficient crossed-field device is collected on the RF circuit so the circuit must be capable of dissipating the spent-beam power. In addition, there may be a significant amount of heating of the cathode by back bombardment so that the cathode may have to be cooled.

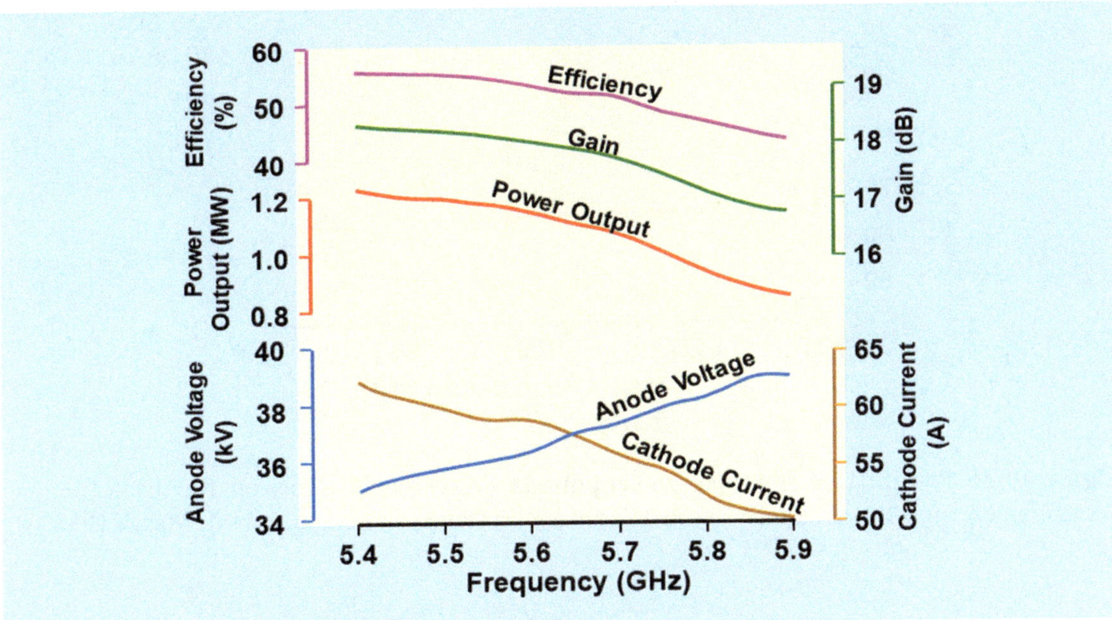

Figure 20-44 CFA performance over a 500-MHz band. (From: *Microwave Tube Manual* by Varian Associates, Air Force Publication Number T.O.00-25-251, October 1979.)

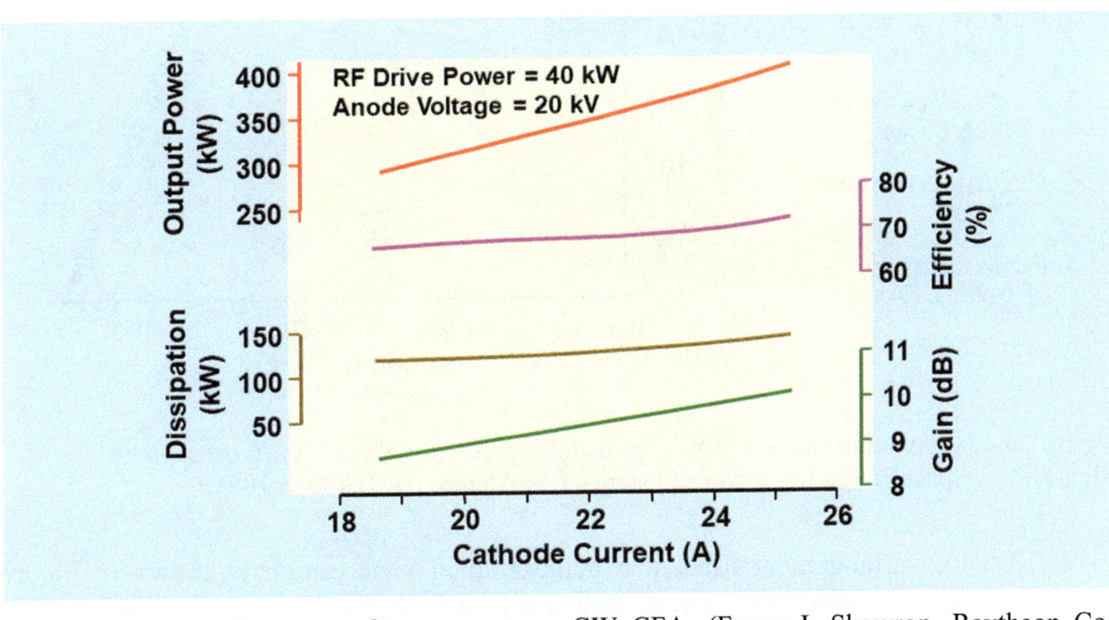

Figure 20-45 Performance of a super-power CW CFA. (From: J. Skowron, Raytheon Co., Workshop on High-Power, Space-Based Microwave Systems, Los Alamos, NM, March 1985.)

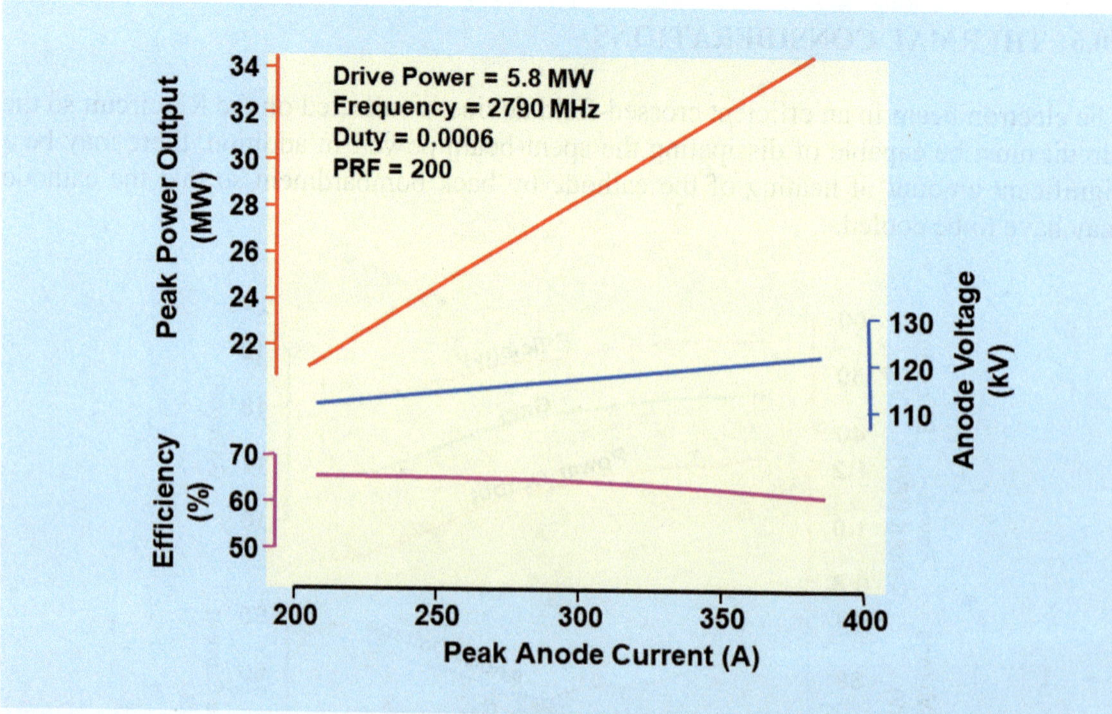

Figure 20-46 Performance of a super-power pulsed CFA. (From: J. Skowron, Raytheon Co., Workshop on High-Power, Space Based Microwave Systems, Los Alamos, NM, March 1985.)

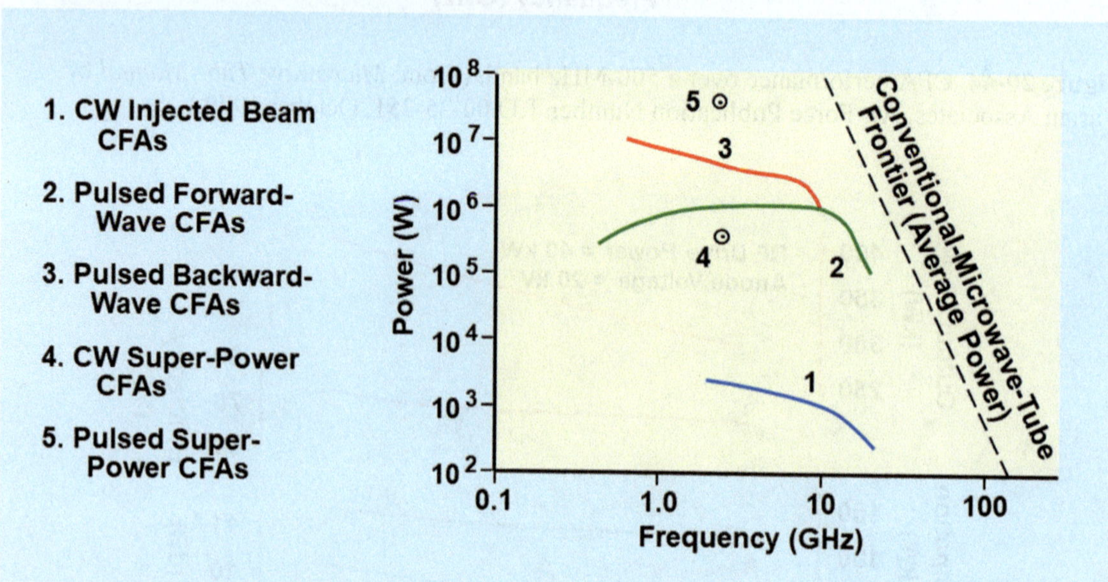

Figure 20-47 State of the art of CFAs. (From: J. Skowron, Raytheon Co., Workshop on High-Power, Space-Based Microwave Systems, Los Alamos, NM, March 1985.)

The cooling arrangement used in a helix-coupled vane circuit is shown in Figure 20-48. In Figure 20-48(a), the thermal path for heat from the electron bombardment of the vane tips is through the entire length of the vanes. For operation at high duty cycles, increased cooling of the vane tips is required and this is provided, as shown in Figure

20-48(b), by injecting coolant into hollow chambers in the vanes. The interception power that is capable of being removed is several kilowatts. The reason for using molybdenum vane tips is that for pulsed operation, the instantaneous heating of the vane tips is so great that copper vane tips are destroyed.

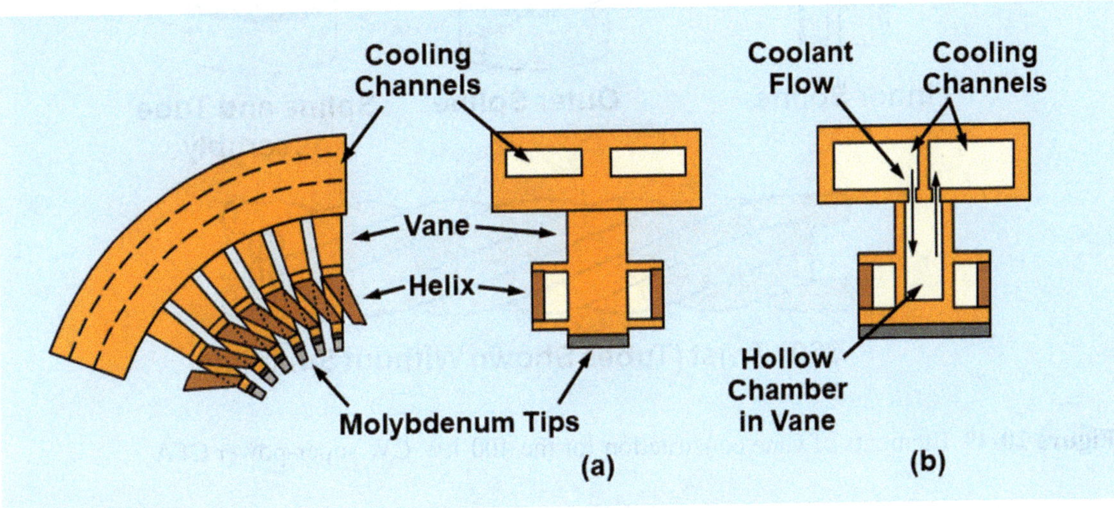

Figure 20-48 Cooling channels in the helix-coupled vane circuit used in forward-wave CFAs.

A cooling arrangement for many CFA circuits, forward-wave and backward-wave, was shown in Figure 20-16. The circuit is fabricated from tubes that are attached to a coolant manifold so coolant is forced through the tubes in parallel. The circuit shown in Figure 20-16 is a forward-wave circuit. By changing the support for the circuit as well as the technique for interconnecting the tubes, the circuit can operate in the backward-wave mode.

The highest dissipation ever achieved in a microwave tube was in the Super Power CFA program carried out by the Raytheon Company [7]. Dissipations of nearly 20 kW/cm^2 were demonstrated. The technique used was a convection process involving a change in phase from a liquid to vapor of the coolant being used. In this process, as the heat is being transferred from the vane to the coolant, it raises the temperature of the coolant to its boiling point, forming a vapor bubble. The bubble then displaces the liquid above, which rushes back to fill the void left by the bubble when it left the hot surface. Although the bubble carries away only a small amount of the total heat being dissipated, it acts somewhat like a pump, carrying hot liquid away from the hot surface and creating intense turbulence within the tubing. For this technique to be used, velocities of approximately 100 feet per second or vortex flow are needed.

The vanes of the CFA were constructed as indicated on Figure 20-49. Copper-clad stainless steel tubes were inserted between the inner and outer splines, short sections of which are shown in Figure 20-49. The resulting spline and tube assembly was then drawn to the final dimension and brazed together. The vane was given a full 360° twist so that each of the six tubes would receive an equal portion of the electron bombardment. (The twist is shown in Figure 20-49 without the splines for clarification.) The twist also gave the water a spiral or vortex rotation, which provided better scrubbing of the bubbles from the wall than could be achieved with straight flow.

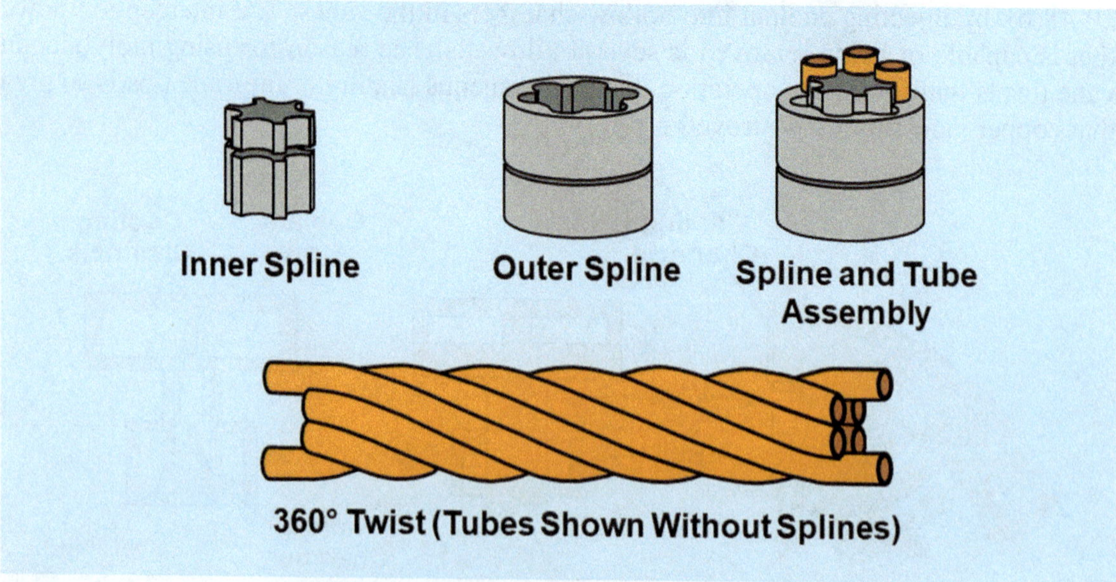

Figure 20-49 Elements of vane construction for the 400-kW CW super-power CFA.

As an example of the way in which power may be divided in a CFA, the power distribution is given in Table 20-1 for the CW CFA, the performance characteristics for which were shown in Figure 20-45. Note that the anode dissipation is over 100 kW and the cathode dissipation is nearly 25 kW.

Table 20.1
Distribution of Power in "Super Power" CW CFA

RF power generated	71.7%	350.0 kW*
Anode dissipation	21.4%	104.5 kW
Cathode dissipation	5.0%	24.6 kW
Transmission line	1.9%	9.3 kW
Totals	100%	488.4 kW

*Note: Drive power is added to power generated for a total output power of 405 kW

Source: J. Skowron (MPTD), Raytheon Co., Workshop on High-Power, Space-Based Microwave Systems, Los Alamos, NM, March 1985.

The severity of the cooling problem becomes apparent when it is realized that the total cross-sectional area of the anode segments ranges from a fraction of a square centimeter in high-frequency devices to a few tens of square centimeters in low-frequency devices. These relatively small areas may result in power densities in excess of a kW/cm^2 that must be dissipated by the anodes.

Table 20-2 contains a listing of the cooling capabilities possible with various techniques. Over 10 kW/cm^2 can be removed by high-velocity water when nucleate boiling occurs on the surfaces being cooled. As a result, this is the cooling technique used in extremely high-power applications.

Shown in Figure 20-50 are the predicted power capabilities of CW CFAs based on anode dissipation capability. The Raytheon QKS 849, which used nucleate boiling to achieve an anode dissipation of 3,600 W/cm^2, is used as a design reference point. Based on this power density, power outputs in excess of 1 MW should be possible at frequencies below about 1,500 MHz.

In pulsed CFAs, electron bombardment of the anode results in a transient temperature rise of the surface during the pulse. The amount of energy that can safely be imparted to the anode is directly related to the heat capacity of the anode material and to the temperature to which the surface may rise.

The calculation of the surface temperature and temperature distribution within an anode during a pulse is relatively straightforward. Because the heat is concentrated in the surface of the anode, a one-dimensional model like that in Figure 20-51 can be used.

Table 20.2

Cooling Capability of Various Media (ΔT = 200°C)

Black body radiation (T ~ 200°C)	0.25 W/cm^2
Air, natural convection to finned forced convection	0.5–110 W/cm^2
Organic oils, natural convection to forced convection, boiling	5–90 W/cm^2
Water, pool boiling	125 W/cm^2
Water, pressurized, nonboiling	155 W/cm^2
Liquid metals, forced convection, nonboiling	450 W/cm^2
Water, high velocity, nucleate boiling Operational Maximum Achieved	 10,000 W/cm^2 20,000 W/cm^2

Source: J. Skowron, (MPTD), Raytheon Co., Workshop on High-Power, Space-Based Microwave Systems, Los Alamos, NM, March 1985.

Shown in Figures 20-52 and 20-53 are temperatures calculated for the case of the 200-μs pulse length, 100 pulses per second, and a peak power density of 67,850 W/cm^2. Notice that the temperature of the face of the anode vane reaches nearly 500°C during the pulse and that the temperature only 0.010 inch into the vane is just 200°C. During the interpulse period, the temperature of the entire vane nearly drops to that of the coolant metal interface. It is because of the high vane surface temperature reached during a pulse that the vane tips of CFAs are sometimes fabricated from a high-temperature metal such as molybdenum.

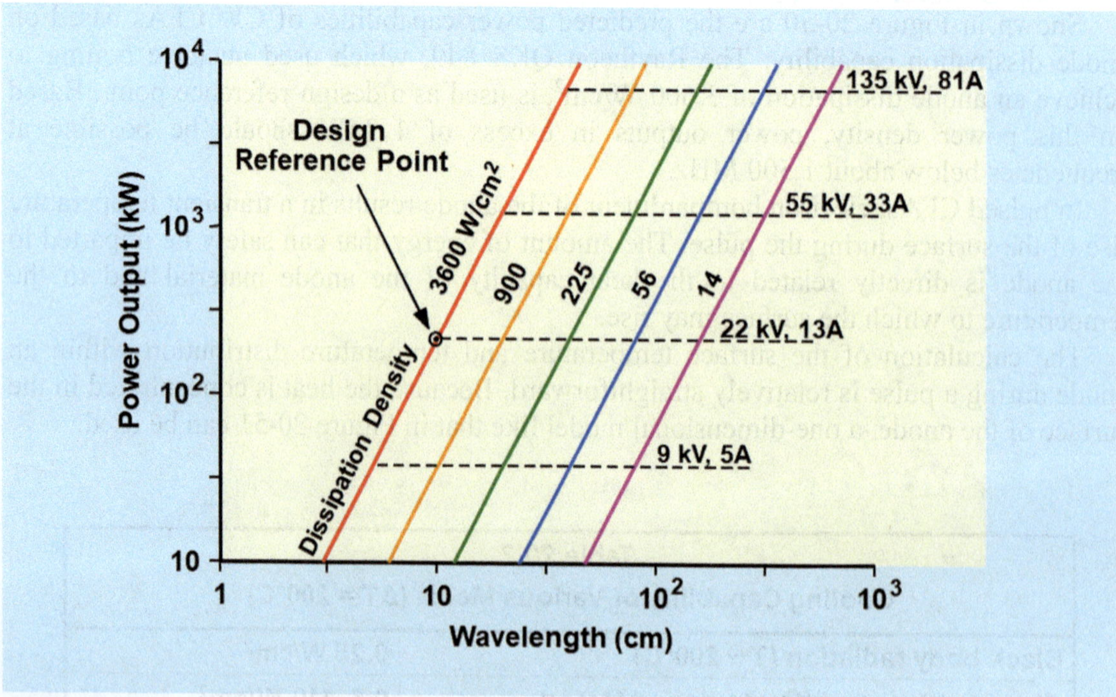

Figure 20-50 Predicted power capabilities of CW CFAs based on anode dissipation power densities. (From: J. Skowron (MPTD), Raytheon Co., Workshop on High-Power, Space-Based Microwave Systems, Los Alamos, NM, March 1985.)

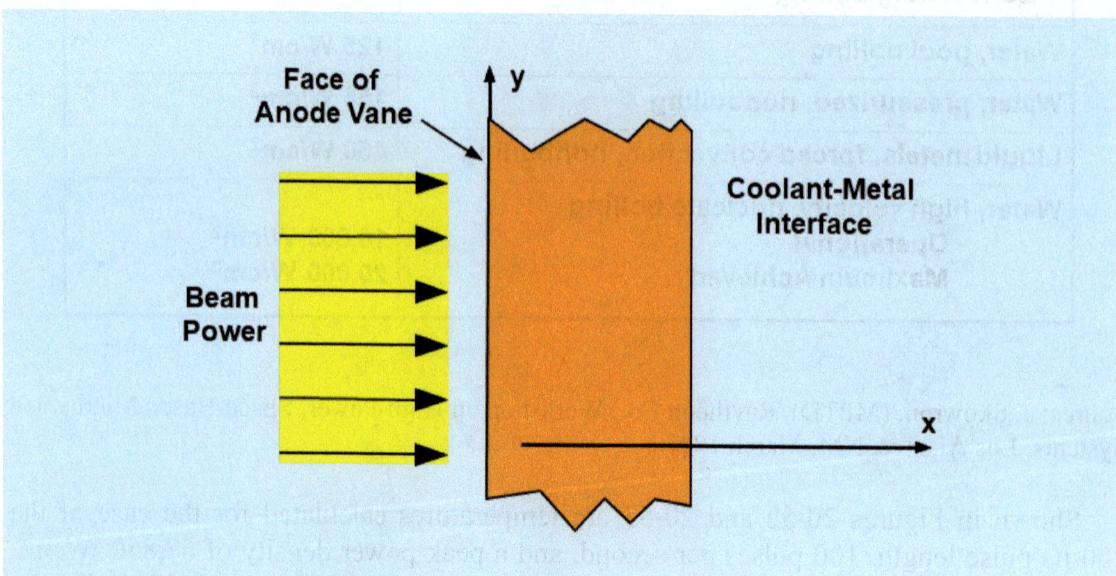

Figure 20-51 Model used for solution of one-dimensional heat flow equation. (From: J. Skowron, *Proc. IEEE*, March 1973. ©1973 IEEE.)

The predicted power capabilities of pulsed CFAs based on the temperature of the vane-tip surface are given in Figure 20-54. For a given surface temperature, as pulse duration is reduced, the peak-power density can be increased before the limiting energy input is reached. At 3,000 MHz, 20 MW has been achieved and is used as the design reference point.

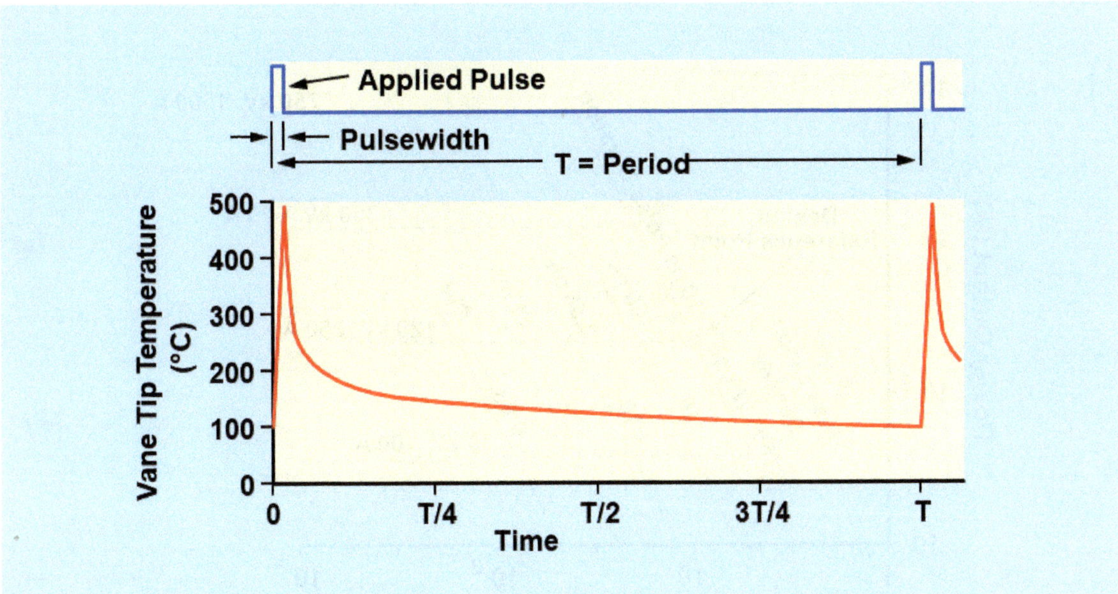

Figure 20-52 Vane tip temperature as a function of time. (From: J. Skowron, *Proc. IEEE*, March 1973. ©1973 IEEE.)

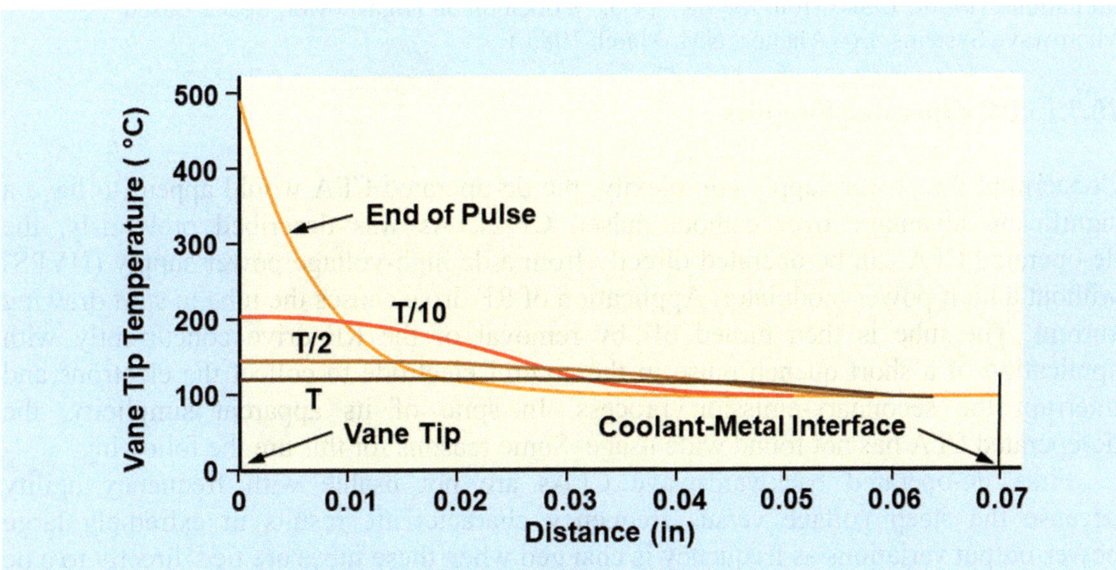

Figure 20-53 Temperature as a function of distance into vane tip. (From: J. Skowron, *Proc. IEEE*, March 1973. ©1973 IEEE.)

20.7 CFA POWER SUPPLY CONSIDERATIONS

The decision concerning whether to use a CFA for a particular application or whether it's a forward-wave device, backward-wave device, pulsed, or dc-operated is based on many factors. Some considerations are the complexity of the power supply [8], the RF power capability, bandwidth, impedance, cost, weight, and size.

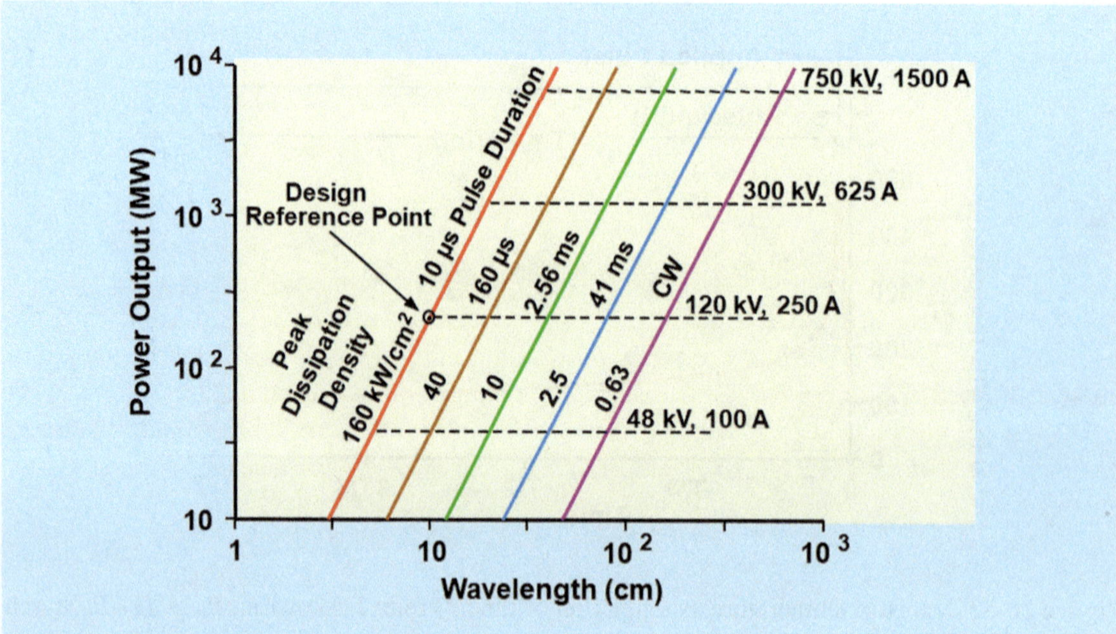

Figure 20-54 Predicted power capabilities of pulsed CFAs based on anode surface temperature limitations. (From: J. Skowron, Raytheon Co., Workshop on High-Power, Space-Based Microwave Systems, Los Alamos, NM, March 1985.)

20.7.1 DC-Operated Supplies

Concerning the power supply complexity, the dc-operated CFA would appear to have a significant advantage over cathode-pulsed CFAs. As was described previously, the dc-operated CFA can be operated directly from a dc high-voltage power supply (HVPS) without a high-power modulator. Application of RF drive causes the tube to start drawing current. The tube is then turned off by removal of the RF drive concurrently with application of a short quench pulse to the control electrode to collect the electrons and interrupt the secondary-emission process. In spite of its apparent simplicity, the dc-operated CFA has not found wide usage. Some reasons for this are the following.

First, dc-operated backward-wave CFAs are not usable with frequency agility because the steep voltage versus frequency characteristic results in extremely large power-output variations as frequency is changed when these tubes are tied directly to a dc supply. For typical backward-wave CFA characteristics, the power output would vary 100% for a 10% change in frequency. A suitable regulator tube that would be capable of overcoming this variation on a pulse-to-pulse basis could also serve as a cathode pulser.

Second, the low dynamic impedance of a CFA sets a limit of only ~1% voltage droop on the HVPS if the RF power-output droop is to be limited to about 10%. A regulator used between the HVPS and the CFA to absorb HVPS droop (and to interrupt arcs) would have to be rated for the full peak current and voltage of the CFA. Thus, the regulator would be just as big as a hard-tube modulator for a cathode-pulsed CFA and would eliminate any advantage of dc operation.

A regulator preceding the HVPS capacitor bank can be considered, but then the CFA is connected directly to the capacitor bank. A CFA arc will draw a very large current that must be promptly interrupted to avoid damage to the CFA. A crowbar can protect the CFA by diverting the energy stored in the capacitor bank, but the power supply must then be recycled to restore operation. As a result, an arc in dc operation may result in interruption of CFA operation for a period of milliseconds to seconds rather than for a single pulse.

20.7.2 Cathode-Pulsing Supplies

If cathode pulsing is used, it doesn't matter very much to the modulator designer whether the voltage of the CFA varies over the band or not. As will be discussed shortly, this conclusion is reached because the types of modulators that would normally be used with cathode-pulsed CFAs can readily absorb the voltage variations of the backward-wave CFA, and they still hold the operating power of the CFA adequately constant. The basis for choice between forward-wave and backward-wave CFAs for cathode pulsing would thus be other characteristics such as availability, efficiency, cost, and life.

For cathode-pulsed CFAs, several types of high-power modulators may be used. For one or a few fixed-pulse widths, low duty cycle, and moderate PRF, a line-type modulator is suitable and has been widely used. Regulation of the CFA operating power, as well as attenuation of HVPS ripple, can be accomplished in several ways. Line-type modulators readily tolerate CFA arcing.

Among the hard-tube modulator alternatives, constant-current operation is attractive because the modulator acts as a regulator and because it limits the effects of CFA arcs, as in the line-type modulator. Besides attenuating power-supply ripple, the regulating effect of a constant-current, hard-tube modulator tends to hold CFA current constant in spite of HVPS voltage droop during the pulse. This makes it possible for a smaller capacitor bank to be used. Therefore, either a line-type modulator or a constant-current, hard-tube modulator is generally preferred for cathode-pulsed CFAs. As noted earlier, both of these types of modulators are able to tolerate pulse-to-pulse CFA voltage variations. The constant-current, hard-tube modulator maintains constant CFA current, while the line-type modulator (being a matched source) maintains constant power to the CFA as its voltage varies.

REFERENCES

[1] G. K. Farney, "Crossed-field microwave devices," Technology Service Corporation/ Crane Division Naval Surface Warfare Center, 1995.

[2] R. R. Warnecke et al., "Magnetron-type traveling-wave amplifier tube," *Proc. IRE*, Vol. 38, No. 5, May 1950, pp. 486–495.

[3] J. F. Skowron, "The continuous cathode (emitting sole) crossed-field amplifier," *Proc. IEEE*, Vol. 61, No. 3, March 1973, pp. 330–336.

[4] W. C. Brown, "Description and operating characteristics of the platinotron-a new microwave tube device," *Proc. IRE*, Vol. 45, 1957, pp. 1209–1222.

[5] NAVSEA Crane, "AEGIS microwave tubes," NAVSURFWARCENDIV.

[6] Litton, "Operation of the crossed-field amplifier (CFA)," Crossed-Field Amplifier Dept. 589, Litton Electron Tube Division, Williamsport, PA.

[7] Raytheon Company, "Ultra high-power amplitron," Vol. 1, CW Amplitron Development, Rome Air Development Center, Griffiss Air Force Base, New York, Technical Report No. RADC-TDR 64-389, May 1965.

[8] T. A. Weil, "Comparison of CFAs for pulsed-radar transmitters," *Microwave Jour.*, No. 3, June 1973.

Chapter 21

Gyrotrons

21.1 INTRODUCTION

Russian scientists at IAP Nizhny Novgorod, R.A.S., operated the first gyrotron in September 1964. The name *gyrotron* was originally used by the Russians for a single-cavity oscillator, now often referred to as a gyro-monotron. The name now refers to a class of devices including both oscillators and amplifiers. As shown in Figure 21-1, there is a one-to-one correspondence between linear-beam (O-type) tubes (klystrons, TWTs, and so forth) and gyro-devices [1].

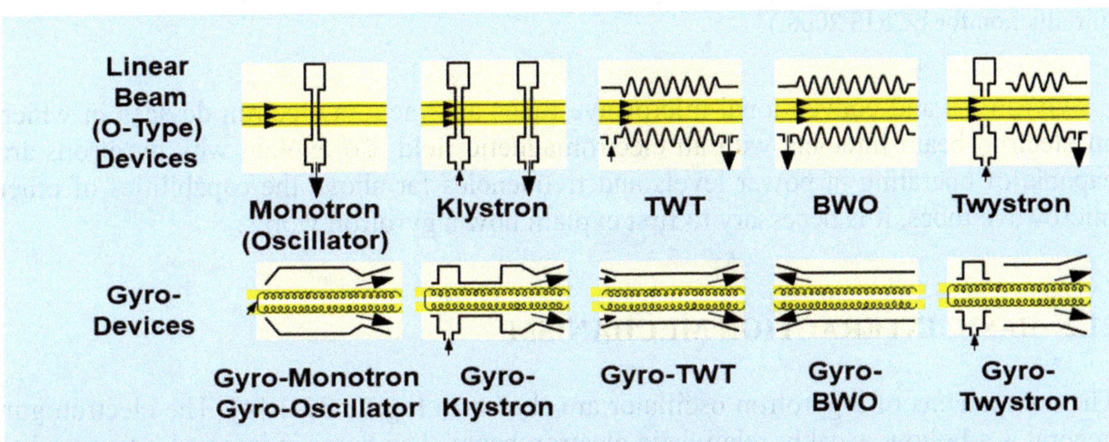

Figure 21-1 Linear-beam devices and corresponding gyro-devices. (From: V. A. Flyagin, et al., *IEEE Trans. Microwave Theory and Techniques,* Vol. 25, No. 6, 1977. © 1977 IEEE.)

Gyrotrons are fast-wave devices that fill an important gap in the spectrum of electromagnetic sources. As indicated in Figure 21-2, the possible operating capabilities of gyrotrons far exceed those of conventional microwave tubes in both frequency and power [2]. The upper limit, indicated by the dashed line, is estimated with the relation that is valid for conventional microwave tubes, which is that the product of power (MW) times frequency squared (GHz^2) is a constant. The value of 5×10^4 shown in the figure is based on the development of a 170-GHz, 2-MW CW gyrotron now in process for the International Thermonuclear Experimental Reactor (ITER) [3].

Either conventional microwave tubes or gyrotrons could operate in the yellow-shaded region between these devices. As gyrotrons develop and the needs arise, it is becoming more likely that gyrotrons will be the devices of choice for operation in this region.

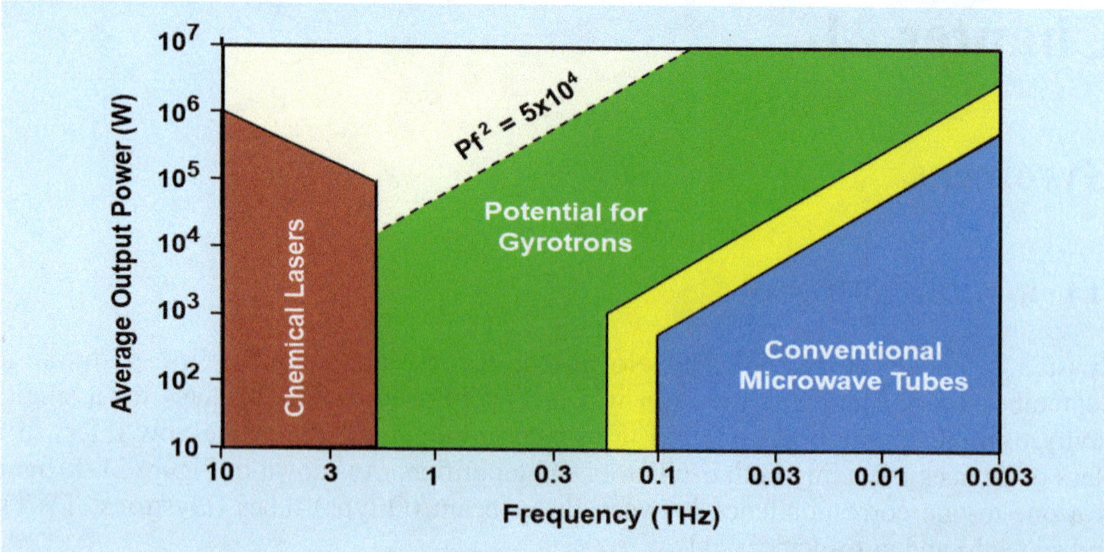

Figure 21-2 The operating region for gyrotrons. (Adapted from: H. Jory et al., Gyrotron Introduction for ECRIS 2008.)

Gyrotrons and conventional microwave tubes are vacuum electron devices in which an electron beam interacts with an electromagnetic field. To explain why gyrotrons are capable of operating at power levels and frequencies far above the capabilities of other microwave tubes, it is necessary to first explain how a gyrotron works.

21.2 BASIC INTERACTION MECHANISM

The components of a gyrotron oscillator are shown in Figure 21-3 [4]. The electron gun generates a hollow, weakly relativistic electron beam. The beam is injected into a cavity, which is immersed in a longitudinal magnetic field. The electrons in the beam spin at the electron cyclotron frequency that is given by

$$\omega_c = \frac{e}{m}B \tag{21-1}$$

where m is the relativistic electron mass and B is the magnetic flux density.

Now, assume, as indicated in Figure 21-3, that there is a transverse electric (TE) electromagnetic field oscillating at ω_c in the cavity. As the beam moves through the cavity, the spinning electrons interact with and amplify the field. In gyrotrons operating at low and modest power levels, the beam and the RF power leave the cavity and travel through a cone-shaped expanding waveguide. The beam is collected on the walls of the waveguide and the RF propagates out of the gyrotron through the output waveguide and window. In high-power gyrotrons, the RF is separated from the electron beam and directed through an output window. The beam continues on to the collector, which may be operating at a depressed potential to recover energy from the spent beam.

Now, the question is: how does the interaction between the electrons take place so that electron energy is transferred to the electromagnetic fields?

The electron gun and beam-focusing fields are designed so that much of the energy of the electrons is in the rotational motion. Because the magnetic field is very large, ω_c is very high and the orbit diameter for the electrons is small. Because the orbit diameter is small, as indicated in Figure 21-4, the thickness of the hollow beam may be several times the diameter of the electron orbit. In effect, the hollow beam contains a large number of small beams, often referred to as beamlets [5].

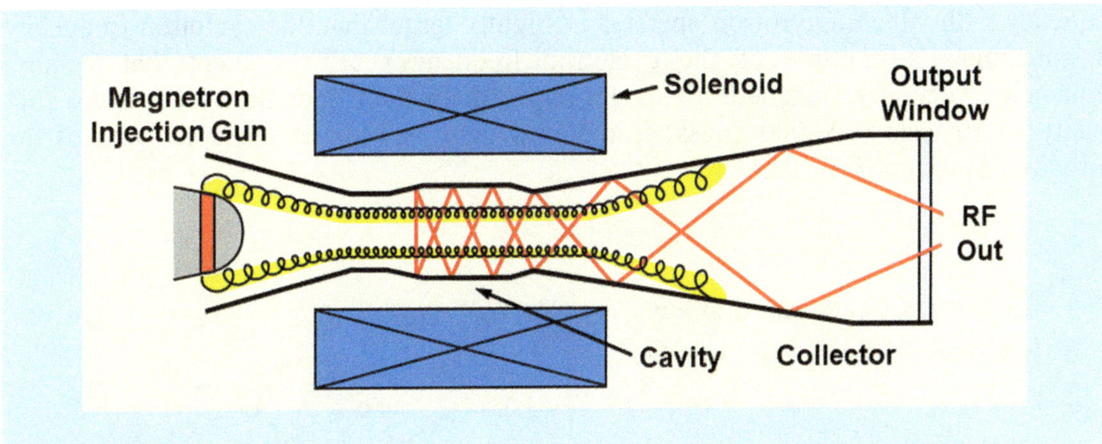

Figure 21-3 Gyrotron oscillator. (Adapted from: A. Grudiev, Hamburg University of Technology, 2002.)

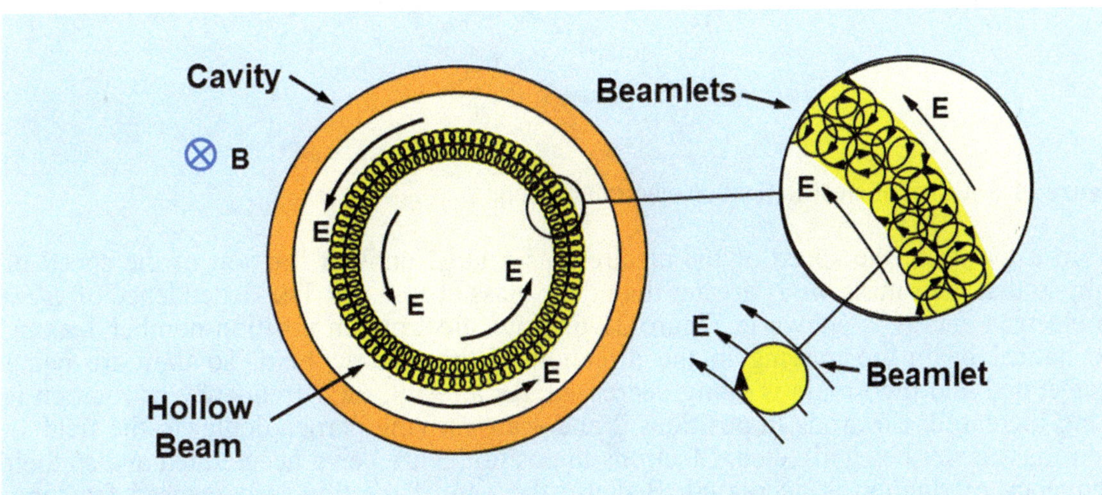

Figure 21-4 Hollow electron beam in a gyrotron cavity. (Adapted from: G. Dohler, *Int. J. Electronics,* 1983.)

As is the case in any oscillator, at the initiation of oscillations, an electromagnetic wave originates from the broadband noise that the electron beam generates. In that noise spectrum, there is a frequency component at the cyclotron frequency. In the cavity, the rotating electrons interact with the circular electric field of the electromagnetic wave. The way in which electrons interact with the wave is at the heart of the operation of all gyro-devices.

The basic operating mechanism may be explained by examining the interaction of the electrons in a single beamlet like that in Figure 21-5 with the electric field. Assume that the electrons in the beamlet are uniformly distributed along a single helical path prior to interaction. Also, assume for this example that the electrons rotate in the counterclockwise direction. The angular rotational frequency of the electrons is the cyclotron frequency, ω_c.

Some values of the frequency of rotation, $f_c = \omega_c/2\pi$, for nonrelativistic electrons are listed in Table 21-1 along with the corresponding magnetic flux densities. Since the frequency with which a gyrotron operates is slightly larger than the cyclotron frequency (or sometimes a multiple of the cyclotron frequency), it is evident that at high frequencies, very large magnetic fields are required for their operation. Note that a flux density of 10 tesla (100,000 gauss) is required for operation at the upper end of the mm-wave spectrum, 300 GHz.

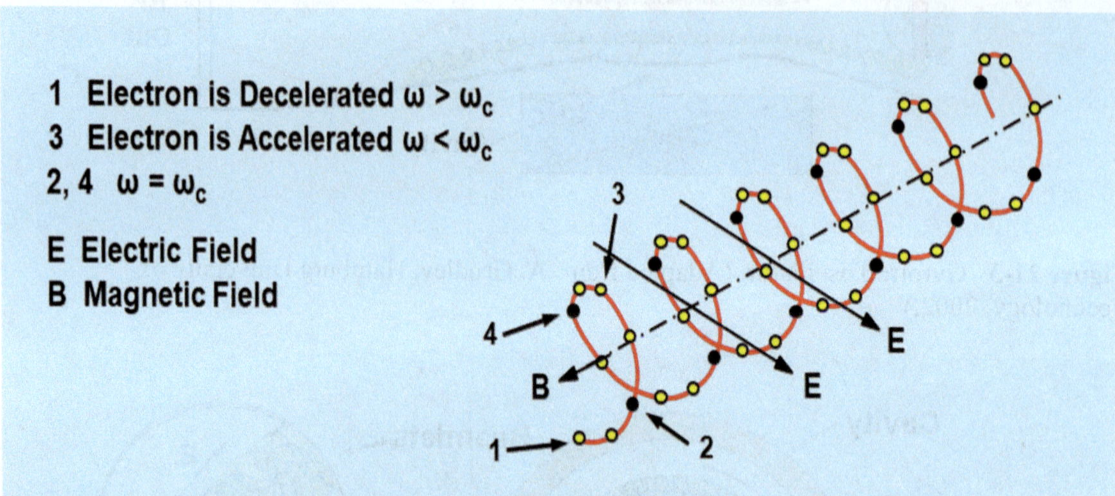

Figure 21-5 Effect of electric field on electron motion.

In a gyrotron, the speed of the electrons is a large enough fraction of the speed of light so that their mass, m, is greater than their mass at rest, m_o. The dependence of m/m_o on electron energy is shown in Figure 21-6. Now, electrons in position number 1 along the helical beam are moving in the direction of the electric field, so they are being decelerated and their mass is being decreased. As a result, their frequency of rotation is being increased. Electrons in positions 2 and 4 are moving perpendicular to the field so their mass is not being affected. Electrons in position 3 are being accelerated and so their frequency of rotation is decreased. Because the rate of rotation is decreased for some electrons and is increased for others, orbital bunching occurs as indicated in Figure 21-7 if the electrons are permitted to drift [6].

Table 21.1

Magnetic Flux Density and Corresponding Cyclotron Frequency		
Flux Density		Cyclotron Frequency
Tesla	Gauss	GHz
0.0357	357	1
0.107	1,071	3
0.357	3,570	10
1.07	10,700	30
3.57	35,700	100
10.7	107,000	300
35.7	357,000	1000

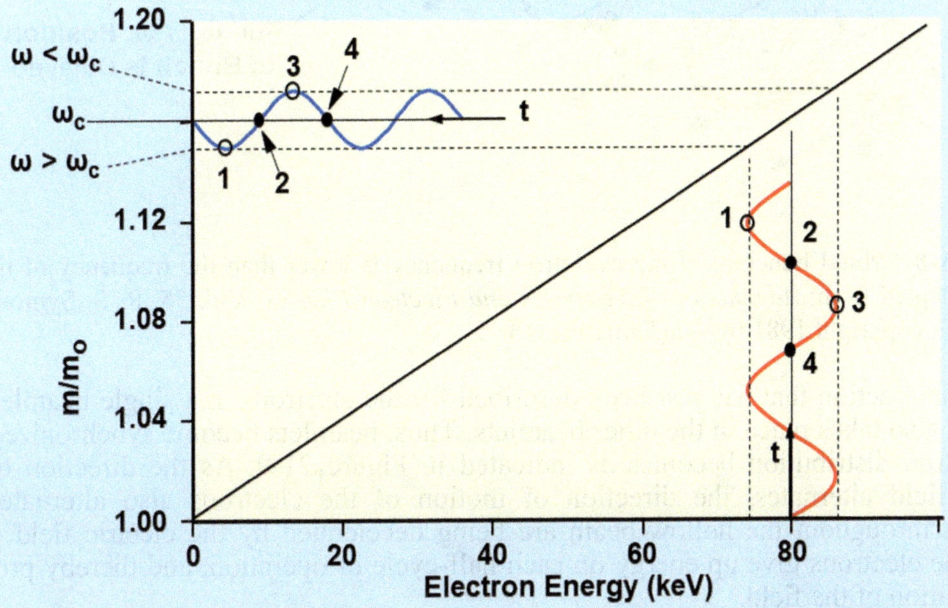

Figure 21-6 Variation of electron mass and frequency of rotation with electron energy.

If the cyclotron frequency is somewhat lower than the frequency of the RF field, then the position of the bunches along the helical orbit is delayed with respect to phase of the applied field as indicated in Figure 21-8. At this point, it should be noted that the bunched electrons are all moving in the direction of the electric field, and so they all give up energy to the field. As the electron bunches rotate in near synchronism with the alternating RF field, they continue to give up energy on each half-cycle of rotation.

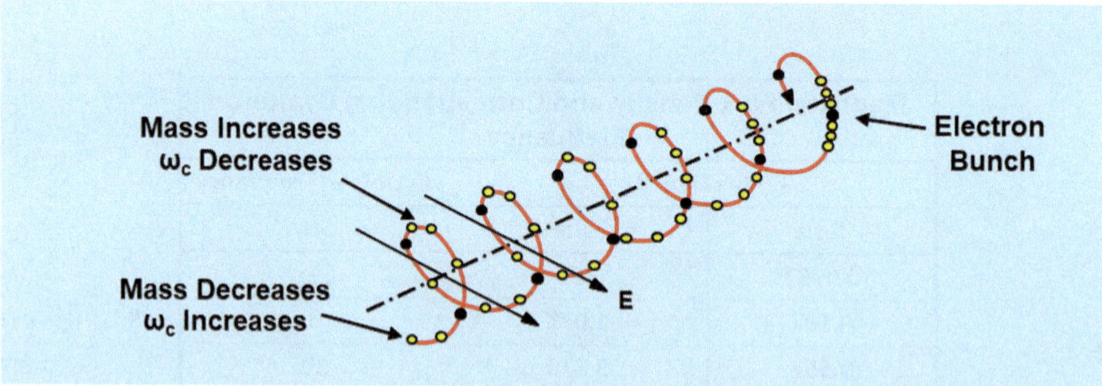

Figure 21-7 Orbital bunching of electrons. (Adapted from: *Advances in Electronics and Electron Physics,* Vol. 55, R. S. Symons and H. R. Jory, copyright 1981 by Academic Press.)

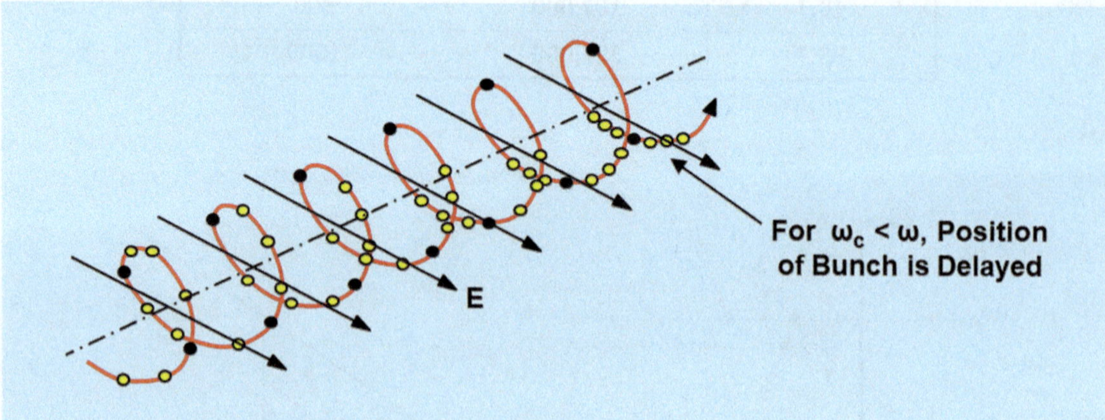

Figure 21-8 Orbital bunching if the cyclotron frequency is lower than the frequency of the RF field. (Adapted from: *Advances in Electronics and Electron Physics,* Vol. 55, R. S. Symons and H. R. Jory, copyright 1981 by Academic Press.)

The interaction that has just been described for the electrons in a single beamlet in a gyrotron also takes place in the other beamlets. Thus, beamlets become synchronized and the electron distribution becomes as indicated in Figure 21-9. As the direction of the electric field alternates, the direction of motion of the electrons also alternates, so electrons throughout the hollow beam are being decelerated by the electric field. As a result, the electrons give up energy on each half-cycle of operation, and thereby produce amplification of the field.

The following is a comparison of some of the characteristics of gyrotrons and conventional microwave tubes.

Conventional Tubes:

1. The RF fields with which the electrons must interact are most intense near the circuit surrounding the beam, so the beam must travel as close to the circuit as possible. This leads to beam interception and heating of the circuit.
2. Circuit dimensions are on the order of the wavelength of the operating frequency so power is severely limited at high frequencies.
3. Cathodes operate space-charge-limited.

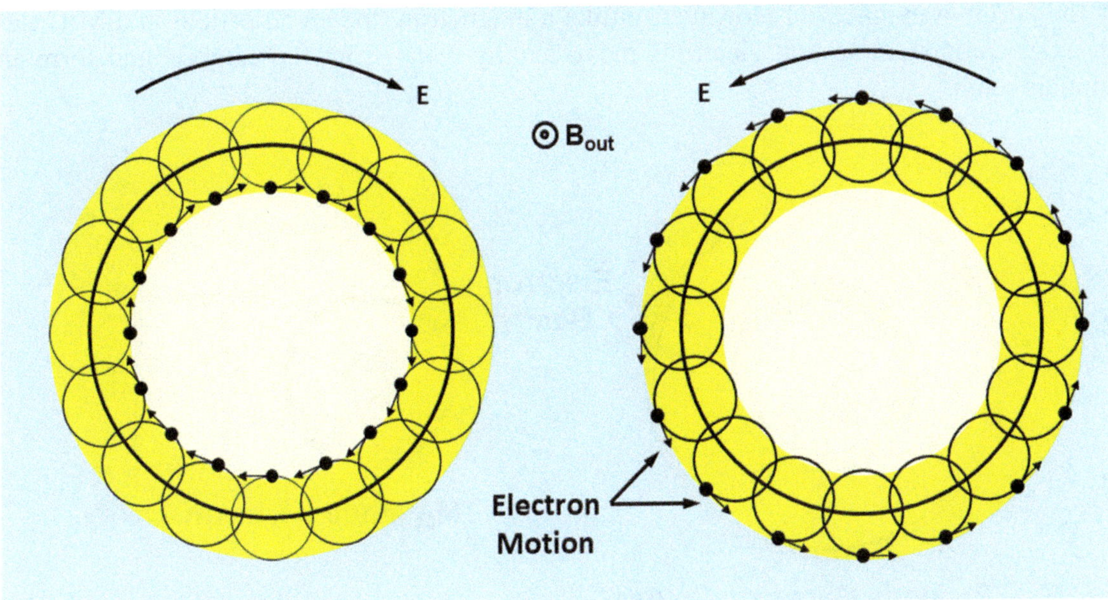

Figure 21-9 Orbiting electrons rotate in synchronism with the alternating electric field and continuously give up energy to the field. (From: Varian brochure *Introduction to Gyrotrons.*)

Gyrotrons:

1. The frequency is not defined by the cavity geometry, but by the cavity magnetic field;
2. The cavity geometry is chosen to excite a TE mode, which is close to cutoff and has low wall losses;
3. The field with which the beam must interact is most intense some distance from the cavity surface. Beam interception problems are far less severe than in linear-beam tubes;
4. Because of the mode structure of the electric field in gyrotrons, the cavities in which interaction takes place can be very large (many wavelengths in diameter);
5. Gyrotron cathodes operate temperature-limited and so they don't have a space-charge cloud near the cathode surface to help correct for nonuniformities in emission.

We can now understand why a gyrotron can operate at power levels and frequencies far above the capabilities of conventional microwave tubes. Figure 21-10 shows a comparison of the sizes of a gyrotron cavity and a klystron cavity for the same frequency of operation. For operation with higher-order modes than the TE_{01} mode, gyrotron cavities are even larger than the TE_{01} cavity.

21.3 MAGNETRON INJECTION GUNS

In Section 21.2 the basic gyrotron interaction process was described. To make this process work properly, an annular electron beam with very specific characteristics is required. To generate this beam, most gyro-devices use an electron gun that is referred to as a magnetron injection gun (MIG). The reference to magnetron arises because most versions of these guns have a cathode, which is similar to the cathode in a magnetron. As shown in Figure 21-11, the cathode operates in a magnetic field that is nearly axial and an electric field that is nearly radial so that electrons follow cycloidal orbits near the cathode as they do in a magnetron. However, unlike a magnetron, the electric field in a MIG has an axial component so that electrons move axially away from the cathode and form an annular beam.

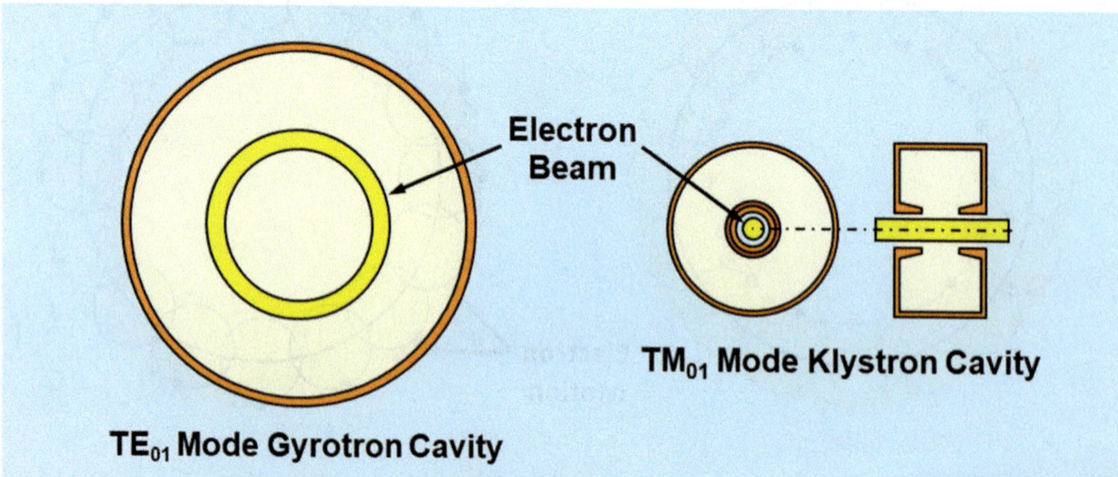

Figure 21-10 Comparison of a gyrotron cavity and a klystron cavity for the same frequency of operation.

21.3.1 MIG Configurations

The type of interaction cavity and the beam required for the desired mode of interaction in that cavity determine, to a large extent, the configuration of the MIG. A MIG can have the conventional design used in most gyrotrons in which the interaction cavity has no center conductor (referred to as a hollow cavity) or it can be a coaxial MIG for use in a gyrotron with a coaxial interaction cavity.

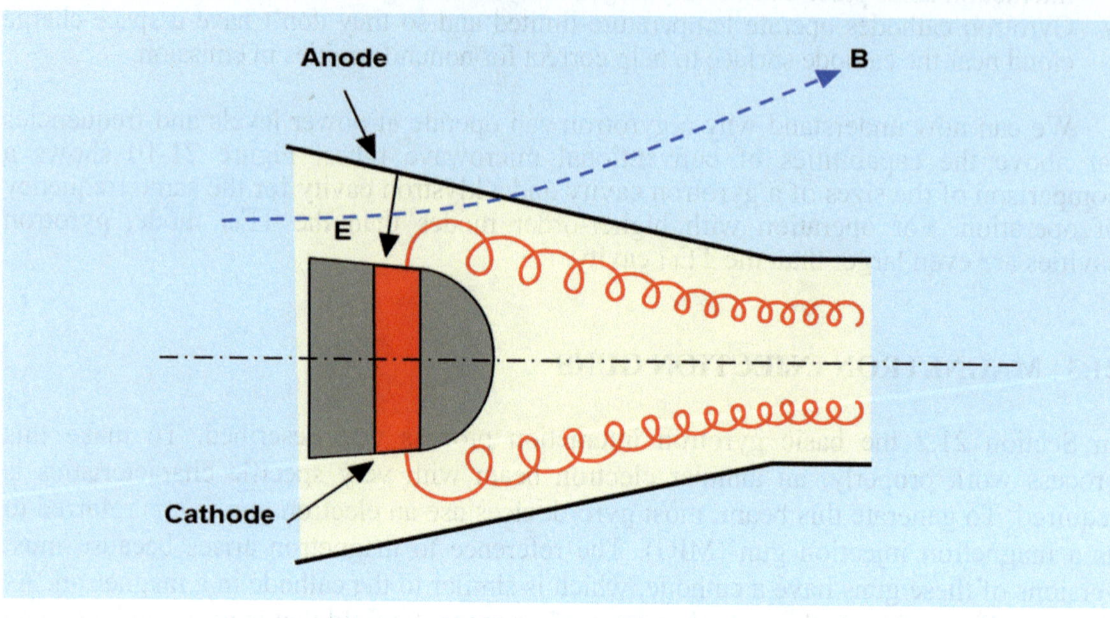

Figure 21-11 Electric and magnetic fields near the cathode in a MIG.

Examples of cathodes for conventional MIG designs are shown in Figure 21-12 [7]. Depending on the slant angle of the cathode surface, the electron, flow can be nonlaminar, as shown in Figure 21-12(a, b) where electron trajectories cross, or "laminar" as in Figure 21-12(c). (Electron flow is never truly laminar because of the emitter characteristics such as surface roughness, work function variations, and thermal emission velocities.)

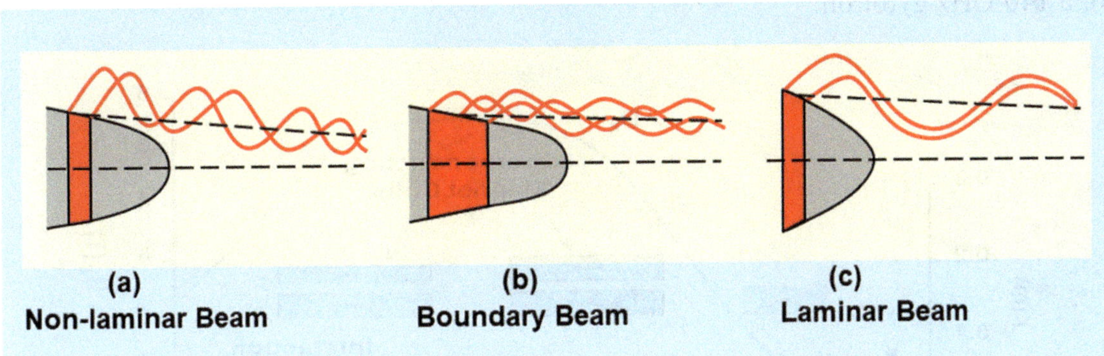

Figure 21-12 Cathodes for conventional MIGs. (Adapted from: *Introduction to the Physics of Gyrotrons*, G. Nusinovich, © 2004 The Johns Hopkins University Press.)

The gun in Figure 21-13 has a cathode with a very steep slant angle (50°) [8]. This angle was critical in minimizing the transverse velocity spread of the electrons. Also, the beam is relatively insensitive to electrode misalignments and field errors. A transverse relative velocity spread of 1.6% and a transverse-to-axial velocity ratio of 1.5 was obtained in simulations for a 6-A, 65-kV beam.

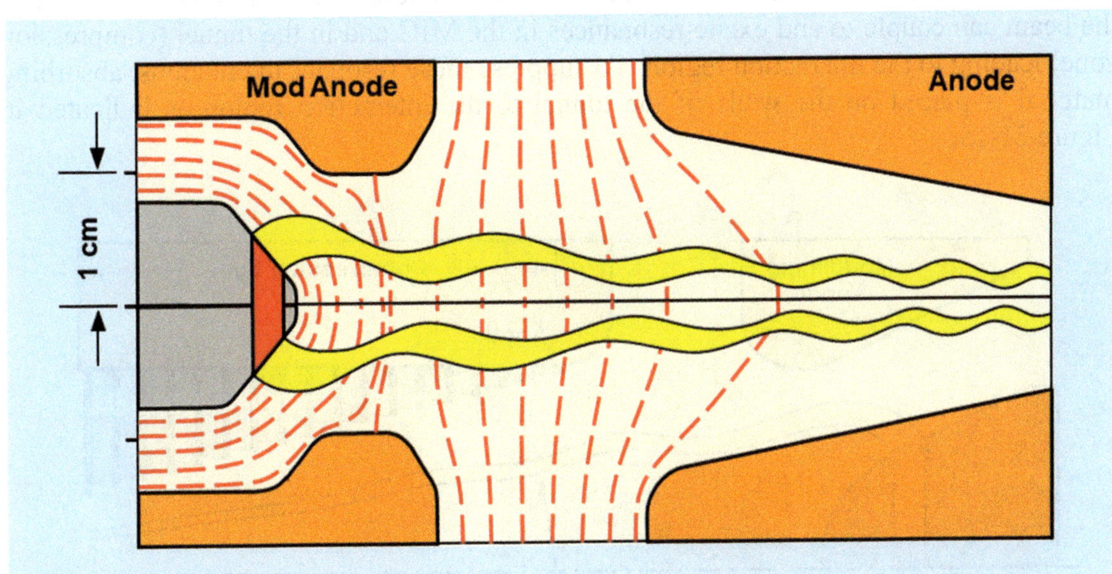

Figure 21-13 MIG design for a 94-GHz gyro-klystron. (Adapted from: Nguyen et al., *IEEE Trans. Plasma Science*, 1998. © 1998 IEEE.)

The MIG in Figure 21-13 was designed for a 94-GHz, 80-kW peak power, 10-kW average power gyro-klystron, the cavities of which operated in the TE_{01} mode. This MIG design employs an optimized double-anode geometry. The modulating anode is used to vary the ratio of the transverse velocity of the electrons to the axial velocity.

The magnetic field in a gyrotron focuses the electron beam and also establishes the frequency of operation. The variation in the field as the beam travels from the MIG to and through the interaction region is enormous. This variation is indicated in Figure 21-14 for a 140-GHz gyrotron.

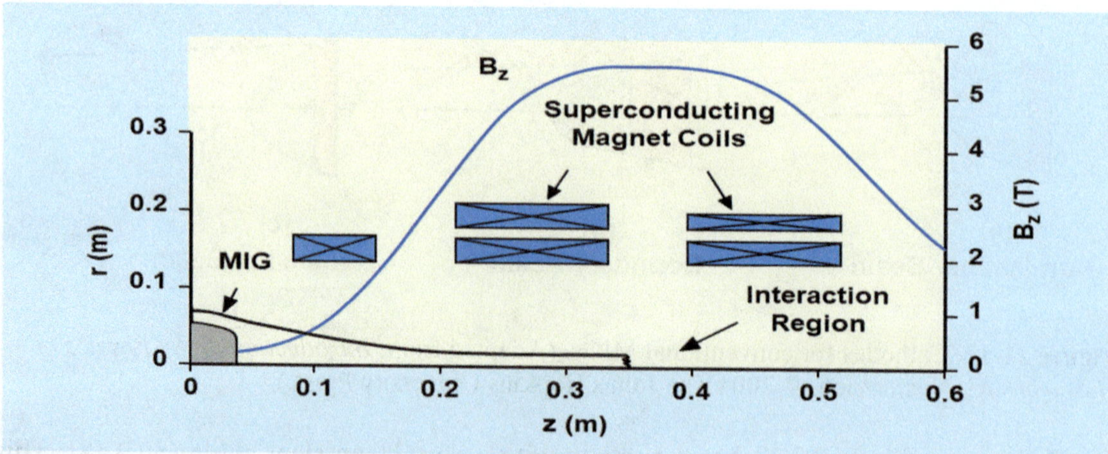

Figure 21-14 Magnetic flux density as a function of axial position for a 140-GHz gyrotron. (Adapted from: *Gyrotrons*, M. V. Kartikeyan et al., © Springer-Verlag 2004.)

The rotation frequency of the electrons varies over an extremely wide range as the beam is compressed and travels from the MIG to the interaction region. Because of this, the beam can couple to and excite resonances in the MIG and in the tunnel (compression zone) leading to the interaction region. To suppress these resonant interactions, absorbing material is placed on the walls of the tunnel to the interaction region as indicated in Figure 21-15.

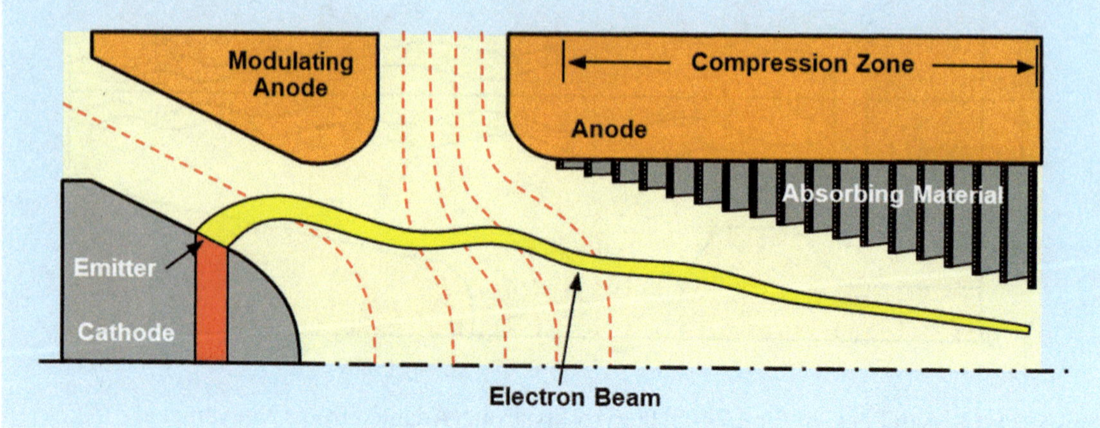

Figure 21-15 Placement of absorbing material for suppressing resonances. (Adapted from: *Gyrotrons*, M. V. Kartikeyan et al., © Springer-Verlag 2004.)

21.3.2 First-Order Design

As is the case with linear-beam tubes, the design of the electron gun depends on numerical simulation to determine beam performance and ensure proper interface with the RF circuit. Also, it is very important to have a good starting design before proceeding with the simulation. This is the value of the Vaughan synthesis procedure for Pierce guns (Chapter 6).

Baird and Lawson [9] provided seven equations for obtaining a first-order design for the MIG configuration shown in Figure 21-16. The following material is from their analysis.

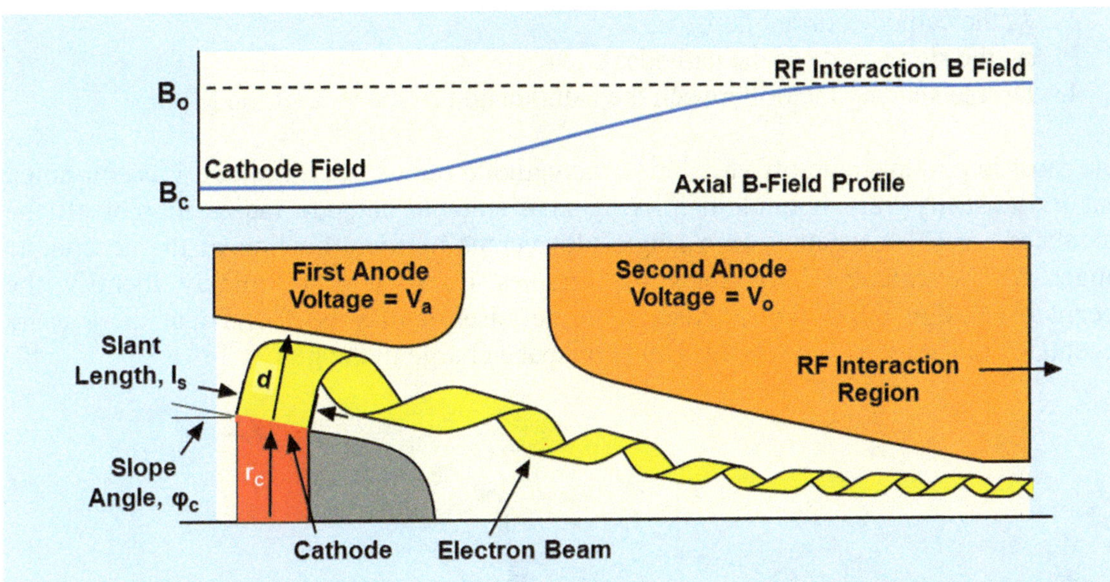

Figure 21-16 MIG configuration used by Baird and Lawson in obtaining a first-order design [9].

There are five MIG design parameters, which are required by the RF interaction region[1]:

1. Beam power;
2. Electron energy;
3. Required cyclotron frequency;
4. Average radius of the electron guiding centers;
5. Transverse-to-longitudinal velocity ratio.

[1] By comparison, there are three Pierce-gun design parameters, which are set by the RF interaction region: beam voltage, beam current or perveance, and beam radius. These are comparable to items 1, 2, and 4. The magnetic field is established by these three quantities and is used only for focusing the beam. In the gyrotron, the magnetic field guides the beam and determines the frequency of operation so it must be specified in the gun design. The only place where the transverse velocities are of considerable importance in linear-beam tube design is in confined flow PPM-focused beams. There, a considerable amount of energy can be expended in transverse velocities, which do not contribute to RF performance and so tube efficiency suffers. In the gyrotron, the efficiency of the interaction depends on the fraction of beam power that is in the transverse motion of the electrons.

These parameters are usually well known and relatively fixed in value. They determine the RF interaction region values of the voltage and current (V_o and I_o), the transverse and longitudinal velocity components $v_{\perp o}$ and v_{zo}, the magnetic field B_o, and the position and size of the orbits (r_{go} = mean guiding center radius and r_{Lo} = Larmor radius). The transverse velocity and the radii are shown in Figure 21-17, which is a cross section of the electron beam downstream from the MIG.

In addition to the design parameters required by the RF interaction region, there are additional variables in the MIG design that can be used to form a satisfactory beam. Baird and Lawson selected the following four variables[2]:

1. r_c, the average cathode radius;
2. J_c, the cathode current density;
3. φ_c, the slope angle of the cathode;
4. D_F, the spacing factor between the cathode and first anode of the MIG.

The most important trade-off variable is the cathode radius, r_c. Baird and Lawson noted that it is usually very helpful in arriving at a suitable cathode radius to plot all the calculated variables together on a single plot versus r_c. Identification of the acceptable ranges of the calculated variables then makes it possible to quickly identify the acceptable design range for r_c. Also, in some cases it may be found that there is no acceptable range for r_c due to a high level of space charge in the gun.

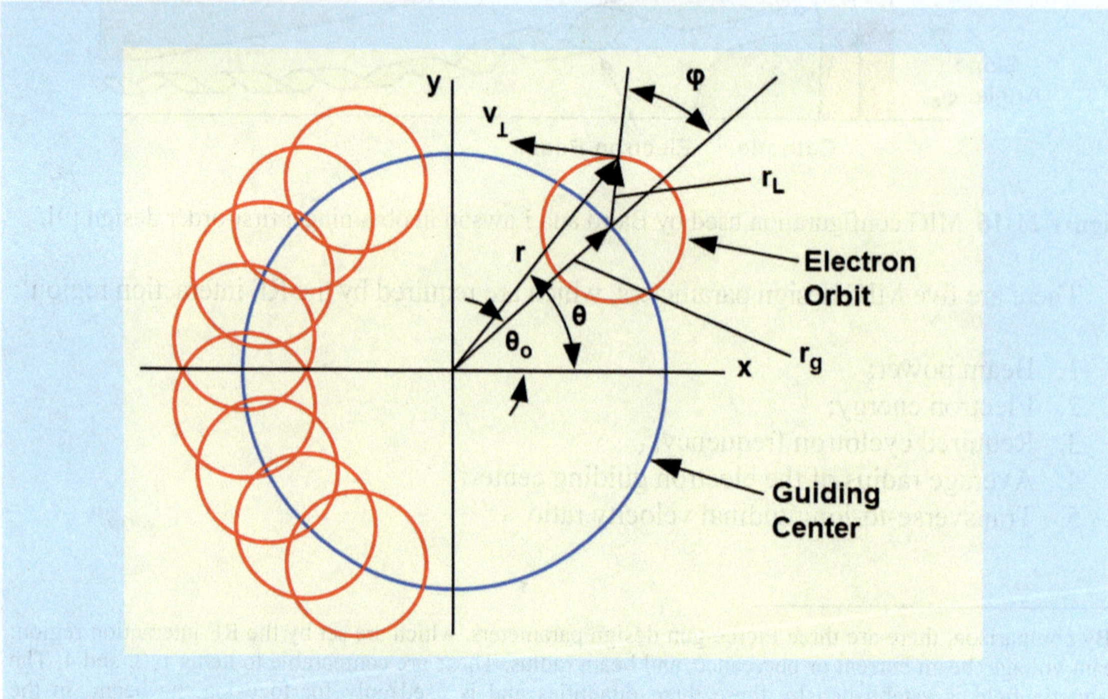

Figure 21-17 Model used for the cross section of the beam in the interaction region [9].

[2]In the Pierce gun design, only two additional design parameters, r_c and J_c, are required.

The starting point in the analysis is the conservation of angular momentum that gives the magnetic field compression ratio.[3] This is the reduction in the beam cross-sectional area that results from the increase in magnetic field as the beam moves from the gun to the interaction region. Conservation of angular-momentum for electron trajectories in cylindrically symmetric dc fields $E(r, z)$ and $B(r, z)$, can be written (assuming the near axis approximation $B_z = B_z(z)$ only) as

$$\left(\gamma m_o r^2 \dot{\theta} - eB_z r^2/2\right) = \text{constant} \tag{21-2}$$

where γ is the relativistic mass factor, $\dot{\theta}$ is the angular velocity, and r is the instantaneous radius of cycling electron orbits.

The constant is found by evaluating the equation at the cathode, where $\dot{\theta} = 0$. Thus,

$$\text{constant} = eB_{zc} r_c^2 / 2 \tag{21-3}$$

where the subscript c indicates that the variable is evaluated at the cathode.

When the angular momentum is interpreted in the beam drift region (see Figure 21-16) where the flux density is B_o, the frequency of rotation $\dot{\theta}$ is the Larmor frequency, $(e/2\gamma m_o)B_o$ and the radius of the electron orbit is the Larmor radius r_{Lo}. In the beam drift region, the second term in the conservation of momentum relation accounts for the rotation of the electron orbit about the gun axis so the value of that quantity is $(e/2)B_o r_{go}^2$.

Now, the momentum equation reduces to

$$B_{zc} r_c^2 = B_o\left(r_{go}^2 - r_{Lo}^2\right) \tag{21-4}$$

By normalizing the distance to r_{Lo} and the magnetic field to B_{zc}, this equation yields Baird and Lawson's equation for the magnetic field compression ratio, F_m,

$$F_m = \frac{R_c^2}{R_g^2 - 1} \tag{21-5}$$

where R_c is the normalized cathode radius.

The emitting area of the cathode in the MIG must be large enough to supply the specified beam current, I_o, at the selected cathode current density, J_o. This leads to the second Baird and Lawson equation for determining the ratio L_s/R_c, where $L_s = l_s/r_{Lo}$ is the normalized slant length of the cathode. This comes from area calculations of a conical segment of the cathode.

By combining the slant length with the magnetic field compression ratio, the third Baird and Lawson design equation for the spread δR_g in the guiding centers in the interaction region of the beam can be determined. This is an important design quantity, which must be controlled in a MIG because a large spread in the guiding centers is usually associated with a decrease in the RF interaction efficiency.

The fourth Baird and Lawson equation relates the slant distance, d, from the cathode to

[3] In linear-beam tubes, Busch's theorem, which is an adaptation of the conservation of momentum, is used, primarily in the electron gun, to find the rate of beam rotation as electrons cross magnetic field lines and enter the main focusing field.

the anode (see Figure 21-17) to the Larmor radius at the cathode. This equation says that the increase in distance from the cathode to the anode along the path d is proportional to r_{Lc}. When $d < 2r_{Lc}$, beam interception on the first anode occurs. As d is made larger than $2r_{LC}$, the clearance between the beam and the first anode increases and a higher voltage on the first anode is required.

The voltage, V_a, which is required on the first anode to produce the specified transverse beam velocity, is estimated from the fifth Baird and Lawson relation. This equation is derived using conservation of angular momentum from the cathode surface to the first peak in the Larmor orbit above the cathode. A relativistic Hull cutoff equation is found, and this establishes the potential at the first beam peak (peak of the first Larmor orbit). Then, using modified coaxial line equations, the voltage on the anode is determined.

The final two design equations provide the most important information required to assess the viability of any given design choices. The sixth relation gives the electric field required at the cathode to achieve the desired beam performance. If this field is above the acceptable design limit of about 100 kV/cm, the MIG is in danger of arcing during operation. (The actual field levels resulting from field enhancements should be limited to 100 kV/cm. The field on the nose of the cathode electrode tends to be the highest stress point.) If the electric field for a given design is too high, the primary remedy is to increase r_c. Thus the sixth relation sets a *lower bound* on acceptable values of the cathode radius.

Baird and Lawson's seventh equation alternatively sets an *upper bound* on r_c. This equation gives the ratio of the beam current density J_c to the space-charge-limiting current density, J_L. As the cathode radius is increased, the space-charge-limiting current decreases very rapidly and the ratio J_c/J_L increases quickly to its maximum value of unity. MIG design, however, is based on negligible space-charge effects. It is therefore mandatory that Jc/JL be kept below about 15%–20% to rely on these design rules. The seventh equation, therefore, imposes an upper bound on r_c, which cannot be exceeded without finding some method for including the effects of space-charge in the design process.

The MIG in Figure 21-18 is an example of the application of Baird and Lawson's analysis to the design of a 600-kW, 70-kV MIG for use at 35 GHz. Even though the gun design has two anodes, full beam voltage was applied to the first anode to create a smoothly varying E-field in the cathode-anode region to better approximate adiabatic flow conditions.

21.3.3 MIG Performance

Gyrotrons share an extremely important characteristic with other electron devices that operate in vacuum. That is, the electron beam must be, and must remain for the life of the device, nearly perfect. For optimum gyrotron performance, there are very stringent requirements on the MIG and on the electron beam that it provides,

First, only the transverse energy of the electrons can be converted to RF power output. If the ratio of transverse to axial velocity of the electrons is α, then the maximum efficiency of conversion of beam power to the RF field is

$$\eta = \frac{\alpha^2}{1 + \alpha^2} \tag{21-6}$$

This means that the transverse to axial velocity ratio should be as high as possible. Ratios between 1.2 and 1.5 are typical [8].

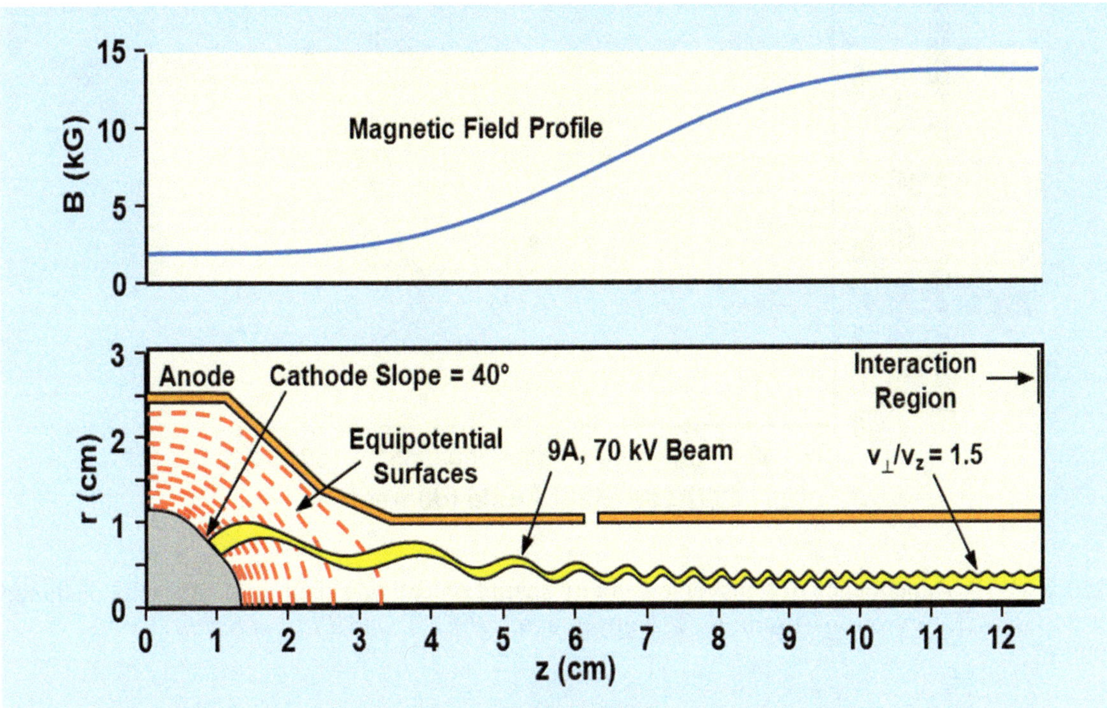

Figure 21-18 An example of the application of Baird and Lawson's analysis to the design of a 600-kW, 70-kV MIG for use at 35 G.

The transverse velocity spread must be as low as possible to help ensure phase-bunching stability. Velocity spread is a function of many factors:

1. MIG cathode slant angle. Figure 21-19 shows an example calculated for a 94-GHz gyrotron [8].
2. The presence of space charge. The cathodes in MIG guns are operated temperature-limited to avoid instabilities that occur in the space-charge cloud that surrounds the cathode in magnetron-type cathodes. Brown [10] reported a reduction in noise by 60 dB or more from a magnetron when cathode temperature was reduced so that operation was temperature limited rather than space-charge-limited (see Chapter 19).
3. Rack velocity noise, shot current, flicker current, and surface roughness.

21.4 BEAM-WAVE INTERACTION

The operating mode that we have discussed so far has been the TE_{01} mode (Figures 21-9, 21-10, and 21-19) and the cavities have been hollow cavities; that is, there is no center conductor in the cavities. In this section we discuss coaxial cavities as well as hollow cavities and modes of operation that are far more complex than the TE_{01} mode. MIGs for supplying beams to hollow and to coaxial cavities were discussed in the previous section.

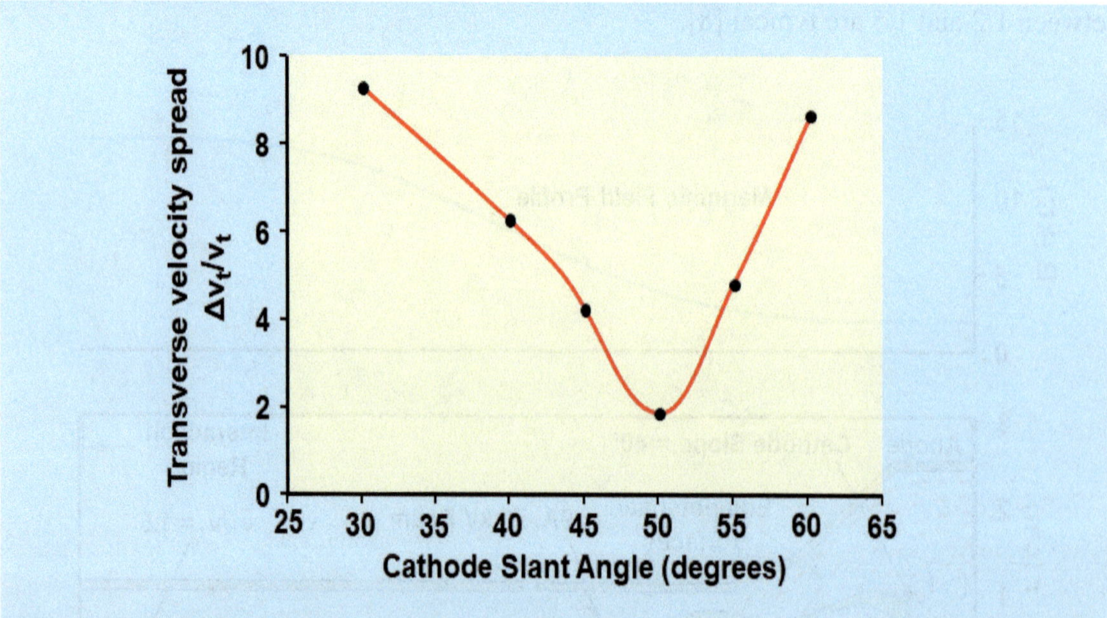

Figure 21-19 Dependence of transverse velocity spread on cathode slant angle for MIG designed for 94-GHz TE_{01} gyrotron. (From: K. T. Nguyen et al., *IEEE Trans. Plasma Science*, 1998. © 1998 IEEE.)

21.4.1 Cavities

Most high-power gyrotrons use highly overmoded cavities. The TE_{01} cavity mode used in the discussion in the previous section has a constant electric field in the azimuthal direction and one maxima in the radial direction. Overmoded cavities can have several cycles of variation in the field in the azimuthal direction and several maxima in the radial direction. Cavities may be hollow or coaxial.

Examples of radial field variations in hollow cavities are shown in Figure 21-20 [11]. Five different waveguide modes are shown. The placement of hollow electron beams for maximum interaction with these modes at the fundamental cyclotron frequency is also shown. In general, TE_{0n} modes ($n > 1$) are used to increase the power handling capacity of the device while retaining the favorable field geometry and mode control properties of these modes. The TE_{52} and TE_{91} modes are used because the fields are localized near the outside wall of the cavity. These are called whispering-gallery modes. With the fields localized near the wall of the cavity, it is possible to use a large radius electron beam and, at the same time, preferentially excite the desired operating mode. These modes, therefore, permit a measure of mode control in a large overmoded cavity with an enlarged (high-power) electron beam.

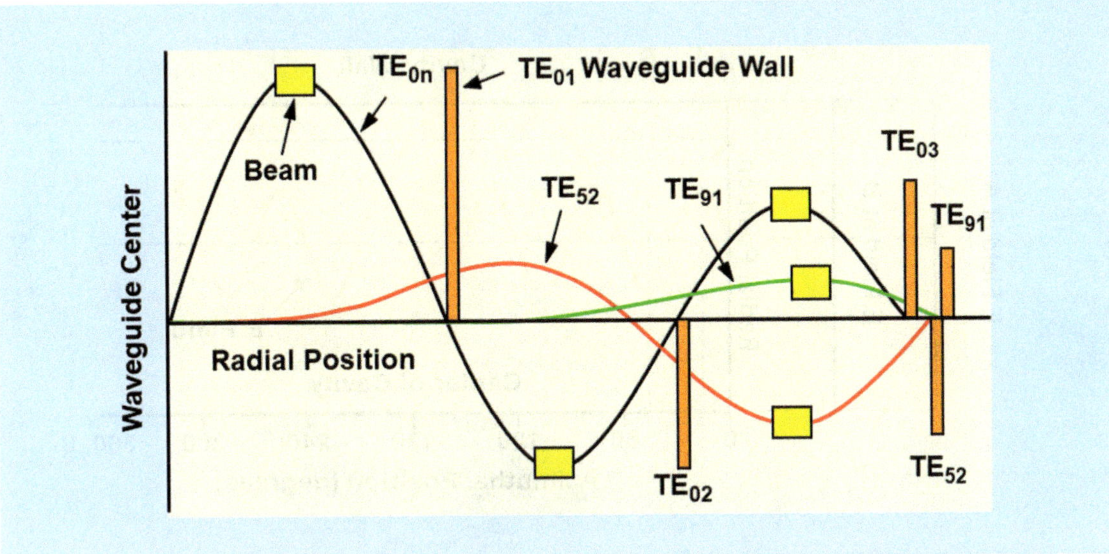

Figure 21-20 Radial field variations for several hollow cavity modes. (Adapted from: J. M. Baird and W. Lawson, *Int. J. Electron,* 1986.)

An example of the field pattern in an overmoded hollow cavity is shown in Figure 21-21 [2]. This is a plot of the magnitude of the square of the electric field as functions of the radial and azimuthal directions for the TE_{62} mode. Electron orbits and electric field lines for maximum interaction with the electrons are shown.

It is helpful in explaining the operation of an overmoded cavity like the cavity with the TE_{62} mode in Figure 21-21 to unroll the field pattern and show it in a linear manner as in Figures 21-22 and 21-24. The field has six cycles of variation in the azimuthal direction and one cycle (two maxima) in the radial direction.

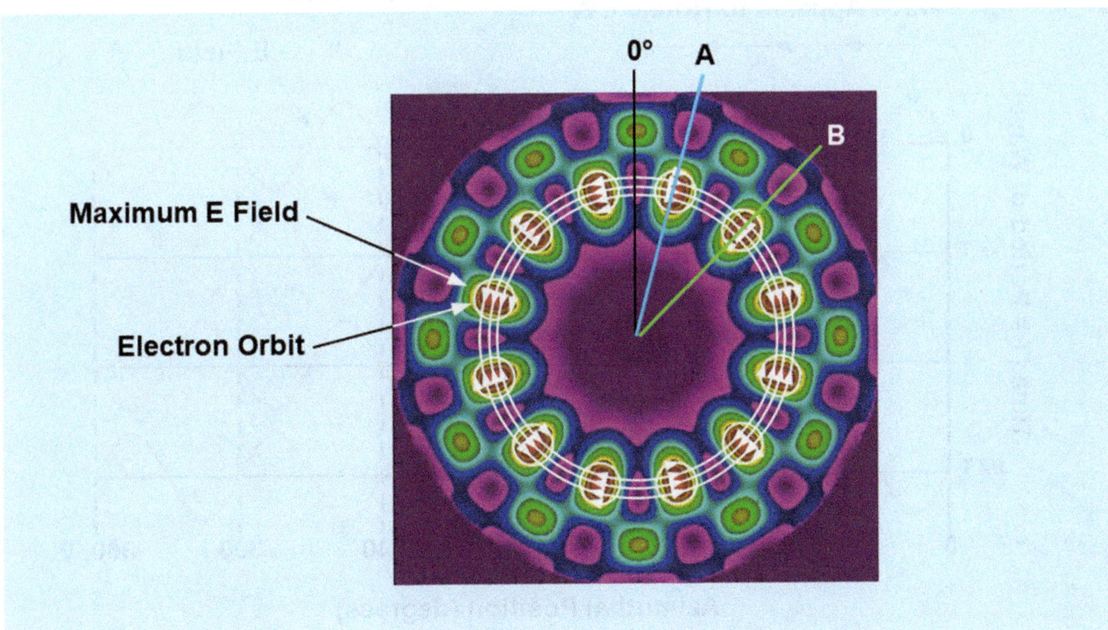

Figure 21-21 The magnitude of the square of the electric field for the TE_{62} mode is shown here with optimum placement of orbits of electrons. (Adapted from: H. Jory, et al. [2].)

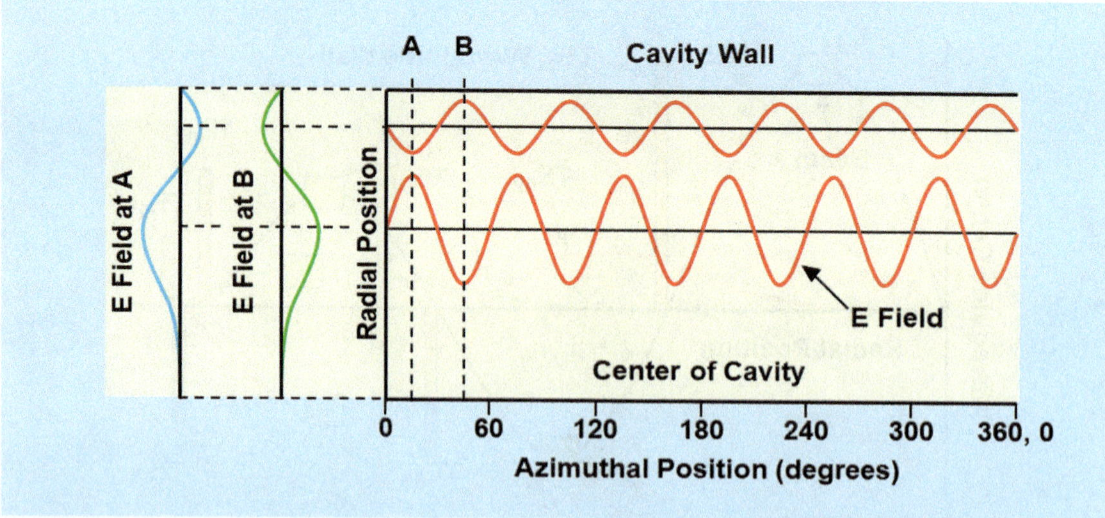

Figure 21-22 E field in Figure 21-21 as functions of radial and azimuthal positions.

As a function of time, the fields vary as shown in Figure 21-23. Electron orbits are shown to rotate in synchronism with the alternating field. If one particular phase of the wave is chosen, such as the phase of the peak field, the wave appears to move to the right (rotate in the clockwise direction in the cavity). Notice that the phase of the field peaks could have been chosen so that the wave would appear to rotate counterclockwise. The TE_{62} mode is referred to as a rotating mode and, in fact, all TE_{mn} modes with $m \neq 0$ are rotating modes.

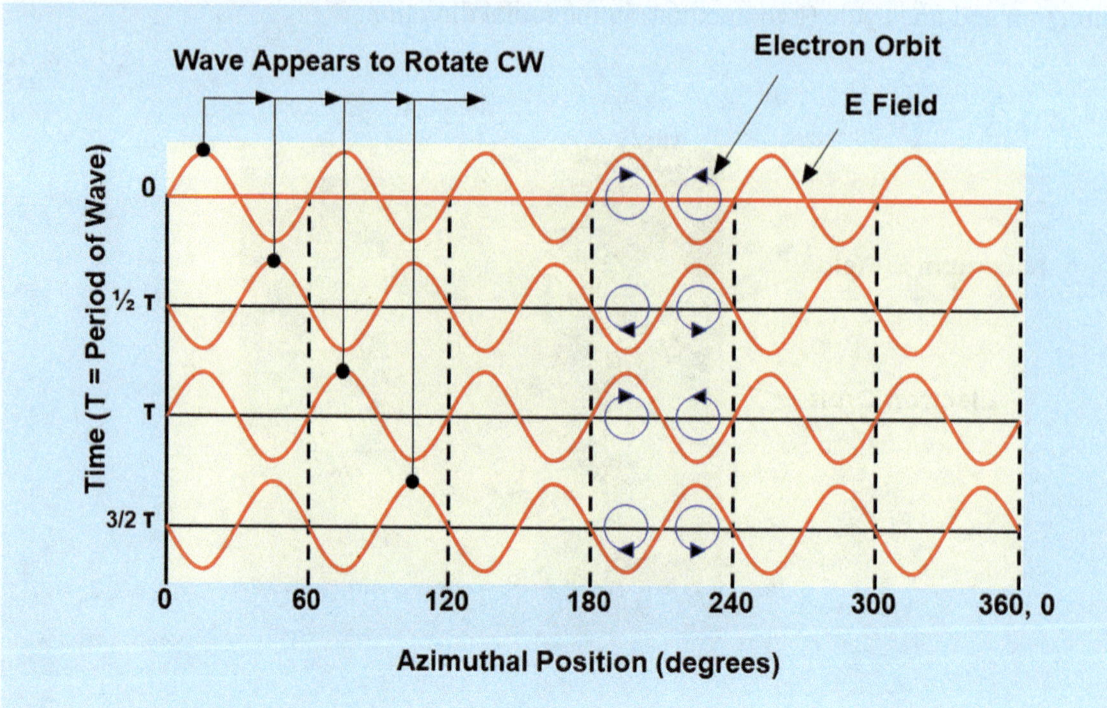

Figure 21-23 E field as a function of time. When one particular phase is selected, the field appears to rotate clockwise (or counterclockwise).

The apparent rotation of the mode in the gyrotron cavity can be compared to the traveling wave in a magnetron. Figure 21-24 depicts an eight-cavity magnetron that has been unrolled [12]. Only the anode segments that make up the capacitive portion of the cavities are shown. As the cavity fields oscillate, the potential across the gaps alternates. For an electron spoke that is rotating about the cathode to continuously interact with the anode field, the spoke must travel at the speed of the traveling anode wave. Thus, this magnetron, like all modern magnetrons, is referred to as a traveling-wave magnetron.

So we see that the magnetron and the TE_{mn} mode gyrotron ($m \neq 0$) are both traveling-wave devices. The difference between them is that in the magnetron, the electrons must travel with the field, whereas in the gyrotron the electrons stay relatively stationary and spin in small orbits in synchronism with the alternating field as the field travels past them.

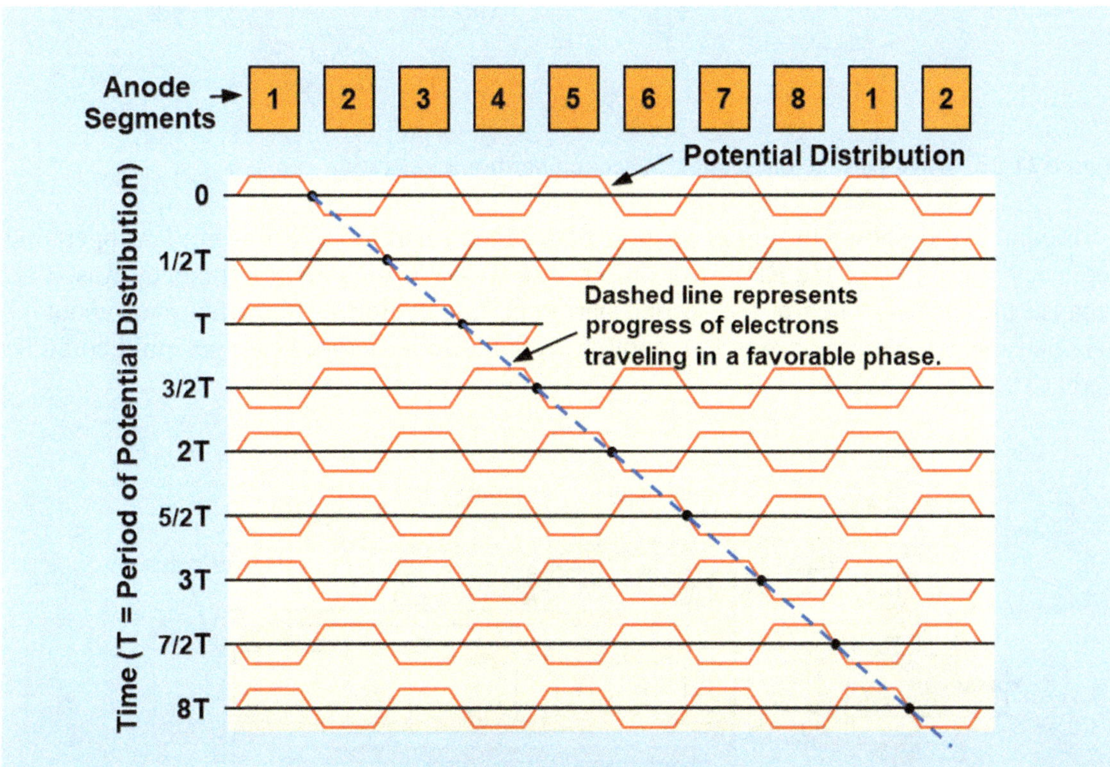

Figure 21-24 Eight-cavity traveling-wave magnetron. (Adapted from: *Theory and Application of Microwaves,* Bronwell and Beam, copyright 1947 by McGraw-Hill, Inc.)

21.4.2 Harmonic Operation

With the proper shape for the RF field, it is possible for harmonic interactions with the electrons to occur. In the example shown in Figure 21-25, the field varies at twice the cyclotron frequency. The direction of the field reverses in the center of the electron orbit. Thus, an electron that is initially decelerated by the field (Figure 21-25(a)) is moving transverse to the field when the field reverses (Figure 21-25(b)) and so its orbital energy is not changed. By the time the field reverses again, the electron has moved another 90° around its orbit and is again in a decelerating field. Thus, during each orbit of the electron, the RF field goes through two complete cycles. For a given operating frequency, the cyclotron frequency is half the value used in the fundamental mode of operation. As a result, the magnetic field is reduced by a factor of 2.

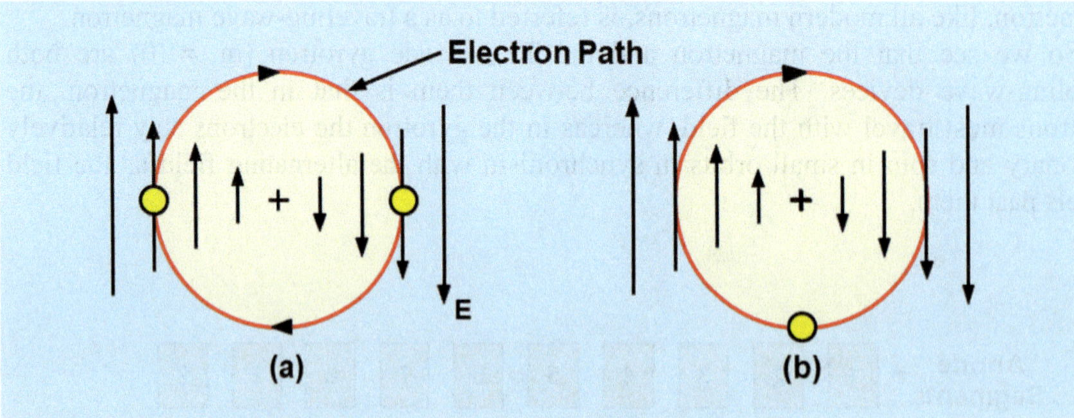

Figure 21-25 Wave particle interaction in second harmonic operation.

Figure 21-26 shows the cross section of a second-harmonic gyrotron. The operating frequency was 34 GHz, the RF power output was 10-kW CW, and the efficiency was 40%. Because the tube operated at the second harmonic, the cyclotron frequency was about 17 GHz and the magnetic field was low enough so that a room-temperature solenoid could be used [13].

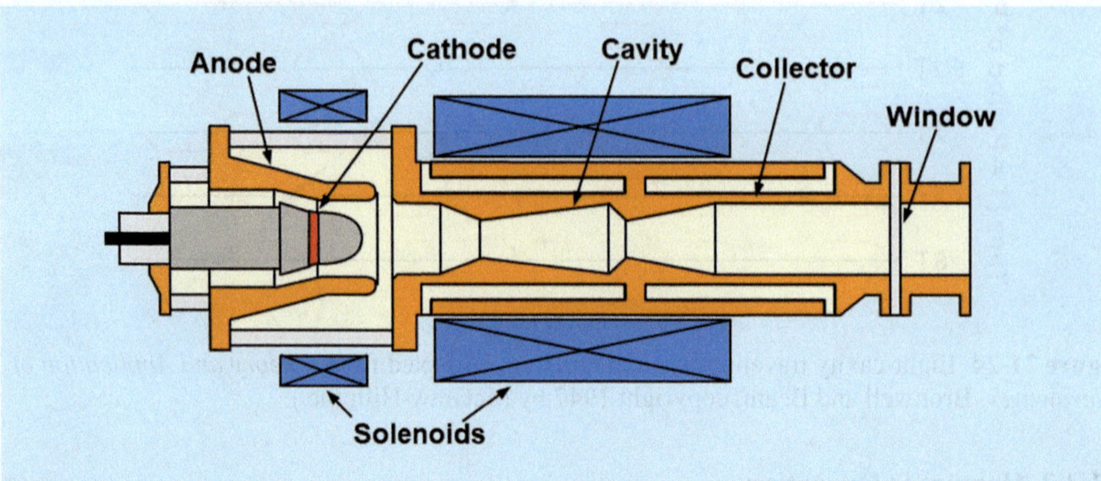

Figure 21-26 Second-harmonic gyrotron with room-temperature solenoid. (From: D. V. Kisel et al., *Radio Engineering and Electronic Physics*, 1974.)

21.5 GYRO-AMPLIFIERS

The gyro-amplifiers to be discussed are shown in Figure 21-27: the gyro-klystron, gyro-twystron, and gyro-TWT. Gyro-amplifiers are being developed for applications requiring phase coherence and instantaneous bandwidth. A primary application is radar, where they are of considerable interest for future high-performance applications [14].

Gyro-amplifiers have larger weight and volume than state-of-the-art coupled-cavity TWTs, but they can provide significantly higher powers than the TWTs. The wide variety of radar applications for which gyro-amplifiers are now being considered are both those existing millimeter-wave radar applications and new radar applications previously not considered due to the absence of powerful amplifiers at the required frequencies.

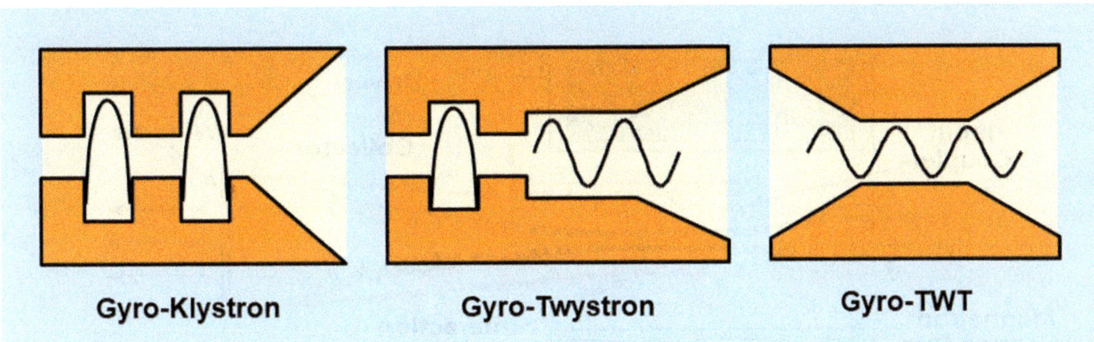

Figure 21-27 Gyro-amplifiers.

The frequency bands in which gyro-amplifiers are expected to find important radar applications are primarily Ka-band (26.5–40 GHz) and W-band (75–110 GHz). Within these bands, radars are generally designed to take advantage of the atmospheric windows at frequencies of approximately 35 GHz and 94 GHz.

The bandwidth requirement of the system determines the tube type:

- Gyro-klystron for < 1% bandwidth;
- Gyro-twystron for 1% to 2% bandwidth;
- Gyro-TWT for bandwidths > 2%.

These devices are described in Sections 21.5.1 through 21.5.3.

21.5.1 Gyro-Klystrons

Conceptually, the only difference between the gyro-klystron and the gyrotron oscillator is the addition of the input cavity. The operation of a gyro-klystron is similar to a conventional klystron except that electron bunching occurs in the azimuthal direction rather than in the axial direction. In a gyro-klystron, a signal to be amplified is fed into an input cavity where the cyclotron bunching process is initiated. Then, the beam is permitted to drift while bunches form. As the beam passes through a second cavity, the amplified signal may be removed or the bunching process may be enhanced and the signal removed in a subsequent cavity. A sketch of a four-cavity gyro-klystron is shown in Figure 21-28 [15].

An example of a gyro-klystron is the CPI VGB-8194. As shown in Figure 21-29, this five-cavity amplifier operates at 93.5 to 94.5 GHz with a 3-dB bandwidth of 700 MHz. The peak RF output power is 100 kW and the average output power is 10 kW. The amplifier is available with a cryogen-free 4-tesla superconducting magnet [16].

Among other applications, the VGB-8194 gyro-klystron is the key component of the NRL WARLOC radar and is what made this development unique [17]. WARLOC is a high average power, coherent W-band (94-GHz) radar operating at the NRL Chesapeake Bay Detachment (CBD) site in Maryland. Developed by the Navy as a research radar for imaging and tracking applications, WARLOC also brings unprecedented sensitivity to the measurement of cloud parameters (including phase, morphology, and particle size and motion) at spatial resolutions unparalleled by contemporary radar systems.

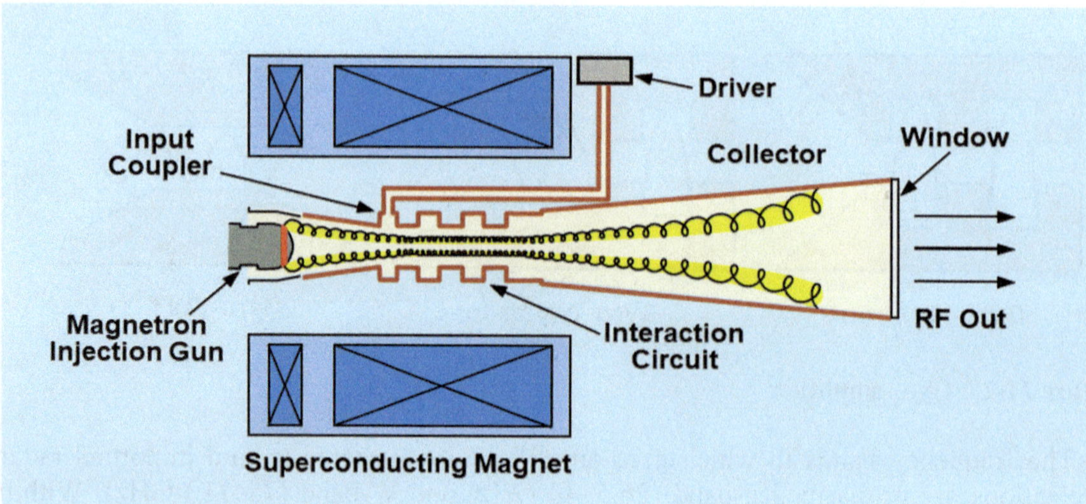

Figure 21-28 Gyro-klystron. (From: M. Blank et al., *IEEE Trans. Plasma Science.* © 2002 IEEE.)

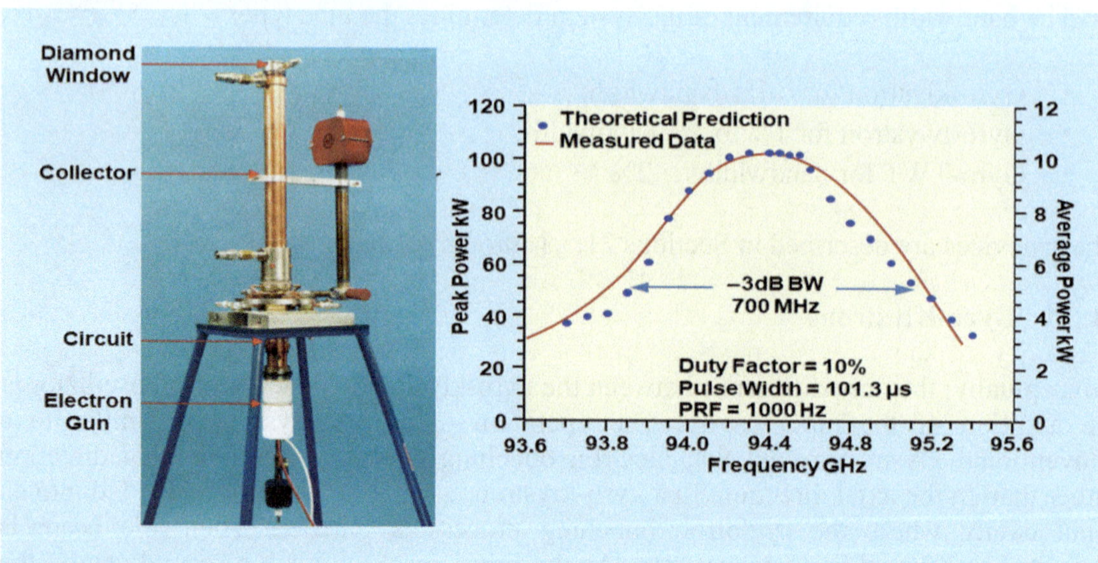

Figure 21-29 CPI VGB-8194 Gyro-klystron. (Courtesy of CPI.)

21.5.2 Gyro-TWTs

There has been substantial interest and development effort in gyro-TWTs. The gyro-TWT is an attractive candidate for use as the transmitter power amplifier in millimeter-wave radars. Gyro-TWTs are capable of a much broader bandwidth than gyro-klystrons. A CPI VBG-8190 gyro-TWT is shown in Figure 21-30.

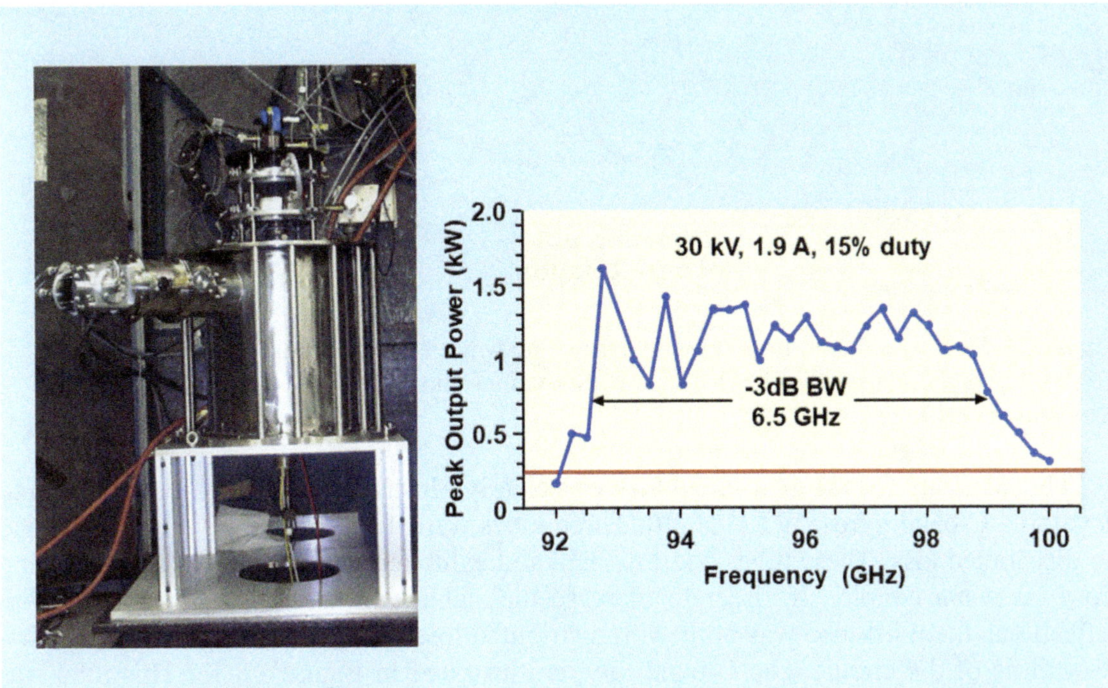

Figure 21-30 CPI VGB – 8190 gyro TWT.

In gyro-TWTs, a nonresonant RF structure is used to produce a traveling-wave interaction. As in other gyro-devices, a spiraling electron beam immersed in an axial magnetic field is used. Traveling waves are launched into the interaction space by an input coupler. Axial phase synchronism is required between the traveling wave and the rotating electrons as indicated in Figure 21-31.

As the electromagnetic wave and the spiraling electron beam move through the interaction space, the kinetic energy of the electron beam is transferred into the electromagnetic fields, creating RF amplification. For typical operation near the cutoff frequency of the interaction waveguide, it is primarily the transverse component of electron motion that interacts with the electromagnetic wave.

An important problem that must be dealt with in gyro-TWT design is maintaining stability and preventing BWOs in the interaction space. At the same time, of course, acceptable performance, including high overall gain, stability to local reflective oscillations, and high average power capability must be achieved [18].

To ensure stability, the majority of this interaction space should exhibit a moderate amount of electromagnetic loss per unit length. This distributed loading approach has been shown to have superior stability characteristics compared to using localized lossy severs. Finally, the downstream part of the interaction space is completed with a short, unloaded cylindrical tunnel in which the final, highest powered portion of the amplification takes place.

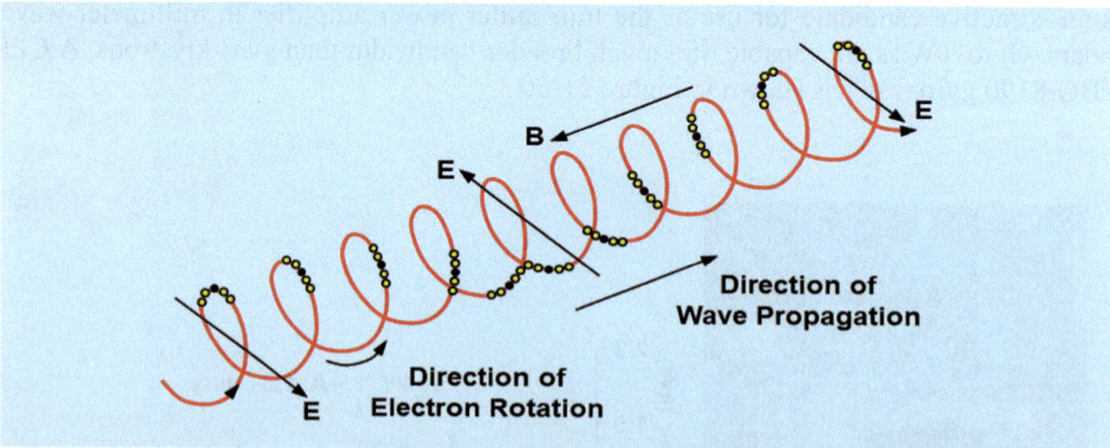

Figure 21-31 Interaction of beamlet with a traveling wave. (Adapted from: *Advances in Electronics and Electron Physics,* Vol. 55, R. S. Symons and H. R. Jory, copyright 1981 by Academic Press.)

The necessity for the distributed loss was clearly illustrated by Varian in a program to develop a C-band gyro-TWT [19, 20]. Three tubes were built, the first two of which had no distributed loss. These tubes had low gain and exhibited very large powers emanating from the *input* coupler. In order to increase the stable gain and reduce problems with reflections, the third tube was built with distributed loss. Kanthal was applied to the first two-thirds of the circuit. The C-band tube is illustrated in Figure 21-32. This tube was basically a large-scale model of a 94-GHz TWT. The development effort for this tube was undertaken in preparation for the development of the 94-GHz gyro-TWT.

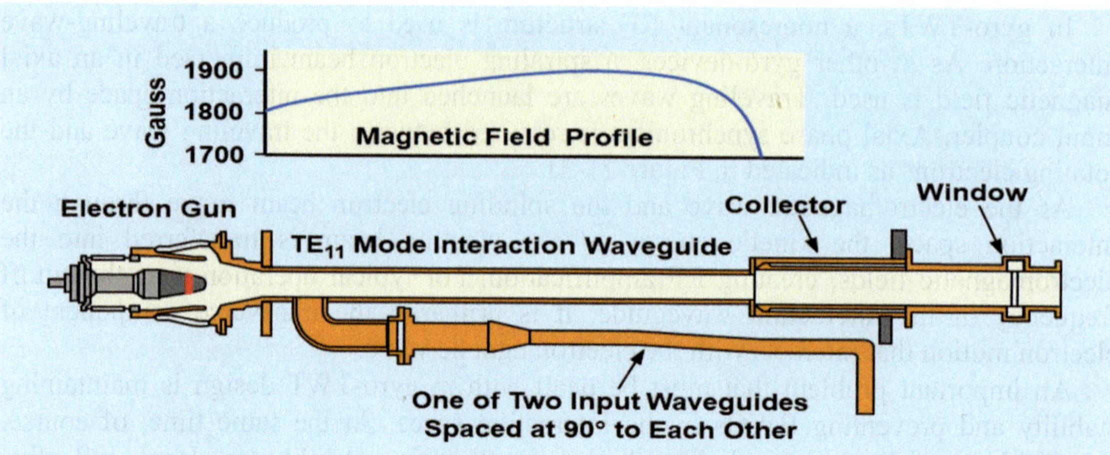

Figure 21-32 C-band gyro-TWT. (Adapted from: P. Ferguson and R. Symons, *Technical Digest, 1980,* IEDM. © 1980 IEEE.)

The interaction waveguide used in the C-band tube was a circular waveguide operating in the TE_{11} mode. Two rectangular to circular waveguide input transitions at 90° to each other were used to match into the two orthogonal TE_{11} modes that can exist in the circular waveguide. The output window was at the end of the collector. Transverse field permanent magnets deflected electrons to the collector wall before they could reach the window.

The tube had two gun solenoids and four main field solenoids. As indicated by the magnetic flux density plot, optimum results were obtained by using a slightly tapered field. The power output characteristics for this device are shown in Figure 21-33. Modulating anode pulsing was used and, at a beam voltage of 65 kV, the power output was 120 kW at an efficiency of 26%. The 3-dB bandwidth was 7.25%.

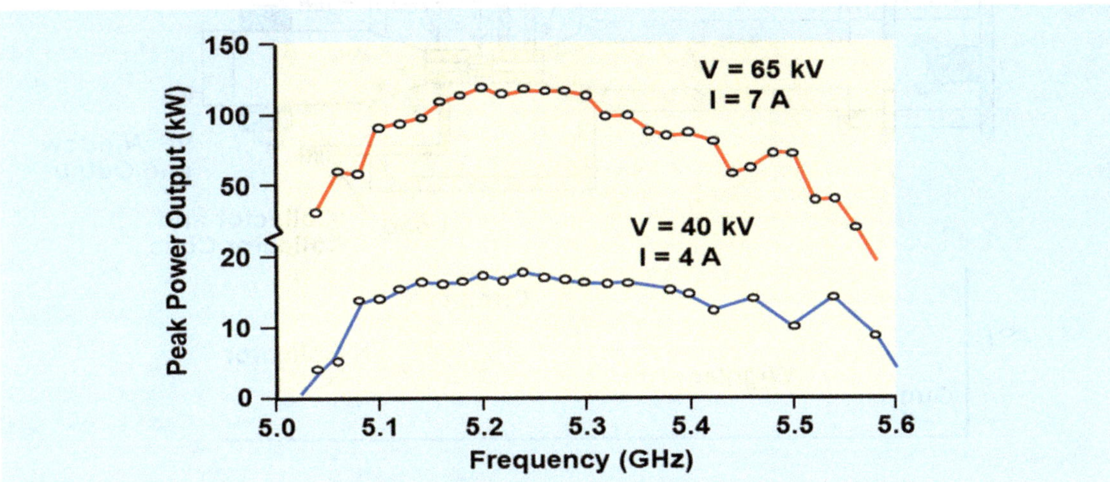

Figure 21-33 Peak power output for a gyro-TWT for two operating conditions. (Adapted from: P. Ferguson and R. Symons, *Technical Digest, 1980,* IEDM. © 1980 IEEE.)

In another early effort to develop a 95-GHz gyro-TWT, Varian used a Pierce gun and a contra-wound helical winding along the beam axis to produce the required hollow electron beam [21]. This winding, along with a straight magnetic field along the axis, provides a force, which is perpendicular to the beam axis and which rotates in azimuth with distance along the beam axis. This force gives the electrons a transverse velocity as they move through this section of the tube. A schematic drawing of the device is shown in Figure 21-34 and the power output is plotted in Figure 21-35. The tube demonstrated output power in excess of 20-kW peak.

Now, CPI's VGB-8190 gyro-TWT has demonstrated the ability to produce 2-kW peak power (0.8 kW average) with a bandwidth of 6.5 GHz around a center frequency of 95.5 GHz. A gyro-TWT similar to this (8-GHz bandwidth, 500-W peak, >250-W average from 92–100 GHz) is used as the driver for twystrons in the Haystack radar [2].

21.5.3 Gyro-Twystrons

The gyro-twystron, like the conventional twystron, is a hybrid device with a modest bandwidth capability. It is derived from the gyro-klystron by using the gyro-klystron in the cavities near the RF input and by replacing the output cavity with a slightly tapered waveguide section as in the gyro-TWT to be described in the next section. The output section is excited by the electron beam, which is bunched by the interaction in the klystron section.

The configuration of the gyro-twystron can prevent the problem of breakdown at high-power levels because the RF power density in the output waveguide can be much smaller than in the gyro-klystron output cavity.

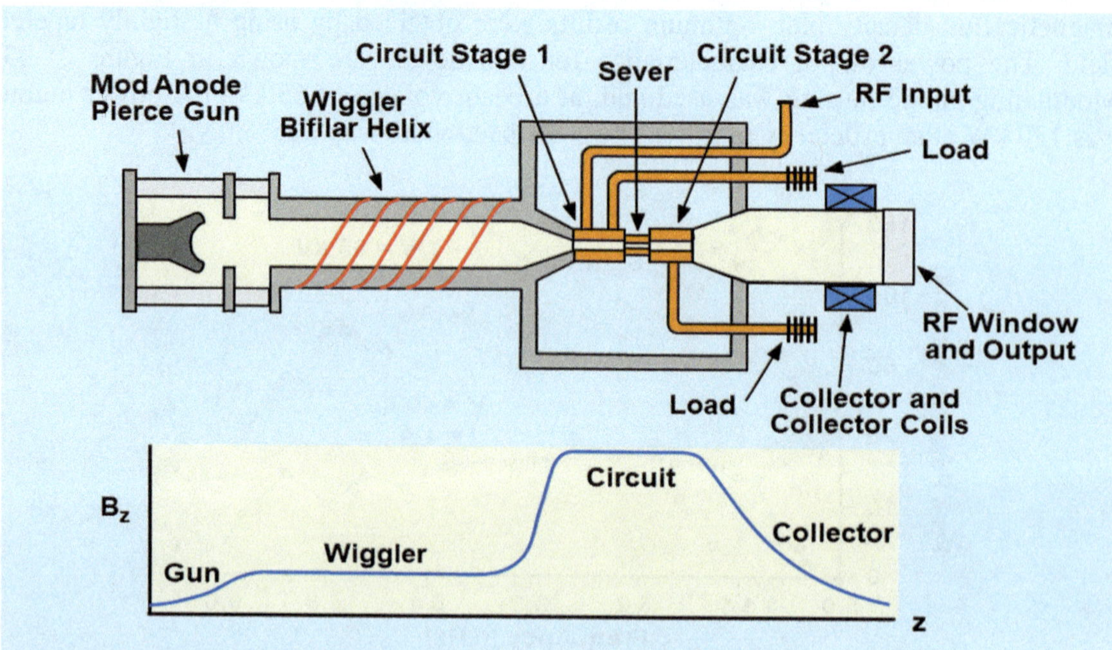

Figure 21-34 The VGT-8195A gyro TWT using a Pierce gun and a wiggler magnet for operation at 95 GHz and powers up to 20-kW peak.

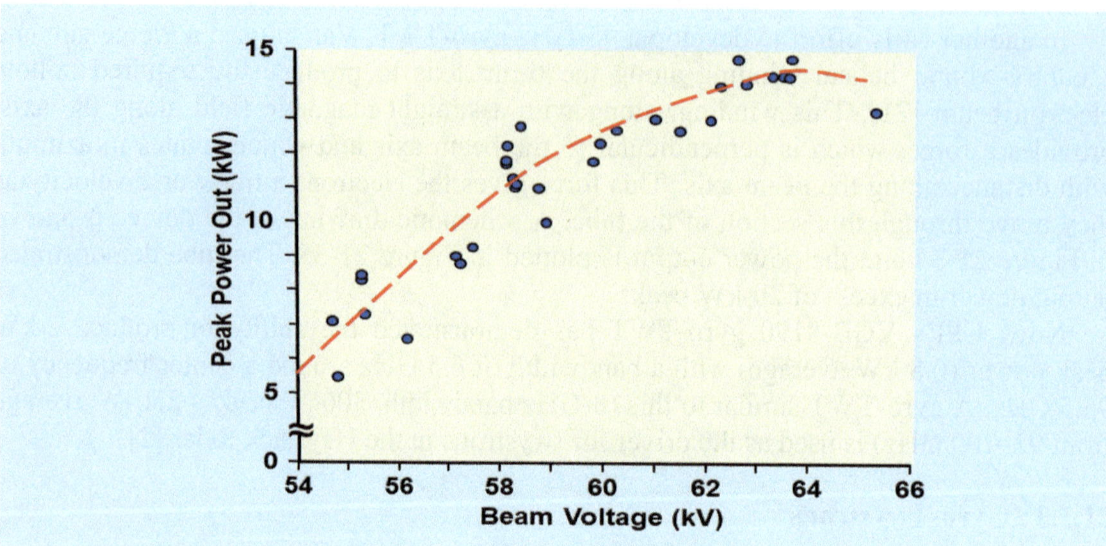

Figure 21-35 Power output of the VGT-8195A gyro-TWT.

An application for gyro-twystrons is the Haystack radar upgrade. The Haystack Ultra-wideband Satellite Imaging Radar (HUSIR) development is being conducted at the Lincoln Space Surveillance Complex by MIT Lincoln Laboratory. For the Haystack upgrade, it is planned that 16 CPI VGB-8193 W-band gyro-twystrons at 55-kW peak RF output and 10% duty will be used to cover four frequencies in the 92–100-GHz band [2]. A VGB-8193 twystron is shown in Figure 21-36 and power output data for two of the HUSIR twystrons are shown in Figure 21-37.

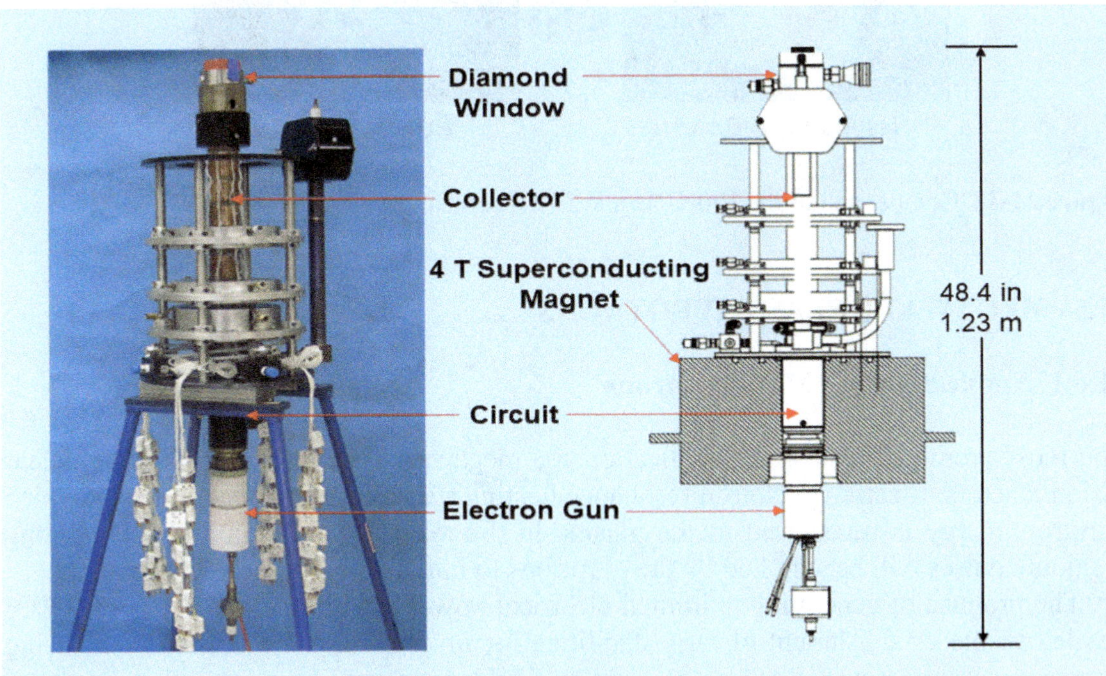

Figure 21-36 VGB-8193 twystron.

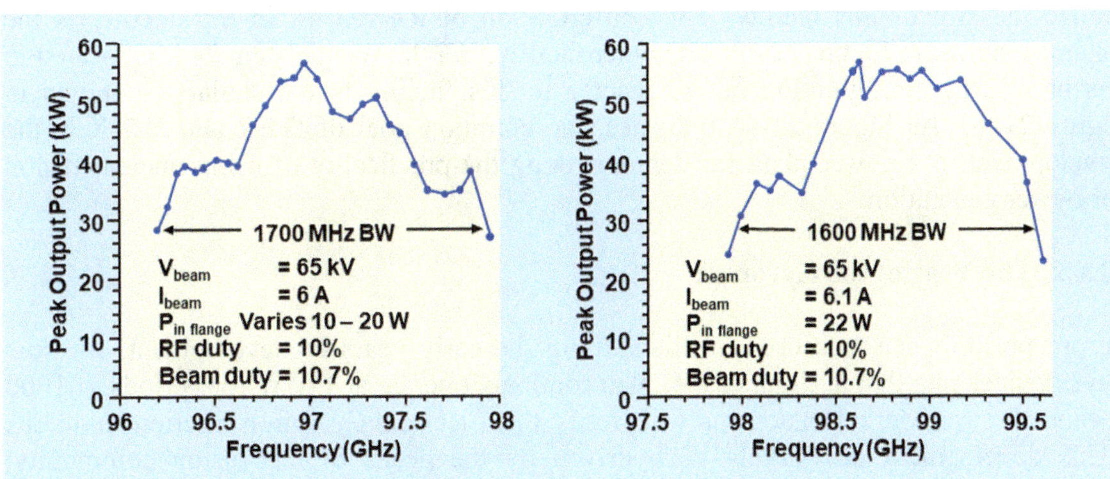

Figure 21-37 Power output for two of the twystrons in HUSIR, the Haystack radar upgrade.

The radar images in Figure 21-38 illustrate the improvement in resolution that is realized by increasing the radar frequency from 10 GHz to 100 GHz.

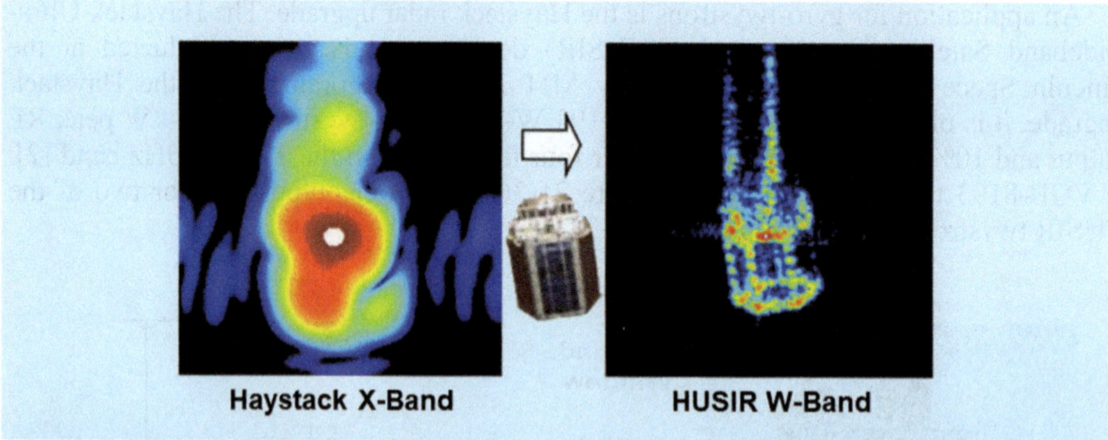

Figure 21-38 Comparison of the 100-GHz and 10-GHz radar images of a satellite.

21.6 MEGAWATT-CLASS GYROTRONS

21.6.1 Applications for MW Gyrotrons

The most promising near-term application for megawatt gyrotrons is in thermonuclear fusion reactors. Electron cyclotron resonance heating (ECRH) is the mechanism by which gyrotron energy is transferred to the plasma in the reactor. One thousand second-long gigajoule pulses will be supplied by the gyrotrons to initiate thermonuclear reactions.

The promise of generating unlimited electrical power with no pollutants such as CO_2 has led to the establishment of large facilities for investigating the feasibility of using thermonuclear reactors for power generation. The largest facility is located in Southern France and is ITER, the international thermonuclear experimental reactor [3]. A drawing of ITER is shown in Figure 21-39. Notice the image of the man near the bottom right to realize the size of this facility. The configuration of the reactor in this facility is the tokamak shown in Figure 21-40. The other facility, W7-X, Wendelstein 7-X, is located in Germany. The configuration of the reactor in this facility is the stellarator shown in Figure 21-40. As Figure 21-41 indicates, the common goal of ITER and W7-X is the development of a power plant for demonstrating the practicality of using nuclear fusion for power generation.

21.6.2 The Year of the Gyrotron

"From performance measured in kilowatts in the early years of development, we now have devices like that in Figure 21-42 that routinely operate at 1 MW for periods of 1000 seconds," says Mark Henderson, who leads the Electron Cyclotron Section at ITER. "This development has largely been driven by the needs of the fusion community, although gyrotrons also have applications as heating devices in industry or as microwave sources in spectroscopy and diagnostics." In view of the timing and successes in bringing MW-class gyrotrons on line, Henderson called 2018 "the year of the gyrotron" [3]!

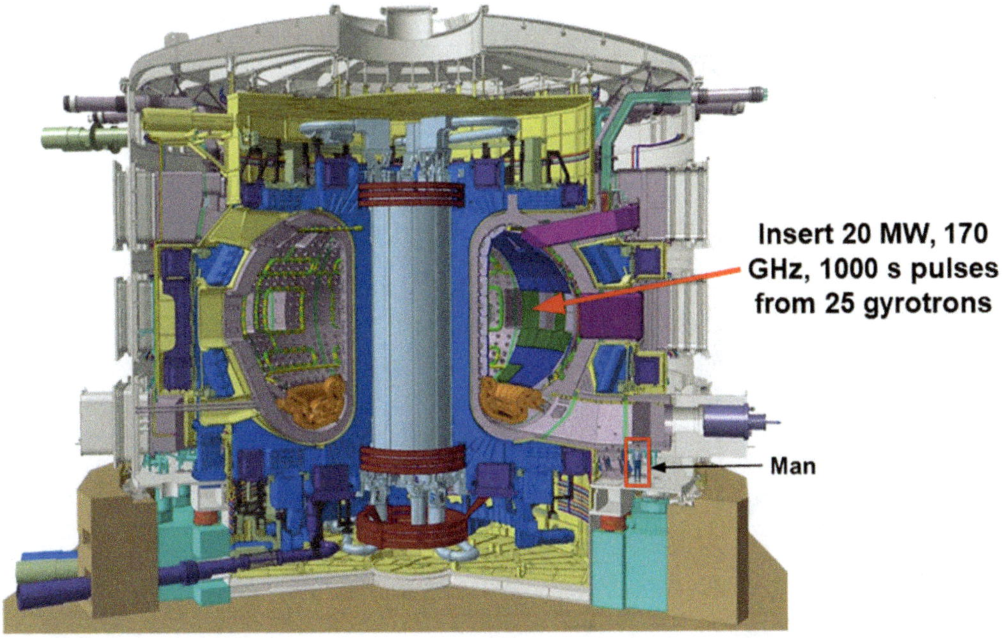

Figure 21-39 ITER, the international thermonuclear experimental reactor.

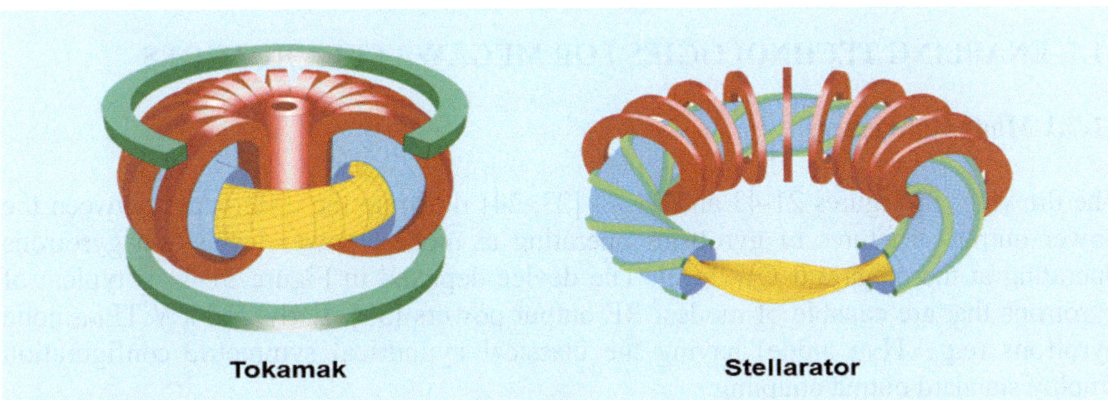

Figure 21-40 Reactors used in ITER (Tokamak) and W7-X (Stellarator).
\

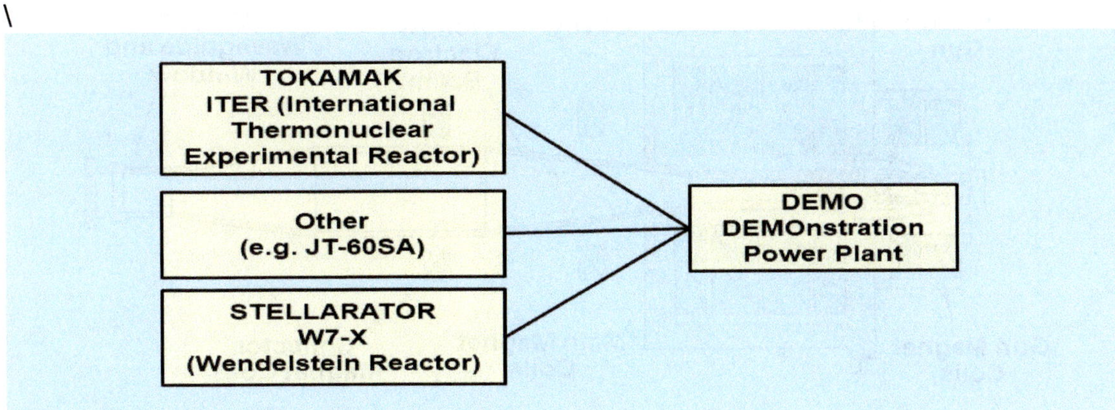

Figure 21-41 The long-term goal is to build a power plant that will illustrate the feasibility of using thermonuclear fusion for power generation.

Figure 21-42 Thales megawatt-class gyrotron for ITER.

21.7 ENABLING TECHNOLOGIES FOR MEGAWATT GYROTRONS

21.7.1 Mode Converters

The drawings in Figures 21-43 and 21-44 [23, 24] illustrate the differences between the power output structures of gyrotrons operating at modest power levels and gyrotrons operating at the megawatt CW level. The device depicted in Figure 21-43 is typical of gyrotrons that are capable of modest RF output powers [33]. Early 200-kW TE_{0n}-mode gyrotrons (e.g., TE_{02} mode) having the classical cylindrical symmetric configuration employ standard output coupling.

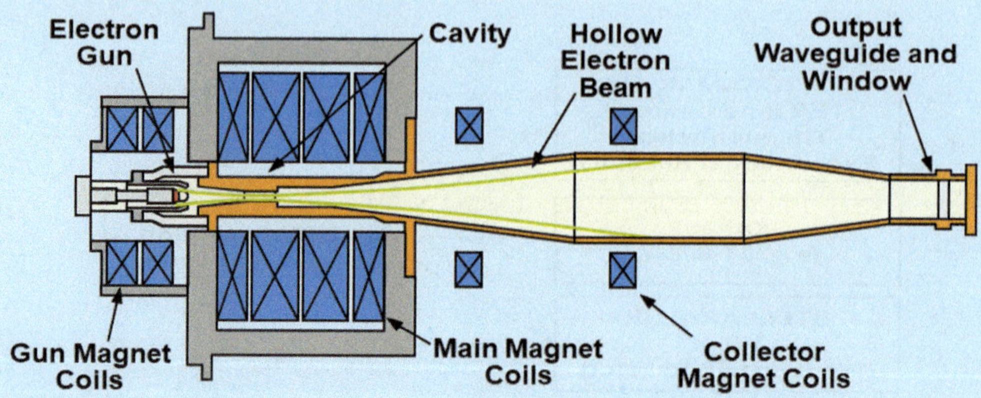

Figure 21-43 Cross section of a gyrotron capable of generating modest RF power levels. (From: Varian, *Introduction to Gyro-Devices.*)

For CW gyrotrons operating at the megawatt level, the quasi-optical mode converter makes it possible to separate the spent electron beam from the outgoing RF power as in Figure 21-44. This makes it possible to increase the collector beam interception area as well as depress the collector.

The operating mode of a gyrotron is a TE_{0n} or TE_{mn} mode in a circular waveguide cavity. For most applications, these modes are not suitable for use because of serious diffraction and polarization losses. To be of use, they must be converted into the fundamental Gaussian mode or lower-order linearly polarized hybrid mode (HE_{11}). The conventional waveguide mode converters that were used with early gyrotrons are in the form of rippled-wall or corrugated circular waveguides and serpentine structures. These devices have high efficiency but are long and have narrow bandwidth, so they are not appropriate for use in high-power gyrotrons.

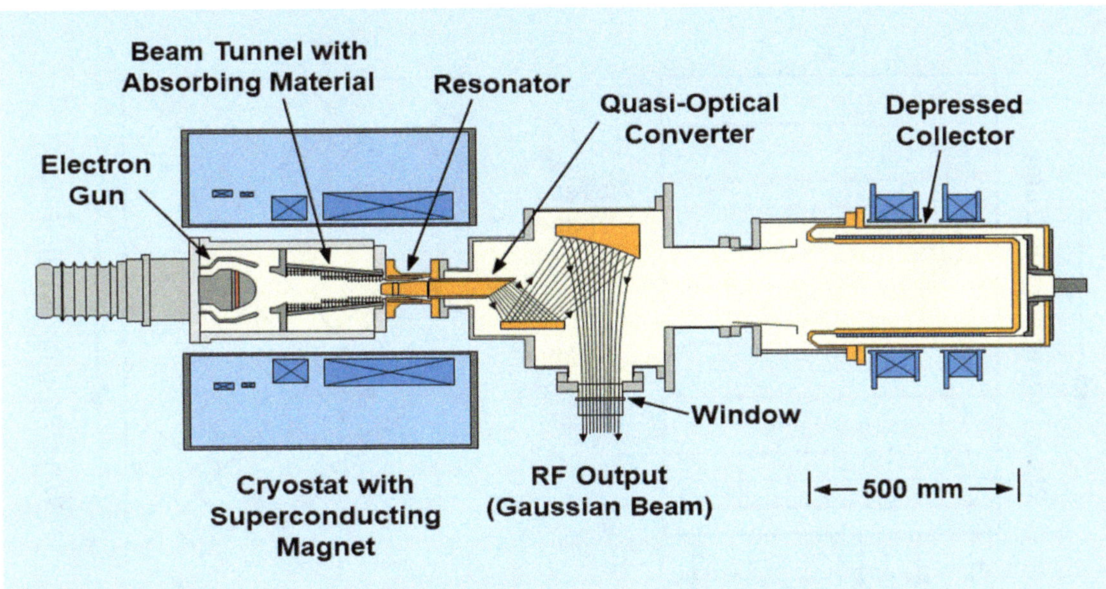

Figure 21-44 Primary components of a high-power gyrotron. (Adapted from: M. Thumm, 2007.)

The short-wavelength, high-output power and the rotating asymmetric high-order cavity modes have led to the need for internal quasi-optical mode conversion concepts, which were first developed in Russia [26] and are known as Vlasov converters. These converters are compact and can be placed between the gyrotron cavity and the output window. The Denisov quasi-optical converter [27], is a modification of the Vlasov converter, which provides much lower diffraction losses and higher power transmission efficiency. This is a highly efficient way to transform the complex circular cavity mode into a linearly polarized Gaussian beam. Efficient conversion is crucial for high-power gyrotrons to avoid severe cooling problems and, possibly, to avoid device destruction. An additional benefit of these quasi-optical converters is that they make it possible to separate the RF beam from the electron beam.

The basic principles of quasi-optical mode converters for circular symmetric (TE_{0n}) modes and rotating asymmetric (TE_{mn}) modes are shown in Figure 21-45. In Figure 21-46, the field patterns are unrolled to simplify the visualization of the operation of the mode launchers that are part of the quasi-optical converters.

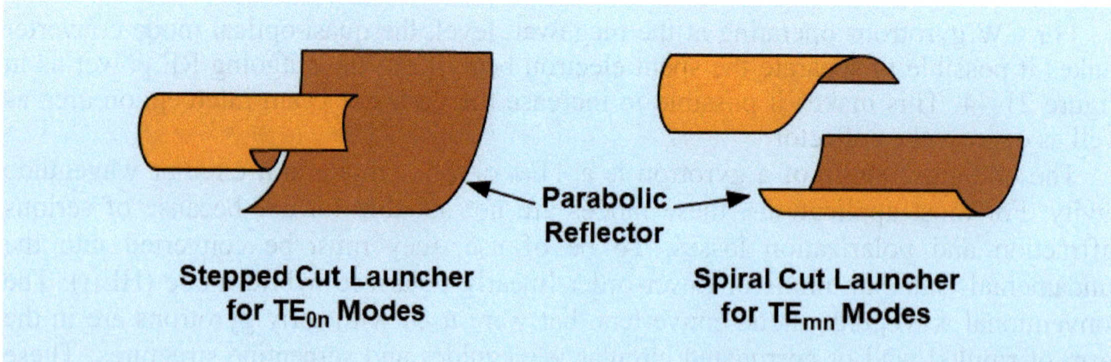

Figure 21-45 Launchers for TE_{0n} and TE_{mn} gyrotron cavity modes. (Adapted from: *Gyrotrons*, M. V. Kartikeyan et al., © Springer-Verlag 2004.)

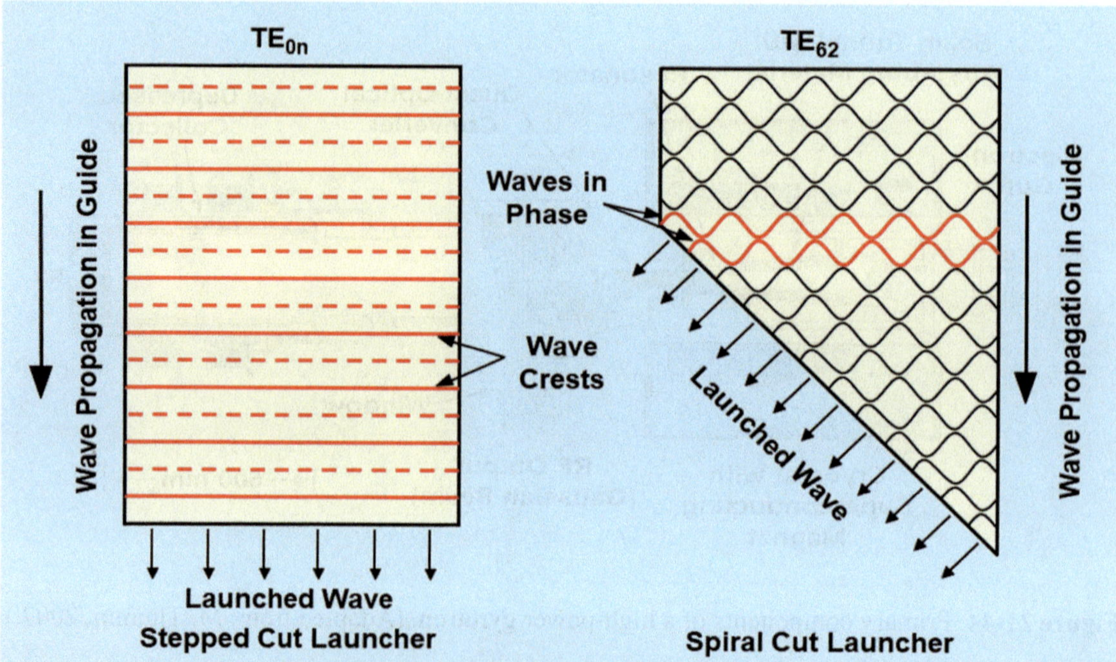

Figure 21-46 Unrolled wave patterns at the launcher for TE_{0n} and TE_{mn} gyrotron cavity modes.

Because the TE_{0n} modes have no variations in the field in the azimuthal direction, all parts of a wave crest arrive at the launcher simultaneously and so the launcher has a square-cut end for launching the wave. In contrast to this, TE_{mn} modes have cyclical variations in the field in the azimuthal direction. By cutting the launcher at the appropriate angle as shown in Figure 21-46, wave fronts, all parts of which are in phase, can be launched. After being reflected by an appropriately shaped mirror (quasi-parabolic) a beam of parallel rays is formed (Figure 21-47).

A quasi-optical mode converter has been theoretically and experimentally investigated at Karlsruhe Institute of Technology (KIT). The converter was developed for the 2-MW, 170-GHz, $TE_{34,19}$ mode coaxial-cavity ITER gyrotron. The converter, shown in Figure 21-48, consists of a launcher, a quasi-elliptical mirror, and two toroidal mirrors [28].

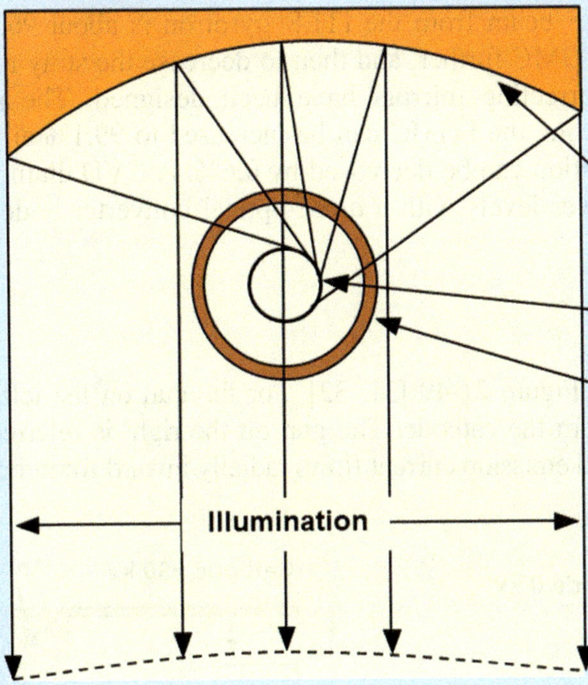

Figure 21-47 Wave propagation from a spiral cut launcher. (Adapted from: *Gyrotrons*, M. V. Kartikeyan et al., © Springer-Verlag 2004.)

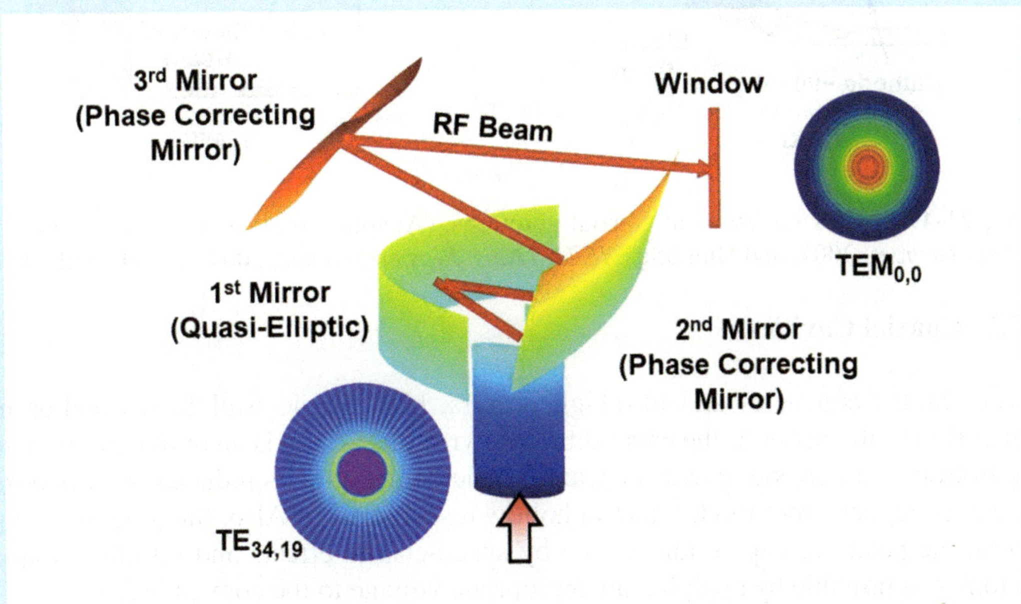

Figure 21-48 High-efficiency quasi-optical mode converter. (Adapted from: J. Jin, IVEC 2010. © 2010 IEEE.)

Theoretical and experimental results show that the TEM_{00} fundamental gaussian mode content (FGMC) of the RF beam from the ITER gyrotron is about 96.7% at the window plane. To improve the FGMC further, and then to decrease the stray radiation in the tube, two adapted phase-correcting mirrors have been designed. The simulation results show that with these mirrors, the FGMC can be increased to 99.1% at the output window plane and the stray radiation can be decreased by 0.6%. A CVD diamond output window that is used at high-power levels with a quasi-optical converter is described in Chapter 22.

21.7.2 Coaxial MIGs

Two coaxial MIGs are shown in Figure 21-49 [31, 32]. For the gun on the left, emission current flows radially outward from the cathode. The gun on the right is referred to as an inverted coaxial MIG (ICMIG) and emission current flows radially inward from the cathode.

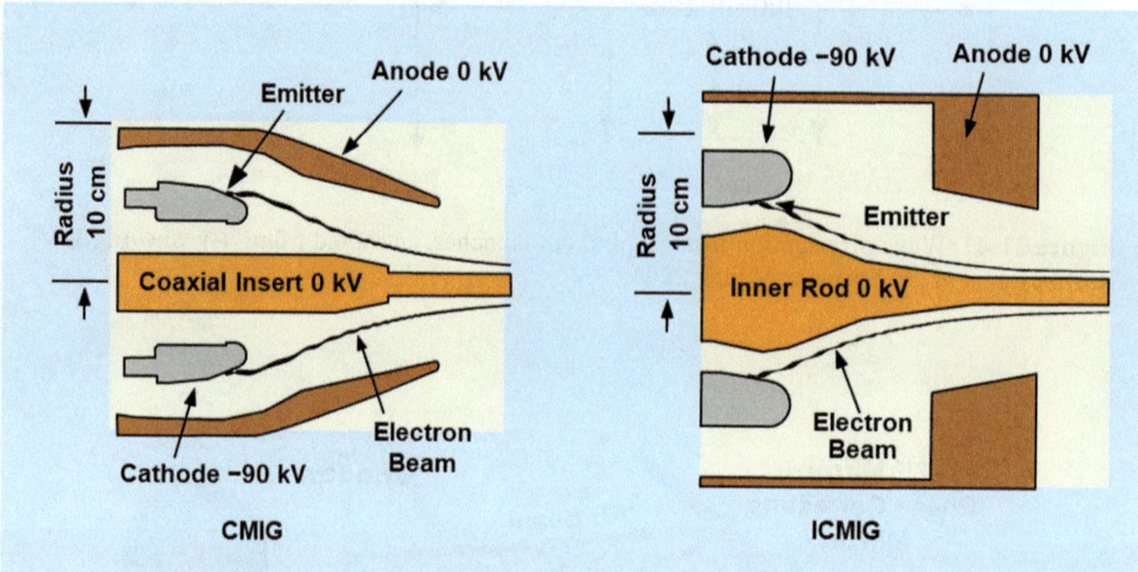

Figure 21-49 MIGs for use with coaxial gyrotrons. (Adapted from: B. Piosczyk, *IEEE Trans. Electron Devices*, 2001, and Dumbrajs, *IEEE Trans. Plasma Science*, 2004. © 2001 and 2004.)

21.7.3 Coaxial Cavities

At very high frequencies and ultrahigh powers, high ohmic wall losses and/or mode competition problems limit the capabilities of gyrotrons. There is an argument to be made for gyrotrons with coaxial cavities because mode competition is reduced, so it is possible to use even higher-order modes than in hollow resonators [x]. Also, the inner rod helps in limiting the potential depression caused by space-charge effects and ultrafast frequency step tuning is possible by applying an appropriate voltage to the coaxial insert.

In a coaxial cavity, ohmic losses can be controlled by corrugating the center conductor as indicated in Figures 21-50 and 21-51. If axial rectangular grooves are placed in the center conductor, which are $1/4\lambda$ deep, the impedance of the surface of the center conductor at the frequency of interest appears to be infinite, so currents and losses are very low.

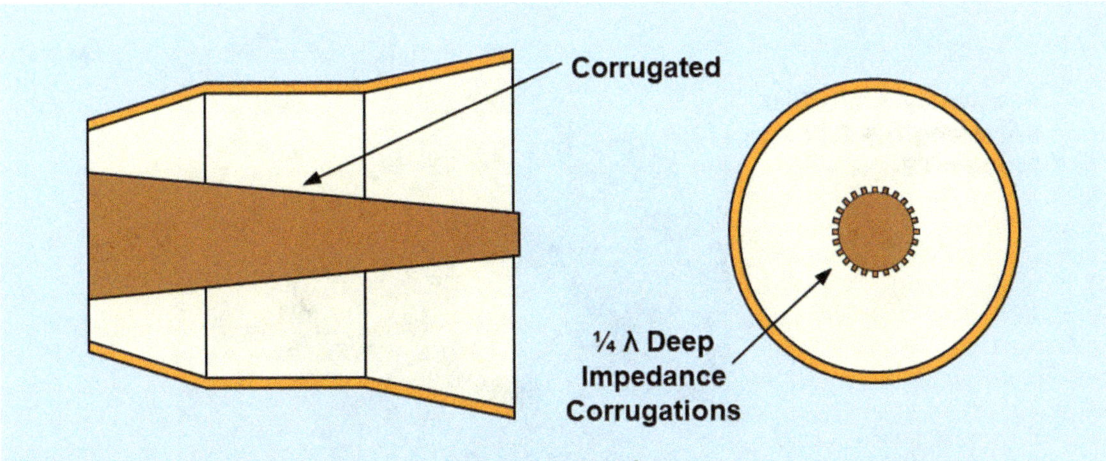

Figure 21-50 Coaxial gyrotron cavity. (From: B. Piosczyk et al., IAEA Technical Meeting on ECRH Physics and Technology for ITER, 2003.)

In either a hollow cavity or a coaxial cavity, the tunnel expands from the cavity to the output window so RF propagates towards the window. The tunnel contracts toward the gun and this prevents propagation of the RF wave toward the gun.

Figure 21-52 shows the transverse field of the $TE_{31,17}$ mode in a high-order mode coaxial cavity [4]. The corrugations in the center conductor are 0.45 mm deep and 0.35 mm wide and the number of grooves is 62.

21.7.4 Step Frequency Tuning

The mode and frequency that are excited in a cavity like that shown in Figure 21-51 are highly dependent on the beam voltage, magnetic field and the voltage on the center conductor. Gantenbein et al. reported on the step tuning of a 1-MW D-band gyrotron. This tube was equipped with a Brewster angle CVD diamond output window so that operation over the frequency range from 111.6 to 165.7 GHz was possible. (Brewster angle windows are described in Chapter 22.) Step tuning was accomplished by varying the magnetic field [22].

21.7.5 Collectors

Collectors [31–36], including depressed collectors, for linear-beam tubes were discussed in Chapter 16. There is usually very little magnetic field in those collectors and the shape of the beam entering the collector is controlled by space-charge forces and electric field configurations so the design of the undepressed collectors is relatively straightforward.

In TWTs, the interaction of the beam with the circuit is relatively weak and the energy spread of the electrons in the spent beam is about 30%–40% of the initial beam energy. This makes it possible to design depressed collectors having as many as 10 stages of depression and to achieve overall efficiencies over 70%. The author (of this book) is not aware of depressed collectors for linear-beam tubes that operate at power levels above 100 kW.

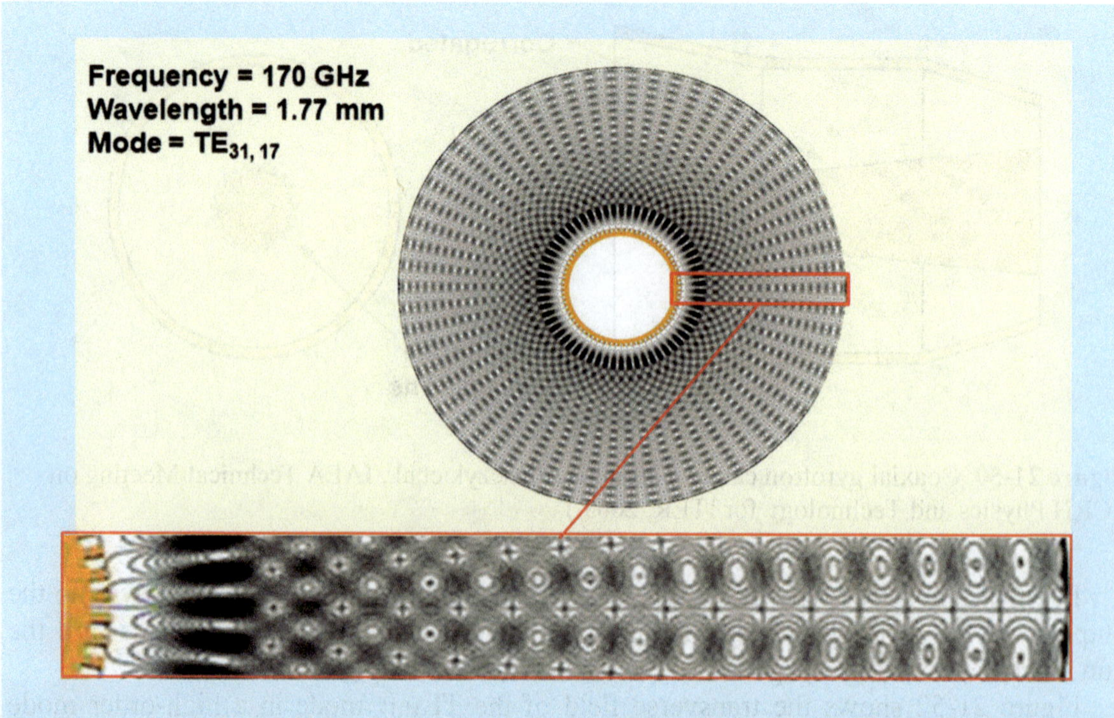

Figure 21-51 High-order mode in a coaxial cavity. (From: A. Grudiev, Hamburg University of Technology, 2010.)

For high-power gyrotrons, on the other hand, depressed collectors are being operated at power levels over 1 MW CW. The design of these collectors is particularly challenging because the magnetic field in the collector may be several tens of gauss and the energy range of the electrons in the spent beam can be 35%–140% of the incident beam energy [8]. Still, single-stage depressed collectors have been used on high-power gyrotrons and have increased the overall efficiency to above 50% [28–30]. Although these experiments were limited to single-stage collectors, they demonstrated the value of depressed collectors in reducing the power loading, X-ray generation, and possibly of using less costly power supplies.

In a high-frequency gyrotron the magnetic flux density in the interaction region is several tens of tesla. In going from the interaction region to the collector, the flux density decreases to several tens of gauss. Electrons follow the expanding flux lines in the collector and land on a ring-shaped band on the collector surface, the width of which can be estimated.

Thus, knowing the beam current and voltage, the power density imparted to the collector surface (the wall loading) can be estimated. In high-power gyrotrons, the diameter of the collector has to be excessively large to keep the wall loading within the technological limit of ~500 W/cm^2 [8]. It should be noted here that in linear-beam tube design, the wall loading limit is ~1 kW/cm^2 (Chapter 16).

If the radius of the collector is to be limited, the width of the band on which the electrons land must be made larger. Auxiliary collector coils can be used to widen the interception band. In the collector in Figure 21-52, these coils are located in the insert in the center of the collector [31]. The second way to reduce loading is to use sweeping coils to move the electron beam axially in the collector. For the collector shown in Figure 21-53 transverse and vertical sweeping coils are used. Transverse sweeping is accomplished with a rotating magnetic field generated by three pairs of coils powered by a three-phase 50-Hz supply. The part of the collector where the coils are located is made from nonmagnetic stainless steel. Two of the coils can be seen in Figure 21-53. The vertical sweeping coils, which are not shown in Figure 21-53, operate at 7 Hz.

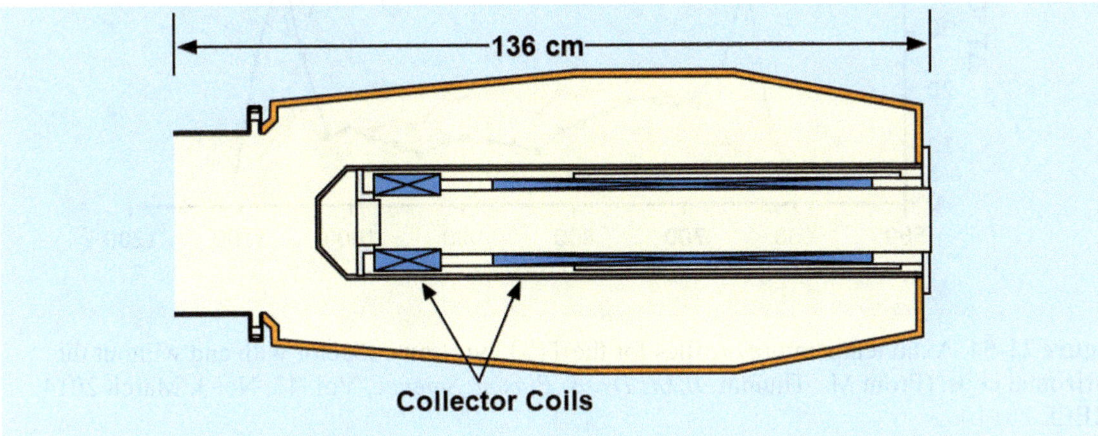

Figure 21-52 Single-stage depressed collector for a 2.2-MW CW gyrotron. (From: B. Piosczyk, Conference on Infrared and Millimeter Waves, 2001.)

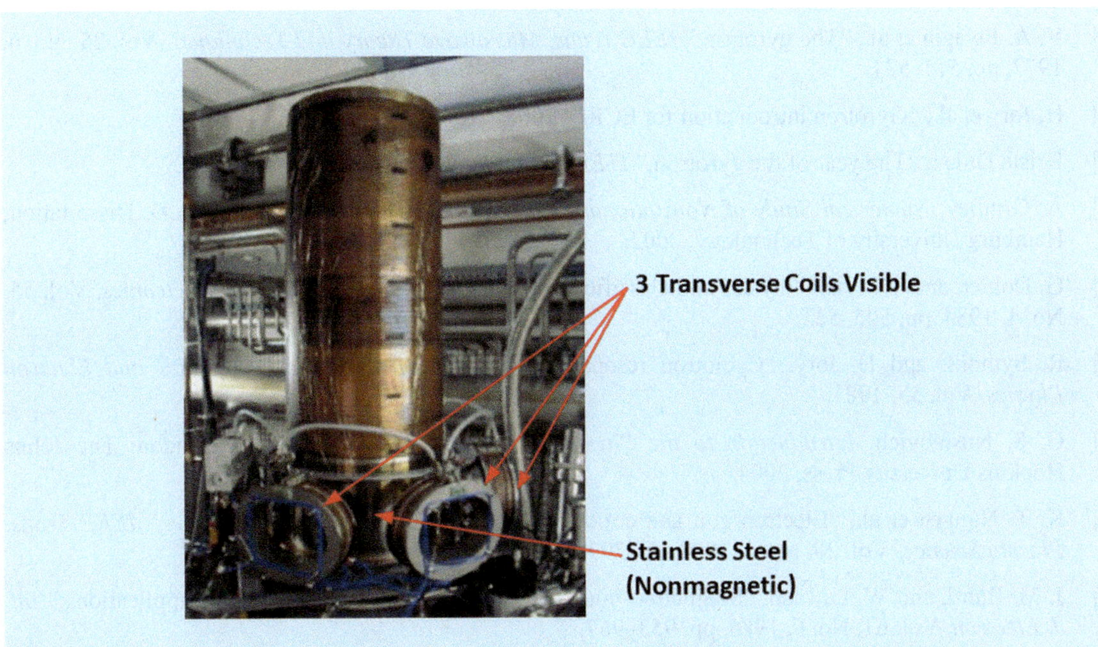

Figure 21-53 TED gyrotron collector with three pairs of transverse sweeping coils. (The vertical coils have been removed.) (From M. Thumm, *IEEE Trans. Plasma Science*, Vol. 42, No.3, March 2014.) © IEEE 2014.

Axial temperature profiles of the TED collector in Figure 21-52 are shown in Figure 21-53. The temperature is much better controlled by using vertical and horizontal coils rather than just the vertical coils.

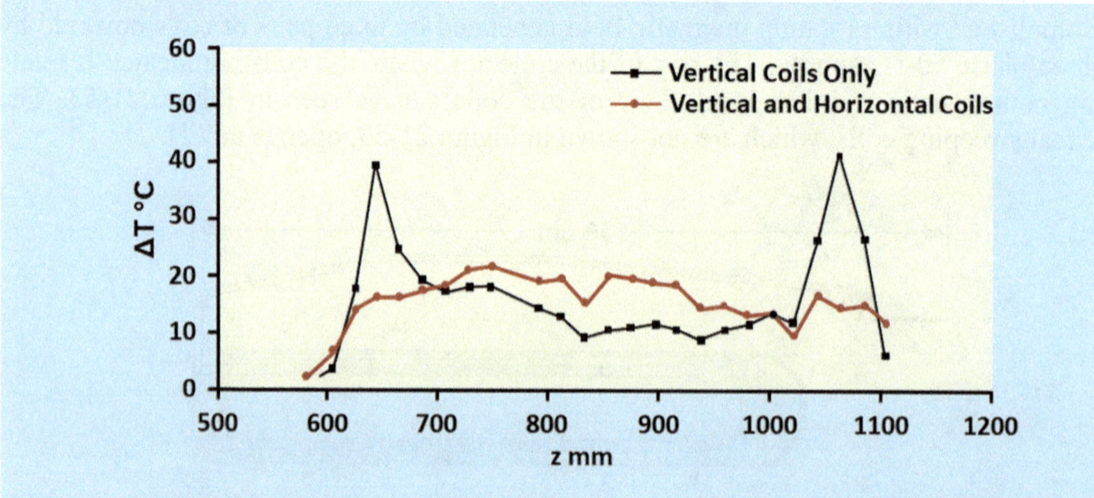

Figure 21-54 Axial temperature profiles for the TED gyrotron collector with and without the horizontal coils. (From M. Thumm, *IEEE Trans. Plasma Science*, Vol. 42, No. 3, March 2014. ©IEEE 2014.)

REFERENCES

[1] V. A. Flyagin et al., "The gyrotron," *IEEE Trans. Microwave Theory and Techniques,* Vol. 25, No. 6, 1977, pp. 514–521.

[2] H. Jory et al., "Gyrotron introduction for ECRIS 2008," 2008.

[3] Krista Dulon, "The year of the gyrotron," *ITER Newsline*, February, 2018.

[4] A. Grudiev, *Numerical Study of Nonstationary Phenomena in Gyro-Oscillators*, Ph.D. Dissertation, Hamburg University of Technology, 2002.

[5] G. Dohler, and W. Friz, "Physics and classification of fast-wave devices," *Int. J. Electronics,* Vol. 55, No. 4, 1983, pp. 505–522.

[6] R. Symons, and H. Jory, "Cyclotron resonance devices," *Advances in Electronics and Electron Physics,* Vol. 55, 1981.

[7] G. S. Nusinovich, *Introduction to the Physics of Gyrotrons*, Baltimore and London: The Johns Hopkins University Press, 2004.

[8] K. T. Nguyen et al., "Electron gun and collector design for 94-GHz gyro-amplifiers," *IEEE Trans. Plasma Science,* Vol. 26, No. 3, 1998, pp. 799–813.

[9] J. M. Baird, and W. Lawson, "Magnetron injection gun (MIG) design for gyrotron applications," *Int. J. Electron,* Vol. 61, No. 6, 1986, pp. 953–967.

[10] W. C. Brown, "The magnetron—a low noise, long life amplifier," *Applied Microwaves*, 1990.

[11] J. M. Baird, "Survey of fast wave tube developments." IEDM 1979, pp. 156–163.

[12] A. B. Bronwell, and R. E. Beam, *Theory and Application of Microwaves*, New York: McGraw-Hill, Inc. 1947.

[13] D. V. Kisel et al., "An experimental study of a gyrotron, operating at the second harmonic of the cyclotron frequency, with optimized distribution of the high frequency field," *Radio Engineering and Electronic Physics,* Vol. 19, No. 4, 1974, pp. 95–100.

[14] K. L. Felch et al., "Characteristics and applications of fast-wave gyrodevices," *Proc. IEEE,* Vol. 87, No. 5, 1999, pp. 752–781.

[15] M. Blank et al., "Development and demonstration of high-average power W-band gyro-amplifiers for radar applications," *IEEE Trans. Plasma Science,* Vol. 30, No. 3, 2002, pp. 865–875.

[16] CPI, "VGB-8194, gyrotron amplifier."

[17] G. J. Linde et al., "WARLOC: A high-power coherent 94 GHz radar," *IEEE Trans. Aerospace and Electronic Systems,* Vol. 44, No. 3, 2008, pp. 1102–1117.

[18] J. P. Calame et al., "Gyrotron-traveling wave-tube circuits based on lossy ceramics," *IEEE Trans. Electron Devices,* Vol. 49, No. 8, 2002, pp. 1469–1477.

[19] R. S. Symons et al., "An Experimental Gyro-TWT," *IEEE Trans. Microwave Theory and Techniques,* Vol. 29, No. 3, 1981, pp. 181–184.

[20] P. Ferguson, and R. Symons, "A C-band gyro-TWT," *Technical Digest, IEDM,* 1980, pp. 310–313.

[21] P. E. Ferguson, and R. S. Symons, "A gyro-TWT with a space-charge limited gun," *Technical Digest, IEDM,* 1981, pp. 198–201.

[22] G. Gantendein, et al., "First operation of a step-frequency tunable 1-mw gyrotron with a diamond Brewster angle output window," *IEEE Trans. Electron Devices,* Vol. 61, No. 6, June 2014.

[23] S. Sabchevski, Quasi-optical converters for high-power gyrotrons: a brief review of physical models, numerical methods and computer codes," *Jour. of Physics,* Vol. 442006, pp. 102–109.

[24] M. K. Thumm, and W. Kasparek, "Passive high-power microwave components," *IEEE Trans. Plasma Science,* Vol. 30, No. 3, 2002, pp. 755–786.

[25] S. N. Vlasov, L. I. Zagryadskaya, and M. I. Petelin, "Transformation of a whispering gallery mode propagating in a circular waveguide into a beam of waves," *Radio Eng. Electron. Phys.,* Vol. 12, 1975.

[26] G. G. Denisov et al., "110 GHz gyrotron with built-in high-efficiency converter," *Int. J. Electronics,* Vol. 72, 1992, pp. 1079–1091.

[27] J. Jin et al., "2.2: Design of phase correcting mirror system for coaxial-cavity ITER gyrotron," *IEEE IVEC,* 2010, pp. 29–30.

[28] B. Piosczyk et al., "Single-stage depressed collectors for gyrotrons," *IEEE Trans. Plasma Science,* Vol. 24, 1996, pp. 579–585.

[29] T. A. Hargreaves et al., "Depressed collector performance on the NRL quasi-optical gyrotron," *Conf. Digest, 15th Int. Conf. on Infrared and Millimeter Waves,* 1990, p. 330.

[30] K. Sakamoto et al., "Major improvements in gyrotron efficiency with beam energy recovery," *Phys. Rev. Lett.,* Vol. 73, 1994, p. 3532.

[31] B. Piosczyk, "A 2.2 MW CW collector for a coaxial cavity gyrotron," *Conf. Digest, 26th Int. Conf. on Infrared and Millimeter Waves,* 2001.

[32] G. Dammertz, "Beam tunnel, resonator, collector and RF window," *Gyrotron oscillators, their principles and practice,* C. Edgecombe, (ed.), London: Taylor and Francis, 1993.

[33] Varian brochure, *Introduction to Gyro-Devices,* 2000.

[34] M. Thumm, "State-of-the-art of high-power gyro-devices and free electron masers update 2006," 2007.

[35] M. Thumm, "History, presence and future of gyrotrons," *IEEE IVEC,* 2009, pp. 37–40.

[36] T. Rzesnicki et al., "2.2-MW record power of the 170-GHz European preprototype coaxial-cavity gyrotron for ITER," *IEEE Trans. Plasma Science,* Vol. 38, No. 6, 2010, pp. 1141–1149.

Chapter 22

Windows

22.1 BACKGROUND

The problem of transmitting RF energy from the vacuum environment inside a tube to an external load is common to all types of vacuum electron devices. The window used for this purpose is one of the most critical elements in a high-power microwave system and is also one of the most difficult elements to design, develop, fabricate, and use.

The number of possible window configurations seems to be limited only by the ingenuity of the designer [1, 2]. Many fairly distinct designs in coaxial lines and waveguides have been used successfully. All of these techniques have been highly dependent on materials and vacuum seal technology. Over the last 80 years, windows have followed a gradual evolution from simple mica membranes in waveguides to high purity alumina, beryllia, sapphire [3-6], synthetic polycrystalline diamond [7], and now, possibly synthetic single-crystal diamond [8].

The characteristics of ideal and real windows are summarized in Figure 22-1. With the ideal window, there are no losses and no impedance mismatches. On the other hand, materials available for use in real windows are all lossy to some extent and so heating is often a problem. All materials available for use in windows have a dielectric constant that is considerably larger than that of the vacuum and/or gas that the window must interface with so reflections occur. These reflections lead to resonances within the structure containing the window (trapped modes) and/or resonances within the window material (ghost modes).

22.2 COAXIAL WINDOWS

In the case of coaxial windows, the mechanical design and fabrication can be difficult. Two vacuum seals are required (inner conductor to ceramic and outer conductor to ceramic). Also, at high frequencies, the center conductor in the ceramic becomes very small and metallization and brazing become difficult. The RF design of coaxial windows is relatively easy because the transmission line uses the TEM mode and the ceramic is a disk, as shown in Figure 22-2.

The impedance of a coaxial line operating in the TEM mode is given by

$$Z_o = \frac{60}{\sqrt{\varepsilon}} \ln \frac{b}{a} \text{ Ohms} \tag{22-1}$$

where ε is the relative dielectric constant of the media between the center and outer conductors. For vacuum or air where $\varepsilon = 1$, the ratio $b/a = 2.3$ for an impedance of 50Ω.

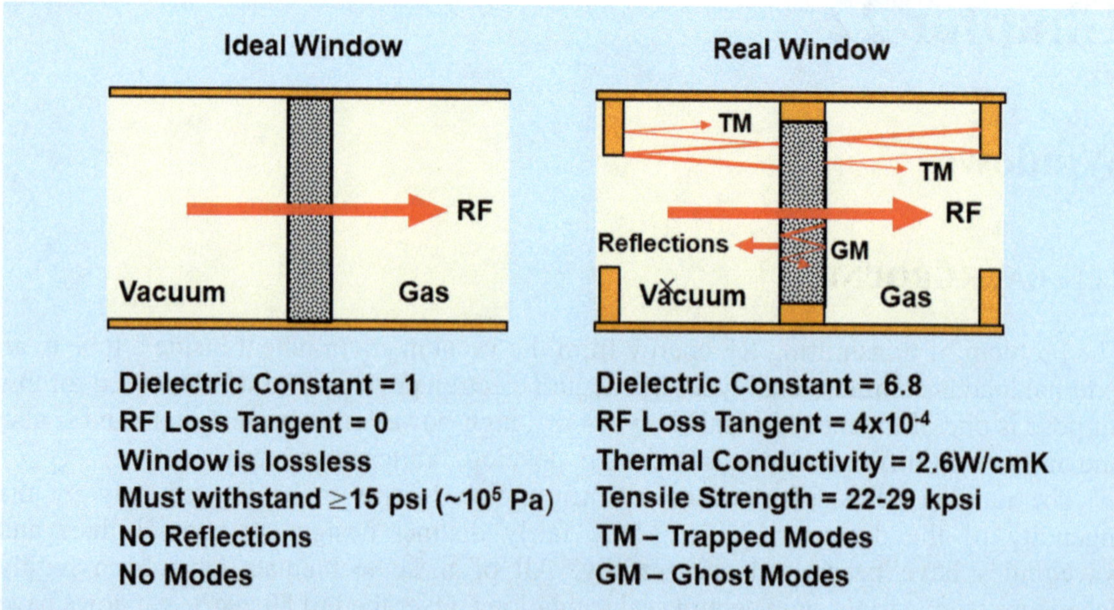

Figure 22-1 Characteristics of ideal and real windows.

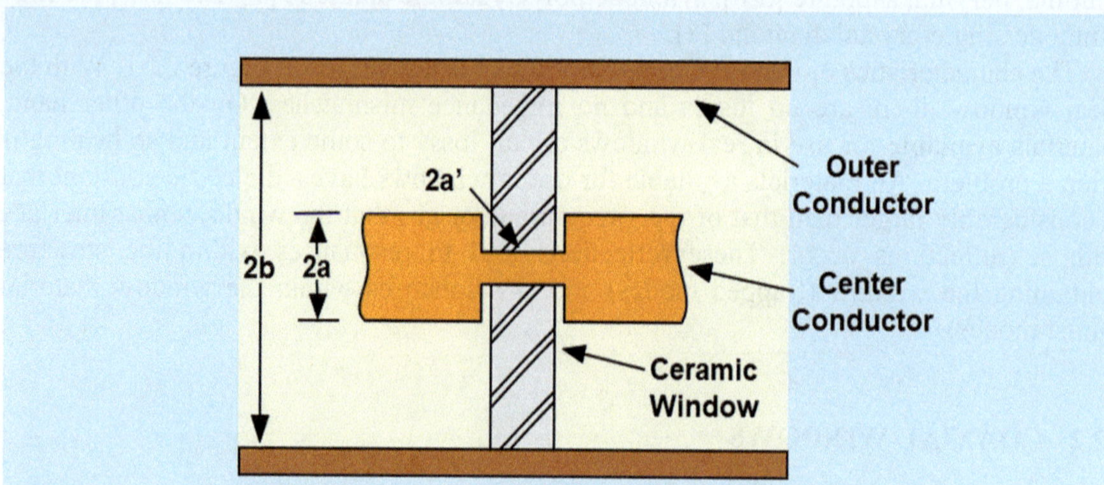

Figure 22-2 Basic configuration of a coaxial window.

The ceramic that is used for the window is nearly always Al_2O_3 or, when high thermal conductivity is required, BeO or aluminum nitride is used. (Because of its toxicity, BeO is replaced by aluminum nitride when machining is required.) The relative dielectric constant of alumina is 9.3 and so, for an impedance of 50Ω, $b/a = 12.7$. This means that the radius of the center conductor in alumina (a') must be about 20% of that in vacuum or air (a). For beryllia, the relative dielectric constant is 6.4 so the radius of the center conductor must be about 30% of that in vacuum or air. Figure 22-3 shows the relative size of conductors for a coaxial window used in the relative dielectric constant of alumina is 9.3 and so, for an impedance of 50Ω, b/a = 12.7 in practice. Notice that the radius of the center conductor in the ceramic is about one-third of the radius in air or vacuum. It should also be noticed that the reduced diameter is extended a short distance from the ceramic. This is done to improve the impedance match.

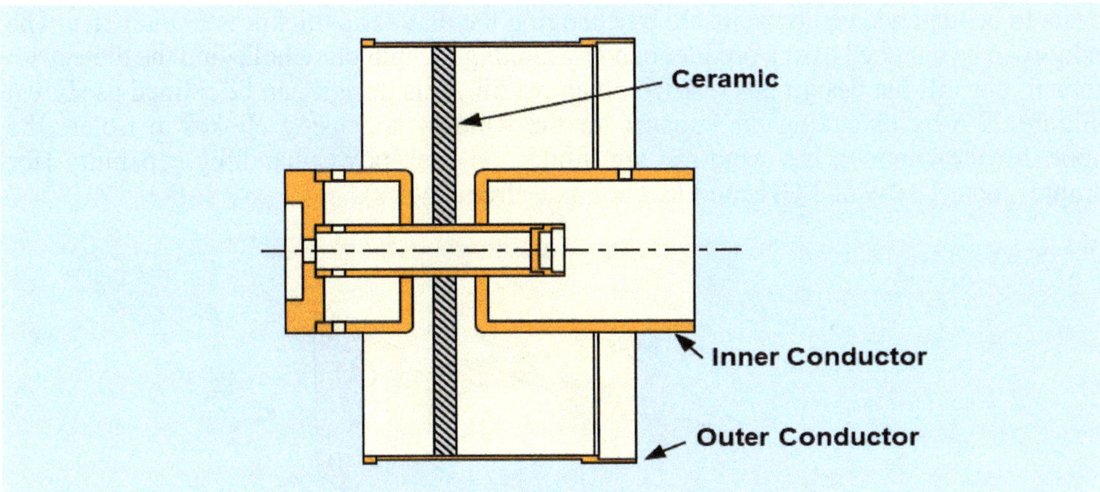

Figure 22-3 Coaxial window for operation at 1.25 GHz. (Adapted from: *Power Klystrons Today,* M. J. Smith and G. Phillips, copyright 1995 by Research Studies Press, Ltd.)

Hughes Electron Dynamics Division has described the integral coax-to-waveguide adapter window shown in Figure 22-4. This was used on a coupled-cavity TWT with a waveguide input. The interface with the customer's system was coaxial. In this case, the design was empirical. Impedance match adjustments could be made by varying the waveguide backwall position, the inner-conductor diameter, and the depth of penetration of the inner conductor in the waveguide. The vacuum seal was made to the outer conductor only.

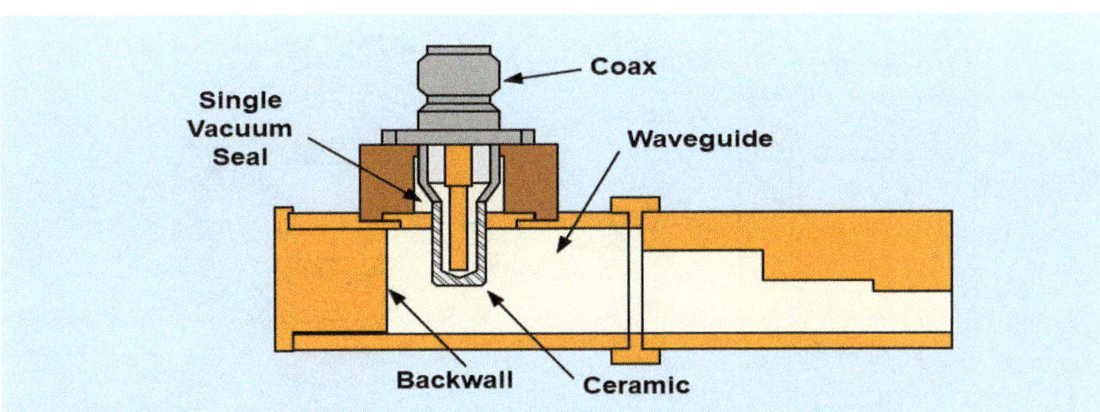

Figure 22-4 Coax window with waveguide adapter. (Courtesy of Hughes Electron Dynamics Division.)

22.3 CONVENTIONAL WAVEGUIDE WINDOWS

The simplest window consists of an inductive iris into which is sealed a capacitive dielectric plate. An example is shown in Figure 22-5. The impedance is adjusted so that it is resonant and the shunt admittance is made zero at the band center. The frequency at which the window appears to be matched can be adjusted by changing the dielectric thickness or diameter. The window can be matched over a broader band by making the inductive hole (and the dielectric) elliptical, but still the design has restricted bandwidth. This design can be refined as shown in Figure 22-6 by extending the support for the window and using chokes to isolate the support for the window. Iris windows are limited on peak power-handling capability (for example, about 1 MW at 3 GHz and less at higher frequencies).

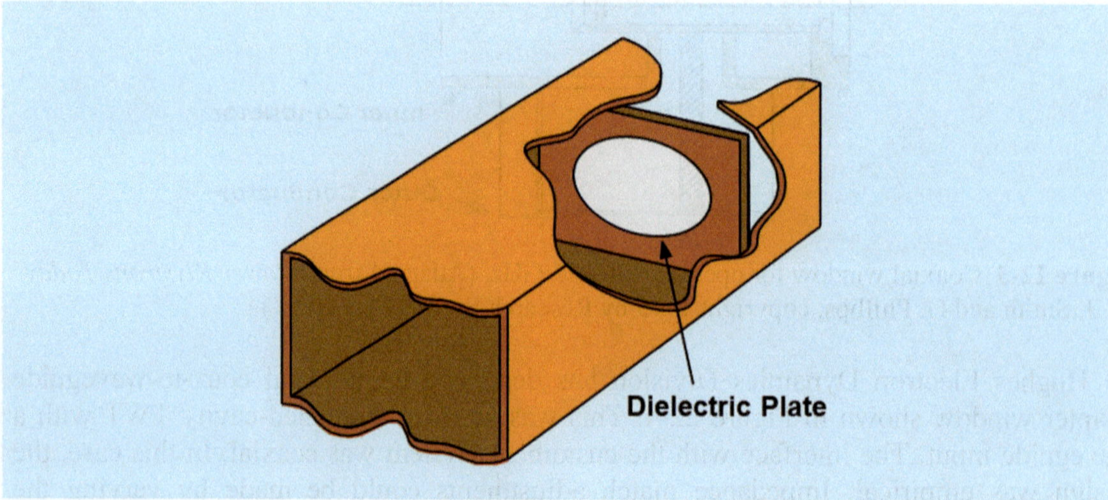

Figure 22-5 Iris window. (Adapted from: *Power Klystrons Today,* M. J. Smith and G. Phillips, copyright 1995 by Research Studies Press, Ltd.)

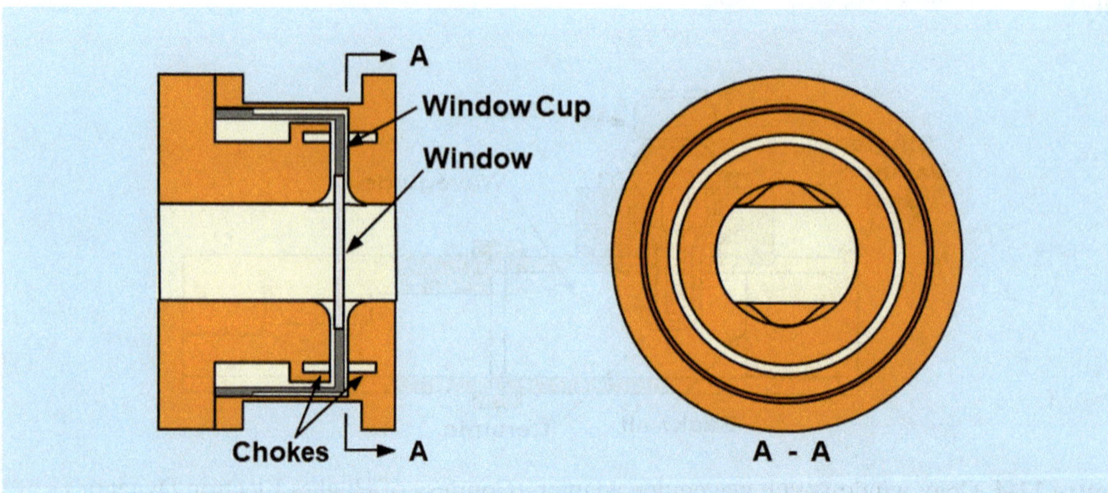

Figure 22-6 Iris window with choke. (Adapted from: *Power Klystrons Today,* M. J. Smith and G. Phillips, copyright 1995 by Research Studies Press, Ltd.)

The two most generally used window designs suitable for medium- and high-power tubes consist of either a rectangular ceramic block in the output waveguide, or a circular ceramic disk (sometimes referred to as a poker chip) in a section of round waveguide that is then adapted to rectangular waveguides on both ends. The window thickness is almost always one-half wavelength or less (although the thickness is somewhat arbitrary, since almost any configuration can be designed for good transmission). The most important factors are the methods of design used, the average and peak power-handling capabilities, the bandwidth available, the elimination of unwanted resonances, and the basic technology of manufacture.

The block window is not used as often as the disk window, but it is easier to design and is useful in illustrating some of the characteristics and problems in design and use. The most common block window is half a wavelength thick and fills the entire cross section of the waveguide. These windows are well matched at the resonant frequency and are suitable for narrow bandwidths without modification (see Figure 22-7). If irises or other matching elements are added, the window can be made into a filter with a bandwidth of 10% to 20%. It is mainly a high-power window, particularly suitable for high average power. The transmission loss is low, on the order of 0.2%, and because it is a filter, the reflection from it is essentially reactive. A VSWR plot for one of these windows is shown in Figure 22-8. The electrical design is straightforward and can easily be adapted to computer approaches where the matching elements are determined automatically for a given bandwidth [7].

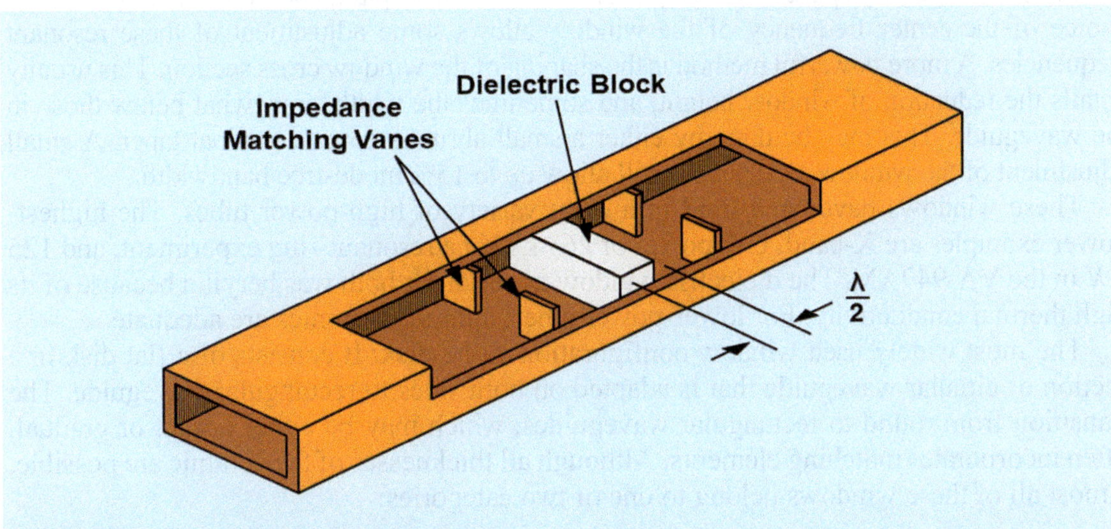

Figure 22-7 Half-wave resonant block window in rectangular waveguide. (Adapted from: *Linear-Beam Tubes,* A. Staprans, Internal Varian Report, September 1968.)

Probably the most severe problem for block windows are ghost modes. Forrer and Jaynes have described a method for determining these resonances, which arise from stored energy in higher-order waveguide modes inside the dielectric [8]. These modes are beyond cutoff in the waveguide and behave as high-Q cavity modes that are loosely coupled to the dominant propagating mode. These resonances, although barely visible on a VSWR plot, are hazardous for high-power operation because of the possibility of high-power absorption and subsequent failure of the window at ghost-mode frequencies.

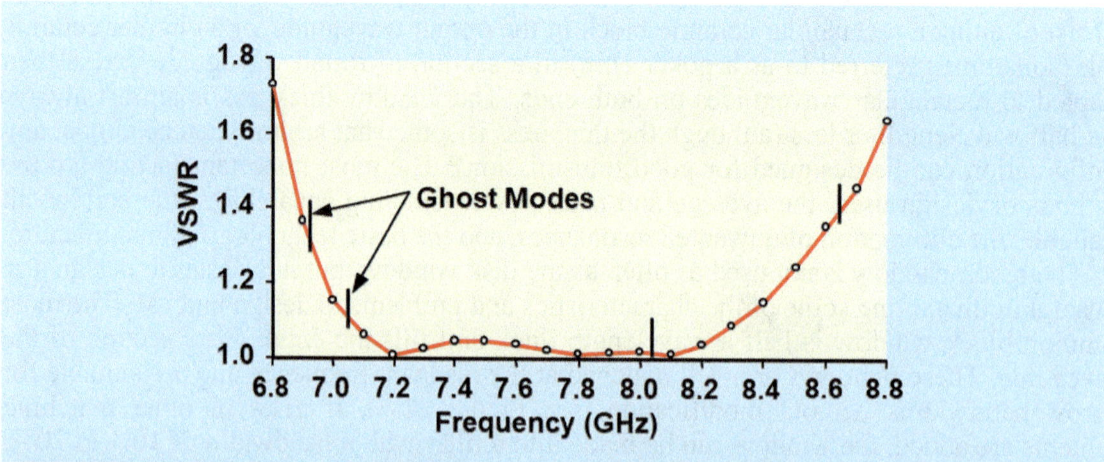

Figure 22-8 Impedance match of a half-wave resonant block window in a rectangular waveguide. (Adapted from: *Linear-Beam Tubes,* A. Staprans, Internal Varian Report, September 1968.)

For a standard window configuration using alumina ceramic, more than 10 such resonances, irregularly spaced, will be present in the ordinary passband of the waveguide. The choice of the center frequency of the window allows some adjustment of these resonant frequencies. A more powerful method is the shaping of the window cross section. This usually entails the reduction of window height, and sometimes the width, somewhat below those in the waveguide. This may be done by either a small abrupt step or a gradual taper. A small adjustment of the window aspect ratio will allow up to 15% mode-free bandwidth.

These windows have been used in a great variety of high-power tubes. The highest-power examples are X-band, CW power of 263 kW in a resonant-ring experiment, and 125 kW in the VA-949AM. The dielectric window material for both was beryllia because of its high thermal conductivity. For lower-power tubes, alumina ceramics are adequate.

The most widely used window configuration is the disk. It consists of a flat disk in a section of circular waveguide that is adapted on both sides to rectangular waveguide. The transition from round to rectangular waveguides, which may be either abrupt or gradual, often incorporates matching elements. Although all thicknesses of the ceramic are possible, almost all of these windows belong to one of two categories:

1. The ceramic is made as thin as mechanical stresses and the fear of electrical breakdown will allow (about one-tenth of a wavelength).
2. The window is made self-resonant (a half-wavelength long (or several half-wavelengths long at mm-wave frequencies) at the center frequency). These windows are sturdier and generally less lossy than thin windows but their use requires attention to technological problems such as modes.

Disk windows are substantially easier to build than rectangular windows. However, trapped-mode resonances are a problem. These resonances are due to modes that propagate in the round part of a waveguide, but are cut off at the transitions to the rectangular waveguides. In addition, ghost modes may also be present in the ceramic. Their density increases with the thickness of the ceramic. Since a thicker dielectric is usually used in round rather than in rectangular windows, the ghost-mode population in round windows is usually higher than in rectangular windows.

The simplest and most commonly used form of the disk window is a thin ceramic in a short circular waveguide section joined abruptly to the rectangular waveguides. This window, which in its basic form was patented by Symons [9], is shown in Figure 22-9. It is an easy window to make and has been used for tubes at all power levels. It can be made exceptionally broadband if trapped modes can be tolerated, as is the case in many low- and medium-power tubes.

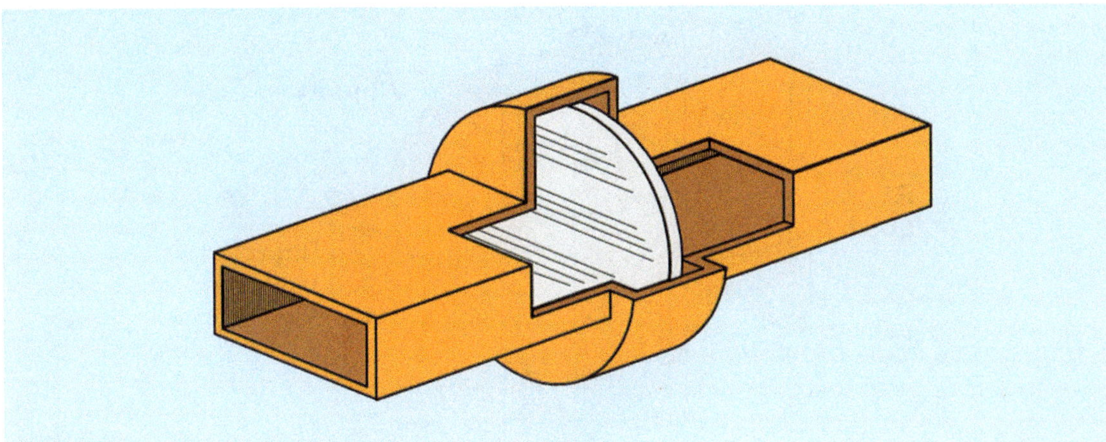

Figure 22-9 Thin-disk, pillbox-type window. (Adapted from: *Linear-Beam Tubes,* A. Staprans, Internal Varian Report, September 1968.)

The VSWR of a broadbanded version of this window is shown in Figure 22-10. Notice that the trapped-mode population is relatively high, although only one of the modes couples very strongly to the propagating wave. By proper adjustment of all dimensions, these windows can be made completely mode-free in a 15% to 20% bandwidth.

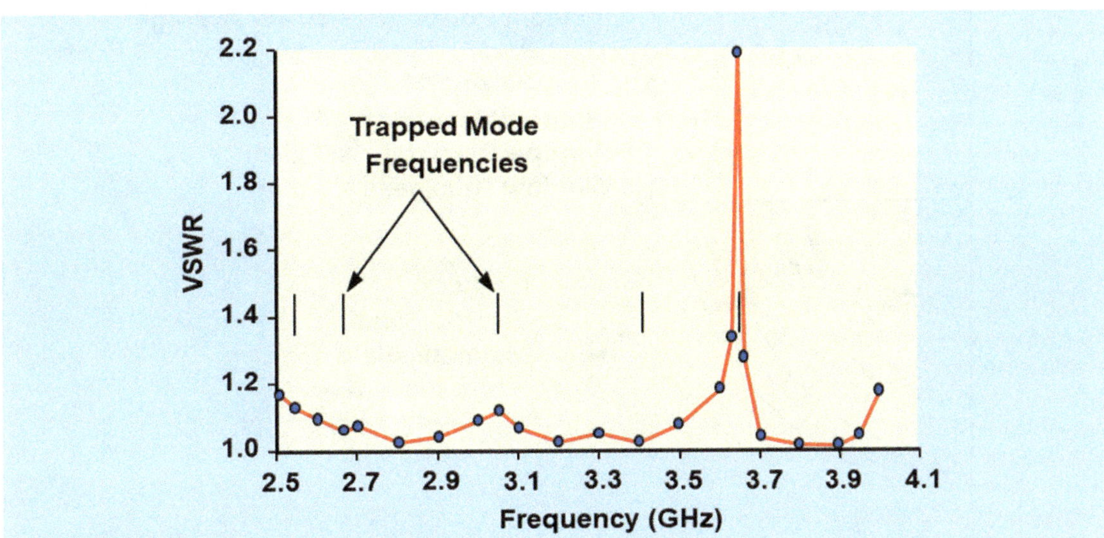

Figure 22-10 Impedance match of a thin-disk, pillbox-type window. (Adapted from: *Linear-Beam Tubes,* A. Staprans, Internal Varian Report, September 1968.)

A liquid cooled version of the thin-disk, pillbox window described by Faillon and Kern

[10] is shown in Figure 22-11. The temperature distribution in the window resulting from the application of 50-kW average at 3 GHz is shown in Figure 22-12. Temperature as a function of radial position is given in Figure 22-13. Heating results in compression at the center of the disk and traction in the outer regions. As Figure 22-15 shows, forces as high as 6 kg/mm^2 result from heating.

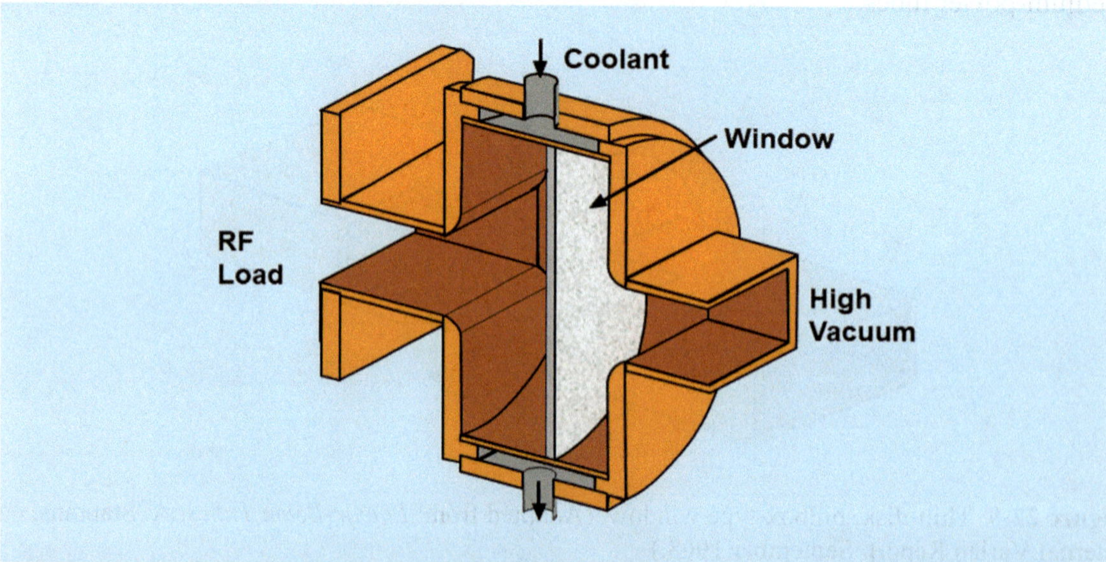

Figure 22-11 Thin-disk, pillbox window. (Adapted from: G. Faillon and P. Kern, LINAC 84.)

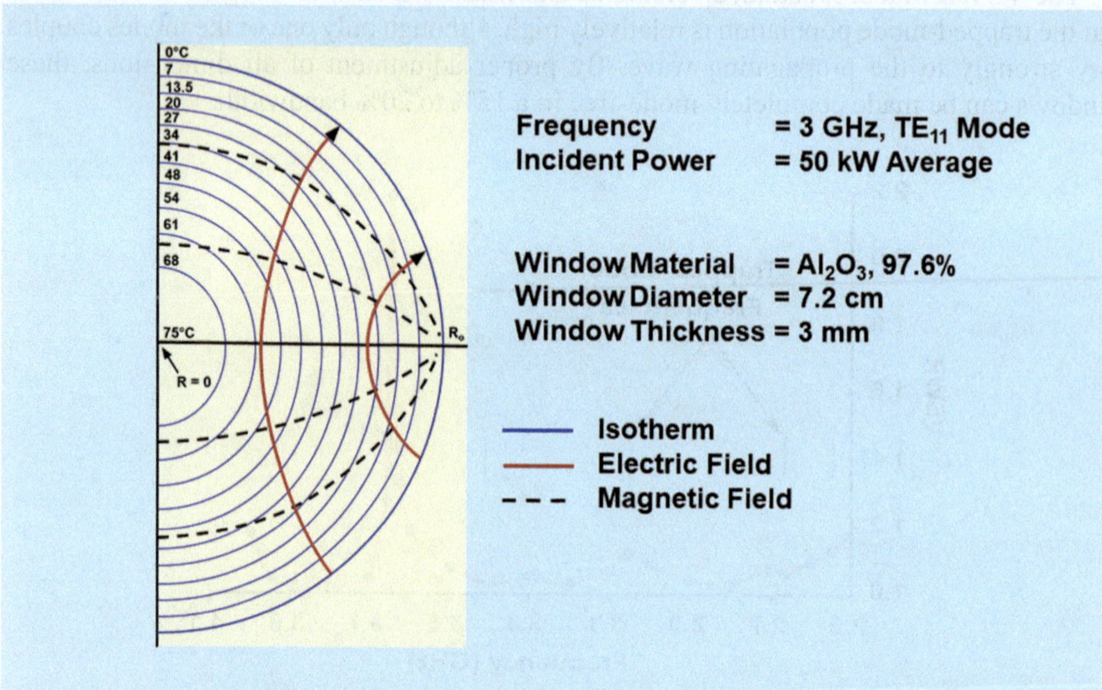

Figure 22-12 Temperature distribution in a thin disk ceramic window. (Adapted from: G. Faillon and P. Kern, LINAC 84.)

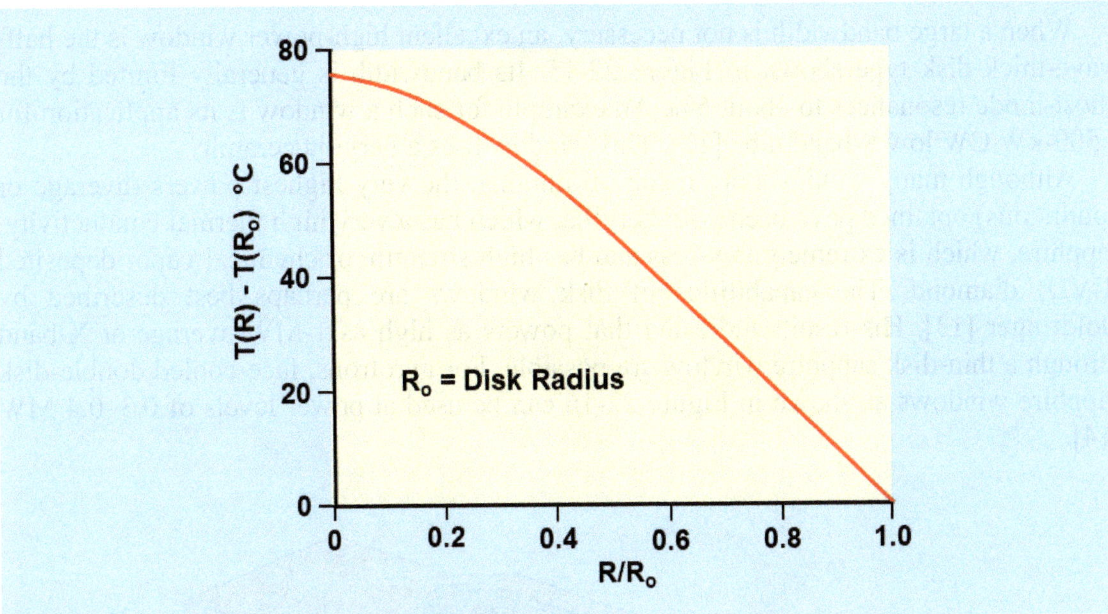

Figure 22-13 Temperature as a function of radius in a pillbox window. (Adapted from: G. Faillon and P. Kern, LINAC 84.)

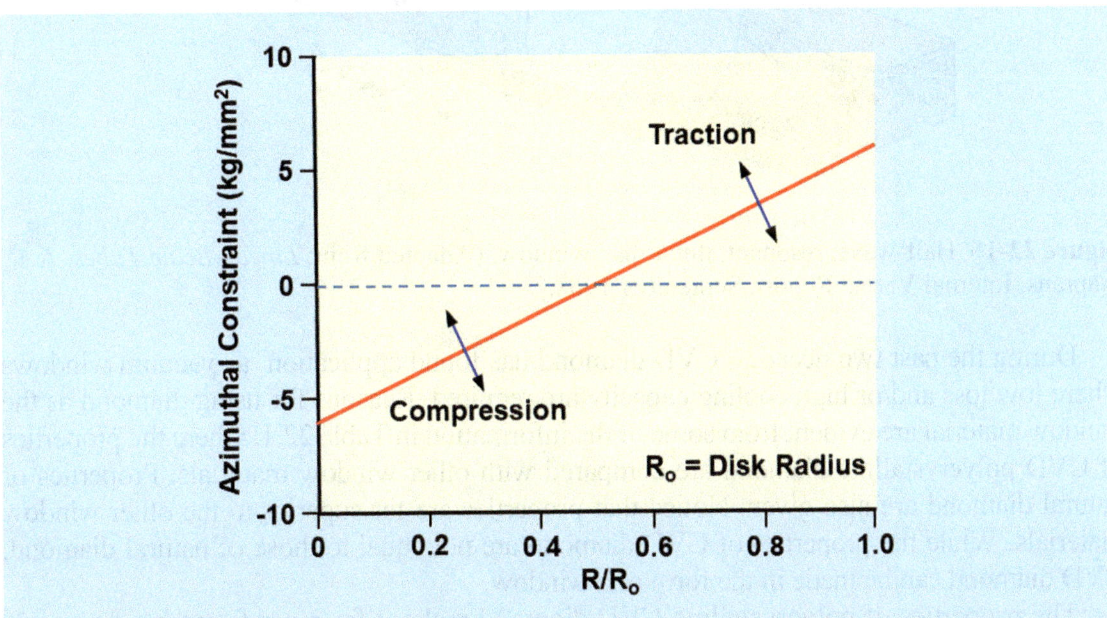

Figure 22-14 Heating results in compression near the center of a disk window and traction in the outer regions. (Adapted from: G. Faillon and P. Kern, LINAC 84.)

Since most output windows rely on cooling at the edge only, the disk window is inferior to the rectangular window for average power because of the longer heat path to the periphery. Nevertheless, by taking advantage of extremely low-loss materials, such as synthetic sapphire, very high average powers have been realized in these windows. The thin-disk window is also suitable for high peak power. One of its many applications was in the 25-MW peak power S-band klystron for the Stanford linear accelerator [11].

When a large bandwidth is not necessary, an excellent high-power window is the half-wave-thick disk type shown in Figure 22-15. Its bandwidth is generally limited by the ghost-mode resonances to about 5%. An example for such a window is its application for a 500-kW CW low S-band tube [12]. This window uses a beryllia ceramic.

Although many windows are made of alumina, the very highest powers (average or continuous) obtained have been with beryllia, which has a very high thermal conductivity, sapphire, which is extremely low-loss and has high strength, or chemical vapor deposited (CVD) diamond. The capabilities of disk windows are perhaps best described by Goldfinger [13]. His results indicated that powers as high as 1-MW average at X-band through a thin-disk sapphire window are possible. For gyrotrons, face-cooled double-disk sapphire windows as shown in Figure 22-16 can be used at power levels of 0.3–0.4 MW [14].

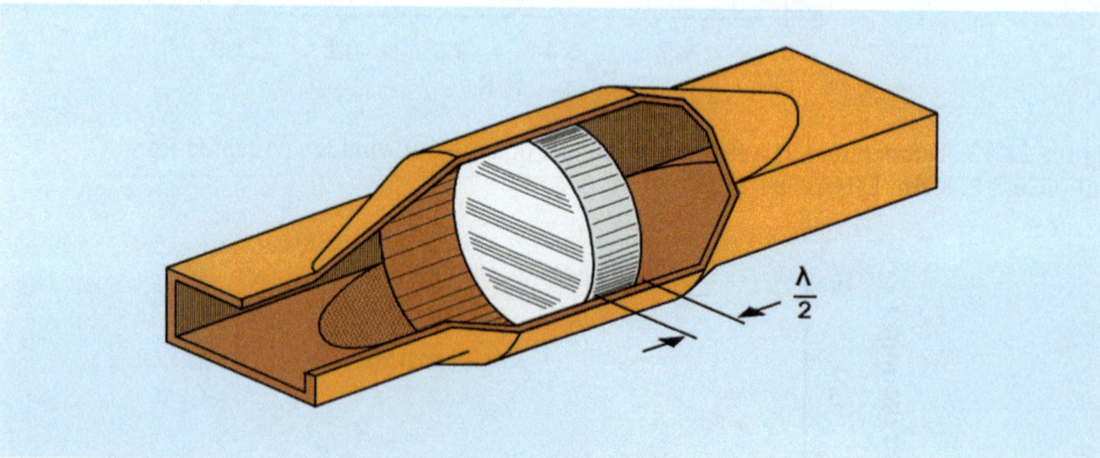

Figure 22-15 Half-wave, resonant, thick-disk window. (Adapted from: *Linear-Beam Tubes,* A. Staprans, Internal Varian Report, September 1968.)

During the past two decades, CVD diamond has found application as vacuum windows where low loss and/or high cooling capacity are required. Reasons for using diamond as the window material are evident from some of the information in Table 22.1, where the properties of CVD polycrystalline diamond are compared with other window materials. Properties of natural diamond are also given. Notice that properties are far superior to the other window materials. While the properties of CVD diamond are not equal to those of natural diamond, CVD diamond can be made in the form of a window.

The properties of polycrystalline CVD diamond make it far more favorable for use in high-power tubes than either BeO or sapphire. The low-loss tangent coupled with the extremely high thermal conductivity make it an excellent material for high-power applications.

As an example of the improved thermal performance of CVD diamond compared to BeO, [15] gives a comparison of calculations of the temperature rise in the center of Ku-band windows made of the two materials. The power transmitted through the windows was 10 kW. The calculation for the BeO window showed 19W of loss and a 3.7°C gradient between the center and the edge of the window. For the CVD diamond, the loss was only 4.71W and the temperature gradient was 0.546°C. The temperature gradient in the CVD diamond was a factor of 6.7 below that in BeO.

It should be noted that the gyro-klystrons, gyro-TWTs and gyro-twystrons manufactured by CPI all use diamond windows. These tubes are discussed in Chapter 21.

Table 22.1
Properties of Diamond Compared to Selected Materials

Material	RF Loss Tangent × 10^{-4}	Dielectric Constant	Thermal Conductivity W/cmK	Tensile Strength kpsi
BeO [15]	4.0	6.8	2.6	22 – 29
Alumina [15]	2 – 8	9.8	0.2	—
Sapphire [15]	2	9.4 – 9.8	0.2 – 0.5	29 – 54
Quartz [15]	0.4 – 3	3.8 – 4.4	0.01 – 0.06	10 – 17
CVD Polycrystaline Diamond [15]	2.0 – 6.8	5.7	12 - 18	33 - 59
Type IIA Natural Diamond [15]	1.4	5.64	20	

Before diamond windows were available for high power-applications, face-cooled double-disk sapphire windows like that shown in Figure 22-16 were used at average power levels up to a few hundred kilowatts. With the promise of the development of gyrotrons capable of generating megawatts of RF power, it became clear that windows capable of operating at the megawatt level must be developed.

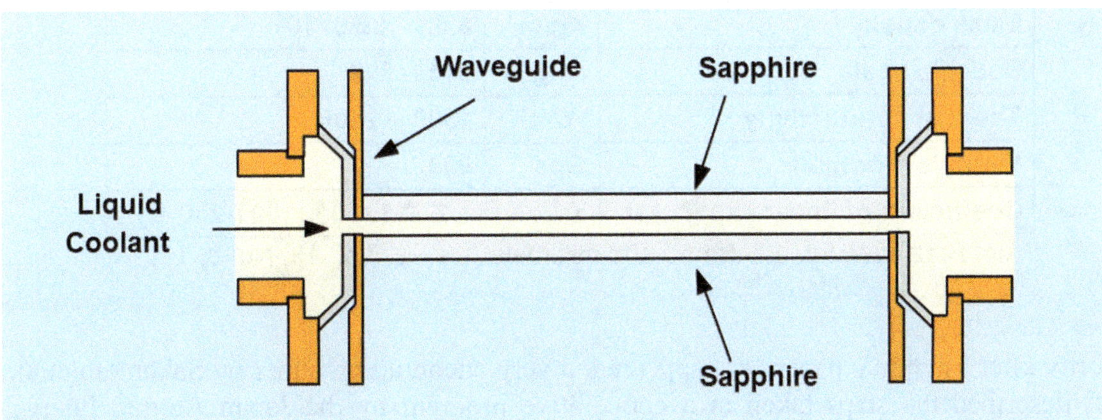

Figure 22-16 Face-cooled double-disk sapphire window. (Using, for example, a low-loss, Fluor-Carbon liquid coolant.)

22.4 MEGAWATT AVERAGE POWER WINDOWS

In Chapter 21, we discussed, the highly successful programs that led to the development of megawatt-class gyrotrons and the naming of 2018 as "the year of the gyrotron." by

Mark Henderson [16]. Dr. Henderson leads the Electron Cyclotron Section at ITER. The reason for Henderson's announcement in February 2018 is that the first energy-generating devices (gyrotrons) of ITER's electron cyclotron resonance heating system will be finalized in 2018. One of the key efforts that made the 1-MW gyrotrons possible was the development of windows that could transmit long pulses at the megawatt power level. Thumm [17] summarized possible approaches to the development of a megawatt-capable millimeter-wave window with emphasis on the CVD diamond. Table 22-2 provides information supporting the encouragement to place an emphasis on the CVD diamond:

1. Extremely high thermal conductivity (five times that of copper);
2. The possibility of very low loss (a loss tangent of 2×10^{-5} was measured in the program to be described);
3. Adequate strength to withstand the rigors of installation and use.

Table 22.2
Properties of CVD diamond

Property	Unit	Value
Diameter	mm	96.1
Thickness	mm	2.23
Roughnesss	μm	0.25 – 0.28
Loss tangent		$0.55 - 2.0\times10^{-4}$
Permittivity		5.67
Mass density	Kg/m^3	$3.52 - 3.56\times10^3$
Specific heat	J/kgK	514 – 542
Thermal conductivity	W/mK	1300 – 2100
Young's modulus	Gpa	800
Coefficient of linear expansion	K^{-1}	$1.2\text{-}\ 2.1\times10^{-6}$ (100T-250°C)

From Braz et al. Int. J. infrared and millimeter waves, Vol. 18, No. 8, 1997

Shortly after Thumm's paper [17] appeared, a very encouraging paper by Sakamoto et al. [18] described the steps taken in a cooperative program by the Japan Atomic Energy Research Institute (JAERI), Karlsrhue Institute of Technology (FIT, formerly FZK) and DeBeers in developing a window for an ITER gyrotron.

One of the key developments that made the 1-MW gyrotrons possible was the use of CVD polycrystalline diamond disk output windows like the ones shown in Figure 22-17.

Figure 22-17 Polycrystalline CVD diamond disks. (Karlsrhue Institute of Technology.)

The main challenge was the creation of diamond disks with diameters of about 100 mm, thicknesses of ~2 mm, and low dielectric loss. It was also necessary to manufacture a flange assembly for attaching the window to a gyrotron and for cooling the window. It was assumed at this point that a single-edge cooled diamond disk would be adequate for operation at cw power levels up to at least a megawatt.

The gyrotron used for these tests was capable of providing powers up to 0.5 megawatt at 170 GHz. The diamond disk and its supporting flange were mounted on the gyrotron and baked at 450°C for two days.

The results of tests at 0.5-MW of the 170 GHz window are shown in Figure 22-20. RF power was applied to the window in multi-second-long pulses. The temperature at the center of the window was measured. The red dots represent measured temperatures with 0.5 MW transmitted through the window. The blue line is the temperature that is determined from the rate of heat input to the disk (from the change in cooling water temperature and the water flow rate). The result is that the loss tangent for this disk is 1.3×10^{-4} and that thermal conductivity is 1,800 W/mK. The temperature rise in the disk was ~150°C. The red line shows the calculated temperature rise of a disk for which the fabrication process had been improved. For this disk, the transmitted power was assumed to be 1 MW. The temperature rise was only 40°C and the loss tangent was 2×10^{-5}. Sakamoto noted that this low loss tangent should permit operation at several megawatts and greatly increase the maximum energy/pulse that the gyrotron could supply. At that time, the energy per pulse was limited to 3.6 MJ. (0.45 MW, 8 s).

In January 2020, Thumm reported that, the Japan 170-GHz ITER gyrotron had set the world record energy per pulse of 2.88 GJ (0.8 MW for 60 min.)

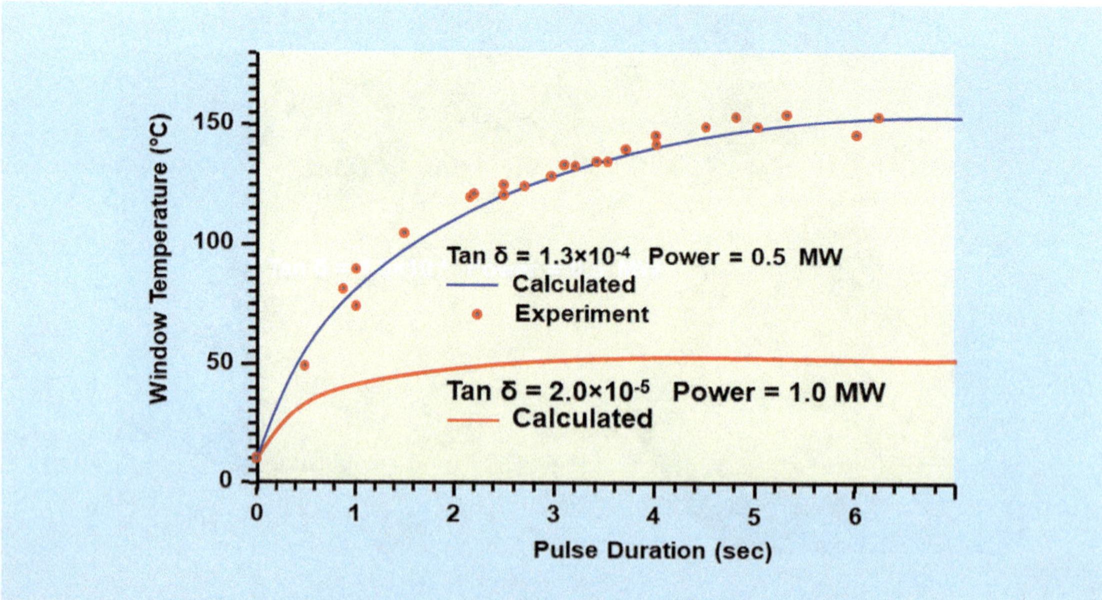

Figure 22-20 Window-center temperature as a function of time while transmitting 0.5 MW of power.

22.4.1 Step Tuning

The diamond disk window in Figure 22-17 was made to operate at a single frequency. In Chapter 21, we discussed requirements for gyrotron operation at several frequencies. An example is operation with a step-tuned gyrotron. It would be possible but slow and inconvenient to use mechanical switching to change the window configuration every time frequency was changed. It would be far more convenient to use an output window having a wide bandwidth.

Tests have been conducted by Grantenbein et al. using an output window consisting of a diamond disk mounted at the Brewster angle in a circular waveguide [21–23]. Figure 22-18 contains a photo and a drawing of a diamond disk in a circular waveguide. The Brewster angle is the angle at which there is complete transmission of a vertically polarized wave impinging on the interface between two dielectrics. In effect, it is the angle that results in an impedance match between dielectrics. The Brewster angle is not a function of frequency so that it should be possible to make very broadband windows.

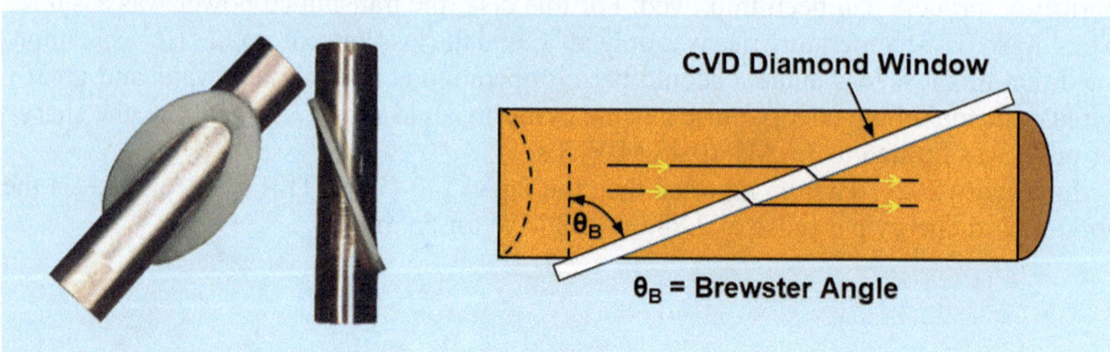

Figure 22-22 Photo and drawing of a CVD diamond Brewster- angle window mounted into a circular waveguide [23].

The Brewster angle is given by

$$\theta_B = arc\,tan\left(\frac{n_1}{n_2}\right)$$

where n_1 and n_2 are the refractive indices of the two dielectrics. The refractive index is proportional to wave velocity, which is proportional to (dielectric constant)$^{0.5}$. For polycrystalline CVD diamond, the dielectric constant is 5.70, so

$$\frac{n_1}{n_2} = \frac{v_1}{v_2} = \frac{\sqrt{5.70}}{1} = 2.39$$

And so

$$\theta_B = 67.3°$$

The power reflection for diamond as a function of angle is given in Figure 22-19. The reflection remains below 1% (–20 dB) between 61 and 71 degrees [24].

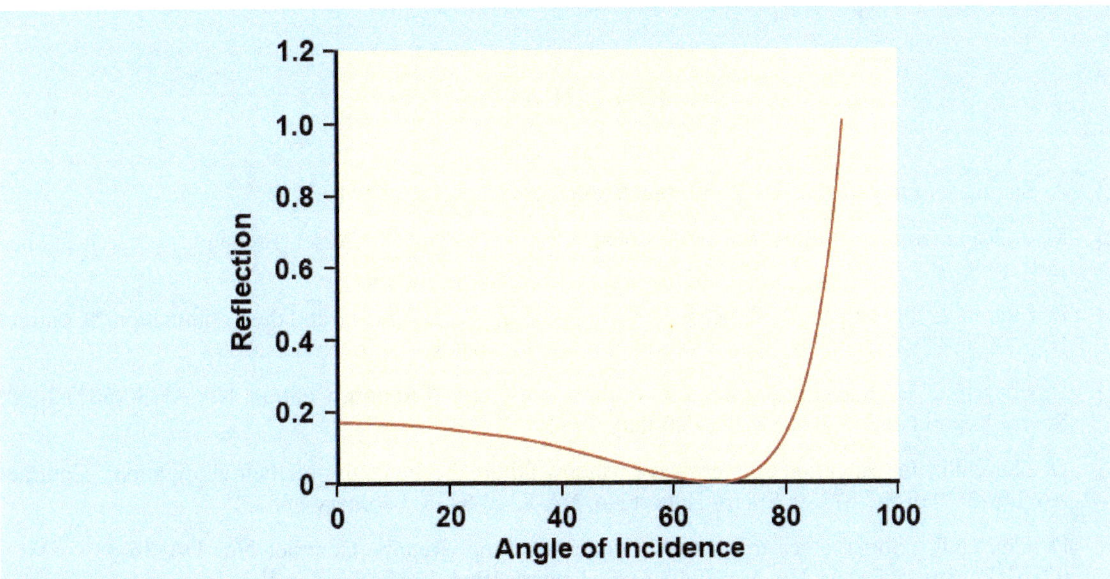

Figure 22-23 Power reflection from CVD diamond disk Brewster angle window [18].

Fabrication of the CVD diamond, Brewster-angle window is not straight forward, primarily because there is a 16:1 difference in expansion coefficients between the diamond and the attached metal. Also, excessive braze temperatures result in carburization of the diamond. Thales manufactured the windows used in [17]. After considerable experimentation, they brazed the windows into OFHC copper cuffs at 800°C [24].

The window was installed in a D-band gyrotron and was tested at the megawatt power level using 3-ms-long pulses. By varying the magnetic field, the gyrotron mode of operation was switched from $TE_{19,6}$ to $TE_{27,9}$ and the frequency of operation was step tuned from 111.6 GHz to 165.7 GHz in 10 frequency increments [21].

Recently, there have been reports of success in fabricating large single-crystal diamond disks [25]. If the physical characteristics of these disks are like the characteristics of type IIa diamond, it should be possible to make broadband windows with even less loss than the one described in [18].

22.5 SCALING OF WINDOWS

The most common failure mode of a window is fracture due to thermal stresses. As a result, the limitation of thermal stresses is one of the most important factors in window design. The average power capability of a window can be scaled to another frequency with reasonable accuracy. If the loss tangent and dielectric constant are held constant and the size of the window is scaled directly with wavelength, the average power-handling capability is inversely proportional to frequency.

At a fixed frequency, the power-handling capability is virtually insensitive to the window cross-section dimensions. This is because, even though dissipation density is decreased for a larger window, a longer path for the heat flux is created. This results in essentially the same maximum temperature at the center of the window regardless of the window size. As a result, there is no advantage from an average-power standpoint in going to a very large window. However, problems associated with operation at high peak-power levels, such as breakdown, will be alleviated in a larger window.

REFERENCES

[1] A. Staprans, *Linear-Beam Tubes*, Internal Varian Report, September 1968.

[2] M. J. Smith, and G. Phillips, *Power Klystrons Today*, Taunton, Somerset, England: Research Studies Ltd., 1995.

[3] D. Churchill, "Investigation of microwave window failure mechanisms and their elimination," Contract No. DA-36-039-SC-78314, Sperry Report No. NA-8240-8182-5, November 1960.

[4] D. Churchill, "High average-power RF window study," Final Report, Contract No. AF 30(602)-2428, Sperry Report No. NA-8240-8299, February 1962.

[5] D. Churchill, "Investigation of microwave window failure mechanisms and their elimination," Contract No. DA-36-039-SC-87389, Sperry Report No. NA-8240-8338, February 1963.

[6] D. Churchill, "Superpower microwave windows," Final Report, Contract No. DA 36-039 AMC-02161(E), Sperry Report No. NA-8240-8372, January 1964.

[7] F. O. Johnson, "Computer synthesis of filter networks as applied to waveguide windows," *Proc. International Conference on Microwave Circuit Theory and Information Theory*, Tokyo, Japan, September 7–11, 1964.

[8] M. P. Forrer, and E. T. Jaynes, "Resonant modes in waveguide windows," *IRE Trans. on Microwave Theory and Techniques*, Vol. MTT-8, No. 1, March 1960, pp. 147–150.

[9] R. S. Symons, and H. E. Schoennauer, "Sealed waveguide window," U.S. Patent 2,958,834, November 1, 1960.

[10] G. Faillon, and P. Kern, A high frequency, 37 MW, # GHz, 5 μs, multicavity klystron for linear accelerators," Linear Accelerator Conference (LINAC 84), 7-11 May, 1984, Darmstadt, Germany.

[11] J. H. Jasberg, and J. V. Lebacqz, "High-power microwave windows," *Proc. 5th International Congress on Microwave Tubes*, Paris, September 1964, pp. 246–250.

[12] R. H. Giebler, "A 500 kW CW klystron at S-band," IEEE IEDM, Washington, D.C., October 1967.

[13] A. Goldfinger, "High-power RF window study," Varian Associates, Eimac Division, Final Report, RADC-TR-66-657, January 1967.

[14] M. V. Kartikeyan, E. Borie, and M. Thumm, *Gyrotrons*, Berlin: Springer, 2003.

[15] B. G. James, "Diamond windows for high-power broadband microwave tubes," Communications and Power Industries, Final Report, Contract No. N00164-94-C-004, December 1997.

[16] M. Henderson, "The year of the gyrotron," Reported by Krista Dulon, *ITER Newsline,* February 2018.

[17] M. Thumm, "Development of output windows for high-power long-pulse gyrotrons and ec wave applications,".*Int. J. Infrared and Millimeter Waves,* Vol. 19, No. 1, January 1998.

[18] K Sakamoto, "Diamond window and its application to ITER gyrotron" *ITER EDA Newsletter*, Vol. 8, No. 1, 1999.

[19] K. Felch et al., "Recent ITER-relevant gyrotron tests," *Jour. of Physics: Conference Series,* Vol. 25, No. 1, 2005, pp. 13–23.

[20] M. Thumm, "State of the art of high-power gyro-devices and free electron masers," *Int. J. Infrared and Millimeter Waves*, Vol. 19, No.31, January2020.

[21] G. Gantendein, et al., "First operation of a step-frequency tunable 1-MW gyrotron with a diamond Brewster angle output window," *IEEE Trans. Electron Devices*, Vol. 61, No. 6, June 2014.

[22] G. Gantendein et al., "Operation of a step-frequency tunable gyrotron with a diamond Brewster angle output window," IEEE IVEC 2013.

[23] G. Aiellio et al. "CVB diamond Brewster window: feasibility study by FEM analyses," http://www.epj-conferences.org.

[24] M Read, "Advanced output coupling for high power gyrotrons," Final Technical Report Grant Number DE-SC0006212, Calabazas Creek Research, August 2016.

[25] M. Schreck, et al., "A 155 carat diamond with 92 mm diameter,", March 22, 2017, University of Augsburg, Germany.

Appendix A

Useful Constants and Conversions

Constants

Boltzmann constant	k	$1.38°10^{-23}$ J/K
electronic charge	e	1.6×10^{-19} C
electron rest mass	m	9.1×10^{-31} kg
electron charge-to-mass ratio	η	1.76×10^{11} C/kg
permittivity of vacuum	ε_o	8.85×10^{-12} F/m
permeability of vacuum	μ_o	$4\pi\times10^{-7}$ H/m
Planck constant	h	6.625×10^{-34} J-s
speed of light in vacuum	c	3×10^{8} m/s
hydrogen atom mass	m_h	1,836×m (electron)

Conversions

1 eV	=	1.6×10^{-19} J
1 tesla	=	10^{4} Gauss
0°C	=	273 K
room temp. (20°C) kT	=	1/40 eV

Appendix B

Vacuum Technology

If a microwave tube is to operate properly, a high or ultrahigh vacuum must be maintained throughout its life. As a result, special techniques, processes, and materials must be used. The emphasis in this appendix is on vacuum technology. Subjects discussed are:

- Units of measurement;
- Ranges of operation;
- Sources of gas;
- Vacuum systems;
- Vacuum pumps;
- Vacuum gauges;
- Bakeout;
- Microwave tube materials;
- Fabrication techniques.

Several excellent books on vacuum technology are listed with the references at the end of this appendix [1–11].

B.1 UNITS OF MEASUREMENT

Many years ago, the most commonly used unit of pressure was the millimeter of mercury (mmHg). Atmospheric pressure is 760 mmHg. Now, the accepted units of pressure are the torr and the newton/m^2 (pascal). The relation between these units is

$$1 \text{ torr} \approx 1 \text{ mmHg} \approx 1.33 \times 10^2 \text{ pascal}$$

Another unit of pressure sometimes used at relatively low vacuum (high pressure) levels is the micron:

$$1 \text{ micron} = 10^{-6} \text{ meters of Hg} = 10^{-3} \text{ mmHg} \approx 10^{-3} \text{ torr}$$

Still another unit of pressure used in European literature is the mbar (millibar):

$$1.33 \text{ mbar} = 1 \text{ torr}$$

In microwave tube work, it is rarely and perhaps never necessary to know pressure to an accuracy better than a factor of two or so. As a result, it is safe to remember that

$$1 \text{ torr} \approx 1 \text{ mmHg} \approx 1 \text{ mbar}$$

B.2 RANGES OF OPERATION

In evacuating a microwave tube, the pressure changes from atmospheric to a value as low as 10^{-8} to 10^{-9} torr or lower. Thus, the pressure is reduced by over eleven orders of magnitude. It is common to divide this very wide pressure variation into four ranges:

- 760 torr to ~1 torr – Rough vacuum
- 1 torr to ~10^{-3} torr – Medium vacuum
- 10^{-3} torr to ~10^{-7} torr – High vacuum
- $< 10^{-7}$ torr – Ultrahigh vacuum

In most electron tubes, the pressure is at the low end of the high vacuum range or the high end of the ultrahigh vacuum range (i.e., from 10^{-7} to 10^{-9} torr). In some high-power klystrons that have been in operation for > 50,000 hrs, the pressure is even lower.

At atmospheric pressure, the gas density is about 2.7×10^{19} molecules/cm^3. As shown in Figure B-1, density drops linearly with decreasing pressure and, at 10^{-8} torr, is about 3×10^{8} molecules/cm^3. While this density appears to be high, 3×10^{8} molecules will form a monolayer on an area of only 10^{-6} cm^2. The time required to form a monolayer is also shown in Figure B-1. At a pressure of 10^{-8} torr, a monolayer is formed in a about 3 min.

Mean free path length is the mean distance traveled by a molecule between collisions. Figure B-1 shows that at 10^{-8} torr, the mean free path length is ~10^{4} m. So the average molecule could pass through a 1m vacuum chamber 10,000 times without colliding with another molecule. At 1 torr however, the mean free path length is ~10^{-4} m so molecules collide with each other more frequently than with the walls of a vacuum system.

At relatively high pressures, where molecules collide with each other far more frequently than with vacuum chamber walls, gas flow is referred to as viscous flow. At low pressures, where molecules collide with vacuum system walls more frequently than with each other, gas flow is referred to as molecular flow. In the viscous flow and molecular flow regions, the analysis of gas flow is relatively straightforward. In the pressure range from ~10^{-2} torr to 10^{-4} torr, a transition from viscous to molecular flow occurs, and gas flow behavior is difficult to analyze.

The effects of viscous and molecular flow in vacuum system design are illustrated in Figure B-2. In the viscous flow region, molecules gradually migrate in the direction of the pump because of the change in pressure, ΔP. With molecular flow, molecules collide only with the walls of the chamber and only occasionally are directed toward the pump. As a result, a large pump orifice is required if a high pump speed is to be achieved. For example, if 1-inch (2.5-cm) tubulation was used to connect the vacuum system to the pump, the speed would be limited to a few liters per second. If a 6-inch (15-cm) diameter orifice was used, the speed could be well over 1,000 liters per second.

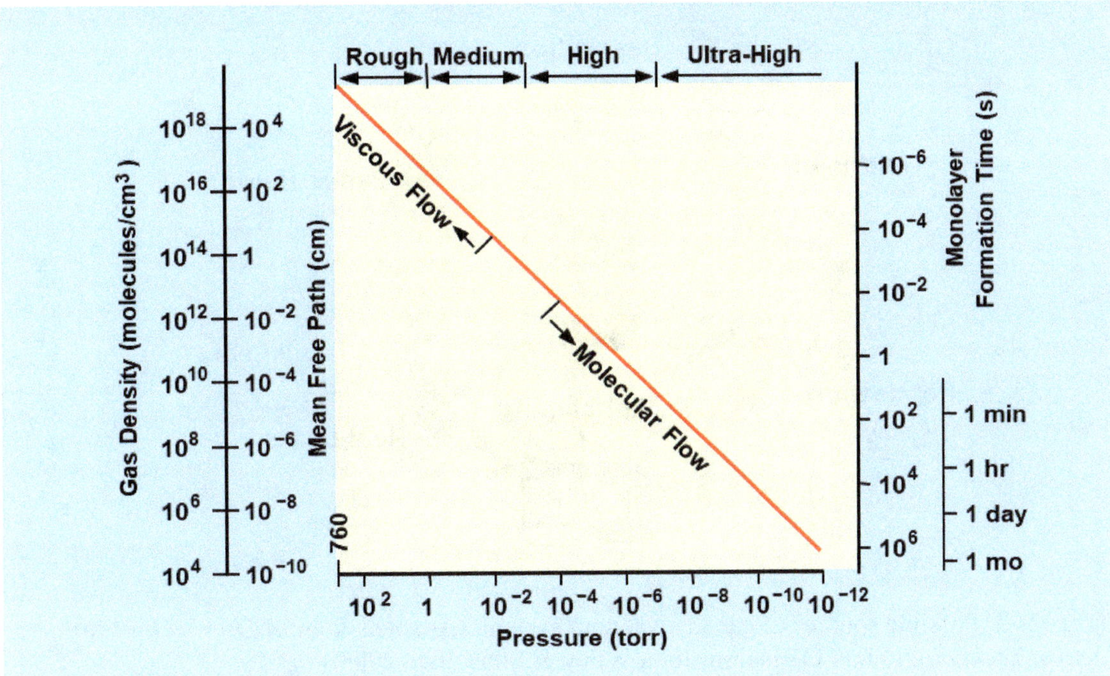

Figure B-1 Relationship of several quantities defining the degree of vacuum.

B.3 SOURCES OF GAS

If the only gas source in a container that is to be evacuated is the gas located in the volume of the container, then the container could be pumped out very quickly. In addition, if the pump being used is a perfect pump; that is, regardless of pressure, the pumping speed is unchanged and the pump contributed no gas to the system, then the ultimate pressure would be unlimited. Unfortunately, there are many gas sources in any container that is to be evacuated and there is no perfect pump. The gas sources are identified in Figure B-3 and described next.

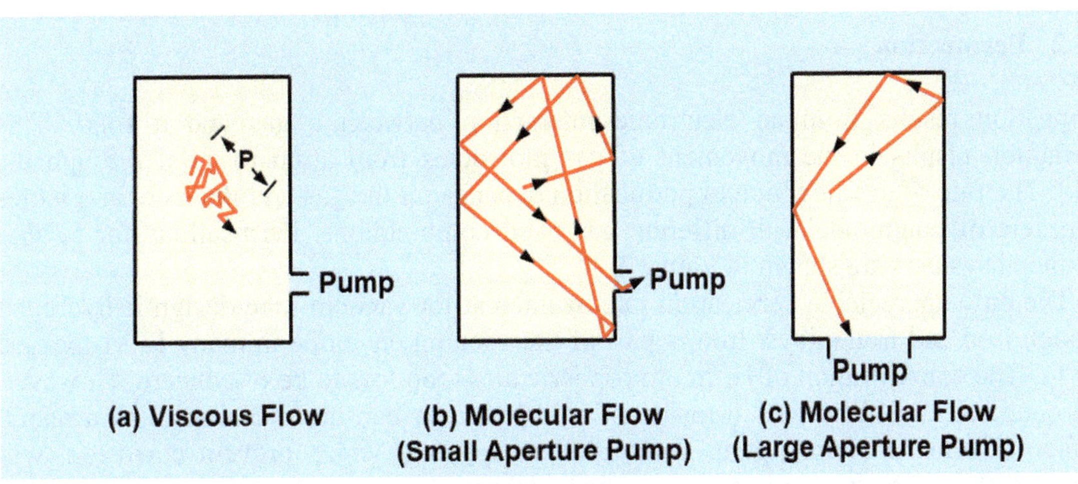

Figure B-2 Comparison of viscous and molecular flow.

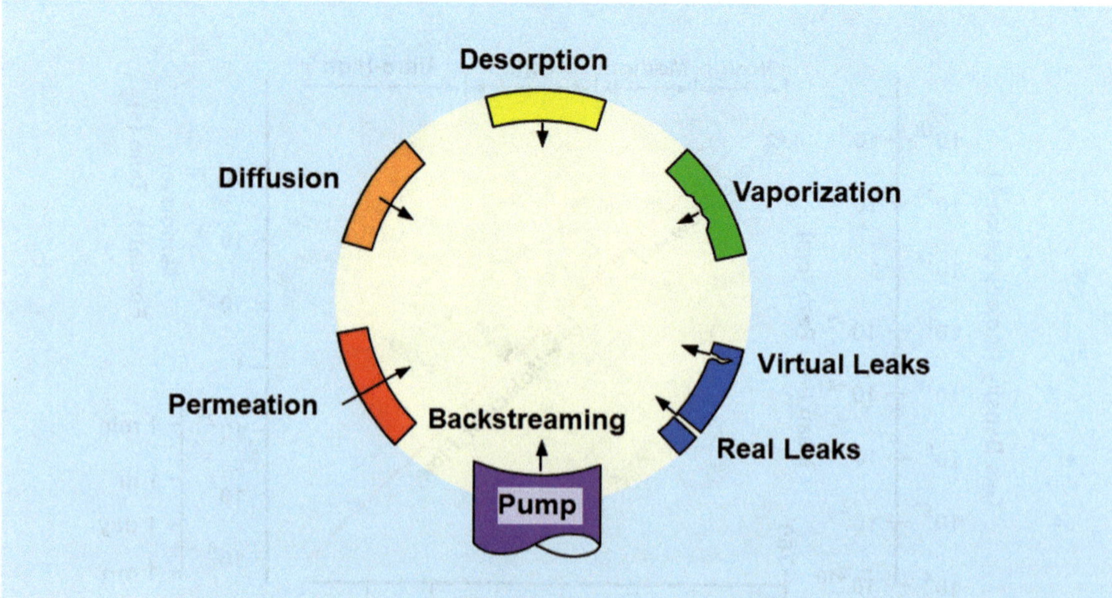

Figure B-3 Possible sources of gas in a vacuum system. (Adapted from: *A User's Guide to Vacuum Technology*, J. F. O'Hanlon, John Wiley & Sons, Inc., 2003.)

B.3.1 Backstreaming

Backstreaming is the transport of gas from a vacuum pump back into the chamber being pumped. Perhaps the best examples of pumps for which backstreaming is sometimes a problem are the oil-sealed mechanical pump and the diffusion pump. The oil in the mechanical pump vaporizes and the oil molecules can diffuse throughout the vacuum system being pumped. In the diffusion pump, oil is the working fluid that produces the pumping action, but the oil vapors can diffuse toward the system being pumped. When either or both of these pumps are used, backstreaming occurs. The backstreaming gas consists of molecules of the pumping fluid and its fractions. A baffle can be used between the pump and the chamber being evacuated to reduce backstreaming, but still some molecules of pumping fluid migrate to the chamber being evacuated.

B.3.2 Permeation

Permeation results from an electronic interaction between a gas and a solid. This interaction results in the movement of gas molecules from atom to atom through the solid. The rate of gas flow due to permeation depends on the gas and the solid and varies by orders of magnitude with different gas-solid combinations. Permeations for several gas-metal systems are shown in Figure B-4.

The only appreciable permeation rate of interest for vacuum-tube design is hydrogen through iron and iron alloys. Iron is part of the vacuum envelope in many PPM-focused TWTs. The concentration of H_2 in atmospheric air is too low to be of concern. However, hydrogen released from the corrosion of iron by water can be of concern. As a result, surface treatments of iron parts, such as electroplating, which prevent corrosion, will minimize the generation and permeation of hydrogen.

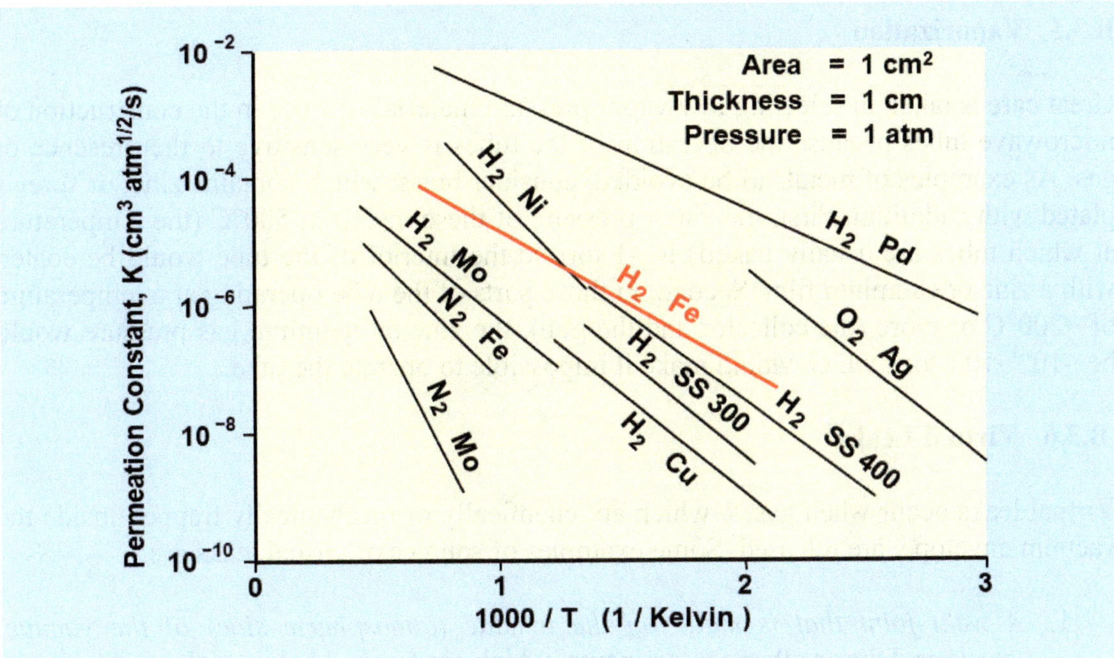

Figure B-4 Permeation rates for several gas-metal systems. (Adapted from: A. Roth, *Vacuum Technology*, Elsevier Science Publishers B.V., 1990.)

B.3.3 Diffusion

In general, diffusion is the movement of one material through another. In high and ultrahigh vacuum chambers, we are concerned mostly about the diffusion of gases through the metallic walls of the chamber to the interior surface of the walls of the chamber. There, the gas is desorbed and contributes to the gas that must be removed from the system. The rate at which gas diffuses through metals is an exponential function of temperature so heat (baking) to accelerate the diffusion process is often used.

B.3.4 Desorption

Gas molecules adhere to surfaces within a vacuum chamber and are said to be adsorbed on the surfaces. The source of this gas can be the atmosphere within the chamber or diffusion or permeation from the walls of the chamber. Gas-free surfaces in vacuum can become covered with gas molecules very quickly. For example, Figure B-1 shows that at a pressure of 10^{-6} torr, the time required to form a monolayer of gas on a gas-free surface is only one second. When adsorbed gas molecules are released from a surface, they are said to be desorbed. The rate of desorption is an exponential function of temperature, so baking is very effective in removing desorbed gas.

Many adsorbed gasses are too tightly bonded to surfaces to be set free by baking. These can be desorbed by electron impact. As a result, many microwave tubes are operated with a pump attached for several hundred hours with the electron beam being used to assist in the outgassing process.

B.3.5 Vaporization

Great care is taken in selecting low-vapor pressure materials for use in the construction of microwave tubes because the operation of the tubes is very sensitive to the presence of gas. As examples of metals to be avoided, consider brass, which contains zinc, or screws plated with cadmium. First, the vapor pressure of these metals at 500°C (the temperature at which tubes are usually baked) is ~1 torr so the interior of the tube would be coated with a zinc or cadmium film. Second, in those parts of the tube operating at a temperature of ~200°C or more (the collector and the gun), the zinc or cadmium gas pressure would be ~10^{-4}–10^{-5} torr, which would make it impossible to operate the tube.

B.3.6 Virtual Leaks

Virtual leaks occur when gases, which are chemically or mechanically trapped inside the vacuum envelope, are released. Some examples of sources of virtual leaks are:

1. *A weld joint that is made on the outside (atmospheric side) of the vacuum envelope.* Dirt or other contaminants, which are trapped between the welded parts on the inside (vacuum side) of the joint, release gas that slowly leaks into the vacuum chamber.
2. *A screw that is used to fasten a part in an electron gun.* The threads of the screw have not been properly vented and so gas that is trapped and not completely driven out during bakeout is slowly released.
3. *An inadequately outgassed part inside the vacuum envelope.* Gas from the part slowly diffuses into the vacuum chamber.
4. *During thermal cycling, a microscopic crack that develops on the inner (vacuum side) of a ceramic insulator.* This crack exposes a tiny gas-filled void and the gas slowly leaks into the vacuum chamber.

Virtual leaks are avoided by following good design practice, eliminating gas traps, and properly processing parts. There is no external test that can detect a virtual leak. The use of a getter or an appendage pump is an inexpensive and straightforward way to prevent gas problems in the event that a virtual leak occurs.

B.3.7 Real Leaks

A real leak results from a pore in the wall of the vacuum envelope. Examples of pores are voids caused by impurities or inclusions in the wall material or cracks caused by thermal or mechanical stress.

The leak rate for air as a function of pore size (radius and length) is shown in Figure B-5. The slope of 1/3 indicates molecular flow. A hole size on the order of at least 3 molecular diameters (~10A or 10^{-7} cm) is required for pore leakage. Smaller holes are likely to be plugged by large molecules. Thus, pore leakage is unlikely to occur below a leak rate of about 10^{-15} torr-liters/sec for a 1-mm-long pore. (1 torr-liter/sec ≈ 1 cm^3/sec at standard temperature and pressure). At this leak rate, the pressure in a dormant TWT with an internal volume of 0.025 liter and without a getter or pump would increase to about 10^{-6} torr in one year. At this pressure, there is high likelihood that proper tube operation would not be possible. Thus, even the smallest pore leak is unacceptable for most applications.

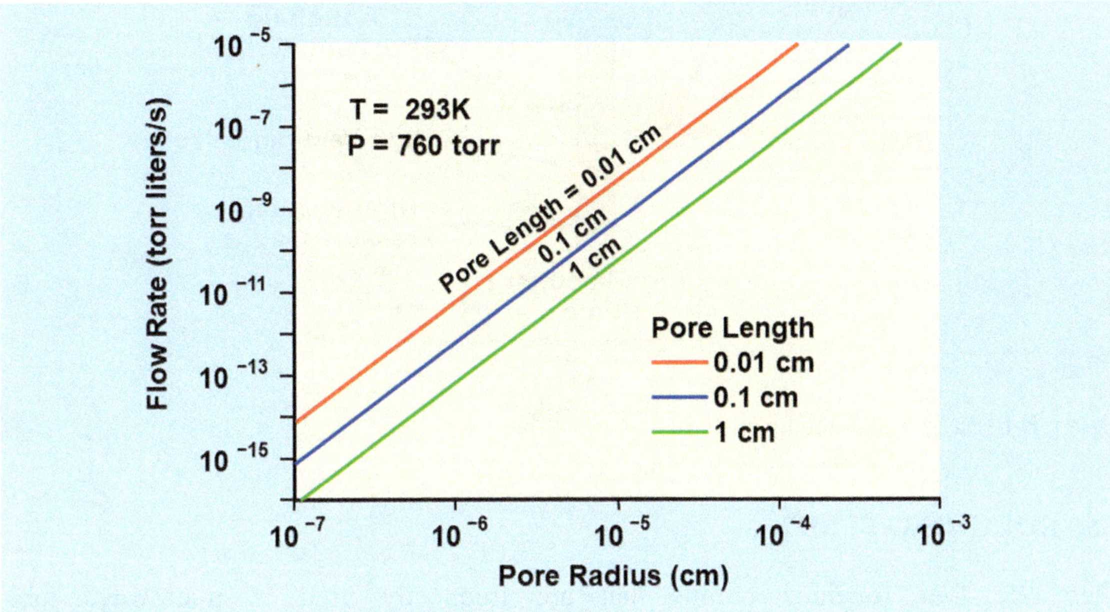

Figure B-5 Flow rate of air as a function of pore size.

B.4 VACUUM SYSTEMS

There is no single vacuum pump that is capable of evacuating a chamber from atmospheric pressure down to the ultrahigh vacuum range. Instead, as shown in Figure B-6, at least two pumps are required. One, the roughing pump or fore pump, evacuates the system from atmospheric pressure down to the 10^{-2}–10^{-4} torr range. The other, the high- vacuum pump, reduces the pressure to the high to ultrahigh vacuum range.

The operation of the system is straightforward. If the entire system is at atmospheric pressure, then, with both valves open, the roughing pump is started. When the pressure in the system has decreased to the pressure at which the high-vacuum pump can start operating (usually ~10^{-2}–10^{-4} torr), then the high-vacuum pump is started and the roughing pressure valve is closed. If all goes well (there are no leaks, for example) then the high-vacuum pump will reduce the pressure to the high to ultrahigh vacuum range.

Very often, the entire system is not started at atmospheric pressure. Instead, initially, the high-vacuum valve is closed and the high-vacuum pump is operating (it was evacuated and started during a previous pump-down cycle). Then, the roughing pump is started, the tube to be evacuated is pumped to the 10^{-2}–10^{-4} torr range, the roughing valve is closed, and the high-vacuum valve is opened.

From the discussion in this section, it is clear that vacuum pumps may be divided into two major categories: roughing pumps and high-vacuum pumps. The pumps most commonly used in the fabrication and operation of microwave tubes are described in Sections B.5 and B.6.

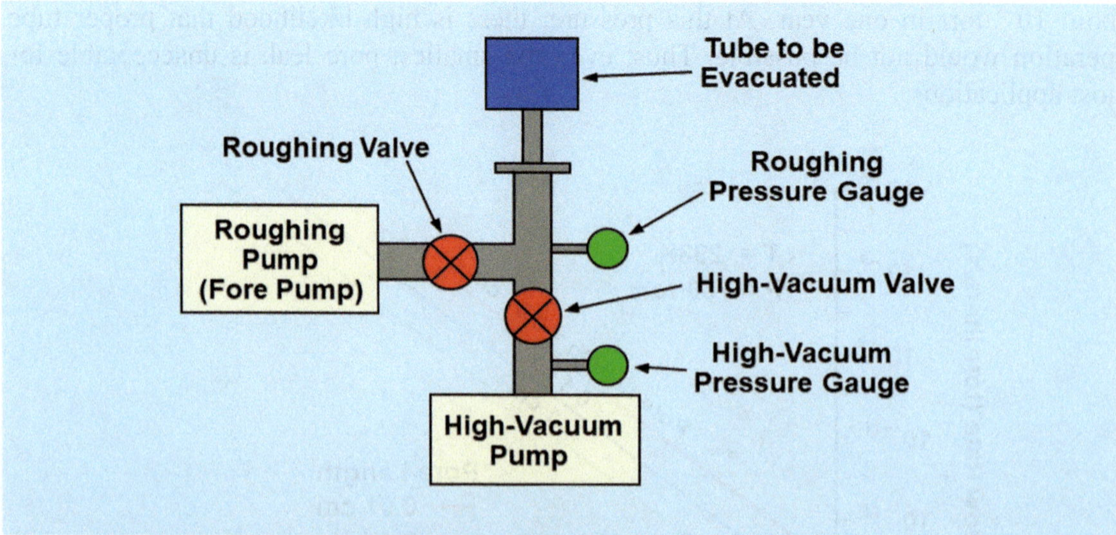

Figure B-6 Basic high-vacuum system.

B.5 ROUGHING PUMPS

There are four roughing pumps that are frequently used in microwave tube manufacturing:

- Oil-filled mechanical pump;
- Scroll pump;
- Sorption pump;
- Venturi pump.

B.5.1 Oil-Filled Mechanical Pumps

An example of the oil-filled mechanical pump is the sliding vane rotary pump, the essential elements of which are illustrated in Figure B-7. A rotating cylinder containing spring-loaded vanes transfers gas from the inlet port to the output port. A high-grade vacuum oil is used in the assembly containing the rotating cylinder and vanes. The oil serves as a lubricant and also provides a vacuum seal between moving parts.

A problem with mechanical pumps is the pumping of condensable vapors such as water. If these vapors condense during the compression cycle, they may emulsify with the pump oil and degrade the lubricating properties. Also, the ultimate pressure capability of the pump is reduced because the vapors are released from the pump oil in the inlet side of the pump. This problem may be avoided with the use of a gas ballast device that operates during the compression cycle. Pumping speed curves for sliding vane rotary pumps are shown in Figure B-8. The ultimate pressure is ~10^{-2}–10^{-4} torr.

B.5.2 Scroll Pumps

Scroll pumps are oil-free and require no lubricant or sealant. The scroll pump uses two plates, each of which has a spiral-shaped pattern impressed in it. One of these plates is stationary and fits inside the other, which moves with an orbital motion. As the movable scroll orbits, moving crescent-shaped gas pockets are formed that compress the gas that they contain. This action is indicated in Figure B-9. First a large pocket of gas is trapped (in a), then as the movable plate orbits (b through h), the pocket becomes smaller and smaller. Finally (in i and j), the gases are expelled at the center of the fixed plate. A pumping speed curve for a Varian triscroll pump is shown in Figure B-10.

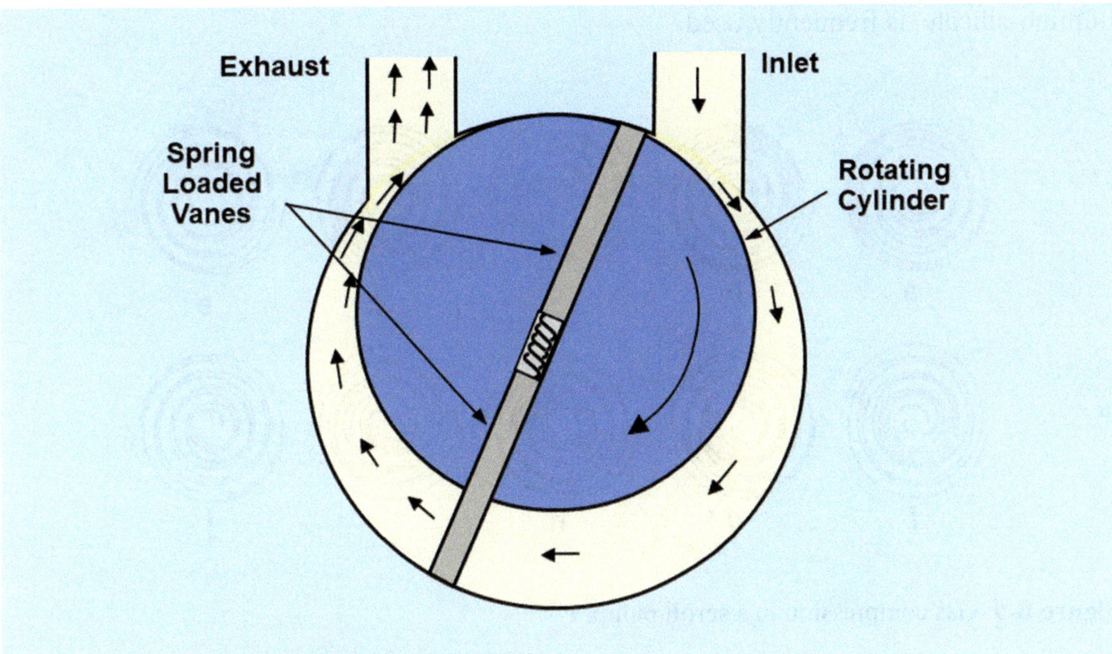

Figure B-7 Essential elements of a sliding vane rotary pump.

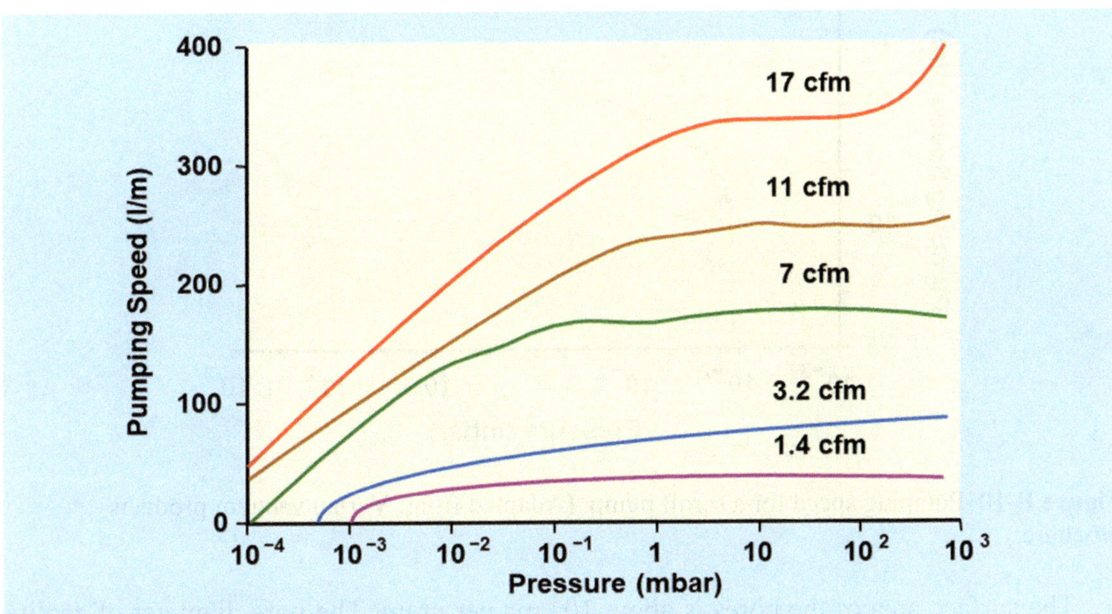

Figure B-8 Pumping speeds of sliding vane rotary pumps. (Adapted from: Varian vacuum products brochure.)

B.5.3 Sorption Pumps

A sorption pump is capable of evacuating a chamber from atmospheric pressure to the medium vacuum range. A typical pump is shown in Figure B-11. The operating principle is the physical adsorption of gases on the surfaces of molecular sieves. Zeolite, an alkali alumino-silicate, is frequently used.

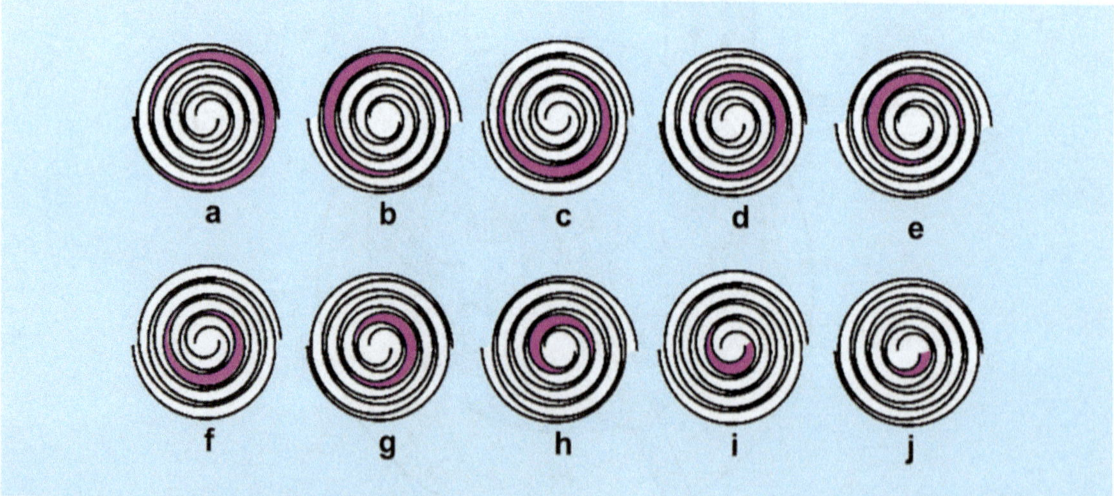

Figure B-9 Gas compression in a scroll pump.

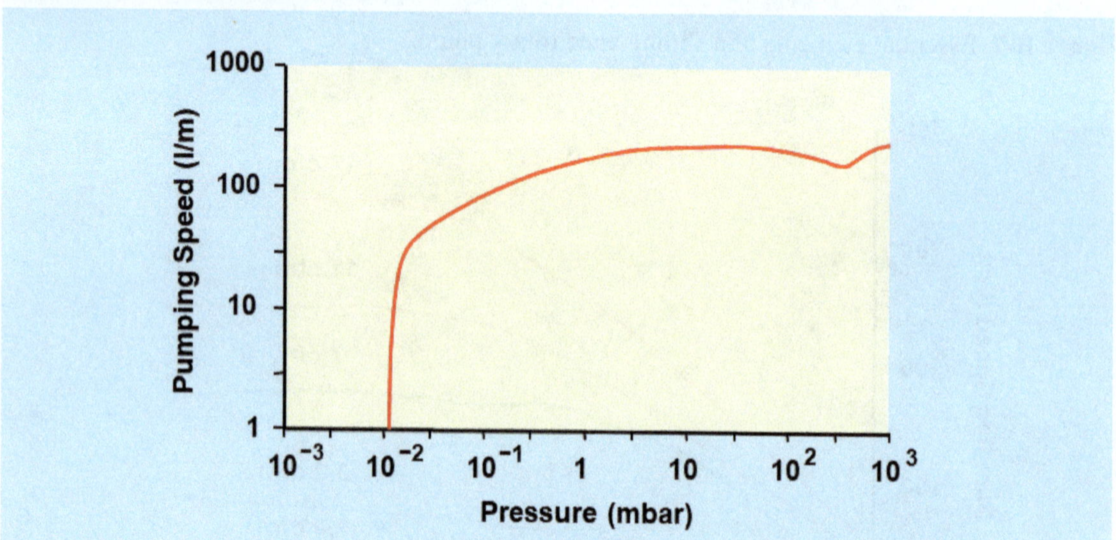

Figure B-10 Pumping speed for a scroll pump. (Adapted from: Varian vacuum products brochure.)

The surface area of the pores is about 10^3 m^2 per gram. The pore diameter of zeolite 13X (about 13Å) is approximately the size of water vapor, oil vapor, and large gas molecules. Large molecules are, therefore, readily adsorbed. Small molecules such as hydrogen, helium, and neon are very poorly adsorbed.

The adsorption of gases is dependent on pressure and temperature, as shown in Figure B-12. Note that the difference between the quantity of nitrogen adsorbed and the quantity of neon or helium adsorbed is at least four orders of magnitude.

In practice, hydrogen and helium are normally present in the gases to be evacuated from a microwave tube. This is because a hydrogen atmosphere is used during brazing processes and an inert gas atmosphere containing helium is frequently used for inert gas welding. Since hydrogen and helium are normally present, it is necessary to be cautious in using a sorption pump. Because the pump operates mostly in the viscous flow range, hydrogen and helium are dragged along with the gases readily pumped by the zeolite. As long as the pumping process takes place fairly quickly, hydrogen and helium can be pulled into the sorption pump, and then a valve can be closed before the hydrogen and helium diffuse back into the chamber being pumped.

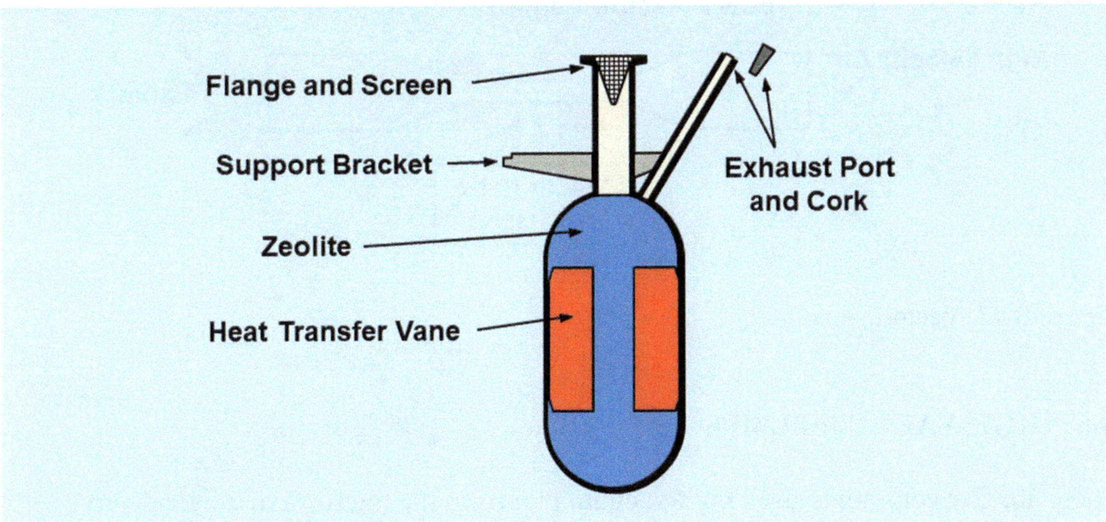

Figure B-11 Typical sorption pump.

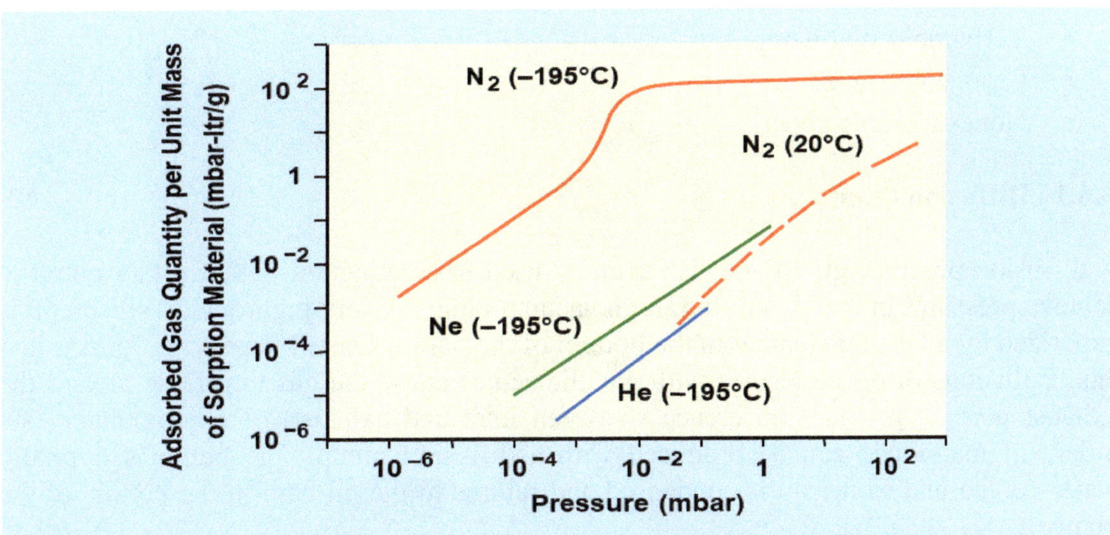

Figure B-12 Adsorption isotherms of zeolite 13X for nitrogen, neon, and helium. (Adapted from: *Product and Vacuum Technology Reference Book* by Leybold Heraeus, Inc.)

Sorption pumps are immersed in liquid nitrogen to achieve large pumping capacities. To rejuvenate zeolite after pumping, the pump is warmed to room temperature. Gas is released from the zeolite and escapes from the pump through an exhaust port. If a great deal of water vapor or other condensables has been pumped, it may be necessary to bake the zeolite for a few hours at 200°C or more.

B.5.4 Venturi Pumps

The operation of the sorption pump can be improved by using a venturi pump (Figure B-13) to remove 50%–80% of the gas in the system before the sorption pump is used.

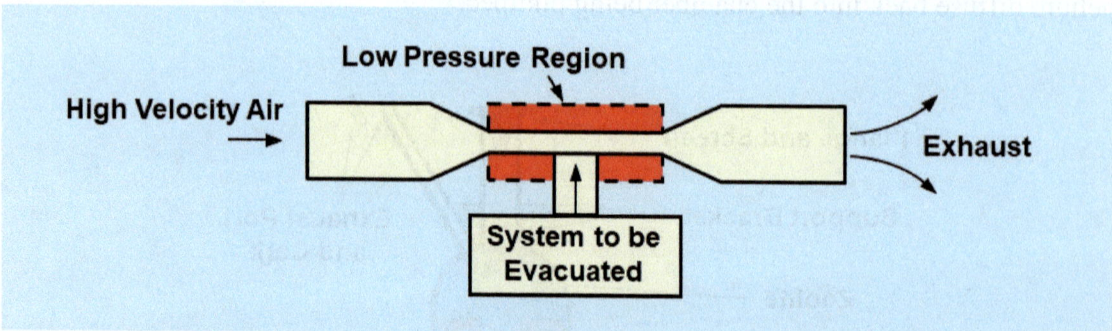

Figure B-13 Venturi pump.

B.6 HIGH-VACUUM PUMPS

There are five commonly used high-vacuum pumps in the microwave tube industry:

- Diffusion pump;
- Ion pump;
- Turbomolecular pump;
- Cryogenic pump;
- Nonevaporable getter.

B.6.1 Diffusion Pumps

A diffusion pump (Figure B-14) is normally used in conjunction with a rotary pump to achieve pressures in the high or ultrahigh vacuum range. A very high-quality silicon oil is vaporized by a heating element in the bottom of the pump. The oil vapors are formed into jets. Collisions of oil molecules with gas molecules cause the gas to diffuse toward the exhaust port. A pressure difference (between inlet and exhaust) of approximately six orders of magnitude can be produced with a diffusion pump. The pump is normally water-cooled and so the oil is condensed and returns to the oil bath in the bottom of the pump.

Typical pumping speed and throughput for a diffusion pump are shown in Figure B-15. Note that speed is constant over a wide range of pressures. At very low pressures, the ultimate vacuum level is limited primarily by diffusion of oil molecules back into the chamber being pumped. This is referred to as backstreaming and was discussed previously in this appendix. A baffle is sometimes used between the pump and the chamber to limit backstreaming. Metallic elements are arranged as indicated in Figure B-16 so that there is no line-of-sight path through the baffle. Baffles are normally cooled, sometimes by liquid nitrogen, to enhance the condensation of oil vapors. With a baffle it is sometimes possible to reduce the ultimate pressure by over two orders of magnitude.

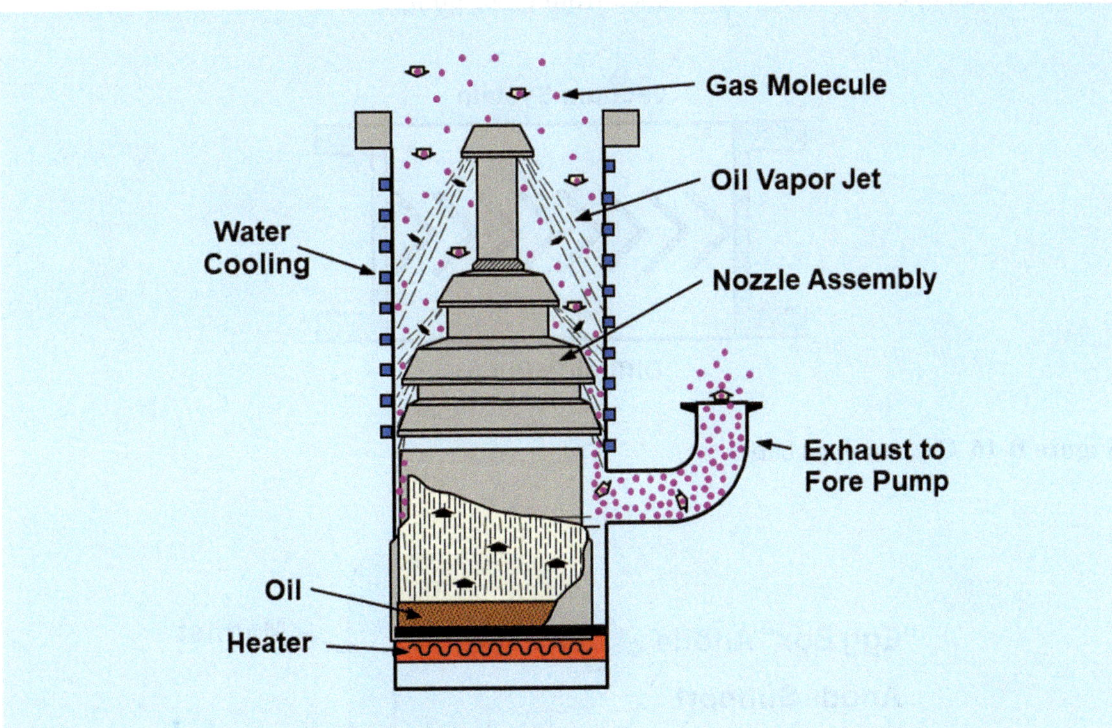

Figure B-14 Oil-filled diffusion pump. (Adapted from: *Product and Vacuum Technology Reference Book* by Leybold Heraeus, Inc.)

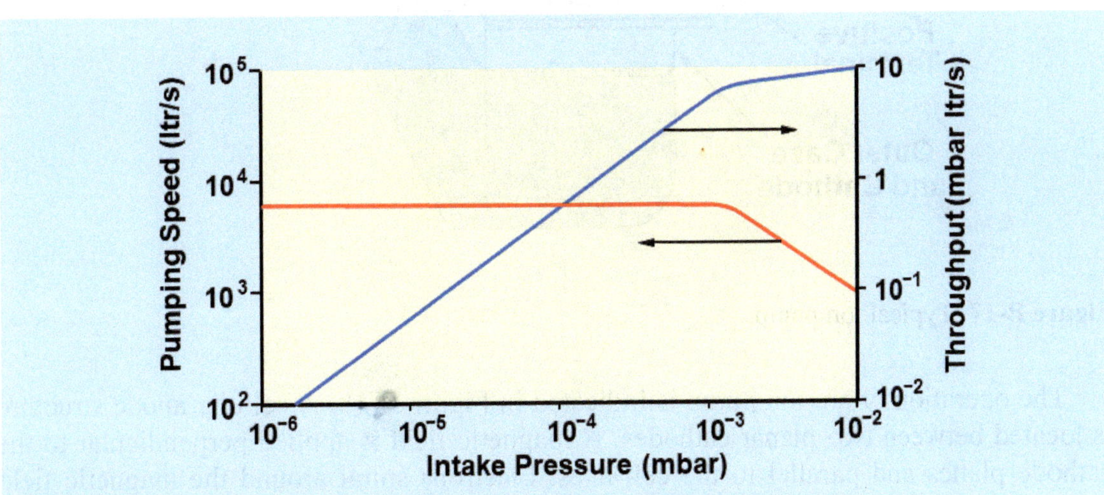

Figure B-15 Typical pumping speed and throughput for a diffusion pump.

B.6.2 Ion Pumps

The configuration of an ion pump is shown in Figure B-17. Ion pumps are capable of evacuating a chamber to the ultrahigh vacuum range. They require a pressure in the low end of the medium vacuum range (~10^{-2} torr) before they can be started. Often, an sorption pump is used to generate this medium vacuum level. Once an ion pump has been started, it can operate without assistance from other pumps.

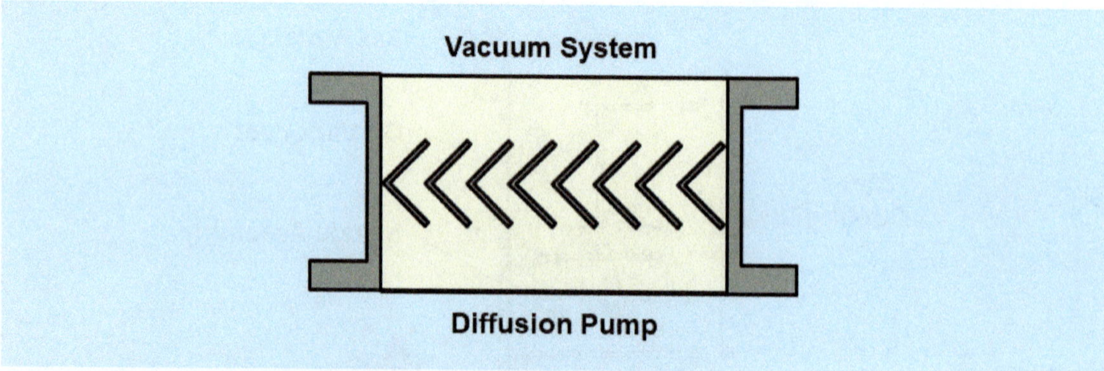

Figure B-16 Chevron-type baffle.

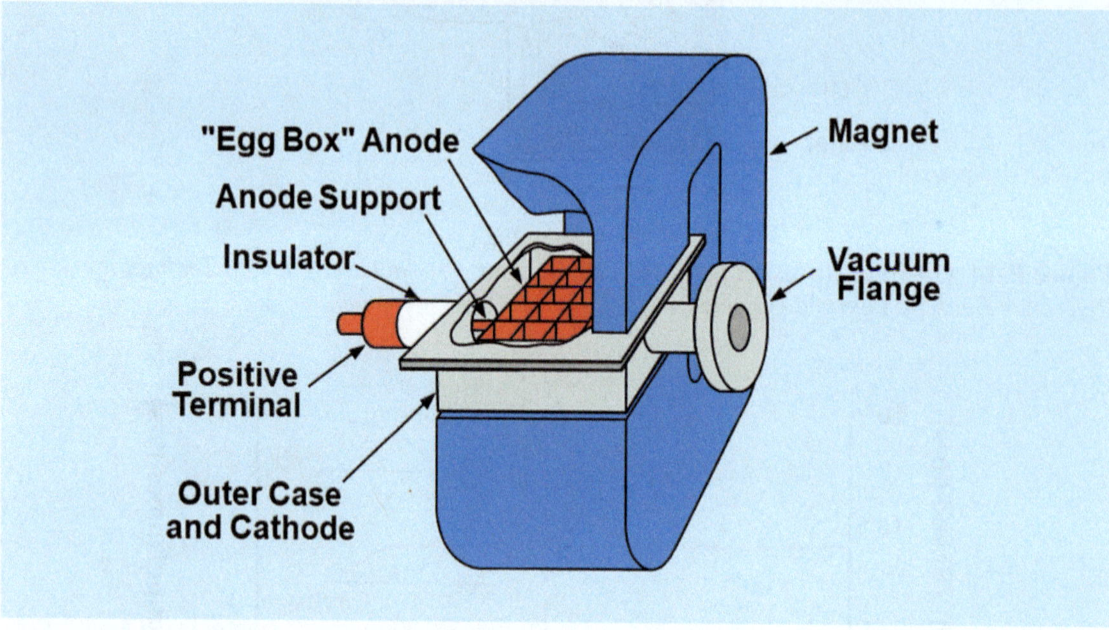

Figure B-17 Typical ion pump.

The operation of an ion pump is indicated in Figure B-18. A cellular anode structure is located between two planar cathodes. A magnetic field is applied perpendicular to the cathode planes and parallel to the cell axes. Electrons spiral around the magnetic field lines and ionize gas molecules. Electrons released during ionization participate in further ionization of other gas molecules. During collisions, electrons give up energy and gradually migrate to the anode. The positive ions are accelerated to the negative electrode, which is fabricated from titanium. The ions interact with the titanium, and at least three pumping actions occur as indicated in Figure B-19:

1. Ions sputter titanium cathode material, which deposits and acts as a getter film, adsorbing reactive gases (such as nitrogen, oxygen, and hydrogen).
2. The energy of some ions is sufficient for them to be buried in the electrode surfaces, where they are adsorbed by ion implantation. This process is effective for all types of ions including those of rare or inert gases.
3. Sputtered titanium from the cathodes buries gas atoms on the target electrode.

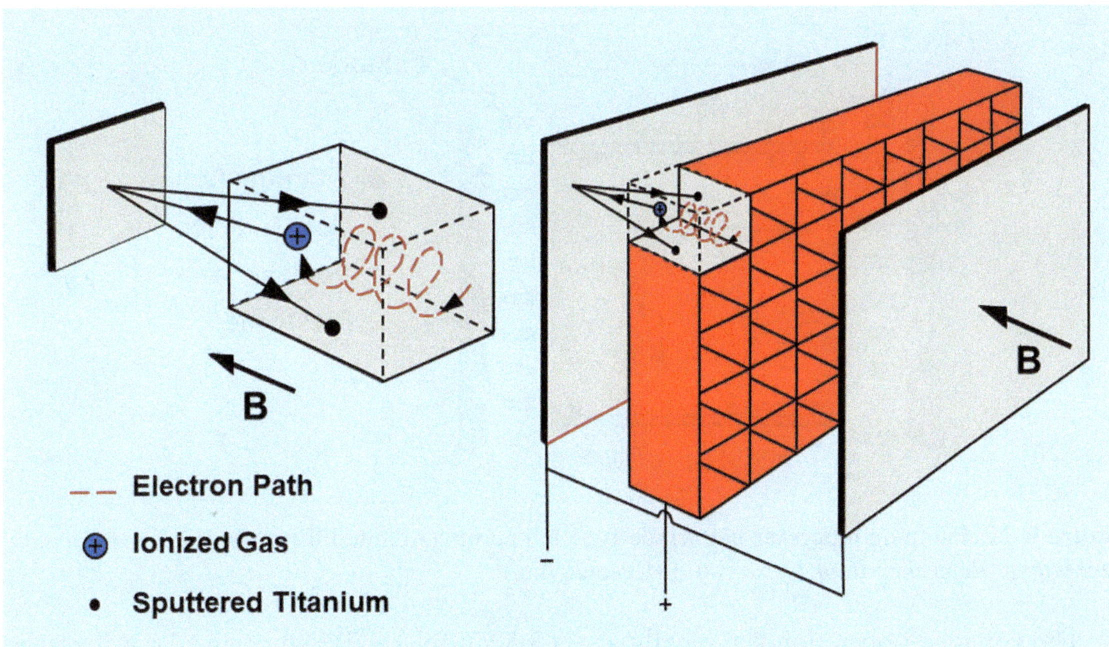

Figure B-18 Operation of diode-type ion pump.

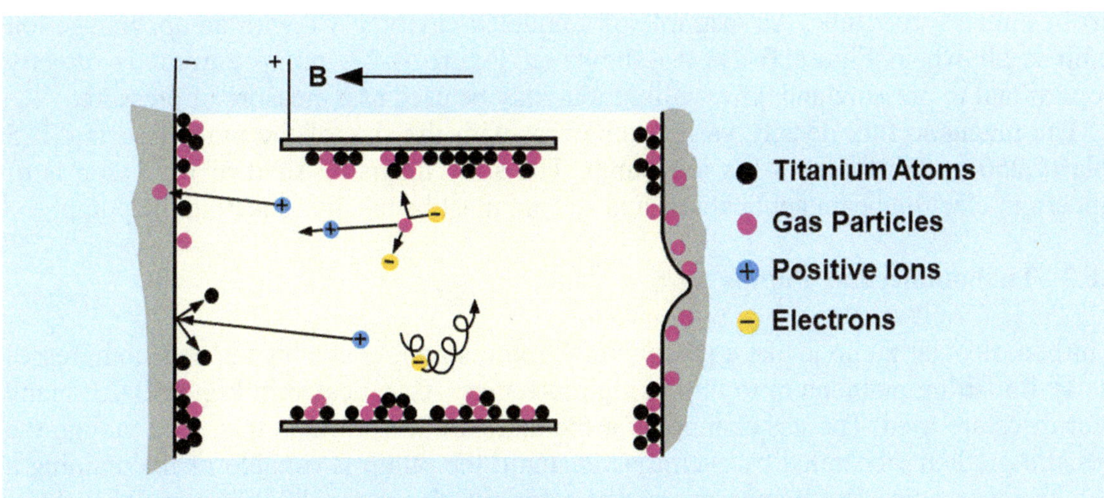

Figure B-19 Pumping processes in a diode-type ion pump. (Adapted from: *Product and Vacuum Technology Reference Book* by Leybold Heraeus, Inc.)

The pumping speed of an ion pump depends on the type of gas being pumped. Compared with air, the speed for other gases is

- Hydrogen — 150% to 200%;
- Oxygen — 80%;
- Methane — 100%;
- Other light hydrocarbons — 80% to 120%;
- Argon — 30%;
- Helium — 28%.

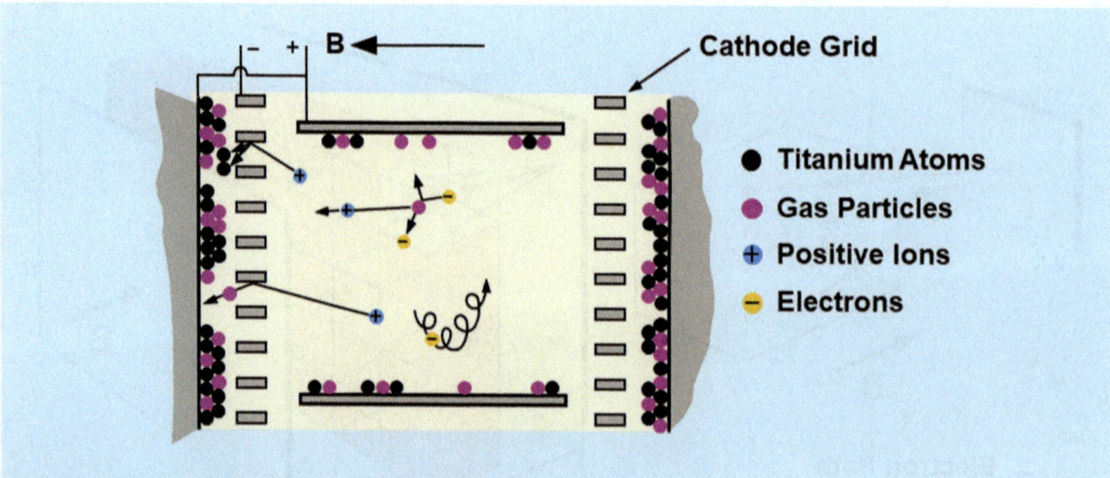

Figure B-20 Pumping processes in a triode-type ion pump. (Adapted from: *Product and Vacuum Technology Reference Book* by Leybold Heraeus, Inc.)

The voltage of operation is typically about 3.5 kV for small pumps and 7 kV for large pumps. Ion pumps as small as 0.2-L/s and as large as several thousand liters per second are available. A small pump is often used as an appendage pump, which is an integral part of a microwave tube. An example of a coupled-cavity TWT with an appendage ion pump is shown in Figure B-21. As shown in Figure B-22, pump current is directly proportional to pressure and, after calibration, may be used as a measure of pressure.

The magnetic flux density varies somewhat with the size of the pump and is 0.125 tesla (1,250 gauss) for an 8-L/s ion pump. The stray magnetic field of the pump is of concern in electron beam applications and so care must be taken in locating the pump.

B.6.3 Turbomolecular Pumps

A turbomolecular pump is like a jet engine. A rotor, spinning at a high speed, compresses gas by imparting momentum to impinging molecules. As indicated in Figure B-23, many rotor discs are used. The gas channels for the discs are in series, so that even though the pressure gradient generated by each disc is small, the pump is capable of maintaining a high-pressure ratio. The pumps operate at rotor speeds up to ~80,000 rpm. They have large compression ratios for hydrogen and ultimate pressures in the 10^{-10} torr range. The fore pressure is ~10^{-2} torr so oil-free backing pumps such as the scroll pump can be used.

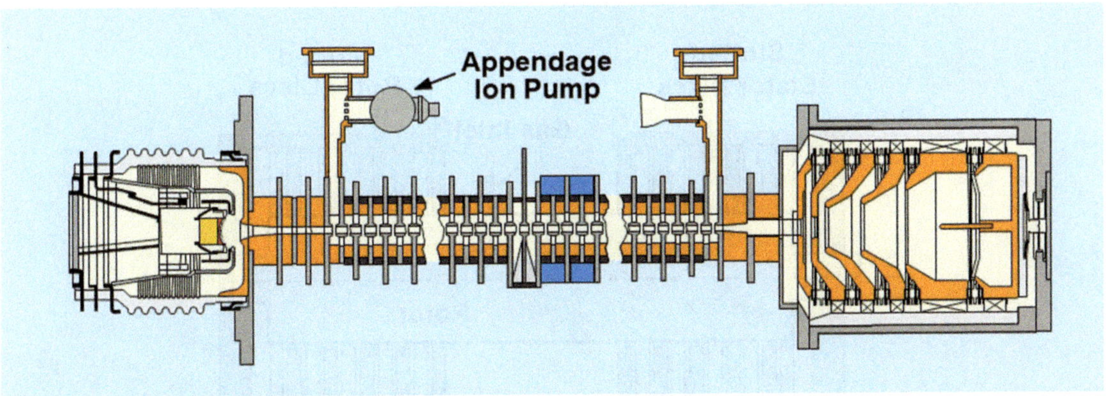

Figure B-21 Coupled-cavity TWT with an appendage pump.

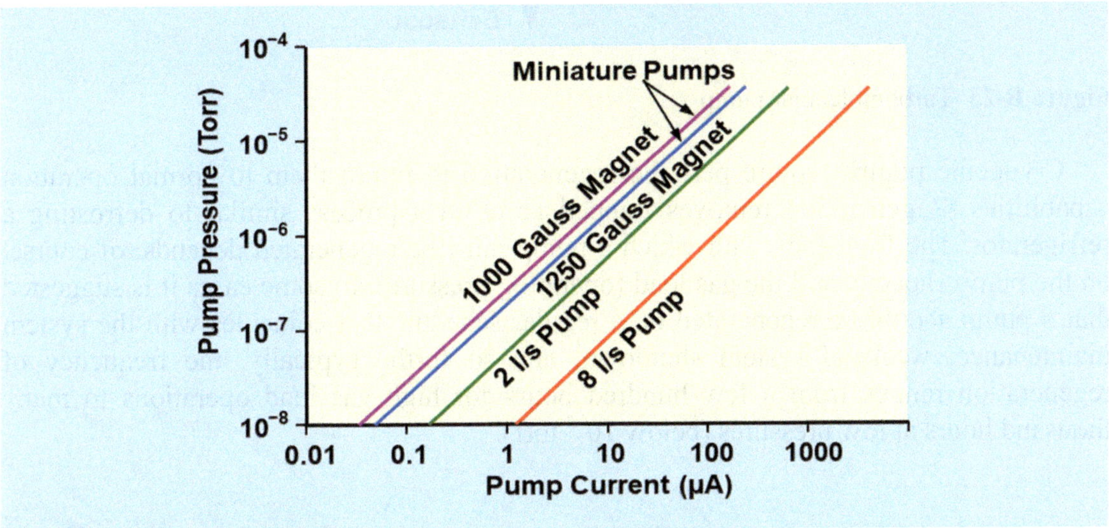

Figure B-22 Ion pump pressure versus current.

B.6.4 Cryogenic Pumps

Cryogenic pumps operate on the principle that gases can be condensed at low temperatures and have extremely low vapor pressures. An example of a cryogenic pump is shown in Figure B-24. This pump operates in the 10^{-3}–10^{-10} torr range.
Pumping speeds for some common gases are:

- Water — 4,000 liters/second;
- Air — 1,500 liters/second;
- Hydrogen — 2,500 liters/second;
- Argon — 1,200 liters/second.

In operation, the 80K array traps water and hydrocarbon vapors. The 15K array condenses nitrogen, oxygen, and argon. The specially processed charcoal in this array traps helium, hydrogen, and neon.

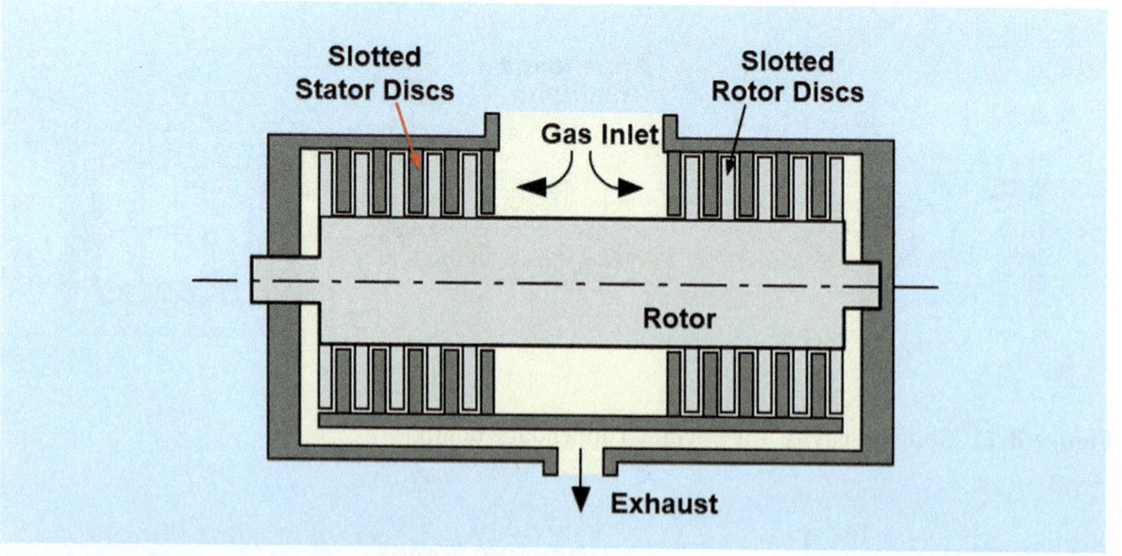

Figure B-23 Turbomolecular pump.

Cryogenic pumps require periodic regeneration to return them to normal operation capabilities. Regeneration removes trapped gases in a process similar to defrosting a refrigerator. The frequency with which a pump must be regenerated depends, of course, on the pump capacity and the gas load (operating pressure). In some cases it is suggested that a pump should be regenerated on a regular schedule that coincides with the system maintenance, weekend system shutdown, and so forth. Typically, the frequency of regeneration ranges from a few hundred hours for high gas load operations to many thousand hours at low pressures (below 10^{-5} torr).

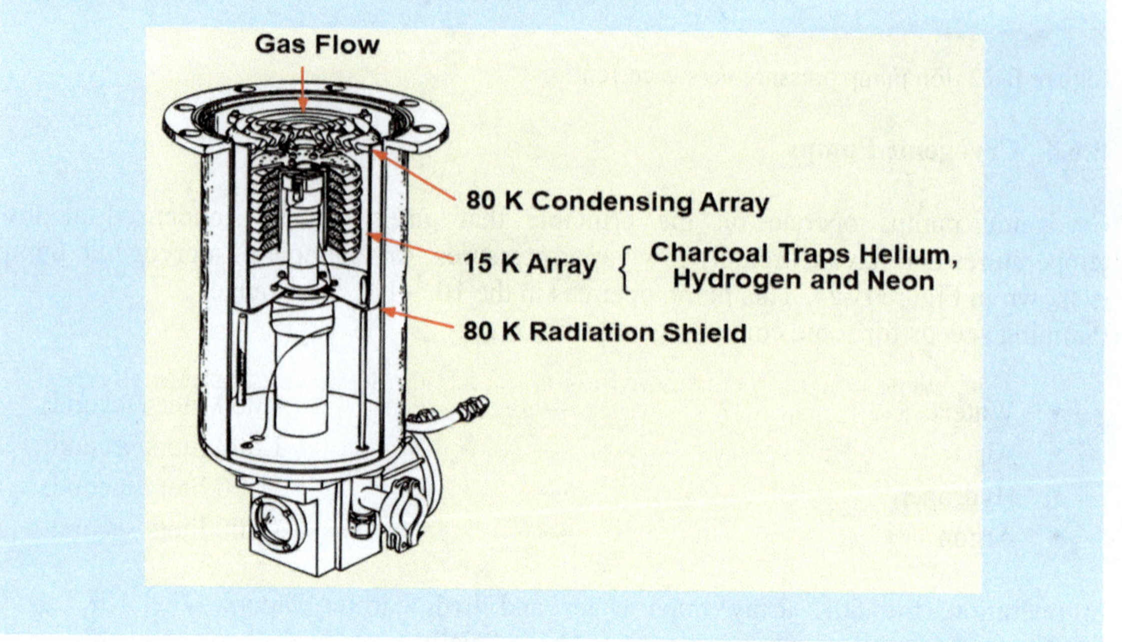

Figure B-24 The Cryo-Torr 8 high-vacuum pump. (Adapted from: CTI-Cryogenics *Cryo-Torr Manual.*)

B.6.5 Nonevaporable Getters

The phenomenon of gettering has long been used in vacuum tubes. In conventional getters, a chemically active metal is evaporated to form a film on the inner surface of the vacuum envelope. Gases that react with the metal film are said to be gettered. Conventional getters have not found application in high-power microwave tubes primarily because the evaporated film may cause electrical leakage or breakdown in high-voltage structures.

Nonevaporable getters are now commonly used in microwave tubes. The ST175 getters available from SAES GETTERS/USA Inc. are particularly attractive for use. These getters utilize a mixture of titanium and molybdenum powders in a weight ratio of 9:1. The material is lightly pressed and sintered to form a structure that is porous and has good mechanical stability. It is possible to use these getters in holders made of nichrome as indicated in Figure B-25.

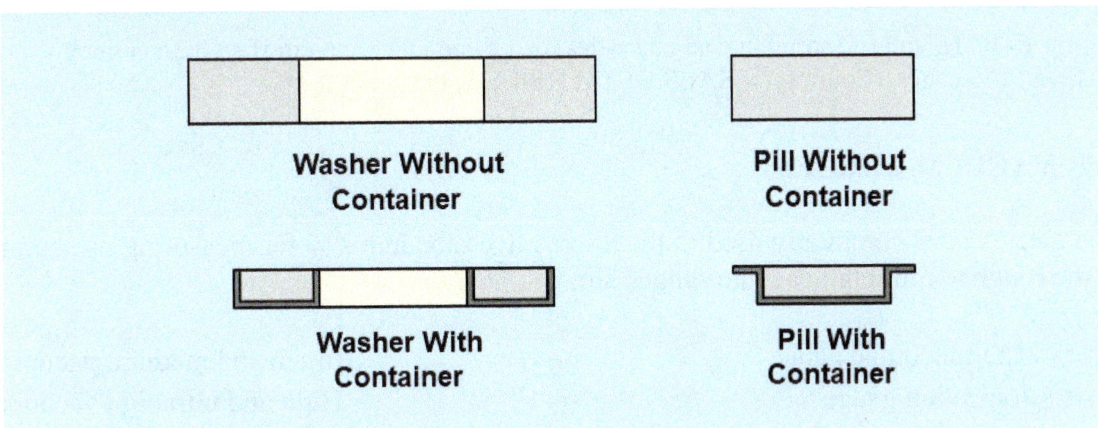

Figure B-25 SAES ST75 getters without and with nichrome containers. (Courtesy of SAES GETTERS/USA, Inc.)

The ST175 can be satisfactorily activated by heating to 500°C for 10 minutes in vacuum ($<10^{-3}$ torr) so these getters can be conveniently used during the bakeout of tubes. As an example, the results of sorption tests for CO and H_2 activated in this manner are shown in Figure B-26.

Some of the microwave tube experience with nonevaporable getters installed in an electron gun is as follows:

1. Appendage ion pump current, which might sometimes be high, is virtually zero.
2. Arcing in the electron gun is greatly reduced.
3. After extended periods of storage, voltage can be applied immediately to a tube without gassing or arcing.
4. The getters are recoverable. After activation and then exposure to air, getters can be reactivated.

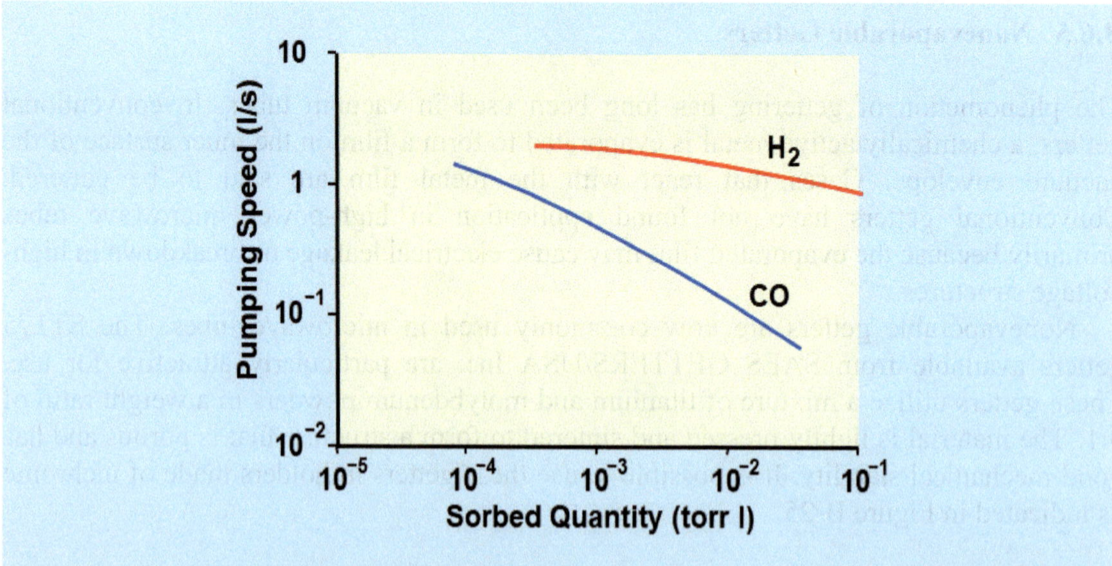

Figure B-26 H_2 and CO sorption characteristics for 1.5-cm OD × 0.6-cm ID × 0.2-cm thick SAES ST175 getter. (Courtesy of SAES GETTERS/USA, Inc.)

B.7 VACUUM GAUGES

The gauges most commonly used in the microwave tube industry for measuring pressures in the rough to ultrahigh vacuum ranges are:

- Thermocouple gauge — Rough and medium vacuum
- Ionization gauge — High and ultrahigh vacuum
- Ion pump current — High and ultrahigh vacuum

The use of ion-pump current as a measure of pressure was discussed in Section B.6.2. Thermocouple and ionization gauges will be described in this section.

B.7.1 Thermocouple Gauge

The thermocouple gauge is one of several gauges that use the thermal conductivity of gas to provide the pressure indication. As indicated in Figure B-27, a filament of wire heated by an electric current is mounted in an envelope attached to the vacuum system. The temperature of the wire depends on heat loss through the gas, which depends on gas pressure. The wire temperature is monitored with a thermocouple, which is calibrated to indicate pressure. Thermocouple gauges operate in the pressure range of ~10 to 10^{-3} torr.

B.7.2 Ionization Gauge

Ionization gauges of the type shown in Figure B-28 are commonly used to measure pressures in the high and ultrahigh vacuum ranges. Electrons, accelerated by a positive grid, produce ions that are collected by an ion collector. The resulting ion current is a measure of gas pressure. These gauges are sometimes referred to as inverted triode ion gauges, or Bayard-Alpert gauges. In these gauges, a very small diameter wire is used as the ion collector. This is to minimize the X-rays produced when electrons strike the positive grid. Some of these X-rays impinge on the collector wire, causing it to release electrons. The resulting flow of electron current away from the collector causes an error in pressure indication.

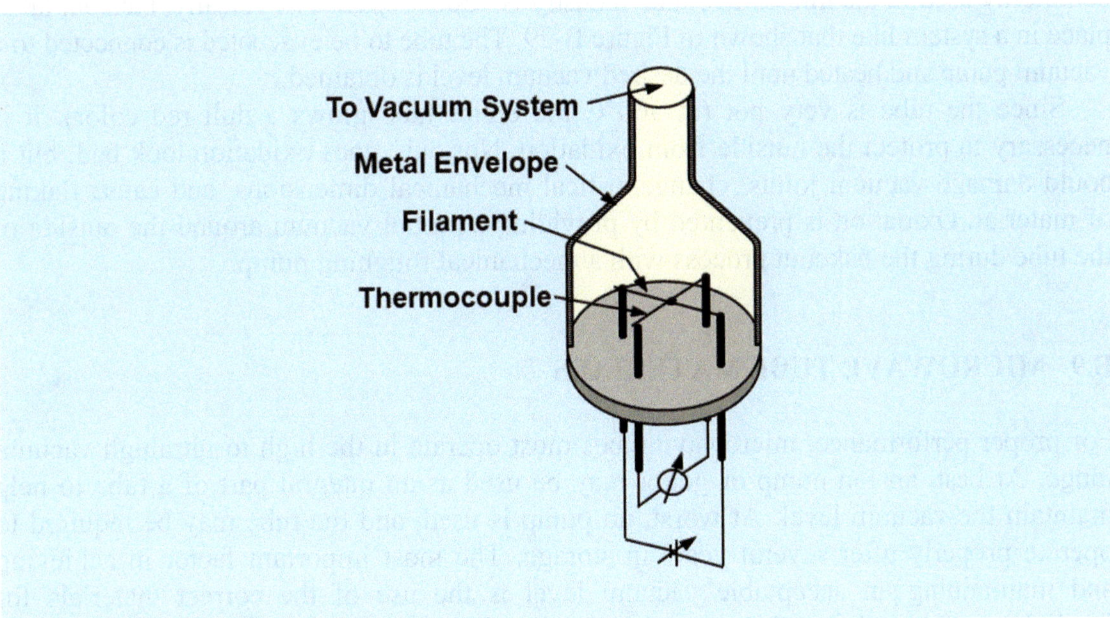

Figure B-27 Thermocouple gauge. (Adapted from: *Handbook of Thin Film Technology* by L. I. Maissel and R. Glang, McGraw-Hill, Inc., 1970.)

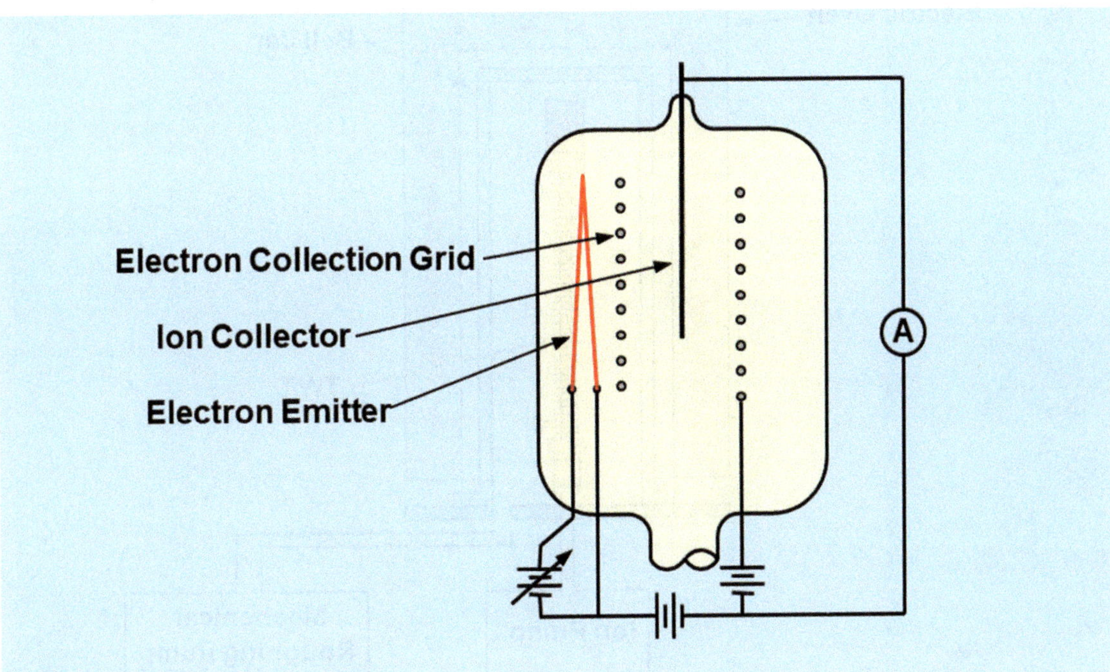

Figure B-28 Ionization gauge.

B.8 BAKEOUT

Bakeout is the process of heating a tube to facilitate evacuation. As pointed out previously in this appendix, the rates of diffusion and the desorption of gases vary exponentially with temperature, so heat is used to aid in the evacuation process. The bakeout temperature commonly used in the microwave tube industry is ~500°C–525°C. Typically, bakeout takes place in a system like that shown in Figure B-29. The tube to be evacuated is connected to a vacuum pump and heated until the desired vacuum level is obtained.

Since the tube is very hot (at 500°C the entire tube glows a dull red color), it is necessary to protect the outside from oxidation. Not only does oxidation look bad, but it could damage vacuum joints, change critical mechanical dimensions, and cause flaking of material. Oxidation is prevented by providing a partial vacuum around the outside of the tube during the bakeout process with a mechanical roughing pump.

B.9 MICROWAVE TUBE MATERIALS

For proper performance, microwave tubes must operate in the high to ultrahigh vacuum range. At best, an ion pump or getter may be used as an integral part of a tube to help maintain the vacuum level. At worst, no pump is used, and the tube may be required to operate properly after several years in storage. The most important factor in achieving and maintaining an acceptable vacuum level is the use of the correct materials for fabricating a tube. Many of the materials commonly used are listed in Figure B-30.

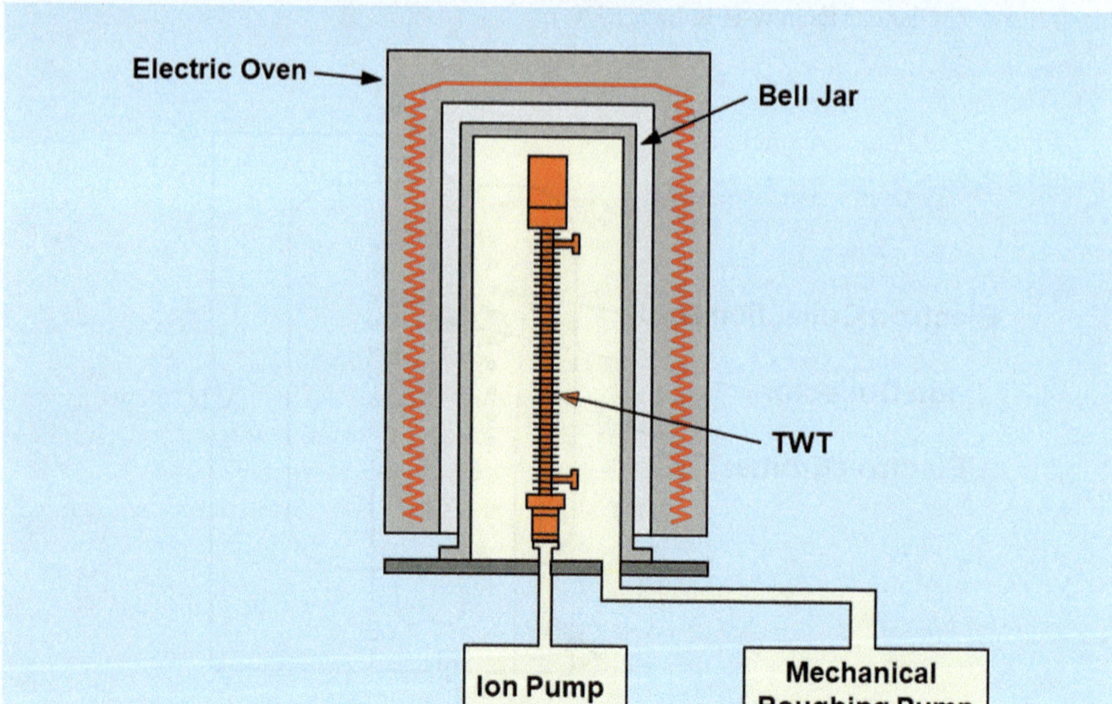

Figure B-29 Configuration of equipment for baking out a TWT.

	Aluminum (O_2)	Barium	Beryllium (O_2)	Boron (Nitride)	Calcium	Carbon	Chromium	Copper	Gold	Iridium	Iron	Manganese	Molybdenum	Nickel	Osmium	Palladium	Platinum	Rhenium	Rhodium	Scandium	Silver	Strontium	Thorium	Titanium	Tungsten	Vanadium	Zirconium
Heater	■																	■							■		
Cathode	■	■	■		■					■				■	■		■			■		■	■		■		
Grids						■		■	■				■												■		
Anode								■																			
Cavities								■																			
Helix								■	■				■					■							■		
Support	■		■	■		■																					
Attenuator						■																					
Collector						■		■																			
Envelope							■	■			■			■													
Brazes								■	■			■	■			■	■		■		■						■
Polepieces											■															■	
Window	■		■			■																					
Getter		■																						■		■	■

Figure B-30 Some materials used inside the vacuum envelope of a microwave tube.

The primary factor considered in selecting a metal for use inside the vacuum envelope is vapor pressure. The charts in Figures B-31, B-32, and B-33 give vapor pressure as a function of temperature for the metallic elements.

As an example of the use of these charts, consider zinc (see Figure B-31). At low temperatures, the vapor pressure is reasonable (10^{-9} torr at 100°C). At a somewhat elevated temperature (400°C), the vapor pressure is only 10^{-1} torr. This shows why brass (which contains zinc) must not be used in a vacuum system that may be heated. If brass was used and baked at 400°C or higher, zinc vapor would permeate the entire system.

The curve for barium in Figure B-33 is significant in considering the operation of cathodes. At a cathode operating temperature of 1,000°C, the vapor pressure of barium is over 1 torr. Thus, excess barium (over ~ monolayer) on the cathode surface evaporates very rapidly. This evaporated barium may eventually deposit on insulating surfaces and cause electrical leakage or breakdown. To remove excess barium, a new cathode is often placed in vacuum and operated at a high temperature before it is placed in a tube.

Nonmetallic materials, such as those shown in Figure B-34, are sometimes used in demountable vacuum systems. These materials severely limit the pressure and temperature at which the system can operate.

When a nonmetallic material is required in a system that must be baked, one of a very small number of materials is selected. If the material is part of the vacuum envelope, then either Al_2O_3 (aluminum oxide), BeO (beryllium oxide), glass or, in special cases, CVD diamond is selected. These materials have extremely low vapor pressures. Inside the vacuum envelope, in addition to Al_2O_3 or BeO, boron nitride or diamond is used.

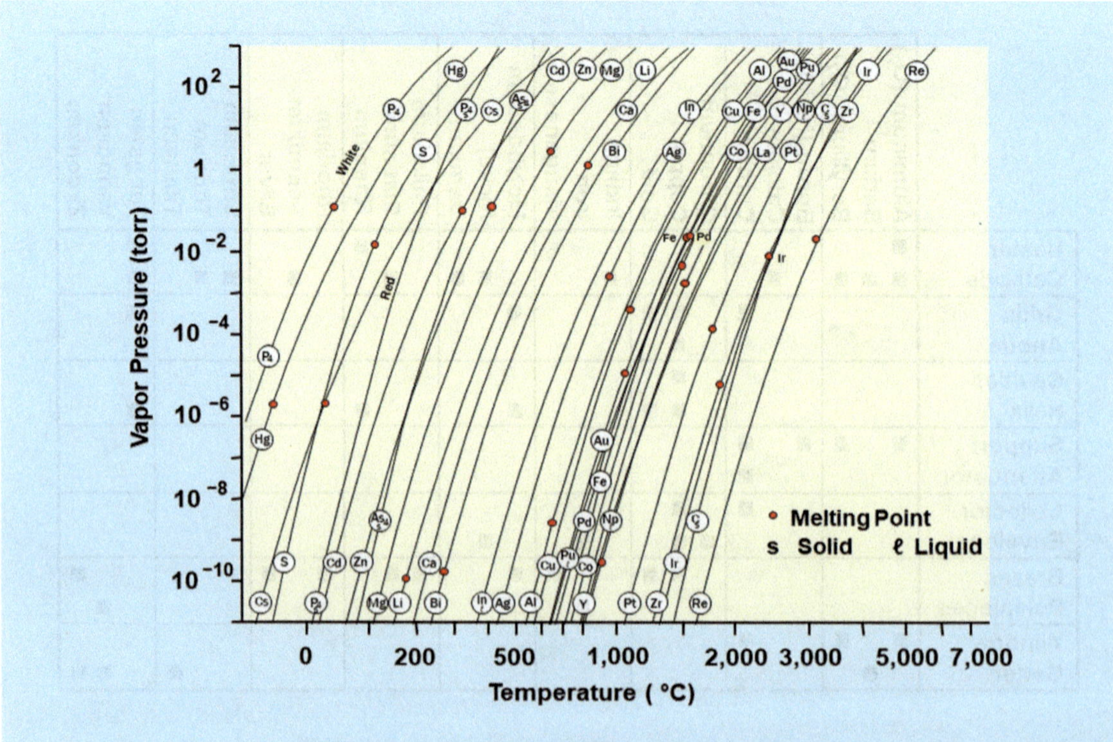

Figure B-31 Vapor pressures of the elements. (Adapted from: R. E. Honig and D. A. Kramer, *RCA Review*, June 1969, copyright 1969 by Radio Corporation of America.)

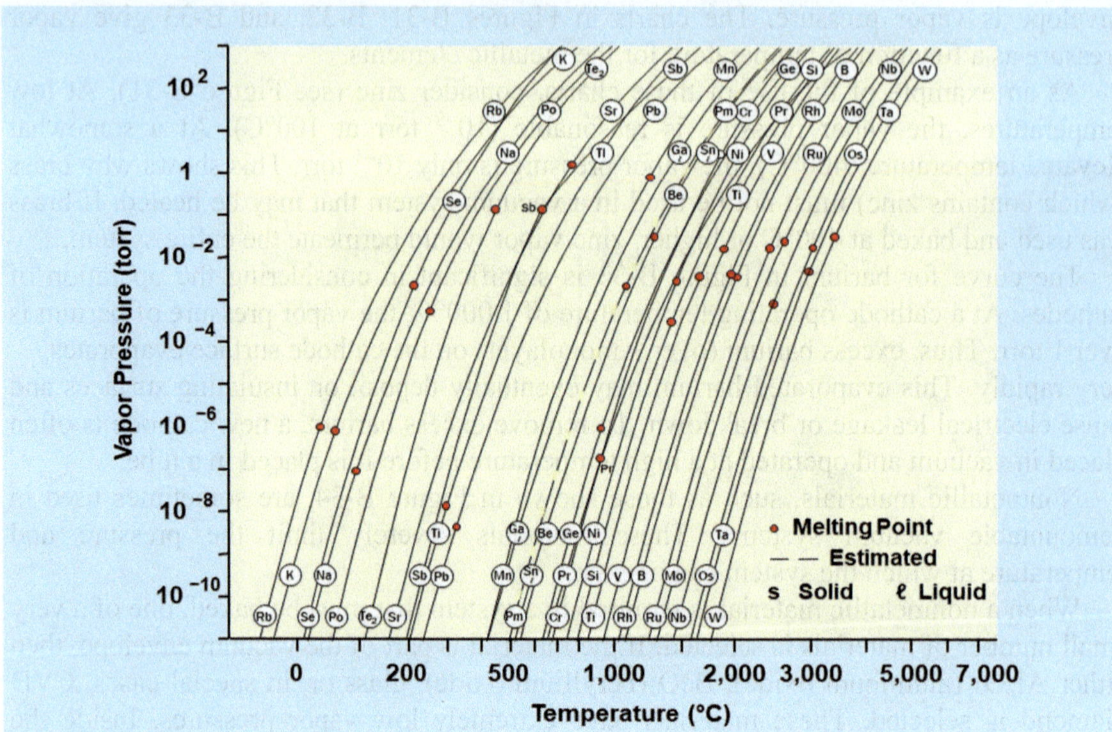

Figure B-32 More vapor pressures of the elements.

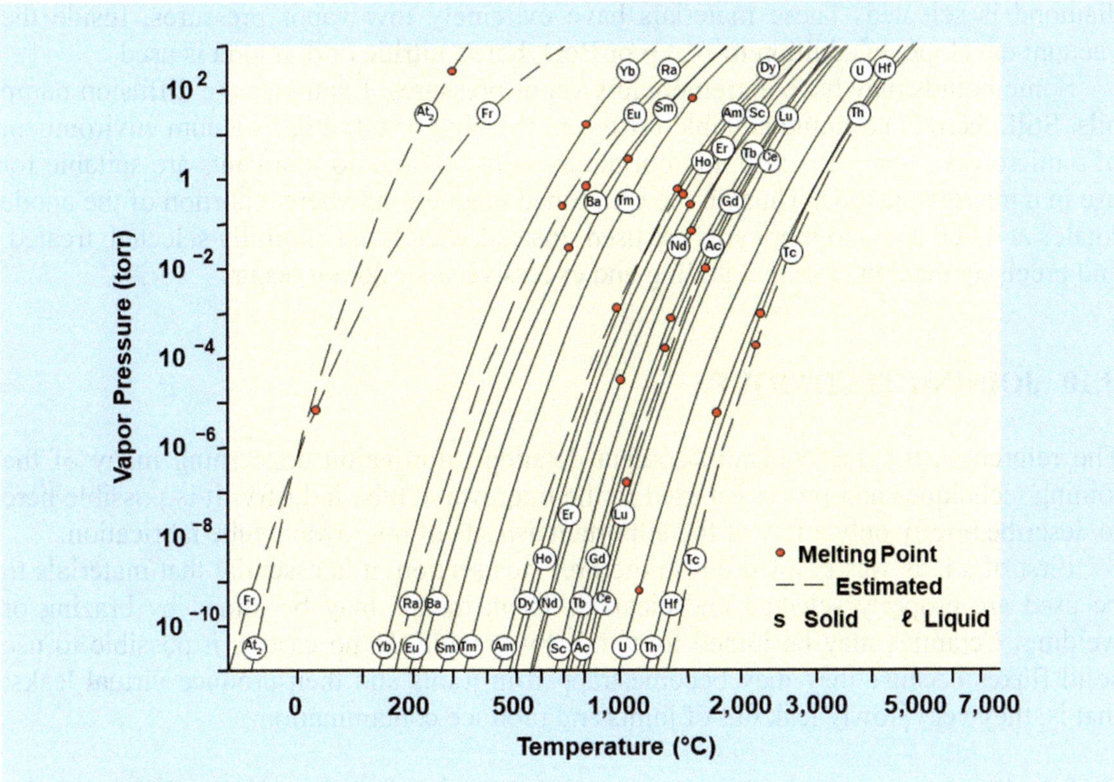

Figure B-33 More vapor pressures of the elements.

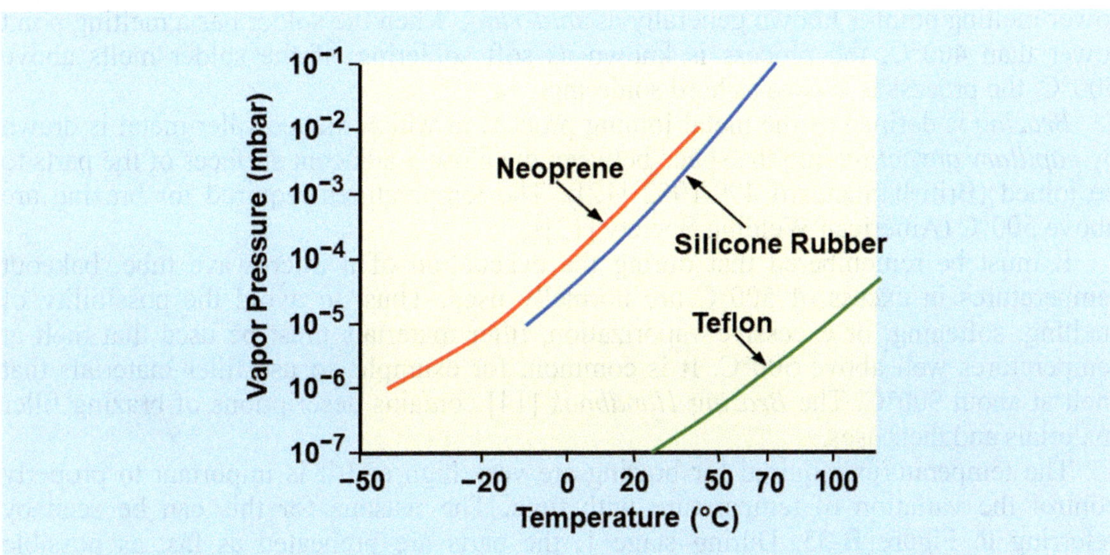

Figure B-34 Vapor pressure of nonmetallic sealing materials. (Adapted from: *Product and Vacuum Technology Reference Book* by Leybold Heraeus, Inc.)

When a nonmetallic material is required in a system that must be baked, one of a very small number of materials is selected. If the material is part of the vacuum envelope, then either Al_2O_3 (aluminum oxide), BeO (beryllium oxide), glass or, in special cases, CVD diamond is selected. These materials have extremely low vapor pressures. Inside the vacuum envelope, in addition to Al_2O_3 or BeO, boron nitride or diamond is used.

Some liquids may have extremely low vapor pressures. Examples are diffusion pump oils. Still, there is no liquid suitable for use in the high or ultrahigh vacuum environment of a microwave tube. It is also appropriate to point out that no lubricants are suitable for use in a microwave tube. Thus, in the spin tuned magnetron, where a portion of the anode rotates at 4,000 rpm, no lubricants are used. Instead, metals are carefully selected, treated, and precisely machined so that galling and excessive wear do not occur.

B.10 JOINING TECHNIQUES

The references for this appendix contain detailed information concerning many of the joining techniques and processes used in the microwave tube industry. It is possible here to describe briefly only a few of the important aspects of microwave tube fabrication.

First of all, as was pointed out in the previous section, it is essential that materials to be used are properly selected and cleaned. Then, metals may be joined by brazing or welding. Ceramics may be joined to metals by brazing. In no case is it possible to use solid fluxes because they may become trapped in joints and then produce virtual leaks; that is, they very slowly leak out of joints and produce contamination.

B.10.1 Brazing

From O'Hanlon [3], "The joining process of two metal parts with a third one having a lower melting point is known generally as *soldering*. When the solder has a melting point lower than 400°C, the process is known as soft soldering; if the solder melts above 500°C, the process is known as hard soldering.

Brazing is defined as the metal joining process in which melted filler metal is drawn by *capillary attraction* into the space between the closely adjacent surfaces of the parts to be joined (British Standard 499/1952 [12]). The temperatures required for brazing are above 500°C (American Welding Society [13]).

It must be remembered that during the evacuation of a microwave tube, bakeout temperatures in excess of 500°C are normally used. Thus, to avoid the possibility of melting, softening, or excessive vaporization, filler materials must be used that melt at temperatures well above 500°C. It is common, for example, to use filler materials that melt at about 900°C. The *Brazing Handbook* [14] contains descriptions of brazing filler materials and their uses.

The temperatures required for brazing are very high and it is important to properly control the variation of temperature with time. The reasons for this can be seen by referring to Figure B-35. During stage 1, the parts are preheated as fast as possible consistent with avoiding warping and thermal shock. Thermal equilibrium is established during stage 2. It is important for all parts to reach the solidus temperature at the same time. Stage 3 is the brazing step. The final temperature is adjusted to obtain the desired viscosity of the filler. The time spent in the liquidus region is minimized. Finally, the parts are cooled as quickly as possible. The most critical portion of cooldown is dropping the temperature below the solidus temperature.

Some of the problems that occur if temperature is not properly controlled during brazing are shown in Figure B-36. The ideal braze joint is the one with the full fillet. If either temperature or time is inadequate, then the filler material may not be completely melted and so the underbrazed condition results. If the parts are not properly cleaned or the joint tolerance is incorrect, then voids may occur that can result in leaks or virtual leaks. If the temperature is too high or the time that the parts are held at a high temperature is too long, then the filler may alloy with the parts being brazed and cause erosion.

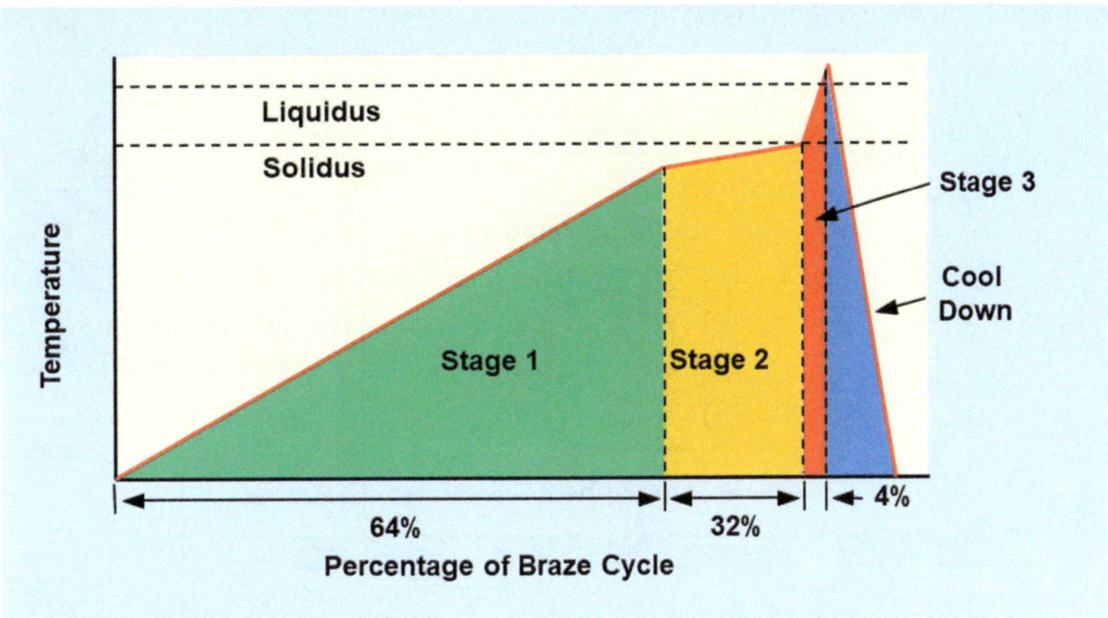

Figure B-35 Temperature–time profile for avoiding brazing problems. (From: R. English, Microwave Tube Manufacture lectures, Crane Division, Naval Surface Warfare Center.)

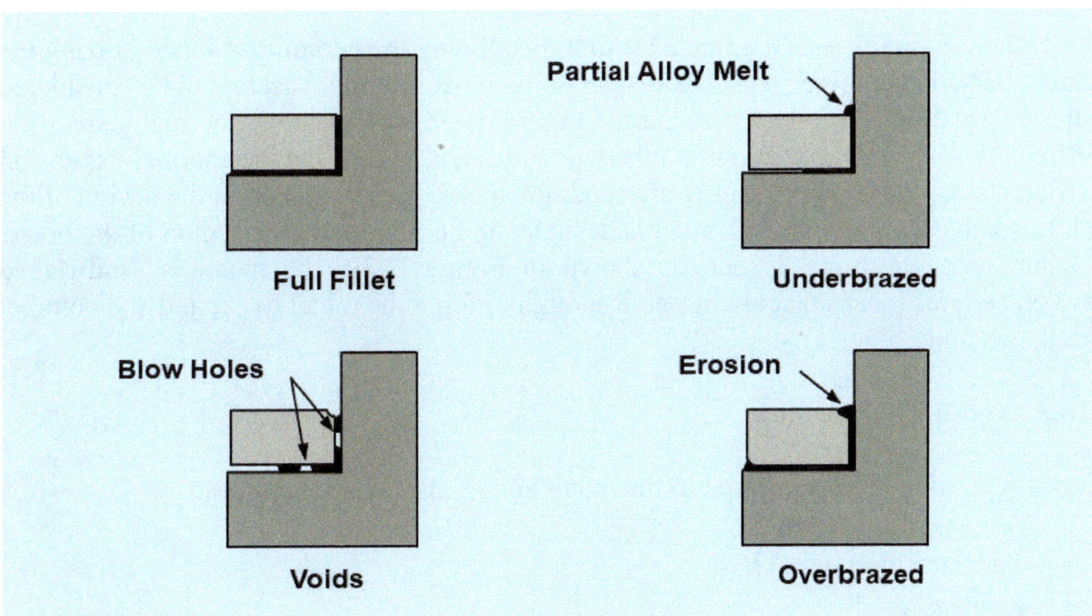

Figure B-36 Examples of braze problems. (From: R. English, Microwave Tube Manufacture lectures, Crane Division, Naval Surface Warfare Center.)

Because of the critical nature of the brazing operation, there are only a few processes that can be used. The most common are oven brazing and induction brazing. Fluxes cannot be used so brazes must be performed in vacuum or in a reducing atmosphere of hydrogen. Hydrogen is often used rather than a vacuum because brazing can be performed more rapidly. In some cases, such as brazing of stainless steel, extremely dry hydrogen must be used to prevent oxidation. An induction heating system for brazing in hydrogen is shown in Figure B-37.

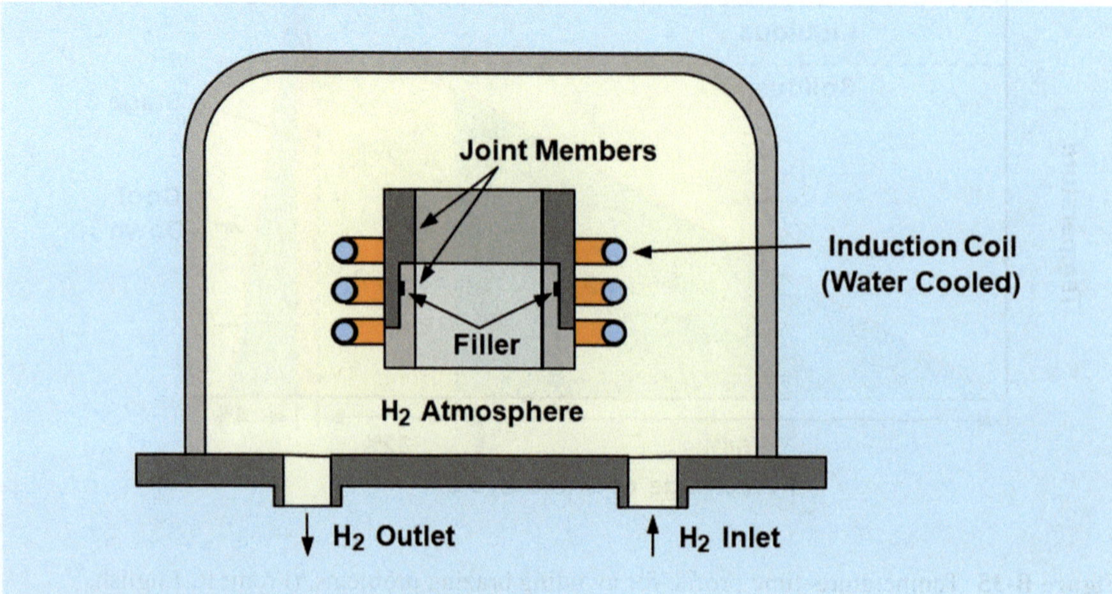

Figure B-37 Brazing by induction heating in hydrogen.

Ceramic-to-metal seals are made by first metallizing the ceramic and then brazing the metallic part to the metallized surface. One of the most important factors to be considered is the thermal expansion of the parts. Figure B-38 shows how thermal expansion coefficients vary with temperature. When ceramics and metals having thermal expansion coefficients that differ significantly are used, great care must be taken in the design of the seal. In some cases the metal element is designed to be flexible in the region of the braze. In a butt ceramic-to-metal seal, as shown in Figure B-39, the metal is sandwiched between ceramic pieces that are massive enough to force the metal to expand and contract with the ceramic.

B.10.2 Welding

Several types of welding are used in the manufacture of microwave tubes:

- Tungsten inert gas (TIG);
- Resistance;
- Electron beam;
- Laser.

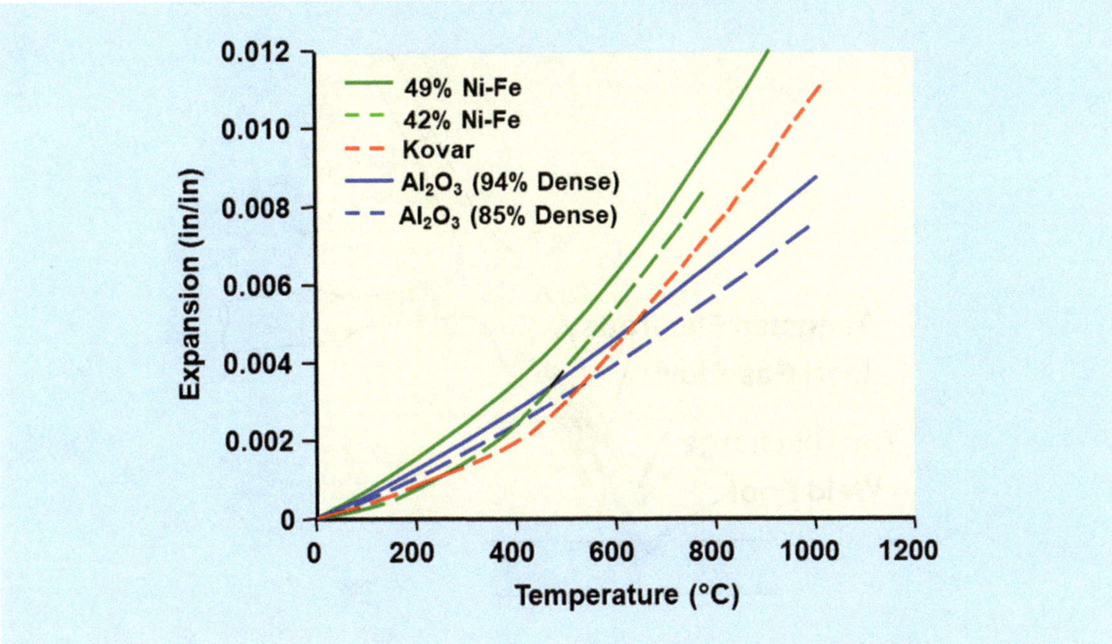

Figure B-38 Thermal expansion characteristics. (Adapted from: *Handbook of Electron Tube and Vacuum Techniques* by Fred Rosebury, copyright 1965 by Addison-Wesley Publishing Company, Inc.)

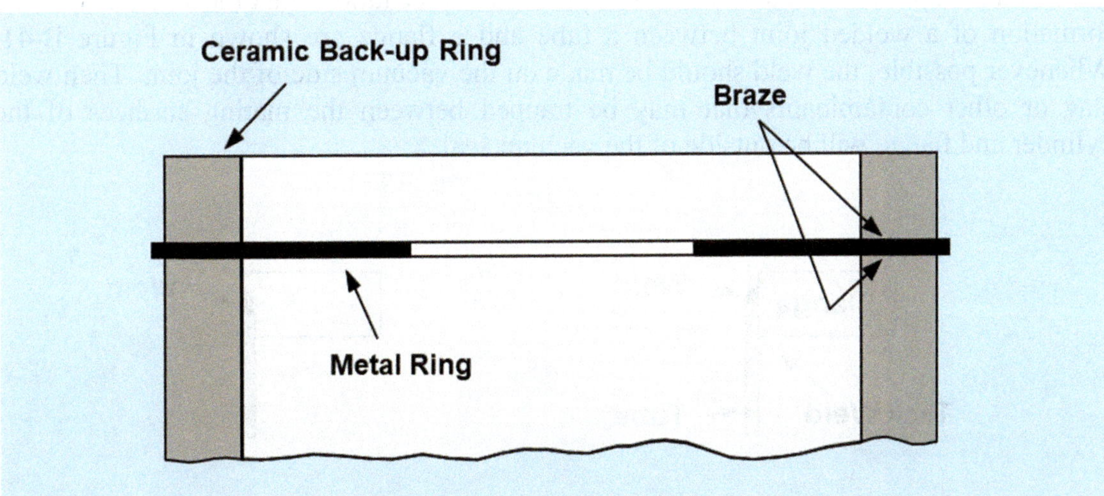

Figure B-39 Butt ceramic-to-metal seal.

TIG welding can be used manually or in automated systems A drawing of a manual inert gas welding torch is given in Figure B-40. With TIG welding, the heat produced by an arc melts the metal of the joint between workpieces and joins them together. Important parameters in the welding process are the current, which controls the depth and speed of the weld, and current types (dc positive or negative or ac). Either argon or helium or a mixture of the two is the gas that is used. The gas must be very pure, with impurities of 50 ppm at a maximum.

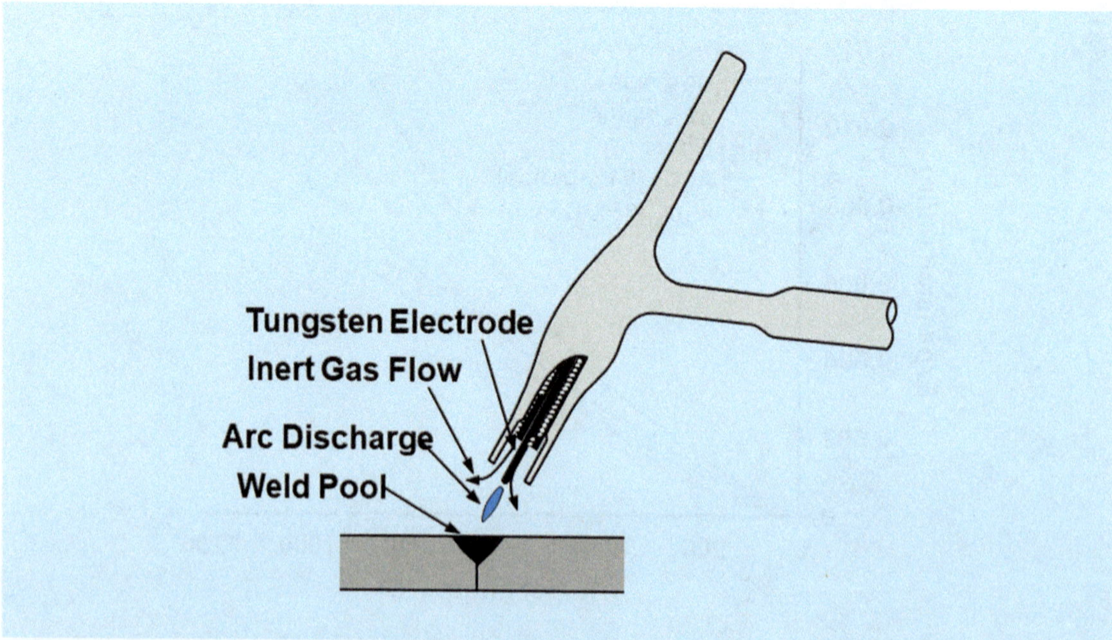

Figure B-40 Schematic of tungsten inert gas (TIG) welding torch. (Adapted from: *Handbook of Thin Film Technology* by L. I. Maissel and R. Glang. ©1970 McGraw-Hill, Inc.)

Care must be taken in designing weld joints if adequate strength is to be obtained and gas and dirt traps are to be avoided. Examples of the precautions that must be taken in the formation of a welded joint between a tube and a flange are shown in Figure B-41. Whenever possible, the weld should be made on the vacuum side of the joint. Then weld slag or other contaminants that may be trapped between the mating surfaces of the cylinder and flange will be outside of the vacuum seal.

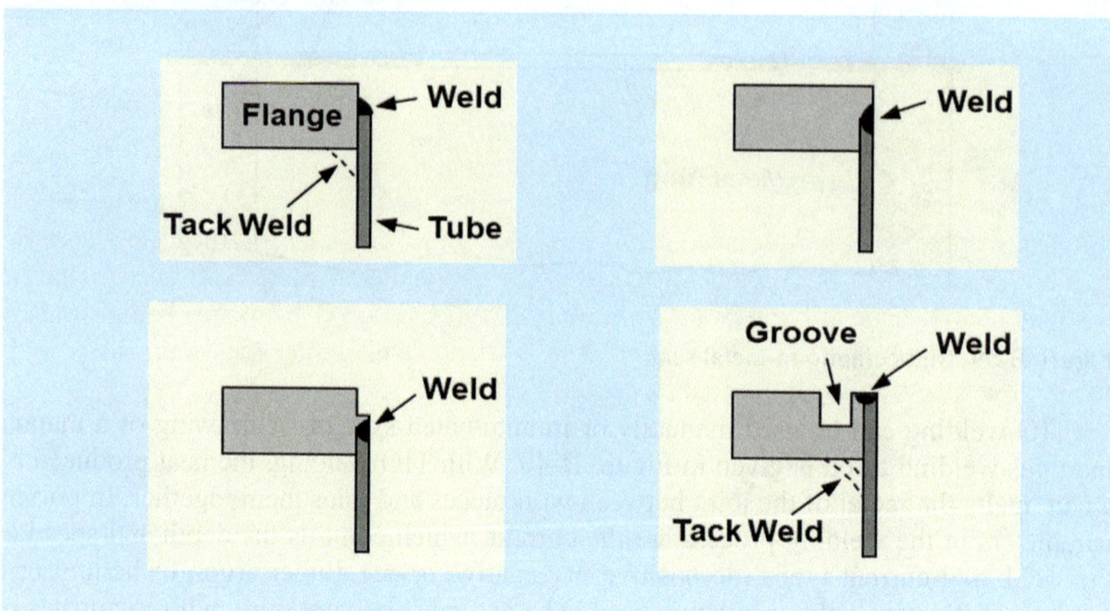

Figure B-41 Common tube-to-flange weld designs. (Adapted from: *Handbook of Thin Film Technology* by L. I. Maissel and R. Glang. ©1970 McGraw-Hill, Inc.)

Resistance welding (sometimes called spot welding) is perhaps the simplest method for joining two metals. Using a device like that depicted in Figure B-42, high pressure is applied to the metal parts to be welded and then a current pulse is passed through the parts, causing them to melt and fuse together.

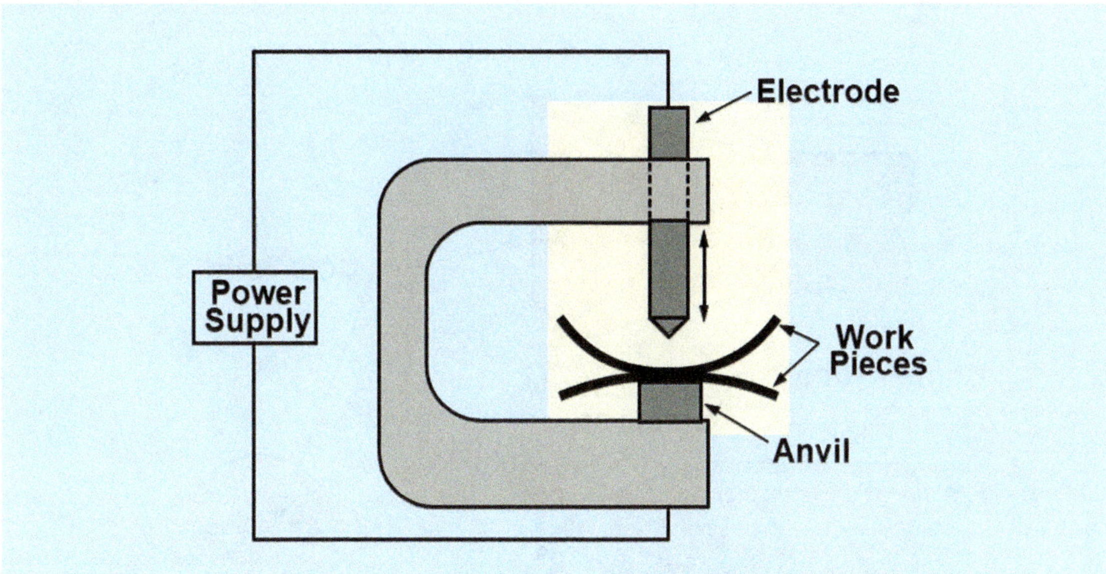

Figure B-42 Apparatus for spot-welding metal parts. (Adapted from: R. English, Microwave Tube Manufacture lectures, Crane Division, Naval Surface Warfare Center.)

Electron beam welding is generally performed in vacuum but can be done in an atmosphere. A beam accelerated to 0.3 – 0.7 times the speed of light is used to form deep narrow welds. The region affected by the weld is small so welding close to parts affected by heat is possible. Disadvantages are that a vacuum requirement is usually required, equipment costs are high, and close-fitting joints are required because of the narrow beam width that is used.

Laser welding is accomplished with either a pulsed or a CW light beam. It is often used for welds of small parts in hard-to-reach locations such as the connection in a TWT of the last turn of a helix to the center conductor of a coaxial connector. It is also used when complex weld patterns are required. Compared with electron beam welding, smaller spot sizes are possible and the region affected by heat is smaller. One disadvantage of laser welding is that a large fraction of the light beam is reflected until the metals being welded have melted. Other disadvantages are that it is difficult to define the beamwidth, and precise alignments and control of gaps are required.

After assembly and processing of a metal-to-ceramic vacuum tube has been completed, it is necessary to separate the tube from the pumping station. (An appendage pump may remain as an integral part of the tube.) The technique universally used to separate the tube from the pumping station is illustrated in Figure B-43. Soft copper tubulation is used to make the vacuum connection to the pumping station. This tubulation is molecularly clean on the internal surfaces because it has been baked along with the microwave tube (at 450°C or higher).

By squeezing the tubulation with a special tool, the internal mating surfaces of the copper can be forced to flow together to form a cold weld. In its crudest form, the tool is similar to a manually operated bolt cutter with the cutting edges replaced by cylindrical steel bars.

Because the manual squeeze-off (pinch-off) operation with the tool that is much like a bolt cutter is highly operator-dependent, an automated system is normally employed.

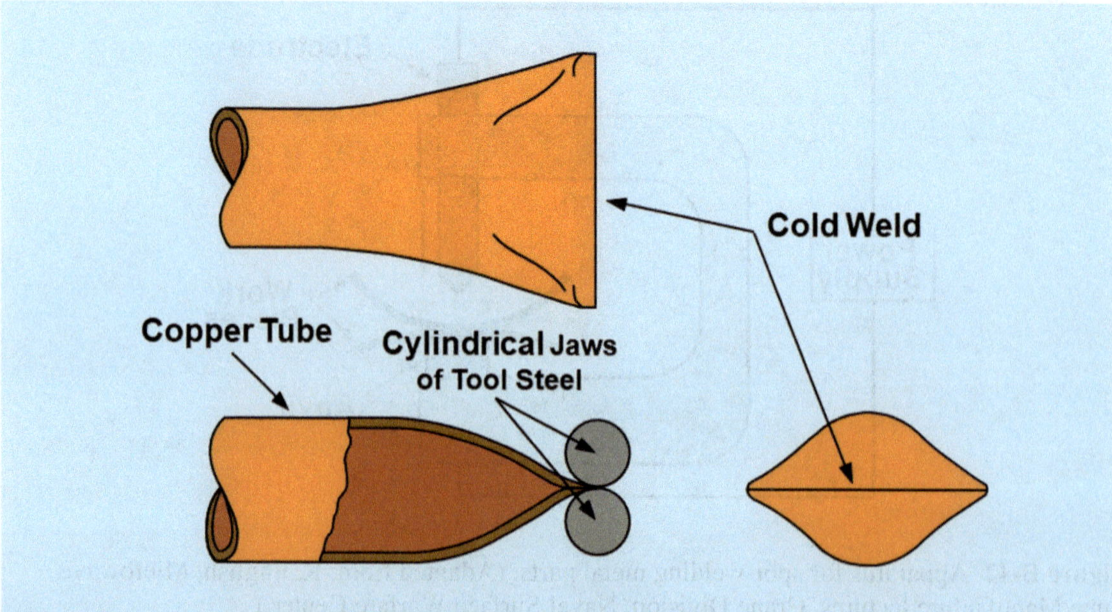

Figure B-43 Method used to squeeze off (pinch off) copper tubulation. (Adapted from: *Power Travelling Wave Tubes* by J. F. Gittins, American Elsevier Publishing Company, Inc., 1965.)

REFERENCES

[1] *Vacuum Physics and Technology, Methods of Experimental Physics*, Vol. 14, New York: Academic Press, 1979.

[2] S. Dushman, and J. M. Lafferty, *Scientific Foundations of Vacuum Techniques*, New York: John Wiley and Sons, 1962.

[3] J. F. O'Hanlon, *A User's Guide to Vacuum Technology*, New York: John Wiley and Sons, 2003.

[4] M. Pirani, and J. Yarwood, *Principles of Vacuum Engineering*, New York: Reinhold, 1963.

[5] R. W. Roberts, and T. A. Vanderslice, *Ultra High Vacuum and Its Applications*, Englewood Cliffs, NJ: Prentice Hall, 1963.

[6] A. Roth, *Vacuum Technology*, Amsterdam, the Netherlands: Elsevier Science Publishers B.V., 1990.

[7] A. Roth, *Vacuum Sealing Techniques*, Woodbury, NY: AIP Press, 1994.

[8] M. Knoll, *Materials and Processes of Electron Devices*, Berlin: Springer-Verlag, 1959.

[9] W. H. Kohl, *Materials and Techniques for Electron Tubes*, New York: Reinhold, 1960.

[10] F. Rosebury, *Handbook of Electron Tube and Vacuum Techniques*, Reading, MA: Addison-Wesley, 1965.

[11] R. English, "Lectures on Microwave Tube Manufacture," Crane Division of the Naval Surface Warfare Center.

[12] British Standard 499/1952.

[13] American Welding Society.

[14] *Brazing Handbook*, Miami, FL: American Welding Society, 1991.

Appendix C

Magnetics

Magnetic fields are used for focusing electron beams in nearly all microwave tubes. These fields may be generated by permanent magnets or electromagnets. In a permanent magnet, electron motion within the material produces the magnetic field. In an electromagnet, current passing through a wire produces a magnetic field surrounding the wire. Ferromagnetic materials, permanent magnets, and electromagnets are discussed in this appendix [1–5].

C.1 MAGNETIC QUANTITIES

In considerations of magnetics and magnetic materials, three quantities are of primary interest:

- Magnetic field intensity, H, is one of the elementary fields in nature.
- Magnetic flux density, B, is the density of magnetic flux lines produced by the magnetic field.
- Magnetic permeability, μ, is the property of a medium that determines how much flux is produced by a certain field level. Flux density is related to field intensity by

$$B = \mu H \tag{C-1}$$

The units commonly used for H, B, and μ are given in the following table along with conversion factors. The value for μ in air (free space), μ_o, is

$$\mu_o = 4\,\pi \times 10^{-7} \text{ H/m} \tag{C-2}$$

It is common to use the term *relative permeability*, μ_r, which is the ratio of the permeability of a material to that of free space. That is,

$$\mu_r = \frac{\mu}{\mu_o} \tag{C-3}$$

Quantity	*MKS Unit (a)*	*CGS Unit (b)*	*Conversion*
H	Amperes/meter	Oersted	(a) = 79.577 × (b)
B	Tesla	Gauss	(a) = 10^{-4} × (b)
μ	Henrys/meter	Gauss/Oersted	(a) = 1.257×10^{-6} × (b)

C.2 MAGNETIC CIRCUITS

It is convenient, when thinking about magnetic circuits, to relate magnetic quantities to analogous quantities in electrical circuits. Figure C-1 indicates how magnetic and electric circuits are related. *MMF* is the magnetomotive force and is analogous to voltage. The source of the magnetomotive force can be either a permanent magnet or an electromagnet. Φ is the magnetic flux and is analogous to the current, and reluctance is analogous to resistance.

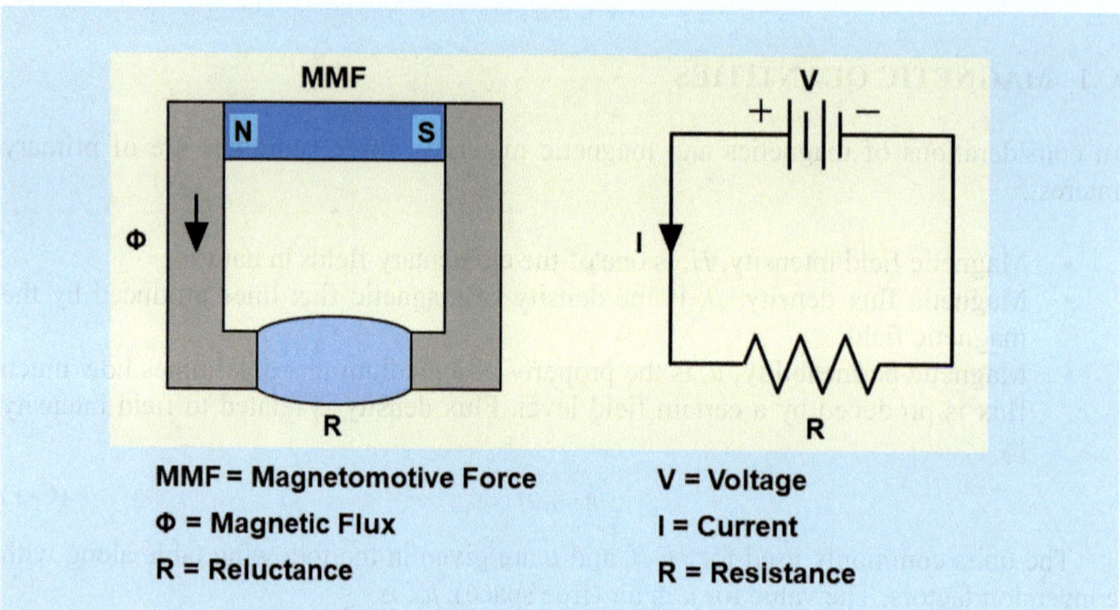

Figure C-1 Comparison of magnetic and electric circuits.

Permeance, denoted by *P*, is the inverse of reluctance, is analogous to electrical conductance, and is often used rather than reluctance. *MMF*, Φ, *P*, and *R* are related to *H* and *B* by

$$MMF = \int H \cdot dl \tag{C-4}$$

$$\Phi = \int B \cdot dA \tag{C-5}$$

$$R = \frac{1}{P} = \frac{MMF}{\Phi} \tag{C-6}$$

Just as resistance is related to the physical characteristics of a medium by $R = \ell / \sigma A$, where σ is conductivity, reluctance is related by $R = \ell / \mu A$ where μ is permeability.

Much can be learned by examining the magnetic circuit in Figure C-1. Consider Figure C-2 and assume that there is no flux leakage outside the magnetic circuit and also assume that the flux density in the gap, G, is uniform. The problem is to find expressions for the flux density, B_g, in the gap and for the operating conditions of the magnet.

Let H_m be the magnetic field strength of the magnet, let H_{Fe} be the magnetic field in iron pole pieces that connect the magnet to the gap, and let H_g be the magnetic field in the gap. In terms of field quantities, the magnetic circuit equation is as follows:

$$\int H\,dl = \int_A^B H_{Fe}\,dl + \int_B^C H_G\,dl + \int_C^D H_{Fe}\,dl + \int_D^A H_m\,dl = 0 \tag{C-7}$$

which, assuming a constant H_{Fe} in the pole pieces with lengths ℓ_1 and ℓ_2, becomes

$$H_m T = H_{Fe}\ell_1 + H_{Fe}\ell_2 + H_g G \tag{C-8}$$

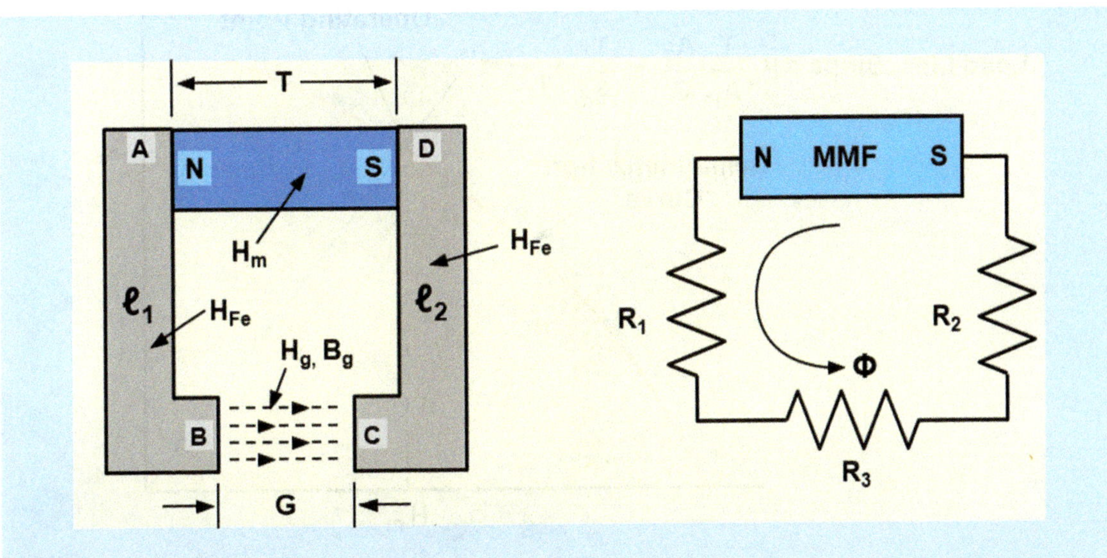

Figure C-2 Simple magnetic circuit.

In terms of circuit quantities, the magnetic circuit equation is

$$MMF = \Phi(R_1 + R_2 + R_3) \tag{C-9}$$

The reluctances, R_1, R_2, and R_3 are given by:

$$R_1 = \frac{\ell_1}{\mu_{Fe} A_1}, \quad R_2 = \frac{\ell_2}{\mu_{Fe} A_2}, \text{ and } R_3 = \frac{G}{\mu_o A_g} \tag{C-10}$$

where μ_{Fe} is the permeability of iron and A is the cross-sectional area. Since $\mu_{Fe} >> \mu_o$ and

$MMF = H_m T$,

$$H_m T = \Phi \frac{G}{\mu_o A_g} = B_g G / \mu_o \tag{C-11}$$

or

$$B_g = \mu_o H_m \frac{T}{G} \tag{C-12}$$

Thus, if H_m is known, the flux density in the gap can be found.

Since magnetic flux is constant, $B_g = B_m A_m / A_g$, and so

$$\frac{B_m}{H_m} = \mu_o \frac{T}{A_m} \frac{A_g}{G} = \frac{T}{A_m} P_g \tag{C-13}$$

where P_g is the permeance of the gap. This relation gives the slope of a line, called the load line, on the B-H characteristic for the magnet material. This is illustrated in Figure C-3. The demagnetizing curve is discussed in the following section.

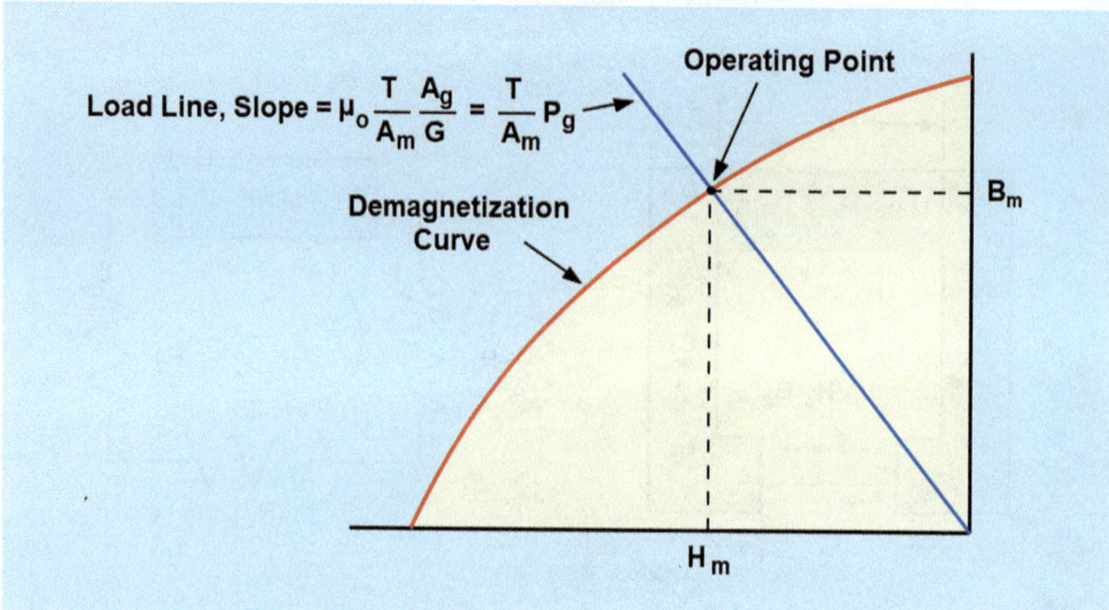

Figure C-3 Illustration of magnetic load line.

C.3 MAGNETIC MATERIALS

Magnetic materials can be divided into three categories. These are:

- $\mu_r < 1$ diamagnetic
- $\mu_r > 1$ paramagnetic
- $\mu_r >> 1$ ferromagnetic

Diamagnetism is present in all materials. The establishment of magnetic flux is opposed by the material so that the relative permeability is less than unity. This effect is explained by the laws of Faraday and Lenz. When a magnetic field is applied, an electric field is produced. This electric field acts on the electron orbits in a material to produce microscopic reaction currents. The reaction currents produce a magnetic field that opposes the applied field. Diamagnetism is an extremely small effect. In most materials, it is overridden by paramagnetism or ferromagnetism.

Paramagnetic materials have a relative permeability slightly greater than unity. In these materials, the application of an external magnetic field aligns permanent magnetic dipoles of atoms, ions, or molecules with the applied field, thus enhancing the field.

In ferromagnetic materials, a specialized condition exists that greatly aids the applied magnetic field. This condition involves the electronic structures of atoms and the distances between atoms. The most stable configuration of the electrons is when their spin magnetic moments are parallel. The result is that complete alignment of the spins occurs within regions called *magnetic domains*. The application of an external field both aligns domains and shifts domain boundaries to produce an enormous field.

C.3.1 Ferromagnetic Materials

Ferromagnetic materials are highly nonlinear. The relation between B and H must be determined experimentally. This is done with a circuit like that in Figure C-4.

From Ampere's law, the magnetic field, H, in a magnetic sample is proportional to the current, I. From Faraday's law, the voltage, V, induced in a winding is proportional to the time rate of change of magnetic flux linking the winding. The integral of voltage is therefore proportional to the flux and the magnetic flux density, B.

A signal proportional to B is applied to the vertical input of an oscilloscope. A signal proportional to H is applied to the horizontal input. The result displayed on the oscilloscope is known as a hysteresis curve and is illustrated in Figure C-5.

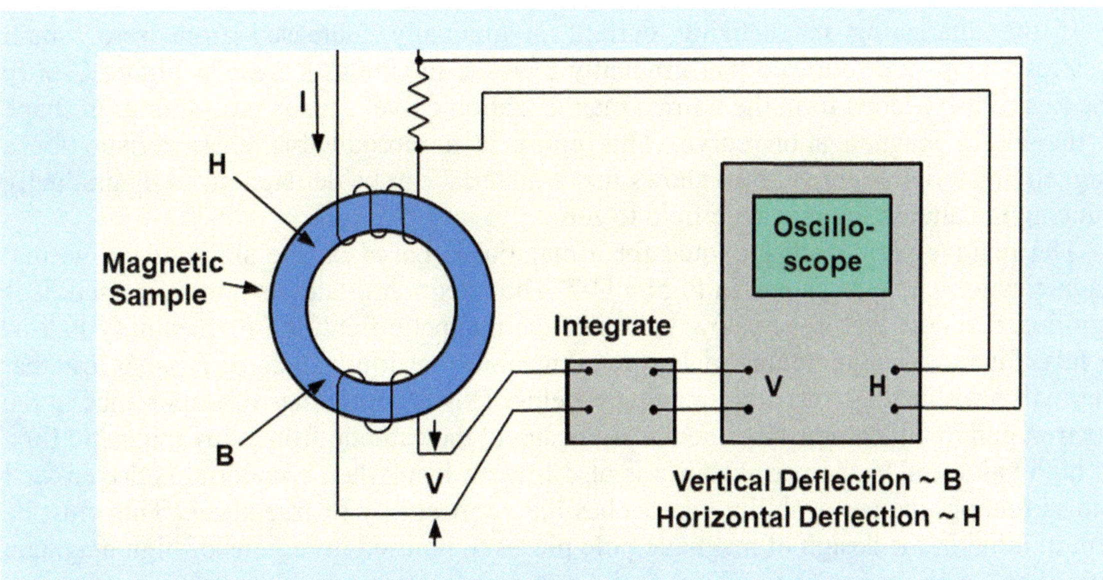

Figure C-4 Circuit for determining the magnetic characteristic of materials.

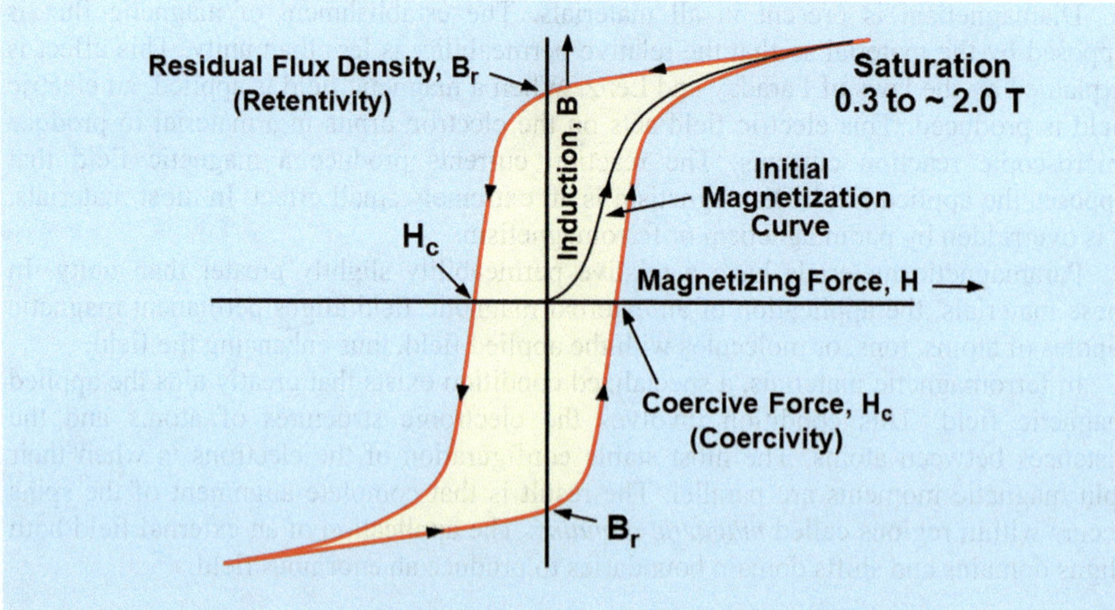

Figure C-5 Hysteresis curve resulting from use of circuit shown in Figure C-4.

When current is increased, the right-hand portion of the hysteresis curve is followed. When current is decreased, the left-hand portion is followed. At very large current excursions, the shape of the hysteresis curve is relatively independent of the current, and a loop known as the saturation loop is traversed.

Several important characteristics of magnetic materials can be illustrated with the hysteresis loop. Some of these are shown in Figure C-5. When a magnetic field is first applied to a demagnetized sample, the initial magnetization curve is followed. When H is zero, there is a residual flux density, B_r, in the material called the *retentivity*. The flux density, B, is reduced to zero when a field, H_c, called the *coercive force* or *coercivity*, is applied.

If the alternating magnetizing current is gradually increased from zero, small hysteresis loops are followed that gradually increase in size as shown in Figure C-6(a). The tips of these loops form the normal magnetization curve. This is very similar in shape to the initial magnetization curve. This curve is a reproducible characteristic for a magnetic material. Figure C-6(b) shows that a material can be demagnetized by gradually reducing the alternating magnetic field to zero.

The relative permeability quoted for a material is based on the slope of the normal magnetization curve as shown in Figure C-7. The maximum value is normally used. It is significant to note that at very low values of the magnetic field, the permeability is low. In fact, for most magnetic materials, the permeability at low values of H is so low that they will not shield against stray magnetic fields. This is sometimes of importance in the electron gun of a tube when it is necessary to shield the cathode from stray magnetic flux. At high values of H, the permeability is also low. In fact, when a material is driven well into saturation, the permeability approaches the permeability of free space. This must be kept in mind in the design of magnetic pole pieces to be used in regions of high magnetic field. The concentration of flux lines at the corners of pole pieces, for example, can cause saturation and distortion of the field pattern.

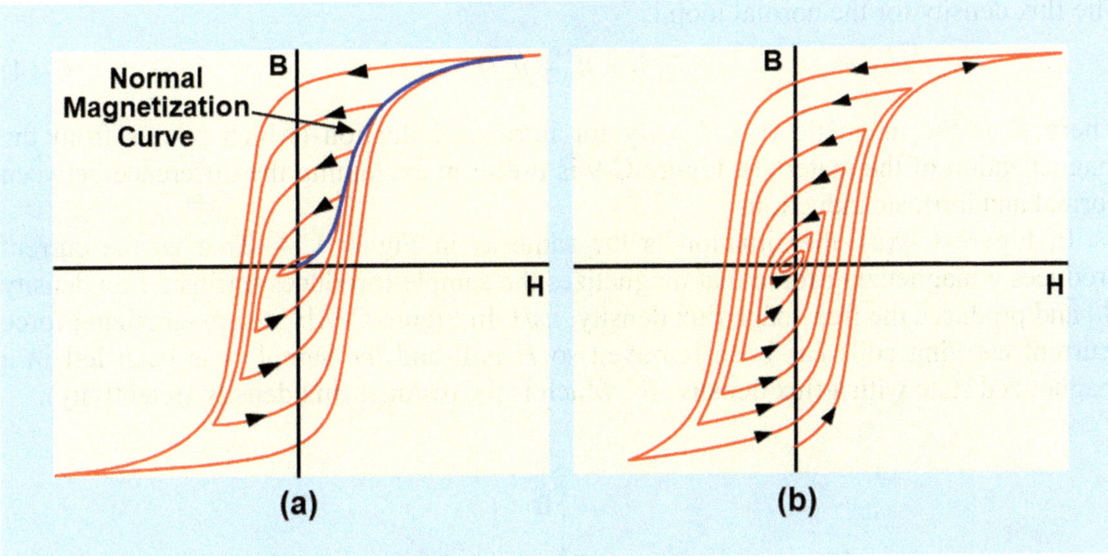

Figure C-6 (a) Magnetization and (b) demagnetization with a gradually changing alternating current.

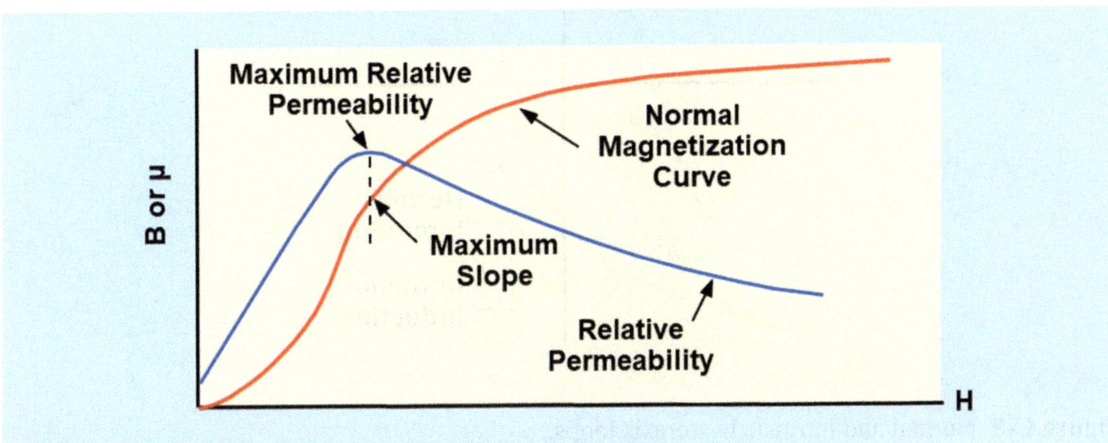

Figure C-7 Relation of relative permeability to normal magnetization curve.

C.3.2 Normal and Intrinsic Hysteresis Curves

The current used for applying a magnetizing force to the sample in the test described in Figure C-4 does two things:

1. Orients the domains and magnetizes the sample;
2. Generates a field that would be there in the absence of the sample.

As a result, to find the true (intrinsic) hysteresis characteristic of the sample, the additional field that must be supplied must be subtracted. The result is that two hysteresis loops shown in Figure C-8 must be considered. These are:

1. The normal induction loop that contains the additional field;
2. The intrinsic induction loop that is a characteristic of only the metal.

The flux density for the normal loop is

$$B = B_i + \mu_o H \tag{C-14}$$

where B_i is the magnetic flux density for intrinsic induction (which results from the magnetization of the material). Figure C-9 is useful in explaining the difference between normal and intrinsic induction.

In Figure C-9(a), the situation is the same as in Figure C-4. That is, the current produces a magnetizing force that magnetizes the sample (produces intrinsic flux density B_i) and produces the free-space flux density, $\mu_o H$. In Figure C-9(b), the magnetizing force (current carrying coil) has been removed so H is 0 and the sample has been left in a magnetized state with a flux density, B_r, which is the residual flux density (retentivity).

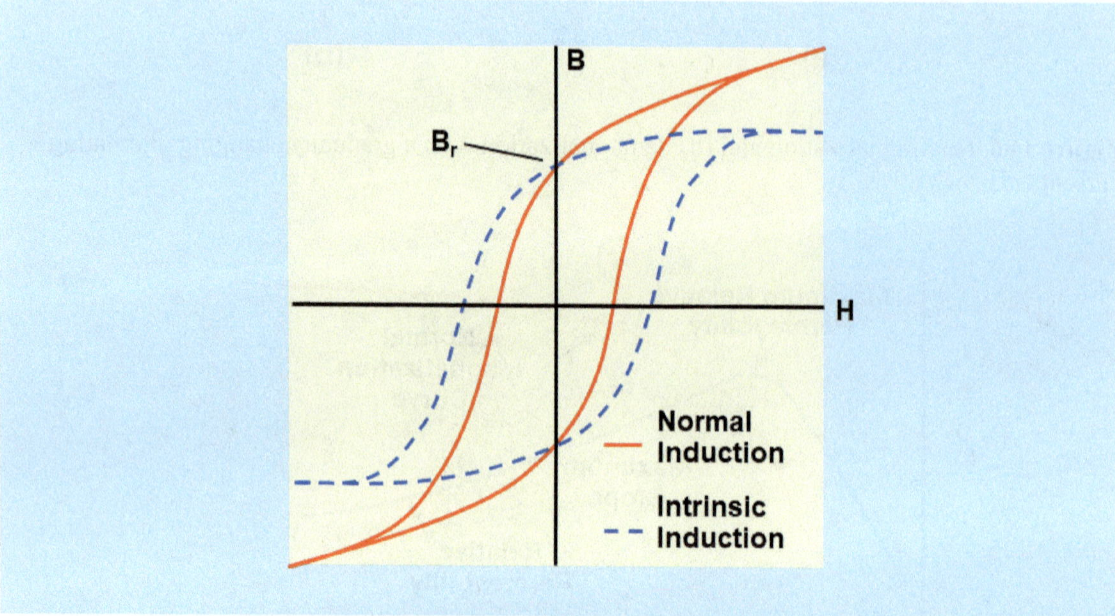

Figure C-8 Normal and intrinsic hysteresis loops.

In Figure C-9(c), there is a gap in the magnetic sample. When the gap is cut, north and south magnetic poles are formed in the faces of the magnet. Within the gap, the magnetic field, H, is from north to south and is in the same direction as the flux density, B. The flux density produced by H in the gap is $\mu_o H$. In the magnet, the field, H, is also from north to south but is in the direction *opposite* to B. The flux density produced by H in the magnet is $-\mu_o H$. Thus, H tends to demagnetize the magnet. This effect is the same as reversing the current through the coil in Figure C-9(a). This reverses the field generated by the coil and moves the operating point on the B–H curve to the left of the B axis as shown in Figure C-10. This is the second quadrant of the hysteresis loop. The curve in this region is called the demagnetization curve. Here, the field is opposite to B so demagnetization occurs. From Figure C-9, it is seen that the flux density in the gap is $B_i + \mu_o H$ so it is the normal B–H curve that is used for magnet design and for determining demagnetization.

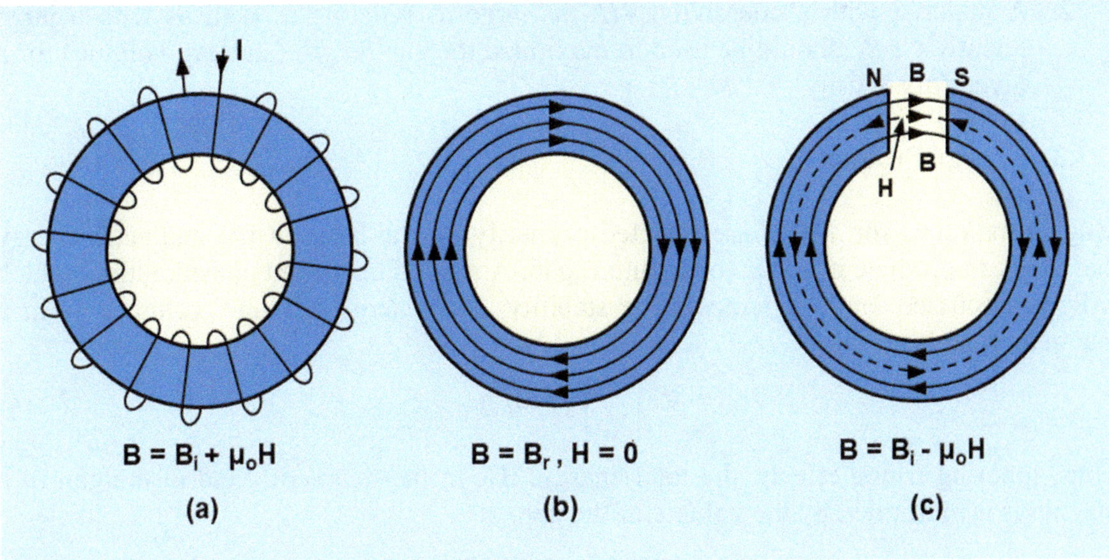

Figure C-9 Normal and intrinsic inductance effects in magnetic material with a gap.

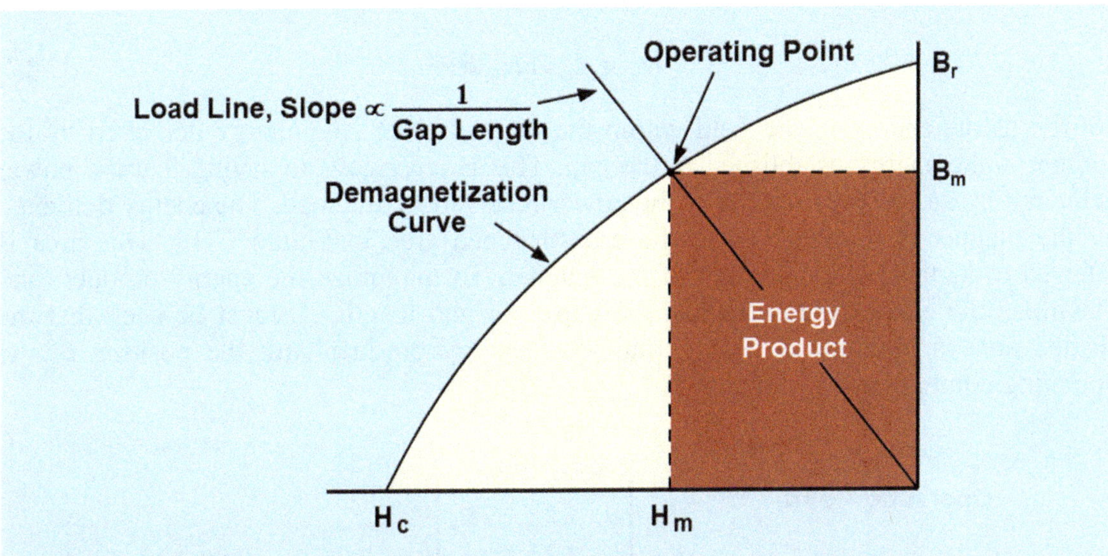

Figure C-10 Demagnetizing curve in the second quadrant of the hysteresis loop.

In Figure C-9, when the magnetizing coil was removed, the sample was left in a magnetized state with a flux density of B_r. Thus, a permanent magnet was formed, but all of the flux was contained within the magnetic material, so there is no way that the flux could be used, for example, to exert a force on electrons. When a gap was cut in the sample, flux was made available for use.

Now, two considerations concerning the design of permanent magnet circuits are immediately evident from Figure C-10:

1. For optimum utilization of the magnet, the gap must be long enough to provide a reasonable gap volume in which interactions with the flux density can take place. On the other hand, the gap must not be so long that the flux density is too low.

2. A material with a coercivity, $-H_c$, as large as possible as well as with a large retentivity, B_r, should be used to maximize the gap length (and gap volume) for a given flux density.

C.3.3 Energy Product

Magnet materials for TWTs are selected primarily on the basis of the magnetic energy that they can produce in an air (or vacuum) gap. A second important consideration, which will be discussed later, is temperature stability. The energy per unit volume, w, in a magnetic field is

$$w = \int \boldsymbol{H} \bullet d\boldsymbol{B} \tag{C-15}$$

Thus, ignoring fringe effects, the total energy, W_g, in the magnetic field in the gap of a magnet is w multiplied by the volume of the gap, or

$$W_g = \ell_g A H_g B \tag{C-16}$$

where A is the cross-sectional area of the gap. Because $H_m \ell_m = H_g \ell_m$,

$$W_g = \ell_m A H_m B \tag{C-17}$$

but this is the energy in the field within the magnet. Thus, the energy delivered by the magnet is the energy established in the gap. This is analogous to saying that the power delivered by an electrical circuit is the power received by the load. The energy delivered by the magnet is proportional to the cross-hatched area in Figure C-10. This area is referred to as the energy product of the magnet. To maximize the energy product (and maximize the energy established in the gap), the gap length, l_g, must be such that the shaded area is maximized. The variation of energy product with the position of the operating point is shown in Figure C-11.

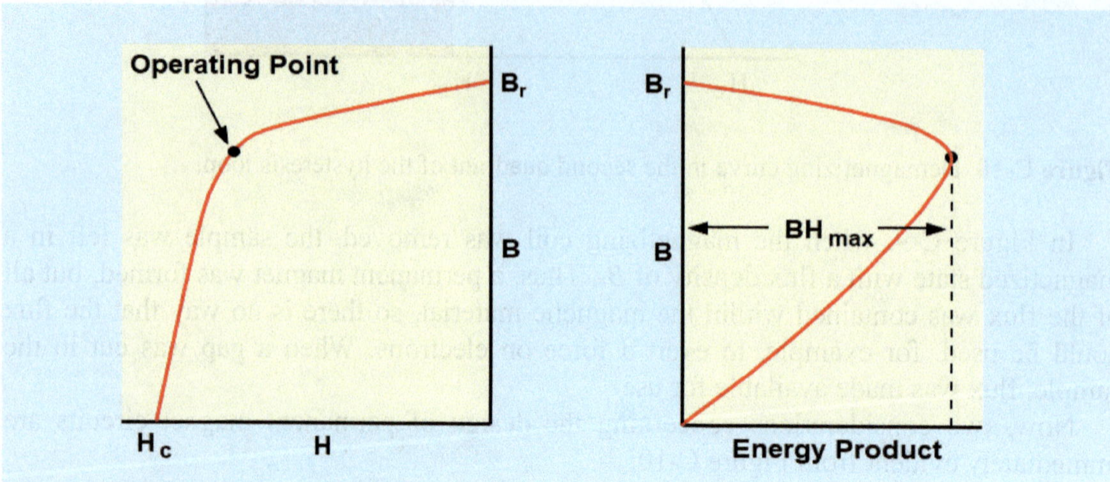

Figure C-11 Demagnetization curve and energy product.

C.3.4 Rare Earth Magnet Materials

Rare earth magnets are the magnets of choice for most modern microwave tube applications. The most common material used is samarium cobalt. The advantage of samarium cobalt over some other materials that have been used in the past is shown by the demagnetization curves in Figure C-12. The energy product for samarium cobalt is approximately four times that of alnico V, which is a magnet material that was in common use in microwave tubes 30 years ago. The coercive force is over an order of magnitude greater than that of alnico V, with the result that the samarium cobalt magnets are extremely stable.

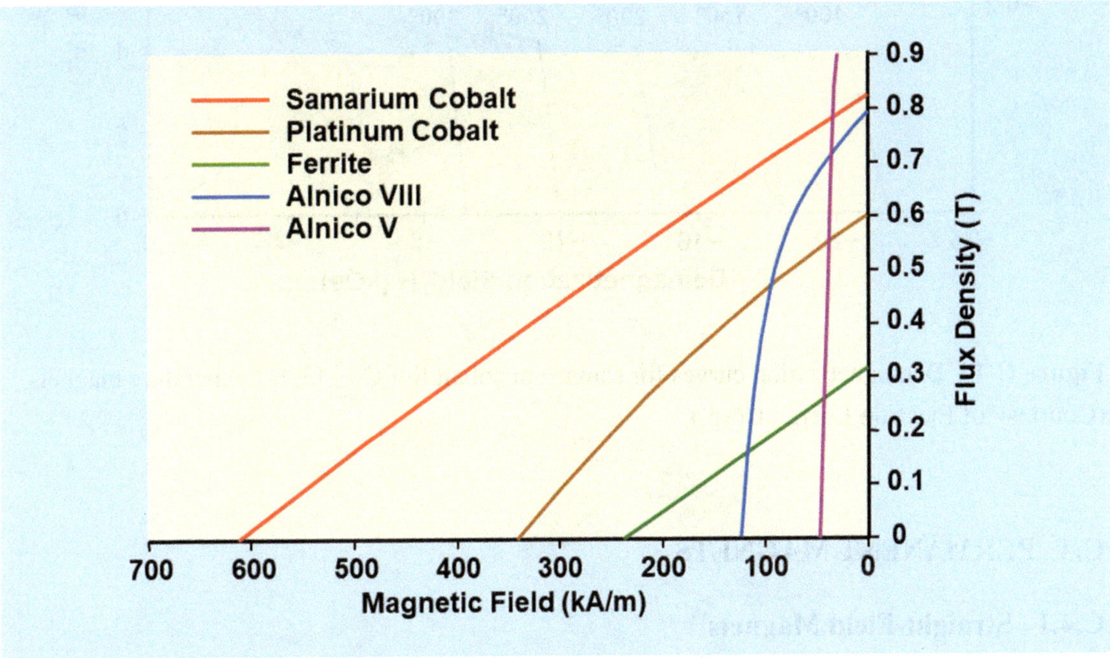

Figure C-12 Comparison of rare earth cobalt magnets with conventional magnets. (Adapted from: *Rare Earth Cobalt Magnets,* Varian Brochure, 1972.)

Permanent magnets on microwave tubes must be able to operate over a wide temperature range with no significant change in characteristics. Conventional samarium cobalt ($SmCo_5$) has a field strength temperature coefficient of about 0.04%/°C. Electron Energy Corporation developed a series of Sm_2Co_{17} magnet grades that are temperature-compensated so that they have a near-zero temperature coefficient. This was done by substituting gadolinium for a portion of the samarium in the alloy [6].

Figure C-13 shows the demagnetization curves at various temperatures for one of a series of Sm_2Co_{17} high-temperature magnets. These magnets have an energy product of 20 MGOe. Notice that the normal demagnetization plots are extremely linear. This simplifies the analysis of magnetic circuits in which these magnets are used.

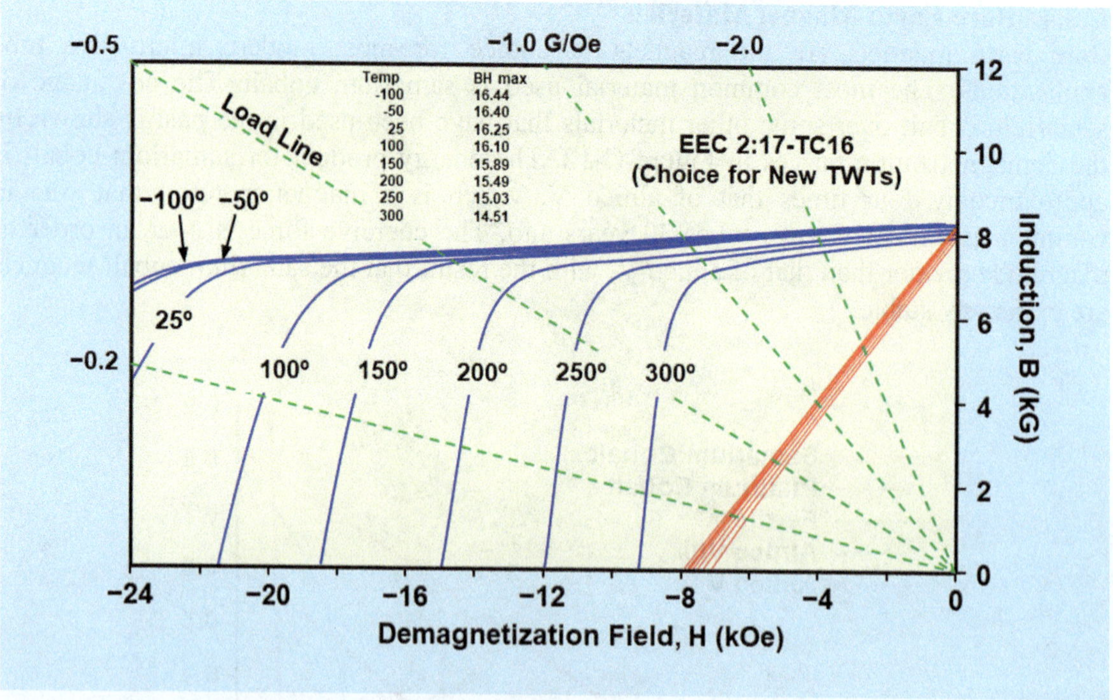

Figure C-13 Demagnetization curves for samarium cobalt Sm_2Co_{17} high-temperature magnets. (Courtesy of Electron Energy Corp.)

C.4 PERMANENT MAGNETS

C.4.1 Straight-Field Magnets

Historically, permanent magnets [5] were sometimes used for electron beam focusing in TWTs and klystrons. These magnets were large and heavy and could produce only limited magnetic fields. Figure C-14 shows the magnet used on the SLAC 2422 klystron [7]. The magnet was barrel-shaped, produced a magnetic flux density of about 0.06T, and was designed for a gap length of 16.5 inches. The inside diameter was 8.5 inches, which provided space for the insertion of lead shielding and for the easy removal of the tube from the magnet. The maximum outside diameter was approximately 18 inches and the total weight was nearly 800 pounds. Permanent magnets were used because they made it possible to eliminate 240 focusing coils, focusing supplies, and water interlocks.

C.4.2 Periodic Permanent Magnets

In the majority of applications of microwave tubes, it is necessary to minimize weight and size. For these applications, periodic permanent magnet (PPM) focusing is used. The weight reduction of the focusing system, when compared with a solenoid or a unidirectional magnet, can be one to two orders of magnitude. The reason for this weight reduction can be explained by first noting that a solenoid or unidirectional magnet establishes a magnetic field in a volume far larger than that required for focusing the electron beam. Since the weight of the magnet is roughly proportional to the magnetic energy stored in the volume surrounding the magnet, much of the weight of the magnet results from the production of unused magnetic field. Figure C-15 [8] is useful in explaining why the PPM weight is so low. Assume first that the magnet in Figure C-15(a) has the most efficient design (magnet weight is minimized) to produce a magnetic field to focus an electron beam as it travels through the magnet. The most efficient design is the one that nearly optimizes the energy product for the magnet.

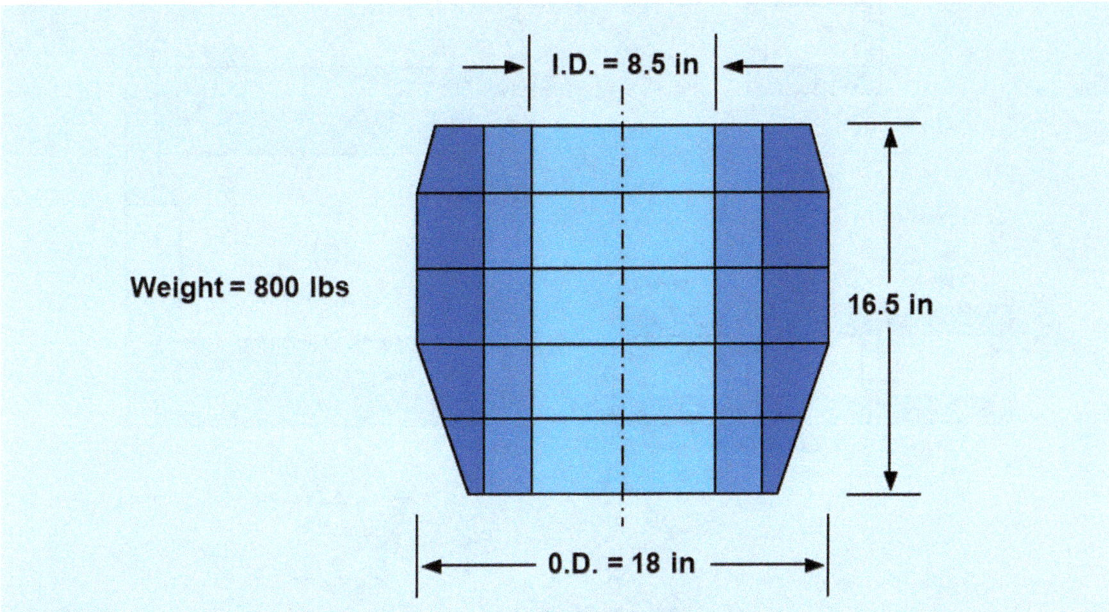

Figure C-14 Permanent magnet used on the SLAC 2422 klystron. (Adapted from: J. V. Lebacqz, SLAC-PUB-87, 1965.)

Next, assume that it is desired to focus the same beam over a distance that is three times the length of the magnet in Figure C-15(a). This may be accomplished with the magnet in Figure C-15(b), but to produce the same magnetic field that was produced by the magnet in Figure C-15(a), it is necessary to increase all dimensions of the magnet by a factor of three. (Remember that the dimensions of a permanent magnet are established to nearly maximize the energy product. A magnet configuration other than that shown in Figure C-15(b) would result in either a reduction in magnetic field or an increase in magnet weight.) As a result, the magnet in Figure C-15(b) weighs 3^3 or 27 times as much as the magnet in (a), and the energy stored in the field surrounding the magnet in Figure C-15(b) is 27 times that for Figure C-15(a). By comparison, if three magnets like that in (a) were stacked with like poles together as shown in Figure C-15(c), then a field sufficient to focus the beam could be produced over the same length as for the magnet in Figure C-15(b). The total weight and energy stored for the magnets in Figure C-15(c) would be one-ninth of the weight and energy stored for the magnet in Figure C-15(b).

If the magnet in Figure C-15(b) was replaced by n smaller magnets aligned like those in Figure C-15(c), the weight of the resulting structure would be $1/n^2$ times that of the larger magnet. In practice, the weight saving is not this great. This is because the field on the axis of the periodic structure does not appear as a square wave with abrupt reversals occurring at magnet interfaces. Instead, the field may be more nearly sinusoidal and the peak value must be increased to produce the same effective field as would be produced by the square wave. As a result, the magnet sizes must be increased to produce the required effective field.

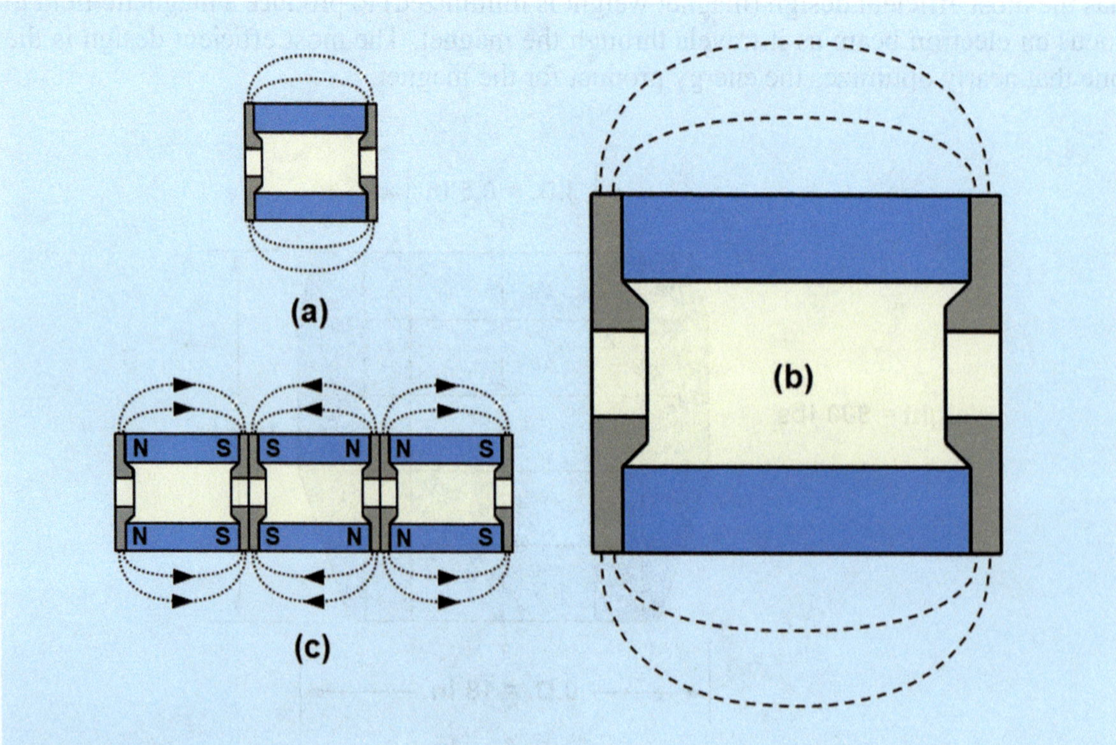

Figure C-15 Size (and weight) reduction resulting from the use of periodic permanent magnets. (Adapted from: J. T. Mendel et al., *Proc. IRE*, May 1954. © 1954 IRE (now IEEE).)

Since the magnet weight is roughly proportional to $1/n^2$, and since it is usually desirable to minimize magnet and tube weight, it is reasonable to ask what limits the number of magnets that are used. Qualitatively, this question can be answered by referring to Figure C-16 in which an electron beam focused by a periodic permanent magnet structure is illustrated. First, the interior dimension of the magnets has to be large enough to surround the electron beam and (not shown in Figure C-16) the RF circuit. Second, each magnet and the gap between pole pieces must be long enough so that magnetic field lines of sufficient strength for focusing penetrate into the electron beam. Third, as shown in Chapter 7, the wrong choice of magnet length and period of the magnet structure can lead to excessive beam ripple or beam divergence. Finally, other considerations to be discussed later in this appendix lead to the use of *double-period focusing* or possibly to *long-period focusing*.

In PPM structures like that in Figure C-16, ring-shaped magnets are used to form magnetic lenses that periodically focus the electron beam traveling along the center of the structure. Periodic magnetic focusing of an electron beam is discussed in Chapter 7.

A problem with the ring-shaped magnets illustrated in Figure C-16 is that each must be split in order to be placed on the tube. This is one of the factors leading to nonuniformities in the magnetic fields produced by these structures. To obtain acceptable tube performance, it is often necessary to compensate for the nonuniformities by placing magnetic shims on the outside of the magnetic assembly.

A technique for avoiding some of the problems of ring-shaped magnets is shown in Figure C-17. Several (often four) pill-shaped magnets are placed between magnetic pole pieces that surround the axis of the tube. The magnets can be magnetized to the desired level and then placed on the tube without being split. Field uniformity is increased (as compared with the ring-shaped magnets). In addition, magnet cost is reduced because magnets of one size can be used on tubes having a wide range of sizes.

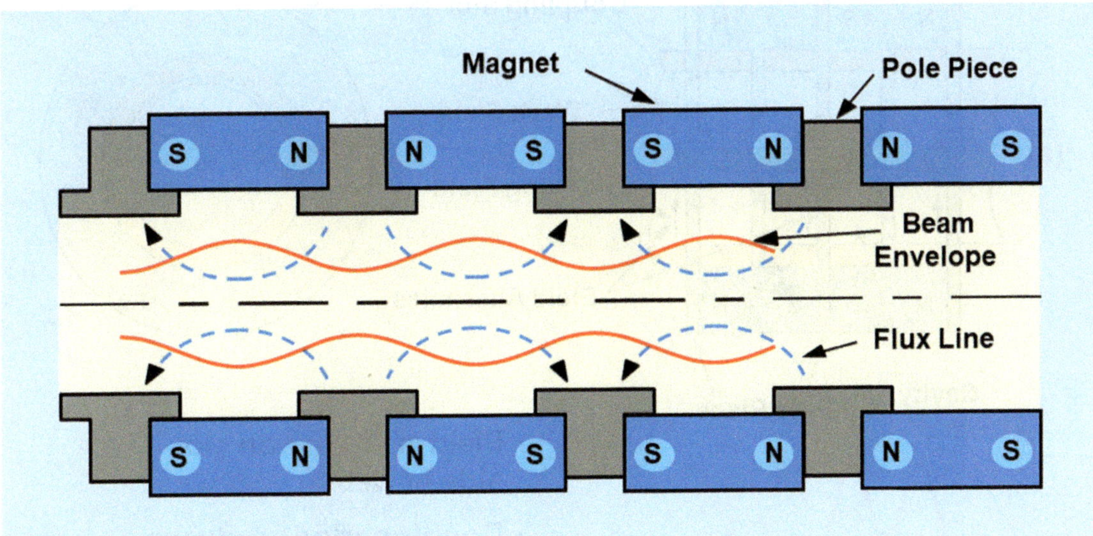

Figure C-16 Periodic permanent magnet structure.

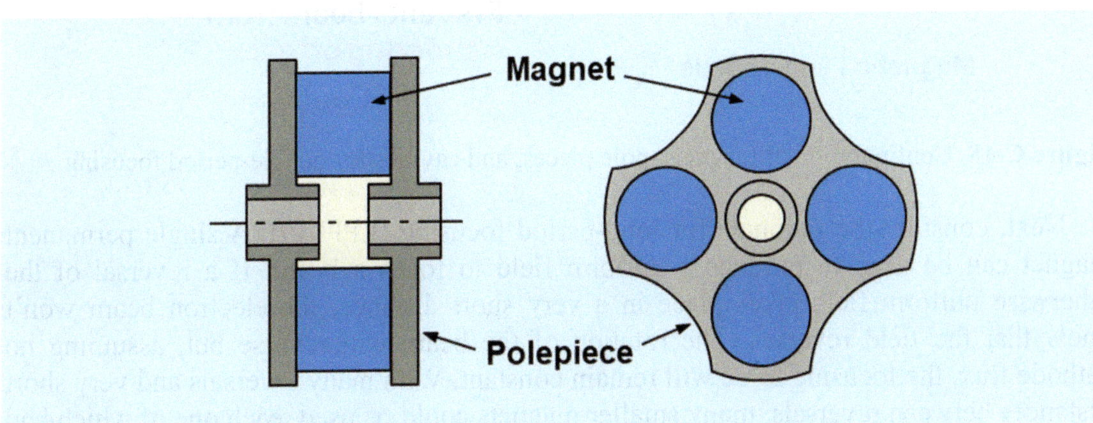

Figure C-17 A PPM configuration using pill-shaped magnets. (Cavity not shown.)

C.4.3 Double-Period and Long-Period Focusing

Double-period focusing is used in coupled-cavity TWTs. In those tubes, as shown in Figure C-18, some cavity walls also serve as pole pieces that concentrate the magnetic flux in the electron beam. Other cavity walls serve as flux guides to shape the magnetic field in the beam.

Slots in the cavity walls are used to couple the RF field from cavity to cavity. Often, the slots in adjacent walls are on opposite sides of the axis. The slots produce asymmetries in the focusing field, which produce transverse forces that deflect the electron beam. By arranging the magnets as shown in Figure C-18, the transverse forces produced by adjacent pole pieces are in opposite directions and so the deflections of the beam are in opposite directions. This prevents the beam from gradually being forced off axis as it moves from pole piece to pole piece.

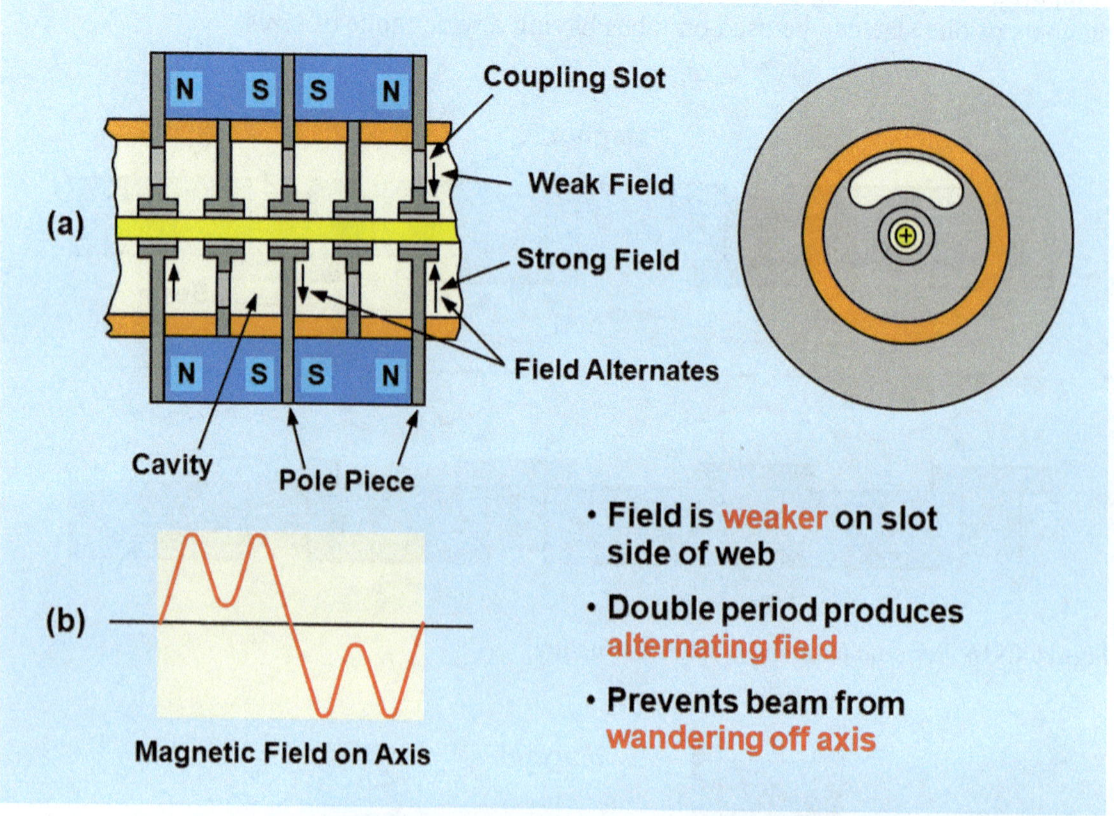

Figure C-18 Configuration of magnets, pole pieces, and cavities for double-period focusing.

Next, consider the rationale for long-period focusing (LPF) [9]. A single permanent magnet can be used to produce a uniform field to focus a beam. If a reversal of the otherwise uniform field takes place in a very short distance, the electron beam won't know that the field reverses. The rotation of the beam will reverse but, assuming no cathode flux, the focusing force will remain constant. With many reversals and very short distances between reversals, many smaller magnets could be used, each one of which had its polarity reversed from the previous magnet. The advantage in doing this would be the reduction in the weight of the magnets. PPM focusing is the limit, in terms of weight reduction but not necessarily in terms of practicality and cost. Long-period focusing is the compromise between minimum weight and minimum cost.

A primary application proposed for long-period focusing is high-frequency TWTs that have very small-diameter electron beams and, with PPM focusing, require many very small magnets. For example, 50 or more small ring magnets are required for a millimeter-wave TWT. The cost of parts and of labor to assemble the parts is very high. Phillips et al. proposed the use of long-period focusing to reduce the parts count, the cost of parts, and the cost of labor to assemble the parts [9]. A portion of an LPF structure and the associated magnetic field produced by it is shown in Figure C-19. With this configuration, compared with PPM focusing, the number of ring magnets is reduced by a factor of 4. In the single period shown, a pair of saturable flux guides and four spacers replace the eight spacers that would be required by PPM focusing. Phillips et al. demonstrated the feasibility of LPF in tests on a 200W TWT.

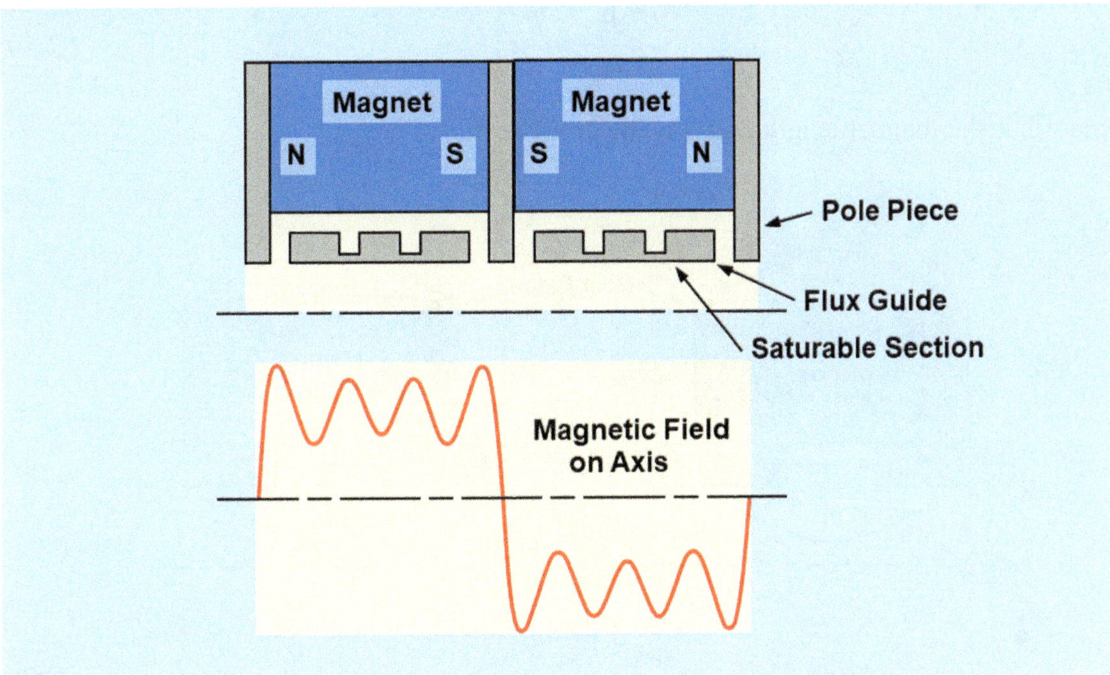

Figure C-19 Example of long-period focusing. (From: R. M. Phillips, et al., *Final Technical Report,* RADC-TR-90-110, June 1990.)

C.5 POLE PIECES

In many of the preceding figures, reference to pole pieces has been made. The pole pieces must be capable of carrying the flux from the magnets to the electron beam. In low-frequency tubes, where dimensions are large and flux densities are modest, the flux densities in the pole pieces are modest. However, in high-frequency devices, where dimensions are small, the flux density in the pole pieces can be high enough so that saturation occurs. When saturation occurs, the flux passes through the pole piece material as if it were made of air and the field available for focusing the beam is degraded. To avoid this problem, the saturation flux density of the pole pieces must be as high as possible and the pole piece must be designed so that the flux density is below the saturation value.

It is tedious but reasonably straightforward to estimate the flux density in the pole pieces and to determine if there is a danger of saturation [10, 11]. The flux density can be estimated if the operating point of the magnet is known and the dimensions of the magnetic circuit are known. To find the operating point of the magnet, the load line must be determined. To plot the load line, it is necessary to know the permeance of the magnet configuration.

As an example, the total permeance in Figure C-20(a) is

$$P_T = P_1 + P_2 + P_3 \tag{C-18}$$

where P_1 is the path between ferrules, P_2 is the leakage path inside the ferrules, and P_3 is the leakage path outside the pole pieces. The load line is given by

$$\frac{B_m}{H_m} = \frac{T}{A} P_T \tag{C-19}$$

where T is the magnet length and A is the cross-sectional area.

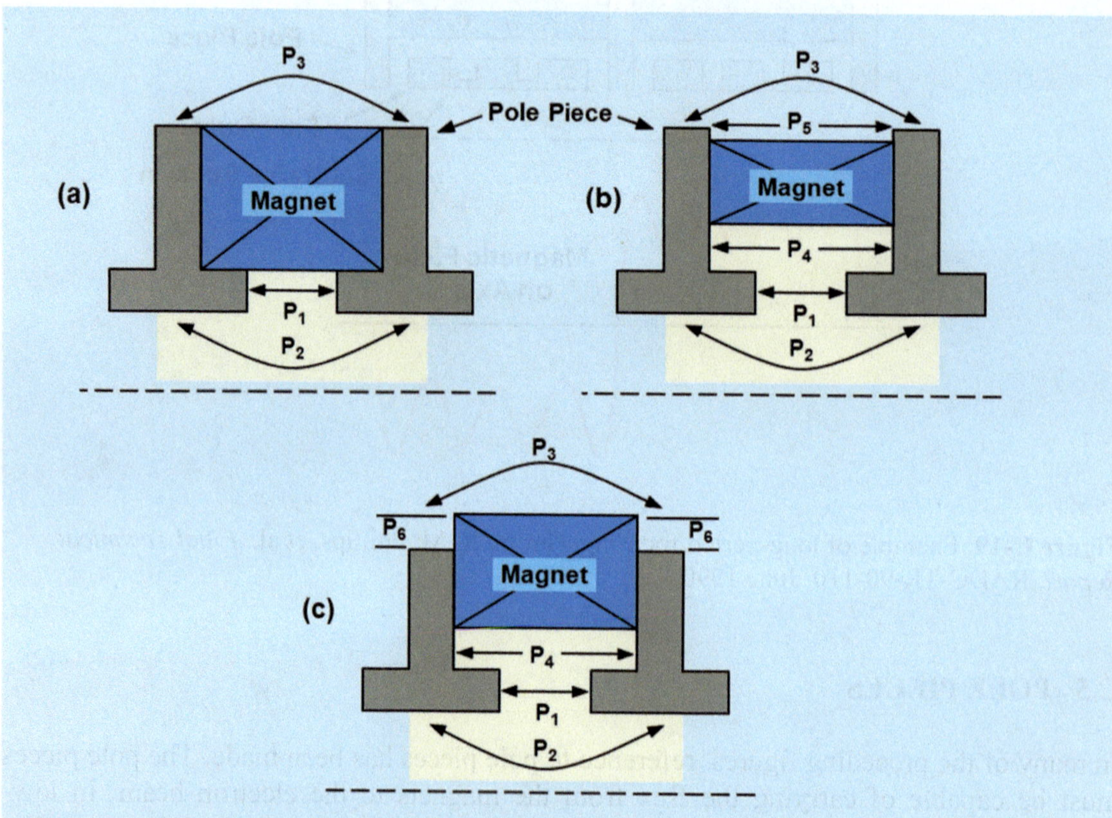

Figure C-20 Some common magnet and pole piece configurations. (Adapted from: M. Santra et al., *IEEE Trans. Electron Devices*, Vol. 56, No. 5, 2009. © 2009 IEEE.)

Permeance depends very much on the position of the magnets relative to the pole pieces as indicated in Figure C-20. The configuration in Figure C-20(a) is representative of those used on helix TWTs. In coupled-cavity TWTs, the configuration is more like that shown in Figure C-20(b) or C-20(c). The pole pieces are the walls of the cavities and the magnets are at a radial distance that is large enough to make room for the cavities. As a result, P_4 is a significant permeance. Permeances P_5 and P_6 depend largely on flux requirements and thermal considerations.

The calculation of some of the permeances is straightforward. For example, referring to the dimensions in Figure C-21,

$$P_1 = \mu_o \frac{\left(r_{f2}^2 - r_{f1}^2\right)}{g} \tag{C-20}$$

On the other hand, permeances P_2 and P_3 are not so easy to determine. To find these, boundary value problems must be solved using Fourier series involving Bessel functions as well as sine and cosine functions.

After the permeances have been determined and the magnet load line has been established, the magnetic field that the pole pieces must support can be found. With the magnetic field and the pole piece dimensions, magnetic flux densities can be calculated. This is when the magnetic characteristics of the pole pieces must be known. As shown in Figure C-22, the saturation flux density for iron, which is commonly used, is about 21 kG. The highest saturation flux density material is vanadium permendur which, in Figure C-22, is shown to be about 22 kG but is sometimes reported to be as high as 23 kG [12].

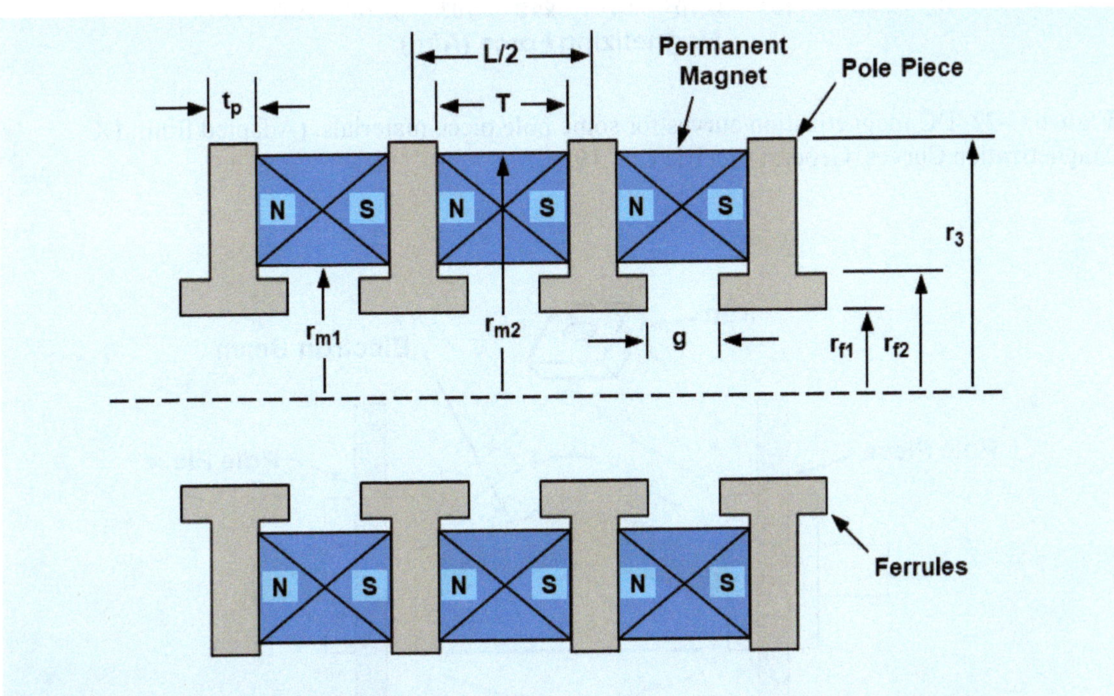

Figure C-21 Dimensions required for the determination of the flux densities in pole pieces. (Adapted from: M. Santra et al., *IEEE Trans. Electron Devices*, Vol. 56, No. 5, 2009. © 2009 IEEE.)

Before leaving the subjects of pole pieces and permanent magnets, we will briefly discuss the effect of magnetic objects on the focusing of an electron beam. In connection with Figure C-16, it was pointed out that ring-shaped magnets have to be broken in order to be placed on a tube. As a result, nonuniformities in the focusing field for the beam occur and magnetic shims are placed on the outside of the tube in an effort to compensate for the nonuniformities.

It is important to realize that magnetic materials of any kind that are brought into close vicinity of a permanent magnet focused tube can affect the focusing of the beam. Magnetic tools are particularly dangerous, but even a small magnetic scrap like that shown in Figure C-23 can affect the beam trajectory and possibly destroy the tube.

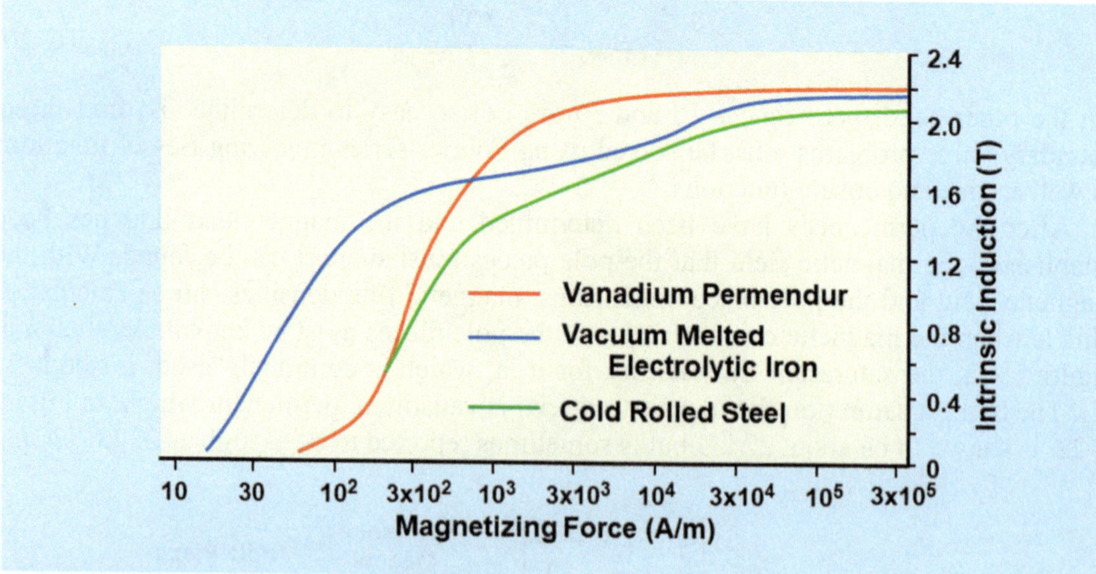

Figure C-22 DC magnetization curves for some pole piece materials. (Adapted from: DC Magnetization Curves, General Electric Co., 1956.)

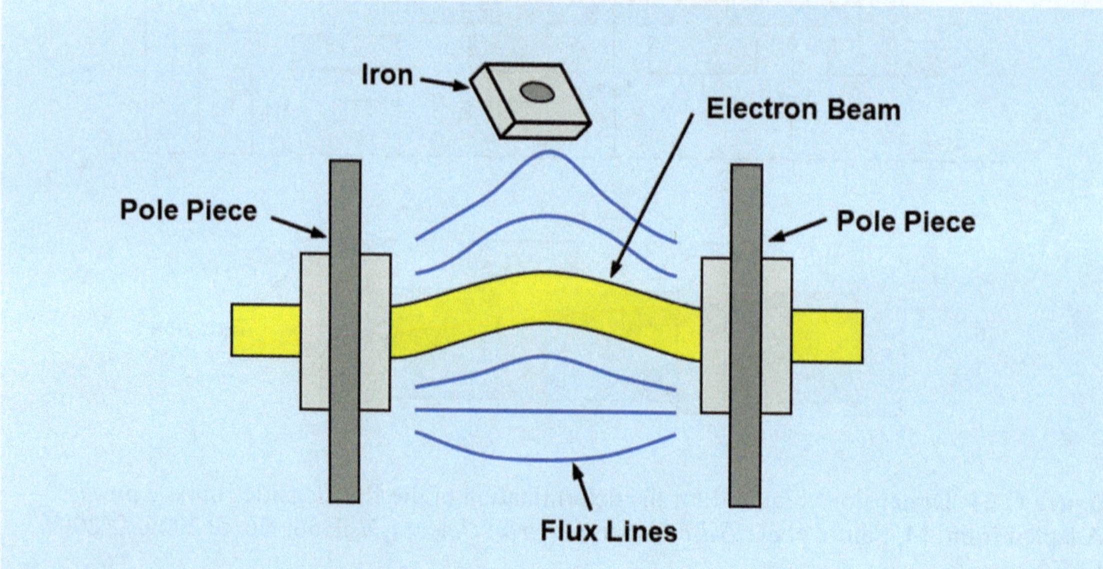

Figure C-23 Distortion of magnetic focusing field by a magnetic scrap.

C.6 ELECTROMAGNETS

A wire carrying a current produces a magnetic field as shown in Figure C-24. The direction of the field lines in relation to the direction of current flow can be remembered with the aid of the familiar right-hand rule. (The thumb points in the direction of conventional current and the fingers point in the direction of the field lines.) For microwave tube applications, the current-carrying wire is formed into a solenoid, as shown in Figure C-25.

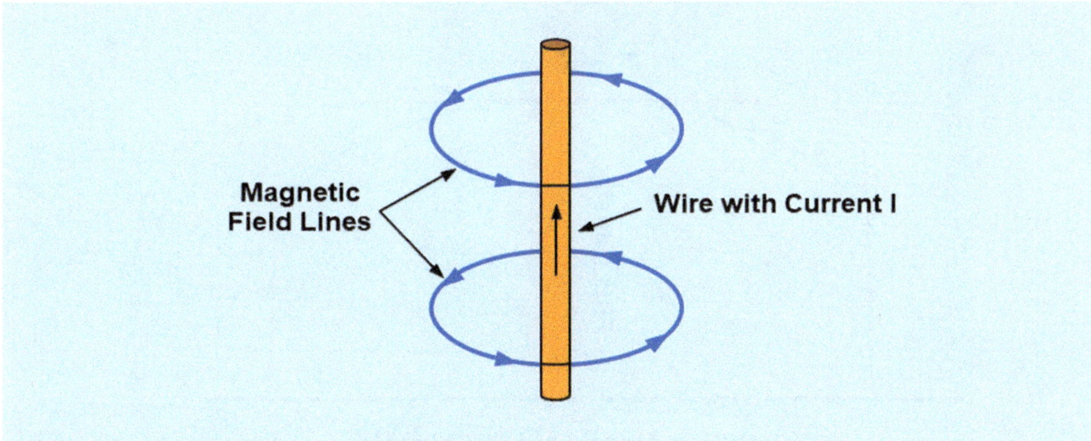

Figure C-24 Magnetic field around a wire carrying a current.

Very close to each turn of the solenoid, the flux lines are nearly circular. However, at greater distances, the flux in the region between adjacent turns cancels, so the flux lines around each pair of turns are nearly elliptical. At still greater distances, there is further cancellation of opposing flux lines and the pattern in the center of Figure C-25 occurs. Near the middle of the solenoid, the flux lines are parallel and the flux density is constant. Near the ends of the solenoid, the flux lines diverge and the flux density decreases as indicated by the dashed line in Figure C-26. This large decrease in field near the ends of the solenoid is not acceptable for most microwave tube applications.

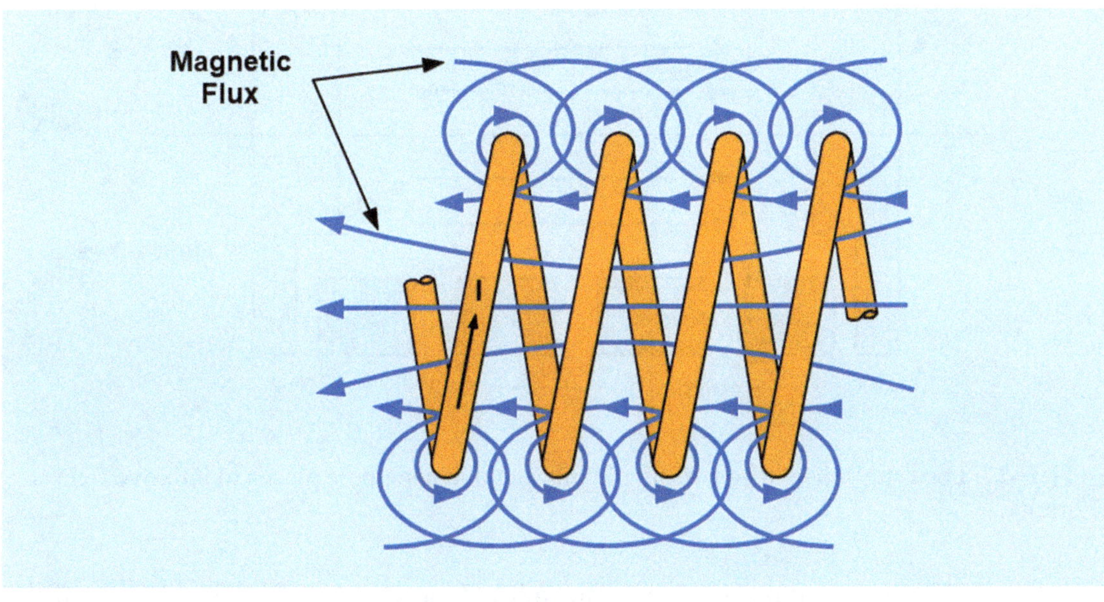

Figure C-25 Portion of a solenoid showing magnetic flux linkages.

To enhance uniformity of the field on axis where the electron beam is to flow, magnetic pole pieces, as shown in Figure C-27, are used. The pole pieces have a high relative permeability ($\mu_r \approx 5{,}000$) and their effect is to strengthen the field near the ends of the solenoid as shown by the solid curve in Figure C-26. The magnetic behavior of the pole pieces was described in the previous section.

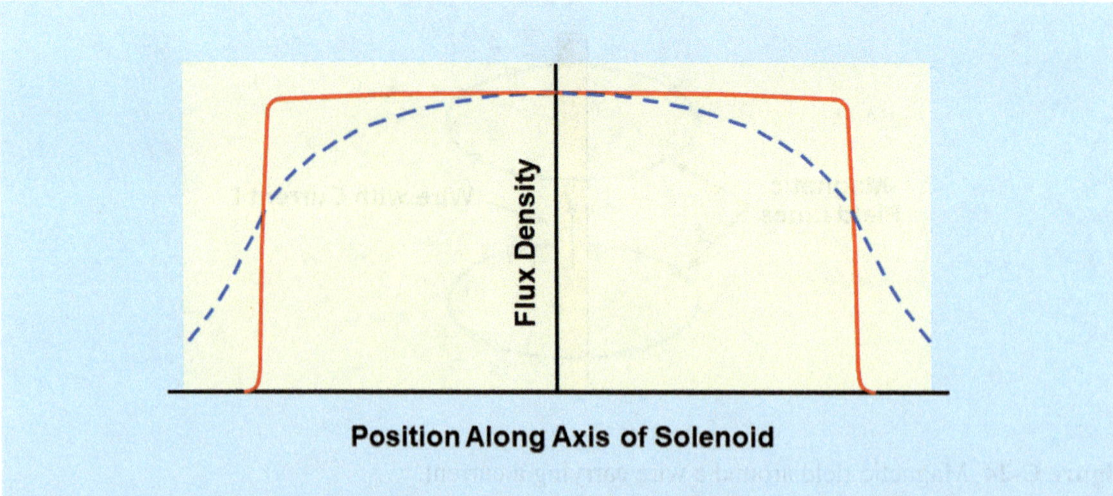

Figure C-26 Solenoid flux density without (dashed line) and with (solid line) magnetic pole pieces.

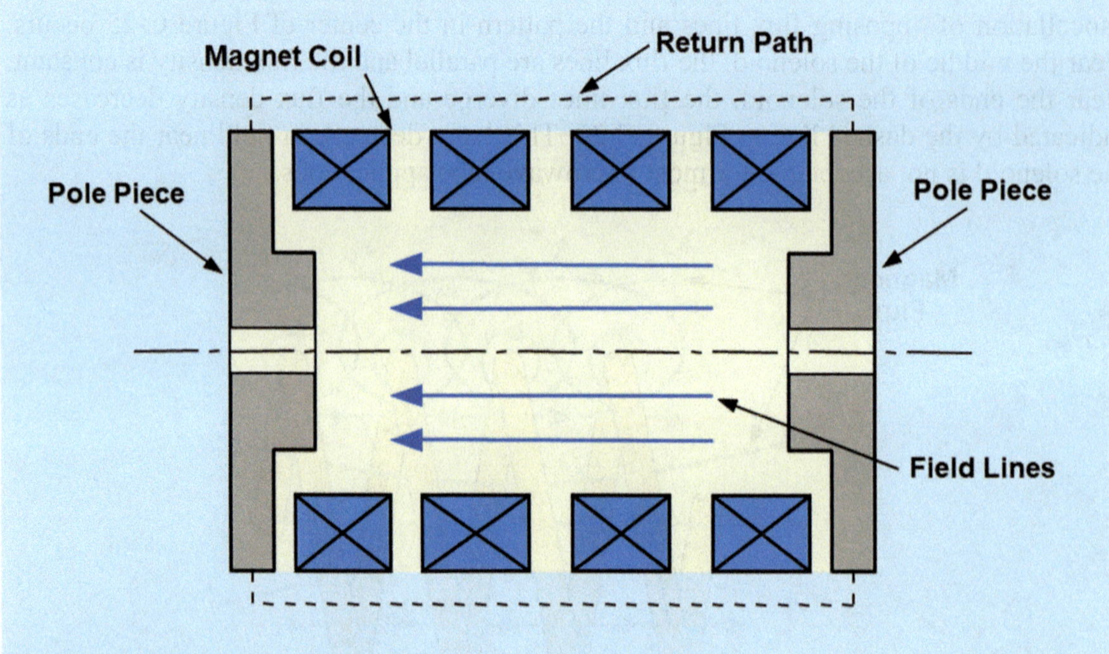

Figure C-27 Focusing solenoid consisting of coils and magnetic pole pieces to improve field uniformity.

Figure C-28 shows a typical solenoid for a TWT or klystron. Individual coils of the solenoid are separated to provide space for RF feedthroughs, cooling lines, and so forth. Normally, a solenoid is surrounded by a magnetic shield, or magnetic bars, as shown in Figure C-28, to provide a return path for flux lines. These magnetic shields do not enhance the magnetic flux density on the axis of the solenoid. Instead, they confine the magnetic flux lines surrounding a solenoid to a localized region near the solenoid. This helps minimize the possibility that a magnetic object, such as a tool, will accidentally be pulled toward the tube being focused by the solenoid, thus causing damage to, or destruction of, the tube. The magnetic shields or bars also help to minimize the interactions of the solenoid with nearby electrical circuits.

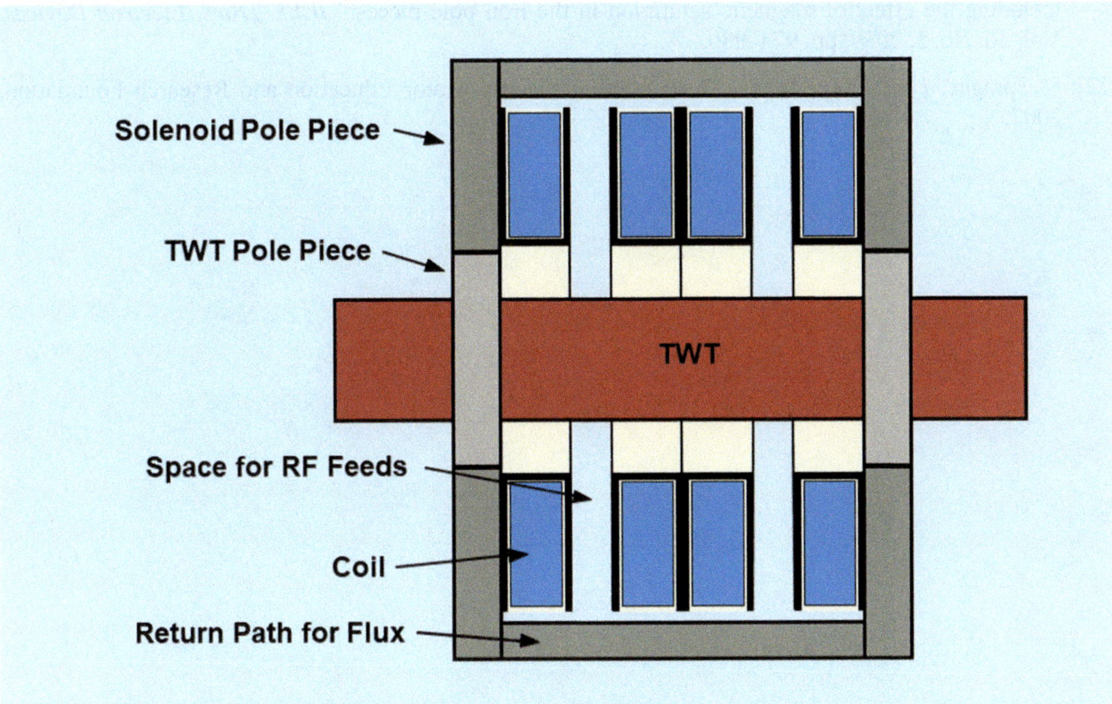

Figure C-28 Typical configuration of a solenoid and a microwave tube.

REFERENCES

[1] W. H. Hayt, Jr., *Engineering Electromagnetics*, New York: McGraw-Hill, 1981.

[2] J. D. Krauss and K. R. Carver, *Electromagnetics*, New York: McGraw-Hill, 1973.

[3] L. M. Magid, *Electromagnetic Fields, Energy, and Waves*, New York: McGraw-Hill, 1972.

[4] C. G. Skitek, and S. V. Marshall, *Electromagnetic Concepts and Applications*, Englewood Cliffs, NJ: Prentice Hall, 1982.

[5] R. J. Parker and R. J. Strudders, *Permanent Magnets and Their Applications*, New York: John Wiley and Sons, 1962.

[6] J. F. Liu and M. H. Walmer, "Thermal stability and performance data for SmCo 2:17 high-temperature magnets on PPM focusing structures," *IEEE Trans. Electron Devices,* Vol. 52, No. 5, 2005, pp. 899–902.

[7] J. V. Lebacqz, *High Power Klystrons*, Vol. SLAC-PUB-87, Stanford Linear Accelerator Center, Stanford, CA, 1965.

[8] J. T. Mendel, C. F. Quate, and W. H. Yocom, "Electron beam focusing with periodic permanent magnet fields," *Proc. IRE*, Vol. 42, No. 5, May 1954, pp. 800–810.

[9] R. M. Phillips, S. Snyder, and M. Baird, "Advanced electron beam focusing (long period focusing)," *Final Technical Report RADC-TR-90-119*, June 1990.

[10] F. Sterzer and W. W. Siekanowicz, "Design of periodic permanent magnets for focusing of electron beams," *RCA Review,* Vol. 18, No. 1, 1957, pp. 39–59.

[11] M. Santra, L. Kumar, and J. Balakrishnan, "An improved analysis of PPM focusing structures including the effect of magnetic saturation in the iron pole pieces," *IEEE Trans. Electron Devices,* Vol. 56, No. 5, 2009, pp. 974–980.

[12] S. Sprague, *Lamination Steels,* Third Edition, Electric Motor Education and Research Foundation, 2007.

Appendix D

Cyclotron, Larmor, and Scallop Frequencies

D.1 CYCLOTRON FREQUENCY

The force, $\boldsymbol{F}$, produced by magnetic flux, $\boldsymbol{B}$, on an electron moving with velocity, $\boldsymbol{u}$, is

$$\boldsymbol{F} = -e\,\boldsymbol{u} \times \boldsymbol{B} \tag{D-1}$$

The quantities $\boldsymbol{F}$, $\boldsymbol{B}$, and $\boldsymbol{u}$ are bold to indicate that they are vectors. The vector product $\boldsymbol{u} \times \boldsymbol{B}$ is maximum when the vectors are perpendicular. For vectors that are not perpendicular, the vector product decreases by the sine of the angle between vectors.

Figure D-1 shows the trajectory of an electron traveling with velocity u_o at a right angle to the direction of the flux density, B, which is constant. The force on the electron is eu_oB. An easy way to determine the direction of the force on the electron is by using the right-hand rule. As is indicated in Figure D-1,

1. Point fingers toward direction of charge velocity, u_o;
2. Rotate fingers toward magnetic field, B;
3. Thumb points to force on positive charge, $F+$;
4. Force on electron is opposite.

The force deflects the trajectory of the electron as shown in Figure D-1 and, because its direction is perpendicular to the trajectory of the electron, it does not change the speed of the electron. Since the speed is constant, the force is constant. The result is that the path of the electron is circular.

The radius of the path can be determined by considering the outward centrifugal force on the electron as it follows the circular path. The magnitude of this force is mu_o^2/r, where r is the radius of the electron path. This outward force equals the inward force produced by the magnetic field so

$$e\,u_o\,B = \frac{m\,u_o^2}{r} \tag{D-2}$$

from which

$$r = \frac{u_o}{\eta B} \tag{D-3}$$

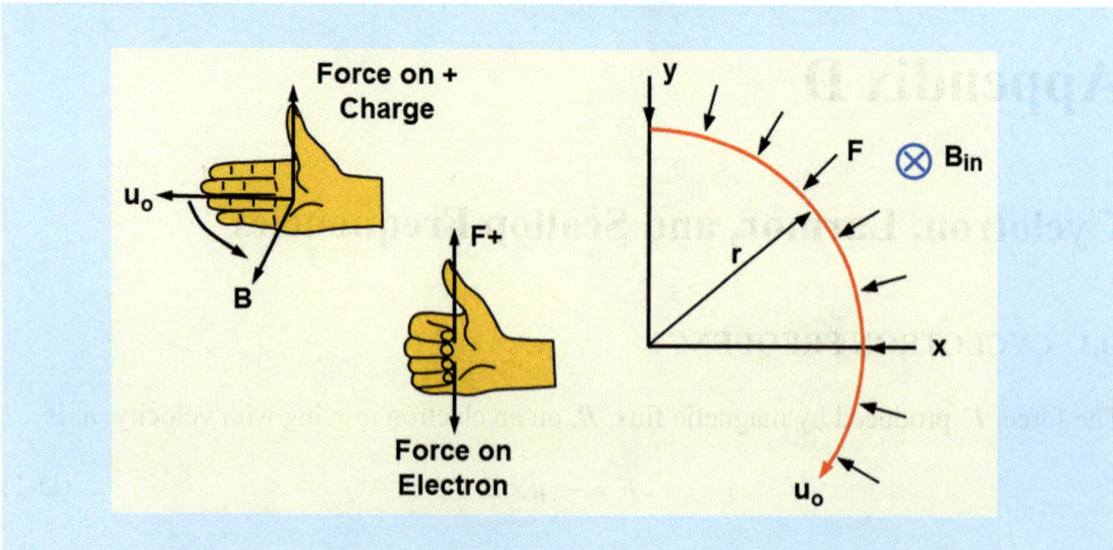

Figure D-1 Deflection of an electron by a magnetic field.

In terms of the voltage, V_o, through which an electron was accelerated to reach velocity u_o, r is

$$r = 3.37\times10^{-6}\frac{V_o^{1/2}}{B}\ \text{m} \tag{D-4}$$

At V_o = 10,000V and B = 0.1T (1,000) G, r = 3.37 mm.

The frequency with which the electron rotates is easy to determine because the path length for one rotation is $2\pi r$ and the velocity is u_o. Thus, the time required for each revolution is

$$t = \frac{2\pi r}{u_o} = \frac{2\pi}{\eta B} \tag{D-5}$$

and the frequency of rotation is

$$f = \frac{1}{t} = \frac{1}{2\pi}\eta B \tag{D-6}$$

Note that this frequency depends only on the electron charge and mass and on the magnetic flux density but not on the electron velocity or the radius of the path. The angular frequency with which the electron rotates is called the *cyclotron frequency*, ω_c, and is given by

$$\omega_c = 2\pi f_c = \eta B \tag{D-7}$$

f_c (MHz)	Flux Density (G)
2.8	1
28	10
280	100
2,800	1,000
28,000	10,000

D.2 BUSCH'S THEOREM

Busch's theorem is a powerful tool used in the analysis of the motion of electrons when axial and radial electric and magnetic fields are present. Busch's theorem can be derived by examining the motion of a ring of electrons passing through an axially symmetric magnetic field as shown in Figure D-2. In cylindrical coordinates, where r is radial position, θ is rotational position, and z is axial position, the equation for the θ directed force on an electron is

$$F_\theta = m\left(2\dot{r}\dot{\theta} + r\ddot{\theta}\right) = e\dot{r}B_z - e\dot{z}B_r \tag{D-8}$$

This can be written as

$$\eta r(\dot{r}B_z - \dot{z}B_r) = \frac{d}{dt}\left(r^2\dot{\theta}\right) \tag{D-9}$$

where B_z and B_r are the axial and radial components of the magnetic flux density.

Multiplying by dt, the following can be obtained:

$$\frac{\eta}{2\pi}(2\pi r dr B_z - 2\pi r dz B_r) = d\left(r^2\dot{\theta}\right) \tag{D-10}$$

To attach physical significance to this equation, consider a ring of electrons moving axially and shrinking radially as shown in Figure D-2. Assume that the ring is crossing magnetic flux lines.

As the ring shrinks, it crosses axial components of the magnetic flux. The total axial flux change, $d\Psi_z$, in distance, dz, is

$$d\Psi_z = 2\pi r dr B_z \tag{D-11}$$

The resulting force on the electrons, $-eu_r x B_z$, is in the counterclockwise ($-\theta$) direction as shown in Figure D-3.

As the ring moves axially, it crosses radial components of flux. The total radial flux change, $d\Psi_r$, in the distance, dz, is

$$d\Psi_r = 2\pi r dz B_r \tag{D-12}$$

The resulting force on the electrons, $-eu_z x B_r$, is in the clockwise ($+\theta$) direction, and this is indicated in Figure D-4.

Now, (D-11) and (D-12) are the quantities on the left side of (D-10) so the net force producing rotation of the electron ring results from the flux change, $d\Psi = d\Psi_z - d\Psi_r$, and the rotation can be found from

$$\frac{\eta}{2\pi} d\Psi = d\left(r^2\dot{\theta}\right) \tag{D-13}$$

When this is integrated, the result is

$$r^2\dot{\theta} - r_o^2\dot{\theta}_o = \frac{\eta}{2\pi}(\Psi - \Psi_o) \tag{D-14}$$

where r_o, and Ψ_o are the radius and rate of rotation of the ring and the flux contained within it at z_o. The quantities r, $\dot{\theta}$, and Ψ are the corresponding values at position z.

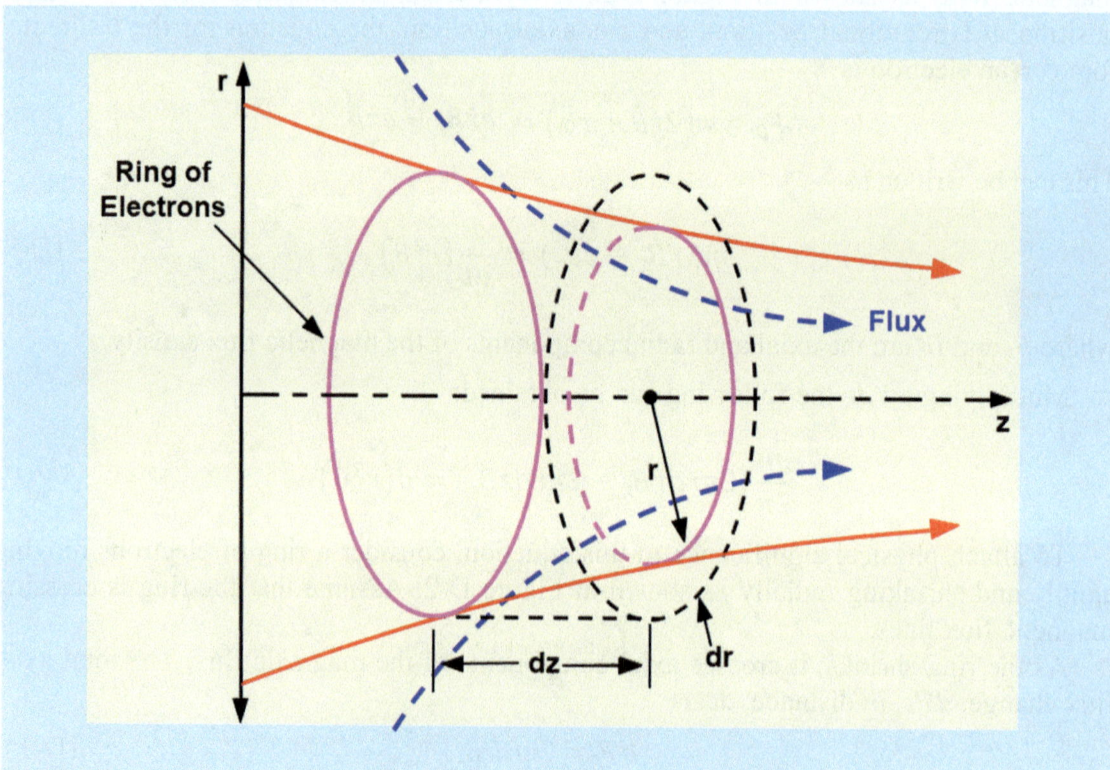

Figure D-2 Ring of electrons crossing axially symmetric magnetic flux lines.

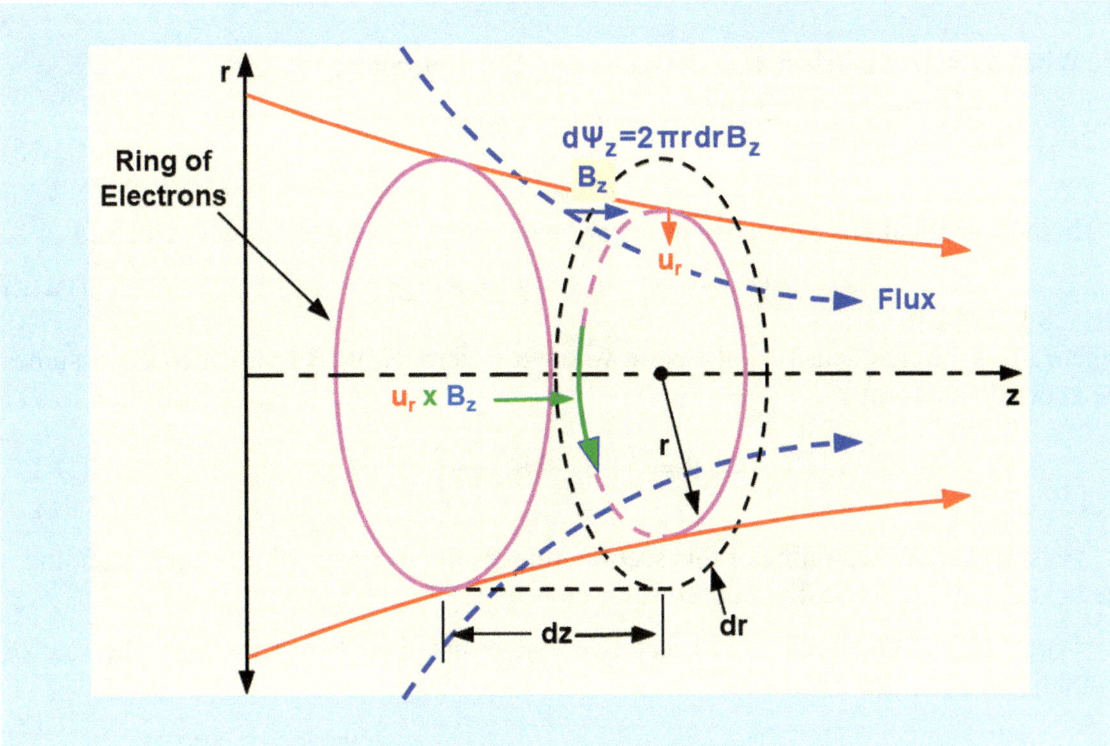

Figure D-3 Counterclockwise rotation results from crossing axial flux, B_z.

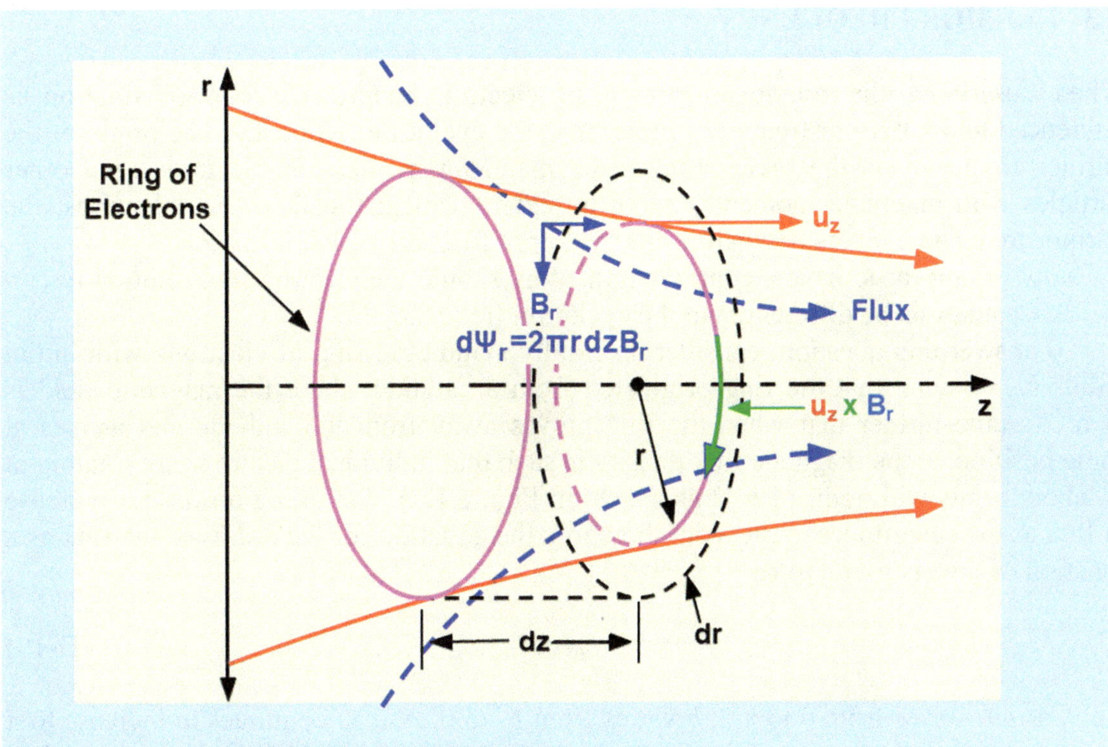

Figure D-4 Clockwise rotation results from crossing radial flux, B_r.

When $\dot{\theta}_o = 0$ at z_o (such as at the cathode), (D-14) becomes

$$\dot{\theta} = \frac{\frac{\eta}{2\pi}(\Psi - \Psi_o)}{r^2} \quad \text{(D-15)}$$

In most applications of interest in linear-beam tubes, Ψ changes slowly with z so

$$\Psi \approx \pi r^2 B_z \quad \text{and} \quad \Psi_o \approx \pi r_o^2 B_{zo} \quad \text{(D-16)}$$

Where B_{zo} is the flux density at the point where $\dot{\theta}$ is zero. Now, Busch's theorem assumes the approximate form

$$\dot{\theta} = \frac{\eta}{2}\left(B_z - B_{zo}\frac{r_o^2}{r^2}\right) \quad \text{(D-17)}$$

As stated at the beginning of this section, this result has great value in understanding the behavior of magnetically focused electron beams.

When B_{zo} is zero, such as at a cathode that is not linked by flux lines, the rate of rotation is the Larmor frequency,

$$\omega_L = \eta B_z/2 \quad \text{(D-18)}$$

which is one half of the cyclotron frequency.

D.3 LARMOR FREQUENCY

When describing the rotational motion of electron beams, the rate of rotation is referenced to the Larmor frequency rather than the cyclotron frequency. The origin of the Larmor frequency is the precession of the magnetic moment of electrons (and other particles with magnetic moments) about an external magnetic field. For electrons, the Larmor frequency, ω_L, is one half of the cyclotron frequency; that is, $\omega_c/2$.

Now we appear to have a contradiction. Why should the electron ring rotation rate in the Busch analysis be different than the cyclotron frequency?

To answer this question, consider an axially symmetric ring of electrons with initial radius b_o. Assume that the ring originated from a cathode where the magnetic field is zero. Assume further that when the ring moves away from the cathode and arrives at some position, z, the magnetic flux density is such that individual electrons are rotating at ω_c about orbits with radii of $b_o/2$ as shown in Figure D-5. Assuming no further increase in flux as the ring moves in the axial direction, the equation for the radius of the ring as a function of time is found to be

$$b = b_o \cos \omega_L t \quad \text{(D-19)}$$

Now, as $\omega_L t$ varies from 0 to $\pi/2$, b varies from b_o to 0. As $\omega_L t$ continues to increase to π radians, b increases from 0 to b_o. So, as indicated in Figure D-5, there is one complete rotation of an electron as ω_L goes through π radians. The actual frequency of rotation of the electron is $2\omega_L = \omega_c$, as it should be. Thus, in this case, the Larmor frequency arises from the coordinate system used to describe the beam shape. As the ring of electrons moves in the axial direction, the shape of the contour formed by the rotating electrons resembles a string of empty sausages, as is shown in Figure D-6.

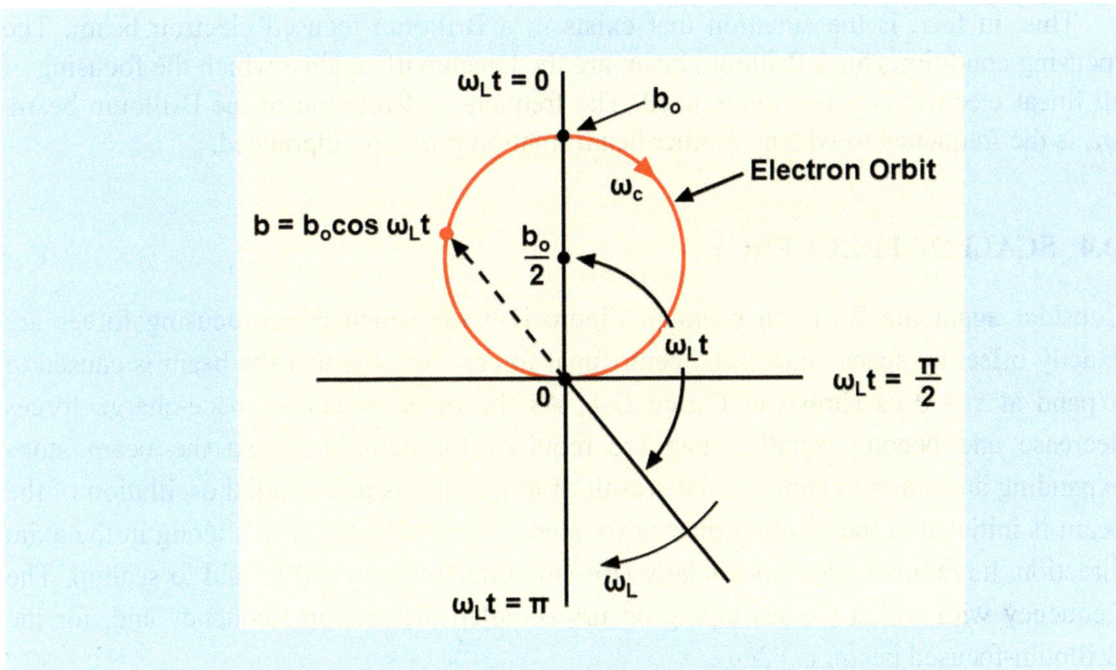

Figure D-5 Offset electron orbit plotted in cylindrical coordinates.

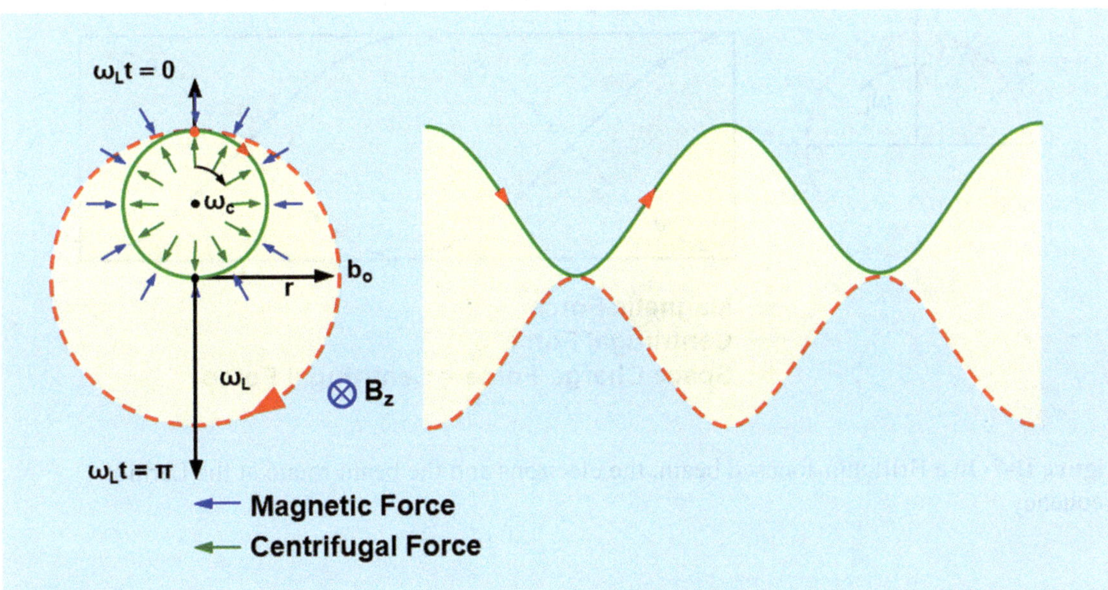

Figure D-6 Cyclotron, ω_c, and Larmor, ω_L, frequencies for a ring of electrons injected into a magnetic field with flux density B_z.

Next, as is shown in Figure D-7, assume that the ring of electrons encloses a beam of electrons. Also assume that the beam produces a radial space-charge force equal to the centrifugal force so the force pushing out on the ring is doubled. The rotational electron speed is not affected so the path length for a single electron orbit is twice that for cyclotron rotation. Thus, all of the electrons (i.e., the beam) rotate at the Larmor frequency, ω_L.

This, in fact, is the situation that exists in a Brillouin-focused electron beam. The focusing conditions for a Brillouin beam are the benchmark against which the focusing of all linear electron beams is referenced. The frequency of rotation of the Brillouin beam, ω_L, is the frequency to which all other beam rotation rates are referenced.

D.4 SCALLOP FREQUENCY

Consider again the Brillouin beam in Figure D-7 for which beam-focusing forces are exactly offset by space-charge and centrifugal forces. Assume that the beam is caused to expand at $z = 0$ as shown in Figure D-8. As the beam expands, space-charge forces decrease and become smaller than are required for equilibrium so the beam stops expanding and starts to contract. The result of this action is that a radial oscillation of the beam is initiated at the electron plasma frequency. As the beam moves along in the axial direction, its radius varies sinusoidally (for small deflections) and is said to scallop. The frequency with which the scalloping occurs is called the scallop frequency and, for the Brillouin-focused beam, is $\sqrt{2}\omega_L$.

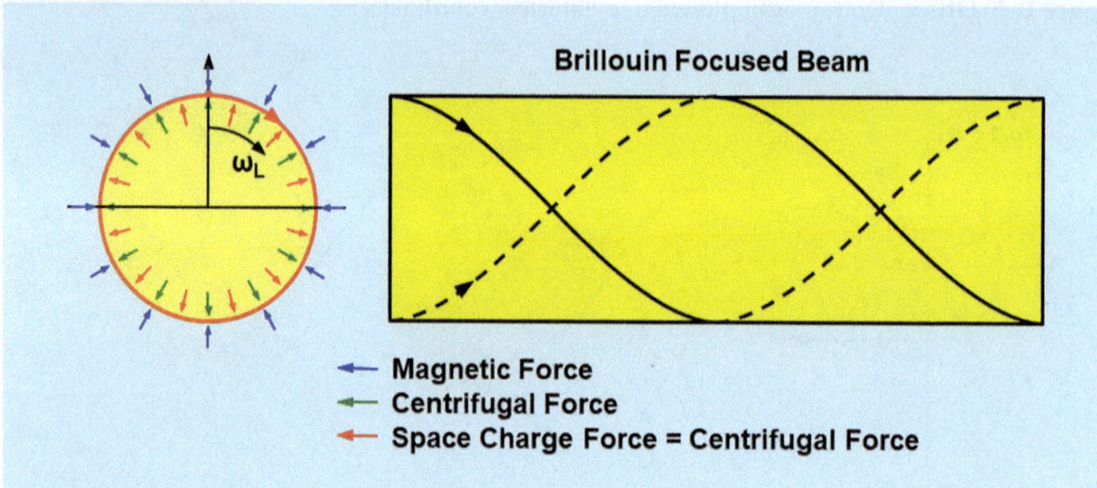

Figure D-7 In a Brillouin-focused beam, the electrons and the beam rotate at the Larmor frequency.

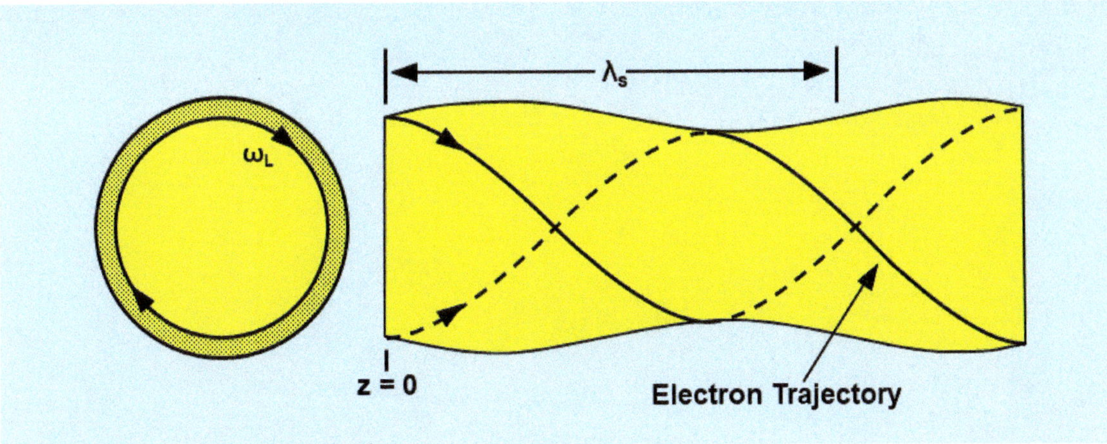

Figure D-8 Dependence of electron dynamics on space charge. ω_L is the Larmor frequency and λ_s is the scallop wavelength.

Appendix E

Breakdown and Protection

One of the most common and perhaps most poorly understood classes of failure mechanisms in microwave tubes centers on electrical breakdown. In the courses from which this book was originally developed, electrical breakdown was, at first, only briefly discussed. Because of the strong interest in breakdown phenomena expressed by those taking the course, that section was developed to cover major aspects of breakdown phenomena of concern to those who use microwave tubes.

Breakdown in microwave tubes may occur inside or outside the vacuum envelope at any of several locations as is indicated here and in Figure E-1.

- Electron gun (between electrodes, or leads, or from electrodes or leads to ground);
- Collector (between electrodes, or leads, or from electrodes or leads to ground);
- High-power portions of the RF structure.

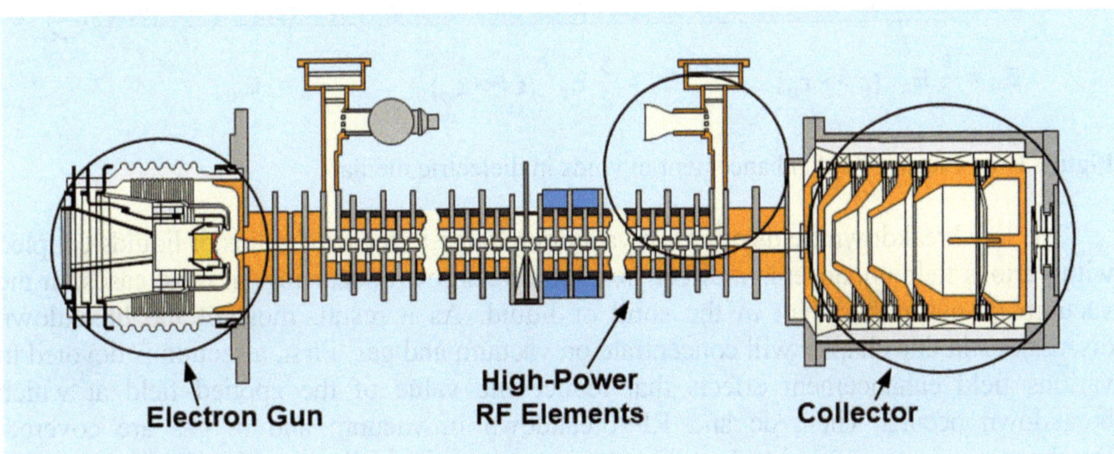

Figure E-1 Regions in a microwave tube where electrical breakdown may occur.

Often, it is not the tube that is at fault. For example, there may be a failure at any one of several locations in the RF system. If excessive RF power is reflected, the tube may be damaged for any one of several reasons. All too often the tube is thought to be at fault and is replaced before the original cause of the difficulty is corrected.

Whether or not breakdown will occur depends on two factors:

1. The applied field level and local field enhancement effects;
2. The breakdown field of the medium (gas, vacuum, liquid, or solid).

Very approximately, the breakdown fields for various media are as follows:

- Gas ~ tens of V/cm to 10^5 V/cm depending on pressure and type;
- Vacuum ~ 0.5–3×10^5 V/cm;
- Liquid ~ 0.5–1.0×10^6 V/cm;
- Solid ~ 0.5–1.0×10^6 V/cm.

At interfaces between solids, liquids and vacuum, or gas, the breakdown field may be significantly degraded. In addition, in a void in a solid or a liquid, the field is normally enhanced. The amount of enhancement depends on the shape and orientation of the void and the dielectric constants of the media, as indicated in Figure E-2.

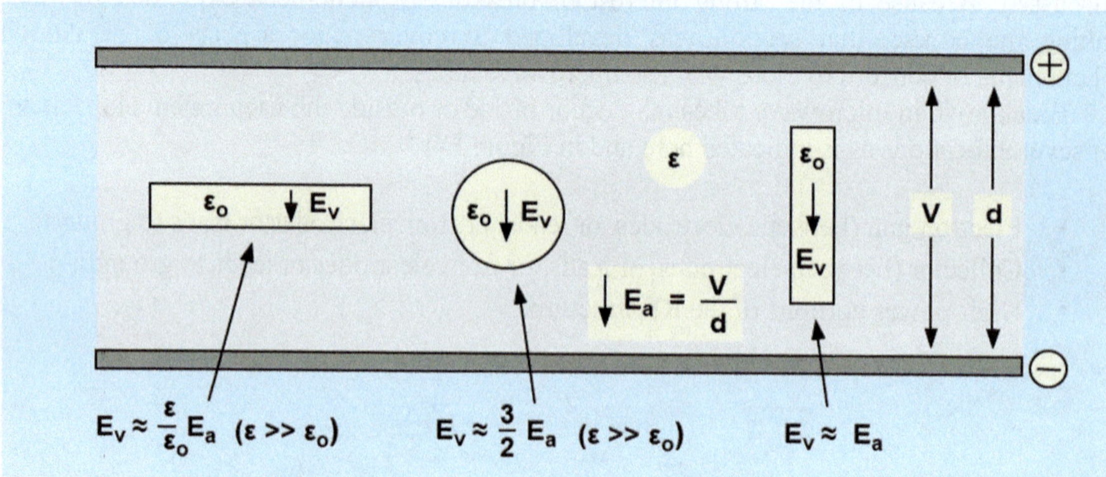

Figure E-2 Electric field enhancement in voids in dielectric media.

The low breakdown fields of gas or vacuum relative to those of solids or liquids coupled with various field enhancement effects result in breakdown occurring, in most cases, in the vacuum or gas rather than in the solid or liquid. As a result, most of the breakdown discussions in this chapter will concentrate on vacuum and gas. First, a section is devoted to various field enhancement effects that reduce the value of the applied field at which breakdown occurs. Then, dc and RF breakdown in vacuum and in gas are covered. Breakdown in or across an insulator is sometimes a cause of failure and is also discussed.

E.1 FIELD ENHANCEMENT

The electric field between parallel plane electrodes is the voltage applied from one electrode to the other divided by the distance by which the electrodes are separated. In practice, it is common to calculate field levels in all sorts of configurations simply by dividing the applied voltage by the electrode separation distance. Fields calculated in this manner are often in error by a factor of from two to ten or more. Worse yet, because the power varies as the square of the field, the calculated power-handling capability is often in error by a factor ranging up to two orders of magnitude or more.

The parallel plane and coaxial conductors shown in Figure E-3 provide a simple case for the comparison of field levels. A potential of 1,000V between electrodes is assumed, and equipotential profiles at 250V, 500V, and 750V are shown. For the parallel-plane electrode case, the potential profiles are equally spaced and the electric field is 1,000V/cm everywhere between the electrodes.

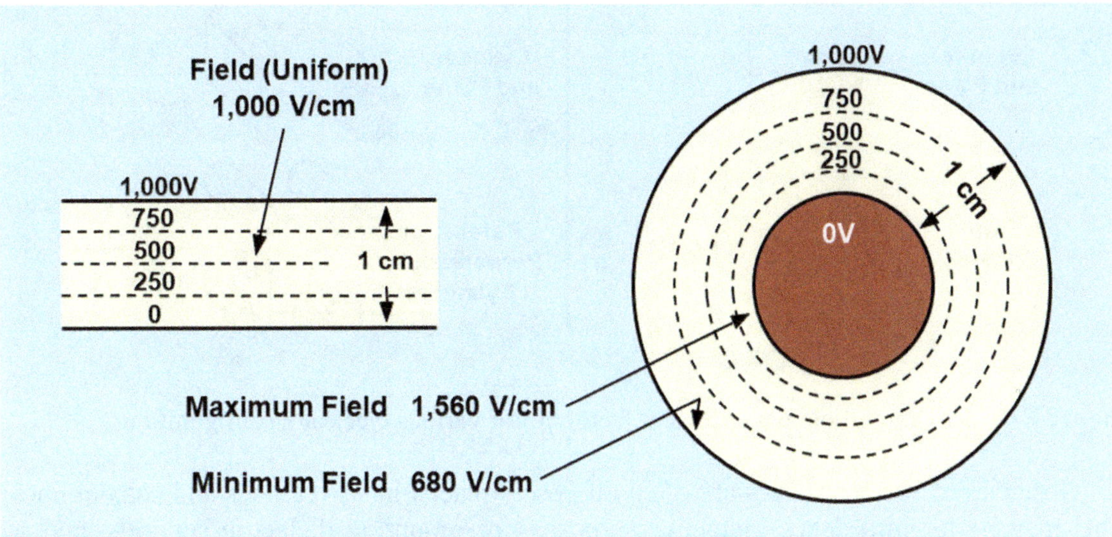

Figure E-3 Potential distributions in parallel plane and 50-Ω coaxial electrodes.

For the coaxial electrode case, the equipotential profiles near the center conductor are more closely spaced than in the parallel plane electrode case. The electric field at the surface of the center conductor is easy to calculate for this configuration and is 1,560 V/cm, or over 50% above that for parallel plane electrodes.

If the maximum field was limited to 1,000 V/cm because of breakdown considerations, then the maximum voltage that could be applied to the coaxial electrodes would be 1,000/1.56, or 641V. In the TE11 mode (for which the potential distributions in Figure E-3 are correct), the maximum RF power capability of the coaxial configuration would be $(1/1.56)^2$, or 40% of that for the parallel-plane electrodes.

The factor by which the electric field exceeds the field for the parallel-plane electrode case is normally called the *field enhancement factor* and is denoted by β. At the center conductor of the 50-Ω coaxial line, the field enhancement factor is 1.56.

Field enhancement factors have been calculated for many electrode configurations. Some of these are given in Figure E-4. In most practical cases of interest, the electrode geometries are close enough to those shown in Figure E-4 to permit a reasonable estimate of the value of β to be made. In some cases, values of β for two or more configurations can be combined to obtain an estimate for a configuration in question. For example, the value of β at the surface of a drop of condensation on the center conductor of a 50-Ω coaxial line is approximately 3 (β for a hemisphere on a plane) times 1.56 (β at the center conductor of a 50-Ω coaxial line), or about 4.7.

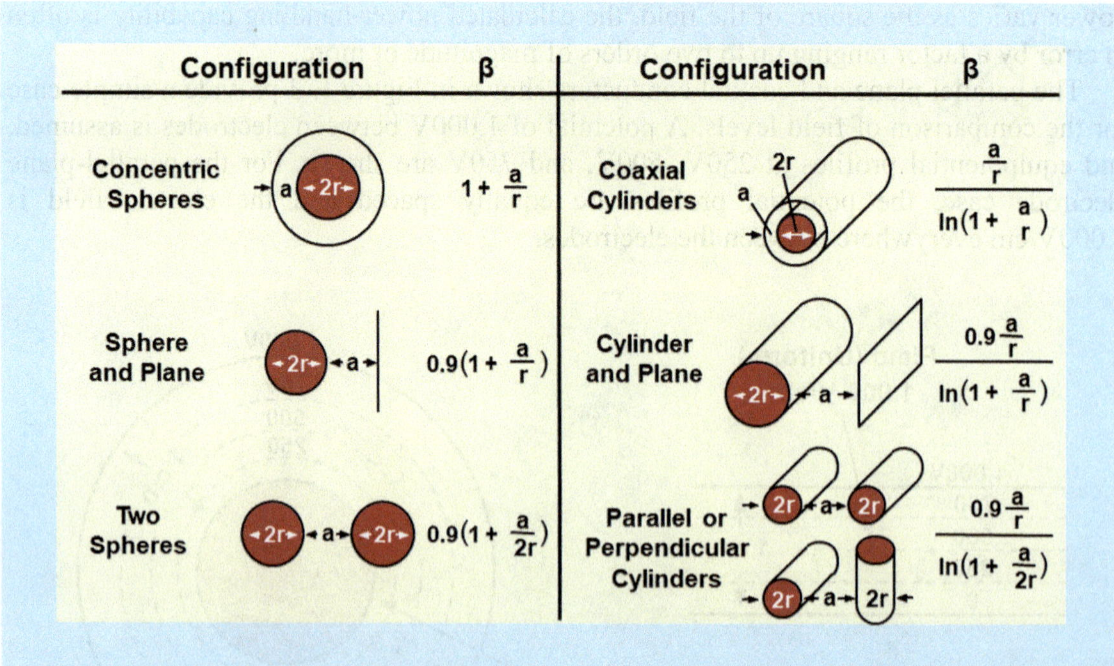

Figure E-4 Maximum field enhancement factor, β, for various electrode configurations.

Obstacles of various shapes are often purposely placed in waveguides and coaxial lines. This may be for impedance matching purposes, for mounting dielectric supports, and so forth. Values of β and power-handling degradation are given in Figure E-5 for some of these obstacle shapes in a rectangular waveguide. Note that the reduction in power capability can easily be an order of magnitude or more.

The importance of field enhancement on electrical breakdown cannot be overemphasized. The most troublesome aspect of field enhancement is that in many cases, its magnitude is either not known or is badly underestimated. The value of β of 1.56 calculated for the center conductor of a 50-Ω coaxial line is virtually meaningless if there are irregularities, such as scratches or machining marks, on the center conductor. These can easily have a β of 10 or more.

E.2 DC BREAKDOWN IN VACUUM

Under ideal conditions, the breakdown field for vacuum exceeds that of all media including liquids and solids, and may be 10 MV/cm or higher. The upper limit results from field emission from the negative electrode. In practice, the field at which breakdown occurs is two to three orders of magnitude below the upper limit. For example, for average gap lengths (~1 cm), breakdown occurs at a few tens to a few hundreds of kV/cm [1–3].

For many vacuum tube applications, Kilpatrick's criterion [4] is used as the guideline to the maximum electric field that can be used in vacuum if breakdown is to be avoided. According to Kilpatrick, breakdown in vacuum varies as electrode spacing raised to the 0.75 power. Kilpatrick refers to field levels below this criterion as the no-spark region and to higher field levels as the possible-spark region.

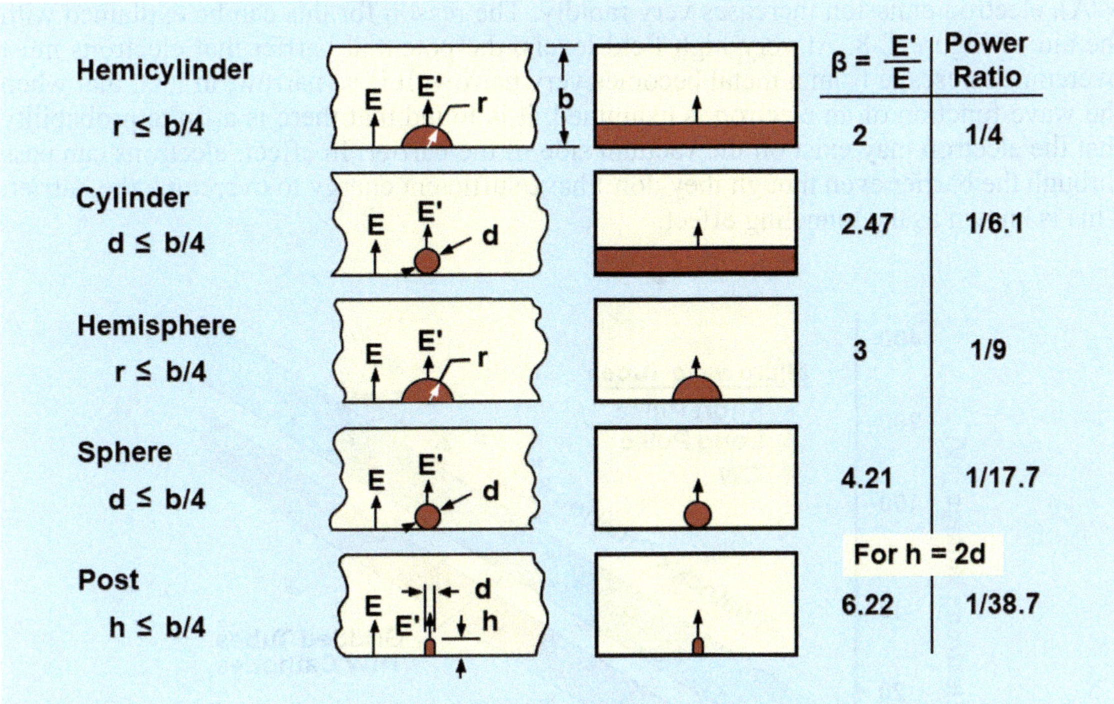

Figure E-5 Values of β and power-handling degradation for various obstacles in rectangular waveguides. (From: R. D. Wegenroth, Wheeler Laboratories Report 953P, June 1960.)

Figure E-6 contains Kilpatrick's criterion and some results from the tube industry [5, 6]. The shaded areas indicate roughly the values of plate hold-off voltages for gridded tubes with cylindrical configurations and with thoriated tungsten (ThW) or oxide-coated cathodes. The data points represent the experimental breakdown characteristics of Pierce guns. The points marked by circles are for dc operation. The triangles and crosses represent long-pulse and short pulse operation, respectively. Other results presented by Staprans [7] are in basic agreement with those given in Figure E-6. Note that for pulse operation, field levels somewhat above the Kilpatrick curve are sometimes used.

The Kilpatrick criterion is based on results for parallel plane electrodes while the tube results in Figure E-6 are for various electrode configurations. Thus, varying degrees of field enhancement contribute to the scatter in the data. Apparently, when oxide-coated cathodes are used, decomposition products form surface layers of reduced work function, greater roughness, different hardness, and so forth, which result in reduced breakdown voltage.

E.2.1 Electrode Phenomena Leading to Breakdown

The mechanisms responsible for breakdown in vacuum at field levels that are orders of magnitude below the field emission limit for a planar surface have been under investigation for decades. The electrode phenomena thought most likely to lead to vacuum breakdown in microwave tubes are shown in Figure E-7. These phenomena and the way in which they cause breakdown are discussed in this section.

To understand how microprotrusions can cause breakdown, it is necessary to investigate field emission phenomena. As was discussed in the chapter on cathodes, as the electric field applied to a cathodic electrode is increased to 10^7–10^8 V/cm (0.1–1.0

V/Å), electron emission increases very rapidly. The reason for this can be explained with the aid of Figure E-8. At very high field levels, the potential barrier that electrons must overcome to escape from a metal becomes very narrow. It is so narrow, in fact, that when the wave function of an electron is examined, it is found that there is a finite probability that the electron may exist on the vacuum side of the barrier. In effect, electrons can pass through the barrier even though they don't have sufficient energy to overcome the barrier. This is known as the tunneling effect.

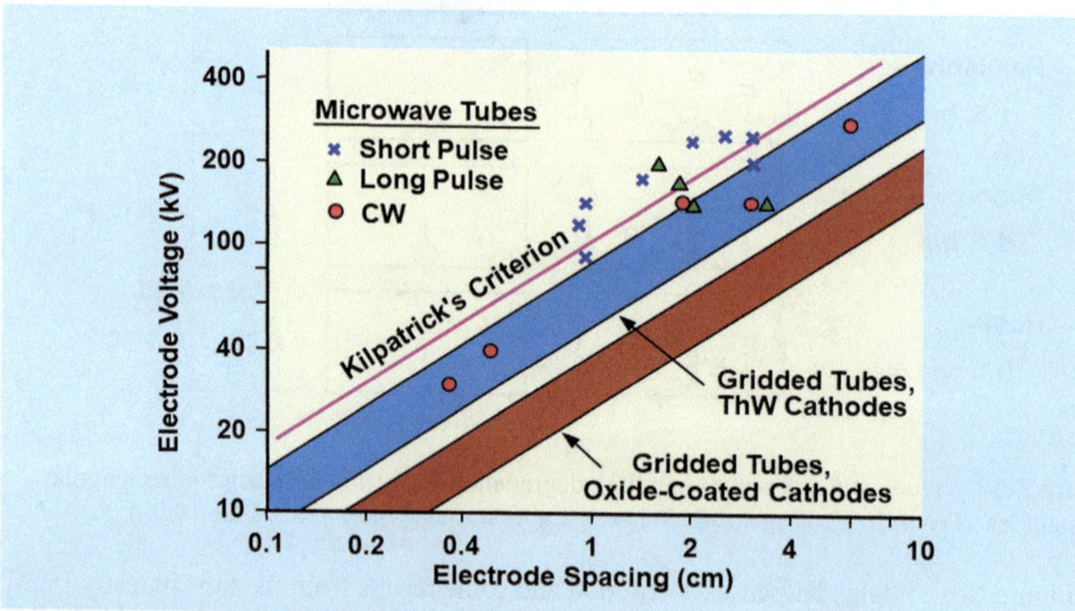

Figure E-6 Kilpatrick's criterion and experimental breakdown characteristics of tubes.

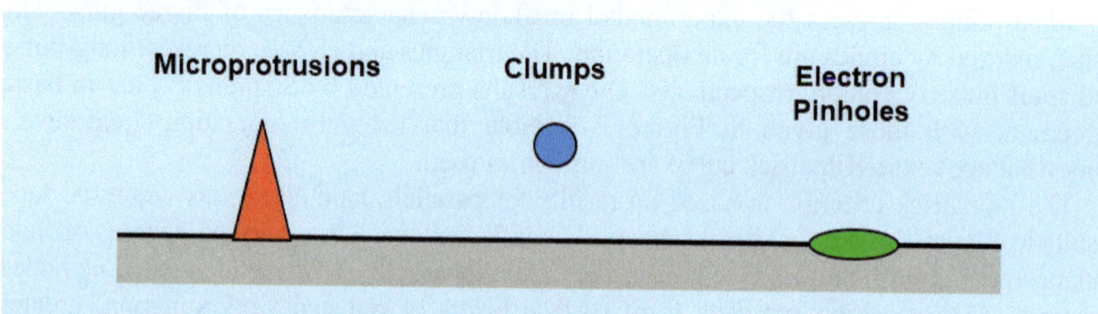

Figure E-7 Electrode phenomena that lead to breakdown.

E.2.1.1 Breakdown Caused by Microprotrusions

When a microprotrusion like that shown in Figure E-9 is present, equipotential surfaces near the tip are crowded together so the electric field is higher than on nearby planar surfaces. If the field enhancement factor, β, at the tip is high enough (~ 100 or more) then field emission can occur at average field levels of ~ 10^5 V/cm.

The relation for current as a function of electric field when field emission is occurring is called the Fowler-Nordheim equation and is

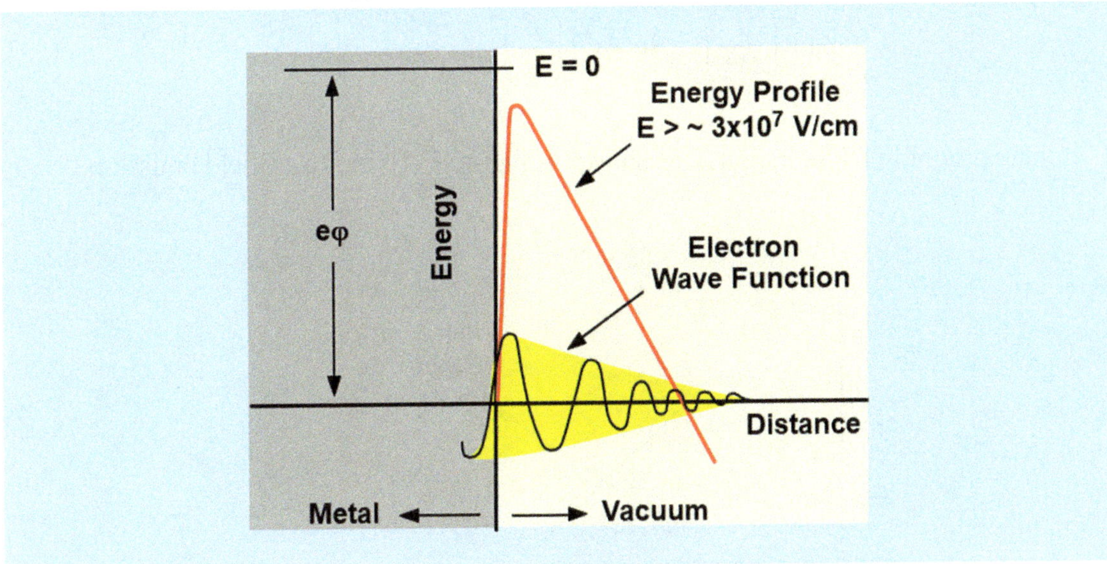

Figure E-8 Electron potential energy profile with a very strong electric field.

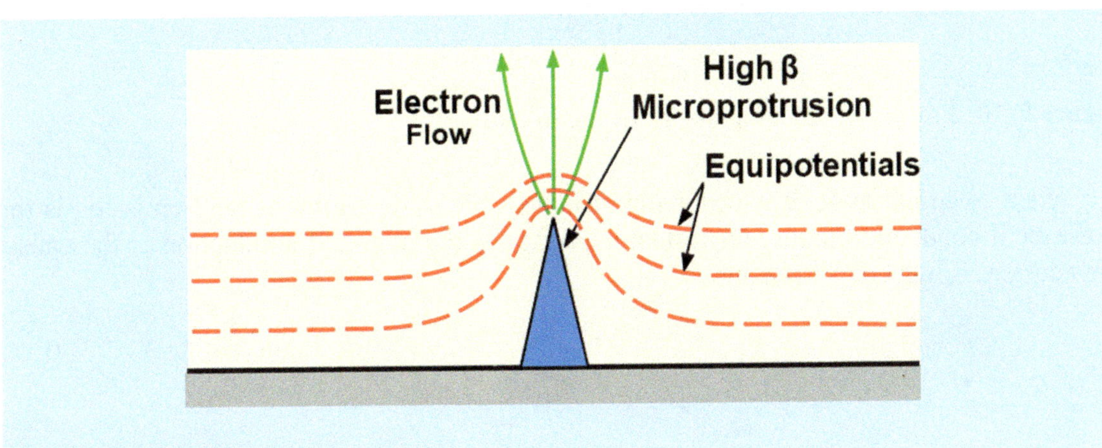

Figure E-9 Equipotential concentration near the tip of a microprotrusion.

$$J_o = C_1 E^2 e^{-C_2/E} \tag{E-1}$$

where C_1 and C_2 are constants that depend on work function.

Letting current density, $J_o = I/A$, where A is emitter area, and $E = \beta V/d$, where β is the field enhancement and d is spacing, then

$$I = A\, C_1 \left(\frac{\beta V}{d}\right)^2 e^{-\frac{C_2 d}{\beta V}} \tag{E-2}$$

or

$$\ln\left(\frac{I}{V^2}\right) = \ln AC_1\left(\frac{\beta}{d}\right)^2 - \frac{dC_2}{\beta}\left(\frac{1}{V}\right) \tag{E-3}$$

So a linear plot of ln I/V^2 versus $1/V$ as shown in Figure E-10 indicates field emission.

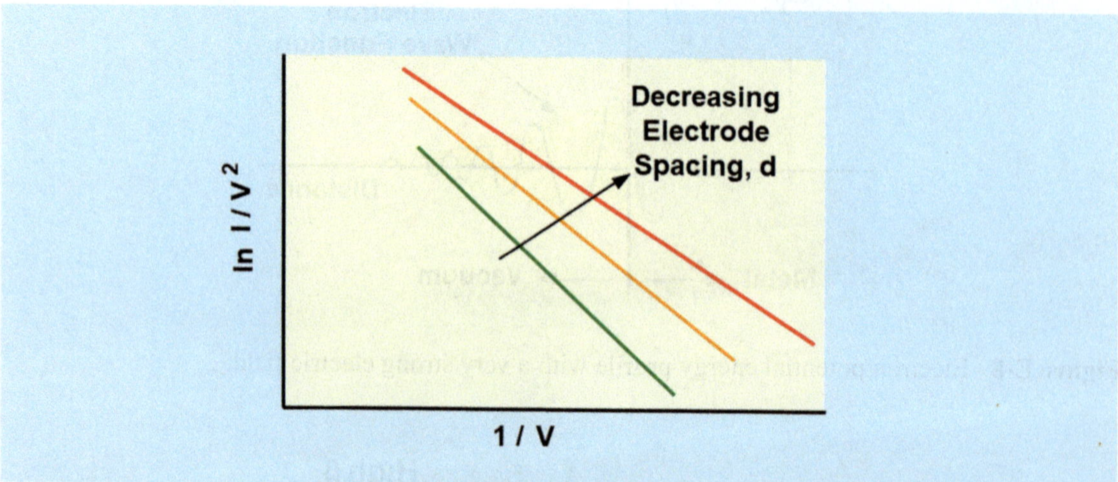

Figure E-10 Fowler-Nordheim plots for field emission.

When electrical leakage is occurring, it is possible to determine if field emission is the cause or if conduction across an insulator surface is the cause. If conduction is the cause, then current is related to voltage by

$$I = CV \tag{E-4}$$

or

$$\frac{I}{V^2} = \frac{C}{V} \tag{E-5}$$

so

$$\ln\frac{I}{V^2} = \ln C + \ln\frac{1}{V} \tag{E-6}$$

Figure E-11 shows an example of a Fowler-Nordheim plot for an electron gun in which it was apparent that electron conduction was causing leakage at low voltages and field emission was occurring at high voltages.

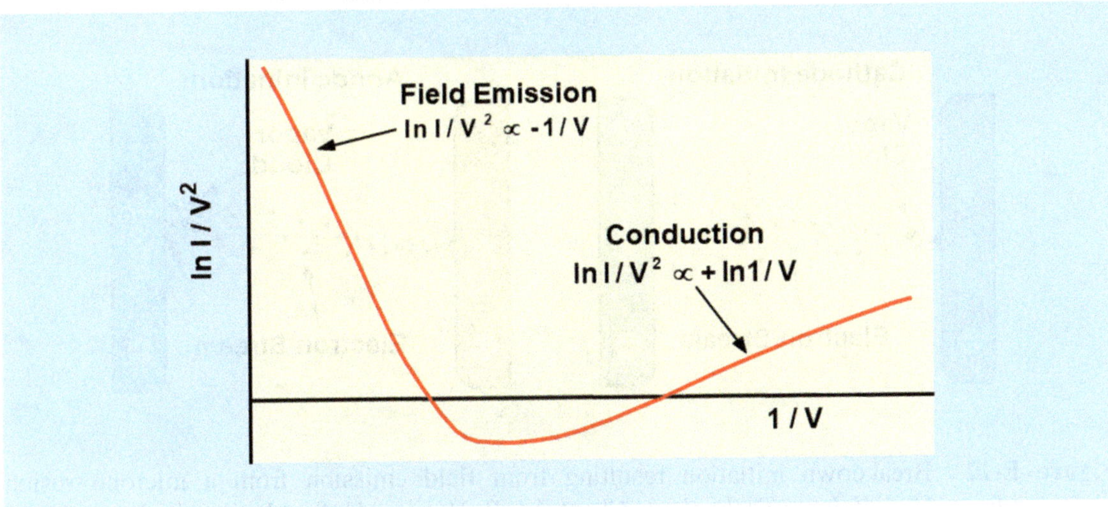

Figure E-11 Fowler-Nordheim plot for an electron gun.

Microprotrusions can originate in several ways including the following:

1. Machining operations on some materials can produce sharp fibers.
2. A freshly polished surface can have filamentary structures lying in the plane of the surface. These can be erected normal to the surface by the strong electromechanical forces associated with a large electric field.
3. It is possible for microprotrusions to grow in the presence of large electric fields. This is a field activated surface diffusion process that is enhanced by heating the negative surface. Published data suggest that this effect is pronounced with electrode materials such as copper and aluminum that give unstable gaps.

For breakdown to be initiated by a microprotrusion, a microplasma must be generated. This is only possible if localized melting and vaporization of electrode material takes place. Outgassing can contribute to this process. As shown in Figure E-12, there are two ways in which this can happen:

1. Cathode-initiated breakdown results from vaporization by current flow and resistive heating of the tip of a microprotrusion;
2. Anode-initiated breakdown occurs if the electron stream from the microprotrusion heats a spot on the anode to the point where melting and vaporization occur.

As indicated in Figure E-13, there can also be in-flight vaporization of a microparticle released from the anode hot spot.

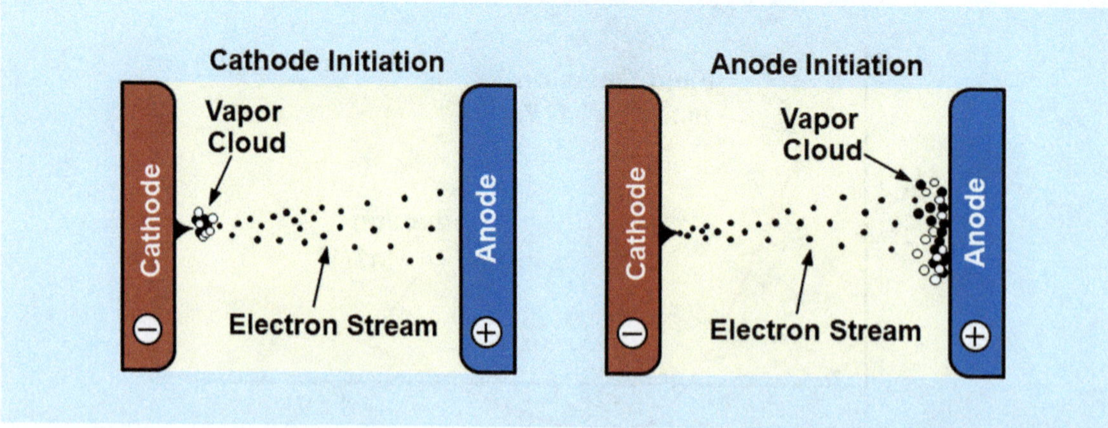

Figure E-12 Breakdown initiation resulting from field emission from a microprotrusion. (Adapted from: *High Voltage Technology*, Chapter 4, R. Hawley, Oxford University Press, 1968.)

Whether breakdown is cathode-initiated or anode-initiated depends on the field enhancement factor of the microprotrusion. Utsumi [8] found the high β emitters ($\beta > \sim 500$) (see Figure E-14) usually lead to cathode-initiated breakdown while low β emitters ($\beta < \sim 30$) lead to anode-initiated breakdown. As shown in Figure E-15, Utsumi also found that electrode separation played a role in determining whether breakdown was anode initiated or cathode-initiated.

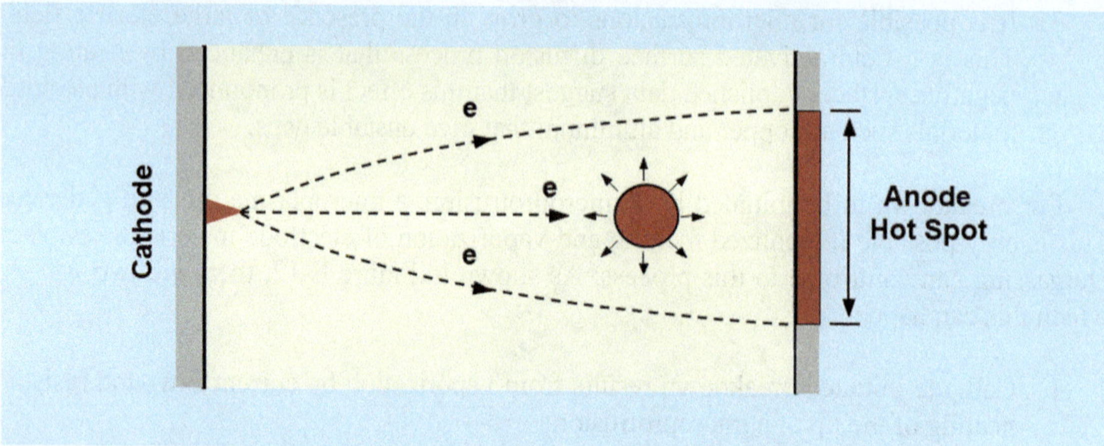

Figure E-13 In-flight vaporization by electron bombardment of an anode microparticle.

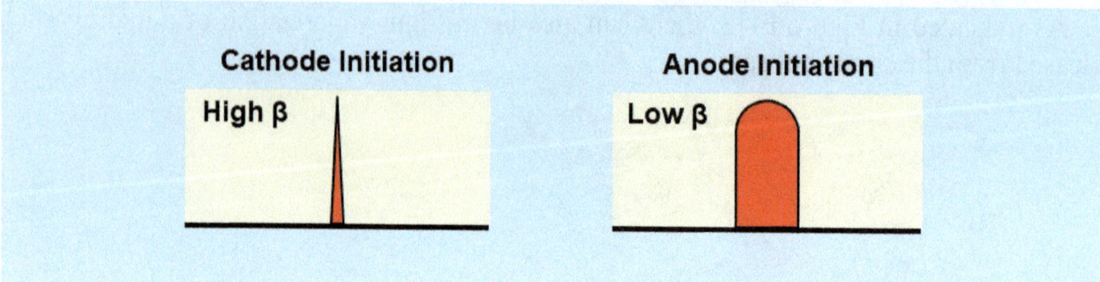

Figure E-14 Emitters with high and low field enhancement factors.

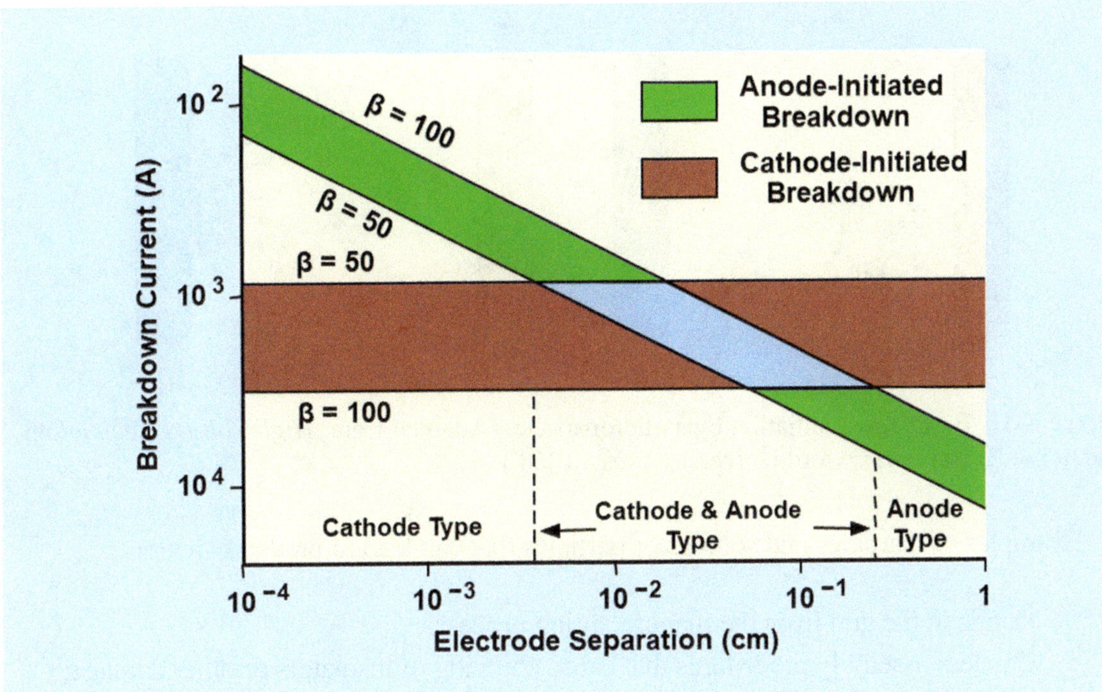

Figure E-15 Electrode dominance in vacuum arc initiation. (From: T. Utsumi, *Technical Report* RADC-TR-66-712, January 1967.)

E.2.1.2 Breakdown Caused by Clumps

Cranberg [9] recognized that in the presence of a high electric field, there are strong electromechanical forces. These forces tend to detach microparticles or clumps of microparticles from electrodes, as shown in Figure E-16. These clumps are charged to the same polarity as the parent electrode so they are also repelled. The resulting charged particles are accelerated to high velocities during the transit to the opposite electrode. On impact, particle energy is dissipated as heat and/or mechanical shock energy. If the impact energy is large enough, there can be sufficient vaporization of the particle and target to form a microplasma and possibly cause breakdown.

For large particles, breakdown can occur between a particle in transit and the electrode surface that it approaches. This can lead to breakdown of the gap.

When a particle strikes a surface, a crater can be produced. The splash rim of material surrounding the crater may have sharp protrusions that field emit and result in breakdown.

A very practical indication of the role of particles in breakdown may be had from a technique sometimes used by the tube industry, apparently to eliminate breakdown caused by particles. An electron gun to be highly stressed is placed in a vacuum chamber, high voltage is applied, and it is vibrated until arcing stops. The process may be repeated with varying orientations of the gun. Particles are observed to fall from the gun during the process, and it is presumed that they are the cause of arcing.

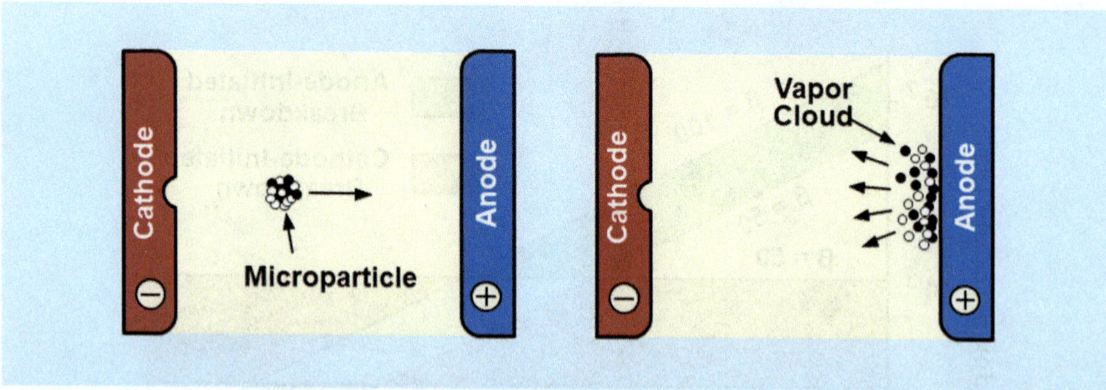

Figure E-16 Breakdown initiation by a microparticle. (Adapted from: *High Voltage Technology*, Chapter 4, R. Hawley, Oxford University Press, 1968.)

Examples of particles and sources of particles that can lead to breakdown are:

- Debris in the gun from the manufacturing process;
- Careless assembly procedures that cause chipping of insulators or other damage;
- Dust from inadequate cleanliness;
- Cathode evaporants that form loose films on various gun surfaces;
- Use of improper getters;
- Particles from heater potting material;
- Debris that enters the gun through the anode aperture;
- Embedded particles that are exposed by chemical etching.

In addition to these examples, as indicted in Figure E-17, a shower of microparticles in Figure E-17(b) can be produced by the impact of a primary microparticle from the opposing electrode in Figure E-17(a). A shower of microparticles can also be produced by an arc in vacuum (to be discussed later in this chapter), and this can, subsequently, produce breakdown.

To find a relation for the conditions under which breakdown may be initiated by a clump, it is necessary to consider the energy, U, acquired by a particle that is accelerated across a gap. That energy is the product of the particle charge, q, and the accelerating voltage, V; that is

$$U = qV \tag{E-7}$$

If A is the cross-sectional area of the particle, the energy density on impact is

$$\frac{U}{A} = \frac{q}{A}V \tag{E-8}$$

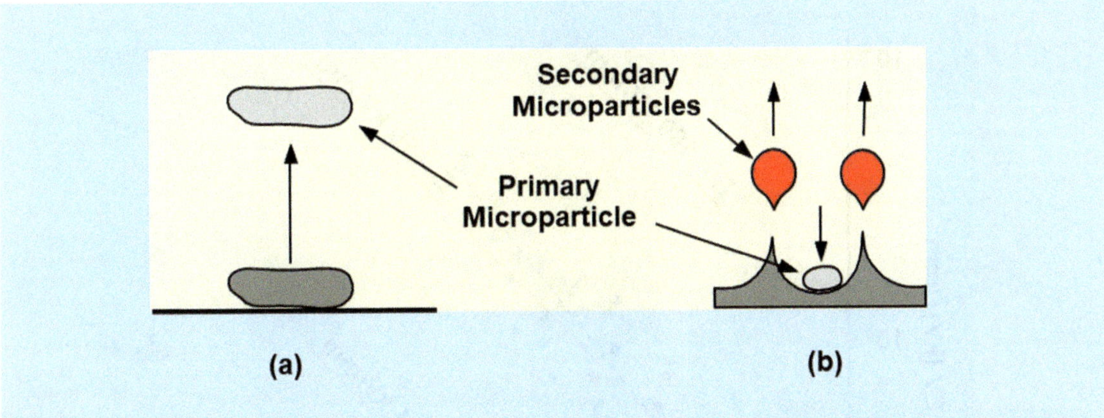

Figure E-17 A primary microparticle can produce secondary microparticles. (Adapted from: *High Voltage Vacuum Insulation*, R. Latham, Academic Press, London, 1995.)

The charge density, q/A, acquired by the particle is proportional to the electric field, E, so

$$\frac{U}{A} \propto EV \tag{E-9}$$

Cranberg showed from experimental results that in the limit when breakdown occurs,

$$E_{\lim} V \approx 10^4 \tag{E-10}$$

where E is in kV/mm and V is in kV.

Figure E-18 shows that the Cranberg limit applies to cathode-modulated Pierce guns. For anode-modulated tubes and dc-operated tubes the EV limit is closer to 10^3 $(\mathrm{kV})^2/\mathrm{mm}$. From Pierce gun design considerations, Durand and Shroff concluded that the relation between $E_{\lim}V$ and V should be

$$E_{\lim} V \sim V^{1.25} \tag{E-11}$$

This predicts that $E_{\lim}$ should increase with voltage. The dashed line in Figure E-18 indicates the average of the $E_{\lim}V$ products for Pierce guns and shows that there is, in fact, a trend for the maximum field to increase with increasing voltage.

It is interesting to observe that Kilpatrick's criterion (Figure E-6) shows that as spacing increases, the average electric field at which breakdown occurs decreases by about a factor of five (from ~ 200 kV/cm to ~ 40 kV/cm) as breakdown voltage increases from 20 kV to 400 kV. Over this same voltage range, (E-11) predicts that $E_{\lim}$ should increase by about a factor of two before breakdown occurs. Of course, we can understand from the Cranberg theory why $E_{\lim}$ can *increase* as the average field decreases. It is $E_{\lim}$ that is responsible for the charging of a clump and it is the gap voltage (and the average gap field) that accelerates the clump. The clump energy necessary to initiate breakdown is proportional to the product of the two fields and thus, as one decreases, the other can increase.

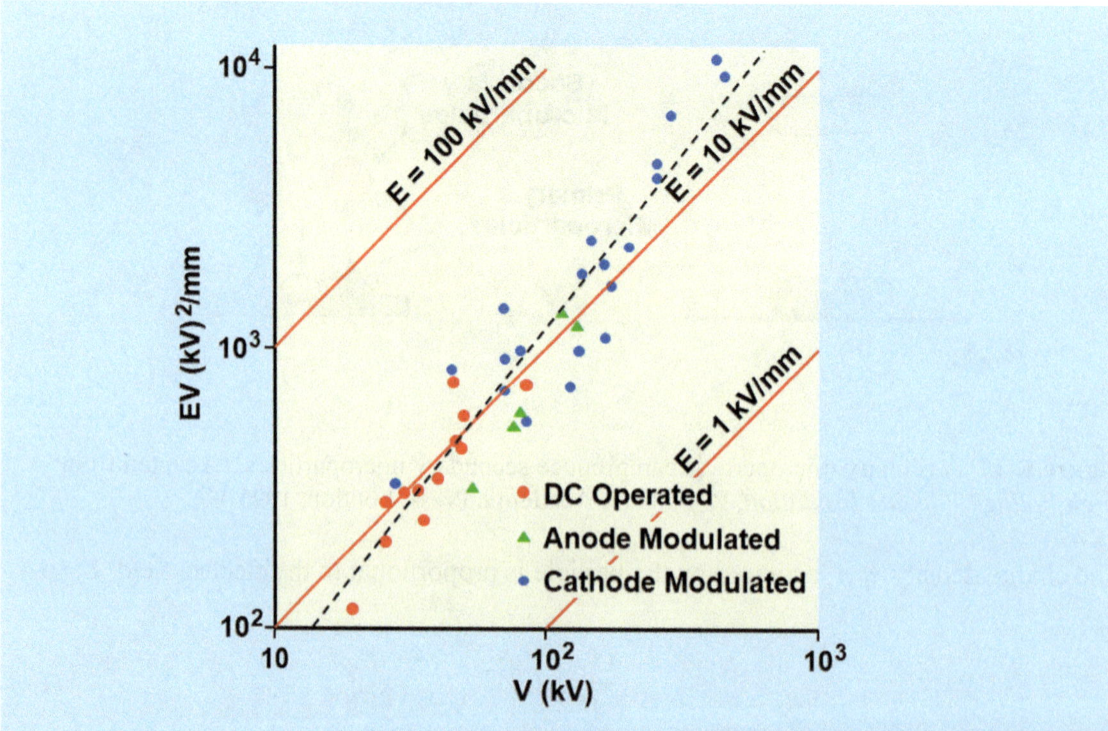

Figure E-18 *EV* versus *V* for Pierce guns. (Adapted from: *High Voltage Vacuum Insulation*, Chapter 11, A. J. Durand and A. M. Shroff, Academic Press London, 1995.)

Pulse length affects the maximum *EV* product as shown in Figure E-19. As pulse length increases, *EV* decreases. Durand and Shroff note that when points corresponding to high values of *EV* are connected, almost all guns are below a limit given by

$$EV = \frac{100}{\text{pulse length}^{0.34}} \tag{E-12}$$

This limit must be taken into consideration during the design of electron guns. It should be noted that the squares in Figure E-19 are the results of breakdown experiments on high-voltage guns from X-band klystrons at SLAC. These results show that it is possible to exceed the limit indicated by (E-12).

In Figures E-18 and E-19, most of the data indicate *EV* limits well below the Cranberg value of 10^4 $(\text{kV})^2$/mm. There is the possibility that an explanation given by Latham may account for some of the low *EV* values. As shown in Figure E-20, a particle may bounce back and forth between two electrodes and gradually acquire enough energy to cause breakdown. As a result, the gap voltage required to produce breakdown may be significantly below that needed for breakdown caused by a single-transit particle.

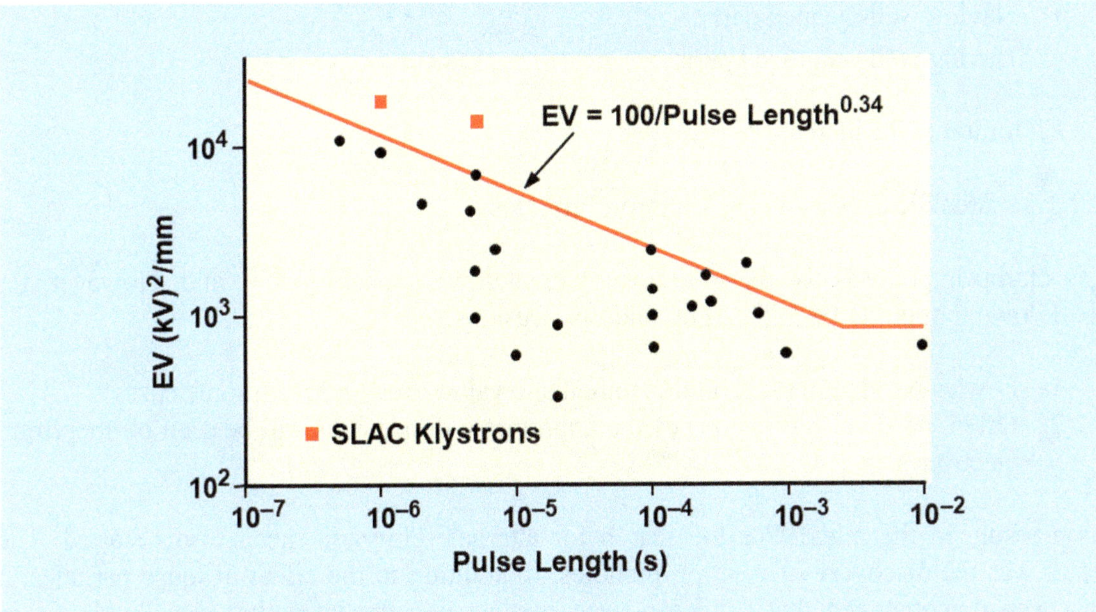

Figure E-19 The effect of pulse length on *EV*. (Adapted from: *High Voltage Vacuum Insulation*, Chapter 11, A. J. Durand and A. M. Shroff, Academic Press, London, 1995.)

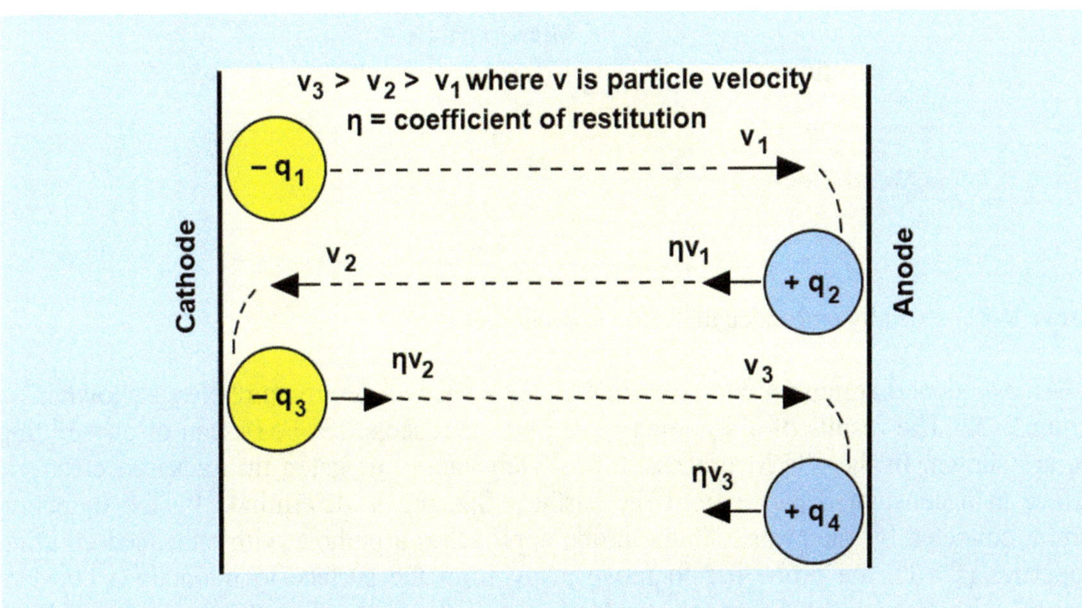

Figure E-20 Particle bouncing and gradual energy accumulation proposed by Latham. (Adapted from: *High Voltage Vacuum Insulation*, R. Latham, Academic Press, London, 1995.)

Durand and Shroff attributed the low *EV* values in Pierce guns to the presence of a cathode. Not only are the cathode and nearby electrodes very hot, there is the deposition of cathode evaporants (especially barium) to contend with. With attention to the following:

- Lowering the temperature of electrodes;

- Having well-cleaned parts;
- Having well-outgassed parts.

the *EV* limit may be increased.

E.2.1.3 Breakdown Caused by Electron Pinholes

The clump hypothesis for the initiation of breakdown appears to be valid. However, for breakdown thought to be initiated by microprotrusions:

1. Fowler-Nordheim plots predict unrealistic values of β (> 250 in some cases);
2. Often, no microprotrusions of the appropriate dimensions can be seen on the virgin electrodes.

As a result, in the mid-1970s the search for alternate emission mechanisms started. The result was the discovery of electron pinholes. In addition to the effect of loose particles, it has been demonstrated that a micron-sized particle like the insulating particle shown in Figure E-21 that is attached to an electrode surface or partly embedded in the surface can be the site for the initiation of breakdown.

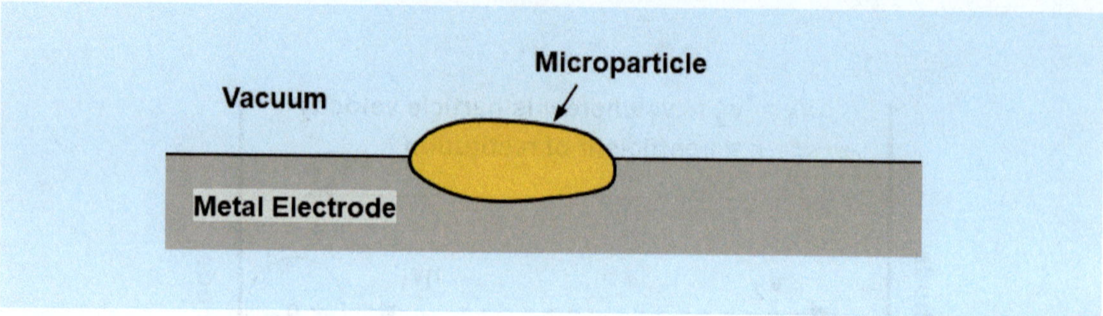

Figure E-21 Partially embedded dielectric microparticle.

An excellent diagram for demonstrating the existence of electron pinholes is shown in Figure E-22. The results of a scanning tunneling microscope (STEM) scan of an emitting site are shown. In the STEM process, a positively biased tungsten tip tracks an electrode surface at a constant distance from the surface. Spacing is determined by the tunneling current collected by the probe. As the probe approaches a pinhole with enhanced emitting properties ($\beta > 1$), the probe has to move away from the surface to maintain a constant current (with fixed bias). From the displacement of the probe from the surface in Figure E-22(a), the emitting properties of the surface Figure E-22(b) can be determined. It has been shown that both the edge and the center of a particle can emit.

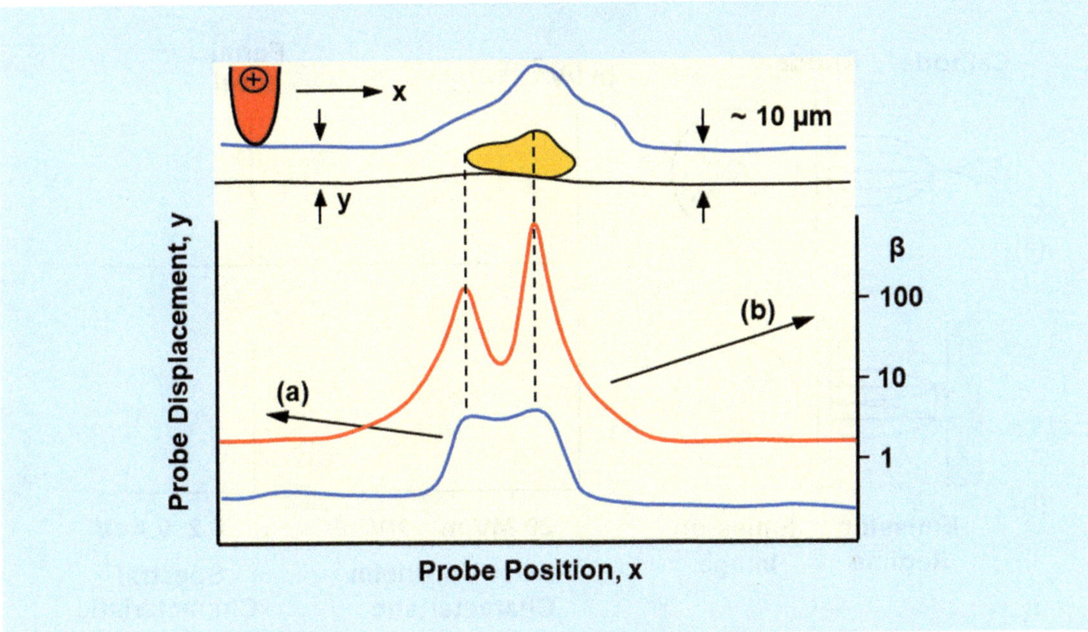

Figure E-22 STEM analysis of an emission site. (Adapted from: *High Voltage Vacuum Insulation*, R. Latham, Academic Press, London, 1995.)

The research of R.V. Latham [10] has provided a great deal of insight into the electron pinhole phenomena that cause this breakdown. The key results of Latham's studies (and the studies of others referenced by Latham) are summarized in Figure E-23, which contrasts the properties of emission from a particulate site with those from a metallic whisker. In summary, the results show the following:

1. The basic emission process is associated with micron-sized particulate structures on the cathode surface. Frequently, such structures appear to be electrically insulated from the metal surface.
2. The material composition of the particles can be very varied, and apparently independent of the substrate cathode metal, with carbon being the single most common emission material, typically accounting for 20% of all sites.
3. In contrast to the linear Fowler-Nordheim characteristic of the metallic whisker, particulate emission sites have nonlinear characteristics, particularly at high fields.
4. Emission images obtained from individual sites indicate that the electrons originate from a number of apparently independent subsites that switch on and off randomly with mean frequencies that are dependent on pressure.
5. The electrons emitted from individual subsites have single-peak spectra that are symmetrical in shape, have large half-widths (typically > 300 meV), and exhibit a strong field-dependent shift from the Fermi level, which is characteristic of a nonmetallic emission mechanism. The key physical implication of these data is that the emission process must involve the creation of hot electrons and that, consequently, the emission regime itself must include a dielectric or semiconducting medium.
6. The emission of electrons is associated with an electroluminescent effect that indicates the involvement of a semiconducting or insulating medium.

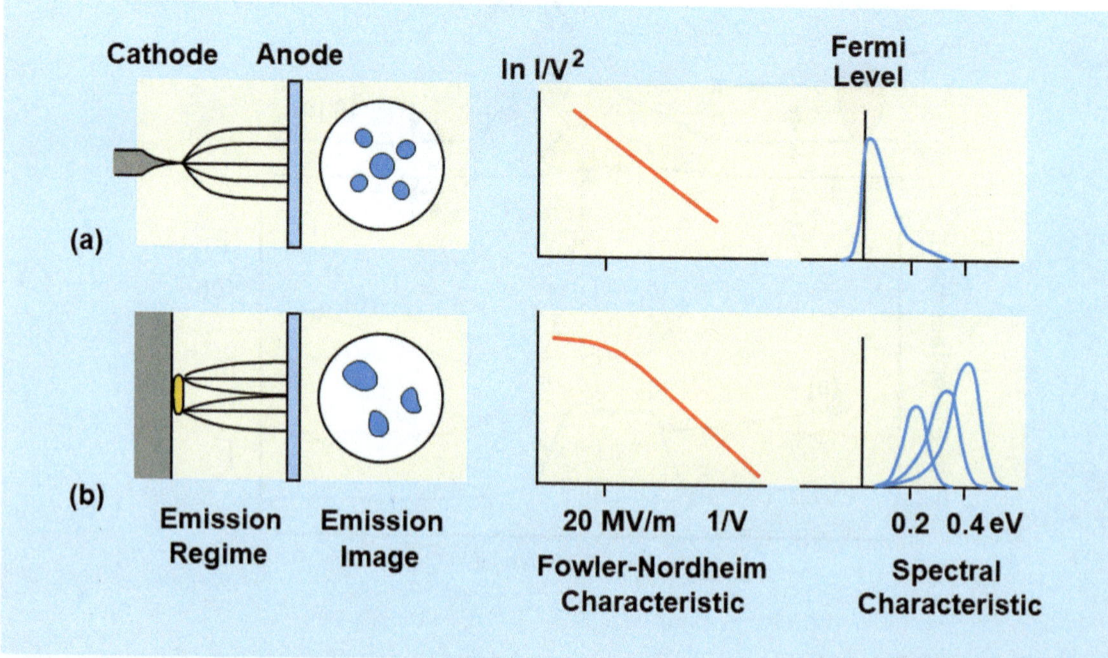

Figure E-23 Comparison of emission properties of (a) a metallic whisker (manmade) and (b) a particulate site. (From: R. V. Latham and N. S. Xu, *Vacuum*, 1991.)

On the basis of the results listed above, two possible models shown in Figure E-24 for the emission sites have been proposed. The M-I-V model involves a metal-insulator-vacuum regime. In the presence of a high electric field, electrons tunnel from the Fermi level of the metal into the conduction band of the insulator. They then move through the insulator under the influence of the penetrating field. Upon approaching the insulator-vacuum interface, the electrons experience very much higher field conditions, which cause them to be heated (accelerated) to temperatures in excess of 1,000°C. As a result, they are able to be quasi-thermionically emitted into the vacuum over the surface potential barrier at an energy that is significantly below the Fermi level. It is also assumed with this model that traps are present in the bulk of the insulator and these serve to control the flow of current at high field levels. This is the cause of the nonlinear Fowler-Nordheim characteristic shown in Figure E-23.

The metallic particle in contact with the insulating surface in the M-I-M-V model in Figure E-24(b) serves to enhance the field at its contact point with the insulator. A conducting channel is formed in the region of the contact point, and electrons are emitted into the metal flake as well as into the vacuum.

E.2.2 Avoiding Breakdown

Four primary factors must be considered in avoiding breakdown:

1. Electrode geometry;
2. Electrode material;
3. Electrode surface preparation;
4. Electrode conditioning.

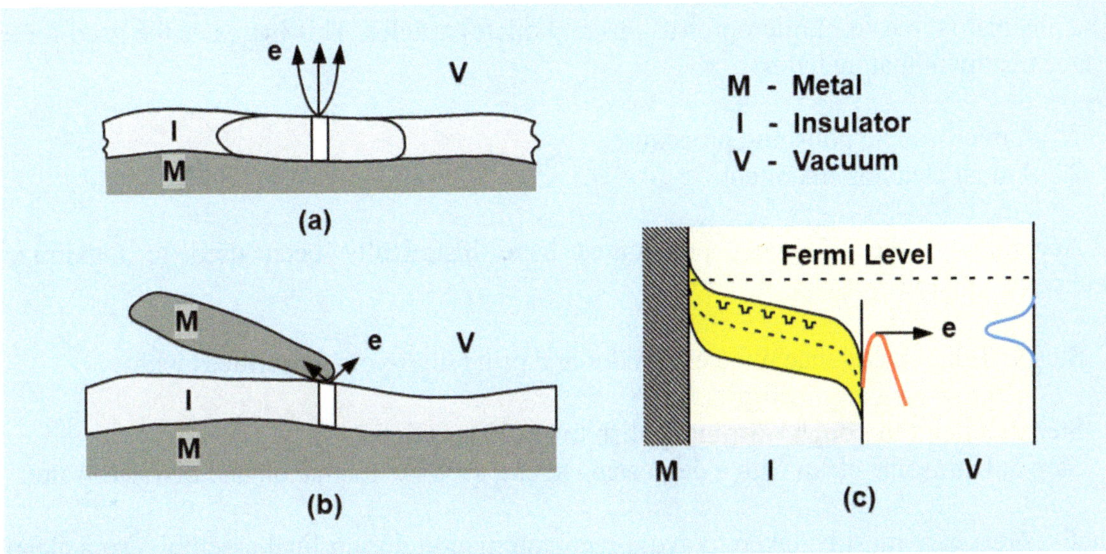

Figure E-24 (a) M-I-V model, (b) M-I-M-V model, and (c) energy diagram of the conducting channel phenomenon. (From: R. V. Latham and N. S. Xu, *Vacuum*, 1991.)

The area and spacing of electrodes along with field enhancement characteristics all have strong effects on breakdown voltage. Obviously, radii should be as large as possible to minimize β. High β regions like slotted screw heads should be avoided or shielded. An increase in area will generally reduce breakdown voltage because more emitting sites and microparticles will be exposed.

For small spacings ($< \sim 0.5$ mm), breakdown tends to be field-dependent. This indicates that breakdown is initiated by microprotrusion or electron pinhole phenomena. For large spacings ($> \sim 2$ mm), breakdown becomes voltage-dependent. This indicates that breakdown becomes dominated by microparticle processes so cleanliness and the absence of particulate contamination are absolutely essential in microwave tube applications where spacings are almost always greater than 2 mm. In any event, electrode spacing should be as large as possible without overly compromising the operation of the device.

It has long been recognized that the choice of electrode material has a pronounced effect on breakdown voltage. It is generally agreed that at one extreme, there are materials like copper and nickel that are particularly susceptible to breakdown. At the other extreme, there are physically hard materials such as stainless steel, titanium, and molybdenum for which breakdown problems are minimized. However, hardness is not the only requirement because tungsten does not perform well.

Other than hardness, a second significant property of good electrode materials is their resistance to corrosion. The surfaces of these materials have a strong insulating oxide film, whereas the surfaces of materials such as copper often have a semiconducting oxide film. While there is still controversy concerning breakdown mechanisms, these characteristics seem to be consistent with clump-initiated breakdown. For example, hard materials are less susceptible than soft materials to damage and the generation of secondary microparticles by impinging clumps. Semiconducting films are more likely to produce clump charging than insulating films. Increased clump charges result in increased clump velocities in the presence of high fields and increased surface damage on impact.

Concerning surface preparation, historically, the ideal electrode surface has been thought to be one that is free from microprotrusions and microparticles. This indicates the need for a surface treatment that includes:

1. A microscopic polishing procedure;
2. A final cleaning treatment.

Accordingly, the following procedures have historically been used to maximize breakdown voltage:

Step 1: Take freshly machined electrodes and grind high-voltage surfaces with progressively finer grits.
Step 2: After a thorough washing, polish to a 0.25-μm finish.
Step 3: Ultrasonic clean with a degreasing agent, acetone, methanol, and distilled water.

Finally, great care must be taken to avoid recontamination during final assembly, so a clean room environment and procedures must be used. Still, surfaces prepared in this way will have a very complex surface microstructure so electrodes are often treated by using either a chemical or an electrochemical polishing process. In addition, it has been shown that heating to ~1,100–1,200°C suppresses electron pinholes. (Of course, this is usually not possible with most microwave tube operations.)

Electrode conditioning is used to safely quench as many as possible of the sources of prebreakdown current. The simplest and most widely used technique is current conditioning, which is particularly suitable for initial treatment of electron guns. A high-voltage supply is used with a current-limiting resistor to increase voltage in small steps so that the prebreakdown current is stabilized in each step. The time profiles of current and voltage for a microdischarge that does not result in breakdown are shown in Figure E-25.

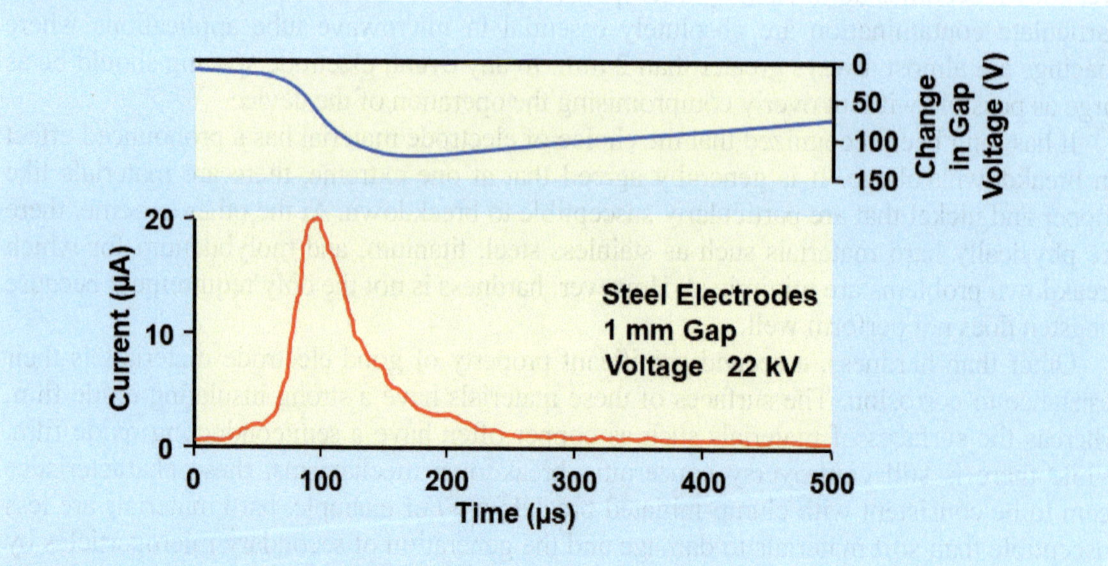

Figure E-25 Microdischarge for 22 kV applied to a 1-mm gap between steel electrodes. (From: High Voltage Technology Seminar, Ion Physics Corporation, 1969.)

As high voltage is gradually applied to the electrodes (Figure E-26), microdischarges are produced that "clean up" the electrodes. That is, microprotrusions are assumed to be thermally blunted or removed by the strong electromechanical forces. With a slowly increasing voltage, microparticles acquire a minimum charge and are not accelerated enough to cause breakdown. Instead, they are removed by continuous bouncing between electrodes or become welded to an electrode surface.

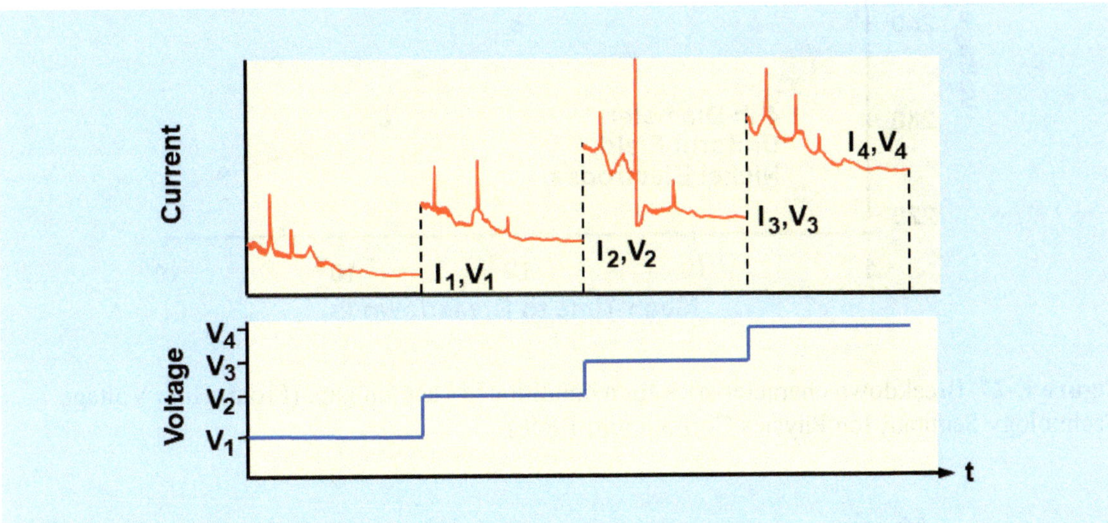

Figure E-26 Voltage and current during conditioning. (Adapted from: *High Voltage Vacuum Insulation: The Physical Basis*, R. V. Latham, Academic Press, 1981.)

In a high-voltage device, this conditioning process must be carried out when voltage is first applied and then may have to be repeated in an accelerated manner each time the device is turned on. It should also be noted that even for a newly conditioned gap, the voltage hold-off capability degrades with time, as shown in Figure E-27.

E.2.3 Vacuum Arcs

Regardless of the initiation mechanism, dc breakdown in a vacuum leads to an arc on the negative electrode (at the current levels encountered under fault conditions in a microwave tube). This arc is normally known as a *vacuum arc*.

In a vacuum arc, the material of the negative electrode is vaporized and ionized by arc spots to form a conducting medium between the electrodes. Vacuum arcs can be established on any reasonably good conductor. The voltage between electrodes during a vacuum arc varies little with electrode spacing for typical (cm) spacings and is essentially the cathode-fall potential. This is because there is no appreciable anode-fall voltage and because the voltage drop in the plasma column is extremely small (~0.01 V/cm). The voltage drop is dependent on the material of the cathode and varies from 8V for mercury to 33V for cold steel. Figure E-28 shows the relationship between the voltage at which a vacuum arc is sustained and the product of the boiling temperature and thermal conductivity of the cathode material. Barium is not included in this figure but is comparable to cadmium in arcing characteristics. As shown in Figure E-29, the arcing voltage is a slowly varying function of current.

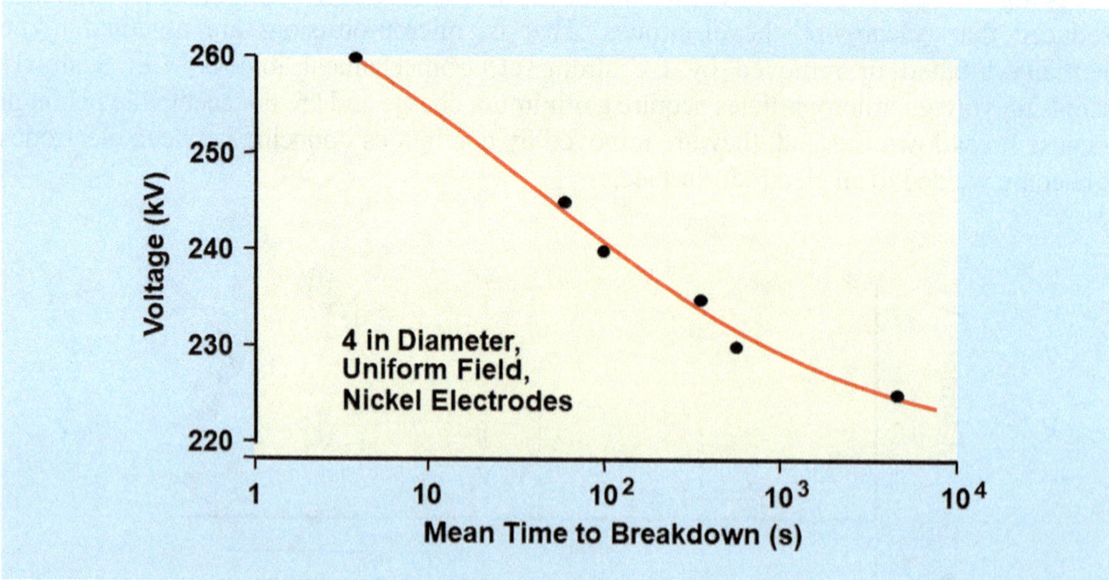

Figure E-27 Breakdown characteristics for a conditioned vacuum gap. (From: High Voltage Technology Seminar, Ion Physics Corporation, 1969.)

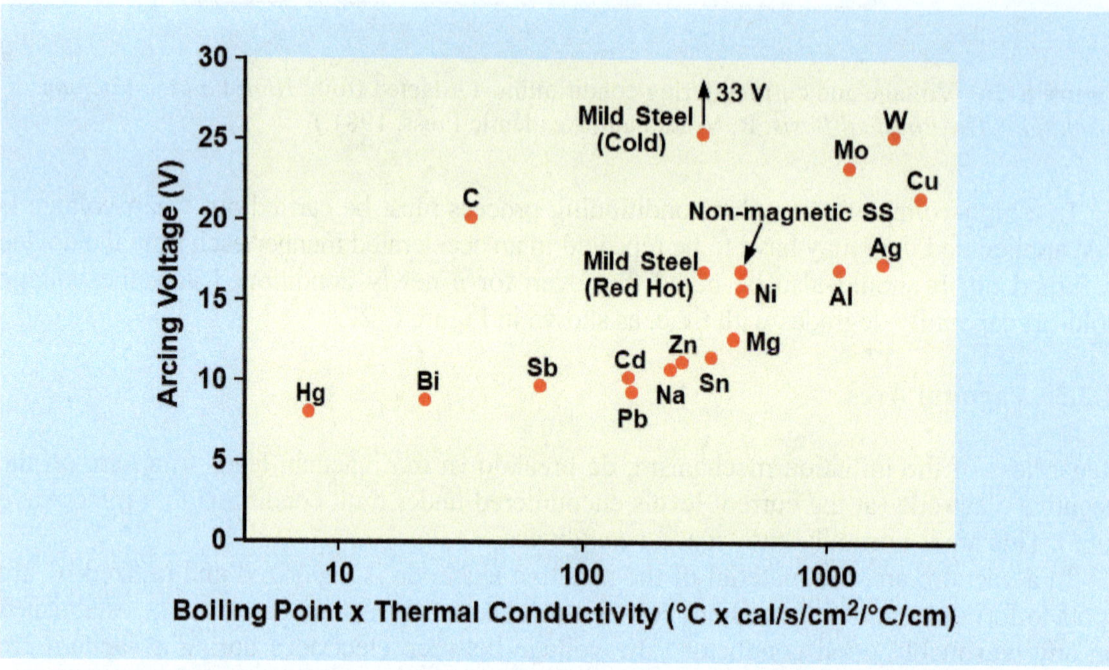

Figure E-28 Voltage at which a vacuum arc is sustained for several commonly used metals. (Adapted from: M. P. Reece, *Proc. IEE*, 1963.)

On a surface consisting of differing metals, a vacuum arc will operate on the low-arcing voltage metals. For example, if a calcium film partially covers a copper surface, a vacuum arc will remove the calcium film before attacking the copper. This results from the relatively low arcing voltage of calcium (about 12V) relative to that of copper (about 20V). Similarly, the evaporation products from thermionic cathodes will be preferentially removed from nearby grid or control electrode surfaces by a vacuum arc. This is because barium, strontium, calcium, and so forth all have similar arcing voltages, which are about one-half (or less) of those of the normal materials of grids and electrodes. As a side note, arcing may produce deposition of material on critical surfaces such as insulators.

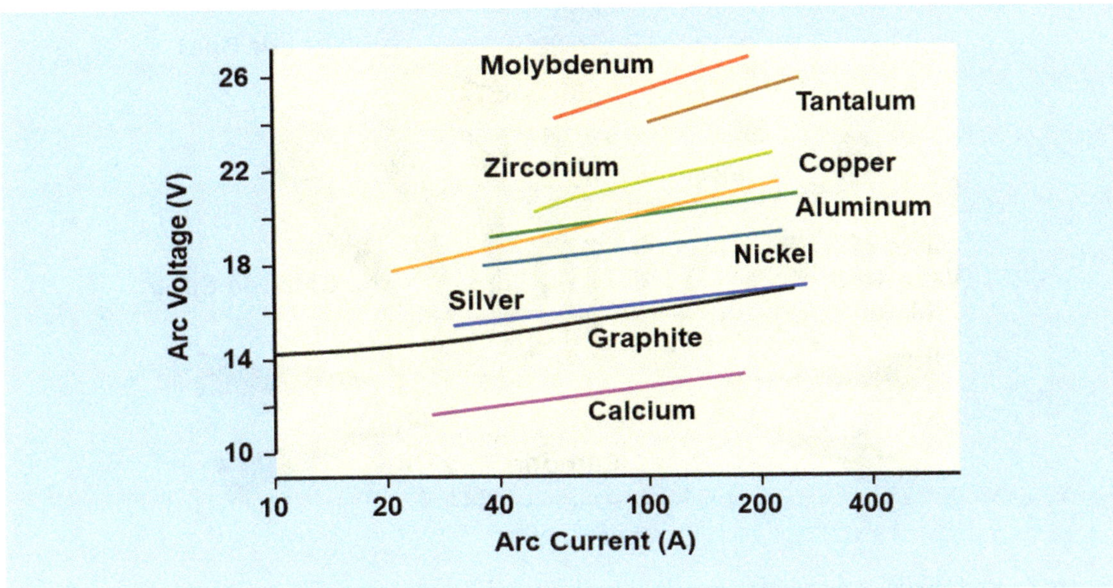

Figure E-29 The dc volt-ampere characteristics of vacuum arcs. (From: W. D. Davis and H.C. Miller, *Jour. App. Phys*., 1969.)

Unlike an arc in gas, which consists of one large spot anchored in position, many very small arc spots exist in a vacuum arc, as indicated in Figure E-30. These spots travel at high speed (meters per second to tens of meters per second) with apparently random motion across the electrode surface. On a macroscopic scale, a surface subjected to prolonged vacuum arcing has the texture of a piece of frosted glass. On a microscopic scale, individual arc tracks somewhat resemble the tracks of insects crawling in random directions across a freshly painted surface.

It has been shown [11] that the cathode spot in the vacuum arc consists of a finite number of cells, each of which passes a current that is restricted to a narrow range of values. For example, Kesaev has shown that the current range of a cell in a mercury arc is approximately 0.08 to 0.1A. If the current through the arc is about 0.1A, then only one cell exists. If the current is raised above 0.1A, the cell divides into two cells. Further increases in current increase the number of cells until the total is equal to the current through the arc divided by the current per cell. If the current through a cell is reduced below about 0.08A, the cell extinguishes in a time that is thought to be less than 0.1 μs.

It is assumed that vacuum arcs on nearly all metals are cellular in nature. Possible exceptions are metals such as tungsten, which are capable of thermionic emission rates in excess of 10^6 A/cm^2 at temperatures slightly above their boiling points (as calculated using the Richardson-Dushman equation and room-temperature values of work functions). At these high temperatures, the computed evaporation rates lead to mass-to-charge ratios that are near the value of 0.1 mg/coulomb that is observed in vacuum-arc experiments. As a result, if it is assumed that the cellular nature of the arc is necessary for the intense emission rates to occur on low-temperature materials, then cells may not necessarily exist on some high boiling point materials.

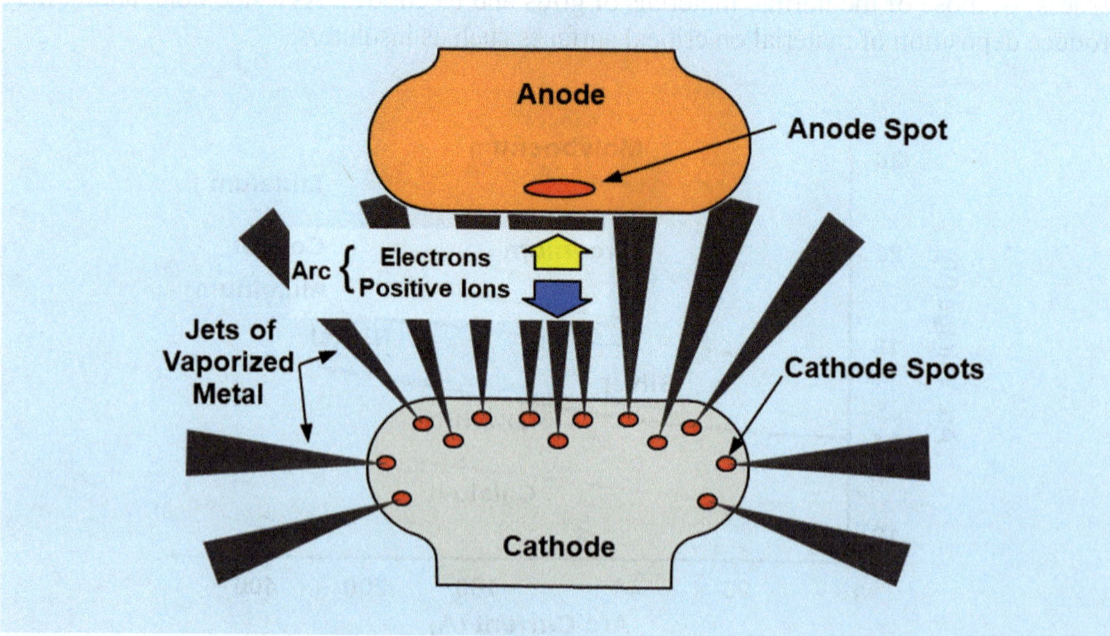

Figure E-30 Details of a vacuum arc. (Adapted from: *Vacuum Arcs, Theory and Application,* J. M. Lafferty (ed.), John Wiley & Sons, 1980.)

As the current through a vacuum arc is reduced, the number of cells decreases and the probability that all cells will extinguish simultaneously increases. As a result, the average arc lifetime becomes shorter and shorter. Figure E-31 shows the interruption characteristics for several metals and shows, for example, that when the current through a tungsten (W) arc approaches 3A, the average lifetime approaches 10 μs.

When a vacuum arc extinguishes, it does so very rapidly. In fact, the current is said to chop, and the chopping time is usually less than 0.1 μs. The chopping characteristics of vacuum arcs are important to the designer of protection circuitry that is used with microwave tubes to prevent damage to the tube in the event that an arc occurs. For example, as the current is driven toward zero by a protection circuit, the arc will extinguish well before the current reaches zero. If there is significant inductance in the protection circuitry, the voltage generated by an arc that chops can be sufficient to cause insulation failure.

The rate at which material is removed from a surface by a vacuum arc varies somewhat from metal to metal but is generally about 0.1 mg per coulomb of charge passed through the arc. Thus, if the time history of the current in a tube during an arc is known, the amount of erosion can be estimated with reasonable accuracy.

Similarly, if energy input to an arc in an electron gun is known, the erosion can be estimated. This is because the arcing voltage is restricted to a reasonably narrow range of values. For example, if a 20J arc occurs, the charge passed through the arc is estimated to be 1C using an average arcing voltage of 20V. For 1C of charge, 0.1 mg of material is removed. For a density of 10 gm/cm^3 (typical for most electron-gun metals), the volume of electrode material removed is 10^{-5} cm^3. For a 1-cm^2 area of electrode material, the thickness removed is only 10^{-5} cm or 0.1 μm. Thus, the amount of material removed by a 1C or a 20J arc is extremely small.

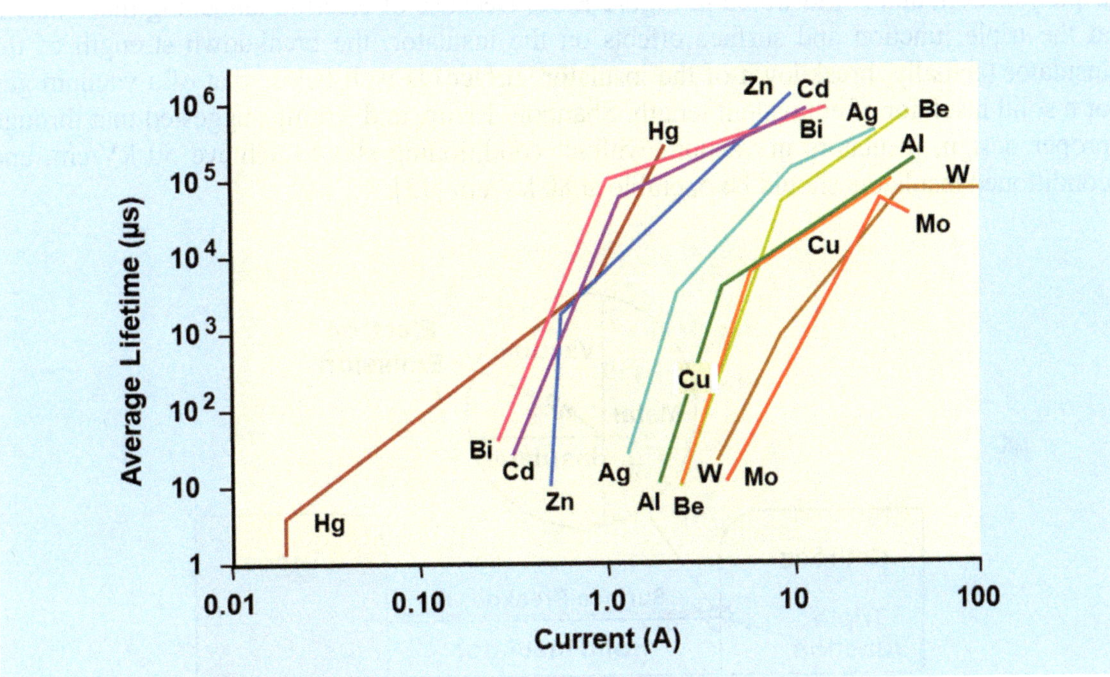

Figure E-31 Average arc lifetime for pure electrode materials. (From: J. D. Cobine and G. A. Farrall, *Jour. App. Phys.*, December 1960.)

Unfortunately, the small amount of material removed by the 20J arc is not the only factor that must be considered. If the surfaces on which the vaporized arc material deposits are critical, such as insulators, leakage currents or insulator breakdown may result. Furthermore, gas molecules on the arced surfaces or nearby surfaces may be driven off by the arc, producing a gas burst. As an example, the release of a monolayer of gas from 1 cm^2 of area into a volume of 1 liter raises the pressure from 10^{-6} to 10^{-4} torr. This gas must be removed (by adsorption, an appendage pump, or a getter) before tube operation can be resumed.

The maximum permissible energy input during an arc in the gun is sometimes specified by the manufacturer. This energy is based largely on the manufacturer's experience. Maximum permissible energy levels from joules to tens of joules are not uncommon. Knowing the maximum permissible energy level, it is then possible for the tube user to design a protection circuit to interrupt the flow of energy before it becomes excessive.

Once a vacuum arc has extinguished, the voltage recovery rate can be extremely rapid. The recovery rate depends on the duration and intensity of the arc; however, a value of 10 kV/μs is not uncommon. Thus, it is often necessary to cause only a very brief interruption of current through a vacuum arc before voltage is reapplied to the electrodes. In fact, this principle has been used in the development of vacuum arc interrupters [12].

E.3 DC BREAKDOWN ON INSULATOR SURFACES

One of the weakest points in a vacuum device from an electrical breakdown point of view is the interface between a metal, an insulator, and a vacuum. This interface is often called the triple junction, and is illustrated in Figure E-32. Because of electron launching mechanisms at the triple junction and surface effects on the insulator, the breakdown strength of the insulator (actually, breakdown of the insulator surface) is well below that of a vacuum gap or a solid insulator of equivalent length. Shannon, Philip, and Trump suggested that through proper design, insulators in vacuum without conditioning should achieve 50 kV/cm, and conditioned insulators should be operable at 80 kV/cm [13].

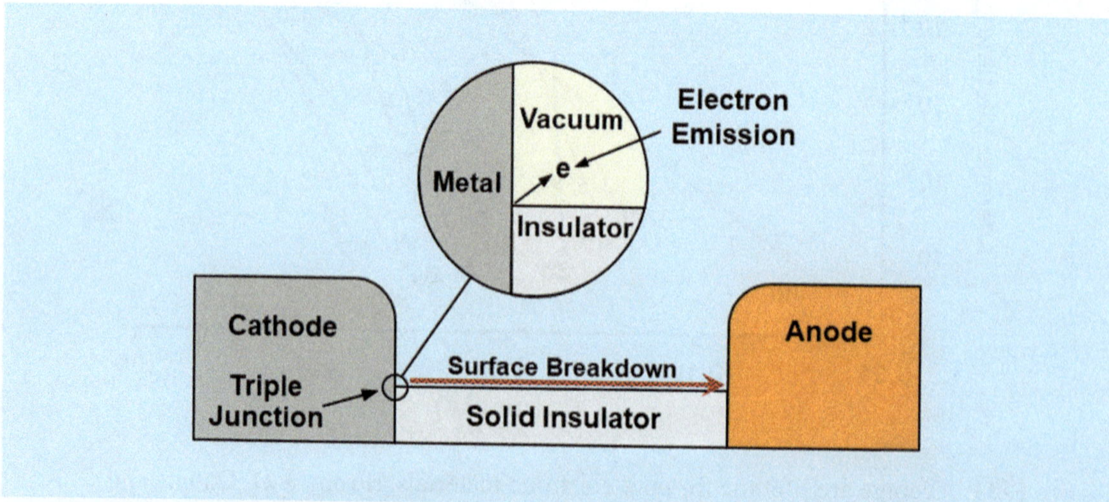

Figure E-32 Triple junction, electron emission, and breakdown across insulator surface.

At the triple junction there are at least two situations that might initiate breakdown. In one, the insulator is brazed to the metal surface. In the process of metallizing the surface of the insulator and brazing it to metal, it is virtually impossible to avoid the formation of a fillet of braze material on the surface of the insulator. This fillet has a razor-sharp edge with a high field enhancement factor. When voltage is applied, the high field can cause micro-protuberances to grow, which further increases in the enhancement factor so field emission and possibly the generation of microparticles occur at relatively low voltages.

The second situation is when the insulator is mechanically held in place but not brazed. Then there are voids between the insulator and the metal along the triple junction. The effect of a void on equipotential surfaces is shown in Figure E-33. The field enhancement factor in these voids may be as high as the relative dielectric constant of the insulator. For aluminum oxide, which is commonly used in these situations, the relative dielectric constant is about 9. This field enhancement factor, coupled with those of microprotuberances, is thought to lead to field emission and possibly microparticle generation.

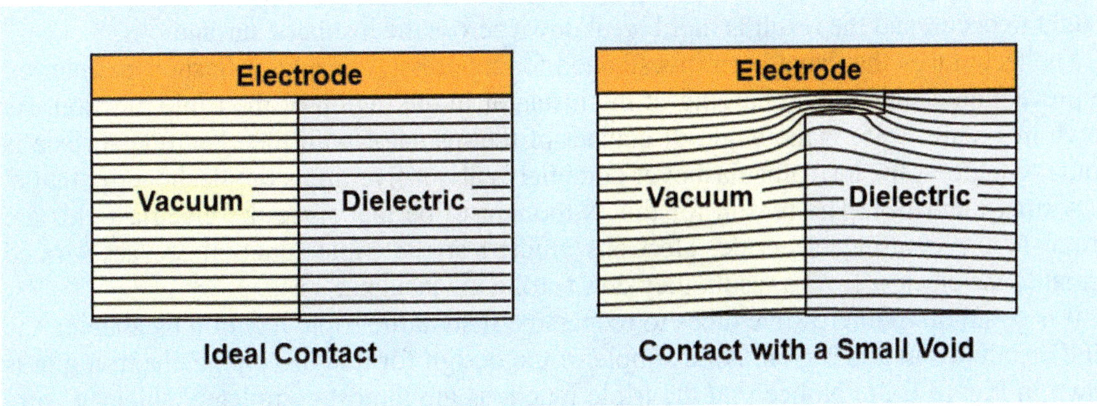

Figure E-33 Effect of a small void on equipotential surfaces.

The second step in insulator flashover is thought to be positive charging of the insulator in the vicinity of the triple junction. This occurs, as indicated in Figure E-34, because electrons that are field emitted from the triple junction strike the insulator surface, causing secondary emission to occur. The secondary emission coefficient for most insulating materials is relatively high (values from 5 to 10 are common, especially if the primary electrons strike the surface at a grazing angle).

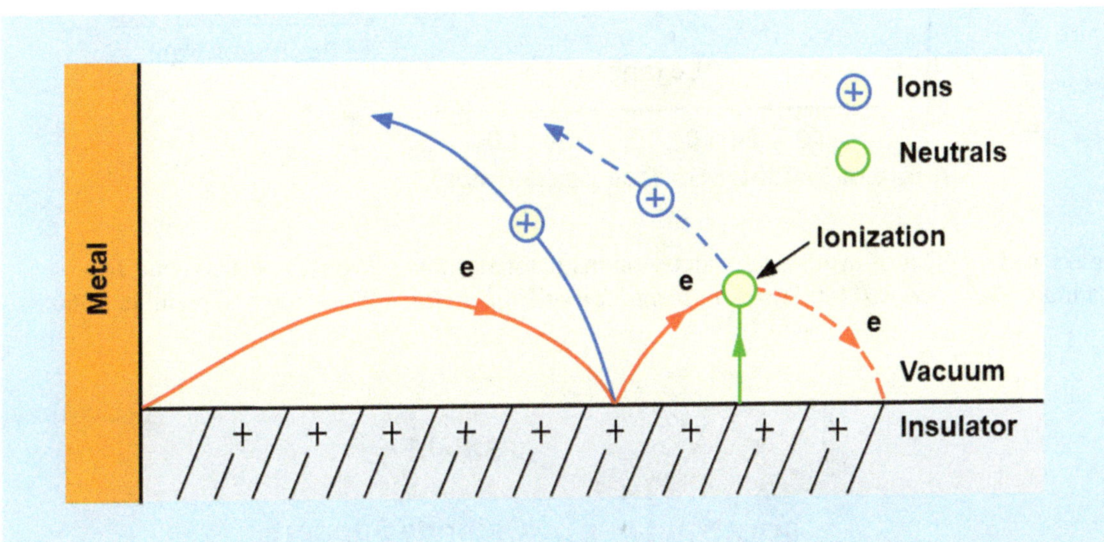

Figure E-34 Events leading to insulator flashover. (Adapted from: *High Voltage Vacuum Insulation: The Physical Basis,* R.V. Latham, Academic Press, 1981.)

The secondary emission process is regenerative in that the secondary electrons are accelerated by the field at the surface to produce additional secondaries. The result is that the insulator surface becomes positively charged in the triple junction region, and this increases the field and the rate of electron emission at the triple junction.

The next phase in the flashover process is thought to be desorption of gas molecules from the insulator surface and subsequent ionization as indicated in Figure E-34. The positive ions are accelerated toward the cathode where they further enhance the field and, as a result, increase electron emission. Other processes possibly involving photon emission are thought to occur, and the result is rapid breakdown across the insulator surface.

The techniques that have been investigated for suppressing insulator flashover center on the prevention of the initial charging of the insulator in the region of the triple junction. As shown in Figure E-35, if the angle of contact of the insulator with the electrode surface is about 45 degrees, the insulator flashover potential is about five times the flashover potential for normal incidence. The reason for this is thought to be that since the electric fields are normal to the emitting electrode, electrons emitted at the triple junction are accelerated normal to the electrode. As a result, they don't strike the insulator surface.

It is common in microwave tubes to reduce the field at the triple junction by shaping the metal electrodes in this region. An example of the design for a high-voltage electron gun is shown in Figure E-36. Notice that the triple junctions are almost completely shielded from equipotential surfaces.

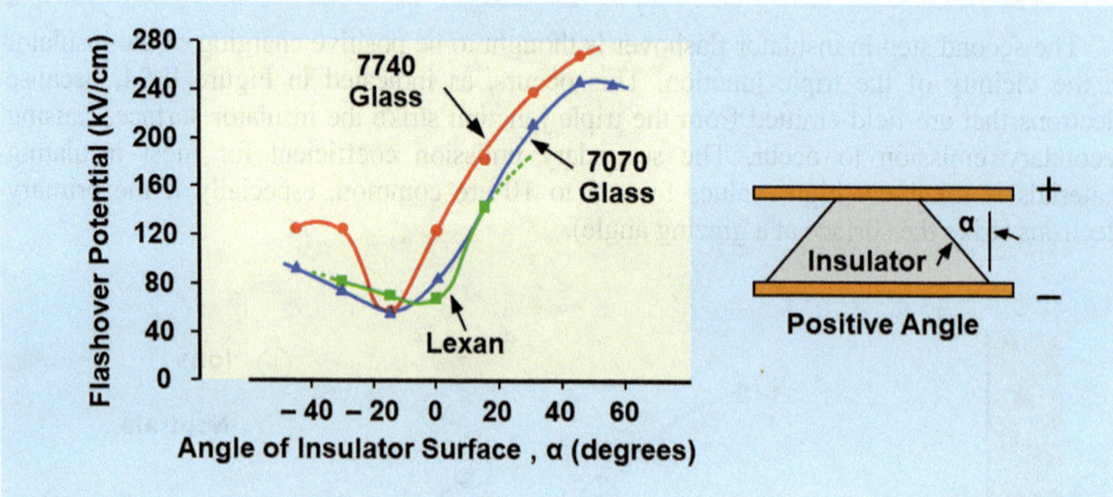

Figure E-35 Effect of insulator geometry on insulator flashover. From A. Watson and J. Shannon, *Proc. Second Int. Symp. on Insulation of High Voltage in Vacuum,* September 1966.

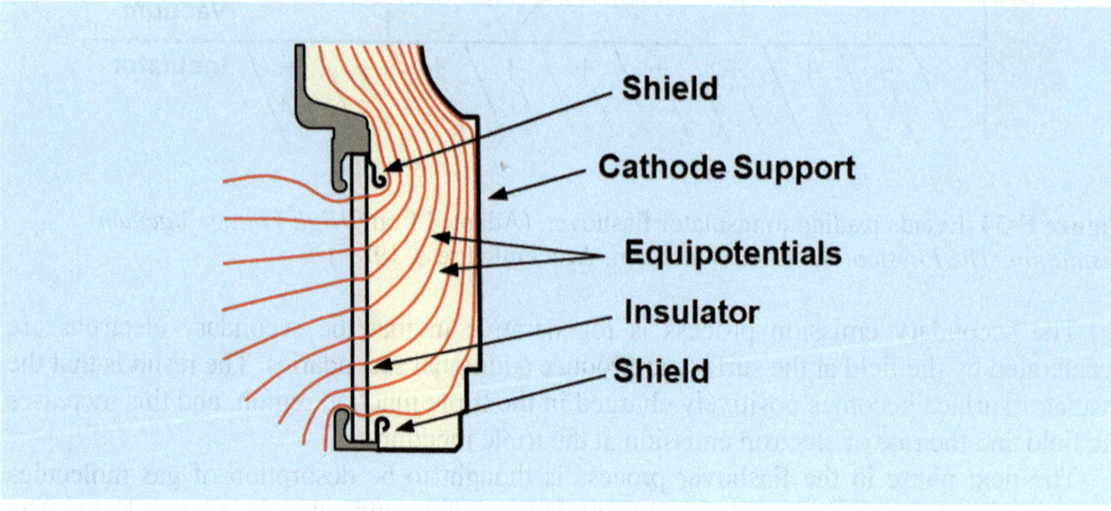

Figure E-36 Electrode configuration for shielding the triple junctions in an electron gun.

E.4 RF BREAKDOWN IN VACUUM

Kilpatrick's criterion, discussed in Section E.2, is often used as a guideline to the maximum RF voltage that can be applied between electrodes. Very often, however, another kind of discharge develops at voltage levels well below the Kilpatrick level. This discharge is called a *multipactor discharge*. In its simplest form, this discharge occurs when electrons move back and forth between two electrodes in synchronism with an RF field. As indicated in Figure E-37, if the secondary emission coefficient of the electrodes is greater than unity (to be more specific, if the product of the secondary emission coefficients of the two electrodes is greater than unity), then the number of electrons involved in the process builds up with time. Many multipacting modes other than that shown in Figure E-37 are possible. These will be described in this section.

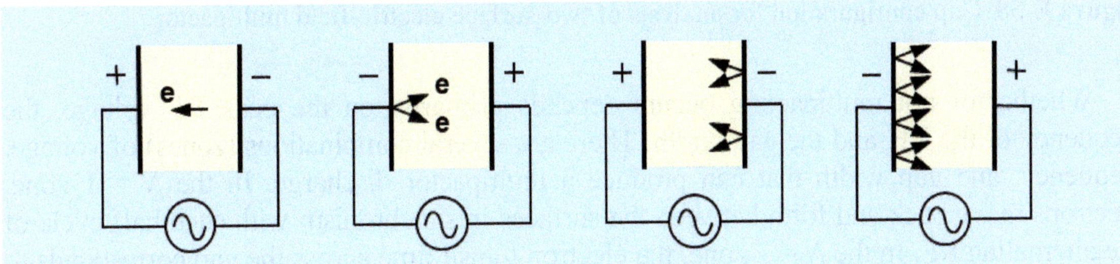

Figure E-37 Illustration of simple two-surface multipactor.

During multipacting, large amounts of energy can be discharged in a small volume with the result that electrodes or RF windows can be damaged or destroyed. Even if excessive damage does not occur, multipacting can load an RF circuit so that the output power is reduced and modulated with a noiselike signal.

The phenomenon of secondary electron resonance, or multipactor, was first recognized and described by Farnsworth over 80 years ago [14]. His primary interest in multipactor was its use in amplifying devices. With the exception of Vaughan's report in 1988 [15], not much has been published about multipactor in microwave tubes. As noted by Preist [16], "There is likewise very little to be heard on the subject of fleas, perhaps for the same reason, that undue familiarity with the subject may be taken to mean that one has problems." Still, as pointed out by Vaughan, multipactor is present in one form or another in nearly all high-power microwave tubes; often it is observable but harmless, occasionally beneficial, and sometimes disastrous. Thus, it is very important for microwave tube engineers to understand multipactor and the approaches and techniques that can be taken to control or eliminate it. In this section, multipacting modes are described along with some specific corrective actions that have been taken to eliminate multipactor.

E.4.1 Two-Surface Multipactor with No Magnetic Field

The theory of two-surface electric-field multipactor has been presented in various reports. The results given here were reported by Vaughan [15]. The configuration that he considered consists of two parallel electrodes driven by an RF voltage as shown in Figure E-38.

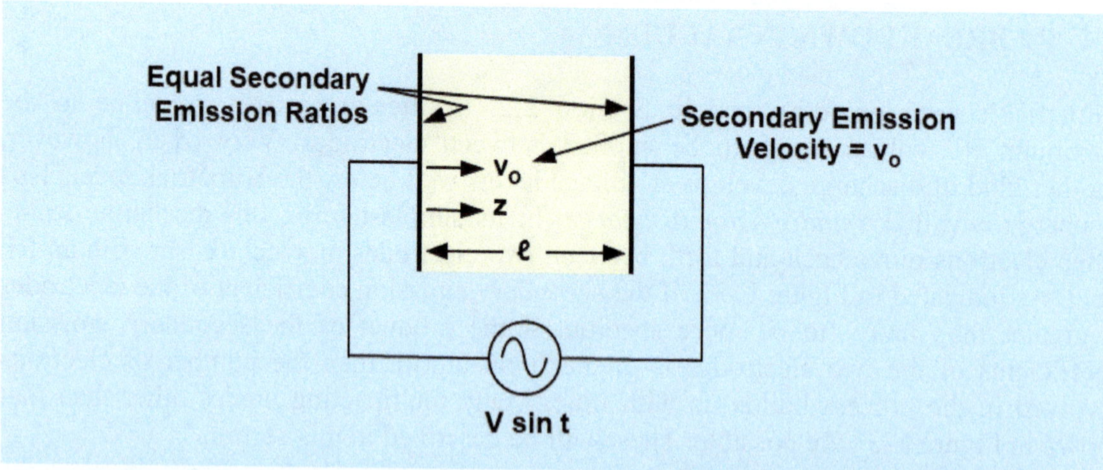

Figure E-38 Gap configuration for analysis of two-surface electric-field multipactor.

Whether or not multipacting occurs depends primarily on the peak RF voltage, the frequency of the RF, and the gap width. There are several combinations (zones) of voltage, frequency, and gap width that can produce a multipactor discharge. In the $N = 1$ zone, electrons move back and forth between the surfaces in synchronism with each half cycle of the alternating RF. In the $N = 3$ zone, the electron transit time across the gap corresponds to 3/2 cycle of the applied RF voltage. This means that electrons start across the gap, stop and reverse direction of travel as the RF voltage reverses polarity, and then stop and reverse direction of travel again when the voltage reverses polarity before finally landing on the opposing electrode. In the $N = 5$ zone, the electron transit time across the gap corresponds to 5/2 cycle of the applied RF voltage.

The conditions that produce three of the multipactor zones are shown in Figure E-39. For each zone, approximate contours of constant impact voltage are indicated by dashed lines. The lowest impact voltage is 100V. This is because, as Figure E-40 shows, the secondary emission coefficient for typical materials drops below unity at impact energy levels below about 100 eV and so multipactor does not occur. At high impact voltages, the energy level at which the secondary emission ratio drops below unity is highly dependent on the angle of incidence of the impacting electrons. As shown in Figure E-40, for normal incidence, the value is about 2,500 eV. The curve for impact at about 55° shows that the secondary emission coefficient drops below unity well above 4,000 eV.

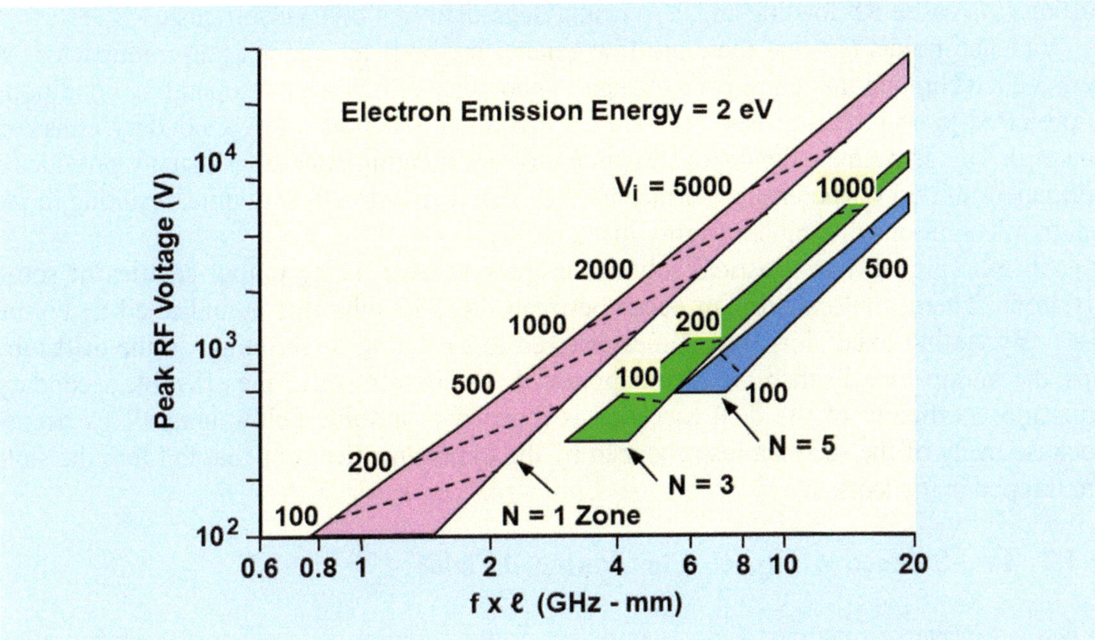

Figure E-39 RF voltages at which two-surface electric-field multipactor discharges may occur. (From: J.R.M. Vaughan, *IEEE Trans. Electron Devices*, July 1988. © 1988 IEEE.)

For some forms of multipactor to be discussed later, the impact angle can approach 90° from normal. The energy of these electrons can be almost unlimited and the secondary emission coefficient will still be above unity.

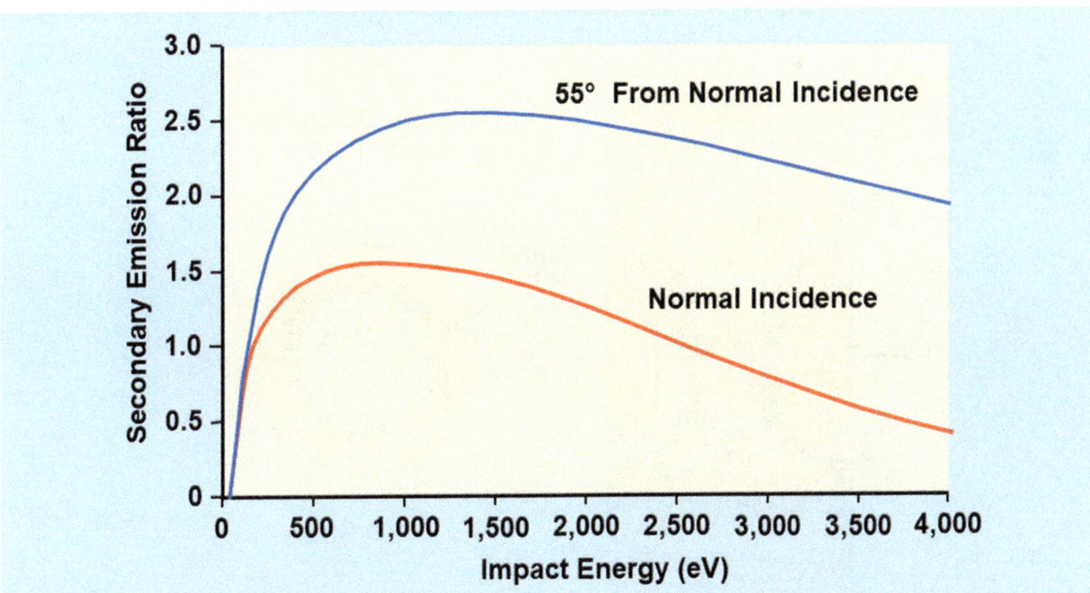

Figure E-40 Typical variations of secondary emission ratio with impact energy for metal surfaces. (From: J. R. M. Vaughan, *IEEE Trans. Electron Devices*, July 1988. © 1988 IEEE.)

The power level of the simple multipactor discussed in the previous paragraphs is usually too low to cause physical damage to the electrodes. However, the power level is sufficient to cause RF loading and, as a result, degradation of tube performance.

Vaughan points out that there are few general methods for suppressing multipactor. In general, a technique that employs a change in geometry or a change in operating conditions is preferred to one that is based on a surface treatment (such as a low secondary emission coating). This is because the change in geometry or operating conditions usually provides a permanent solution, whereas the surface treatment may deteriorate with time, resulting in the return of conditions favorable to a multipactor.

An example of a geometrical solution is the one used in the output cavities of some klystrons. There, multipactor can occur between the drift tube tips as indicated in Figure E-41. By cutting axial slots (sometimes referred to as fluting or serrating) in the drift tube tips, the multipactor discharge can be suppressed. The reason is that the effective secondary emission coefficient of the drift tube tips is reduced to a value below unity. This occurs because many of the secondaries produced by the impacting electrons that fall into the slots are trapped in the slots.

E.4.2 Two-Surface Multipactor in Combined Fields

When a transverse magnetic field is imposed on the parallel-plate electrode configuration considered in Section E.4.1, the electrons are deflected from straight-line paths to the curved trajectories shown in Figure E-42. The most significant aspect of the change in trajectory shape is that the electrons can impact on the electrodes at an angle that can approach 90° from normal incidence. Thus, referring back to Figure E-40, it is possible to see that the energies of the electrons can be tens of thousands of electron volts and the secondary emission coefficient of the surface will still remain above unity. Thus, the crossed-field multipactor can be a very high-voltage and high-power discharge.

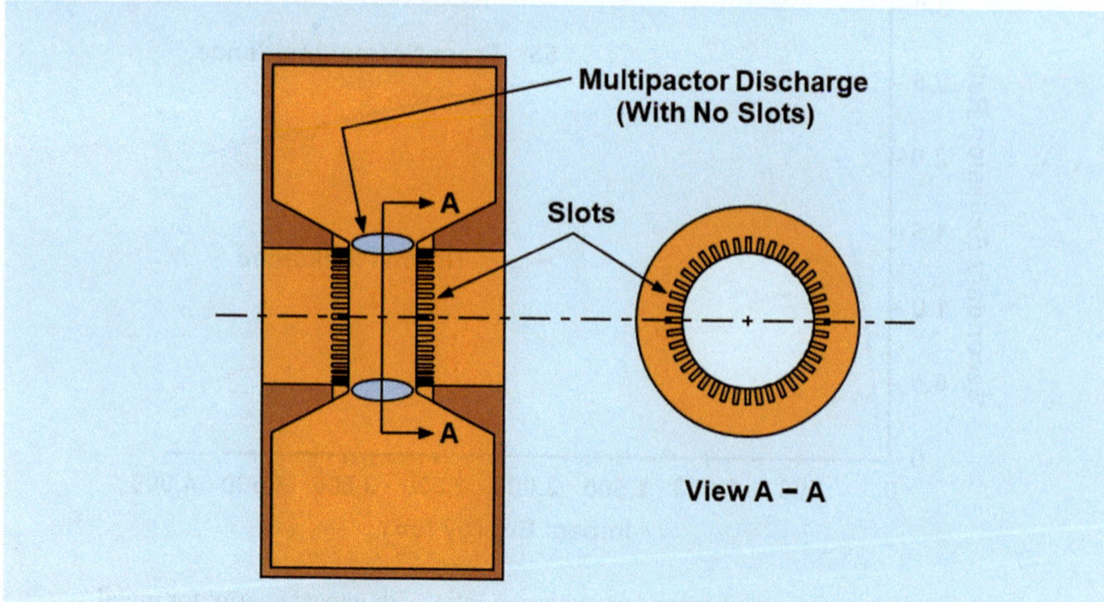

Figure E-41 Use of slots in drift tube tips to suppress multipactor.

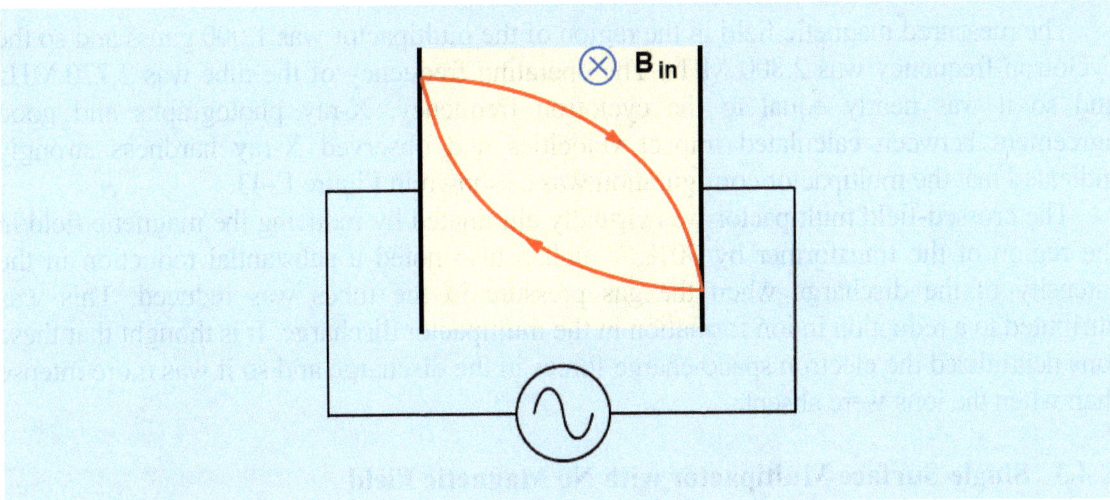

Figure E-42 Illustration of the effect of a transverse magnetic field on electron trajectories.

Maximum secondary emission will occur when electron impact is tangential to the electrode surfaces and this is when we would expect the crossed-field discharge to be most intense. This will occur when an electron reaches the top of a cycloidal orbit in one-half cycle of the applied frequency and this, in turn, occurs when the applied frequency is equal to the cyclotron frequency; that is, when

$$\omega = \omega_c = \eta B \tag{E-13}$$

Vaughan [17] reported the case of crossed-field multipactor in the impedance transformer of a magnetron, the configuration of which is shown in Figure E-43. It was thought that electrons generated by the multipactor at point A and missing the transformer surface at point B continued on to strike the window at point C, causing window failure.

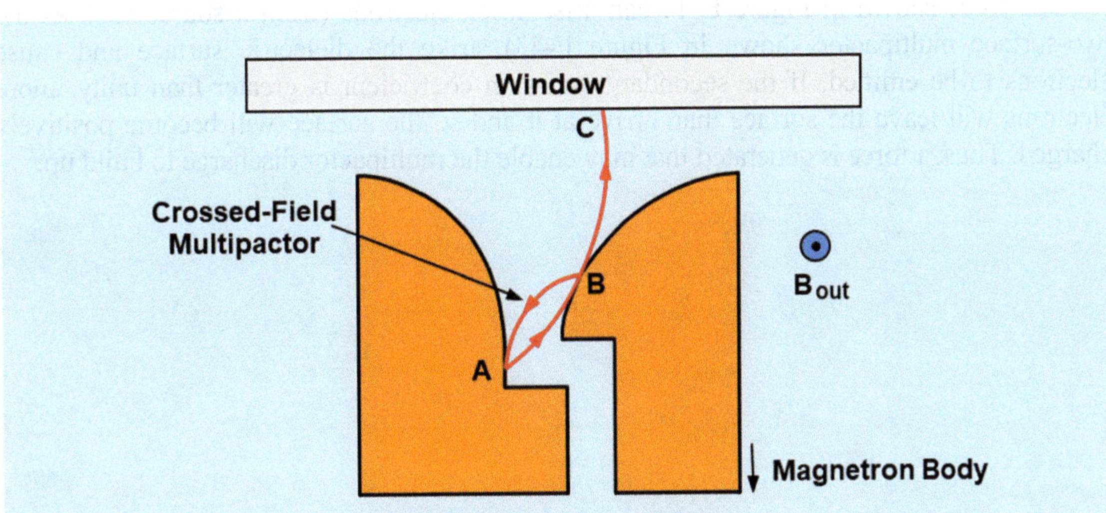

Figure E-43 Possible crossed-field multipactor in magnetron impedance transformer. (From: J. R. M. Vaughan, *IEEE Trans. Electron Devices*, July 1988. © 1988 IEEE.)

The measured magnetic field in the region of the multipactor was 1,000 gauss and so the cyclotron frequency was 2,800 MHz. The operating frequency of the tube was 2,770 MHz and so it was nearly equal to the cyclotron frequency. X-ray photographs and good agreement between calculated impact velocities and observed X-ray hardness strongly indicated that the multipactor configuration was as shown in Figure E-43.

The crossed-field multipactor was virtually eliminated by reducing the magnetic field in the region of the transformer by 90%. Vaughan also noted a substantial reduction in the intensity of the discharge when the gas pressure in the tubes was reduced. This was attributed to a reduction in ion formation in the multipactor discharge. It is thought that these ions neutralized the electron space-charge forces in the discharge and so it was more intense than when the ions were absent.

E.4.3 Single-Surface Multipactor with No Magnetic Field

In the absence of a magnetic field, a multipactor discharge can occur on a single dielectric surface as indicated in Figure E-44 [16]. A dielectric is required because there must be a component of the electric field parallel to the surface. Electrons emitted from the surface (indicated by the vectors labeled with v_o in Figure E-44) have a component of velocity perpendicular to the surface. The electric field accelerates the electrons in the direction parallel to the surface and, in the absence of any force toward the surface, the electrons drift away, as indicated by trajectories 1 and 2.

If there is a force toward the dielectric surface, then electrons follow trajectories like 3 and 4 in Figure E-44 and strike the surface. For electrons with sufficient energy and a surface with a secondary emission coefficient greater than unity, one or more secondaries will be emitted for each primary. Each of these can follow a trajectory like that shown in Figure E-45 and produce additional secondaries. As a result, a multipactor discharge can build up on the dielectric surface and cause heating and failure of the dielectric.

The force shown in Figure E-44 can arise when electrons (from a source such as the two-surface multipactor shown in Figure E-43) strike the dielectric surface and cause electrons to be emitted. If the secondary emission coefficient is greater than unity, more electrons will leave the surface than arrive at it and so the surface will become positively charged. Thus, a force is generated that may enable the multipactor discharge to build up.

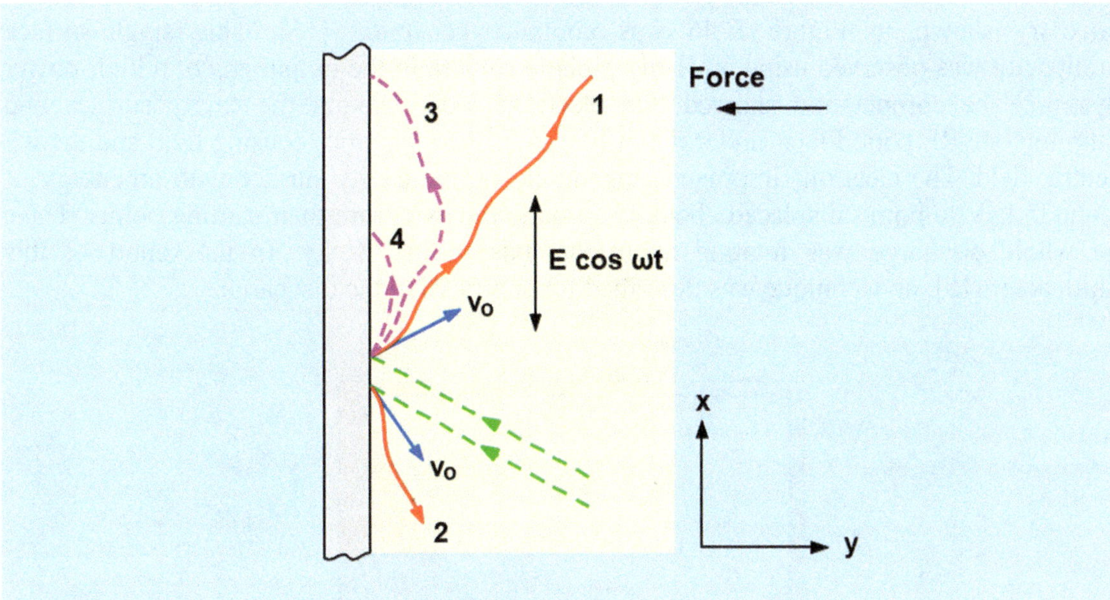

Figure E-44 Electron motions near a dielectric surface with no magnetic field. (From: D. H. Preist, *Proc. 4th International Congress on Microwave Tubes*, 1963.)

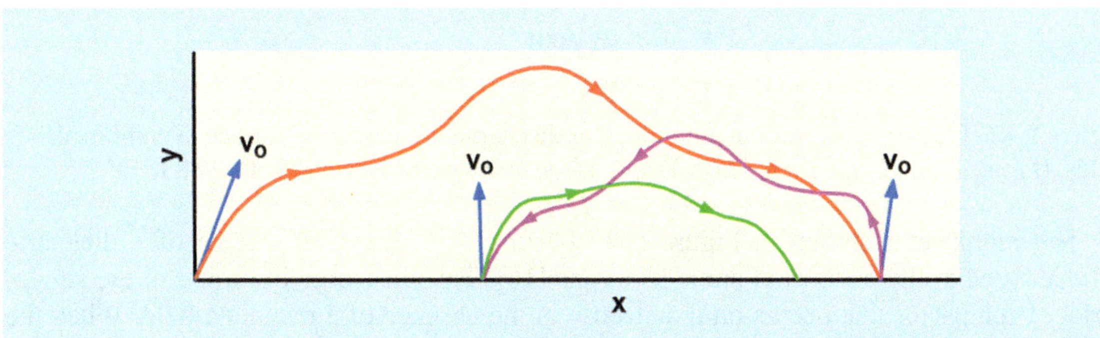

Figure E-45 Electron motions during single-surface multipactor with no magnetic field. (From: D. H. Preist, *Proc. 4th International Congress on Microwave Tubes*, 1963.)

It should be noted that, unlike the two-surface multipactor discharges, there is no requirement with the single-surface discharge for a definite transit angle. Thus, the single-surface multipactor can exist over a large range of electric field strengths as long as the field is sufficient to accelerate electrons to impact velocities high enough so that the secondary emission coefficient is greater than unity.

The secondary emission coefficients of all of the dielectric materials used in microwave tubes is greater than unity. To suppress multipactor, it has been necessary to coat the dielectrics with low secondary yield films. Titanium-based films have been successfully used for this purpose [16, 18].

E.4.4 Single-Surface Multipactor in Combined Fields

An example of single-surface multipactor in combined fields on a metallic surface is shown in Figure E-46 and an example on a dielectric surface is given in Figure E-47. The electron trajectory shown in Figure E-46 was computer-generated [15]. This single-surface multipactor was observed using an X-ray pinhole camera in the output gap of a high-power klystron. The computation showed that electrons were leaving the cavity surface and returning one RF period later under the influence of the magnetic focusing field and the RF electric field. The electrons impinged tangentially on the cavity surface with an energy of about 35 keV to points displaced about 35° around the axis from their starting points. Thus, the whole discharge was rotating about the axis of the cavity. In the report of this multipactor [15], no technique was described for suppressing the discharge.

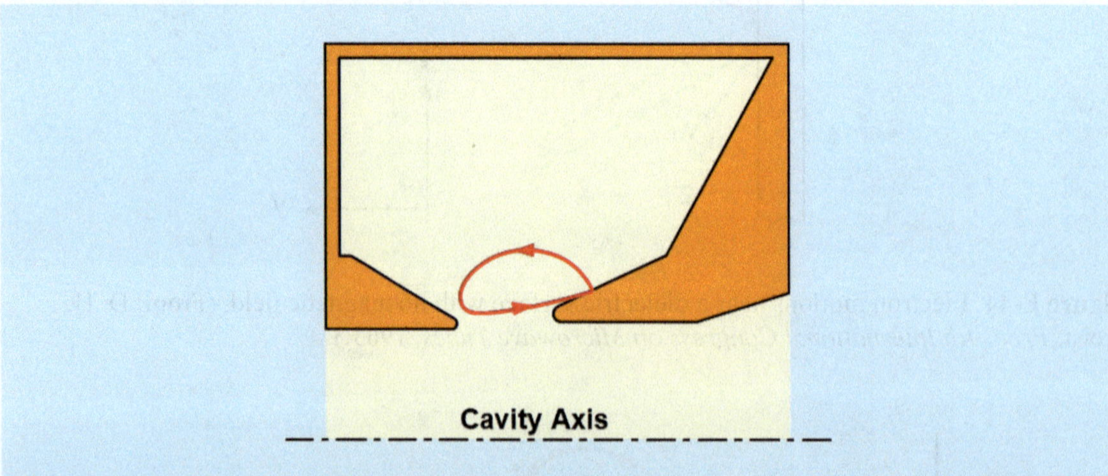

Figure E-46 Electron trajectory in a multipactor discharge on a metallic surface in combined fields. (From: J. R. M. Vaughan, *IEEE Trans. Electron Devices*, July 1988. © 1988 IEEE.)

The multipactor shown in Figure E-47 occurred on the inner surface of a dielectric cylinder used in the cavities of some klystrons [16]. This discharge is similar to the single-surface multipactor that occurs on a dielectric in the absence of a magnetic field. When the magnetic field is present, the force driving the electrons toward the surface of the dielectric is provided by the magnetic field in conjunction with the component of the initial electron velocity normal to the surface. The most intense discharge occurs when the cyclotron frequency is equal to the radio frequency. Magnetic fields stronger than the cyclotron field tend to prevent this form of multipactor.

E.5 RF BREAKDOWN OF INSULATORS

The multipactor discharges on dielectrics described in the previous section are discharges that occur on a surface. They may produce heating of the surface and that, in turn, may result in cracking and insulation failure. Vaughan [19] described another cause of window cracking that results from arcing to a conductive deposit on the vacuum side of the window. This problem was eliminated by preventing the formation of the conductive deposit.

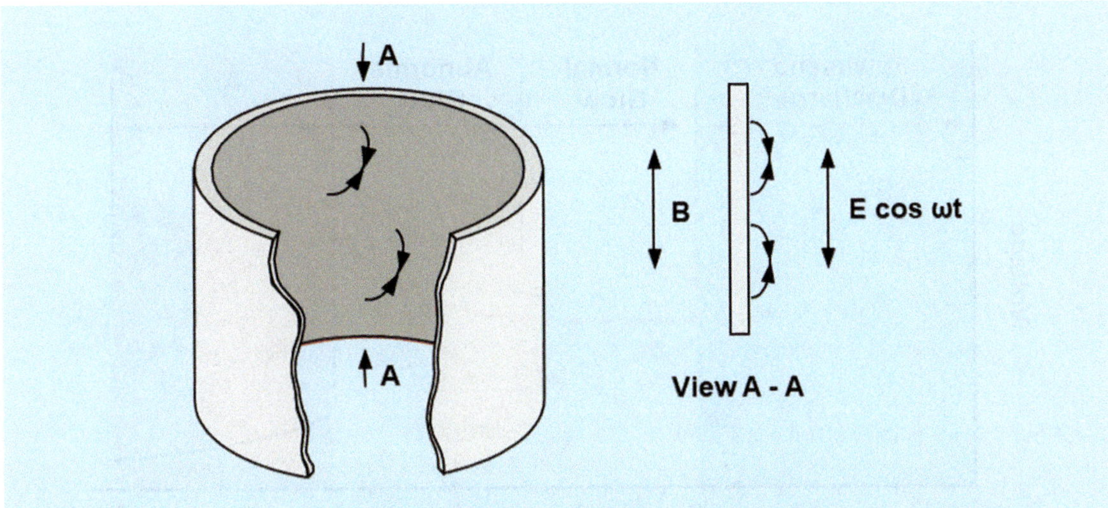

Figure E-47 Electron motions in a multipactor discharge on a dielectric surface in aligned fields. (From: D. H. Preist, *Proc. 4th International Congress on Microwave Tubes*, 1963.)

Vaughan [19] has also described window punctures that resulted from the combined actions of an RF electric field and dc window charging on electrons deposited on the surface of a window. The window configuration was that shown in Figure E-43, and the source of electrons was the crossed-field multipactor that occurred in the magnetron impedance transformer. The windows were generally observed to be charged positively between 5 and 12 kV on the *outside*. As a result, there was an electrostatic field of about 50 kV/cm through the window. Vaughan postulated that electrons striking the window formed a small cavity in the window surface and that the field at the tip of this cavity enhanced the electrostatic field in the window. The resulting dc force attracting electrons into the window, coupled with RF forces (produced by the RF electric field component parallel to the window surface) causing the electrons to oscillate sideways, produced a disintegrating action that resulted in electrons burrowing through the window. The result was the formation of a pinhole with a diameter on the order of 10 μm. This failure mode was eliminated by suppressing the crossed-field multipactor, which was the source of the electrons that initiated the failure process.

E.6 DC BREAKDOWN IN GAS

Electrical discharges in gases are extremely complex. Several different types of discharges can exist, depending on the current and voltage applied to the gas. These various discharge types are indicated in Figure E-48. Current and relative voltages shown in Figure E-48 are representative values only and depend on factors such as gas type and the area of the electrodes used during testing [20–22].

When breakdown occurs, the discharge rapidly increases in intensity. Depending on circuit conditions (voltage and impedance), either a glow discharge or an arc may develop.

The breakdown voltage depends on the ease with which the gas is ionized and on the effectiveness of the ions in producing electrons when they strike the negative electrode. As a result, the breakdown voltage is a function of gas type and pressure, and electrode spacing and material.

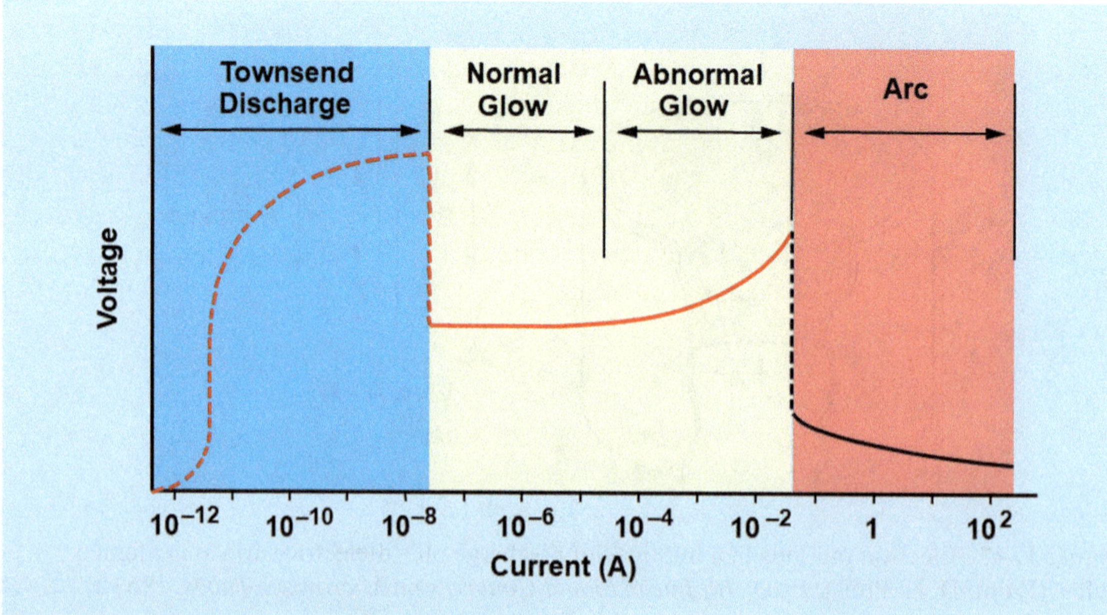

Figure E-48 Typical volt-ampere characteristics of a gas discharge. (Adapted from: *Principles and Applications of Electron Devices,* P. D. Ankrum.)

The breakdown voltage is normally plotted in the manner shown in Figure E-49. Breakdown voltage plots of this type are referred to as Paschen characteristics because of Paschen's law, which states that, for parallel plane electrodes in a particular gas at a particular temperature, the breakdown voltage is a function of the product of the gas pressure and the electrode spacing.

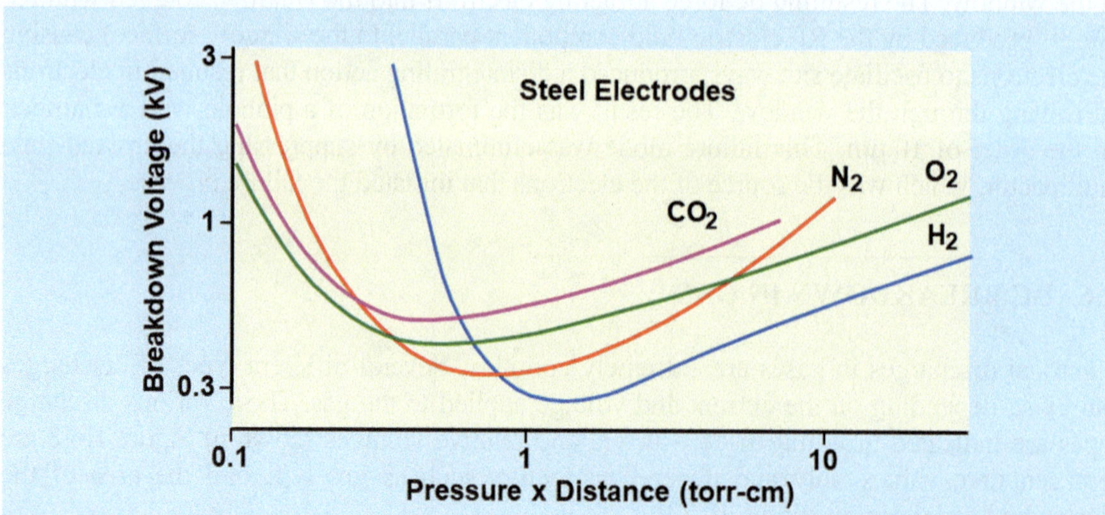

Figure E-49 Paschen characteristics for some molecular gases. (From: M. J. Schonhuber, *IEEE Trans. P.A.S.*, February 1969. © 1969 IEEE.)

The shapes of the Paschen characteristics may be explained as follows: to the left of the minimum, the breakdown voltage increases as pressure and density decrease (with fixed electrode spacing). This is because the gas molecules are farther apart, and more electrons travel through the gas without producing ionization. In the limit, under high-vacuum conditions, there is essentially no ionization. Breakdown occurs because of electrode phenomena. DC breakdown in a vacuum is discussed in Section E.2.

To the right of the Paschen minimum, the breakdown voltage increases as pressure and gas density increase. In this case, gas molecules are closer together, and so a large electric field is required to give the electrons enough energy in the short distances between collisions to produce ionization. Thus, the breakdown voltage is high. At atmospheric pressure and above, the breakdown voltage for a 1-cm gap can be several tens and even hundreds of thousands of volts.

At intermediate pressures and electrode spacings, optimum conditions for ionization occur, and so there is a minimum in the Paschen curve. The minimum breakdown voltage for most gases is a few hundred volts and occurs at pressure × distance (torr-cm) near unity.

Breakdown voltages for several gases are listed in Table E-1. The values given were obtained with a 1-cm gap at the Paschen minimum (the critical pressure-spacing product) and at atmospheric pressure. It should be noted that the freons and sulfur hexafluoride have high-breakdown voltages. These gases are referred to as electronegative gases. Their molecules require one or two electrons to fill the outer shell. As a result, they can easily capture free electrons, and this results in the high-breakdown voltages for these gases.

Electronegative gases are expensive and tend to be somewhat hazardous, and so they are often mixed with more common gases such as air or dry nitrogen. As shown in Figure E-50, even a small amount of sulfur hexafluoride (SF_6) in nitrogen significantly increases breakdown voltage.

The power supply or modulator for a microwave tube is normally capable of delivering a short-circuit current of at least several amperes and sometimes several kiloamperes. As a result, when breakdown occurs, a glow discharge rapidly transitions to an arc. Depending on the arc voltage and arc duration, damage can be significant.

Table 25.1
Breakdown Voltages for Several Commonly Used Gases

Gas	Volts (dc) (Paschen Min)	Kilovolts/cm (dc) (Atm. Pressure)
Air	315	33
Argon	280	4.8
Carbon Dioxide	430	28
Freon 14	480	32
Freon 115	430	90
Freon 116	500	—
Helium	189	1.63
Hydrogen	292	17
Nitrogen	265	32
Oxygen	440	—
Sulfur Hexafluoride	520	89

Source: W. G. Dunbar, "High Voltage Design Guide," Vol. IV, AFWAL-TR-82-2057.

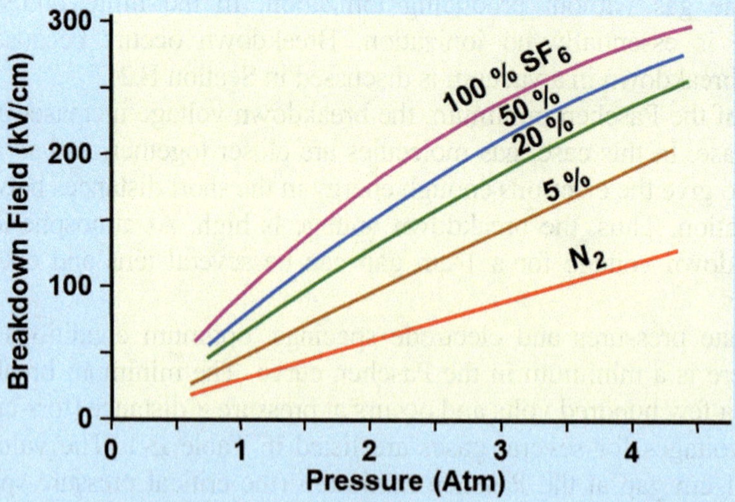

Figure E-50 Effect on breakdown voltage of mixing SF6 with dry nitrogen. (From: N. H. Malik and A. H. Qureshi, *IEEE Trans. Elec. Ins.*, October 1980. © 1980 IEEE.)

The arc voltage depends on the voltage gradient of the arc column, which is largely dependent on gas type, arc current, and pressure. The arc-column voltage gradients for several common gases at atmospheric pressure are given in Figure E-51. The combination of the column gradient of a few tens of volts/cm with the electrode drops results in arc voltages in roughly the 50- to 100V range for arcs of average length (a few centimeters).

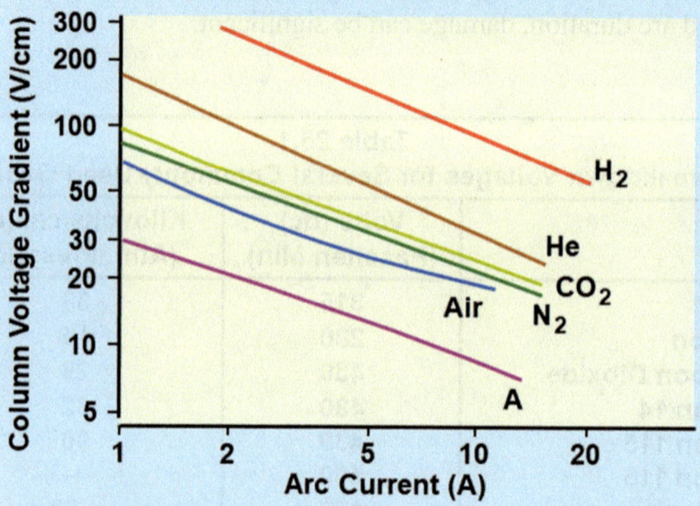

Figure E-51 Arc-column voltage gradient at atmospheric pressure. (Adapted from: *Gaseous Conductors,* J. D. Cobine, Dover Publications, 1958.)

At the end of Section E.3, estimates are made of the energy and charge levels at which excessive damage may occur during a dc arc inside the vacuum envelope of a tube. Unfortunately, it is not so easy to make comparable estimates for dc arcs outside the vacuum envelope. If the arc occurs between bare electrodes, it is possible that a substantial amount of energy (hundreds of joules or more) can be dissipated without causing excessive damage. If, however, the arc occurs in the vicinity of an insulating material, it is possible that the material may be damaged with only a few joules of energy dissipation.

E.7 RF BREAKDOWN IN GAS

In the previous section, it was shown that the dc breakdown field for air at atmospheric pressure is about 30 kV/cm. At low frequencies (up to tens of Hz), the breakdown field remains at 30 kV/cm. As frequency is increased to the audio range and then into the RF range, the breakdown field drops and becomes highly dependent on frequency and electrode spacing, as well as pressure [21].

The unified plot given in Figure E-52 was developed by Woo [23] for 50-Ω coaxial transmission lines and gives breakdown power as functions of frequency, pressure, and spacing under laboratory conditions. This plot appears complex at first glance, but it is relatively straightforward once a person becomes familiar with it.

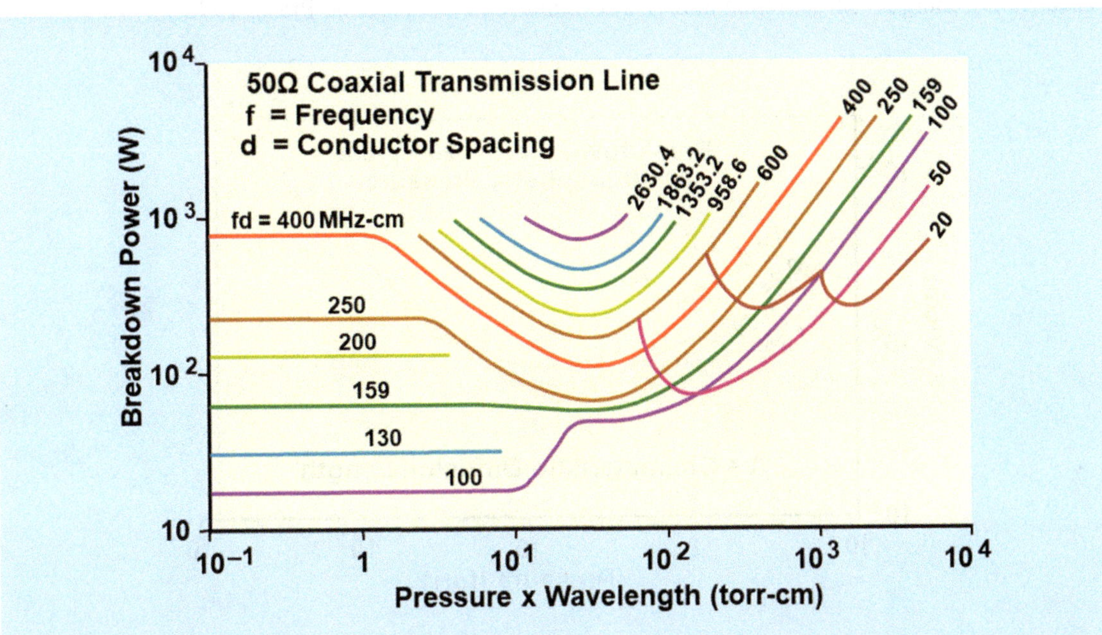

Figure E-52 Unified plot for RF breakdown of air in 50-Ω coaxial transmission lines. (From: R. Woo, *Proc. IEEE*, February 1969. © 1969 IEEE.)

The actual value of the electric field at which breakdown occurs has been determined for selected frequencies in the microwave range. Two such selected frequencies are shown in Figures E-53 and E-54. In both of these cases (992 MHz and 9.4 GHz), extrapolations to atmospheric pressure indicate that the breakdown field is about 20 kV/cm, or comparable to the dc breakdown field level of about 30 kV/cm. The characteristic diffusion length in Figures E-53 and E-54 is a measure of the size and shape of the cavity in which the measurements were made.

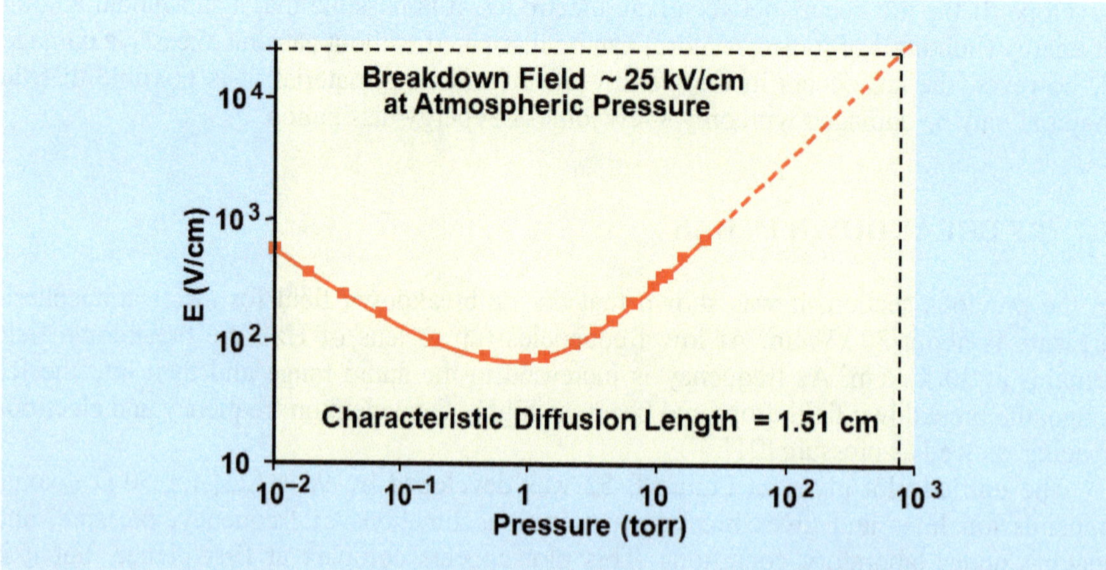

Figure E-53 Breakdown in air at 992 MHz. (Adapted from: *Gaseous Electronics*, Volume I, "Electric Discharge," M. H. Hirsh and H. J. Oskam (eds.), Academic Press, 1978.)

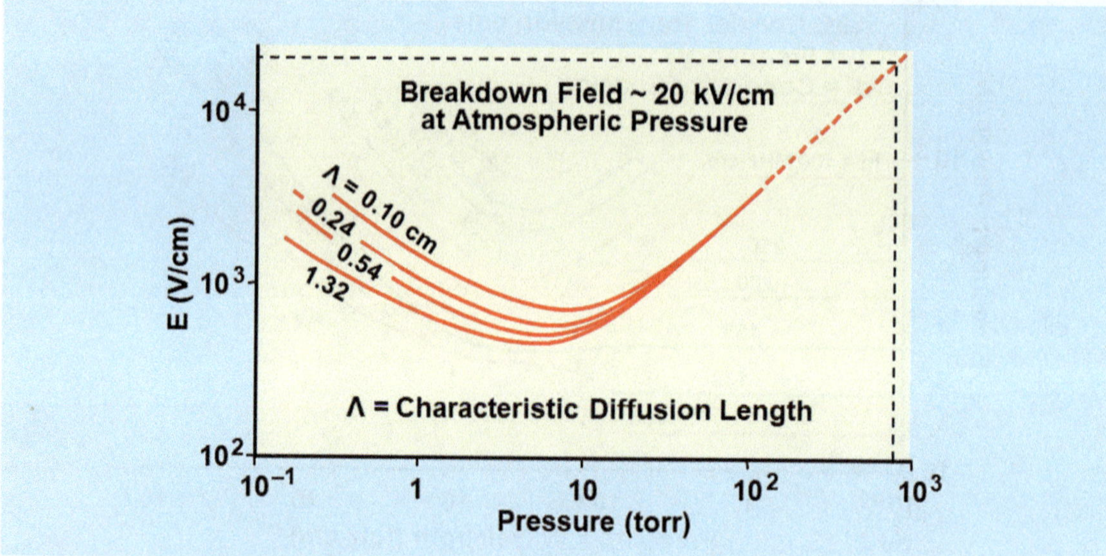

Figure E-54 Breakdown in air at 9.4 GHz. (Adapted from: *Gaseous Electronics*, Volume I, "Electric Discharge," M. H. Hirsh and H. J. Oskam (eds.), Academic Press, 1978.)

The information in Figures E-52, E-53, and E-54 would lead one to believe that extremely high-power levels can be transmitted through waveguides and coaxial lines of relatively modest size. For example, an extrapolation of Woo's unified plots to atmospheric pressure in a 7/8-inch coaxial line and a frequency of 1 GHz indicates that the breakdown power is about 200 kW. Still, as shown in Figure E-55, the power ratings for a standard 7/8-inch coaxial line are 43 kW peak and only about 1,250W average. Some of the degradation results from field enhancement associated with supporting elements for the center conductor. More important, for the average power case, heating, especially of the center conductor and gas, is a major factor limiting the power rating.

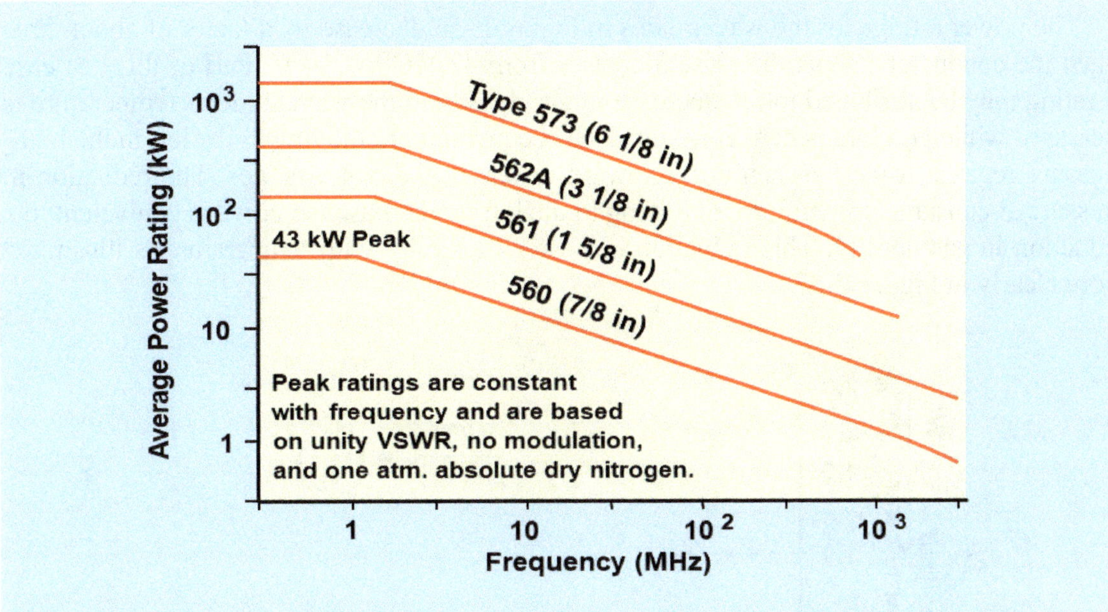

Figure E-55 Average power ratings for standard rigid lines. (Adapted from: *Microwave Engineers Handbook*, T. Saad (ed.), Artech House, 1971.)

A situation similar to that in coaxial lines exists in waveguides as shown in Figure E-56. The ratings in the left-hand plot are for a conductor temperature rise of 42.4°C, and those for the right-hand plot are for a temperature rise of 110°C. Based on a breakdown field of 20 kV/cm, it would be expected that WR 90, which has a small dimension of about 1 cm, should be able to transmit megawatts of power. Instead, the rating is in the kilowatt range.

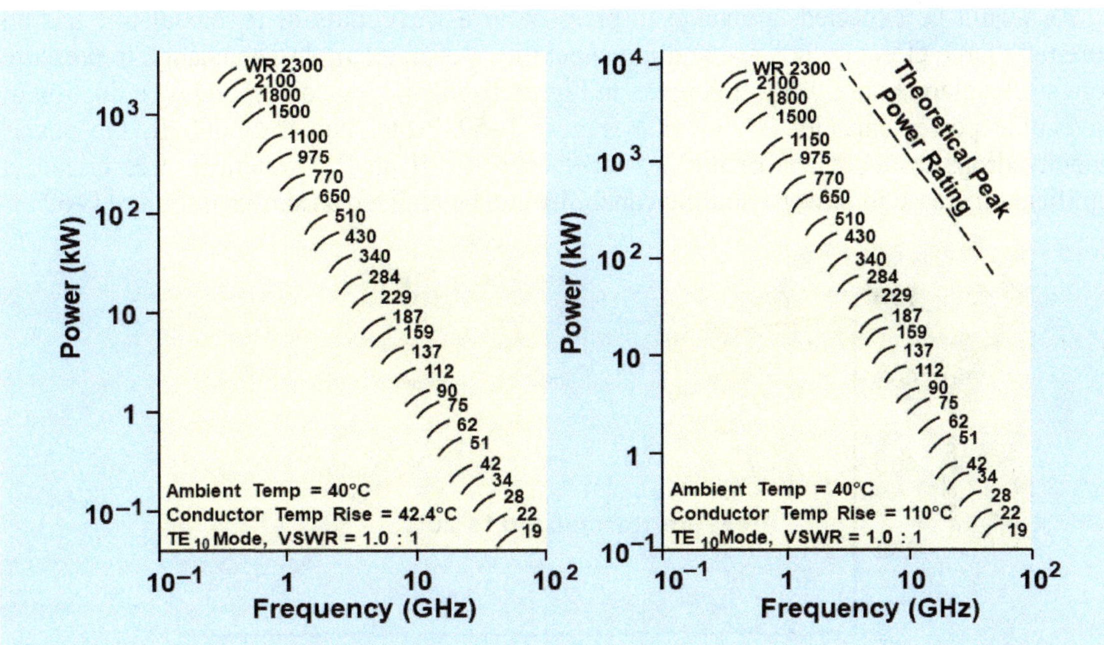

Figure E-56 Average power rating for copper rectangular waveguide. (Adapted from: *Microwave Engineers Handbook*, T. Saad (ed.), Artech House, 1971.)

The power ratings for the waveguides in Figure E-56 decrease by a factor of about three when the conductor temperature rise increases from 42.4°C to 110°C. Part of this decrease in rating may be attributed to the decrease in gas density in the waveguide as temperature is increased while pressure is held constant. This is equivalent to moving to the left in the high-pressure region on the Paschen curve toward lower breakdown voltages. The reduction in pressure at constant temperature in moving to the left on the Paschen curve is equivalent to a reduction in gas density. This reduction in power capability with temperature is illustrated more clearly in Figure E-57.

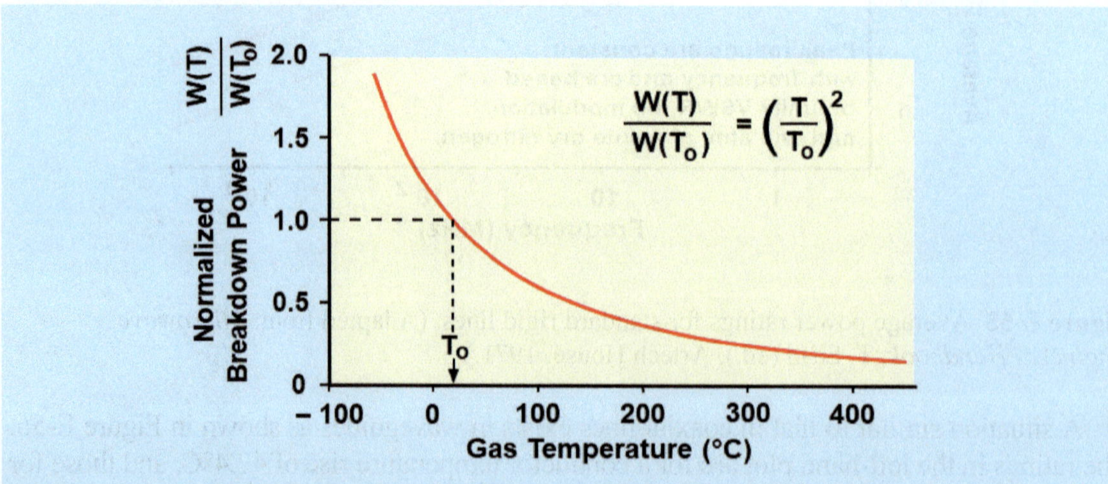

Figure E-57 Waveguide breakdown power versus temperature at fixed pressure. (From: R. H. Stone and R. M. White, *Final Report TIS R61 ELM 205,* General Electric Co., January 1961.)

As would be expected, a change in pressure in a waveguide or a coaxial line has an extremely large effect on power-handling capability. As an example, the change in pressure from sea level to about 50,000 ft is given in Figure E-58, and the corresponding reduction in waveguide power capacity is shown in Figure E-59. Note that at 50,000 ft, the power capacity drops to about 3% of the sea level capacity. If an RF system is pressurized, a significant increase in power-handling capability can be realized as shown in Figure E-60.

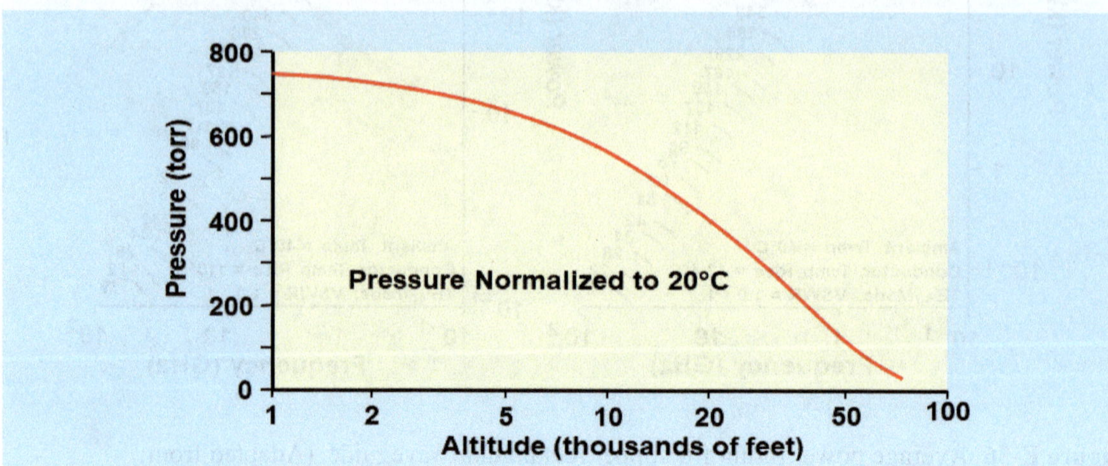

Figure E-58 Pressure as a function of altitude.

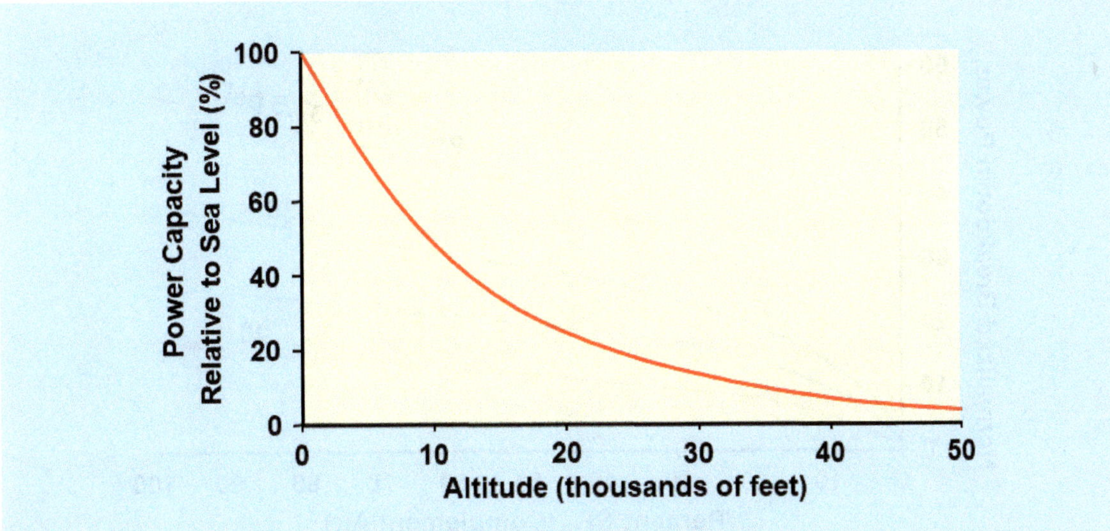

Figure E-59 Waveguide power capacity versus altitude. (From: *Microwave Tube Manual*, Varian, Air Force Publication No. T.O.00-25-251, October 1979.)

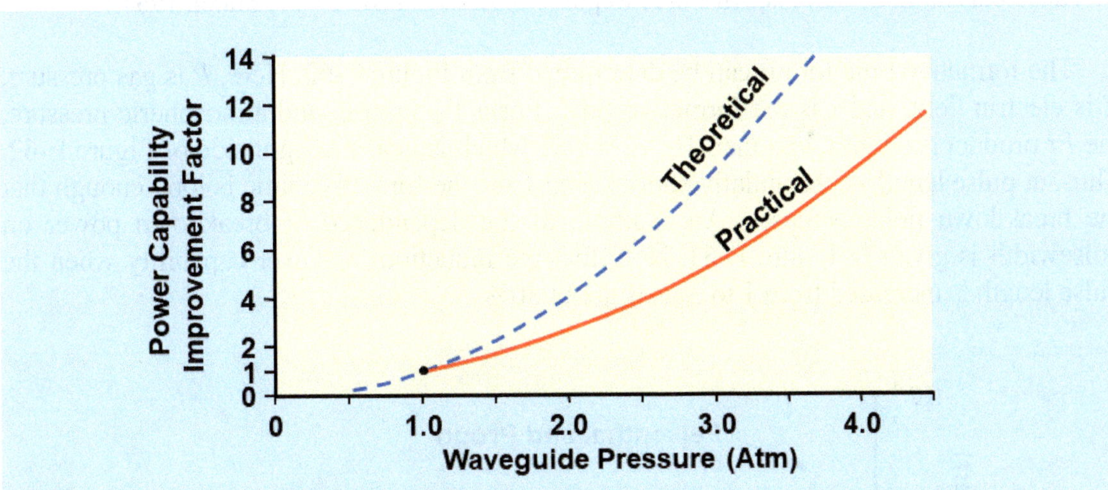

Figure E-60 Improvement in power capability with pressurization. (From: *Microwave Tube Manual*, Varian, Air Force Publication No. T.O.00-25-251, October 1979.)

In Section E.7, the use of electronegative gases for increasing breakdown voltage was discussed. As shown in Figure E-61, similar improvements are possible in RF components. The improvement in power-handling capability with pure sulfur hexafluoride (SF_6) at a pressure of 20 psi is a factor of about 16. Even a small percentage of SF_6 mixed with air produces a significant increase in power capability. Other electronegative gases such as freon 12 and freon 114 produce similar improvements in power-handling capability.

In pulsed RF systems, the peak power capability increases for at least two reasons. The first has already been pointed out in connection with Figure E-55. That is, the heating effect is decreased. The other reason is based on the fact that it takes time for a gas to break down. First, there is the time required for the breakdown process to start. This is called the formative time. Second, there is the time required for the discharge to develop to the point where appreciable RF power is absorbed or reflected.

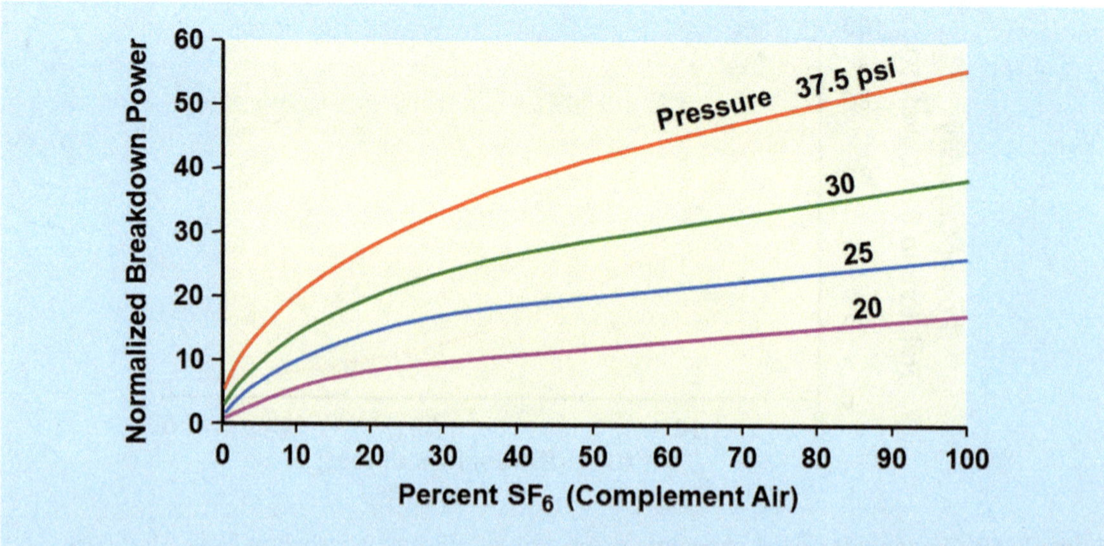

Figure E-61 Breakdown power for mixtures of SF6 and air normalized to air at 20 psi. (From: V. C. Vannicola and L. L. Stevens, Air Force Report No. RADC-TDR-64-67, March 1964.)

The formative time for air can be determined from Figure E-62. Here, P is gas pressure, E is electric field, and τ is the formative time. For a 1-μs pulse, and atmospheric pressure, the $P\tau$ product is slightly less than 10^{-3} torr-sec, which is near the right side of Figure E-62. Thus, at pulse lengths substantially shorter than 1 μs, the formative time is long enough that the breakdown field increases. An example of the dependence of breakdown power on pulsewidth is given in Figure E-63. Note that the reduction in power capability when the pulse length is increased from 1 to 3 μs is about 30%.

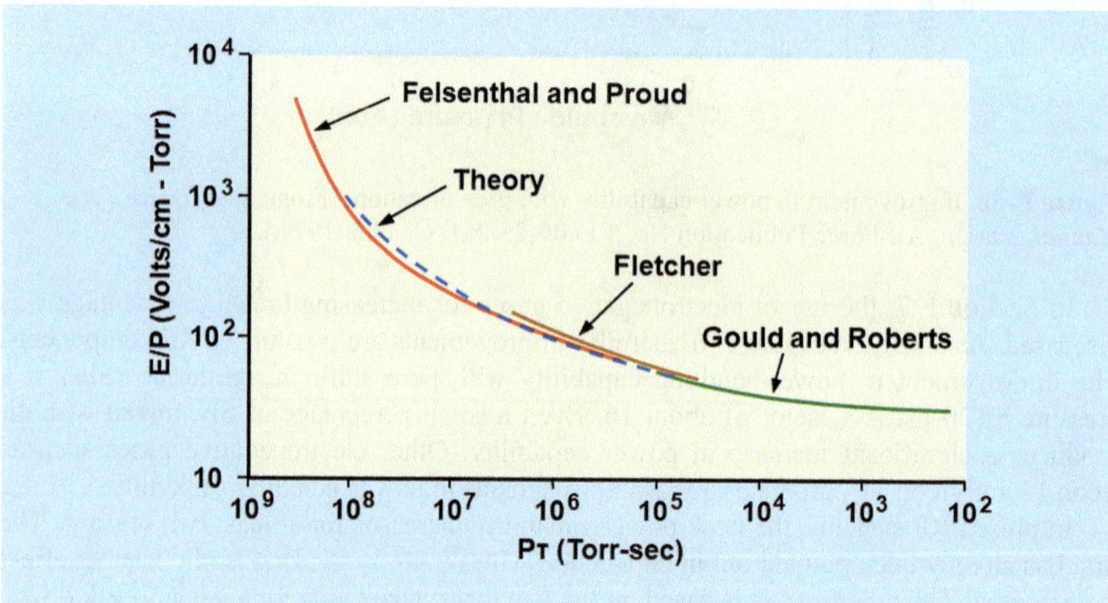

Figure E-62 Formative time in air. (From: P. Felsenthal and J. M. Proud, *Phys. Rev.*, 1965.)

VSWR in the RF output system of a tube may cause breakdown to occur. The reason for this is, of course, that the reflected wave adds to the forward wave to produce regions of increased field level. A VSWR of 1.5:1 reduces the breakdown power by 35%, and 2:1 reduces breakdown power by 50%. If a system must be designed for operation with a significant VSWR, then the normal waveguide or coaxial line rating must be reduced to prevent breakdown as indicated by Figure E-64.

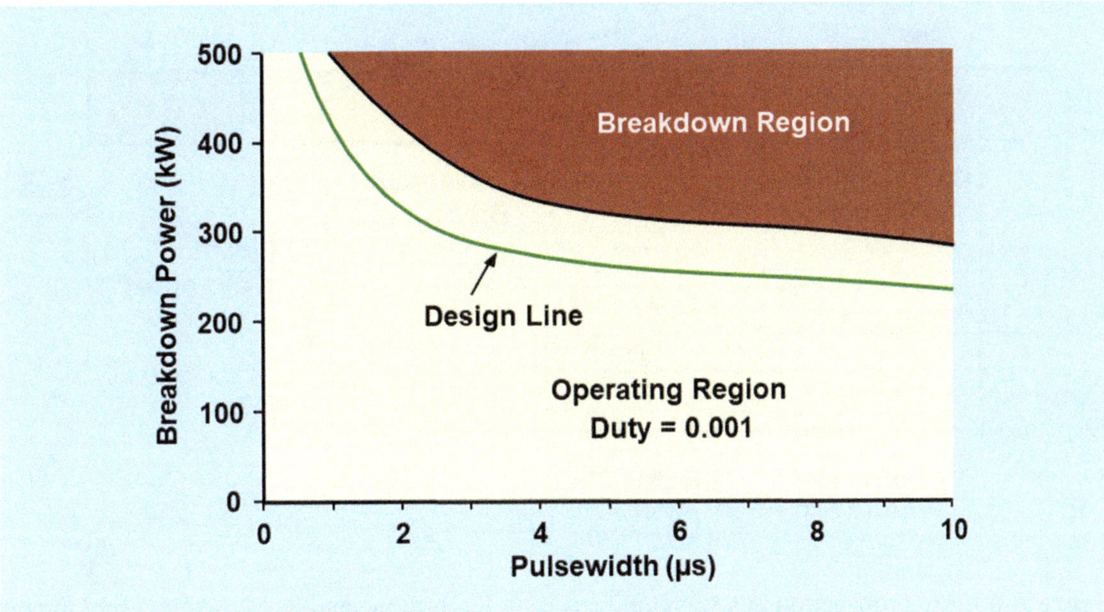

Figure E-63 Effect of pulsewidth on breakdown power in waveguide. (From: *Microwave Tube Manual*, Varian, Air Force Publication No. T.O.00-25-251, October 1979.)

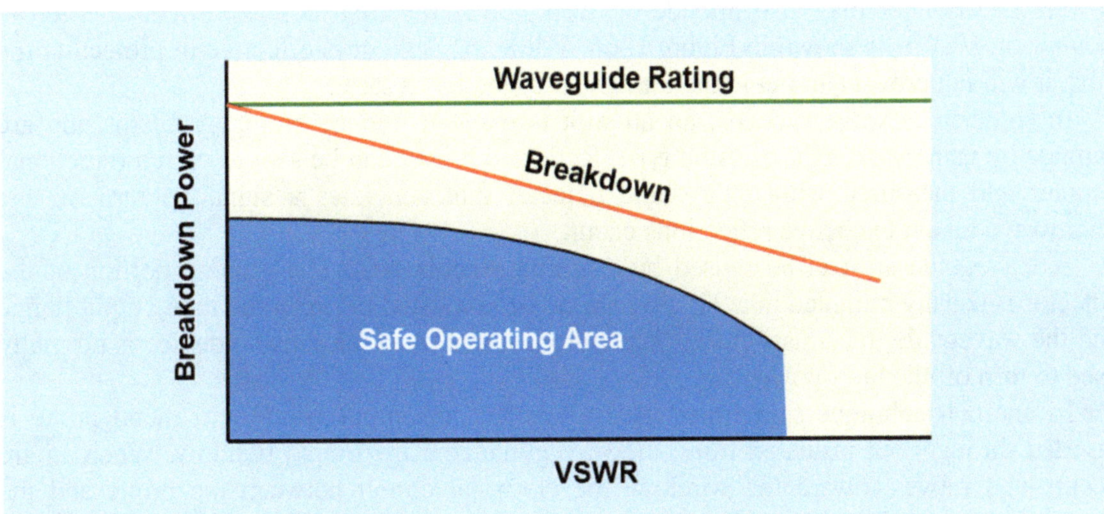

Figure E-64 Effect of VSWR on RF power system rating. (From: *Microwave Tube Manual*, Varian, Air Force Publication No. T.O.00-25-251, October 1979.)

If breakdown does occur in the RF output system connected to a microwave tube, it is probable that the resulting arc will reflect most of the incident power back toward the tube. As a result, the VSWR will be very high and may produce breakdown at other points along the RF system, as is indicated in Figure E-65. In this case, an arc at plane I produces a standing wave and causes the electric field to exceed the breakdown level of a critical element such as the output window of a tube located at plane IV.

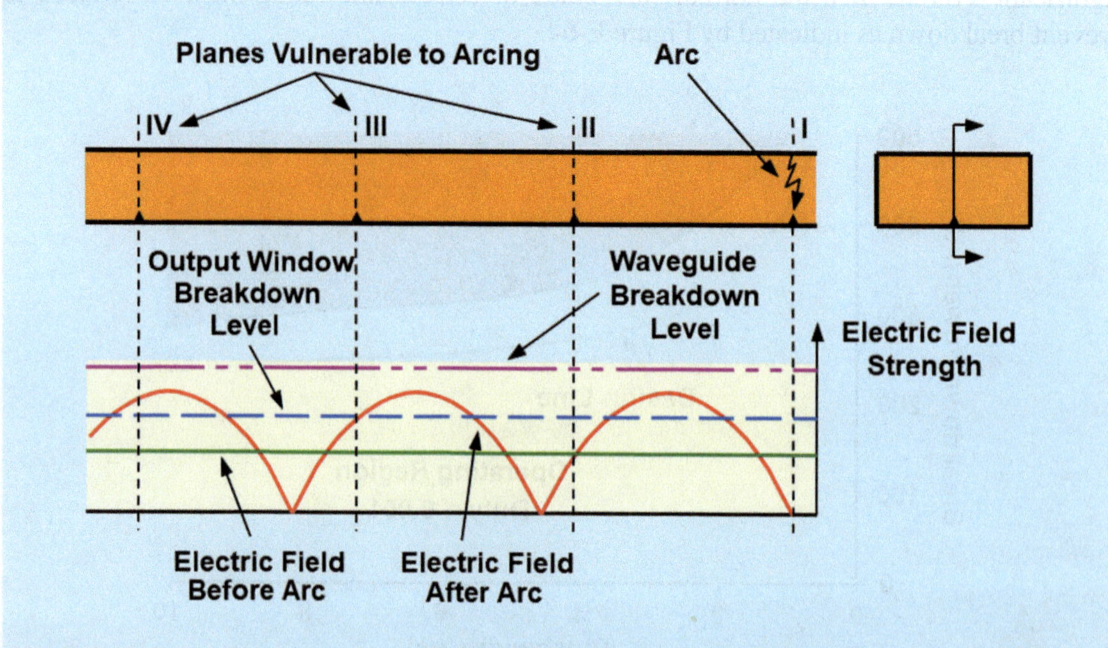

Figure E-65 Arc propagation in a waveguide. (From: R. H. Stone and R. M. White, *Final Report TIS R61 ELM 205,* General Electric Co., January 1961.)

An isolator is often used to protect a microwave tube from excessive reflected RF power. The isolator may also impede or even stop a traveling arc. The effect of a ferrite isolator on VSWR is shown in Figure E-66. While the isolator is effective in protecting the tube, it will not prevent arcing in the remainder of the RF system.

In some microwave systems, an attempt is made to reduce arcing problems and arc damage by using a detector of some type. Reflected power can be sensed with a directional coupler and measured with a threshold detector that generates a signal to turn off the microwave tube if excessive reflections occur.

A waveguide arc can be sensed by detecting visible arcs. The sensing portion of the detector is usually mounted in a 90° waveguide bend located between the tube output flange and the waveguide transmission line. The output signal from the sensing device is normally used to turn off the high-power tube.

In another technique sometimes used to protect an output window, a metal probe is inserted through, but insulated from, the waveguide near the output window. When an arc occurs and moves toward the window, the electrical circuit between the probe and the waveguide is closed so that a relay can be activated to turn off the tube.

E.8 FAULT DETECTION AND TUBE PROTECTION

There are several types of faults that occur in microwave tubes. Two that often occur are

1. Excess body (interception) current;
2. Excess reflected RF power.

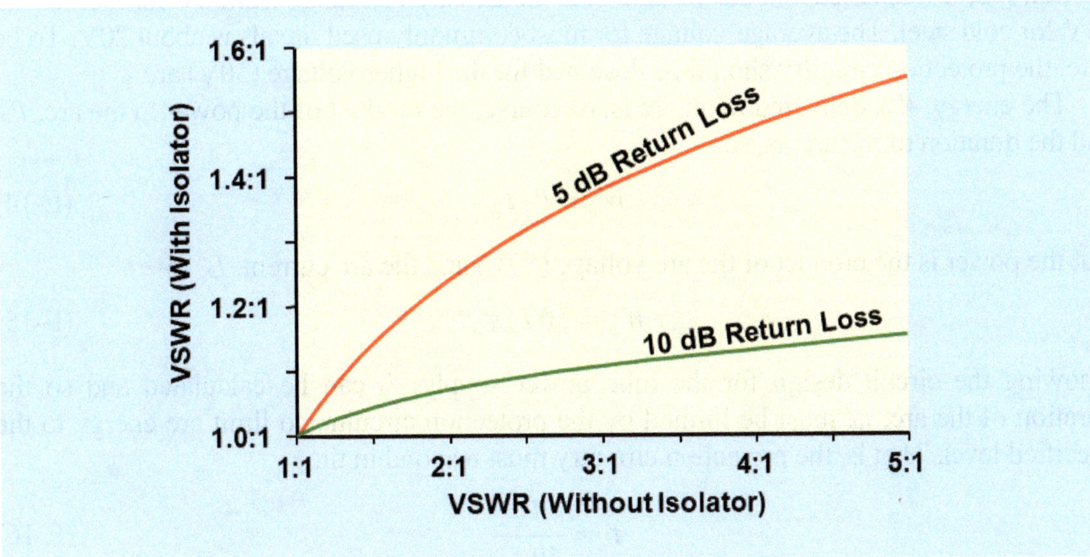

Figure E-66 Effect of ferrite isolator on VSWR. (From: *Microwave Tube Manual*, Varian, Air Force Publication No. T.O.00-25-251, October 1979.)

Excess body current can result from an arc in the gun or from any one of many factors that upset the electron beam dynamics and cause too much of the beam to strike the RF circuit or the anode of the electron gun. Excess reflected RF power is normally caused by a problem in the RF circuit attached to the output of the tube. Even though the tube is often not at fault, excess reflected power may cause it to fail and so the tube must be protected.

E.8.1 Excess Body Current

The most common location for an arc in a microwave tube is in the electron gun and the arc that is formed is a vacuum arc. There is also the possibility that an arc may occur outside the vacuum envelope in the gun region. When an arc occurs, excess current flows to the gun anode (or to the anode lead outside the vacuum envelope) and, if the anode is connected to the body of the tube (as it often is), then the current returns to the power supply through the tube body and so the current is referred to as body current.

Body current caused by an arc is usually distinguishable from body current caused by beam interception because it is usually much larger in magnitude and occurs much more rapidly. In designing circuits that will protect a tube from excess body current, it is necessary to design for the fast-rising, large-amplitude current of the arc as well as for the (usually) more gradual increase of beam interception current to an unacceptable level. The more difficult aspect of the circuit design is usually the arc protection part and that is what we will concentrate on here. In designing arc detection and tube protection circuits, the two arc characteristics that must be known are

1. Arc voltage;
2. Maximum permissible arc energy.

Early in this section it was shown that the arc voltage for an arc in air is about 50V. In Section E.2, it was noted that the voltage of a vacuum arc can range from 8V for mercury to 33V for cold steel. The average voltage for most commonly used metals is about 20V. To be safe, the protection circuitry should be designed for the higher voltage (50V) arc.

The energy, W_a, delivered to an arc is, of course, the product of the power to the arc, P_a, and the duration of the arc, τ_a, so

$$W_a = P_a \tau_a \tag{E-14}$$

But the power is the product of the arc voltage (50V) and the arc current, I_a, so

$$W_a = 50\, I_a \tau_a \tag{E-15}$$

Knowing the circuit design for the tube power supply, I_a can be calculated and so the duration of the arc, τ_a, must be limited by the protection circuitry to limit arc energy to the specified level. That is, the protection circuitry must respond in time

$$\tau_a < \frac{W_a}{50\, I_a} \tag{E-16}$$

For example, if the maximum arc energy specification is 2 joules and the short-circuit current from the power supply is 1,000A (this high current is entirely possible when a supply using large filter capacitors is used), then

$$\tau_a < 40 \text{ µs} \tag{E-17}$$

Thus, the time response requirement on the protection circuitry is usually very modest. In fact, arc energies can easily be limited to a small fraction of a joule by limiting short-circuit current and using a switching circuit that reacts within a few microseconds.

E.8.2 Excess Reflected RF Power

Figure E-67 shows some of the elements of an RF system that can cause excessive reflected RF power. In the cases shown, an impedance mismatch occurs because of a mechanical problem in the RF system. Arcing in the output RF system can also produce a very large impedance mismatch and almost total power reflection.

If the power reflection occurs on the output side of the isolator and the isolator is not defective, then the reflected power will not reach the tube, and thus, of course, the tube will be protected. Usually, a directional coupler (which is not shown in Figure E-67) is used to measure reflected power and turn off the system in the event of excessive reflected power.

If the power reflection occurs on the input side of the isolator, or the isolator is defective (or there is no isolator and no directional coupler for measuring reflected power), then excessive power may reach the microwave tube. This may cause failure of the output window or failure of internal tube elements such as the attenuator in a traveling-wave tube. In some systems an optical detector is placed in the RF output line near the tube window to shut the system off if an arc occurs.

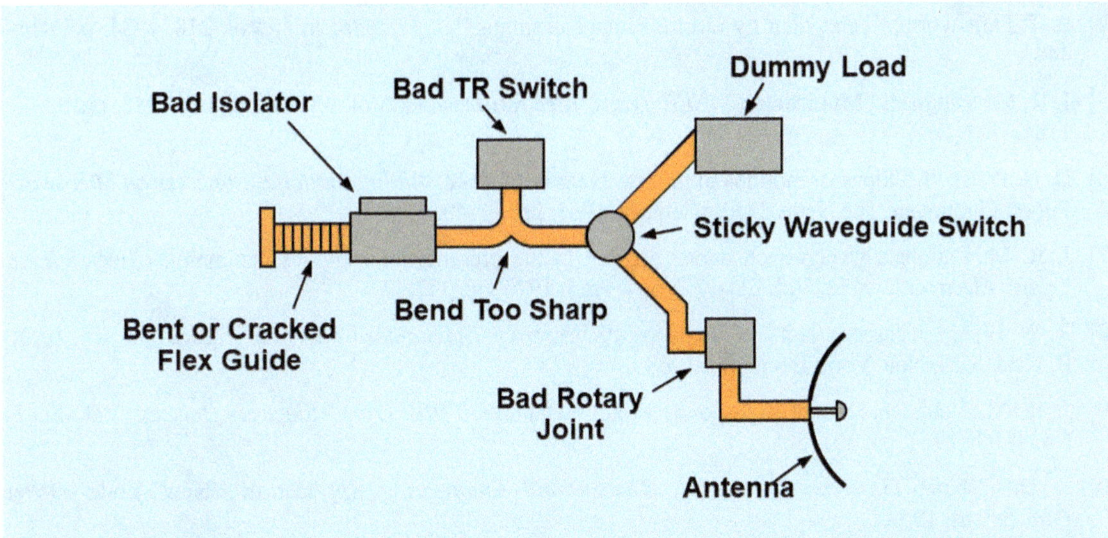

Figure E-67 Common problem areas in RF systems. (From: *Microwave Tube Manual*, Varian, Air Force Publication No. T.O.00-25-251, October 1979.)

REFERENCES

[1] J. M. Lafferty, *Vacuum Arcs, Theory and Applications*, New York: John Wiley and Sons, 1980.

[2] R. V. Latham, *High Voltage Vacuum Insulation: The Physical Basis*, New York: Academic Press, 1981.

[3] R. V. Latham, *High Voltage Vacuum Insulation: Basic Concepts and Technical Practice*, New York: Academic Press, 1995.

[4] W. D. Kilpatrick, "Criterion for vacuum sparking designed to include both RF and dc," *Rev. Sci. Inst.*, Vol. 28, No. 10, 1957, pp. 824–826.

[5] Machlett Laboratories, "Machlett pulse tubes for high voltage, high power, video and RF pulsing," June 1964.

[6] A. Staprans, "Voltage breakdown limitations of electron guns for high-power microwave tubes," *Proc. Second International Symposium on Insulation of Voltages in Vacuum*, Cambridge, MA, 1966.

[7] A. Staprans, "Electron gun breakdown," *High Voltage Workshop*, Monterey, CA, February 1985.

[8] T. Utsumi, "Physical processes in electrical breakdown in vacuum," *Technical Report* RADC-TR-66-712, January 1967.

[9] L. Cranberg, "Initiation of electrical breakdown in vacuum," *Jour. of App. Phys.*, Vol. 23, No. 5, May 1952, pp. 518–522.

[10] R. V. Latham, and N. S. Xu, "Electron pin holes: the limiting defect for insulating high voltages by vacuum, a basis for new cold cathode electron sources," *Vacuum*, Vol. 42, No. 18, 1991, pp. 1173–1181.

[11] I. G. Kesaev, *Cathode Processes in the Mercury Arc*, New York: Consultants Bureau, 1964.

[12] A. S. Gilmour, Jr., and D. L. Lockwood, "The interruption of vacuum arcs at high dc voltages," *IEEE Trans. Electron Devices*, Vol. ED-22, No. 4, April 1975, pp. 173–180.

[13] J. P. Shannon, S. F. Philip, and J. G. Trump, "Insulation of high voltages across solid insulators in vacuum," *Jour. of Vacuum Science and Technology*, Vol. 2, No. 5, September-October 1965, pp. 234–238.

[14] P. T. Farnsworth, "Television by electron image scanning," *J. Franklin Inst.*, Vol. 218, 1934, pp. 411–444.

[15] J. R. M. Vaughan, "Multipactor," *IEEE Trans. Electron Devices*, Vol. 35, No. 7, July 1988, pp. 1172–1180.

[16] D. H. Preist, "Multipactor motions in microwave tubes," *Proc. 4th International Congress on Microwave Tubes*, Eindhoven, The Netherlands: Centrex, 1963, pp. 34–40.

[17] J. R. M. Vaughan, "Magnetron anode erosion in the presence of cathodes containing oxides," *IEEE Trans. Electron Devices*, Vol. ED-17, No. 4, April 1970, pp. 377.

[18] R. W. Bierce, J. Jasberg, and J. V. Lebacqz, "Klystrons," *The Stanford Two Mile Accelerator,* Ch. 10, R. B. Neal, (ed.), New York: Benjamin, 1968.

[19] J. R. M. Vaughan, "Some high-power window failures," *IRE Trans. Electron Devices*, Vol. ED-8, No. 4, July 1961, pp. 302–308.

[20] J. D. Cobine, *Gaseous Conductors, Theory and Engineering Applications*, New York: Dover Publications, 1958.

[21] M. N. Hirsh and H. J. Oskam, *Gaseous Electronics, Volume 1, Electrical Discharges*, New York: Academic Press, 1978.

[22] P. D. Ankrum, *Principles and Applications of Electron Devices*, Scranton, PA: International Textbook Company, 1959.

[23] R. Woo, "RF voltage breakdown in coaxial transmission lines," *Proc. IEEE*, Vol. 57, No. 2, February 1969, pp. 254–256.

Glossary

Absolute zero: The lowest temperature theoretically possible, is equal to zero degrees Kelvin or –273°C.

Admittance: A measure of how well alternating current flows in a conductor. The real part of admittance is conductance and imaginary part is susceptance. It is the reciprocal of resistance.

AM near-carrier noise: AM noise and spurious outputs at frequencies close to the carrier frequency, usually 100 Hz to 500 kHz relative to the carrier. Components not related to the presence of the carrier (signal) are usually caused by ripple on power supply voltages. Interactions in the TWT are usually responsible for components that exist only when a carrier (signal) is present. (*See also* near-carrier noise.)

Amplitude pulling factor: The amount of gain change experienced when the VSWR of the input source or the load is changed.

Amplitude pushing factor: The amount of gain change experienced as the voltage on a particular tube element is varied, usually expressed in dB/volt.

AM/AM conversion: *amplitude modulation to amplitude modulation conversion* describes a type of distortion in microwave amplifiers. It is the change in the output RF voltage produced by variations in input signal level, usually expressed in dB/dB. An ideal amplifier has an AM/AM conversion of 1 dB/dB. As the amplifier approaches saturation, the amplification process becomes nonlinear and gain compression occurs.

AM/PM conversion: *Amplitude modulation to phase modulation conversion* describes a type of distortion in microwave amplifiers. It is the change in phase angle of the output RF voltage produced by variations in input signal level, usually expressed in degrees/dB. An ideal amplifier has an AM/PM conversion of 0 degrees/dB.

AM/PM transfer: *Amplitude modulation to phase modulation transfer* describes a type of distortion in microwave amplifiers. It is the change in phase of the output RF voltage on one signal produced by variations in input signal level on a signal at another frequency, usually expressed in degrees/dB.

Amplitude shift keying (ASK): A form of digital modulation in which the amplitude of the carrier signal is switched between two states.

***b*:** Pierce's velocity parameter, b is a measure of actual beam velocity relative to synchronous beam velocity.

Backstreaming: (1) A condition in which electrons reflected from the collector or secondary electrons from the collector travel backward toward the electron gun. This is an undesirable effect, distorting the primary electron beam and any modulation that may be present. (2) A condition in which gas particles from a vacuum pump travel backward into the chamber being pumped.

Backward-wave oscillator (BWO): A wideband voltage tunable oscillator related to a traveling wave tube. It uses a broadband circuit similar to a TWT. Regeneration exists because the backward-wave mode on the circuit propagates RF energy in a direction opposite to that of the electron stream. Oscillation occurs when the loop gain is equal to or greater than unity. High-power TWTs can experience undesired backward-wave oscillations when the backward-wave mode has sufficient regeneration.

Barium: The primary emitting constituent in a dispenser cathode.

Beam efficiency: (also called electronic efficiency) The RF output power divided by the beam power.

Beam power: The product of cathode voltage and cathode current.

Body current: The part of the electron current from the gun that does not arrive at the collector. This current is intercepted by the RF circuit and other electrodes.

Body-helix protection: A combination of circuit elements in the TWT power supply, ranging from a simple resistive network to a complex crowbar device. The circuitry is designed to prevent damage to the TWT as the result of arcing or unusually high body current.

Brillouin field: *See* Brillouin focusing.

Brillouin flow: Electron flow resulting from the use of Brillouin focusing.

Brillouin focusing: A magnetic focusing scheme in which the magnetic field that provides focusing is parallel to the direction of electron flow and the electron gun is shielded from the magnetic field. The magnetic field required to focus an ideal beam is called the *Brillouin field.*

Bunching: The result of the interaction of an electron beam with an RF field. Electrons which are accelerated by the field move toward those that have been decelerated, forming a high-density region usually referred to as a *bunch.*

C: Pierce's gain parameter, C is a measure of the effectiveness of the circuit in a TWT in interacting with the electron beam.

Cathode loading: The current density at the emitting surface of the cathode, usually expressed in amperes per square centimeter.

Cavity: A microwave-resonant circuit, usually consisting of a hollow metallic enclosure in which RF fields can be established at one or more frequencies.

Centipede structure: A fundamental forward wave coupled-cavity structure. Adjacent cavities are coupled by a multiplicity of dual coupling loops.

Characteristic impedance: The impedance of a transmission line as measured at the driving point if the line was infinitely long.

SCharge density: The amount of electric charge contained in a unit of volume, usually expressed in coulombs per cubic meter.

Circuit efficiency: The efficiency with which the slow wave circuit of a microwave tube delivers the RF power generated in the tube to the output connector.

Cloverleaf structure: A fundamental forward wave coupled-cavity structure. Each cavity has the shape of a four-leaf clover. Adjacent cavities are magnetically coupled by radial slots and are rotated 45°.

Coaxial magnetron injection gun (CMIG): An electron gun used in a coaxial gyrotron.

Cold match: The input or output RF impedance match of a TWT obtained when the TWT is not operating (the TWT is electrically cold).

Collector: The electrode that collects the spent electron beam after it has been used to provide power for microwave amplification or oscillation. The kinetic energy in the spent electron beam is converted to thermal energy (heat) in the collector or is recovered and returned to the power supply by the collector.

Collector efficiency: The efficiency with which the collector recovers power from the spent beam.

Confined flow: A type of beam focusing similar to Brillouin flow except that a small amount of magnetic flux is caused to thread the cathode. More field is required than for Brillouin flow, but the beam is less affected by RF bunching (less scalloping is introduced when RF is applied to the tube and body current is affected less by the application of RF drive). Confined flow is sometimes referred to as *immersed flow*.

Confocal mirrors: For identical mirrors, the radius of curvature of the mirrors is the distance between the mirrors.

Contrawound helix: A helix slow-wave structure where two helices, wound in opposite directions, are superimposed into a single structure. This circuit offers substantially higher power than a conventional helix with some sacrifice of bandwidth.

Control grid: An electrode, usually a gridlike structure, mounted between the cathode and the anode to control the flow of electrons. An appropriate negative voltage (with respect to the cathode) reduces the electron flow (beam current) to zero or cutoff and an appropriate positive (or, in some cases, less negative) voltage allows current to flow. This electrode is usually some sort of mesh structure.

Coupled-cavity tube: A TWT with a slow-wave structure made up of a number of cavities electrically coupled by means of coupling holes, slots, or loops. This circuit is capable of very high-power operation.

Crossed-field device: A microwave tube in which the electric field between the cathode and anode is perpendicular to the magnetic field. The resulting average motion of the electrons emitted from the cathode is nearly perpendicular to both the electric and magnetic fields and parallel to the direction of propagation of energy on a nearby slow wave circuit.

Current density: The current per unit area. For electrons in an electron beam, the current density is the beam current divided by the cross-sectional area of the beam. The current density at the cathode surface (cathode loading) is equal to the current divided by the surface area of the cathode and is usually expressed in amperes per square centimeter.

Chemical vapor deposited diamond (CVD): Used as the output window for RF power on high-power high frequency microwave tubes and gyrotrons.

***d*:** Pierce's loss parameter, d is a measure of the reduction in gain that occurs in a TWT as the result of circuit loss.

dBc: The level of a spurious output relative to the main signal (carrier) expressed in decibels below the carrier power level.

dBm: Power level in decibels above a milliwatt; 60 dBm is a power level of 60 dB above a milliwatt and is a kilowatt.

dBW: Power level in decibels above a watt; 60 dBW is a power level of 60 dB above a watt and is a megawatt.

Denison mode converter: A modification of the Vlasov converter, which is compact and can be placed between the gyrotron cavity and the output window.

Density modulation: The process of controlling the density of electrons emitted from a cathode, usually by a grid or other control electrode located near the cathode.

Depressed collector: Collector depression is the process of applying a negative potential (with respect to the tube body or ground) to the collector to reduce the velocity of electrons entering the collector. This reduces the kinetic energy of the electrons and so less energy is converted to heat and X-rays when the electrons impinge on the electrodes of the collector. The energy recovered from the electrons is returned to the power supply and so the conversion efficiency is greatly improved. A collector designed to operate in this mode is called a *depressed collector*. Depressed collectors can have one or more depressed stages. Multiple-stage depressed collectors use velocity sorting techniques to direct the high-velocity electrons to stages having the greatest depression and slow electrons to the stages having the least depression.

Dispenser cathode: A cathode that has a body made of porous tungsten. The pores contain the active materials that enhance electron emission. When the cathode is heated to its operating temperature, the material in the pores migrates toward the emitting surface, continuously replenishing the supply of active material at the emitting surface.

Dispersion: A term used to describe the variation in phase velocity of an RF wave with frequency. A nondispersive circuit propagates RF energy at a phase velocity that is independent of frequency.

Drift tube: A section of metal tubing in which an electron beam drifts without being accelerated or decelerated by applied electric fields.

Drive: A term to indicate the RF input or RF signal to an electronic device.

Dual mode: Usually refers to devices capable of operating efficiently in both a low power CW mode and a high-power pulse mode.

Duty cycle: The fraction of the time during which a device is operating. The term *duty cycle* is usually used with pulsed operation and is the product of pulsewidth and pulse repetition frequency.

Dynamic range for linear operation: The range of operation of an amplifier in which the power output is nearly proportional to the power input. The lower limit is the input drive level at which the signal input is equal to the noise input plus the noise figure. The upper limit is the input drive level that produces 1 dB or less of gain compression.

Dynamic velocity taper (DVT): A variation in the pitch of the helix of a TWT to improve efficiency and linearity. (*See also* tapered velocity.)

Efficiency: The RF output power divided by the sum of all power provided by the power supplies. (*See also* overall efficiency).

Electronic efficiency: The efficiency with which dc beam power is converted to RF power at the desired frequency within a microwave tube.

Electronic power conditioner (EPC): A power supply/modulator usually associated with space-qualified TWTAs. In addition to providing power, EPCs often include protection circuits, circuits for receiving commands, and circuits for providing telemetry data.

Energy product: A measure of the magnetic energy produced in the gap in a permanent magnet. The energy product is the product of B and H at the magnet operating point.

Equalizer: A passive device providing selective loss and/or phase shift over an operating band such that the net gain and/or phase shift of the equalized amplifier matches a required profile.

External Q: A measure of the effect on the Q of a resonant circuit by an external circuit attached to that resonant circuit. External Q is the reciprocal of the difference between the reciprocals of the loaded and the unloaded Q.

Ferrule: A drift tube tip in the cavity of a klystron or coupled-cavity TWT.

Fine grain: A term used to refer to small variations in gain or phase of an amplifier.

Focus electrode: An element in the electron gun that is used to focus the electrons into a well-defined beam. It is sometimes called the *beam-forming electrode* (BFE).

Folded helix circuit: A helix slow-wave structure where two helices, wound in the same direction, are superimposed into a single structure. This circuit is sometimes called a *bifilar helix* and offers substantially higher power than a conventional helix with some sacrifice of bandwidth.

Folded waveguide circuit: Another name for a coupled-cavity circuit.

Frequency shift keying (FSK): A form of digital modulation in which the frequency of the carrier signal is switched between two states.

Fundamental backward wave circuit: A slow-wave circuit used in coupled-cavity TWTs. The fundamental mode of the circuit is a backward-wave mode. The circuit is operated in a forward space-harmonic mode.

Fundamental forward-wave circuit: A slow-wave circuit used in high-power coupled-cavity TWTs. The fundamental mode of the circuit is a forward-wave mode.

Fundamental Gaussian mode content (FGMC): The fraction of the mode content in the output beam from a gyrotron that is Gaussian.

Gain: The ratio of output voltage, current, or power to the input voltage, current, or power, respectively, in an amplifier stage, receiver, or system. Power gain is usually expressed in dB.

Gain compression: The decrease in gain of an amplifier between the small signal gain and the gain at saturation. Typically, gain compression is 6 to 8 dB in TWTs.

Gated emission: Emission from a cathode that is controlled (gated), usually with a grid adjacent to the cathode.

Getter: A device that, after activation, absorbs or adsorbs gases within an electron tube.

Grid: An electrode mounted between the cathode and the anode to control the flow of electrons. When used to control the amount of electron current in a TWT or klystron, the grid is called a *control grid.* The grid electrode may be a cylindrical-shaped ring or a section of wire screen or mesh.

Group delay: The distortion that results when the time delay of a signal being processed is not constant as a function of frequency. Areas of interest are linear delay, parabolic delay, and the ripple component. The linear component is the difference in delay between two frequencies within the specified bandwidth. Parabolic delay is determined by fitting the equation of a parabola to a plot of delay versus frequency. Ripple is the maximum peak-to-peak variation relative to a plot of the smoothed curve.

Group velocity: The velocity at which energy advances along an RF circuit. For a coaxial transmission line with no dielectric, the group velocity is the speed of light. For waveguide, the group velocity is less than the speed of light. The group velocity cannot be greater than the speed of light.

Gyrotron: An oscillator or amplifier that is capable of generating extremely high powers at frequencies of tens of gigahertz to terahertz.

Gyro-monotron or gyro-oscillator: a gyrotron oscillator.

Gyro-klystron: A gyrotron amplifier.

Gyro-twystron: A gyrotron amplifier.

Gyro-TWT: A gyrotron amplifier.

Harmonic drive: The inclusion of phase-conditioned harmonic power in the input RF signal to reduce harmonic power and harmonic capture, and to improve efficiency at the low end of the operating band of a TWT.

Harmonic interaction: The effect of the harmonic content of the RF input signal on the beam modulation. This is generally undesirable and usually reduces the fundamental power output.

Heater: An electric heating element for raising an indirectly heated cathode to the operating temperature.

Hot match: The RF impedance match at the input or output of a TWT, obtained when the TWT is operating (electrically hot).

Hub: *See* Ferrule or magnetron.

Hull cutoff condition: The combination of perpendicular electric and magnetic fields that causes an electron in a smooth bore magnetron to just graze anode.

Hysteresis loop: *See* intrinsic hysteresis loop and normal hysteresis loop.

Inverted magnetron injection gun (ICMIG): An electron gun used on coaxial gyrotrons.

Immersed flow: *See* Confined flow.

Inline slot structure: A fundamental backward-wave coupled-cavity structure in which adjacent cavities are electromagnetically coupled by two slots spaced 180° apart in the wall between the cavities. The slots in the walls of adjacent cavity are aligned.

Insertion loss: The reduction in signal strength that results when a passive device is installed (inserted) into an RF transmission line. Insertion loss is caused by power being absorbed in the device and by reflections from discontinuities at the interfaces between the device and the transmission line. Usually expressed in decibels.

Intercepting grid: A control grid that is not shielded from electrons emitted by the cathode. The grid intercepts approximately 10% to 15% of the emission current when a positive potential is applied.

Intercept point: On a plot of intermodulation distortion data as a function of RF input drive, the carrier data at small signal drive levels has a slope of 1:1, third-order intermodulation products have a slope of 3:1, and fifth-order intermodulation products have a slope of 5:1. Extrapolating the slopes for the carriers and the third-order products produces an intercept point. The output power at this point is called the *third-order intercept point.* Similarly, the fifth-order intercept point is the point at which the carrier slope intercepts the fifth-order product slope. (*See also* Intermodulation distortion.)

Interdigital line: A slow-wave circuit composed of a comblike structure with alternate segments being connected together at one end and remaining segments connected together at the opposite end.

Intermodulation distortion: Distortion resulting from the production of frequencies that are the sum of, and the difference between, frequencies contained in the applied waveform.

Intrinsic hysteresis loop: The magnetization characteristic of a ferromagnetic metal.

Intrinsic efficiency: Efficiency calculated by dividing the RF output power from a microwave tube by the sum of all power provided with the exception of heater power and that provided to the solenoid for focusing the beam.

Ion oscillations: Fluctuations in microwave tube operation resulting from ion effects in the electron beam. These fluctuations are commonly observed as variations in RF power output or as RF signals removed from the carrier frequency by the frequency of ion oscillations in the electron beam.

Iridium: A transition metal sometimes used with dispenser cathodes for enhancing emission.

Klystrode: The trademarked name for an IOT that CPI uses.

Klystron: A velocity-modulated microwave tube that uses the interaction between an electron beam and the RF energy in cavities to provide signal amplification. In a klystron, the RF interaction takes place at discrete locations along the electron beam. Common types of klystrons are the reflex klystron (an oscillator having only one cavity), two-cavity klystron amplifiers and oscillators, and multicavity klystron amplifiers.

Leakage current: Undesirable flow of current between electrodes often resulting from conductive paths through or over the surface of an insulator or from unwanted electron emission from electrode surfaces.

Linear-beam tube: A microwave tube (usually a klystron or TWT) in which a long cylindrical electron beam is used.

Linearizer: A device that improves the linearity of the AM/AM conversion curve of an amplifier (causing the slope of power output versus power input to be nearly 1 dB/dB over a wide range of drive levels). Linearizers are usually designed to also reduce the AM/PM conversion (causing the phase shift to remain nearly constant over a wide range of drive levels). The use of a linearizer reduces the intermodulation distortion.

Loaded Q: The Q of a resonant circuit when there is external coupling to the circuit.

Loss buttons: In coupled-cavity TWTs, lossy elements used to provide frequency selective loss to inhibit the excitation of higher order modes.

Magnetron: A crossed-field microwave oscillator containing a concentric cathode and anode containing resonant cavities. An axial magnetic field causes electrons to orbit between the cathode and the anode. The RF voltages across gaps in the resonators modulate the velocities of the electrons. This causes the orbiting electrons to form into spokes that rotate around the tube axis. As the spokes rotate past the resonator gaps, they induce currents that excite the cavities. The RF voltages build up to large levels. High-power output is obtained at high efficiency.

Metal-ceramic: A term applied to tube assemblies that employ ceramics as a part of their vacuum walls. Electrical isolation between elements is achieved by using ceramic insulators. The assembly is integrated by brazing the ceramic and metal parts.

Micron: A measure of pressure. One micron is 10^{-3} torr.

Mode interference: *See* interfering mode.

M-type tube: A crossed-field tube.

Multimode: Having the capability of operating efficiently at more than one power level. (*See also* dual mode.)

Multioctave: Capable of operating satisfactorily over a frequency range of two or more octaves.

Multipactor: An electron-RF field interaction in which electrons oscillate back and forth in synchronism with an RF voltage applied between electrodes. Secondary electrons produced when primary electrons strike the electrodes cause the number of oscillating electrons to increase with time. Multipactor is usually an undesirable effect and can occur across cavity or drift tube gaps, in output waveguides, or on the surface of the ceramic output window.

Multistage collector: A collector with several segments, each successive segment being depressed more than the preceding segment. This enhances the efficiency of the collector in recovering power from the spent electron beam.

Near-carrier noise: AM noise, PM noise, and spurious outputs at frequencies close to the carrier frequency, usually 100 Hz to 500 kHz relative to the carrier. Components not related to the presence of the carrier (signal) are usually caused by ripple on power supply voltages. Interactions in the TWT are usually responsible for components that exist only when a carrier (signal) is present.

Noise figure: The ratio of the signal-to-noise on the output of a device to the signal-to-noise on the input. It is important because it indicates the amount of noise the amplifier contributes to the signal, and it is an absolute indicator of the sensitivity of the device. Noise figure is usually expressed in decibels, and is abbreviated NF or F.

Noise power: The noise generated by a device or amplifier when measured at the output port when the input port is terminated and no RF drive is applied.

Noise power ratio (NPR): A measure of intermodulation when an infinite number of carriers is being amplified. The input to the amplifier is a well-defined bandwidth of random noise with a deep notch in the center. Intermodulation in the amplifier produces intermodulation products that partially fill the notch. The ratio of noise power in the passband to the noise power in the notch, measured at the output of the amplifier, is the NPR.

Nonintercepting grid: A control grid that is mechanically shielded from the cathode by a shadow grid in such a way that few electrons are intercepted by the control grid, even when a positive voltage is applied to the control grid (with respect to the cathode).

Normal hysteresis loop: The magnetization characteristic of a ferromagnetic metal plus the field that would be there in the absence of the metal.

Octave: The interval between two frequencies having a ratio of 2:1.

Osmium: A transition metal sometimes used with dispenser cathodes for enhancing emission.

Outgassing: A term used to describe the release of various gasses from the internal surfaces of a vacuum device during processing and testing.

O-type tube: A linear-beam tube.

Overall efficiency: The ratio of RF output power from a microwave tube to the total of all power inputs to the tube.

Overdrive: An input signal level greater than that required for saturation, resulting in decreasing output power.

Oxide-coated cathode: A cathode with an emissive surface consisting of oxides of alkaline-earth metals (usually barium, calcium, and strontium).

Penultimate cavity: The next to the last cavity of a multicavity klystron.

Perveance: A numerical constant relating current and voltage in a space-charge–limited diode. The relation for perveance, usually denoted by P, is

$$P = \frac{I}{V^{3/2}}$$

Perveance depends only on the geometry of the diode. When the voltage is the cathode-to-anode voltage in the electron gun, the perveance is called the *gun perveance*. When the voltage is the cathode-to-circuit voltage, the perveance is called the *beam perveance*.

Phase linearity: A term referring to the degree of deviation from a straight line of the phase versus frequency characteristic of a device.

Phase-pulling factor: The amount by which a change in input or output VSWR changes the phase shift in the TWT.

Phase-pushing factor: The amount by which a change in the voltage on an electrode changes the phase shift in the TWT, usually expressed in degrees/volt.

Phase tracking: The closeness or similarity of the phase characteristics of a number of devices. This is an important consideration when power combining the outputs from two or more devices.

Phase velocity: The velocity at which phase fronts for an electromagnetic wave propagate. For a coaxial transmission line with no dielectric, the magnitude of the phase velocity is the speed of light. For a waveguide, the magnitude of the phase velocity is greater than the speed of light. Slow-wave circuits are designed to provide a phase velocity having a magnitude much less than the speed of light.

PM near-carrier noise: PM (phase modulation) noise and spurious output at frequencies close to the carrier frequency, usually 100 Hz to 500 kHz relative to the carrier. Components not related to the presence of the carrier (signal) are usually caused by ripple on power supply voltages. Interactions in the TWT are usually responsible for components that exist only when a carrier (signal) is present. (*See also* near-carrier noise.)

Pole piece: A section of ferromagnetic material used to shape the magnetic flux distribution from a solenoid (electromagnet) or permanent magnet.

Power combining: A scheme whereby the outputs from two or more amplifiers are combined to provide a greater output power than available from a single amplifier.

Power curve: A plot of output power versus input drive or input frequency. (For power curve versus input drive, *see also* AM/AM conversion.)

Phase shift keying (PSK): A form of digital modulation in which the phase of the carrier signal is switched between two states.

Periodic permanent magnet focusing (PPM): A method of focusing a TWT where permanent magnets of opposite polarity are placed side by side along the length of the tube.

Pulse-up ratio: The ratio, usually expressed in dB, between the CW power level and the pulse-power level in a dual mode device.

***QC*:** Pierce's space charge parameter, QC is a measure of the strength of space charge forces causing debunching relative to the strength of the circuit field that is bunching the beam.

Quadrature phase shift keying (QPSK): A form of digital modulation in which the phase of the carrier signal is switched between four states.

Quasi-optical: Basic elements are optical but effects of finite wavelength are accounted for.

Reike diagram: A Smith chart based load diagram for microwave oscillators, particularly klystrons and magnetrons.

Residual gas: The gas remaining in an evacuated vacuum tube.

RF efficiency: (also called beam efficiency) The RF output power divided by the beam power.

Rhenium: A transition metal sometimes used with dispenser cathodes for enhancing emission.

Ring-bar tube: A TWT with a slow-wave structure composed of ringlike segments connected by straight straps or bars. This device is capable of higher power levels than a conventional helix tube at a significant reduction in bandwidth.

Ring-loop tube: Similar to a ring-bar tube, except that the bars connecting the ringlike segments are replaced by curved straps.

Samarium cobalt: A high-energy product magnet material commonly used on microwave tubes.

Saturated gain: The gain of an amplifier operating at saturation.

Saturated power output: The power output of an amplifier operating at saturation.

Saturation: A term used to describe that point on the RF power output versus RF power input characteristic where an increase in input power does not produce an increase in output power.

Screen grid: A grid structure placed between the control grid and anode to reduce the capacitive coupling between those two electrodes. In a multimode electron gun, it is the element used to control beam current from the edge of the cathode.

Secondary emission: Emission of electrons resulting from the bombardment of a surface by high energy electrons.

Serrodyne: An operating mode for a TWT in which the input signal is translated to a new frequency at the output. A sawtooth voltage is superimposed on the dc voltage between the cathode and the slow wave circuit, causing the phase shift in the TWT to change in a linear fashion as a function of time. The amplitude of the sawtooth voltage can be adjusted to provide 360° of phase shift during each cycle, causing the output of the TWT to be offset by the frequency of the sawtooth waveform (the flyback time of the sawtooth must be very short so the TWT thinks that the phase is being offset continuously).

Shadow grid: A grid structure placed between the cathode and control grid and electrically connected to the cathode. This element shields the control grid from interception.

Shot noise: Noise power generation in a vacuum tube resulting from the random nature of the emission of electrons.

Single-stage collector: A TWT or klystron collector that has only one element for collecting electrons. The stage may be operated at ground potential or may be depressed. (*See also* depressed collector, multistage collector.)

Slow-wave circuit: Any structure that slows the phase velocity of an RF wave below the velocity of light to establish synchronism or near synchronism between the velocities of the wave and an electron beam.

Small-signal gain: The gain of an amplifier operating at small-signal levels.

Space charge: The electrical charge of electrons in a vacuum environment (as opposed to electrons within the crystal structure of a metal).

Space-charge–limited operation: Operation of a cathode at an emission density high enough so that the cathode current is limited by the space-charge field of the electron cloud near the cathode surface.

Spent electron beam: The electron beam exiting from the interaction region of a linear-beam tube.

Spherical diode: A diode constructed to duplicate the spherical geometry of a typical electron gun. Used for the evaluation and analysis of electron emission, analysis of electron gun designs, and for realistic cathode life testing.

Stagger tuning: A method of tuning the cavities of a klystron, also called *broadband tuning*. One or more of the cavities is tuned to a frequency(s) above and one or more is tuned to a frequency(s) below the center of the frequency band to provide approximately uniform amplification over a band of frequencies.

Staggered-slot structure: A fundamental backward-wave coupled-cavity structure in which adjacent cavities are electromagnetically coupled by one opening or slot in the wall between the cavities. The slots are rotated 180° from cavity wall to cavity wall.

Synchronous tuning: The tuning of all cavities of a klystron to the same frequency.

Synchronous voltage: The beam voltage in a TWT at which the electron beam has the same velocity as the phase velocity of the RF wave on the circuit.

Tapered termination: A gradual increase in the amount of loss applied to a slow-wave structure to control reflections within a TWT.

Tapered velocity: A change in the pitch of a helix, the height of cavities, or some other axial dimensions of a slow-wave circuit to change the phase velocity of the RF wave. This is done because the extraction of kinetic energy from the electron beam causes the beam to slow down near the output of the TWT. The taper helps to maintain synchronism between the RF wave and the electron beam, which improves the efficiency of the TWT. (*See also* DVT.)

Temperature-limited operation: Operation of a cathode under conditions such that the cathode current depends on cathode temperature.

Thermionic emission: The emission of electrons from a surface as the result of heating of the surface.

Tetrode: A vacuum tube having four elements, usually a cathode, control grid, screen grid, and anode. In multimode TWTs, *tetrode* is a term describing the electron gun (cathode, shadow grid, control grid, and screen grid).

Torr: A unit of pressure. Atmospheric pressure is 760 torr. Most linear-beam microwave tubes operate at pressures below 10^{-7} torr.

Transfer curve: A curve or family of curves for which output power is plotted as a function of input drive power at one or more beam voltages. Sometimes referred to as a *gain curve*. (*See also* AM/AM transfer.)

Triode: A three-element vacuum tube having a cathode, a control grid, and an anode.

Traveling-wave tube (TWT): A microwave tube using a broadband circuit in which an electron beam interacts continuously with a guided electromagnetic field to amplify microwave frequencies.

Traveling-wave tube amplifier (TWTA): A combination of a power supply, a modulator (for pulsed systems), and a traveling-wave tube, often packaged in a common enclosure.

Twystron: A high-power microwave tube having an input circuit consisting of klystron cavities and a cloverleaf coupled-cavity structure for the output circuit. A twystron has a nearly flat power output characteristic over a bandwidth as high as 10% or more.

Vacuum envelope: Any structure containing or capable of containing a high vacuum environment. The body structure of a microwave tube is usually referred to as the *vacuum envelope*.

Velocity modulation: The modulation of the velocities of electrons in a microwave tube as a result of their interaction with RF electric fields.

Velocity resynchronization: Any method for changing the axial velocity of an RF wave or of an electron beam to improve the synchronism between that wave and an electron beam. (*See also* Tapered velocity, Velocity jump.)

Velocity step: *See* tapered velocity.

Velocity taper: *See* tapered velocity.

Vlasov mode converter: A smooth surface circular waveguide with a launching cut and reflectors that form a Gaussian-like beam in a gyrotron.

Work function: The minimum energy, usually expressed in electron volts (eV), required for an electron to be thermionically emitted from a cathode.

Whispering gallery mode: An electromagnetic wave mode in a gyrotron, TE_{mn} mode, where $m >> n$.

About the Author

Dr. A. S. Gilmour, Jr. received a B.E.E., an M.S.E.E., and a Ph.D. from Cornell University. He was an assistant professor of electrical engineering at Cornell when he received a two-year appointment as professor in residence at the Cornell Aeronautical Laboratory in Buffalo, New York. He was then named head of the Wave Electronics Section at the Laboratory. This was followed by a two-year assignment as manager of the Electronics Department of Sanders Associates in Williamsville, New York. He was then appointed professor of electrical and computer engineering at the State University of New York at Buffalo. He served as chairman for several years and is now professor emeritus.

Dr. Gilmour has been actively involved in the research, design, development and application of microwave tubes for over six decades. His klystron and TWT work resulted in a nearly perfect Brillouin electron beam. He made the first detailed measurements of the RF current and velocity distributions in Brillouin beams. He has participated in the design or improvement of transmitters for a large number of military microwave systems. He was instrumental in the establishment of a NASA/USAF program to develop high-efficiency TWTs for electronic counter measures systems. During the past three decades he has worked extensively with government personnel to correct deficient tube and system designs in order to reduce failure rates and enhance performance. He participated in the design and development of high-power microwave systems for the Strategic Defense Initiative. At the University at Buffalo, he compiled and published the *Conference Records of the Power Modulator Symposia* (over 5,000 pages)

Dr. Gilmour has presented nearly 200 courses on microwave devices and related power systems to well over 3,000 engineers and scientists. Nearly 100 of those courses were focused on RF vacuum electron devices (microwave tubes). The locations for most of these courses were in house. Some of the courses were open to the public and were attended by scientists and engineers from Europe and Asia as well as the United States.

In the process of preparing the lecture material for the vacuum electron device courses, Dr. Gilmour wrote the following four books. These were published by Artech House.

1. *Microwave Tubes,* 490 pages, 1986.
2. *Principles of Traveling Wave Tubes,* 625 pages, 1994. This book was also translated and published in Russia.
3. *Klystrons, Traveling Wave Tubes, Magnetrons, Crossed-Field Amplifiers, and Gyrotrons,* 860 pages, 2011. This book was also translated and published in China.
4. *Microwave and Millimeter Wave Vacuum Electron Devices* (this book), 900 pages with over 1,000 color figures, 2020.

Dr. Gilmour was a member of the Program Committees for the fourteenth through twentieth Power Modulator Symposia and was chairman in 1988. He was responsible for hosting the

Twelfth Power Modulator Symposium in Buffalo, New York. For nearly a decade, he served on Program Committees for High Voltage Workshops and was chairman in 1990. He was a member of the Committee on Pulsed Power Sources of the Air Force Studies Board, Assembly of Engineering, National Research Council, National Academy of Sciences. He served on the Program Committees for International Pulsed Power Conferences for over a decade.

Dr. Gilmour is a fellow member of the Institute of Electrical and Electronic Engineers. His background includes certification as an electromagnetic compatibility engineer and as a professional engineer in the State of New York.

He has received several awards, including the 2018 IEEE John R. Pierce Award for Excellence in Vacuum Electronics. The citation for this award is "For contributions to the development of high efficiency traveling wave tubes and for educating generations of scientists and engineers worldwide through his books and courses on vacuum electronics".

Index

G

Q

R

Artech House Microwave Library

Behavioral Modeling and Linearization of RF Power Amplifiers, John Wood

Chipless RFID Reader Architecture, Nemai Chandra Karmakar, Prasanna Kalansuriya, Randika Koswatta, and Rubayet E-Azim

Control Components Using Si, GaAs, and GaN Technologies, Inder J. Bahl

Design of Linear RF Outphasing Power Amplifiers, Xuejun Zhang, Lawrence E. Larson, and Peter M. Asbeck

Design Methodology for RF CMOS Phase Locked Loops, Carlos Quemada, Guillermo Bistué, and Iñigo Adin

Design of CMOS Operational Amplifiers, Rasoul Dehghani

Design of RF and Microwave Amplifiers and Oscillators, Second Edition, Pieter L. D. Abrie

Digital Filter Design Solutions, Jolyon M. De Freitas

Discrete Oscillator Design Linear, Nonlinear, Transient, and Noise Domains, Randall W. Rhea

Distortion in RF Power Amplifiers, Joel Vuolevi and Timo Rahkonen

Distributed Power Amplifiers for RF and Microwave Communications, Narendra Kumar and Andrei Grebennikov

Electric Circuits: A Primer, J. C. Olivier

Electronics for Microwave Backhaul, Vittorio Camarchia, Roberto Quaglia, and Marco Pirola, editors

EMPLAN: Electromagnetic Analysis of Printed Structures in Planarly Layered Media, Software and User's Manual, Noyan Kinayman and M. I. Aksun

An Engineer's Guide to Automated Testing of High-Speed Interfaces, Second Edition, José Moreira and Hubert Werkmann

Envelope Tracking Power Amplifiers for Wireless Communications, Zhancang Wang

Essentials of RF and Microwave Grounding, Eric Holzman

Frequency Measurement Technology, Ignacio Llamas-Garro, Marcos Tavares de Melo, and Jung-Mu Kim

FAST: Fast Amplifier Synthesis Tool—Software and User's Guide, Dale D. Henkes

Feedforward Linear Power Amplifiers, Nick Pothecary

Filter Synthesis Using Genesys S/Filter, Randall W. Rhea

Foundations of Oscillator Circuit Design, Guillermo Gonzalez

Frequency Synthesizers: Concept to Product, Alexander Chenakin

Fundamentals of Nonlinear Behavioral Modeling for RF and Microwave Design, John Wood and David E. Root, editors

Generalized Filter Design by Computer Optimization, Djuradj Budimir

Handbook of Dielectric and Thermal Properties of Materials at Microwave Frequencies, Vyacheslav V. Komarov

Handbook of RF, Microwave, and Millimeter-Wave Components, Leonid A. Belov, Sergey M. Smolskiy, and Victor N. Kochemasov

High-Efficiency Load Modulation Power Amplifiers for Wireless Communications, Zhancang Wang

High-Linearity RF Amplifier Design, Peter B. Kenington

High-Speed Circuit Board Signal Integrity, Second Edition, Stephen C. Thierauf

Integrated Microwave Front-Ends with Avionics Applications, Leo G. Maloratsky

Intermodulation Distortion in Microwave and Wireless Circuits, José Carlos Pedro and Nuno Borges Carvalho

Introduction to Modeling HBTs, Matthias Rudolph

An Introduction to Packet Microwave Systems and Technologies, Paolo Volpato

Introduction to RF Design Using EM Simulators, Hiroaki Kogure, Yoshie Kogure, and James C. Rautio

Introduction to RF and Microwave Passive Components, Richard Wallace and Krister Andreasson

Klystrons, Traveling Wave Tubes, Magnetrons, Crossed-Field Amplifiers, and Gyrotrons, A. S. Gilmour, Jr.

Lumped Elements for RF and Microwave Circuits, Inder Bahl

Lumped Element Quadrature Hybrids, David Andrews

Microstrip Lines and Slotlines, Third Edition, Ramesh Garg, Inder Bahl, and Maurizio Bozzi

Microwave Circuit Modeling Using Electromagnetic Field Simulation, Daniel G. Swanson, Jr. and Wolfgang J. R. Hoefer

Microwave Component Mechanics, Harri Eskelinen and Pekka Eskelinen

Microwave Differential Circuit Design Using Mixed-Mode S-Parameters, William R. Eisenstadt, Robert Stengel, and Bruce M. Thompson

Microwave Engineers' Handbook, Two Volumes, Theodore Saad, editor

Microwave Filters, Impedance-Matching Networks, and Coupling Structures, George L. Matthaei, Leo Young, and E. M. T. Jones

Microwave Imaging Methods and Applications, Matteo Pastorino and Andrea Randazzo

Microwave Material Applications: Device Miniaturization and Integration, David B. Cruickshank

Microwave Materials and Fabrication Techniques, Second Edition, Thomas S. Laverghetta

Microwave Materials for Wireless Applications, David B. Cruickshank

Microwave and Millimeter-Wave Vacuum Electron Devices: Inductive Output Tubes, Klystrons, Traveling-Wave Tubes, Magnetrons, Crossed-Field Amplifiers, and Gyrotrons, A. S. Gilmour, Jr.

Microwave Mixer Technology and Applications, Bert Henderson and Edmar Camargo

Microwave Mixers, Second Edition, Stephen A. Maas

Microwave Network Design Using the Scattering Matrix, Janusz A. Dobrowolski

Microwave Power Amplifier Design with MMIC Modules, Howard Hausman

Microwave Radio Transmission Design Guide, Second Edition, Trevor Manning

Microwave and RF Semiconductor Control Device Modeling, Robert H. Caverly

Microwave Transmission Line Circuits, William T. Joines, W. Devereux Palmer, and Jennifer T. Bernhard

Microwaves and Wireless Simplified, Third Edition, Thomas S. Laverghetta

Modern Microwave Circuits, Noyan Kinayman and M. I. Aksun

Modern Microwave Measurements and Techniques, Second Edition, Thomas S. Laverghetta

Modern RF and Microwave Filter Design, Protap Pramanick and Prakash Bhartia

Neural Networks for RF and Microwave Design, Q. J. Zhang and K. C. Gupta

Noise in Linear and Nonlinear Circuits, Stephen A. Maas

Nonlinear Microwave and RF Circuits, Second Edition, Stephen A. Maas

On-Wafer Microwave Measurements and De-Embedding, Errikos Lourandakis

Parameter Extraction and Complex Nonlinear Transistor Models, Günter Kompa

Passive RF Component Technology: Materials, Techniques, and Applications, Guoan Wang and Bo Pan, editors

Practical Analog and Digital Filter Design, Les Thede

Practical Microstrip Design and Applications, Günter Kompa

Practical Microwave Circuits, Stephen Maas

Practical RF Circuit Design for Modern Wireless Systems, Volume I: Passive Circuits and Systems, Les Besser and Rowan Gilmore

Practical RF Circuit Design for Modern Wireless Systems, Volume II: Active Circuits and Systems, Rowan Gilmore and Les Besser

Principles of RF and Microwave Design, Matthew A. Morgan

Production Testing of RF and System-on-a-Chip Devices for Wireless Communications, Keith B. Schaub and Joe Kelly

Q Factor Measurements Using MATLAB , Darko Kajfez

QMATCH: Lumped-Element Impedance Matching, Software and User's Guide, Pieter L. D. Abrie

Radio Frequency Integrated Circuit Design, Second Edition, John W. M. Rogers and Calvin Plett

Reflectionless Filters, Matthew A. Morgan

RF Bulk Acoustic Wave Filters for Communications, Ken-ya Hashimoto

RF Design Guide: Systems, Circuits, and Equations, Peter Vizmuller

RF Linear Accelerators for Medical and Industrial Applications, Samy Hanna

RF Measurements of Die and Packages, Scott A. Wartenberg

The RF and Microwave Circuit Design Handbook, Stephen A. Maas

RF and Microwave Coupled-Line Circuits, Rajesh Mongia, Inder Bahl, and Prakash Bhartia

RF and Microwave Oscillator Design, Michal Odyniec, editor

RF Power Amplifiers for Wireless Communications, Second Edition, Steve C. Cripps

RF Systems, Components, and Circuits Handbook, Ferril A. Losee

Scattering Parameters in RF and Microwave Circuit Analysis and Design, Janusz A. Dobrowolski

The Six-Port Technique with Microwave and Wireless Applications, Fadhel M. Ghannouchi and Abbas Mohammadi

Solid-State Microwave High-Power Amplifiers, Franco Sechi and Marina Bujatti

Spin Transfer Torque Based Devices, Circuits, and Memory, Brajesh Kumar Kaushik and Shivam Verma

Stability Analysis of Nonlinear Microwave Circuits, Almudena Suárez and Raymond Quéré

Substrate Noise Coupling in Analog/RF Circuits, Stephane Bronckers, Geert Van der Plas, Gerd Vandersteen, and Yves Rolain

System-in-Package RF Design and Applications, Michael P. Gaynor

Technologies for RF Systems, Terry Edwards

Terahertz Metrology, Mira Naftaly, editor

TRAVIS 2.0: Transmission Line Visualization Software and User's Guide, Version 2.0, Robert G. Kaires and Barton T. Hickman

Understanding Microwave Heating Cavities, Tse V. Chow Ting Chan and Howard C. Reader

Understanding Quartz Crystals and Oscillators, Ramón M. Cerda

Vertical GaN and SiC Power Devices, Kazuhiro Mochizuki

The VNA Applications Handbook, Gregory Bonaguide and Neil Jarvis

For further information on these and other Artech House titles, including previously considered out-of-print books now available through our In-Print-Forever® (IPF®) program, contact:

Artech House Publishers
685 Canton Street
Norwood, MA 02062
Phone: 781-769-9750
Fax: 781-769-6334
e-mail: artech@artechhouse.com

Artech House Books
16 Sussex Street
London SW1V 4RW UK
Phone: +44 (0)20 7596 8750
Fax: +44 (0)20 7630 0166
e-mail: artech-uk@artechhouse.com

Find us on the World Wide Web at: www.artechhouse.com